Chapter 4   Exponents and Polynomials

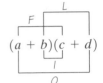

Rules of Exponents

1. $x^m \cdot x^n = x^{m+n}$ **product rule**

2. $\dfrac{x^m}{x^n} = x^{m-n}, x \neq 0$ **quotient rule**

3. $(x^m)^n = x^{m \cdot n}$ **power rule**

4. $x^0 = 1, x \neq 0$ **zero exponent rule**

5. $x^{-m} = \dfrac{1}{x^m}, x \neq 0$ **negative exponent rule**

6. $\left(\dfrac{ax}{by}\right)^m = \dfrac{a^m x^m}{b^m y^m}, b \neq 0, y \neq 0$ **expanded power rule**

7. $\left(\dfrac{a}{b}\right)^{-m} = \left(\dfrac{b}{a}\right)^m, a \neq 0, b \neq 0$ **a fraction raised to a negative exponent rule**

FOIL method (*First, Outer, Inner, Last*) of multiplying binomials:
$(a + b)(c + d) = ac + ad + bc + bd$

$$(a + b)(c + d)$$

Product of the sum and difference of the same two terms:
$(a + b)(a - b) = a^2 - b^2$

Squares of binomials: $(a + b)^2 = a^2 + 2ab + b^2$
$(a - b)^2 = a^2 - 2ab + b^2$

Chapter 5   Factoring

If $a \cdot b = c$, then a and b are **factors** of c.
Difference of two squares: $a^2 - b^2 = (a + b)(a - b)$
Sum of two cubes: $a^3 + b^3 = (a + b)(a^2 - ab + b^2)$
Difference of two cubes: $a^3 - b^3 = (a - b)(a^2 + ab + b^2)$

To Factor a Polynomial

1. If all the terms of the polynomial have a greatest common factor other than 1, factor it out.
2. If the polynomial has two terms, determine if it is a difference of two squares or a sum or a difference of two cubes. If so, factor using the appropriate formula.
3. If the polynomial has three terms, factor the trinomial using one of the procedures discussed.
4. If the polynomial has more than three terms, try factoring by grouping.

5. As a final step, examine your factored polynomial to see if the terms in any factors listed have a common factor. If you find a common factor, factor it out at this point.

Quadratic equation: $ax^2 + bx + c = 0, a \neq 0$.

Zero-factor Property: If $ab = 0$, then $a = 0$ or $b = 0$.

To Solve a Quadratic Equation by Factoring

1. Write the equation in standard form with the squared term positive. This will result in one side of the equation being 0.
2. Factor the side of the equation that is not 0.
3. Set each factor containing a variable equal zero and solve each equation.
4. Check the solution found in step 3 in the original equation.

Pythagorean Theorem: $a^2 + b^2 = c^2$

Chapter 6   Rational Expressions and Equations

To Simplify Rational Expressions

1. Factor both the numerator and denominator as completely as possible.
2. Divide out any factors common to both the numerator and denominator.

To Multiply Rational Expressions

1. Factor all numerators and denominators completely.
2. Divide out common factors.
3. Multiply the numerators together and multiply the denominators together.

To Add or Subtract Two Rational Expressions

1. Determine the least common denominator (LCD).
2. Rewrite each fraction as an equivalent fraction with the LCD.
3. Add or subtract numerators while maintaining the LCD.

4. When possible, factor the remaining numerator and simplify the fraction.

To Solve Rational Expressions

1. Determine the LCD of all fractions in the equation.
2. Multiply both sides of the equation by the LCD. This will result in every term in the equation being multiplied by the LCD.
3. Remove any parentheses and combine like terms on each side of the equation.
4. Solve the equation.
5. Check your solution in the original equation.

Variation

Direct Variation: $y = kx$

Inverse Variation: $y = \dfrac{k}{x}$

Annotated Instructor's Edition

Elementary Algebra
Algebra
For College Students
Sixth Edition

Allen R. Angel
Monroe Community College

with assistance from
Richard Semmler
Northern Virginia Community College

Donna R. Petrie
Monroe Community College

PEARSON
Prentice
Hall

PEARSON EDUCATION, INC.
Upper Saddle River, New Jersey 07458

Senior Acquisitions Editor: Paul Murphy
Editor in Chief: Christine Hoag
Project Manager: Ann Heath
Media Project Manager, Developmental Math: Audra J. Walsh
Vice President/Director of Production and Manufacturing: David W. Riccardi
Executive Managing Editor: Kathleen Schiaparelli
Senior Managing Editor: Linda Mihatov Behrens
Production Editor: Elm Street Publishing Services, Inc.
Production Assistant: Nancy Bauer
Assistant Managing Editor, Math Media Production: John Matthews
Manufacturing Buyer: Michael Bell
Manufacturing Manager: Trudy Pisciotti
Executive Marketing Manager: Eilish Collins Main
Marketing Assistant: Annett Uebel
Development Editor: Don Gecewicz
Supplements Coordinator: Liz Covello
Editor in Chief, Development: Carol Trueheart
Editorial Assistant/Supplements Editor: Kerri-Ann O'Donnell
Art Director/Cover Designer: John Christiana
Interior Designer: Jonathan Boylan
Art Editor: Thomas Benfatti
Creative Director: Carole Anson
Director of Creative Services: Paul Belfanti
Director, Image Resource Center: Melinda Reo
Manager, Rights and Permissions: Zina Arabia
Interior Image Specialist: Beth Brenzel
Cover Image Specialist: Karen Sanatar
Image Coordinator: Charles Morris
Photo Researcher: Sheila Norman
Cover Photo: © Galen Rowell/CORBIS
Art Studio: Scientific Illustrators
Compositor: Preparé, Inc.

© 2004, 2000, 1996, 1992, 1988, 1985 by Pearson Education, Inc.
Pearson Education, Inc.
Upper Saddle River, NJ 07458

Printed in the United States of America
10 9 8 7 6 5 4 3 2

ISBN 0-13-140024-X   ISBN 0-13-140023-1 (Student edition)

Pearson Education LTD., *London*
Pearson Education Australia PTY, Limited, *Sydney*
Pearson Education Singapore, Pte. Ltd
Pearson Education North Asia Ltd, *Hong Kong*
Pearson Education Canada, Ltd., *Toronto*
Pearson Educación de Mexico, S.A. de C.V.
Pearson Education — Japan, *Tokyo*
Pearson Education Malaysia, Pte. Ltd

To my wife, Kathy
and my sons, Robert and Steven

Contents

6   Rational Expressions and Equations   366

7   Graphing Linear Equations   439

8   Systems of Linear Equations   508

Preface

This book was written for college students and other adults who have never been exposed to algebra or those who have been exposed but need a refresher course. My primary goal was to write a book that students can read, understand, and enjoy. To achieve this goal I have used short sentences, clear explanations, and many detailed, worked-out examples. I have tried to make the book relevant to college students by using practical applications of algebra throughout the text.

Features of the Text

Full-Color Format   Color is used pedagogically in the following ways:

- Important definitions and procedures are color screened.
- Color screening or color type is used to make other important items stand out.
- Artwork is enhanced and clarified with use of multiple colors.
- The full-color format allows for easy identification of important features by students.
- The full-color format makes the text more appealing and interesting to students.

Readability   One of the most important features of the text is its readability. The book is very readable, even for those with weak reading skills. Short, clear sentences are used and more easily recognized, and easy-to-understand language is used whenever possible.

Accuracy   Accuracy in a mathematics text is essential. To ensure accuracy in this book, mathematicians from around the country have read the pages carefully for typographical errors and have checked all the answers.

Connections   Many of our students do not thoroughly grasp new concepts the first time they are presented. In this text we encourage students to make connections. That is, we introduce a concept, then later in the text briefly reintroduce it and build upon it.

Often an important concept is used in many sections of the text. Students are reminded where the material was seen before, or where it will be used again. This also serves to emphasize the importance of the concept. Important concepts are also reinforced throughout the text in the Cumulative Review Exercises and Cumulative Review Tests.

Chapter Opening Application   Each chapter begins with a real-life application related to the material covered in the chapter. By the time students complete the chapter, they should have the knowledge to work the problem.

A Look Ahead   This feature at the beginning of each chapter gives students a preview of the chapter and also indicates where this material will be used again in other chapters of the book. This material helps students see the connections between various topics in the book and the connection to real-world situations.

The Use of Icons   At the beginning of each chapter and of each section, a variety of icons are illustrated. These icons are provided to tell students where they may be able to get extra help if needed. There are icons for the *Student's Solution Manual*, ; the *Student's Study Guide*, ; *CDs and videotapes*, ; *Math Pro 4/5 Software*, ; the *Prentice Hall Tutor Center*, ; and the *Angel Website*, . Each of these items will be discussed shortly.

Keyed Section Objectives   Each section opens with a list of skills that the student should learn in that section. The objectives are then keyed to the appropriate portions of the sections with red numbers such as **1**.

Problem Solving   Polya's five-step problem-solving procedure is discussed in Section 1.2. Throughout the book problem solving and Polya's problem-solving procedure are emphasized.

Practical Applications   Practical applications of algebra are stressed throughout the text. Students need to learn how to translate application problems into algebraic symbols. The problem-solving approach used throughout this text gives students ample practice in setting up and solving application problems. The use of practical applications motivates students.

Detailed, Worked-Out Examples   A wealth of examples have been worked out in a step-by-step, detailed manner. Important steps are highlighted in color, and no steps are omitted until after the student has seen a sufficient number of similar examples.

Now Try Exercise   In each section, students are asked to work exercises that parallel the examples given in the text. These Now Try Exercises make the students *active*, rather than passive, learners and they reinforce the concepts as students work the exercises. Through these exercises students have the opportunity to immediately apply what they have learned. Now Try Exercises are indicated in green type such as 35, in the exercise sets.

Study Skills Section   Many students taking this course have poor study skills in mathematics. Section 1.1, the first section of this text, discusses the study skills needed to be successful in mathematics. This section should be very beneficial for your students and should help them to achieve success in mathematics.

Helpful Hints   The Helpful Hint boxes offer useful suggestions for problem solving and other varied topics. They are set off in a special manner so that students will be sure to read them.

Helpful Hints—Study Tips   This is a new feature. These Helpful Hint—Study Tips boxes offer valuable information on items related to studying and learning the material.

Avoiding Common Errors   Errors that students often make are illustrated. The reasons why certain procedures are wrong are explained, and the correct procedure for working the problem is illustrated. These Avoiding Common Errors boxes will help prevent your students from making those errors we see so often.

Mathematics in Action   This new feature stresses the need for and the uses of mathematics in real-life situations. Examples of the use of mathematics in many professions, and how we use mathematics daily without ever giving it much thought, are given. This can be a motivational feature for your students and can give them a better appreciation of mathematics.

Using Your Calculator   The Using Your Calculator boxes, placed at appropriate locations in the text, are written to reinforce the algebraic topics presented in the section and to give the student pertinent information on using a scientific calculator to solve algebraic problems.

Using Your Graphing Calculator   Using Your Graphing Calculator boxes are placed at appropriate locations throughout the text. They reinforce the algebraic topics taught and sometimes offer alternate methods of working problems. This book is designed to give the instructor the option of using or not using a graphing calculator in his or her course. Some of the Using Your Graphing Calculator boxes contain graphing calculator exercises, whose answers appear in the answer section of the book. The illustrations shown in the Using Your Graphing Calculator boxes are from a Texas Instruments 83 Plus calculator. The Using Your Graphing Calculator boxes are written assuming that the student has no prior graphing calculator experience.

Exercise Sets

The exercise sets are broken into three main categories: Concept/Writing Exercises, Practice the Skills, and Problem Solving. Many exercise sets also contain Challenge Problems and/or Group Activities. Each exercise set is graded in difficulty. The early problems help develop the student's confidence, and then students are eased gradually into the more difficult problems. A sufficient number and variety of examples are given in each section for the student to successfully complete even the more difficult exercises. The number of exercises in each section is more than ample for student assignments and practice.

Concept/Writing Exercises   Most exercise sets include exercises that require students to write out the answers in words. These exercises improve students' understanding and comprehension of the material. Many of these exercises involve problem solving and conceptualization and help develop better reasoning and critical thinking skills. Writing exercises are indicated by the symbol ⟍.

Problem Solving Exercises   These exercises have been added to help students become better thinkers and problem solvers. Many of these exercises involve real-life applications of algebra. It is important for students to be able to apply what they learn to real-life situations. Many problem solving exercises help with this.

Challenge Problems   These exercises, which are part of many exercise sets, provide a variety of problems. Many were written to stimulate student thinking.

Others provide additional applications of algebra or present material from future sections of the book so that students can see and learn the material on their own before it is covered in class. Others are more challenging than those in the regular exercise set.

Video Icon Exercises   The exercises that are worked out in detail on the videotapes are marked with the video icon, 📼. This will prove helpful for your students.

Cumulative Review Exercises   All exercise sets (after the first two) contain questions from previous sections in the chapter and from previous chapters. These cumulative review exercises will reinforce topics that were previously covered and help students retain the earlier material, while they are learning the new material. For the students' benefit the Cumulative Review Exercises are keyed to the section where the material is covered, using brackets, such as *[3.4]*.

Group Activities   Many exercise sets have group activity exercises that lead to interesting group discussions. Many students learn well in a cooperative learning atmosphere, and these exercises will get students talking mathematics to one another.

Chapter Summary   At the end of each chapter is a chapter summary that includes a glossary and important chapter facts.

Chapter Review Exercises   At the end of each chapter are review exercises that cover all types of exercises presented in the chapter. The review exercises are keyed, using color numbers and brackets, to the sections where the material was first introduced.

Chapter Practice Tests   The comprehensive end-of-chapter practice test will enable the students to see how well they are prepared for the actual class test. The Instructor's Test Manual includes several forms of each chapter test that are similar to the student's practice test. Multiple choice tests are also included in the Instructor's Test Manual.

Cumulative Review Tests   These tests, which appear at the end of each chapter after the first, test the students' knowledge of material from the beginning of the book to the end of that chapter. Students can use these tests for review, as well as for preparation for the final exam. These exams, like the cumulative review exercises, will serve to reinforce topics taught earlier. The answers to the Cumulative Review Test questions directly follow the test so that students can quickly check their work. After each answer the section and

objective numbers where that material was covered are given using brackets, such as [See 4.2, obj. 5].

Answers   The *odd answers* are provided for the exercise sets. *All answers* are provided for the Using Your Graphing Calculator Exercises, Cumulative Review Exercises, Review Exercises, Practice Tests, and Cumulative Review Tests. *Answers* are not provided for the Group Activity exercises since we want students to reach agreement by themselves on the answers to these exercises.

National Standards

Recommendations of the *Curriculum and Evaluation Standards for School Mathematics*, prepared by the National Council of Teachers of Mathematics (NCTM), and *Crossroads in Mathematics: Standards for Introductory College Mathematics Before Calculus*, prepared by the American Mathematical Association of Two Year Colleges (AMATYC), are incorporated into this edition.

Prerequisite

This text assumes no prior knowledge of algebra. However, a working knowledge of arithmetic skills is important. Fractions are reviewed early in the text, and decimals are reviewed in Appendix A.

Modes of Instruction

The format and readability of this book lends itself to many different modes of instruction. The constant reinforcement of concepts will result in greater understanding and retention of the material by your students.

The features of the text and the large variety of supplements available make this text suitable for many types of instructional modes including:

- lecture
- distance learning
- self-paced instruction
- modified lecture
- cooperative or group study
- learning laboratory

Changes in the Sixth Edition

When I wrote the sixth edition I considered many letters and reviews I got from students and faculty alike. I would like to thank all of you who made suggestions for improving the sixth edition. I would also like to thank the many instructors and students who wrote to inform me of how much they enjoyed, appreciated, and learned from the text. Some of the changes made in the sixth edition of the text include:

- Chapter 3, Formulas and Applications of Algebra, has been rewritten. Many of the exercises have been shortened and additional explanations have been added where necessary. Section 3.5 has been reorganized for greater clarity. The real-life application problems have been updated.

- Addition and subtraction of fractions in Chapter 1 have been greatly enhanced.

- Solving equations containing fractions is now introduced in Chapter 2. Many examples and exercises containing fractions were added.

- There is expanded coverage of order of operations with nested parentheses.

- The identity properties and inverse properties have been added to this edition.

- The Pythagorean Theorem has been moved earlier, to Chapter 5, at the request of many instructors.

- Direct and Inverse Variation has been added as the last section in Chapter 6.

- There is now an entire section dedicated to Applications of Quadratic Equations.

- A greater variety of exercises has been added to exercise sets throughout the book. In general, the exercise sets have been greatly enhanced.

- There is more emphasis on geometry than in the previous edition. More sections have examples and exercises that relate to geometry.

- In selected sections, more difficult exercises have been added at the end of exercise sets.

- Rational numbers have been explained more completely.

- A brief introduction to complex numbers has been added to Chapter 10 for those instructors who wish to introduce this topic to their students.

- The book has a new design with the purpose of making the exercise sets flow more smoothly so that exercises will be easy to spot and identify.

- The Cumulative Review Tests now have the answers directly following the test so that students can get immediate feedback. In addition, the section and objective numbers where the material was discussed are given after the answer.

- A Look Ahead has replaced the Preview and Perspectives. The information provided gives students an overview of the chapter and how it relates to other material in the book and to real-life situations.

- A new feature called Mathematics in Action has been added. This feature stresses the need for and the importance of mathematics in real life. This may be motivational for your students.

- More and new Helpful Hints and Avoiding Common Errors have been added where appropriate.

- Helpful Hint—Study Tips have been added. These reinforce and expand upon the Study Skills for Success in Mathematics covered in Section 1.1.

- Application problems throughout the book have been updated and made more interesting.

- Using Your Graphing Calculator boxes now show keystrokes and screens from a Texas Instruments 83 Plus calculator.

- A fraction raised to a negative exponent is covered more completely.

- The balance bars have been removed from the explanations in Chapter 2.

- When factoring by grouping, the common factor is now placed on the left for consistency with other factoring problems.

- Perpendicular lines are now introduced in the text rather than in the exercise set.

- Chapter 9, Roots and Radicals, has been rewritten and reorganized for greater clarity and understanding. The material also now flows more smoothly.

- Variables other than x and y are used more often in examples and exercises.

- More photos have been added to the text to make it more attractive and interesting for students.

- The basic colors used in the text have been softened to make the text easier to read.

- A brief introduction to metric units of measurement is now presented in the Scientific Notation section.

Supplements to the Sixth Edition

For this edition of the book the author has personally coordinated the development of the *Student's Solution*

Manual and the *Instructor's Solution Manual*. Experienced mathematics professors who have prior experience in writing supplements, and whose works have been of superior quality, have been carefully selected for authoring the supplements.

For Instructors

Printed Supplements

Annotated Instructor's Edition (0-13-140024-X)

- Contains all of the content found in the current edition.
- Answers to exercises are printed on the same text page (graphed answers are in a special graphing section at the back of the text).
- Teaching Tips throughout the text are placed at key points in the margin.

NEW! Instructor's Examples Manual
(0-13-141761-4)

- Provides similar examples for every example presented in the text. Instructors will find this very convenient when they wish to work additional examples on the chalkboard.

Instructor's Solutions Manual (0-13-140026-6)

- Solutions to even-numbered section exercises.
- Solutions to every exercise found in the Chapter Reviews, Chapter Tests, and Cumulative Review Tests.

Instructor's Test Manual (0-13-140027-4)

- Two free-response Pretests per chapter.
- Eight Chapter Tests per chapter (3 multiple choice, 5 free response).
- Two Cumulative Review Tests (one multiple choice, one free response) every two chapters.
- Eight Final Exams (3 multiple choice, 5 free response).
- Twenty additional exercises per section for added test exercises if needed.

Media Supplements

NEW! TestGen-EQ with QuizMaster CD-ROM
(Windows/Macintosh) (0-13-140035-5)

- Algorithmically driven, text-specific testing program.
- Networkable for administering tests and capturing grades online.

- Edit and add your own questions to create a nearly unlimited number of tests and worksheets.
- Use the new "Function Plotter" to create graphs.
- Tests can be easily exported to HTML so they can be posted to the Web for student practice.
- Includes an e-mail function for network users, enabling instructors to send a message to a specific student or an entire group.
- Network-based reports and summaries for a class or student and for cumulative or selected scores available.

 NEW! MathPro5
(Instructor Version) (0-13-140030-4)

- Online customizable tutorial, diagnostic, and assessment program.
- Text-specific at the learning objective level.
- Diagnostic option identifies student skills, and provides individual learning plan and tutorial reinforcement.
- Integration of TestGen-EQ allows for testing to operate within the tutorial environment.
- Course management tracking of tutorial and testing activity.

 MathPro Explorer 4.0
(Network Version IBM/Mac)
(0-13-140028-2)

- Enables instructors to create either customized or algorithmically generated practice quizzes from any section of a chapter.
- Includes e-mail function for network users enabling instructors to send a message to a specific student or to an entire group.
- Network-based reports and summaries for a class or student and for cumulative or selected scores.

Companion Website (*www.prenhall.com/angel*)

- Create a customized online syllabus with Syllabus Manager.
- Assign quizzes or monitor student self-quizzes by having students e-mail results.
- Destination links provide additional opportunities to explore related sites.

For Students

Printed Supplements

 Student's Solutions Manual
SSM (0-13-140025-8)

- Solutions to odd-numbered section exercises.
- Solutions to every (even and odd) exercise found in the Cumulative Review Exercises, Chapter Tests, and Cumulative Review Tests.

 **Student's Study Guide**
Study Guide (0-13-141760-6)

- Includes additional worked-out examples and additional exercises, practice tests, and answers.
- Emphasizes important concepts and includes information to help students study and succeed in mathematics.

Media Supplements

 NEW! MathPro5 (Student Version)
MathPro 4/5 (0-13-140034-7)

- Online customizable tutorial, diagnostic, and assessment program.
- Text-specific at the learning objective level.
- Algorithmically driven for a virtually unlimited number of practice problems with immediate feedback.
- Includes "Watch" screen videoclips.
- Step-by-step solutions.
- Summary of progress.

 MathPro 4.0 Explorer Student Version
MathPro 4/5 (IBM/Mac) (0-13-140033-9)

- Available on CD-Rom for stand-alone use or can be networked in the school laboratory.
- Text-specific tutorial exercises and instructions at the objective level.
- Algorithmically generated Practice Problems.
- Includes "Watch" screen videoclips.

 Videotape Series
CD/Video (0-13-140031-2)

- Keyed to each section of the text.
- Step-by-step solutions to exercises from each section of the text. Exercises from the text that are worked in the videos are marked with a video icon.

 NEW! Digitized Lecture Videos on CD-ROM
CD/Video (0-13-140032-0)

- The entire set of Angel, *Elementary Algebra*, Sixth Edition, lecture videotapes in digital form.
- Convenient access anytime to video tutorial support from a computer at home or on campus.
- Available shrinkwrapped with the text or stand-alone.

 NEW! Prentice Hall Mathematics
PH Math Tutor Center **Tutor Center**

- Staffed with developmental math instructors and open 5 days a week, 7 hours per day.
- Obtain help for examples and exercises in Angel, *Elementary Algebra*, Sixth Edition, via toll-free telephone, fax, or e-mail.
- The Prentice Hall Mathematics Tutor Center is accessed through a registration number that may be bundled with a new text or purchased separately with a used book. Visit *http://www.prenhall.com/tutorcenter* to learn more.

 Companion Website
prenhall.com/Angel (*www.prenhall.com/angel*)

- Practice problems and quizzes with instant feedback.
- Graphing calculator keystroke instructions.
- Destination links provide additional opportunities to explore related sites.

Acknowledgments

Writing a textbook is a long and time-consuming project. Many people deserve thanks for encouraging and assisting me with this project. Most importantly, my special thanks goes to my wife Kathy and sons, Robert and Steven. Without their constant encouragement and understanding, this project would not have become a reality. I would also like to thank my daughter-in-law, Kathy, for her support.

I would like to thank Richard Semmler of Northern Virginia Community College and Larry Clar and Donna Petrie of Monroe Community College for their conscientiousness and attention to detail they provided in checking pages, artwork, and answers. Special thanks to Richard who has been involved in all aspects of the project.

I want to thank Lauri Semarne for also reading the pages and checking the answers. Mitchel Levy of

Broward Community College also deserves my thanks for offering valuable suggestions on the exercise sets and for helping with the Cumulative Review Tests.

I would like to thank my editors at Prentice Hall, Paul Murphy and Ann Heath, my developmental editor Don Gecewicz, and my project editor Phyllis Crittenden, of Elm Street Publishing Services, Inc., for their many valuable suggestions and conscientiousness with this project.

I want to thank those who worked with me on the many supplements for this book. A partial listing follows:

Student's and Instructor's Solutions Manuals—
Doreen Kelly, *Mesa Community College*
Instructor's Test Manual—Kelli Hammer, *Broward Community College*

I would like to thank the following reviewers and proofreaders for their thoughtful comments and suggestions:

Frances Alvarado, *University of Texas–Pan American*
Jose Alvarado, *University of Texas–Pan American*
Ben Anderson, *Darton College (GA)*
Sharon Berrian, *Northwest Shoals Community College*
Dianne Bolen, *Northeast Mississippi Community College*
Julie Bonds, *Sonoma State University*
Clark Brown, *Mojave Community College (AZ)*
Connie Buller, *Metropolitan Community College (NE)*

Lisa DeLong Cuneo, *Pennsylvania State University–Dubois*
Stephan Delong, *Tidewater Community College*
William Echols, *Houston Community College (TX)*
Dale Felkins, *Arkansas Technical University*
Reginald Fulwood, *Palm Beach Community College (FL)*
Abdollah Hajikandi, *State University of New York–Buffalo*
Richard Hobbs, *Mission College (CA)*
Joe Howe, *St. Charles Community College*
Mary Johnson, *Inver Hills Community College (MN)*
Mike Kirby, *Tidewater Community College*
Mitchel Levy, *Broward Community College (FL)*
Mitzi Logan, *Pitt Community College (NC)*
Constance Meade, *College of Southern Idaho*
Lynnette Meslinsky, *Erie Community College (NY)*
Elizabeth Morrison, *Valencia Community College (FL)*
Elsie Newman, *Owens Community College*
Charlotte Newsom, *Tidewater Community College*
Charles Odion, *Houston Community College (TX)*
Behnaz Rouhani, *Athens Technical College (GA)*
Brian Sanders, *Modesto Junior College (CA)*
Rebecca Schantz, *Prairie State College (IL)*
Cristela Sifuentez, *University of Texas–Pan American*
Fereja Tahir, Illinois Central College
Burnette Thompson, Jr., *Houston Community College (TX)*
Ronald Yates, *Community College of Southern Nevada*

Focus on Pedagogy

The Angel series is well-known and widely respected for its realistic, practical approach to algebra. The Angel approach offers thoughtful pedagogy, integrated, up-to-date real-world examples and data, and superior exercise sets.

A Look Ahead In this chapter we provide the building blocks for this course and all other mathematics courses you will take. For many students, Section 1.1, Study Skills for Success in Mathematics, may be the most important section in the book. *Read it carefully and follow the advice given.* If you follow this advice carefully, you will greatly enhance your chance of success in this course.

In Section 1.2, we introduce a 5-step problem-solving procedure that we will use throughout the book. Other important topics covered in this chapter are fractions and the structure of the real number system. It is essential that you understand *addition, subtraction, multiplication, and division of real numbers* covered in Sections 1.6 through 1.8 before you go on to the next chapter.

1.1 STUDY SKILLS FOR SUCCESS IN MATHEMATICS

SSM   Study Guide   CD/Video

MathPro 4/5   PH Math Tutor Center   prenhall.com/Angel

1 Understand the goals of this text.
2 Learn proper study skills.
3 Prepare for and take exams.
4 Learn to manage time.
5 Purchase a calculator.

This section is extremely important. Take the time to read it carefully and follow the advice given. For many of you this section may be the most important section of the text.

Most of you taking this course fall into one of three categories: (1) those who did not take algebra in high school, (2) those who took algebra in high school but did not understand the material, or (3) those who took algebra in high school and were successful but have been out of school for some time and need to take the course again. Whichever the case, you will need to acquire study skills for mathematics courses.

Before we discuss study skills, we will present the goals of this text. These goals may help you realize why certain topics are covered in the text and why they are covered as they are.

▲ Page 2

A Look Ahead

Every chapter begins with **A Look Ahead** to give students an overview of the chapter and how it relates to other material in the book and to real-life situations.

Study Skills for Success In Mathematics (Section 1.1)

Following the study skills presented in this section greatly increases a student's chance for success in this and all other mathematics courses.

In-text Examples

A wealth of **in-text examples** illustrate the concept being presented and provide a step-by-step annotated solution.

Now Try Exercises

Now Try Exercises appear after selected examples to reinforce important concepts. Now Try Exercises provide students with immediate practice and make the student an active learner.

---

178 • Chapter 3 • Formulas and Applications of Algebra

EXAMPLE 10   **Perimeter of Rectangle** The formula for the perimeter of a rectangle is $P = 2l + 2w$. Solve this formula for the length, l.

Solution   We must get l all by itself on one side of the equation. We begin by removing the $2w$ from the right side of the equation to isolate the term containing the l.

$$P = 2l + 2w$$
$$P - 2w = 2l + 2w - 2w \qquad \text{Subtract } 2w \text{ from both sides.}$$
$$P - 2w = 2l$$
$$\frac{P - 2w}{2} = \frac{2l}{2} \qquad \text{Divide both sides by 2.}$$
$$\frac{P - 2w}{2} = l \quad \left(\text{or} \quad l = \frac{P}{2} - w\right)$$

NOW TRY EXERCISE 45

EXAMPLE 11   **Simple Interest Formula** We used the simple interest formula, $i = prt$, in Example 1. Solve the simple interest formula for the principal, p.

Solution   We must isolate the p. Since p is multiplied by both r and t, we divide both sides of the equation by rt.

▲ Page 178

xvi

The Angel series is designed to help students see the important information that they need to learn concepts and topics.

Definitions, Procedures and Important Facts

Definitions, **Procedures** and **Important Facts** are presented in boxes throughout the text to make it easy for students to focus on this material when studying or preparing for quizzes and tests.

Page 21 ▶

Multiplication Symbols

If a and b represent any two mathematical quantities, then each of the following may be used to indicate the product of a and b ("a times b").

$$ab \quad a \cdot b \quad a(b) \quad (a)b \quad (a)(b)$$

Examples

3 times 4 may be written:	3 times x may be written:	x times y may be written:
	$3x$	xy
$3(4)$	$3(x)$	$x(y)$
$(3)4$	$(3)x$	$(x)y$
$(3)(4)$	$(3)(x)$	$(x)(y)$

Now we will introduce the term *factors*, which we shall be using throughout the text. Below we define factors.

DEFINITION | The numbers or variables that are multiplied in a multiplication problem are called **factors**.

If $a \cdot b = c$, then a and b are *factors* of c.

For example, in $3 \cdot 5 = 15$, the numbers 3 and 5 are factors of the product 15. In $2 \cdot 15 = 30$, the numbers 2 and 15 are factors of the product 30. Note that 30 has many other factors. Since $5 \cdot 6 = 30$, the numbers 5 and 6 are also factors of 30. Since $3x$ means 3 times x, both the 3 and the x are factors of $3x$.

Helpful Hints

Helpful Hints offer useful suggestions for problem solving and various other topics.

HELPFUL HINT | From Examples 2 and 3, we can see that when a factor is moved from the denominator to the numerator or from the numerator to the denominator, the sign of the *exponent* changes.

$$x^{-4} = \frac{1}{x^4} \qquad \frac{1}{x^{-4}} = x^4$$

$$3^{-5} = \frac{1}{3^5} \qquad \frac{1}{3^{-5}} = 3^5$$

▲Page 253

Helpful Hints— Study Tips

Helpful Hints—Study Tips reinforce and expand upon the Study Skills for success in Mathematics covered in Section 1.1

HELPFUL HINT

STUDY TIP

Here are some suggestions if you find you are having some difficulty with application problems.
1. Instructor—Make an appointment to see your instructor. Make sure you have read the material in the book and attempted all the homework problems. Go with specific questions for your instructor.
2. Videotapes—Find out if the videotapes that accompany this book are available at your college. If so, view the videotapes that go with this chapter. Using the pause control, you can watch the videotapes at your own pace.
3. Student's Study Guide—If a copy of the Student's Study Guide is available, you may wish to read the material related to this chapter.

▲Page 206

Avoiding Common Errors

Avoiding Common Errors boxes illustrate common mistakes, explain why certain procedures are wrong, and show the correct methods for working the problem.

AVOIDING COMMON ERRORS | An expression raised to the zero power is not equal to 0; it is equal to 1.

CORRECT	INCORRECT
$x^0 = 1$	$x^0 = 0$
$5^0 = 1$	$5^0 = 0$

▲Page 245

Focus on Problem Solving

The sixth editions of the Angel series continue to place a strong focus on problem solving. Angel's exemplary approach to Problem Solving helps students learn to solve problems with confidence. In the process, the Angel texts help students understand *why* they are working on a certain operation while teaching them *how* to perform it. Problem solving is introduced early and incorporated as a theme throughout the texts.

Five-Step Problem-Solving Procedure

The in-text examples demonstrate how to solve each exercise based on Polya's five-step problem-solving procedure: **Understand, Translate, Carry Out, Check,** and **State Answer.**

Guidelines for Problem Solving

1. **Understand the problem.**
   - Read the problem *carefully* at least twice. In the first reading, get a general overview of the problem. In the second reading, determine **(a)** exactly what you are being asked to find and **(b)** what information the problem provides.
   - Make a list of the given facts. Determine which are pertinent to solving the problem.
   - Determine whether you can substitute smaller or simpler numbers to make the problem more understandable.
   - If it will help you organize the information, list the information in a table.
   - If possible, make a sketch to illustrate the problem. Label the information given.

2. **Translate the problem to mathematical language.**
   - This will generally involve expressing the problem in terms of an algebraic expression or equation. (We will explain how to express application problems as equations in Chapter 3.)
   - Determine whether there is a formula that can be used to solve the problem.

3. **Carry out the mathematical calculations necessary to solve the problem.**

4. **Check the answer obtained in step 3.**
   - Ask yourself, "Does the answer make sense?" "Is the answer reasonable?" If the answer is not reasonable, recheck your method for solving the problem and your calculations.
   - Check the solution in the original problem if possible.

5. **Make sure you have answered the question.**
   - State the answer clearly.

▲ Page 9

Problem Solving

17. ***Commissions*** Barbara Riedell earns a 5% commission on appliances she sells. Her sales last week totaled $9400. Find her week's earnings.

18. ***Empire State Building*** May 1, 1931, was the opening day of the Empire State Building. It stands 1454 feet or 443 meters high. Use this information to determine the approximate number of feet in a meter.

19. ***Sales Tax*** **a)** The sales tax in Jefferson County is 7%. What was the sales tax that Jack Mayleben paid on a used car that cost $16,700 before tax?
   b) What is the total cost of the car including tax?

20. ***Checking Account*** The balance in Lois Heater's checking account is $312.60. She purchased five compact disks at $17.11 each including tax. If she pays by check, what is the new balance in her checking account?

21. ***Buying a Computer*** Scott Borden wants to purchase a computer that sells for $950. He can either pay the total amount at the time of purchase or agree to pay the store $200 down and $33 a month for 24 months.
   a) If he pays the down payment and monthly charge, how much will he pay for the computer?
   b) How much money can he save by paying the total amount at the time of purchase?

24. ***Energy Values*** The following table gives the approximate energy values of some foods and the approximate energy consumption of some activities, in kilojoules (kJ).
   Determine how long it would take for you to use up the energy from the following.
   a) a hamburger by running
   b) a chocolate milkshake by walking
   c) a glass of skim milk by cycling

▲ Page 17

Problem-Solving Exercises

Problem Solving exercises are designed to help students become better thinkers.

Focus on Real-World Applications & Data

Each chapter begins with an illustrated, real-world application to motivate students and encourage them to see algebra as an important part of their daily lives. Problems that are based on real data from a broad range of subjects appear throughout the text, in the end-of-chapter material, and in the exercise sets.

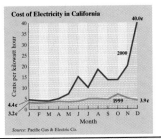

28. Electricity Cost Using the graph shown, determine the approximate difference in the cost of electricity for a family that used 1500 kilowatt hours of electricity in December, 1999, versus December, 2000, if they received their electricity from Pacific Gas & Electric Company.

Cost of Electricity in California

Source: Pacific Gas & Electric Co.

leap) year?

b) If water costs $5.20 per 1000 gallons, how much additional money per year is being spent on the water bill?

32. Tire Pressure When Sandra Hakanson's car tire pressure is 28 pounds per square inch (psi), her car averages 17.3 miles per gallon (mpg) of gasoline. If her tire pressure is increased to 32 psi, it averages 18.0 mpg.

a) How much farther will she travel on a gallon of gas if her tires are inflated to the higher pressure?

b) If she drives an average of 12,000 miles per year, how many gallons of gasoline will she save in a year by increasing her tire pressure from 28 to 32 psi?

c) If gasoline costs $1.40 per gallon, how much money will she save in a year

33. Taxi Ride A taxicab charges $2 upon a customer's entering the taxi, then 30 cents for each $\frac{1}{4}$ mile traveled and 20 cents for each 30 seconds stopped in traffic. David Lopez takes a taxi ride for a distance of 3 miles where the taxi spends 90 seconds stopped in traffic. De-

Real-World Applications

An abundance of wonderfully updated, **real-world applications** give students needed practice with practical applications of algebra. **Real data** is used and real-world situations emphasize the relevance of the material being covered to students' everyday lives.

◀ Page 18

Chapter-Opening Applications

New chapter-opening applications emphasize the role of mathematics in everyday life, and in the workplace, giving students an applied, real-world introduction to the chapter material.

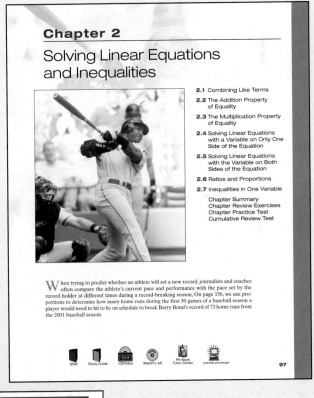

Chapter 2

Solving Linear Equations and Inequalities

2.1 Combining Like Terms
2.2 The Addition Property of Equality
2.3 The Multiplication Property of Equality
2.4 Solving Linear Equations with a Variable on Only One Side of the Equation
2.5 Solving Linear Equations with the Variable on Both Sides of the Equation
2.6 Ratios and Proportions
2.7 Inequalities in One Variable

Chapter Summary
Chapter Review Exercises
Chapter Practice Test
Cumulative Review Test

When trying to predict whether an athlete will set a new record, journalists and coaches often compare the athlete's current pace and performance with the pace set by the record holder at different times during a record-breaking season, On page 156, we use proportions to determine how many home runs during the first 50 games of a baseball season a player would need to hit to be on schedule to break Barry Bond's record of 73 home runs from the 2001 baseball season.

SSM   Study Guide   CD/Video   MathPro 4/5   PH Math Tutor Center   prenhall.com/Angel

97

Page 97 ▶

Mathematics in Action

The Air You Breathe

The Environmental Protection Agency (EPA) has identified indoor air quality as the number one environmental health concern of today. It is largely poor outdoor air quality that creates substandard indoor air quality, particularly in urban areas. Contaminants such as dust, pollen, and automobile pollutants often make their way directly into buildings. With more and more chemicals being used, people are increasingly suffering from chemical sensitivities to formaldehyde, pesticides, ozone, cleaning solvents, fiberglass, asbestos, lead, and radon. Allergies to molds are now more widespread than ever. No building or workplace is immune.

For workplaces involved with pharmaceuticals and biotechnology, the issue of contaminant-free air becomes even more crucial. One type of filter used in these setting is the HEPA (High Efficiency Particulate Air), which was originally developed to remove radioactive contaminants from the air in the development of the first atomic bomb.

The science of filtration involves the trapping of objects ranging from coal dust to viruses. The size of the particles being filtered is typically expressed in microns, where 1 micron (which is short for 1 micrometer) = 1 millionth of a meter = 10^{-6} meters = 0.000001 meters. The measurement refers to the diameter of the particle.

Mathematics In Action

Mathematics In Action emphasizes the need for and the importance of mathematics in real-life.

◀Page 267

Focus on Exercises

Exercise sets throughout the text have been greatly enhanced. Each exercise set progresses in difficulty to help students gain confidence and succeed with more difficult exercises. The end of each exercise set also includes a set of Challenge Problems.

Concept/Writing Exercises

Concept/Writing exercises encourage students to analyze and write about the concepts they are learning.

▼ Page 180

180 • Chapter 3 • Formulas and Applications of Algebra

Exercise Set 3.1

Concept/Writing Exercises

1. What is a formula?
2. What does it mean to *evaluate a formula*?
3. Write the simple interest formula, then indicate what each letter in the formula represents.
4. What is a quadrilateral?
5. What is the relationship between the radius and the diameter of a circle?
6. Is π equal to 3.14? Explain your answer.
7. By using any formula for area, explain why area is measured in square units.
8. By using any formula for volume, explain why volume is measured in cubic units.

Practice the Skills Exercises

Practice the Skills exercises cover all types of exercises presented in the chapter.

Practice the Skills

Simplify each fraction. If a fraction is already simplified, so state.

21. $\dfrac{3}{12}$

22. $\dfrac{40}{10}$

23. $\dfrac{10}{15}$

24. $\dfrac{19}{25}$

25. $\dfrac{17}{17}$

26. $\dfrac{9}{21}$

27. $\dfrac{36}{76}$

28. $\dfrac{16}{72}$

29. $\dfrac{40}{264}$

30. $\dfrac{60}{105}$

31. $\dfrac{12}{25}$

32. $\dfrac{80}{124}$

▲ Page 28

Problem Solving

17. **Commissions** Barbara Riedell earns a 5% commission on appliances she sells. Her sales last week totaled $9400. Find her week's earnings.

18. **Empire State Building** May 1, 1931, was the opening day of the Empire State Building. It stands 1454 feet or 443 meters high. Use this information to determine the approximate number of feet in a meter.

19. **Sales Tax a)** The sales tax in Jefferson County is 7%. What was the sales tax that Jack Mayleben paid on a used car that cost $16,700 before tax?
    b) What is the total cost of the car including tax?

20. **Checking Account** The balance in Lois Heater's checking account is $312.60. She purchased five compact disks at $17.11 each including tax. If she pays by check, what is the new balance in her checking account?

21. **Buying a Computer** Scott Borden wants to purchase a computer that sells for $950. He can either pay the total amount at the time of purchase or agree to pay the store $200 down and $33 a month for 24 months.
    a) If he pays the down payment and monthly charge, how much will he pay for the computer?
    b) How much money can he save by paying the total amount at the time of purchase?

24. **Energy Values** The following table gives the approximate energy values of some foods and the approximate energy consumption of some activities, in kilojoules (kJ).
    Determine how long it would take for you to use up the energy from the following.
    a) a hamburger by running
    b) a chocolate milkshake by walking
    c) a glass of skim milk by cycling

Problem-Solving Exercises

Problem-Solving exercises are designed to help students become better thinkers.

▲ Page 17

Challenge Problems

Challenge Problems stimulate student interest with exercises that are conceptually and computationally more demanding.

Group Activity

Group Activities provide students with opportunities for collaborative learning.

Cumulative Review Exercises

Cumulative Review Exercises reinforce previously covered topics. These exercises are keyed to sections where the material is explained.

Challenge Problems

111. *Cereal Box* A cereal box is to be made by folding the cardboard along the dashed lines as shown in the figure on the right.

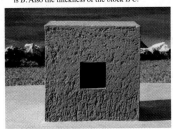

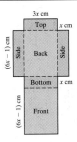

a) Using the formula

volume = length · width · height

write an equation for the volume of the box.

b) Find the volume of the box when $x = 7$ cm.

c) Write an equation for the surface area of the box.

d) Find the surface area when $x = 7$ cm.

Group Activity

112. *Square Face on Cube* Consider the following photo. The front of the figure is a square with a smaller black square painted on the center of the larger square. Suppose the length of one side of the larger square is A, and length of one side of the smaller (the black square) is B. Also the thickness of the block is C.

a) Group member one: Determine an expression for the surface area of the black square.

b) Group member two: Determine an expression for the surface area of the larger square (which includes the smaller square).

c) Group member three: Determine the surface area of the larger square minus the black square (the purple area shown).

d) As a group, write an expression for the volume of the entire solid block.

e) As a group, determine the volume of the entire solid block if its length is 1.5 feet and its width is 0.8 feet.

Cumulative Review Exercises

[1.9] **113.** Evaluate. $\left[4\left(12 \div 2^2 - 3\right)^2\right]^2$.

[2.6] **114.** *Horses* A stable has four Morgan and six Arabian horses. Find the ratio of Arabians to Morgans.

115. *Emptying Pool* It takes 3 minutes to siphon 25 gallons of water out of a swimming pool. How long will it take to empty a 13,500-gallon swimming pool by siphoning? Write a proportion that can be used to solve the problem, and then find the desired value.

[2.7] **116.** Solve $2(x - 4) \geq 3x + 9$.

▲ Page 184

To the Student

Algebra is a course that cannot be learned by observation. To learn algebra you must become an active participant. You must read the text, pay attention in class, and, most importantly, you must work the exercises. The more exercises you work, the better.

The text was written with you in mind. Short, clear sentences are used, and many examples are given to illustrate specific points. The text stresses useful applications of algebra. Hopefully, as you progress through the course, you will come to realize that algebra is not just another math course that you are required to take, but a course that offers a wealth of useful information and applications.

This text makes full use of color. The different colors are used to highlight important information. Important procedures, definitions, and formulas are placed within colored boxes.

The boxes marked **Helpful Hints** should be studied carefully, for they stress important information. The boxes marked **Avoiding Common Errors** should also be studied carefully. These boxes point out errors that students commonly make, and provide the correct procedures for doing these problems.

Ask your professor early in the course to explain the policy on when the calculator may be used. Pay particular attention to the **Using Your Calculator** boxes. You should also read the **Using Your Graphing Calculator** boxes even if you are not using a graphing calculator in class. You may find the information presented here helps you better understand the algebraic concepts.

Other questions you should ask your professor early in the course include: What supplements are available for use? Where can help be obtained when the professor is not available? Supplements that may be available, and the icons used to represent them in the text, include: Student's Solution Manual, ▌; Student's Study Guide, ▌; Math Pro tutorial software, ◉; and CDs and videotapes, 🔒 (including a videotape on study skills needed for success in mathematics). Other sources of help include the Prentice Hall Tutor Center, 📱; and the Angel Website, 🖥. These items are discussed under the heading of Supplements in Section 1.1.

You may wish to form a study group with other students in your class. Many students find that working in small groups provides an excellent way to learn the material. By discussing and explaining the concepts and exercises to one another you reinforce your own understanding. Once guidelines and procedures are determined by your group, make sure to follow them.

One of the first things you should do is to read Section 1.1, Study Skills for Success in Mathematics. Read this section slowly and carefully, and pay particular attention to the advice and information given. Occasionally, refer back to this section. This could be the most important section of the book. Carefully read the material on doing your homework and on attending class.

At the end of all Exercise Sets (after the first two) are **Cumulative Review Exercises**. You should work these problems on a regular basis, even if they are not assigned. These problems are from earlier sections and chapters of the text, and they will refresh your memory and reinforce those topics. If you have a problem when working these exercises, read the appropriate section of the text or study your notes that correspond to that material. The section of the text where the Cumulative Review Exercise was introduced is indicated in brackets, [ ], to the left of the exercise. After reviewing the material, if you still have a problem, make an appointment to see your professor. Working the Cumulative Review Exercises throughout the semester will also help prepare you to take your final exam.

At the end of each chapter are a **Chapter Summary**, **Chapter Review Exercises**, and a **Chapter Practice Test**. Before each examination you should review this material carefully and take the Practice Test. If you do well on the Practice Test, you should do well on the class test. The questions in the Review Exercises are marked to indicate the section in which that material was first introduced. If you have a problem with a Review Exercise question, reread the section indicated.

You may also wish to take the **Cumulative Review Test** that appears at the end of every chapter.

In the back of the text there is an **answer section** which contains the answers to the *odd-numbered* exercises, including the Challenge Problems. Answers to *all* Using Your Graphing Calculator Exercises, Cumulative Review Exercises, Chapter Review Exercises, and Chapter Practice Tests are provided. Answers to the Group Activity exercises are not provided, for we wish students to reach agreement by themselves on answers to these exercises. The answers should be used only to check your work. The answers to the Cumula-

tive Review Test exercises are provided directly after the test for immediate feedback. After each answer the section and objective number where that type of exercise was covered is provided.

I have tried to make this text as clear and error free as possible. No text is perfect, however. If you find an error in the text, or an example or section that you believe can be improved, I would greatly appreciate hearing from you. If you enjoy the text, I would also appreciate hearing from you. You can contact me at *www.prenhall.com/angel.*

Allen R. Angel

Chapter 1

Real Numbers

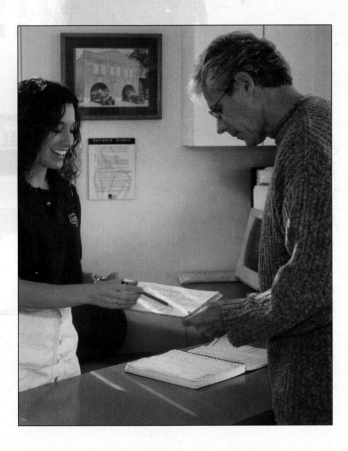

C overage provided by medical insurance policies differs according to the type of policy held by an individual. In addition to yearly premiums, certain health maintenance plans require a co-payment for each visit made to a doctor's office. Other types of policies require that the individual pay a certain dollar amount in medical expenses each year, after which the insurance company pays a large percentage of the remaining costs. On pages 11 and 12 we use problem-solving techniques developed by a famous mathematician, George Polya, to determine the portion of a medical bill a person is responsible for and how much of that bill the insurance company will pay.

SSM   Study Guide   CD/Video   MathPro 4/5   PH Math Tutor Center   prenhall.com/Angel

A Look Ahead

In this chapter we provide the building blocks for this course and all other mathematics courses you will take. For many students, Section 1.1, Study Skills for Success in Mathematics, may be the most important section in the book. *Read it carefully and follow the advice given.* If you follow this advice carefully, you will greatly enhance your chance of success in this course.

In Section 1.2, we introduce a 5-step problem-solving procedure that we will use throughout the book. Other important topics covered in this chapter are fractions and the structure of the real number system. It is essential that you understand *addition, subtraction, multiplication, and division of real numbers* covered in Sections 1.6 through 1.8 before you go on to the next chapter.

1.1 STUDY SKILLS FOR SUCCESS IN MATHEMATICS

SSM

Study Guide

CD/Video

MathPro 4/5

PH Math Tutor Center

prenhall.com/Angel

1 Understand the goals of this text.
2 Learn proper study skills.
3 Prepare for and take exams.
4 Learn to manage time.
5 Purchase a calculator.

This section is extremely important. Take the time to read it carefully and follow the advice given. For many of you this section may be the most important section of the text.

Most of you taking this course fall into one of three categories: (1) those who did not take algebra in high school, (2) those who took algebra in high school but did not understand the material, or (3) those who took algebra in high school and were successful but have been out of school for some time and need to take the course again. Whichever the case, you will need to acquire study skills for mathematics courses.

Before we discuss study skills, we will present the goals of this text. These goals may help you realize why certain topics are covered in the text and why they are covered as they are.

1 Understand the Goals of This Text

The goals of this text include:

1. Presenting traditional algebra topics

2. Preparing you for more advanced mathematics courses

3. Building your confidence in, and your enjoyment of, mathematics

4. Improving your reasoning and critical thinking skills

5. Increasing your understanding of how important mathematics is in solving real-life problems

6. Encouraging you to think mathematically, so that you will feel comfortable translating real-life problems into mathematical equations, and then solving the problems.

It is important to realize that this course is the foundation for more advanced mathematics courses. A thorough understanding of algebra will make it easier for you to succeed in other mathematics courses and in life.

Have a Positive Attitude

You may be thinking to yourself, "I hate math," or "I wish I did not have to take this class." You may have heard of "math anxiety" and feel you fit this category. The first thing to do to be successful in this course is to change your attitude to a more positive one. You must be willing to give this course, and yourself, a fair chance.

Based on past experiences in mathematics, you may feel that this is difficult. However, mathematics is something you need to work at. Many of you are more mature now than when you took previous mathematics courses. Your maturity and desire to learn are extremely important and can make a tremendous difference in your ability to succeed in mathematics. I believe you can be successful in this course, but you also need to believe it.

Prepare for and Attend Class

To be prepared for class, you need to do your homework. If you have difficulty with the homework, or some of the concepts, write down questions to ask your instructor. If you were given a reading assignment, read the appropriate material carefully before class. If you were not given a reading assignment, spend a few minutes previewing any new material in the textbook before class. At this point you don't have to understand everything you read. Just get a feeling for the definitions and concepts that will be discussed. This quick preview will help you understand what your instructor is explaining during class.

After the material is explained in class, read the corresponding sections of the text slowly and carefully, word by word.

You should plan to attend every class. Most instructors agree that there is an inverse relationship between absences and grades. That is, the more absences you have, the lower your grade will be. Every time you miss a class, you miss important information. If you must miss a class, contact your instructor ahead of time, and get the reading assignment and homework. If possible, before the next class, try to copy a friend's notes to help you understand the material you missed.

To be successful in this course, you must thoroughly understand the material in this first chapter, especially fractions and adding and subtracting real numbers. If you are having difficulty with these topics, see your instructor for help.

In algebra and other mathematics courses, the material you learn is cumulative. The new material is built on material that was presented previously. You must understand each section before moving on to the next section, and each chapter before moving on to the next chapter. Therefore, do not let yourself fall behind. Seek help as soon as you need it—do not wait! Make sure that you do all your homework assignments completely and study the text carefully. You will greatly increase your chance of success in this course by following the study skills presented in this section.

While in class, pay attention to what your instructor is saying. If you don't understand something, ask your instructor to repeat the material. If you have read the assigned material before class and have questions that have not been answered, ask your instructor. If you don't ask questions, your instructor will not know that you have a problem understanding the material.

In class, take careful notes. Write numbers and letters clearly, so that you can read them later. Make sure your x's do not look like y's and vice versa. It is not necessary to write down every word your instructor says. Copy the major points and the examples that do not appear in the text. You should not be taking notes so frantically that you lose track of what your instructor is saying. It is a mistake to believe that you can copy material in class without understanding it and then figure it out later.

Read the Text

A mathematics text is not a magazine. Mathematics textbooks should be read slowly and carefully, word by word. If you don't understand what you are reading, reread the material. When you come across a new concept or definition, you may wish to underline or highlight it, so that it stands out. Then it will be easier to find later. When you come across an example, read and follow it line by line. Don't just skim it. Then work out the example on another sheet of paper. Also, work the **Now Try Exercises** that appear in the text next to many examples. The **Now Try Exercises** are designed so that you have the opportunity to immediately apply new ideas. Make notes of anything you don't understand to ask your instructor.

This textbook has special features to help you. I suggest that you pay particular attention to these highlighted features, including the Avoiding Common Errors boxes, the Helpful Hint boxes, and important procedures and definitions identified by color. The **Avoiding Common Errors** boxes point out the most common errors made by students. Read and study this material very carefully and make sure that you understand what is explained. If you avoid making these common errors, your chances of success in this and other mathematics classes will be increased greatly. The **Helpful Hints** offer many valuable techniques for working certain problems. They may also present some very useful information or show an alternative way to work a problem.

Do the Homework

Two very important commitments that you must make to be successful in this course are attending class and doing your homework regularly. Your assignments must be worked conscientiously and completely. Do your homework as soon as possible, so the material presented in class will be fresh in your mind. Research has shown that for mathematics courses, studying and doing homework shortly after learning the material improves retention and performance. Mathematics cannot be learned by observation. You need to practice what was presented in class. It is through doing homework that you truly learn the material. While working homework you will become aware of the types of problems that you need further help with. If you do not work the assigned exercises, you will not know what questions to ask in class.

When you do your homework, make sure that you write it neatly and carefully. List the exercise number next to each problem and work each problem step by step. Then you can refer to it later and understand what is written. Pay particular attention to copying signs and exponents correctly.

Don't forget to check the answers to your homework assignments. This book contains the answers to the odd-numbered exercises in the back of the book. In addition, the answers to all the Cumulative Review Exercises, Chapter Review Exercises, and Chapter Practice Tests are in the back of the book. The answers to the Cumulative Review Test questions follow the questions themselves. In addition, the section and objective number where the material was first introduced is provided in brackets. Answers to the Group Activity Exercises are not provided for we want you to arrive at the answers as a group.

Ask questions in class about homework problems you don't understand. You should not feel comfortable until you understand all the concepts needed to work every assigned problem successfully.

Study for Class

Study in the proper atmosphere, in an area where you will not be constantly disturbed, so that your attention can be devoted to what you are reading. The area where you study should be well ventilated and well lit. You should have sufficient desk space to spread out all your materials. Your chair should be comfortable. You should try to minimize distractions while you are studying. You should not study for hours on end. Short study breaks are a good idea.

Before you begin studying, make sure that you have all the materials you need (pencils, markers, calculator, etc.). You may wish to highlight the important points covered in class or in the book.

It is recommended that students study and do homework for at least two hours for each hour of class time. Some students require more time than others. It is important to spread your studying time out over the entire week rather than studying during one large block of time.

When studying, you should not only understand how to work a problem but also know *why* you follow the specific steps you do to work the problem. If you do not have an understanding of why you follow the specific process, you will not be able to transfer the process to solve similar problems.

This book has **Cumulative Review Exercises** at the end of every section after Section 1.2. Even if these exercises are not assigned for homework, I urge you to work them as part of your studying process. These exercises reinforce material presented earlier in the course, and you will be less likely to forget the material if you review it repeatedly throughout the course. They will also help prepare you for the final exam. If you forget how to work one of the Cumulative Review Exercises, turn to the section indicated in brackets next to the problem and review that section. Then try the problem again.

3   Prepare for and Take Exams

If you study a little bit each day you should not need to cram the night before an exam. Begin your studying early. If you wait until the last minute, you may not have time to seek the help you may need if you find you cannot work a problem.

To prepare for an exam:

1. Read your class notes.
2. Review your homework assignments.
3. Study formulas, definitions, and procedures you will need for the exam.
4. Read the Avoiding Common Errors boxes and Helpful Hint boxes carefully.
5. Read the summary at the end of each chapter.
6. Work the review exercises at the end of each chapter. If you have difficulties, restudy those sections. If you still have trouble, seek help.
7. Work the chapter practice test.
8. Rework quizzes previously given if the material covered in the quizzes will be included on the test.
9. If your exam is a cumulative exam, work the Cumulative Review Test.

Prepare for Midterm and Final Exam

When studying for a comprehensive midterm or final exam follow the procedures discussed for preparing for an exam. However, also:

1. Study all your previous tests and quizzes carefully. Make sure that you have learned to work the problems that you may have previously missed.
2. Work the Cumulative Review Tests at the end of each chapter. These tests cover the material from the beginning of the book to the end of that chapter.
3. If your instructor has given you a worksheet or practice exam, make sure that you complete it. Ask questions about any problems you do not understand.
4. Begin your studying process early so that you can seek all the help you need in a timely manner.

Take an Exam

Make sure you get sufficient sleep the night before the test. If you studied properly, you should not have to stay up late preparing for a test. Arrive at the exam site early so that you have a few minutes to relax before the exam. If you rush into the exam, you will start out nervous and anxious. After you are given the exam, you should do the following:

1. Carefully write down any formulas or ideas that you want to remember.
2. Look over the entire exam quickly to get an idea of its length. Also make sure that no pages are missing.
3. Read the test directions carefully.
4. Read each question carefully. Answer each question completely, and make sure that you have answered the specific question asked.
5. Work the questions you understand best first; then go back and work those you are not sure of. Do not spend too much time on any one problem or you may not be able to complete the exam. Be prepared to spend more time on problems worth more points.
6. Attempt each problem. You may get at least partial credit even if you do not obtain the correct answer. If you make no attempt at answering the question, you will lose full credit.
7. Work carefully step by step. Copy all signs and exponents correctly when working from step to step, and make sure to copy the original question from the test correctly.
8. Write clearly so that your instructor can read your work. If your instructor cannot read your work, you may lose credit. Also, if your writing is not clear, it is easy to make a mistake when working from one step to the next. When appropriate, make sure that your final answer stands out by placing a box around it.
9. If you have time, check your work and your answers.
10. Do not be concerned if others finish the test before you or if you are the last to finish. Use any extra time to check your work.

Stay calm when taking your test. Do not get upset if you come across a problem you can't figure out right away. Go on to something else and come back to that problem later.

4 Learn to Manage Time

As mentioned earlier, it is recommended that students study and do homework for at least two hours for each hour of class time. Finding the necessary time to study is not always easy. Below are some suggestions that you may find helpful.

1. Plan ahead. Determine when you will study and do your homework. Do not schedule other activities for these periods. Try to space these periods evenly over the week.
2. Be organized, so that you will not have to waste time looking for your books, your pen, your calculator, or your notes.
3. If you are allowed to use a calculator, use it for tedious calculations.
4. When you stop studying, clearly mark where you stopped in the text.
5. Try not to take on added responsibilities. You must set your priorities. If your education is a top priority, as it should be, you may have to reduce time spent on other activities.
6. If time is a problem, do not overburden yourself with too many courses. Consider taking fewer credits. If you do not have sufficient time to study, your understanding and all your grades may suffer.

Use Supplements

This text comes with a large variety of supplements. Find out from your instructor early in the semester which supplements are available and might be beneficial for you to use. Supplements should not replace reading the text, but should be used to enhance your understanding of the material. If you miss a class, you may want to review the videotape on the topic you missed before attending the next class.

The supplements are listed on each chapter opening page, and on each page with a section head. The icons used for the supplements are: ▮ for the Student's Solution Manual (SSM); ▯ for the Student's Study Guide; 🔒 for CDs and videotapes (which show about 20 minutes of lecture per section and include the worked out solutions to the exercises marked with this icon); 💿 for Math Pro (tutorial and practice test) software; 📱 for the Prentice Hall Mathematics Tutor (The tutor center offers individual tutoring on examples, exercises, and problems contained in the text. Students can access the tutor center by phone, fax, or email once your instructor has initiated the process with the publisher.); and 🖥 for the Allen R. Angel textbook Web site at Prentice Hall. The Web site includes additional exercises, practice quizzes that can be graded, graphing calculator instructions for all brands of graphing calculators, and chapter projects.

Seek Help

Be sure to get help as soon as you need it! Do not wait! In mathematics, one day's material is usually based on the previous day's material. So, if you don't understand the material today, you may not be able to understand the material tomorrow.

Where should you seek help? There are often a number of resources on campus. Try to make a friend in the class with whom you can study. Often, you can help one another. You may wish to form a study group with other students in your class. Discussing the concepts and homework with your peers will reinforce your own understanding of the material.

You should know your instructor's office hours, and you should not hesitate to seek help from your instructor when you need it. Make sure that you have read the assigned material and attempted the homework before meeting with your instructor. Come prepared with specific questions to ask.

There are often other sources of help available. Many colleges have a mathematics lab or a mathematics learning center, where tutors are available. Ask your instructor early in the semester where and when tutoring is available. Arrange for a tutor as soon as you need one.

TEACHING TIP
You may want to suggest that students exchange phone numbers with a study partner.

5 Purchase a Calculator

TEACHING TIP
You may wish to recommend a type of calculator for students to purchase and a place to purchase one at a good price.

I strongly urge you to purchase a scientific or graphing calculator as soon as possible. Ask your instructor if he or she recommends a particular calculator for this or a future mathematics class. Also ask your instructor if you may use a calculator in class, on homework, and on tests. If so, you should use your calculator whenever possible to save time.

If a calculator contains a │ LOG │ key or │ SIN │ key, it is a scientific calculator. You *cannot* use the square root key │ √ │ to identify scientific calculators since both scientific calculators and nonscientific calculators may have this key. You should pay particular attention to the Using Your Calculator boxes in this book. The boxes explain how to use your calculator to solve problems. If you are using a graphing calculator, pay particular attention to the Using Your Graphing Calculator boxes. You may also need to use the reference manual that comes with your calculator at various times.

A Final Word

You can be successful at mathematics if you attend class regularly, pay attention in class, study your text carefully, do your homework daily, review regularly, and seek help as soon as you need it. Good luck in your course.

Exercise Set 1.1

Do you know:

1.–10., 15., 17. Answers will vary.

1. your professor's name and office hours?
2. your professor's office location and telephone number?
3. where and when you can obtain help if your professor is not available?
4. the name and phone number of a friend in your class?

5. what supplements are available to assist you in learning?
6. if your instructor is recommending the use of a particular calculator?
7. when you can use your calculator in this course?

If you do not know the answer to questions 1–7, you should find out as soon as possible.

8. What are your reasons for taking this course?
9. What are your goals for this course?
10. Are you beginning this course with a positive attitude? It is important that you do!
11. List the things you need to do to prepare properly for class.
12. Explain how a mathematics text should be read.
13. For each hour of class time, how many hours outside of class are recommended for studying and doing homework?

14. When studying, you should not only understand how to work a problem, but also why you follow the specific steps you do. Why is this important?
15. Two very important commitments that you must make to be successful in this course are **a)** doing homework regularly and completely and **b)** attending class regularly. Explain why these commitments are necessary.
16. Write a summary of the steps you should follow when taking an exam. See page 5.
17. Have you given any thought to studying with a friend or a group of friends? Can you see any advantages in doing so? Can you see any disadvantages in doing so?

11. Do all the homework carefully and completely; preview the new material to be covered in class.   **12.** Read slowly and carefully; do not just skim the text.   **13.** At least 2 hours of study and homework time for each hour of class time is generally recommended.   **14.** so that you will be able to solve similar types of problems

1.2 PROBLEM SOLVING

SSM

Study Guide

CD/Video

MathPro 4/5

PH Math Tutor Center

prenhall.com/Angel

1 Learn the five-step problem-solving procedure.
2 Solve problems involving bar, line, and circle graphs.
3 Solve problems involving statistics.

1 Learn the Five-Step Problem-Solving Procedure

One of the main reasons we study mathematics is that we can use it to solve many real-life problems. Throughout the book, we will be problem solving. To solve most real-life problems mathematically, we need to be able to express the problem in mathematical symbols. This is an important part of the problem-solving procedure that we will present shortly. In Chapter 3, we will also spend a great deal of time explaining how to express real-life applications mathematically.

We will now give the general five-step **problem-solving procedure** that was developed by George Polya and presented in his book *How to Solve It.* You can approach any problem by following this general procedure.

George Polya

Guidelines for Problem Solving

1. **Understand the problem.**
   - Read the problem *carefully* at least twice. In the first reading, get a general overview of the problem. In the second reading, determine **(a)** exactly what you are being asked to find and **(b)** what information the problem provides.
   - Make a list of the given facts. Determine which are pertinent to solving the problem.
   - Determine whether you can substitute smaller or simpler numbers to make the problem more understandable.
   - If it will help you organize the information, list the information in a table.
   - If possible, make a sketch to illustrate the problem. Label the information given.

2. **Translate the problem to mathematical language.**
   - This will generally involve expressing the problem in terms of an algebraic expression or equation. (We will explain how to express application problems as equations in Chapter 3.)
   - Determine whether there is a formula that can be used to solve the problem.

3. **Carry out the mathematical calculations necessary to solve the problem.**

4. **Check the answer obtained in step 3.**
   - Ask yourself, "Does the answer make sense?" "Is the answer reasonable?" If the answer is not reasonable, recheck your method for solving the problem and your calculations.
   - Check the solution in the original problem if possible.

5. **Make sure you have answered the question.**
   - State the answer clearly.

In step 2 we use the words *algebraic expression*. An **algebraic expression**, sometimes just referred to as an **expression**, is a general term for any collection of numbers, letters (called variables), grouping symbols such as parentheses ( ) or brackets [ ], and **operations** (such as addition, subtraction, multiplication, and division). In this section we will not be using variables, so we will discuss their use later.

Examples of Expressions

$$3 + 4, \qquad 6(12 \div 3), \qquad (2)(7)$$

The following examples show how to apply the guidelines for problem solving. We will sometimes provide the steps in the examples to illustrate the five-step procedure. However, in some problems it may not be possible or necessary to list every step in the procedure. In some of the examples, we use decimal numbers and percents. If you need to review procedures for adding, subtracting, multiplying, or dividing decimal numbers, or if you need a review of percents, read Appendix A before going on.

EXAMPLE 1 **Transportation** Chicago's O'Hare airport is the busiest in the world with about 65 million passengers arriving and departing annually. The airport express bus operates between the airport and downtown, a distance of 19 miles. A particular airport express bus makes 8 round trips daily between the airport and downtown and carries an average of 12 passengers per trip (each way). The fare each way is $17.50.

a) What are the bus's receipts from one day's operation?

b) If the one-way fare is increased by 10%, determine the new fare.

Solution   **a)** Understand the problem   A careful reading of the problem shows that the task is to find the bus's total receipts from one day's operation. Make a list of all the information given and determine which information is needed to find the total receipts.

Information Given	Pertinent to Solving the Problem
65 million passengers arrive/depart annually	no
19 miles from airport to downtown	no
8 round trips daily	yes
12 passengers per trip (each way)	yes
$17.50 fare (each way)	yes

To find the total receipts, it is not necessary to know the number of passengers who use the airport or the distance between the airport and downtown. Solving this problem involves realizing that the total receipts depend on the number of one-way trips per day, the average number of passengers per trip, and the one-way cost per passenger. The product of these three numbers will yield the total daily receipts. For the 8 round trips daily, there are 2 × 8 or 16 one-way trips daily.

Translate the problem into mathematical language

$$\begin{pmatrix} \text{receipts} \\ \text{for one} \\ \text{day} \end{pmatrix} = \begin{pmatrix} \text{number of} \\ \text{one-way} \\ \text{trips per day} \end{pmatrix} \times \begin{pmatrix} \text{number of} \\ \text{passengers} \\ \text{per trip} \end{pmatrix} \times \begin{pmatrix} \text{cost per} \\ \text{passenger} \\ \text{each way} \end{pmatrix}$$

Carry out the calculations

$$= 16 \times 12 \times \$17.50 = \$3360.00$$

We could also have used 8 round trips and a fare of $35.00 per person to obtain the answer. Can you explain why?

Check the answer   The answer $3360.00 is a reasonable answer based on the information given.

Answer the question asked   The receipts for one day's operation are $3360.00.

b) Understand   If the fare is increased by 10%, the new fare becomes 10% greater than $17.50. Thus you need to add 10% of $17.50 to $17.50 to obtain the answer. When performing calculations, numbers given in percent are usually changed to decimal numbers.

Translate   new fare = original fare + 10% of original fare

Carry Out   new fare = $17.50 + 0.10($17.50)

                              = $17.50 + $1.75 = $19.25

Check   The answer $19.25, which is a little larger than $17.50, seems reasonable.

Answer   When increased by 10% the new fare is $19.25.

EXAMPLE 2 **Processor Speed** In February, 2001, the fastest Intel processor, the Pentium 4, could perform about 1.5 billion operations per second (1.5 gigahertz, symbolized 1.5 GHz). How many operations could it perform in 0.3 seconds?

Solution *Understand* We are given the name of the processor, a speed of about 1.5 billion (1,500,000,000) operations per second, and 0.3 second. To determine the answer to this problem, the name of the processor, Pentium 4, is not needed.

TEACHING TIP
Before discussing the solution to Example 2 ask, "How many could it perform in 2 seconds? $\frac{1}{2}$ second? $\frac{1}{4}$ second? $\frac{1}{10}$ second? Between which 2 numbers should our answer be? Why?

To obtain the answer, will we need to multiply or divide? Often a fairly simple problem seems more difficult because of the numbers involved. When very large or very small numbers make the problem seem confusing, try substituting commonly used numbers in the problem to determine how to solve the problem. Suppose the problem said that the processor can perform 6 operations per second. How many operations can it perform in 2 seconds? The answer to this question should be more obvious. It is 6×2 or 12. Since we multiplied to obtain this answer, we also will need to multiply to obtain the answer to the given problem.

Translate
number of operations in 0.3 seconds $= 0.3$(number of operations per second)

Carry Out
$$= 0.3(1,500,000,000)$$
$$= 450,000,000 \qquad \textit{From a calculator}$$

Check The answer, 450,000,000 operations, is less than the 1,500,000,000 operations per second, which makes sense because the processor is operating for less than a second.

Answer In 0.3 second, the processor can perform about 450,000,000 operations.

EXAMPLE 3 **Medical Insurance** Beth Rechsteiner's medical insurance policy is similar to that of many workers. Her policy requires that she pay the first $100 of medical expenses each calendar year (called a deductible). After the deductible is paid, she pays 20% of the medical expenses (called a co-payment) and the insurance company pays 80%. (There is a maximum co-payment of $600 that she must pay each year. After that, the insurance company pays 100% of the fee schedule.) On January 1, Beth sprained her ankle playing tennis. She went to the doctor's office for an examination and X rays. The total bill of $325 was sent to the insurance company.

a) How much of the bill will Beth be responsible for?

b) How much will the insurance company be responsible for?

Solution **a)** *Understand* First we list all the *relevant* given information.

Given Information

$100 deductible

20% co-payment after deductible

80% paid by insurance company after deductible

$325 doctor bill

All the other information is not needed to solve the problem. Beth will be responsible for the first $100 and 20% of the remaining balance. The insurance company will be responsible for 80% of the balance after the deductible. Before we can find what Beth owes, we need to first find the balance of the bill after the deductible. The balance of the bill after the deductible is $325 - $100 = $225.

Translate   Beth's responsibility = deductible + 20% of balance of bill after the deductible

Carry Out   Beth's responsibility = $100 + 20\%(225)$

$$= 100 + 0.20(225)$$
$$= 100 + 45$$
$$= 145$$

Check and Answer   The answer appears reasonable. Beth will owe the doctor $145.

b) The insurance company will be responsible for 80% of the balance after the deductible.

insurance company's responsibility = 80% of balance after deductible

$$= 0.80(225)$$
$$= 180$$

Thus the insurance company is responsible for $180. This checks because the sum of Beth's responsibility and the insurance company's responsibility is equal to the doctor's bill.

$$\$145 + \$180 = \$325$$

We could have also found the answer to part **b)** by subtracting Beth's responsibility from the total amount of the bill, but to give you more practice with percents we decided to show the solution as we did.

NOW TRY EXERCISE 33

2 Solve Problems Involving Bar, Line, and Circle Graphs

Problem solving often involves understanding and reading graphs and sets of data (or numbers). We will be using bar, line, and circle (or pie) graphs and sets of data, throughout the book. We will illustrate a number of such graphs in this section and explain how to interpret them. To work Example 4, you must interpret bar and circle graphs and work with data.

EXAMPLE 4   **Global E-Commerce**   The following information was provided by Forrester Research Inc. (forrester.com). The graphs provide information about global e-commerce sales (sales made over the Internet) and includes both business-to-business and business-to-consumer sales. It is estimated that e-commerce sales could reach $6.8 trillion in 2004.

Figure 1.1 is a **bar graph** that shows worldwide e-commerce sales. Figure 1.2 is a **circle graph** that shows the breakdown of e-commerce sales by selected regions.

a) Using the graph in Figure 1.1, estimate the worldwide e-commerce sales in 2004.

b) If the worldwide e-commerce sales in 2004 are $6.8 trillion, use Figure 1.2 to estimate the e-commerce sales from North America, Asia/Pacific, Western Europe, and the rest of the world.

Solution   **a)** Using the bar above 2004 in Figure 1.1, we estimate that the e-commerce sales are about $6.8 trillion. Notice that we are using the front face of the bar to get our estimate. Whenever we are working with a three-dimensional bar graph, as in the case here, we will use the front face to determine the reading.

c) Understand   The circle graph in Figure 1.2 indicates that about 50.9% of the world's e-commerce sales in 2004 is from North America, 24.3% is from Asia/Pacific, 22.6% is from Western Europe, and 2.2% is from the rest of the world. The sum of these

Worldwide E-commerce Sales*

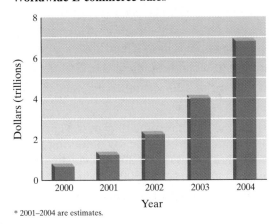

* 2001–2004 are estimates.

FIGURE 1.1

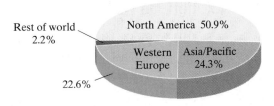

Estimated Breakdown of Global E-commerce
(Sales by region in 2004)

Rest of world 2.2%

North America 50.9%

Western Europe

Asia/Pacific 24.3%

22.6%

FIGURE 1.2

percents is 100%. (Notice that 50.9% of the total area of the circle is North America, 24.3% is Asia/Pacific, 22.6% is Western Europe, and 2.2% is the rest of the world.) To determine the approximate amount of e-commerce sales in North America, we need to find 50.9% of the total e-commerce sales. To do this, we multiply as follows.

Translate

$$\binom{\text{amount of e-commerce}}{\text{sales in North America}} = \binom{\text{percent of total e-commerce}}{\text{sales from North America}}\binom{\text{total e-commerce}}{\text{sales}}$$

Carry Out

Amount of e-commerce sales in North America $= 0.509(\$6.8 \text{ trillion})$

$= \$3.4612 \text{ trillion}$

Thus about $3.4612 trillion of e-commerce sales are expected to occur in North America in 2004.

To find the amounts of e-commerce sales in the other areas of the world, we do similar calculations.

amount of e-commerce sales in Asia/Pacific $= 0.243(\$6.8 \text{ trillion})$

$= \$1.6524 \text{ trillion}$

amount of e-commerce sales in Western Europe $= 0.226(\$6.8 \text{ trillion})$

$= \$1.5368 \text{ trillion}$

amount of e-commerce sales in the rest of the world $= 0.022(\$6.8 \text{ trillion})$

$= \$0.1496 \text{ trillion}$

Check   If we add the four amounts, we obtain the $6.8 trillion total. Therefore our answer is correct.

$\$3.4612 \text{ trillion} + \$1.6524 \text{ trillion} + \$1.5368 \text{ trillion} + \$0.1496 \text{ trillion} = \$6.8 \text{ trillion}$

Answer   In 2004, the e-commerce sales is estimated to be as follows: North America, $3.4612 trillion; Asia/Pacific, $1.6524 trillion; Western Europe, $1.5368 trillion; the rest of the world, $0.1496 trillion.

NOW TRY EXERCISE 23

In Example 5 we will use the symbol ≈, which is read "**is approximately equal to.**" If, for example, the answer to a problem is 34.12432, we may write the answer as ≈34.1.

EXAMPLE 5

The Super Bowl   The **line graph** in Figure 1.3 shows the cost of a 30-second commercial during Super Bowls from 1967 to 2002. The advertising prices are set by the TV network.

a) Estimate the cost of 30-second commercials in 1975 and 2001.

b) How much more was the cost for a 30-second commercial in 2001 than in 1975?

c) How many times greater was the cost of a 30-second commercial in 2001 than in 1975?

Solution

a) When reading a line graph where the line has some thickness, as in Figure 1.3, we will use the center of the line to make our estimate. By observing the dashed lines on the graph, we can estimate that the cost of a 30-second commercial was about $375,000 in 1975 and about $2.3 million (or $2,300,000) in 2001.

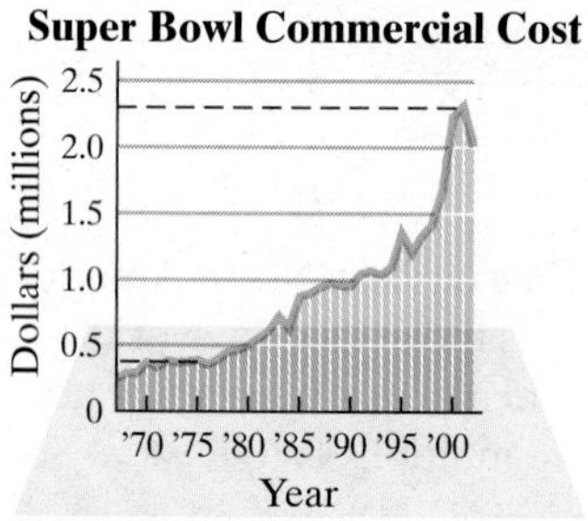

b) We use the problem-solving procedure to answer the question.

Understand   To determine how much more the cost of a 30-second commercial was in 2001 than in 1975, we need to subtract.

Translate   difference in cost = cost in 2001 − cost in 1975

Carry Out   = $2,300,000 − $375,000 = $1,925,000

Answer and Check   The answer appears reasonable. The cost was $1,925,000 more in 2001 than in 1975.

Super Bowl Commercial Cost

Source: NFL research
Note: Prices are in 2001 dollars.

FIGURE 1.3

c) Understand   If you examine parts b) and c), they may appear to ask the same question, but they do not. In Section 1.1 we indicated that it is important to read a mathematics book carefully, word by word. The two parts are different in that part b) asks "how much more was the cost" whereas part c) asks "how many <u>times</u> greater." To determine the number of times greater the cost was in 2001 than in 1975, we need to divide the cost in 2001 by the cost in 1975, as shown below.

Translate   number of times greater = $\dfrac{\text{cost in 2001}}{\text{cost in 1975}}$

Carry Out   number of times greater = $\dfrac{2,300,000}{375,000} \approx 6.13$

Check and Answer   By observing the graph, we see that the answer is reasonable. The cost of a 30-second commercial during the Super Bowl in 2001 was about 6.13 times the cost in 1975.

3 Solve Problems Involving Statistics

Because understanding statistics is so important in our society, we will now discuss certain statistical topics and use them in solving problems.

The *mean* and *median* are two **measures of central tendency**, which are also referred to as *averages*. An average is a value that is representative of a set of data (or numbers). If you take a statistics course you will study these averages in more detail, and you will be introduced to other averages.

The **mean** of a set of data is determined by adding all the values and dividing the sum by the number of values. For example, to find the mean of 6, 9, 3, 12, 12, we do the following.

$$\text{mean} = \frac{6 + 9 + 3 + 12 + 12}{5} = \frac{42}{5} = 8.4$$

We divided the sum by 5 since there are five values. The mean is the most commonly used average and it is generally what is thought of when we use the word *average*.

Another average is the median. The **median** is the value in the middle of a set of **ranked data**. The data may be ranked from smallest to largest or largest to smallest. To find the median of 6, 9, 3, 12, 12, we can rank the data from smallest to largest as follows.

$$3, 6, \boxed{9,} 12, 12$$

↑
Middle value

The value in the middle of the ranked set of data is 9. Therefore, the median is 9. Note that half the values will be above the median and half will be below the median.

If there is an even number of pieces of data, the median is halfway between the two middle pieces. For example, to find the median of 3, 12, 5, 12, 17, 9, we can rank the data as follows.

$$3, 5, \boxed{9, 12,} 12, 17$$

↑
Middle values

TEACHING TIP
After discussing mean and median, ask students, "Which average would you like me to use to determine your grade, mean or median? Why?"

Since there are six pieces of data (an even number), we find the value halfway between the two middle pieces, the 9 and the 12. To find the median, we add these values and divide the sum by 2.

$$\text{median} = \frac{9 + 12}{2} = \frac{21}{2} = 10.5$$

Thus, the median is 10.5. Note that half the values are above and half are below 10.5.

EXAMPLE 6 **The Mean Grade** Alfonso Ramirez's first six exam grades are 90, 87, 76, 84, 78, and 62.

a) Find the mean for Alfonso's six grades.

b) If one more exam is to be given, what is the minimum grade that Alfonso can receive to obtain at least a B average (a mean average of 80 or better)?

c) Is it possible for Alfonso to obtain an A average (90 or better)? Explain.

Solution **a)** To obtain the mean, we add the six grades and divide by 6.

$$\text{mean} = \frac{90 + 87 + 76 + 84 + 78 + 62}{6} = \frac{477}{6} = 79.5$$

b) We will show the problem-solving steps for this part of the example.

Understand The answer to this part may be found in a number of ways. For the mean average of seven exams to be 80, the total points for the seven exams must be 7(80) or 560. Can you explain why? The minimum grade needed can be found by subtracting the sum of the first six grades from 560.

Translate

minimum grade needed
on the seventh exam = 560 − sum of first six exam grades

Carry Out
$$= 560 - (90 + 87 + 76 + 84 + 78 + 62)$$
$$= 560 - 477$$
$$= 83$$

Check   We can check to see that a seventh grade of 83 gives a mean of 80 as follows.

$$\text{mean} = \frac{90 + 87 + 76 + 84 + 78 + 62 + 83}{7} = \frac{560}{7} = 80$$

Answer   A seventh grade of 83 or higher will result in at least a B average.

c) We can use the same reasoning as in part **b)**. For a 90 average, the total points that Alfonso will need to attain is $90(7) = 630$. Since his total points are 477, he will need $630 - 477$ or 153 points to obtain an A average. Since the maximum number of points available on most exams is 100, Alfonso would not be able to obtain an A in the course.

NOW TRY EXERCISE 41

✻

1. understand, translate, calculate, check, state answer   2. a collection of numbers, letters, grouping symbols, and operations
3. substitute smaller or larger numbers   4. divide the sum of the data by the number of pieces of data

Exercise Set 1.2

5. Rank the data. The median is the value in the middle.

Concept/Writing Exercises

1. Outline the five-step problem-solving procedure.
2. What is an expression?
3. If a problem is difficult to solve because the numbers in the problem are very large or very small, what can you do that may help make the problem easier to solve?
4. Explain how to find the mean of a set of data.
5. Explain how to find the median of a set of data.
6. What measure of central tendency do we generally think of as "the average"? the mean
7. Consider the set of data 2, 3, 5, 6, 30. Without doing any calculations, can you determine whether the mean or the median is greater? Explain your answer. mean

8. Consider the set of data 4, 101, 102, 103. Without doing any calculations, determine which is greater, the mean or the median. Explain your answer. median
9. To get a grade of B, a student must have a mean average of 80. Pat Mast has a mean average of 79 for 10 quizzes. He approaches his teacher and asks for a B, reasoning that he missed a B by only one point. What is wrong with his reasoning? missed by 10 points
10. Consider the set of data 3, 3, 3, 4, 4, 4. If one 4 is changed to 5, which of the following will change, the mean and/or the median? Explain. mean

Practice the Skills

In this exercise set, use a calculator as needed to save time.

🔒 11. *Test Grades* Jenna Webber's test grades are 78, 97, 59, 74, and 74. For Jenna's grades, determine the **a)** mean and **b)** median. a) 76.4 b) 74

12. *Bowling Scores* Eric Flemming's bowling scores for five games were 161, 131, 187, 163, and 145. For Eric's games, determine the **a)** mean and **b)** median. a) 157.4 b) 161

13. *Grocery Bills* Liz Kaster's monthly grocery bills for the first five months of 2003 were $204.83, $153.85, $210.03, $119.76, and $128.38. For Liz's grocery bills, determine the **a)** mean and **b)** median. a) $163.37 b) 153.85

14. *Electric Bills* The Foxes' electric bills for January through June, 2002, were $96.56, $108.78, $87.23, $85.90, $79.55, and $65.88. For these bills, determine the **a)** mean and **b)** median. a) ≈$87.32 b) ≈$86.57

15. *Dry Summers* The following figure shows the 10 driest summers in the Southeast from 1895 through 2001. Determine **a)** the mean and **b)** the median inches of rainfall for the 10 years shown. a) 11.593 b) 11.68

Driest Summers in the Southeast
The average summer rainfall in the Southeast is 15.61 inches.

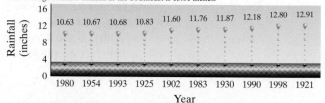

Source: Gloria Forthun, Southeast Regional Climate Center
Records are from the 1895 through 2001. The Southeast is Va., N.C., S.C., Ga., Fla., and Ala.

16. **Homes for Sale** Eight homes are for sale in a community. The sale prices are $124,100, $175,900, $142,300, $164,800, $146,000, $210,000, $112,200, and $153,600.

Determine **a)** the mean and **b)** the median sale price of the eight homes. **a)** $153,612.50 **b)** $149,800

Problem Solving

🔒 17. **Commissions** Barbara Riedell earns a 5% commission on appliances she sells. Her sales last week totaled $9400. Find her week's earnings. $470

18. **Empire State Building** May 1, 1931, was the opening day of the Empire State Building. It stands 1454 feet or 443 meters high. Use this information to determine the approximate number of feet in a meter. ≈3.28

19. **Sales Tax a)** The sales tax in Jefferson County is 7%. What was the sales tax that Jack Mayleben paid on a used car that cost $16,700 before tax? $1169

    b) What is the total cost of the car including tax? $17,869

20. **Checking Account** The balance in Lois Heater's checking account is $312.60. She purchased five compact disks at $17.11 each including tax. If she pays by check, what is the new balance in her checking account? $227.05

21. **Buying a Computer** Scott Borden wants to purchase a computer that sells for $950. He can either pay the total amount at the time of purchase or agree to pay the store $200 down and $33 a month for 24 months.

    a) If he pays the down payment and monthly charge, how much will he pay for the computer? $992

    b) How much money can he save by paying the total amount at the time of purchase? $42

22. **Parking Lot** The Midtown Parking Lot charges $1.50 for each hour of parking or part thereof. Alfredo Irizarry parks his car in the garage from 9:00 A.M. to 5:00 P.M., 5 days a week.

    a) What is his weekly cost for parking? $60

    b) How much money would he save by paying a weekly parking rate of $35.00? $25

23. **Military** The chart below shows where women have enlisted in the United States military as of January, 2001. If the total number of women who have enlisted is approximately 91,600, determine approximately how many more women enlisted in the army than the navy. ≈19,236

**Where Women
Have Enlisted**

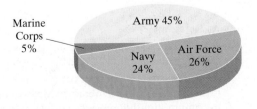

Source: U.S. Department of Defense

See Exercise 24. b)

24. **Energy Values** The following table gives the approximate energy values of some foods and the approximate energy consumption of some activities, in kilojoules (kJ). Determine how long it would take for you to use up the energy from the following.

    a) a hamburger by running 19.375 min

    b) a chocolate milkshake by walking 88 min

    c) a glass of skim milk by cycling 10 min

Energy Value, Food	(kJ)	Energy Consumption, Activity	(kJ/min)
Chocolate milkshake	2200	Walking	25
Fried egg	460	Cycling	35
Hamburger	1550	Swimming	50
Strawberry shortcake	1440	Running	80
Glass of skim milk	350		

🔒 25. **Gas Mileage** When the odometer in Tribet LaPierre's car reads 16,741.3, he fills his gas tank. The next time he fills his tank it takes 10.5 gallons, and his odometer reads 16,935.4. Determine the number of miles per gallon that his car gets. ≈18.49 mpg

26. **Jet Ski** The rental cost of a jet ski from Don's Ski Rental is $10.00 per 15 minutes, and the rental cost from Carol's Ski Rental is $25 per half hour. Suppose you plan to rent a jet ski for 3 hours.

    a) Which is the better deal? Don's

    b) How much will you save? $30

27. *Buying Tires* Eric Weiss purchased four tires by mail order. He paid $62.30 plus $6.20 shipping and handling per tire. There was no sales tax on this purchase. When he received the tires, Eric had to pay $8.00 per tire for mounting and balancing. At a local tire store, his total cost for the four tires with mounting and balancing would have been $425 plus 8% sales tax. How much did Eric save by purchasing the tires through the mail? $153

28. *Electricity Cost* Using the graph shown, determine the approximate difference in the cost of electricity for a family that used 1500 kilowatt hours of electricity in December, 1999, versus December, 2000, if they received their electricity from Pacific Gas & Electric Company. $541.50

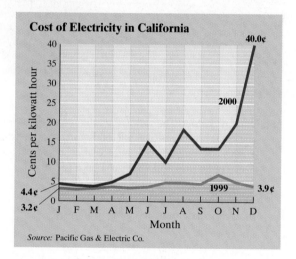

Cost of Electricity in California

Source: Pacific Gas & Electric Co.

29. *Income Taxes* The federal income tax rate schedule for a *joint return* in 2001 is illustrated in the following table.

Adjusted Gross Income	Taxes
$0 − $45,200	15% of income
$45,200 − $109,250	$6780.00 + 27.5% in excess of $45,200
$109,250 − $166,500	$24,393.75 + 30.5% in excess of $109,250
$166,500 − $297,350	$41,855.00 + 35.5% in excess of $166,500
$297,350 and up	$88,306.75 + 39.1% in excess of $297,350

a) If the Donovins' adjusted gross income in 2001 was $34,612 determine their taxes. $5191.80

b) If the Ortegas' 2001 adjusted gross income was $53,710 determine their taxes. $9120.25

30. b) ≈1.47 feet per second

30. *Conversions* a) What is 1 mile per hour equal to in feet per hour? One mile contains 5280 feet. 5280 feet per hour

b) What is 1 mile per hour equal to in feet per second?

c) What is 60 miles per hour equal to in feet per second? 88 feet per second

31. *Leaky Faucet* A faucet that leaks 1 ounce of water per minute wastes 11.25 gallons in a day.

a) How many gallons of water are wasted in a (non-leap) year? 4106.25 gal

b) If water costs $5.20 per 1000 gallons, how much additional money per year is being spent on the water bill? ≈$21.35

32. *Tire Pressure* When Sandra Hakanson's car tire pressure is 28 pounds per square inch (psi), her car averages 17.3 miles per gallon (mpg) of gasoline. If her tire pressure is increased to 32 psi, it averages 18.0 mpg.

a) How much farther will she travel on a gallon of gas if her tires are inflated to the higher pressure? 0.7 mi

b) If she drives an average of 12,000 miles per year, how many gallons of gasoline will she save in a year by increasing her tire pressure from 28 to 32 psi? ≈26.97 gal

c) If gasoline costs $1.40 per gallon, how much money will she save in a year ≈$37.76

33. *Taxi Ride* A taxicab charges $2 upon a customer's entering the taxi, then 30 cents for each $\frac{1}{4}$ mile traveled and 20 cents for each 30 seconds stopped in traffic. David Lopez takes a taxi ride for a distance of 3 miles where the taxi spends 90 seconds stopped in traffic. Determine David's cost of the taxi ride. $6.20

34. *Insurance* Drivers under the age of 25 who pass a driver education course generally have their auto insurance premium decreased by 10%. Most insurers will offer this 10% deduction until the driver reaches 25. A particular driver education course costs $70. Andre DePue, who just turned 18, has auto insurance that costs $630 per year.

a) Excluding the cost of the driver education course, how much would Andre save in auto insurance premiums, from the age of 18 until the age of 25, by taking the driver education course? $441

b) What would be his net savings after the cost of the course? $371

35. *Child Care* The chart on the top of the next page shows average rates for care of one child in various cities. Determine

a) the difference for child care using a day-care center in Austin and in Santa Monica for 20 weeks. $7420

b) the average number of hours of evening babysitting (ages 14 to 18) you can obtain in Minneapolis if the maximum monthly amount you wish to spend is $132. 22 hours

Average Rates For	Austin	Minneapolis	New York City	Santa Monica, Calif.
Day-care center[1]	**$109**/week	**$135**/week	**$350**/week	**$480**/week
Nanny (full time)	**$550**/week	**$575**/week	**$650**/week	**$700**/week
Nanny (part time)	**$12**/hour	**$12**/hour	**$14**/hour	**$13**/hour
Evening babysitter (18 or older)	**$12**/hour	**$10**/hour	**$13**/hour	**$12**/hour
Evening babysitter (ages 14 to 18)	**$7**/hour	**$6**/hour	**$8**/hour	**$7**/hour

Note: [1] For a preschooler.
Source: Money magazine Sept, 2001

36. **Baseball Salaries** Alex Rodriguez, of the Texas Rangers, was the highest paid professional baseball player in 2001, earning $25.2 million. Roger Clemens, of the New York Yankees, was the highest paid pitcher, earning $15.5 million. In 2001, Rodriguez batted 632 times and Clemens (in the regular season) pitched 220.1 innings. Determine approximately how much more Clemens received per inning pitched than Rodriguez did per at bat. ≈$30,549.12

37. **Balance** Consider the figure shown. Assuming the green and red blocks have the same weight, where should a single green block, ■, be placed to make the scale balanced? Explain how you determined your answer.

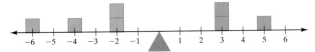

38. **Exams** Andy Gilfillan's mean average on six exams is 78. Find the sum of his scores. 468

39. **Hotel Cost** Lisa Davis, a consultant, stayed one night each in 8 different hotels on business. The total amount of the 8 hotel bills was $1470.72. Determine the mean cost, in dollars, of her hotel stays. $183.84

40. **Construct Data** Construct a set of five pieces of data with a mean of 70 and no two values the same.

41. **Test Grades** A mean of 60 on all exams is needed to pass a course. On his first five exams Lamond Paine's grades are 50, 59, 67, 80, and 56.
   a) What is the minimum grade that Lamond can receive on the sixth exam to pass the course? 48
   b) An average of 70 is needed to get a C in the course. Is it possible for Lamond to get a C? If so, what is the minimum grade that Lamond can receive on the sixth exam? no

42. **Earnings** Consider the following data provided in a press release on July 18, 2002, by the U.S. Census Bureau. The data shows the average lifetime earnings of individuals with different educational backgrounds.

   Assume the average person works 40 years, for 40 hours per week, for 48 weeks a year (disregard paid holidays and vacations).

Average Lifetime Earnings	
Education	**Earnings**
Professional degree	$4.4 million
Doctorate degree	$3.4 million
Master's degree	$2.5 million
Bachelor's degree	$2.1 million
Associate degree	$1.6 million
High school diploma	$1.2 million

   a) Determine the number of hours worked in a lifetime. 76,800 hr
   b) Determine the average hourly wage of a person who has received a high school diploma. ≈$15.63
   c) Determine the average hourly wage for a person who has received a professional degree. ≈$57.29

43. **Life Span** In October, 2001, the U.S. Census Bureau stated that the average life span of Americans had increased slightly to 76.9 years. Do you feel the average the bureau is using is the mean or median? Explain. 43. Answers will vary

37. On the 3 on the right   40. One example is 50, 60, 70, 80, 90

Challenge Problem

44. **Reading Meters** The figure on the top left of the following page shows how to read an electric (or gas) meter.
   1. Start with the dial on the right. Use the smaller number on the dial (except when the dial is between 9 and 0, then use the 9). Notice the arrows above the meters indicate the direction the dial is moving (clockwise, then counterclockwise).
   2. If the pointer is directly on a number, check the dial to the *right* to make sure it has passed 0 and is headed towards 1. If the dial to the right has not passed 0, use the next lower number. The number on the meters on the top left of the next page is 16064.

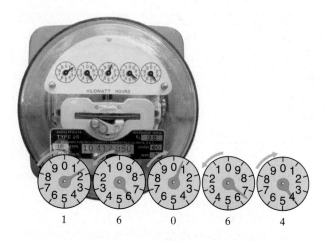

1   6   0   6   4

Source: Southern California Edison, *Understanding Your Electricity Bill*

Suppose your previous month's reading was as shown on the left, and this month's meter reading is as shown below.

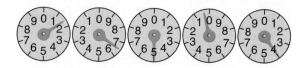

a) Determine this month's meter reading. 16504

b) Determine your electrical cost for this month by first subtracting last month's meter reading from this month's meter reading (measured in kilowatt hours), and then multiplying the difference by the cost per kilowatt hour of electricity. Assume electricity costs 24.3 cents per kilowatt hour. ≈$106.92

1.3 FRACTIONS

1   Learn multiplication symbols and recognize factors.
2   Simplify fractions.
3   Multiply fractions.
4   Divide fractions.
5   Add and subtract fractions.
6   Change mixed numbers to fractions and vice versa.

Students taking algebra for the first time often ask, "What is the difference between arithmetic and algebra?" When doing arithmetic, all the quantities used in the calculations are known. In algebra, however, one or more of the quantities are unknown and must be found.

EXAMPLE 1   **Flour Needed for a Recipe**   Mrs. Clark has 2 cups of flour. A recipe calls for 3 cups of flour. How many additional cups does she need?

Solution   The answer is 1 cup.

Although elementary, this is an example of an algebraic problem. The unknown quantity is the number of additional cups of flour needed.

An understanding of decimal numbers (see Appendix A) and fractions is essential to success in algebra. You will need to know how to simplify a fraction and how to add, subtract, multiply, and divide fractions. We will review these topics in this section. We will also explain the meaning of factors.

1 Learn Multiplication Symbols and Recognize Factors

In algebra we often use letters called **variables** to represent numbers. Letters commonly used as variables are *x*, *y*, and *z*, but other letters can be used as variables. Variables are usually shown in italics. So that we do not confuse the variable *x* with the multiplication sign, we often use different notation to indicate multiplication.

Multiplication Symbols

If a and b represent any two mathematical quantities, then each of the following may be used to indicate the product of a and b ("a times b").

$$ab \quad a\cdot b \quad a(b) \quad (a)b \quad (a)(b)$$

Examples

3 times 4 may be written:	3 times x may be written:	x times y may be written:
	$3x$	xy
$3(4)$	$3(x)$	$x(y)$
$(3)4$	$(3)x$	$(x)y$
$(3)(4)$	$(3)(x)$	$(x)(y)$

Now we will introduce the term *factors*, which we shall be using throughout the text. Below we define factors.

DEFINITION

The numbers or variables that are multiplied in a multiplication problem are called **factors**.

If $a \cdot b = c$, then a and b are *factors* of c.

For example, in $3 \cdot 5 = 15$, the numbers 3 and 5 are factors of the product 15. In $2 \cdot 15 = 30$, the numbers 2 and 15 are factors of the product 30. Note that 30 has many other factors. Since $5 \cdot 6 = 30$, the numbers 5 and 6 are also factors of 30. Since $3x$ means 3 times x, both the 3 and the x are factors of $3x$.

2 Simplify Fractions

Now we have the necessary information to discuss **fractions**. The top number of a fraction is called the **numerator**, and the bottom number is called the **denominator**. In the fraction $\frac{3}{5}$, the 3 is the numerator and the 5 is the denominator.

HELPFUL HINT

Consider the fraction $\frac{3}{5}$. There are equivalent methods of expressing this fraction, as illustrated below.

$$\frac{3}{5} = 3 \div 5 = 5\overline{)3}$$

In general, $\frac{a}{b} = a \div b = b\overline{)a}$

Now we will discuss how to simplify a fraction.

A fraction is **simplified** (or **reduced to its lowest terms**) when the numerator and denominator have no common factors other than 1. To simplify a fraction, follow these steps.

To Simplify a Fraction

1. Find the largest number that will divide (without remainder) both the numerator and the denominator. This number is called the **greatest common factor** (GCF).
2. Then divide both the numerator and the denominator by the greatest common factor.

If you do not remember how to find the greatest common factor of two or more numbers, read Appendix B.

EXAMPLE 2 Simplify **a)** $\dfrac{10}{25}$ **b)** $\dfrac{6}{18}$.

Solution **a)** The largest number that divides both 10 and 25 is 5. Therefore, 5 is the greatest common factor. Divide both the numerator and the denominator by 5 to simplify the fraction.

$$\frac{10}{25} = \frac{10 \div 5}{25 \div 5} = \frac{2}{5}$$

b) Both 6 and 18 can be divided by 1, 2, 3, and 6. The largest of these numbers, 6, is the greatest common factor. Divide both the numerator and the denominator by 6.

$$\frac{6}{18} = \frac{6 \div 6}{18 \div 6} = \frac{1}{3}$$

Note in Example **2b)** that both the numerator and denominator could have been written with a factor of 6. Then the common factor 6 could be divided out.

$$\frac{6}{18} = \frac{1 \cdot \cancel{6}}{3 \cdot \cancel{6}} = \frac{1}{3}$$

NOW TRY EXERCISE 23 When you work with fractions you should simplify your answers.

3 Multiply Fractions

To multiply two or more fractions, multiply their numerators together and multiply their denominators together.

To Multiply Fractions

$$\frac{a}{b} \cdot \frac{c}{d} = \frac{ac}{bd}$$

EXAMPLE 3 Multiply $\dfrac{3}{13}$ by $\dfrac{5}{11}$.

Solution $\dfrac{3}{13} \cdot \dfrac{5}{11} = \dfrac{3 \cdot 5}{13 \cdot 11} = \dfrac{15}{143}$

Before multiplying fractions, to help avoid having to simplify an answer, we often divide both a numerator and a denominator by a common factor.

EXAMPLE 4 Multiply **a)** $\dfrac{8}{17} \cdot \dfrac{5}{16}$ **b)** $\dfrac{27}{40} \cdot \dfrac{16}{9}$.

Solution **a)** Since the numerator 8 and the denominator 16 can both be divided by the common factor 8, we divide out the 8 first. Then we multiply.

$$\frac{8}{17} \cdot \frac{5}{16} = \frac{\overset{1}{\cancel{8}}}{17} \cdot \frac{5}{\underset{2}{\cancel{16}}} = \frac{1 \cdot 5}{17 \cdot 2} = \frac{5}{34}$$

b)

$$\frac{27}{40}\cdot\frac{16}{9}=\frac{\overset{3}{27}}{40}\cdot\frac{16}{\underset{1}{9}} \qquad \textit{Divide both 27 and 9 by 9.}$$

$$=\frac{\overset{3}{27}}{\underset{5}{40}}\cdot\frac{\overset{2}{16}}{\underset{1}{9}} \qquad \textit{Divide both 40 and 16 by 8.}$$

$$=\frac{3\cdot2}{5\cdot1}=\frac{6}{5}$$

NOW TRY EXERCISE 51

The numbers $0, 1, 2, 3, 4,\ldots$ are called **whole numbers**. The three dots after the 4 which is called an *ellipsis*, indicate that the whole numbers continue indefinitely in the same manner. Thus the numbers 468 and 5043 are also whole numbers. Whole numbers will be discussed further in Section 1.4. To multiply a whole number by a fraction, write the whole number with a denominator of 1 and then multiply.

EXAMPLE 5 **Lawn Mower Engine** Some engines run on a mixture of gas and oil. A particular lawn mower engine requires a mixture of $\frac{5}{64}$ gallon of oil for each gallon of gasoline used. A lawn care company wishes to make a mixture for this engine using 12 gallons of gasoline. How much oil must be used?

Solution We must multiply 12 by $\frac{5}{64}$ to determine the amount of oil that must be used. First we write 12 as $\frac{12}{1}$, then we divide both 12 and 64 by their greatest common factor, 4, as follows.

$$12\cdot\frac{5}{64}=\frac{12}{1}\cdot\frac{5}{64}=\frac{\overset{3}{12}}{1}\cdot\frac{5}{\underset{16}{64}}=\frac{3\cdot5}{1\cdot16}=\frac{15}{16}$$

Thus, $\frac{15}{16}$ gallon of oil must be added to the 12 gallons of gasoline to make the proper mixture.

4 Divide Fractions

To divide one fraction by another, invert the divisor (the second fraction if written with ÷) and proceed as in multiplication.

To Divide Fractions

$$\frac{a}{b}\div\frac{c}{d}=\frac{a}{b}\cdot\frac{d}{c}=\frac{ad}{bc}$$

Sometimes, rather than being asked to obtain the answer to a problem by adding, subtracting, multiplying, or dividing, you may be asked to evaluate an expression. To **evaluate** an expression means to obtain the answer to the problem using the operations given.

EXAMPLE 6 Evaluate **a)** $\frac{3}{5}\div\frac{5}{6}$ **b)** $\frac{3}{8}\div12$.

Solution **a)** $\frac{3}{5}\div\frac{5}{6}=\frac{3}{5}\cdot\frac{6}{5}=\frac{3\cdot6}{5\cdot5}=\frac{18}{25}$

b) Write 12 as $\frac{12}{1}$. Then invert the divisor and multiply.

NOW TRY EXERCISE 59

$$\frac{3}{8} \div 12 = \frac{3}{8} \div \frac{12}{1} = \frac{\overset{1}{\cancel{3}}}{8} \cdot \frac{1}{\underset{4}{\cancel{12}}} = \frac{1}{32}$$

5 Add and Subtract Fractions

Fractions that have the same (or a common) *denominator can be added or subtracted.* To add (or subtract) fractions with the same denominator, add (or subtract) the numerators and keep the common denominator.

To Add and Subtract Fractions

$$\frac{a}{c} + \frac{b}{c} = \frac{a+b}{c} \quad \text{or} \quad \frac{a}{c} - \frac{b}{c} = \frac{a-b}{c}$$

EXAMPLE 7 Add **a)** $\frac{6}{15} + \frac{2}{15}$ **b)** $\frac{8}{13} - \frac{5}{13}$

Solution **a)** $\frac{6}{15} + \frac{2}{15} = \frac{6+2}{15} = \frac{8}{15}$ **b)** $\frac{8}{13} - \frac{5}{13} = \frac{8-5}{13} = \frac{3}{13}$

To add (or subtract) fractions with unlike denominators, we must first rewrite each fraction with the same, or a common, denominator. The smallest number that is divisible by two or more denominators is called the **least common denominator** or **LCD**. *If you have forgotten how to find the least common denominator review Appendix B now.*

EXAMPLE 8 Add $\frac{1}{2} + \frac{1}{5}$.

Solution We cannot add these fractions until we rewrite them with a common denominator. Since the lowest number that both 2 and 5 divide into (without remainder) is 10, we will first rewrite both fractions with the least common denominator of 10.

$$\frac{1}{2} = \frac{1}{2} \cdot \frac{5}{5} = \frac{5}{10} \quad \text{and} \quad \frac{1}{5} = \frac{1}{5} \cdot \frac{2}{2} = \frac{2}{10}$$

Now we add.

$$\frac{1}{2} + \frac{1}{5} = \frac{5}{10} + \frac{2}{10} = \frac{7}{10}$$

TEACHING TIP
Illustrate why fractions must have common denominators by asking "If you have a quarter and a dime, can you tell me the value of your money in just quarters? In just dimes? Is there a unit you could use to describe the value of your money?" Then show how to represent the situation in fraction form.

Note that multiplying both the numerator and denominator by the same number is the same as multiplying by 1. Thus the value of the fraction does not change.

EXAMPLE 9 How much larger is $\frac{3}{4}$ inch than $\frac{2}{3}$ inch?

Solution To find how much larger, we need to subtract $\frac{2}{3}$ inch from $\frac{3}{4}$ inch.

$$\frac{3}{4} - \frac{2}{3}$$

The least common denominator is 12. Therefore, we rewrite both fractions with a denominator of 12.

$$\frac{3}{4} = \frac{3}{4} \cdot \frac{3}{3} = \frac{9}{12} \quad \text{and} \quad \frac{2}{3} = \frac{2}{3} \cdot \frac{4}{4} = \frac{8}{12}$$

Now we subtract.

$$\frac{3}{4} - \frac{2}{3} = \frac{9}{12} - \frac{8}{12} = \frac{1}{12}$$

NOW TRY EXERCISE 75 Therefore, $\frac{3}{4}$ inch is $\frac{1}{12}$ inch greater than $\frac{2}{3}$ inch.

AVOIDING COMMON ERRORS

It is important that you remember that dividing out a common factor in the numerator of one fraction and the denominator of a different fraction can be performed only when multiplying fractions. *This process cannot be performed when adding or subtracting fractions.*

CORRECT
MULTIPLICATION PROBLEMS

$$\frac{\overset{1}{\cancel{3}}}{5} \cdot \frac{1}{\cancel{3}_{1}}$$

$$\frac{\overset{2}{\cancel{8}} \cdot 3}{\underset{1}{4}}$$

INCORRECT
ADDITION PROBLEMS

$$\frac{\overset{1}{\cancel{3}}}{5} + \frac{1}{\cancel{3}_{1}}$$

$$\frac{\overset{2}{\cancel{8}} + 3}{\underset{1}{4}}$$

---

6 Change Mixed Numbers to Fractions and Vice Versa

Consider the number $5\frac{2}{3}$. This is an example of a **mixed number**. A mixed number consists of a whole number followed by a fraction. The mixed number $5\frac{2}{3}$ means $5 + \frac{2}{3}$. We can change $5\frac{2}{3}$ to a fraction as follows.

TEACHING TIP
You may wish to teach your students that $5\frac{2}{3} = \frac{3 \cdot 5 + 2}{3} = \frac{17}{3}$

$$5\frac{2}{3} = \boxed{5} + \frac{2}{3} = \boxed{\frac{15}{3}} + \frac{2}{3} = \frac{15 + 2}{3} = \frac{17}{3}$$

Notice that we expressed the whole number, 5, as a fraction with a denominator of 3, then added the fractions.

EXAMPLE 10 Change $7\frac{3}{8}$ to a fraction.

Solution

$$7\frac{3}{8} = \boxed{7} + \frac{3}{8} = \boxed{\frac{56}{8}} + \frac{3}{8} = \frac{56 + 3}{8} = \frac{59}{8}$$

TEACHING TIP
You may wish to teach your students that $43 \div 6 = 7$, remainder 1, and $\frac{43}{6} = 7\frac{1}{6}$

Now consider the fraction $\frac{17}{3}$. This fraction can be converted to a mixed number as follows.

$$\frac{17}{3} = \boxed{\frac{15}{3}} + \frac{2}{3} = \boxed{5} + \frac{2}{3} = 5\frac{2}{3}$$

Notice we wrote $\frac{17}{3}$ as a sum of two fractions, each with the denominator of 3. The first fraction being added, $\frac{15}{3}$, is the equivalent of the largest integer that is less than $\frac{17}{3}$.

EXAMPLE 11 Change $\frac{43}{6}$ to a mixed number.

Solution

$$\frac{43}{6} = \boxed{\frac{42}{6}} + \frac{1}{6} = \boxed{7} + \frac{1}{6} = 7\frac{1}{6}$$

Notice in Example 11 the fraction $\frac{43}{6}$ is simplified because the greatest common divisor of the numerator and denominator is 1. Do not confuse simplifying a fraction with changing a fraction with a value greater than 1 to a mixed number. The fraction $\frac{43}{6}$ can be converted to the mixed number $7\frac{1}{6}$. However, $\frac{43}{6}$ is a simplified fraction.

Now we will work examples that contain mixed numbers.

EXAMPLE 12   **Plumbing**   To repair a plumbing leak, a coupling $\frac{1}{2}$ inch long is glued to a piece of plastic pipe that is $2\frac{9}{16}$ inches long. How long is the combined length? See Figure 1.4.

Solution   Understand and Translate   We need to add $2\frac{9}{16}$ inches and $\frac{1}{2}$ inch to obtain the combined lengths. We will place the two numbers one above the other. After we write both fractions with a common denominator, we add the numbers.

Carry Out

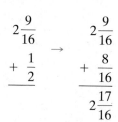

$$
2\frac{9}{16} \qquad \qquad 2\frac{9}{16}
$$
$$
\underline{+\ \frac{1}{2}} \quad \rightarrow \quad \underline{+\ \frac{8}{16}}
$$
$$
2\frac{17}{16}
$$

FIGURE 1.4

Since $2\frac{17}{16} = 2 + \frac{17}{16} = 2 + 1\frac{1}{16} = 3\frac{1}{16}$, the sum is $3\frac{1}{16}$.

Check and Answer   The answer appears reasonable. Thus the total length is $3\frac{1}{16}$ inches.   ✳

EXAMPLE 13   **Growing Taller**   The graph in Figure 1.5 shows the height of Kelly on January 1, 2002, and on January 1, 2003. How much had Kelly grown during this time period?

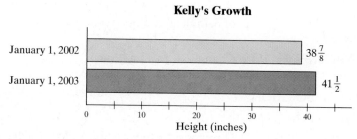

Kelly's Growth

FIGURE 1.5

Solution   Understand and Translate   To find the increase, we need to subtract the height on January 1, 2002, from the height on January 1, 2003, we will subtract vertically.

Carry Out

$$
41\frac{1}{2} \qquad \qquad 41\frac{4}{8}
$$
$$
\underline{-38\frac{7}{8}} \quad \rightarrow \quad \underline{-38\frac{7}{8}}
$$

Since we wish to subtract $\frac{7}{8}$ from $\frac{4}{8}$, and $\frac{7}{8}$ is greater than $\frac{4}{8}$, we write $41\frac{4}{8}$ as $40\frac{12}{8}$. To get $40\frac{12}{8}$, we take 1 unit from the number 41 and write it as $\frac{8}{8}$. This gives $40 + 1 + \frac{4}{8} = 40 + \frac{8}{8} + \frac{4}{8} = 40 + \frac{12}{8} = 40\frac{12}{8}$. Now we subtract as follows.

$$
\begin{array}{c}
41\dfrac{1}{2} \\
-38\dfrac{7}{8}
\end{array}
\rightarrow
\begin{array}{c}
41\dfrac{4}{8} \\
-38\dfrac{7}{8}
\end{array}
\rightarrow
\begin{array}{c}
40\dfrac{12}{8} \\
-38\dfrac{7}{8} \\
\hline
2\dfrac{5}{8}
\end{array}
$$

Check and Answer   By examining the graph, we see that the answer is reasonable. Thus, Kelly grew by $2\frac{5}{8}$ inches during this time period.   ✳

Although it is not necessary to change mixed numbers to fractions when adding or subtracting mixed numbers, it is necessary to change mixed numbers to fractions when multiplying or dividing mixed numbers. We illustrate this procedure in Example 14.

EXAMPLE 14   **Cutting Strips**   A rectangular piece of material 3 feet wide by $12\frac{1}{2}$ feet long is cut into five equal strips, as shown in Figure 1.6. Find the dimensions of each strip.

Solution   Understand and Translate   We know from the diagram that one side will have a width of 3 feet. To find the length of the strips, we need to divide $12\frac{1}{2}$ by 5.

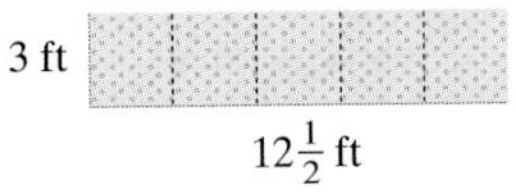

3 ft

$12\frac{1}{2}$ ft

Carry Out $12\dfrac{1}{2} \div 5 = \dfrac{25}{2} \div \dfrac{5}{1} = \dfrac{\overset{5}{\cancel{25}}}{2} \cdot \dfrac{1}{\underset{1}{\cancel{5}}} = \dfrac{5}{2} = 2\dfrac{1}{2}$

FIGURE 1.6

Check and Answer   If you multiply $2\frac{1}{2}$ times 5 you obtain the original length, $12\frac{1}{2}$. Thus the calculation was correct. The dimensions of each strip will be 3 feet by $2\frac{1}{2}$ feet.   ✳

NOW TRY EXERCISE 95

---

1. b) x, y, z **2.** numbers or variables that are multiplied together   **3.** $5(x), (5)x, (5)(x), 5x, 5 \cdot x$ **4. a)** numerator   **b)** denominator
5. divide out the common factors   **6. a)** ellipsis   **b)** numbers continue in the same manner

Exercise Set 1.3

7. a) the smallest number divisible by the two denominators   **b)** Answers will vary.

Concept/Writing Exercises

🖉 **1. a)** What are variables?   letters that represent numbers

  b) What letters are often used to represent variables?

🖉 **2.** What are factors?

  3. Show five different ways that "5 times x" may be written.

  4. In a fraction, what is the name of the **a)** top number and **b)** bottom number?

🖉 **5.** Explain how to simplify a fraction.

🖉 **6. a)** What are the three dots in 4, 5, 6, 7, … called?

  b) What do the three dots following the 7 mean?

🖉 **7. a)** What is the least common denominator of two or more fractions?

  b) Write two fractions and then give the LCD of the two fractions you wrote.

🖉 **8.** What is the LCD of the fractions $\frac{3}{8}$ and $\frac{7}{10}$? Explain.   40

In Exercises 9 and 10, which part a) or b) shows simplifying a fraction? Explain.

🖉 **9. a)** $\dfrac{\overset{1}{\cancel{2}}}{5} \cdot \dfrac{1}{\underset{2}{\cancel{4}}}$ **b)** $\dfrac{\overset{1}{\cancel{2}}}{\underset{2}{\cancel{4}}}$ b)

🖉 **10. a)** $\dfrac{\overset{1}{\cancel{4}}}{\underset{3}{\cancel{12}}}$ **b)** $\dfrac{7}{\underset{3}{\cancel{12}}} \cdot \dfrac{\overset{1}{\cancel{4}}}{5}$ a)

In Exercises 11 and 12, one of the procedures a) or b) is incorrect. Determine which is the incorrect procedure and explain why.

11. a) $\dfrac{\overset{1}{4}}{5} + \dfrac{3}{\underset{2}{8}}$ **b)** $\dfrac{\overset{1}{4}}{5} \cdot \dfrac{3}{\underset{2}{8}}$ a) is incorrect

12. a) $\dfrac{4}{\underset{3}{15}} \cdot \dfrac{\overset{1}{5}}{7}$ **b)** $\dfrac{4}{\underset{3}{15}} + \dfrac{\overset{1}{5}}{7}$ b) is incorrect

In Exercises 13 and 14 indicate any parts where a common factor can be divided out as a first step in evaluating each expression. Explain your answer.

13. a) $\dfrac{4}{5} + \dfrac{1}{4}$ **b)** $\dfrac{4}{5} - \dfrac{1}{4}$ **c)** $\dfrac{4}{5} \cdot \dfrac{1}{4}$ **d)** $\dfrac{4}{5} \div \dfrac{1}{4}$ c)

14. a) $6 + \dfrac{5}{12}$ **b)** $6 \cdot \dfrac{5}{12}$ **c)** $6 - \dfrac{5}{12}$ **d)** $6 \div \dfrac{5}{12}$ b)

15. Explain how to multiply fractions.

16. Explain how to divide fractions.

17. Explain how to add or subtract fractions.

18. Give an example of a mixed number. $3\dfrac{1}{2}$

19. Is the fraction $\dfrac{24}{5}$ simplified? Explain your answer.   yes

20. Is the fraction $\dfrac{20}{3}$ simplified? Explain your answer.   yes

15. divide out common factors, multiply numerators, multiply denominators   **16.** invert the divisor, then multiply
17. Write fractions with common denominator, add or subtract numerators, keep common denominator

Practice the Skills

Simplify each fraction. If a fraction is already simplified, so state.

21. $\dfrac{3}{12}$ $\frac{1}{4}$ **22.** $\dfrac{40}{10}$ 4 **23.** $\dfrac{10}{15}$ $\frac{2}{3}$ **24.** $\dfrac{19}{25}$ simplified

25. $\dfrac{17}{17}$ 1 **26.** $\dfrac{9}{21}$ $\frac{3}{7}$ 🔒 **27.** $\dfrac{36}{76}$ $\frac{9}{19}$ **28.** $\dfrac{16}{72}$ $\frac{2}{9}$

29. $\dfrac{40}{264}$ $\frac{5}{33}$ **30.** $\dfrac{60}{105}$ $\frac{4}{7}$ **31.** $\dfrac{12}{25}$ simplified   **32.** $\dfrac{80}{124}$ $\frac{20}{31}$

Convert each mixed number to a fraction.

33. $2\dfrac{3}{5}$ $\frac{13}{5}$ **34.** $5\dfrac{1}{3}$ $\frac{16}{3}$ 🔒 **35.** $2\dfrac{13}{15}$ $\frac{43}{15}$ **36.** $6\dfrac{5}{12}$ $\frac{77}{12}$

37. $4\dfrac{3}{4}$ $\frac{19}{4}$ **38.** $6\dfrac{2}{9}$ $\frac{56}{9}$ **39.** $4\dfrac{13}{19}$ $\frac{89}{19}$ **40.** $3\dfrac{3}{32}$ $\frac{99}{32}$

Write each fraction as a mixed number.

41. $\dfrac{7}{4}$ $1\frac{3}{4}$ **42.** $\dfrac{17}{5}$ $3\frac{2}{5}$ **43.** $\dfrac{15}{4}$ $3\frac{3}{4}$ **44.** $\dfrac{9}{2}$ $4\frac{1}{2}$

45. $\dfrac{110}{20}$ $5\frac{1}{2}$ **46.** $\dfrac{67}{13}$ $5\frac{2}{13}$ 🔒 **47.** $\dfrac{32}{7}$ $4\frac{4}{7}$ **48.** $\dfrac{72}{14}$ $5\frac{1}{7}$

Find each product or quotient. Simplify the answer.

49. $\dfrac{2}{3} \cdot \dfrac{4}{5}$ $\frac{8}{15}$ **50.** $\dfrac{3}{5} \cdot \dfrac{5}{7}$ $\frac{3}{7}$ **51.** $\dfrac{5}{12} \cdot \dfrac{4}{15}$ $\frac{1}{9}$ **52.** $\dfrac{1}{2} \cdot \dfrac{12}{15}$ $\frac{2}{5}$

53. $\dfrac{3}{4} \div \dfrac{1}{2}$ $\frac{3}{2}$ or $1\frac{1}{2}$ **54.** $\dfrac{15}{16} \cdot \dfrac{4}{3}$ $\frac{5}{4}$ or $1\frac{1}{4}$ **55.** $\dfrac{3}{8} \div \dfrac{3}{4}$ $\frac{1}{2}$ **56.** $\dfrac{3}{8} \cdot \dfrac{10}{11}$ $\frac{15}{44}$

57. $\dfrac{5}{12} \div \dfrac{4}{3}$ $\frac{5}{16}$ **58.** $\dfrac{15}{4} \cdot \dfrac{2}{3}$ $\frac{5}{2}$ or $2\frac{1}{2}$ **59.** $\dfrac{10}{3} \div \dfrac{5}{9}$ 6 **60.** $\dfrac{12}{5} \div \dfrac{3}{7}$ $\frac{28}{5}$ or $5\frac{3}{5}$

61. $\left(2\dfrac{1}{5}\right)\left(\dfrac{7}{8}\right)$ $\frac{77}{40}$ or $1\frac{37}{40}$ **62.** $\dfrac{28}{13} \cdot \dfrac{2}{7}$ $\frac{8}{13}$ 🔒 **63.** $5\dfrac{3}{8} \div 1\dfrac{1}{4}$ $\frac{43}{10}$ or $4\frac{3}{10}$ **64.** $4\dfrac{4}{5} \div \dfrac{8}{15}$ 9

Add or subtract. Simplify each answer.

65. $\dfrac{1}{4} + \dfrac{3}{4}$ 1

66. $\dfrac{3}{8} + \dfrac{2}{8}$ $\dfrac{5}{8}$

67. $\dfrac{5}{12} - \dfrac{1}{12}$ $\dfrac{1}{3}$

68. $\dfrac{18}{36} - \dfrac{1}{36}$ $\dfrac{17}{36}$

69. $\dfrac{8}{17} + \dfrac{2}{34}$ $\dfrac{9}{17}$

70. $\dfrac{3}{7} + \dfrac{17}{35}$ $\dfrac{32}{35}$

71. $\dfrac{4}{5} + \dfrac{6}{15}$ $\dfrac{6}{5}$ or $1\dfrac{1}{5}$

72. $\dfrac{5}{6} - \dfrac{3}{4}$ $\dfrac{1}{12}$

73. $\dfrac{1}{6} - \dfrac{1}{18}$ $\dfrac{1}{9}$

74. $\dfrac{11}{28} + \dfrac{1}{7}$ $\dfrac{15}{28}$

75. $\dfrac{5}{12} - \dfrac{1}{8}$ $\dfrac{7}{24}$

76. $\dfrac{5}{8} - \dfrac{4}{7}$ $\dfrac{3}{56}$

77. $\dfrac{7}{12} - \dfrac{2}{9}$ $\dfrac{13}{36}$

78. $\dfrac{3}{7} + \dfrac{5}{12}$ $\dfrac{71}{84}$

79. $\dfrac{5}{9} - \dfrac{4}{15}$ $\dfrac{13}{45}$

80. $\dfrac{1}{32} + \dfrac{5}{12}$ $\dfrac{43}{96}$

81. $6\dfrac{1}{3} - 3\dfrac{1}{5}$ $\dfrac{47}{15}$ or $3\dfrac{2}{15}$

82. $2\dfrac{3}{8} + \dfrac{1}{4}$ $\dfrac{21}{8}$ or $2\dfrac{5}{8}$

83. $5\dfrac{3}{4} - \dfrac{1}{3}$ $\dfrac{65}{12}$ or $5\dfrac{5}{12}$

84. $2\dfrac{1}{3} + 1\dfrac{1}{8}$ $\dfrac{83}{24}$ or $3\dfrac{11}{24}$

Problem Solving

85. *Height Gain* The following graph shows Kim Brugger's height, in inches, on her 8th and 12th birthdays. How much had Kim grown in the 4 years? $8\dfrac{15}{16}$ in.

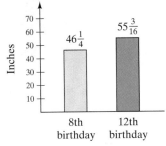

86. *Road Paving* The following graph shows the progress of the Davenport Paving Company in paving the Memorial Highway. How much of the highway was paved from June through August? $6\dfrac{7}{8}$ mi

Highway Paved in Selected Months

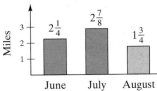

In many problems you will need to subtract a fraction from 1, where 1 represents "the whole" or the "total amount." Exercises 87–90 are answered by subtracting the given fraction from 1.

87. *Online Employees* In 2001 approximately $\dfrac{25}{36}$ of all employees in the U.S. were online. What fraction of all employees in the U.S. in 2001 were not online? $\dfrac{11}{36}$

88. *Global Warming* The probability that an event does not occur may be found by subtracting the probability that the event does occur from 1. If the probability that global warming is occurring is $\dfrac{7}{9}$, find the probability that global warming is not occurring. $\dfrac{2}{9}$

89. *Truck Sales* Use the following graph to determine the approximate fraction of sales for the pickup truck, minivan and SUV market that were for imported vehicles in 2001. $\dfrac{11}{50}$

U.S. Sales in 2001 for Pickup Trucks, Minivans, and SUVs

Source: J.D. Power and Associates

90. *Union Membership* The following graph shows the approximate fraction of U.S. workers who belonged to unions in 2001. Determine the fraction of U.S. workers who did not belong to a union in 2001. $\dfrac{22}{25}$

U.S. Union Membership, 2001

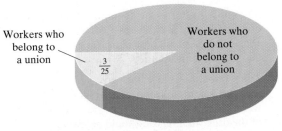

Source: Bureau of Labor Statistics

91. *Truck Weight* A flatbed tow truck weighing $4\dfrac{1}{2}$ tons is carrying two cars. One car weighs $1\dfrac{1}{6}$ tons, the other weighs $1\dfrac{3}{4}$ tons. What is the total weight of the tow truck and the two cars? $7\dfrac{5}{12}$ tons

92. Wood Cut A length of $3\frac{1}{16}$ inches is cut from a piece of wood $16\frac{3}{4}$ inches long. What is the length of the remaining piece of wood? $13\frac{11}{16}$ in.

93. Albino's Growth An albino python at Cypress Gardens, Florida, was 3 feet, $3\frac{1}{4}$ inches when born. Its present length is 15 feet $2\frac{1}{2}$ inches. How much has it grown since birth? 11 ft, $11\frac{1}{4}$ in. or $143\frac{1}{4}$ in. or ≈ 11.94 ft

94. Pants Inseam The inseam on a new pair of pants is 30 inches. If Sean Leland's inseam is $28\frac{3}{8}$ inches by how much will the pants need to be shortened? $1\frac{5}{8}$ in.

95. Wood Cut Dawn Foster cuts a piece of wood measuring $3\frac{1}{8}$ inches into two equal pieces. How long is each piece? $1\frac{9}{16}$ in.

96. Baking Turkey The instructions on a turkey indicate that a 12- to 16-pound turkey should bake at 325°F for about 22 minutes per pound. Donna Draus is planning to bake a $13\frac{1}{2}$-pound turkey. Approximately how long should the turkey be baked? 297 min or 4 hr 57 min

97. Chopped Onions A recipe for pot roast calls for $\frac{1}{4}$ cup chopped onions for each pound of beef. For $5\frac{1}{2}$ pounds of beef, how many cups of chopped onions are needed? $1\frac{3}{8}$ cups

98. Fencing Rick O'Shea wants to fence in his backyard as shown. The three sides to be fenced measure $16\frac{2}{3}$ yards, $22\frac{2}{3}$ yards, and $14\frac{1}{8}$ yards.

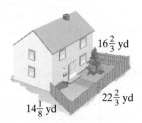

16$\frac{2}{3}$ yd

22$\frac{2}{3}$ yd

14$\frac{1}{8}$ yd

a) How much fence will Rick need? $53\frac{11}{24}$ yd

b) If Rick buys 60 yards of fence, how much will be left over? $6\frac{13}{24}$ yd

99. Shampoo A bottle of shampoo contains 15 fluid ounces. If Tierra Bentley uses $\frac{3}{8}$ of an ounce each time she washes her hair, how many times can Tierra wash her hair using this bottle? 40 times

100. Drug Amount A nurse must give $\frac{1}{16}$ milligram of a drug for each kilogram of patient weight. If Mr. Duncan weighs (or has a mass of) 80 kilograms, find the amount of the drug Mr. Duncan should be given. 5 mg

101. Windows An insulated window for a house is made up of two pieces of glass, each $\frac{1}{4}$-inch thick, with a 1-inch space between them. What is the total thickness of this window? $1\frac{1}{2}$ in.

102. Cream Pie A Boston cream pie weights $1\frac{5}{16}$ pounds. If the pie is to be divided equally among 6 people, how much will each person get? $\frac{7}{32}$ lb

103. Cutting Wood A 28-inch length of wood is to be cut into $4\frac{2}{3}$ inch strips. How many whole strips can be made? Disregard loss of wood due to cuts made. 6

104. Fasten Bolts A mechanic wishes to use a bolt to fasten a piece of wood $4\frac{1}{2}$ inches thick to a metal tube $2\frac{1}{3}$ inches thick. If the thickness of the nut is $\frac{1}{8}$ inch, find the length of the shaft of the bolt so that the nut fits flush with the end of the bolt (see the figure below). $6\frac{23}{24}$ in.

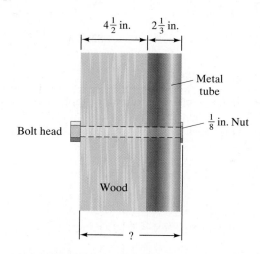

$4\frac{1}{2}$ in. $2\frac{1}{3}$ in.

Metal tube

Bolt head

$\frac{1}{8}$ in. Nut

Wood

?

105. Computer Cabinet Marcinda James is considering purchasing a mail order computer. The catalog describes the computer as $7\frac{1}{2}$ inches high and the monitor as $14\frac{3}{8}$ inches high. Marcinda is hoping to place the monitor on top of the computer and to place the computer and monitor together in the opening where the computer is shown.

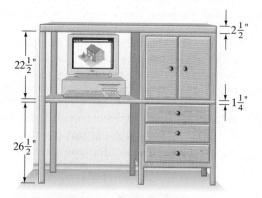

$2\frac{1}{2}$"

$22\frac{1}{2}$"

$1\frac{1}{4}$"

$26\frac{1}{2}$"

a) Will there be sufficient room to do this? yes

b) If so, how much extra space will she have? $\frac{5}{8}$ in.

c) Find the total height of the computer desk $52\frac{3}{4}$ in.

106. Soda If five 2-liter bottles of soda are split evenly among 30 people, how many liters of soda will each person get? $\frac{1}{3}$ liter

Challenge Problems

107. Add or subtract the following fractions using the rule discussed in this section. Your answer should be a single fraction, and it should contain the symbols given in the exercise.

a) $\dfrac{*}{a} + \dfrac{?}{a} \quad \dfrac{*+?}{a}$ **b)** $\dfrac{\odot}{?} - \dfrac{\square}{?} \quad \dfrac{\odot-\square}{?}$

c) $\dfrac{\triangle}{\square} + \dfrac{4}{\square} \quad \dfrac{\triangle+4}{\square}$ **d)** $\dfrac{x}{3} - \dfrac{2}{3} \quad \dfrac{x-2}{3}$

e) $\dfrac{12}{x} - \dfrac{4}{x} \quad \dfrac{8}{x}$

108. Multiply the following fractions using the rule discussed in this section. Your answer should be a single fraction

and it should contain the symbols given in the exercise.

a) $\dfrac{\triangle}{a} \cdot \dfrac{\square}{b} \quad \dfrac{\triangle\square}{ab}$ **b)** $\dfrac{6}{3} \cdot \dfrac{\triangle}{\square} \quad \dfrac{2\triangle}{\square}$

c) $\dfrac{x}{a} \cdot \dfrac{y}{b} \quad \dfrac{xy}{ab}$ **d)** $\dfrac{3}{8} \cdot \dfrac{4}{y} \quad \dfrac{3}{2y}$

e) $\dfrac{3}{x} \cdot \dfrac{x}{y} \quad \dfrac{3}{y}$

109. *Drug Dosage* An allopurinal pill comes in 300-milligram doses. Dr. Highland wants a patient to get 450 milligrams each day by cutting the pills in half and taking $\frac{1}{2}$ pill three times a day. If she wants to prescribe enough pills for a 6-month period (assume 30 days per month), how many pills should she prescribe?   270 pills

Group Activity

110. a) Flakes, 2 cups; milk, 6 tbsp   **b)** Flakes, 2 cups; milk, $7\frac{1}{3}$ tbsp   **c)** Flakes, 2 cups; milk, $\frac{1}{2}$ cup or 8 tbsp
d) Flakes, 2 cups; milk, $8\frac{2}{3}$ tbsp   **e)** Flakes are same, milk is different—$\frac{1}{3}$ cup is not twice 2 tbsp

Discuss and answer Exercise 110 as a group.

110. *Potatoes* The following table gives the amount of each ingredient recommended to make 2, 4, and 8 servings of instant mashed potatoes.

Servings	2	4	8
Water	$\frac{2}{3}$ cup	$1\frac{1}{3}$ cups	$2\frac{2}{3}$ cups
Milk	2 tbsp	$\frac{1}{3}$ cup	$\frac{2}{3}$ cup
Butter*	1 tbsp	2 tbsp	4 tbsp
Salt†	$\frac{1}{4}$ tsp	$\frac{1}{2}$ tsp	1 tsp
Potato flakes	$\frac{2}{3}$ cup	$1\frac{1}{3}$ cup	$2\frac{2}{3}$ cups

* or margarine
† Less salt can be used if desired.

Determine the amount of potato flakes and milk needed to make 6 servings by the different methods described. When working with milk, 16 tbsp = 1 cup.

a) Group member 1: Determine the amount of potato flakes and milk needed to make 6 servings by multiplying the amount for 2 servings by 3.

b) Group member 2: Determine the amounts by adding the amounts for 2 servings to the amounts for 4 servings.

c) Group member 3: Determine the amounts by finding the average (mean) of 4 and 8 servings.

d) As a group, determine the amount by subtracting the amount for 2 servings from the amount for 8 servings.

e) As a group, compare your answers from parts **a)** through **d)**. Are they all the same? If not, can you explain why? (This might be a little tricky.)

Cumulative Review Exercises

[1.1] **111.** What is your instructor's name and office hours?

[1.2] **112.** What is the mean of 9, 8, 15, 32, 16?   16

113. What is the median of 9, 8, 15, 32, 16?   15

[1.3] **114.** What are variables?   letters used to represent numbers

---

1.4 THE REAL NUMBER SYSTEM

SSM   Study Guide   CD/Video

MathPro 4/5   PH Math Tutor Center   prenhall.com/Angel

1   Identify sets of numbers.

2   Know the structure of the real numbers.

We will be talking about and using various types of numbers throughout the text. This section introduces you to some of those numbers and to the structure of the real number system.

1 Identify Sets of Numbers

A **set** is a collection of **elements** listed within braces. The set $\{a, b, c, d, e\}$ consists of five elements, namely *a, b, c, d,* and *e*. A set that contains no elements is called an **empty set** (or **null set**). The symbols { } or ∅ are used to represent the empty set.

There are many different sets of numbers. Two important sets are the natural numbers and the whole numbers. The whole numbers were introduced earlier.

Natural numbers: $\{1, 2, 3, 4, 5, \ldots\}$

Whole numbers: $\{0, 1, 2, 3, 4, 5, \ldots\}$

An aid in understanding sets of numbers is the real number line (Fig. 1.7).

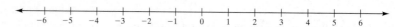

FIGURE 1.7

The real number line continues indefinitely in both directions. The numbers to the right of 0 are positive and those to the left of 0 are negative. Zero is neither positive nor negative (Fig. 1.8).

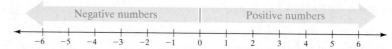

FIGURE 1.8

Figure 1.9 illustrates the natural numbers marked on a number line. The natural numbers are also called the **positive integers** or the **counting numbers**.

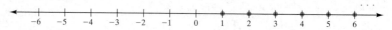

FIGURE 1.9

Another important set of numbers is the integers.

Integers: $\{\ldots, -5, -4, -3, -2, -1, 0, 1, 2, 3, 4, 5, \ldots\}$

The integers consist of the negative integers, 0, and the positive integers. The integers are marked on the number line in Figure 1.10.

FIGURE 1.10

Can you think of any numbers that are not integers? You probably thought of "fractions" or "decimal numbers." Fractions and certain decimal numbers belong to the set of rational numbers. The set of **rational numbers** consists of all the numbers that can be expressed as a quotient (or a ratio) of two integers, with the denominator not 0.

Rational numbers: {quotient of two integers, denominator not 0}

All integers are rational numbers since they can be written with a denominator of 1. For example, $3 = \frac{3}{1}$, $-12 = \frac{-12}{1}$, and $0 = \frac{0}{1}$. All fractions containing integers in the numerator and denominator (with the denominator not 0) are rational numbers. For example, the fraction $\frac{3}{5}$ is a quotient of two integers and is a rational number.

When a fraction that is a ratio of two integers is converted to a decimal number by dividing the numerator by the denominator, the quotient will always be either a *terminating decimal number*, such as 0.3 and 3.25, or a *repeating decimal number* such as 0.3333 . . . and 5.2727. . . . The three dots at the end of a number like 0.3333 . . . indicate that the numbers continue to repeat in the same manner indefinitely. All terminating decimal numbers and all repeating decimal numbers are rational numbers which can be expressed as a quotient of two integers. For example, $0.3 = \frac{3}{10}$, $3.25 = \frac{325}{100}$, $0.3333 \ldots = \frac{1}{3}$ and $5.2727 \ldots = \frac{522}{99}$. Some rational numbers are illustrated on the number line in Figure 1.11.

FIGURE 1.11

Most of the numbers that we use are rational numbers. However, some numbers are not rational. Numbers such as the square root of 2, written $\sqrt{2}$, are not rational numbers. Any number that can be represented on the number line that is not a rational number is called an **irrational number**. Irrational numbers are nonterminating, non-repeating decimal numbers. Thus, for example, $\sqrt{2}$ cannot be expressed exactly as a decimal number. Irrational numbers can only be *approximated* by decimal numbers. $\sqrt{2}$ is *approximately* 1.41. Thus we may write $\sqrt{2} \approx 1.41$. Some irrational numbers are illustrated on the number line in Figure 1.12. Rational and irrational numbers will be discussed further in later chapters.

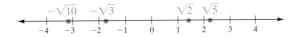

FIGURE 1.12

2 Know the Structure of the Real Numbers

Notice that many different types of numbers can be illustrated on a number line. Any number that can be represented on the number line is a **real number**.

Real numbers: {all numbers that can be represented on a real number line}

The symbol $\mathbb{R}$ is used to represent the set of real numbers. All the numbers mentioned thus far are real numbers. The natural numbers, the whole numbers, the integers, the rational numbers, and the irrational numbers are all real numbers. There are some types of numbers that are not real numbers, but these numbers are not discussed in this chapter. Figure 1.13 on page 34 illustrates the relationships between the various sets of numbers within the set of real numbers.

In Figure 1.13a, we can see that when we combine the rational numbers and the irrational numbers we get the real numbers. When we combine the integers with the noninteger rational numbers (such as $\frac{1}{2}$ and 0.42), we get the rational numbers. When we combine the whole numbers and the negative integers, we get the integers.

Consider the natural number 5. If we follow the branches in Figure 1.13a upward, we see that the number 5 is also a whole number, an integer, a rational number, and a real number. Now consider the number $\frac{1}{2}$. It belongs to the noninteger rational numbers. If we follow the branches upward, we can see that $\frac{1}{2}$ is also a rational number and a real number.

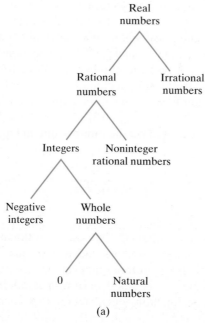

Real Numbers

Rational numbers	Irrational numbers
(Integers and noninteger rational numbers)	(Certain* square roots)
-12 4 0	$\sqrt{2}$ $\sqrt{5}$
$\frac{3}{8}$ $\frac{1}{3}$ -1.24	π $\sqrt{12}$
$-1\frac{3}{5}$ -2.463	

*Other higher roots like $\sqrt[3]{2}$ and $\sqrt[4]{5}$ are also irrational numbers.

(a)   (b)

FIGURE 1.13

EXAMPLE 1   Consider the following set of numbers.

$$\left\{ -2, -0.8, 4\frac{1}{2}, -59, \sqrt{3}, 0, 9, -\frac{4}{7}, -2.9, \sqrt{7}, -\sqrt{5} \right\}$$

List the elements of the set that are

a) natural numbers.   **b)** whole numbers.   **c)** integers.

d) rational numbers.   **e)** irrational numbers.   **f)** real numbers.

Solution   We will list the elements from left to right as they appear in the set. However, the elements may be listed in any order.

a) 9 **b)** $0, 9$ **c)** $-2, -59, 0, 9$ **d)** $-2, -0.8, 4\frac{1}{2}, -59, 0, 9, -\frac{4}{7}, -2.9$

NOW TRY EXERCISE 51   **e)** $\sqrt{3}, \sqrt{7}, -\sqrt{5}$ **f)** $-2, -0.8, 4\frac{1}{2}, -59, \sqrt{3}, 0, 9, -\frac{4}{7}, -2.9, \sqrt{7}, -\sqrt{5}$ ✳

2. the empty or null set   **3.** Answers will vary.   **4.** { } and ∅   **5.** 0 is an element of the whole numbers, but not of the natural numbers   **6.** the set of counting numbers, the set of positive integers   **7. a)** quotient of two integers, denominator not 0

Exercise Set 1.4

b) can be written with a denominator of 1

Concept/Writing Exercises

8. a)–c) All natural numbers are whole numbers, rational numbers, and real numbers.

1. What is a set?   a collection of elements

2. What is a set that contains no elements called?

3. Describe a set that is an empty set.

4. Give two symbols used to represent the empty set.

5. How do the sets of natural numbers and whole numbers differ?

6. What are two other names for the set of natural numbers?

7. a) What is a rational number?

     b) Explain why every integer is a rational number.

8. Explain why the natural number 7 is also a

     a) whole number.

     b) rational number.

     c) real number.

9. Is 0 a member of the set of

     a) integers?   Yes

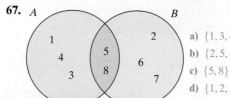

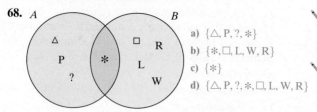

b) positive integers? No

c) negative integers? No

d) rational numbers? Yes

10. Write a paragraph or two describing the structure of the real number system. Explain how whole numbers, counting numbers, integers, rational numbers, irrational numbers, and real numbers are related.
Answers will vary.

Practice the Skills

In Exercise 11–15, list each set of numbers.

11. Integers. $\{\ldots, -3, -2, -1, 0, 1, 2, 3, \ldots\}$

12. Counting numbers. $\{1, 2, 3, 4, \ldots\}$

13. Whole numbers. $\{0, 1, 2, 3, \ldots\}$

14. Positive integers. $\{1, 2, 3, 4, \ldots\}$

15. Negative integers. $\{\ldots, -3, -2, -1\}$

16. Natural numbers. $\{1, 2, 3, 4, \ldots\}$

In Exercises 17–48, indicate whether each statement is true or false.

17. 0 is a whole number. T

18. -1 is a negative integer. T

19. -7.3 is a real number. T

20. $\frac{3}{5}$ is an integer. F

21. 0.6 is an integer. F

22. 0 is an integer. T

23. $\sqrt{2}$ is a rational number. F

24. $\sqrt{3}$ is a real number. T

25. $-\frac{1}{5}$ is a rational number. T

26. $-2\frac{1}{3}$ is a rational number. T

27. 0 is a rational number. T

28. -2.4 is a rational number. T

29. $4\frac{5}{8}$ is an irrational number. F

30. 0 is not a positive number. T

31. $-\frac{5}{3}$ is an irrational number. F

32. Every counting number is a rational number. T

33. The symbol $\varnothing$ is used to represent the empty set. T

34. Every integer is positive. F

35. Every real number is a rational number. F

36. Every negative integer is a real number. T

37. Every rational number is a real number. T

38. Every negative number is a negative integer. F

39. Some real numbers are not rational numbers. T

40. Some rational numbers are not real numbers. F

41. When zero is added to the set of counting numbers, the set of whole numbers is formed. T

42. All real numbers can be represented on a number line. T

43. The symbol $\mathbb{R}$ is used to represent the set of real numbers. T

44. Any number to the left of zero on a number line is a negative number. T

45. Every number greater than zero is a positive integer. F

46. Irrational numbers cannot be represented on a number line. F

47. When the negative integers, the positive integers, and 0 are combined, the integers are formed. T

48. The natural numbers, counting numbers, and positive integers are different names for the same set of numbers. T

49. We generally think of house addresses as being integers greater than 0. Have you ever seen a house number that was not an integer greater than 0? There are quite a few such addresses in cities and towns across America. The house numbers 0 and $2\frac{1}{2}$ appear on Legare Street in Charleston, SC. Considering the numbers 0 and $2\frac{1}{2}$, list those that are

a) integers 0

b) rational numbers $0, 2\frac{1}{2}$

c) real numbers $0, 2\frac{1}{2}$

50. Some of the older hotels in Europe have elevators that list negative numbers for floors below the lobby level. For example, a floor might be designated as -2. In the United States and in many countries, floor number 13 is omitted because of superstition. Considering the numbers -2 and 13, list those that are

a) positive integers 13

b) rational numbers $-2, 13$

c) real numbers $-2, 13$

d) whole numbers 13

51. Consider the following set of numbers.

$$\left\{ -\frac{5}{7}, 0, -2, 3, 6\frac{1}{4}, \sqrt{7}, -\sqrt{3}, 1.63, 77 \right\}$$

List the numbers that are

a) positive integers. $3, 77$

b) whole numbers. $0, 3, 77$

c) integers. $0, -2, 3, 77$

d) rational numbers. $-\frac{5}{7}, 0, -2, 3, 6\frac{1}{4}, 1.63, 77$

e) irrational numbers. $\sqrt{7}, -\sqrt{3}$

f) real numbers. $-\frac{5}{7}, 0, -2, 3, 6\frac{1}{4}, \sqrt{7}, -\sqrt{3}, 1.63, 77$

52. Consider the following set of numbers.

$$\left\{ -6, 7, 12.4, -\frac{9}{5}, -2\frac{1}{4}, \sqrt{3}, 0, 9, \sqrt{7}, 0.35 \right\}$$

List the numbers that are

a) positive integers. $7, 9$

b) whole numbers. $7, 0, 9$

c) integers. $-6, 7, 0, 9$

d) rational numbers. $-6, 7, 12.4, -\frac{9}{5}, -2\frac{1}{4}, 0, 9, 0.35$

e) irrational numbers. $\sqrt{3}, \sqrt{7}$

f) real numbers. $-6, 7, 12.4, -\frac{9}{5}, -2\frac{1}{4}, \sqrt{3}, 0, 9, \sqrt{7}, 0.35$

Problem Solving

55. $\sqrt{2}, \sqrt{3}, \sqrt{5}$ **59.** $-13, -5, -1$ **61.** $\sqrt{2}, \sqrt{3}, -\sqrt{5}$

Give three examples of numbers that satisfy the given conditions.   **53–64.** Answers will vary

53. An integer but not a negative integer. $0, 1, 2$

54. A real number but not an integer. $\frac{1}{2}, -\frac{1}{2}, 0.6$

55. An irrational number and a positive number.

56. A real number and a rational number. $1, \frac{1}{2}, -\frac{3}{5}$

57. A rational number but not an integer. $-\frac{2}{3}, \frac{1}{2}, 6.3$

58. An integer and a rational number. $-5, 0, 4$

59. A negative integer and a rational number.

60. A negative integer and a real number. $-1, -2, -3$

61. A real number but not a positive rational number.

62. A rational number but not a negative number. $1.5, 3, 6\frac{1}{4}$

63. A real number but not an irrational number. $-7, 1, 5$

64. A negative number but not a negative integer. $-\frac{1}{2}, -\frac{5}{8}, -\sqrt{5}$

Three dots inside a set indicate that the set continues in the same manner. For example, $\{1, 2, 3, \ldots, 84\}$ is the set of natural numbers from 1 up to and including 84. In Exercises 65 and 66, determine the number of elements in each set.

65. $\{8, 9, 10, 11, \ldots, 94\}$ 87

66. $\{-4, -3, -2, -1, 0, 1, \ldots, 64\}$ 69

Challenge Problems

*The diagrams in Exercises 67 and 68 are called Venn diagrams (named after the English mathematician John Venn). Venn diagrams are used to illustrate sets. For example, in the diagrams, circle A contains all the elements in set A, and circle B contains all the elements in set B. For each diagram, determine **a)** set A, **b)** set B, **c)** the set of elements that belong to both set A and set B, and **d)** the set of elements that belong to either set A or set B.*

67.

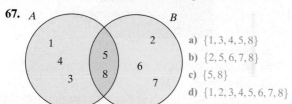

a) $\{1, 3, 4, 5, 8\}$
b) $\{2, 5, 6, 7, 8\}$
c) $\{5, 8\}$
d) $\{1, 2, 3, 4, 5, 6, 7, 8\}$

68.

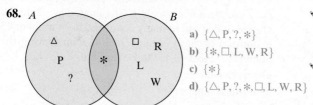

a) $\{\triangle, P, ?, *\}$
b) $\{*, \square, L, W, R\}$
c) $\{*\}$
d) $\{\triangle, P, ?, *, \square, L, W, R\}$

69. Consider the sets $A = \{1, 2, 3, 4\}$ and $B = \{1, 2, 3, 4, \ldots\}$.
a) Explain the difference between set A and set B.
b) How many elements are in set A? 4
c) How many elements are in set B? infinite number
d) Set A is an example of a *finite set*. Can you guess the name given to a set like set B? infinite set

70. How many decimal numbers are there
a) between 1.0 and 2.0;
b) between 1.4 and 1.5? Explain your answer.

71. How many fractions are there
a) between 1 and 2;
b) between $\frac{1}{3}$ and $\frac{1}{5}$? Explain your answer.

69. a) set B continues beyond 4   **70. a)** and **b)** an infinite number   **71. a)** and **b)** an infinite number

Group Activity

Discuss and answer Exercise 72 as a group.

72. Set A **union** set B, symbolized $A \cup B$, consists of the set of elements that belong to set A or set B (or both sets). Set A **intersection** set B, symbolized $A \cap B$, consists of the set of elements that both set A and set B have in common. Note that the elements that belong to both sets are listed only once in the union of the sets.

Consider the pairs of sets below.

Group member 1: $A = \{2, 3, 4, 6, 8, 9\}$ $B = \{1, 2, 3, 5, 7, 8\}$
Group member 2: $A = \{a, b, c, d, g, i, j\}$ $B = \{b, c, d, h, m, p\}$
Group member 3: $A = \{red, blue, green, yellow\}$ $B = \{pink, orange, purple\}$

a) Group member 1: Find the union and intersection of the sets marked Group member 1.

b) Group member 2: Find the union and intersection of the sets marked Group member 2.

c) Group member 3: Find the union and intersection of the sets marked Group member 3.

d) Now as a group, check each other's work. Correct any mistakes.

e) As a group, using group member 1's sets, construct a Venn diagram like those shown in Exercises 67 and 68.

72. a) $\cup : \{1, 2, 3, 4, 5, 6, 7, 8, 9\}, \cap : \{2, 3, 8\}$ **b)** $\cup : \{a, b, c, d, g, h, i, j, m, p\}, \cap : \{b, c, d\}$

Cumulative Review Exercises

c) $\cup : \{red, blue, green, yellow, pink, orange, purple\}, \cap : \varnothing$

[1.3] 73. Convert $5\frac{2}{5}$ to a fraction. $\frac{27}{5}$

74. Write $\frac{16}{3}$ as a mixed number. $5\frac{1}{3}$

75. Add $\frac{3}{5} + \frac{5}{8}$. $\frac{49}{40}$ or $1\frac{9}{40}$

76. Multiply $\left(\frac{5}{9}\right)\left(4\frac{2}{3}\right)$. $\frac{70}{27}$ or $2\frac{16}{27}$

1.5 INEQUALITIES

SSM   Study Guide   CD/Video

MathPro 4/5   PH Math Tutor Center   prenhall.com/Angel

1   Determine which is the greater of two numbers.

2   Find the absolute value of a number.

1   Determine Which Is the Greater of Two Numbers

TEACHING TIP
Point out that the arrowhead on the number line pointing to larger numbers is similar to the greater than symbol and the arrowhead pointing to smaller numbers is similar to the less than symbol.

The number line, which shows numbers increasing from left to right, can be used to explain inequalities (see Fig. 1.14). When comparing two numbers, **the number to the right on the number line is the greater number, and the number to the left is the lesser number.** The symbol $>$ is used to represent the words "is greater than." The symbol $<$ is used to represent the words "is less than."

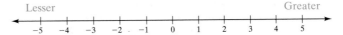

Lesser   Greater

$-5 \quad -4 \quad -3 \quad -2 \quad -1 \quad 0 \quad 1 \quad 2 \quad 3 \quad 4 \quad 5$

FIGURE 1.14

The statement that the number 3 is greater than the number 2 is written $3 > 2$. Notice that 3 is to the right of 2 on the number line. The statement that the number 0 is greater than the number -1 is written $0 > -1$. Notice that 0 is to the right of -1 on the number line.

Instead of stating that 3 is greater than 2, we could state that 2 is less than 3, written $2 < 3$. Notice that 2 is to the left of 3 on the number line. The statement that the number -1 is less than the number 0 is written $-1 < 0$. Notice that -1 is to the left of 0 on the number line.

EXAMPLE 1   Insert either $>$ or $<$ in the shaded area between each pair of numbers to make a true statement.

a) $-4 \;\blacksquare\; -2$ **b)** $-\frac{3}{2} \;\blacksquare\; 2.5$ **c)** $\frac{1}{2} \;\blacksquare\; \frac{1}{4}$ **d)** $-2 \;\blacksquare\; 4$

Solution The points given are shown on the number line (Fig. 1.15).

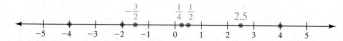

FIGURE 1.15

a) $-4 < -2$; notice that -4 is to the left of -2.

b) $-\dfrac{3}{2} < 2.5$; notice that $-\dfrac{3}{2}$ is to the left of 2.5.

c) $\dfrac{1}{2} > \dfrac{1}{4}$; notice that $\dfrac{1}{2}$ is to the right of $\dfrac{1}{4}$.

d) $-2 < 4$; notice that -2 is to the left of 4.

EXAMPLE 2 Insert either $>$ or $<$ in the shaded area between each pair of numbers to make a true statement.

a) $-1 \ \blacksquare \ -2$ **b)** $-1 \ \blacksquare \ 0$ **c)** $-2 \ \blacksquare \ 2$ **d)** $-4.09 \ \blacksquare \ -4.9$

Solution The numbers given are shown on the number line (Fig. 1.16).

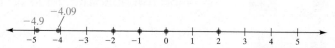

FIGURE 1.16

a) $-1 > -2$; notice that -1 is to the right of -2.

b) $-1 < 0$; notice that -1 is to the left of 0.

c) $-2 < 2$; notice that -2 is to the left of 2.

NOW TRY EXERCISE 29 **d)** $-4.09 > -4.9$; notice that -4.09 is to the right of -4.9.

2 Find the Absolute Value of a Number

The concept of absolute value can be explained with the help of the number line shown in Figure 1.17. The **absolute value** of a number can be considered the distance between the number and 0 on a number line. Thus, the absolute value of 3, written $|3|$, is 3 since it is 3 units from 0 on a number line. Similarly, the absolute value of -3, written $|-3|$, is also 3 since -3 is 3 units from 0.

$$|3| = 3 \quad \text{and} \quad |-3| = 3$$

FIGURE 1.17

Since the absolute value of a number measures the distance (without regard to direction) of a number from 0 on the number line, *the absolute value of every number will be either positive or zero.*

Number	Absolute Value of Number
6	$\lvert 6 \rvert = 6$
-6	$\lvert -6 \rvert = 6$
0	$\lvert 0 \rvert = 0$
$-\dfrac{1}{2}$	$\left\lvert -\dfrac{1}{2} \right\rvert = \dfrac{1}{2}$

The negative of the absolute value of a nonzero number will always be a negative number. For example,

$$-\lvert 2 \rvert = -(2) = -2 \quad \text{and} \quad -\lvert -3 \rvert = -(3) = -3$$

EXAMPLE 3   Insert either $>$, $<$, or $=$ in each shaded area to make a true statement.

a) $\lvert 3 \rvert \; \blacksquare \; 3$ **b)** $\lvert -2 \rvert \; \blacksquare \; \lvert 2 \rvert$ **c)** $-2 \; \blacksquare \; \lvert -4 \rvert$ **d)** $\lvert -5 \rvert \; \blacksquare \; 0$ **e)** $\left\lvert -\dfrac{4}{9} \right\rvert \; \blacksquare \; \lvert -18 \rvert$

Solution   **a)** $\lvert 3 \rvert = 3$.   **b)** $\lvert -2 \rvert = \lvert 2 \rvert$, since both $\lvert -2 \rvert$ and $\lvert 2 \rvert$ equal 2.

c) $-2 < \lvert -4 \rvert$, since $\lvert -4 \rvert = 4$.   **d)** $\lvert -5 \rvert > 0$, since $\lvert -5 \rvert = 5$.

e) $\left\lvert -\dfrac{4}{9} \right\rvert < \lvert -18 \rvert$, since $\left\lvert -\dfrac{4}{9} \right\rvert = \dfrac{4}{9}$ and $\lvert -18 \rvert = 18$.

NOW TRY EXERCISE 57

We will use absolute value in Sections 1.6 and 1.7 to add and subtract real numbers. The concept of absolute value is very important in higher-level mathematics courses. If you take a course in intermediate algebra, you will learn a more formal definition of absolute value.

1. a) and b)

2. the distance between the number and 0 on a number line   **3. a)** 4 is 4 units from 0.   **b)** -4 is 4 units from 0.   **c)** 0 is 0 units from 0.
4. Yes, 0. The absolute value of 0 is 0.   **7.** No, $-3 > -4$ but $\lvert -3 \rvert < \lvert -4 \rvert$.   **8.** No, $\lvert -3 \rvert < \lvert -4 \rvert$ but $-3 > -4$.
9. No, $\lvert -4 \rvert > \lvert -3 \rvert$ but $-4 < -3$.   **10.** No, $-4 < -3$ but $\lvert -4 \rvert > \lvert -3 \rvert$.

Exercise Set 1.5

Concept/Writing Exercises

1. a) Draw a number line.

b) Mark the numbers -2 and -4 on your number line.

c) Is -2 less than -4 or is -2 greater than -4? Explain. $>$

In parts **d)** and **e)** write a correct statement using -2 and -4 and the symbol given.

d) $<$ $-4 < -2$

e) $>$ $-2 > -4$

2. What is the absolute value of a number?

3. a) Explain why the absolute value of 4, $\lvert 4 \rvert$, is 4.

b) Explain why the absolute value of -4, $\lvert -4 \rvert$, is 4.

c) Explain why the absolute value of 0, $\lvert 0 \rvert$, is 0.

4. Are there any real numbers whose absolute value is not a positive number? Explain your answer.

5. Suppose a and b represent any two real numbers. If $a > b$ is true, will $b < a$ also be true? Explain and give some examples using specific numbers for a and b.   yes

6. Will $\lvert a \rvert - \lvert a \rvert = 0$ always be true for any real number a? Explain.   yes

7. Suppose a and b represent any two real numbers. Suppose $a > b$ is true. Will $\lvert a \rvert > \lvert b \rvert$ also be true? Explain and give an example to support your answer.

8. Suppose a and b represent any two real numbers. Suppose $\lvert a \rvert < \lvert b \rvert$ is true. Will $a < b$ also be true? Explain and give an example to support your answer.

9. Suppose a and b represent any two real numbers. Suppose $\lvert a \rvert > \lvert b \rvert$ is true. Will $a > b$ also be true? Explain and give an example to support your answer.

10. Suppose a and b represent any two real numbers. Suppose $a < b$ is true. Will $\lvert a \rvert < \lvert b \rvert$ also be true? Explain and give an example to support your answer.

Practice the Skills

Evaluate.

11. $\lvert 7 \rvert$ 7   **12.** $\lvert -6 \rvert$ 6   **13.** $\lvert -15 \rvert$ 15   **14.** $\lvert -10 \rvert$ 10   **15.** $\lvert 0 \rvert$ 0

16. $\lvert 54 \rvert$ 54   **17.** $-\lvert -5 \rvert$ -5 **18.** $-\lvert 92 \rvert$ -92 **19.** $-\lvert 21 \rvert$ -21 **20.** $-\lvert -34 \rvert$ -34

Insert either < or > in each shaded area to make a true statement.

21. $5 \;\blacksquare\; 2 \quad >$

22. $4 \;\blacksquare\; -2 \quad >$

23. $-6 \;\blacksquare\; 0 \quad <$

24. $-6 \;\blacksquare\; -4 \quad <$

🔒 **25.** $\dfrac{1}{2} \;\blacksquare\; -\dfrac{2}{3} \quad >$

26. $\dfrac{3}{5} \;\blacksquare\; \dfrac{4}{5} \quad <$

27. $0.7 \;\blacksquare\; 0.8 \quad <$

28. $-0.2 \;\blacksquare\; -0.4 \quad >$

29. $-\dfrac{1}{2} \;\blacksquare\; -1 \quad >$

30. $-0.1 \;\blacksquare\; -0.9 \quad >$

31. $3 \;\blacksquare\; -3 \quad >$

32. $-\dfrac{3}{4} \;\blacksquare\; -1 \quad >$

🔒 **33.** $-2.1 \;\blacksquare\; -2 \quad <$

34. $-1.83 \;\blacksquare\; -1.82 \quad <$

35. $\dfrac{4}{5} \;\blacksquare\; -\dfrac{4}{5} \quad >$

36. $-9 \;\blacksquare\; -12 \quad >$

37. $-\dfrac{3}{8} \;\blacksquare\; \dfrac{3}{8} \quad <$

38. $-4.09 \;\blacksquare\; -5.3 \quad >$

39. $0.49 \;\blacksquare\; 0.43 \quad >$

40. $-1.0 \;\blacksquare\; -0.7 \quad <$

41. $5 \;\blacksquare\; -7 \quad >$

42. $0.001 \;\blacksquare\; 0.002 \quad <$

43. $-0.006 \;\blacksquare\; -0.007 \quad >$

44. $\dfrac{1}{2} \;\blacksquare\; -\dfrac{1}{2} \quad >$

45. $\dfrac{5}{8} \;\blacksquare\; 0.6 \quad >$

46. $2.7 \;\blacksquare\; \dfrac{10}{3} \quad <$

47. $-\dfrac{4}{3} \;\blacksquare\; -\dfrac{1}{3} \quad <$

48. $\dfrac{9}{2} \;\blacksquare\; \dfrac{7}{2} \quad >$

49. $-\dfrac{1}{2} \;\blacksquare\; -\dfrac{3}{2} \quad >$

50. $-0.4 \;\blacksquare\; -0.5 \quad >$

51. $0.3 \;\blacksquare\; \dfrac{1}{3} \quad <$

52. $\dfrac{9}{20} \;\blacksquare\; 0.42 \quad >$

53. $\dfrac{13}{15} \;\blacksquare\; \dfrac{8}{9} \quad <$

54. $-\dfrac{17}{30} \;\blacksquare\; -\dfrac{16}{20} \quad >$

55. $-(-6) \;\blacksquare\; -(-5) \quad >$

56. $-\left(-\dfrac{12}{13}\right) \;\blacksquare\; \dfrac{7}{8} \quad >$

Insert either <, >, or = in each shaded area to make a true statement.

57. $3 \;\blacksquare\; |-2| \quad >$

58. $|-8| \;\blacksquare\; |-7| \quad >$

59. $|-4| \;\blacksquare\; \dfrac{2}{3} \quad >$

60. $|-4| \;\blacksquare\; -3 \quad >$

61. $|0| \;\blacksquare\; |-4| \quad <$

62. $|-2.1| \;\blacksquare\; |-1.8| \quad >$

🔒 **63.** $4 \;\blacksquare\; \left|-\dfrac{9}{2}\right| \quad <$

64. $-5 \;\blacksquare\; |5| \quad <$

65. $\left|-\dfrac{4}{5}\right| \;\blacksquare\; \left|-\dfrac{5}{4}\right| \quad <$

66. $\left|\dfrac{2}{5}\right| \;\blacksquare\; |-0.40| \quad =$

67. $|-4.6| \;\blacksquare\; \left|-\dfrac{23}{5}\right| \quad =$

68. $\left|-\dfrac{8}{3}\right| \;\blacksquare\; |-3.5| \quad <$

Insert either >, <, or = in each shaded area to make a true statement.

69. $\dfrac{2}{3} + \dfrac{2}{3} + \dfrac{2}{3} + \dfrac{2}{3} \;\blacksquare\; 4 \cdot \dfrac{2}{3} \quad =$

70. $\dfrac{2}{3} \cdot \dfrac{2}{3} \;\blacksquare\; \dfrac{2}{3} + \dfrac{2}{3} \quad <$

🔒 **71.** $\dfrac{1}{2} \cdot \dfrac{1}{2} \;\blacksquare\; \dfrac{1}{2} \div \dfrac{1}{2} \quad <$

72. $5 \div \dfrac{2}{3} \;\blacksquare\; \dfrac{2}{3} \div 5 \quad >$

73. $\dfrac{5}{8} - \dfrac{1}{2} \;\blacksquare\; \dfrac{5}{8} \div \dfrac{1}{2} \quad <$

74. $2\dfrac{1}{3} \cdot \dfrac{1}{2} \;\blacksquare\; 2\dfrac{1}{3} + \dfrac{1}{2} \quad <$

Arrange the numbers from smallest to largest.

75. $|-6|, -|-8|, \dfrac{3}{5}, \dfrac{4}{9}, 0.38 \quad -|-8|, 0.38, \dfrac{4}{9}, \dfrac{3}{5}, |-6|$

76. $-\dfrac{3}{4}, -|0.6|, -\dfrac{5}{9}, -1.74, |-1.9| \quad -1.74, -\dfrac{3}{4}, -|0.6|, -\dfrac{5}{9}, |-1.9|$

77. $\dfrac{2}{3}, 0.6, |-2.6|, \dfrac{19}{25}, \dfrac{5}{12} \quad \dfrac{5}{12}, 0.6, \dfrac{2}{3}, \dfrac{19}{25}, |-2.6|$

78. $-|-5|, |-9|, \left|-\dfrac{12}{5}\right|, 2.7, \dfrac{7}{12} \quad -|-5|, \dfrac{7}{12}, \left|-\dfrac{12}{5}\right|, 2.7, |-9|$

Problem Solving

79. What numbers are 4 units from 0 on a number line?
$4, -4$

80. What numbers are 5 units from 0 on the number line?
$5, -5$

In Exercises 81–88, give three real numbers that satisfy all the stated criteria. If no real numbers satisfy the criteria, so state and explain why.

81. greater than 4 and less than 6

82. less than -2

🔒 **83.** less than -2 and greater than -6

84. less than 4 and greater than 6 not possible

81. Answers will vary; $4\dfrac{1}{2}, 5, 5.5$.   **82.** Answers will vary; $-3, -4, -5$.   **83.** Answers will vary; $-3, -4, -5$.

85. Answers will vary: 4, 5, 6. **86.** Answers will vary: −4, −5, −6. **87.** Answers will vary: 3, 4, 5.

85. greater than −3 and greater than 3

86. less than −3 and less than 3

87. greater than |−2| and less than |−6|

88. greater than |−3| and less than |3| not possible

89. a) Consider the word *between*. What does this word mean?   does not include the endpoints

   b) List three real numbers between 4 and 6   4.1, 5, 5½

   c) Is the number 4 between the numbers 4 and 6? Explain.   no

   d) Is the number 5 between the numbers 4 and 6? Explain.   yes

   e) Is it true or false that the real numbers between 4 and 6 are the real numbers that are both greater than 4 and less than 6? Explain.   true

90. The following graph shows the U.S. population from 1790 to 2000.

U.S. Population 1790-2000

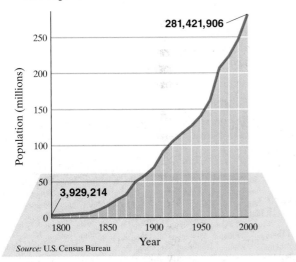

Source: U.S. Census Bureau

   a) Estimate the ten-year period when the U.S. population was first greater than 100 million people.   1920–1930

91. The following graph shows the average cost of a gallon of gasoline in the United States from January, 1999, through June, 2000. The graph also shows the components that make up the price of a gallon of gas.

b) Estimate the ten-year period when the U.S. population was first greater than 200 million people.   1970–1980

c) Estimate the ten year periods when the U.S. population was greater than 50 million and less than 150 million people.   1880–1950

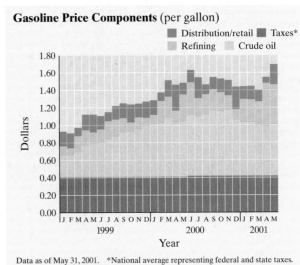

Gasoline Price Components (per gallon)

Data as of May 31, 2001.   *National average representing federal and state taxes.
Source: Energy Information Administration

   a) During this time period, when were the taxes less than 50 cents?   all the time

   b) During this time period, when was the cost of a gallon of gasoline greater than $1.50 for the first time?   March, 2000

   c) During this time period, when was the cost of a gallon of gas less than $1.00?   January–March, 1999

   d) During this time period, when was the cost of a gallon of gasoline greater than or equal to $1.40 but less than or equal to $1.60?   March, 2000–May, 2001, except June 2000

Challenge Problems   **93.** greater than   **96. a)** x **b)** $-x$ **c)** $x, -x$

92. A number greater than 0 and less than 1 (or between 0 and 1) is multiplied by itself. Will the product be less than, equal to, or greater than the original number selected? Explain why this is always true.   less than

93. A number between 0 and 1 is divided by itself. Will the quotient be less than, equal to, or greater than the original number selected? Explain why this is always true.

94. What two numbers can be substituted for x to make $|x| = 3$ a true statement?   3, −3

95. Are there any values for x that would make $|x| = -|x|$ a true statement?   yes, 0

96. a) To what is $|x|$ equal if x represents a real number greater than or equal to 0?   **b)** To what is $|x|$ equal if x represents a real number less than 0?   **c)** Fill in the following shaded areas to make a true statement.

   c) $|x| = \begin{cases} \blacksquare, & x \geq 0 \\ \blacksquare, & x < 0 \end{cases}$

Group Activity

Discuss and answer Exercise 97 as a group.

97. a) Group member 1: Draw a number line and mark points on the line to represent the following numbers.

$$|-2|, \quad -|3|, \quad -\left|\frac{1}{3}\right|$$

b) Group member 2: Do the same as in part **a)**, but on your number line mark points for the following numbers.

$$|-4|, \quad -|2|, \quad \left|-\frac{3}{5}\right|$$

c) Group member 3: Do the same as in parts **a)** and **b)**, but mark points for the following numbers.

$$|0|, \quad \left|\frac{16}{5}\right|, \quad -|-3|$$

d) As a group, construct one number line that contains all the points listed in parts **a)**, **b)**, and **c)**.

(a)–(d)

Cumulative Review Exercises

[1.3] **98.** Subtract $1\frac{2}{3} - \frac{3}{8}$. $\frac{31}{24}$ or $1\frac{7}{24}$

[1.4] **99.** List the set of whole numbers. $\{0, 1, 2, 3, \ldots\}$

100. List the set of counting numbers. $\{1, 2, 3, 4, \ldots\}$

101. Consider the following set of numbers.

$$\left\{ 5, -2, 0, \frac{1}{3}, \sqrt{3}, -\frac{5}{9}, 2.3 \right\}$$

List the numbers that are

a) natural numbers.   5

b) whole numbers.   5, 0

c) integers.   5, −2, 0

d) rational numbers. $5, -2, 0, \frac{1}{3}, -\frac{5}{9}, 2.3$

e) irrational numbers. $\sqrt{3}$

f) real numbers. $5, -2, 0, \frac{1}{3}, \sqrt{3}, -\frac{5}{9}, 2.3$

1.6 ADDITION OF REAL NUMBERS

SSM

Study Guide

CD/Video

MathPro 4/5

PH Math Tutor Center

prenhall.com/Angel

1   Add real numbers using a number line.

2   Add fractions.

3   Identify opposites.

4   Add using absolute values.

5   Add using calculators.

There are many practical uses for negative numbers. A submarine diving below sea level, a bank account that has been overdrawn, a business spending more than it earns, and a temperature below zero are some examples.

The four basic **operations** of arithmetic are addition, subtraction, multiplication, and division. In the next few sections we will explain how to add, subtract, multiply, and divide numbers. We will consider both positive and negative numbers. In this section we discuss the operation of addition.

1   Add Real Numbers Using a Number Line

To add numbers, we make use of a number line. Represent the first number to be added (first *addend*) by an arrow starting at 0. The arrow is drawn to the right if the number is positive. If the number is negative, the arrow is drawn to the left. From the tip of the first arrow, draw a second arrow to represent the second addend. The second arrow is drawn to the right or left, as just explained. The sum of the two numbers is found at the tip of the second arrow. Note that with the exception of 0, *any number without a sign in front of it is positive.* For example, 3 means +3 and 5 means +5.

EXAMPLE 1   Evaluate $3 + (-4)$ using a number line.

Solution   *Always begin at 0.* Since the first addend, the 3, is positive, the first arrow starts at 0 and is drawn 3 units to the right (Fig. 1.18).

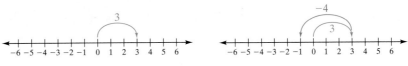

FIGURE 1.18   FIGURE 1.19

The second arrow starts at 3 and is drawn 4 units to the left, since the second addend is negative (Fig. 1.19). The tip of the second arrow is at -1. Thus

$$3 + (-4) = -1$$

EXAMPLE 2   Evaluate $-4 + 2$ using a number line.

Solution   Begin at 0. Since the first addend is negative, -4, the first arrow is drawn 4 units to the left. From there, since 2 is positive, the second arrow is drawn 2 units to the right. The second arrow ends at -2 (Fig. 1.20).

FIGURE 1.20

$$-4 + 2 = -2$$

EXAMPLE 3   Evaluate $-3 + (-2)$ using a number line.

Solution   Start at 0. Since both numbers being added are negative, both arrows will be drawn to the left (Fig. 1.21).

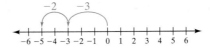

FIGURE 1.21

$$-3 + (-2) = -5$$

In Example 3, we can think of the expression $-3 + (-2)$ as combining a *loss* of 3 and a *loss* of 2 for a total *loss* of 5, or -5.

EXAMPLE 4   Add $5 + (-5)$ using a number line.

Solution   The first arrow starts at 0 and is drawn 5 units to the right. The second arrow starts at 5 and is drawn 5 units to the left. The tip of the second arrow is at 0. Thus, $5 + (-5) = 0$ (Fig. 1.22).

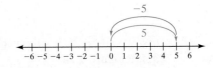

FIGURE 1.22

$$5 + (-5) = 0$$

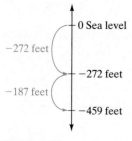

FIGURE 1.23

EXAMPLE 5   **Depth of a Submarine**   A submarine dives 272 feet. Later it dives an additional 187 feet. Find the depth of the submarine (assume that depths below sea level are indicated by negative numbers).

Solution   A vertical number line (Fig. 1.23) may help you visualize this problem.

NOW TRY EXERCISE 117 $-272 + (-187) = -459$ feet   ✳

2   Add Fractions

We introduced fractions, and explained how to add positive fractions, in Section 1.3. To add fractions, where one or more of the fractions is negative, we use the same general procedure discussed in Section 1.3. Whenever the denominators are not the same, we must obtain a common denominator. Once we obtain the common denominator, we obtain the answer by adding the numerators while keeping the common denominator. For example, suppose after obtaining a common denominator, we have $-\frac{19}{29} + \frac{13}{29}$. To obtain the numerator of the answer, we may add $-19 + 13$ on the number line to obtain -6. The denominator of the answer is the common denominator, 29. Thus the answer is $-\frac{6}{29}$. We show these calculations as follows.

$$-\frac{19}{29} + \frac{13}{29} = \frac{-19 + 13}{29} = \frac{-6}{29} = -\frac{6}{29}$$

Let's look at one more example. Suppose after obtaining a common denominator, we have $-\frac{7}{40} + \left(-\frac{5}{40}\right)$. To obtain the numerator of the answer, we may add $-7 + (-5)$ on the number line to obtain -12. The denominator of the answer is 40. Thus, the answer before being simplified is $-\frac{12}{40}$. The final answer simplifies to $-\frac{3}{10}$. We show these calculations as follows.

$$-\frac{7}{40} + \left(-\frac{5}{40}\right) = \frac{-7 + (-5)}{40} = \frac{-12}{40} = -\frac{3}{10}$$

EXAMPLE 6   Add $-\dfrac{5}{12} + \dfrac{4}{5}$.

Solution   First, we determine that the least common denominator is 60. Changing each fraction to a fraction with a denominator of 60 yields

$$-\frac{5}{12} \cdot \boxed{\frac{5}{5}} + \frac{4}{5} \cdot \boxed{\frac{12}{12}}$$

$$\text{or} \quad -\frac{25}{60} + \frac{48}{60}$$

Now the fractions being added have a common denominator. To add these fractions, we keep the common denominator and add the numerators to get

$$-\frac{5}{12} + \frac{4}{5} = -\frac{25}{60} + \frac{48}{60} = \frac{-25 + 48}{60}$$

Now we add $-25 + 48$ on the number line to get the numerator of the fraction, 23; see Figure 1.24.

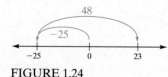

FIGURE 1.24

Thus $\dfrac{-25 + 48}{60} = \dfrac{23}{60}$ and $-\dfrac{5}{12} + \dfrac{4}{5} = \dfrac{23}{60}$.   ✳

EXAMPLE 7   Add $-\dfrac{7}{8} + \left(-\dfrac{3}{40}\right)$.

Solution   The least common denominator (or LCD) is 40. Rewriting each fraction with the LCD gives the following.

$$-\frac{7}{8} + \left(-\frac{3}{40}\right) = -\frac{7}{8} \cdot \boxed{\frac{5}{5}} + \left(-\frac{3}{40}\right)$$

$$= -\frac{35}{40} + \left(-\frac{3}{40}\right) = \frac{-35 + (-3)}{40}$$

Now we add $-35 + (-3)$ to get the numerator of the fraction, -38; see Figure 1.25.

FIGURE 1.25

NOW TRY EXERCISE 83   Thus $-\dfrac{7}{8} + \left(-\dfrac{3}{40}\right) = \dfrac{-35 + (-3)}{40} = -\dfrac{38}{40} = -\dfrac{19}{20}$.

HELPFUL HINT

In Example 6, Figure 1.24, we illustrated that starting at 0 and moving 25 units to the left, then moving 48 units to the right, ends at a positive 23. Thus $-25 + 48 = 23$.

When the numbers are large, you are not expected to actually mark and count the units. For example, you will not need to count 48 units to the right from -25 to obtain the answer 23.

In objective 3, we will show you how to obtain $-25 + 48$ without having to draw number lines. We present adding on the number line here to help you understand the concept of addition of signed numbers, and to help you in determining, without doing any calculations, whether the sum of two signed numbers will be a positive number, a negative number, or zero.

---

3 Identify Opposites

Now let's consider **opposites**, or **additive inverses**.

DEFINITION

Any two numbers whose sum is zero are said to be **opposites** (or **additive inverses**) of each other. In general, if we let a represent any real number, then its opposite is $-a$ and $a + (-a) = 0$.

In Example 4 the sum of 5 and -5 is 0. Thus -5 is the opposite of 5 and 5 is the opposite of -5.

EXAMPLE 8   Find the opposite of each number. **a)** 3 **b)** -4 **c)** $-\dfrac{7}{8}$

Solution   **a)** The opposite of 3 is -3, since $3 + (-3) = 0$.

b) The opposite of -4 is 4, since $-4 + 4 = 0$.

NOW TRY EXERCISE 15   **c)** The opposite of $-\frac{7}{8}$ is $\frac{7}{8}$, since $-\frac{7}{8} + \frac{7}{8} = 0$.

4 Add Using Absolute Values

Now that we have had some practice adding signed numbers on a number line, we give a rule (in two parts) for using absolute value to add signed numbers. Remember that the absolute value of a nonzero number will always be positive. The first part of the rule follows.

> **To add real numbers with the same sign** (either both positive or both negative), add their absolute values. The sum has the same sign as the numbers being added.

EXAMPLE 9 Add $4 + 8$.

Solution Since both numbers have the same sign, both positive, we add their absolute values: $|4| + |8| = 4 + 8 = 12$. Since both numbers being added are positive, the sum is positive. Thus $4 + 8 = 12$. ✳

EXAMPLE 10 Add $-6 + (-9)$.

Solution Since both numbers have the same sign, both negative, we add their absolute values: $|-6| + |-9| = 6 + 9 = 15$. Since both numbers being added are negative, their sum is negative. Thus $-6 + (-9) = -15$. ✳

The sum of two positive numbers will always be positive and the sum of two negative numbers will always be negative.

> **To add two signed numbers with different signs** (one positive and the other negative), subtract the smaller absolute value from the larger absolute value. The answer has the sign of the number with the larger absolute value.

EXAMPLE 11 Add $10 + (-6)$.

Solution The two numbers being added have different signs, so we subtract the smaller absolute value from the larger: $|10| - |-6| = 10 - 6 = 4$. Since $|10|$ is greater than $|-6|$ and the sign of 10 is positive, the sum is positive. Thus, $10 + (-6) = 4$. ✳

EXAMPLE 12 Add $12 + (-18)$.

Solution The numbers being added have different signs, so we subtract the smaller absolute value from the larger: $|-18| - |12| = 18 - 12 = 6$. Since $|-18|$ is greater than $|12|$ and the sign of -18 is negative, the sum is negative. Thus, $12 + (-18) = -6$. ✳

EXAMPLE 13 Add $-21 + 20$

Solution The two numbers being added have different signs, so we subtract the smaller absolute value from the larger: $|-21| - |20| = 21 - 20 = 1$. Since $|-21|$ is greater than $|20|$, the sum is negative. Therefore, $-21 + 20 = -1$. ✳

NOW TRY EXERCISE 45 Now let's look at some examples that contain fractions and decimal numbers.

EXAMPLE 14 Add $-\dfrac{3}{5} + \dfrac{4}{7}$

Solution We can write each fraction with the least common denominator, 35.

$$-\frac{3}{5} + \frac{4}{7} = -\frac{3}{5} \cdot \frac{7}{7} + \frac{4}{7} \cdot \frac{5}{5}$$

$$= \frac{-21}{35} + \frac{20}{35} = \frac{-21 + 20}{35}$$

Since $|-21|$ is greater than $|20|$, the final answer will be negative. In Example 13 we found that $-21 + 20 = -1$. Thus, we can write

$$\frac{-21}{35} + \frac{20}{35} = \frac{-21 + 20}{35} = \frac{-1}{35} = -\frac{1}{35}$$

Thus $-\frac{3}{5} + \frac{4}{7} = -\frac{1}{35}$. ✳

EXAMPLE 15 Add $-24.23 + (-17.96)$.

Solution Since both numbers have the same sign, both negative, we add their absolute values: $|-24.23| + |-17.96| = 24.23 + 17.96$.

$$\begin{array}{r} 24.23 \\ + 17.96 \\ \hline 42.19 \end{array}$$

Since two negative numbers are being added, the sum is negative. Therefore, $-24.23 + (-17.96) = -42.19$. ✳

NOW TRY EXERCISE 57

The sum of two signed numbers with different signs may be either positive or negative. The sign of the sum will be the same as the sign of the number with the larger absolute value.

HELPFUL HINT Architects often make a scale model of a building before starting construction of the building. This "model" helps them visualize the project and often helps them avoid problems.

Mathematicians also construct models. A mathematical *model* may be a physical representation of a mathematical concept. It may be as simple as using tiles or chips to represent specific numbers. For example, below we use a model to help explain addition of real numbers. This may help some of you understand the concepts better.

We let a red chip represent $+1$ and a green chip represent -1.

● = +1   ● = -1

If we add $+1$ and -1, or a red and a green chip, we get 0.
Now consider the addition problem $3 + (-5)$. We can represent this as

● ● ● + ● ● ● ● ●
‾3‾   ‾‾‾-5‾‾‾

If we remove 3 red chips and 3 green chips, or three zeros, we are left with 2 green chips, which represents a sum of -2. Thus, $3 + (-5) = -2$,

⊘ ⊘ ⊘ + ⊘ ⊘ ⊘ ● ●

(continued on the next page)

Now consider the problem $-4 + (-2)$. We can represent this as

$$\underbrace{\bullet\ \bullet\ \bullet\ \bullet}_{-4} + \underbrace{\bullet\ \bullet}_{-2}$$

Since we end up with 6 green chips, and each green chip represents -1, the sum is -6. Therefore, $-4 + (-2) = -6$.

EXAMPLE 16   **Net Profit or Loss**   The B.J. Donaldson Printing Company had a loss of $4000 for the first 6 months of the year and a profit of $29,500 for the second 6 months of the year. Find the net profit or loss for the year.

Solution   Understand and Translate   This problem can be represented as $-4000 + 29,500$. Since the two numbers being added have different signs, subtract the smaller absolute value from the larger.

Carry Out $|29{,}500| - |-4000| = 29{,}500 - 4{,}000 = 25{,}500$

Thus, $-4000 + 29{,}500 = 25{,}500$.

Check and Answer   The answer is reasonable. Thus, the net profit for the year was $25,500.

NOW TRY EXERCISE 121

5   Add Using Calculators

Throughout the book we will provide information about calculators in the Using Your Calculator boxes. Some will be for scientific calculators; some will be for graphing calculators, also called graphers; and some will be for both. To the left are pictures of a scientific calculator (on left) and a graphing calculator (on right). Graphing calculators can do everything that a scientific calculator can do and more. Ask your instructor if he or she is recommending a particular calculator for this course. If you plan on taking additional mathematics courses, you may want to consider purchasing the calculator that will be used in those courses.

It is important that you understand the procedures for adding, subtracting, multiplying, and dividing real numbers *without* using a calculator. *You should not need to rely on a calculator to work problems.* If you are permitted to use a calculator, you can, however, use the calculator to help save time on difficult calculations. If you have an understanding of the basic concepts, you should be able to tell if you have made an error entering information on the calculator if the answer shown does not seem reasonable.

Following is the first of many Using Your Calculator boxes and the first of many Using Your Graphing Calculator boxes. Note that *no new material* will be presented in the boxes. The boxes are provided to help you use your calculator, if you are using one in this course. Do not be concerned if you do not know what all the calculator keys do. Your instructor will tell you which calculator keys you will need to know how to use.

In Using Your Graphing Calculator boxes, when we show keystrokes or graphing screens (called windows), they will be for the Texas Instruments TI-83 or Texas Instruments TI-83 Plus calculators. For the material covered in this book, the keystrokes used on both calculators, and the graphs obtained will be the same. You should read your graphing calculator manual for more detailed instructions.

TEACHING TIP
Point out that scientific calculators know the rules for order of operation which we will discuss in Section 1.9. Then say, "Check if your calculator is a scientific calculator by entering $10 - 4 \times 2 =$. Did you get 2 or 12? If your answer is 12, the calculator is not a scientific calculator."

Using Your Calculator

Entering Negative Numbers

Most scientific calculators contain a $\boxed{+/-}$ key, which is used to enter a negative number. To enter the number -5, press 5 $\boxed{+/-}$ and a -5 will be displayed. Now we show how to evaluate some addition problems on a scientific calculator.

Addition of Real Numbers

EVALUATE	KEYSTROKES*	ANSWER DISPLAYED
$-9 + 24$	9 $\boxed{+/-}$ $\boxed{+}$ 24 $\boxed{=}$	15
$15 + (-22)$	15 $\boxed{+}$ 22 $\boxed{+/-}$ $\boxed{=}$	-7
$-30 + (-16)$	30 $\boxed{+/-}$ $\boxed{+}$ 16 $\boxed{+/-}$ $\boxed{=}$	-46

*With some scientific calculators the negative sign is entered before the number, as is done with graphing calculators.

Using Your Graphing Calculator

Entering Negative Numbers

Graphing calculators have two keys that look similar, as shown below.

$\boxed{(-)}$ $\boxed{-}$

↑   ↑

Used to make a number negative   *Used to subtract*

Addition of Real Numbers

EVALUATE	KEYSTROKES	ANSWER DISPLAYED
$-9 + 24$	$\boxed{(-)}$ 9 + 24 $\boxed{ENTER}$	15
$15 + (-22)$	15 + $\boxed{(-)}$ 22 $\boxed{ENTER}$	-7
$-30 + (-16)$	$\boxed{(-)}$ 30 + $\boxed{(-)}$ 16 $\boxed{ENTER}$	-46

To make a number negative on a graphing calculator, you first press the $\boxed{(-)}$ key, then enter the number.

NOW TRY EXERCISE 97

Exercise Set 1.6

1. addition, subtraction, multiplication, division
2. a) two numbers that sum to 0   **b)** One example is 3 and -3.   **6. b)** negative

Concept/Writing Exercises

1. What are the four basic operations of arithmetic?

2. a) What are opposites or additive inverses?

    b) Give an example of two numbers that are opposites.

3. a) Are the numbers $-\frac{2}{3}$ and $\frac{3}{2}$ opposites? Explain.   no

    b) If the numbers in part **a)** are not opposites, what number is the opposite of $-\frac{2}{3}$? $\frac{2}{3}$

4. If we add two negative numbers, will the sum be a positive number or a negative number? Explain.   negative

5. If we add a positive number and a negative number, will the sum be positive, negative, or can it be either? Explain.   can be either

6. a) If we add $-24{,}692$ and $30{,}519$, will the sum be positive or negative? Without performing any calculations, explain how you determined your answer.   positive

    b) Repeat part **a)** for the numbers $24{,}692$ and $-30{,}519$.

    c) Repeat part **a)** for the numbers $-24{,}692$ and $-30{,}519$.   negative

7., 8., 9. a) and c), 10. a) and c) Answers will vary.

7. Explain in your own words how to add two numbers with like signs.

8. Explain in your own words how to add two numbers with unlike signs.

9. Mr. Dabskic charged $175 worth of goods on his charge card. He later made a payment of $93.

   a) Explain why the balance remaining on his card may be found by the addition $-175 + 93$.

   b) Find the sum of $-175 + 93$. $\;-82$

c) In part b), you should have obtained a sum of -82. Explain why this -82 indicates that Mr. Dabskic owes $82 on his credit card.

10. Mrs. Goldstein owed $163 on her credit card. She charged another item costing $56.

   a) Explain why the new balance on her credit card may be found by the addition $-163 + (-56)$.

   b) Find the sum of $-163 + (-56)$. $\;-219$

   c) In part b), you should have obtained a sum of -219. Explain why this -219 indicates that Mrs. Goldstein owes $219 on her credit card.

In Exercises 11 and 12, are the calculations shown correct? If not, explain why not.

11. $\dfrac{-5}{12} + \dfrac{9}{12} = \dfrac{-5+9}{12} = \dfrac{4}{12} = \dfrac{1}{3}$ correct

12. $\dfrac{-6}{70} + \left(\dfrac{-9}{70}\right) = \dfrac{-6+(-9)}{70} = \dfrac{-15}{70} = -\dfrac{3}{14}$ correct

Practice the Skills

Write the opposite of each number.

13. 9 -9

14. -7 7

15. -28 28

16. 3 -3

17. 0 0

18. -5 5

19. $\dfrac{5}{3}$ $-\dfrac{5}{3}$

20. $-\dfrac{1}{4}$ $\dfrac{1}{4}$

21. $2\dfrac{3}{5}$ $-2\dfrac{3}{5}$

22. -1 1

23. 3.72 -3.72

24. -0.721 0.721

Add.

25. $5 + 6$ 11

26. $-8 + 2$ -6

27. $4 + (-3)$ 1

28. $8 + (-7)$ 1

29. $-4 + (-2)$ -6

30. $-3 + (-5)$ -8

31. $6 + (-6)$ 0

32. $-8 + 8$ 0

33. $-4 + 4$ 0

34. $-6 + 9$ 3

35. $-8 + (-2)$ -10

36. $6 + (-5)$ 1

37. $-6 + 6$ 0

38. $-7 + 3$ -4

39. $-8 + (-5)$ -13

40. $0 + (-3)$ -3

41. $0 + 0$ 0

42. $0 + (-0)$ 0

43. $-6 + 0$ -6

44. $-9 + 13$ 4

45. $18 + (-9)$ 9

46. $-7 + 7$ 0

47. $-33 + (-31)$ -64

48. $-27 + (-9)$ -36

49. $-42 + (-9)$ -51

50. $56 + (-14)$ 42

51. $6 + (-27)$ -21

52. $-16 + 9$ -7

53. $-35 + 40$ 5

54. $-12 + 17$ 5

55. $-4.2 + 6.5$ 2.3

56. $-8.3 + 5.7$ -2.6

57. $-9.7 + (-5.4)$ -15.1

58. $10.34 + (-8.57)$ 1.77

59. $180 + (-200)$ -20

60. $-33 + (-92)$ -125

61. $-67 + 28$ -39

62. $183 + (-183)$ 0

63. $184 + (-93)$ 91

64. $-19 + 176$ 157

65. $-452 + 312$ -140

66. $-94 + (-98)$ -192

67. $-26 + (-79)$ -105

68. $49 + (-63)$ -14

69. $-24.6 + (-13.9)$ -38.5

70. $80.5 + (-90.4)$ -9.9

71. $106.3 + (-110.9)$ -4.6

72. $-124.7 + (-19.3)$ -144.0

Add.

73. $\dfrac{3}{5} + \dfrac{1}{7}$ $\dfrac{26}{35}$

74. $\dfrac{5}{8} + \dfrac{3}{5}$ $\dfrac{49}{40}$

75. $\dfrac{5}{12} + \dfrac{6}{7}$ $\dfrac{107}{84}$

76. $\dfrac{2}{9} + \dfrac{3}{10}$ $\dfrac{47}{90}$

77. $-\dfrac{8}{11} + \dfrac{4}{5}$ $\dfrac{4}{55}$

78. $-\dfrac{4}{9} + \dfrac{5}{27}$ $-\dfrac{7}{27}$

79. $-\dfrac{7}{10} + \dfrac{11}{90}$ $-\dfrac{26}{45}$

80. $\dfrac{8}{9} + \left(-\dfrac{1}{3}\right)$ $\dfrac{5}{9}$

81. $\dfrac{9}{25} + \left(-\dfrac{3}{50}\right)$ $\dfrac{3}{10}$

82. $\dfrac{3}{20} + \left(-\dfrac{9}{100}\right)$ $\dfrac{3}{50}$

83. $-\dfrac{7}{30} + \left(-\dfrac{5}{6}\right)$ $-\dfrac{16}{15}$

84. $-\dfrac{1}{15} + \left(-\dfrac{5}{6}\right)$ $-\dfrac{9}{10}$

85. $-\dfrac{4}{5} + \left(-\dfrac{5}{75}\right)$ $-\dfrac{13}{15}$

86. $-\dfrac{9}{24} + \dfrac{5}{7}$ $\dfrac{19}{56}$

87. $\dfrac{5}{36} + \left(-\dfrac{5}{24}\right)$ $-\dfrac{5}{72}$

88. $-\dfrac{9}{40} + \dfrac{4}{15}$ $\dfrac{1}{24}$

89. $-\dfrac{5}{12} + \left(-\dfrac{3}{10}\right)$ $-\dfrac{43}{60}$

90. $\dfrac{7}{16} + \left(-\dfrac{5}{24}\right)$ $\dfrac{11}{48}$

91. $-\dfrac{13}{14} + \left(-\dfrac{7}{42}\right)$ $-\dfrac{23}{21}$

92. $-\dfrac{11}{27} + \left(-\dfrac{7}{18}\right)$ $-\dfrac{43}{54}$

In Exercises 93–108, **a)** determine by observation whether the sum will be a positive number, zero, or a negative number; **b)** find the sum using your calculator; and **c)** examine your answer to part **b)** to see whether it is reasonable and makes sense.

93. $587 + (-197)$ 390

94. $-140 + (-629)$ −769

95. $-84 + (-289)$ −373

96. $-647 + 352$ −295

97. $-947 + 495$ −452

98. $762 + (-762)$ 0

99. $-496 + (-804)$ −1300

100. $-354 + 1090$ 736

101. $-375 + 263$ −112

102. $1127 + (-84)$ 1043

103. $-1833 + (-2047)$ −3880

104. $-426 + 572$ 146

105. $3124 + (-2013)$ 1111

106. $-9095 + (-647)$ −9742

107. $-1025 + (-1025)$ −2050

108. $7513 + (-4361)$ 3152

Indicate whether each statement is true or false.

109. The sum of two negative numbers is always a negative number. T

110. The sum of a negative number and a positive number is sometimes a negative number. T

111. The sum of two positive numbers is never a negative number. T

112. The sum of a positive number and a negative number is always a negative number. F

113. The sum of a positive number and a negative number is always a positive number. F

114. The sum of a number and its opposite is always equal to zero. T

Problem Solving   **120.** 69 feet below sea level or −69 feet

Write an expression that can be used to solve each problem and then solve.

115. *Credit Card* Mr. Peter owed $94 on his bank credit card. He charged another item costing $183. Find the amount that Mr. Peter owed the bank. $277

116. *Charge Card* Mrs. Chu charged $142 worth of goods on her charge card. Find her balance after she made a payment of $87. $55

117. *Football* A football team lost 18 yards on one play and then lost 3 yards on the following play. What was the total loss in yardage? 21 yd

118. *Income Tax* Mrs. Poweski paid $1823 in federal income tax. When she was audited, Mrs. Poweski had to pay an additional $471. What was her total tax? $2294

119. *Drilling for Water* A company is drilling a well. During the first week they drilled 27 feet, and during the second week they drilled another 34 feet before they struck water. How deep is the well? 61 ft

120. *Death Valley* The Duncans are at a point 267 feet below sea level in Death Valley, California. They proceed to climb up a mountain a vertical distance of 198 feet. What is their vertical distance in terms of sea level?

121. *High Mountain* The Guinness Book of World Records, 2002 edition, lists Mauna Kea in Hawaii as the tallest mountain in the world when measured from its base to its peak. The base of Mauna Kea is 19,684 feet below sea level. The total height of the mountain from its base to its peak is 33,480 feet. How high is the peak of Mauna Kea above sea level? 13,796 ft

122. *Coffee Bar* The Frenches opened a coffee bar. Their income and expenses for their first three months of operation are shown in the following graph.
   a) Find the net profit or loss (the sum of income and expenses) for the first month. loss of $7310
   b) Find the net profit or loss for the second month.
   c) Find the net profit or loss for the third month.

122. **b)** gain of $5230   **c)** gain of $2440

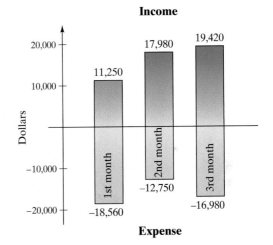

123. The following graph shows the profit or loss for the U.S. Postal Service for the years 1990 through 2001.

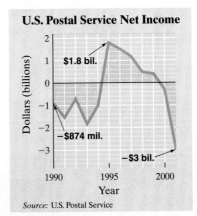

Source: U.S. Postal Service

a) Determine the net gain or net loss for the Postal Service in 2001. −$3 billion

123. b) 1999, $0.4 billion; 2000, − $ 0.3 billion; 2001, − $ 3 billion; net loss, − $ 2.9 billion

b) Estimate the net income or loss for the U.S. Postal Service for each of the years 1999, 2000, and 2001. Then estimate the net gain or loss from 1999 through 2001 by adding the three net incomes.

124. The following chart shows percent changes from 1996 to 2000 for selected crimes in the United States.

a) Determine the percent change in auto thefts from 1998 through 2000 by adding the individual percents for the years 1998, 1999, and 2000. −13.4%

b) Determine the percent change in U.S. violent crimes from 1998 through 2000. −13.0%

Percent change from previous year for U.S. crimes		
Year	Violent crime	Auto theft
1996	−6.5%	−5.2%
1997	−3.2%	−2.9%
1998	−6.4%	−8.4%
1999	−6.7%	−7.7%
2000	0.1%	2.7%

Source: FBI

Challenge Problems

Evaluate each exercise by adding the numbers from left to right. We will discuss problems like this shortly.

125. $(-4) + (-6) + (-12)$ −22

126. $5 + (-7) + (-8)$ −10

127. $29 + (-46) + 37$ 20

128. $4 + (-5) + 6 + (-8)$ −3

129. $(-12) + (-10) + 25 + (-3)$ 0

130. $(-4) + (-2) + (-15) + (-27)$ −48

131. $\frac{1}{2} + \left(-\frac{1}{3}\right) + \frac{1}{5}$ $\frac{11}{30}$

132. $-\frac{3}{8} + \left(-\frac{2}{9}\right) + \left(-\frac{1}{2}\right)$ $-\frac{79}{72}$

Find the following sums. Explain how you determined your answer. (Hint: Pair small numbers with large numbers from the ends inward.)

133. $1 + 2 + 3 + \cdots + 10$ 55

134. $1 + 2 + 3 + \cdots + 20$ 210

Cumulative Review Exercises

[1.3] 135. Multiply $\left(\frac{3}{5}\right)\left(1\frac{2}{3}\right)$. 1

136. Subtract $3 - \frac{5}{16}$. $\frac{43}{16}$ or $2\frac{11}{16}$

[1.4] 137. List the set of counting numbers. $\{1, 2, 3, 4, \ldots\}$

[1.5] *Insert either* $<, >,$ *or* $=$ *in each shaded area to make a true statement.*

138. $|-3| \blacksquare 2$ >

139. $8 \blacksquare |-7|$ >

1.7 SUBTRACTION OF REAL NUMBERS

SSM   Study Guide   CD/Video   MathPro 4/5   PH Math Tutor Center   prenhall.com/Angel

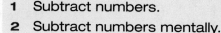

1 Subtract numbers.

2 Subtract numbers mentally.

3 Evaluate expressions containing more than two numbers.

1 Subtract Numbers

Any subtraction problem can be rewritten as an addition problem using the additive inverse.

To Subtract Real Numbers

In general, if a and b represent any two real numbers, then

$$a - b = a + (-b)$$

This rule says that to subtract b from a, add the opposite or additive inverse of b to a.

EXAMPLE 1   Evaluate $9 - (+4)$.

Solution   We are subtracting a positive 4 from 9. To accomplish this, we add the opposite of $+4$, which is -4, to 9.

$$9 - (+4) = 9 + (-4) = 5$$

Subtract   Positive 4   Add   Negative 4

We evaluated $9 + (-4)$ using the procedures for *adding* real numbers presented in Section 1.6.

Often in a subtraction problem, when the number being subtracted is a positive number, the $+$ sign preceding the number being subtracted is not shown. For example, in the subtraction $9 - 4$,

$$9 - \boxed{4} \text{ means } 9 - \boxed{(+4)}$$

Thus, to evaluate $9 - 4$, we must add the opposite of 4, which is -4, to 9.

$$9 - 4 = 9 + (-4) = 5$$

Subtract   Positive 4   Add   Negative 4

This procedure is illustrated in Example 2.

EXAMPLE 2   Evaluate $5 - 3$.

Solution   We must subtract a positive 3 from 5. To change this problem to an addition problem, we add the opposite of 3, which is -3, to 5.

$$\overbrace{5 - 3}^{\substack{\text{Subtraction} \\ \text{problem}}} = \overbrace{5 + (-3)}^{\substack{\text{Addition} \\ \text{problem}}} = 2$$

Subtract Positive 3 Add Negative 3

EXAMPLE 3   Evaluate

a) $4 - 9$ **b)** $-4 - 2$

Solution   **a)** Add the opposite of 9, which is -9, to 4.

$$4 - 9 = 4 + (-9) = -5$$

NOW TRY EXERCISE 15   **b)** Add the opposite of 2, which is -2, to -4.

$$-4 - 2 = -4 + (-2) = -6$$

EXAMPLE 4   Evaluate $4 - (-2)$.

Solution   We are asked to subtract a negative 2 from 4. To do this, we add the opposite of -2, which is 2, to 4.

$$4 - (-2) = 4 + 2 = 6$$

Subtract   Negative 2   Add   Positive 2

HELPFUL HINT

By examining Example 4, we see that

$$4 - (-2) = 4 + 2$$

Two negative   Plus
signs together

Whenever we subtract a negative number, we can replace the two negative signs with a plus sign.

EXAMPLE 5 Evaluate

a) $7 - (-5)$ **b)** $-15 - (-12)$

Solution **a)** Since we are subtracting a negative number, adding the opposite of -5, which is 5, to 7 will result in the two negative signs being replaced by a plus sign.

$$7 - (-5) = 7 + 5 = 12$$

NOW TRY EXERCISE 19   **b)** $-15 - (-12) = -15 + 12 = -3$ ✳

HELPFUL HINT

We will now indicate how we may illustrate subtraction using colored chips. Remember from the preceding section that a red chip represents $+1$ and a green chip -1.

● $= +1$ ● $= -1$

Consider the subtraction problem $2 - 5$. If we change this to an addition problem, we get $2 + (-5)$. We can then add, as was done in the preceding section. The figure below shows that $2 + (-5) = -3$.

🌑🌑 + 🌑🌑●●●

Now consider $-2 - 5$. This means $-2 + (-5)$, which can be represented as follows:

●● + ●●●●●

Thus, $-2 - 5 = -7$.
Now consider the problem $-3 - (-5)$. This can be rewritten as $-3 + 5$, which can be represented as follows:

🌑🌑🌑 + 🌑🌑🌑●●

Thus, $-3 - (-5) = 2$.
Some students still have difficulty understanding why when you subtract a negative number you obtain a positive number. Let's look at the problem $3 - (-2)$. This time we will look at it from a slightly different point of view. Let's start with 3:

● ● ●

From this we wish to subtract a negative 2. To the $+3$ shown above we will add two zeros by adding two $+1 - 1$ combinations. Remember, $+1$ and -1 sum to 0.

● ● ● + ● ● + ● ●
  $+3$ 0 0

Now we can subtract or "take away" the two -1's as shown:

● ● ● + ● 🌑 + ● 🌑

From this we see that we are left with $3 + 2$ or 5. Thus, $3 - (-2) = 5$.

EXAMPLE 6   Subtract 12 from 3.

Solution
$$3 - 12 = 3 + (-12) = -9$$

HELPFUL HINT

Example 6 asked us to "subtract 12 from 3." Some of you may have expected this to be written as $12 - 3$ since you may be accustomed to getting a positive answer. However, the correct way of writing this is $3 - 12$. Notice that the number following the word "from" is our starting point. That is where the calculation begins. For example:

Subtract 2 from 7 means $7 - 2$.   From 7, subtract 2 means $7 - 2$.

Subtract 5 from -1 means $-1 - 5$.   From -1, subtract 5 means $-1 - 5$.

Subtract -4 from -2 From -2, subtract -4
means $-2 - (-4)$.   means $-2 - (-4)$.

Subtract -3 from 6 means $6 - (-3)$.   From 6, subtract -3 means $6 - (-3)$.

Subtract a from b means $b - a$.   From a, subtract b means $a - b$.

EXAMPLE 7   Subtract 5 from 5.

Solution
$$5 - 5 = 5 + (-5) = 0$$

EXAMPLE 8   Subtract -6.48 from 4.25.

Solution
$$4.25 - (-6.48) = 4.25 + 6.48 = 10.73$$

NOW TRY EXERCISE 65   Now we will perform subtraction problems that contain fractions.

EXAMPLE 9   Subtract $\dfrac{5}{9} - \dfrac{13}{15}$

Solution   Begin by changing the subtraction problem to an addition problem.

$$\frac{5}{9} - \frac{13}{15} = \frac{5}{9} + \left(-\frac{13}{15}\right)$$

Now rewrite the fractions with the LCD, 45, and add the fractions as was done in the last section.

$$\frac{5}{9} + \left(-\frac{13}{15}\right) = \frac{5}{9} \cdot \frac{5}{5} + \left(-\frac{13}{15}\right) \cdot \frac{3}{3}$$

$$= \frac{25}{45} + \left(-\frac{39}{45}\right) = \frac{25 + (-39)}{45} = \frac{-14}{45} = -\frac{14}{45}$$

Thus $\dfrac{5}{9} - \dfrac{13}{15} = -\dfrac{14}{45}$.

EXAMPLE 10   Subtract $-\dfrac{7}{18}$ from $-\dfrac{9}{15}$.

Solution   This problem is written $-\frac{9}{15} - \left(-\frac{7}{18}\right)$.

We can simplify this as follows.

$$-\frac{9}{15} - \left(-\frac{7}{18}\right) = -\frac{9}{15} + \frac{7}{18}.$$

The LCD of 15 and 18 is 90. Rewriting the fractions with a common denominator gives

$$-\frac{9}{15} \cdot \frac{6}{6} + \frac{7}{18} \cdot \frac{5}{5} = -\frac{54}{90} + \frac{35}{90} = \frac{-54 + 35}{90}$$

$$= \frac{-19}{90} = -\frac{19}{90}.$$

NOW TRY EXERCISE 87

Let us now look at some applications that involve subtraction.

EXAMPLE 11 **A Checkbook Balance** Tom Duncan's checkbook indicated a balance of $237 before he wrote a check for $364. Find the balance in his checkbook.

Solution Understand and Translate We can obtain the balance by subtracting 364 from 237.

Carry Out
$$237 - 364 = 237 + (-364) = -127$$

Check and Answer The negative indicates a deficit, which is what we expect. Therefore, Tom is overdrawn by $127.

EXAMPLE 12 **Temperature Difference** On February 1, 2002, the low temperature for the day in Honolulu, Hawaii, was 72° F. On the same day, the low temperature in Delta Junction, Alaska (the end of the Alaska Highway, about 80 miles southeast of Fairbanks), was −23° F. Find the difference in their temperatures.

Solution Understand and Translate The word "difference" in the example title indicates subtraction. We can obtain the difference in their temperatures by subtracting as follows.

Carry Out
$$72 - (-23) = 72 + 23 = 95$$

Check and Answer Therefore the temperature in Honolulu is 95° greater than the temperature in Delta Junction.

EXAMPLE 13 **Measuring Rainfall** At Beverly Broomell's house a rain gauge is placed in the yard and is left untouched for 2 days. Suppose that on the first day $2\frac{1}{4}$ inches of rain fall. On the second day, no rain falls, but $\frac{3}{8}$ inch of the first day's rainfall evaporates.

How much water remains in the gauge after the second day?

Solution Understand and Translate From the first amount, $2\frac{1}{4}$ inches, we must subtract $\frac{3}{8}$ inch.

Carry Out We begin by changing the subtraction problem to an addition problem. We then change the mixed number to a fraction, and then rewrite each fraction with the LCD 8.

$$2\frac{1}{4} - \frac{3}{8} = 2\frac{1}{4} + \left(-\frac{3}{8}\right)$$

$$= \frac{9}{4} + \left(-\frac{3}{8}\right)$$

$$= \frac{9}{4} \cdot \frac{2}{2} + \left(-\frac{3}{8}\right)$$

$$= \frac{18}{8} + \left(-\frac{3}{8}\right)$$

$$= \frac{18 + (-3)}{8} = \frac{15}{8} \quad \text{or} \quad 1\frac{7}{8}$$

Check and Answer   Thus, after the second day there was $1\frac{7}{8}$ inches of water in the gauge. Based upon the numbers given in the problem, the answer seems

NOW TRY EXERCISE 135   reasonable.

EXAMPLE 14   Evaluate.

a) $15 + (-4)$ b) $-16 - 3$ c) $19 + (-14)$

d) $7 - (-9)$ e) $-9 - (-3)$ f) $8 - 13$

Solution   Parts **a)** and **b)** are addition problems, whereas the other parts are subtraction problems. We can rewrite each subtraction problem as an addition problem to evaluate.

a) $15 + (-4) = 11$ b) $-16 - 3 = -16 + (-3) = -19$

c) $19 + (-14) = 5$ d) $7 - (-9) = 7 + 9 = 16$

e) $-9 - (-3) = -9 + 3 = -6$ f) $8 - 13 = 8 + (-13) = -5$

2 Subtract Numbers Mentally

In the previous examples, we changed subtraction problems to addition problems. We did this because we know how to add real numbers. After this chapter, when we work out a subtraction problem, we will not show this step. *You need to practice and thoroughly understand how to add and subtract real numbers. You should understand this material so well that, when asked to evaluate an expression like $-4 - 6$, you will be able to compute the answer mentally. You should understand that $-4 - 6$ means the same as $-4 + (-6)$, but you should not need to write the addition to find the value of the expression, -10.*

Let's evaluate a few subtraction problems without showing the process of changing the subtraction to addition.

EXAMPLE 15   Evaluate.

a) $-7 - 5$ b) $4 - 12$ c) $18 - 25$ d) $-20 - 12$

Solution   a) $-7 - 5 = -12$ b) $4 - 12 = -8$

c) $18 - 25 = -7$ d) $-20 - 12 = -32$

In Example 15 **a)**, we may have reasoned that $-7 - 5$ meant $-7 + (-5)$, which is -12, but we did not need to show it.

EXAMPLE 16   Evaluate $-\frac{3}{5} - \frac{7}{8}$

Solution   The least common denominator is 40. Write each fraction with the lcd, 40.

NOW TRY EXERCISE 79

$$-\frac{3}{5}\cdot\frac{8}{8} - \frac{7}{8}\cdot\frac{5}{5} = -\frac{24}{40} - \frac{35}{40} = \frac{-24 - 35}{40} = -\frac{59}{40} = -1\frac{19}{40}$$ ✳

Notice in Example 16, when we had $-\dfrac{24}{40} - \dfrac{35}{40}$, we could have written $\dfrac{-24 + (-35)}{40}$,

but at this time we elected to write it as $\dfrac{-24 - 35}{40}$. Since $-24 - 35$ is -59, the

answer is $-\dfrac{59}{40}$ or $-1\dfrac{19}{40}$.

---

3 Evaluate Expressions Containing More Than Two Numbers

In evaluating expressions involving more than one addition and subtraction, work from left to right unless parentheses or other grouping symbols appear.

EXAMPLE 17   Evaluate.

a) $-5 - 13 - 4$ b) $-3 + 1 - 7$ c) $8 - 10 + 2$

Solution   We work from left to right.

a)	b)	c)
$-5 - 13 - 4$	$-3 + 1 - 7$	$8 - 10 + 2$
$= -18 - 4$	$= -2 - 7$	$= -2 + 2$
$= -22$	$= -9$	$= 0$

NOW TRY EXERCISE 119

✳

TEACHING TIP
Stress that when an expression contains several additions and/or subtractions you evaluate the expression from left to right.

After this section you will generally not see an expression like $3 + (-4)$. Instead, the expression will be written as $3 - 4$. Recall that $3 - 4$ means $3 + (-4)$ by our definition of subtraction. **Whenever we see an expression of the form $a + (-b)$, we can write the expression as $a - b$.** For example, $12 + (-15)$ can be written as $12 - 15$ and $-6 + (-9)$ can be written as $-6 - 9$.

As discussed earlier, **whenever we see an expression of the form $a - (-b)$, we can rewrite it as $a + b$.** For example $6 - (-13)$ can be rewritten as $6 + 13$ and $-12 - (-9)$ can be rewritten as $-12 + 9$. Using both of these concepts, the expression $9 + (-12) - (-8)$ may be simplified to $9 - 12 + 8$.

EXAMPLE 18   a) Evaluate $-5 - (-9) + (-12) + (-3)$.

b) Simplify the expression in part a).

c) Evaluate the simplified expression in part b).

Solution   a) We work from left to right. The shading indicates the additions being performed to get to the next step.

$$-5 - (-9) + (-12) + (-3) = -5 + 9 + (-12) + (-3)$$
$$= 4 + (-12) + (-3)$$
$$= -8 + (-3)$$
$$= -11$$

b) The expression simplifies as follows:

$$-5 - (-9) + (-12) + (-3) = -5 + 9 - 12 - 3$$

c) Evaluate the simplified expression from left to right. Begin by adding $-5 + 9$ to obtain 4.

$$-5 + 9 - 12 - 3 = 4 - 12 - 3$$
$$= -8 - 3$$
$$= -11$$

When you come across an expression like the one in Example 18 **a)**, you should simplify it as we did in part **b)** and then evaluate the simplified expression.

NOW TRY EXERCISE 125

Using Your Calculator

Subtraction on a Scientific Calculator

In the Using Your Calculator box on page 49, we indicated that the $+/-$ key is usually pressed after a number is entered to make the number negative. Following are some examples of subtraction on a scientific calculator.

EVALUATE	KEYSTROKES	ANSWER DISPLAYED
$-5 - 8$	5 $+/-$ $-$ 8 $=$	-13
$2 - (-7)$	2 $-$ 7 $+/-$ $=$	9

Subtraction on a Graphing Calculator

In the Using Your Graphing Calculator box on page 49 we mentioned that on a graphing calculator we press the $(-)$ key before the number is entered to make the number negative. The $-$ key on a graphing calculator is used to perform subtraction. Following are some examples of subtraction on a graphing calculator.

EVALUATE	KEYSTROKE	DISPLAY
$-5 - 8$	$(-)$ 5 $-$ 8 ENTER	-13
$2 - (-7)$	2 $-$ $(-)$ 7 ENTER	9

Exercise Set 1.7

1. $2 - 7$ 3. $\Box - *$ 4. $\odot - ?$ 5. c) -9 6. a) $a + b$ 6. c) 8 7. a) $a - b$ 7. c) -2 8. left to right

Concept/Writing Exercises

1. Write an expression that illustrates 7 subtracted from 2.
2. Write an expression that illustrates -6 subtracted from -4. $-4 - (-6)$
3. Write an expression that illustrates $*$ subtracted from $\Box$.
4. Write an expression that illustrates ? subtracted from $\odot$.
5. a) Explain in your own words how to subtract a number b from a number a. Answers will vary.
   b) Write an expression using addition that can be used to subtract 14 from 5. $5 + (-14)$
   c) Evaluate the expression you determined in part **b)**.

6. a) Express the subtraction $a - (-b)$ in a simplified form.
   b) Write a simplified expression that can be used to evaluate $-4 - (-12)$. $-4 + 12$
   c) Evaluate the simplified expression obtained in part **b)**.
7. a) Express the subtraction $a - (+b)$ in a simplified form.
   b) Simplify the expression $7 - (+9)$. $7 - 9$
   c) Evaluate the simplified expression obtained in part **b)**.
8. When we add three or more numbers without parentheses, how do we evaluate the expression?

9. a) Simplify $3 - (-6) + (-5)$ by eliminating two signs next to one another and replacing them with a single sign. (See Example 18b).) Explain how you determined your answer. $3 + 6 - 5$

b) Evaluate the simplified expression obtained in part **a)**.

10. a) Simplify $-12 + (-5) - (-4)$. Explain how you determined your answer. $-12 - 5 + 4$

b) Evaluate the simplified expression obtained in part **a)**.

9. b) 4   **10. b)** -13

In Exercises 11 and 12, are the following calculations correct? If not, explain why.

11. $\dfrac{4}{9} - \dfrac{3}{7} = \dfrac{28}{63} - \dfrac{27}{63} = \dfrac{28 - 27}{63} = \dfrac{1}{63}$ Correct

12. $-\dfrac{5}{12} - \dfrac{7}{9} = -\dfrac{15}{36} - \dfrac{28}{36} = \dfrac{-15 - 28}{36} = -\dfrac{43}{36}$ Correct

Practice the Skills

Evaluate.

13. $12 - 5$ 7
14. $-1 - 6$ -7
15. $8 - 9$ -1
16. $3 - 3$ 0

17. $-4 - 2$ -6
18. $-7 - (-4)$ -3
19. $-4 - (-3)$ -1
20. $-3 - 3$ -6

21. $-4 - 4$ -8
22. $7 - (-7)$ 14
23. $0 - 7$ -7
24. $9 - (-9)$ 18

25. $8 - 8$ 0
26. $9 - (-3)$ 12
27. $-3 - 1$ -4
28. $-5 - (-3)$ -2

29. $5 - 3$ 2
30. $4 - 9$ -5
31. $6 - (-3)$ 9
32. $6 - 10$ -4

33. $0 - (-5)$ 5
34. $-6 - 6$ -12
35. $-9 - 11$ -20
36. $-4 - (-2)$ -2

37. $-4 - (-4)$ 0
38. $14 - 7$ 7
39. $-8 - (-12)$ 4
40. $9 - 9$ 0

41. $-6 - (-2)$ -4
42. $18 - (-4)$ 22
43. $-9 - 2$ -11
44. $-25 - 16$ -41

45. $-35 - (-8)$ -27
46. $37 - 40$ -3
47. $-90 - 60$ -150
48. $-52 - 37$ -89

49. $-45 - 37$ -82
50. $-50 - (-40)$ -10
51. $70 - (-70)$ 140
52. $130 - (-90)$ 220

53. $42.3 - 49.7$ -7.4
54. $81.3 - 92.5$ -11.2
55. $-7.85 - (-3.92)$ -3.93
56. $-12.43 - (-9.57)$ -2.86

57. Subtract 9 from -20. -29
58. Subtract -3 from -10. -7
59. Subtract 8 from -8. -16

60. Subtract 5 from -20. -25
61. Subtract -3 from -5. -2
62. Subtract 15 from -5. -20

63. Subtract -4 from 9.   13
64. Subtract -23 from -23.   0
65. Subtract 24 from 13. -11

66. Subtract -11 from -5.   6
67. Subtract -12.4 from -6.3.   6.1
68. Subtract 17.3 from -9.8. -27.1

Evaluate.

69. $\dfrac{4}{5} - \dfrac{5}{6}$ $-\dfrac{1}{30}$
70. $\dfrac{5}{9} - \dfrac{3}{8}$ $\dfrac{13}{72}$
71. $\dfrac{8}{15} - \dfrac{7}{45}$ $\dfrac{17}{45}$
72. $\dfrac{5}{12} - \dfrac{7}{8}$ $-\dfrac{11}{24}$

73. $-\dfrac{1}{4} - \dfrac{2}{3}$ $-\dfrac{11}{12}$
74. $-\dfrac{7}{10} - \dfrac{5}{12}$ $-\dfrac{67}{60}$
75. $-\dfrac{4}{15} - \dfrac{3}{20}$ $-\dfrac{5}{12}$
76. $-\dfrac{5}{4} - \dfrac{7}{11}$ $-\dfrac{83}{44}$

77. $\dfrac{3}{8} - \dfrac{6}{48}$ $\dfrac{1}{4}$
78. $-\dfrac{5}{6} - \dfrac{3}{32}$ $-\dfrac{89}{96}$
79. $-\dfrac{7}{12} - \dfrac{5}{40}$ $-\dfrac{17}{24}$
80. $\dfrac{17}{18} - \dfrac{13}{20}$ $\dfrac{53}{180}$

81. $\dfrac{3}{16} - \left(-\dfrac{5}{8}\right)$ $\dfrac{13}{16}$
82. $-\dfrac{4}{9} - \left(-\dfrac{3}{5}\right)$ $\dfrac{7}{45}$
83. $-\dfrac{5}{12} - \left(-\dfrac{3}{8}\right)$ $-\dfrac{1}{24}$
84. $\dfrac{5}{20} - \left(-\dfrac{1}{8}\right)$ $\dfrac{3}{8}$

85. Subtract $\dfrac{7}{9}$ from $\dfrac{4}{7}$. $-\dfrac{13}{63}$

86. Subtract $\dfrac{7}{15}$ from $\dfrac{5}{8}$. $\dfrac{19}{120}$

87. Subtract $-\dfrac{3}{10}$ from $-\dfrac{5}{12}$. $-\dfrac{7}{60}$

88. Subtract $-\dfrac{5}{16}$ from $-\dfrac{9}{10}$. $-\dfrac{47}{80}$

*In Exercises 89–106, **a)** determine by observation whether the difference will be a positive number, zero, or a negative number; **b)** find the difference using your calculator; and **c)** examine your answer to part **b)** to see whether it is reasonable and makes sense.*

89. $378 - 279$ 99
90. $483 - 569$ -86
91. $-482 - 137$ -619
92. $178 - (-377)$ 555

93. $843 - (-745)$ 1588
94. $864 - (-762)$ 1626
95. $-408 - (-604)$ 196
96. $-623 - 111$ -734

97. $-1024 - (-576)$ -448
98. $-104.7 - 27.6$ -132.3
99. $165.7 - 49.6$ 116.1
100. $-40.2 - (-12.6)$ -27.6

101. Subtract 364 from 295. -69
102. Subtract -433 from -932. -499
103. Subtract 647 from -1023. -1670

104. Subtract 2432 from -4120. -6552
105. Subtract -7.62 from -7.62.   0
106. Subtract 36.7 from -103.2. -139.9

Evaluate.

107. $7 + 5 - (+8)$ 4

108. $15 - (+9) - (+5)$ 1

109. $-6 + (-6) + 6$ −6

110. $9 - 4 + (-2)$ 3

111. $-13 - (+5) + 3$ −15

112. $7 - (+4) - (-3)$ 6

113. $-9 - (-3) + 4$ −2

114. $15 + (-7) - (-3)$ 11

115. $5 - (-9) + (-1)$ 13

116. $12 + (-5) - (-4)$ 11

117. $17 + (-8) - (+14)$ −5

118. $-7 + 6 - 3$ −4

119. $-36 - 5 + 9$ −32

120. $45 - 3 - 7$ 35

121. $-2 + 7 - 9$ −4

122. $-2 - 7 - 13$ −22

🔒 **123.** $25 - 19 + 3$ 9

124. $-4 - 1 + 5$ 0

125. $-4 - 6 + 5 - 7$ −12

126. $-9 - 3 - (-4) + 5$ −3

🔒 **127.** $17 + (-3) - 9 - (-7)$ 12

128. $32 + 5 - 7 - 12$ 18

129. $-9 + (-7) + (-5) - (-3)$ −18

130. $6 - 9 - (-3) + 12$ 12

Problem Solving

131. *Lands' End*   The Lands' End catalog department had 300 ladies' blue cardigan sweaters in stock on December 1. By December 9, they had taken orders for 343 of the sweaters.
   a) How many sweaters were on back order?   43
   b) If they wanted 100 sweaters in addition to those already ordered, how many sweaters would they need to back order?   143

132. *Leadville, Co*   According to the *Guinness Book of World Records*, the city with the greatest elevation in the United States is Leadville, Colorado at 10,152 feet. The city with the lowest elevation in the United States, at 184 feet below sea level, is Calipatria, California. What is the difference in the elevation of these cities?   10,336 ft

Leadville, Colorado

133. *Drilling*   The Jacksons, who live near Myrtle Beach, South Carolina, have a house at an elevation of 42 feet above sea level. They hire the RL Schlicter Drilling Company to dig a well. After hitting water, the drilling company informs the Jacksons that they had to dig 58 feet to hit water. How deep is the well with regards to sea level?   16 feet below sea level

134. *Death Valley*   A medical supply package is dropped into Death Valley, California, from a helicopter 1605 feet above sea level. The package lands at a location in Death Valley 267 feet below sea level. What vertical distance did the package travel?   1872 feet

🔒 **135.** *Temperature Change*   The greatest change in temperature ever recorded within a 24-hour period occurred at Browning, Montana, on January 23, 1916. The temperature fell from 44°F to −56°F. How much did the temperature drop?   100°F

136. *Trains*   Two trains start at the same station at the same time. The Amtrak travels 68 miles in 1 hour. The Pacific Express travels 80 miles in 1 hour.
   a) If the two trains travel in opposite directions, how far apart will they be in 1 hour?   148 mi
   b) If the two trains travel in the same direction, how far apart will they be in 1 hour?   12 mi

137. *Golf*   As of this writing, Tiger Woods is the youngest person to win the Masters golf tournament, which is held each April in Augusta, Georgia. The chart below shows his scores at the Masters for various years.

Tiger Woods at the Masters	
Date*	**Score (above or below par)**
2002	−12
2001	−16
2000	−4
1999	+1
1998	−3
1997	−18
1995	+5
* Tiger Woods did not make the cut to finish the tournament in 1996 *Source: USA Today*, April 5, 2001, page 5E	

a) If par for the Masters is 288 strokes, determine his score in 2002. 276

b) What was the difference in his strokes for the Masters between 2002 and 2000? 8 strokes less

c) What was his average stroke score for the Masters for 2000, 2001, and 2002 if par for the course was 288 strokes each year? ≈ 277.33

138. *Sales Descline* The chart on the right shows the total monthly change in sales, in percent, for the Ford Motor Company from January, 2000 through July 2001.

a) Estimate the change in sales between June, 2001, and July, 2001. a decrease of about 6%

b) Estimate the change in sales between April, 2000, and December, 2000. a decrease of about 27%

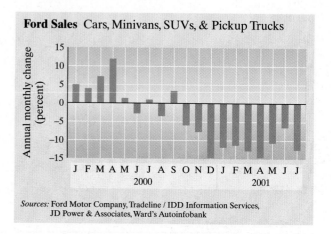

Ford Sales Cars, Minivans, SUVs, & Pickup Trucks

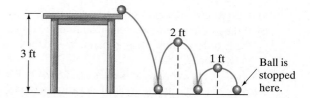

Sources: Ford Motor Company, Tradeline / IDD Information Services, JD Power & Associates, Ward's Autoinfobank

Challenge Problems

Find each sum.

139. $1 - 2 + 3 - 4 + 5 - 6 + 7 - 8 + 9 - 10$ −5

140. $1 - 2 + 3 - 4 + 5 - 6 + \cdots + 99 - 100$ −50

141. Consider a number line. **141. a)** 7 units

a) What is the distance, in units, between −2 and 5?

b) Write a subtraction problem to represent this distance (the distance is to be positive). $5 - (-2)$

142. *Stock* Jody Coffee buys a stock for $50. Will the stock be worth more if it decreases by 10% and then increases by 10%, or if it increases by 10% and then decreases by 10%, or will the value be the same either way? Same

143. *Rolling Ball* A ball rolls off a table and follows the path indicated in the figure. Suppose the maximum height reached by the ball on each bounce is 1 foot less than on the previous bounce.

a) Determine the total vertical distance traveled by the ball. 9 ft

b) If we consider the ball moving in a downward direction as negative, and the ball moving in an upward direction as positive, what was the net vertical distance traveled (from its starting point) by the ball? −3 ft

145. The set of rational numbers together with the set of irrational numbers form the set of real numbers.

Cumulative Review Exercises

[1.4] 144. List the set of integers.
$$\{\ldots, -3, -2, -1, 0, 1, 2, 3, \ldots\}$$

145. Explain the relationship between the set of rational numbers, the set of irrational numbers, and the set of real numbers.

[1.5] *Insert either* $>, <,$ *or* $=$ *in each shaded area to make the statement true.*

146. $|-3|$ ▦ -5 >

147. $|-6|$ ▦ $|-7|$ <

[1.7] 148. Evaluate $\dfrac{4}{5} - \dfrac{3}{8}$. $\dfrac{17}{40}$

1.8 MULTIPLICATION AND DIVISION OF REAL NUMBERS

SSM Study Guide CD/Video

MathPro 4/5 PH Math Tutor Center prenhall.com/Angel

1 Multiply numbers.

2 Divide numbers.

3 Remove negative signs from denominators.

4 Evaluate divisions involving 0.

1 Multiply Numbers

TEACHING TIP
You may wish to mention that multiplication and division of decimal numbers are covered in Appendix A.

The following rules are used in determining the sign of the product when two numbers are multiplied.

The Sign of the Product of Two Real Numbers

1. The product of two numbers with **like** signs is a **positive** number.

2. The product of two numbers with **unlike** signs is a **negative** number.

By this rule, the product of two positive numbers or two negative numbers will be a positive number. The product of a positive number and a negative number will be a negative number.

EXAMPLE 1   Evaluate.

a) $4(-5)$ **b)** $(-6)(7)$ **c)** $(-9)(-3)$

Solution   **a)** Since the numbers have unlike signs, the product is negative.

$$4(-5) = -20$$

b) Since the numbers have unlike signs, the product is negative.

$$(-6)(7) = -42$$

c) Since the numbers have like signs, both negative, the product is positive.

$$(-9)(-3) = 27$$

EXAMPLE 2   Evaluate.

a) $(-8)(5)$ **b)** $(-4)(-8)$ **c)** $4(-9)$ **d)** $0(6)$ **e)** $0(-2)$ **f)** $-3(-6)$

Solution   **a)** $(-8)(5) = -40$ **b)** $(-4)(-8) = 32$ **c)** $4(-9) = -36$

d) $0(6) = 0$ **e)** $0(-2) = 0$ **f)** $-3(-6) = 18$

NOW TRY EXERCISE 29   *Note that zero multiplied by any real number equals zero.*

HELPFUL HINT

At this point some students begin confusing problems like $-2 - 3$ with $(-2)(-3)$ and problems like $2 - 3$ with $2(-3)$. If you do not understand the difference between problems like $-2 - 3$ and $(-2)(-3)$, make an appointment to see your instructor as soon as possible.

Subtraction Problems	Multiplication Problems
$-2 - 3 = -5$	$(-2)(-3) = 6$
$2 - 3 = -1$	$(2)(-3) = -6$

EXAMPLE 3   Evaluate

a) $\left(\dfrac{-1}{8}\right)\left(\dfrac{-3}{5}\right)$ **b)** $\left(\dfrac{3}{20}\right)\left(\dfrac{-3}{10}\right)$.

Solution   **a)** $\left(\dfrac{-1}{8}\right)\left(\dfrac{-3}{5}\right) = \dfrac{(-1)(-3)}{8(5)} = \dfrac{3}{40}$ **b)** $\left(\dfrac{3}{20}\right)\left(\dfrac{-3}{10}\right) = \dfrac{3(-3)}{20(10)} = -\dfrac{9}{200}$

NOW TRY EXERCISE 41

Sometimes you may be asked to perform more than one multiplication in a given problem. When this happens, the sign of the final product can be determined by counting the number of *negative* numbers being multiplied. *The product of an even number of negative numbers will always be positive. The product of an odd number of negative numbers will always be negative.* Can you explain why?

EXAMPLE 4 Evaluate

a) $(-4)(3)(-2)(-1)$ **b)** $(-3)(2)(-1)(-2)(-4)$.

Solution **a)** Since there are three negative numbers (an odd number of negatives), the product will be negative, as illustrated.

$$(-4)(3)(-2)(-1) = (-12)(-2)(-1)$$
$$= (24)(-1)$$
$$= -24$$

b) Since there are four negative numbers (an even number), the product will be positive.

$$(-3)(2)(-1)(-2)(-4) = (-6)(-1)(-2)(-4)$$
$$= (6)(-2)(-4)$$
$$= (-12)(-4)$$
$$= 48$$

NOW TRY EXERCISE 35

2 Divide Numbers

The rules for dividing numbers are very similar to those used in multiplying numbers.

The Sign of the Quotient of Two Real Numbers

1. The quotient of two numbers with **like** signs is a **positive** number.

2. The quotient of two numbers with **unlike** signs is a **negative** number.

Therefore, the quotient of two positive numbers or two negative numbers will be a positive number. The quotient of a positive number and a negative number will be a negative number.

EXAMPLE 5 Evaluate.

a) $\dfrac{10}{-5}$ **b)** $\dfrac{-45}{5}$ **c)** $\dfrac{-36}{-6}$

Solution **a)** Since the numbers have unlike signs, the quotient is negative.

$$\frac{10}{-5} = -2$$

b) Since the numbers have unlike signs, the quotient is negative.

$$\frac{-45}{5} = -9$$

c) Since the numbers have like signs, both negative, the quotient is positive.

$$\frac{-36}{-6} = 6$$

EXAMPLE 6 Evaluate **a)** $-16 \div (-2)$ **b)** $\dfrac{-2}{3} \div \dfrac{-5}{7}$

Solution **a)** Since the numbers have like signs, both negative, the quotient is positive.

$$\frac{-16}{-2} = 8$$

b) Invert the *divisor*, $\dfrac{-5}{7}$, and then multiply.

$$\frac{-2}{3} \div \frac{-5}{7} = \left(\frac{-2}{3}\right)\left(\frac{7}{-5}\right) = \frac{-14}{-15} = \frac{14}{15}$$

HELPFUL HINT

For multiplication and division of two real numbers:

$$\left.\begin{array}{l} (+)(+) = + \\[2mm] (-)(-) = + \end{array}\right\} \quad \left.\begin{array}{l} \dfrac{(+)}{(+)} = + \\[2mm] \dfrac{(-)}{(-)} = + \end{array}\right\} \quad \text{\textit{Like signs give positive products and quotients.}}$$

$$\left.\begin{array}{l} (+)(-) = - \\[2mm] (-)(+) = - \end{array}\right\} \quad \left.\begin{array}{l} \dfrac{(+)}{(-)} = - \\[2mm] \dfrac{(-)}{(+)} = - \end{array}\right\} \quad \text{\textit{Like signs give negative products and quotients.}}$$

3 Remove Negative Signs from Denominators

We now know that the quotient of a positive and a negative number is a negative number. The fractions $-\frac{3}{4}, \frac{-3}{4}$, and $\frac{3}{-4}$ all represent the same negative number, negative three-fourths.

If a and b represent any real numbers, $b \neq 0$, then

$$\frac{a}{-b} = \frac{-a}{b} = -\frac{a}{b}$$

In mathematics we generally do not write a fraction with a negative sign in the denominator. When a negative sign appears in a denominator, we can move it to the numerator or place it in front of the fraction. For example, the fraction $\frac{5}{-7}$ should be written as either $-\frac{5}{7}$ or $\frac{-5}{7}$. Fractions also can be written using a slash, /. For example, the fraction $-\frac{5}{7}$ may be written $-5/7$ or $-(5/7)$.

EXAMPLE 7 Evaluate $\dfrac{2}{5} \div \left(\dfrac{-8}{15}\right)$.

Solution

$$\frac{2}{5} \div \left(\frac{-8}{15}\right) = \frac{\overset{1}{2}}{\underset{1}{5}} \cdot \left(\frac{\overset{3}{15}}{\underset{4}{-8}}\right) = \frac{1(3)}{1(-4)} = \frac{3}{-4} = -\frac{3}{4}$$

NOW TRY EXERCISE 73

The operations on real numbers are summarized in Table 1.1.

TABLE 1.1 Summary of Operations on Real Numbers				
Signs of Numbers	**Addition**	**Subtraction**	**Multiplication**	**Division**
Both Numbers Are Positive	Sum Is Always Positive	Difference May Be Either Positive or Negative	Product Is Always Positive	Quotient Is Always Positive
Examples				
6 and 2	$6 + 2 = 8$	$6 - 2 = 4$	$6 \cdot 2 = 12$	$6 \div 2 = 3$
2 and 6	$2 + 6 = 8$	$2 - 6 = -4$	$2 \cdot 6 = 12$	$2 \div 6 = \frac{1}{3}$
One Number Is Positive and the Other Number Is Negative	Sum May Be Either Positive or Negative	Difference May Be Either Positive or Negative	Product Is Always Negative	Quotient Is Always Negative
Examples				
6 and -2	$6 + (-2) = 4$	$6 - (-2) = 8$	$6(-2) = -12$	$6 \div (-2) = -3$
-6 and 2	$-6 + 2 = -4$	$-6 - 2 = -8$	$-6(2) = -12$	$-6 \div 2 = -3$
Both Numbers Are Negative	Sum Is Always Negative	Difference May Be Either Positive or Negative	Product Is Always Positive	Quotient Is Always Positive
Examples				
-6 and -2	$-6 + (-2) = -8$	$-6 - (-2) = -4$	$-6(-2) = 12$	$-6 \div (-2) = 3$
-2 and -6	$-2 + (-6) = -8$	$-2 - (-6) = 4$	$-2(-6) = 12$	$-2 \div (-6) = \frac{1}{3}$

4 Evaluate Divisions Involving 0

Now let's look at divisions involving the number 0. What is $\frac{0}{1}$ equal to? Note that $\frac{6}{3} = 2$ because $3 \cdot 2 = 6$. We can follow the same procedure to determine the value of $\frac{0}{1}$. Suppose that $\frac{0}{1}$ is equal to some number, which we will designate by ?.

$$\text{If} \quad \frac{0}{1} = \boxed{?} \quad \text{then} \quad 1 \cdot \boxed{?} = 0$$

Since only $1 \cdot 0 = 0$, the ? must be 0. Thus, $\frac{0}{1} = 0$. Using the same technique, we can show that zero divided by any nonzero number is zero.

> If a represents any real number except 0, then
>
> $$0 \div a = \frac{0}{a} = 0$$

Now what is $\frac{1}{0}$ equal to?

$$\text{If} \quad \frac{1}{0} = \boxed{?} \quad \text{then} \quad 0 \cdot \boxed{?} = 1$$

But since 0 multiplied by any number will be 0, there is no value that can replace ?. We say that $\frac{1}{0}$ is **undefined**. Using the same technique, we can show that any real number, except 0, divided by 0 is undefined.

TEACHING TIP
It may be helpful to use a table and a calculator to divide 1 by numbers closer and closer to 0. Then have students evaluate $\frac{1}{0}$ on their calculator.

> If a represents any real number except 0, then
>
> $$a \div 0 \quad \text{or} \quad \frac{a}{0} \quad \text{is } \textbf{undefined}$$

What is $\frac{0}{0}$ equal to?

$$\text{If} \quad \frac{0}{0} = \boxed{?} \quad \text{then} \quad 0 \cdot \boxed{?} = 0$$

But since the product of any number and 0 is 0, the $\boxed{?}$ can be replaced by any real number. Therefore the quotient $\frac{0}{0}$ cannot be determined, and so there is no answer. Therefore, we will not use it in this course.*

Summary of Division Involving 0

If a represents any real number except 0, then

$$\frac{0}{a} = 0 \qquad \frac{a}{0} \text{ is undefined}$$

EXAMPLE 8 Indicate whether each quotient is 0 or undefined.

a) $\dfrac{0}{2}$ **b)** $\dfrac{5}{0}$ **c)** $\dfrac{0}{-4}$ **d)** $\dfrac{-2}{0}$

Solution The answer to parts **a)** and **c)** is 0. The answer to parts **b)** and **d)** is undefined. ✳

NOW TRY EXERCISE 95

Using Your Calculator

Multiplication and Division on a Scientific Calculator

Below we show how numbers may be multiplied and divided on a calculator.

EVALUATE	KEYSTROKES	ANSWER DISPLAYED
$6(-23)$	6 $\boxed{\times}$ 23 $\boxed{+/-}$ $\boxed{=}$	-138
$\dfrac{-240}{-16}$	240 $\boxed{+/-}$ $\boxed{\div}$ 16 $\boxed{+/-}$ $\boxed{=}$	15

Multiplication and Division on a Graphing Calculator

EVALUATE	KEYSTROKES	ANSWER DISPLAYED
$6(-23)$	6 $\boxed{\times}$ $\boxed{(-)}$ 23 $\boxed{\text{ENTER}}$	-138
$\dfrac{-240}{-16}$	$\boxed{(-)}$ 240 $\boxed{\div}$ $\boxed{(-)}$ 16 $\boxed{\text{ENTER}}$	15

Since a positive number multiplied by a negative number will be negative, to obtain the product of $6(-23)$, you can multiply, $(6)(23)$ and write a negative sign before the answer. Since a negative number divided by a negative number is positive, $\frac{-240}{-16}$ could have been found by dividing $\frac{240}{16}$.

---

*At this level, some professors prefer to call $\frac{0}{0}$ *indeterminate* while others prefer to call $\frac{0}{0}$ *undefined*. In higher-level mathematics courses, $\frac{0}{0}$ is sometimes referred to as the *indeterminate form*.

1.–2. like signs: product or quotient is positive; unlike signs: product or quotient is negative   **3.** even number of negative numbers:

Exercise Set 1.8

product is positive, odd number: negative   **5.** $-\frac{a}{b}$ or $\frac{-a}{b}$

7. a) $3 - 5$, subtraction; $3(-5)$, multiplication   **8. a)** $-4 - 2$, subtraction; $(-4)(-2)$, multiplication   **9. a)** $x - y$, subtraction; $x(-y)$,

Concept/Writing Exercises

multiplication   **b)** 7   **c)** 10   **d)** -3 **10. a)** -24 **b)** 24   **c)** -11 **d)** 5

1. State the rules used to determine the sign of the product of two real numbers.

2. State the rules used to determine the sign of the quotient of two real numbers.

3. When multiplying three or more real numbers, explain how to determine the sign of the product of the numbers.

4. What is the product of 0 and any real number? 0

5. How do we generally rewrite a fraction of the form $\frac{a}{-b}$, where a and b represent any positive real numbers?

6. **a)** What is $\frac{0}{a}$ equal to where a is any nonzero real number? 0

b) What is $\frac{a}{0}$ equal to where a is any nonzero real number? undefined

7. **a)** Explain the difference between $3 - 5$ and $3(-5)$.

b) Evaluate $3 - 5$ and $3(-5)$. $-2, -15$

8. **a)** Explain the difference between $-4 - 2$ and $(-4)(-2)$.

b) Evaluate $-4 - 2$ and $(-4)(-2)$. $-6, 8$

9. **a)** Explain the difference between $x - y$ and $x(-y)$ where x and y represent any real numbers. If x is 5 and y is -2, find the value of **b)** $x - y$, **c)** $x(-y)$, and **d)** $-x - y$.

10. If x is -8 and y is 3, find the value of **a)** xy, **b)** $x(-y)$, **c)** $x - y$, and **d)** $-x - y$.

Determine the sign of each product. Explain how you determined the sign.

11. $(8)(4)(-5)$ negative

12. $(-9)(-12)(20)$ positive

13. $(-102)(-16)(24)(19)$ positive

14. $(1054)(-92)(-16)(-37)$ negative

15. $(-40)(-16)(30)(50)(-13)$ negative

16. $(-1)(3)(-462)(-196)(-312)$ positive

Practice the Skills

Find each product.

17. $(-5)(-4)$ 20

18. $-4(2)$ -8

19. $6(-3)$ -18

20. $6(-2)$ -12

21. $(-2)(4)$ -8

22. $(-3)(2)$ -6

23. $0(-5)$ 0

24. $-1(8)$ -8

25. $6(7)$ 42

26. $-9(-4)$ 36

27. $(7)(-8)$ -56

28. $7(-7)$ -49

29. $(-5)(-6)$ 30

30. $-2(5)$ -10

31. $0(3)(8)$ 0

32. $5(-4)(2)$ -40

33. $(21)(-1)(4)$ -84

34. $2(8)(-1)(-3)$ 48

35. $-1(-3)(3)(-8)$ -72

36. $(2)(-4)(-5)(-1)$ -40

37. $(-4)(5)(-7)(1)$ 140

38. $(-3)(2)(5)(3)$ -90

39. $(-1)(3)(0)(-7)$ 0

40. $(-6)(6)(4)(-4)$ 576

Find each product.

41. $\left(\frac{-1}{2}\right)\left(\frac{3}{5}\right)$ $-\frac{3}{10}$

42. $\left(\frac{1}{3}\right)\left(\frac{-3}{5}\right)$ $-\frac{1}{5}$

43. $\left(\frac{-5}{9}\right)\left(\frac{-7}{15}\right)$ $\frac{7}{27}$

44. $\left(\frac{4}{5}\right)\left(\frac{-3}{10}\right)$ $-\frac{6}{25}$

45. $\left(\frac{6}{-3}\right)\left(\frac{4}{-2}\right)$ 4

46. $\left(\frac{9}{-10}\right)\left(\frac{6}{-7}\right)$ $\frac{27}{35}$

47. $\left(\frac{-3}{8}\right)\left(\frac{5}{6}\right)$ $-\frac{5}{16}$

48. $\left(\frac{9}{10}\right)\left(\frac{7}{-8}\right)$ $-\frac{63}{80}$

Find each quotient.

49. $\frac{10}{5}$ 2

50. $25 \div 5$ 5

51. $-16 \div (-4)$ 4

52. $\frac{-18}{9}$ -2

53. $\frac{-36}{-9}$ 4

54. $\frac{30}{-6}$ -5

55. $\frac{36}{-2}$ -18

56. $\frac{-15}{-1}$ 15

57. $\frac{-19}{-1}$ 19

58. $-15/(-3)$ 5

59. $40/(-4)$ -10

60. $\frac{-6}{-1}$ 6

61. $\frac{-42}{7}$ -6

62. $\frac{-25}{-5}$ 5

63. $\frac{36}{-4}$ -9

64. $\frac{-10}{10}$ -1

65. $-64 \div (-8)$ 8

66. $-64/(-4)$ 16

67. Divide 0 by 4.   0

68. Divide 26 by -13. -2

69. Divide 30 by -10. -3

70. Divide -30 by -5.   6

71. Divide -120 by 30. -4

72. Divide -25 by -5.   5

Find each quotient.

73. $\dfrac{3}{12} \div \left(\dfrac{-5}{8}\right)$ $-\dfrac{2}{5}$

74. $(-3) \div \dfrac{5}{19}$ $-\dfrac{57}{5}$

75. $\dfrac{-5}{12} \div (-3)$ $\dfrac{5}{36}$

76. $\dfrac{-4}{9} \div \left(\dfrac{-6}{7}\right)$ $\dfrac{14}{27}$

🔒 **77.** $\dfrac{-15}{21} \div \left(\dfrac{-15}{21}\right)$ 1

78. $\dfrac{6}{15} \div \left(\dfrac{7}{30}\right)$ $\dfrac{12}{7}$

79. $(-12) \div \dfrac{5}{12}$ $-\dfrac{144}{5}$

80. $\dfrac{-16}{3} \div \left(\dfrac{5}{-9}\right)$ $\dfrac{48}{5}$

Evaluate.

🔒 **81.** $-4(8)$ -32

82. $\dfrac{-18}{-2}$ 9

83. $\dfrac{100}{-5}$ -20

84. $-50 \div (-10)$ 5

85. $-7(2)$ -14

86. $-5(-7)$ 35

87. $27 \div (-3)$ -9

88. Divide -120 by -10.   12

89. $-100 \div 5$ -20

90. $4(-2)(-1)(-5)$ -40

91. Divide 60 by -60. -1

92. $(6)(1)(-3)(4)$ -72

Indicate whether each quotient is 0 or undefined.

93. $0 \div 8$ 0

94. $0 \div (-7)$ 0

95. $\dfrac{5}{0}$ undefined

96. $\dfrac{-2}{0}$ undefined

97. $\dfrac{0}{1}$ 0

98. $\dfrac{6}{0}$ undefined

99. 8 divided by 0   undefined

100. 0 divided by 12   0

*Exercises 101–116, **a)** determine by observation whether the product or quotient will be a positive number, zero, a negative number; or undefined; **b)** find the product or quotient on your calculator (an error message indicates that the quotient is undefined); **c)** examine your answer in part **b)** to see whether it is reasonable and makes sense.*

101. $92(-38)$ -3496

102. $-168 \div 42$ -4

103. $-240/15$ -16

104. $190/10$ 19

🔒 **105.** $243 \div (-27)$ -9

106. $(323)(-115)$ $-37,145$

107. $(-49)(-126)$ 6174

108. $(1530)(0)$ 0

109. $0 \div 5335$ 0

110. $-86.4 \div (-36)$ 2.4

111. $8.2 \div 0$ undefined

112. $-37.74 \div 37$ -1.02

113. $8 \div (2.5)$ 3.2

114. $(1.1)(9.72)(6.3)$ 67.3596

115. $(-3.0)(4.2)(-18)$ 226.8

116. $-288.86/1.43$ -202

Indicate whether each statement is true or false.

117. The product of two negative numbers is a negative number.   F

118. The product of a positive number and a negative number is a negative number.   T

119. The quotient of two numbers with unlike signs is a positive number.   F

120. The quotient of two negative numbers is a positive number.   T

121. The product of an even number of negative numbers is a positive number.   T

122. The product of an odd number of negative numbers is a negative number.   T

123. Zero divided by 1 is 1.   F

124. Six divided by 0 is 0.   F

125. Zero divided by 1 is undefined.   F

126. Five divided by 0 is undefined.   T

127. The product of 0 and any real number is 0.   T

128. Division by 0 does not result in a real number.   T

129. 60 yards loss or −60 yards   136. a) get $\frac{1}{\text{number}}$ as a decimal   c) and d) get error

Problem Solving

129. *Football* A high school football team is penalized four times, each time with a loss of 15 yards, or −15 yards. Find the total loss due to penalties.

130. *Submarine Dive* A submarine is at a depth of −160 feet (160 feet below sea level). It dives to 3 times that depth. Find its new depth.   −480 ft

131. *Credit Card* Leona De Vito's balance on her credit card is −$450 (she owes $450). She pays back $\frac{1}{3}$ of this balance.
   a) How much did she pay back? $150
   b) What is her new balance?   −$300

132. *Money Owed* Dominike Jason owes a friend $300. After he makes four payments of $30 each, how much will he still owe? $180

133. *Stock Loss* If a stock loses $1\frac{1}{2}$ points each day for three successive days, how much has it lost in total? $4\frac{1}{2}$ points

134. *Wind Chill* On Monday in Minneapolis the wind chill was −30°F. On Tuesday the wind chill was only $\frac{1}{3}$ of what it was on Monday. What was the wind chill on Tuesday?   −10°F

135. *Employment* The following chart shows the change in nonfarm payroll employment (seasonally adjusted) for the United States from November, 2000 (N), through October, 2001 (O).

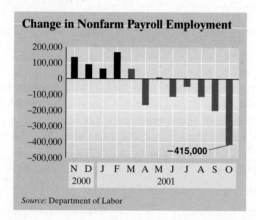

Source: Department of Labor

a) How many times larger was the change in October than the change in September?   ≈2.075

b) How many times larger was the change in October than the change in July?   ≈8.3

136. *Reciprocals* Most calculators have a *reciprocal key*, $\boxed{1/x.}$

   a) Press a number key between 1 and 9 on your calculator. Then press the $\boxed{1/x}$ key. Indicate what happened when you pressed this key.

   b) Press the $\boxed{1/x}$ key a second time. What happens now?   get number

   c) What do you think will happen if you enter 0 and then press the $\boxed{1/x}$ key? Explain your answer.

   d) Enter 0 and then press the $\boxed{1/x}$ key to see whether your answer to part **c)** was correct. If not, explain why.

137. *Heart Rate* The Johns Hopkins Medical Letter states that to find a person's *target heart rate* in beats per minute, follow this procedure. Subtract the person's age from 220, then multiply this difference by 60% and 75%. The difference multiplied by 60% gives the lower limit and the difference multiplied by 75% gives the upper limit.

   a) Find the target heart rate rate range of a 50-year-old.   102–128 beats per minute

   b) Find you own target heart rate.   Answers will vary.

Challenge Problems

We will learn in the next section that $2^3 = 2 \cdot 2 \cdot 2$ *and* $x^n = \underbrace{x \cdot x \cdot x \cdots \cdot x}_{n \text{ factors of } x}.$

Use this information to evaluate each expression.

138. 3^4 81   **139.** $(-2)^3$ −8   **140.** $\left(\dfrac{2}{3}\right)^3$ $\dfrac{8}{27}$ **141.** 1^{100} 1   **142.** $(-1)^{81}$ −1

143. Will the product of $(-1)(-2)(-3)(-4)\cdots(-10)$ be a positive number or a negative number? Explain how you determined your answer.   positive, 10 negative numbers

144. Will the product of $(1)(-2)(3)(-4)(5)(-6)\cdots(33)(-34)$ be a positive number or a negative number? Explain how you determined your answer.   negative, 17 negative numbers

Group Activity

Discuss and answer Exercise 145 as a group, according to the instructions.

145. a) Each member of the group is to do this procedure separately. At this time do not share your number with the other members of your group.

  1. Choose a number between 2 and 10.

  2. Multiply your number by 9.

  3. Add the two digits in the product together.

  4. Subtract 5 from the sum.

  5. Now choose the corresponding letter of the alphabet that corresponds with the difference found. For example, 1 is a, 2 is b, 3 is c, and so on.

  6. Choose a *one-word* country that starts with that letter.

  7. Now choose a *one-word* animal that starts with the last letter of the country selected.

  8. Finally, choose a color that starts with the last letter of the animal chosen.

b) Now share your final answer with the other members of your group. Did you all get the same answer?

c) Most people will obtain the answer *orange*. As a group, write a paragraph or two explaining why.

Country will start with D. Most students will select Denmark. They will most likely select kangaroo which leads to orange.

Cumulative Review Exercises

[1.3] **146.** Find the quotient $\frac{5}{7} \div \frac{1}{5}$. $\frac{25}{7}$ or $3\frac{4}{7}$

[1.7] **147.** Subtract -18 from -20. -2

  148. Evaluate $6 - 3 - 4 - 2$ -3

  149. Evaluate $5 - (-2) + 3 - 7$ 3

[1.8] **150.** Evaluate $-40 \div (-8)$ 5

1.9 EXPONENTS, PARENTHESES, AND THE ORDER OF OPERATIONS

1 Learn the meaning of exponents.

2 Evaluate expressions containing exponents.

3 Learn the difference between $-x^2$ and $(-x)^2$.

4 Learn the order of operations.

5 Learn the use of parentheses.

6 Evaluate expressions containing variables.

SSM   Study Guide   CD/Video   MathPro 4/5   PH Math Tutor Center   prenhall.com/Angel

1 Learn the Meaning of Exponents

To understand certain topics in algebra, you must understand exponents. Exponents are introduced in this section and are discussed in more detail in Chapter 4.

In the expression 4^2, the 4 is called the **base**, and the 2 is called the **exponent**. The number 4^2 is read "4 squared" or "4 to the second power" and means

$$4 \cdot 4 = 4^2 \leftarrow exponent$$

2 factors of 4

The number 4^3 is read "4 cubed" or "4 to the third power" and means

$$4 \cdot 4 \cdot 4 = 4^3$$

3 factors of 4

In general, the number b to the nth power, written b^n, means

$$\underbrace{b \cdot b \cdot b \cdot \,\cdots\, \cdot b}_{n \text{ factors of } b} = b^n$$

Thus, $b^4 = b \cdot b \cdot b \cdot b$ or $bbbb$ and $x^3 = x \cdot x \cdot x$ or xxx.

2 Evaluate Expressions Containing Exponents

Now let's evaluate some expressions that contain exponents.

EXAMPLE 1 Evaluate. **a)** 3^2 **b)** 2^5 **c)** 1^5 **d)** $(-6)^2$ **e)** $(-2)^3$ **f)** $\left(\dfrac{2}{3}\right)^2$

Solution **a)** $3^2 = 3 \cdot 3 = 9$

b) $2^5 = 2 \cdot 2 \cdot 2 \cdot 2 \cdot 2 = 32$

c) $1^5 = 1 \cdot 1 \cdot 1 \cdot 1 \cdot 1 = 1$ (1 raised to any power equals 1; why?)

d) $(-6)^2 = (-6)(-6) = 36$

e) $(-2)^3 = (-2)(-2)(-2) = -8$

f) $\left(\dfrac{2}{3}\right)^2 = \left(\dfrac{2}{3}\right)\left(\dfrac{2}{3}\right) = \dfrac{4}{9}$

NOW TRY EXERCISE 29

It is not necessary to write exponents of 1. Thus, when writing xxy, we write x^2y and not x^2y^1. **Whenever we see a letter or number without an exponent, we always assume that the letter or number has an exponent of 1.**

Examples of Exponential Notation

a) $xyxx = x^3y$ **b)** $xyzzy = xy^2z^2$

c) $3aabbb = 3a^2b^3$ **d)** $5xyyyy = 5xy^4$

e) $4 \cdot 4rrs = 4^2r^2s$ **f)** $5 \cdot 5 \cdot 5mmn = 5^3m^2n$

Notice in parts **a)** and **b)** that the order of the factors does not matter.

HELPFUL HINT Note that $x + x + x + x + x + x = 6x$ and $x \cdot x \cdot x \cdot x \cdot x \cdot x = x^6$. Be careful that you do not get addition and multiplication confused.

---

3 Learn the Difference between $-x^2$ and $(-x)^2$

An exponent refers only to the number or letter that directly precedes it unless parentheses are used to indicate otherwise. For example, in the expression $3x^2$, only the x is squared. In the expression $-x^2$ only the x is squared. We can write $-x^2$ as $-1x^2$ because any real number may be multiplied by 1 without affecting its value.

$$-x^2 = -1x^2$$

By looking at $-1x^2$ we can see that only the x is squared, not the -1. If the entire expression $-x$ were to be squared, we would need to use parentheses and write $(-x)^2$. Note the difference in the following two examples:

$$-x^2 = -(x)(x)$$
$$(-x)^2 = (-x)(-x)$$

Group Activity

Discuss and answer Exercise 145 as a group, according to the instructions.

145. a) Each member of the group is to do this procedure separately. At this time do not share your number with the other members of your group.

    1. Choose a number between 2 and 10.
    2. Multiply your number by 9.
    3. Add the two digits in the product together.
    4. Subtract 5 from the sum.
    5. Now choose the corresponding letter of the alphabet that corresponds with the difference found. For example, 1 is a, 2 is b, 3 is c, and so on.

    6. Choose a *one-word* country that starts with that letter.
    7. Now choose a *one-word* animal that starts with the last letter of the country selected.
    8. Finally, choose a color that starts with the last letter of the animal chosen.

b) Now share your final answer with the other members of your group. Did you all get the same answer?

c) Most people will obtain the answer *orange*. As a group, write a paragraph or two explaining why.

Country will start with D. Most students will select Denmark. They will most likely select kangaroo which leads to orange.

Cumulative Review Exercises

[1.3] **146.** Find the quotient $\dfrac{5}{7} \div \dfrac{1}{5}$. $\dfrac{25}{7}$ or $3\dfrac{4}{7}$

[1.7] **147.** Subtract -18 from -20. -2

    148. Evaluate $6 - 3 - 4 - 2$ -3

    149. Evaluate $5 - (-2) + 3 - 7$ 3

[1.8] **150.** Evaluate $-40 \div (-8)$ 5

1.9 EXPONENTS, PARENTHESES, AND THE ORDER OF OPERATIONS

SSM   Study Guide   CD/Video

MathPro 4/5   PH Math Tutor Center   prenhall.com/Angel

1   Learn the meaning of exponents.
2   Evaluate expressions containing exponents.
3   Learn the difference between $-x^2$ and $(-x)^2$.
4   Learn the order of operations.
5   Learn the use of parentheses.
6   Evaluate expressions containing variables.

1   Learn the Meaning of Exponents

To understand certain topics in algebra, you must understand exponents. Exponents are introduced in this section and are discussed in more detail in Chapter 4.

In the expression 4^2, the 4 is called the **base**, and the 2 is called the **exponent**. The number 4^2 is read "4 squared" or "4 to the second power" and means

$$\underbrace{4 \cdot 4}_{2 \text{ factors of } 4} = 4^2 \leftarrow \text{exponent}$$

The number 4^3 is read "4 cubed" or "4 to the third power" and means

$$\underbrace{4 \cdot 4 \cdot 4}_{3 \text{ factors of } 4} = 4^3$$

In general, the number b to the nth power, written b^n, means

$$\underbrace{b \cdot b \cdot b \cdot \,\cdots\, \cdot b}_{n \text{ factors of } b} = b^n$$

Thus, $b^4 = b \cdot b \cdot b \cdot b$ or $bbbb$ and $x^3 = x \cdot x \cdot x$ or xxx.

2 Evaluate Expressions Containing Exponents

Now let's evaluate some expressions that contain exponents.

EXAMPLE 1 Evaluate. **a)** 3^2 **b)** 2^5 **c)** 1^5 **d)** $(-6)^2$ **e)** $(-2)^3$ **f)** $\left(\dfrac{2}{3}\right)^2$

Solution
a) $3^2 = 3 \cdot 3 = 9$
b) $2^5 = 2 \cdot 2 \cdot 2 \cdot 2 \cdot 2 = 32$
c) $1^5 = 1 \cdot 1 \cdot 1 \cdot 1 \cdot 1 = 1$ (1 raised to any power equals 1; why?)
d) $(-6)^2 = (-6)(-6) = 36$
e) $(-2)^3 = (-2)(-2)(-2) = -8$
f) $\left(\dfrac{2}{3}\right)^2 = \left(\dfrac{2}{3}\right)\left(\dfrac{2}{3}\right) = \dfrac{4}{9}$

NOW TRY EXERCISE 29

It is not necessary to write exponents of 1. Thus, when writing xxy, we write x^2y and not x^2y^1. **Whenever we see a letter or number without an exponent, we always assume that the letter or number has an exponent of 1.**

Examples of Exponential Notation

a) $xyxx = x^3y$ **b)** $xyzzy = xy^2z^2$

c) $3aabbb = 3a^2b^3$ **d)** $5xyyyy = 5xy^4$

e) $4 \cdot 4rrs = 4^2r^2s$ **f)** $5 \cdot 5 \cdot 5mmn = 5^3m^2n$

Notice in parts **a)** and **b)** that the order of the factors does not matter.

HELPFUL HINT

Note that $x + x + x + x + x + x = 6x$ and $x \cdot x \cdot x \cdot x \cdot x \cdot x = x^6$. Be careful that you do not get addition and multiplication confused.

---

3 Learn the Difference between $-x^2$ and $(-x)^2$

TEACHING TIP
It may be helpful to verbalize $-x^2$ as the opposite of x^2.

An exponent refers only to the number or letter that directly precedes it unless parentheses are used to indicate otherwise. For example, in the expression $3x^2$, only the x is squared. In the expression $-x^2$ only the x is squared. We can write $-x^2$ as $-1x^2$ because any real number may be multiplied by 1 without affecting its value.

$$-x^2 = -1x^2$$

By looking at $-1x^2$ we can see that only the x is squared, not the -1. If the entire expression $-x$ were to be squared, we would need to use parentheses and write $(-x)^2$. Note the difference in the following two examples:

$$-x^2 = -(x)(x)$$
$$(-x)^2 = (-x)(-x)$$

Consider the expressions -3^2 and $(-3)^2$. How do they differ?

$$-3^2 = -(3)(3) = -9$$
$$(-3)^2 = (-3)(-3) = 9$$

HELPFUL HINT The expression $-x^2$ is read "negative x squared," or "the opposite of x squared." The expression $(-x)^2$ is read "negative x, quantity squared."

EXAMPLE 2 Evaluate. **a)** -5^2 **b)** $(-5)^2$ **c)** -2^3 **d)** $(-2)^3$

Solution **a)** $-5^2 = -(5)(5) = -25$ **b)** $(-5)^2 = (-5)(-5) = 25$
c) $-2^3 = -(2)(2)(2) = -8$ **d)** $(-2)^3 = (-2)(-2)(-2) = -8$ ✳

EXAMPLE 3 Evaluate. **a)** -2^4 **b)** $(-2)^4$

Solution **a)** $-2^4 = -(2)(2)(2)(2) = -16$ **b)** $(-2)^4 = (-2)(-2)(-2)(-2) = 16$ ✳

NOW TRY EXERCISE 105

Using Your Calculator

Use of $\boxed{x^2}$, $\boxed{y^x}$, and $\boxed{\wedge}$ Keys

The $\boxed{x^2}$ key is used to square a value. For example, to evaluate 5^2, we would do the following.

	KEYSTROKES	ANSWER DISPLAYED
Scientific Calculator	5 $\boxed{x^2}$	25
Graphing Calculator	5 $\boxed{x^2}$ $\boxed{\text{ENTER}}$	25

To evaluate $(-5)^2$ on a calculator, we would do the following.

	KEYSTROKES	ANSWER DISPLAYED
*Scientific Calculator	5 $\boxed{+/-}$ $\boxed{x^2}$	25
Graphing Calculator	$\boxed{(}$ $\boxed{(-)}$ 5 $\boxed{)}$ $\boxed{x^2}$ $\boxed{\text{ENTER}}$	25

To raise a value to a power greater than 2, we use the $\boxed{y^x}$ * or $\boxed{\wedge}$ key. To use these keys you enter the number, then press either the $\boxed{y^x}$ or $\boxed{\wedge}$ key, then enter the exponent. Following we show how to evaluate 2^5 and $(-2)^5$.

	EVALUATE	KEYSTROKES	ANSWER DISPLAYED
Scientific Calculator	2^5	2 $\boxed{y^x}$ 5 $\boxed{=}$	32
*Scientific Calculator	$(-2)^5$	2 $\boxed{+/-}$ $\boxed{y^x}$ 5 $\boxed{=}$ **	−32
Graphing Calculator	2^5	2 $\boxed{\wedge}$ 5 $\boxed{\text{ENTER}}$	32
Graphing Calculator	$(-2)^5$	$\boxed{(}$ $\boxed{(-)}$ 2 $\boxed{)}$ $\boxed{\wedge}$ 5 $\boxed{\text{ENTER}}$	−32

Possibly the easiest way to raise negative numbers to a power may be to raise the positive number to the power and then write a negative sign before the final answer if needed. *A negative number raised to an odd power will be negative, and a negative number raised to an even power will be positive.* Can you explain why this is true?

* Some newer scientific calculators have a $\boxed{(-)}$ key. To evaluate a negative expression raised to a power on these calculators, follow the instructions for the graphing calculator.

** Some calculators use a $\boxed{x^y}$ key instead of a $\boxed{y^x}$ key.

4 Learn the Order of Operations

Now that we have introduced exponents we can present the **order of operations**. Can you evaluate $2 + 3 \cdot 4$? Is it 20? Or is it 14? To answer this, you must know the order of operations to follow when evaluating a mathematical expression. You will often have to evaluate expressions containing multiple operations.

Order of Operations
To Evaluate Mathematical Expressions,
Use the Following Order

1. First, evaluate the information within **parentheses** ( ), brackets [ ], or braces { }. These are **grouping symbols**, for they group information together. A fraction bar, —, also serves as a grouping symbol. If the expression contains nested parentheses (one pair of parentheses within another pair), evaluate the information in the innermost parentheses first.

2. Next, evaluate all **exponents**.

3. Next, evaluate all **multiplications** or **divisions** in the order in which they occur, working from left to right.

4. Finally, evaluate all **additions** or **subtractions** in the order in which they occur, working from left to right.

Some students remember the word PEMDAS or the phrase "Please Excuse My Dear Aunt Sally" to help them remember the order of operations. PEMDAS helps them remember the order: **P**arentheses, **E**xponents, **M**ultiplication, **D**ivision, **A**ddition, **S**ubtraction. Remember, this does not imply multiplication before division or addition before subtraction.

We can now evaluate $2 + 3 \cdot 4$. Since multiplications are performed before additions,

$$2 + 3 \cdot 4 \quad \text{means} \quad 2 + (3 \cdot 4) = 2 + 12 = 14$$

Using Your Calculator

We now know that $2 + 3 \cdot 4$ means $2 + (3 \cdot 4)$ and has a value of 14. What will a calculator display if you key in the following?

$$2 \boxed{+} 3 \boxed{\times} 4 \boxed{=}$$

The answer depends on your calculator. *Scientific and graphing calculators* will evaluate an expression following the rules just stated.

	KEYSTROKES	ANSWER DISPLAYED
Scientific Calculator	$2 \boxed{+} 3 \boxed{\times} 4 \boxed{=}$	14
Graphing Calculator	$2 \boxed{+} 3 \boxed{\times} 4 \boxed{\text{ENTER}}$	14

Nonscientific calculators will perform operations in the order they are entered.

		ANSWER DISPLAYED
Nonscientific Calculator	$2 \boxed{+} 3 \boxed{\times} 4 \boxed{=}$	20

Remember that in algebra, unless otherwise instructed by parentheses, we always perform multiplications and divisions before additions and subtractions. In this course you should be using either a scientific or graphing calculator.

5 Learn the Use of Parentheses

Parentheses or brackets may be used (1) to change the order of operations to be followed in evaluating an algebraic expression or (2) to help clarify the understanding of an expression.

To evaluate the expression $2 + 3 \cdot 4$, we would normally perform the multiplication, $3 \cdot 4$, first. If we wished to have the addition performed before the multiplication, we could indicate this by placing parentheses around $2 + 3$:

$$(2 + 3) \cdot 4 = 5 \cdot 4 = 20$$

Consider the expression $1 \cdot 3 + 2 \cdot 4$. According to the order of operations, multiplications are to be performed before additions. We can rewrite this expression as $(1 \cdot 3) + (2 \cdot 4)$. Note that the order of operations was not changed. The parentheses were used only to help clarify the order to be followed.

Sometimes it may be necessary to use more than one set of parentheses to indicate the order to be followed in an expression. As indicated in step 1 of the Order of Operations box, when one set of parentheses is within another set of parentheses, we call this **nested parentheses**. For example, the expression $6(2 + 3(4 + 1))$ has nested parentheses. Often, to make an expression with nested parentheses easier to follow, brackets, [ ], or braces, { }, are used in place of multiple parentheses. Thus, we could write the expression $6(2 + 3(4 + 1))$ as $6[2 + 3(4 + 1)]$. Whenever we are given an expression with nested parentheses, we always evaluate the numbers in the *innermost parentheses first*. Color shading is used in the following examples to indicate the order in which the expression is evaluated.

$$6[2 + 3(4 + 1)] = 6[2 + 3(5)] = 6[2 + 15] = 6[17] = 102$$

$$4[3(6 - 4) \div 6] = 4[3(2) \div 6] = 4[6 \div 6] = 4[1] = 4$$

$$\{2 + [(8 \div 4)^2 - 1]\}^2 = [2 + (2^2 - 1)]^2 = [2 + (4 - 1)]^2 = (2 + 3)^2 = 5^2 = 25$$

Now we will work some examples, but before we do, read the following Helpful Hint.

HELPFUL HINT If parentheses are not used to change the order of operations, multiplications and divisions are always performed before additions and subtractions. When a problem has only multiplications and divisions, work from left to right. Similarly, when a problem has only additions and subtractions, work from left to right.

EXAMPLE 4 Evaluate $6 + 3 \cdot 5^2 - 4$.

Solution Colored shading is used to indicate the order in which the expression is to be evaluated.

$$6 + 3 \cdot 5^2 - 4 \quad \text{Exponent}$$
$$= 6 + 3 \cdot 25 - 4 \quad \text{Multiply}$$
$$= 6 + 75 - 4 \quad \text{Add/subtract, left to right}$$
$$= 81 - 4$$
$$= 77$$

EXAMPLE 5 Evaluate $6 + 3[(32 \div 4^2) + 5]$.

Solution

$$6 + 3[(32 \div \boxed{4^2}) + 5] \quad \textit{Exponent}$$
$$= 6 + 3[(\boxed{32 \div 16}) + 5] \quad \textit{Divide}$$
$$= 6 + 3[\boxed{2 + 5}] \quad \textit{Add}$$
$$= 6 + \boxed{3[7]} \quad \textit{Multiply}$$
$$= 6 + 21$$
$$= 27$$

EXAMPLE 6 Evaluate $(8 \div 2) + 7(5 - 2)^2$.

Solution

$$\boxed{(8 \div 2)} + 7\boxed{(5 - 2)}^2 \quad \textit{Parentheses}$$
$$= 4 + 7\boxed{(3)^2} \quad \textit{Exponent}$$
$$= 4 + \boxed{7 \cdot 9} \quad \textit{Multiply}$$
$$= 4 + 63$$

NOW TRY EXERCISE 81
$$= 67$$

EXAMPLE 7 Evaluate $-8 - 81 \div 9 \cdot 2^2 + 7$.

Solution

$$-8 - 81 \div 9 \cdot \boxed{2^2} + 7 \quad \textit{Exponent}$$
$$= -8 - \boxed{81 \div 9} \cdot 4 + 7 \quad \textit{Multiply/divide, left to right}$$
$$= -8 - \boxed{9 \cdot 4} + 7 \quad \textit{Multiply}$$
$$= \boxed{-8 - 36} + 7 \quad \textit{Add/subtract, left to right}$$
$$= -44 + 7$$
$$= -37$$

EXAMPLE 8 Evaluate. **a)** $-4^2 + 6 \div 3$ **b)** $(-4)^2 + 6 \div 3$

Solution

a)
$$- \boxed{4^2} + 6 \div 3 \quad \textit{Exponent}$$
$$= -16 + \boxed{6 \div 3} \quad \textit{Divide}$$
$$= -16 + 2$$
$$= -14$$

b)
$$\boxed{(-4)^2} + 6 \div 3 \quad \textit{Exponent}$$
$$= 16 + \boxed{6 \div 3} \quad \textit{Divide}$$
$$= 16 + 2$$
$$= 18$$

EXAMPLE 9 Evaluate $\dfrac{3}{8} - \dfrac{2}{5} \cdot \dfrac{1}{12}$.

Solution First perform the multiplication.

$$\frac{3}{8} - \left(\frac{\overset{1}{2}}{5} \cdot \frac{1}{\underset{6}{12}} \right) \quad \textit{Multiply}$$

$$= \frac{3}{8} - \frac{1}{30} \quad \textit{Subtract}$$

$$= \frac{45}{120} - \frac{4}{120}$$

$$= \frac{41}{120}$$

NOW TRY EXERCISE 91

EXAMPLE 10   Write the following statements as mathematical expressions using parentheses and brackets and then evaluate: Subtract 3 from 15. Divide this difference by 2. Multiply this quotient by 4.

Solution

$$15 - 3$$ *Subtract 3 from 15.*

$$(15 - 3) \div 2$$ *Divide by 2.*

$$4[(15 - 3) \div 2]$$ *Multiply the quotient by 4.*

Now evaluate.

$$4[(15 - 3) \div 2]$$
$$= 4[\,12 \div 2\,]$$
$$= 4(6)$$
$$= 24$$

As shown in Example 10, sometimes brackets are used in place of parentheses to help avoid confusion. If only parentheses had been used, the preceding expression would appear as $4((15 - 3) \div 2)$.

Using Your Calculator

Using Parentheses

When evaluating an expression on a calculator where the order of operations is to be changed, you will need to use parentheses. If you are not sure whether they are needed, it will not hurt to add them. Consider $\frac{8}{4 - 2}$. Since we wish to divide 8 by the difference $4 - 2$, we neeed to use parentheses.

EVALUATE	KEYSTROKES	ANSWER DISPLAYED
$\dfrac{8}{4 - 2}$	8 $\div$ (4 $-$ 2) $=$ *	4

What would you obtain if you evaluated 8 $\div$ 4 $-$ 2 $=$ on a scientific calculator? Why would you get that result?

To evaluate $\left(\frac{2}{5}\right)^2$ on a scientific calculator, we press the following keys.

EVALUATE	KEYSTROKES	ANSWER DISPLAYED
$\left(\dfrac{2}{5}\right)^2$	2 $\div$ 5 $=$ x^2 **	.16
	or (2 $\div$ 5) x^2	.16

What would you obtain if you evaluated 2 $\div$ 5 x^2 on a scientific calculator? Why?

* If using a graphing calculator, replace $=$ with ENTER . Everything else remains the same.

** On a graphing calculator, replace $=$ with ENTER and press ENTER after x^2 .

---

6   Evaluate Expressions Containing Variables

Now we will evaluate some expressions for given values of the variables.

EXAMPLE 11   Evaluate $5x - 4$ when $x = 3$.

Solution   Substitute 3 for x in the expression.

$$5x - 4 = 5(3) - 4 = 15 - 4 = 11$$

EXAMPLE 12   Evaluate **a)** x^2 and **b)** $-x^2$ when $x = 3$.

Solution   Substitute 3 for x.

a) $x^2 = 3^2 = 3(3) = 9$

b) $-x^2 = -3^2 = -(3)(3) = -9$

EXAMPLE 13   Evaluate **a)** y^2 and **b)** $-y^2$ when $y = -4$.

Solution   Substitute -4 for y.

a) $y^2 = (-4)^2 = (-4)(-4) = 16$

b) $-y^2 = -(-4)^2 = -(-4)(-4) = -16$

Note that $-x^2$ will always be a negative number for any nonzero value of x, and $(-x)^2$ will always be a positive number for any nonzero value of x. Can you explain why? See Exercise 6 on page 80.

AVOIDING COMMON ERRORS

The expression $-x^2$ means $-(x^2)$. When asked to evaluate $-x^2$ for any real number x, many students will incorrectly treat $-x^2$ as $(-x)^2$. For example, to evaluate $-x^2$ when $x = 5$,

CORRECT

$-5^2 = -(5^2) = -(5)(5)$
$= -25$

INCORRECT

$-5^2 = (-5)(-5)$
$= 25$

EXAMPLE 14   Evaluate $(4x + 1) + 2x^2$ when $x = \dfrac{1}{4}$.

Solution   Substitute $\frac{1}{4}$ for each x in the expression, then evaluate using the order of operations.

$$(4x + 1) + 2x^2 = \left[4\left(\frac{1}{4}\right) + 1\right] + 2\left(\frac{1}{4}\right)^2 \quad \textit{Substitute}$$

$$= [1 + 1] + 2\left(\frac{1}{4}\right)^2 \quad \textit{Multiply}$$

$$= 2 + 2\left(\frac{1}{16}\right) \quad \textit{Add, exponent}$$

$$= 2 + \frac{1}{8} \quad \textit{Multiply}$$

$$= 2\frac{1}{8}$$

EXAMPLE 15   Evaluate $-y^2 + 3(x + 2) - 5$ when $x = -3$ and $y = -2$.

Solution   Substitute -3 for each x and -2 for each y, then evaluate using the order of operations.

$$-y^2 + 3(x + 2) - 5 = -(-2)^2 + 3(-3 + 2) - 5 \quad \textit{Substitute}$$

$$= -(-2)^2 + 3(-1) - 5 \quad \textit{Parentheses}$$

$$= -(4) + 3(-1) - 5 \quad \textit{Exponent}$$

$$= -4 - 3 - 5 \quad \textit{Multiply}$$

$$= -7 - 5 \quad \textit{Subtract, left to right}$$

$$= -12$$

Using Your Calculator

Evaluating Expressions on a Scientific Calculator

Later in this course you will need to evaluate an expression like $3x^2 - 2x + 5$ for various values of x. Below we show how to evaluate such expressions.

EVALUATE	KEYSTROKES

a) $3x^2 - 2x + 5$, for $x = 4$

$3(4)^2 - 2(4) + 5$

3 $\boxed{\times}$ 4 $\boxed{x^2}$ $\boxed{-}$ 2 $\boxed{\times}$ 4 $\boxed{+}$ 5 $\boxed{=}$ 45

b) $3x^2 - 2x + 5$, for $x = -6$

$3(-6)^2 - 2(-6) + 5$

3 $\boxed{\times}$ 6 $\boxed{+/-}$ $\boxed{x^2}$ $\boxed{-}$ 2 $\boxed{\times}$ 6 $\boxed{+/-}$ $\boxed{+}$ 5 $\boxed{=}$ 125

c) $-x^2 - 3x - 5$, for $x = -2$

$-(-2)^2 - 3(-2) - 5$

1 $\boxed{+/-}$ $\boxed{\times}$ 2 $\boxed{+/-}$ $\boxed{x^2}$ $\boxed{-}$ 3 $\boxed{\times}$ 2 $\boxed{+/-}$ $\boxed{-}$ 5 $\boxed{=}$ -3

Remember in part **c)** that $-x^2 = -1x^2$.

Using Your Graphing Calculator

Evaluating Expressions on a Graphing Calculator

All graphing calculators have a procedure for evaluating expressions. To do so you will generally need to enter the value to be used for the variable and the expression to be evaluated. After the $\boxed{\text{ENTER}}$ key is pressed, the graphing calculator displays the answer. The procedure varies from calculator to calculator. Below we show how an expression is evaluated on a TI-83 Plus. On this calculator the store key—$\boxed{\text{STO} \blacktriangleright}$ —is used to store a value. Stored values and expressions are separated using a colon, which is obtained by pressing $\boxed{\text{ALPHA}}$ followed by $\boxed{\cdot}$.

EVALUATE

$3x^2 - 2x + 5$ for $x = -6$

KEYSTROKES ON TI-83 PLUS

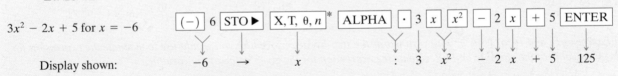

Display shown: $-6 \quad \rightarrow \quad x \quad : \quad 3 \quad x^2 \quad - \quad 2 \quad x \quad + \quad 5 \quad 125$

Notice that to obtain an x^2 on the display, we press the $\boxed{\text{X,T, }\theta\text{, }n}$ key to select the variable x, then we press the $\boxed{x^2}$ key, which is used to square the variable or number selected.

A nice feature of graphing calculators is that to evaluate an expression for different values of a variable you do not have to re-enter the expression each time. For example, on the TI-83 Plus if you press $\boxed{2^{nd}}$ followed by $\boxed{\text{ENTER}}$ it displays the expression again. Then you just go back and change the value stored for x to the new value and press $\boxed{\text{ENTER}}$ to evaluate the expression with the new value of the variable.

Each brand of calculator uses different keys and procedures. We have given just a quick overview. Please read the manual that comes with your graphing calculator for a complete explanation of how to evaluate expressions.

*This key can be used to generate any of these letters (θ is a Greek letter). From this point on, in displays of keystrokes, to generate an x we will just show $\boxed{x}$ rather than $\boxed{\text{X,T, }\theta\text{, }n}$.

6. a) $-x^2$ means $-(x^2)$

6. b) $(-x)^2$ means $(-x)(-x)$, product positive if $x < 0$ or $x > 0$

Exercise Set 1.9

7. parentheses, exponents, multiplication or division, addition or subtraction

Concept/Writing Exercises

1. In the expression a^b, what is the a called and what is the b called? base, exponent

2. Explain the meaning of

   a) 3^2 2 factors of 3

   b) 5^4 4 factors of 5

   c) x^n n factors of x

3. a) What is the exponent on a number or letter that has no exponent illustrated? 1

   b) In the expression $5x^3y^2z$, what is the exponent on the 5, the x, the y, and the z? 1, 3, 2, 1

4. Write a simplified expression for the following.

   a) $y + y + y + y$ 4y

   b) $y \cdot y \cdot y \cdot y$ y^4

5. Write a simplified expression for the following.

   a) $x + x + x + x + x$ 5x

   b) $x \cdot x \cdot x \cdot x \cdot x$ x^5

6. a) Explain why $-x^2$ will always be a negative number for any nonzero real number selected for x.

   b) Explain why $(-x)^2$ will always be a positive number for any nonzero real number selected for x.

7. List the order of operations to be followed when evaluating a mathematical expression.

8. When an expression has only additions and subtractions or only multiplications or divisions, how is it evaluated? left to right

9. If you evaluate $4 + 5 \times 2$ on a calculator and obtain an answer of 18, is the calculator a scientific calculator? Explain. no

10. List two reasons why parentheses are used in an expression. to change order or to clarify

11. Determine the results obtained on a scientific calculator if the following keys are pressed.

    a) 20 $\div$ 5 $-$ 3 $=$ 1

    b) 20 $\div$ (5 $-$ 3) $=$ 10

    c) Which keystrokes, a) or b), are used to evaluate $\frac{20}{5 - 3}$? Explain b)

12. Determine the results obtained on a scientific calculator if the following keys are pressed.

    a) 15 $-$ 10 $\div$ 5 $=$ 13

    b) (15 $-$ 10) $\div$ 5 $=$ 1

    c) Which keystrokes, a) or b), are used to evaluate $\frac{15 - 10}{5}$? Explain. b)

In Exercises 13 and 14, a) write in your own words the step-by-step procedure you would use to evaluate the expression, and b) evaluate the expression.

13. $[10 - (16 \div 4)]^2 - 6^3$ b) -180

14. $[(8 \cdot 3) - 4^2]^2 - 5$ b) 59

In Exercises 15 and 16, a) write in your own words the step-by-step procedure you would use to evaluate the expression for the given value of the variable, and b) evaluate the expression for the given value of the variable.

15. $-4x^2 + 3x - 6$ when $x = 5$ b) -91

16. $-5x^2 - 2x + 8$ when $x = -2$ b) -8

Practice the Skills

Evaluate.

17. 5^2 25
18. 2^3 8
19. 1^7 1
20. 3^3 27
21. -5^2 -25
22. 7^3 343
23. $(-3)^2$ 9
24. -6^3 -216
25. $(-1)^3$ -1
26. 2^5 32
27. $(-8)^2$ 64
28. 5^3 125
29. $(-6)^2$ 36
30. $(-3)^3$ -27
31. 4^1 4
32. -7^2 -49
33. $(-4)^4$ 256
34. -1^4 -1
35. -2^4 -16
36. $3^2(4)^2$ 144
37. $\left(\frac{3}{4}\right)^2$ $\frac{9}{16}$
38. $\left(\frac{5}{8}\right)^3$ $\frac{125}{512}$
39. $\left(-\frac{1}{2}\right)^5$ $-\frac{1}{32}$
40. $\left(-\frac{2}{3}\right)^4$ $\frac{16}{81}$
41. $5^2 \cdot 3^2$ 225
42. $(-1)^4(2)^4$ 16
43. $4^3 \cdot 3^2$ 576
44. $(-2)^3(-1)^9$ 8

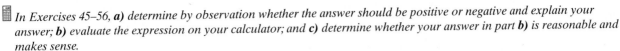 *In Exercises 45–56,* ***a)*** *determine by observation whether the answer should be positive or negative and explain your answer;* ***b)*** *evaluate the expression on your calculator; and* ***c)*** *determine whether your answer in part* ***b)*** *is reasonable and makes sense.*

45. 7^3 343

46. 4^6 4096

47. 6^4 1296

48. -2^5 -32

49. $(-3)^5$ -243

50. 10^3 1000

51. $(-5)^4$ 625

52. $(1.3)^3$ 2.197

53. $(4.6)^4$ 447.7456

54. $(-3.3)^3$ -35.937

55. $-\left(\dfrac{7}{8}\right)^2$ -0.765625

56. $\left(-\dfrac{3}{4}\right)^3$ -0.421875

Evaluate.

57. $3 + 2 \cdot 6$ 15

58. $7 - 5^2 + 2$ -16

59. $6 - 6 + 8$ 8

60. $(8^2 \div 4) - (20 - 4)$ 0

61. $1 + 3 \cdot 2^2$ 13

62. $8 \cdot 3^2 + 1 \cdot 5$ 77

63. $-3^3 + 27$ 0

64. $(-2)^3 + 8 \div 4$ -6

65. $(4 - 3) \cdot (5 - 1)^2$ 16

66. $-10 - 6 - 3 - 2$ -21

67. $3 \cdot 7 + 4 \cdot 2$ 29

68. $[1 - (4 \cdot 5)] + 6$ -13

69. $5 - 2(7 + 5)$ -19

70. $8 + 3(6 + 4)$ 38

71. $-32 - 5(7 - 10)^2$ -77

72. $-40 - 3(4 - 8)^2$ -88

73. $\dfrac{3}{4} + 2\left(\dfrac{1}{5}\right)^2$ $\dfrac{83}{100}$

74. $-\dfrac{2}{3} - 3\left(\dfrac{3}{4}\right)^2$ $-\dfrac{113}{48}$

75. $4^2 - 3 \cdot 4 - 6$ -2

76. $-2[-5 + (7 + 4)]$ -12

77. $(6 \div 3)^3 + 4^2 \div 8$ 10

78. $4 + (4^2 - 13)^4 - 3$ 82

79. $[-8(-2 + 5)]^2$ 576

80. $(-3)^2 + 6^2 \div 2^2 + 5$ 23

81. $(3^2 - 1) \div (3 + 1)^2$ $\dfrac{1}{2}$

82. $2[3(8 - 2^2) - 6]$ 12

83. $[4 + ((5 - 2)^2 \div 3)^2]^2$ 169

84. $(20 \div 5 \cdot 5 \div 5 - 5)^2$ 1

85. $2.5 + 7.56 \div 2.1 + (9.2)^2$ 90.74

86. $(8.4 + 3.1)^2 - (3.64 - 1.2)$ 129.81

87. $2[1.55 + 5(3.7)] - 3.35$ 36.75

88. $\dfrac{1}{2} + \dfrac{3}{4} \cdot \dfrac{5}{6}$ $\dfrac{9}{8}$ or $1\dfrac{1}{8}$

89. $\left(\dfrac{2}{5} + \dfrac{3}{8}\right) - \dfrac{3}{20}$ $\dfrac{5}{8}$

90. $\left(\dfrac{5}{6} \cdot \dfrac{4}{5}\right) + \left(\dfrac{2}{3} \cdot \dfrac{5}{8}\right)$ $\dfrac{13}{12}$ or $1\dfrac{1}{12}$

91. $\dfrac{3}{4} - 4 \cdot \dfrac{5}{40}$ $\dfrac{1}{4}$

92. $\dfrac{1}{8} - \dfrac{1}{4} \cdot \dfrac{3}{2} + \dfrac{3}{5}$ $\dfrac{7}{20}$

93. $\dfrac{4}{5} + \dfrac{3}{4} \div \dfrac{1}{2} - \dfrac{2}{3}$ $\dfrac{49}{30}$

94. $\dfrac{12 - (4 - 6)^2}{6 + 4^2 \div 2^2}$ $\dfrac{4}{5}$

95. $\dfrac{5 - [3(6 \div 3) - 2]}{5^2 - 4^2 \div 2}$ $\dfrac{1}{17}$

96. $\dfrac{[(7 - 3)^2 - 4]^2}{9 - 16 \div 8 - 4}$ 48

97. $\dfrac{-[4 - (6 - 12)^2]}{[(9 \div 3) + 4]^2 + 2^2}$ $\dfrac{32}{53}$

98. $\dfrac{[(5 - (3 - 7)) - 2]^2}{2[(16 \div 2^2) - (8 \cdot 4)]}$ $-\dfrac{49}{56}$

99. $\{5 - 2[4 - (6 \div 2)]^2\}^2$ 9

100. $\{-6 - [3(16 \div 4^2)]\}^2$ 81

101. $-\{4 - [-3 - (2 - 5)]^2\}$ -4

102. $3\{4[(3 - 4)^2 - 3]^3 - 1\}$ -99

Evaluate ***a)*** x^2, ***b)*** $-x^2$, *and* ***c)*** $(-x)^2$ *for the following values of x.*

103. 3 $9, -9, 9$

104. 8 $64, -64, 64$

105. -4 $16, -16, 16$

106. -5 $25, -25, 25$

107. 6 $36, -36, 36$

108. 7 $49, -49, 49$

109. $-\dfrac{1}{3}$ $\dfrac{1}{9}, -\dfrac{1}{9}, \dfrac{1}{9}$

110. $\dfrac{3}{4}$ $\dfrac{9}{16}, -\dfrac{9}{16}, \dfrac{9}{16}$

Evaluate each expression for the given value of the variable or variables.

111. $x + 6; x = -2$ 4

112. $2x - 4x + 5; x = 3$ -1

113. $5z - 2; z = 4$ 18

114. $3(x - 2); x = 5$ 9

115. $a^2 - 6; a = -3$ 3

116. $b^2 - 8; b = 5$ 17

117. $-4x^2 - 2x + 1; x = -1$ -1

118. $2r^2 - 5r + 3; r = 1$ 0

119. $3p^2 - 6p - 4; p = 2$ -4

120. $-w^2 - 5w + 3; w = 4$ -33

121. $-x^2 - 2x + 5; x = \dfrac{1}{2}$ $\dfrac{15}{4}$

122. $2x^2 - 4x - 10; x = \dfrac{3}{4}$ $-\dfrac{95}{8}$

123. $4(3x + 1)^2 - 6x; x = 5$ 994

124. $3n^2(2n - 1) + 5; n = -4$ -427

125. $-3s + 2t; s = 3, t = 4$ -1

126. $7m + 4n^2 - 5; m = 6, n = 7$ 233

127. $r^2 - s^2; r = -2, s = -3$ -5

128. $p^2 - q^2; p = 5, q = -3$ 16

129. $2(x + 2y) + 4x - 3y; x = 2; y = -3$ 9

130. $4(x + y)^2 + 2(x + y) + 3; x = 2, y = 4$ 159

131. $6x^2 + 3xy - y^2; x = 2, y = -3$ -3

132. $3(x - 4)^2 - (3y - 4)^2; x = -1, y = -2$ -25

Problem Solving

Write the following statements as mathematical expressions using parentheses and brackets, and then evaluate.

133. Multiply 6 by 3. From this product, subtract 4. From this difference, subtract 2. $[(6 \cdot 3) - 4] - 2; 12$

134. Add 4 to 9. Divide this sum by 2. Add 10 to this quotient. $[(9 + 4) \div 2] + 10; \frac{33}{2}$

135. Divide 18 by 3. Add 9 to this quotient. Subtract 8 from this sum. Multiply this difference by 9.

136. Multiply 6 by 3. To this product, add 27. Divide this sum by 8. Multiply this quotient by 10.

🔒 **137.** Add $\frac{4}{5}$ to $\frac{3}{7}$ Multiply this sum by $\frac{2}{3}$. $\left(\frac{4}{5} + \frac{3}{7}\right) \cdot \frac{2}{3}; \frac{86}{105}$

138. Multiply $\frac{3}{8}$ by $\frac{4}{5}$. To this product, add $\frac{7}{120}$. From this sum, subtract $\frac{1}{60}$. $\left[\left(\frac{3}{8} \cdot \frac{4}{5}\right) + \frac{7}{120}\right] - \frac{1}{60}; \frac{41}{120}$

139. For what value or values of x does $-(x^2) = -x^2$?

140. For what value or values of x does $x = x^2$? $0, 1$

141. **Sales Tax** If the sales tax on an item is 7%, the sales tax on an item costing d dollars can be found by the expression $0.07d$. Determine the sales tax on a compact disk that costs $15.99. $1.12

142. **Road Trip** If a car travels at 60 miles per hour, the distance it travels in t hours is $60t$. Determine how far a car traveling at 60 miles per hour travels in 2.5 hours. 150 mi

143. **Car Cost** If the sales tax on an item is 7%, then the total cost of an item c, including sales tax, can be found by the expression $c + 0.07c$. Find the total cost of a car that costs $15,000. $16,050

144. **Bicycle Business** The profit or loss of a business, in dollars, can be found by the expression $-x^2 + 60x - 100$ where x is the number of bicycles (from 0 to 30) sold in a week. Determine the profit or loss if 20 bicycles are sold in a week. $700 profit

✏️ **145.** In the Using Your Calculator box on page 77 we showed that to evaluate $\left(\frac{2}{5}\right)^2$ we press the following keys:

$$2 \boxed{\div} 5 \boxed{=} \boxed{x^2}$$

We obtained an answer of .16. Indicate what a scientific (or graphing) calculator would display if the following keys are pressed. (On a graphing calculator, press $\boxed{\text{ENTER}}$ instead of $\boxed{=}$ and end part **b)** with $\boxed{\text{ENTER}}$). Explain your reason for each answer. Check your answer on your calculator.

a) $2 \boxed{\div} 5 \boxed{x^2} \boxed{=}$ .08

b) $\boxed{(} 2 \boxed{\div} 5 \boxed{)} \boxed{x^2}$ .16

✏️ **146.** We will discuss using zero as an exponent in Section 6.1. On your calculator find the value of 4^0 by using your $\boxed{y^x}$, $\boxed{x^y}$, or $\boxed{\wedge}$ key and record its value. Evaluate a few other numbers raised to the zero power. Can you make any conclusions about a real number (other than 0) raised to the zero power? equals 1

135. $9\{[(18 \div 3) + 9] - 8\}; 63$

136. $10\{[(6 \cdot 3) + 27] \div 8\}; \frac{225}{4}$

139. all real numbers

Challenge Problems

147. **Grass Growth** The rate of growth of grass in inches per week depends on a number of factors, including rainfall and temperature. For a certain region of the country, the growth per week can be approximated by the expression $0.2R^2 + 0.003RT + 0.0001T^2$, where R is the weekly rainfall, in inches, and T is the average weekly temperature, in degrees Fahrenheit. Find the amount of growth of grass for a week in which the rainfall is 2 inches and the average temperature is 70°F. 1.71 in.

Insert one pair of parentheses to make each statement true.

148. $14 + 6 \div 2 \times 4 = 40$ $(14 + 6) \div 2 \times 4$

150. $24 \div 6 \div 2 + 2 = 1$ $24 \div 6 \div (2 + 2)$

149. $12 - 4 - 6 + 10 = 24$ $12 - (4 - 6) + 10$

Group Activity

*Discuss and answer Exercises 151–154 as a group, according to the instructions. Each question has four parts. For parts **a**), **b**), and **c**), simplify the expression and write the answer in exponential form. Use the knowledge gained in parts **a**)–**c**) to answer part **d**. (General rules that may be used to solve exercises like these will be discussed in Chapter 6.)*

a) *Group member 1: Do part **a**) of each exercise.*

b) *Group member 2: Do part **b**) of each exercise.*

c) *Group member 3: Do part **c**) of each exercise.*

d) *As a group, answer part **d**) of each exercise. You may need to make up other examples like parts **a**)–**c**) to help you answer part **d**).*

151. a) $2^2 \cdot 2^3$ 2^5 **b)** $3^2 \cdot 3^3$ 3^5 **c)** $2^3 \cdot 2^4$ 2^7 **d)** $x^m \cdot x^n$ x^{m+n}

152. a) $\dfrac{2^3}{2^2}$ $2^1 = 2$ **b)** $\dfrac{3^4}{3^2}$ 3^2 **c)** $\dfrac{4^5}{4^3}$ 4^2 **d)** $\dfrac{x^m}{x^n}$ x^{m-n}

153. a) $\left(2^3\right)^2$ 2^6 **b)** $\left(3^3\right)^2$ 3^6 **c)** $\left(4^2\right)^2$ 4^4 **d)** $\left(x^m\right)^n$ x^{mn}

154. a) $(2x)^2$ $2^2 x^2$ **b)** $(3x)^2$ $3^2 x^2$ **c)** $(4x)^3$ $4^3 x^3$ **d)** $(ax)^m$ $a^m x^m$

Cumulative Review Exercises

[1.2] **155.** **House Occupancy** The graph shows the number of occupants in various houses selected at random in a neighborhood.

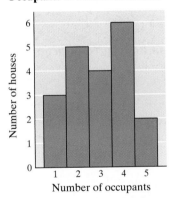

Occupants in Selected Houses

a) How many houses have three occupants?   4

b) Make a chart showing the number of houses that have one occupant, two occupants, three occupants, and so on.

c) How many occupants in total are there in all the houses?   59

d) Determine the mean number of occupants in all the houses surveyed.   2.95

156. **Taxi Cost** Yellow Cab charges $2.40 for the first $\frac{1}{2}$ mile plus 20 cents for each additional $\frac{1}{8}$ mile or part thereof. Find the cost of a 3-mile trip. $6.40

155. b)	Occupants	Houses
	1	3
	2	5
	3	4
	4	6
	5	2

[1.8] *Evaluate.*

157. $(-2)(-4)(6)(-1)(-3)$ 144

158. $\left(\dfrac{-5}{7}\right) \div \left(\dfrac{-3}{14}\right)$ $\dfrac{10}{3}$ or $3\dfrac{1}{3}$

1.10 PROPERTIES OF THE REAL NUMBER SYSTEM

SSM   Study Guide   CD/Video

MathPro 4/5   PH Math Tutor Center   prenhall.com/Angel

1   Learn the commutative property.

2   Learn the associative property.

3   Learn the distributive property.

Here, we introduce various properties of the real number system. We will use these properties throughout the text.

1 Learn the Commutative Property

The **commutative property of addition** states that the order in which any two real numbers are added does not matter.

> **Commutative Property of Addition**
>
> If a and b represent any two real numbers, then
>
> $$a + b = b + a$$

Notice that the commutative property involves a change in *order*. For example,

$$4 + 3 = 3 + 4$$
$$7 = 7$$

The **commutative property of multiplication** states that the order in which any two real numbers are multiplied does not matter.

> **Commutative Property of Multiplication**
>
> If a and b represent any two real numbers, then
>
> $$a \cdot b = b \cdot a$$

For example,

$$6 \cdot 3 = 3 \cdot 6$$
$$18 = 18$$

The commutative property **does not hold** *for subtraction or division.* For example, $4 - 6 \neq 6 - 4$ and $6 \div 3 \neq 3 \div 6$.

2 Learn the Associative Property

The **associative property of addition** states that, in the addition of three or more numbers, parentheses may be placed around any two adjacent numbers without changing the results.

> **Associative Property of Addition**
>
> If a, b, and c represent any three real numbers, then
>
> $$(a + b) + c = a + (b + c)$$

Notice that the associative property involves a change of *grouping*. For example,

$$(3 + 4) + 5 = 3 + (4 + 5)$$
$$7 + 5 = 3 + 9$$
$$12 = 12$$

In this example the 3 and 4 are grouped together on the left, and the 4 and 5 are grouped together on the right.

The **associative property of multiplication**, states that, in the multiplication of three or more numbers, parentheses may be placed around any two adjacent numbers without changing the results.

Associative Property of Multiplication

If a, b, and c represent any three real numbers, then

$$(a \cdot b) \cdot c = a \cdot (b \cdot c)$$

For example,

$$(6 \cdot 2) \cdot 4 = 6 \cdot (2 \cdot 4)$$
$$12 \cdot 4 = 6 \cdot 8$$
$$48 = 48$$

Since the associative property involves a change of grouping, when the associative property is used, the content within the parentheses changes.

The associative property **does not hold** *for subtraction or division.* For example, $(4 - 1) - 3 \neq 4 - (1 - 3)$ and $(8 \div 4) \div 2 \neq 8 \div (4 \div 2)$.

Often when we add numbers we group the numbers so that we can add them easily. For example, when we add $70 + 50 + 30$ we may first add the $70 + 30$ to get 100. We are able to do this because of the commutative and associative properties. Notice that

$$
\begin{aligned}
(70 + 50) + 30 &= 70 + (50 + 30) && \textit{Associative property of addition} \\
&= 70 + (30 + 50) && \textit{Commutative property of addition} \\
&= (70 + 30) + 50 && \textit{Associative property of addition} \\
&= 100 + 50 && \textit{Addition facts} \\
&= 150
\end{aligned}
$$

Notice in the second step that the same numbers remained in parentheses but the order of the numbers changed, $50 + 30$ to $30 + 50$. Since this step involved a change in order (and not grouping), this is the commutative property of addition.

3 Learn the Distributive Property

A very important property of the real numbers is the **distributive property of multiplication over addition**. We often shorten the name to the **distributive property**.

Distributive Property

If a, b, and c represent any three real numbers, then

$$a(b + c) = ab + ac$$

For example, if we let $a = 2$, $b = 3$, and $c = 4$, then

$$2(3 + 4) = (2 \cdot 3) + (2 \cdot 4)$$
$$2 \cdot 7 = 6 + 8$$
$$14 = 14$$

Therefore, we may either add first and then multiply, or multiply first and then add. Another example of the distributive property is

$$2(x + 3) = 2 \cdot x + 2 \cdot 3 = 2x + 6$$

The distributive property can be expanded in the following manner:

$$\boxed{a}\,(b + c + d + \cdots + n) = \boxed{a}\,b + \boxed{a}\,c + \boxed{a}\,d + \cdots + \boxed{a}\,n$$

For example, $3(x + y + 5) = 3x + 3y + 15$.

The distributive property will be discussed in more detail in Chapter 2.

HELPFUL HINT

The *commutative property* changes *order*.

The *associative property* changes *grouping*.

The *distributive property* involves *two operations*, usually multiplication and addition.

EXAMPLE 1 Name each property illustrated.

a) $4 + (-2) = -2 + 4$ **b)** $5(r + s) = 5 \cdot r + 5 \cdot s = 5r + 5s$

c) $x \cdot y = y \cdot x$ **d)** $(-12 + 3) + 4 = -12 + (3 + 4)$

Solution

a) Commutative property of addition

b) Distributive property

c) Commutative property of multiplication

NOW TRY EXERCISE 29   **d)** Associative property of addition ✳

HELPFUL HINT

Do not confuse the distributive property with the associative property of multiplication. Make sure you understand the difference.

Distributive Property	**Associative Property of Multiplication**
$3(4 + x) = 3 \cdot 4 + 3 \cdot x$	$3(4 \cdot x) = (3 \cdot 4)x$
$\qquad = 12 + 3x$	$\qquad = 12x$

For the distributive property to be used, within the parentheses, there must be two *terms*, separated by a plus or minus sign as in $3(4 + x)$.

4 Learn the Identity Properties

Now we will discuss the **identity properties**. When the number 0 is added to any real number, the real number is unchanged. For example, $5 + 0 = 5$ and $0 + 5 = 5$. For this reason we call 0 the **identity element of addition**. When any real number is multiplied by 1, the real number is unchanged. For example, $7 \cdot 1 = 7$ and $1 \cdot 7 = 7$. For this reason we call 1 the **identity element of multiplication**. The identity property for addition states that when 0 is added to any real number, the sum is the real number you started with. The identity property for multiplication states that when any real number is multiplied by 1, the product is the real number you started with. These properties are summarized below.

Identity Properties

If a represents any real number, then

$$a + 0 = a \quad \text{and} \quad 0 + a = a \qquad \textit{Identity Property of Addition}$$

and

$$a \cdot 1 = a \quad \text{and} \quad 1 \cdot a = a \qquad \textit{Identity Property of Multiplication}$$

We often use the identity properties without realizing we are using them. For example, when we reduce $\frac{15}{50}$, we may do the following:

$$\frac{15}{50} = \frac{3 \cdot 5}{10 \cdot 5} = \frac{3}{10} \cdot \frac{5}{5} = \frac{3}{10} \cdot 1 = \frac{3}{10}$$

When we showed that $\frac{3}{10} \cdot 1 = \frac{3}{10}$, we used the identity property of multiplication.

5   Learn the Inverse Properties

The last properties we will discuss in this chapter are the **inverse properties**. On page 45 we indicated that numbers like 3 and -3 were opposites or *additive inverses* because $3 + (-3) = 0$ and $-3 + 3 = 0$. Any two numbers whose sum is 0 are called *additive inverses* of each other. In general, for any real number a its additive inverse is $-a$.

We also have multiplicative inverses. Any two numbers whose product is 1 are called *multiplicative inverses* of each other. For example, because $4 \cdot \frac{1}{4} = 1$ and $\frac{1}{4} \cdot 4 = 1$, the numbers 4 and $\frac{1}{4}$ are *multiplicative inverses* (or *reciprocals*) of each other. In general, for any real number a, its multiplicative inverse is $\frac{1}{a}$. The inverse properties are summarized below.

Inverse Properties

If a represents any real number, then

$$a + (-a) = 0 \quad \text{and} \quad -a + a = 0 \qquad \textit{Inverse Property of Addition}$$

and

$$a \cdot \frac{1}{a} = 1 \qquad \text{and} \qquad \frac{1}{a} \cdot a = 1 \, (a \neq 0) \quad \textit{Inverse Property of Multiplication}$$

We often use the inverse properties without realizing we are using them. For example, to evaluate the expression $6x + 2$, when $x = \frac{1}{6}$ we may do the following:

$$6x + 2 = 6\left(\frac{1}{6}\right) + 2 = 1 + 2 = 3$$

When we multiplied $6\left(\frac{1}{6}\right)$ and replaced it with 1, we used the inverse property of multiplication. We will be using both the identity and inverse properties throughout the book, although we may not specifically refer to them by name.

EXAMPLE 2   Name each property illustrated.

a) $2(x + 6) = (2 \cdot x) + (2 \cdot 6) = 2x + 12$ b) $3x \cdot 1 = 3x$

c) $(3 \cdot 6) \cdot 5 = 3 \cdot (6 \cdot 5)$ d) $y \cdot \dfrac{1}{y} = 1$

e) $2a + (-2a) = 0$ f) $3y + 0 = 3y$

Solution   a) Distributive property

b) Identity property of multiplication

c) Associative property of multiplication

d) Inverse property of multiplication

e) Inverse property of addition

NOW TRY EXERCISE 55   f) Identity property of addition

Mathematics in Action

Mathematics, a Powerful Language

According to a special report to the U.S. Department of Education, a mathematics education will affect a person's lifetime income more than will knowledge in any other discipline. Why is this? It is because mathematics is the language that has proved the most efficient for analyzing, controlling, and creating the resources people need or want. Think about what *you* need or want. Everything from health insurance to the logistics of fast food, from downloadable audio files to mind-bending special effects in the movies, from computerized fashion design to barcode reading at the supermarket register, would be unimaginable without the application of mathematical techniques and thinking.

At a more personal level, but still of great importance in your day-to-day life, your quality of living is determined in many instances by decisions you make that depend on mathematical understanding and calculation. Figuring out how best to finance a big purchase, plan a trip, and prepare a recipe for a family gathering all benefit from mathematical knowledge, not to mention larger issues like whether a politician is distorting facts by applying dubious statistical techniques.

In so many ways and at so many levels, the mastery of mathematics allows you to shape your own life rather than have other people shape it for you.

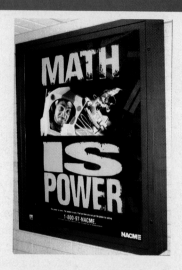

Mathematics doesn't care whether you were born in Los Angeles or Taiwan or whether you speak with a Mississippi, Maine, or Zapateca accent. Nothing stands in the way of developing your own mathematical style and strengths but your own willingness to make the needed effort.

As you read through the examples and exercises in this book, never lose sight of the big picture— how each bit of knowledge and skill you gain and each problem you solve, incrementally brings you closer to the goal of being able to choose how you want to work and live in a world where mathematics is key.

Exercise Set 1.10

Concept/Writing Exercises

1. Explain the commutative property of addition and give an example of it. $3 + 4 = 4 + 3$

2. Explain the commutative property of multiplication and give an example of it. $5(6) = 6(5)$

3. Explain the associative property of addition and give an example of it. $(2 + 3) + 4 = 2 + (3 + 4)$

4. Explain the associative property of multiplication and give an example of it. $(2 \cdot 3) \cdot 4 = 2 \cdot (3 \cdot 4)$

5. **a)** Explain the difference between $x + (y + z)$ and $x(y + z)$. Answers will vary

   b) Find the value of $x + (y + z)$ when $x = 4, y = 5$, and $z = 6$. 15

   c) Find the value of $x(y + z)$ when $x = 4, y = 5$, and $z = 6$. 44

6. Explain the distributive property and give an example of it. $2(3 + 4) = 2(3) + 2(4)$

7. Explain how you can tell the difference between the associative property of multiplication and the distributive property. Answers will vary

8. **a)** Write the associative property of addition using $x + (y + z)$. $x + (y + z) = (x + y) + z$

   b) Write the distributive property using $x(y + z)$.

9. What number is the additive identity element? 0

10. What number is the multiplicative identity element? 1

8. b) $x(y + z) = xy + xz$

Practice the Skills

*In Exercises 11–22, for the given expression, determine **a)** the additive inverse, and **b)** the multiplicative inverse.*

11. 6 a) -6 b) $\dfrac{1}{6}$

12. 5 a) -5 b) $\dfrac{1}{5}$

13. -3 a) 3 b) $-\dfrac{1}{3}$

14. -7 a) 7 b) $-\dfrac{1}{7}$

15. x a) $-x$ b) $\dfrac{1}{x}$

16. y a) $-y$ b) $\dfrac{1}{y}$

17. 1.6 a) -1.6 b) $\dfrac{1}{1.6}$ or 0.625

18. -0.125 a) 0.125 b) $-\dfrac{1}{0.125}$ or -8

19. $\dfrac{1}{5}$ a) $-\dfrac{1}{5}$ b) 5

20. $\dfrac{1}{8}$ a) $-\dfrac{1}{8}$ b) 8

21. $-\dfrac{3}{5}$ a) $\dfrac{3}{5}$ b) $-\dfrac{5}{3}$

22. $-\dfrac{2}{9}$ a) $\dfrac{2}{9}$ b) $-\dfrac{9}{2}$

Practice the Skills

Name each property illustrated.

23. $(x + 3) + 5 = x + (3 + 5)$ associative, addition

24. $3 + y = y + 3$ commutative, addition

25. $6(x + 7) = 6x + 42$ distributive

26. $1(x + 3) = (1)(x) + (1)(3) = x + 3$ distributive

27. $5 \cdot y = y \cdot 5$ commutative, multiplication

28. $x \cdot y = y \cdot x$ commutative, multiplication

29. $p \cdot (q \cdot r) = (p \cdot q) \cdot r$ associative, multiplication

30. $2(x + 4) = 2x + 8$ distributive

31. $4(d + 3) = 4d + 12$ distributive

32. $3 + (4 + t) = (3 + 4) + t$ associative, addition

33. $0 + 3y = 3y$ identity property for addition

34. $3z \cdot 1 = 3z$ identity property for multiplication

35. $2y \cdot \dfrac{1}{2y} = 1$ inverse property for multiplication

36. $-4x + 4x = 0$ inverse property for addition

In Exercises 37–58 the name of a property is given followed by part of an equation. Complete the equation, to the right of the equal sign, to illustrate the given property.

37. commutative property of addition
$x + 6 = \quad 6 + x$

38. commutative property of addition
$-3 + 4 = \quad 4 + (-3)$

39. associative property of multiplication
$-6 \cdot (4 \cdot 2) = \quad (-6 \cdot 4) \cdot 2$

40. associative property of addition
$-5 + (6 + 8) = \quad (-5 + 6) + 8$

41. distributive property
$1(x + y) = \quad x + y$

42. distributive property
$4(x + 3) = \quad 4x + 12$

43. commutative property of multiplication
$x \cdot y = \quad y \cdot x$

44. distributive property
$6(x + y) = \quad 6x + 6y$

45. commutative property of addition
$4x + 3y = \quad 3y + 4x$

46. associative property of addition
$(3 + r) + s = \quad 3 + (r + s)$

47. associative property of addition
$(a + b) + 3 = \quad a + (b + 3)$

48. commutative property of multiplication
$(x + 2)3 = \quad 3(x + 2)$

49. associative property of addition
$(3x + 4) + 6 = \quad 3x + (4 + 6)$

50. commutative property of addition
$3(x + y) = \quad 3(y + x)$

51. commutative property of multiplication
$3(m + n) = \quad (m + n)3$

52. associative property of multiplication
$(3x)y = \quad 3(xy)$

53. distributive property
$4(x + y + 3) \quad 4x + 4y + 12$

54. distributive property
$3(x + y + 2) = \quad 3x + 3y + 6$

55. inverse property of addition
$3n + (-3n) = \quad 0$

56. identity property for addition
$0 + 2x = \quad 2x$

57. identity property for multiplication
$\left(\dfrac{5}{2}n\right)(1) = \quad \dfrac{5}{2}n$

58. inverse property for multiplication
$\left(\dfrac{x}{2}\right)\left(\dfrac{2}{x}\right) = \quad 1$

Problem Solving

Indicate whether the given processes are commutative. That is, does changing the order in which the actions are done result in the same final outcome? Explain each answer.

59. Putting sugar and then cream in coffee; putting cream and then sugar in coffee. yes

60. Brushing your teeth and then washing your face; washing your face and then brushing your teeth. yes

61. Applying suntan lotion and then sunning yourself; sunning yourself and then applying suntan lotion. no

62. Putting on your socks and then your shoes; putting on your shoes and then your socks. no

63. Getting your hands dirty and then washing your hands. Washing your hands and then getting your hands dirty. no

64. Writing on the blackboard and then erasing the blackboard; erasing the blackboard and then writing on the blackboard. no

In Exercises 65–70, indicate whether the given processes are associative. For a process to be associative, the final outcome must be the same when the first two actions are performed first or when the last two actions are performed first. Explain each answer.

65. Brushing your teeth, washing your face, and combing your hair. yes

66. In a store, buying cereal, soap, and dog food. yes

67. Putting on a shirt, a tie, and a sweater. no

68. A coffee machine dropping the cup, dispensing the coffee, and then adding the sugar. no

69. Starting a car, moving the shift lever to drive, and then stepping on the gas. no

70. Putting cereal, milk, and sugar in a bowl. yes

71. The commutative property of addition is $a + b = b + a$. Explain why $(3 + 4) + x = x + (3 + 4)$ also illustrates the commutative property of addition. treat $(3 + 4)$ as one value

72. The commutative property of multiplication is $a \cdot b = b \cdot a$. Explain why $(3 + 4) \cdot x = x \cdot (3 + 4)$ also illustrates the commutative property of multiplication. treat $(3 + 4)$ as one value

Challenge Problems

73. Consider $x + (3 + 5) = x + (5 + 3)$. Does this illustrate the commutative property of addition or the associative property of addition? Explain. commutative

74. Consider $x + (3 + 5) = (3 + 5) + x$. Does this illustrate the commutative property of addition or the associative property of addition? Explain. commutative

75. Consider $x + (3 + 5) = (x + 3) + 5$. Does this illustrate the commutative property of addition? Explain. no, associative property of addition

76. The commutative property of multiplication is $a \cdot b = b \cdot a$. Explain why $(3 + 4) \cdot (5 + 6) = (5 + 6) \cdot (3 + 4)$ also illustrates the commutative property of multiplication. treat $(3 + 4)$ as one value and $(5 + 6)$ as a second value

Cumulative Review Exercises

[1.3] **77.** Add $2\frac{3}{5} + \frac{2}{3}$. $\frac{49}{15}$ or $3\frac{4}{15}$

78. Subtract $3\frac{5}{8} - 2\frac{3}{16}$. $\frac{23}{16}$ or $1\frac{7}{16}$

[1.9] Evaluate.

79. $12 - 24 \div 8 + 4 \cdot 3^2$ 45

80. $-4x^2 + 6xy + 3y^2$ when $x = 2, y = -3$ -25

CHAPTER SUMMARY

Key Words and Phrases

1.2
Approximately equal to
Bar graph
Circle (or pie) graph
Expression
Line graph
Measures of central
   tendency
Mean
Median
Operations
Problem-solving
   procedure
Ranked data

1.3
Denominator
Evaluate
Factors

Fraction
Greatest common factor
Least common
   denominator
Mixed number
Numerator
Reduced to lowest terms
Variables
Whole numbers

1.4
Counting numbers
Elements of a set
Empty (or null) set
Integers
Irrational numbers
Natural numbers
Negative integers

Positive integers
Rational numbers
Real numbers
Set

1.5
Absolute value

1.6
Additive inverses
Operations
Opposites

1.8
Undefined

1.9
Base
Exponent

Grouping symbols
Nested parentheses
Order of operations
Parentheses

1.10
Associative properties
Commutative properties
Distributive property
Identity element for
   addition
Identity element for
   multiplication
Identity properties
Inverse properties
Reciprocal

IMPORTANT FACTS

Problem–Solving Procedure

1. Understand the question.
2. Translate into mathematical language.
3. Carry out the calculations.
4. Check the answer.
5. Answer the question asked.

Fractions

$$\frac{a}{c} + \frac{b}{c} = \frac{a+b}{c} \qquad \frac{a}{c} - \frac{b}{c} = \frac{a-b}{c}$$

$$\frac{a}{b} \cdot \frac{c}{d} = \frac{ac}{bd} \qquad \frac{a}{b} \div \frac{c}{d} = \frac{a}{b} \cdot \frac{d}{c} = \frac{ad}{bc}$$

Sets of Numbers

Natural numbers: $\{1, 2, 3, 4, \ldots\}$

Whole numbers: $\{0, 1, 2, 3, 4, \ldots\}$

Integers: $\{\ldots, -3, -2, -1, 0, 1, 2, 3, \ldots\}$

Rational numbers: {quotient of two integers, denominator not 0}

Irrational numbers: {real numbers that are not rational numbers}

Real numbers: {all numbers that can be represented on a number line}

Operations on the Real Numbers

To *add real numbers with the same sign*, add their absolute values. The sum has the same sign as the numbers being added.

To *add real numbers with different signs*, subtract the smaller absolute value from the larger absolute value. The answer has the sign of the number with the larger absolute value.

(continued on the next page)

To *subtract b from a*, add the opposite of *b* to *a*.

$$a - b = a + (-b)$$

The *products* and *quotients* of numbers with *like signs* will be *positive*. The *products* and *quotients* of numbers with *unlike signs* will be *negative*.

Division Involving 0	Exponents
If *a* represents any real number except 0, then $$\frac{0}{a} = 0$$ $$\frac{a}{0} \text{ is undefined}$$	$$b^n = \underbrace{b \cdot b \cdot b \cdot \cdots \cdot b}_{n \text{ factors of } b}$$

Order of Operations

1. Evaluate expressions within parentheses.
2. Evaluate all expressions with exponents.
3. Perform multiplications or divisions working left to right.
4. Perform additions or subtractions working left to right.

Properties of the Real Number System

Property	Addition	Multiplication
Commutative	$a + b = b + a$	$ab = ba$
Associative	$(a + b) + c = a + (b + c)$	$(ab)c = a(bc)$
Identity	$a + 0 = a$ and $0 + a = a$	$a \cdot 1 = a$ and $1 \cdot a = a$
Inverse	$a + (-a) = 0$ and $-a + a = 0$	$a \cdot \frac{1}{a} = 1$ and $\frac{1}{a} \cdot a = 1, a \neq 0,$
Distributive		$a(b + c) = ab + ac$

Chapter Review Exercises

[1.2] Solve.

1. **State Fair** Paul Vaupel sells hot dogs and sausages at state fairs. On day 1 of a 4-day fair, he sold 162 hot dogs. On day 2 he sold 187 hot dogs. On day 3 he sold 196 hot dogs, and on day 4 he sold 95 hot dogs. If he purchased 60 dozen hot dogs for the fair, how many did he have left over? 80

2. **Inflation** Assume that the rate of inflation is 5% for the next two years. What will be the cost of goods two years from now, adjusted for inflation, if the goods cost $500.00 today? $551.25

3. **Cost Increase** The cost of a car increases by 20% and then decreases by 20%. Is the resulting price of the car greater than, less than, or equal to the original price of the car? Explain. less than

4. **Fax Machine** Krystal Streeter wants to purchase a fax machine that sells for $300. She can either pay the total amount at the time of purchase, or she can agree to pay the store $30 down and $25 a month for 12 months. How much money can she save by paying the total amount at the time of purchase? $30

5. *Test Grades* On Kristen Reid's first five exams her grades were 75, 79, 86, 88, and 64. Find the **a)** mean and **b)** median of her grades. a) 78.4 b) 79

6. *Temperatures* The high temperatures, in degrees Fahrenheit, on July 1 in the last five years in Honolulu, Hawaii, were 76, 79, 84, 82, and 79. Find the **a)** mean and **b)** median of the temperatures. a) 80°F b) 79°F

7. *Prescriptions* In November, 2001, the average U.S. prescription cost $45.79. The following piece of art shows where the money goes.

  a) How much money from the average prescription is profit for the drug manufacturer? $8.33

  b) If a pharmacist pays $60 for a drug, and the percents in the art apply, determine the cost at which the pharmacist will sell the drug. $73.20

U.S. Prescription Cost

Source: National Association of Chain Drug Stores, University of Wisconsin Sonderegger Research Center

8. *Gas Reserve* The following bar graph shows proven oil reserves, in billions of barrels, as of November, 2001.

  a) What is the difference, in billions of barrels, between the oil reserves of the United States and Canada? 17.1 billion

  b) How many times greater are the oil reserves in the Middle East than in North America? ≈12.5

  c) What are the world's total oil reserves? 1028.8 billion barrels

Proven Conventional Oil Reserves (billions of barrels)

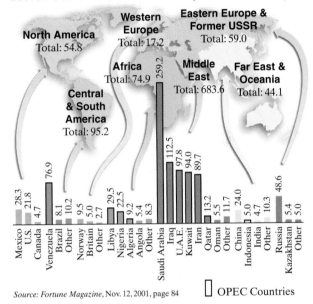

Source: Fortune Magazine, Nov. 12, 2001, page 84   ☐ OPEC Countries

[1.3] *Perform each indicated operation. Simplify your answers.*

9. $\dfrac{3}{5} \cdot \dfrac{5}{6}$ $\dfrac{1}{2}$

10. $\dfrac{2}{5} \div \dfrac{10}{9}$ $\dfrac{9}{25}$

11. $\dfrac{5}{12} \div \dfrac{3}{5}$ $\dfrac{25}{36}$

12. $\dfrac{5}{6} + \dfrac{1}{3}$ $\dfrac{7}{6}$ or $1\dfrac{1}{6}$

13. $\dfrac{3}{8} - \dfrac{1}{9}$ $\dfrac{19}{72}$

14. $2\dfrac{1}{3} - 1\dfrac{1}{5}$ $\dfrac{17}{15}$ or $1\dfrac{2}{15}$

[1.4] **15.** List the set of natural numbers. {1, 2, 3, …}
16. List the set of whole numbers. {0, 1, 2, 3, …}
17. List the set of integers. {… −3, −2, −1, 0, 1, 2, 3, …}
18. Describe the set of rational numbers.
19. Consider the following set of numbers.

$$\left\{ 3, -5, -12, 0, \dfrac{1}{2}, -0.62, \sqrt{7}, 426, -3\dfrac{1}{4} \right\}$$

List the numbers that are
a) positive integers. 3, 426
b) whole numbers. 3, 0, 426
c) integers. 3, −5, −12, 0, 426
d) rational numbers. $3, -5, -12, 0, \frac{1}{2}, -0.62, 426, -3\frac{1}{4}$
e) irrational numbers. $\sqrt{7}$
f) real numbers. $3, -5, -12, 0, \frac{1}{2}, -0.62, \sqrt{7}, 426, -3\frac{1}{4}$

20. Consider the following set of numbers.

$$\left\{ -2.3, -8, -9, 1\dfrac{1}{2}, \sqrt{2}, -\sqrt{2}, 1, -\dfrac{3}{17} \right\}$$

List the numbers that are
a) natural numbers. 1
b) whole numbers. 1
c) negative integers. −8, −9
d) integers. −8, −9, 1
e) rational numbers. $-2.3, -8, -9, 1\frac{1}{2}, 1, -\frac{3}{17}$
f) real numbers. $-2.3, -8, -9, 1\frac{1}{2}, \sqrt{2}, -\sqrt{2}, 1, -\frac{3}{17}$

18. {quotient of two integers, denominator not 0}

[1.5] *Insert either* $<$, $>$, *or* $=$ *in each shaded area to make a true statement.*

21. $-7 \blacksquare -5$ $<$
22. $-2.6 \blacksquare -3.6$ $>$
23. $0.50 \blacksquare 0.509$ $<$
24. $4.6 \blacksquare 4.06$ $>$

25. $-3.2 \blacksquare -3.02$ $<$
26. $5 \blacksquare |-3|$ $>$
27. $-9 \blacksquare |-7|$ $<$
28. $|-2.5| \blacksquare \left|\dfrac{5}{2}\right|$ $=$

[1.6–1.7] *Evaluate.*

29. $-9 + (-5)$ -14
30. $-6 + 6$ 0
31. $0 + (-3)$ -3
32. $-10 + 4$ -6

33. $-8 - (-2)$ -6
34. $-2 - (-4)$ 2
35. $4 - (-4)$ 8
36. $12 - 12$ 0

37. $7 - (-7)$ 14
38. $2 - 7$ -5
39. $0 - (-4)$ 4
40. $-7 - 5$ -12

41. $\dfrac{4}{3} - \dfrac{3}{4}$ $\dfrac{7}{12}$
42. $\dfrac{1}{2} + \dfrac{3}{5}$ $\dfrac{11}{10}$
43. $\dfrac{5}{9} - \dfrac{3}{4}$ $-\dfrac{7}{36}$
44. $-\dfrac{5}{7} + \dfrac{3}{8}$ $-\dfrac{19}{56}$

45. $-\dfrac{5}{12} - \dfrac{5}{6}$ $-\dfrac{5}{4}$
46. $-\dfrac{6}{7} + \dfrac{5}{12}$ $-\dfrac{37}{84}$
47. $\dfrac{2}{9} - \dfrac{3}{10}$ $-\dfrac{7}{90}$
48. $\dfrac{7}{15} - \left(-\dfrac{7}{60}\right)$ $\dfrac{7}{12}$

Evaluate.

49. $9 - 4 + 3$ 8
50. $-5 + 7 - 6$ -4
51. $-5 - 4 - 3$ -12
52. $-2 + (-3) - 2$ -7

53. $7 - (+4) - (-3)$ 6
54. $6 - (-2) + 3$ 11

[1.8] *Evaluate.*

55. $-3(9)$ -27
56. $(-8)(-5)$ 40
57. $(-4)(-5)(-6)$ -120
58. $\left(\dfrac{3}{5}\right)\left(\dfrac{-2}{7}\right)$ $-\dfrac{6}{35}$

59. $\left(\dfrac{10}{11}\right)\left(\dfrac{3}{-5}\right)$ $-\dfrac{6}{11}$
60. $\left(\dfrac{-5}{8}\right)\left(\dfrac{-3}{7}\right)$ $\dfrac{15}{56}$
61. $0\left(\dfrac{4}{9}\right)$ 0
62. $(-4)(-6)(-2)(-3)$ 144

Evaluate.

63. $15 \div (-3)$ -5
64. $12 \div (-2)$ -6
65. $-20 \div 5$ -4
66. $0 \div 4$ 0

67. $90 \div (-9)$ -10
68. $-4 \div \left(\dfrac{-4}{9}\right)$ 9
69. $\dfrac{28}{-3} \div \left(\dfrac{9}{-2}\right)$ $\dfrac{56}{27}$
70. $\dfrac{14}{3} \div \left(\dfrac{-6}{5}\right)$ $-\dfrac{35}{9}$

Indicate whether each quotient is 0 or undefined.

71. $0 \div 5$ 0
72. $0 \div (-6)$ 0
73. $8 \div 0$ undefined

74. $-4 \div 0$ undefined
75. $\dfrac{8}{0}$ undefined
76. $\dfrac{0}{-5}$ 0

[1.6–1.8, 1.9] *Evaluate.*

77. $-5(3 - 8)$ 25
78. $2(4 - 8)$ -8
79. $(3 - 6) + 4$ 1

80. $(-4 + 3) - (2 - 6)$ 3
81. $[6 + 3(-2)] - 6$ -6
82. $(-4 - 2)(-3)$ 18

83. $[12 + (-4)] + (6 - 8)$ 6
84. $9[3 + (-4)] + 5$ -4
85. $-4(-3) + [4 \div (-2)]$ 10

86. $(-3 \cdot 4) \div (-2 \cdot 6)$ 1
87. $(-3)(-4) + 6 - 3$ 15
88. $[-2(3) + 6] - 4$ -4

[1.9] *Evaluate.*

89. 7^2 49
90. 9^3 729
91. 3^4 81
92. $(-3)^3$ -27

93. $(-1)^9$ -1
94. $(-2)^5$ -32
95. $\left(\dfrac{-4}{5}\right)^2$ $\dfrac{16}{25}$
96. $\left(\dfrac{2}{5}\right)^3$ $\dfrac{8}{125}$

97. $5^3 \cdot (-2)^2$ 500
98. $(-2)^4\left(\dfrac{1}{2}\right)^2$ 4
99. $\left(-\dfrac{2}{3}\right)^2 \cdot 3^3$ 12
100. $(-4)^3(-2)^3$ 512

Evaluate.

101. $3 + 5 \cdot 4$ 23

102. $4 \cdot 6 + 4 \cdot 2$ 32

103. $(3 - 7)^2 + 6$ 22

104. $10 - 36 \div 4 \cdot 3$ −17

105. $6 - 3^2 \cdot 5$ −39

106. $[6 - (3 \cdot 5)] + 5$ −4

107. $\dfrac{4 + 5^2 \div 5}{6 - (-3 + 2)}$ $\dfrac{9}{7}$

108. $\dfrac{6^2 - 4 \cdot 3^2}{-[6 - (3 - 4)]}$ 0

109. $3[9 - (4^2 + 3)] \cdot 2$ −60

110. $(-3^2 + 4^2) + (3^2 \div 3)$ 10

111. $2^3 \div 4 + 6 \cdot 3$ 20

112. $(4 \div 2)^4 + 4^2 \div 2^2$ 20

113. $(8 - 2^2)^2 - 4 \cdot 3 + 10$ 14

114. $4^3 \div 4^2 - 5(2 - 7) \div 5$ 9

115. $-\{-4[27 \div 3^2 - 2(4 - 2)]\}$ −4

116. $2\{4^3 - 6[4 - (2 - 4)] - 3\}$ 50

Evaluate each expression for the given values.

117. $6x - 6; x = 5$ 24

118. $6 - 4x; x = -5$ 26

119. $2x^2 - 5x + 3; x = 6$ 45

120. $5y^2 + 3y - 2; y = -1$ 0

121. $-x^2 + 2x - 3; x = 2$ −3

122. $-x^2 + 2x - 3; x = -2$ −11

123. $-3x^2 - 5x + 5; x = 1$ −3

124. $-x^2 - 8x - 12y; x = -3, y = -2$ 39

🖩 *[1.6–1.9]* **a)** *Use a calculator to evaluate each expression, and* **b)** *check to see whether your answer is reasonable.*

125. $278 + (-493)$ −215

126. $324 - (-29.6)$ 353.6

127. $\dfrac{-17.28}{6}$ −2.88

128. $(-62)(-1.9)$ 117.8

129. $(-3)^6$ 729

130. $-(4.2)^3$ −74.088

[1.10] *Name each indicated property.*

131. $(7 + 4) + 9 = 7 + (4 + 9)$ associative, addition

132. $6 \cdot x = x \cdot 6$ commutative, multiplication

133. $4(x + 3) = 4x + 12$ distributive

134. $(x + 4)3 = 3(x + 4)$ commutative, multiplication

135. $6x + 3x = 3x + 6x$ commutative, addition

136. $(x + 7) + 4 = x + (7 + 4)$ associative, addition

137. $r + 0 = r$ identity, addition

138. $\dfrac{1}{n} \cdot n = 1$ inverse, multiplication

Chapter Practice Test

1. *Shopping* While shopping, Faith Healy purchases two half-gallons of milk for $1.30 each, one Boston cream pie for $4.75, and three 2-liter bottles of soda for $1.10 each.

   a) What is her total bill before tax? $10.65

   b) If there is a 7% sales tax on the bottles of soda, how much is the sales tax? $0.23

   c) How much is her total bill including tax? $10.88

   d) How much change will she receive from a $50 bill? $39.12

2. *Health Care* The graph on the right shows how health care costs have risen from 1997 through 2002.

   a) Estimate the average cost, per employee, for employers in 2002. ≈ $4400

   b) Estimate the difference in the average cost, per employee, for employers from 1997 to 2002. ≈ $1400

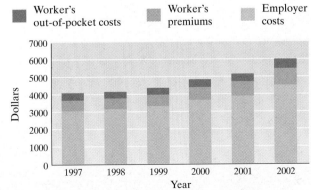

Rising Healthcare Costs

Worker's out-of-pocket costs Worker's premiums Employer costs

Source: Hewitt Associates

3. **Population Information** The following graph illustrates information about the U.S. population in 1900 and in 2000 according to the U.S. Census Bureau.

a) Determine approximately how many *households* there were in the United States in 2000? Explain how you determined your answer. ≈ 108.5 million

b) The median age in 2000 was 35.3 years. Explain what this means. Half the population was above and half was below this age.

The U.S.A.

1900 | 2000

Population

| 76 mil. | 281 million |

Average family size (people)

| 4.76 | 2.59 |

Median age

| 22.9 | 35.3 |

4. Consider the following set of numbers.

$$\left\{-6, 42, -3\frac{1}{2}, 0, 6.52, \sqrt{5}, \frac{5}{9}, -7, -1\right\}$$

List the numbers that are

a) natural numbers. 42

b) whole numbers. 42, 0

c) integers. $-6, 42, 0, -7, -1$

d) rational numbers. $-6, 42, -3\frac{1}{2}, 0, 6.52, \frac{5}{9}, -7, -1$

e) irrational numbers. $\sqrt{5}$

f) real numbers. $-6, 42, -3\frac{1}{2}, 0, 6.52, \sqrt{5}, \frac{5}{9}, -7, -1$

Insert either $<$, $>$, or $=$ in each shaded area to make a true statement.

5. $-9 \,\blacksquare\, -12$ $>$

6. $|-3| \,\blacksquare\, |-2|$ $>$

Evaluate.

7. $-7 + (-8)$ -15

8. $-6 - 5$ -11

9. $15 - 12 - 17$ -14

10. $(-4 + 6) - 3(-2)$ 8

11. $(-4)(-3)(2)(-1)$ -24

12. $\left(\dfrac{-2}{9}\right) \div \left(\dfrac{-7}{8}\right)$ $\dfrac{16}{63}$

13. $\left(-18 \cdot \dfrac{1}{2}\right) \div 3$ -3

14. $-\dfrac{3}{8} - \dfrac{4}{7}$ $-\dfrac{53}{56}$

15. $-6(-2 - 3) \div 5 \cdot 2$ 12

16. $\left(\dfrac{3}{5}\right)^3$ $\dfrac{27}{125}$

17. Write $2 \cdot 2 \cdot 5 \cdot 5yyzzz$ in exponential form. $2^2 5^2 y^2 z^3$

18. Write $2^2 3^3 x^4 y^2$ as a product of factors. $2 \cdot 2 \cdot 3 \cdot 3 \cdot 3xxxxyy$

Evaluate the expression for the given values.

19. $5x^2 - 8; x = -4$ 72

20. $6x - 3y^2 + 4; x = 3, y = -2$ 10

21. $-x^2 - 6x + 3; x = -2$ 11

22. $-x^2 + xy + y^2; x = 1, y = -2$ 1

Name each indicated property.

23. $x + 3 = 3 + x$ commutative, addition

24. $4(x + 9) = 4x + 36$ distributive

25. $(2 + x) + 4 = 2 + (x + 4)$ associative, addition

Chapter 2

Solving Linear Equations and Inequalities

When trying to predict whether an athlete will set a new record, journalists and coaches often compare the athlete's current pace and performance with the pace set by the record holder at different times during a record-breaking season, On page 156, we use proportions to determine how many home runs during the first 50 games of a baseball season a player would need to hit to be on schedule to break Barry Bond's record of 73 home runs from the 2001 baseball season.

SSM

Study Guide

CD/Video

MathPro 4/5

PH Math Tutor Center

prenhall.com/Angel

A Look Ahead

When many students describe algebra they use the words "solving equations." Solving equations is an important part of algebra. The major emphasis of this chapter is to teach you how to solve linear equations. We will be using principles learned in this chapter throughout the book.

To be successful in solving linear equations, you need to have a thorough understanding of adding, subtracting, multiplying, and dividing real numbers. This material was discussed in Chapter 1. The material presented in the first four sections of this chapter are the building blocks for solving linear equations. In Section 2.5 we combine the material presented previously to solve a variety of linear equations.

In Section 2.6 we discuss ratios and proportions and how to set up and solve them. For many students, proportions may be the most common types of equations used to solve real-life applications. In Section 2.7 we discuss solving linear inequalities, which is an extension of solving linear equations.

2.1 COMBINING LIKE TERMS

SSM

Study Guide

CD/Video

MathPro 4/5

PH Math Tutor Center

prenhall.com/Angel

1 Identify terms.
2 Identify like terms.
3 Combine like terms.
4 Use the distributive property.
5 Remove parentheses when they are preceded by a plus or minus sign.
6 Simplify an expression.

1 Identify Terms

In Section 1.3 and other sections of the text, we indicated that letters called **variables** are used to represent numbers. A variable can represent a variety of different numbers.

As was indicated in Chapter 1, an **expression** (sometimes referred to as an **algebraic expression**) is a collection of numbers, variables, grouping symbols, and operation symbols.

Examples of Expressions

$$5, \quad x^2 - 6, \quad 4x - 3, \quad 2(x + 5) + 6, \quad \frac{x + 3}{4}$$

When an algebraic expression consists of several parts, the parts that are *added* are called the **terms** of the expression. Consider the expression $2x - 3y - 5$. The expression can be written as $2x + (-3y) + (-5)$, and so the expression $2x - 3y - 5$ has three terms: $2x$, $-3y$, and -5. The expression $3x^2 + 2xy + 5(x + y)$ also has three terms: $3x^2$, $2xy$, and $5(x + y)$.

When listing the terms of an expression, it is not necessary to list the $+$ sign at the beginning of a term.

TEACHING TIP
Emphasize that it is important to list the minus sign when identifying the term.

Expression	Terms
$-2x + 3y - 8$	$-2x,\ 3y,\ -8$
$3y^2 - 2x + \dfrac{1}{2}$	$3y^2,\ -2x,\ \dfrac{1}{2}$

$$7 + x + 4 - 5x \qquad\qquad 7,\ x,\ 4,\ -5x$$

$$3(x - 1) - 4x + 2 \qquad\qquad 3(x - 1),\ -4x,\ 2$$

$$\frac{x + 4}{3} - 5x + 3 \qquad\qquad \frac{x + 4}{3},\ -5x,\ 3$$

NOW TRY EXERCISE 1

The numerical part of a term is called its **numerical coefficient** or simply its **coefficient**. In the term $6x$, the 6 is the numerical coefficient. Note that $6x$ means the variable x is multiplied by 6.

TEACHING TIP
Point out that the language of mathematics often assumes the reader knows certain things by the absence of a symbol. For instance, $3x$ is assumed to have a positive sign associated with it because no sign is given.

Term	Numerical Coefficient
$3x$	3
$-\dfrac{1}{2}x$	$-\dfrac{1}{2}$
$4(x - 3)$	4
$\dfrac{2x}{3}$	$\dfrac{2}{3}$, since $\dfrac{2x}{3}$ means $\dfrac{2}{3}x$
$\dfrac{x + 4}{3}$	$\dfrac{1}{3}$, since $\dfrac{x + 4}{3}$ means $\dfrac{1}{3}(x + 4)$

Whenever a term appears without a numerical coefficient, we assume that the numerical coefficient is 1.

Examples

TEACHING TIP
Point out that variables are used when different numbers can be used in a situation and constants are used when only one number can be used. For instance, if you are charged a monthly fee of $9.95 for internet service and an hourly fee of $1.25, your charge is represented by $1.25x + 9.95$.

$$x \text{ means } 1x \qquad\qquad -x \text{ means } -1x$$
$$x^2 \text{ means } 1x^2 \qquad\qquad -x^2 \text{ means } -1x^2$$
$$xy \text{ means } 1xy \qquad\qquad -xy \text{ means } -1xy$$
$$(x + 2) \text{ means } 1(x + 2) \qquad -(x + 2) \text{ means } -1(x + 2)$$

If an expression has a term that is a number (without a variable), we refer to that number as a **constant term**, or simply a **constant**. In the expression $x^2 + 3x - 4$, the -4 is a constant term, or a constant.

2 Identify Like Terms

Like terms are terms that have the same variables with the same exponents. Some examples of like terms and unlike terms follow. Note that if two terms are like terms, only their numerical coefficients may differ.

Like Terms	Unlike Terms	
$3x,\ -4x$	$3x,\ 2$	*(One term has a variable, the other is a constant.)*
$4y,\ 6y$	$3x,\ 4y$	*(Variables differ)*
$5,\ -6$	$x,\ 3$	*(One term has a variable, the other is a constant.)*
$3(x + 1),\ -2(x + 1)$	$2x,\ 3xy$	*(Variables differ)*
$3x^2,\ 4x^2$	$3x,\ 4x^2$	*(Exponents differ)*
$5ab,\ 2ab$	$4a,\ 2ab$	*(Variables differ)*

EXAMPLE 1   Identify any like terms.

a) $2x + 3x + 4$ b) $2x + 3y + 2$ c) $x + 3 + y - \dfrac{1}{2}$ d) $x + 3x^2 - 4x^2$

Solution   a) $2x$ and $3x$ are like terms.

b) There are no like terms.

c) 3 and $-\dfrac{1}{2}$ are like terms.

d) $3x^2$ and $-4x^2$ are like terms.

EXAMPLE 2   Identify any like terms.

a) $5x - x + 6$ b) $3 - 2x + 4x - 6$ c) $12 + x^2 - x + 7$

Solution   a) $5x$ and $-x$ (or $-1x$) are like terms.

b) 3 and -6 are like terms; $-2x$ and $4x$ are like terms.

c) 12 and 7 are like terms.

3   Combine Like Terms

We often need to simplify expressions by combining like terms. **To combine like terms** means to add or subtract the like terms in an expression. To combine like terms, we can use the procedure that follows.

To Combine Like Terms

1. Determine which terms are like terms.

2. Add or subtract the coefficients of the like terms.

3. Multiply the number found in step 2 by the common variable(s).

Examples 3 through 8 illustrate this procedure.

EXAMPLE 3   Combine like terms: $5x + 4x$.

Solution $5x$ and $4x$ are like terms with the common variable x. Since $5 + 4 = 9$, then $5x + 4x = 9x$.

EXAMPLE 4   Combine like terms: $\dfrac{3}{5}x - \dfrac{2}{3}x$.

Solution   Since $\dfrac{3}{5} - \dfrac{2}{3} = \dfrac{9}{15} - \dfrac{10}{15} = -\dfrac{1}{15}$, then $\dfrac{3}{5}x - \dfrac{2}{3}x = -\dfrac{1}{15}x$.

EXAMPLE 5   Combine like terms: $5.23a - 7.45a$.

Solution   Since $5.23 - 7.45 = -2.22$, then $5.23a - 7.45a = -2.22a$.

EXAMPLE 6   Combine like terms: $3x + x + 5$.

Solution   The $3x$ and x are like terms.

$$3x + x + 5 = 3x + 1x + 5 = 4x + 5$$

Because of the commutative property of addition, the order of the terms in the answer is not critical. Thus $5 + 4x$ is also an acceptable answer to Example 6. When writing answers, we generally list the terms containing variables in alphabetical order from left to right, and list the constant term on the right.

The commutative and associative properties of addition will be used to rearrange the terms in Examples 7 and 8.

EXAMPLE 7 Combine like terms: $3b + 6a - 5 - 2a$

Solution The only like terms are $6a$ and $-2a$.

$$3b + 6a - 5 - 2a = 6a - 2a + 3b - 5 \qquad \textit{Rearrange terms.}$$
$$= 4a + 3b - 5 \qquad \textit{Combine like terms.}$$

EXAMPLE 8 Combine like terms: $-2x^2 + 3y - 4x^2 + 3 - y + 5$.

Solution $-2x^2$ and $-4x^2$ are like terms.

$3y$ and $-y$ are like terms.

3 and 5 are like terms.

Grouping the like terms together gives

$$-2x^2 + 3y - 4x^2 + 3 - y + 5 = -2x^2 - 4x^2 + 3y - y + 3 + 5$$
$$= -6x^2 + 2y + 8$$

NOW TRY EXERCISE 29

4 Use the Distributive Property

We introduced the distributive property in Section 1.10. Because this property is so important, we will study it again. But before we do, let's go back briefly to the subtraction of real numbers. Recall from Section 1.7 that

$$6 - 3 = 6 + (-3)$$

In general,

> For any real numbers a and b,
>
> $$a - b = a + (-b)$$

We will use the fact that $a + (-b)$ means $a - b$ in discussing the distributive property.

TEACHING TIP
Before discussing Example 9a, have students rewrite the expression $2(x + 4)$ as an addition problem: $2(x + 4) = (x + 4) + (x + 4) = 2x + 8$. Then ask, "How would we simplify this problem using the Distributive Property?"

Distributive Property

For any real numbers a, b, and c,

$$a(b + c) = ab + ac$$

EXAMPLE 9 Use the distributive property to remove parentheses.

a) $2(x + 4)$ **b)** $-2(w + 4)$

Solution **a)** $2(x + 4) = 2x + 2(4) = 2x + 8$

b) $-2(w + 4) = -2w + (-2)(4) = -2w + (-8) = -2w - 8$

Note in part **b)** that, instead of leaving the answer $-2w + (-8)$, we wrote it as $-2w - 8$, which is the proper form of the answer.

EXAMPLE 10 Use the distributive property to remove parentheses.

a) $3(x - 2)$ **b)** $-2(4x - 3)$

Solution **a)** By the definition of subtraction, we may write $x - 2$ as $x + (-2)$.

$$3(x - 2) = 3[x + (-2)] = 3x + 3(-2)$$
$$= 3x + (-6)$$
$$= 3x - 6$$

b) $-2(4x - 3) = -2[4x + (-3)] = -2(4x) + (-2)(-3) = -8x + 6$ ✳

The distributive property is used often in algebra, so you need to understand it well. You should understand it so well that you will be able to simplify an expression using the distributive property without having to write down all the steps that we listed in working Examples 9 and 10. Study closely the Helpful Hint that follows.

HELPFUL HINT

With a little practice, you will be able to eliminate some of the intermediate steps when you use the distributive property. When using the distributive property, there are eight possibilities with regard to signs. Study and learn the eight possibilities that follow.

Positive Coefficient

a) $2(x) = 2x$

$2(x +3) = 2x+6$

$2(+3) = +6$

b) $2(x) = 2x$

$2(x -3) = 2x-6$

$2(-3) = -6$

c) $2(-x) = -2x$

$2(-x+3) = -2x+6$

$2(+3) = +6$

d) $2(-x) = -2x$

$2(-x-3) = -2x-6$

$2(-3) = -6$

Negative Coefficient

e) $(-2)(x) = -2x$

$-2(x +3) = -2x-6$

$(-2)(+3) = -6$

f) $(-2)(x) = -2x$

$-2(x -3) = -2x+6$

$(-2)(-3) = +6$

g) $(-2)(-x) = 2x$

$-2(-x+3) = 2x-6$

$(-2)(+3) = -6$

h) $(-2)(-x) = 2x$

$-2(-x-3) = 2x+6$

$(-2)(-3) = +6$

TEACHING TIP
Have students write $4(x + y + z)$ as an addition problem and then simplify. After they compare the original expression with the simplified one ask, "Can the Distributive Property be used when we have 3 terms?"

The distributive property can be expanded as follows:

$$a(b + c + d + \cdots + n) = ab + ac + ad + \cdots + an$$

Examples of the Expanded Distributive Property

$$3(x + y + z) = 3x + 3y + 3z$$
$$2(x + y - 3) = 2x + 2y - 6$$

EXAMPLE 11 Use the distributive property to remove parentheses.

a) $4(x - 3)$ **b)** $-2(2x - 4)$ **c)** $-\dfrac{1}{2}(4r + 5)$ **d)** $-2(3x - 2y + 4z)$

Solution   **a)** $4(x - 3) = 4x - 12$ **b)** $-2(2x - 4) = -4x + 8$

c) $-\dfrac{1}{2}(4r + 5) = -2r - \dfrac{5}{2}$ **d)** $-2(3x - 2y + 4z) = -6x + 4y - 8z$ ✳

The distributive property can also be used from the right, as in Example 12.

EXAMPLE 12   Use the distributive property to remove parentheses from the expression $(2x - 8y)4$.

Solution   We distribute the 4 on the right side of the parentheses over the terms within the parentheses.

$$(2x - 8y)4 = 2x(4) - 8y(4)$$
$$= 8x - 32y$$

NOW TRY EXERCISE 83   ✳

Example 12 could have been rewritten as $4(2x - 8y)$ by the commutative property of multiplication, and then the 4 could have been distributed from the left to obtain the same answer, $8x - 32y$.

5   Remove Parentheses When They Are Preceded by a Plus or Minus Sign

In the expression $(4x + 3)$, how do we remove parentheses? Recall that the coefficient of a term is assumed to be 1 if none is shown. Therefore, we may write

$$(4x + 3) = 1(4x + 3)$$
$$= 1(4x) + (1)(3)$$
$$= 4x + 3$$

Note that $(4x + 3) = 4x + 3$. **When no sign or a plus sign precedes parentheses, the parentheses may be removed without having to change the expression inside the parentheses.**

Examples

$$(x + 3) = x + 3$$
$$(2x - 3) = 2x - 3$$
$$+(2x - 5) = 2x - 5$$
$$+(x + 2y - 6) = x + 2y - 6$$

Now consider the expression $-(4x + 3)$. How do we remove parentheses in this expression? Here, the coefficient in front of the parentheses is -1, so each term within the parentheses is multiplied by -1.

$$-(4x + 3) = -1(4x + 3)$$
$$= -1(4x) + (-1)(3)$$
$$= -4x + (-3)$$
$$= -4x - 3$$

Thus, $-(4x + 3) = -4x - 3$. **When a negative sign precedes parentheses, the signs of all the terms within the parentheses are changed when the parentheses are removed.**

Examples

$$-(x + 4) = -x - 4$$
$$-(-2x + 3) = 2x - 3$$
$$-(5x - y + 3) = -5x + y - 3$$
$$-(-4c - 3d - 5) = 4c + 3d + 5$$

NOW TRY EXERCISE 77

6 Simplify an Expression

Combining what we learned in the preceding discussions, we have the following procedure for **simplifying an expression**.

> **To Simplify an Expression**
>
> 1. Use the distributive property to remove any parentheses.
> 2. Combine like terms.

EXAMPLE 13 Simplify $6 - (2x + 3)$.

Solution

$$6 - (2x + 3) = 6 - 2x - 3 \quad \text{Use the distributive property.}$$
$$= -2x + 3 \quad \text{Combine like terms.}$$

Note: $3 - 2x$ is the same as $-2x + 3$; however, we generally write the term containing the variable first. ✳

EXAMPLE 14 Simplify $-\left(\frac{2}{3}x - \frac{1}{4}\right) + 3x$

Solution

$$-\left(\frac{2}{3}x - \frac{1}{4}\right) + 3x = -\frac{2}{3}x + \frac{1}{4} + 3x \quad \text{Distributive Property}$$

$$= -\frac{2}{3}x + 3x + \frac{1}{4} \quad \text{Rearrange terms}$$

$$= -\frac{2}{3}x + \frac{9}{3}x + \frac{1}{4} \quad \text{Write x terms with the LCD, 3}$$

$$= \frac{7}{3}x + \frac{1}{4} \quad \text{Combine like terms} \quad ✳$$

Notice in Example 14 that $\frac{7}{3}x$ and $\frac{1}{4}$ could not be combined because they are not like terms.

EXAMPLE 15 Simplify $\frac{3}{4}x + \frac{1}{2}(3x - 5)$

Solution

$$\frac{3}{4}x + \frac{1}{2}(3x - 5) = \frac{3}{4}x + \frac{1}{2}(3x) + \frac{1}{2}(-5) \quad \text{Distributive Property}$$

$$= \frac{3}{4}x + \frac{3}{2}x - \frac{5}{2}$$

$$= \frac{3}{4}x + \frac{6}{4}x - \frac{5}{2} \quad \text{Write x terms with the LCD, 4}$$

$$= \frac{9}{4}x - \frac{5}{2} \quad \text{Combine like terms} \quad ✳$$

EXAMPLE 16   Simplify $3(2a - 5) - 3(b - 6) - 4a$

Solution $3(2a - 5) - 3(b - 6) - 4a = 6a - 15 - 3b + 18 - 4a$ *Distributive property*

$$= 6a - 4a - 3b - 15 + 18 \quad \textit{Rearrange terms.}$$

$$= 2a - 3b + 3 \quad \textit{Combine like terms.}$$

NOW TRY EXERCISE 103

HELPFUL HINT

TEACHING TIP
Point out that although there is more than one way to write the solution, it is easier to compare answers if we all use the same style. Then ask students to identify the style used in the book.

Keep in mind the difference between the concepts of *term* and *factor*. When two or more expressions are **multiplied**, each expression is a **factor** of the product. For example, since $4 \cdot 3 = 12$, the 4 and the 3 are factors of 12. Since $3 \cdot x = 3x$, the 3 and the x are factors of $3x$. Similarly, in the expression $5xyz$, the $5, x, y$, and z are all factors.

In an expression, the parts that are **added** are the **terms** of the expression. For example, the expression $2x^2 + 3x - 4$, has three terms, $2x^2, 3x$, and -4. Note that the terms of an expression may have factors. For example, in the term $2x^2$, the 2 and the x^2 are factors because they are multiplied.

1. a) the parts that are added   **c)** $6xy, 3x, -y, -9$ **2. a)** have same variable with same exponents   **3. a)** the parts that are multiplied
b) multiplied together   **5. a)** numerical coefficient or coefficient   **6.** remove any parentheses and combine like terms   **7. a)** The signs of all terms inside the parentheses change when the parentheses are removed.

8. a) Parentheses may be removed without changing the expression.

Exercise Set 2.1

Concept/Writing Exercises

1. a) What are the terms of an expression?
   b) What are the terms of $3x - 4y - 5$? $3x, -4y, -5$
   c) What are the terms of $6xy + 3x - y - 9$?

2. a) What are like terms? Determine whether the following are like terms. If not, explain why.
   b) $3x, 4y$ no   **c)** $7, -2$ yes
   d) $5x^2, 2x$ no   **e)** $4x, -5xy$ no

3. a) What are the factors of an expression?
   b) Explain why 3 and x are factors of $3x$.
   c) Explain why $5, x$, and y are all factors of the expression $5xy$.   multiplied together

4. Consider the expression $2x - 5$.
   a) What is the x called?   a variable
   b) What is the -5 called?   a constant
   c) What is the 2 called?   coefficient

5. a) What is the name given to the numerical part of a term? List the coefficient of the following terms.
   b) $4x$ 4   **c)** x 1   **d)** $-x$ -1
   e) $\dfrac{3x}{5}$ $\dfrac{3}{5}$ **f)** $\dfrac{4}{7}(3t - 5)$ $\dfrac{4}{7}$

6. What does it mean to simplify an expression?

7. a) When a minus sign precedes an expression within parentheses, explain how to remove parentheses.
   b) Write $-(x - 8)$ without parentheses. $-x + 8$

8. a) When no sign or a plus sign precedes an expression within parentheses, explain how to remove parentheses.
   b) Write $+(x - 8)$ without parentheses. $x - 8$

Practice the Skills

Combine like terms when possible. If not possible, rewrite the expression as is.

9. $5x + 3x$ $8x$
10. $3x + 6$ $3x + 6$
11. $4x - 5x$ $-x$
12. $4x + 3y$ $4x + 3y$
13. $y + 3 + 4y$ $5y + 3$
14. $-2x - 3x$ $-5x$
15. $-2x + 5x$ $3x$
16. $4x - 7x + 4$ $-3x + 4$
17. $2 - 6x + 5$ $-6x + 7$
18. $-7 - 4m - 6$ $-4m - 13$
19. $-2w - 3w + 5$ $-5w + 5$
20. $5x + 2y + 3 + y$ $5x + 3y + 3$
21. $-x + 2 - x - 2$ $-2x$
22. $8x - 2y - 1 - 3x$ $5x - 2y - 1$
23. $3 + 6x - 3 - 6x$ 0
24. $y - 2y + 5$ $-y + 5$
25. $5 + 2x - 4x + 6$ $-2x + 11$
26. $5s - 3s - 2s$ 0
27. $4r - 6 - 6r - 2$ $-2r - 8$
28. $-6t + 5 + 2t - 9$ $-4t - 4$
29. $2 - 3x - 2x + y$ $-5x + y + 2$
30. $7x - 3 - 2x$ $5x - 3$
31. $-2x + 4x - 3$ $2x - 3$
32. $4 - x + 4x - 8$ $3x - 4$

39. $2x^2 + 4x + 8y^2$ **45.** $-3n^2 - 2n + 13$ **47.** $21.72x - 7.11$ **51.** $5w^3 + 2w^2 + w + 3$ **53.** $-7z^3 - z^2 + 2z$ **57.** $4a^2 + 3ab + b^2$

33. $b + 4 + \dfrac{3}{5}$ $b + \dfrac{23}{5}$

34. $\dfrac{3}{4}x + 2 + x$ $\dfrac{7}{4}x + 2$

35. $5.1n + 6.42 - 4.3n$ $0.8n + 6.42$

36. $13.4x + 1.2x + 8.3$ $14.6x + 8.3$

37. $\dfrac{1}{2}a + 3b + 1$ $\dfrac{1}{2}a + 3b + 1$

38. $x + \dfrac{1}{2}y - \dfrac{3}{8}y$ $x + \dfrac{1}{8}y$

39. $2x^2 + 3y^2 + 4x + 5y^2$

40. $-4x^2 - 3.1 - 5.2$ $-4x^2 - 8.3$

41. $-x^2 + 2x^2 + y$ $x^2 + y$

42. $1 + x^2 + 6 - 3x^2$ $-2x^2 + 7$ 🔒 **43.** $2x - 7y - 5x + 2y$ $-3x - 5y$

44. $3x - 7 - 9 + 4x$ $7x - 16$

45. $4 - 3n^2 + 9 - 2n$

46. $9x + y - 2 - 4x$ $5x + y - 2$

47. $-19.36 + 40.02x + 12.25 - 18.3x$

48. $52x - 52x - 63.5 - 63.5$ -127 **49.** $\dfrac{3}{5}x - 3 - \dfrac{7}{4}x - 2$ $-\dfrac{23}{20}x - 5$

50. $\dfrac{1}{2}y - 4 + \dfrac{3}{4}x - \dfrac{1}{5}y$ $\dfrac{3}{4}x + \dfrac{3}{10}y - 4$

51. $5w^3 + 2w^2 + w + 3$

52. $4p^2 - 3p^2 + 2p - 5p$ $p^2 - 3p$

53. $2z - 5z^3 - 2z^3 - z^2$

54. $5ab - 3ab$ $2ab$

55. $6x^2 - 6xy + 3y^2$ $6x^2 - 6xy + 3y^2$

56. $x^2 - 3xy - 2xy + 6$ $x^2 - 5xy + 6$

57. $4a^2 - 3ab + 6ab + b^2$

58. $4b^2 - 8bc + 5bc + c^2$ $4b^2 - 3bc + c^2$

Use the distributive property to remove parentheses.

59. $5(x + 2)$ $5x + 10$

60. $3(x - 6)$ $3x - 18$

61. $5(x + 4)$ $5x + 20$

62. $-2(y + 8)$ $-2y - 16$

63. $-2(x - 4)$ $-2x + 8$

64. $2(-y + 5)$ $-2y + 10$

65. $-\dfrac{1}{2}(2x - 4)$ $-x + 2$

66. $-4(x + 6)$ $-4x - 24$

67. $1(-4 + x)$ $x - 4$

68. $4(m - 6)$ $4m - 24$

69. $\dfrac{4}{5}(s - 5)$ $\dfrac{4}{5}s - 4$

70. $5(x - y + 5)$ $5x - 5y + 25$

71. $-0.3(3x + 5)$ $-0.9x - 1.5$

72. $-(x - 3)$ $-x + 3$

73. $\dfrac{1}{3}(3r - 12)$ $r - 4$

74. $-2(x + y - z)$ $-2x - 2y + 2z$

75. $0.7(2x + 0.5)$ $1.4x + 0.35$

76. $-(x + 4y)$ $-x - 4y$

77. $-(-x + y)$ $x - y$

78. $(3x + 4y - 6)$ $3x + 4y - 6$ 🔒 **79.** $-(2x + 4y - 8)$ $-2x - 4y + 8$

80. $-3(2a + 3b - 7)$

81. $1.1(3.1x - 5.2y + 2.8)$

82. $-4(-2m - 3n + 8)$

83. $2\left(3x - 2y + \dfrac{1}{4}\right)$ $6x - 4y + \dfrac{1}{2}$

84. $2\left(-\dfrac{1}{2}x + 4y + 3\right)$ $-x + 8y + 6$

85. $(x + 3y - 9)$ $x + 3y - 9$

86. $(-p + 2q - 3)$ $-p + 2q - 3$

87. $-3(-x + 2y + 4)$ $3x - 6y - 12$

88. $2.3(1.6x + 5.1y - 4.1)$

80. $-6a - 9b + 21$ **81.** $3.41x - 5.72y + 3.08$ **82.** $8m + 12n - 32$ **88.** $3.68x + 11.73y - 9.43$

Simplify. **116.** $-a - 2b + 10$ **117.** $0.2x - 4y - 2.8$ **118.** $-3x - 10y + 30$ **119.** $-6x + 7y$

89. $3(x - 5) - x$ $2x - 15$

90. $2 + (x - 3)$ $x - 1$

91. $-2(3 - x) + 7$ $2x + 1$

92. $-(3x - 3) + 5$ $-3x + 8$

93. $6x + 2(4x + 9)$ $14x + 18$

94. $3(x + y) + 2y$ $3x + 5y$

🔒 **95.** $2(x - y) + 2x + 3$ $4x - 2y + 3$

96. $6 + (x - 5) + 3x$ $4x + 1$

97. $4(2c - 3) - 3(c - 4)$ $5c$

98. $4 + (2y + 2) + y$ $3y + 6$

99. $8x - (x - 3)$ $7x + 3$

100. $-(x - 5) - 3x + 4$ $-4x + 9$

101. $2(x - 3) - (x + 3)$ $x - 9$

102. $3y - (2x + 2y) - 6x$ $-8x + y$

103. $4(x - 1) + 2(3 - x) - 4$ $2x - 2$

104. $4(x + 3) - 2x$ $2x + 12$

105. $-(3s + 4) - (s + 2)$ $-4s - 6$

106. $6 - 2(x + 3) + 5x$ $3x$

107. $-3(x + 1) + 5x + 6$ $2x + 3$

108. $-(x + 2) + 3x - 6$ $2x - 8$

109. $4(m + 3) - 4m - 12$ 0

110. $-3(a + 2b) + 3(a + 2b)$ 0

111. $0.4 + (y + 5) + 0.6 - 2$ $y + 4$

112. $4 - (2 - x) + 3x$ $4x + 2$

113. $4 + (3x - 4) - 5$ $3x - 5$

114. $2y - 6(y - 2) + 3$ $-4y + 15$

🔒 **115.** $4(x + 2) - 3(x - 4) - 5$ $x + 15$

116. $6 - (a - 5) - (2b + 1)$

117. $-0.2(6 - x) - 4(y + 0.4)$

118. $-5(2y - 8) - 3(1 + x) - 7$

119. $-6x + 7y - (3 + x) + (x + 3)$ **120.** $3(t - 2) - 2(t + 4) - 6$ $t - 20$ **121.** $\dfrac{1}{2}(x + 3) + \dfrac{1}{3}(3x + 6)$ $\dfrac{3}{2}x + \dfrac{7}{2}$

122. $\dfrac{2}{3}(r - 2) - \dfrac{1}{2}(r + 4)$ $\dfrac{1}{6}r - \dfrac{10}{3}$

Problem Solving

If □ + □ + □ + ⊙ + ⊙ can be represented as 3□ + 2⊙, write an expression to represent each of the following.

123. □ + ⊖ + ⊖ + □ + ⊖   2□ + 3⊖
124. ⊗ + ☺ + ⊗ + ☺ + ☺ + ☺   2⊗ + 4☺
125. $x + y + \triangle + \triangle + x + y + y$ $2x + 3y + 2\triangle$
126. $2 + x + 2 + \ominus + \ominus + 2 + y$ $x + y + 2\ominus + 6$

In Exercises 127 and 128, consider the following. The positive factors of 6 are 1, 2, 3, and 6 since

$$1 \cdot 6 = 6$$
$$2 \cdot 3 = 6$$
$$\uparrow \;\; \uparrow$$
factors

127. List all the positive factors of 12.   1, 2, 3, 4, 6, 12.

128. List all the positive factors of 16.   1, 2, 4, 8, 16

Combine like terms.

129. $3\triangle + 5\square - \triangle - 3\square$ $2\triangle + 2\square$

130. $8☺ - 4▣ - 2▢ - 3☺$ $5☺ - 6▢$

Challenge Problems

Simplify.   **131.** $22x^2 - 25y^2 - 4x + 3$ **132.** $9x^2 - 7x - 8$ **133.** $6x^2 + 5y^2 + 3x + 7y$

131. $4x^2 + 5y^2 + 6(3x^2 - 5y^2) - 4x + 3$

133. $x^2 + 2y - y^2 + 3x + 5x^2 + 6y^2 + 5y$

132. $2x^2 - 4x + 8x^2 - 3(x + 2) - x^2 - 2$

134. $2[3 + 4(x - 5)] - [2 - (x - 3)]$ $9x - 39$

Cumulative Review Exercises

[1.5] Evaluate.

135. $|-7|$ 7
136. $-|-16|$ -16
[1.7] **137.** Evaluate $-4 - 3 - (-6)$ -1

✎ *[1.9]* **138.** Write a paragraph explaining the order of operations.   Answers will vary.

139. Evaluate $-x^2 + 5x - 6$ when $x = -1$. -12

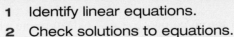

2.2 THE ADDITION PROPERTY OF EQUALITY

SSM   Study Guide   CD/Video

MathPro 4/5   PH Math Tutor Center   prenhall.com/Angel

1 Identify linear equations.
2 Check solutions to equations.
3 Identify equivalent equations.
4 Use the addition property to solve equations.
5 Solve equations by doing some steps mentally.

1 Identify Linear Equations

A statement that shows two algebraic expressions are equal is called an **equation**. For example, $4x + 3 = 2x - 4$ is an equation. In this chapter we learn to solve **linear equations** in one variable.

DEFINITION

A **linear equation** in one variable is an equation that can be written in the form

$$ax + b = c$$

where a, b, and c are real numbers and $a \neq 0$.

Examples of Linear Equations

$$x + 4 = 7$$
$$2x - 4 = 6$$

2 Check Solutions to Equations

The **solution to an equation** is the number or numbers that when substituted for the variable or variables make the equation a true statement. For example, the solution to $x + 4 = 7$ is 3. We will shortly learn how to find the solution to an equation, or to **solve an equation**. But before we do this we will learn how to *check* the solution to an equation.

The solution to an equation may be **checked** by substituting the value that is believed to be the solution for the variable in the original equation. If the substitution results in a true statement, your solution is correct. If the substitution results in a false statement, then either your solution or your check is incorrect, and you need to go back and find your error. Try to check all your solutions. Checking the solutions will improve your arithmetic and algebra skills.

When we show the check of a solution we shall use the $\overset{?}{=}$ notation. This notation is used when we are questioning whether a statement is true. For example, if we use

$$2 + 3 \overset{?}{=} 2(3) - 1$$

we are asking "Does $2 + 3 = 2(3) - 1$?"

To check whether 3 is the solution to $x + 4 = 7$, we substitute 3 for each x in the equation.

Check: $x = 3$

$$x + 4 = 7$$
$$3 + 4 \overset{?}{=} 7$$
$$7 = 7 \quad \textit{True}$$

Since the check results in a true statement, 3 is a solution.

EXAMPLE 1

Consider the equation $2x - 4 = 6$. Determine whether 3 is a solution.

Solution

To determine whether 3 is a solution to the equation, we substitute 3 for x.

Check: $x = 3$

$$2x - 4 = 6$$
$$2(3) - 4 \overset{?}{=} 6$$
$$6 - 4 \overset{?}{=} 6$$
$$2 = 6 \quad \textit{False}$$

Since we obtained a false statement, 3 is not a solution.

Now check to see if 5 is a solution to the equation in Example 1. Your check should show that 5 is a solution.

We can use the same procedures to check more complex equations, as shown in Examples 2 and 3.

EXAMPLE 2 Determine whether 18 is a solution to the following equation.

$$3x - 2(x + 3) = 12$$

Solution To determine whether 18 is a solution, we substitute 18 for each x in the equation. If the substitution results in a true statement, then 18 is a solution.

Check: $x = 18$

$$3x - 2(x + 3) = 12$$
$$3(18) - 2(18 + 3) \stackrel{?}{=} 12$$
$$3(18) - 2(21) \stackrel{?}{=} 12$$
$$54 - 42 \stackrel{?}{=} 12$$
$$12 = 12 \quad \textit{True}$$

Since we obtain a true statement, 18 is a solution.

EXAMPLE 3 Determine whether $-\dfrac{3}{2}$ is a solution to the following equation.

$$3(n + 3) = 6 + n$$

Solution In this equation n is the variable. Substitute $-\dfrac{3}{2}$ for each n in the equation.

Check: $n = -\dfrac{3}{2}$

$$3(n + 3) = 6 + n$$
$$3\left(-\frac{3}{2} + 3\right) \stackrel{?}{=} 6 + \left(-\frac{3}{2}\right)$$
$$3\left(-\frac{3}{2} + \frac{6}{2}\right) \stackrel{?}{=} \frac{12}{2} - \frac{3}{2}$$
$$3\left(\frac{3}{2}\right) \stackrel{?}{=} \frac{9}{2}$$
$$\frac{9}{2} = \frac{9}{2} \quad \textit{True}$$

Thus, $-\dfrac{3}{2}$ is a solution.

NOW TRY EXERCISE 21

Using Your Calculator

Checking Solutions

Calculators can be used to check solutions to equations. For example, to check whether $\frac{-10}{3}$ is a solution to the equation $2x + 3 = 5(x + 3) - 2$, we perform the following steps.

1. Substitute $\frac{-10}{3}$ for each x as shown below.

$$2x + 3 = 5(x + 3) - 2$$

$$2\left(\frac{-10}{3}\right) + 3 \stackrel{?}{=} 5\left(\frac{-10}{3} + 3\right) - 2$$

2. Evaluate each side of the equation separately using your calculator. If you obtain the same value on both sides, your solution checks.

Scientific Calculator

To evaluate the left side of the equation, $2\left(\frac{-10}{3}\right) + 3$, press the following keys:

2 ⊠ ((10 +/− ÷ 3)) + 3 = −3.6666667

To evaluate the right side of the equation, $5\left(\frac{-10}{3} + 3\right) - 2$, press the following keys:

5 ⊠ ((10 +/− ÷ 3 + 3)) − 2 = −3.6666667

Since both sides give the same value, the solution checks. Note that because calculators differ in their electronics, sometimes the last digit of a calculation will differ.

Graphing Calculator

Left side of equation: 2 (((−) 10 ÷ 3)) + 3 ENTER −3.666666667

Right side of the equation: 5 (((−) 10 ÷ 3 + 3)) − 2 ENTER −3.666666667

Since both sides give the same solution, the solution checks.

3 Identify Equivalent Equations

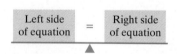

FIGURE 2.1

TEACHING TIP
Begin by asking "If I have a balanced scale and add 3 pounds to one side of the scale, what must I do to keep the scale balanced?"

Now that we know how to check a solution to an equation we will discuss solving equations. Complete procedures for solving equations will be given shortly. For now, you need to understand that **to solve an equation, it is necessary to get the variable alone on one side of the equal sign. We say that we isolate the variable**. To isolate the variable, we make use of two properties: the addition and multiplication properties of equality. Look first at Figure 2.1.

Think of an equation as a balanced statement whose left side is balanced by its right side. When solving an equation, we must make sure that the equation remains balanced at all times. That is, both sides must always remain equal. **We ensure that an equation always remains equal by doing the same thing to both sides of the equation**. For example, if we add a number to the left side of the equation, we must add exactly the same number to the right side. If we multiply the right side of the equation by some number, we must multiply the left side by the same number.

When we add the same number to both sides of an equation or multiply both sides of an equation by the same nonzero number, we do not change the solution to the equation, just the form. Two or more equations with the same solution are called **equivalent equations**. The equations $2x - 4 = 2$, $2x = 6$, and $x = 3$ are equivalent, since the solution to each is 3.

Check: $x = 3$

$$2x - 4 = 2 \qquad\qquad 2x = 6 \qquad\qquad x = 3$$
$$2(3) - 4 \stackrel{?}{=} 2 \qquad 2(3) \stackrel{?}{=} 6 \qquad 3 = 3 \quad \text{True}$$
$$6 - 4 \stackrel{?}{=} 2 \qquad\quad 6 = 6 \quad \text{True}$$
$$2 = 2 \quad \text{True}$$

When solving an equation, we use the addition and multiplication properties to express a given equation as simpler equivalent equations until we obtain the solution.

4 Use the Addition Property to Solve Equations

Now we are ready to define the **addition property of equality**.

Addition Property of Equality

If $a = b$, then $a + c = b + c$ for any real numbers a, b, and c.

This property means that the same number can be added to both sides of an equation without changing the solution. **The addition property is used to solve equations of the form $x + a = b$.** To isolate the variable x in equations of this form, add the opposite or additive inverse of a, $-a$, to both sides of the equation.

To isolate the variable when solving equations of the form $x + a = b$, **we use the addition property to eliminate the number on the same side of the equal sign as the variable**. Study the following examples carefully.

Equation	To Solve, Use the Addition Property to Eliminate the Number
$x + 8 = 10$	8
$x - 7 = 12$	-7
$5 = x - 12$	-12
$-4 = x + 9$	9

TEACHING TIP
Point out that the Addition Property of Equality is used to make an equivalent equation. When used intelligently, it can be used to solve an equation. Have students use the property on $x + 3 = 10$ and $x - 4 = 5$ to **a)** make an equivalent equation which does not give the solution. **b)** make the equivalent equation which gives the solution.

Now let's work some examples.

EXAMPLE 4 Solve the equation $x - 4 = -3$.

Solution To isolate the variable, x, we must eliminate the -4 from the left side of the equation. To do this we add 4, the opposite of -4, to *both sides* of the equation.

$$x - 4 = -3$$
$$x - 4 + 4 = -3 + 4 \quad \text{Add 4 to both sides.}$$
$$x + 0 = 1$$
$$x = 1$$

Note how the process helps to isolate x.

Check:
$$x - 4 = -3$$
$$1 - 4 \stackrel{?}{=} -3$$
$$-3 = -3 \quad \text{True}$$

In Example 5, we will not show the check. Space limitations prevent us from showing all checks. However, *you should check all of your answers.*

EXAMPLE 5 Solve the equation $x + 5 = 9$.

Solution To solve this equation, we must isolate the variable, x. Therefore, we must eliminate the 5 from the left side of the equation. To do this, we add -5, the opposite of 5, to *both sides* of the equation.

$$x + 5 = 9$$
$$x + 5 + (-5) = 9 + (-5) \quad \text{Add } -5 \text{ to both sides.}$$
$$x + 0 = 4$$
$$x = 4$$

In Example 5 we added -5 to both sides of the equation. From Section 1.7 we know that $5 + (-5) = 5 - 5$. Thus, we can see that adding a negative 5 to both sides of the equation is equivalent to subtracting a 5 from both sides of the equation. According to the addition property, the same number may be *added* to both sides of an equation. **Since subtraction is defined in terms of addition, the addition property also allows us to *subtract* the same number from both sides of the equation.** Thus, Example 5 could have also been worked as follows:

$$x + 5 = 9$$
$$x + 5 - 5 = 9 - 5 \quad \text{Subtract 5 from both sides.}$$
$$x + 0 = 4$$
$$x = 4$$

In this text, unless there is a specific reason to do otherwise, rather than adding a negative number to both sides of the equation, we will subtract a number from both sides of the equation.

EXAMPLE 6 Solve the equation $k + 7 = -3$.

Solution We must isolate the variable k.

$$k + 7 = -3$$
$$k + 7 - 7 = -3 - 7 \quad \text{Subtract 7 from both sides.}$$
$$k + 0 = -10$$
$$k = -10$$

Check:
$$k + 7 = -3$$
$$-10 + 7 \stackrel{?}{=} -3$$
$$-3 = -3 \quad \text{True}$$

NOW TRY EXERCISE 51

HELPFUL HINT Remember that our goal in solving an equation is to get the variable alone on one side of the equation. To do this, we add or subtract **the number on the same side of the equation as the variable** to both sides of the equation.

(continued on the next page)

Equation	Must Eliminate	Number to Add (or Subtract) to (or from) Both Sides of the Equation	Correct Results	Solution
$x - 5 = 8$	-5	add 5	$x - 5 + 5 = 8 + 5$	$x = 13$
$x - 3 = -12$	-3	add 3	$x - 3 + 3 = -12 + 3$	$x = -9$
$2 = x - 7$	-7	add 7	$2 + 7 = x - 7 + 7$	$9 = x$ or $x = 9$
$x + 12 = -5$	$+12$	subtract 12	$x + 12 - 12 = -5 - 12$	$x = -17$
$6 = x + 4$	$+4$	subtract 4	$6 - 4 = x + 4 - 4$	$2 = x$ or $x = 2$
$13 = x + 9$	$+9$	subtract 9	$13 - 9 = x + 9 - 9$	$4 = x$ or $x = 4$

Notice that under the *Correct Results* column, when the equation is simplified by combining terms, the x will become isolated because the sum of a number and its opposite is 0, and $x + 0$ equals x.

EXAMPLE 7 Solve the equation $6 = x - 9$.

Solution The variable x is on the right side of the equation. To isolate the x, we must eliminate the -9 from the right side of the equation. This can be accomplished by adding 9 to both sides of the equation.

$$6 = x - 9$$
$$6 + 9 = x - 9 + 9 \quad \text{Add 9 to both sides.}$$
$$15 = x + 0$$
$$15 = x$$

Thus, the solution is 15.

EXAMPLE 8 Solve the equation $-6.25 = y + 12.78$.

Solution The variable is on the right side of the equation. Subtract 12.78 from both sides of the equation to isolate the variable.

$$-6.25 = y + 12.78$$
$$-6.25 - 12.78 = y + 12.78 - 12.78 \quad \text{Subtract 12.78 from both sides.}$$
$$-19.03 = y + 0$$
$$-19.03 = y$$

NOW TRY EXERCISE 67   The solution is -19.03.

AVOIDING COMMON ERRORS

When solving an equation, our goal is to get the variable alone on one side of the equal sign. Consider the equation $x + 3 = -4$. How do we solve it?

CORRECT	WRONG
Remove the 3 from the left side of the equation.	Remove the -4 from the right side of the equation.
$x + 3 = -4$	$x + 3 = -4$
$x + 3 - 3 = -4 - 3$	$x + 3 + 4 = -4 + 4$
$x = -7$	$x + 7 = 0$
Variable is now isolated.	*Variable is **not** isolated.*

Remember, use the addition property to *remove the number that is on the same side of the equation as the variable.*

5 Solve Equations by Doing Some Steps Mentally

Consider the following two problems.

a)
$$x - 5 = 12$$
$$x - 5 + 5 = 12 + 5$$
$$x + 0 = 12 + 5$$
$$x = 17$$

b)
$$15 = x + 3$$
$$15 - 3 = x + 3 - 3$$
$$15 - 3 = x + 0$$
$$12 = x$$

Note how the number on the same side of the equal sign as the variable is transferred to the opposite side of the equal sign when the addition property is used. Also note that the sign of the number changes when transferred from one side of the equal sign to the other.

When you feel comfortable using the addition property of equality, you may wish to do some of the steps mentally to reduce some of the written work. For example, the preceding two problems may be shortened as follows:

SHORTENED FORM

a)
$$x - 5 = 12$$
$$x - 5 + 5 = 12 + 5$$ ← Do this step mentally.
$$x = 12 + 5$$
$$x = 17$$

$$x - 5 = 12$$
$$x = 12 + 5$$
$$x = 17$$

SHORTENED FORM

b)
$$15 = x + 3$$
$$15 - 3 = x + 3 - 3$$ ← Do this step mentally.
$$15 - 3 = x$$
$$12 = x$$

$$15 = x + 3$$
$$15 - 3 = x$$
$$12 = x$$

NOW TRY EXERCISE 55

1. a statement that shows two algebraic expressions are equal   2. a) the number(s) that make the equation a true statement
b) to find the solutions to an equation   3. Substitute the value in the equation. Then determine if it results in a true statement

Exercise Set 2.2

5. two or more equations with the same solution

Concept/Writing Exercises

1. What is an equation?
2. a) What is meant by the "solution to an equation"?
   b) What does it mean to "solve an equation"?
3. Explain how the solution to an equation may be checked.
4. In your own words, explain the addition property of equality. Answers will vary
5. What are equivalent equations?
6. To solve an equation we "isolate the variable."
   a) Explain what this means.
   b) Explain how to isolate the variable in the equations discussed in this section. Answers will vary
7. When solving the equation $x - 4 = 6$, would you add 4 to both sides of the equation or subtract 6 from both sides of the equation? Explain. Add 4

9. one example is $x + 2 = 1$ 10. all have the same solution, 1

8. When solving the equation $6 = x + 2$, would you subtract 6 from both sides of the equation or subtract 2 from both sides of the equation? Explain. Subtract 2
9. Give an example of a linear equation in one variable.
10. Explain why the following three equations are equivalent.
$$2x + 3 = 5, \quad 2x = 2, \quad x = 1$$
11. Explain why the addition property allows us to subtract the same quantity from both sides of an equation.
12. To solve the equation $x - \square = \triangle$ for x, do we add $\square$ to both sides of the equation or do we subtract $\triangle$ from both sides of the equation? Explain. Add $\square$.

6. a) get the variable by itself on one side of the equation
11. subtraction is defined in terms of addition

Practice the Skills

13. Is $x = 2$ a solution of $4x - 3 = 5$ yes
14. Is $x = -6$ a solution of $2x + 1 = x - 5$? yes
15. Is $x = -3$ a solution of $2x - 5 = 5(x + 2)$? no
16. Is $x = 1$ a solution of $2(x - 3) = -3(x + 1)$? no
17. Is $p = 0$ a solution of $3p - 4 = 2(p + 3) - 10$? yes
18. Is $k = -1$ a solution of $-3(k - 3) = -4k + 3 - 5k$ yes

19. Is $x = 3.4$ a solution of $3(x + 2) - 3(x - 1) = 9$? yes
20. Is $x = \frac{1}{2}$ a solution of $x + 3 = 3x + 2$? yes
21. Is $x = \frac{1}{2}$ a solution of $4x - 4 = 2x - 2$? no
22. Is $x = \frac{1}{2}$ a solution of $3x + 4 = 2x + 9$? no
23. Is $x = \frac{11}{2}$ a solution of $3(x + 2) = 5(x - 1)$? yes
24. Is $h = 3$ a solution of $-(h - 5) - (h - 6) = 3h - 4$? yes

Solve each equation and check your solution.

25. $x + 5 = 9$ 4
26. $x - 4 = 13$ 17
27. $x + 1 = -6$ -7
28. $x - 4 = -8$ -4
29. $x + 4 = -5$ -9
30. $x - 16 = 36$ 52
31. $x + 9 = 52$ 43
32. $9 + n = 9$ 0
33. $-6 + w = 9$ 15
34. $3 = 7 + t$ -4
35. $27 = x + 16$ 11
36. $50 = x - 25$ 75
37. $-18 = -14 + x$ -4
38. $7 + x = -50$ -57
39. $9 + x = 4$ -5
40. $x + 29 = -29$ -58
41. $4 + x = -9$ -13
42. $9 = x - 3$ 12
43. $7 + r = -23$ -30
44. $a - 5 = -9$ -4
45. $8 = 8 + v$ 0
46. $9 + x = 12$ 3
47. $-4 = x - 3$ -1
48. $-13 = 4 + x$ -17
49. $12 = 16 + x$ -4
50. $62 = z - 15$ 77
51. $15 + x = -5$ -20
52. $-20 = 4 + x$ -24
53. $-10 = -10 + x$ 0
54. $8 = 8 + x$ 0
55. $5 = x - 12$ 17
56. $-12 = 20 + c$ -32
57. $-50 = x - 24$ -26
58. $-29 + x = -15$ 14
59. $43 = 15 + p$ 28
60. $-25 = 74 + x$ -99
61. $40.2 + x = -5.9$ -46.1
62. $-27.23 + x = 9.77$ 37
63. $-37 + x = 9.5$ 46.5
64. $7.2 + x = 7.2$ 0
65. $x - 8.77 = -17$ -8.23
66. $6.1 + x = 10.2$ 4.1
67. $9.32 = x + 3.75$ 5.57
68. $139 = x - 117$ 256

Problem Solving

69. Do you think the equation $x + 1 = x + 2$ has a real number as a solution? Explain. (We will discuss equations like this in Section 2.5) no

70. Do you think the equation $x + 4 = x + 4$ has more than one real number as a solution? If so, how many solutions does it have? Explain. yes, infinite number

Challenge Problems

We can solve equations that contain unknown symbols. Solve each equation for the symbol indicated by adding (or subtracting) a symbol to (or from) both sides of the equation. Explain each answer. (Remember that to solve the equation you want to isolate the symbol you are solving for on one side of the equation.)

71. $x - \triangle = \square$, for x $x = \square + \triangle$
72. $\square + \odot = \triangle$, for $\odot$ $\odot = \triangle - \square$
73. $\odot = \square + \triangle$, for $\square$ $\square = \odot - \triangle$
74. $\square = \triangle + \odot$, for $\odot$ $\odot = \square - \triangle$

Group Activity

Discuss and answer Exercise 75 as a group.

75. Consider the equation $2(x + 3) = 2x + 6$.
  a) Group member 1: Determine whether 4 is a solution to the equation. yes
  b) Group member 2: Determine whether -2 is a solution to the equation. yes
  c) Group member 3: Determine whether 0.3 is a solution to the equation. yes
  d) Each group member: Select a number not used in parts **a)–c)** and determine whether that number is a solution to the equation. yes
  e) As a group, write what you think is the solution to the equation $2(x + 3) = 2x + 6$ and write a paragraph explaining your answer. all real numbers

Cumulative Review Exercises

[1.9] Evaluate.

76. $3x + 4(x - 3) + 2$ when $x = 4$ 18

77. $6x - 2(2x + 1)$ when $x = -3$ -8

[2.1] Simplify.

78. $4x + 3(x - 2) - 5x - 7$ $2x - 13$

79. $-(2t + 4) + 3(4t - 5) - 3t$ $7t - 19$

2.3 THE MULTIPLICATION PROPERTY OF EQUALITY

SSM   Study Guide   CD/Video

MathPro 4/5   PH Math Tutor Center   prenhall.com/Angel

1. Identify reciprocals.
2. Use the multiplication property to solve equations.
3. Solve equations of the form $-x = a$.
4. Do some steps mentally when solving equations.

1 Identify Reciprocals

In Section 1.10 we introduced the **reciprocal** (or multiplative inverse) of a number. Recall that two numbers are reciprocals of each other when their product is 1. Some examples of numbers and their reciprocals follow.

Number	Reciprocal	Product
2	$\dfrac{1}{2}$	$(2)\left(\dfrac{1}{2}\right) = 1$
$-\dfrac{3}{5}$	$-\dfrac{5}{3}$	$\left(-\dfrac{3}{5}\right)\left(-\dfrac{5}{3}\right) = 1$
-1	-1	$(-1)(-1) = 1$

The reciprocal of a positive number is a positive number and the reciprocal of a negative number is a negative number. Note that 0 has no reciprocal. Why?

In general, if a represents any nonzero number, its reciprocal is $\dfrac{1}{a}$. For example, the reciprocal of 3 is $\dfrac{1}{3}$ and the reciprocal of -2 is $\dfrac{1}{-2}$ or $-\dfrac{1}{2}$. The reciprocal of $-\dfrac{3}{5}$ is $\dfrac{1}{-\dfrac{3}{5}}$, which can be written as $1 \div \left(-\dfrac{3}{5}\right)$. Simplifying, we get $\left(\dfrac{1}{1}\right)\left(-\dfrac{5}{3}\right) = -\dfrac{5}{3}$. Thus, the reciprocal of $-\dfrac{3}{5}$ is $-\dfrac{5}{3}$.

2 Use the Multiplication Property to Solve Equations

In Section 2.2 we used the addition property of equality to solve equations of the form $x + a = b$, where a and b represent real numbers. In this section we use the multiplication property of equality to solve equations of the form $ax = b$, where a and b represent real numbers.

It is important that you recognize the difference between equations like $x + 2 = 8$ and $2x = 8$. In $x + 2 = 8$ the 2 is a *term* that is being added to x, so we use the addition property to solve the equation. In $2x = 8$ the 2 is a *factor* of $2x$. The 2 is the coefficient multiplying the x, so we use the multiplication property to solve the equation. The multiplication property of equality is used to solve linear equations where the coefficient of the x-term is a number other than 1.

Now we present the *multiplication property of equality*.

Multiplication Property of Equality

If $a = b$, then $a \cdot c = b \cdot c$ for any real numbers a, b, and c.

The multiplication property means that both sides of an equation can be multiplied by the same nonzero number without changing the solution. **The multiplication property can be used to solve equations of the form $ax = b$.** We can isolate the variable in equations of this form by multiplying both sides of the equation by the reciprocal of a, which is $\frac{1}{a}$. Doing so makes the numerical coefficient of the variable, x, become 1, which can be omitted when we write the variable. By following this process, we say that we *eliminate* the coefficient from the variable.

Equation	To Solve, Use the Multiplication Property to Eliminate the Coefficient
$4x = 9$	4
$-5x = 20$	-5
$15 = \dfrac{1}{2}x$	$\dfrac{1}{2}$
$7 = -9x$	-9

Now let's work some examples.

EXAMPLE 1   Solve the equation $3x = 6$.

Solution   To isolate the variable, x, we must eliminate the 3 from the left side of the equation. To do this, we multiply both sides of the equation by the reciprocal of 3, which is $\frac{1}{3}$.

$$3x = 6$$
$$\frac{1}{3} \cdot 3x = \frac{1}{3} \cdot 6 \qquad \textit{Multiply both sides by } \frac{1}{3}.$$
$$\frac{1}{\cancel{3}} \cdot \cancel{3}x = \frac{1}{\cancel{3}} \cdot \cancel{6}^{2} \qquad \textit{Divide out the common factors.}$$
$$1x = 2$$
$$x = 2$$

Notice in Example 1 that $1x$ is replaced by x in the next step. Usually we do this step mentally.

EXAMPLE 2   Solve the equation $\dfrac{x}{2} = 4$.

Solution   Since dividing by 2 is the same as multiplying by $\frac{1}{2}$, the equation $\frac{x}{2} = 4$ is the same as $\frac{1}{2}x = 4$. We will therefore multiply both sides of the equation by the reciprocal of $\frac{1}{2}$, which is 2.

$$\frac{x}{2} = 4$$
$$\overset{1}{\cancel{2}}\left(\frac{x}{\cancel{2}}\right) = 2 \cdot 4 \qquad \textit{Multiply both sides by 2.}$$
$$x = 2 \cdot 4$$
$$x = 8$$

NOW TRY EXERCISE 15

EXAMPLE 3 Solve the equation $\frac{2}{3}x = 6$.

Solution The reciprocal of $\frac{2}{3}$ is $\frac{3}{2}$. We multiply both sides of the equation by $\frac{3}{2}$.

$$\frac{2}{3}x = 6$$

$$\frac{3}{2} \cdot \frac{2}{3}x = \frac{3}{2} \cdot 6 \qquad \textit{Multiply both sides by } \frac{3}{2}.$$

$$1x = 9$$

$$x = 9$$

We will show a check of this solution.

Check: $$\frac{2}{3}x = 6$$

$$\frac{2}{3}(9) \stackrel{?}{=} 6$$

NOW TRY EXERCISE 49 $$6 = 6 \qquad \textit{True} \qquad \text{✳}$$

In Example 1, we multiplied both sides of the equation $3x = 6$ by $\frac{1}{3}$ to isolate the variable. We could have also isolated the variable by dividing both sides of the equation by 3, as follows:

$$3x = 6$$

$$\frac{\overset{1}{\cancel{3}}x}{\underset{1}{\cancel{3}}} = \frac{\overset{2}{\cancel{6}}}{\underset{1}{\cancel{3}}} \qquad \textit{Divide both sides by 3.}$$

$$x = 2$$

We can do this because dividing by 3 is equivalent to multiplying by $\frac{1}{3}$. **Since division can be defined in terms of multiplication $\left(\frac{a}{b} \text{ means } a \cdot \frac{1}{b}\right)$, the multiplication property also allows us to divide both sides of an equation by the same nonzero number.** This process is illustrated in Examples 4 through 6.

EXAMPLE 4 Solve the equation $8w = 3$.

Solution In this equation w is the variable. To solve the equation we divide both sides of the equation by 8.

$$8w = 3$$

$$\frac{8w}{8} = \frac{3}{8} \qquad \textit{Divide both sides by 8.}$$

$$w = \frac{3}{8} \qquad \qquad \text{✳}$$

EXAMPLE 5 Solve the equation $-15 = -3z$.

Solution In this equation the variable, z, is on the right side of the equal sign. To isolate z, we divide both sides of the equation by -3.

$$-15 = -3z$$

$$\frac{-15}{-3} = \frac{-3z}{-3} \qquad \textit{Divide both sides by } -3.$$

$$5 = z \qquad \qquad \text{✳}$$

EXAMPLE 6 Solve the equation $0.24x = 1.20$.

Solution We begin by dividing both sides of the equation by 0.24 to isolate the variable x.

$$0.24x = 1.20$$

$$\frac{0.24x}{0.24} = \frac{1.20}{0.24} \qquad \textit{Divide both sides by 0.24.}$$

$$x = 5$$

NOW TRY EXERCISE 35 Working problems involving decimal numbers on a calculator will probably save you time.

HELPFUL HINT When solving an equation of the form $ax = b$, we can isolate the variable by

1. multiplying both sides of the equation by the reciprocal of a, $\frac{1}{a}$, as was done in Examples 1, 2, and 3, or

2. dividing both sides of the equation by a, as was done in Examples 4, 5, and 6.

Either method may be used to isolate the variable. However, if the equation contains a fraction, or fractions, you will arrive at a solution more quickly by multiplying by the reciprocal of a. This is illustrated in Examples 7 and 8.

EXAMPLE 7 Solve the equation $-2x = \dfrac{3}{5}$.

Solution Since this equation contains a fraction, we will isolate the variable by multiplying both sides of the equation by $-\frac{1}{2}$, which is the reciprocal of -2.

$$-2x = \frac{3}{5}$$

$$\left(-\frac{1}{2}\right)(-2x) = \left(-\frac{1}{2}\right)\left(\frac{3}{5}\right) \qquad \textit{Multiply both sides by } -\frac{1}{2}.$$

$$1x = \left(-\frac{1}{2}\right)\left(\frac{3}{5}\right)$$

$$x = -\frac{3}{10}$$

In Example 7, if you wished to solve the equation by dividing both sides of the equation by -2, you would have to divide the fraction $\frac{3}{5}$ by -2.

EXAMPLE 8 Solve the equation $-6 = -\dfrac{3}{5}x$.

Solution Since this equation contains a fraction, we will isolate the variable by multiplying both sides of the equation by the reciprocal of $-\frac{3}{5}$, which is $-\frac{5}{3}$.

$$-6 = -\frac{3}{5}x$$

$$\left(-\frac{5}{3}\right)(-6) = \left(-\frac{5}{3}\right)\left(-\frac{3}{5}x\right) \qquad \textit{Multiply both sides by } -\frac{5}{3}.$$

NOW TRY EXERCISE 63 $$10 = x$$

In Example 8, the equation was written as $-6 = -\frac{3}{5}x$. This equation is equivalent to the equations $-6 = \frac{-3}{5}x$ and $-6 = \frac{3}{-5}x$. Can you explain why? All three equations have the same solution, 10.

3 Solve Equations of the Form $-x = a$

When solving an equation, we may obtain an equation like $-x = 7$. This is not a solution since $-x = 7$ means $-1x = 7$. The solution to an equation is of the form $x = $ some number. When an equation is of the form $-x = 7$, we can solve for x by multiplying both sides of the equation by -1, as illustrated in the following example.

EXAMPLE 9 Solve the equation $-x = 7$.

Solution $-x = 7$ means that $-1x = 7$. We are solving for x, not $-x$. We can multiply both sides of the equation by -1 to isolate x on the left side of the equation.

$$-x = 7$$
$$-1x = 7$$
$$(-1)(-1x) = (-1)(7) \qquad \text{\textit{Multiply both sides by} } -1.$$
$$1x = -7$$
$$x = -7$$

Check:
$$-x = 7$$
$$-(-7) \stackrel{?}{=} 7$$
$$7 = 7 \quad \textit{True}$$

Thus, the solution is -7.

Example 9 may also be solved by dividing both sides of the equation by -1. Try this now and see that you get the same solution. Whenever we have the opposite (or negative) of a variable equal to a quantity, as in Example 9, we can solve for the variable by multiplying (or dividing) both sides of the equation by -1.

EXAMPLE 10 Solve the equation $-x = -5$.

Solution
$$-x = -5$$
$$-1x = -5$$
$$(-1)(-1x) = (-1)(-5) \qquad \text{\textit{Multiply both sides by} } -1.$$
$$1x = 5$$
$$x = 5$$

NOW TRY EXERCISE 23

HELPFUL HINT

For any real number a, if $-x = a$, then $x = -a$.

Examples

$$-x = 7 \qquad\qquad -x = -2$$
$$x = -7 \qquad\qquad x = -(-2)$$
$$\qquad\qquad\qquad x = 2$$

4 Do Some Steps Mentally When Solving Equations

When you feel comfortable using the multiplication property, you may wish to do some of the steps mentally to reduce some of the written work. Now we present two examples worked out in detail, along with their shortened form.

EXAMPLE 11 Solve the equation $-3x = -21$.

Solution

$$-3x = -21$$

$$\frac{-3x}{-3} = \frac{-21}{-3} \quad \longleftarrow \quad \boxed{\text{Do this step mentally.}}$$

$$x = \frac{-21}{-3}$$

$$x = 7$$

SHORTENED FORM

$$-3x = -21$$

$$x = \frac{-21}{-3}$$

$$x = 7$$

EXAMPLE 12 Solve the equation $\dfrac{1}{3}x = 9$.

Solution

$$\frac{1}{3}x = 9$$

$$3\left(\frac{1}{3}x\right) = 3(9) \quad \longleftarrow \quad \boxed{\text{Do this step mentally.}}$$

$$x = 3(9)$$

$$x = 27$$

SHORTENED FORM

$$\frac{1}{3}x = 9$$

$$x = 3(9)$$

$$x = 27$$

NOW TRY EXERCISE 47

In Section 2.2 we discussed the addition property and in this section we discussed the multiplication property. It is important that you understand the difference between the two. The following Helpful Hint should be studied carefully.

HELPFUL HINT

The **addition property** is used to solve equations of the form $x + a = b$. The *addition property* is used when a number is *added to or subtracted from* a variable.

$$x + 3 = -6$$

$$x + 3 - 3 = -6 - 3$$

$$x = -9$$

$$x - 5 = -2$$

$$x - 5 + 5 = -2 + 5$$

$$x = 3$$

The **multiplication property** is used to solve equations of the form $ax = b$. It is used when a variable is *multiplied* or *divided by* a number.

$$3x = 6$$

$$\frac{3x}{3} = \frac{6}{3}$$

$$x = 2$$

$$\frac{x}{2} = 4$$

$$2\left(\frac{x}{2}\right) = 2(4)$$

$$x = 8$$

$$\frac{2}{5}x = 12$$

$$\left(\frac{5}{2}\right)\left(\frac{2}{5}x\right) = \left(\frac{5}{2}\right)(12)$$

$$x = 30$$

2. Division is defined in terms of multiplication.   **4.** Divide by 3.   **5.** Divide by −2.   **8.** By a, to isolate the variable x.

Exercise Set 2.3

Concept/Writing Exercises

1. In your own words, explain the multiplication property of equality.   Answers will vary

2. Explain why the multiplication property allows us to divide both sides of an equation by a nonzero quantity.

3. **a)** If $-x = a$, where a represents any real number, what does x equal? $-a$
   b) If $-x = 5$, what is x? -5
   c) If $-x = -5$, what is x? 5

4. When solving the equation $3x = 5$, would you divide both sides of the equation by 3 or by 5? Explain.

5. When solving the equation $-2x = 5$, would you divide both sides of the equation by -2 or by 5? Explain.

6. When solving the equation $\dfrac{x}{2} = 3$, what would you do to isolate the variable? Explain.   Multiply both sides by 2.

7. When solving the equation $4 = \dfrac{x}{3}$, what would you do to isolate the variable? Explain.   Multiply both sides by 3.

8. When solving the equation $ax = b$ for x, you would divide both sides of the equation by a or b? Explain.

Practice the Skills

Solve each equation and check your solution.

9. $4x = 12$ 3

10. $5x = 50$ 10

11. $\dfrac{x}{2} = 4$ 8

12. $\dfrac{y}{5} = 3$ 15

🔒 13. $-4x = 12$ -3

14. $8 = 16y$ $\dfrac{1}{2}$

15. $\dfrac{x}{4} = -2$ -8

16. $\dfrac{x}{3} = -3$ -9

17. $\dfrac{x}{5} = 1$ 5

18. $-7x = 49$ -7

19. $-27n = 81$ -3

20. $16 = -4y$ -4

21. $-7 = 3r$ $-\dfrac{7}{3}$

22. $\dfrac{x}{8} = -3$ -24

23. $-x = -11$ 11

24. $-x = 9$ -9

25. $10 = -y$ -10

26. $-9 = \dfrac{k}{6}$ -54

27. $-\dfrac{w}{3} = -13$ 39

28. $5 = \dfrac{z}{12}$ 60

29. $4 = -12x$ $-\dfrac{1}{3}$

30. $12y = -15$ $-\dfrac{5}{4}$

🔒 31. $-\dfrac{x}{3} = -2$ 6

32. $-\dfrac{a}{8} = -7$ 56

33. $43t = 26$ $\dfrac{26}{43}$

34. $-24x = -18$ $\dfrac{3}{4}$

35. $-4.2x = -8.4$ 2

36. $-3.88 = 1.94y$ -2

37. $7x = -7$ -1

38. $3x = \dfrac{3}{5}$ $\dfrac{1}{5}$

39. $5x = -\dfrac{3}{8}$ $-\dfrac{3}{40}$

40. $-2b = -\dfrac{4}{5}$ $\dfrac{2}{5}$

41. $15 = -\dfrac{x}{4}$ -60

42. $\dfrac{c}{9} = 0$ 0

43. $-\dfrac{b}{4} = -60$ 240

44. $-x = -\dfrac{5}{9}$ $\dfrac{5}{9}$

45. $\dfrac{x}{5} = -7$ -35

46. $-3r = 0$ 0

47. $5 = \dfrac{x}{4}$ 20

48. $-3 = \dfrac{x}{-5}$ 15

🔒 49. $\dfrac{3}{5}d = -30$ -50

50. $\dfrac{2}{7}x = 7$ $\dfrac{49}{2}$

51. $\dfrac{y}{-2} = 0$ 0

52. $-6x = \dfrac{5}{2}$ $-\dfrac{5}{12}$

53. $\dfrac{-7}{8}w = 0$ 0

54. $-x = \dfrac{5}{8}$ $-\dfrac{5}{8}$

55. $\dfrac{1}{5}x = 4.5$ 22.5

56. $5 = -\dfrac{3}{5}s$ $-\dfrac{25}{3}$

🔒 57. $-4 = -\dfrac{2}{3}z$ 6

58. $-9 = \dfrac{-5}{3}n$ $\dfrac{27}{5}$

59. $-1.4x = 28.28$ -20.2

60. $-0.42x = -2.142$ 5.1

61. $-4w = \dfrac{7}{12}$ $-\dfrac{7}{48}$

62. $6x = \dfrac{8}{3}$ $\dfrac{4}{9}$

63. $\dfrac{2}{3}x = 6$ 9

64. $-\dfrac{1}{4}x = \dfrac{3}{4}$ -3

Problem Solving

65. **a)** Explain the difference between $5 + x = 10$ and $5x = 10$.   Answers will vary
   b) Solve $5 + x = 10$. $x = 5$
   c) Solve $5x = 10$. $x = 2$

66. **a)** Explain the difference between $3 + x = 6$ and $3x = 6$.   Answers will vary
   b) Solve $3 + x = 6$. $x = 3$
   c) Solve $3x = 6$. $x = 2$

67. Consider the equation $\frac{2}{3}x = 4$. This equation could be solved by multiplying both sides of the equation by $\frac{3}{2}$, the reciprocal of $\frac{2}{3}$, or by dividing both sides of the equation by $\frac{2}{3}$. Which method do you feel would be easier? Explain your answer. Find the solution to the equation. *multiply by $\frac{3}{2}$; 6*

68. Consider the equation $4x = \frac{3}{5}$. Would it be easier to solve this equation by dividing both sides of the equa-

tion by 4 or by multiplying both sides of the equation by $\frac{1}{4}$, the reciprocal of 4? Explain your answer. Find the solution to the problem. *multiply by $\frac{1}{4}$; $\frac{3}{20}$*

69. Consider the equation $\frac{3}{7}x = \frac{4}{5}$. Would it be easier to solve this equation by dividing both sides of the equation by $\frac{3}{7}$ or by multiplying both sides of the equation by $\frac{7}{3}$, the reciprocal of $\frac{3}{7}$? Explain your answer. Find the solution to the equation. *multiply by $\frac{7}{3}$; $\frac{28}{15}$*

Challenge Problems

70. Consider the equation $\square\odot = \triangle$.
 a) To solve for $\odot$, what symbol do we need to isolate? $\odot$
 b) How would you isolate the symbol you specified in part **a)**? *Divide both sides by $\square$.*
 c) Solve the equation for $\odot$. $\odot = \dfrac{\triangle}{\square}$.

71. Consider the equation. $\smiley = \triangle\boxdot$
 a) To solve for $\boxdot$, what symbol do we need to isolate? $\boxdot$
 b) How would you isolate the symbol you specified in part **a)**? *Divide both sides by $\triangle$.*

 c) Solve the equation for $\square$. $\square = \dfrac{\smiley}{\triangle}$

72. Consider the equation $\# = \dfrac{\smiley}{\triangle}$.

 a) To solve for $\smiley$, what symbol do we need to isolate? $\smiley$
 b) How would you isolate the symbol you specified in part **a)**? *Multiply both sides by $\triangle$.*
 c) Solve the equation for $\smiley$. $\smiley = \#\triangle$

Cumulative Review Exercises

76. $-11x + 38$

[1.7] **73.** Subtract -4 from -8. -4
 74. Evaluate $6 - (-3) - 5 - 4$. 0
[1.9] **75.** Evaluate $4^2 - 2^3 \cdot 6 \div 3 + 6$ 6

[2.1] **76.** Simplify $-(x + 3) - 5(2x - 7) + 6$.
[2.2] **77.** Solve the equation $-48 = x + 9$. -57

2.4 SOLVING LINEAR EQUATIONS WITH A VARIABLE ON ONLY ONE SIDE OF THE EQUATION

1 Solve linear equations with a variable on only one side of the equal sign.

2 Solve equations containing decimal numbers or fractions.

SSM   Study Guide   CD/Video

MathPro 4/5   PH Math Tutor Center   prenhall.com/Angel

1 Solve Linear Equations with a Variable on Only One Side of the Equal Sign

In this section, we discuss how to solve linear equations using *both* the addition and multiplication properties of equality when a variable appears on only one side of the equal sign. In Section 2.5, we will discuss how to solve linear equations using both properties when a variable appears on both sides of the equal sign.

The general procedure we use to solve equations is to "isolate the variable." That is, get the variable, x, alone on one side of the equal sign.

No one method is the "best" to solve all linear equations. But the following general procedure can be used to solve linear equations when the variable appears on only one side of the equation.

> **To Solve Linear Equations with a Variable on Only One Side of the Equal Sign**
>
> **1.** If the equation contains fractions, multiply **both** sides of the equation by the least common denominator (LCD). This will eliminate the fractions from the equation.

(continued on the next page)

2. Use the distributive property to remove parentheses.

3. Combine like terms on the same side of the equal sign.

4. Use the addition property to obtain an equation with the term containing the variable on one side of the equal sign and a constant on the other side. This will result in an equation of the form $ax = b$.

5. Use the multiplication property to isolate the variable. This will give a solution of the form $x = \dfrac{b}{a}$ $\left(\text{or } 1x = \dfrac{b}{a}\right)$.

6. Check the solution in the original equation.

When solving an equation, you should always check your solution, as is indicated in step 6. To conserve space, we will not show all checks.

When solving an equation, remember that our goal is to isolate the variable on one side of the equation.

Consider the equation $2x + 4 = 10$ which contains no fractions or parentheses, and no like terms on the same side of the equal sign. Therefore, we start with step 4, using the addition property. Remember that the addition property allows us to add (or subtract) the same quantity to (or from) both sides of an equation without changing its solution. Here we subtract 4 from both sides of the equation to isolate the term containing the variable.

Equation

$2x + 4 = 10$

$2x + 4 - 4 = 10 - 4$ *Addition property*

or $2x = 6$ *x-term is now isolated*

Notice how the term containing the variable, $2x$, is now by itself on one side of the equal sign. Now we use the multiplication property, step 5, to isolate the variable x. Remember that the multiplication property allows us to multiply or divide both sides of the equation by the same nonzero number without changing its solution. Here we divide both sides of the equation by 2, the coefficient of the term containing the variable, to obtain the solution, 3.

$2x = 6$

$\dfrac{\overset{1}{2}x}{\underset{1}{2}} = \dfrac{\overset{3}{6}}{\underset{1}{2}}$ *Multiplication property*

$x = 3$ *x is now isolated*

The solution to the equation $2x + 4 = 10$ is 3. Now let's work some examples.

EXAMPLE 1   Solve the equation $3x - 6 = 15$.

Solution   We will follow the procedure outlined for solving equations. Since the equation contains no fractions nor parentheses, and since there are no like terms to be combined, we start with step 4.

Step 4 $3x - 6 = 15$

$3x - 6 + 6 = 15 + 6$ *Add 6 to both sides.*

$3x = 21$

Step 5
$$\frac{3x}{3} = \frac{21}{3}$$ *Divide both sides by 3.*
$$x = 7$$

Step 6   Check:
$$3x - 6 = 15$$
$$3(7) - 6 \overset{?}{=} 15$$
$$21 - 6 \overset{?}{=} 15$$
$$15 = 15 \qquad \textit{True}$$

Since the check is true, the solution is 7. Note that after completing step 4 we obtain $3x = 21$, which is an equation of the form $ax = b$. After completing step 5, we obtain the answer in the form $x =$ some real number.   ✳

HELPFUL HINT

When solving an equation that does not contain fractions, **the addition property (step 4) is to be used before the multiplication property (step 5)**. If you use the multiplication property before the addition property, it is still possible to obtain the correct answer. However, you will usually have to do more work, and you may end up working with fractions. What would happen if you tried to solve Example 1 using the multiplication property before the addition property?

EXAMPLE 2   Solve the equation $-2r - 6 = -3$.

Solution
$$-2r - 6 = -3$$

Step 4
$$-2r - 6 + 6 = -3 + 6 \qquad \textit{Add 6 to both sides.}$$
$$-2r = 3$$

Step 5
$$\frac{-2r}{-2} = \frac{3}{-2} \qquad \textit{Divide both sides by }-2.$$
$$r = -\frac{3}{2}$$

Step 6   Check:
$$-2r - 6 = -3$$
$$-2\left(-\frac{3}{2}\right) - 6 \overset{?}{=} -3$$
$$3 - 6 \overset{?}{=} -3$$
$$-3 = -3 \qquad \textit{True}$$

The solution is $-\frac{3}{2}$.   ✳

NOW TRY EXERCISE 19

Note that checks are always made with the *original* equation. In some of the following examples, the check will be omitted to save space. You should check all of your answers.

EXAMPLE 3   Solve the equation $16 = 4x + 6 - 2x$.

Solution   Again we must isolate the variable x. Since the right side of the equation has two like terms containing the variable x, we will first combine these like terms.

$$16 = 4x + 6 - 2x$$

Step 3
$$16 = 2x + 6 \qquad \textit{Like terms were combined.}$$

Step 4 $16 - 6 = 2x + 6 - 6$ *Subtract 6 from both sides.*

$$10 = 2x$$

Step 5 $\dfrac{10}{2} = \dfrac{2x}{2}$ *Divide both sides by 2.*

$$5 = x$$

The preceding solution can be condensed as follows.

$$16 = 4x + 6 - 2x$$

$16 = 2x + 6$ *Like terms were combined.*

$10 = 2x$ *6 was subtracted from both sides.*

$5 = x$ *Both sides were divided by 2.*

EXAMPLE 4   Solve the equation $2(x + 4) - 5x = -3$.

Solution $2(x + 4) - 5x = -3$

Step 2 $2x + 8 - 5x = -3$ *Distributive property was used.*

Step 3 $-3x + 8 = -3$ *Like terms were combined.*

Step 4 $-3x + 8 - 8 = -3 - 8$ *Subtract 8 from both sides.*

$$-3x = -11$$

Step 5 $\dfrac{-3x}{-3} = \dfrac{-11}{-3}$ *Divide both sides by −3.*

$$x = \frac{11}{3}$$

The solution to Example 5 can be condensed as follows:

$$2(x + 4) - 5x = -3$$

$2x + 8 - 5x = -3$ *The distributive property was used.*

$-3x + 8 = -3$ *Like terms were combined.*

$-3x = -11$ *8 was subtracted from both sides.*

$x = \dfrac{11}{3}$ *Both sides were divided by −3.*

EXAMPLE 5   Solve the equation $2t - (t + 2) = 6$.

Solution $2t - (t + 2) = 6$

$2t - t - 2 = 6$ *The distributive property was used.*

$t - 2 = 6$ *Like terms were combined.*

NOW TRY EXERCISE 79 $t = 8$ *2 was added to both sides.*

2   Solve equations containing decimal numbers or fractions

In Chapter 3, we will be solving many equations that contain decimal numbers. To solve such equations, we may follow the same procedure as outlined earlier. Example 6 illustrates two methods to solve an equation that contains decimal numbers.

EXAMPLE 6   Solve the equation $x + 1.24 - 0.07x = 4.96$.

Solution   We will work this example using two methods. In method 1, we work with decimal numbers throughout the solving process. In method 2, we multiply both sides of the equation by a power of 10 to change the decimal numbers to whole numbers.

Method 1
$$x + 1.24 - 0.07x = 4.96$$
$$0.93x + 1.24 = 4.96 \quad \textit{Like terms were combined,}$$
$$\textit{1x} - 0.07x = 0.93x.$$
$$0.93x + 1.24 - 1.24 = 4.96 - 1.24 \quad \textit{Subtract 1.24 from both sides.}$$
$$0.93x = 3.72$$
$$\frac{0.93x}{0.93} = \frac{3.72}{0.93} \quad \textit{Divide both sides by 0.93.}$$
$$x = 4$$

Method 2   Some students prefer to eliminate the decimal numbers from the equation by multiplying both sides of the equation by 10 if the decimal numbers are given in tenths, by 100 if the decimal numbers are given in hundredths, and so on. In Example 6, since the decimal numbers are in hundredths, you can eliminate the decimals from the equation by multiplying both sides of the equation by 100. This alternate method would give the following.

$$x + 1.24 - 0.07x = 4.96$$
$$100(x + 1.24 - 0.07x) = 100(4.96) \quad \textit{Multiply both sides of equation by 100}$$
$$100(x) + 100(1.24) - 100(0.07x) = 496 \quad \textit{Distributive property}$$
$$100x + 124 - 7x = 496$$
$$93x + 124 = 496 \quad \textit{Like terms were combined}$$
$$93x = 372 \quad \textit{124 was subtracted from both sides}$$
$$x = 4 \quad \textit{Both sides were divided by 93}$$

NOW TRY EXERCISE 41   Study both methods provided to see which method you prefer.   ✳

Now let's look at solving equations that contain fractions. There will be various times throughout the course when we will need to solve equations containing fractions. The first step in solving equations containing fractions is to multiply both sides of the equation by the LCD to eliminate the fractions from the equations. Examples 7–9 illustrate the procedure.

EXAMPLE 7   Solve $\dfrac{x - 3}{5} = 7$.

Solution   The LCD of the fraction in this question is 5. Step 1 in the procedure box tells us to multiply both sides of the equation by the LCD. This step will eliminate fractions from the equation.

$$\frac{x - 3}{5} = 7$$

Step 1
$$5\left(\frac{x - 3}{5}\right) = 5 \cdot 7 \quad \textit{Multiply both sides by the LCD, 5}$$
$$x - 3 = 35$$

Step 4
$$x = 38 \quad \textit{3 was added to both sides}$$

Step 6   Check:

$$\frac{x - 3}{5} = 7$$

$$\frac{38 - 3}{5} \stackrel{?}{=} 7$$

$$\frac{35}{5} \stackrel{?}{=} 7$$

$$7 = 7 \qquad \textit{True}$$

NOW TRY EXERCISE 45 Thus, the solution is 38.

In Example 7, $\dfrac{x - 3}{5} = 7$ could also be expressed as $\dfrac{1}{5}(x - 3) = 7$. Had we been given the equation in this form, we would have begun the solution the same way, by multiplying both sides of the equation by the LCD, 5.

EXAMPLE 8   Solve $\dfrac{d}{2} + 3d = 14$.

Solution   Step 1 tells us to multiply both sides of the equation by the LCD, 2. This step will eliminate fractions from the equation.

Step 1 $$2\left(\frac{d}{2} + 3d\right) = 14 \cdot 2 \qquad \textit{Multiply both sides by the LCD, 2}$$

Step 2 $$2\left(\frac{d}{2}\right) + 2 \cdot 3d = 14 \cdot 2 \qquad \textit{Distributive Property}$$

$$d + 6d = 28$$

Step 3 $$7d = 28 \qquad \textit{Like terms were combined}$$

Step 5 $$d = 4 \qquad \textit{Both sides were divided by 7}$$

Step 6 Check: $$\frac{d}{2} + 3d = 14$$

$$\frac{4}{2} + 3(4) \stackrel{?}{=} 14$$

$$2 + 12 \stackrel{?}{=} 14$$

NOW TRY EXERCISE 89 $$14 = 14 \qquad \textit{True}$$

EXAMPLE 9   Solve the equation $\dfrac{1}{5}x - \dfrac{3}{8}x = \dfrac{1}{10}$.

Solution   The LCD of 5, 8, and 10 is 40. Multiply both sides of the equation by 40 to eliminate fractions from the equation.

$$\frac{1}{5}x - \frac{3}{8}x = \frac{1}{10}$$

Step 1 $$40\left(\frac{1}{5}x - \frac{3}{8}x\right) = 40\left(\frac{1}{10}\right) \qquad \textit{Multiply both sides by the LCD, 40}$$

Step 2 $$40\left(\frac{1}{5}x\right) - 40\left(\frac{3}{8}x\right) = 40\left(\frac{1}{10}\right) \qquad \textit{Distributive property}$$

$$8x - 15x = 4$$

Step 3 $$-7x = 4 \qquad \textit{Like terms were combined}$$

Step 5 $$x = -\frac{4}{7}$$ *Both sides were divided by −7*

Step 6 Check: Substitute $-\frac{4}{7}$ for each x in the equation.

$$\frac{1}{5}x - \frac{3}{8}x = \frac{1}{10}$$

$$\frac{1}{5}\left(-\frac{4}{7}\right) - \frac{3}{8}\left(-\frac{4}{7}\right) \stackrel{?}{=} \frac{1}{10}$$ *Substitute $-\frac{4}{7}$ for each x*

$$-\frac{4}{35} + \frac{3}{14} \stackrel{?}{=} \frac{1}{10}$$ *Divide out common factors, then multiply fractions*

$$-\frac{8}{70} + \frac{15}{70} \stackrel{?}{=} \frac{7}{70}$$ *Write each fraction with the LCD, 70*

$$\frac{7}{70} = \frac{7}{70}$$ *True*

NOW TRY EXERCISE 99

HELPFUL HINT

In Example 9, we multiplied both sides of the equation by the LCD, 40. When solving equations containing fractions, multiplying both sides of the equation by *any* common denominator will eventually lead to the correct answer (if you don't make a mistake), but you may have to work with larger numbers. In Example 9, if you multiplied both sides of the equation by 80, 120, or 160, for example, you would eventually obtain the answer $-\frac{4}{7}$. When solving equations containing fractions, you should multiply both sides of the equation by the LCD. But if you mistakenly multiply both sides of the equation by a different common denominator to clear fractions, you will still end up with the correct answer. To show that other common denominators may be used, solve Example 9 now by multiplying both sides of the equation by the common denominator 80 instead of the LCD, 40.

When checking solutions to equations that contain fractions, you may sometimes want to perform the check using a calculator. When checking a solution using a calculator, work with each side of the equation separately. Below we show the steps used to evaluate the left side of the equation in Example 9 for $x = -\frac{4}{7}$ using a scientific calculator.*

$$\frac{1}{5}x - \frac{3}{8}x = \frac{1}{10}$$

$$\frac{1}{5}\left(-\frac{4}{7}\right) - \frac{3}{8}\left(-\frac{4}{7}\right) = \frac{1}{10}$$

Evaluate the left side of equation

$$1 \;\boxed{\div}\; 5 \;\boxed{\times}\; 4 \;\boxed{+/-}\; \boxed{\div}\; 7 \;\boxed{-}\; 3 \;\boxed{\div}\; 8 \;\boxed{\times}\; 4 \;\boxed{+/-}\; \boxed{\div}\; 7 \;\boxed{=}\; 0.1.$$

NOW TRY EXERCISE 99 Since the right side of the equation, $\frac{1}{10} = 0.1$, the answer checks.

HELPFUL HINT

Some of the most commonly used terms in algebra are "evaluate," "simplify," "solve," and "check." Make sure you understand what each term means and when each term is used.

(continued on the next page)

---

*Keystrokes may differ on some scientific calculators. Read the instruction manual for your calculator.

Evaluate: To *evaluate an expression* means to find its numerical value.

Evaluate $16 \div 2^2 + 36 \div 4$

$= 16 \div 4 + 36 \div 4$

$= 4 + 36 \div 4$

$= 4 + 9$

$= 13$

Evaluate $-x^2 + 3x - 2$ when $x = 4$

$= -4^2 + 3(4) - 2$

$= -16 + 12 - 2$

$= -4 - 2$

$= -6$

Simplify: To *simplify an expression* means to perform the operations and combine like terms.

Simplify $3(x - 2) - 4(2x + 3)$

$3(x - 2) - 4(2x + 3) = 3x - 6 - 8x - 12$

$= -5x - 18$

Note that when you simplify an expression containing variables you do not generally end up with just a numerical value unless all the variable terms happen to add to zero.

Solve: To *solve an equation* means to find the value or the values of the variable that make the equation a true statement.

Solve $2x + 3(x + 1) = 18$

$2x + 3x + 3 = 18$

$5x + 3 = 18$

$5x = 15$

$x = 3$

Check: To *check the proposed solution to an equation*, substitute the value in the original equation. If this substitution results in a true statement, then the answer checks. For example, to check the solution to the equation just solved, we substitute 3 for x in the equation.

Check $2x + 3(x + 1) = 18$

$2(3) + 3(3 + 1) \stackrel{?}{=} 18$

$6 + 3(4) \stackrel{?}{=} 18$

$6 + 12 \stackrel{?}{=} 18$

$18 = 18$ *True*

Since we obtained a true statement, the 3 checks.

It is important to realize that expressions may be evaluated or simplified (depending on the type of problem) and equations are solved and then checked.

Exercise Set 2.4

10. Evaluate   **12.** Multiply both sides of the equation by the LCD.

Concept/Writing Exercises

1. Does the equation $x + 3 = 2x + 5$ contain a variable on only one side of the equation? Explain. no

2. Does the equation $2x - 4 = 3$ contain a variable on only one side of the equation? Explain. yes

3. If $1x = \dfrac{1}{3}$, what does x equal? $x = \dfrac{1}{3}$

4. If $1x = -\dfrac{3}{5}$, what does x equal? $x = -\dfrac{3}{5}$

5. If $-x = \dfrac{1}{2}$ what does x equal? $x = -\dfrac{1}{2}$

6. If $-x = \dfrac{7}{8}$, what does x equal? $x = -\dfrac{7}{8}$

7. If $-x = -\dfrac{3}{5}$, what does x equal? $x = \dfrac{3}{5}$

8. If $-x = -\dfrac{4}{9}$, what does x equal? $x = \dfrac{4}{9}$

9. Do you evaluate or solve an equation? Explain. solve

10. Do you evaluate or solve an expression? Explain.

11. a) In your own words, write the general procedure for solving an equation where the variable appears on only one side of the equal sign. Answers will vary

b) Refer to pages 123 and 124 to see whether you omitted any steps.

12. When solving equations that contain fractions, what is the first step in the process of solving the equation?

13. a) Explain, in a step-by-step manner, how to solve the equation $2(3x + 4) = -4$. Answers will vary

b) Solve the equation by following the steps you listed in part **a)**. -2

14. a) Explain, step-by-step, how to solve the equation $4x - 2(x + 3) = 4$. Answers will vary

b) Solve the equation by following the steps you listed in part **a)**. 5

Practice the Skills

Solve each equation. You may wish to use a calculator to solve equations containing decimal numbers.

15. $3x + 6 = 12$ 2

16. $2x - 4 = 8$ 6

17. $-4w - 5 = 11$ -4

18. $-4x + 6 = 20$ $-\frac{7}{2}$

19. $5x - 6 = 19$ 5

20. $6 - 3x = 18$ -4

21. $5x - 2 = 10$ $\frac{12}{5}$

22. $-5k - 4 = -19$ 3

23. $-2t + 9 = 21$ -6

24. $20 = 2d + 6$ 7

25. $12 - x = 9$ 3

26. $-3x - 3 = -12$ 3

27. $8 + 3x = 19$ $\frac{11}{3}$

28. $-2x + 7 = -10$ $\frac{17}{2}$

29. $16x + 5 = -14$ $-\frac{19}{16}$

30. $19 = 25 + 4x$ $-\frac{3}{2}$

31. $-4.2 = 3x + 25.8$ -10

32. $-24 + 16x = -24$ 0

33. $7r - 16 = -2$ 2

34. $-2w + 4 = -8$ 6

35. $60 = -5s + 9$ $-\frac{51}{5}$

36. $15 = 7x + 1$ 2

37. $-2x - 7 = -13$ 3

38. $-2 - x = -12$ 10

39. $2.3x - 9.34 = 6.3$ 6.8

40. $x + 0.05x = 21$ 20

41. $x + 0.07x = 16.05$ 15

42. $-2.7 = -1.3 + 0.7x$ -2

43. $28.8 = x + 1.40x$ 12

44. $8.40 = 2.45x - 1.05x$ 6

45. $\dfrac{x - 4}{6} = 9$ 58

46. $\dfrac{m - 6}{5} = 2$ 16

47. $\dfrac{d + 3}{7} = 9$ 60

48. $\dfrac{1}{5}(x + 2) = -3$ -17

49. $\dfrac{1}{3}(t - 5) = -6$ -13

50. $\dfrac{2}{3}(n - 3) = 8$ 15

51. $\dfrac{3}{4}(x - 5) = -12$ -11

52. $\dfrac{x + 4}{7} = \dfrac{3}{7}$ -1

53. $\dfrac{1}{4} = \dfrac{z + 1}{4}$ 0

54. $\dfrac{4x + 5}{6} = \dfrac{7}{2}$ 4

55. $\dfrac{3}{4} = \dfrac{4m - 5}{6}$ $\frac{19}{8}$

56. $\dfrac{5}{6} = \dfrac{5t - 4}{2}$ $\frac{17}{15}$

57. $4(n + 2) = 8$ 0

58. $3(x - 2) = 12$ 6

59. $5(3 - x) = 15$ 0

60. $-2(x - 3) = 26$ -10

61. $-4 = -(x + 5)$ -1

62. $-3(2 - 3x) = 9$ $\frac{5}{3}$

63. $12 = 4(x - 3)$ 6

64. $-2(x + 8) - 5 = 1$ -11

65. $22 = -(3x - 4)$ -6

66. $-2 = 5(3x + 1) - 12x$ $-\frac{7}{3}$

67. $-3r + 4(r + 2) = 11$ 3

68. $9 = -2(a - 3)$ $-\frac{3}{2}$

69. $x - 3(2x + 3) = 11$ -4

70. $3(4 - x) + 5x = 9$ $-\frac{3}{2}$

71. $5x + 3x - 4x - 7 = 9$ 4

72. $4(x + 2) = 13$ $\frac{5}{4}$

73. $0.7(x - 3) = 1.4$ 5

74. $21 + (c - 9) = 24$ 12

75. $2.5(4q - 3) = 0.5$ 0.8

76. $0.1(2.4x + 5) = 1.7$ 5

77. $3 - 2(x + 3) + 2 = 1$ -1

78. $2(3x - 4) - 4x = 12$ 10

79. $1 + (x + 3) + 6x = 6$ $\frac{2}{7}$

80. $5x - 2x + 7x = -81$ -8.1

81. $4.85 - 6.4x + 1.11 = 22.6$ -2.6

82. $5.76 - 4.24x - 1.9x = 27.864$ -3.6

83. $7 = 8 - 5(m + 3)$ $-\frac{14}{5}$

84. $-4 = 3 - 6(t - 5)$ $\dfrac{37}{6}$

85. $10 = \dfrac{2s + 4}{5}$ 23

86. $12 = \dfrac{4d - 1}{3}$ $\dfrac{37}{4}$

87. $x + \dfrac{2}{3} = \dfrac{3}{5}$ $-\dfrac{1}{15}$

88. $n - \dfrac{1}{4} = \dfrac{1}{2}$ $\dfrac{3}{4}$

89. $\dfrac{t}{4} - t = \dfrac{3}{2}$ -2

90. $n - \dfrac{n}{3} = \dfrac{1}{2}$ $\dfrac{3}{4}$

91. $\dfrac{3}{7} = \dfrac{3t}{4} + 1$ $-\dfrac{16}{21}$

92. $\dfrac{5}{8} = \dfrac{5t}{6} + 2$ $-\dfrac{33}{20}$

93. $\dfrac{1}{2}r + \dfrac{1}{5}r = 7$ 10

94. $\dfrac{2}{8} + \dfrac{3}{4} = \dfrac{w}{5}$ 5

95. $\dfrac{x}{3} - \dfrac{3x}{4} = \dfrac{1}{12}$ $-\dfrac{1}{5}$

96. $\dfrac{x}{4} - \dfrac{x}{6} = \dfrac{1}{4}$ 3

97. $\dfrac{1}{2}x + 4 = \dfrac{1}{6}$ $-\dfrac{23}{3}$

98. $\dfrac{4}{5} + n = \dfrac{1}{3}$ $-\dfrac{7}{15}$

 99. $\dfrac{4}{5}s - \dfrac{3}{4}s = \dfrac{1}{10}$ 2

100. $\dfrac{2}{3} = \dfrac{1}{5}(t + 2)$ $\dfrac{4}{3}$

101. $\dfrac{4}{9} = \dfrac{1}{3}(n - 7)$ $\dfrac{25}{3}$

102. $-\dfrac{3}{8} = \dfrac{1}{8} - \dfrac{2x}{7}$ $\dfrac{7}{4}$

103. $-\dfrac{3}{5} = -\dfrac{1}{9} - \dfrac{3}{4}x$ $\dfrac{88}{135}$

104. $-\dfrac{3}{5} = -\dfrac{1}{6} - \dfrac{5}{4}m$ $\dfrac{26}{75}$

Problem Solving

105. a) Explain why it is easier to solve the equation $3x + 2 = 11$ by first subtracting 2 from both sides of the equation rather than by first dividing both sides of the equation by 3. You will not have to work with fractions.
b) Solve the equation. 3

106. a) Explain why it is easier to solve the equation $5x - 3 = 12$ by first adding 3 to both sides of the equation rather than by first dividing both sides of the equation by 5. You will not have to work with fractions.
b) Solve the equation. 3

Challenge Problems

For exercises 107–109, solve the equation.

107. $3(x - 2) - (x + 5) - 2(3 - 2x) = 18$ $\dfrac{35}{6}$

109. $4[3 - 2(x + 4)] - (x + 3) = 13$ -4

110. $\ddot{\smile} = \dfrac{@ + \triangledown}{\square}$

108. $-6 = -(x - 5) - 3(5 + 2x) - 4(2x - 4)$ $\dfrac{4}{5}$

110. Solve the equation $\square\ddot{\smile} - \triangledown = @$ for $\ddot{\smile}$.

Group Activity

In Chapter 3 we will discuss procedures for writing application problems as equations. Let's look at an application now.

Birthday Party John Logan purchased 2 large chocolate bars and a birthday card. The birthday card cost $3. The total cost was $9. What was the price of a single chocolate bar?

This problem can be represented by the equation $2x + 3 = 9$, which can be used to solve the problem. Solving the equation we find that x, the price of a single chocalate bar, is $3.

*For Exercises 111 and 112, each group member should do parts **a)** and **b)**. Then do part **c)** as a group.*

a) *Obtain an equation that can be used to solve the problem.*
b) *Solve the equation and answer the question.*
c) *Compare and check each other's work.*

111. Stationary Eduardo Verner purchased three boxes of stationery. He also purchased wrapping paper and thank-you cards. If the wrapping paper and thank-you cards together cost $6, and the total he paid was $42, find the cost of a box of stationery.
a) $3x + 6 = 42$ **b)** $12

112. Candies Mahandi Ison purchased three rolls of peppermint candies and the local newspaper. The newspaper cost 50 cents. He paid $2.75 in all. What did a roll of candies cost?
a) $3x + 0.5 = 2.75$ **b)** $0.75

Exercise Set 2.4

10. Evaluate   12. Multiply both sides of the equation by the LCD.

Concept/Writing Exercises

1. Does the equation $x + 3 = 2x + 5$ contain a variable on only one side of the equation? Explain. no

2. Does the equation $2x - 4 = 3$ contain a variable on only one side of the equation? Explain. yes

3. If $1x = \dfrac{1}{3}$, what does x equal? $x = \dfrac{1}{3}$

4. If $1x = -\dfrac{3}{5}$, what does x equal? $x = -\dfrac{3}{5}$

5. If $-x = \dfrac{1}{2}$ what does x equal? $x = -\dfrac{1}{2}$

6. If $-x = \dfrac{7}{8}$, what does x equal? $x = -\dfrac{7}{8}$

7. If $-x = -\dfrac{3}{5}$, what does x equal? $x = \dfrac{3}{5}$

8. If $-x = -\dfrac{4}{9}$, what does x equal? $x = \dfrac{4}{9}$

9. Do you evaluate or solve an equation? Explain. solve

10. Do you evaluate or solve an expression? Explain.

11. a) In your own words, write the general procedure for solving an equation where the variable appears on only one side of the equal sign. Answers will vary

   b) Refer to pages 123 and 124 to see whether you omitted any steps.

12. When solving equations that contain fractions, what is the first step in the process of solving the equation?

13. a) Explain, in a step-by-step manner, how to solve the equation $2(3x + 4) = -4$. Answers will vary

   b) Solve the equation by following the steps you listed in part a). −2

14. a) Explain, step-by-step, how to solve the equation $4x - 2(x + 3) = 4$. Answers will vary

   b) Solve the equation by following the steps you listed in part a). 5

Practice the Skills

Solve each equation. You may wish to use a calculator to solve equations containing decimal numbers.

15. $3x + 6 = 12$ 2
16. $2x - 4 = 8$ 6
17. $-4w - 5 = 11$ −4
18. $-4x + 6 = 20$ $-\dfrac{7}{2}$

19. $5x - 6 = 19$ 5
20. $6 - 3x = 18$ −4
21. $5x - 2 = 10$ $\dfrac{12}{5}$
22. $-5k - 4 = -19$ 3

23. $-2t + 9 = 21$ −6
24. $20 = 2d + 6$ 7
25. $12 - x = 9$ 3
26. $-3x - 3 = -12$ 3

27. $8 + 3x = 19$ $\dfrac{11}{3}$
28. $-2x + 7 = -10$ $\dfrac{17}{2}$
29. $16x + 5 = -14$ $-\dfrac{19}{16}$
30. $19 = 25 + 4x$ $-\dfrac{3}{2}$

31. $-4.2 = 3x + 25.8$ −10
32. $-24 + 16x = -24$ 0
33. $7r - 16 = -2$ 2
34. $-2w + 4 = -8$ 6

35. $60 = -5s + 9$ $-\dfrac{51}{5}$
36. $15 = 7x + 1$ 2
37. $-2x - 7 = -13$ 3
38. $-2 - x = -12$ 10

39. $2.3x - 9.34 = 6.3$ 6.8
40. $x + 0.05x = 21$ 20
41. $x + 0.07x = 16.05$ 15
42. $-2.7 = -1.3 + 0.7x$ −2

43. $28.8 = x + 1.40x$ 12
44. $8.40 = 2.45x - 1.05x$ 6
45. $\dfrac{x - 4}{6} = 9$ 58
46. $\dfrac{m - 6}{5} = 2$ 16

47. $\dfrac{d + 3}{7} = 9$ 60
48. $\dfrac{1}{5}(x + 2) = -3$ −17
49. $\dfrac{1}{3}(t - 5) = -6$ −13
50. $\dfrac{2}{3}(n - 3) = 8$ 15

51. $\dfrac{3}{4}(x - 5) = -12$ −11
52. $\dfrac{x + 4}{7} = \dfrac{3}{7}$ −1
53. $\dfrac{1}{4} = \dfrac{z + 1}{4}$ 0
54. $\dfrac{4x + 5}{6} = \dfrac{7}{2}$ 4

55. $\dfrac{3}{4} = \dfrac{4m - 5}{6}$ $\dfrac{19}{8}$
56. $\dfrac{5}{6} = \dfrac{5t - 4}{2}$ $\dfrac{17}{15}$
57. $4(n + 2) = 8$ 0
58. $3(x - 2) = 12$ 6

59. $5(3 - x) = 15$ 0
60. $-2(x - 3) = 26$ −10
61. $-4 = -(x + 5)$ −1
62. $-3(2 - 3x) = 9$ $\dfrac{5}{3}$

63. $12 = 4(x - 3)$ 6
64. $-2(x + 8) - 5 = 1$ −11
65. $22 = -(3x - 4)$ −6

66. $-2 = 5(3x + 1) - 12x$ $-\dfrac{7}{3}$
67. $-3r + 4(r + 2) = 11$ 3
68. $9 = -2(a - 3)$ $-\dfrac{3}{2}$

69. $x - 3(2x + 3) = 11$ −4
70. $3(4 - x) + 5x = 9$ $-\dfrac{3}{2}$
71. $5x + 3x - 4x - 7 = 9$ 4

72. $4(x + 2) = 13$ $\dfrac{5}{4}$
73. $0.7(x - 3) = 1.4$ 5
74. $21 + (c - 9) = 24$ 12

75. $2.5(4q - 3) = 0.5$ 0.8
76. $0.1(2.4x + 5) = 1.7$ 5
77. $3 - 2(x + 3) + 2 = 1$ −1

78. $2(3x - 4) - 4x = 12$ 10
79. $1 + (x + 3) + 6x = 6$ $\dfrac{2}{7}$
80. $5x - 2x + 7x = -81$ −8.1

81. $4.85 - 6.4x + 1.11 = 22.6$ −2.6
82. $5.76 - 4.24x - 1.9x = 27.864$ −3.6
83. $7 = 8 - 5(m + 3)$ $-\dfrac{14}{5}$

84. $-4 = 3 - 6(t - 5)$ $\frac{37}{6}$

85. $10 = \frac{2s + 4}{5}$ 23

86. $12 = \frac{4d - 1}{3}$ $\frac{37}{4}$

87. $x + \frac{2}{3} = \frac{3}{5}$ $-\frac{1}{15}$

88. $n - \frac{1}{4} = \frac{1}{2}$ $\frac{3}{4}$

89. $\frac{t}{4} - t = \frac{3}{2}$ -2

90. $n - \frac{n}{3} = \frac{1}{2}$ $\frac{3}{4}$

91. $\frac{3}{7} = \frac{3t}{4} + 1$ $-\frac{16}{21}$

92. $\frac{5}{8} = \frac{5t}{6} + 2$ $-\frac{33}{20}$

93. $\frac{1}{2}r + \frac{1}{5}r = 7$ 10

94. $\frac{2}{8} + \frac{3}{4} = \frac{w}{5}$ 5

95. $\frac{x}{3} - \frac{3x}{4} = \frac{1}{12}$ $-\frac{1}{5}$

96. $\frac{x}{4} - \frac{x}{6} = \frac{1}{4}$ 3

97. $\frac{1}{2}x + 4 = \frac{1}{6}$ $-\frac{23}{3}$

98. $\frac{4}{5} + n = \frac{1}{3}$ $-\frac{7}{15}$

 99. $\frac{4}{5}s - \frac{3}{4}s = \frac{1}{10}$ 2

100. $\frac{2}{3} = \frac{1}{5}(t + 2)$ $\frac{4}{3}$

101. $\frac{4}{9} = \frac{1}{3}(n - 7)$ $\frac{25}{3}$

102. $-\frac{3}{8} = \frac{1}{8} - \frac{2x}{7}$ $\frac{7}{4}$

103. $-\frac{3}{5} = -\frac{1}{9} - \frac{3}{4}x$ $\frac{88}{135}$

104. $-\frac{3}{5} = -\frac{1}{6} - \frac{5}{4}m$ $\frac{26}{75}$

Problem Solving

105. a) Explain why it is easier to solve the equation $3x + 2 = 11$ by first subtracting 2 from both sides of the equation rather than by first dividing both sides of the equation by 3. You will not have to work with fractions.
b) Solve the equation. 3

106. a) Explain why it is easier to solve the equation $5x - 3 = 12$ by first adding 3 to both sides of the equation rather than by first dividing both sides of the equation by 5. You will not have to work with fractions.
b) Solve the equation. 3

Challenge Problems

For exercises 107–109, solve the equation.

107. $3(x - 2) - (x + 5) - 2(3 - 2x) = 18$ $\frac{35}{6}$
109. $4[3 - 2(x + 4)] - (x + 3) = 13$ -4

110. ☻ $= \dfrac{@ + \triangledown}{\square}$

108. $-6 = -(x - 5) - 3(5 + 2x) - 4(2x - 4)$ $\frac{4}{5}$
110. Solve the equation $\square$☻ $- \triangledown = @$ for ☻.

Group Activity

In Chapter 3 we will discuss procedures for writing application problems as equations. Let's look at an application now.

Birthday Party John Logan purchased 2 large chocolate bars and a birthday card. The birthday card cost $3. The total cost was $9. What was the price of a single chocolate bar?

This problem can be represented by the equation $2x + 3 = 9$, which can be used to solve the problem. Solving the equation we find that x, the price of a single chocolate bar, is $3.

*For Exercises 111 and 112, each group member should do parts **a)** and **b)**. Then do part **c)** as a group.*

a) *Obtain an equation that can be used to solve the problem.*
b) *Solve the equation and answer the question.*
c) *Compare and check each other's work.*

111. *Stationary* Eduardo Verner purchased three boxes of stationery. He also purchased wrapping paper and thank-you cards. If the wrapping paper and thank-you cards together cost $6, and the total he paid was $42, find the cost of a box of stationery.
a) $3x + 6 = 42$ **b)** $12

112. *Candies* Mahandi Ison purchased three rolls of peppermint candies and the local newspaper. The newspaper cost 50 cents. He paid $2.75 in all. What did a roll of candies cost?
a) $3x + 0.5 = 2.75$ **b)** $0.75

Cumulative Review Exercises

[1.9] **113.** Evaluate $[5(2 - 6) + 3(8 \div 4)^2]^2$.   64

      114. Evaluate $-2x^2 + 3x - 12$ when $x = 5$.   −47

[2.2] **115.** To solve an equation, what do you need to do to the variable?   Isolate the variable on one side of the equation.

[2.3] **116.** To solve the equation $7 = -4x$, would you add 4 to both sides of the equation or divide both sides of the equation by −4? Explain your answer.   Divide both sides by −4.

2.5 SOLVING LINEAR EQUATIONS WITH THE VARIABLE ON BOTH SIDES OF THE EQUATION

SSM

Study Guide

CD/Video

MathPro 4/5

PH Math Tutor Center

prenhall.com/Angel

1   Solve equations with the variable on both sides of the equal sign.

2   Solve equations containing decimal numbers or fractions.

3   Identify identities and contradictions.

---

1   Solve Equations with the Variable on Both Sides of the Equal Sign

The equation $4x + 6 = 2x + 4$ contains the variable x on both sides of the equal sign. To solve equations of this type, we must use the appropriate properties to rewrite the equation with all terms containing the variable on only one side of the equal sign and all terms not containing the variable on the other side of the equal sign. This will allow us to isolate the variable, which is our goal. Following is a general procedure, similar to the one outlined in Section 2.4, that can be used to solve linear equations with the variable on both sides of the equal sign.

> **To Solve Linear Equations with the Variable on Both Sides of the Equal Sign**
>
> 1. If the equation contains fractions, multiply **both** sides of the equation by the least common denominator. This will eliminate fractions from the equation.
>
> 2. Use the distributive property to remove parentheses.
>
> 3. Combine like terms on the same side of the equal sign.
>
> 4. Use the addition property to rewrite the equation with all terms containing the variable on one side of the equal sign and all terms not containing the variable on the other side of the equal sign. It may be necessary to use the addition property twice to accomplish this goal. You will eventually get an equation of the form $ax = b$.
>
> 5. Use the multiplication property to isolate the variable. This will give a solution of the form $x =$ some number.
>
> 6. Check the solution in the original equation.

The steps listed here are basically the same as the steps listed in the boxed procedure on pages 123 and 124, except that in step 4 you may need to use the addition property more than once to obtain an equation of the form $ax = b$.

      Remember that our goal in solving an equation is to isolate the variable, that is, to get the variable alone on one side of the equation.

      Consider the equation $3x + 4 = x + 12$ which contains no fractions or parentheses, and no like terms on the same side of the equal sign. Therefore, we start with step 4, the addition property. We will use the addition property twice in order to obtain an equation where the variable appears on only one side of the equal sign. We begin by subtracting x from both sides of the equation to get all the terms containing the variable on the left side of the equation. This will give the following:

Equation

$$3x + 4 = x + 12$$

$$3x - x + 4 = x - x + 12 \quad \textit{Addition property}$$

$$\text{or} \quad 2x + 4 = 12 \qquad \textit{Variable appears only on left side of equal sign}$$

TEACHING TIP
Point out that once the x is subtracted from both sides of the equation the resulting equation is just like those we solved in the previous section.

Notice that the variable, x, now appears on only one side of the equation. However, $+4$ still appears on the same side of the equal sign as the $2x$. We use the addition property a second time to get the term containing the variable by itself on one side of the equation. Subtracting 4 from both sides of the equation gives $2x = 8$, which is an equation of the form $ax = b$.

Equation

$$2x + 4 = 12$$

$$2x + 4 - 4 = 12 - 4 \qquad \textit{Addition property}$$

$$2x = 8 \qquad \textit{x-term is now isolated}$$

Now that the $2x$ is by itself on one side of the equation, we can use the multiplication property, step 5, to isolate the variable and solve the equation for x. We divide both sides of the equation by 2 to isolate the variable and solve the equation.

$$2x = 8$$

$$\frac{\overset{1}{\cancel{2}}x}{\underset{1}{\cancel{2}}} = \frac{\overset{4}{\cancel{8}}}{\underset{1}{\cancel{2}}} \qquad \textit{Multiplication property}$$

$$x = 4 \qquad \textit{x is now isolated}$$

The solution to the equation is 4.

EXAMPLE 1 Solve the equation $4x + 6 = 2x + 4$.

Solution Our goal is to get all terms with the variable on one side of the equal sign and all terms without the variable on the other side. The terms with the variable may be collected on either side of the equal sign. Many methods can be used to isolate the variable. We will illustrate two. In method 1, we will isolate the variable on the left side of the equation. In method 2, we will isolate the variable on the right side of the equation. In both methods, we will follow the steps given in the box on page 133. Since this equation does not contain fractions or parentheses, and there are no like terms on the same side of the equal sign, we begin with step 4.

Method 1: Isolating the variable on the left

$$4x + 6 = 2x + 4$$

Step 4 $\quad 4x - 2x + 6 = 2x - 2x + 4 \quad$ *Subtract 2x from both sides.*

$$2x + 6 = 4$$

Step 4 $\quad 2x + 6 - 6 = 4 - 6 \qquad$ *Subtract 6 from both sides.*

$$2x = -2$$

Step 5
$$\frac{2x}{2} = \frac{-2}{2}$$ 　　　　　*Divide both sides by 2.*
$$x = -1$$

Method 2:　Isolating the variable on the right

$$4x + 6 = 2x + 4$$

Step 4
$$4x - 4x + 6 = 2x - 4x + 4$$ 　　*Subtract 4x from both sides.*
$$6 = -2x + 4$$

Step 4
$$6 - 4 = -2x + 4 - 4$$ 　　*Subtract 4 from both sides.*
$$2 = -2x$$

Step 5
$$\frac{2}{-2} = \frac{-2x}{-2}$$ 　　　　*Divide both sides by −2.*
$$-1 = x$$

The same answer is obtained whether we isolate the variable on the left or right side. However, we need to divide both sides of the equation by a negative number in method 2.

Step 6　Check:
$$4x + 6 = 2x + 4$$
$$4(-1) + 6 \stackrel{?}{=} 2(-1) + 4$$
$$-4 + 6 \stackrel{?}{=} -2 + 4$$
$$2 = 2 \qquad \text{True}$$ ✳

NOW TRY EXERCISE 19

TEACHING TIP
Since there is more than one way to arrive at the correct answer, you may wish to suggest that students choose a style to use when solving equations. Some wish to solve so the variable is isolated on the left side of the equations. Others may wish to avoid dividing by a negative number so the variable is isolated on whichever side has a larger quantity of the variable. You may wish to indicate which method you prefer.

EXAMPLE 2　Solve the equation $2x - 3 - 5x = 13 + 4x - 2$.

Solution　We will choose to collect the terms containing the variable on the right side of the equation in order to create a positive coefficient of x. Since there are like terms *on the same side of the equal sign*, we will begin by combining these like terms.

Step 3
$$2x - 3 - 5x = 13 + 4x - 2$$
$$-3x - 3 = 4x + 11$$ 　　　*Like terms were combined.*

Step 4
$$-3x + 3x - 3 = 4x + 3x + 11$$ 　　*Add 3x to both sides.*
$$-3 = 7x + 11$$

Step 4
$$-3 - 11 = 7x + 11 - 11$$ 　　*Subtract 11 from both sides.*
$$-14 = 7x$$

Step 5
$$\frac{-14}{7} = \frac{7x}{7}$$ 　　　　*Divide both sides by 7.*
$$-2 = x$$

Step 6　Check:
$$2x - 3 - 5x = 13 + 4x - 2$$
$$2(-2) - 3 - 5(-2) \stackrel{?}{=} 13 + 4(-2) - 2$$
$$-4 - 3 + 10 \stackrel{?}{=} 13 - 8 - 2$$
$$-7 + 10 \stackrel{?}{=} 5 - 2$$
$$3 = 3 \qquad \text{True}$$

Since the check is true, the solution is -2. ✳

The solution to Example 2 could be condensed as follows:

$$2x - 3 - 5x = 13 + 4x - 2$$
$$-3x - 3 = 4x + 11 \qquad \textit{Like terms were combined.}$$
$$-3 = 7x + 11 \qquad \textit{3x was added to both sides.}$$
$$-14 = 7x \qquad \textit{11 was subtracted from both sides.}$$
$$-2 = x \qquad \textit{Both sides were divided by 7.}$$

We solved Example 2 by moving the terms containing the variable to the right side of the equation. Now rework the problem by moving the terms containing the variable to the left side of the equation. You should obtain the same answer.

EXAMPLE 3 Solve the equation $2(p + 3) = -3p + 10$.

Solution
$$2(p + 3) = -3p + 10$$

Step 2 $\qquad 2p + 6 = -3p + 10 \qquad$ *Distributive property was used.*

Step 4 $\quad 2p + 3p + 6 = -3p + 3p + 10 \qquad$ *Add 3p to both sides.*

$$5p + 6 = 10$$

Step 4 $\quad 5p + 6 - 6 = 10 - 6 \qquad$ *Subtract 6 from both sides.*

$$5p = 4$$

Step 5 $\qquad \dfrac{5p}{5} = \dfrac{4}{5} \qquad$ *Divide both sides by 5.*

$$p = \dfrac{4}{5}$$

The solution to Example 4 could be condensed as follows:

$$2(p + 3) = -3p + 10$$
$$2p + 6 = -3p + 10 \qquad \textit{Distributive property was used.}$$
$$5p + 6 = 10 \qquad \textit{3p was added to both sides.}$$
$$5p = 4 \qquad \textit{6 was subtracted from both sides.}$$
$$p = \dfrac{4}{5} \qquad \textit{Both sides were divided by 5.}$$

NOW TRY EXERCISE 27

HELPFUL HINT

After the distributive property was used in Example 3, we obtained the equation $2p + 6 = -3p + 10$. Then we had to decide whether to collect terms with the variable on the left or the right side of the equal sign. If we wish the sum of the terms containing a variable to be positive, we use the addition property to eliminate the variable, with the *smaller* numerical coefficient from one side of the equation. Since -3 is smaller than 2, we added $3p$ to both sides of the equation. This eliminated $-3p$ from the right side of the equation and resulted in the sum of the variable terms on the left side of the equation, $5p$, being positive.

EXAMPLE 4 Solve the equation $2(x - 5) + 3 = 3x + 9$.

Solution
$$2(x - 5) + 3 = 3x + 9$$

Step 2 $\quad 2x - 10 + 3 = 3x + 9 \qquad$ *Distributive property was used.*

Step 3 $\qquad 2x - 7 = 3x + 9 \qquad$ *Like terms were combined.*

Step 4 $\qquad -7 = x + 9 \qquad$ *2x was subtracted from both sides.*

Step 4 $\qquad -16 = x \qquad$ *9 was subtracted from both sides.*

EXAMPLE 5   Solve the equation $7 - 2x + 5x = -2(-3x + 4)$.

Solution
$$7 - 2x + 5x = -2(-3x + 4)$$

Step 2	$7 - 2x + 5x = 6x - 8$	*Distributive property was used.*
Step 3	$7 + 3x = 6x - 8$	*Like terms were combined.*
Step 4	$7 = 3x - 8$	*3x was subtracted from both sides.*
Step 4	$15 = 3x$	*8 was added to both sides.*
Step 5	$5 = x$	*Both sides were divided by 3.*

NOW TRY EXERCISE 63   The solution is 5.   ✳

2 Solve Equations Containing Decimal Numbers or Fractions

Now we will solve an equation that contains decimal numbers. As explained in the previous section, equations containing decimal numbers may be solved by a number of different procedures. We will illustrate two procedures for solving Example 6.

EXAMPLE 6   Solve the equation $5.74x + 5.42 = 2.24x - 9.28$.

Solution   **Method 1**   We first notice that there are no like terms on the same side of the equal sign that can be combined. We will elect to collect the terms containing the variable on the left side of the equation.

$$5.74x + 5.42 = 2.24x - 9.28$$

Step 4 $5.74x - 2.24x + 5.42 = 2.24x - 2.24x - 9.28$ *Subtract 2.24x from both sides.*

$$3.50x + 5.42 = -9.28$$

Step 4 $3.50x + 5.42 - 5.42 = -9.28 - 5.42$ *Subtract 5.42 from both sides.*

$$3.50x = -14.7$$

Step 5
$$\frac{3.50x}{3.50} = \frac{-14.7}{3.50}$$ *Divide both sides by 3.5.*

$$x = -4.20$$

Method 2   In the previous section, we introduced a procedure to eliminate decimal numbers from equations. If the equation contains decimals given in tenths, multiply both sides of the equation by 10. If the equation contains decimals given in hundredths, multiply both sides of the equation by 100, and so on. Since the given equation has numbers given in hundredths, we will multiply both sides of the equation by 100.

$$5.74x + 5.42 = 2.24x - 9.28$$
$$100(5.74x + 5.42) = 100(2.24x - 9.28)$$ *Multiply both sides by 100.*
$$100(5.74x) + 100(5.42) = 100(2.24x) - 100(9.28)$$ *Distributive property.*
$$574x + 542 = 224x - 928$$

Step 4 $574x + 542 - 542 = 224x - 928 - 542$ *Subtract 542 from both sides.*

$$574x = 224x - 1470$$

Step 4 $574x - 224x = 224x - 224x - 1470$ *Subtract 224x from both sides.*

$$350x = -1470$$

Step 5
$$\frac{350x}{350} = \frac{-1470}{350}$$ *Divide both sides by 350.*
$$x = -4.20$$

NOW TRY EXERCISE 25

Notice we obtain the same answer using either method. You may use either method to solve equations of this type. ✳

Now let's work some equations that contain fractions on both sides of the equal sign.

EXAMPLE 7   Solve the equation $\frac{1}{2}a = \frac{3}{4}a + \frac{1}{5}$.

Solution   Step 1   In this equation we are solving for a. The least common denominator is 20. Begin by multiplying both sides of the equation by the LCD.

$$\frac{1}{2}a = \frac{3}{4}a + \frac{1}{5}$$

Step 1 $20\left(\frac{1}{2}a\right) = 20\left(\frac{3}{4}a + \frac{1}{5}\right)$ *Multiply both sides by the LCD, 20.*

Step 2 $10a = \overset{5}{20}\left(\frac{3}{\underset{1}{4}}a\right) + \overset{4}{20}\left(\frac{1}{\underset{1}{5}}\right)$ *Distributive property*

$$10a = 15a + 4$$

Step 4 $-5a = 4$ *15a was subtracted from both sides.*

Step 5 $a = -\frac{4}{5}$ *Both sides were divided by −5.*

Step 6   Check:
$$\frac{1}{2}a = \frac{3}{4}a + \frac{1}{5}$$

$$\frac{1}{2}\left(-\frac{4}{5}\right) \overset{?}{=} \frac{3}{4}\left(-\frac{4}{5}\right) + \frac{1}{5}$$

$$-\frac{2}{5} \overset{?}{=} -\frac{3}{5} + \frac{1}{5}$$

$$-\frac{2}{5} = -\frac{2}{5}$$ *True*

NOW TRY EXERCISE 51

The solution to the equation is $-\frac{4}{5}$. ✳

HELPFUL HINT   The equation in Example 7, $\frac{1}{2}a = \frac{3}{4}a + \frac{1}{5}$ could have been given as $\frac{a}{2} = \frac{3a}{4} + \frac{1}{5}$ because $\frac{1}{2}a$ is the same as $\frac{a}{2}$, and $\frac{3}{4}a$ is the same as $\frac{3a}{4}$. You would solve the equation $\frac{a}{2} = \frac{3a}{4} + \frac{1}{5}$ the same way you solved the equation in Example 7. You would begin by multiplying both sides of the equation by the LCD, 20.

EXAMPLE 8   Solve the equation $\dfrac{x}{4} + 3 = 2(x - 2)$.

Solution   Begin by multiplying both sides of the equation by the LCD, 4.

$$\frac{x}{4} + 3 = 2(x - 2)$$

$$4\left(\frac{x}{4} + 3\right) = 4[2(x - 2)] \quad \text{Multiply both sides by the LCD, 4}$$

$$4\left(\frac{x}{4}\right) + 4(3) = 4[2(x - 2)] \quad \text{Distributive Property (used on left)}$$

$$x + 12 = 8(x - 2)$$

$$x + 12 = 8x - 16 \quad \text{Distributive Property (used on right)}$$

$$12 = 7x - 16 \quad \text{x was subtracted from both sides.}$$

$$28 = 7x \quad \text{16 was added to both sides.}$$

$$4 = x \quad \text{Both sides were divided by 7.}$$

A check will show that 4 is the solution.

HELPFUL HINT   Notice the equation in Example 8 had *two terms* on the left side of the equal sign, $\dfrac{x}{4}$ and 3. The equation had only *one term* on the right side of the equal sign, $2(x - 2)$. Therefore, after we multiplied both sides of the equation by 4, the next step was to use the distributive property on the left side of the equation.

EXAMPLE 9   Solve the equation $\dfrac{1}{2}(2x + 3) = \dfrac{2}{3}(x - 6) + 4$.

Notice this equation contains one term on the left side of the equal sign and two terms on the right side of the equal sign.

Solution   Multiply both sides of the equation by the LCD, 6.

$$\frac{1}{2}(2x + 3) = \frac{2}{3}(x - 6) + 4$$

$$\overset{3}{6} \cdot \frac{1}{2}(2x + 3) = 6\left[\frac{2}{3}(x - 6) + 4\right] \quad \text{Multiply both sides by the LCD, 6.}$$

$$3(2x + 3) = \overset{2}{6} \cdot \frac{2}{3}(x - 6) + 6 \cdot 4 \quad \text{Distributive property (used on right)}$$

$$6x + 9 = 4(x - 6) + 24 \quad \text{Distributive property (used on left)}$$

$$6x + 9 = 4x - 24 + 24 \quad \text{Distributive property (used on right)}$$

$$6x + 9 = 4x \quad \text{Like items were combined.}$$

$$9 = -2x \quad \text{6x was subtracted from both sides.}$$

$$-\frac{9}{2} = x \quad \text{Both sides were divided by } -2.$$

Check: Substitute $-\dfrac{9}{2}$ for x in the equation.

$$\frac{1}{2}(2x + 3) = \frac{2}{3}(x - 6) + 4$$

$$\frac{1}{2}\left[2\left(-\frac{9}{2}\right) + 3\right] \overset{?}{=} \frac{2}{3}\left(-\frac{9}{2} - 6\right) + 4$$

$$\frac{1}{2}(-9 + 3) \overset{?}{=} \frac{2}{3}\left(-\frac{9}{2} - \frac{12}{2}\right) + 4$$

$$\frac{1}{2}(-6) \overset{?}{=} \frac{2}{3}\left(-\frac{21}{2}\right) + 4$$

$$-3 \overset{?}{=} -7 + 4$$

$$-3 = -3 \qquad\qquad \textit{True}$$

Thus the solution is $-\dfrac{9}{2}$.

In Example 9, the given equation could have been represented as $\dfrac{2x + 3}{2} = \dfrac{2(x - 6)}{3} + 4$. If you were given the equation in this form, the first step

NOW TRY EXERCISE 75

in determining the solution would have been to multiply both sides of the equation by the LCD, 6. Because this is just another way of writing the equation in Example 9, the answer would be $-\frac{9}{2}$.

We will discuss solving equations containing fractions further in Section 6.6.

3 Identify Identities and Contradictions

Thus far all the equations we have solved have had a single value for a solution. Equations of this type are called **conditional equations**, for they are only true under specific conditions. Some equations, as in Example 10, are true for infinitely many values of x. Equations that are true for infinitely many values of x are called **identities**. A third type of equation, as in Example 11, has no solution and is called a **contradiction**.

EXAMPLE 10 Solve the equation $2x + 6 = 2(x + 3)$.

Solution
$$2x + 6 = 2(x + 3)$$
$$2x + 6 = 2x + 6$$

Since the same expression appears on both sides of the equal sign, the statement is true for infinitely many values of x. If we continue to solve this equation further, we might obtain

$$2x = 2x \qquad\qquad \textit{6 was subtracted from both sides.}$$
$$0 = 0 \qquad\qquad \textit{2x was subtracted from both sides.}$$

Note: The solution process could have been stopped at $2x + 6 = 2x + 6$. Since one side is identical to the other side, the equation is true for infinitely many values of x. *The solution to this equation is all real numbers.* **When solving an equation, like the equation in Example 10, that is always true, write your answer as "all real numbers".**

EXAMPLE 11 Solve the equation $-3x + 4 + 5x = 4x - 2x + 5$.

Solution
$$-3x + 4 + 5x = 4x - 2x + 5$$
$$2x + 4 = 2x + 5 \qquad\qquad \textit{Like terms were combined.}$$

$$2x - 2x + 4 = 2x - 2x + 5 \quad \textit{Subtract 2x from both sides.}$$
$$4 = 5 \quad \textit{False}$$

When solving a equation, if you obtain an obviously false statement, as in this example, the equation has *no solution.* No value of x will make the equation a true statement. **When solving an equation like the equation in Example 11, that is never true, write your answer as "*no solution*".** An answer left blank may be marked wrong.

NOW TRY EXERCISE 31

HELPFUL HINT

Some students start solving equations correctly but do not complete the solution. Sometimes they are not sure that what they are doing is correct and they give up for lack of confidence. You must have confidence in yourself. As long as you follow the procedure on page 133, you should obtain the correct solution even if it takes quite a few steps. Remember two important things: (1) your goal is to isolate the variable, and (2) whatever you do to one side of the equation you must also do to the other side. That is, you must treat both sides of the equation equally.

1., 9. a), 10. a) Answers will vary   **2.** an equation that is true only under specific conditions

Exercise Set 2.5

Concept/Writing Exercises

1. a) In your own words, write the general procedure for solving an equation that does not contain fractions where the variable appears on both sides of the equation.

b) Refer to page 133 to see whether you omitted any steps.

2. What is a conditional equation?

3. a) What is an identity?

b) What is the solution to the equation $3x + 5 = 3x + 5$?

4. When solving an equation, how will you know if the equation is an identity?

5. Explain why the equation $x + 5 = x + 5$ must be an identity? Both sides are identical.

6. a) What is a contradiction?

b) What is the solution to a contradiction? no solution

7. When solving an equation, how will you know if the equation has no solution? will obtain a false statement

8. Explain why the equation $x + 5 = x + 4$ must be a contradiction. subtract x from both sides; $5 \neq 4$

9. a) Explain, step-by-step, how to solve the equation $4(x + 3) = 6(x - 5)$.

b) Solve the equation by following the steps you listed in part **a)**. 21

10. a) Explain, step-by-step, how to solve the equation $4x + 3(x + 2) = 5x - 10$.

b) Solve the equation by following the steps you listed in part **a)**. -8

3. a) an equation that is true for infinitely many values of the variable. **b)** all real numbers   **4.** the same expression will appear on both sides of the equation   **6. a)** an equation that has no solution

Practice the Skills

Solve each equation.   **26.** 0.7

11. $3x = -2x + 15$ 3

12. $x + 4 = 2x - 7$ 11

13. $-4x + 10 = 6x$ 1

14. $3a = 4a + 8$ -8

15. $5x + 3 = 6$ $\frac{3}{5}$

16. $-6x = 2x + 16$ -2

17. $21 - 6p = 3p - 2p$ 3

18. $8 - 3x = 4x + 50$ -6

19. $2x - 4 = 3x - 6$ 2

20. $-5x = -4x + 9$ -9

21. $6 - 2y = 9 - 8y + 6y$ no solution

22. $-4 + 2y = 2y - 6 + y$ 2

23. $9 - 0.5x = 4.5x + 8.5$ 0.1

24. $124.8 - 9.4x = 4.8x + 32.5$ 6.5

25. $0.62x - 0.65 = 9.75 - 2.63x$ 3.2

26. $8.71 - 2.44x = 11.02 - 5.74x$

27. $5x + 3 = 2(x + 6)$ 3

28. $x - 14 = 3(x + 2)$ -10

29. $x - 25 = 12x + 9 + 3x$ $\quad -\frac{17}{7}$

30. $5y + 6 = 2y + 3 - y$ $\quad -\frac{3}{4}$

31. $2(x - 2) = 4x - 6 - 2x$ $\quad$ no solution

32. $4r = 10 - 2(r - 4)$ $\quad 3$

33. $-(w + 2) = -6w + 32$ $\quad \frac{34}{5}$

34. $7(-3m + 5) = 3(10 - 6m)$ $\quad \frac{5}{3}$

35. $5 - 3(2t - 5) = 3t + 13$ $\quad \frac{7}{9}$

36. $4(2x - 3) = -2(3x + 16)$ $\quad -\frac{10}{7}$

37. $\frac{a}{5} = \frac{a - 3}{2}$ $\quad 5$

38. $\frac{b}{16} = \frac{b - 6}{4}$ $\quad 8$

39. $6 - \frac{x}{4} = \frac{x}{8}$ $\quad 16$

40. $\frac{n}{10} = 9 - \frac{n}{5}$ $\quad 30$

41. $\frac{5}{2} - \frac{x}{3} = 3x$ $\quad \frac{3}{4}$

42. $\frac{x}{4} - 3 = -2x$ $\quad \frac{4}{3}$

43. $\frac{5}{8} + \frac{1}{4}a = \frac{1}{2}a$ $\quad \frac{5}{2}$

44. $\frac{3}{4}x + \frac{1}{2} = \frac{1}{2}x$ $\quad -2$

45. $0.1(x + 10) = 0.3x - 4$ $\quad 25$

46. $5(3.2x - 3) = 2(x - 4)$ $\quad \frac{1}{2}$

47. $2(x + 4) = 4x + 3 - 2x + 5$ $\quad$ all real numbers

48. $3(y - 1) + 9 = 8y + 6 - 5y$ $\quad$ all real numbers

49. $5(3n + 3) = 2(5n - 4) + 6n$ $\quad 23$

50. $-4(-3z - 5) = -(10z + 8) - 2z$ $\quad -\frac{7}{6}$

51. $-(3 - p) = -(2p + 3)$ $\quad 0$

52. $12 - 2x - 3(x + 2) = 4x + 6 - x$ $\quad 0$

53. $-(x + 4) + 5 = 4x + 1 - 5x$ $\quad$ all real numbers

54. $18x + 3(4x - 9) = -6x + 81$ $\quad 3$

55. $35(2x - 1) = 7(x + 4) + 3x$ $\quad \frac{21}{20}$

56. $10(x - 10) + 5 = 5(2x - 20)$ $\quad$ no solution

57. $0.4(x + 0.7) = 0.6(x - 4.2)$ $\quad 14$

58. $0.5(6x - 8) = 1.4(x - 5) - 0.2$ $\quad -2$

59. $\frac{3}{5}x + 4 = \frac{1}{5}x + 5$ $\quad \frac{5}{2}$

60. $\frac{3}{5}x - 2 = x + \frac{1}{3}$ $\quad -\frac{35}{6}$

61. $\frac{3}{4}(2x - 4) = 4 - 2x$ $\quad 2$

62. $\frac{1}{5}(w + 2) = w + 4$ $\quad -\frac{9}{2}$

63. $-(x - 5) + 2 = 3(4 - x) + 5x$ $\quad -\frac{5}{3}$

64. $3(x - 4) = 2(x - 8) + 5x$ $\quad 1$

65. $3(x - 6) - 4(3x + 1) = x - 22$ $\quad 0$

66. $-2(-3x + 5) + 6 = 4(x - 2)$ $\quad -2$

67. $5 + 2x = 6(x + 1) - 5(x - 3)$ $\quad 16$

68. $4 - (6x + 6) = -(-2x + 10)$ $\quad 1$

69. $7 - (-y - 5) = 2(y + 3) - 6(y + 1)$ $\quad -\frac{12}{5}$

70. $12 - 6x + 3(2x + 3) = 2x + 5$ $\quad 8$

71. $\frac{1}{2}(2d + 4) = \frac{1}{3}(4d - 4)$ $\quad 10$

72. $\frac{3}{5}(x - 6) = \frac{2}{3}(3x - 5)$ $\quad -\frac{4}{21}$

73. $\frac{3(2r - 5)}{5} = \frac{3r - 6}{4}$ $\quad \frac{10}{3}$

74. $\frac{3(x - 4)}{4} = \frac{5(2x - 3)}{3}$ $\quad \frac{24}{31}$

75. $\frac{2}{7}(5x + 4) = \frac{1}{2}(3x - 4) + 1$ $\quad 30$

76. $\frac{5}{12}(x + 2) = \frac{2}{3}(2x + 1) + \frac{1}{6}$ $\quad 0$

77. $\frac{a - 5}{2} = \frac{3a}{4} + \frac{a - 25}{6}$ $\quad 4$

78. $\frac{a - 7}{3} = \frac{a + 5}{2} - \frac{7a - 1}{6}$ $\quad 5$

79. **a)** Answers will vary, one example is $x + x + 1 = x + 2$.

80. **a)** Answers will vary, one example is $x + 2 = x + x + 1$.

81. **a)** Answers will vary, one example is $x + x + 1 = 2x + 1$

82. **a)** Answers will vary, one example is $2x + 1 = x + x + 1$.

Problem Solving

79. **a)** Construct a *conditional equation* containing three terms on the left side of the equal sign and two terms on the right side of the equal sign.

b) Explain how you know your answer to part **a)** is a conditional equation. has a single solution

c) Solve the equation. $x = 1$

80. **a)** Construct a *conditional equation* containing two terms on the left side of the equal sign and three terms on the right side of the equal sign.

b) Explain how you know your answer to part **a)** is a conditional equation. has a single solution

c) Solve the equation. $x = 1$

81. **a)** Construct an *identity* containing three terms on the left side of the equal sign and two terms on the right side of the equal sign.

b) Explain how you know your answer to part **a)** is an identity.

c) What is the solution to the equation?

82. **a)** Construct an *identity* containing two terms on the left side of the equal sign and three terms on the right side of the equal sign.

b) Explain how you know your answer to part **a)** is an identity.

c) What is the solution to the equation?

81. b) and **82. b)** when simplified, both sides are the same   **81. c)** and **82. c)** all real numbers

83. a) Construct a *contradiction* containing three terms on the left side of the equal sign and two terms on the right side of the equal sign.

b) Explain how you know your answer to part **a)** is a contradiction. Can never be true

c) What is the solution to the equation? no solution

83. a) Answers will vary, one example is $x + x + 1 = 2x + 2$

84. a) Construct a *contradiction* containing three terms on the left side of the equal sign and four terms on the right side of the equal sign.

b) Explain how you know your answer to part **a)** is a contradiction. Can never be true

c) What is the solution to the equation? no solution

84. a) Answers will vary, one example is
$x + x + 2 = x + x + 2 + 1$

Challenge Problems   86. no solution   87. all real numbers

85. Solve the equation $5* - 1 = 4* + 5$ for $*$. $* = 6$
86. Solve the equation $2\triangle - 4 = 3\triangle + 5 - \triangle$ for $\triangle$.
87. Solve the equation $3\odot - 5 = 2\odot - 5 + \odot$ for $\odot$.

88. Solve $-2(x + 3) + 5x = 3(4 - 2x) - (x + 2)$. $\frac{8}{5}$
89. Solve $4 - [5 - 3(x + 2)] = x - 3$. -4

Group Activity

Discuss and answer Exercises 90 as a group. In the next chapter we will be discussing procedures for writing application problems as equations. Let's get some practice now.

90. Chocolate Bars Consider the following word problem. Mary Kay purchased two large chocolate bars. The total cost of the two chocolate bars was equal to the cost of one chocolate bar plus $6. Find the cost of one chocolate bar.

a) Each group member: Represent this problem as an equation with the variable x. $2x = x + 6$

b) Each group member: Solve the equation you determined in Part a). $6

c) As a group, check your equation and your answer to make sure that it makes sense.

91. a) 4 **b)** 7 **c)** 0   **92.** ≈0.131687243   **93.** Factors are expressions that are multiplied; Terms are expressions that are added

Cumulative Review Exercises

[1.5] **91.** Evaluate **a)** $|4|$ **b)** $|-7|$ **c)** $|0|$.

[1.9] **92.** Evaluate $\left(\frac{2}{3}\right)^5$ on your calculator.

[2.1] **93.** Explain the difference between factors and terms.

94. Simplify $2(x - 3) + 4x - (4 - x)$. $7x - 10$

[2.4] **95.** Solve $2(x - 3) + 4x - (4 - x) = 0$. $\frac{10}{7}$

96. Solve $(x + 4) - (4x - 3) = 16$. -3

2.6 RATIOS AND PROPORTIONS

 SSM
 Study Guide
 CD/Video
 MathPro 4/5
 PH Math Tutor Center
 prenhall.com/Angel

1 Understand ratios.
2 Solve proportions using cross-multiplication.
3 Solve applications.
4 Use proportions to change units.
5 Use proportions to solve problems involving similar figures.

1 Understand Ratios

A **ratio** is a quotient of two quantities. Ratios provide a way to compare two numbers or quantities. The ratio of the number a to the number b may be written

$$a \text{ to } b, \quad a:b, \quad \text{or} \quad \frac{a}{b}$$

where a and b are called the **terms of the ratio**.

EXAMPLE 1 **Taste Test** In a number of different shopping malls, a beverage company was performing a taste test using three beverages, A, B, and C, respectively. The purpose of the test was to determine which of the three new beverages should be released to the public. Participants in the Mall of America were asked to sample all three and to indicate which one was their favorite. The results of the survey are indicated on the graph in Figure 2.2.

a) Find the ratio of the number of people who selected B to those who selected A.

b) Find the ratio of the number who selected C to the total number in the survey.

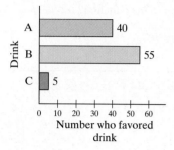

FIGURE 2.2

Solution We will use our five-step problem-solving procedure.

a) Understand and Translate The ratio we are seeking is

Number who selected B : Number who selected A

Carry Out We substitute the appropriate values into the ratio. This gives

$$55:40$$

Now we simplify by dividing each number in the ratio by 5, the greatest number that divides both terms in the ratio. This gives

$$11:8$$

Check and Answer Our division is correct. The ratio is $11:8$.

b) We use the same procedure as in part a). Five people selected drink C. There were $40 + 55 + 5$ or 100 people surveyed. Thus the ratio is $5:100$, which simplifies to

$$1:20$$

NOW TRY EXERCISE 17

The answer in Example 1, part **a)** could also have been written $\frac{11}{8}$ or 11 to 8. The answer in part **b)** could also have been written $\frac{1}{20}$ or 1 to 20.

EXAMPLE 2 **Cholesterol Level** There are two types of cholesterol: low-density lipoprotein, (LDL—considered the harmful type of cholesterol) and high-density lipoprotein (HDL—considered the healthful type of cholesterol). Some doctors recommend that the ratio of low- to high-density cholesterol be less than or equal to $4:1$. Mr. Suarez's cholesterol test showed that his low-density cholesterol measured 167 milligrams per deciliter, and his high-density cholesterol measured 40 milligrams per deciliter. Is Mr. Suarez's ratio of low- to high-density cholesterol less than or equal to the recommended $4:1$ ratio?

Solution Understand We need to determine if Mr. Suarez's low- to high-density cholesterol is less than or equal to $4:1$.

Translate Mr. Suarez's low- to high-density cholesterol is $167:40$. To make the second term equal to 1, we divide both terms in the ratio by the second term, 40.

Carry Out
$$\frac{167}{40}:\frac{40}{40}$$

or $4.175:1$

Check and Answer Our division is correct. Therefore, Mr. Suarez's ratio is not less than or equal to the desired $4:1$ ratio.

EXAMPLE 3 **Oil-Gas mixture** Some power equipment, such as chainsaws and blowers, use a gas–oil mixture to run the engine. The instructions on a particular chainsaw indicate that 5 gallons of gasoline should be mixed with 40 ounces of special oil to obtain the proper gas–oil mixture. Find the ratio of gasoline to oil in the proper mixture.

Solution **Understand** To express these quantities in a ratio, both quantities must be in the same units. We can either convert 5 gallons to ounces or 40 ounces to gallons.

Translate Let's change 5 gallons to ounces. Since there are 128 ounces in 1 gallon, 5 gallons of gas equals 5(128) or 640 ounces. The ratio we are seeking is

ounces of gasoline : ounces of oil

Carry Out   640 : 40

or   16 : 1   *Divide both terms by 40 to simplify.*

Check and Answer Our simplification is correct. The correct ratio of gas to oil for this chainsaw is 16 : 1. ✳

EXAMPLE 4   **Gear Ratio**   The *gear ratio* of two gears is defined as

$$\text{gear ratio} = \frac{\text{number of teeth on the driving gear}}{\text{number of teeth on the driven gear}}$$

Find the gear ratio of the gears shown in Figure 2.3.

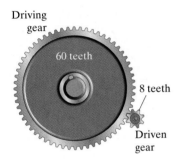

Driving gear — 60 teeth / 8 teeth / Driven gear

FIGURE 2.3

Solution   **Understand and Translate**   To find the gear ratio we need to substitute the appropriate values.

Carry Out $\text{gear ratio} = \frac{\text{number of teeth on driving gear}}{\text{number of teeth on driven gear}} = \frac{60}{8} = \frac{15}{2}$

Thus, the gear ratio is 15 : 2. Gear ratios are generally given as some quantity to 1. If we divide both terms of the ratio by the second term, we will obtain a ratio of some number to 1. Dividing both 15 and 2 by 2 gives a gear ratio of 7.5 : 1.

Check and Answer   The gear ratio is 7.5 : 1. This means that as the driving gear goes around once the driven gear goes around 7.5 times. (A typical first gear ratio on a passenger car may be 3.545 : 1). ✳

NOW TRY EXERCISE 27

2 Solve Proportions Using Cross-Multiplication

A **proportion** is a special type of equation. It is a statement of equality between two ratios. One way of denoting a proportion is $a : b = c : d$, which is read "a is to b as c is to d." In this text we write proportions as

$$\frac{a}{b} = \frac{c}{d}$$

The a and d are referred to as the **extremes**, and the b and c are referred to as the **means** of the proportion. In Sections 2.4 and 2.5 we solved equations containing fractions by multiplying both sides of the equation by the LCD to eliminate fractions. For example, for the proportion

$$\frac{x}{3} = \frac{35}{15}$$

$$15\left(\frac{x}{3}\right) = 15\left(\frac{35}{15}\right)$$ *Multiply both sides by the LCD, 15*

$$5x = 35$$

$$x = 7$$

Another method that can be used to solve proportions is **cross-multiplication**. This process of cross-multiplication gives the same results as multiplying both sides of the equation by the LCD. However, many students prefer to use cross-multiplication because they do not have to determine the LCD of the fractions, and then multiply both sides of the equation by the LCD.

Cross-Multiplication

$$\text{If } \frac{a}{b} = \frac{c}{d}, \text{ then } ad = bc.$$

Note that *the product of the means is equal to the product of the extremes.*

If any three of the four quantities of a proportion are known, the fourth quantity can easily be found.

EXAMPLE 5 Solve $\dfrac{x}{3} = \dfrac{35}{15}$ for x by cross-multiplying.

Solution

$$\frac{x}{3} = \frac{35}{15}$$

$$x \cdot 15 = 3 \cdot 35$$

$$15x = 105$$

$$x = \frac{105}{15} = 7$$

Check:

$$\frac{x}{3} = \frac{35}{15}$$

$$\frac{7}{3} \stackrel{?}{=} \frac{35}{15}$$

$$\frac{7}{3} = \frac{7}{3} \quad \textit{True}$$

Before we introduced cross-multiplication, we solved the proportion $\dfrac{x}{3} = \dfrac{35}{15}$ by multiplying both sides of the equation by 15. In Example 5, we solved the same proportion using cross-multiplication. Notice we obtained the same solution, 7, in each case. When you solve an equation using cross-multiplication, you are in effect, multiplying both sides of the equation by the product of the two denominators, and then dividing out the common factors. However, this process is not shown.

EXAMPLE 6 Solve $\dfrac{-8}{3} = \dfrac{64}{x}$ for x by cross-multiplying.

Solution

$$\frac{-8}{3} = \frac{64}{x}$$

$$-8 \cdot x = 3 \cdot 64$$

$$-8x = 192$$

$$\frac{-8x}{-8} = \frac{192}{-8}$$

$$x = -24$$

Check:

$$\frac{-8}{3} = \frac{64}{x}$$

$$\frac{-8}{3} \stackrel{?}{=} \frac{64}{-24}$$

$$\frac{-8}{3} \stackrel{?}{=} \frac{8}{-3}$$

$$\frac{-8}{3} = \frac{-8}{3} \quad \textit{True}$$

NOW TRY EXERCISE 41

3 Solve Applications

Often, practical problems can be solved using proportions. To solve such problems, use the five-step problem-solving procedure we have been using throughout the book. Below we give that procedure with more specific directions for translating problems into proportions.

To Solve Problems Using Proportions

1. Understand the problem.

2. Translate the problem into mathematical language.

   a) First, represent the unknown quantity by a variable (a letter).

   b) Second, set up the proportion by listing the given ratio on the left side of the equal sign, and the unknown and the other given quantity on the right side of the equal sign. When setting up the right side of the proportion, the same respective quantities should occupy the same respective positions on the left and the right. For example, an acceptable proportion might be

   $$\text{Given ratio}\left\{\frac{\text{miles}}{\text{hour}} = \frac{\text{miles}}{\text{hour}}\right.$$

3. Carry out the mathematical calculations necessary to solve the problem.

   a) Once the proportion is correctly written, drop the units and cross-multiply.

   b) Solve the resulting equation.

4. Check the answer obtained in step 3.

5. Make sure you have answered the question.

   Note that the two ratios* must have the same units. For example, if one ratio is given in miles/hour and the second ratio is given in feet/hour, one of the ratios must be changed before setting up the proportion.

EXAMPLE 7 **Applying Fertilizer**   A 30-pound bag of fertilizer will cover an area of 2500 square feet.

a) How many pounds are needed to cover an area of 16,000 square feet?

b) How many bags of fertilizer are needed?

Solution **a)** Understand   The given ratio is 30 pounds per 2500 square feet. The unknown quantity is the number of pounds necessary to cover 16,000 square feet.

Translate   Let x = number of pounds.

$$\text{Given ratio}\left\{\frac{30 \text{ pounds}}{2500 \text{ square feet}} = \frac{x \text{ pounds}}{16,000 \text{ square feet}}\right. \begin{array}{l} \longleftarrow \textit{Unknown} \\ \longleftarrow \textit{Given quantity} \end{array}$$

Note how the weight and the area are given in the same relative positions.

---

*Strictly speaking, a quotient of two quantities with different units, such as $\frac{6 \, miles}{1 \, hour}$, is called a *rate*. However, few books make the distinction between ratios and rates when discussing proportions.

Carry Out

$$\frac{30}{2500} = \frac{x}{16{,}000}$$

$$30(16{,}000) = 2500x \quad \textit{Cross-multiply.}$$

$$480{,}000 = 2500x \quad \textit{Solve.}$$

$$\frac{480{,}000}{2500} = x$$

$$192 = x$$

Check   Using a calculator, we determine that both ratios in the proportion, 30/2500 and 192/16,000, have a value of 0.012. Thus, the answer 192 pounds, checks.

Answer   The amount of fertilizer needed to cover an area of 16,000 square feet is 192 pounds.

b) Since each bag weighs 30 pounds, the number of bags is found by division.

$$192 \div 30 = 6.4 \text{ bags}$$

The number of bags needed is therefore 7, since one must purchase whole bags.

NOW TRY EXERCISE 65

EXAMPLE 8   **Charity Luncheon**   Each year in Tampa, Florida, the New York Yankees host a charity luncheon with the proceeds going to support the Tampa Boys and Girls Clubs. At the luncheon, the guests meet and get autographs from members of the team. If a particular player signs, on the average, 33 autographs in 4 minutes, how much time must be allowed for him to sign 350 autographs?

Solution   The unknown quantity is the time needed for the player to sign 350 autographs. We are given that, on the average, he signs 33 autographs in 4 minutes. We will use this given ratio in setting up our proportion.

Translate   We will let x represent the time to sign 350 autographs.

$$\text{Given ratio} \begin{cases} \dfrac{33 \text{ autographs}}{4 \text{ minutes}} = \dfrac{350 \text{ autographs}}{x \text{ minutes}} \end{cases}$$

Carry Out

$$\frac{33}{4} = \frac{350}{x}$$

$$33x = 4(350)$$

$$33x = 1400$$

$$x = \frac{1400}{33} \approx 42.4$$

Check and Answer   Using a calculator, we can determine that both ratios in the proportion, $\dfrac{33}{4}$ and $\dfrac{350}{42.4}$, have approximately the same value of 8.25. Thus, about 42.4 minutes would be needed for the player to sign the 350 autographs.

EXAMPLE 9   **Drug Dosage**   A doctor asks a nurse to give a patient 250 milligrams of the drug simethicone. The drug is available only in a solution whose concentration is 40 milligrams of simethicone per 0.6 milliliter of solution. How many milliliters of solution should the nurse give the patient?

Solution Understand and Translate We can set up the proportion using the medication on hand as the given ratio and the number of milliliters needed to be given as the unknown.

$$\text{Given ratio (medication on hand)} \begin{cases} \dfrac{40 \text{ milligrams}}{0.6 \text{ milliliter}} = \dfrac{250 \text{ milligrams}}{x \text{ milliliters}} \end{cases}$$

Desired medication ← ←Unknown

Carry Out

$$\frac{40}{0.6} = \frac{250}{x}$$
$$40x = 0.6(250) \quad \text{Cross-multiply.}$$
$$40x = 150 \quad \text{Solve.}$$
$$x = \frac{150}{40} = 3.75$$

Check and Answer The nurse should administer 3.75 milliliters of the simethicone solution.

HELPFUL HINT

When you are setting up a proportion, it does not matter which unit in the given ratio is in the numerator and which is in the denominator as long as the units in the other ratio are *in the same relative position*. For example,

$$\frac{60 \text{ miles}}{1.5 \text{ hours}} = \frac{x \text{ miles}}{4.2 \text{ hours}} \quad \text{and} \quad \frac{1.5 \text{ hours}}{60 \text{ miles}} = \frac{4.2 \text{ hours}}{x \text{ miles}}$$

will both give the same answer of 168 (try it and see). When setting up the proportion, set it up so that it makes the most sense to you. Notice that when setting up a proportion containing different units, the same units should not be multiplied by themselves during cross multiplication.

Correct

$$\frac{\text{miles}}{\text{hour}} = \frac{\text{miles}}{\text{hour}}$$

Incorrect

$$\frac{\text{miles}}{\text{hour}} \diagdown \frac{\text{hour}}{\text{miles}}$$

4 Use Proportions to Change Units

Proportions can also be used to convert from one quantity to another. For example, you can use a proportion to convert a measurement in feet to a measurement in meters, or to convert from pounds to kilograms. The following examples illustrate converting units.

EXAMPLE 10 **Feet to Miles** There are 5280 feet in 1 mile. What is the distance, in miles, of 18,362 feet?

Solution Understand and Translate We know that 1 mile is 5280 feet. We use this known fact in one ratio of our proportion. In the second ratio, we set the quantities with the same units in the same respective positions. The unknown quantity is the number of miles, which we will call x.

$$\text{Known ratio} \begin{cases} \dfrac{1 \text{ mile}}{5280 \text{ feet}} = \dfrac{x \text{ miles}}{18,362 \text{ feet}} \end{cases}$$

Note that both numerators contain the same units, and both denominators contain the same units.

Carry Out Now drop the units and solve for x by cross-multiplying.

$$\frac{1}{5280} = \frac{x}{18{,}362}$$

$$1(18{,}362) = 5280x \qquad \textit{Cross-multiply.}$$

$$18{,}362 = 5280x \qquad \textit{Solve.}$$

$$\frac{18362}{5280} = \frac{5280x}{5280}$$

$$3.48 \approx x$$

NOW TRY EXERCISE 75   **Check and Answer**   Thus, 18,362 feet equals about 3.48 miles. ✳

EXAMPLE 11   **Exchanging Currency**   When people travel to a foreign country they often need to exchange currency. Donna Boccio visited Cancun, Mexico. She stopped by a local bank and was told that $1 U.S. could be exchanged for 9.696 pesos.

a) How many pesos would she get if she exchanged $150 U.S.?

b) Later that same day, Donna went to the city market where she purchased a ceramic figurine. The price she negotiated for the figurine was 245 pesos. Using the exchange rate given, determine the cost of the figurine in U.S. dollars.

Solution   **a)** Understand   We are told that $1 U.S. can be exchanged for 9.696 Mexican pesos. We use this known fact for one ratio in our proportion. In the second ratio, we set the quantities with same units in the same respective positions.

Translate   The unknown quantity is the number of pesos, which we shall call x.

$$\textit{Given Ratio} \left\{ \frac{\$1 \text{ U.S.}}{9.696 \text{ pesos}} = \frac{\$150 \text{ U.S.}}{x \text{ pesos}} \right.$$

Note that both numerators contain U.S. dollars and both denominators contain pesos.

Carry Out
$$\frac{1}{9.696} = \frac{150}{x}$$

$$1x = 9.696(150)$$

$$x = 1454.4$$

Check and Answer   Thus, $150 U.S. could be exchanged for 1454.4 Mexican pesos.

b) Understand and Translate   We use the same given ratio that we used in part **a)**. Now we must find the equivalent in U.S. dollars of 245 Mexican pesos. Let's call the equivalent U.S. dollars x.

$$\textit{Given Ratio} \left\{ \frac{\$1 \text{ U.S.}}{9.696 \text{ pesos}} = \frac{\$x \text{ U.S.}}{245 \text{ pesos}} \right.$$

Carry Out
$$\frac{1}{9.696} = \frac{x}{245}$$

$$1(245) = 9.696x$$

$$245 = 9.696x$$

$$25.27 \approx x$$

Check and Answer   The cost of the figurine in U.S. dollars is $25.27. ✳

HELPFUL HINT

Some of the problems we have just worked using proportions could have been done without using proportions. However, when working problems of this type, students often have difficulty in deciding whether to multiply or divide to obtain the correct answer. By setting up a proportion, you may be better able to understand the problem and have more success in obtaining the correct answer.

5 Use Proportions to Solve Problems Involving Similar Figures

Proportions can also be used to solve problems in geometry and trigonometry. The following examples illustrate how proportions may be used to solve problems involving **similar figures**. Two figures are said to be similar when their corresponding angles are equal and their corresponding sides are in proportion. Two similar figures will have the same shape.

EXAMPLE 12 The figures to the left are similar. Find the length of the side indicated by the x.

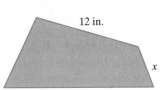

Solution We set up a proportion of corresponding sides to find the length of side x.

Lengths from smaller figure   *Lengths from larger figure*

5 inches and 12 inches are corresponding sides of similar figures.

2 inches and x are corresponding sides of similar figures.

$$\frac{5}{2} = \frac{12}{x}$$

$$5x = 24$$

$$x = \frac{24}{5} = 4.8$$

TEACHING TIP
Point out that scaled drawings are used by architects.

Thus, the side indicated by x is 4.8 inches in length.

Note in Example 12 that the proportion could have also been set up as

$$\frac{5}{12} = \frac{2}{x}$$

because one pair of corresponding sides is in the numerators and another pair is in the denominators.

EXAMPLE 13 Triangles ABC and $AB'C'$ are similar triangles. Use a proportion to find the length of side AB'.

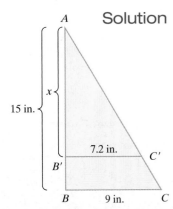

Solution We set up a proportion of corresponding sides to find the length of side AB'. We will let x represent the length of side AB'. One proportion we can use is

$$\frac{\text{length of } AB}{\text{length of } BC} = \frac{\text{length of } AB'}{\text{length of } B'C'}$$

Now we insert the proper values and solve for the variable x.

$$\frac{15}{9} = \frac{x}{7.2}$$

$$(15)(7.2) = 9x$$

$$108 = 9x$$

$$12 = x$$

NOW TRY EXERCISE 53 Thus, the length of side AB' is 12 inches.

Mathematics in Action

Qualifying Ratios

When you apply for a mortgage loan, you may feel that you are being judged as to whether you are worthy of receiving money to purchase a home. In a sense you are, however, the judgment is being made largely by plugging numbers into a couple of simple ratios, called *qualifying ratios*. One of these is the *housing ratio* and the other the *debt ratio*.

If the housing ratio used by the lender is 27, for example, this means that your monthly housing expenses—monthly mortgage principal, interest payments, property taxes and homeowner's insurance, and other fees—should not exceed 27 percent of your gross monthly income. So suppose your household gross income is $6500 and the monthly housing expenses are $1700, the housing ratio is 26.2 (from $\frac{1700}{6500} = 0.262 = 26.2\%$) and you pass the test. By the way, notice how ratios, decimals, and percentages get used interchangeably.

If the lender's debt ratio is 38, this means that your monthly housing expenses plus long-term debt should not exceed 38 percent of your gross monthly income. Continuing with the same example as above, if you are making monthly payments of $900 on a recreational vehicle, the debt ratio would be 40 (from $\frac{1700 + 900}{6500} = 0.40 = 40\%$) and the lender might refuse you the mortgage loan on that basis.

Don't lose heart! The housing ratio and the debt ratio are not the only criteria applied. Even if your ratios are running a bit high, the lender could make a judgment in your favor based on factors like long-term job stability and credit record.

You can find a number of online calculators on the Internet by using a search engine such as *www.google.com* and searching on **mortgage calculator**. You will find interactive calculators in all colors, sizes, and shapes. If you study them carefully, you will discover that the numbers they ask for almost invariably are the figures needed to calculate the housing ratio and the debt ratio.

3. c to $d, c : d, \frac{c}{d}$ **4.** a statement of equality between two ratios

5. need a given ratio and one of the two parts of a second ratio

6. corresponding angles equal and corresponding sides in proportion

Exercise Set 2.6

Concept/Writing Exercises

1. What is a ratio?   a quotient of two quantities

2. In the ratio $a : b$, what are the a and b called?   terms

3. List three ways to write the ratio of c to d.

4. What is a proportion?

5. As you have learned, proportions can be used to solve a wide variety of problems. What information is needed for a problem to be set up and solved using a proportion?

6. What are similar figures?

7. Must similar figures be the same size? Explain.   no

8. Must similar figures have the same shape? Explain.   yes

In Exercises 9–12, is the proportion set up correctly? Explain.

9. $\dfrac{\text{gal}}{\text{min}} = \dfrac{\text{gal}}{\text{min}}$ yes

10. $\dfrac{\text{sq ft}}{\text{lb}} = \dfrac{\text{sq ft}}{\text{lb}}$ yes

11. $\dfrac{\text{ft}}{\text{sec}} = \dfrac{\text{sec}}{\text{ft}}$ no

12. $\dfrac{\text{tax}}{\text{cost}} = \dfrac{\text{cost}}{\text{tax}}$ no

Practice the Skills

The results of a mathematics examination are 6 As, 4 Bs, 9 Cs, 3 Ds, and 2 Fs. Write the following ratios in lowest terms.

13. A's to C's $2:3$

14. B's to total grades $1:6$

15. D's to A's $1:2$

16. Grades better than C to total grades $5:12$

17. Total grades to D's $8:1$

18. Grades better than C to grades less than C $2:1$

Determine the following ratios. Write each ratio in lowest terms.

19. 7 gallons to 4 gallons $7:4$

20. 50 dollars to 60 dollars $5:6$

21. 5 ounces to 15 ounces $1:3$

22. 18 minutes to 24 minutes $3:4$

23. 3 hours to 30 minutes $6:1$

24. 6 feet to 4 yards $1:2$

25. 26 ounces to 4 pounds $13:32$

26. 7 dimes to 12 nickels $7:6$

Find each gear ratio in lowest terms. (See Example 4.)

27. Driving gear, 40 teeth; driven gear, 5 teeth $8:1$

28. Driving gear, 30 teeth; driven gear, 8 teeth
$15:4$ or $3.75:1$

31. a) $1.13:0.38$ **b)** $\approx2.97:1$

32. a) $281:249$ **b)** $\approx1.13:1$

In Exercises 29–32, **a)** *Determine the indicated ratio, and* **b)** *write the ratio as some quantity to 1.*

29. *Summer Olympics* At the 2000 summer Olympics in Sydney, Australia, 199 nations were represented. At the summer Olympics in 1984 in Los Angeles, California, 140 nations were represented. What is the ratio of the number of nations represented in the 2000 summer Olympics to the number of nations represented in 1984 summer Olympics? **a)** $199:140$ **b)** $\approx1.42:1$

30. *Mail Letter* In January 2002, the cost to mail a one-ounce letter was 34 cents and the cost to mail a two-ounce letter was 57 cents. What is the ratio of the cost to mail a one-ounce letter to the cost to mail a two-ounce letter? **a)** $34:57$ **b)** $\approx0.60:1$

31. *Armed Forces* In January, 2000, there were about 1.13 million U.S. armed forces worldwide of which about 0.38 million were in the army. What is the ratio of the total armed forces to those in the army?

32. *Population* The United States population in 1990 was about 249 million, and the population in 2000 was about 281 million. What is the ratio of the U.S. population in 2000 to the U.S. population in 1990?

Exercises 33–36 show graphs. For each exercise, find the indicated ratio.

33. *Farm Size*

  a) Determine the ratio of the average size farm in 2000 to the average size farm in 1940. $434:174$ or $217:87$

  b) Determine the ratio of the average size farm in 1970 to the average size farm in 2000. $374:434$ or $187:217$

34. *Number of Farms*

  a) Estimate the ratio of farms in the U.S. in 1940 to those in the U.S. in 2000. $6.3:2.2$

  b) Estimate the ratio of farms in the U.S. in 2000 to those in the U.S. in 1970. $2.2:3.0$ or $1.1:1.5$

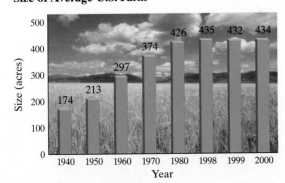

Size of Average U.S. Farm

Source: U.S. Department of Agriculture

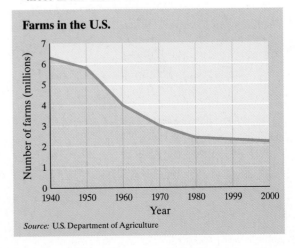

Farms in the U.S.

Source: U.S. Department of Agriculture

35. *Favorite Doughnut*

a) Determine the ratio of people whose favorite doughnut is glazed to people whose favorite doughnut is filled. 40:32 or 5:4

b) Determine the ratio of people whose favorite doughnut is frosted to people whose favorite doughnut is plain. 15:11

Favorite Doughnut Flavors

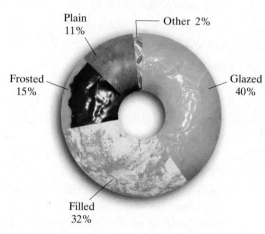

Plain 11%
Other 2%
Frosted 15%
Glazed 40%
Filled 32%

Source: The Heller Research Group

36. *Spam*

a) Determine the ratio of junk email for Financial to the junk email for Health. 32:3

b) Determine the ratio of junk mail for Products to all junk email. 38:100 or 19:50

Junk E-mail (or Spam) Feb. 12–18, 2001

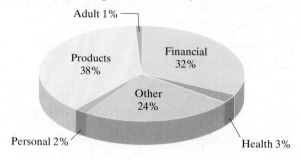

Adult 1%
Products 38%
Financial 32%
Other 24%
Personal 2%
Health 3%

Source: Brightmail Inc.

Solve each proportion for the variable by cross-multiplying.

37. $\dfrac{3}{x} = \dfrac{5}{20}$ 12

38. $\dfrac{x}{8} = \dfrac{24}{48}$ 4

39. $\dfrac{5}{3} = \dfrac{75}{a}$ 45

40. $\dfrac{x}{3} = \dfrac{90}{30}$ 9

41. $\dfrac{90}{x} = \dfrac{-9}{10}$ −100

42. $\dfrac{8}{12} = \dfrac{n}{6}$ 4

43. $\dfrac{15}{45} = \dfrac{x}{-6}$ −2

44. $\dfrac{y}{6} = \dfrac{7}{42}$ 1

45. $\dfrac{3}{z} = \dfrac{-1.5}{27}$ −54

46. $\dfrac{3}{12} = \dfrac{-1.4}{z}$ −5.6

47. $\dfrac{9}{12} = \dfrac{x}{8}$ 6

48. $\dfrac{2}{20} = \dfrac{x}{200}$ 20

The following figures are similar. For each pair, find the length of the side indicated by x.

49.

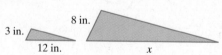

3 in.
8 in.
12 in.
x
32 in.

50.

2 ft
1.8 ft
0.8 ft
x
0.72 ft

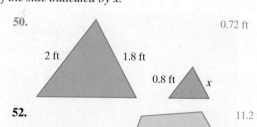

51.

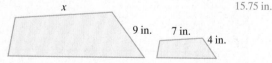

x
9 in.
7 in.
4 in.
15.75 in.

52.

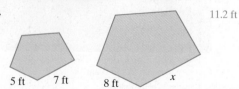

5 ft
7 ft
8 ft
x
11.2 ft

53.

19.5 in.

16 in. 26 in.

12 in.

x

54.

5.6 in.

14 in.

20 in.

x

8 in.

Problem Solving

57. 361.1 mi

In Exercises 55–74, write a proportion that can be used to solve the problem. Then solve the equation to obtain the answer.

55. *Washing Clothes* A bottle of liquid Tide contains 100 fluid ounces. If one wash load requires 4 ounces of the detergent, how many washes can be done with one bottle of Tide?   25 loads

56. *Laying Cable* A telephone cable crew is laying cable at a rate of 42 feet an hour. How long will it take them to lay 252 feet of cable?   6 hr

57. *Car Mileage* A 2002 Ford Mustang with the 4.6 liter engine is rated to get 23 miles per gallon (highway driving). How far can it travel on a full tank of gas, 15.7 gallons?

58. *Painting the House* A gallon of paint covers 825 square feet. How much paint is needed to cover a house with a surface area of 5775 square feet?   7 gal

59. *Model Train* A model train set is in a ratio of 1:20. That is, one foot of the model represents 20 feet of the original train. If a caboose is 30 feet long, how long should the model be?   1.5 ft

60. *Spreading Fertilizer* If a 40-pound bag of fertilizer covers 5000 square feet, how many pounds of fertilizer are needed to cover an area of 26,000 square feet?   208 lb

61. *Insecticide application* The instructions on a bottle of liquid insecticide say "use 3 teaspoons of insecticide per gallon of water." If your sprayer has an 8-gallon capacity, how much insecticide should be used to fill the sprayer?   24 tsp

62. *Property Tax* The property tax in the city of Hendersonville, North Carolina, is $9.475 per $1000 of assessed value. If the Estever's house is assessed at $145,000, how much property tax will they owe? $1373.88

63. *Blue Heron* The photograph shows a blue heron. If the blue heron, that measures 3.5 inches in the photo is actually 3.75 feet tall, approximately how long is its beak if it measures 0.4 inches in the photo?   ≈0.43 ft

64. *Onion Soup* A recipe for 6 servings of French onion soup requires $1\frac{1}{2}$ cups of thinly sliced onions. If the recipe were to be made for 15 servings, how many cups of onions would be needed?   3.75 cups

65. *Maps* On a map, 0.5 inch represents 22 miles. What will be the length on a map that corresponds to a distance of 55 miles?   1.25 in.

66. *Steven King Novel* Karen Estes is currently reading a Steven King novel. If she reads 72 pages in 1.3 hours, how long will it take her to read the entire 656 page novel?   ≈11.84 hours

67. *Wall Street Bull* Suppose the famous bull by the New York Stock Exchange (see photo on page 156) is a replica of a real bull in a ratio of 2.95 to 1. That is, the metal bull is 2.95 times greater than the regular bull. If the length of the Wall Street bull is 28 feet long, approximately how long is the bull that served as its model?   ≈9.49 ft

See Exercise 67.

68. Flood When they returned home from vacation, the Duncans had a foot of water in their basement. They contacted their fire department, which sent equipment to pump out the water. After the pump had been on for 30 minutes, 3 inches of water had been removed. How long, from the time they started pumping, will it take to remove all the water from the basement? 2 hr

69. Drug Dosage A nurse must administer 220 micrograms of atropine sulfate. The drug is available in solution form. The concentration of the atropine sulfate solution is 400 micrograms per milliliter. How many milliliters should be given? 0.55 ml

70. Dosage by body Surface A doctor asks a nurse to administer 0.7 gram of meprobamate per square meter of body surface. The patient's body surface is 0.6 square meter. How much meprobamate should be given? 0.42 g

71. Swimming Laps Jason Abbott swims 3 laps in 2.3 minutes. Approximately how long will it take him to swim 30 laps if he continues to swim at the same rate? 23 min

72. Reading a Novel Mary read 40 pages of a novel in 30 minutes. If she continues reading at the same rate, how long will it take her to read the entire 760-page book? 570 min or 9 hr 30 min

73. Prader-Willi Syndrome It is estimated that each year in the United States about 1 in every 12,000 (1 : 12,000) people is born with a genetic disorder called Prader-Willi syndrome. If there were approximately 4,063,000 births in the United States in 2000, approximately how may children were born with Prader-Willi syndrome?

74. U.S. Population In the United States in 2000, the birth rate was 14.5 per one thousand people. In the United States in 2000, there were approximately 4,063,000 births. What was the U.S. population in 2000? ≈280,207,000

73. ≈339 children

In Exercises 75–86, use a proportion to make the conversion. Round your answers to two decimal places.

75. Convert 78 inches to feet. 6.5 ft

76. Convert 22,704 feet to miles (5280 feet = 1 mile). 4.3 mi

77. Convert 26.1 square feet to square yards (9 square feet = 1 square yard). 2.9 sq yd

78. Convert 146.4 ounces to pounds. 9.15 lb

79. Newborn One inch equals 2.54 centimeters. Find the length of a newborn, in inches, if it measures 50.8 centimeters. 20 in.

80. Distance One mile equals approximately 1.6 kilometers. Find the distance, in kilometers, from San Diego, California, to San Francisco, California—a distance of 520 miles. 832 km

81. Home Run Record Barry Bonds, who plays for the San Francisco Giant's baseball team, holds the record for the most home runs, 73, in a 162 game season. In the first 50 games of a season, how many home runs would a player need to hit to be on schedule to break Bond's record? ≈23

82. Topsoil A 40 pound bag of topsoil covers 12 square feet (one inch deep). How many pounds of the top soil are needed to cover 350 square feet (one inch deep)?

83. Gold If gold is selling for $408 per 480 grains (a troy ounce), what is the cost per grain? $0.85

84. Interest on Savings Jim Chao invests a certain amount of money in a savings account. If he earned $110.52 in 180 days, how much interest would he earn in 500 days assuming the interest rate stays the same? $307

85. Statistics In a statistics course, we find that for one particular set of scores 15 points equals 3.75 standard deviations. How many points equals 1 standard deviation?

86. Currency Exchange When Mike Weatherbee visited the United States from Canada, he exchanged $13.50 Canadian for $10 U.S.. If he exchanges his remaining

82. ≈1166.67 lb   85. 4 points

87. Yes, her ratio is 2.12 : 1   89. must increase   90. must decrease

$600 Canadian for U.S. dollars, how much more in dollars will he receive? $444.44

The Peace Bridge connecting the United States and Canada

87. *Cholesterol* Mrs. Ruff's low-density cholesterol level is 127 milligrams per deciliter (mg/dL). Her high-density cholesterol level is 60 mg/dL. Is Mrs. Ruff's ratio of low-to high-density cholesterol level less than or equal to the 4 : 1 recommended level? (See Example 2.)

88. *Cholesterol*

a) Another ratio used by some doctors when measuring cholesterol level is the ratio of total cholesterol to high-density cholesterol.* Is this ratio increased or decreased if the total cholesterol remains the same but the high-density level is increased? Explain. decreased

b) Doctors recommend that the ratio of total cholesterol to high-density cholesterol be less than or equal to 4.5 : 1. If Mike's total cholesterol is 220 mg/dL and his high-density cholesterol is 50 mg/dL, is his ratio less than or equal to 4.5 : 1? Explain. Yes, ratio is 4.4 : 1

89. For the proportion $\frac{a}{b} = \frac{c}{d}$, if a increases while b and d stay the same, what must happen to c? Explain.

90. For the proportion $\frac{a}{b} = \frac{c}{d}$, if a and c remain the same while d decreases, what must happen to b? Explain.

92. $\frac{1}{3}$ cup flour, $\frac{2}{3}$ tsp nutmeg, $\frac{2}{3}$ tsp cinnamon; $\frac{1}{6}$ tsp salt, $1\frac{1}{3}$ tbsp butter, 1 cup sugar   93. 0.625 cc

Challenge Problems

91. *Wear on Tires* A new Goodyear tire has a tread of about 0.34 inches. After 5000 miles the tread is about 0.31 inches. If the legal minimum amount of tread for a tire is 0.06 inches, how many more miles will the tires last? (Assume no problems with the car or tires and that the tires wear at an even rate.) ≈41,667 mi

92. *Apple Pie* The recipe for the filling for an apple pie calls for

12 cups sliced apples	$\frac{1}{4}$ teaspoon salt
$\frac{1}{2}$ cup flour	2 tablespoons butter
1 teaspoon nutmeg	or margarine
1 teaspoon cinnamon	$1\frac{1}{2}$ cups sugar

Determine the amount of each of the other ingredients that should be used if only 8 cups of apples are available.

93. *Insulin* Insulin comes in 10-cubic-centimeter (cc) vials labeled in the number of units of insulin per cubic centimeter. Thus a vial labeled U40 means there are 40 units of insulin per cubic centimeter of fluid. If a patient needs 25 units of insulin, how many cubic centimeters of fluid should be drawn up into a syringe from the U40 vial?

94. *Slope* An important concept, which we will discuss in Chapter 4, is *slope*. The slope of a line may be defined as a *ratio* of the vertical change to the horizontal change between any two points on a line. In the following figures, the vertical change is found using the red dashed lines, and the horizontal change is found using the green dashed lines.

a) Determine the slope of Figure 2.4a. 2
b) Determine the slope of Figure 2.4b. $\frac{2}{3}$
c) Determine the slope of Figure 2.4c. −1

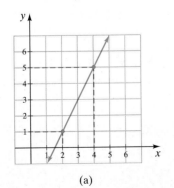

(a)

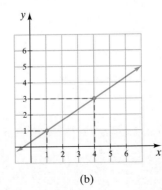

(b)

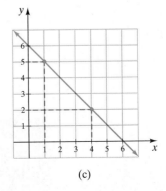

(c)

FIGURE 2.4

*Total cholesterol includes both low- and high-density cholesterol, plus other types of cholesterol.

 Group Activity

Discuss and answer Exercises 95 and 96 as a group.

95. a) Each group member: Find the ratio of your height to your arm span (finger tips to finger tips) when your arms are extended horizontally outward. You will need help from your group in getting these measurements.

b) If a box were to be drawn about your body with your arms extended, would the box be a square or a rectangle? If a rectangle, would the longer length be your arm span or your height measurement? Explain.

c) Compare these results with other members of your group.

d) What one ratio would you use to report the height to arm span for your group as a whole? Explain.

96. A special ratio in mathematics is called the *golden ratio*. Do research in a history of mathematics book or on the Internet, and as a group write a paper that explains what the golden ratio is and why it is important.

Cumulative Review Exercises

[1.11] Name each illustrated property.

97. $x + 3 = 3 + x$ commutative, addition

98. $3(xy) = (3x)y$ associative, multiplication

99. $2(x - 3) = 2x - 6$ distributive property

[2.5] **100.** Solve $-(2x + 6) = 2(3x - 6)$ $\frac{3}{4}$

101. Solve $3(4x - 3) = 6(2x + 1) - 15$ all real numbers

2.7 INEQUALITIES IN ONE VARIABLE

 SSM
 Study Guide
 CD/Video

1 Solve linear inequalities.

2 Solve linear inequalities that have all real numbers as their solution, or have no solution.

 MathPro 4/5
 PH Math Tutor Center
 prenhall.com/Angel

1 Solve Linear Inequalities

The is-greater-than symbol, $>$, and is-less-than symbol, $<$, were introduced in Section 1.5. The symbol $\geq$ means is greater than or equal to and $\leq$ means is less than or equal to. A mathematical statement containing one or more of these symbols is called an **inequality**. The direction of the symbol is sometimes called the **sense** or **order of the inequality**.

Examples of Inequalities in One Variable
$$x + 3 < 5 \qquad x + 4 \geq 2x - 6 \qquad 4 > -x + 3$$

To solve an inequality, we must get the variable by itself on one side of the inequality symbol. To do this, we make use of properties very similar to those used to solve equations. Here are four properties used to solve inequalities. Later in this section, we will introduce two additional properties.

Properties Used to Solve Inequalities

For real numbers, a, b, and c:

1. If $a > b$, then $a + c > b + c$.

2. If $a > b$, then $a - c > b - c$.

(continued on the next page)

> **3.** If $a > b$ and $c > 0$, then $ac > bc$.
>
> **4.** If $a > b$ and $c > 0$, then $\dfrac{a}{c} > \dfrac{b}{c}$.

Property 1 says the same number may be added to both sides of an inequality. Property 2 says the same number may be subtracted from both sides of an inequality. Property 3 says the same *positive* number may be used to multiply both sides of an inequality. Property 4 says the same *positive* number may be used to divide both sides of an inequality. When any of these four properties is used, *the direction of the inequality symbol does not change.*

EXAMPLE 1   Solve the inequality $x - 4 > 7$, and graph the solution on a number line.

Solution   To solve this inequality, we need to isolate the variable, x. Therefore, we must eliminate the -4 from the left side of the inequality. To do this, we add 4 to both sides of the inequality.

$$x - 4 > 7$$
$$x - 4 + 4 > 7 + 4 \quad \text{Add 4 to both sides.}$$
$$x > 11$$

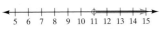

FIGURE 2.5

The solution is all real numbers greater than 11. We can illustrate the solution on a number line by placing an open circle at 11 on a number line and drawing an arrow to the right (Fig. 2.5).

The open circle at the 11 indicates that the 11 is *not* part of the solution. The arrow going to the right indicates that all the values greater than 11 are solutions to the inequality.

EXAMPLE 2   Solve the inequality $2x + 6 \leq -2$, and graph the solution on a number line.

Solution   To isolate the variable, we must eliminate the $+6$ from the left side of the inequality. We do this by subtracting 6 from both sides of the inequality.

$$2x + 6 \leq -2$$
$$2x + 6 - 6 \leq -2 - 6 \quad \text{Subtract 6 from both sides.}$$
$$2x \leq -8$$
$$\frac{2x}{2} \leq \frac{-8}{2} \quad \text{Divide both sides by 2.}$$
$$x \leq -4$$

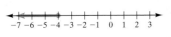

FIGURE 2.6

The solution is all real numbers less than or equal to -4. We can illustrate the solution on a number line by placing a closed, or darkened, circle at -4 and drawing an arrow to the left (Fig. 2.6).

The darkened circle at -4 indicates that -4 *is* a part of the solution. The arrow going to the left indicates that all the values less than -4 are also solutions to the inequality.

NOW TRY EXERCISE 21

Notice in properties 3 and 4 that we specified that $c > 0$. What happens when an inequality is multiplied or divided by a negative number? Examples 3 and 4 illustrate that **when an inequality is multiplied or divided by a negative number, the direction of the inequality symbol changes.**

EXAMPLE 3   Multiply both sides of the inequality $8 > -4$ by -2.

Solution

$$8 > -4$$
$$-2(8) < -2(-4) \qquad \textit{Change the direction of the inequality symbol.}$$
$$-16 < 8$$

EXAMPLE 4   Divide both sides of the inequality $8 > -4$ by -2.

Solution

$$8 > -4$$
$$\frac{8}{-2} < \frac{-4}{-2} \qquad \textit{Change the direction of the inequality symbol.}$$
$$-4 < 2$$

TEACHING TIP
Stress that the change of direction of the inequality symbol occurs in the step where you multiply by a negative number (not the step where you simplify the multiplication).

Now we state two additional properties, used when an inequality is multiplied or divided by a negative number.

Additional Properties Used to Solve Inequalities

5. If $a > b$ **and** $c < 0$, then $ac < bc$.

6. If $a > b$ **and** $c < 0$, then $\dfrac{a}{c} < \dfrac{b}{c}$.

EXAMPLE 5   Solve the inequality $-2x > 6$, and graph the solution on a number line.

Solution   To isolate the variable, we must eliminate the -2 on the left side of the inequality. To do this, we can divide both sides of the inequality by -2. When we do this, however, we must remember to *change the direction* of the inequality symbol.

$$-2x > 6$$
$$\frac{-2x}{-2} < \frac{6}{-2} \qquad \begin{array}{l}\textit{Divide both sides by } -2\textit{, and change}\\ \textit{the direction of the inequality symbol.}\end{array}$$
$$x < -3$$

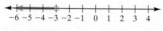

FIGURE 2.7

The solution is all real numbers less than -3. The solution is graphed on a number line in Figure 2.7.

EXAMPLE 6   Solve the inequality $4 \geq -5 - x$, and graph the solution on a number line. We will illustrate two methods that can be used to solve this inequality.

Solution   Method 1:
$$4 \geq -5 - x$$
$$4 + 5 \geq -5 + 5 - x \qquad \textit{Add 5 to both sides.}$$
$$9 \geq -x$$
$$-1(9) \leq -1(-x) \qquad \begin{array}{l}\textit{Multiply both sides by } -1\textit{, and change}\\ \textit{the direction of the inequality symbol.}\end{array}$$
$$-9 \leq x$$

TEACHING TIP
Suggest that students adopt the style of always giving their solution to an inequality with the variable on the left side.

The inequality $-9 \leq x$ can also be written $x \geq -9$.

Method 2:
$$4 \geq -5 - x$$
$$4 + x \geq -5 - x + x \qquad \textit{Add } x \textit{ to both sides.}$$
$$4 + x \geq -5$$

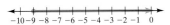

FIGURE 2.8

NOW TRY EXERCISE 25

$$4 - 4 + x \geq -5 - 4 \qquad \textit{Subtract 4 from both sides.}$$
$$x \geq -9$$

The solution is graphed on a number line in Figure 2.8. Other methods could also be used to solve this problem.

Notice in Example 6, Method 1, we wrote $-9 \leq x$ as $x \geq -9$. Although the solution $-9 \leq x$ is correct, it is customary to write the solution to an inequality with the variable on the left. One reason we write the variable on the left is that it often makes it easier to graph the solution on the number line. How would you graph $-3 > x$? How would you graph $-5 \leq x$? If you rewrite these inequalities with the variable on the left side, the answer becomes clearer.

$$-3 > x \quad \text{means} \quad x < -3$$
$$-5 \leq x \quad \text{means} \quad x \geq -5$$

Notice that you can change an answer from an is-greater-than statement to an is-less-than statement or from an is-less-than statement to an is-greater-than statement. When you change the answer from one form to the other, remember that the inequality symbol must point to the letter or number to which it was pointing originally.

HELPFUL HINT

$a > x$ means $x < a$ *Note that both inequality symbols point to x.*
$a < x$ means $x > a$ *Note that both inequality symbols point to a.*

EXAMPLES

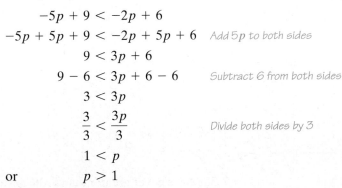

$-3 > x$ means $x < -3$

$-5 \leq x$ means $x \geq -5$

TEACHING TIP
If you prefer to use brackets and parentheses in place of open and closed circles, you may wish to discuss this with your students now.

Now let's solve inequalities where the variable appears on both sides of the inequality symbol. To solve these inequalities, we use the same basic procedure that we used to solve equations. However, we must remember that whenever we multiply or divide both sides of an inequality by a negative number, we must change the direction of the inequality symbol.

EXAMPLE 7   Solve the inequality $-5p + 9 < -2p + 6$, and graph the solution on a number line.

Solution   This equation uses the variable p. The variable used does not affect the procedure for solving the inequality.

TEACHING TIP
After working Example 7, have your students rework the example starting by adding $2p$ to both sides of the equation. Ask which method they prefer and why.

$$-5p + 9 < -2p + 6$$
$$-5p + 5p + 9 < -2p + 5p + 6 \qquad \textit{Add 5p to both sides}$$
$$9 < 3p + 6$$
$$9 - 6 < 3p + 6 - 6 \qquad \textit{Subtract 6 from both sides}$$
$$3 < 3p$$
$$\frac{3}{3} < \frac{3p}{3} \qquad \textit{Divide both sides by 3}$$
$$1 < p$$

or $p > 1$

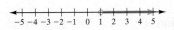

FIGURE 2.9

The solution is graphed in Figure 2.9.

EXAMPLE 8   Solve the inequality $\frac{1}{2}x + 3 \le -\frac{1}{3}x + 7$, and graph the solution on a number line.

Solution   Since the inequality contains fractions, we begin by multiplying both sides of the inequality by the LCD, 6, to eliminate the fractions.

$$\frac{1}{2}x + 3 \le -\frac{1}{3}x + 7$$

$$6\left(\frac{1}{2}x + 3\right) \le 6\left(-\frac{1}{3}x + 7\right) \quad \text{\textit{Multiply both sides by the LCD, 6}}$$

$$3x + 18 \le -2x + 42 \quad \text{\textit{Distributive property}}$$

$$5x + 18 \le 42 \quad \text{\textit{2x was added to both sides}}$$

$$5x \le 24 \quad \text{\textit{18 was subtracted from both sides}}$$

$$x \le \frac{24}{5} \quad \text{\textit{Both sides were divided by 5}}$$

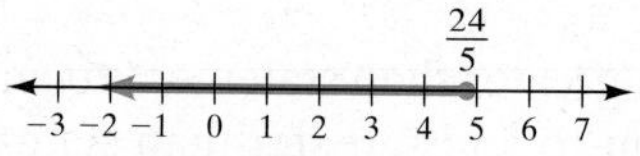

FIGURE 2.10

NOW TRY EXERCISE 53   The solution is graphed in Figure 2.10.

2 Solve Linear Inequalities That Have All Real Numbers as Their Solution, or Have No Solution

In Examples 9 and 10 we illustrate two special types of inequalities. Example 9 is an inequality that is true for all real numbers, and Example 10 is an inequality that is never true for any real number.

EXAMPLE 9   Solve the inequality $2(x + 3) \le 5x - 3x + 8$, and graph the solution on a number line.

Solution
$$2(x + 3) \le 5x - 3x + 8$$
$$2x + 6 \le 5x - 3x + 8 \quad \text{\textit{Distributive property was used.}}$$
$$2x + 6 \le 2x + 8 \quad \text{\textit{Like terms were combined.}}$$
$$2x - 2x + 6 \le 2x - 2x + 8 \quad \text{\textit{Subtract 2x from both sides.}}$$
$$6 \le 8$$

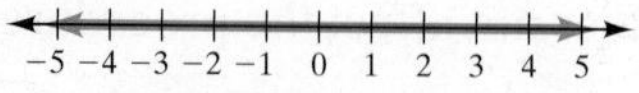

FIGURE 2.11   Since 6 is always less than or equal to 8, the solution is **all real numbers** (Fig. 2.11).

EXAMPLE 10   Solve the inequality $4(x + 1) > x + 5 + 3x$, and graph the solution on a number line.

Solution
$$4(x + 1) > x + 5 + 3x$$
$$4x + 4 > x + 5 + 3x \quad \text{\textit{Distributive property was used.}}$$
$$4x + 4 > 4x + 5 \quad \text{\textit{Like terms were combined.}}$$
$$4x - 4x + 4 > 4x - 4x + 5 \quad \text{\textit{Subtract 4x from both sides.}}$$
$$4 > 5$$

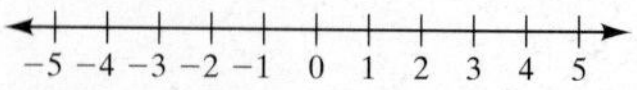

FIGURE 2.12

NOW TRY EXERCISE 43   Since 4 is never greater than 5, the answer is **no solution** (Fig. 2.12). There is no real number that makes the statement true.

Exercise Set 2.7

Concept/Writing Exercises

1. List the four inequality symbols given in this section and write how each is read. $>, \geq, <, \leq$

2. Explain the difference between $>$ and $\geq$.

3. Are the following statements true or false? Explain.

   a) $3 > 3$ false   **b)** $3 \geq 3$ true

4. If $a < b$ is a true statement, must $b > a$ also be a true statement? Explain. yes

5. When solving an inequality, under what conditions will it be necessary to change the direction of the inequality symbol?

6. List the six rules used to solve inequalities.

7. When solving an inequality, if you obtain the result $3 < 5$, what is the solution? all real numbers

8. When solving an inequality, if you obtain the result $4 \geq 2$, what is the solution? all real numbers

9. When solving an inequality, if you obtain the result $5 < 2$, what is the solution? no solution

10. When solving an inequality, if you obtain the result $-4 \geq -2$, what is the solution? no solution

2. $\geq$ includes "is equal to" **5.** when multiplying or dividing by a negative number **6.** If $a > b$ then $a + c > b + c$ and $a - c > b - c$. If $c > 0$ then $ac > bc$ and $a/c > b/c$. If $a > b$ and $c < 0$ then $ac < bc$ and $a/c < b/c$.

Practice the Skills

Solve each inequality, and graph the solution on a number line.

11. $x + 2 > 6$ $x > 4$,

12. $x - 5 > -1$ $x > 4$,

13. $x + 9 \geq 6$ $x \geq -3$,

14. $4 - x \geq 3$ $x \leq 1$,

15. $-x + 3 < 8$ $x > -5$,

16. $7 < 3 + w$ $w > 4$,

17. $8 \leq 2 - r$ $r \leq -6$,

18. $2x < 4$ $x < 2$,

19. $-2x < 3$ $x > -\frac{3}{2}$,

20. $-12 \geq -3b$ $b \geq 4$,

21. $2x + 3 \leq 5$ $x \leq 1$,

22. $-4x - 3 > 5$ $x < -2$,

23. $6n - 12 < -12$ $n < 0$,

24. $7x - 4 \leq 9$ $x \leq \frac{13}{7}$,

25. $4 - 6x > -5$ $x < \frac{3}{2}$,

26. $8 < 4 - 2x$ $x < -2$,

27. $15 > -9x + 50$ $x > \frac{35}{9}$,

28. $3x - 4 < 5$ $x < 3$,

29. $6 < 3x + 10$ $x > -\frac{4}{3}$,

30. $-3x > 2x + 10$ $x < -2$,

31. $6s + 2 \leq 6s - 9$ no solution,

32. $-2x - 4 \leq -5x + 12$ $x \leq \frac{16}{3}$,

33. $x - 4 \leq 3x + 8$ $x \geq -6$,

34. $-4n - 6 > 4n - 20$ $n < \frac{7}{4}$,

35. $-x + 4 < -3x + 6$ $x < 1$,

36. $2(x - 3) < 4x + 10$ $x > -8$,

37. $6(2m - 4) \geq 2(6m - 12)$ all real numbers,

38. $-2(w + 3) \leq 4w + 5$ $w \geq -\frac{11}{6}$,

39. $x + 3 < x + 4$ all real numbers,

40. $x + 5 \geq x - 2$ all real numbers,

41. $6(3 - x) < 2x + 12$ $x > \frac{3}{4}$,

42. $2(3 - x) + 4x < -6$ $x < -6$,

43. $4x - 4 < 4(x - 5)$ no solution,

44. $-2(-5 - x) > 3(x + 2) + 4 - x$

45. $5(2x + 3) \geq 6 + (x + 2) - 2x$ $x \geq -\frac{7}{11}$,

46. $-3(-2x + 12) < -4(x + 2) - 6$

47. $1.2x + 3.1 < 3.5x - 3.8$ $x > 3$,

48. $-5.3r - 6.7 \geq 2.3 - 6.5r$ $r \geq 7.5$,

49. $1.2(m - 3) \geq 4.6(2 - m) + 1.7$ $m \geq 2.5$,

50. $-4.6(4 - x) < 2.4(x - 3) - 0.2$ $x < 5$,

51. $\frac{x}{3} \leq \frac{x}{4} + 4$ $x \leq 48$,

52. $\frac{x}{5} - 2 \leq \frac{x}{6}$ $x \leq 60$,

53. $t + \frac{1}{6} > \frac{2}{3}t$ $t > -\frac{1}{2}$,

54. $\frac{3}{5}r - 9 < \frac{3}{8}r$ $r < 40$,

55. $\frac{1}{8}(4 - r) \leq \frac{1}{4}$ $r \geq 2$,

56. $5 - \frac{1}{6}x < \frac{2}{3}x$ $x > 6$,

57. $\frac{2}{3}(t + 2) \leq \frac{1}{4}(2t - 6)$ $t \leq -17$,

58. $\frac{3}{4}(n - 4) \geq \frac{2}{3}(n - 4)$ $n \geq 4$,

44. no solution,   **46.** $x < \frac{11}{5}$,

Problem Solving

59. *Chicago Temperatures* The following chart shows the average high and low monthly temperatures in Chicago over a 126-year period (*Source:* Richard Koeneman/WGN-TV meteorologist). Notice that the months are not listed in order.

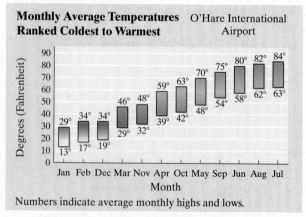

Numbers indicate average monthly highs and lows.

a) In what months was the average high temperature >65°F? May, Sept., June, Aug., July

b) In what months was the average high temperature ≤59°F? Jan., Feb., Dec., Mar., Nov., Apr.

c) In what months was the average low temperature <29°F? Jan., Feb., Dec.

d) In what months was the average low temperature ≥58°F? June, Aug., July

60. *Popular Parks* The following graph indicates the 5 most visited sites in the United States National Park system during the year 2000.

Most Visited Parks in the U.S.

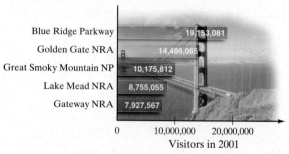

Source: National Parks Service

a) In which sites was the number of visits ≥ 12,000,000? Blue Ridge Pkwy, Golden Gate

b) In which sites was the number of visits ≥ 10,000,000 but ≤ 15,000,000? Golden Gate, Smoky Mt.

c) In which sites was the number of visits > 8,000,000 but ≤ 10,175,812? Smoky Mt., Lake Mead

d) In which sites was the number of visits ≥ 14,486,065 and ≤ 14,486,065? Golden Gate

61. The inequality symbols discussed so far are <, ≤, >, and ≥. Can you name an inequality symbol that we have not mentioned in this section? ≠

62. Reproduced below is a portion of the Florida Individual and Joint Intangible Tax Return for 2001.

TAX CALCULATION WORKSHEET

Instructions: Determine which column applies based on filing status. Complete *only* the applicable column.	Individual	Joint
6A. Enter Total Intangible Assets from Schedule A, Line 5	$	$
6B. Multiply by Tax Rate	× 0.001	× 0.001
6C. Gross Tax	$	$
6D. Subtract Personal Exemption	− $20.00	− $40.00
6E. Enter Total Tax Due Carry Amount to Schedule A, Line 6	$	$

*(Complete only **one** column below)*

Use the Tax Calculation Worksheet to determine your total tax due (line 6E) if your total taxable assets from Schedule A, line 5 are as follows.

a) $30,000 and your filing status is individual $10
b) $175,000 and your filing status is individual $155
c) $200,000 and your filing status is joint $160
d) $300,000 and your filing status is joint $260

63. Consider the inequality $xy > 6$, where x and y represent real numbers. Explain why we *cannot* do the following step:

$$\frac{xy}{y} > \frac{6}{y} \quad \text{Divide both sides by y.}$$

Don't know that y is positive. If negative, must reverse sign of inequality.

Challenge Problems

64. Solve the following inequality.

$$3(2 - x) - 4(2x - 3) \le 6 + 2x - 4x \qquad x \ge \frac{4}{3}$$

65. Solve the following inequality.

$$6x - 6 > -4(x + 3) + 5(x + 6) - x \qquad x > 4$$

Cumulative Review Exercises

[1.9] **66.** Evaluate $-x^2$ for $x = 3$. -9
67. Evaluate $-x^2$ for $x = -5$. -25
[2.5] **68.** Solve $4 - 3(2x - 4) = 5 - (x + 3)$. $\frac{14}{5}$

[2.6] **69.** *Electric Bill* The Milford Electric Company charges $0.174 per kilowatt-hour of electricity. The Vega's monthly electric bill was $87 for the month of July. How many kilowatt-hours of electricity did the Vega's use in July? 500 kWh

CHAPTER SUMMARY
Key Words and Phrases

2.1
Combine like terms
Constant
Expression
Factor
Like terms
Numerical coefficient
Simplify an expression
Term
Variable

2.2
Addition property of equality
Check an equation
Equation
Equivalent equations
Isolate the variable
Linear equation
Solution to an equation
Solve an equation

2.3
Multiplication property of equality
Reciprocal

2.5
Conditional equations
Contradiction
Identity

2.6
Cross-multiplication

Extremes
Means
Proportion
Ratio
Similar figures
Slope
Terms of a ratio

2.7
Inequality
Sense (or order) of the inequality

(continued on the next page)

IMPORTANT FACTS

Distributive property	$a(b + c) = ab + ac$
Addition property	If $a = b$, then $a + c = b + c$.
Multiplication property	If $a = b$, then $a \cdot c = b \cdot c, c \neq 0$.
Cross-multiplication	If $\dfrac{a}{b} = \dfrac{c}{d}$, then $ad = bc$.

Properties used to solve inequalities

1. If $a > b$, then $a + c > b + c$.
2. If $a > b$, then $a - c > b - c$.

3. If $a > b$ and $c > 0$, then $ac > bc$.
4. If $a > b$ and $c > 0$, then $\dfrac{a}{c} > \dfrac{b}{c}$.

5. If $a > b$ and $c < 0$, then $ac < bc$.
6. If $a > b$ and $c < 0$, then $\dfrac{a}{c} < \dfrac{b}{c}$.

Chapter Review Exercises

[2.1] Use the distributive property to simplify.

1. $3(x + 4)$ $3x + 12$
2. $3(x - 2)$ $3x - 6$
3. $-2(x + 4)$ $-2x - 8$
4. $-(x + 2)$ $-x - 2$
5. $-(m + 3)$ $-m - 3$
6. $-4(4 - x)$ $-16 + 4x$
7. $5(5 - p)$ $25 - 5p$
8. $6(4x - 5)$ $24x - 30$
9. $-5(5x - 5)$ $-25x + 25$
10. $4(-x + 3)$ $-4x + 12$
11. $\frac{1}{2}(2x + 4)$ $x + 2$
12. $-(3 + 2y)$ $-3 - 2y$
13. $-(x + 2y - z)$ $-x - 2y + z$
14. $-3(2a - 5b + 7)$ $-6a + 15b - 21$

Simplify.

15. $7x - 3x$ $4x$
16. $5 - 3y + 3$ $-3y + 8$
17. $1 + 3x + 2x$ $5x + 1$
18. $-2x - x + 3y$ $-3x + 3y$
19. $4m + 2n + 4m + 6n$ $8m + 8n$
20. $9x + 3y + 2$ $9x + 3y + 2$
21. $6x - 2x + 3y + 6$ $4x + 3y + 6$
22. $x + 8x - 9x + 3$ 3
23. $-4x^2 - 8x^2 + 3$ $-12x^2 + 3$
24. $-2(3a^2 - 4) + 6a^2 - 8$ 0
25. $2x + 3(x + 4) - 5$ $5x + 7$
26. $4(3 - 2b) - 2b$ $-10b + 12$
27. $6 - (-x + 6) - x$ 0
28. $2(2x + 5) - 10 - 4$ $4x - 4$
29. $-6(4 - 3x) - 18 + 4x$ $22x - 42$
30. $4y - 3(x + y) + 6x^2$ $6x^2 - 3x + y$
31. $\frac{1}{4}d + 2 - \frac{3}{5}d + 5$ $-\frac{7}{20}d + 7$
32. $3 - (x - y) + (x - y)$ 3
33. $\frac{5}{6}x - \frac{1}{3}(2x - 6)$ $\frac{1}{6}x + 2$
34. $\frac{2}{3} - \frac{1}{4}n - \frac{1}{3}(n + 2)$ $-\frac{7}{12}n$

[2.2–2.5] Solve.

35. $6x = 6$ 1
36. $x + 6 = -7$ -13
37. $x - 4 = 7$ 11
38. $\frac{x}{3} = -9$ -27
39. $2x + 4 = 8$ 2
40. $14 = 3 + 2x$ $\frac{11}{2}$
41. $4c + 3 = -21$ -6
42. $4 - 2a = 10$ -3
43. $-x = -12$ 12
44. $3(x - 2) = 6$ 4
45. $-12 = 3(2x - 8)$ 2
46. $4(6 + 2x) = 0$ -3

67. and **70.** all real numbers   **68.** -4

47. $-6n + 2n + 6 = 0$ $\frac{3}{2}$

48. $-3 = 3w - (4w + 6)$ -3

49. $6 - (2n + 3) - 4n = 6$ $-\frac{1}{2}$

50. $4x + 6 - 7x + 9 = 18$ -1

51. $4 + 3(x + 2) = 10$ 0

52. $-3 + 3x = -2(x + 1)$ $\frac{1}{5}$

53. $8.4r - 6.3 = 6.3 + 2.1r$ 2

54. $19.6 - 21.3t = 80.1 - 9.2t$ -5

55. $0.35(c - 5) = 0.45(c + 4)$ -35.5

56. $-2.3(x - 8) = 3.7(x + 4)$ 0.6

57. $\frac{p}{3} + 2 = \frac{1}{4}$ $-\frac{21}{4}$

58. $\frac{d}{6} + \frac{1}{7} = 2$ $\frac{78}{7}$

59. $\frac{3}{5}(r - 6) = 3r$ $-\frac{3}{2}$

60. $\frac{2}{3}w = \frac{1}{7}(w - 2)$ $-\frac{6}{11}$

61. $9x - 6 = -3x + 30$ 3

62. $-(w + 2) = 2(3w - 6)$ $\frac{10}{7}$

63. $2x + 6 = 3x + 9 - 3$ 0

64. $-5a + 3 = 2a + 10$ -1

65. $3x - 12x = 24 - 9x$ no solution

66. $5p - 2 = -2(-3p + 6)$ 10

67. $4(2x - 3) + 4 = 8x - 8$

68. $4 - c - 2(4 - 3c) = 3(c - 4)$

69. $2(x + 7) = 6x + 9 - 4x$ no solution

70. $-5(3 - 4x) = -6 + 20x - 9$

71. $4(x - 3) - (x + 5) = 0$ $\frac{17}{3}$

72. $-2(4 - x) = 6(x + 2) + 3x$ $-\frac{20}{7}$

73. $\frac{x + 3}{2} = \frac{x}{2}$ no solution

74. $\frac{x}{6} = \frac{x - 4}{2}$ 6

75. $\frac{1}{5}(3s + 4) = \frac{1}{3}(2s - 8)$ 52

76. $\frac{2(2t - 4)}{5} = \frac{3t + 6}{4} - \frac{3}{2}$ 32

77. $\frac{2}{5}(2 - x) = \frac{1}{6}(-2x + 2)$ 7

78. $\frac{x}{4} + \frac{x}{6} = \frac{1}{2}(x + 3)$ -18

[2.6] *Determine the following ratios. Write each ratio in lowest terms.*

79. 12 feet to 20 feet $3:5$

80. 80 ounces to 12 pounds $5:12$

81. 32 ounces to 2 pounds $1:1$

Solve each proportion.

82. $\frac{x}{4} = \frac{8}{16}$ 2

83. $\frac{5}{20} = \frac{x}{80}$ 20

84. $\frac{3}{x} = \frac{15}{45}$ 9

85. $\frac{20}{45} = \frac{15}{x}$ $\frac{135}{4}$

86. $\frac{6}{5} = \frac{-12}{x}$ -10

87. $\frac{b}{6} = \frac{8}{-3}$ -16

88. $\frac{-4}{9} = \frac{-16}{x}$ 36

89. $\frac{x}{-15} = \frac{30}{-5}$ 90

The following pairs of figures are similar. For each pair, find the length of the side indicated by x.

90.

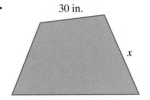

30 in.   40 in.   6 in. x 8 in.

91.

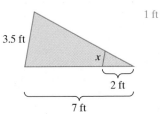

3.5 ft   1 ft x 2 ft   7 ft

[2.7] *Solve each inequality, and graph the solution on a number line.*

92. $3x + 4 \geq 10$ $x \geq 2$,

93. $-4a - 6 > 4a - 14$ $a < 1$,

94. $5 - 3r \leq 2r + 15$ $r \geq -2$,

95. $2(x + 4) \leq 2x - 5$ no solution,

96. $2(x + 3) > 6x - 4x + 4$ all real numbers,

97. $x + 6 > 9x + 30$ $x < -3$,

98. $x - 2 \leq -4x + 7$ $x \leq \frac{9}{5}$,

99. $-(x + 2) < -2(-2x + 5)$ $x > \frac{8}{5}$,

100. $\frac{x}{2} < \frac{2}{3}(x + 3)$ $x > -12$,

101. $\frac{3}{10}(t - 2) \leq \frac{3}{4}(4 + 2t)$ $t \geq -3$,

[2.6] Set up a proportion and solve each problem.

102. Boat Trip A ship travels 40 miles in 1.8 hours. If it travels at the same rate, how long will it take for it to travel 140 miles? 6.3 hr

103. 240 calories

103. Cake If a 4-ounce piece of cake has 160 calories, how many calories does a 6-ounce piece of that cake have?

104. Copy Machine If a copy machine can copy 20 pages per minute, how many pages can be copied in 22 minutes? 440 pages

105. Map Scale If the scale of a map is 1 inch to 60 miles, what distance on the map represents 380 miles? $6\frac{1}{3}$ in.

106. Model Car Bryce Winston builds a model car to a scale of 1 inch to 1.5 feet. If the completed model is 10.5 inches, what is the size of the actual car? 15.75 ft

107. Money Exchange If one U.S. dollar can be exchanged for 9.165 Mexican pesos, find the value of 1 peso in terms of U.S. dollars. ≈$0.109

108. Catsup If a machine can fill and cap 80 bottles of catsup in 50 seconds, how many bottles of catsup can it fill and cap in 2 minutes? 192 bottles

Chapter Practice Test

21. all real numbers, ![number line]

Use the distributive property to simplify.

1. $-3(4 - 2x)$ $6x - 12$

2. $-(x + 3y - 4)$ $-x - 3y + 4$

Simplify.

3. $5x - 8x + 4$ $-3x + 4$

4. $4 + 2x - 3x + 6$ $-x + 10$

5. $-y - x - 4x - 6$ $-5x - y - 6$

6. $a - 2b + 6a - 6b - 3$ $7a - 8b - 3$

7. $2x^2 + 3 + 2(3x - 2)$ $2x^2 + 6x - 1$

Solve exercises 8–16.

8. $2.4x - 3.9 = 3.3$ 3

9. $\frac{5}{6}(x - 2) = x - 3$ 8

10. $6m - (4 - 2m) = 0$ $\frac{1}{2}$

11. $3w + 2(2w - 6) = 4(3w - 3)$ 0

12. $2x - 3(-2x + 4) = -13 + x$ $-\frac{1}{7}$

13. $3x - 4 - x = 2(x + 5)$ no solution

14. $-3(2x + 3) = -2(3x + 1) - 7$ all real numbers

15. $\frac{9}{x} = \frac{3}{-15}$ -45

16. $\frac{1}{7}(2x - 5) = \frac{3}{8}x - \frac{5}{7}$ 0

17. What do we call an equation that has

   a) exactly one solution,   conditional equation

   b) no solution,   contradiction

   c) all real numbers as its solution?   identity

Solve 18–21, and graph the solution on a number line.

18. $2x - 4 < 4x + 10$ $x > -7$, ![number line]

19. $3(x + 4) \geq 5x - 12$ $x \leq 12$, ![number line]

20. $4(x + 3) + 2x < 6x - 3$ no solution, ![number line]

21. $-(x - 2) - 3x = 4(1 - x) - 2$

22. The following figures are similar. Find the length of side x. $\frac{32}{3} = 10\frac{2}{3}$ ft

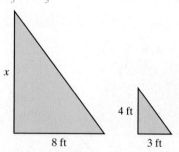

23. Insecticide If 6 gallons of insecticide can treat 3 acres of land, how many gallons of insecticide are needed to treat 75 acres? 150 gal

24. Gas Station Profit Assume a gas station owner makes a profit of 40 cents per gallon of gasoline sold. How many gallons of gasoline would he have to sell in a year to make a profit of $20,000 from gasoline sales? 50,000 gal

25. Travel Time While traveling, you notice that you traveled 25 miles in 35 minutes. If your speed does not change, how long will it take you to travel 125 miles? 175 min or 2 hr 55 min

Cumulative Review Test

Take the following test and check your answers with those that appear at the end of the test. Review any questions that you answered incorrectly. The section and objective where the material was covered are indicated after the answer.

1. Multiply $\dfrac{52}{15} \cdot \dfrac{10}{13}$ $\frac{8}{3}$

2. Divide $\dfrac{5}{24} \div \dfrac{2}{9}$ $\frac{15}{16}$

3. Insert $<$, $>$, or $=$ in the shaded area to make a true statement: $|-2|$ ▨ 1. $>$

4. Evaluate $-5 - (-4) + 12 - 8.$ 3

5. Subtract -6 from $-7.$ -1

6. Evaluate $20 - 6 \div 3 \cdot 2.$ 16

7. Evaluate $3[6 - (4 - 3^2)] - 30.$ 3

8. Evaluate $-2x^2 - 6x + 8$ when $x = -2.$ 12

9. Name the illustrated property.

$$(x + 4) + 6 = x + (4 + 6)$$

Simplify.

10. $8x + 2y + 4x - y$ $12x + y$

11. $9 - \dfrac{2}{3}x + 16 + \dfrac{3}{4}x$ $\frac{1}{12}x + 25$

Solve.

12. $6x + 2 = 10$ $\frac{4}{3}$

13. $-6x - 5x + 6 = 28$ -2

14. $4(x - 2) = 5(x - 1) + 3x + 2$ $-\frac{5}{4}$

15. $\dfrac{40}{30} = \dfrac{3}{x}$ 2.25

16. $\dfrac{3}{4}n - \dfrac{1}{5} = \dfrac{2}{3}n$ $\frac{12}{5}$

Solve, and graph the solution on a number line.

17. $x - 3 > 7$ $x > 10,$ ←———⊕——→
　　　　　　　　　　　　　　10

18. $2x - 7 \le 3x + 5$ $x \ge -12,$ ←———●———→
　　　　　　　　　　　　　　−12

19. *Fertilizer* A 36-pound bag of fertilizer can fertilize an area of 5000 square feet. How many pounds of fertilizer will Marisa Neilson need to fertilize her 22,000-square-foot lawn? 158.4 lb

20. *Earnings* If Samuel earns $10.50 after working for 2 hours scrubbing boats at the marina, how much does he earn after 8 hours? $42

Answers to Cumulative Review Test

1. $\dfrac{8}{3}$; [Sec. 1.3, Obj. 3]　**2.** $\dfrac{15}{16}$; [Sec. 1.3, Obj. 4]　**3.** $>$; [Sec. 1.5, Obj. 2]　**4.** 3; [Sec. 1.7, Obj. 3]　**5.** -1; [Sec. 1.7, Obj. 1]

6. 16; [Sec. 1.9, Obj. 4]　**7.** 3; [Sec. 1.9, Obj. 5]　**8.** 12; [Sec. 1.9, Obj. 6]　**9.** Associative property of addition; [Sec. 1.10, Obj. 2]

10. $12x + y$; [Sec. 2.1, Obj. 3]　**11.** $\dfrac{1}{12}x + 25$; [Sec. 2.1, Obj. 3]　**12.** $\dfrac{4}{3}$; [Sec. 2.4, Obj. 1]　**13.** -2; [Sec. 2.4, Obj. 1]

14. $-\dfrac{5}{4}$; [Sec. 2.5, Obj. 1]　**15.** 2.25; [Sec. 2.6, Obj. 2]　**16.** $\dfrac{12}{5}$; [Sec. 2.5, Obj. 2]　**17.** $x > 10,$ ←——⊕————→; [Sec. 2.7,
　　　　　　　　　　　　　　　　　　　　　　　　　　　　　　　　　　　　　　10

Obj. 1]　**18.** $x \ge -12,$ ←——●————→; [Sec. 2.7, Obj. 1]　**19.** 158.4 pounds; [Sec. 2.6, Obj. 3]　**20.** $42; [Sec. 2.6, Obj. 3]
　　　　　　　　　　　　−12

Chapter 3

Formulas and Applications of Algebra

Physical fitness has become an important part of people's daily lives. We find people using paths in local parks and nature trails to jog, bike, and rollerblade. Exercising with friends can be motivating and can add to the enjoyment. On page 221 we solve an equation based on the distance formula to determine how long it will take a person biking along the paths in Griffith Park in Los Angeles, California to meet a friend who had started running earlier in the day.

| SSM | Study Guide | CD/Video | MathPro 4/5 | PH Math Tutor Center | prenhall.com/Angel |

A Look Ahead

The major goal of this chapter is to teach you the terminology and techniques to write real-life applications as equations. The equations are then solved using the techniques taught in Chapter 2. For mathematics to be relevant it must be useful. In this chapter we explain and illustrate many real-life applications of algebra. This is an important topic and we want you to learn it well and feel comfortable applying mathematics to real-life situations. Thus, we cover the material in this chapter very slowly. You need to have confidence in your work, and you need to do all your homework. The more problems you attempt, the better you will become at setting up and solving application (or word) problems.

This chapter begins with a discussion of formulas. We explain how to evaluate a formula and how to solve for a variable in a formula. Most mathematics and science courses use a wide variety of formulas. Formulas are also used in many other disciplines, including the arts, business and economics, medicine, and technology.

3.1 FORMULAS

SSM

Study Guide

CD/Video

MathPro 4/5

PH Math Tutor Center

prenhall.com/Angel

1 Use the simple interest formula.
2 Use geometric formulas.
3 Solve for a variable in a formula.

A **formula** is an equation commonly used to express a specific relationship mathematically. For example, the formula for the area of a rectangle is

$$\text{area} = \text{length} \cdot \text{width} \quad \text{or} \quad A = lw$$

To **evaluate a formula**, substitute the appropriate numerical values for the variables and perform the indicated operations.

1 Use the Simple Interest Formula

A formula commonly used in banking is the **simple interest formula**.

Simple Interest Formula

$$\text{interest} = \text{principal} \cdot \text{rate} \cdot \text{time} \quad \text{or} \quad i = prt$$

This formula is used to determine the simple interest, i, earned on some savings accounts, or the simple interest an individual must pay on certain loans. In the simple interest formula $i = prt$, p is the principal (the amount invested or borrowed), r is the interest rate in decimal form, and t is the amount of time of the investment or loan.

EXAMPLE 1 **Auto Loan** To buy a car, Darcy Betts borrowed $10,000 from a bank for 3 years. The bank charged 5% simple annual interest for the loan. How much interest will Darcy owe the bank?

Solution Understand and Translate Since the bank charged simple interest, we use the simple interest formula to solve the problem. We are given that the rate, r, is 5%, or 0.05 in decimal form. The principal, p, is $10,000 and the time, t, is 3 years. We substitute these values in the simple interest formula and solve for the interest, i.

$$i = prt$$

Carry Out
$$i = 10,000(0.05)(3)$$
$$i = 1500$$

Check There are various ways to check this problem. First ask yourself "Is the answer realistic?" $1500 is a realistic answer. The interest on $10,000 for 1 year at 5% is $500. Therefore for 3 years, an interest of $1500 is correct.

NOW TRY EXERCISE 91

Answer   Darcy will pay $1500 interest. After 3 years, when she repays the loan, she will pay the principal, $10,000, plus the interest, $1500 for a total of $11,500. ✳

EXAMPLE 2

Savings Account   John Starmack invests $4000 in a savings account that earns simple interest for 2 years. If the interest earned from the account is $500, find the rate.

Solution

Understand and Translate   We use the simple interest formula, $i = prt$. We are given the principal, p, the time, t, and the interest, i. We are asked to find the rate, r. We substitute the given values in the simple interest formula and solve the resulting equation for r.

TEACHING TIP
After discussing Example 2, have students calculate how long it would take for the investment to double, that is earn $5000 in interest.

Carry Out

$$i = prt$$
$$500 = 4000(r)(2)$$
$$500 = 8{,}000r$$
$$\frac{500}{8000} = \frac{8000r}{8000}$$
$$0.0625 = r$$

Check and Answer   The simple interest rate of 0.0625 or 6.25% per year is realistic. If we substitute $P = \$4000$, $r = 0.0625$ and $t = 2$, we obtain the interest, $i = \$500$. Thus the answer checks. The simple interest rate is 6.25%. ✳

2 Use Geometric Formulas

The **perimeter**, P, is the sum of the lengths of the sides of a figure. Perimeters are measured in the same common unit as the sides. For example, perimeter may be measured in centimeters, inches, or feet. The **area**, A, is the total surface within the figure's boundaries. Areas are measured in square units. For example, area may be measured in square centimeters, square inches, or square feet. Table 3.1

TABLE 3.1 Formulas for Areas and Perimeters of Quadrilaterals and Triangles*

Figure	Sketch	Area	Perimeter
Square		$A = s^2$	$P = 4s$
Rectangle		$A = lw$	$P = 2l + 2w$
Parallelogram		$A = lh$	$P = 2l + 2w$
Trapezoid		$A = \frac{1}{2}h(b + d)$	$P = a + b + c + d$
Triangle		$A = \frac{1}{2}bh$	$P = a + b + c$

*See Appendix C for additional information on geometry and geometric figures.

gives the formulas for finding the areas and perimeters of triangles and quadrilaterals. **Quadrilateral** is a general name for a four-sided figure.

In Table 3.1, the letter h is used to represent the *height* of the figure. In the figure of the trapezoid, the sides b and d are called the *bases* of the trapezoid. In the triangle, the side labeled b is called the *base* of the triangle.

EXAMPLE 3

Building an Exercise Area   Dr. Hal Balmer, a veterinarian, decides to fence in a large rectangular area in the yard behind his office for exercising dogs that are boarded overnight. The part of the yard to be fenced in will be 40 feet long and 23 feet wide (see Figure 3.1).

a) How much fencing is needed?

b) How large, in square feet, will the fenced in area be?

Solution

a) Understand   To find the amount of fencing required, we need to find the perimeter of the rectangular area to be fenced in. To find the perimeter, P, substitute 40 for the length, l, and 23 for the width, w, in the perimeter formula, $P = 2l + 2w$.

$$P = 2l + 2w$$

Carry Out

$$P = 2(40) + 2(23) = 80 + 46 = 126$$

Check and Answer   By looking at Figure 3.1, we can see that a perimeter of 126 feet is a reasonable answer. Thus, 126 feet of fencing will be needed to fence in the area for the dogs to exercise.

b) To find the fenced in area, substitute 40 for the length and 23 for the width in the formula for the area of a rectangle. Both the length and width are measured in feet. Since we are multiplying an amount measured in feet by a second amount measured in feet, the answer will be in square feet (or ft^2).

$$A = lw$$
$$= 40(23) = 920 \text{ square feet (or } 920 \text{ ft}^2)$$

Based upon the data given, an area of 920 ft^2 is reasonable. The area to be fenced in will be 920 square feet.   ✳

FIGURE 3.1

40 ft   23 ft

TEACHING TIP
Point out that area is in square units because the formulas involve multiplication of 2 linear dimensions. For instance,
(3 meters)(4 meters) =
(3)(4)(meters)(meters) =
12 meters^2

EXAMPLE 4

Panoramic Photos   Cathy Panic recently purchased a new camera that can take panoramic photos, like the one shown, in addition to regular photos. A panoramic photo has a perimeter of 27 inches and a length of 10 inches. Find the width of a panoramic photo.

Solution **Understand and Translate** The perimeter, P, is 27 inches and the length, l, is 10 inches. Substitute these values into the formula for the perimeter of a rectangle and solve for the width, w.

$$P = 2l + 2w$$
$$27 = 2(10) + 2w$$

Carry Out
$$27 = 20 + 2w$$
$$27 - 20 = 20 - 20 + 2w \quad \text{Subtract 20 from both sides.}$$
$$7 = 2w$$
$$\frac{7}{2} = \frac{2w}{2} \quad \text{Divide both sides by 2.}$$
$$\frac{7}{2} = w$$
$$3.5 = w$$

Check and Answer By looking at the panoramic photo, and from other panoramic photos you may be familiar with, you should realize that dimensions of a length of 10 inches and a width of 3.5 inches is reasonable. The answer is, the width of the photo is 3.5 inches.

NOW TRY EXERCISE 95

EXAMPLE 5 Karin Wagner owns a small sailboat, and she will need to replace a triangular sail shortly because of wear. When ordering the sail, she needs to specify the base and height of the sail. She measures the base and finds that it is 5 feet (see Fig. 3.2). She also remembers that the sail has an area of 30 square feet. She does not want to have to take the sail down to find its height, so she uses algebra to find its height. Find the height of Karin's sail.

Solution **Understand and Translate** We use the formula for the area of a triangle given in Table 3.1.

$$A = \frac{1}{2}bh$$
$$30 = \frac{1}{2}(5)h$$

Carry Out
$$2 \cdot 30 = 2 \cdot \frac{1}{2}(5)h \quad \text{Multiply both sides by 2.}$$
$$60 = 5h$$
$$\frac{60}{5} = \frac{5h}{5} \quad \text{Divide both sides by 5.}$$
$$12 = h$$

FIGURE 3.2

Check and Answer The height of the triangle is 12 feet. By looking at Figure 3.2, and by your knowledge of sailboat sails, you may realize that a sail 12 feet tall and 5 feet wide at the base is reasonable. Thus the height of the sail is 12 feet.

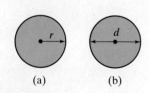

(a)   (b)

FIGURE 3.3

Another figure that we see and use daily is the circle. The **circumference**, C, is the length (or perimeter) of the curve that forms a circle. The **radius**, r, is the line segment from the center of the circle to any point on the circle (Fig. 3.3a). The

diameter of a circle is a line segment through the center whose endpoints both lie on the circle (Fig. 3.3b). *Note that the length of the diameter is twice the length of the radius.*

The formulas for both the area and the circumference of a circle are given in Table 3.2.

TEACHING TIP
Have students write a formula for circumference based on diameter.

TABLE 3.2 Formulas for Circles

Circle	Area	Circumference
	$A = \pi r^2$	$C = 2\pi r$

The value of **pi**, symbolized by the Greek lowercase letter π, is an irrational number which cannot be exactly expressed as a decimal number or a numerical fraction. Pi is *approximately* 3.14.

Using Your Calculator

Scientific and graphing calculators have a key for finding the value of π. If you press the $\boxed{\pi}$ key, your calculator may display 3.1415927. This is only an approximation of π. If you own a scientific or graphing calculator, use the $\boxed{\pi}$ key when evaluating expressions containing π. If your calculator does not have a $\boxed{\pi}$ key, use 3.14 to approximate it. *When evaluating an expression containing π, we will use the $\boxed{\pi}$ key on a calculator to obtain the answer.* The final answer displayed in the text or answer section may therefore be slightly different (and more accurate) than yours if you use 3.14 for π.

EXAMPLE 6 **Helicopter Landing Pad** The University Medical Center has a circular landing pad, for medical helicopters to land, near the entrance to the emergency room. Determine the area and circumference of the circular landing pad if its diameter is 40 feet.

Solution The radius is half its diameter, so $r = \dfrac{40}{2} = 20$ feet.

TEACHING TIP
After discussing Example 6, have students give the dimensions of a square which would have about the same area as the landing area. Then ask, "Why do helicopters land on circular pads?"

$$A = \pi r^2 \qquad\qquad C = 2\pi r$$
$$A = \pi(20)^2 \qquad\qquad C = 2\pi(20)$$
$$A = \pi(400) \qquad\qquad C \approx 125.66 \text{ feet}$$
$$A \approx 1256.64 \text{ square feet}$$

NOW TRY EXERCISE 99

To obtain our answer of 1256.64, we used the $\boxed{\pi}$ key on a calculator and rounded our final answer to the nearest hundredth. If you do not have a calculator with a $\boxed{\pi}$ key and use 3.14 for π, your answer for the area would be 1256. ✳

Table 3.3 on the next page gives formulas for finding the volume of certain **three-dimensional figures**. **Volume** is measured in cubic units, such as cubic centimeters or cubic feet.

TABLE 3.3 Formulas for Volumes of Three-Dimensional Figures		
Figure	**Sketch**	**Volume**
Rectangular solid		$V = lwh$
Right circular cylinder		$V = \pi r^2 h$
Right circular cone		$V = \dfrac{1}{3}\pi r^2 h$
Sphere		$V = \dfrac{4}{3}\pi r^3$

TEACHING TIP
Point out that volume is in cubic units because the formulas involve multiplication of 3 linear dimensions. For instance,
(2 meters)(3 meters)(4 meters) = (2)(3)(4)(meters)(meters)(meters) = 24 meters³

EXAMPLE 7 **Wilson**   In the movie *Castaway* with Tom Hanks, Tom has a volleyball friend that he names Wilson. Wilson has a diameter of approximately 8.6 inches.* Find the volume of air in the volleyball.

Solution   Understand and Translate   Table 3.3 gives the formula for the volume of a sphere. The formula involves the radius. Since the diameter is 8.6 inches, its radius is $\dfrac{8.6}{2} = 4.3$ inches.

$$V = \frac{4}{3}\pi r^3$$

Carry Out
$$V = \frac{4}{3}\pi(4.3)^3 = \frac{4}{3}\pi(79.507) \approx 333.04$$

---

*Officially, a volleyball may have a circumference no less than 25 inches and no greater than 27 inches.

Check and Answer   A cubic foot occupies a space of 12 in. × 12 in. × 12 in. or 1728 cubic inches. Since a volleyball would fit inside a box 1 foot by 1 foot by 1 foot, and the answer is less than 1728 cubic inches, the answer is reasonable. The volume of the volleyball is about 333 cubic inches. ✳

EXAMPLE 8   **The Height of a Gas Can**   The gas can shown in Figure 3.4 is a right circular cylinder. Find the height of the can if it has a radius of 8 inches and a volume of 4021 cubic inches.

Solution   Understand and Translate   We are given that $V = 4021$ and $r = 8$. Substitute these values into the formula for the volume of a right circular cylinder and solve for the height, h.

$$V = \pi r^2 h$$
$$4021 = \pi(8)^2 h$$

Carry Out

$$4021 = \pi(64)h$$
$$\frac{4021}{64\pi} = \frac{\pi(64)h}{64\pi} \qquad \textit{Divide both sides by } 64\pi$$
$$20 \approx h$$

FIGURE 3.4

Answer   The gas can is about 20 inches tall.
NOW TRY EXERCISE 25   ✳

EXAMPLE 9   **Diagonals of Quadrilateral**   A quadrilateral is a polygon* that has four sides (Fig. 3.5). Notice it has two diagonals. The number of diagonals, d, in a polygon of n sides is given by the formula $d = \frac{1}{2}n^2 - \frac{3}{2}n$.

a) How many diagonals does a pentagon (five sides) have?
b) How many diagonals does an octagon (eight sides) have?

FIGURE 3.5

Solution   a) $d = \dfrac{1}{2}n^2 - \dfrac{3}{2}n$ b) $d = \dfrac{1}{2}n^2 - \dfrac{3}{2}n$

$\qquad = \dfrac{1}{2}(5)^2 - \dfrac{3}{2}(5)$ $\qquad = \dfrac{1}{2}(8)^2 - \dfrac{3}{2}(8)$

$\qquad = \dfrac{1}{2}(25) - \dfrac{3}{2}(5)$ $\qquad = \dfrac{1}{2}(64) - 12$

$\qquad = \dfrac{25}{2} - \dfrac{15}{2} = \dfrac{10}{2} = 5$ $\qquad = 32 - 12 = 20$

NOW TRY EXERCISE 77   A pentagon has 5 diagonals and an octagon has 20 diagonals.   ✳

3 Solve for a Variable in a Formula

Often in this course and in other mathematics and science courses, you will be given an equation or formula solved for one variable and have to solve it for a different variable. We will now learn how to do this. This material will reinforce what you learned about solving equations in Chapter 2. We will use the procedures learned here to solve problems in many other sections of the text.

To solve for a variable in a formula, treat each of the quantities, except the one for which you are solving, as if they were constants. Then solve for the desired variable by isolating it on one side of the equation, as you did in Chapter 2.

*A polygon is a closed figure made up of straight line segments (such as a triangle or a square). Polygons are discussed in Appendix C.

EXAMPLE 10 **Perimeter of Rectangle** The formula for the perimeter of a rectangle is $P = 2l + 2w$. Solve this formula for the length, l.

Solution We must get l all by itself on one side of the equation. We begin by removing the $2w$ from the right side of the equation to isolate the term containing the l.

TEACHING TIP
After discussing Example 10, have students solve for w.

$$P = 2l + 2w$$
$$P - 2w = 2l + 2w - 2w \qquad \text{Subtract } 2w \text{ from both sides.}$$
$$P - 2w = 2l$$
$$\frac{P - 2w}{2} = \frac{2l}{2} \qquad \text{Divide both sides by 2.}$$
$$\frac{P - 2w}{2} = l \quad \left(\text{or} \quad l = \frac{P}{2} - w\right)$$

NOW TRY EXERCISE 45

EXAMPLE 11 **Simple Interest Formula** We used the simple interest formula, $i = prt$, in Example 1. Solve the simple interest formula for the principal, p.

Solution We must isolate the p. Since p is multiplied by both r and t, we divide both sides of the equation by rt.

$$i = prt$$
$$\frac{i}{rt} = \frac{prt}{rt}$$
$$\frac{i}{rt} = p$$

NOW TRY EXERCISE 41

When we discuss graphing, in Chapter 7, we will need to solve many equations for the variable y. Also, when graphing an equation on a graphing calculator, you will need to solve the equation for y before you can graph it. The procedure to solve an equation for y is illustrated in Example 12.

EXAMPLE 12 **a)** Solve the equation $2x + 3y = 12$ for y.
b) Find the value of y when $x = 6$.

Solution **a)** Begin by isolating the term containing the variable y.

$$2x + 3y = 12$$
$$2x - 2x + 3y = 12 - 2x \qquad \text{Subtract } 2x \text{ from both sides.}$$
$$3y = 12 - 2x$$
$$\frac{3y}{3} = \frac{12 - 2x}{3} \qquad \text{Divide both sides by 3.}$$
$$y = \frac{12 - 2x}{3} \quad \left(\text{or} \quad y = \frac{12}{3} - \frac{2x}{3} = 4 - \frac{2}{3}x\right)$$

b) To find the value of y when x is 6, substitute 6 for x in the equation solved for y in part **a)**.

$$y = \frac{12 - 2x}{3}$$
$$y = \frac{12 - 2(6)}{3} = \frac{12 - 12}{3} = \frac{0}{3} = 0$$

NOW TRY EXERCISE 61   We see that when $x = 6$, $y = 0$.

Some formulas contain fractions. When a formula contains a fraction, we can eliminate the fraction by multiplying both sides of the equation by the least common denominator, as illustrated in Example 13. We use the Multiplication Property of Equality, as explained in Section 2.3.

EXAMPLE 13   The formula for the area of a triangle is $A = \dfrac{1}{2}bh$. Solve this formula for h.

Solution   We begin by multiplying both sides of the equation by the LCD, 2, to eliminate the fraction. We then isolate the variable h.

$$A = \frac{1}{2}bh$$

$$\boxed{2} \cdot A = \boxed{2} \cdot \frac{1}{2}bh \quad \text{\textit{Multiply both sides by 2.}}$$

$$2A = bh$$

$$\frac{2A}{b} = \frac{bh}{b} \quad \text{\textit{Divide both sides by }} b.$$

$$\frac{2A}{b} = h$$

NOW TRY EXERCISE 43   Thus $h = \dfrac{2A}{b}$.

In Chapter 7, we will introduce the point-slope form of a linear equation. In that chapter you will need to work with equations like the one in Example 14.

EXAMPLE 14   Solve the equation $y - \dfrac{1}{3} = \dfrac{1}{4}(x - 6)$ for y.

Solution   Multiply both sides of the equation by the LCD, 12.

$$y - \frac{1}{3} = \frac{1}{4}(x - 6)$$

$$\boxed{12}\left(y - \frac{1}{3}\right) = \boxed{12} \cdot \frac{1}{4}(x - 6) \quad \text{\textit{Multiply both sides by 12.}}$$

$$12y - 4 = 3(x - 6) \qquad \text{\textit{Distributive property used on left}}$$

$$12y - 4 = 3x - 18 \qquad \text{\textit{Distributive property used on right}}$$

$$12y = 3x - 14 \qquad \text{\textit{Add 4 to both sides}}$$

$$y = \frac{3x - 14}{12} \qquad \text{\textit{Divided both sides by 12.}}$$

The answer $y = \dfrac{3x - 14}{12}$ can be expressed as $y = \dfrac{3}{12}x - \dfrac{14}{12}$ or $y = \dfrac{1}{4}x - \dfrac{7}{6}$.

NOW TRY EXERCISE 73

Exercise Set 3.1

Concept/Writing Exercises

1. What is a formula?

2. What does it mean to *evaluate a formula*?

3. Write the simple interest formula, then indicate what each letter in the formula represents. $i = prt$

4. What is a quadrilateral? a four-sided figure

5. What is the relationship between the radius and the diameter of a circle? $d = 2r$

6. Is π equal to 3.14? Explain your answer. no

7. By using any formula for area, explain why area is measured in square units.

8. By using any formula for volume, explain why volume is measured in cubic units.

1. an equation used to express a relationship mathematically 2. to substitute values and perform the indicated operations 7. When you multiply a unit by the same unit, you get a square unit. 8. When you multiply the same unit three times, you get a cubic unit.

Practice the Skills

Use the formula to find the value of the variable indicated. Use a calculator to save time and where necessary, round your answer to the nearest hundredth.

9. $P = 4s$ (perimeter of a square); find P when $s = 6$. 24

10. $A = lw$ (area of a rectangle); find A when $l = 12$ and $w = 8$. 96

11. $A = s^2$ (area of a square); find A when $s = 7$. 49

12. $c = 2.54i$ (to change inches to centimeters); find c when $i = 12$. 30.48

13. $P = 2l + 2w$ (perimeter of a rectangle); find P when $l = 8$ and $w = 5$. 26

14. $f = 1.47m$ (to change speed from mph to ft/sec); find f when $m = 60$. 88.2

15. $A = \pi r^2$ (area of a circle); find A when $r = 5$. 78.54

16. $p = i^2 r$ (formula for finding electrical power); find r when $p = 2000$ and $i = 4$. 125

17. $z = \dfrac{x - m}{s}$ (statistics formula for finding the z-score); find z when $x = 100$, $m = 80$, and $s = 10$. 2

18. $A = \dfrac{1}{2}bh$ (area of a triangle); find b when $A = 30$ and $h = 10$ 6

19. $V = \dfrac{1}{3}Bh$ (volume of a cone); find h when $V = 60$ and $B = 12$. 15

20. $P = 2l + 2w$ (perimeter of a rectangle); find l when $P = 28$ and $w = 6$. 8

21. $A = \dfrac{m + n}{2}$ (mean of two values); find n when $A = 36$ and $m = 16$. 56

22. $A = P(1 + rt)$ (banking formula to find the amount in an account); find r when $A = 1050$, $t = 1$, and $P = 1000$. 0.05

23. $F = \dfrac{9}{5}C + 32$ (for converting Celsius temperature to Fahrenheit); find F when $C = 15$. 59

24. $V = \dfrac{4}{3}\pi r^3$ (volume of a sphere); find V when $r = 8$. 2144.66

25. $V = \pi r^2 h$ (volume of a cylinder); find h when $V = 678.24$ and $r = 6$. 6.00

26. $C = \dfrac{5}{9}(F - 32)$ (for converting Fahrenheit temperature to Celsius); find F when $C = 68$. 154.4

27. $B = \dfrac{703w}{h^2}$ (for finding body mass index); find w when $B = 24$ and $h = 61$. 127.03

28. $F = \dfrac{1}{2}mg^2$ (for finding force of attraction); Find m when $F = 6000$ and $g = 32$. 11.72

29. $S = C + rC$ (for determining selling price when an item is marked up); find S when $C = 160$ and $r = 0.12$ (or 12%). 179.2

30. $S = R - rR$ (for determining sale price when an item is discounted); find R when $S = 92$ and $r = 0.08$ (or 8%). 100

In Exercises 31–36, use Tables 3.1, 3.2, and 3.3 to find the formula for the area or volume of the figure. Then determine either the area or volume.

31. 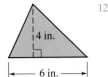 12 in.^{2}

4 in.

6 in.

32. 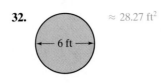 ≈ 28.27 ft^2

6 ft

33. 4 cm 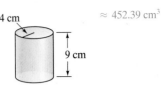 ≈ 452.39 cm^3

9 cm

34. 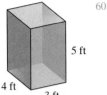 60 ft^3

5 ft

4 ft

3 ft

35. 4 ft 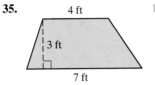 16.5 ft^2

3 ft

7 ft

36. 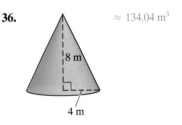 ≈ 134.04 m^3

8 m

4 m

In Exercises 37–60, solve for the indicated variable.

37. $P = 4s$, for s $s = P/4$

38. $A = lw$, for w $w = A/l$

39. $d = rt$, for t $t = d/r$

40. $C = \pi d$, for d $d = C/\pi$

41. $V = lwh$, for l $l = V/(wh)$

42. $i = prt$, for t $t = i/(pr)$

43. $A = \dfrac{1}{2}bh$, for b $b = 2A/h$

44. $E = IR$, for I $I = E/R$

45. $P = 2l + 2w$, for w $w = (P - 2l)/2$

46. $PV = KT$, for T $T = PV/K$

47. $5 - 2t = m$, for t $t = (-m + 5)/2$

48. $3m + 2n = 25$, for n $n = (-3m + 25)/2$

49. $y = mx + b$, for b $b = y - mx$

50. $A = P + Prt$, for r $r = (A - P)/(Pt)$

51. $y = mx + b$, for x $x = (y - b)/m$

52. $d = a + b + c$, for b $b = d - a - c$

53. $ax + by = c$, for y $y = (-ax + c)/b$

54. $ax + by + c = 0$, for y $y = (-ax - c)/b$

55. $V = \pi r^2 h$, for h $h = V/(\pi r^2)$

56. $V = \dfrac{1}{3}\pi r^2 h$, for h $h = 3V/(\pi r^2)$

57. $A = \dfrac{m + d}{2}$, for m $m = 2A - d$

58. $A = \dfrac{m + 2d}{3}$, for d $d = \dfrac{3A - m}{2}$

59. $R = \dfrac{I + 3w}{2}$, for w $w = \dfrac{2R - I}{3}$

60. $A = \dfrac{a + b + c}{3}$, for c $c = 3A - a - b$

*In Exercises 61–76, **a)** solve each equation for y, then **b)** find the value of y for the given value of x.*

61. $3x + y = 5$, $x = 2$ **a)** $y = -3x + 5$ **b)** -1

62. $6x + 2y = -12$, $x = -3$ **a)** $y = -3x - 6$ **b)** 3

63. $4x = 6y - 8$, $x = 10$ **a)** $y = (2x + 4)/3$ **b)** 8

64. $-2y + 6x = -10$, $x = 0$ **a)** $y = 3x + 5$ **b)** 5

65. $5y = -12 + 3x$, $x = 4$ **a)** $y = (3x - 12)/5$ **b)** 0

66. $15 = 3y - x$, $x = 3$ **a)** $y = (x + 15)/3$ **b)** 6

67. $-3x + 5y = -10$, $x = 4$ **a)** $y = (3x - 10)/5$ **b)** $2/5$

68. $3x - 2y = -18$, $x = -1$ **a)** $y = (3x + 18)/2$ **b)** $15/2$

69. $15 - 3x = -6y$, $x = 0$ **a)** $y = (x - 5)/2$ **b)** $-5/2$

70. $-12 = -2x - 3y$, $x = -2$ **a)** $y = (-2x + 12)/3$ **b)** $16/3$

71. $-8 = -x - 2y$, $x = -4$ **a)** $y = (-x + 8)/2$ **b)** 6

72. $2x + 5y = 20$, $x = -5$ **a)** $y = (-2x + 20)/5$ **b)** 6

73. $y + 3 = -\dfrac{1}{3}(x - 4)$, $x = 6$ $y = \dfrac{-x - 5}{3}, -\dfrac{11}{3}$

74. $y - 3 = \dfrac{2}{3}(x + 4)$, $x = 3$ $y = \dfrac{2x + 17}{3}, \dfrac{23}{3}$

75. $y - \dfrac{1}{5} = 2\left(x + \dfrac{1}{3}\right)$, $x = 4$ $y = \dfrac{30x + 13}{15}, \dfrac{133}{15}$

76. $y + 5 = \dfrac{3}{4}\left(x + \dfrac{1}{2}\right)$, $x = 0$ $y = \dfrac{6x - 37}{8}, -\dfrac{37}{8}$

Use the formula in Example 9, $d = \frac{1}{2}n^2 - \frac{3}{2}n$, to find the number of diagonals in a figure with the given number of sides.

77. 10 sides 35

78. 6 sides 9

Use the formula $C = \frac{5}{9}(F - 32)$ to find the Celsius temperature (C) equivalent to the given Fahrenheit temperature (F).

🔒 **79.** $F = 50°$ $\ \ C = 10°$

80. $F = 86°$ $\ \ C = 30°$

Use the formula $F = \frac{9}{5}C + 32$, to find the Fahrenheit temperature (F) equivalent to the given Celsius temperature (C).

81. $C = 25°$ $\ \ F = 77°$

82. $C = 10°$ $\ \ F = 50°$

In chemistry the ideal gas law is $P = KT/V$ where P is pressure, T is temperature, V is volume, and K is a constant. Find the missing quantity.

83. $T = 20, K = 2, V = 1$ $\ \ P = 40$

84. $T = 30, P = 3, K = 0.5$ $\ \ V = 5$

85. $P = 80, T = 100, V = 5$ $\ \ K = 4$

86. $P = 100, K = 2, V = 6$ $\ \ T = 300$

Problem Solving

🔒 **87.** Consider the formula for the area of a square, $A = s^2$. If the length of the side of a square, s, is doubled, what is the change in its area? 4 times as large

88. Consider the formula for the volume of a cube, $V = s^3$. If the length of the side of a cube, s, is doubled, what is the change in its volume? 8 times as large

The sum of the first n even numbers can be found by the formula $S = n^2 + n$. Find the sum of the numbers indicated.

89. First 6 even numbers 42

90. First 10 even numbers 110

In Exercises 91–94, use the simple interest formula.

91. *Auto Loan* Thang Tran decided to borrow $6000 from Citibank to help pay for a car. His loan was for 3 years at a simple interest rate of 8%. How much interest will Thang pay? $1440.

92. *Simple Interest Loan* Danielle Maderi lent her brother $4000 for a period of 2 years. At the end of the 2 years, her brother repaid the $4000 plus $640 interest. What simple interest rate did her brother pay? 8%

93. *Savings Account* Kate Lynch invested a certain amount of money in a savings account paying 3% simple interest per year. When she withdrew her money at the end of 3 years, she received $450 in interest. How much money did Kate place in the savings account? $5000

94. *Savings Account* Peter Ostroushko put $6000 in a savings account earning $3\frac{1}{2}$% simple interest per year. When he withdrew his money, he received $840 in interest. How long had he left his money in the account? 4 yr

Use the formula given in Tables 3.1, 3.2, and 3.3 to work Exercises 95–108.

95. *Triangular Table* To display some books at a convention, Cynthia Lennon used a triangular table top whose

sides were 12 feet, 8 feet, and 5 feet. Find the perimeter of the table top. 25 ft

96. *Calculator Screen* The screen of the Texas Instruments rectangular display screen (or window) is 2.5 inches by 1.5 inches. Find the area of the window. 3.75 in.²

97. *Yield Sign* A yield traffic sign is triangular with a base of 36 inches and a height of 31 inches. Find the area of the sign. 558 in.²

99. ≈7.07 ft²

98. *Fencing* Milt McGowen has a rectangular lot that measures 100 feet by 60 feet. If Milt wants to fence in his lot, how much fencing will he need? 320 ft

99. *Living Room Table* A round living room table top has a diameter of 3 feet. Find the area of the table top.

100. *Swimming Pool* A circular above-ground swimming pool has a diameter of 24 feet. Determine the circumference of the pool. ≈75.40 ft

🔒 **101.** *Kite* Below we show a kite. Determine the area of the kite. 3 ft²

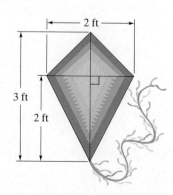

2 ft

3 ft

2 ft

102. *Trapezoidal Sign* Canter Martin made a sign to display at a baseball game. The sign was in the shape of a trapezoid. Its bases are 4 feet and 3 feet, and its height is 2 feet. Find the area of the sign. 7 ft^2

103. *Jacuzzi* The inside of a circular jacuzzi is 8 feet in diameter. If the water inside the jacuzzi is 3 feet deep, determine, in cubic feet, the volume of water in the jacuzzi. $\approx 150.80 \text{ ft}^3$

104. *Banyan Tree* The largest banyan tree in the continental United States is at the Edison House in Fort Myers, Florida. The circumference of the aerial roots of the tree is 390 feet. Find the *diameter* of the aerial roots to the nearest tenth of a foot. 124.1 ft

105. *Amphitheatre* The seats in an amphitheater are inside a trapezoidal area as shown in the figure.

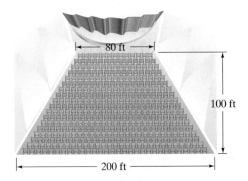

The bases of the trapezoidal area are 80 feet and 200 feet, and the height is 100 feet. Find the area of the floor occupied by seats. $14{,}000 \text{ ft}^2$

106. *Oil Drum* Roberto Sanchez has an empty oil drum that he uses for storage. The oil drum is 4 feet high and has a diameter of 24 inches. Find the volume of the drum in cubic feet. 12.57 ft^3

107. *Ice Cream* Find the volume of an ice cream cone (cone only) if its diameter is 3 inches and its height is 5 inches. $\approx 11.78 \text{ in.}^3$

108. *Rockefeller Center* An ice skating rink at Rockefeller Center in New York City is almost rectangular (the corners are curved). The length is 150 feet and the width is 120 feet. If ice fills the rink to a height of 0.25 ft, determine the volume of ice in the rink (assume the rink is rectangular). 4500 ft^3

109. *Body Mass Index* A person's body mass index (BMI) is found by multiplying a person's weight, w, in pounds by 703, then dividing this product by the square of the person's height, h, in inches.

a) Write a formula to find the BMI. $B = \dfrac{703w}{h^2}$

b) Brandy Belmont is 5 feet 3 inches tall and weighs 135 pounds. Find her BMI. ≈ 23.91

110. *Body Mass Index* Refer to Exercise 109. Mario Guzza's weight is 162 pounds, and he is 5 feet 7 inches tall. Find his BMI. ≈ 25.37

Challenge Problems

111. ***Cereal Box*** A cereal box is to be made by folding the cardboard along the dashed lines as shown in the figure on the right.

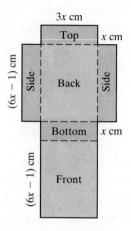

a) Using the formula

$$volume = length \cdot width \cdot height$$

write an equation for the volume of the box.

b) Find the volume of the box when $x = 7$ cm.

c) Write an equation for the surface area of the box.

d) Find the surface area when $x = 7$ cm. 2590 cm²

111. a) $V = 18x^3 - 3x^2$
b) 6027 cm³ **c)** $S = 54x^2 - 8x$

 ## Group Activity

112. ***Square Face on Cube*** Consider the following photo. The front of the figure is a square with a smaller black square painted on the center of the larger square. Suppose the length of one side of the larger square is A, and length of one side of the smaller (the black square) is B. Also the thickness of the block is C.

a) Group member one: Determine an expression for the surface area of the black square. B^2

b) Group member two: Determine an expression for the surface area of the larger square (which includes the smaller square). A^2

c) Group member three: Determine the surface area of the larger square minus the black square (the purple area shown). $A^2 - B^2$

d) As a group, write an expression for the volume of the entire solid block. A^2C

e) As a group, determine the volume of the entire solid block if its length is 1.5 feet and its width is 0.8 feet. 1.8 ft³

Cumulative Review Exercises

[1.9] 113. Evaluate. $\left[4(12 \div 2^2 - 3)^2\right]^2$. 0

[2.6] 114. ***Horses*** A stable has four Morgan and six Arabian horses. Find the ratio of Arabians to Morgans. 3:2

115. ***Emptying Pool*** It takes 3 minutes to siphon 25 gallons of water out of a swimming pool. How

long will it take to empty a 13,500-gallon swimming pool by siphoning? Write a proportion that can be used to solve the problem, and then find the desired value. 1620 min or 27 hr

[2.7] 116. Solve $2(x - 4) \geq 3x + 9$. $x \leq -17$

3.2 CHANGING APPLICATION PROBLEMS INTO EQUATIONS

1 Translate phrases into mathematical expressions.
2 Write expressions involving percent.
3 Express the relationship between two related quantities.
4 Write expressions involving multiplication.
5 Translate applications into equations.

SSM Study Guide CD/Video

MathPro 4/5 PH Math Tutor Center prenhall.com/Angel

1 Translate Phrases into Mathematical Expressions

HELPFUL HINT

STUDY TIP

It is important that you prepare for the remainder of the chapter carefully. Make sure you read the book and work the examples carefully. *Attend class every day, and most of all, work all the exercises assigned to you.*

As you read through the examples in the rest of the chapter, think about how they can be expanded to other, similar problems. For example, in Example 1a) we will state that the distance, d, increased by 12 miles, can be represented by $d + 12$. You can generalize this to other, similar problems. For example, a weight, w, increased by 15 pounds, can be represented as $w + 15$.

One practical advantage of knowing algebra is that you can use it to solve everyday problems involving mathematics. For algebra to be useful in solving everyday problems, you must first be able to *translate application problems into mathematical language*. The purpose of this section is to help you take an application problem, also referred to as a *word* or *verbal problem*, and write it as a mathematical equation.

TEACHING TIP
Point out that "a number decreased by 5" is not the same as "5 decreased by a number." Use a specific number to illustrate the difference between the two expressions.

Often the most difficult part of solving an application problem is translating it into an equation. Before you can translate a problem into an equation, you must understand the meaning of certain words and phrases and how they are expressed mathematically. Table 3.4 is a list of selected words and phrases and the operations they imply. We used the variable x. However, any variable could have been used.

TABLE 3.4

Word or Phrase	Operation	Statement	Algebraic Form
Added to More than Increased by The sum of	Addition	7 *added to* a number 5 *more than* a number A number *increased by* 3 *The sum of* a number and 4	$x + 7$ $x + 5$ $x + 3$ $x + 4$
Subtracted from Less than Decreased by The difference between	Subtraction	6 *subtracted from* a number 7 *less than* a number A number *decreased by* 5 The *difference between* a number and 9	$x - 6$ $x - 7$ $x - 5$ $x - 9$
Multiplied by The product of Twice a number, 3 times a number, etc. Of, when used with a percent or fraction	Multiplication	A number *multiplied by* 6 *The product of* 4 and a number *Twice a number* 20% *of* a number	$6x$ $4x$ $2x$ $0.20x$
Divided by The quotient of One-half of a number, one third, etc.	Division	A number *divided by* 8 *The quotient of* a number and 6 *One-seventh of a number*	$\dfrac{x}{8}$ $\dfrac{x}{6}$ $\dfrac{x}{7}$

Often a statement contains more than one operation. The following chart provides some examples of this.

Statement	Algebraic Form
Four more than twice a number	$\underbrace{2x}_{\text{Twice a number}} + 4$
Five less than 3 times a number	$\underbrace{3x}_{\text{Three times a number}} - 5$
Three times the sum of a number and 8	$3\underbrace{(x + 8)}_{\text{The sum of a number and 8}}$
Twice the difference between a number and 4	$2\underbrace{(x - 4)}_{\text{The difference between a number and 4}}$

To give you more practice with the mathematical terms, we will also convert some algebraic expressions into statements. Often an algebraic expression can be written in several different ways. Following is a list of some of the possible statements that can be used to represent the given algebraic expression.

TEACHING TIP
Point out the importance of a comma. Three times a number, decreased by 4 is written as $3x - 4$. But, three times a number decreased by 4 is written as $3(x - 4)$.

Algebraic	Statements
$2x + 3$	Three more than twice a number The sum of twice a number and 3 Twice a number, increased by 3 Three added to twice a number
$3x - 4$	Four less than 3 times a number Three times a number, decreased by 4 The difference between 3 times a number and 4 Four subtracted from 3 times a number

EXAMPLE 1 Express each statement as an algebraic expression.
a) The distance, d, increased by 12 miles
b) Eight less than twice the area, a
c) Four pounds more than 5 times the weight, w
d) Twice the sum of the height, h, and 3 feet

Solution a) $d + 12$ b) $2a - 8$ c) $5w + 4$ d) $2(h + 3)$ ✳

EXAMPLE 2 Write three different statements to represent the following expressions.
a) $5x - 2$ b) $2x + 7$

Solution a) 1. Two less than 5 times a number
         2. Five times a number, decreased by 2
         3. The difference between 5 times a number and 2
     b) 1. Seven more than twice a number
         2. Two times a number, increased by 7
         3. The sum of twice a number and 7   ✳

EXAMPLE 3   Write a statement to represent each expression.
a) $3x - 4$ **b)** $3(x - 4)$

Solution   **a)** One of many possible statements is 4 less than 3 times a number.
b) The expression within parentheses may be written "the difference between a number and 4." Therefore, the entire expression may be written as 3 times the

NOW TRY EXERCISE 37   difference between a number and 4.   ✳

2   Write Expressions Involving Percent

TEACHING TIP
Have students use the distributive property to rewrite $c + 0.06c$.

Since percents are used so often, you must have a clear understanding of how to write expressions involving percent. Whenever we perform a calculation involving percent, we generally change it to a decimal or a fraction first.

EXAMPLE 4   Express each phrase as an algebraic expression.
a) The cost of a pair of boots, c, increased by 6%
b) The population in the town of Brooksville, p, decreased by 12%

Solution   **a)** When shopping we may see a "25% off" sales sign. We assume that this means 25% off *the original cost*, even though this is not stated. This question asks for the cost increased by 6%. We assume that this means the cost increased by 6% of the original cost, and write

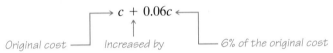

$$c + 0.06c$$

Original cost ⎯⎯ Increased by ⎯⎯ 6% of the original cost

Thus, the answer is $c + 0.06c$.
b) Using the same reasoning as in part **a)** the answer is $p - 0.12p$.   ✳

AVOIDING COMMON
ERRORS

In Example 4**a)** we asked you to represent a cost, c, increased by 6%. Note, the answer is $c + 0.06c$. Often, students write the answer to this question as $c + 0.06$. It is important to realize that a percent of a quantity must always be a percent multiplied by some number or letter. Some phrases involving the word percent and the correct and incorrect interpretations follow.

PHRASE	CORRECT	INCORRECT
A $7\frac{1}{2}$% sales tax on c dollars	$0.075c$	~~0.075~~
The cost, c, increased by a $7\frac{1}{2}$% sales tax	$c + 0.075c$	~~$c + 0.075$~~
The cost, c, reduced by 25%	$c - 0.25c$	~~$c - 0.25$~~

3   Express the Relationship between Two Related Quantities

Sometimes in a problem, two numbers are related to each other in a certain way. We often represent the simplest, or most basic, number that needs to be expressed as a variable, and the other as an expression containing that variable. Some examples follow.

Statement	One Number	Second Number
Two numbers differ by 5	x	$x + 5$
Mike's age now and Mike's age in 8 years	x	$x + 8$
One number is 6 times the other number	x	$6x$
One number is 12% less than the other	x	$x - 0.12x$

TEACHING TIP
Show students that "the sum of two numbers is 10" can be represented by $x + y = 10$. One number is x and the other is y. Then show that by solving the equation for y, y can be described in terms of x.

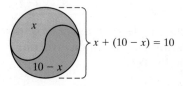

$x + (10 - x) = 10$

FIGURE 3.6

Note that often more than one pair of expressions can be used to represent the two numbers. For example, "two numbers differ by 5" can also be expressed as x and $x - 5$. Let's now look at two more statements.

Statement	One Number	Second Number
The sum of two numbers is 10.	x	$10 - x$
A 25-foot length of wood is cut in two pieces.	x	$25 - x$

It may not be obvious why in "the sum of two numbers is 10" the two numbers are represented as x and $10 - x$. Suppose that one number is 2; what is the other number? Since the sum is 10, the second number must be $10 - 2$ or 8. Suppose that one number is 6; the second number must be $10 - 6$, or 4. In general, if the first number is x, the second number must be $10 - x$. Note that the sum of x and $10 - x$ is 10 (Fig. 3.6).

Consider the statement, "a 25-foot length of wood is cut in two pieces." If we call one length x, then the other length must be $25 - x$. For example, if one length is 6 feet, the other length must be $25 - 6$ or 19 feet (Fig. 3.7).

FIGURE 3.7

EXAMPLE 5 For each relationship, select a variable to represent one quantity and state what that quantity represents. Then express the second quantity in terms of the variable selected.

a) The Bisons scored 12 points more than the Chicklets.

b) Sheila is walking 1.4 times faster than Jim.

c) Bill and Mary share $75.

d) Kim has 7 more than 5 times the amount Sylvia has.

e) The length of a rectangle is 3 feet less than 4 times its width.

f) At Delphi Corporation, the number of employees increased by 12% from 2002 to 2003.

g) The profit of a business, in percent, is shared between two partners, Luigi and Juan.

Solution When answering these questions, we must first decide which quantity we will let the variable represent. In general, if one quantity is given in terms of another quantity, we let the variable represent the basic (or the original quantity) on which the second quantity is based. For example, suppose we are given that "Paul is 6 years

See Example 5 f).

older than George." Since Paul's age is given in terms of George's age, we let the variable represent George's age.

The second thing we need to decide is what letter to use as the variable. Although x is often used to represent the variable, other letters may be used. For example, in "Paul is 6 years older than George," if we let x represent George's age, then Paul's age is $x + 6$. If we let g represent George's age, then Paul's age is $g + 6$. Both answers are correct. It is a matter of preference as to which letter you select. Now let's answer the questions.

a) Let c be the number of points scored by the Chicklets. Then $c + 12$ is the number of points scored by the Bisons.

b) Let j be Jim's speed. Then Sheila's speed is $1.4j$.

c) We are not told how much of the $75 each receives. In this case, we can let the variable represent either person. Let a represent the amount Bill receives. Then the amount Mary receives is $75 - a$. For example, if Bill received $20, then Mary receives $75 - 20$ or $55.

d) Let s represent the amount Sylvia has. Then $5s + 7$ is the amount Kim has.

e) Since the length is given in terms of the width, we will let the variable represent the width. If w represents the width, then the length is $4w - 3$.

f) The number of employees in 2003 is based upon the number of employees in 2002. Therefore, let n be the number of employees in 2002. This number of employees increased by 12% is $n + 0.12n$. Thus the number of employees at Delphi Corporation in 2003 is $n + 0.12n$.

g) The total to be shared is 100%, but we do not know how much each partner gets. We can let the variable represent the percent that either Luigi or Juan gets. If p represents the percent that Luigi gets, than $100 - p$ is the percent Juan gets. Notice that the total profit is 100%. [That is, $p + (100 - p) = 100$.]

NOW TRY EXERCISE 27

4 Write Expressions Involving Multiplication

Consider the statement "the cost of 3 items at $5 each." How would you represent this quantity using mathematical symbols? You would probably reason that the cost would be 3 times $5 and write $3 \cdot 5$ or $3(5)$.

Now consider the statement "the cost of x items at $5 each." How would you represent this statement using mathematical symbols? If you use the same reasoning, you might write $x \cdot 5$ or $x(5)$. Another way to write this product is $5x$. Thus, the cost of x items at $5 each could be represented as $5x$.

Finally, consider the statement "the cost of x items at y dollars each." Following the reasoning used in the previous two illustrations, you might write $x \cdot y$ or $x(y)$. Since these products can be written as xy, the cost of x items at y dollars each can be represented as xy.

EXAMPLE 6

Write each statement as an algebraic expression.
a) The cost of purchasing x pens at $2 each
b) A 5% commission on x dollars in sales
c) The dollar amount earned in h hours if a person earns $6.50 per hour
d) The number of calories in x chocolate mints if each chocolate mint has 55 calories
e) The increase in population in n years for a city growing by 300 persons per year
f) The distance traveled in t hours when 55 miles are traveled each hour

Solution   **a)** We can reason like this: one pen would cost $1(2)$ dollars, two pens would cost $2(2)$ dollars, three pens $3(2)$, four pens $4(2)$, and so on. Continuing this reasoning process, we can see that x pens would cost $x(2)$ or $2x$ dollars.

b) A 5% commission on $1 sales would be $0.05(1)$, on $2 sales $0.05(2)$, on $3 sales $0.05(3)$, on $4 sales $0.05(4)$, and so on. Therefore, the commission on sales of x dollars would be $0.05(x)$ or $0.05x$.

c) $6.50h$

d) $55x$

e) $300n$

NOW TRY EXERCISE 9   **f)** $55t$ ✳

EXAMPLE 7   **Cost of Movie**   The cost for seeing a movie at an AMC Movie Theater is $6.50 for adults and $4.25 for children. Write an algebraic expression to represent the total income received by the theater if x adults and y children pay for admission.

Solution   For x adults, the theater receives $6.50x$ dollars.
For y children, the theater receives $4.25y$ dollars.

NOW TRY EXERCISE 33   For x adults and y children, the theater receives $6.50x + 4.25y$ dollars.   ✳

EXAMPLE 8   Write an algebraic expression for each statement.

a) The number of ounces in x pounds

b) The number of cents in a dimes and b nickels

c) The number of seconds in x hours, y minutes, and z seconds (3600 seconds = 1 hour)

Solution   **a)** Since each pound contains 16 ounces, x pounds is $16 \cdot x$ or $16x$ ounces.

b) Since a dimes is $10a$ cents and b nickels is $5b$ cents, the answer is $10a + 5b$.

c) $3600x + 60y + z$ ✳

Some terms that we will be using are consecutive integers, consecutive even integers, and consecutive odd integers. **Consecutive integers** are integers that differ by 1 unit. For example, the integers 6 and 7 are consecutive integers. Two consecutive integers may be represented as x and $x + 1$. **Consecutive even integers** are even integers that differ by 2 units. For example, 6 and 8 are consecutive even integers. **Consecutive odd integers** also differ by 2 units. For example, 7 and 9 are consecutive odd integers. Two consecutive even integers, or two consecutive odd integers, may be represented as x and $x + 2$.

5 Translate Applications into Equations

The word *is* in an application problem often means *is equal to* and is represented by an equal sign. Some examples of statements written as equations follow.

Statement	Equation
Six more than twice a number *is* 4.	$2x + 6 = 4$
A number decreased by 4 *is* 3 more than twice the number.	$x - 4 = 2x + 3$
The product of two consecutive integers *is* 56.	$x(x + 1) = 56$
The sum of a number and the number increased by 4 *is* 60.	$x + (x + 4) = 60$
Twice the difference of a number and 3 *is* the sum of the number and 20.	$2(x - 3) = x + 20$
A number increased by 15% *is* 120.	$x + 0.15x = 120$
The sum of two consecutive odd integers *is* 24.	$x + (x + 2) = 24$

Now let's translate some equations into statements. Some examples of equations written as statements follow. We will write only two statements for each equation, but remember there are other ways these equations can be written.

Equation	Statements
$3x - 4 = 4x + 3$	Four less than 3 times a number *is* 3 more than 4 times the number. Three times a number, decreased by 4 *is* 4 times the number, increased by 3.
$3(x - 2) = 6x - 4$	Three times the difference between a number and 2 *is* 4 less than 6 times the number. The product of 3 and the difference between a number and 2 *is* 6 times the number, decreased by 4.

EXAMPLE 9 Write two statements to represent the equation $x - 4 = 3x - 6$.

Solution **1.** A number decreased by 4 is 6 less than 3 times the number.

2. The difference between a number and 4 is the difference between 3 times the number and 6.

EXAMPLE 10 Write a statement to represent the equation $x + 2(x - 4) = 6$.

Solution The sum of a number and twice the difference between the number and 4 *is* 6.

EXAMPLE 11 **Translate Words into Equations** Write each problem as an equation.

a) One number is 4 less than twice the other. Their sum is 14.

b) For two consecutive integers, the sum of the smaller and 3 times the larger is 23.

Solution **a)** First, we express the two numbers in terms of the variable.

$$\text{Let } x = \text{one number}$$
$$\text{then } 2x - 4 = \text{second number}$$

Now we write the equation using the information given.

$$\text{first number} + \text{second number} = 14$$
$$x + (2x - 4) = 14$$

b) First, we express the two consecutive integers in terms of the variable.

$$\text{Let } x = \text{smaller consecutive integer}$$
$$\text{then } x + 1 = \text{larger consecutive integer}$$

Now we write the equation using the information given.

$$\text{smaller} + 3 \text{ times the larger} = 23$$
$$x + 3(x + 1) = 23$$

NOW TRY EXERCISE 77

HELPFUL HINT

If you examine Example 11 a) the word *is* is used twice, once in each sentence. However, only one equal sign appears in the equations. When the word *is* appears more than once, as in Example 11 a), generally one of them is being used to express the relationship between the numbers, and the other *is* will represent the equal sign in the equation. When you come across this situation, read the question carefully to determine which *is* will represent the equal sign in the equation.

EXAMPLE 12 **Translate Words into Equations** Write the following problem as an equation. One train travels 3 miles more than twice the distance another train travels. The total distance traveled by both trains is 800 miles.

Solution First express the distance traveled by each train in terms of the variable.

$$\text{Let } x = \text{distance traveled by one train}$$
$$\text{then } 2x + 3 = \text{distance traveled by second train}$$

Now write the equation using the information given.

$$\text{distance of train 1} + \text{distance of train 2} = \text{total distance}$$
$$x + (2x + 3) = 800$$

EXAMPLE 13 **Translate Words into Equations** Write the following problem as an equation. Lori Soushon is 4 years older than 3 times the age of her son Ron. The difference in Lori's age and Ron's age is 26 years.

Solution Since Lori's age is given in terms of Ron's age, we will let the variable represent Ron's age.

$$\text{Let } x = \text{Ron's age}$$
$$\text{then } 3x + 4 = \text{Lori's age}$$

We are told that the difference in Lori's age and Ron's age is 26 years. The word *difference* indicates subtraction.

$$\text{Lori's age} - \text{Ron's age} = 26$$
$$(3x + 4) - x = 26$$

HELPFUL HINT

In a written expression certain other words may be used in place of *is* to represent the equal sign. Some of these are "will be," "was," and "yields." For example, "When 4 is added to a number, the sum *will be* 3 times the number" can be expressed as $x + 4 = 3x$.

"Six subtracted from a number *was* $\frac{1}{2}$ the number" can be expressed as $x - 6 = \frac{1}{2}x$.

"Five added to a number *yields* 3 times the number" can be expressed as $x + 5 = 3x$.

Generally, the word *is* is used to represent the equal sign, but in some cases you may need to study the wording of the question to determine where the equal sign is to be placed.

EXAMPLE 14 **Translate Words into Equations** Express each problem as an equation.

a) George Devenney rented a tiller for x days at a cost of $22 per day. The cost of renting the tiller *was* $88.

TEACHING TIP
Have students work in pairs to
a) write 3 equations for another
pair to translate into sentences.
b) write 3 sentences for another
pair to translate into equations.

b) The distance Scott Borden traveled for x days at 600 miles per day *was* 1500 miles.

c) The population in the town of Rush is increasing by 500 people per year. The increase in population in t years *is* 2500.

d) The number of cents in d dimes *is* 120.

Solution

a) The cost of renting the tiller for x days is $22x$. Therefore, the equation is $22x = 88$.

b) The distance traveled at 600 miles per day for x days is $600x$. Therefore, the equation is $600x = 1500$.

c) The increase in the population in t years is $500t$. Therefore, the equation is $500t = 2500$.

NOW TRY EXERCISE 85

d) The number of cents in d dimes is $10d$. Therefore, the equation is $10d = 120$.

HELPFUL HINT

It is important that you understand this section and work all your assigned homework problems. You will use the material learned in this section in the next two sections, and throughout the book.

In the examples in this section, we used different letters to represent the variable. Often the letter x is used to represent the variable, but other letters might be used. For example, when discussing an expression or equation involving distance, you may use x to represent the distance, or you may choose to use d, or another variable to represent the distance. Thus to represent the expression "the distance increased by 20 miles," you may write $x + 20$ or $d + 20$. Both are correct.

If your teacher, or the exercise, does not indicate which letter to use to represent the variable, you may select the letter you wish to use. If the answer appendix has an answer as $d + 20$ and your answer is $x + 20$, your answer would be correct, if x and d represent the same quantity.

---

2. subtracted from, less than, decreased by, difference between
3. multiplied by, product of, twice, three times.

Exercise Set 3.2

Concept/Writing Exercises

1. Give four phrases that indicate the operation of addition. added to, more than, increased by, sum

2. Give four phrases that indicate the operation of subtraction.

3. Give four phrases that indicate the operation of multiplication.

4. Give four phrases that indicate the operation of division. divided by, quotient of, one-half, one-third.

5. Explain why $c + 0.25$ *does not* represent the cost of an item increased by 25 percent. need $0.25c$

6. Explain why $c - 0.10$ *does not* represent the cost of an item decreased by 10 percent. need $0.10c$

Practice the Skills

7. *Age* LeDawn Webb is n years old now. Write an expression that represents her age in 7 years. $n + 7$

8. *Age* Hannah Whitlock is t years old. Write an expression that represents David Alevy's age if he is 4 times as old as Hannah $4t$

9. *Pens* At a sale, a Dr. Grip pen costs \$4. Write an expression that represents the cost of purchasing x Dr. Grip pens. $4x$

10. *New Price* An item that costs r dollars is increased by \$6. Write an expression that represents the new price. $r + 6$

11. *Motorcycle* Melissa Blum is selling her motorcycle. She was asking x dollars for the motorcycle but has cut the price in half. Write an expression that represents the new price. $x/2$

12. *Shoes* Keri Goldberg is buying some shoes. The store is having a two-for-one sale, when for each pair of shoes you purchase you are given a second pair free. Write an expression that represents the number of pairs of shoes Keri will get if she purchases y pairs. $2y$

13. *Growing Giraffe* In one year, a giraffe's height, h, increased by 0.8 feet. Write an expression that represents the giraffe's present height. $h + 0.8$

14. *Speed-Reading* John Debruzzi used to read p words per minute. After taking a speed-reading course, his speed increased by 60 words per minute. Write an expression that represents his new reading speed. $p + 60$

15. *Stock Price* In 2002, the price of General Electric's stock, p, dropped 8%. Write an expression for the price of the stock after it dropped. $p - 0.08p$

16. *Recycling Tires* Each year t tires are disposed of in the United States. Only 7% of all tires disposed of are recycled. Write an expression that represents the number of tires that are recycled. $0.07t$

17. *Vitamins* In a Solaray vitamin capsule, the number of milligrams of riboflavin is 5 milligrams fewer than one-tenth the number of milligrams of vitamin C. If n represents the number of milligrams of vitamin C, write an expression for the number of milligrams of riboflavin. $\frac{1}{10}n - 5$

18. *Population* In 2001, China had the greatest population while India had the second greatest population. If p represents the population of India, in millions, write an expression for the population of China if China had 40 million more than 1.2 times the population of India. $1.2p + 40$

19. *Promotion* Assume that last year Jose Rivera had a salary of m dollars. This year he received a promotion and his new salary is $16,000 plus eight-ninths of his previous salary. Write an expression that represents his new salary. $\frac{8}{9}m + 16,000$

20. *Roller Coaster* On a specific drop on a roller coaster in Universal Studios, the speed of a cart coming down the drop is 12 miles per hour greater than 8 times the speed, s, of the cart climbing up another part of the roller coaster. Write an expression that represents the speed of the cart coming down the specific drop. $8s + 12$

21. *Truck Rental* Bob Melina rented a truck for a trip. He paid a daily fee of $45 and a mileage fee of 40 cents a mile. Write an expression that represents his total cost when he travels x miles in one day. $45 + 0.40x$

22. *Population Increase* The city of Clarkville has a population of 4000. If the population increases by 300 people per year, write an expression that represents the population in n years. $4000 + 300n$

23. *Money* Carolyn Curley found that she had x quarters in her handbag. Write an expression that represents this quantity of money in cents. $25x$

24. *Height* Dennis De Valeriz's height is x feet and y inches. Write an expression that represents his height in inches. $12x + y$

25. *Weight* Kim Wager's weight is x pounds and y ounces. Write an expression that represents her weight in ounces. $16x + y$

26. *Newborn* Jill's newborn baby is m minutes and s seconds old. Write an expression that represents the baby's age in seconds. $60m + s$

27. *Sales Staff* The sales staff at the Prentice Hall Publishing Company increased by 4% from 2001 to 2002. If n represents the number of people on the sales staff at Prentice Hall in 2001, write an expression for the number of people on the sales staff at Prentice Hall in 2002. $n + 0.04n$

28. *Nuclear Reactors* The number of reactors in western Europe is 137 less than 2.4 times the number of reactors in the United States. If r represents the number of reactors in the United States, write an expression for the number of reactors in western Europe. $2.4r - 137$

29. **Manatees** A particular manatee (also referred to as a sea cow) lost 2% of its weight over the winter months. If it originally weighed p pounds, write an expression for its new weight. $p - 0.02p$

30. **Taxes** Through some careful tax planning, Gil French was able to reduce his federal income tax from 2002 to 2003 by 18%. If his 2002 tax was t dollars, write an expression that represents his 2003 tax. $t - 0.18t$

31. **Counting Calories** Each slice of white bread has 110 calories and each teaspoon of strawberry preserves has 80 calories. If Donna Contoy makes a strawberry preserve sandwich (2 pieces of bread) and uses x teaspoons of preserves, write an expression that represents the number of calories the sandwich will contain. $220 + 80x$

32. **Growing Las Vegas** Las Vegas, Nevada, according to the U.S. Census Bureau, had the greatest population growth from 1990 to 2000 of any major city in the United States. Las Vegas's population in 2000 was 38,156 less than twice its population in 1990. If p represents Las Vegas's population in 1990, write an expression for Las Vegas's 2000 population. $2p - 38{,}156$

33. **Cholesterol** An average chicken egg contains about 275 milligrams (mg) of cholesterol and an ounce of chicken contains about 25 mg of cholesterol. Write an expression that represents the amount of cholesterol in x chicken eggs and y ounces of chicken. $275x + 25y$

See Exercise 32, Las Vegas at Night

34. **Reading Food Labels** According to U.S. guidelines, each gram of carbohydrates contains 4 calories, each gram of protein contains 4 calories, and each gram of fat contains 9 calories. Write an expression that represents the number of calories in a serving of a product that contains x grams of carbohydrates, y grams of protein, and z grams of fat. $4x + 4y + 9z$

Write each mathematical expression as a statement. (There are many acceptable answers.)

35. $x - 3$ three less than a number

36. $x + 5$ five more than a number

37. $4x + 1$ one more than 4 times a number

38. $3x - 4$ three times a number, decreased by 4

39. $6x - 7$ seven less than 6 times a number

40. $7x - 6$ the difference between 7 times a number and 6

41. $4x - 2$ four times a number, decreased by 2

42. $5 - x$ the difference between 5 and a number

43. $2 - 3x$ three times a number, subtracted from 2

44. $4 + 6x$ the sum of 4 and 6 times a number

45. $2(x - 1)$ twice the difference between a number and 1

46. $3(x + 2)$ three times the sum of a number and 2

In Exercises 47–74, a variable is given to represent one quantity. Express the quantity specified in terms of the variable given. For example, if given the statement "Mike is 2 years older than 3 times Don's age, d," then Mike's age would be represented as $3d + 2$. If given the statement "Paul's 2003 salary was 8% greater than his 2002 salary, s," then Paul's 2003 salary would be represented as $s + 0.08s$. See Example 5 for additional illustrations.

47. **Age** Judy is 5 years older than Scott's age, s. Write an expression for Judy's age. $s + 5$

48. **Age** Lois Heater's son is one-third as old as Lois, l. Write an expression for her son's age. $\frac{1}{3}l$

49. **Teaching Career** Traci has been teaching for 6 years less than Ben, b. Write an expression for the time Traci has been teaching. $b - 6$

50. **Money** One hundred dollars is divided between Romayne and Jim, j. Write an expression for Romayne's amount. $100 - j$

51. **Soil** A total of six hundred pounds of soil is put onto two trucks. If a pounds of soil is placed onto one truck, write an expression for the amount of soil placed onto the other truck. $600 - a$

52. *Television Shopping* A Sony television costs 1.4 times what an RCA television costs, r. Write an expression for the cost of the Sony television. $1.4r$

53. *Profits* Monica and Julia share in the profits, in percent, in a toy store. Write an expression for the amount Julia receives if m is the amount Monica receives. $100 - m$

54. *Calories* The calories in a serving of mixed nuts is 280 calories less than twice the number of calories in a serving of cashew nuts, c. Write an expression for the number of calories in a serving of mixed nuts. $2c - 280$

55. *Number Employed* At the Elten-Mark Company, the number of women employed is 6 less than two-thirds the number of men employed, m. Write an expression for the number of women employed. $\frac{2}{3}m - 6$

56. *Land Area* The largest state in area in the United States is Alaska and the smallest is Rhode Island. The area of Alaska is 462 square miles more than 479 times the area of Rhode Island, r. Write an expression for the area of Alaska. $479r + 462$

57. *Home Runs* In professional baseball, the all-time home run-leader is Hank Aaron, and Babe Ruth is second. Hank Aaron had 673 less home runs than twice what Babe Ruth had, r. Write an expression for the number of home runs Hank Aaron had. $2r - 673$

58. *Life Expectancy* In 2001, the country with the highest average life expectancy was Japan, and the country with the lowest average life expectancy was Botswana. The average life expectancy in Japan was 9.3 years greater than twice that in Botswana, b. Write an expression for the average life expectancy in Japan. (Source: United Nations). $2b + 9.3$

59. *Crowded Museum* In 2000, the number of visitors to the National Air and Space Museum (part of the Smithsonian) was 2.7 million less than twice the number who visited the Louvre, in Paris, p. Write an expression for the number of visitors to the Air and Space Museum in 2000. (Source: USA Today) $2p - 2.7$

60. *Movie Ticket* The average price of a movie ticket in the United States in April 2001 was 70 cents less than 30 times the price in 1928, p. Write an expression for the price, in cents, of a movie ticket in April 2001. (Source: Money Magazine) $30p - 70$

61. *Travel to Work* The average travel time to work in New York State (the highest average) is 15 minutes less than 3 times the average travel time to work in North

Dakota (the lowest average), n. Write an expression for the average travel time to work in New York State. (Source: USA Today) $3n - 15$

62. *Broadway Shows* As of September 2, 2001, the longest running Broadway show was *Cats* and the second longest running was *A Chorus Line*. *A Chorus Line* had 16,318 less performances than three times the number of performances than *Cats, c*, had. Write an expression for the number of performances that *A Chorus Line* had. $3c - 16{,}318$

63. *Median Income* According to the Census Bureau, in the year 2000, New Jersey was the state with the greatest median household income. The U.S. average median income was $67,109 less than twice the New Jersey median income, n. Write an expression for the U.S. average (median) income. $2n - 67{,}109$

64. *Stock Price* The price of Enron stock in January 2002 was $0.60 less than $\frac{1}{130}$ times its January 2001 price, p. Write an expression for the price of Enron stock in January 2002. $\frac{1}{130}p - 0.60$

65. *Sales Increase* Mike Sutton is a sales representative for a medical supply company. His 2003 sales increased by 20% over his 2002 sales, s. Write an expression for his 2003 sales. $s + 0.20s$

66. *Electricity Use* George Young's electricity use in 2003 decreased by 12% from his 2002 electricity use, e. Write an expression for his 2003 electricity use. $e - 0.12e$

67. *Salary Increase* Karen Moreau, an engineer, had a salary increase of 15% over last year's salary, s. Write an expression for this year's salary. $s + 0.15s$

68. *Ladies' Golf* Karrie Webb was the ladies' leading professional golf money winner in both 1999 and 2000. Her earnings increased by about 18% from 1999 to 2000. If w represents Karrie Webb's 1999 earnings, write an expression for her 2000 earnings. $w + 0.18w$

69. *Pizza Parlor* The number of customers to Family Fun Pizza Parlor dropped by 12% from September to October. If f represents the number of customers in September, write an expression for the number of customers in October. $f - 0.12f$

70. *Flu Cases* The number of cases of flu in Archville decreased by 2% from the previous year. If f represents the number of cases of flu in Archville in the previous year, write an expression for the number of flu cases this year. $f - 0.02f$

71. *Car Cost* The cost of a new car purchased in Collier County included a 7% sales tax. If c represents the cost of the car before tax, write an expression for the total cost, including the sales tax. $c + 0.07c$

72. *Shirt Sale* At a 25% off everything sale, Bill Winchief purchased a new shirt. If c represents the pre-sale cost, write an expression for the sale price of the shirt. $c - 0.25c$

73. Pollution The pollution level in Detroit decreased by 50%. If p represents the pollution level before the decrease, write an expression for the new pollution level.

74. Grades The number of students earning a grade of A in this course increased by 100%. If n represents the number of students who received an A before the increase, write an expression for the number of students who now receive an A. $n + 1.00n$ or $2n$

73. $p - 0.50p$

79. $2x - 8 = 12$ **80.** $x + 2(x + 1) = 29$ **83.** $s + (2s - 4) = 890$ **85.** $c + 0.07c = 32,600$ **86.** $c - 0.25c = 195$
87. $c + 0.15c = 42.50$ **90.** $(2l - 16.9) - l = 31.4$

In Exercises 75–92, read the exercise carefully. Then select a letter to represent one of the quantities and state exactly what the letter (or variable) represents. Then write an equation to represent the problem. For example, if given "Steve's age is twice Gail's age, and the sum of their ages is 20," we might let g *represent Gail's age, and then Steve's age would be 2g. Since the sum of their ages is 20, the equation we are seeking is 2g + g = 20. Since different letters may be used to represent the variable, other answers are possible. For instance, 2a + a = 20 would also be acceptable if* a *was used to represent Gail's age. See Examples 11 through 14 for further illustrations.*

75. Two Numbers One number is 4 times another. The sum of the two numbers is 20. $x + 4x = 20$

76. Age Marie is 6 years older than Denise. The sum of their ages is 48. $x + (x + 6) = 48$

77. Consecutive Integers The sum of two consecutive integers is 41. $x + (x + 1) = 41$

78. Even Integers The product of two consecutive even integers is 72. $x(x + 2) = 72$

79. Numbers Twice a number, decreased by 8 is 12.

80. Consecutive Integers For two consecutive integers, the sum of the smaller and twice the larger is 29.

81. Numbers One-fifth of the sum of a number and 10 is 150. $\frac{1}{5}(x + 10) = 150$

82. Jogging David Ostrow jogs 5 times as far as Jennifer Freer. The total distance traveled by both people is 8 miles. $f + 5f = 8$

83. Amtrak An Amtrak train travels 4 miles less than twice the distance traveled by a Southern Pacific train. The total distance traveled by both trains is 890 miles.

84. Wagon Ride On a wagon ride the number of girls was 8 less than twice the number of boys. The total number of boys and girls on the wagon was 24. $b + (2b - 8) = 24$

85. New Car Carlotta Diaz bought a new car. The cost of the car plus a 7% sales tax was $32,600.

86. Sport Coat David Gillespie purchased a sport coat at a 25% off sale. He paid $195 for the sport coat.

87. Cost of Meal Beth Rechsteiner ate at a steakhouse. The cost of the meal plus a 15% tip was $42.50.

88. Video Cassette At the Better Buy Warehouse, Anne Long purchased a video cassette recorder that was reduced by 10% for $208. $c - 0.10c = 208$

89. Salary In 2000, the metropolitan area with the highest average annual salary was San Jose, California, and the metropolitan area with the second highest average annual salary was San Francisco. The average salary in San Jose was about 1.28 times the average salary in San Francisco. The difference between the average salary in San Jose and San Francisco was $16,762. (Source: Bureau of Labor Statistics) $1.28f - f = 16,762$

90. Life Expectancy Women lived much longer in 2000 than they did in 1900. The life expectancy for women in the United States, in 2000 was 16.9 years less than twice their life expectancy in 1900. The difference in their life expectancies was 31.4 years. (Source: U.S. Census Bureau)

91. Laser Surgery The number of laser vision surgeries was 0.55 million more in 2000 than in 1999. The total laser surgeries in 1999 and 2000 sum to 2.45 million. (Source: *U.S. News and World Report*) $s + (s + 0.55) = 2.45$

92. Railroad In a narrow gauge railway, the distance between the tracks is about 64% of the distance between the tracks in a standard gauge railroad. The difference in the distances between the tracks in a standard and narrow gauge railroad is about 1.67 feet. $s - 0.64s = 1.67$

In Exercises 93–104, express each equation as a statement. (There are many acceptable answers.)

93. $x + 2 = 5$

94. $x - 5 = 2x$

95. $3x - 1 = 2x + 4$

96. $x - 3 = 2x + 3$

97. $4(x - 1) = 6$

98. $4x + 6 = 2(x - 3)$

99. $5x + 6 = 6x - 1$

100. $x - 3 = 2(x + 1)$

101. $x + (x + 4) = 8$

102. $x + (2x + 1) = 5$

103. $2x + (x + 3) = 5$

104. $2x - (x + 3) = 6$

105. Explain why the cost of purchasing x items at 6 dollars each is represented as $6x$.

106. Explain why the cost of purchasing x items at y dollars each is represented as xy.

93. Two more than a number is 5. **94.** Five less than a number is twice the number. **95.** Three times a number, decreased by 1, is 4 more than twice the number. **96.** Three less than a number is 3 more than twice the number. **97.** Four times the difference between a number and 1 is 6. **98.** Six more than 4 times a number is twice the difference between the number and 3. **99.** Six more than 5 times a number is the difference between 6 times the number and 1.

Challenge Problems

107. *Time*

a) Write an algebraic expression for the number of seconds in d days, h hours, m minutes, and s seconds.
$86{,}400d + 3600h + 60m + s$

b) Use the expression found in part **a)** to determine the number of seconds in 4 days, 6 hours, 15 minutes, and 25 seconds. $368{,}125$ sec

108. *Tolls* At the time of this writing, the toll for southbound traffic on the Golden Gate Bridge is $1.50 per vehicle *axle* (there is no toll for northbound traffic, car pools are free on weekends).

a) If the number of 2-, 3-, 4-, 5-, and 6-axle vehicles are represented with the letters r, s, t, u, and v, respectively, write an *expression* that represents the daily revenue of the Golden Gate Bridge Authority.

b) Write an *equation* that can be used to determine the daily revenue, d. $d = 3r + 4.50s + 6t + 7.50u + 9v$

100. Three less than a number is twice the sum of the number and 1. **101.** The sum of a number and the number increased by 4 is 8. **102.** The sum of a number and 1 more than twice the number is 5. **103.** The sum of twice a number and the number increased by 3 is 5. **104.** The difference between twice a number and 3 more than the number is 6. **105.** and **106.** Answers will vary. **108. a)** $3r + 4.50s + 6t + 7.50u + 9v$

Group Activity

Exercises 109 and 110 will help prepare you for the next section, where we set up and solve application problems. Discuss and work each exercise as a group. For each exercise, write down the quantity you are being asked to find and represent this quantity with a variable. Then write an equation containing your variable that can be used to solve the problem. Do not solve the equation.

109. *Water Usage* An average bath uses 30 gallons of water and an average shower uses 6 gallons of water per minute. How long a shower would result in the same water usage as a bath? $30 = 6t$

110. *Salary Plans* An employee has a choice of two salary plans. Plan A provides a weekly salary of $200 plus a 5% commission on the employee's sales. Plan B provides a weekly salary of $100 plus an 8% commission on the employee's sales. What must be the weekly sales for the two plans to give the same weekly salary? $200 + 0.05s = 100 + 0.08s$

Cumulative Review Exercises

111. Evaluate $3[(4 - 16) \div 2] + 5^2 - 3$. $\quad$ 4

112. Solve the proportion $\dfrac{3.6}{x} = \dfrac{10}{7}$. $\quad$ 2.52

113. Solve the inequality $2x - 4 > 3$ and graph the solution on a number line. $\quad x > \frac{7}{2}$,

114. $P = 2l + 2w$; find l when $P = 40$ and $w = 5$. $\quad$ 15

$\qquad$ **115.** Solve $3x - 2y = 6$ for y. Then find the value of y when x has a value of 6. $\quad y = \dfrac{3x - 6}{2}$ or $y = \dfrac{3}{2}x - 3$; 6

3.3 SOLVING APPLICATION PROBLEMS

SSM $\qquad$ Study Guide $\qquad$ CD/Video

MathPro 4/5 $\quad$ PH Math
Tutor Center $\quad$ prenhall.com/Angel

1 $\quad$ Use the problem solving procedure.
2 $\quad$ Set up and solve number application problems.
3 $\quad$ Set up and solve application problems involving money.
4 $\quad$ Set up and solve applications concerning percent.

1 $\quad$ Use the Problem-Solving Procedure

1 Gallon

2 cups
1 cup

Many types of application problems can be solved using algebra. In this section we introduce several types. In Sections 3.4 and 3.5 we introduce additional types of applications. They are also presented in many other sections and exercise sets throughout the book. Your instructor may not have time to cover all the applications given in this book. If not, you may still wish to spend some time on your own reading those problems just to get a feel for the types of applications presented.

To be prepared for this section, you must understand the material presented in Section 3.2. The best way to learn how to set up an application or word problem is to practice. The more problems you study and attempt, the easier it will become to solve them.

We often translate problems into mathematical terms without realizing it. For example, if you need 3 cups of milk for a recipe and the measuring cup holds only 2 cups, you reason that you need 1 additional cup of milk after the initial 2 cups. You may not realize it, but when you do this simple operation, you are using algebra.

Let x = number of additional cups of milk needed

Thought process: $(\text{initial 2 cups}) + \left(\begin{array}{c}\text{number of} \\ \text{additional cups}\end{array}\right) = \text{total milk needed}$

Equation to represent problem: $2 + x = 3$

When you solve for x, you get 1 cup of milk.

You probably said to yourself: Why do I have to go through all this when I know that the answer is $3 - 2$ or 1 cup? When you perform this subtraction, you have mentally solved the equation $2 + x = 3$.

$$2 + x = 3$$
$$2 - 2 + x = 3 - 2 \qquad \textit{Subtract 2 from both sides.}$$
$$x = 3 - 2$$
$$x = 1$$

This illustration was very basic. However, most of the problems in this chapter will not be solvable by observation and will require the use of algebra.

The general problem-solving procedure given in Section 1.2 can be used to solve all types of verbal problems. Below, we present the **five-step problem-solving procedure** again so you can easily refer to it. We have included some additional information under steps 1 and 2, since in this section we are going to emphasize translating application problems into equations.

> ### Problem-Solving Procedure for Solving Applications
>
> 1. **Understand the problem.**
>   Identify the quantity or quantities you are being asked to find.
> 2. **Translate the problem into mathematical language (express the problem as an equation).**
>   a) Choose a variable to represent one quantity, *and write down exactly what it represents*. Represent any other quantity to be found in terms of this variable.
>   b) Using the information from step a), write an equation that represents the application.
> 3. **Carry out the mathematical calculations (solve the equation).**
> 4. **Check the answer (using the *original* application).**
> 5. **Answer the question asked.**

TEACHING TIP
Point out that when solving word problems, you must answer the question asked.

Sometimes we will combine two steps in the problem-solving procedure when it helps to clarify the explanation. We may not show the check of a problem to save space. Even if we do not show a check, you should check the problem yourself and make sure your answer is reasonable and makes sense.

Let's now set up and solve some application problems using this procedure. We will first work application problems that do not contain percents, then we will work application problems that contain percents.

2 Set Up and Solve Number Application Problems

The examples presented under this objective involve information and data, but do not contain percents. When we work the examples we will follow the 5-step problem-solving procedure.

EXAMPLE 1 **An Unknown Number** Two subtracted from 4 times a number is 10. Find the number.

Solution Understand To solve this problem, we need to express the statement given as an equation. We are asked to find the unknown number. We use the information learned in the previous section to write the equation.

Translate Let x = the unknown number. Now write the equation.

$$4x - 2 = 10$$

Carry Out
$$4x = 12$$
$$x = 3$$

Check   Substitute 3 for the number in the original problem, two subtracted from 4 times a number is 10.

$$4(3) - 2 \overset{?}{=} 10$$
$$10 = 10 \quad \textit{True}$$

Answer   Since the solution checks, the unknown number is 3.   ✳

EXAMPLE 2   **Number Problem**   The sum of two numbers is 26. Find the two numbers if the larger number is 2 less than three times the smaller number.

Solution   Understand   This problem involves finding two numbers. When finding two numbers, if a second number is expressed in terms of a first number, we generally let the variable represent the first number. Then we represent the second number as an expression containing the variable used for the first number. In this example, we are given that "the larger number is 2 less than three times the smaller number." Notice that the larger number is expressed in terms of the smaller number. Therefore, we will let the variable represent the smaller number.

Translate   Let x = smaller number
                          then $3x - 2$ = larger number

The sum of the two numbers is 26. Therefore, we write the equation

$$\text{smaller number} + \text{larger number} = 26$$
$$x + (3x - 2) = 26$$

Carry Out   Now we solve the equation.

$$4x - 2 = 26$$
$$4x = 28$$
$$x = 7$$

The smaller number is 7. Now we find the larger number.

$$\text{larger number} = 3x - 2$$
$$= 3(7) - 2 \quad \textit{Substitute 7 for x.}$$
$$= 19$$

The larger number is 19.

Check   The sum of the two numbers is 26.

$$7 + 19 \overset{?}{=} 26$$
$$26 = 26 \quad \textit{True}$$

NOW TRY EXERCISE 5   Answer   The two numbers are 7 and 19.   ✳

HELPFUL HINT   When reading a word problem, ask yourself "how many answers are required?" In Example 2, the question asked for the two numbers. The answer is 7 and 19. It is important that you read the question and identify what you are being asked to find. If the question had asked "Find the *smaller* of the two numbers if the larger number is 2 less than three times the smaller number," then the answer would have been only the 7. If the question had asked to find the *larger* of the two numbers, then the answer would have been only 19. *Make sure you answer the question asked in the problem.*

EXAMPLE 3 **Sneakers** Candice *Cotton* and Thomas *Johnson* formed a corporation that manufactures sneakers. Their corporation, called CoJo, will manufacture 1200 pairs of sneakers this year. Each year after this year they plan to increase production by 550 pairs until their annual production reaches 4500 pairs. How long will it take them to reach their production goal?

Solution

Understand We are asked to find the *number of years* that it will take for their production to reach 4500 pairs a year. Next year their production will increase by 550 pairs. In two years, their production will increase by 2(550) over the present year's production. In n years, their production will increase by $n(550)$ or $550n$. We will use this information when we write the equation to solve the problem.

Translate

Let n = number of years

then $550n$ = increase in production over n years

$$(\text{present production}) + \begin{pmatrix} \text{increased production} \\ \text{over } n \text{ years} \end{pmatrix} = \text{future production}$$

$$1200 + 550n = 4500$$

Carry Out

$$550n = 3300$$

$$n = \frac{3300}{550}$$

$$n = 6 \text{ years}$$

Check and Answer As a check, let's list the number of pairs produced this year and for the next 6 years.

This year	Next year	Year 2	Year 3	Year 4	Year 5	Year 6
↓	↓	↓	↓	↓	↓	↓
1200	1750	2300	2850	3400	3950	4500

NOW TRY EXERCISE 19   Thus, in 6 years they will produce 4500 pairs of sneakers per year.   ✳

3   Set Up and Solve Application Problems Involving Money

EXAMPLE 4 **Truck Rental** Armondo Rodriguez is moving. He plans to rent a truck for one day to make the local move. The cost of renting the truck is $60 a day plus 40 cents a mile. Find the maximum distance Armondo can drive if he has only $92.

Solution

RENT-A-TRUCK

Understand The total cost of renting the truck consists of two parts, a fixed cost of $60 per day, and a variable cost of 40 cents per mile. We need to determine the *number of miles* that Armondo can drive so that the total rental cost is $92. Since the fixed cost is given in dollars, we will write the variable cost, or mileage cost, in dollars also.

Translate

Let x = number of miles

then $0.40x$ = cost of driving x miles

daily cost + mileage cost = total cost

$$60 + 0.40x = 92$$

Carry Out

$$0.40x = 32$$

$$\frac{0.40x}{0.40} = \frac{32}{0.40}$$

$$x = 80$$

Check   The cost of driving 80 miles at 40 cents a mile is $80(0.40) = \$32$. Adding the \$32 to the daily cost of \$60 gives \$92, so the answer checks.

Answer   Armondo can drive a maximum of 80 miles.   ✳

EXAMPLE 5

Security Systems   Jose Alvarado plans to install a security system in his house. He has narrowed down his choices to two security system dealers: Moneywell and Doile. Moneywell's system costs \$3580 to install and their monitoring fee is \$20 per month. Doile's equivalent system costs only \$2620 to install, but their monitoring fee is \$32 per month. Assuming that their monthly monitoring fees do not change, in how many months would the total cost of Moneywell's and Doile's system be the same?

Solution

Understand   Doile's system has a smaller initial cost (\$2620 vs. \$3580); however, their monthly monitoring fees are greater (\$32 vs. \$20). We are asked to find the number of months after which the total cost of the two systems will be the same.

Translate

Let n = number of months

then $20n$ = monthly monitoring cost for Moneywell's system for n months

and $32n$ = monthly monitoring cost for Doile's system for n months

$$\text{total cost of Moneywell} = \text{total cost of Doile}$$

$$\binom{\text{initial}}{\text{cost}} + \binom{\text{monthly cost}}{\text{for } n \text{ months}} = \binom{\text{initial}}{\text{cost}} + \binom{\text{monthly cost}}{\text{for } n \text{ months}}$$

$$3580 + 20n = 2620 + 32n$$

Carry Out

$$960 + 20n = 32n$$
$$960 = 12n$$
$$80 = n$$

Check and Answer   The total cost would be the same in 80 months or about 6.7 years. We will leave the check of this answer for you. If the security system was kept for less than about 6.7 years, Doile would be the less expensive. After about 6.7 years, Moneywell would be the less expensive.   ✳

NOW TRY EXERCISE 31

EXAMPLE 6

Mortgages   Kristen Schwartz is buying her first home and has narrowed down her choices for a mortgage to two banks, Bank of America and Sun Trust. Bank of America informs her that her monthly mortgage payment of principal plus interest will be \$650 per month. Sun Trust informs her that her monthly payment will be \$637 per month, plus she will need to pay \$1500 in fees (called points) to obtain the mortgage. How long would it take for the total cost of both mortgages to be the same?

Solution

Understand   The monthly payments with Sun Trust are less than Bank of America, but Sun Trust has a \$1500 fee that Bank of America does not have. We need to determine the number of months when the lower payments would make up for the extra \$1500 fee.

Translate   Let x = number of months

then $650x$ = monthly payments in x months with Bank of America

and $637x$ = monthly payments in x months with Sun Trust

Now set up the equation that can be used to solve the problem.

Bank of America   Sun Trust

monthly payments $=$ monthly payments $+$ fees

$$650x = 637x + 1500$$
$$13x = 1500$$
$$x \approx 115.4$$

Check   We can check this answer by determining the total costs in 115.4 months for both banks.

Bank of America   Sun Trust

$650x$ $637x + 1500$

$650(115.4) = 75,010$ $637(115.4) + 1500 = 75,009.80$

Notice the amounts are about the same. There is a small round-off error.

Answer   In about 115.4 months (or about 9.62 years) the total cost of both mortgages would be the same. If Kristen is planning on keeping the house less than this time, the Bank of America loan would be less expensive. If Kristen is planning on keeping the house for more than this time, the Sun Trust would be less expensive.

NOW TRY EXERCISE 37   ✳

4   Set Up and Solve Applications Concerning Percent

Now we'll look at some application problems that involve percent. Remember that a percent is always a percent of something. Thus if the cost of an item, c, is increased by 8%, we would represent the new cost as $c + 0.08c$, and not $c + 0.08$. See the Avoiding Common Errors box on page 187.

EXAMPLE 7   **Water Bike Rental**   At a beachfront hotel, the cost for a water bike rental is $20 per half hour, which includes a $7\frac{1}{2}$% sales tax. Find the cost of the rental before tax.

Solution   **Understand**   We are asked to find the cost of the water bike rental before tax. The cost of the rental before tax plus the tax on the water bike must equal $20.

Translate   Let $x =$ cost of the rental before tax

then $0.075x =$ tax on the rental

Carry Out

(cost of the water bike rental before tax) $+$ (tax on the rental) $= 20$

$$x + 0.075x = 20$$
$$1.075x = 20$$
$$x = \frac{20}{1.075}$$
$$x \approx 18.60$$

Check and Answer   A check will show that if the cost of the rental is $18.60, the cost of the rental including a $7\frac{1}{2}$% tax is $20.

NOW TRY EXERCISE 41   ✳

EXAMPLE 8   **Salary Increase**   In 2003, Cho Chi, an insurance salesman, had a salary increase of 18% over his 2002 salary. If his 2003 salary was $43,000, determine his 2002 salary.

Solution   Understand   We can represent Cho's 2003 salary in terms of his 2002 salary. Therefore, we will select a variable to represent Cho's 2002 salary. Once we obtain an expression for Cho's 2003 salary, we can set the expression equal to $43,000 and solve the equation to obtain the answer.

Translate   Let s = Cho's 2002 salary

then $s + 0.18s$ = Cho's 2003 salary

$$\text{Cho's 2003 salary} = \$43{,}000$$
$$s + 0.18s = 43{,}000$$

Carry Out $$1.18s = 43{,}000$$

$$s = \frac{43{,}000}{1.18}$$

$$s = 36{,}440.68$$

NOW TRY EXERCISE 43   Check and Answer   Since s represents Cho's 2002 salary, and it is less than his 2003 salary, our answer is reasonable. Cho's 2002 salary was $36,440.68.   ✳

EXAMPLE 9   **Salary Plans**   Jacqueline Johnson recently graduated from college and has accepted a position selling medical supplies and equipment. During her first year, she is given a choice of salary plans. Plan 1 is a $450 weekly base salary plus a 3% commission of weekly sales. Plan 2 is a straight 10% commission of weekly sales.

a) Jacqueline must select one of the plans but is not sure of the dollar sales needed for her weekly salary to be the same under the two plans. Can you determine it?

b) If Jacqueline is certain that she can make $8000 in sales per week, which plan should she select?

Solution   **a)** Understand   We are asked to find the *dollar sales* that will result in Jacqueline receiving the same total salary from both plans. To solve this problem, we write expressions to represent the salary from each of the plans. We then obtain the desired equation by setting the salaries from the two plans equal to one another.

Translate   Let x = dollar sales

then $0.03x$ = commission from plan 1 sales

and $0.10x$ = commission from plan 2 sales

$$\text{salary from plan 1} = \text{salary from plan 2}$$
$$\text{base salary} + 3\% \text{ commission} = 10\% \text{ commission}$$
$$450 + 0.03x = 0.10x$$

Carry Out $$450 = 0.07x$$

or $$0.07x = 450$$

$$\frac{0.07x}{0.07} = \frac{450}{0.07}$$

$$x \approx 6428.57$$

NOW TRY EXERCISE 53

Check   We will leave it up to you to show that sales of $6428.57 result in Jacqueline receiving the same weekly salary from both plans.

Answer   Jacqueline's weekly salary will be the same from both plans if she sells $6428.57 worth of medical supplies and equipment.

b) If Jacqueline's sales are $8000, she would earn more weekly by working on straight commission, Plan 2. Check this out yourself by computing Jacqueline's salary under both plans and comparing them.

HELPFUL HINT

STUDY TIP

Here are some suggestions if you find you are having some difficulty with application problems.

1. **Instructor**—Make an appointment to see your instructor. Make sure you have read the material in the book and attempted all the homework problems. Go with specific questions for your instructor.
2. **Videotapes**—Find out if the videotapes that accompany this book are available at your college. If so, view the videotapes that go with this chapter. Using the pause control, you can watch the videotapes at your own pace.
3. **Student's Study Guide**—If a copy of the Student's Study Guide is available, you may wish to read the material related to this chapter.
4. **Tutoring**—If your college learning center offers free tutoring, as many colleges do, you may wish to take advantage of tutoring.
5. **Study Group**—Form a study group with classmates. Exchange phone numbers and e-mail addresses. You may be able to help one another.
6. **Student's Solutions Manual**—If you get stuck on an exercise you may want to use the Student's Solutions Manual to help you understand a problem. Do not use the Solutions Manual in place of working the exercises. In general, the Solutions Manual should be used only to check your work.
7. **Web Site**—If a computer is available, visit the Prentice Hall-Allen Angel Web site at prenhall.com/angel and study the material that relates to this chapter. You will find additional worked out examples and exercises.
8. **Math Pro**—If Math Pro is available, it can provide you with valuable help in understanding the material. Math Pro provides unlimited practice and gives you immediate feedback on your work, including an immediate review of your work via integrated video. Check with your instructor to determine if Math Pro is available.
9. **Prentice Hall Mathematics Tutor**—Once the program has been initiated by your instructor, you can get individual tutoring by phone, fax, or e-mail.

It is important that you keep trying! Remember, the more you practice, the better you will become at solving application problems.

Mathematics in Action

Dynamic Pricing
Many of the applications in the book have to do with buying things. The assumption is that purchasable items *have* a price. They may go on sale or be discounted for various reasons, but if one person walks into a store and a different person calls to purchase an item, it will be sold to both of them for the same amount of money.

That whole world of fixed pricing is changing rapidly thanks to the Internet. Buying airline tickets over the Internet has become an incredible game of tag, with the prospective flyer rushing from site to site hoping to arrive at one of the sites at just the right moment. A ticket hunter can return to a site 15 minutes after he or she was quoted a price and find that the price has already moved up or down. What

(continued on the next page)

the airlines are practicing is something called *dynamic pricing*.

From minute to minute the price of a seat changes depending on factors such as the rate at which the plane is filling up, how close it is to the departure date, whether you are dealing with the airline or a service company and, possibly, the prices of the tickets already sold.

A new breed of dynamic pricing software is being developed that can help Internet merchants go far beyond the airline model. These programs take into account such factors as the customer's age, income, and geographical location, assuming that information is available from a database somewhere. Other factors include sales history to that customer, other sites the customer has visited, time of year, what others have been willing to pay for an item, how much of the item is in inventory, and what competitors are charging for the same or similar items.

Designing software like this, using it, and even figuring out if it has been effective, requires the collaboration of marketing experts, programmers, psy-

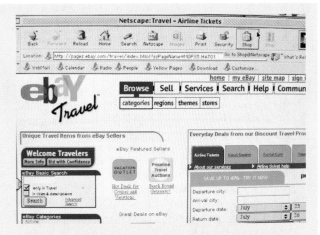

chologists, accountants, and people capable of sophisticated mathematical analysis. It is like some humongous applications problem that can result in millions of dollars in additional revenue if done successfully.

As a person who may be involved in retailing some day, this analytical approach to merchandising may excite you. As a present customer, stay on your toes!

Exercise Set 3.3

Concept/Writing Exercises

1. Outline the five-step problem-solving procedure we use. **1.** and **2.** Answers will vary.

2. If you are having difficulty with some of the material in this section, list some of the things you may be able to do to help you. See the Helpful Hint on page 206.

Practice the Skills/Problem Solving

7. 8, 19 **8.** 12, 31 **10.** 72, 73

*Exercises 3–24, involve finding a number or numbers. Read Examples 1–3, then set up an equation that can be used to solve the problem. Solve the equation and **answer the question asked**. Use a calculator where you feel it is appropriate.*

3. *Consecutive Numbers* The sum of two consecutive integers is 85. Find the numbers. 42, 43

4. *Consecutive Integers* The sum of two consecutive integers is 107. Find the numbers. 53, 54

5. *Odd Integers* The sum of two consecutive odd integers is 104. Find the numbers. 51, 53

6. *Even Integers* The sum of two consecutive even integers is 146. Find the numbers. 72, 74

7. *Sum of Numbers* One number is 3 more than twice a second number. Their sum is 27. Find the numbers.

8. *Sum of Numbers* One number is 5 less than 3 times a second number. Their sum is 43. Find the numbers.

9. *Numbers* The larger of two integers is 8 less than twice the smaller. When the smaller number is subtracted from the larger, the difference is 17. Find the two numbers. 25, 42

10. *Facing Pages* The sum of the two facing page numbers in an open book is 145. What are the page numbers?

11. *Life Expectancy* The life expectancy for men in the United States in 2000 was 19 years less than twice the life expectancy for men in 1900. If the difference in the life expectancies between 2000 and 1900 was 27.3 years, determine the life expectancy for United States men in 2000. (Source: U.S. Census Bureau) 73.6 yr

12. **Work Week** The average work week in 2000 in the United States was 136 hours less than 3 times what it was in 1900. If the sum of the average work weeks in 1900 and 2000 was 104 hours, what was the average work week in 1900? (Source: U.S. Labor Bureau) 60 hr

13. **DVD's** In 2002, Electronics City sold 20 more than twice as many DVD players as it sold in 2001. If a total of 3260 DVD players were sold in these two years, how many were sold in 2002? (Source: U.S. News and World Report) 2180

14. **U.S. Presidents** As of 2001, there were 43 United States presidents. Of the 43 presidents, the number who served as a general was 21 less than the number who did not serve as a general. How many U.S. presidents served as generals*? 11

George Washington by Gilbert Stuart. Museum of the City of New York

15. **Grandma's Gifts** Grandma gave some baseball cards to Richey and some to Erin. She gave 3 times the amount to Erin as she did to Richey. If the total amount she gave to both of them was 260 cards, how many cards did she give to Richey? 65

16. **Ski Shop** The Alpine Valley Ski Shop sells both cross-country and downhill skis. During a year they sold 6 times as many downhill skis as cross-country skis. Determine the number of pairs of cross-country skis sold if the difference in the number of pairs of downhill and cross-country skis sold is 1800. 360 pairs

17. **Animal Art** Many cities are displaying unusual replicas of animals or other items that they eventually sell to raise money for local charities. Joseph Murray designed and built a horse for display. It took him 1.4 hours more than twice the number of hours to attach the baseball gloves to the horse than to design the horse. If the total time it took him to design and attach the gloves to the horse was 32.6 hours, how long did it take him to attach the gloves to the horse? 22.2 hr

The author, Allen R. Angel, is shown in this photo.

18. **Candle Shop** A candle shop makes 60 candles per week. They plan to increase the number of candles they make by 8 per week until they reach a production of 132 candles per week. How many weeks will it take for them to reach their production schedule? 9 weeks

19. **Collecting Frogs** Mary Shapiro collects ceramic and stuffed frogs. She presently has 624 frogs. She wishes to add 6 a week to her collection until her collection reaches a total of 1000 frogs. How long will it take Mary's frog collection to reach 1000 frogs? ≈62.7 weeks

20. **Five Stars** There are only a limited number of officers who have been awarded the rank of five stars (either as generals or admirals). The number of generals who received five stars is 11 less than 4 times the number of admirals who received five stars. The difference between the number of five star generals and admirals is 1. Determine the number of five star admirals and the number of five star generals. admirals: 4, generals: 5

21. **Population** The town of Dover currently has a population of 6500. If its population is increasing at a rate of 1200 people per year, how long will it take for the population to reach 20,600? 11.75 yr

22. **Circuit Boards** The FGN Company produces circuit boards. They now have 4600 employees nationwide. They wish to reduce the number of employees by 250 per year through retirements, until their total employment is 2200. How long will this take? 9.6 yr

23. **Computers** The CTN Corporation has a supply of 3600 computers. They wish to ship 120 computers each week until their supply of computers drops to 2000. How long will this take? ≈13.3 weeks

24. **Lecture Hall** A class in a large lecture hall at American University currently has 280 students. If 4 students drop out of the course each week, how long will it take before the class is down to 228 students? 13 weeks

---

*The eleven U.S. presidents who served as generals were: George Washington, Andrew Jackson, William H. Harrison, Zachary Taylor, Franklin Pierce, Andrew Johnson, Ulysses S. Grant, Rutherford B. Hayes, James Garfield, Benjamin Harrison, and Dwight D. Eisenhower.

*Exercises 25–38 involve money. Read Examples 4–6, then set up an equation that can be used to solve the problem. Solve the equation and **answer the question asked**.*

25. Truck Rental Lori Sullivan rents a truck for one day and pays $50 per day plus 30 cents a mile. How far can Lori drive in one day if she has only $92? 140 mi

26. Gym Membership At Goldies Gym there is a one-time membership fee of $300 plus dues of $40 per month. If Carlos Manieri has spent a total of $700 for Goldies Gym, how long has he been a member? 10 months

27. Copy Machine Yamil Bermadez purchased a copy machine for $2100 and takes out a one-year maintenance protection plan for which he pays 2 cents per copy made. If he spends a total of $2462 in a year, which includes the cost of the machine and the copies made, determine the number of copies he made. 18,100 copies

28. Calling Plan With an AT&T calling plan, you pay $4.95 per month plus 7 cents per minute of talk time. In a specific month, the total cost for monthly fee and talk time was $29.45. Determine the number of minutes of talk time. 350 min

29. Calling Plan With a particular cellular calling plan, there is a monthly charge of $25.95, which entitles the caller to 500 free minutes. After the 500 free minutes are used, the caller pays 40 cents per minute, or part thereof. If Anke Braun's phone bill for one month was $61.95, determine the number of minutes above the 500 that Anke used. 90 min

30. Bridge Toll Mary Choi pays a monthly fee of $25 that allows her to cross the bridge, in either direction, 60 times a month. If she crosses more than 60 times in a month, she pays 25 cents each time she crosses the bridge. During one month, Mary paid a total of $38.25 to cross the bridge. How many extra times (above 60) did Mary cross the bridge that month? 53

31. Washing Machines Scott Montgomery is considering two washing machines, a Kenmore and a Neptune. The Neptune costs $454 while the Kenmore costs $362. The energy guides indicate that the Kenmore will cost an estimated $84 per year to operate and the Neptune will cost an estimated $38 per year to operate. How long will it be before the total cost is the same for both washing machines? 2 yr

32. Buying a Building Jason O'Connor is considering buying the building where he has his office. His monthly rent for his office is $890. If he makes a $31,350 down payment on the building, his monthly mortgage payments would be $560. How many months would it take for his down payment plus his monthly mortgage payments to equal the amount he spends on rent? 95 months

33. Salaries Brooke Mills is being recruited by a number of high-tech companies. Data Technology Corporation has offered her an annual salary of $40,000 per year plus a $2400 increase per year. Nuteck has offered her an annual salary of $49,600 per year plus a $800 increase per year. In how many years will the salaries from the companies be the same? 6 yr

34. Racquet Club The Coastline Racquet Club has two payment plans for its members. Plan 1 has a monthly fee of $20 plus $8 per hour court rental time. Plan 2 has no monthly fee, but court time is $16.25 per hour. If court time is rented in 1-hour intervals, how many hours would you have to play per month so that plan 1 becomes a better buy? 3 hr (more than 2.42)

35. Printers Brandy Hawk will purchase one of two ink jet printers, a Hewlett-Packard or a Lexmark. The Hewlett-Packard costs $149 and the Lexmark costs $99. Suppose, because of the price of the ink cartridges, the cost of printing a page on the Hewlett-Packard is $0.02 per page and the cost of printing a page on the Lexmark is $0.03 per page. How many pages would need to be printed for the two printers to have the same total cost? 5000 pages

36. Internet Provider An Internet service provider has one plan that charges a monthly fee of $20.00, which provides 250 hours of free Internet service. It charges $0.50 per hour for all hours in excess of the 250 hours. If, under this plan, a monthly bill is $38.00, how many extra hours of Internet time was used? 36 extra hours

37. Mortgages The Juan Garcia family is purchasing a house. They are considering two banks, First Union and Kensington, for their mortgage. With First Union, their monthly mortgage payments would be $980 per month and there are no extra fees. With Kensington, their monthly mortgage payment would be $910, but there is a one-time application fee of $2000. How many months would it take for the total cost to be the same for both banks? ≈28.6 mo

38. *Mortgages* Dennis Williams is considering two banks, Citibank and Huntington Bank, for a mortgage. His mortgage payment with Citibank would be $1025 per month, plus he would pay a one time fee of $1500. His mortgage payment with Huntington Bank would be $967, plus he would pay a one time fee of $2500. How long would it take for the total cost to be the same for both banks? ≈17.2 mo

*Exercises 39–60 involve percents. Read Examples 7–9, then set up an equation that can be used to solve the problem. Solve the equation and **answer the questions asked**.*

39. *Financial Planner* Dan Rinn, a financial planner, charges an annual fee of 1% of the client's assets that he is managing. If his annual fee for managing Judy Mooney's retirement portfolio is $620.00, how much money is Dan managing for Judy? $62,000

40. *Warehouse Sales* At the Buyrite Warehouse, for a yearly fee of $60 you save 8% of the price of all items purchased in the store. How much would Mary need to spend during the year so that her savings equal the yearly fee? $750

41. *Airfare* The airfare for a flight from Amarillo, Texas, to New Orleans cost $280, which includes a 7% sales tax. What is the cost of the flight before tax? $261.68

42. *New Car* Yoliette Fournier purchased a new car. The cost of the car, including a 7.5% sales tax was $24,600. What was the cost of the car before tax? $22,883.72

43. *Salary Increase* Zhen Tong just received a job offer that will pay him a 30% greater salary than his present job does. If the salary at his new job will be $30,200, determine his present salary. $23,230.77

44. *New Headquarters* Tarrach and Associates plan on increasing the size of its headquarters by 20%. If its new headquarters is to be 14,200 square feet, determine the size of its present headquarters. ≈11,833.33 sq ft

45. *Downsizing* Marty and Betty McKane's children are grown and have left home, so they have decided to downsize their house. Their new house has 18% fewer square feet than their former house. If their new house has an area of 2200 square feet, determine the size of their former house. ≈2682.93 sq ft

46. *Retirement Income* Ray and Mary Burnham have decided to retire. They estimate their annual income after retirement will be reduced by 15% from their pre-retirement income. If they estimate their retirement income to be $42,000, determine their pre-retirement income. $49,411.76

47. *Autographs* A tennis star was hired to sign autographs at a convention. She was paid $3000 plus 3% of all admission fees collected at the door. The total amount she received for the day was $3750. Find the total amount collected at the door. $25,000

48. $32.49

51. $6000

48. *Sale* At a 1-day 20% off sale, Jane Demsky purchased a hat for $25.99. What is the regular price of the hat?

49. *Wage Cut* A manufacturing plant is running at a deficit. To avoid layoffs, the workers agree on a temporary wage cut of 2%. If the average salary in the plant after the wage cut is $38,600, what was the average salary before the wage cut? $39,387.76

50. *Teachers* During the 2002 contract negotiations, the city school board approved a 4% pay increase for its teachers effective in 2003. If Dana Frick, a first-grade teacher, projects his 2003 annual salary to be $36,400, what is his present salary? $35,000

51. *Sales Volume* Bruce Gregory receives a weekly salary of $350. He also receives a 6% commission on the total dollar volume of all sales he makes. What must his dollar volume be in a week, if he is to make a total of $710?

52. *Lawnmower Ride* Mona Fabricant just got a new rider lawnmower. With the new mower, it takes her 15% less time to mow her lawn than with her previous mower. If it takes her 1.1 hours to mow her lawn with the new lawn mower, how long did it take with the previous lawn mower? ≈1.29 hr

53. *Salary Plans* Vince McAdams, a salesman, is given a choice of two salary plans. Plan 1 is a weekly salary of $600 plus 2% commission of sales. Plan 2 is a straight commission of 10% of sales. How much in sales must Vince make in a week for both plans to result in the same salary? $7500

54. **Salary Plans** Becky Schwartz, a saleswoman, is offered two salary plans. Plan 1 is $400 per week salary plus a 2% commission of sales. Plan 2 is a $250 per week salary plus a 16% commission of sales. How much would Becky need to make in sales for the salary to be the same from both plans? $1071.43

55. **Financial Planning** Belen Poltorade, a financial planner, is offering her customers two financial plans for managing their assets. With plan 1 she charges a planning fee of $1000 plus 1% of the assets she will manage for the customers. With plan 2 she charges a planning fee of $500 plus 2% of the assets she will manage. How much in customer assets would result in both plans having the same total fees? $50,000

56. **Art Show** Bill Rush is an artist. He is going to put on a show of his artwork. Bill is negotiating with an organization to rent its building for his show. The organization has offered Bill two rental plans. Plan 1 is a rental fee of $500 for the week plus 3% of the dollar sales he makes. Plan 2 is $100 for the week plus 15% of the dollar sales he makes. What dollar sales would result in both plans having the same total cost? $3333.33

57. **Membership Fees** The Holiday Health Club has reduced its annual membership fee by 10%. In addition, if you sign up on a Monday, they will take an additional $20 off the already reduced price. If Jorge Sanchez purchases a year's membership on a Monday and pays $250, what is the regular membership fee? $300

58. **Eating Out** After Linda Kodama is seated in a restaurant, she realizes that she has only $30. From this $30 she must pay a 7% tax and she wishes to leave a 15% tip. What is the maximum price for a meal that she can afford to pay? $24.59

59. **Estate** Phil Dodge left an estate valued at $140,000. In his will, he specified that his wife will get 25% more of his estate than his daughter. How much will his wife receive? $77,777.78

60. **Charitable Giving** Charles Ford made a $200,000 cash contribution to two charities, the American Red Cross and the United Way. The amount received by the American Red Cross was 30% greater than the amount received by the United Way. How much did the United Way receive? $86,956.52

Challenge Problems

61. **Average Value** To find the *average* of a set of values, you find the sum of the values and divide the sum by the number of values.

   a) If Paul Lavenski's first three test grades are 74, 88, and 76, write an equation that can be used to find the grade that Paul must get on his fourth exam to have an 80 average.

   b) Solve the equation from part a) and determine the grade Paul must receive.

   a) $80 = \dfrac{74 + 88 + 76 + x}{4}$ b) 82

For Exercises 62 and 63, set up an equation that can be used to solve the problem, then solve the equation and answer the question asked.

62. **Basketball** At a basketball game Boston College scored 78 points. The team made 12 free throws (1 point each). They also made 4 times as many 2-point field goals as 3-point field goals (field goals made from more than 18 feet from the basket). How many 2-point field goals and how many 3-point field goals did they make?
   3 pointers, 6; 2 pointers, 24

63. **Driver Education** A driver education course costs $45 but saves those under age twenty five 10% of their annual insurance premiums until they reach age twenty five. Scott Day has just turned 18, and his insurance costs $600 per year.

   a) How long will it take for the amount saved from insurance to equal the price of the course?

   b) Including the cost of the course, when Scott turns 25, how much will he have saved? $375

   63. a) 0.75 yr or 9 mo

Group Activity

Discuss and answer Exercise 64 as a group.

64. Remodeling Costs

a) Each member of the group use the graph on the right to make up your own verbal problem that can be solved algebraically. There are many different types of problems that can be created. Solve your problem algebraically.

b) Share the problem you made up with all the other members of your group.

c) For each problem given to you, set up an equation that may be used to solve the problem. Solve the equation and answer the question asked.

d) Compare your answers to each problem with all members of your group. If any member's answers do not match, work together to determine the correct answer.

Remodeling Costs and Return

National average cost of five of the most popular home remodeling projects and the average percent of cost returned when the home is sold.

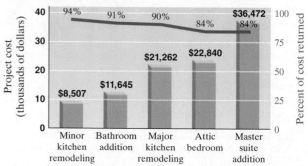

Source: National Association of Home Builders

Cumulative Review Exercises

[1.9] **65.** Evaluate $\dfrac{1}{4} + \dfrac{3}{4} \div \dfrac{1}{2} - \dfrac{1}{3}$. $\quad \dfrac{17}{12}$

Name each indicated property.

[1.10] **66.** $(x + y) + 5 = x + (y + 5)$

67. $xy = yx$

68. $x(x + y) = x^2 + xy$

66. associative property of addition   **67.** commutative property of multiplication
68. distributive property

[2.6] **69. Chicken Barbecue** At the firefighter's annual chicken barbecue, the chef estimates that he will need $\frac{1}{2}$ pound of coleslaw for each 5 people. If he expects 560 residents to attend, how many pounds of coleslaw will he need? 56 lb

[3.1] **70.** Solve the formula $A = \dfrac{1}{2}bh$ for b $b = \dfrac{2A}{h}$

3.4 GEOMETRIC PROBLEMS

1 Solve geometric problems.

   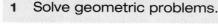
SSM   Study Guide   CD/Video

MathPro 4/5   PH Math Tutor Center   prenhall.com/Angel

1 Solve Geometric Problems

This section serves two purposes. One is to reinforce the geometric formulas introduced in Section 3.1. The second is to reinforce procedures for setting up and solving verbal problems discussed in Sections 3.2 and 3.3. The more practice you have at setting up and solving the verbal problems, the better you will become at solving them.

EXAMPLE 1   **Building a Patio**   Catherine Moushon is planning on building a rectangular concrete patio in the back of her house. The length of the patio is to be 8 feet longer than the width (Figure 3.8). The total amount of wood to be used to build the frame for the patio (called a form) is 56 feet. Find the dimensions of the patio that Catherine plans to build.

Solution

FIGURE 3.8

Understand We are asked to find the dimensions of the patio that Catherine plans to build. Since the amount of wood that will be used to make the frame is 56 feet, the perimeter is 56 feet. Since the length is given in terms of the width, we will let the variable represent the width. Then we can express the length in terms of the variable selected for the width. To solve this problem, we use the formula for the perimeter of a rectangle, $P = 2l + 2w$, where $P = 56$ feet.

Translate

Let w = width of the patio
then $w + 8$ = length of the patio

$$P = 2l + 2w$$
$$56 = 2(w + 8) + 2w$$

Carry Out

$$56 = 2w + 16 + 2w$$
$$56 = 4w + 16$$
$$40 = 4w$$
$$10 = w$$

The width is 10 feet. Since the length is 8 feet greater than the width, the length is $10 + 8 = 18$ feet.

Check We will check the solution by substituting the appropriate values in the perimeter formula.

$$P = 2l + 2w$$
$$56 \stackrel{?}{=} 2(18) + 2(10)$$
$$56 = 56 \qquad \textit{True}$$

NOW TRY EXERCISE 23

Answer The width of the patio will be 10 feet and the length will be 18 feet. ✳

EXAMPLE 2

Corner Lot A triangle that contains two sides of equal length is called an **isosceles triangle**. In isosceles triangles, the angles opposite the two sides of equal length have equal measures. Mr. and Mrs. Harmon Katz have a corner lot that is an isosceles triangle. Two angles of their triangular lot are the same and the third angle is 30° greater than the other two. Find the measure of all three angles (see Fig. 3.9).

Solution

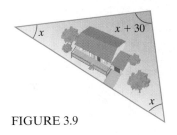

FIGURE 3.9

Understand To solve this problem, you must know that the sum of the angles of any triangle measures 180°. We are asked to find the measure of each of the three angles, where the two smaller angles have the same measure. We will let the variable represent the measure of the smaller angles, and then we will express the larger angle in terms of the variable selected for the smaller angles.

Translate

Let x = the measure of each smaller angle
then $x + 30$ = the measure of larger angle

sum of the 3 angles $= 180$
$$x + x + (x + 30) = 180$$

Carry Out

$$3x + 30 = 180$$
$$3x = 150$$
$$x = \frac{150}{3} = 50°$$

The two smaller angles are each 50°. The larger angle is $x + 30°$ or $50° + 30° = 80°$.

Check and Answer Since $50° + 50° + 80° = 180°$, the answer checks. The two smaller angles are each 50° and the larger angle is 80°. ✳

NOW TRY EXERCISE 11

Recall from Section 3.1 that a quadrilateral is a four-sided figure. Quadrilaterals include squares, rectangles, parallelograms, and trapezoids. The sum of the measures of the angles of any quadrilateral is 360°. We will use this information in Example 3.

EXAMPLE 3   **Water Trough**   Sarah Fuqua owns horses and uses a water trough whose ends are trapezoids. The measure of the two bottom angles of the trapezoid are the same, and the measure of the two top angles are the same. The bottom angles measure 15° less than twice the measure of the top angles. Find the measure of each angle.

Solution   Understand   To help visualize the problem, we draw a picture of the trapezoid, as in Figure 3.10. We use the fact that the sum of the measures of the four angles of a quadrilateral is 360°.

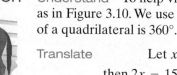

FIGURE 3.10

Translate   Let x = the measure of each of the two smaller angles

then $2x - 15$ = the measure of each of the two larger angles

$$\left(\begin{array}{c}\text{measure of the}\\\text{two smaller angles}\end{array}\right) + \left(\begin{array}{c}\text{measure of the}\\\text{two larger angles}\end{array}\right) = 360$$

$$x + x + (2x - 15) + (2x - 15) = 360$$

Carry Out $$x + x + 2x - 15 + 2x - 15 = 360$$

$$6x - 30 = 360$$

$$6x = 390$$

$$x = 65$$

Each smaller angle is 65°. Each larger angle is $2x - 15 = 2(65) - 15 = 115°$.

Check and Answer   Since $65° + 65° + 115° + 115° = 360°$, the answer checks.

NOW TRY EXERCISE 27   Each smaller angle is 65° and each larger angle is 115°.   ✸

EXAMPLE 4   **Fenced-In Area**   Richard Jeffries recently started an ostrich farm. He is separating the ostriches by fencing in three equal areas, as shown in Figure 3.11. The length of the fenced-in area, l, is to be 30 feet greater than the width and the total amount of fencing available is 660 feet. Find the length and width of the fenced-in area.

Solution   Understand   The fencing consists of four pieces of fence of length w, and two pieces of fence of length l.

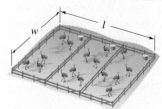

FIGURE 3.11

Translate   Let w = width of fenced-in area

then $w + 30$ = length of fenced-in area

$$\left(\begin{array}{c}\text{4 pieces of fence}\\\text{of length } w\end{array}\right) + \left(\begin{array}{c}\text{two pieces of fence}\\\text{of length } w + 30\end{array}\right) = 660$$

$$4w + 2(w + 30) = 660$$

Carry Out $$4w + 2w + 60 = 660$$

$$6w + 60 = 660$$

$$6w = 600$$

$$w = 100$$

Since the width is 100 feet, the length is $w + 30$ or $100 + 30$ or 130 feet.

Check and Answer   Since $4(100) + 2(130) = 660$, the answer checks. The

NOW TRY EXERCISE 35   width of the ostrich farm is 100 feet and the length is 130 feet.   ✸

Exercise Set 3.4

2. area increases ninefold 4. circumference is tripled 6. volume is 27 times as great
7. a triangle with two equal sides

Concept/Writing Exercises

1. In the equation $A = l \cdot w$, what happens to the area if the length is doubled and the width is halved? Explain your answer. The area remains the same.

2. In the equation $A = s^2$, what happens to the area if the length of a side, s, is tripled? Explain your answer.

3. In the equation $V = l \cdot w \cdot h$, what happens to the volume if the length, width, and height are all doubled? Explain your answer. volume is 8 times as great

4. In the equation $C = 2\pi r$, what happens to the circumference if the radius is tripled?

5. In the equation $A = \pi r^2$, what happens to the area if the radius is tripled? area is 9 times as great

6. In the equation $V = \frac{4}{3}\pi r^3$, what happens to the volume if the radius is tripled? Explain your answer.

7. What is an isosceles triangle?

8. What is a quadrilateral? a four-sided figure

9. What is the sum of the measures of the angles of a triangle? $180°$

10. What is the sum of the measures of the angles of a quadrilateral? $360°$

Practice the Skills/Problem Solving

Solve the following geometric problems.*

11. **Isosceles Triangle** In an isosceles triangle, one angle is $42°$ greater than the other two equal angles. Find the measure of all three angles. See Example 2. $46°, 46°, 88°$

12. **Triangular Building** This building in New York City, referred to as the Flat Iron Building, is in the shape of an isosceles triangle. If the shortest side of the building is 50 feet shorter than the two longer sides, and the perimeter around the building is 196 feet, determine the length of the three sides of the building. 82 feet, 82 feet, 32 feet

13. **A Special Triangle** An **equilateral triangle** is a triangle that has three sides of the same length. The perimeter of an equilateral triangle is 28.5 inches. Find the length of each side. 9.5 in.

14. **Equilateral Triangle** The perimeter of an equilateral triangle is 48.6 centimeters. Find the length of each side. See Exercise 13. 16.2 cm

15. **Complementary Angles** Two angles are **complementary angles** if the sum of their measures is $90°$. Angle A and angle B are complementary angles, and angle A is $21°$ more than twice angle B. Find the measures of angle A and angle B. $A = 67°, B = 23°$

Complementary Angles

16. **Complementary Angles** Angles A and B are complementary angles, and angle B is $14°$ less than angle A. Find the measures of angle A and angle B. See Exercise 15. $A = 52°, B = 38°$

17. **Supplementary Angles** Two angles are **supplementary angles** if the sum of their measures is $180°$. Angle A and angle B are supplementary angles, and angle B is $8°$ less than three times angle A. Find the measures of angle A and angle B. $A = 47°, B = 133°$

Supplementary Angles

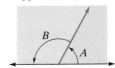

18. **Supplementary Angles** Angles A and B are supplementary angles and angle A is $2°$ more than 4 times angle B. Find the measures of angle A and angle B. See Exercise 17. $A = 144.4°, B = 35.6°$

19. **Vertical Angles** When two lines cross, the opposite angles are called **vertical angles**. Vertical angles have equal measures. Determine the measures

———
*See Appendix C for more material on geometry.

of the vertical angles indicated in the following figure. 88°

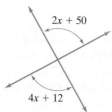

20. Vertical Angles A pair of vertical angles is indicated in the following figure. Determine the measure of the vertical angles indicated. See Exercise 19. 65°

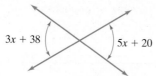

21. Unknown Angles One angle of a triangle is 10° greater than the smallest angle, and the third angle is 30° less than twice the smallest angle. Find the measures of the three angles. 50°, 60°, 70°

22. Unknown Angles One angle of a triangle is 20° larger than the smallest angle, and the third angle is 6 times as large as the smallest angle. Find the measures of the three angles. 20°, 40°, 120°

23. Dimensions of Rectangle The length of a rectangle is 8 feet more than its width. What are the dimensions of the rectangle if the perimeter is 48 feet?

24. Dimensions of Rectangle The perimeter of a rectangle is 120 feet. Find the length and width of the rectangle if the length is twice the width. $l = 40$ ft, $w = 20$ ft

25. Tennis Court The length of a regulation tennis court is 6 feet greater than twice its width. The perimeter of the court is 228 feet. Find the length and width of the court.

23. $l = 16$ ft, $w = 8$ ft
25. $l = 78$ ft, $w = 36$ ft

26. Sandbox Mrs. Christine O'Connor is planning to build a sandbox for her daughter. She has 26 feet of lumber with which to build the perimeter. What should be the dimensions of the rectangular sandbox if the length is to be 3 feet longer than the width? 5 ft by 8 ft

27. Parallelogram In a parallelogram the opposite angles have the same measures. Each of the two larger angles in a parallelogram is 20° less than 3 times the smaller angles. Find the measure of each angle.

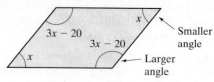

28. Parallelogram The two smaller angles of a parallelogram have equal measures, and the two larger angles each measure 27° less than twice each smaller angle. Find the measure of each angle.

29. Quadrilateral The measure of one angle of a quadrilateral is 10° greater than the smallest angle; the third angle is 14° greater than twice the smallest angle; and the fourth angle is 21° greater than the smallest angle. Find the measures of the four angles of the quadrilateral. 63°, 73°, 140°, 84°

30. Quadrilateral The measure of one angle of a quadrilateral is twice the smallest angle; the third angle is 20° greater than the smallest angle; and the fourth angle is 20° less than twice the smallest angle. Find the measures of the four angles of the quadrilateral. 60°, 120°, 80°, 100°

31. Building a Bookcase A bookcase is to have four shelves, including the top, as shown. The height of the bookcase is to be 3 feet more than the width. Find the width and height of the bookcase if only 30 feet of lumber is available. $w = 4$ ft; $h = 7$ ft

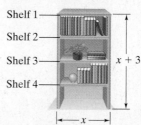

32. Bookcase A bookcase is to have four shelves as shown. The height of the bookcase is to be 2 feet more than the width, and only 20 feet of lumber is available. What should be the width and height of the bookcase? $w = 2\frac{2}{3}$ ft, $h = 4\frac{2}{3}$ ft

27. smaller angles, 50°; larger angles, 130°
28. smaller angles, 69°; larger angles, 111°

33. *Bookcase* What should be the width and height of the bookcase in Exercise 32 if the height is to be twice the width? $w = 2.5$ ft, $h = 5$ ft

34. *Storage Shelves* Carlotta plans to build storage shelves as shown. She has only 45 feet of lumber for the entire unit and wishes the width to be 3 times the height. Find the width and height of the unit. $w = 9$ ft, $h = 3$ ft

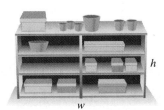

35. *Fenced-In Area* An area is to be fenced in along a straight river bank as illustrated. The length of the fenced-in area is to be 4 feet greater than the width, and the total amount of fencing to be used is 64 feet. Find the width and length of the fenced-in area. $w = 10$ ft, $l = 14$ ft

See Exercise 35.

36. *Gardening* Trina Zimmerman is placing a border around and within a garden where she intends to plant flowers (see the figure). She has 60 feet of bordering, and the length of the garden is to be 2 feet greater than the width. Find the length and width of the garden. The red shows the location of all the bordering in the figure.

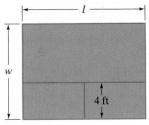

$l = 12$ ft, $w = 10$ ft

Challenge Problems

37. One way to express the area of the figure on the right is $(a + b)(c + d)$. Can you determine another expression, using the area of the four rectangles, to represent the area of the figure? $ac + ad + bc + bd$

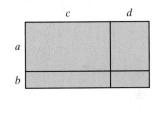

38. a)

f) 20.25 in.^{2}, 40.50 in.^{2}, 40.50 in.^{2}, 81.00 in.^{2}

![icon] **Group Activity**

Discuss and answer Exercise 38 as a group.

38. Consider the four pieces shown. Two are squares and two are rectangles.

  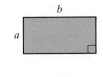

a) Individually, rearrange and place the four pieces together to form one square.

b) The area of the square you constructed is $(a + b)^2$. Write another expression for the area of the square by adding the four individual areas. $a^2 + 2ab + b^2$

c) Compare your answers. If each member of the group did not get the same answers to parts **a)** and **b)**, work together to determine the correct answer.

d) Answer the following question as a group. If b is twice the length of a, and the perimeter of the square you created is 54 inches, find the length of a and b. $a = 4.5$ in., $b = 9$ in.

e) Use the values of a and b found in part **d)** to find the area of the square you created. 182.25 in.^{2}

f) Use the values of a and b found in part **d)** to find the areas of the four individual pieces that make up the large square.

g) Does the sum of the areas of the four pieces found in part **f)** equal the area of the large square found in part **e)**? Is this what you expected? Explain. yes

Cumulative Review Exercises

Insert either >, <, *or* = *in each shaded area to make the statement true.*

[1.5] **39.** $-|-6|$ ▦ $|-4|$ <

40. $|-3|$ ▦ $-|3|$ >

[1.7] **41.** Evaluate $-6 - (-2) + (-4)$. -8

[2.1] **42.** Simplify $-6y + x - 3(x - 2) + 2y$.
$-2x - 4y + 6$

[3.1] **43.** Solve $2x + 3y = 9$ for y; then find the value of y when $x = 3$.
$y = \dfrac{-2x + 9}{3}$ or $y = -\dfrac{2}{3}x + 3, 1$

3.5 MOTION, MONEY, AND MIXTURE PROBLEMS

SSM · Study Guide · CD/Video

MathPro 4/5 · PH Math Tutor Center · prenhall.com/Angel

1 Solve motion problems involving only one rate.
2 Solve motion problems involving two rates.
3 Solve money problems.
4 Solve mixture problems.

We now discuss three additional types of applications: motion, money, and mixture problems. These problems are grouped in the same section because, as you will learn shortly, you use the same general multiplication procedure to solve them. We begin by discussing motion problems.

1 Solve Motion Problems Involving Only One Rate

A **motion problem** is one in which an object is moving at a specified rate for a specified period of time. A car traveling at a constant speed, a swimming pool being filled or drained (the water is added or removed at a specified rate), and spaghetti being cut on a conveyor belt (conveyor belt moving at a specified speed) are all motion problems.

The formula often used to solve motion problems follows.

> **Motion Formula**
>
> $$\text{amount} = \text{rate} \cdot \text{time}$$

The amount can be a measure of many different quantities, depending on the rate. For example, if the rate is measuring *distance* per unit time, the amount will be *distance*. If the rate is measuring *volume* per unit time, the amount will be *volume*; and so on.

EXAMPLE 1

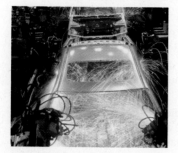

Robotic Welders In an auto plant in Detroit, Michigan, a robot is making welds on cars. If the robot can make 8 welds per minute, how many welds can the robot make in 20 minutes?

Understand and Translate The amount in this example is the number of welds that can be made. Therefore, the formula we will use is number of welds = rate · time. We are given the rate, 8 welds per minute, and the time is 20 minutes. We are asked to find the number of welds.

$$\text{number of welds} = \text{rate} \cdot \text{time}$$

Carry Out

$$= 8 \cdot 20 = 160$$

NOW TRY EXERCISE 3

TEACHING TIP
After discussing the solution to Example 1 ask, "How many welds will be made in an hour?" Be sure students understand that the time units must be converted to minutes.

Answer   Thus, in 20 minutes 160 welds would be made.

Let's look at the units in Example 1. The rate is given in welds per minute and the time is given in minutes. If we analyze the units (a process called *dimensional analysis*), we see that the answer is given in welds.

$$\text{number of welds} = \text{rate} \cdot \text{time}$$
$$= \frac{\text{welds}}{\cancel{\text{minutes}}} \cdot \cancel{\text{minutes}}$$
$$= \text{welds}$$

When the *amount* in the rate formula is *distance*, we often refer to the formula as the **distance formula**.

Distance Formula

$$\text{distance} = \text{rate} \cdot \text{time} \quad \text{or} \quad d = r \cdot t$$

Example 2 illustrates the use of the distance formula.

EXAMPLE 2   **The Alaska Pipeline**   The Alaskan Pipeline extends from Prudhoe Bay, Alaska, to Valdez, Alaska. The distance, or length, of the pipeline is 800.3 miles. Oil flows through the pipeline at an average rate of 5.4 miles per hour. How long will it take oil entering the pipeline in Prudhoe Bay to exit the pipeline in Valdez?

Solution   Understand and Translate   Since we are given a distance of 800.3 miles, we will use the distance formula. We are given the distance and rate and need to solve for the time, t.

$$\text{distance} = \text{rate} \cdot \text{time}$$
$$800.3 = 5.4t$$

Carry Out
$$\frac{800.3}{5.4} = t$$
$$148.2 \approx t$$
$$\text{or} \quad t \approx 148.2 \text{ hours}$$

NOW TRY EXERCISE 9

Answer   It will take approximately 148.2 hours, or about 6.18 days, for the oil to go from Prudhoe Bay to Valdez through the Alaskan Pipeline.

HELPFUL HINT   When working motion problems, the units must be consistent with each other. If you are given a problem where the units are not consistent, you will need to change one of the quantities so that the units will agree before you substitute the values into the formula. For example, if the rate is in feet per second and the distance is given in inches, you will either need to change the distance to feet or the rate to inches per second.

---

2   Solve Motion Problems Involving Two Rates

Now we will look at some motion problems that involve *two rates*, such as two trains traveling at different speeds. In these problems, we generally begin by letting the variable represent one of the unknown quantities, and then we represent the second unknown quantity in terms of the first unknown quantity. For example, suppose that one train travels 20 miles per hour faster than another train. We might let r represent the rate of the slower train and $r + 20$ represent the rate of the faster train.

To solve problems of this type that use the distance formula, we generally add the two distances, or subtract the smaller distance from the larger, or set the two distances equal to each other, depending on the information given in the problem.

Often when working problems involving two different rates, we construct a table, like the following one, to organize the information. The formula above the table shows how the distance in the last column is calculated.

Rate × Time = Distance			
Item	**Rate**	**Time**	**Distance**
Item 1			distance 1
Item 2			distance 2

Depending upon the information given in the problem, we set up one of three types of equations, as indicated below, to solve the problem.

$$\text{distance 1} + \text{distance 2} = \text{total distance}$$
$$\text{distance 1} - \text{distance 2} = \text{difference in distance}$$
$$(\text{or distance 2} - \text{distance 1} = \text{difference in distance})$$
$$\text{distance 1} = \text{distance 2}$$

Examples 3 through 5 illustrate the procedure used.

EXAMPLE 3

Camping Trip   The Justinger family decides to go on a camping trip. They will travel by canoes on the Erie Canal in New York state. They start in Tonawanda (near Buffalo) heading toward Rochester. Mike and Danny, the teenage boys, are in one canoe while Paul and Maryanne, the father and mother, are in a second canoe. About noon, Mike and Danny decide to paddle faster than their parents to get to their campsite before their parents so they can set up camp. While the parents paddle the canoe at a leisurely speed of 2 miles per hour, the boys paddle the canoe at 4 miles per hour. In how many hours will the boys and their parents be 5 miles apart?

Solution   Understand and Translate   We are asked to find the time it takes for the canoes to become separated by 5 miles. We will construct a table to aid us in setting up the problem.

Let t = time when canoes are 5 miles apart

We draw a sketch to help visualize the problem (Fig. 3.12). When the two canoes are 5 miles apart, each has traveled for the same number of hours, t.

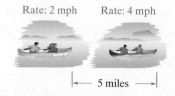

Rate: 2 mph   Rate: 4 mph

|← 5 miles →|

Canoe	Rate	Time	Distance
Parents	2	t	$2t$
Sons	4	t	$4t$

FIGURE 3.12

Since the canoes are traveling in the same direction, the distance between them is found by subtracting the distance traveled by the slower canoe from the distance traveled by the faster canoe.

$$\left(\begin{array}{c}\text{distance traveled}\\\text{by faster canoe}\end{array}\right) - \left(\begin{array}{c}\text{distance traveled}\\\text{by slower canoe}\end{array}\right) = 5 \text{ miles}$$

$$4t - 2t = 5$$

Carry Out
$$2t = 5$$
$$t = 2.5$$

Answer   After 2.5 hours the two canoes will be 5 miles apart.

EXAMPLE 4   **Sewer Pipe**   Two construction crews are 20 miles apart working toward each other. Both are laying sewer pipe in a straight line that will eventually be connected together. Both crews will work the same hours. One crew has better equipment and more workers and can lay a greater length of pipe per day. The faster crew lays 0.4 mile of pipe per day more than the slower crew, and the two pipes are connected after 10 days. Find the rate at which each crew lays pipe.

Solution   Understand and Translate   We are asked to find the two rates. We are told that both crews work for 10 days.

$$\text{Let } r = \text{rate of slower crew}$$
$$\text{then } r + 0.4 = \text{rate of faster crew}$$

We make a sketch (Fig. 3.13) and set up a table of values.

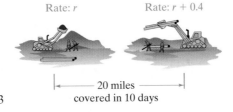

Rate: r Rate: $r + 0.4$

Crew	Rate	Time	Distance
Slower	r	10	$10r$
Faster	$r + 0.4$	10	$10(r + 0.4)$

20 miles covered in 10 days

FIGURE 3.13

The total distance covered by both crews is 20 miles.

$$\left(\begin{array}{c}\text{distance covered}\\\text{by slower crew}\end{array}\right) + \left(\begin{array}{c}\text{distance covered}\\\text{by faster crew}\end{array}\right) = 20 \text{ miles}$$

$$10r + 10(r + 0.4) = 20$$

Carry Out
$$10r + 10r + 4 = 20$$
$$20r + 4 = 20$$
$$20r = 16$$
$$\frac{20r}{20} = \frac{16}{20}$$
$$r = 0.8$$

Answer   The slower crew lays 0.8 mile of pipe per day and the faster crew lays
NOW TRY EXERCISE 19 $r + 0.4$ or $0.8 + 0.4 = 1.2$ miles of pipe per day.

EXAMPLE 5   **Jogging in Griffith Park**   Griffith Park in Los Angeles is the largest municipal park and urban wilderness area in the United States. It has many trails and activities that can be used for recreational purposes. In the park, Connie Buller starts rollerblading at 6 miles per hour at Crystal Springs Drive, going toward the Los Angeles Zoo. Her friend Richard Zucker plans to meet her on the trail. Richard starts at the same point $\frac{1}{2}$ hour after Connie and starts biking in the same direction. Richard rides at 10 miles per hour.

a) How long after Richard starts riding will they meet?

b) How far from their starting point will they be when they meet?

Solution

a) Understand and Translate   Since Richard will travel faster, he will cover the same distance in less time. When they meet, each has traveled the same distance, but Connie will have been on the trail for $\frac{1}{2}$ hour more than Richard. The question asks, how long after Richard starts riding will they meet? Since the rate is given in miles per hour, the time will be in hours.

$$\text{Let } t = \text{time Richard biking}$$

$$\text{then } t + \frac{1}{2} = \text{time Connie is rollerblading}$$

Make a sketch (Fig. 3.14) and set up a table.

Richard   Connie

Traveler	Rate	Time	Distance
Richard	10	t	$10t$
Connie	6	$t + \dfrac{1}{2}$	$6\left(t + \dfrac{1}{2}\right)$

Rate: 10 mph   Rate: 6 mph

FIGURE 3.14   Time: t Time: $t + \frac{1}{2}$

$$\text{Richard's distance} = \text{Connie's distance}$$

$$10t = 6\left(t + \frac{1}{2}\right)$$

Carry Out

$$10t = 6t + 3$$

$$10t - 6t = 6t - 6t + 3$$

$$4t = 3$$

$$t = \frac{3}{4}$$

Answer   Thus, they will meet 3/4 hour after Richard starts biking.

b) To find the distance traveled, we will use Richard rate and time.

$$d = r \cdot t$$

$$= 10 \cdot \frac{3}{4} = \frac{15}{2} = 7.5 \text{ miles}$$

NOW TRY EXERCISE 25   Connie and Richard will meet 7.5 miles from their starting point.   ✳

3 Solve Money Problems

Now we will work some examples that involve money. We place these problems here because they are solved using a procedure very similar to the procedure used to solve motion problems with two rates. One type of money problem involves interest. When working with interest problems, we often let the variable represent one amount of money, then we represent the second amount in terms of the variable. For example, if we know the total amount invested in two accounts is $20,000, we might let x represent the amount in one account, then $20,000 - x$ would be the amount invested in the other account. Recall from Section 3.1 that the simple interest formula is *interest = principal · rate · time*. When solving an interest problem we can use a table, as illustrated below, just as we did when working with motion problems.

Principal × Rate × Time = Interest				
Account	**Principal**	**Rate**	**Time**	**Interest**
Account 1				Interest 1
Account 2				Interest 2

After determining the interest columns on the right, we generally use one of the following formulas, depending on the question, to determine the answer.

$$\text{Interest 1} + \text{Interest 2} = \text{Total Interest}$$
$$\text{Interest 1} - \text{Interest 2} = \text{Difference in interest}$$
$$(\text{or Interest 2} - \text{Interest 1} = \text{Difference in interest})$$
$$\text{Interest 1} = \text{Interest 2}$$

Do you see the similarities with the motion problems? Now let's work an example.

EXAMPLE 6 **Investments** Mitch Levy just had a certificate of deposit mature and he now has $15,000 to invest. He is considering two investments. One is a loan he can make to another party through the Gibraltar Mortgage Company. This investment pays him 11% simple interest for a year. A second investment, which is more secure, is a 1 year certificate of deposit that pays 5%. Mitch decides that he wants to place some money in each investment, but he needs to earn a total of $1500 interest in 1 year from the two investments. How much money should Mitch put in each investment?

Solution Understand and Translate We use the simple interest formula that was introduced in Section 3.1 to solve this problem: interest = principal · rate · time.

$$\text{Let } x = \text{amount to be invested at 5\%}$$
$$\text{then } 15,000 - x = \text{amount to be invested at 11\%}$$

Account	**Principal**	**Rate**	**Time**	**Interest**
CD	x	0.05	1	$0.05x$
Loan	$15,000 - x$	0.11	1	$0.11(15,000 - x)$

Since the sum of the interest from the two investments is $1500, we write the equation

$$\left(\begin{array}{c}\text{interest from}\\ \text{5\% CD}\end{array}\right) + \left(\begin{array}{c}\text{interest from}\\ \text{11\% investment}\end{array}\right) = \text{total interest}$$
$$0.05x + 0.11(15,000 - x) = 1500$$

Carry Out
$$0.05x + 0.11(15,000) - 0.11(x) = 1500$$
$$0.05x + 1650 - 0.11x = 1500$$
$$-0.06x + 1650 = 1500$$
$$-0.06x = -150$$
$$x = \frac{-150}{-0.06} = 2500$$

Check and Answer Thus, $2500 should be invested at 5% interest. The amount to be invested at 11% is

$$15,000 - x = 15,000 - 2500 = 12,500$$

NOW TRY EXERCISE 35 The total amount invested is $15,000, which checks with the information given. ✳

In Example 6, we let x represent the amount invested at 5%. If we had let x represent the amount invested at 11%, the answer would not have changed. Rework Example 6 now, letting x represent the amount invested at 11%.

In other types of problems involving money, we generally set up a similar table. For example, if your annual payments for rent include two different amounts, you might develop a table like the one that follows.

Number of Months × Rent = Amount paid			
Rent	**Number of Months**	**Rent**	**Amount paid**
lower rent			Amount paid at lower rent
higher rent			Amount paid at higher rent

Then we use the amount paid, just as we did with distance and interest, to answer the question.

Now let's look at another example involving money.

EXAMPLE 7 **Art Fair** Mary Gallagher sells both small and large paintings at an art fair. The small paintings sell for $50 each and large paintings sell for $175 each. By the end of the day, Mary lost track of the number of paintings of each size she sold. However, by looking at her receipts she knows that she sold a total of 14 paintings for a total of $1200. Determine the number of small and the number of large paintings she sold that day.

Solution Understand and Translate We are asked to find the number of paintings of each size sold.

$$\text{Let } x = \text{number of small paintings sold}$$

$$\text{then } 14 - x = \text{number of large paintings sold}$$

The income received from the sale of the small paintings is found by multiplying the number of small paintings sold by the cost of a small painting. The income received from the sale of the large paintings is found by multiplying the number of large paintings sold by the cost of a large painting. The total income received for the day is the sum of the income from the small paintings and the large paintings.

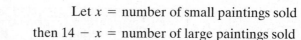

$$\left(\begin{array}{c}\text{Cost of}\\\text{Painting}\end{array}\right) \times \left(\begin{array}{c}\text{Number of}\\\text{Paintings}\end{array}\right) = \left(\begin{array}{c}\text{Income from}\\\text{Paintings}\end{array}\right)$$

Painting	Cost	Number of Paintings	Income from Paintings
Small	50	x	$50x$
Large	175	$14 - x$	$175(14 - x)$

$$\begin{pmatrix} \text{income from} \\ \text{small paintings} \end{pmatrix} + \begin{pmatrix} \text{income from} \\ \text{large paintings} \end{pmatrix} = \text{total income}$$

$$50x + 175(14 - x) = 1200$$

Carry Out

$$50x + 2450 - 175x = 1200$$
$$-125x + 2450 = 1200$$
$$-125x = -1250$$
$$x = \frac{-1250}{-125} = 10$$

Check and Answer   Ten small paintings and $14 - 10$ or 4 large paintings were sold.

Check

$$\begin{aligned} \text{income from 10 small paintings} &= \quad 500 \\ \text{income from 4 large paintings} &= \quad \underline{700} \\ \text{total} &= 1200 \quad \textit{True} \end{aligned}$$

NOW TRY EXERCISE 45

4 Solve Mixture Problems

Now we will work some mixture problems. Any problem in which two or more quantities are combined to produce a different quantity or a single quantity is separated into two or more different quantities may be considered a **mixture problem**. Mixture problems are familiar to everyone, as we can see in the everyday examples that follow.

When solving mixture problems, we often let the variable represent one unknown quantity, and then we represent a second unknown quantity in terms of the first unknown quantity. For example, if we know that when two solutions are mixed they make a total of 80 liters, we may represent the number of liters of one of the solutions as x and the number of liters of the second solution as $80 - x$. Note that when we add x and $80 - x$ we get 80, the total amount.

We generally solve mixture problems by using the fact that the amount (or value) of one part of the mixture plus the amount (or value) of the second part of the mixture is equal to the total amount (or value) of the total mixture.

As we did with motion problems involving two rates, we will use a table to help analyze the problem.

When we construct a table for mixture problems, our table will generally have three rows instead of two as with motion and money problems. One row will be for each of the two individual items being mixed, and the third row will be for the mixture of the two items. When working with solutions, we use the formula, *Amount of substance in the solution = strength of solution (in percent) × quantity of solution*. When we are mixing two quantities and are interested in the *composition* of the mixture, we generally use the following table or a variation of the table.

Strength (in percent) × Quantity = Amount of Substance			
Solution	**Strength**	**Quantity**	**Amount of Substance**
Solution 1			Amount of substance in solution 1
Solution 2			Amount of substance in solution 2
Mixture			Amount of substance in mixture

When using this table, we generally use the following formula to solve the problem.

$$\begin{pmatrix} \text{Amount of Substance} \\ \text{in Solution 1} \end{pmatrix} + \begin{pmatrix} \text{Amount of Substance} \\ \text{in Solution 2} \end{pmatrix} = \begin{pmatrix} \text{Amount of Substance} \\ \text{in Mixture} \end{pmatrix}$$

Let us now look at an example of a mixture problem where two solutions are combined.

EXAMPLE 8

Mixing Acid Solutions   Mr. Dave Lumsford, a chemistry instructor, needs a 10% acetic acid solution for a chemistry experiment. After checking the store room, he finds that there are only 5% and 20% acetic acid solutions available. Since there is not sufficient time to order the 10% solution, Mr. Lumsford decides to make the 10% solution by combining the 5% and 20% solutions. How many liters of the 5% solution must he add to 8 liters of the 20% solution to get a solution that is 10% acetic acid?

Solution

Understand and Translate   We are asked to find the number of liters of the 5% acetic acid solution to mix with 8 liters of the 20% acetic acid solution.

Let x = number of liters of 5% acetic acid solution

Let's draw a sketch of the solution (Fig. 3.15)

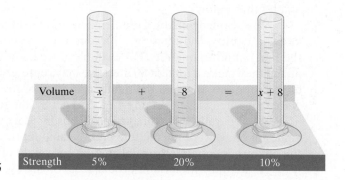

FIGURE 3.15

The amount of acid in a given solution is found by multiplying the percent strength by the number of liters.

Solution	Strength	Liters	Amount of Acetic Acid
5%	0.05	x	$0.05x$
20%	0.20	8	$0.20(8)$
Mixture	0.10	$x + 8$	$0.10(x + 8)$

$$\begin{pmatrix} \text{amount of acid} \\ \text{in 5% solution} \end{pmatrix} + \begin{pmatrix} \text{amount of acid} \\ \text{in 20% solution} \end{pmatrix} = \begin{pmatrix} \text{amount of acid} \\ \text{in 10% mixture} \end{pmatrix}$$

$$0.05x + 0.20(8) = 0.10(x + 8)$$

TEACHING TIP
Have students work in pairs to
write a motion problem and a
mixture problem. Then have them
give their problems to another
pair of students to solve.

Carry Out

$$0.05x + 1.6 = 0.10x + 0.8$$
$$0.05x + 0.8 = 0.10x$$
$$0.8 = 0.05x$$
$$\frac{0.8}{0.05} = x$$
$$16 = x$$

Answer Sixteen liters of 5% acetic acid solution must be added to the 8 liters of 20% acetic acid solution to get a 10% acetic acid solution. The total number of liters that will be obtained is 16 + 8 or 24.

NOW TRY EXERCISE 57

When we are combining two items and are interested in the *value* of the mixture, the following table, or a variation of it, is often used.

Price (per unit) × Quantity = Value of Item			
Item	**Price**	**Quantity**	**Value of Item**
item 1			value of item 1
item 2			value of item 2
mixture			value of mixture

When we use this table, we generally use the following formula to solve the problem.

value of item 1 + value of item 2 = value of mixture

Now let's look at a mixture problem where we discuss the value of the mixture.

EXAMPLE 9

TEACHING TIP
Before discussing the solution to
Example 9 ask, "Would anybody
buy the mix at $8.00 per pound?
Explain. Would anybody sell the
mix at $3.00 per pound? Explain.
What does Becky's price of $6 per
pound for the coffee mix tell you
about the mix?"

Mixing Coffee Beans Becky Bugos owns a coffee shop in Santa Fe, New Mexico. In her shop are many varieties of coffee. One, an orange-flavored coffee, sells for $7 per pound, and a second, a hazelnut coffee, sells for $4 per pound. One day, by mistake, she mixed some orange-flavored beans with some hazelnut beans, and she found that some of her customers liked the blend when they sampled the coffee. So Becky decided to make and sell a blend of the two coffees.

a) How much of the orange-flavored coffee should she mix with 12 pounds of the hazelnut coffee to get a mixture that sells for $6 per pound?

b) How much of the mixture will be produced?

Solution **a)** Understand and Translate We are asked to find the number of pounds of orange flavored coffee.

Let x = number of pounds of orange-flavored coffee

We make a sketch of the situation (Fig. 3.16), then construct a table.

FIGURE 3.16

The value of the coffee is found by multiplying the number of pounds by the price per pound.

Coffee	Price per Pound	Number of Pounds	Value of Coffee
Orange-flavored	7	x	$7x$
Hazelnut	4	12	$4(12)$
Mixture	6	$x + 12$	$6(x + 12)$

$$\left(\begin{array}{c}\text{value of}\\\text{orange-flavored coffee}\end{array}\right) + \left(\begin{array}{c}\text{value of}\\\text{hazelnut coffee}\end{array}\right) = \text{value of mixture}$$

Carry Out

$$7x + 4(12) = 6(x + 12)$$
$$7x + 48 = 6x + 72$$
$$x + 48 = 72$$
$$x = 24 \text{ pounds}$$

Answer   Thus, 24 pounds of the orange-flavored coffee must be mixed with 12 pounds of the hazelnut coffee to make a mixture worth $6 per pound.

b) The number of pounds of the mixture is

NOW TRY EXERCISE 49

$$x + 12 = 24 + 12 = 36 \text{ pounds}$$

Exercise Set 3.5

Practice the Skills / Problem Solving

Set up an equation that can be used to solve each problem. Solve the equation, and answer the question. Use a calculator when you feel it is appropriate.

1. **Average Speed** On her way from Omaha, Nebraska, to Kansas City, Kansas, Peg Hovde traveled 150 miles in 3 hours. What was her average speed? 50 mph

2. **Water Usage** An article found on the Internet states that a typical shower uses 30 gallons of water and lasts for 6 minutes. How much water is typically used per minute? 5 gal/min

3. **Lasers** Lasers have many uses, from eye surgery to cutting through steel doors. One manufacturer of lasers produces a laser that is capable of cutting through steel at a rate of 0.2 centimeters per minute. During one of their tests, they went through a steel door in 12 minutes. How thick was the door? 2.4 cm

4. **Making Copies** Paul Murphy is at Mailboxes, Etc. making copies of an advertisement. While the copies are being run, he counts the copies made in 2.5 minutes and finds that 100 copies were made. At what rate is the copy machine running? 40 copies/min

5. **Laying Tile** Mary Ann Tuerk is laying tile. She can lay 30 square feet of tile per hour. How long will it take her to tile a room that is 420 square feet? 14 hr

6. **Free Fall** While swimming in the ocean, Elyse's glasses fell off her head. If the glasses fall at a rate of 4 feet per second, how long will it take for the glasses to fall 70 feet to the sand at the bottom? 17.5 sec

7. **IV's** Janette Rider is in a hospital recovering from minor surgery. The nurse must administer 1500 cubic centimeters of an intravenous fluid that contains an antibiotic over a 6 hour period. What is the flow rate of the fluid? 250 cm³/hr

8. **Cement** A conveyer belt at a cement plant is transporting 600 pounds of crushed stone per minute. Find the time it will take for the conveyer belt to transport 20,000 pounds of crushed stone. ≈ 33.33 min

9. **Canoe Trip** On a canoe trip, Jody Fry paddles at a speed of 4 feet per second. How long would it take her to paddle a mile (5280 feet)? 1320 sec or 22 min

10. *Film* A production line at a factory checks and packages 620 rolls of film in an hour. How long will it take to check and pack 2170 rolls of film? 3.5 hr

11. *Auto Race* The record speed for a 500-mile auto race was obtained at the Michigan 500 on August 9, 1990. Al Unser Jr. completed the race in about 2.635 hours. Find his average speed during the race. ≈189.75 mph

12. *Typing Speed* Using a word processor on his computer, Howie Sorkin typed an average of 700 words in 14 minutes. Determine his typing speed.
50 words per minute

Solve the following motion problems that involve two rates.

13. *Walkie-Talkies* Willie and Shanna Johnston have walkie-talkies that have a range of 16.8 miles. Willie and Shanna start at the same point and walk in opposite directions. If Willie walks at 3 miles per hour and Shanna walks at 4 miles per hour, how long will it take before they are out of range? 2.4 hr

14. *Blue Angels* At a Navy Blue Angel air show assume that two F/A-18 Hornet jets travel toward each other both at a speed of 1000 miles per hour. After they pass each other, if they were to keep flying in the same direction and speed, how long would it take for them to be 500 miles apart? 0.25 hr or 15 min

15. *Product Testing* The Goodyear Tire Company is testing a new tire. The tires are placed on a machine that can simulate the tires riding on a road. The machine is first set to 60 miles per hour and the tires run at this speed for 7.2 hours. The machine is then set to a second speed and runs at this speed for 6.8 hours. After this 14-hour period, the machine indicates the tires have traveled the equivalent of 908 miles. Find the second speed to which the machine was set. 70 mph

16. *Ski Lifts* On Whistler Mountain, people must use two different ski lifts to get to the top of the mountain. Assume that the change occurs exactly half way up the mountain. The first lift travels at 4 miles per hour for 0.2 hours. The second lift travels for 0.3 hours to reach the top of the mountain. If the total distance traveled up the mountain is 1.2 miles, find the average speed of the second ski lift. ≈1.33 mph

See Exercise 16.

17. *Navigating O'Hare Airport* Sadie Bragg and Dale Ewen are in Chicago's O'Hare airport walking between terminals. Sadie walks on the moving walkway (like a flat escalator moving along the floor). Her speed (relative to the ground) is 220 feet per minute. Dale starts walking at the same time and walks alongside the walkway at a speed of 100 feet per minute. How long have they been walking when Sadie is 600 feet ahead of Dale? 5 min

18. *Earthquakes* Earthquakes generate two circular waves, *p*-waves and *s*-waves, that travel outwards (see the figure). The *p*-waves travel faster, but generally the *s*-waves do the most damage. Suppose the *p*-waves have a velocity of 3.6 miles per second and *s*-waves have a velocity of 1.8 miles per second. How long after the earthquake will *p*-waves and *s*-waves be 80 miles apart? ≈44.4 sec

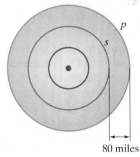

80 miles

19. *Disabled Boat* Two Coast Guard cutters are 225 miles apart traveling toward each other, one from the east and the other from the west, searching for a disabled boat. Because of the current, the eastbound cutter travels 5 miles per hour faster than the westbound cutter. If the two cutters pass each other after 3 hours, find the average speed of each cutter. 35 mph, 40 mph

20. *Walk in the Snow* Kathy Huet goes for a walk in the snow. She walks for 1.2 hours at 4 miles per hour. Then she turns around and returns using the same path. If her return trip takes her 1.5 hours, determine her average speed on her walk back. 3.2 mph

21. *Ironman Triathlon* A triathlon consists of three parts: swimming, cycling, and running. One of the more famous triathlons is the Hawaiian Ironman Triathlon held in Kailua-Kona, Hawaii (the island of Hawaii). Participants in the Ironman must swim a certain distance, cycle a certain distance, and then run for a certain distance. The 2001 women's winner was Natascha Badmann, age 34, from Switzerland. She swam at an average of about 2.38 miles per hour for about 1.01 hours, then she cycled at an average of 21.17 miles per hour for about 5.29 hours. Finally, she ran at an average of 8.32 miles per hour for about 3.15 hours.

a) Estimate the distance that Natascha swam. 2.4 mi
b) Estimate the distance that Natascha cycled. 112.0 mi
c) Estimate the distance that Natascha ran. 26.2 mi
d) Estimate the total distance covered during the triathlon. 140.6 mi
e) Estimate the winning time of the triathlon.*
   9.45 hr

22. *Paving Road* Two crews are laying blacktop on a road. They start at the same time at opposite ends of a 12-mile road and work toward one another. One crew lays blacktop at an average rate of 0.75 mile a day faster than the other crew. If the two crews meet after 3.2 days, find the rate of each crew.

23. *Sailing* Two sailboats are 9.8 miles apart and sailing toward each other. The larger boat, the *Pythagoras*, sails 4 miles per hour faster than the smaller boat, the *Apollo*. The two boats pass each other after 0.7 hour. Find the speed of each boat.

24. *Beach Clean-up* On Earth Day, two groups of people walk a 7-mile stretch of Myrtle Beach, cleaning up the beach. One group, headed by Auturo Perez, starts at one end, and the other group, headed by Jane Ivanov, starts at the other end of the beach. They start at the same time and walk toward each other. Auturo's group is traveling at a rate of 0.5 miles per hour faster than Jane's group, and they meet in 2 hours. Find the speed of each group. Jane's, 1.5 mph; Auturo's, 2 mph

25. *Bank Robbers* A motorboat carrying a bank robber leaves Galveston, Texas, and goes through the Gulf of Mexico heading toward Cozumel, Mexico. One-half hour after the boat leaves Galveston, the Coast Guard gets news of the bank robber's route and sends out a Coast Guard cutter to capture the bank robber. The motorboat is traveling at 25 miles per hour, and the Coast Guard cutter is traveling at 35 miles per hour.

a) How long will it take for the Coast Guard to catch the bank robber? 1.25 hr
b) How far from shore will the boats be when they meet? 43.75 mi

26. *Mountain Climbing* Serge and Francine Saville go mountain climbing together. Francine begins climbing the mountain 30 minutes before Serge and averages 18 feet per minute. When Serge begins climbing, he averages 20 feet per minute. How far up the mountain will they meet? 5400 ft

27. *Long Pass* Phil Cheifetz is the quarterback for his high school football team and Peter Kerwin an end. From the start of a play, Pete runs down the field at about 25 feet per second. Two seconds after the play starts, Phil throws the ball, at about 50 feet per second, to Pete who is still running.

a) How long after Phil throws the ball will Pete catch it? 2 sec
b) How far will Pete be from where the ball was thrown? 100 ft

28. *Exercising* Dien and Phuong Vu belong to a health club and exercise together regularly. They start running on two treadmills at the same time. Dien's machine is set for 6 miles per hour and Phuong's machine is set for 4 miles per hour. When they finish, they compare the distances and find that together they have run a total of 11 miles. How long had they run? 1.1 hr

29. *Traffic Jam* Betty Truitt drives for a number of hours at 70 miles per hour. Then the traffic slows and she drives at 50 miles per hour. She travels at 50 miles per hour for 0.5 hours longer than she traveled at 70 miles per hour. The difference in the distance traveled at 50 miles per hour and 70 miles per hour is 5 miles. Determine how long Betty traveled at 50 miles per hour. 1.5 hr

22. 1.5 mi/day, 2.25 mi/day   **23.** *Apollo*, 5 mph; *Pythagoras*, 9 mph

---

*The record winning time for the triathalon by a woman was by Californian Paula Newby-Frasier, an eight-time winner who in 1992 completed the course in 8:55:28. The actual distances are: swim, 2.4 mi; cycle, 112 mi; run 26 mi, 385 yds.

35. $7000 at 7%, $2400 at 5%
37. $2400 at 6%, $3600 at 4% **38.** $5769.23 at 7%, $6730.77 at 6% **39.** $2000 at 4%, $8000 at 5% **40.** $12,000 at 7%, $8000 at 5%

30. _Salt Mine_ At a salt mine the ore must travel on two different conveyer belts to be loaded onto a train. The second conveyer belt travels at a rate of 0.6 feet per second faster than the first conveyer belt. The ore travels 180 seconds on the first belt and 160 seconds on the second belt. If the total distance traveled by the ore is 1116 feet, determine the speed of the second belt.   3.6 ft/sec

31. _Flight Speed_ An airplane is scheduled to leave San Diego, California, at 9 A.M. and arrive in Cleveland, Ohio, at 1 P.M. Because of mechanical problems on the plane, the plane is delayed by 0.2 hours. To arrive at Cleveland at its originally scheduled time, the plane will need to increase its planned speed by 30 miles per hour. Find the plane's planned speed and its increased speed.   570 mph, 600 mph

32. _Distance Driven_ Yajun Yang started driving to the shopping mall at an average speed of 30 miles per hour. A short while later, he realized he had left his credit card on the kitchen counter. He turned around and headed back to the house, driving at 20 miles per hour (more traffic returning). If it took him a total of 0.6 hours to leave and then return home, how far had Yajun driven before he turned around?   7.2 mi

33. _Fixing Road_ Both a road leading to a bridge and the bridge must be replaced. The engineer estimates that it will take 20 days for the road crew to tear up and clear the road and an additional 60 days for the same road crew to dismantle and clear the bridge. The rate for clearing the road is 1.2 feet per day faster than the rate for clearing the bridge, and the total distance cleared is 124 feet. Find the rate for clearing the road and the rate for clearing the bridge.   road, 2.45 ft/day; bridge, 1.25 ft/day

34. _Mail Delivery_ Rich Poorman delivers the mail to residents along a 10.5-mile route. It normally takes him 5 hours to do the whole route. One Friday he needs to leave work early, so he gets his friend Keri Goldberg to help him. Rich will start delivering at one end and Keri will start at the other end 1 hour

later, and they will meet somewhere along his route. If Keri covers 1.6 miles per hour, how long after Keri begins will the two meet?   ≈2.27 hr

Solve the following money problems.

35. _Simple Interest_ Paul and Donna Petrie invested $9400, part at 5% simple interest and the rest at 7% simple interest for a period of 1 year. How much did they invest at each rate if their total annual interest from both investments was $610? (Use interest = principal · rate · time.)

36. _Simple Interest_ Jerry Correa invested $7000, part at 8% simple interest and the rest at 5% simple interest for a period of 1 year. If he received a total annual interest of $476 from both investments, how much did he invest at each rate? $4200 at 8%, $2800 at 5%

37. _Simple Interest_ Aleksandra Tomich invested $6000, part at 6% simple interest and part at 4% simple interest for a period of 1 year. How much did she invest at each rate if each account earned the same interest?

38. _Simple Interest_ Aimee Calhoun invested $12,500, part at 7% simple interest and part at 6% simple interest for a period of 1 year. How much was invested at each rate if each account earned the same interest?

39. _Simple Interest_ Míng Wang invested $10,000, part at 4% and part at 5% simple interest for a period of one year. How much was invested in each account if the interest earned in the 5% account was $320 greater than the amount invested in the 4% account?

40. _Simple Interest_ Sharon Sledge invested $20,000, part at 5% and part at 7% simple interest for a period of one year. How much was invested in each account if the interest earned in the 7% account was $440 greater than the amount invested in the 5% account?

41. _Rate Increase_ During the year the Public Service Commission approved a rate increase for the General Telephone Company. For a one-line household, the basic service rate increased from $17.10 to $18.40. While preparing her income tax return, Patricia Burgess found that she paid a total of $207.80 for basic telephone service for the year. In what month did the rate increase take effect?   November

42. Cable TV Violet Kokola knows that her subscription rate for the basic tier of cable television increased from $18.20 to $19.50 at some point during the calendar year. She knew that during the calendar year she paid a total of $230.10 to the cable company. Determine the month of the rate increase.

43. Wages Mihály Sarett holds two part-time jobs. One job, at Home Depot, pays $6.50 an hour and the second job, at a veterinary clinic, pays $7.00 per hour. Last week Mihály worked a total of 18 hours and earned $122.00. How many hours did Mihály work at each job?

44. Hall of Fame At the Baseball Hall of Fame adult admission is $9.50 and junior admission (7–12 years of age) is $4.00. During one day, a total of 2000 adult and junior admissions were collected, and $14,710 in admission fees was collected. How many adult admissions were collected? Adults: 1220

45. Computer Systems Comp U.S.A. is having a sale on two different computer systems. One system sells for $1320 and the other system sells for $1550. If a total of 200 of the systems were sold and the receipts from the 200 systems were $282,400, how many of the $1550 computer systems were sold?

46. Ticket Sales At a movie theatre the cost of an evening show was $7.50 and the cost at a matinee was $4.75. On one day there was one matinee and one evening showing of Harry Potter and the Sorcerer's Stone. On that day a total of 310 adult tickets were sold, which resulted in ticket sales of $2022.50. How many adults went to the matinee and how many went to the evening show?

47. Stock Purchase Suppose General Electric stock is selling at $74 a share and PepsiCo stock is selling at $35 a share. Mike Moussa has a maximum of $8000 to invest. He wishes to purchase five times as many shares of PepsiCo as of General Electric. Only whole shares of stock can be purchased.
a) How many shares of each will he purchase?
b) How much money will be left over? $32

48. Stock Purchase Suppose Wal Mart stock is selling at $59 a share and Mattel stock is selling at $28 a share. Amy Waller has a maximum of $6000 to invest. She wishes to purchase four times as many shares of Wal Mart as of Mattel. Only whole shares of stock can be purchased.
a) How many shares of each will she purchase?
b) How much money will be left over? $192

Solve the following mixture problems.

49. Grass Seed Scott's Family grass seed sells for $2.45 per pound and Scott's Spot Filler grass seed sells for $2.10 per pound. How many pounds of each should be mixed to get a 10-pound mixture that sells for $2.20 per pound? 2.86 lb Family; 7.14 lb Spot Filler

50. Nut Shop Jean Valjean owns a nut shop where walnuts cost $6.80 per pound and almonds cost $6.40 per pound. Jean gets an order that specifically requests a 30-pound mixture of walnuts and almonds that will cost $6.65 per pound. How many pounds of each type of nut should Jean mix to get the desired mixture?

51. Pumping Gas Ken Yoshimoto owns a gas station. He is out of premium gasoline and wants to make 500 gallons of premium by combining regular and premium plus gasoline. The regular gasoline costs $1.20 per gallon and the premium plus costs $1.35 per gallon. If the premium is to cost $1.26 per gallon, how much of each should he mix to make the 500 gallons of premium gasoline? 300 gal regular, 200 gal premium plus

52. Bird Food At Agway Gardens, bird food is sold in bulk. In one barrel are sunflower seeds that sell for $1.80 per pound. In a second barrel is cracked corn that sells for $1.40 per pound. If the store makes bags of a mixture of the two by mixing 2.5 pounds of the sunflower seeds with 1 pound of the cracked corn, what should be the cost per pound of the mixture?

53. Bulk Candies At a grocery store certain candies are sold in bulk from barrels. The Good and Plenty cost $2.49 per pound and Sweet Treats cost $2.89 per pound. If Jim Strange takes 3 scoops of the Good and Plenty and mixes it with 5 scoops of the Sweet Treats, how much per pound should the mixture sell for? Assume each scoop contained the same weight of candy.

54. Starbucks Ruth Cordeff runs a coffee house where chocolate almond coffee beans sell for $7.00 per pound and hazelnut coffee beans sell for $6.10 per pound. One customer asks Ruth to make a 6 pound mixture of the chocolate almond beans and the hazelnut coffee beans. How many pounds of each should be used if the mixture is to cost $6.40 per pound?

55. Beef Wellington Chef Ramon marinates his beef, which he uses in making beef Wellington, overnight in a wine that is a blend of two red wines. To make his blend, he mixes 5 liters of a wine that is 12% alcohol by volume with 2 liters of a wine that is 9% alcohol by volume. Determine the alcohol content of the mixture. ≈11.1%

56. *Pharmacy* Susan Staples, a pharmacist, has a 60% solution of the drug sodium iodite. She also has a 25% solution of the same drug. She gets a prescription calling for a 40% solution of the drug. How much of each solution should she mix to make 0.5 liter of the 40% solution? 0.21 L, 60%; 0.29 L, 25%

57. *Sulfuric Acid* In chemistry class, Todd Corbin has 1 liter of a 20% sulfuric acid solution. How much of a 12% sulfuric acid solution must he mix with the 1 liter of 20% solution to make a 15% sulfuric acid solution? $1\frac{2}{3}$ L

58. *Paint* Nick Pappas has two cans of white paint, both of which contain a small percent of a yellow pigment. One can contains a 2% yellow pigment and the other can contains a 5% yellow pigment. Nick wants to mix the two paints to get paint with a 4% yellow pigment. How much of the 5% yellow pigment paint should be mixed with 0.4 gallons of the 2% yellow pigment paint to get the desired paint? 0.8 gal

59. *Clorox* Clorox bleach, used in washing machines, is 5.25% sodium hypochlorite by weight. Swimming pool shock treatments are 10.5% sodium hypochlorite by weight. The instructions on the Clorox bottle say to add 8 ounces (1 cup) of Clorox to a quart of water. How much swimming pool shock treatment should Willie Williams add to a quart of water to get the same amount of sodium hypochlorite in the mixture as when 1 cup of Clorox is added to a quart of water? 4 oz

60. *Mouthwash* The label on the Listerine Cool Mint Antiseptic mouthwash says that it is 21.6% alcohol by volume. The label on the Scope Original Mint mouthwash says that it is 15.0% alcohol by volume. If Hans mixes 6 ounces of the Listerine with 4 ounces of the Scope, what is the percent alcohol content of the mixture? 18.96%

61. *Milk* Chuck Levy knows that reduced fat milk has 2% milkfat by weight and low fat milk has 1% milkfat by weight, but he does not know the milkfat content of whole milk. Chuck also knows a piece of trivia; that is, if you mix 4 gallons of whole milk with 5 gallons of low fat milk you obtain 9 gallons of reduced fat milk. Use this information to find the milkfat content of whole milk. 3.25%

62. *Orange Juice* Mary Ann Terwilliger has made 6 quarts of an orange juice punch for a party. The punch contains 12% orange juice. She feels that she may need

more punch, but she has no more orange juice so she adds $\frac{1}{2}$ quart of water to the punch. Find the percent of orange juice in the new mixture. ≈11.1%

63. *Salt Concentration* Suppose the dolphins at Sea World must be kept in salt water with an 0.8% salt content. After a week of warm weather, the salt content has increased to 0.9% due to water evaporation. How much water with 0% salt content must be added to 50,000 gallons of the 0.9% salt water to lower the salt concentration to 0.8%? 6250 gallons

64. *Hawaiian Punch* The label on a 12-ounce can of frozen concentrate Hawaiian Punch indicates that when the can of concentrate is mixed with 3 cans of cold water the resulting mixture is 10% juice. Find the percent of pure juice in the concentrate. 40% pure juice

65. *Antifreeze* Prestone antifreeze contains 12% ethylene glycose, and Xeres antifreeze contains 9% ethylene glycose. Nina's radiator needs antifreeze to be added, so Nina pours the remainder of a container of Prestone and 1 gallon of Xeres antifreeze into her radiator. If the mixture contained 10% ethylene glycose, how much Prestone antifreeze was added? 0.5 gal

66. *Calcium Requirements* The label on Orange Sweet orange juice with calcium added indicates that one serving satisfies 12% of a person's daily requirement of calcium. The label on Tree Top orange juice indicates that one serving satisfies 2% of a person's daily requirement. How many servings of Tree Top orange juice must be mixed with 0.3 serving of Orange Sweet to obtain a mixture which satisfies 8% of a person's daily requirement of calcium? 0.2 serving

Challenge Problems

67. *Fat Albert* The home base of the Navy's Blue Angels is at the Naval Air Station in Pensacola, Florida. They spend winters at the Naval Air Facility (NAF) in El Centro, California. Assume they fly at about 900 miles per hour when they fly from Pensacola to El Centro in their F/A-18 Hornets. On every trip, their C-130 transport (affectionately called Fat Albert) leaves before them carrying supplies and support personnel. The C-130 generally travels at about 370 miles per hour. If the Blue Angels are making

their trip from Pensacola to El Centro, how long before the Hornets leave should Fat Albert leave if it is to arrive 3 hours before the Hornets? The flying distance between Pensacola and El Centro is 1720 miles. ≈5.74 hr

68. *Filling Radiator* The radiator of Mark Jillian's 2003 Chevrolet holds 16 quarts. It is now filled with a 20% antifreeze solution. How many quarts must Mark drain and replace with pure antifreeze for the radiator to contain a 50% antifreeze solution? 6 qt

Group Activity

Discuss and answer Exercises 69 and 70 as a group.

69. *Race Horse* According to the *Guinness Book of World Records*, the fastest race horse speed recorded was by a horse called Big Racket on February 5, 1945, in Mexico City, Mexico. Big Racket ran a $\frac{3}{4}$-mile race in 62.41 seconds. Find Big Racket's speed in miles per hour. Round your answer to the nearest hundredth.
43.26 mph

70. *Garage Door Opener* An automatic garage door opener is designed to begin to open when a car is 100 feet from the garage. At what rate will the garage door have to open if it is to raise 6 feet by the time a car traveling at 4 miles per hour reaches it? (1 mile per hour ≈ 1.47 feet per second.) ≈ 0.35 ft/sec

Cumulative Review Exercises

[1.3] **71. a)** Divide $2\frac{3}{4} \div 1\frac{5}{8}$. $\frac{22}{13}$ or $1\frac{9}{13}$

   b) Add $2\frac{3}{4} + 1\frac{5}{8}$. $\frac{35}{8}$ or $4\frac{3}{8}$

[2.5] **72.** Solve the equation $6(x - 3) = 4x - 18 + 2x$.
   all real numbers

[2.6] **73.** Solve the proportion $\frac{6}{x} = \frac{72}{9}$. $\frac{3}{4}$ or 0.75

[2.7] **74.** Solve the inequality $3x - 4 \leq -4x + 3(x - 1)$.
   $x \leq \frac{1}{4}$

CHAPTER SUMMARY

Key Words and Phrases

3.1
Area
Circle
Circumference
Diameter
Evaluate a formula
Formula
Perimeter
Pi (π)
Polygon
Quadrilateral

Radius
Simple interest formula
Three-dimensional
   figure
Triangle
Volume

3.2
Consecutive even
   integer
Consecutive integer

Consecutive odd integer
Translate application
   problems into
   mathematical
   language

3.3
Problem-solving
   procedure

3.4
Complementary angles

Equilateral triangle
Isosceles triangle
Supplementary angle
Vertical angles

3.5
Mixture problems
Money problems
Motion problems

IMPORTANT FACTS

Simple interest formula: $i = prt$

Distance formula: $d = rt$

   The sum of the measures of the angles in any triangle is $180°$.

   The sum of the measures of the angles of a quadrilateral is $360°$.

Problem-Solving Procedure for Solving Application Problems
   1. Understand the problem.
      Identify the quantity or quantities you are being asked to find.
   2. Translate the problem into mathematical language (express the problem as an equation).

(continued on the next page)

a) Choose a variable to represent one quantity, *and write down exactly what it represents.* Represent any other quantity to be found in terms of this variable.

b) Using the information from part **a)**, write an equation that represents the application.

3. Carry out the mathematical calculations (solve the equation).

4. Check the answer (using the original application).

5. Answer the question asked.

Chapter Review Exercises

[3.1] Use the formula to find the value of the indicated variable. Use a calculator to save time and round your answer to the nearest hundredth when necessary.

1. $C = 2\pi r$ (circumference of a circle); find C when $r = 6$
37.70

2. $P = 2l + 2w$ (perimeter of a rectangle); find P when $l = 4$ and $w = 5$. 18

3. $A = \dfrac{1}{2}bh$ (area of a triangle); find A when $b = 8$ and $h = 12$. 48

4. $K = \dfrac{1}{2}mv^2$ (energy formula); find m when $K = 200$ and $v = 4$. 25

5. $y = mx + b$ (slope-intercept form); find b when $y = 15, m = 3,$ and $x = -2$. 21

6. $P = \dfrac{f}{1 + i}$ (investment formula); find f when $P = 4716.98$ and $i = 0.06$. 5000

*In Exercises 7–10 **a)** solve each equation for y; then **b)** find the value of y for the given value of x.*

7. $2x = 2y + 4, x = 10$ **a)** $y = x - 2$ **b)** 8

8. $6x + 3y = -9, x = 12$ **a)** $y = -2x - 3$ **b)** -27

9. $5x - 2y = 16, x = 2$ **a)** $y = \dfrac{5}{2}x - 8$ **b)** -3

10. $2x = 3y + 12, x = -6$ **a)** $y = \dfrac{2}{3}x - 4$ **b)** -8

Solve for the indicated variable.

11. $A = lw$ for w $w = \dfrac{A}{l}$

12. $A = \frac{1}{2}bh$, for h $h = 2A/b$

13. $i = prt$, for t $t = i/(pr)$

14. $P = 2l + 2w$, for w $w = (P - 2l)/2$

15. $V = \pi r^2 h$, for h $h = V/(\pi r^2)$

16. $V = \dfrac{1}{3}Bh$ for h. $h = \dfrac{3V}{B}$

17. *Simple Interest* How much interest will Tom Proietti pay if he borrows $600 for 2 years at 9% simple interest? (Use $i = prt$.) $108

18. *Perimeter* The perimeter of a rectangle is 16 inches. Find the length of the rectangle if the width is 2 inches. 6 in.

19. Express $x + (x + 5) = 9$ as a statement.

20. Express $x + (2x - 1) = 10$ as a statement.

[3.2, 3.3] Solve each problem.

21. *Numbers* One number is 8 more than the other. Find the two numbers if their sum is 74. 33 and 41

22. *Consecutive Integers* The sum of two consecutive integers is 237. Find the two integers. 118 and 119

23. *Numbers* The larger of two integers is 3 more than 5 times the smaller integer. Find the two numbers if the smaller subtracted from the larger is 31. 38 and 7

24. *New Car* Shaana recently purchased a new car. What was the cost of the car before tax if the total cost including a 7% tax was $23,260? $21,738.32

25. *Bagels* A bakery currently makes and ships 520 bagels per month to various outlets. They wish to increase the production and shipment of bagels by 20 per month until reaching a production and shipment level of 900 bagels. How long will this take? 19 months

26. *Salary Comparison* In Ron Gigliotti's present position as a salesman he receives a base salary of $500 per week plus a 3% commission on all sales he makes. He is considering moving to another company where he would sell the same goods. His base salary would be only $400 per week, but his commission would be 8% on all sales he makes. What weekly dollar sales would he have to make for the total salaries from each company to be the same? $2000

19. The sum of a number and the number increased by 5 is 9.

20. The sum of a number and twice the number decreased by 1 is 10.

27. *Sales Price* During the first week of a going-out-of-business sale, all prices were reduced by 20%. During the second week of the sale, all prices that still cost more than $100 were reduced by an additional $25. During the second week of the sale, Kathy Golladay purchased a camcorder for $495. What was the original price of the camcorder? $650

28. *Mortgages* The Burds are considering two banks for a mortgage, First Federal and Internet Bank. The monthly mortgage payments with First Federal would be $889 per month plus a one-time application fee of $900. The monthly payment with Internet Bank would be $826 plus a one-time fee of $1200. How many months would it take for the total payments for both banks to be the same? ≈4.76 months

[3.4] *Solve each problem.*

29. *Unknown Angles* One angle of a triangle measures 10° greater than the smallest angle, and the third angle measures 10° less than twice the smallest angle. Find the measures of the three angles. 45°, 55°, 80°

30. *Unknown Angles* One angle of a trapezoid measures 10° greater than the smallest angle; a third angle measures five times the smallest angle; and the fourth angle measures 20° greater than four times the smallest angle. Find the measure of the four angles. 30°, 40°, 150°, 140°

31. *Garden* Jackie Donofrio has a rectangular garden whose length is 4 feet longer than its width. The perimeter of the garden is 70 feet. Find the width and length of the garden? $w = 15.5$ ft, $l = 19.5$ ft

32. *Designing a House* Iram Hafeez is designing a house he plans to build. The basement will be rectangular with two areas. He has placed poles in the ground and attached string to the poles to mark off the two rooms (see the figure at the top of the next column). The length of the basement is to be 30 feet greater than the width, and a total of 310 feet of string was used to mark off the rooms. Find the width and length of the basement. $w = 50$ ft, $l = 80$ ft

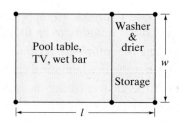

See Exercise 32.

[3.5] *Solve each problem.*   **34.** 6.5 mph

33. *Swimming Pool* Corey Christensen is filling a swimming pool using a hose. After 3.5 hours, the pool has a volume of 105 gallons. Find the rate of flow of the water. 30 gal/hr

34. *Marathon Runner* Dave Morris completed the 26-mile Boston Marathon in 4 hours. Find his average speed.

35. *Jogging* Two joggers follow the same route. Harold Lowe jogs at 8 kilometers per hour and Susan Karney Fackert at 6 kilometers per hour. If they leave at the same time, how long will it take for them to be 4 kilometers apart? 2 hr

36. *Trains Leaving* Two trains going in opposite directions leave from the same station on parallel tracks. One train travels at 50 miles per hour and the other at 60 miles per hour. How long will it take for the trains to be 440 miles apart? 4 hr

37. *Pittsburgh Incline* The Duquesne Incline in Pittsburgh, Pennsylvania, is shown below. The two cars start at the same time at opposite ends of the incline and travel toward each other at the same speed. The length of the incline is 400 feet and the time it takes for the cars to be at the halfway point is about 22.73 seconds. Determine, in feet per second, the speed the cars travel. ≈8.8 ft/sec

38. *Savings Accounts* Tatiana wishes to place part of $12,000 into a savings account earning 8% simple interest and part into a savings account earning $7\frac{1}{4}$% simple interest. How much should she invest in each if she wishes to earn $900 in interest for the year?

39. *Savings Accounts* Aimee Tait invests $4000 into two savings accounts. One account pays 3% simple interest and the other account pays 3.5% simple interest. If the interest earned in the account paying 3.5% simple interest is $94.50 more than the interest earned in the 3% account, how much was invested in each account?

38. $4000 at 8%, $8000 at $7\frac{1}{4}$%   **39.** $700 at 3%, $3300 at 3.5%

40. *Holiday Punch* Marcie Waderman is having a holiday party at her house. She made 2 gallons of a punch solution that contains 2% alcohol. How much pure punch must Marcie add to the punch to reduce the alcohol level to 1.5%? ≈0.67 gal

41. *Wind Chimes* Alan Carmell makes and then sells wind chimes. He makes two types, a smaller one that sells for $8 and a larger one that sells for $20. At an arts and crafts show he sells a total of 30 units, and his total receipts were $492. How many of each type of chime did he sell? small: 9, larger: 21

42. *Acid Solution* Bruce Kennan, a chemist, wishes to make 2 liters of an 8% acid solution by mixing a 10% acid solution and a 5% acid solution. How many liters of each should he use? 1.2 L of 10%, 0.8 L of 5%

[3.1–3.5] Solve each problem.

43. *Numbers* The sum of two consecutive odd integers is 208. Find the two integers. 103 and 105

44. *TV* What is the cost of a television set before tax if the total cost, including a 6% tax, is $477? $450

45. *Medical Supplies* Mr. Chang sells medical supplies. He receives a weekly salary of $300 plus a 5% commission on the sales he makes. If Mr. Chang earned $900 last week, what were his sales in dollars? $12,000

46. *Triangle* One angle of a triangle is 8° greater than the smallest angle. The third angle is 4° greater than twice the smallest angle. Find the measure of the three angles of the triangle. 42°, 50°, 88°

47. *Increase Staff* The Darchelle Leggett Company plans to increase its number of employees by 25 per year. If the company now has 427 employees, how long will it take before they have 627 employees? 8 years

48. *Parallelogram* The two larger angles of a parallelogram each measure 40° greater than the two smaller angles. Find the measure of the four angles.

49. *Copy Centers* Two copy centers across the street from one another are competing for business and both have made special offers. Under Copy King's plan, for a monthly fee of $20 each copy made in that month costs only 4 cents. King Kopie charges a monthly fee of $25 plus 3 cents a copy. How many copies made in a month would result in both copy centers charging the same amount? 500 copies

50. *Swim to Shore* Sisters Kathy and Chris Walter decide to swim to shore from a rowboat. Kathy starts swimming two minutes before Chris and averages 50 feet per minute. When Chris starts swimming, she averages 60 feet per minute.

   a) How long after Chris begins swimming will they meet? 10 min

   b) How far will they be from the rowboat when they meet? 600 ft

51. *Butcher* A butcher combined ground beef that cost $3.50 per pound with ground beef that cost $4.10 per pound. How many pounds of each were used to make 80 pounds of a mixture that sells for $3.65 per pound?

52. *Speed Traveled* Two brothers who are 230 miles apart start driving toward each other at the same time. The younger brother travels 5 miles per hour faster than the older brother, and the brothers meet after 2 hours. Find the speed traveled by each brother.

53. *Acid Solution* How many liters of a 30% acid solution must be mixed with 2 liters of a 12% acid solution to obtain a 15% acid solution? 0.4 L

48. 70°, 70°, 110°, 110° **51.** 60 lb of $3.50, 20 lb of $4.10 **52.** older brother, 55 mph; younger brother, 60 mph

Chapter Practice Test

1. *Simple Interest* Liz Wood took out a $12,000 3-year simple interest loan. If the interest she paid on the loan was $3240, find the simple interest rate (use $i = prt$).

2. Use $P = 2l + 2w$ to find P when $l = 6$ feet and $w = 3$ feet. 18 ft

3. Use $A = P + Prt$ to find A when $P = 100, r = 0.15$, and $t = 3$. 145

4. Use $A = \dfrac{m + n}{2}$ to find n when $A = 79$ and $m = 73$.

5. Use $C = 2\pi r$ to find r when $C = 50$. ≈7.96

6. a) Solve $4x = 3y + 9$ for y. $y = \frac{4}{3}x - 3$

   b) Find y when $x = 12$. 13

1. 9% **4.** 85

11. $500 - n$ **14.** 56 and 102 **16.** $2500 **18.** Peter: $40,000, Julie: $80,000

In Exercises 7 and 8, solve for the indicated variable.

7. $P = IR$, for R $R = P/I$

8. $A = \dfrac{a + b}{3}$, for a $a = 3A - b$

9. Find the area of the trapezoid shown.   28 sq ft

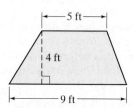

10. Find the area of the skating rink shown. The ends of the rink are semicircular. $\approx 3106.86 \text{ ft}^2$

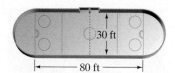

11. *Money* Five hundred dollars was divided between Boris and Monique. If Monique received n dollars, write an expression for the amount Boris received.

12. *Restaurants* The money Sally Sestini earned in one year from her second restaurant was $6000 more than twice what she earned from her first restaurant, f. Write an expression for what Sally earned from her second restaurant. $2f + 6000$

13. Express $x + (x + 4) = 9$ as a verbal statement.
The sum of a number and the number increased by 4 is 9.

In Exercises 14–25, set up an equation that can be used to solve the problem. Solve the problem and answer the question asked.

14. *Integers* The sum of two integers is 158. Find the two integers if the larger is 10 less than twice the smaller.

15. *Consecutive Integers* The sum of two consecutive integers is 43. Find the two integers.   21 and 22

16. *Lawn Furniture* Tim Kent purchased a set of lawn furniture. The cost of the furniture, including a 6% tax, was $2650. Find the cost of the furniture before tax.

17. *Eating Out* Mark Sullivan has only $40. He wishes to leave a 15% tip and must pay 7% tax. Find the price of the most expensive meal that he can order. $32.79

18. *Business Venture* Two friends form a very successful business. Since Julie Burgmeier invested twice as much money in the business as Peter Ancona, she receives

twice the profit that Peter does. If the profit for the year was $120,000, how much will each receive?

19. *Snow Plowing* William Echols is going to hire a service to plow his driveway whenever the snow totals 3 inches or more. Elizabeth Suco offers a service charging an annual fee of $80, plus $5 each time she plows. For the same service, Jon Wilkins charges an annual fee of $50, plus $10 each time he plows. How many times would the snow need to be plowed for the cost of both plans to be the same?   6 times

20. *Mortgages* Mike and Beverly Zwick are considering two banks for a mortgage, the Bank of Washington and First Trust. With the Bank of Washington, their monthly mortgage payment would be $980 per month plus $1500 in fees. With First Trust, their monthly mortgage payment would be $1025, but there are no additional fees. How long would it take for the total amount paid to be the same with both banks? ≈ 33.3 months

21. *Triangle* A triangle has a perimeter of 75 inches. Find the three sides if one side is 15 inches larger than the smallest side, and the third side is twice the smallest side.   15 in., 30 in., 30 in.

22. *American Flag* Kim Martino's American flag has a perimeter of 28 feet. Find the dimensions of the flag if the length is 4 feet less than twice its width.

23. *Laying Cable* Ellis and Harlene Matza are digging a shallow 67.2-foot-long trench to lay electrical cable to a new outdoor light fixture they just installed. They start digging at the same time at opposite ends of where the trench is to go, and dig toward each other. Ellis digs at a rate of 0.2 feet per minute faster than Harlene, and they meet after 84 minutes. Find the speed that each digs.   Harlene: 0.3 ft/min; Ellis: 0.5 ft/min

24. *Bulk Candy* A candy shop sells candy in bulk. In one bin is Jelly Belly candy, which sells for $2.20 per pound, and in a second bin is Kits, which sells for $2.75 per pound. How much of each type should be mixed to obtain a 3 pound mixture, which sells for $2.40 per pound?

25. *Salt Solution* How many liters of 20% salt solution must be added to 60 liters of 40% salt solution to get a solution that is 35% salt?   20 L

22. width: 6 ft, length: 8 ft **24.** Jelly Belly: ≈ 1.91 lb, Kits: ≈ 1.09 lb

Cumulative Review Test

Take the following test and check your answers with those that appear at the end of the test. Review any questions that you answered incorrectly. The section and objective where the material was covered is indicated after the answer.

1. ***Social Security*** The following circle graph shows what the typical retiree receives in social security, as a percent of their total income.

Where Social Security Recipients Get Their Income

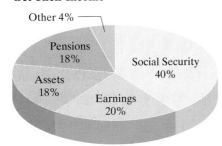

Source: Newsweek

If Emily receives $40,000 per year, and her income is typical of all social security recipients, how much is she receiving in social security? $16,000

2. ***Filing Electronically*** Each year more and more of us file our income taxes electronically. The following graph shows the increase in the number of tax forms filed electronically from 1998 through 2001.

Tax Returns Filed Electronically

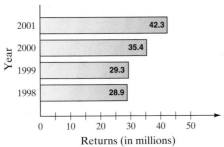

Returns (in millions)

Source: Internal Revenue Service

In Exercises 10–12, solve the equation.

10. $4x - 6 = x + 12$ 6

11. $6r = 2(r + 3) - (r + 5)$ $\frac{1}{5}$

12. $2(x + 5) = 3(2x - 4) - 4x$ no solution

13. ***Gas Needed*** If Lisa Shough's car can travel 50 miles on 2 gallons of gasoline, how many gallons of gas will it need to travel 225 miles?   9 gal

14. Solve the inequality $3x - 4 \le -1$ and graph the solution on a number line.

a) How many more tax forms were filed electronically in 2001 than in 2000?   6.9 million

b) How many times greater was the number of tax forms filed electronically in 2001 than in 2000? ≈ 1.19

3. ***Carbon Dioxide Levels*** David Warner, an environmentalist, was checking the level of carbon dioxide in the air. On five readings, he got the following results.

Test	Carbon Dioxide (parts per million)
1	5
2	6
3	8
4	12
5	5

a) Find the mean level of carbon dioxide detected.   7.2 ppm

b) Find the median level of carbon dioxide detected.   6 ppm

4. Evaluate $\frac{5}{12} \div \frac{3}{4} \cdot \frac{5}{9}$

5. How much larger is $\frac{2}{3}$ inch than $\frac{3}{8}$ inch? $\frac{7}{24}$ in.

6. a) List the set of natural numbers.

   b) List the set of whole numbers. $\{0, 1, 2, 3, \ldots\}$

   c) What is a rational number?

7. a) Evaluate $|-9|$.   9

   b) Which is greater, $|-5|$ or $|-3|$? Explain.

8. Evaluate $2 - 6^2 \div 2 \cdot 2$. -34

9. Simplify $4(2x - 3) - 2(3x + 5) - 6$. $2x - 28$

14. $x \le 1$, $\xleftarrow{\hspace{1.5em}\bullet\hspace{1.5em}}$ 17. $w = \dfrac{P - 2l}{2}$
                      1

15. If $A = \pi r^2$, find A when $r = 6$. ≈ 113.10

16. Consider the equation $4x + 8y = 16$.

    a) Solve the equation for y. $y = -\frac{1}{2}x + 2$

    b) Find y when $x = -4$.   4

17. Solve the formula $P = 2l + 2w$ for w.

18. ***Sum of Numbers*** The sum of two numbers is 29. Find the two numbers if the larger is 11 greater than twice the smaller.   6, 23

19. *Calling Plan* Lori Sypher is considering two cellular telephone plans. Plan A has a monthly charge of $19.95 plus 35 cents per minute. Plan B has a monthly charge of $29.95 plus 10 cents per minute. How long would Lori need to talk in a month for the two plans to have the same total cost? 40 min

20. *Quadrilateral* One angle of a quadrilateral measures 5° larger than the smallest angle; the third angle measures 50° larger than the smallest angle; and the fourth angle measures 25° greater than 4 times the smallest angle. Find the measure of each angle of the quadrilateral. 40°, 45°, 90°, 185°

Answers to Cumulative Review Test

1. $16,000 [Sec. 1.2, Obj. 2] **2. a)** 6.9 million **b)** $\approx$1.19 [Sec. 1.2, Obj. 2] **3. a)** 7.2 parts per million **b)** 6 parts per million [Sec. 1.2, Obj. 3] **4.** $\frac{5}{9}$ [Sec. 1.3, Obj. 4] **5.** $\frac{7}{24}$ inch [Sec. 1.3, Obj. 5] **6. a)** $\{1, 2, 3, 4, \ldots\}$ **b)** $\{0, 1, 2, 3, \ldots\}$ **c)** a quotient of two integers where the denominator is not 0 [Sec. 1.4, Obj. 1] **7. a)** 9 **b)** $|-5|$ [Sec. 1.5, Obj. 2] **8.** -34 [Sec. 1.9, Obj. 5] **9.** $2x - 28$ [Sec. 2.1, Obj. 6] **10.** 6 [Sec. 2.5, Obj. 1] **11.** $\frac{1}{5}$ [Sec. 2.5, Obj. 1] **12.** No solution [Sec. 2.5, Obj. 3] **13.** 9 gallons [Sec. 2.6, Obj. 3] **14.** $x \leq 1$, ⟵————•———⟶ [Sec. 2.7, Obj. 1] **15.** $\approx$113.10 [Sec. 3.1, Obj. 2] $\quad\quad\quad\quad\quad\quad\quad\quad 1$

16. a) $y = -\frac{1}{2}x + 2$ **b)** 4 [Sec. 3.1, Obj. 2] **17.** $w = \frac{P - 2l}{2}$ [Sec. 3.1, Obj. 3] **18.** 6, 23 [Sec. 3.3, Obj. 2]
19. 40 minutes [Sec. 3.3, Obj. 3] **20.** 40°, 45°, 90°, 185° [Sec. 3.4, Obj. 1]

Chapter 4

Exponents and Polynomials

Most areas of science and technology deal with very small and very large numbers. In our daily lives, we now hear more and more terminology that indicates small and large quantities. For example, our computer may have a 40-gigabyte hard drive, and the time that it takes computers to perform calculations may be expressed in microseconds. Scientific notation is a convenient way to work with small and large quantities. In Exercise 85 on page 270 we ask you to use scientific notation to determine how long it takes light from the sun to reach us here on earth, given the sun's distance from earth and the speed that light travels.

SSM

Study Guide

CD/Video

MathPro 4/5

PH Math
Tutor Center

prenhall.com/Angel

A Look Ahead

In this chapter we discuss exponents and polynomials. In Sections 4.1 and 4.2 we discuss the rules of exponents. When discussing scientific notation in Section 4.3, we use the rules of exponents to solve real-life application problems that involve very large or very small numbers. A knowledge of scientific notation may also help you in science and other courses.

In Sections 4.4 through 4.6 we explain how to add, subtract, multiply, and divide polynomials. To be successful with this material, you must understand the rules of exponents presented in the first two sections of this chapter. *To understand factoring, which is covered in Chapter 5, you need to understand polynomials, especially multiplication of polynomials.* As you will learn, factoring polynomials is the reverse of multiplying polynomials. We will be working with polynomials throughout the book.

4.1 EXPONENTS

SSM Study Guide CD/Video

MathPro 4/5 PH Math Tutor Center prenhall.com/Angel

1 Review exponents.
2 Learn the rules of exponents.
3 Simplify an expression before using the expanded power rule.

1 Review Exponents

To use polynomials, we need to expand our knowledge of exponents. Exponents were introduced in Section 1.9. Let's review the fundamental concepts. In the expression x^n, x is referred to as the **base** and n is called the **exponent**. x^n is read "x to the nth power."

$$x^2 = \underbrace{x \cdot x}_{2 \ factors \ of \ x}$$

$$x^4 = \underbrace{x \cdot x \cdot x \cdot x}_{4 \ factors \ of \ x}$$

$$x^m = \underbrace{x \cdot x \cdot x \cdots \cdot x}_{m \ factors \ of \ x}$$

EXAMPLE 1 Write $xxxxyyy$ using exponents.

Solution

$$\underbrace{x \, x \, x \, x}_{\substack{4 \ factors \\ of \ x}} \, \underbrace{y \, y \, y}_{\substack{3 \ factors \\ of \ y}} = x^4 y^3$$

Remember, when a term containing a variable is given without a numerical coefficient, the numerical coefficient of the term is assumed to be 1. For example $x = 1x$ and $x^2 y = 1x^2 y$.

Also recall that when a variable or numerical value is given without an exponent, the exponent of that variable or numerical value is assumed to be 1. For example, $x = x^1$, $xy = x^1 y^1$, $x^2 y = x^2 y^1$, and $2xy^2 = 2^1 x^1 y^2$.

2 Learn the Rules of Exponents

Now we will learn the rules of exponents.

EXAMPLE 2   Multiply $x^4 \cdot x^3$.

Solution

$$\underbrace{\overbrace{x \cdot x \cdot x \cdot x}^{x^4} \cdot \overbrace{x \cdot x \cdot x}^{x^3}} = x^7$$

Example 2 illustrates that when multiplying expressions with the same base we keep the base and *add* the exponents. This is the **product rule for exponents**.

TEACHING TIP
It may be helpful to verbalize some of the items in Example 3 as
a) The product of 2 factors of three and 1 factor of three is how many factors of three? **b)** The product of 4 factors of two and 2 factors of two is how many factors of two?

> ### Product Rule for Exponents
>
> $$x^m \cdot x^n = x^{m+n}$$

In Example 2, we showed that $x^4 \cdot x^3 = x^7$. This problem could also be done using the product rule: $x^4 \cdot x^3 = x^{4+3} = x^7$.

EXAMPLE 3   Multiply each expression using the product rule.

a) $3^2 \cdot 3$ **b)** $2^4 \cdot 2^2$ **c)** $x \cdot x^4$ **d)** $x^3 \cdot x^6$ **e)** $y^4 \cdot y^7$

Solution   **a)** $3^2 \cdot 3 = 3^2 \cdot 3^1 = 3^{2+1} = 3^3$ or 27   **b)** $2^4 \cdot 2^2 = 2^{4+2} = 2^6$ or 64

c) $x \cdot x^4 = x^1 \cdot x^4 = x^{1+4} = x^5$ **d)** $x^3 \cdot x^6 = x^{3+6} = x^9$

NOW TRY EXERCISE 17   **e)** $y^4 \cdot y^7 = y^{4+7} = y^{11}$

AVOIDING COMMON ERRORS

Note in Example 3**a)** that $3^2 \cdot 3^1$ is 3^3 and not 9^3. When multiplying powers of the same base, *do not multiply the bases.*

CORRECT	INCORRECT
$3^2 \cdot 3^1 = 3^3$	$3^2 \cdot 3^1 = 9^3$

Example 4 will help you understand the **quotient rule for exponents**.

EXAMPLE 4   Divide $x^5 \div x^3$.

Solution

$$\frac{x^5}{x^3} = \frac{\cancel{x} \cdot \cancel{x} \cdot \cancel{x} \cdot x \cdot x}{\cancel{x} \cdot \cancel{x} \cdot \cancel{x}} = \frac{1x^2}{1} = x^2$$

When dividing expressions with the same base, keep the base and *subtract* the exponent in the denominator from the exponent in the numerator.

TEACHING TIP
When discussing Example 4, point out that when dividing out common factors, a numerator or denominator is never completely removed. If all factors are divided out of the denominator (numerator), the denominator (numerator) equals 1.

> ### Quotient Rule for Exponents
>
> $$\frac{x^m}{x^n} = x^{m-n}, \qquad x \neq 0$$

In Example 4 we showed that $x^5/x^3 = x^2$. This problem could also be done using the quotient rule: $x^5/x^3 = x^{5-3} = x^2$.

EXAMPLE 5   Divide each expression using the quotient rule.

a) $\dfrac{3^5}{3^2}$ **b)** $\dfrac{6^4}{6}$ **c)** $\dfrac{x^{12}}{x^5}$ **d)** $\dfrac{y^{10}}{y^8}$ **e)** $\dfrac{z^8}{z}$

Solution   **a)** $\dfrac{3^5}{3^2} = 3^{5-2} = 3^3$ or 27   **b)** $\dfrac{6^4}{6} = \dfrac{6^4}{6^1} = 6^{4-1} = 6^3$ or 216

c) $\dfrac{x^{12}}{x^5} = x^{12-5} = x^7$ **d)** $\dfrac{y^{10}}{y^8} = y^{10-8} = y^2$

e) $\dfrac{z^8}{z} = \dfrac{z^8}{z^1} = z^{8-1} = z^7$

NOW TRY EXERCISE 23

AVOIDING COMMON ERRORS

Note in Example 5**a)** that $3^5/3^2$ is 3^3 and not 1^3. When dividing powers of the same base, *do not divide out the bases.*

CORRECT   INCORRECT

$\dfrac{3^3}{3^1} = 3^2$ or 9 $\dfrac{3^3}{3^1} \ne 1^2$

The answer to Example 5**c)**, x^{12}/x^5, is x^7. We obtained this answer using the quotient rule. This answer could also be obtained by dividing out the common factors in both the numerator and denominator as follows.

$$\frac{x^{12}}{x^5} = \frac{(x \cdot x \cdot x \cdot x \cdot x) \cdot x \cdot x \cdot x \cdot x \cdot x \cdot x \cdot x}{(x \cdot x \cdot x \cdot x \cdot x)} = x^7$$

We divided out the product of five x's, which is x^5. We can indicate this process in shortened form as follows.

$$\frac{x^{12}}{x^5} = \frac{x^5 \cdot x^7}{x^5} = x^7$$

In this section, to simplify an expression when the numerator and denominator have the same base and the exponent in the denominator is greater than the exponent in the numerator, we divide out common factors. For example, x^5/x^{12} can be simplified by dividing out the common factor, x^5, as follows.

$$\frac{x^5}{x^{12}} = \frac{x^5}{x^5 \cdot x^7} = \frac{1}{x^7}$$

We will now simplify some expressions by dividing out common factors.

EXAMPLE 6   Simplify by dividing out a common factor in both the numerator and denominator.

a) $\dfrac{x^9}{x^{12}}$ **b)** $\dfrac{y^4}{y^9}$

TEACHING TIP
After discussing Example 6, help students build their mathematical intuition by having them try to write the expressions in 6**a)** and 6**b)** without using fractions. Have them save their predictions to be checked when discussing negative exponents in the next section.

Solution   **a)** Since the numerator is x^9, we write the denominator with a factor of x^9. Since $x^9 \cdot x^3 = x^{12}$, we rewrite x^{12} as $x^9 \cdot x^3$.

$$\frac{x^9}{x^{12}} = \frac{x^9}{x^9 \cdot x^3} = \frac{1}{x^3}$$

b) $\dfrac{y^4}{y^9} = \dfrac{y^4}{y^4 \cdot y^5} = \dfrac{1}{y^5}$

NOW TRY EXERCISE 27

In the next section, we will show another way to evaluate expressions like $\dfrac{x^9}{x^{12}}$ by using the negative exponent rule.

Example 7 leads us to our next rule, the **zero exponent rule**.

EXAMPLE 7 Divide $\dfrac{x^3}{x^3}$.

Solution By the quotient rule,

$$\frac{x^3}{x^3} = x^{3-3} = x^0$$

However,

$$\frac{x^3}{x^3} = \frac{1x^3}{1x^3} = \frac{1 \cdot x \cdot x \cdot x}{1 \cdot x \cdot x \cdot x} = \frac{1}{1} = 1$$

Since $x^3/x^3 = x^0$ and $x^3/x^3 = 1$, then x^0 must equal 1.

Zero Exponent Rule

$$x^0 = 1, \qquad x \neq 0$$

By the zero exponent rule, any real number, except 0, raised to the zero power equals 1. Note that 0^0 is undefined.

EXAMPLE 8 Simplify each expression. Assume $x \neq 0$.
a) 3^0 **b)** x^0 **c)** $3x^0$ **d)** $(3x)^0$ **e)** $4x^2y^3z^0$

Solution **a)** $3^0 = 1$

b) $x^0 = 1$

c) $3x^0 = 3(x^0)$ *Remember, the exponent refers only to the immediately*
$\quad\quad = 3 \cdot 1 = 3$ *preceding symbol unless parentheses are used.*

d) $(3x)^0 = 1$

e) $4x^2y^3z^0 = 4x^2y^3 \cdot 1 = 4x^2y^3$

AVOIDING COMMON ERRORS

An expression raised to the zero power is not equal to 0; it is equal to 1.

CORRECT	INCORRECT
$x^0 = 1$	$x^0 = 0$
$5^0 = 1$	$5^0 = 0$

The **power rule** will be explained with the aid of Example 9.

EXAMPLE 9 Simplify $(x^3)^2$.

TEACHING TIP
After discussing Example 9, have students simplify the following
a) $(x^3)^3$ **b)** $(x^3)^4$ **c)** $(x^3)^5$
Then have them write a procedure for simplifying these kinds of expressions.

Solution
$$(x^3)^2 = x^3 \cdot x^3 = x^{3+3} = x^6$$
2 factors of x^3

Power Rule for Exponents

$$(x^m)^n = x^{m \cdot n}$$

The power rule indicates that when we raise an exponential expression to a power, we keep the base and *multiply* the exponents. Example 9 could also be simplified using the power rule: $(x^3)^2 = x^{3 \cdot 2} = x^6$.

EXAMPLE 10 Simplify. **a)** $(x^3)^5$ **b)** $(3^4)^2$ **c)** $(y^5)^7$

Solution **a)** $(x^3)^5 = x^{3 \cdot 5} = x^{15}$ **b)** $(3^4)^2 = 3^{4 \cdot 2} = 3^8$ **c)** $(y^5)^7 = y^{5 \cdot 7} = y^{35}$ ✳

HELPFUL HINT

Students often confuse the product and power rules. Note the difference carefully.

Product Rule	Power Rule
$x^m \cdot x^n = x^{m+n}$	$(x^m)^n = x^{m \cdot n}$
$2^3 \cdot 2^5 = 2^{3+5} = 2^8$	$(2^3)^5 = 2^{3 \cdot 5} = 2^{15}$

NOW TRY EXERCISE 51

Example 11 will help us in explaining the **expanded power rule**. As the name suggests, this rule is an expansion of the power rule.

EXAMPLE 11 Simplify $\left(\dfrac{ax}{by}\right)^4$.

Solution

$$\left(\frac{ax}{by}\right)^4 = \frac{ax}{by} \cdot \frac{ax}{by} \cdot \frac{ax}{by} \cdot \frac{ax}{by}$$

$$= \frac{a \cdot a \cdot a \cdot a \cdot x \cdot x \cdot x \cdot x}{b \cdot b \cdot b \cdot b \cdot y \cdot y \cdot y \cdot y} = \frac{a^4 \cdot x^4}{b^4 \cdot y^4} = \frac{a^4 x^4}{b^4 y^4}$$ ✳

Expanded Power Rule for Exponents

$$\left(\frac{ax}{by}\right)^m = \frac{a^m x^m}{b^m y^m}, \qquad b \neq 0, y \neq 0$$

The expanded power rule illustrates that every factor within parentheses is raised to the power outside the parentheses when the expression is simplified.

EXAMPLE 12 Simplify each expression.

a) $(4x)^2$ **b)** $(-x)^3$ **c)** $(5xy)^3$ **d)** $\left(\dfrac{-3y}{2z}\right)^2$

Solution **a)** $(4x)^2 = 4^2 x^2 = 16x^2$ **b)** $(-x)^3 = (-1x)^3 = (-1)^3 x^3 = -1x^3 = -x^3$

c) $(5xy)^3 = 5^3 x^3 y^3 = 125x^3 y^3$ **d)** $\left(\dfrac{-3y}{2z}\right)^2 = \dfrac{(-3)^2 y^2}{2^2 z^2} = \dfrac{9y^2}{4z^2}$ ✳

3 Simplify an Expression before Using the Expanded Power Rule

Whenever we have an expression raised to a power, it helps to simplify the expression in parentheses before using the expanded power rule. This procedure is illustrated in Examples 13 and 14.

EXAMPLE 13   Simplify $\left(\dfrac{9x^3y^2}{3xy^2}\right)^3$.

Solution   We first simplify the expression within parentheses by dividing out common factors.

$$\left(\frac{9x^3y^2}{3xy^2}\right)^3 = \left(\frac{9}{3}\cdot\frac{x^3}{x}\cdot\frac{y^2}{y^2}\right)^3 = (3x^2)^3$$

Now we use the expanded power rule to simplify further.

$$(3x^2)^3 = 3^3(x^2)^3 = 27x^6$$

TEACHING TIP
After discussing Example 13, have students verify that the result is the same when the expression within the parentheses is not simplified first. Then ask, "Which method do you prefer?"

Thus, $\left(\dfrac{9x^3y^2}{3xy^2}\right)^3 = 27x^6$.

HELPFUL HINT

STUDY TIP

Be very careful when writing exponents. Since exponents are generally smaller than regular text, take your time and write them clearly, and position them properly. If exponents are not written clearly it is very easy to confuse exponents such as 2 and 3, or 1 and 4, or 0 and 6. If you write down or carry an exponent from step to step incorrectly, you will obtain an incorrect answer.

EXAMPLE 14   Simplify $\left(\dfrac{25x^4y^3}{5x^2y^7}\right)^4$.

Solution   Begin by simplifying the expression within parentheses.

$$\left(\frac{25x^4y^3}{5x^2y^7}\right)^4 = \left(\frac{25}{5}\cdot\frac{x^4}{x^2}\cdot\frac{y^3}{y^7}\right)^4 = \left(\frac{5x^2}{y^4}\right)^4$$

Now use the expanded power rule to simplify further.

$$\left(\frac{5x^2}{y^4}\right)^4 = \frac{5^4(x^2)^4}{(y^4)^4} = \frac{625x^8}{y^{16}}$$

NOW TRY EXERCISE 91   Thus, $\left(\dfrac{25x^4y^3}{5x^2y^7}\right)^4 = \dfrac{625x^8}{y^{16}}$.

AVOIDING COMMON ERRORS

Students sometimes make errors in simplifying expressions containing exponents. One of the most common errors follows. Study this error carefully to make sure you do not make the same mistake.

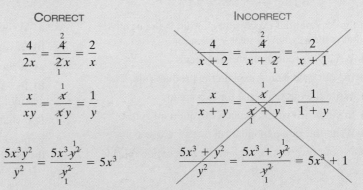

CORRECT

$$\frac{4}{2x} = \frac{\overset{2}{\cancel{4}}}{\underset{1}{\cancel{2}}x} = \frac{2}{x}$$

$$\frac{x}{xy} = \frac{\overset{1}{\cancel{x}}}{\underset{1}{\cancel{x}}y} = \frac{1}{y}$$

$$\frac{5x^3y^2}{y^2} = \frac{5x^3\overset{1}{\cancel{y^2}}}{\underset{1}{\cancel{y^2}}} = 5x^3$$

INCORRECT

$$\frac{4}{x+2} = \frac{\overset{2}{\cancel{4}}}{x+\underset{1}{\cancel{2}}} = \frac{2}{x+1}$$

$$\frac{x}{x+y} = \frac{\overset{1}{\cancel{x}}}{\underset{1}{\cancel{x}}+y} = \frac{1}{1+y}$$

$$\frac{5x^3+y^2}{y^2} = \frac{5x^3+\overset{1}{\cancel{y^2}}}{\underset{1}{\cancel{y^2}}} = 5x^3+1$$

(continued on the next page)

The simplifications on the right side are not correct because only common *factors* can be divided out (remember, factors are multiplied together). In the first denominator on the right, $x + 2$, the x and 2 are terms, not factors, since they are being added. Similarly, in the second denominator, $x + y$, the x and the y are terms, not factors, since they are being added. Also, in the numerator $5x^3 + y^2$, the $5x^3$ and y^2 are terms, not factors. No common factors can be divided out in the fractions on the right.

EXAMPLE 15   Simplify $(3y^3z^2)^4(2y^4z)$

Solution   First simplify $(3y^3z^2)^4$ by using the expanded power rule.

$$(3y^3z^2)^4 = 3^4y^{3\cdot4}z^{2\cdot4} = 81y^{12}z^8$$

Now use the product rule to simplify further.

$$(3y^3z^2)^4(2y^4z) = (81y^{12}z^8)(2y^4z^1)$$
$$= 81 \cdot 2 \cdot y^{12} \cdot y^4 \cdot z^8 \cdot z^1$$
$$= 162y^{12+4}z^{8+1}$$
$$= 162y^{16}z^9$$

NOW TRY EXERCISE 125   Thus, $(3y^3z^2)^4(2y^4z) = 162y^{16}z^9$.

Summary of the Rules of Exponents Presented in This Section

1. $x^m \cdot x^n = x^{m+n}$ **product rule**

2. $\dfrac{x^m}{x^n} = x^{m-n}, \quad x \neq 0$ **quotient rule**

3. $x^0 = 1, \quad x \neq 0$ **zero exponent rule**

4. $(x^m)^n = x^{m\cdot n}$ **power rule**

5. $\left(\dfrac{ax}{by}\right)^m = \dfrac{a^mx^m}{b^my^m}, \quad b \neq 0, \quad y \neq 0$ **expanded power rule**

Exercise Set 4.1

3. a) $\dfrac{x^m}{x^n} = x^{m-n}, x \neq 0$ 6. a) $\left(\dfrac{ax}{by}\right)^m = \dfrac{a^mx^m}{b^my^m}$

Concept/Writing Exercises

1. In the exponential expression c^r, what is the c called? What is the r called? *c:* base, *r:* exponent

2. a) Write the product rule for exponents. $x^m \cdot x^n = x^{m+n}$
   b) In your own words, explain the product rule.

3. a) Write the quotient rule for exponents.
   b) In your own words, explain the quotient rule.

4. a) Write the zero exponent rule. $x^0 = 1, x \neq 0$
   b) In your own words, explain the zero exponent rule.

5. a) Write the power rule for exponents. $(x^m)^n = x^{m\cdot n}$
   b) In your own words, explain the power rule.

6. a) Write the expanded power rule for exponents.
   b) In your own words, explain the expanded power rule.

7. For what value of x is $x^0 \neq 1$? 0

8. Explain the difference between the product rule and the power rule. Give an example of each.

2. b), 3. b), 4. b), 5. b), 6. b), 8. Answers will vary.

Practice the Skills

Simplify.

9. $x^5 \cdot x^4$ x^9

10. $x^6 \cdot x$ x^7

11. $z^4 \cdot z$ z^5

12. $x^4 \cdot x^2$ x^6

13. $3^2 \cdot 3^3$ 243

14. $4^2 \cdot 4^3$ 1024

15. $y^3 \cdot y^2$ y^5

16. $x^3 \cdot x^4$ x^7

17. $z^3 \cdot z^5$ z^8

18. $2^2 \cdot 2^2$ 16

19. $y^6 \cdot y$ y^7

20. $x^4 \cdot x^4$ x^8

Simplify.

21. $\dfrac{6^2}{6}$ 6

22. $\dfrac{x^4}{x^3}$ x

23. $\dfrac{x^{10}}{x^3}$ x^7

24. $\dfrac{y^5}{y}$ y^4

25. $\dfrac{3^5}{3^2}$ 27

26. $\dfrac{4^5}{4^3}$ 16

27. $\dfrac{y^4}{y^6}$ $\dfrac{1}{y^2}$

28. $\dfrac{a^7}{a^9}$ $\dfrac{1}{a^2}$

29. $\dfrac{c^4}{c^4}$ 1

30. $\dfrac{3^4}{3^4}$ 1

31. $\dfrac{a^3}{a^7}$ $\dfrac{1}{a^4}$

32. $\dfrac{x^9}{x^{13}}$ $\dfrac{1}{x^4}$

Simplify.

33. x^0 1

34. 5^0 1

35. $3x^0$ 3

36. $-4x^0$ -4

37. $4(5d)^0$ 4

38. $-2(4x)^0$ -2

39. $-3(-4y)^0$ -3

40. $-(-x)^0$ -1

41. $5x^3yz^0$ $5x^3y$

42. $-5xy^2z^0$ $-5xy^2$

43. $-5r(st)^0$ $-5r$

44. $-3(a^2b^5c^3)^0$ -3

Simplify.

45. $(x^4)^2$ x^8

46. $(x^5)^3$ x^{15}

47. $(x^5)^5$ x^{25}

48. $(y^5)^2$ y^{10}

49. $(x^3)^1$ x^3

50. $(x^3)^2$ x^6

51. $(x^4)^3$ x^{12}

52. $(x^5)^4$ x^{20}

53. $(n^6)^3$ n^{18}

54. $(2w^2)^3$ $8w^6$

55. $(1.3x)^2$ $1.69x^2$

56. $(-3x)^2$ $9x^2$

57. $(-3x^3)^3$ $-27x^9$

58. $(xy)^4$ x^4y^4

59. $(3a^2b^4)^3$ $27a^6b^{12}$

60. $(4x^3y^2)^3$ $64x^9y^6$

Simplify.

61. $\left(\dfrac{x}{3}\right)^2$ $\dfrac{x^2}{9}$

62. $\left(\dfrac{2}{x}\right)^3$ $\dfrac{8}{x^3}$

63. $\left(\dfrac{y}{x}\right)^4$ $\dfrac{y^4}{x^4}$

64. $\left(\dfrac{3}{y}\right)^4$ $\dfrac{81}{y^4}$

65. $\left(\dfrac{6}{x}\right)^3$ $\dfrac{216}{x^3}$

66. $\left(\dfrac{4m}{n}\right)^3$ $\dfrac{64m^3}{n^3}$

67. $\left(\dfrac{3x}{y}\right)^3$ $\dfrac{27x^3}{y^3}$

68. $\left(\dfrac{3s}{t^2}\right)^2$ $\dfrac{9s^2}{t^4}$

69. $\left(\dfrac{4p}{5}\right)^2$ $\dfrac{16p^2}{25}$

70. $\left(\dfrac{3x^4}{y}\right)^3$ $\dfrac{27x^{12}}{y^3}$

71. $\left(\dfrac{2y^3}{x}\right)^4$ $\dfrac{16y^{12}}{x^4}$

72. $\left(\dfrac{-4x^2}{5}\right)^2$ $\dfrac{16x^4}{25}$

Simplify.

73. $\dfrac{x^6y}{xy^3}$ $\dfrac{x^5}{y^2}$

74. $\dfrac{x^3y^5}{x^7y}$ $\dfrac{y^4}{x^4}$

75. $\dfrac{10x^3y^8}{2xy^{10}}$ $\dfrac{5x^2}{y^2}$

76. $\dfrac{5x^{12}y^2}{10xy^9}$ $\dfrac{x^{11}}{2y^7}$

77. $\dfrac{3ab}{27a^3b^4}$ $\dfrac{1}{9a^2b^3}$

78. $\dfrac{30y^5z^3}{5yz^6}$ $\dfrac{6y^4}{z^3}$

79. $\dfrac{35x^4y^9}{15x^9y^{12}}$ $\dfrac{7}{3x^5y^3}$

80. $\dfrac{6m^3n^9}{9m^7n^{12}}$ $\dfrac{2}{3m^4n^3}$

81. $-\dfrac{36xy^7z}{12x^4y^5z}$ $-\dfrac{3y^2}{x^3}$

82. $\dfrac{4x^4y^7z^3}{32x^5y^4z^9}$ $\dfrac{y^3}{8xz^6}$

83. $-\dfrac{6x^2y^7z}{3x^5y^9z^6}$ $-\dfrac{2}{x^3y^2z^5}$

84. $-\dfrac{25x^4y^{10}}{30x^3y^7z}$ $-\dfrac{5xy^3}{6z}$

Simplify.

85. $\left(\dfrac{10x^4}{5x^6}\right)^3$ $\dfrac{8}{x^6}$

86. $\left(\dfrac{4x^4}{8x^8}\right)^3$ $\dfrac{1}{8x^{12}}$

87. $\left(\dfrac{6y^6}{2y^3}\right)^3$ $27y^9$

88. $\left(\dfrac{25s^4t}{5s^6t^4}\right)^3$ $\dfrac{125}{s^6t^9}$

89. $\left(\dfrac{9a^2b^4}{3a^7b^9}\right)^0$ 1

90. $\left(\dfrac{16y^6}{24y^{10}}\right)^3$ $\dfrac{8}{27y^{12}}$

91. $\left(\dfrac{x^4y^3}{x^2y^5}\right)^2$ $\dfrac{x^4}{y^4}$

92. $\left(\dfrac{2x^7y^2}{4xy}\right)^3$ $\dfrac{x^{18}y^3}{8}$

93. $\left(\dfrac{9y^2z^7}{18y^9z}\right)^4$ $\dfrac{z^{24}}{16y^{28}}$

94. $\left(\dfrac{y^7z^5}{y^8z^4}\right)^{10}$ $\dfrac{z^{10}}{y^{10}}$

95. $\left(\dfrac{4xy^5}{y}\right)^3$ $64x^3y^{12}$

96. $\left(\dfrac{-64xy^6}{32xy^9}\right)^4$ $\dfrac{16}{y^{12}}$

Simplify.

97. $(5xy^4)^2$ $25x^2y^8$

98. $(4ab^3)^3$ $64a^3b^9$

99. $(3ab^3)(b)$ $3ab^4$

100. $(6xy^5)(3x^2y^4)$ $18x^3y^9$

101. $(-2xy)(3xy)$ $-6x^2y^2$

102. $(-2x^4y^2)(5x^2y)$ $-10x^6y^3$

103. $(5x^2y)(3xy^5)$ $15x^3y^6$

104. $(-5xy)(-2xy^6)$ $10x^2y^7$

105. $(-3p^2q)^2(-p^2q)$ $-9p^6q^3$

106. $(2c^3d^2)^2(3cd)^0$ $4c^6d^4$

107. $(5r^3s^2)^2(5r^3s^4)^0$ $25r^6s^4$

108. $(3x^2)^4(2xy^5)$ $162x^9y^5$

Simplify.

109. $(-x)^2$ x^2

110. $(2xy^4)^3$ $8x^3y^{12}$

111. $\left(\dfrac{x^5y^5}{xy^5}\right)^3$ x^{12}

112. $(2x^2y^5)(3x^5y^4)^3$ $54x^{17}y^{17}$

113. $(2.5x^3)^2$ $6.25x^6$

114. $(-3a^2b^3c^4)^3$ $-27a^6b^9c^{12}$

115. $\dfrac{x^7y^2}{xy^6}$ $\dfrac{x^6}{y^4}$

116. $(xy^4)(xy^4)^3$ x^4y^{16}

117. $\left(-\dfrac{m^4}{n^3}\right)^3$ $-\dfrac{m^{12}}{n^9}$

118. $\left(-\dfrac{12x}{16x^7y^2}\right)^2$ $\dfrac{9}{16x^{12}y^4}$

119. $(-6x^3y^2)^3$ $-216x^9y^6$

120. $(3x^6y)^2(4xy^8)$ $36x^{13}y^{10}$

121. $(-2x^4y^2z)^3$ $-8x^{12}y^6z^3$

122. $\left(\dfrac{x}{3}\right)^2$ $\dfrac{x^2}{9}$

123. $(9r^4s^5)^3$ $729r^{12}s^{15}$

124. $(5x^4z^{10})^2(2x^2z^8)$ $50x^{10}z^{28}$

125. $(4x^2y)(3xy^2)^3$ $108x^5y^7$

126. $\dfrac{x^2y^6}{x^4y}$ $\dfrac{y^5}{x^2}$

127. $(7.3x^2y^4)^2$ $53.29x^4y^8$

128. $\left(\dfrac{-3x^3}{4}\right)^3$ $-\dfrac{27x^9}{64}$

129. $(x^7y^5)(xy^2)^4$ $x^{11}y^{13}$

130. $(4c^3d^2)(2c^5d^3)^2$ $16c^{13}d^8$

131. $\left(\dfrac{-x^4z^7}{x^2z^5}\right)^4$ x^8z^8

132. $(x^4y^6)^3(3x^2y^5)$ $3x^{14}y^{23}$

Study the Avoiding Common Errors box on page 247. Simplify the following expressions by dividing out common factors. If the expression cannot be simplified by dividing out common factors, so state.

133. $\dfrac{x+y}{x}$

134. $\dfrac{xy}{x}$ y

135. $\dfrac{y^2+3}{y}$

136. $\dfrac{x+4}{2}$

137. $\dfrac{6yz^4}{yz^2}$ $6z^2$

138. $\dfrac{a^2+b^2}{a^2}$

139. $\dfrac{x}{x+1}$

140. $\dfrac{x^4}{x^2y}$ $\dfrac{x^2}{y}$

133., 135., 136., 138., 139. cannot be simplified

Problem Solving

141. What is the value of x^2y if $x = 4$ and $y = 2$? 32

142. What is the value of xy^2 if $x = -3$ and $y = -4$? -48

143. What is the value of $(xy)^0$ if $x = 2$ and $y = 4$? 1

144. What is the value of $(xy)^0$ if $x = -5$ and $y = 3$? 1

145. Consider the expression $(-x^5y^7)^9$. When the expanded power rule is used to simplify the expression, will the *sign* of the simplified expression be positive or negative? Explain how you determined your answer. negative, odd exponent

146. Consider the expression $(-9x^4y^6)^8$. When the expanded power rule is used to simplify the expression, will the *sign* of the simplified expression be positive or negative? Explain how you determined your answer. positive, even exponent

Write an expression for the total area of the figure or figures shown.

147. 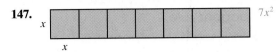 $7x^2$

x

x

148. 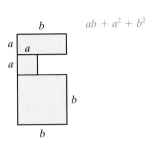 $x^2 + y^2$

y

y x

149. 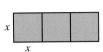 x $3x^2 + 4xy$

x

x

y
y

150. b $ab + a^2 + b^2$

a a

a

b

b

Challenge Problems

Simplify.

151. $\left(\dfrac{3x^4y^5}{6x^6y^8}\right)^3\left(\dfrac{9x^7y^8}{3x^3y^5}\right)^2$ $\dfrac{9x^2}{8y^3}$

152. $(3yz^2)^2\left(\dfrac{2y^3z^5}{10y^6z^4}\right)^0(4y^2z^3)^3$ $576y^8z^{13}$

 ## Group Activity

Discuss and answer Exercise 153 as a group, according to the instructions.

153. In the next section we will be working with negative exponents. To prepare for that work, use the expression $\dfrac{3^2}{3^3}$ to work parts **a)** through **c)**. Work parts **a)** through **d)** individually, then part **e)** as a group.

  a) Divide out common factors in the numerator and denominator and determine the value of the expression. $\frac{1}{3}$

  b) Use the quotient rule on the given expression and write down your results. 3^{-1}

  c) Write a statement of equality using the results of part **a)** and **b)** above. $\frac{1}{3} = 3^{-1}$

  d) Repeat parts **a)** through **c)** for the expression $\dfrac{2^3}{2^4}$.

  e) As a group, compare your answers to parts **a)** through **d)**, then write an exponential expression for $\dfrac{1}{x^m}$. x^{-m}

153. d) $\dfrac{1}{2}, 2^{-1}, \dfrac{1}{2} = 2^{-1}$

Cumulative Review Exercises

156. all real numbers

[1.9] **154.** Evaluate $3^4 \div 3^3 - (5 - 8) + 7$. 13

[2.1] **155.** Simplify $-4(x - 3) + 5x - 2$. $x + 10$

[2.5] **156.** Solve the equation
$$2(x + 4) - 3 = 5x + 4 - 3x + 1.$$

[3.1] **157. a)** Use the formula $P = 2l + 2w$ to find the length of the sides of the rectangle shown if the perimeter of the rectangle is 26 inches.
  4 in., 4 in., 9 in., 9 in.

$x + 5$

x

  b) Solve the formula $P = 2l + 2w$ for w.
$$w = \dfrac{P - 2l}{2}$$

4.2 NEGATIVE EXPONENTS

SSM

Study Guide

CD/Video

MathPro 4/5

PH Math
Tutor Center

prenhall.com/Angel

1 Understand the negative exponent rule.
2 Simplify expressions containing negative exponents.

1 Understand the Negative Exponent Rule

One additional rule that involves exponents is the negative exponent rule. You will need to understand negative exponents to be successful with scientific notation in the next section.

The negative exponent rule will be developed using the quotient rule illustrated in Example 1.

EXAMPLE 1 Simplify x^3/x^5 by **a)** using the quotient rule and **b)** dividing out common factors.

Solution **a)** By the quotient rule,

$$\frac{x^3}{x^5} = x^{3-5} = x^{-2}$$

b) By dividing out common factors,

$$\frac{x^3}{x^5} = \frac{\cancel{x} \cdot \cancel{x} \cdot \cancel{x}}{\cancel{x} \cdot \cancel{x} \cdot \cancel{x} \cdot x \cdot x} = \frac{1}{x^2}$$

In Example 1, we see that x^3/x^5 is equal to both x^{-2} and $1/x^2$. Therefore, x^{-2} must equal $1/x^2$. That is, $x^{-2} = 1/x^2$. This is an example of the **negative exponent rule**.

TEACHING TIP
Have students check the predictions they made earlier (see p. 244).

Negative Exponent Rule

$$x^{-m} = \frac{1}{x^m}, \qquad x \neq 0$$

TEACHING TIP
Have students use their knowledge of the quotient rule to simplify x^0/x^7. Then have them simplify $1/x^7$.

When a variable or number is raised to a negative exponent, the expression may be rewritten as 1 divided by the variable or number to that positive exponent.

Examples

$$x^{-6} = \frac{1}{x^6} \qquad\qquad 4^{-2} = \frac{1}{4^2} = \frac{1}{16}$$

$$y^{-7} = \frac{1}{y^7} \qquad\qquad 5^{-3} = \frac{1}{5^3} = \frac{1}{125}$$

AVOIDING COMMON ERRORS

Students sometimes believe that a negative exponent automatically makes the value of the expression negative. This is not true.

EXPRESSION	CORRECT	INCORRECT	ALSO INCORRECT
3^{-2}	$\dfrac{1}{3^2} = \dfrac{1}{9}$	$\cancel{-3^2}$	$\cancel{-\dfrac{1}{3^2}}$
x^{-3}	$\dfrac{1}{x^3}$	$\cancel{-x^3}$	$\cancel{-\dfrac{1}{x^3}}$

To help you see that the negative exponent rule makes sense, consider the following sequence of exponential expressions and their corresponding values.

$$2^3 = 8, \quad 2^2 = 4, \quad 2^1 = 2, \quad 2^0 = 1, \quad 2^{-1} = \frac{1}{2^1} \text{ or } \frac{1}{2}, \quad 2^{-2} = \frac{1}{2^2} \text{ or } \frac{1}{4}, \quad 2^{-3} = \frac{1}{2^3} \text{ or } \frac{1}{8}$$

Note that each time the exponent decreases by 1, the value of the expression is halved. For example, when we go from 2^3 to 2^2, the value of the expression goes from 8 to 4. If we continue decreasing the exponents beyond $2^0 = 1$, the next exponent in the pattern is -1. And if we take half of 1 we get $\frac{1}{2}$. This pattern illustrates that $x^{-m} = \frac{1}{x^m}$.

2 Simplify Expressions Containing Negative Exponents

Generally, when you are asked to simplify an exponential expression **your final answer should contain no negative exponents.** You may simplify exponential expressions using the negative exponent rule and the rules of exponents presented in the previous section. The following examples indicate how exponential expressions containing negative exponents may be simplified.

EXAMPLE 2 Use the negative exponent rule to write each expression with a positive exponent. Simplify the expressions further when possible.

a) x^{-3} **b)** y^{-4} **c)** 3^{-2} **d)** 5^{-1} **e)** -5^{-3} **f)** $(-5)^{-3}$

Solution **a)** $x^{-3} = \frac{1}{x^3}$ **b)** $y^{-4} = \frac{1}{y^4}$

c) $3^{-2} = \frac{1}{3^2} = \frac{1}{9}$ **d)** $5^{-1} = \frac{1}{5}$

e) $-5^{-3} = -\frac{1}{5^3} = -\frac{1}{125}$ **f)** $(-5)^{-3} = \frac{1}{(-5)^3} = \frac{1}{-125} = -\frac{1}{125}$

EXAMPLE 3 Use the negative exponent rule to write each expression with a positive exponent.

a) $\frac{1}{x^{-2}}$ **b)** $\frac{1}{4^{-1}}$

Solution First use the negative exponent rule on the denominator. Then simplify further.

a) $\frac{1}{x^{-2}} = \frac{1}{1/x^2} = \frac{1}{1} \cdot \frac{x^2}{1} = x^2$ **b)** $\frac{1}{4^{-1}} = \frac{1}{1/4} = \frac{1}{1} \cdot \frac{4}{1} = 4$

HELPFUL HINT

From Examples 2 and 3, we can see that when a factor is moved from the denominator to the numerator or from the numerator to the denominator, the sign of the *exponent* changes.

$$x^{-4} = \frac{1}{x^4} \qquad \frac{1}{x^{-4}} = x^4$$

$$3^{-5} = \frac{1}{3^5} \qquad \frac{1}{3^{-5}} = 3^5$$

Now let's look at additional examples that combine two or more of the rules presented so far.

EXAMPLE 4   Simplify.   **a)** $(z^{-5})^4$ **b)** $(4^2)^{-3}$

Solution   **a)** $(z^{-5})^4 = z^{(-5)(4)}$ *By the power rule*

$$= z^{-20}$$

$$= \frac{1}{z^{20}} \quad \textit{By the negative exponent rule}$$

b) $(4^2)^{-3} = 4^{(2)(-3)}$ *By the power rule*

$$= 4^{-6}$$

$$= \frac{1}{4^6} \quad \textit{By the negative exponent rule}$$

NOW TRY EXERCISE 25

EXAMPLE 5   Simplify.   **a)** $x^3 \cdot x^{-5}$ **b)** $3^{-4} \cdot 3^{-7}$

Solution   **a)** $x^3 \cdot x^{-5} = x^{3+(-5)}$ *By the product rule*

$$= x^{-2}$$

$$= \frac{1}{x^2} \quad \textit{By the negative exponent rule}$$

b) $3^{-4} \cdot 3^{-7} = 3^{-4+(-7)}$ *By the product rule*

$$= 3^{-11}$$

$$= \frac{1}{3^{11}} \quad \textit{By the negative exponent rule}$$

NOW TRY EXERCISE 51

AVOIDING COMMON ERRORS

What is the sum of $3^2 + 3^{-2}$? Look carefully at the correct solution.

CORRECT	INCORRECT

$$3^2 + 3^{-2} = 9 + \frac{1}{9}$$

$$= 9\frac{1}{9}$$

$\cancel{3^2 + 3^{-2} = 0}$

Note that $3^2 \cdot 3^{-2} = 3^{2+(-2)} = 3^0 = 1$.

EXAMPLE 6   Simplify.   **a)** $\dfrac{z^{-6}}{z^{12}}$ **b)** $\dfrac{5^{-7}}{5^{-4}}$

Solution   **a)** $\dfrac{z^{-6}}{z^{12}} = z^{-6-12}$ *By the quotient rule*

$$= z^{-18}$$

$$= \frac{1}{z^{18}} \quad \textit{By the negative exponent rule}$$

b) $\dfrac{5^{-7}}{5^{-4}} = 5^{-7-(-4)}$ *By the quotient rule*

$$= 5^{-7+4}$$

$$= 5^{-3}$$

$$= \frac{1}{5^3} \text{ or } \frac{1}{125} \quad \textit{By the negative exponent rule}$$

NOW TRY EXERCISE 81

The following Helpful Hint should be read carefully.

HELPFUL HINT	Consider a division problem where a variable has a negative exponent in either its numerator or its denominator, such as in Example 6a). Another way to simplify such an expression is to move the variable with the negative exponent from the numerator to the denominator, or from the denominator to the numerator, and change the sign of the exponent. For example,

$$\frac{x^{-4}}{x^5} = \frac{1}{x^5 \cdot x^4} = \frac{1}{x^{5+4}} = \frac{1}{x^9}$$

$$\frac{y^3}{y^{-7}} = y^3 \cdot y^7 = y^{3+7} = y^{10}$$

Now consider a division problem where a number or variable has a negative exponent in both its numerator and its denominator, such as in Example 6b). Another way to simplify such an expression is to move the variable with the more negative exponent from the numerator to the denominator, or from the denominator to the numerator, and change the sign of the exponent from negative to positive. For example,

$$\frac{x^{-8}}{x^{-3}} = \frac{1}{x^8 \cdot x^{-3}} = \frac{1}{x^{8-3}} = \frac{1}{x^5} \quad \textit{Note that } -8 < -3.$$

$$\frac{y^{-4}}{y^{-7}} = y^7 \cdot y^{-4} = y^{7-4} = y^3 \quad \textit{Note that } -7 < -4.$$

EXAMPLE 7 Simplify.   **a)** $7x^4(6x^{-9})$ **b)** $\dfrac{16r^3 s^{-3}}{8rs^2}$ **c)** $\dfrac{2x^2 y^5}{8x^7 y^{-3}}$

Solution   **a)** $7x^4(6x^{-9}) = 7 \cdot 6 \cdot x^4 \cdot x^{-9} = 42x^{-5} = \dfrac{42}{x^5}$

b) $\dfrac{16r^3 s^{-3}}{8rs^2} = \dfrac{16}{8} \cdot \dfrac{r^3}{r} \cdot \dfrac{s^{-3}}{s^2}$

$$= 2 \cdot r^2 \cdot \frac{1}{s^5} = \frac{2r^2}{s^5}$$

c) $\dfrac{2x^2 y^5}{8x^7 y^{-3}} = \dfrac{2}{8} \cdot \dfrac{x^2}{x^7} \cdot \dfrac{y^5}{y^{-3}}$

$$= \frac{1}{4} \cdot \frac{1}{x^5} \cdot y^8 = \frac{y^8}{4x^5}$$

NOW TRY EXERCISE 119

In Example 7**b)**, the variable with the negative exponent, s^{-3}, was moved from the numerator to the denominator. In Example 7**c)**, the variable with the negative exponent, y^{-3}, was moved from the denominator to the numerator. In each case, the sign of the exponent was changed from negative to positive when the variable factor was moved.

EXAMPLE 8 Simplify $(5x^{-3})^{-2}$.

Solution Begin by using the expanded power rule.

$$(5x^{-3})^{-2} = 5^{-2} x^{(-3)(-2)}$$
$$= 5^{-2} x^6$$
$$= \frac{1}{5^2} x^6$$
$$= \frac{x^6}{25}$$

AVOIDING COMMON ERRORS

Can you explain why the simplification on the right is incorrect?

CORRECT

$$\frac{x^3 y^{-2}}{w} = \frac{x^3}{wy^2}$$

INCORRECT

$$\frac{x^3 + y^{-2}}{w} = \frac{x^3}{w + y^2}$$

The simplification on the right is incorrect because in the numerator $x^3 + y^{-2}$ the y^{-2} *is not a factor*; it is a term. We will learn how to simplify expressions like this when we study complex fractions in Section 6.5.

EXAMPLE 9 Simplify $\left(\dfrac{2}{3}\right)^{-2}$

Solution By the expanded power rule, we may write

$$\left(\frac{2}{3}\right)^{-2} = \frac{2^{-2}}{3^{-2}} = \frac{\frac{1}{2^2}}{\frac{1}{3^2}} = \frac{1}{2^2} \cdot \frac{3^2}{1} = \frac{3^2}{2^2} = \frac{9}{4}$$

If we examine the results of Example 7, we see that

$$\left(\frac{2}{3}\right)^{-2} = \frac{3^2}{2^2} = \left(\frac{3}{2}\right)^2.$$

This example illustrates that $\left(\dfrac{a}{b}\right)^{-m} = \left(\dfrac{b}{a}\right)^m$ when $a \neq 0$ and $b \neq 0$. Thus, for example, $\left(\dfrac{3}{4}\right)^{-5} = \left(\dfrac{4}{3}\right)^5$ and $\left(\dfrac{5}{9}\right)^{-3} = \left(\dfrac{9}{5}\right)^3$. We can summarize this information as follows.

A Fraction Raised to a Negative Exponent Rule

For a fraction of the form $\dfrac{a}{b}$, $a \neq 0$ and $b \neq 0$, $\left(\dfrac{a}{b}\right)^{-m} = \left(\dfrac{b}{a}\right)^m$.

EXAMPLE 10 Simplify **a)** $\left(\dfrac{3}{4}\right)^{-3}$ **b)** $\left(\dfrac{x^2}{y^3}\right)^{-4}$

Solution We use the above rule to simplify.

NOW TRY EXERCISE 95

a) $\left(\dfrac{3}{4}\right)^{-3} = \left(\dfrac{4}{3}\right)^3 = \dfrac{4^3}{3^3} = \dfrac{64}{27}$ **b)** $\left(\dfrac{x^2}{y^3}\right)^{-4} = \left(\dfrac{y^3}{x^2}\right)^4 = \dfrac{y^{3 \cdot 4}}{x^{2 \cdot 4}} = \dfrac{y^{12}}{x^8}$

EXAMPLE 11 Simplify **a)** $\left(\dfrac{x^2 y^{-3}}{z^4}\right)^{-5}$ **b)** $\left(\dfrac{2x^{-3} y^2 z}{x^2}\right)^2$

Solution **a)** We will work part **a)** using two different methods. In method 1, we begin by using the expanded power rule. In method 2, we use a fraction raised to a negative exponent rule before we use the expanded power rule. You may use either method.

Method 1

$$\left(\frac{x^2 y^{-3}}{z^4}\right)^{-5} = \frac{x^{2(-5)} y^{(-3)(-5)}}{z^{4(-5)}} \qquad \text{Expanded power rule}$$

$$= \frac{x^{-10} y^{15}}{z^{-20}} \qquad \text{Multiply exponents}$$

$$= \frac{y^{15} z^{20}}{x^{10}} \qquad \text{Negative exponent rule}$$

Method 2
$$\left(\frac{x^2y^{-3}}{z^4}\right)^{-5} = \left(\frac{z^4}{x^2y^{-3}}\right)^5 \qquad \left(\frac{a}{b}\right)^{-m} = \left(\frac{b}{a}\right)^m$$

$$= \left(\frac{y^3z^4}{x^2}\right)^5 \qquad \textit{Simplify expression within parentheses}$$

$$= \frac{y^{3\cdot5}z^{4\cdot5}}{x^{2\cdot5}} \qquad \textit{Expanded power rule}$$

$$= \frac{y^{15}z^{20}}{x^{10}} \qquad \textit{Multiply exponents}$$

b) First simplify the expression within parentheses, then square the results. To simplify, we note that $\dfrac{x^{-3}}{x^2}$ becomes $\dfrac{1}{x^5}$.

$$\left(\frac{2x^{-3}y^2z}{x^2}\right)^2 = \left(\frac{2y^2z}{x^5}\right)^2 = \frac{2^2y^{2\cdot2}z^{1\cdot2}}{x^{5\cdot2}} = \frac{4y^4z^2}{x^{10}}$$

Summary of Rules of Exponents

1. $x^m \cdot x^n = x^{m+n}$ **product rule**

2. $\dfrac{x^m}{x^n} = x^{m-n}, \qquad x \neq 0$ **quotient rule**

3. $x^0 = 1, \qquad x \neq 0$ **zero exponent rule**

4. $(x^m)^n = x^{m\cdot n}$ **power rule**

5. $\left(\dfrac{ax}{by}\right)^m = \dfrac{a^mx^m}{b^my^m}, \qquad b \neq 0, y \neq 0$ **expanded power rule**

6. $x^{-m} = \dfrac{1}{x^m}, \qquad x \neq 0$ **negative exponent rule**

7. $\left(\dfrac{a}{b}\right)^{-m} = \left(\dfrac{b}{a}\right)^m, \qquad a \neq 0, b \neq 0$ **a fraction raised to a negative exponent rule**

Exercise Set 4.2

Concept/Writing Exercises   4. no, a^6/b^2

1. In your own words, describe the negative exponent rule. Answers will vary.

2. Is the expression x^{-2} simplified? Explain. If not simplified, then simplify. no, $1/x^2$

3. Is the expression x^5y^{-3} simplified? Explain. If not simplified, then simplify. no, x^5/y^3

4. Can the expression a^6b^{-2} be simplified to $1/a^6b^{-2}$? If not, what is the correct simplification? Explain.

5. Can the expression $(y^4)^{-3}$ be simplified to $1/y^7$? If not, what is the correct simplification? Explain. no, $1/y^{12}$

6. Are the following expressions simplified? If an expression is not simplified, explain why and then simplify.

a) $\dfrac{5}{x^3}$ yes   b) n^{-5} no, $\dfrac{1}{n^5}$

c) $\dfrac{a^{-4}}{2}$ no, $\dfrac{1}{2a^4}$ d) $\dfrac{x^{-4}}{x^4}$ no, $\dfrac{1}{x^8}$

7. a) Identify the terms in the numerator of the expression $\dfrac{x^5y^2}{z^3}$.   only one term, x^5y^2

b) Identify the factors in the numerator of the expression. x^5, y^2

8. a) Identify the terms in the numerator of the expression $\dfrac{x^{-4}y^3}{z^5}$. only one term, $x^{-4}y^3$

   b) Identify the factors in the numerator of the expression. x^{-4}, y^3

9. Describe what happens to the exponent of a factor when the factor is moved from the numerator to the denominator of a fraction. the sign changes

10. Describe what happens to the exponent of a factor when the factor is moved from the denominator to the numerator of a fraction. the sign changes

Practice the Skills

Simplify.

11. x^{-6} $\dfrac{1}{x^6}$
12. y^{-5} $\dfrac{1}{y^5}$
13. 5^{-1} $\dfrac{1}{5}$
14. 5^{-2} $\dfrac{1}{25}$

15. $\dfrac{1}{x^{-3}}$ x^3
16. $\dfrac{1}{b^{-4}}$ b^4
17. $\dfrac{1}{x^{-1}}$ x
18. $\dfrac{1}{y^{-4}}$ y^4

19. $\dfrac{1}{6^{-2}}$ 36
20. $\dfrac{1}{5^{-3}}$ 125
21. $(x^{-2})^3$ $\dfrac{1}{x^6}$
22. $(m^{-5})^{-2}$ m^{10}

23. $(y^{-5})^4$ $\dfrac{1}{y^{20}}$
24. $(a^5)^{-4}$ $\dfrac{1}{a^{20}}$
25. $(x^4)^{-2}$ $\dfrac{1}{x^8}$
26. $(x^{-9})^{-2}$ x^{18}

27. $(2^{-2})^{-3}$ 64
28. $(2^{-3})^2$ $\dfrac{1}{64}$
29. $y^4 \cdot y^{-2}$ y^2
30. $x^{-3} \cdot x^1$ $\dfrac{1}{x^2}$

31. $x^7 \cdot x^{-5}$ x^2
32. $d^{-3} \cdot d^{-4}$ $\dfrac{1}{d^7}$
33. $3^{-2} \cdot 3^4$ 9
34. $6^{-3} \cdot 6^6$ 216

35. $\dfrac{r^5}{r^6}$ $\dfrac{1}{r}$
36. $\dfrac{x^2}{x^{-1}}$ x^3
37. $\dfrac{p^0}{p^{-3}}$ p^3
38. $\dfrac{x^{-2}}{x^5}$ $\dfrac{1}{x^7}$

39. $\dfrac{x^{-7}}{x^{-3}}$ $\dfrac{1}{x^4}$
40. $\dfrac{z^{-11}}{z^{-12}}$ z
41. $\dfrac{3^2}{3^{-1}}$ 27
42. $\dfrac{4^2}{4^{-1}}$ 64

43. 3^{-3} $\dfrac{1}{27}$
44. x^{-7} $\dfrac{1}{x^7}$
45. $\dfrac{1}{z^{-9}}$ z^9
46. $\dfrac{1}{3^{-3}}$ 27

47. $(p^{-4})^{-6}$ p^{24}
48. $(x^{-3})^{-4}$ x^{12}
49. $(y^{-2})^{-3}$ y^6
50. $z^9 \cdot z^{-12}$ $\dfrac{1}{z^3}$

51. $x^3 \cdot x^{-7}$ $\dfrac{1}{x^4}$
52. $x^{-3} \cdot x^{-5}$ $\dfrac{1}{x^8}$
53. $x^{-8} \cdot x^{-7}$ $\dfrac{1}{x^{15}}$
54. $4^{-3} \cdot 4^3$ 1

55. -4^{-2} $-\dfrac{1}{16}$
56. $(-4)^{-2}$ $\dfrac{1}{16}$
57. $-(-4)^{-2}$ $-\dfrac{1}{16}$
58. -4^{-3} $-\dfrac{1}{64}$

59. $(-4)^{-3}$ $-\dfrac{1}{64}$
60. $-(-4)^{-3}$ $\dfrac{1}{64}$
61. $(-6)^{-2}$ $\dfrac{1}{36}$
62. -6^{-2} $-\dfrac{1}{36}$

63. $\dfrac{x^{-5}}{x^5}$ $\dfrac{1}{x^{10}}$
64. $\dfrac{y^6}{y^{-8}}$ y^{14}
65. $\dfrac{n^{-5}}{n^{-7}}$ n^2
66. $\dfrac{3^{-4}}{3}$ $\dfrac{1}{243}$

67. $\dfrac{2^{-3}}{2^{-3}}$ 1
68. $(5a^2b^3)^0$ 1
69. $(2^{-1} + 3^{-1})^0$ 1
70. $(3^{-1} + 4^2)^0$ 1

71. $\dfrac{2}{2^{-5}}$ 64
72. $(z^{-5})^{-9}$ z^{45}
73. $(x^{-4})^{-2}$ x^8
74. $(x^{-7})^0$ 1

75. $(x^0)^{-2}$ 1
76. $(3^{-2})^{-1}$ 9
77. $2^{-3} \cdot 2$ $\dfrac{1}{4}$
78. $6^4 \cdot 6^{-2}$ 36

79. $6^{-4} \cdot 6^2$ $\dfrac{1}{36}$
80. $\dfrac{z^{-3}}{z^{-7}}$ z^4
81. $\dfrac{x^{-1}}{x^{-4}}$ x^3
82. $\dfrac{r^6}{r}$ r^5

83. $(4^2)^{-1}$ $\dfrac{1}{16}$
84. $(4^{-2})^{-2}$ 256
85. $\dfrac{5}{5^{-2}}$ 125
86. $\dfrac{x^6}{x^7}$ $\dfrac{1}{x}$

87. $\dfrac{3^{-4}}{3^{-2}}$ $\quad \dfrac{1}{9}$

88. $x^{-10} \cdot x^{8}$ $\quad \dfrac{1}{x^2}$

89. $\dfrac{7^{-1}}{7^{-1}}$ $\quad 1$

90. $2x^{-1}y$ $\quad \dfrac{2y}{x}$

91. $(6x^2)^{-2}$ $\quad \dfrac{1}{36x^4}$

92. $(3z^3)^{-2}$ $\quad \dfrac{1}{9z^6}$

🔒 **93.** $3x^{-2}y^2$ $\quad \dfrac{3y^2}{x^2}$

94. $5x^4y^{-1}$ $\quad \dfrac{5x^4}{y}$

95. $\left(\dfrac{1}{2}\right)^{-2}$ $\quad 4$

96. $\left(\dfrac{3}{5}\right)^{-2}$ $\quad \dfrac{25}{9}$

97. $\left(\dfrac{5}{4}\right)^{-3}$ $\quad \dfrac{64}{125}$

98. $\left(\dfrac{3}{5}\right)^{-3}$ $\quad \dfrac{125}{27}$

99. $\left(\dfrac{x^2}{y}\right)^{-2}$ $\quad \dfrac{y^2}{x^4}$

100. $\left(\dfrac{c^4}{d^2}\right)^{-2}$ $\quad \dfrac{d^4}{c^8}$

101. $-\left(\dfrac{r^4}{s}\right)^{-4}$ $\quad -\dfrac{s^4}{r^{16}}$

102. $-\left(\dfrac{m^3}{n^4}\right)^{-5}$ $\quad -\dfrac{n^{20}}{m^{15}}$

103. $7a^{-3}b^{-4}$ $\quad \dfrac{7}{a^3b^4}$

104. $(3x^2y^3)^{-2}$ $\quad \dfrac{1}{9x^4y^6}$

105. $(x^5y^{-3})^{-3}$ $\quad \dfrac{y^9}{x^{15}}$

106. $2w(3w^{-5})$ $\quad \dfrac{6}{w^4}$

107. $(4y^{-2})(5y^{-3})$ $\quad \dfrac{20}{y^5}$

108. $2x^5(3x^{-6})$ $\quad \dfrac{6}{x}$

109. $4x^4(-2x^{-4})$ $\quad -8$

110. $(9x^5)(-3x^{-7})$ $\quad -\dfrac{27}{x^2}$

🔒 **111.** $(4x^2y)(3x^3y^{-1})$ $\quad 12x^5$

112. $(2x^{-3}y^{-2})(x^4y^0)$ $\quad \dfrac{2x}{y^2}$

113. $(5y^2)(4y^{-3}z^5)$ $\quad \dfrac{20z^5}{y}$

114. $(3y^{-2})(5x^{-1}y^3)$ $\quad \dfrac{15y}{x}$

115. $\dfrac{12c^9}{4c^4}$ $\quad 3c^5$

116. $\dfrac{8z^{-4}}{32z^{-2}}$ $\quad \dfrac{1}{4z^2}$

117. $\dfrac{36x^{-4}}{9x^{-2}}$ $\quad \dfrac{4}{x^2}$

118. $\dfrac{12x^{-2}y^0}{2x^3y^2}$ $\quad \dfrac{6}{x^5y^2}$

119. $\dfrac{3x^4y^{-2}}{6y^3}$ $\quad \dfrac{x^4}{2y^5}$

120. $\dfrac{16x^{-7}y^{-2}}{4x^5y^2}$ $\quad \dfrac{4}{x^{12}y^4}$

🔒 **121.** $\dfrac{32x^4y^{-2}}{4x^{-2}y^0}$ $\quad \dfrac{8x^6}{y^2}$

122. $\dfrac{21x^{-3}z^2}{7xz^{-3}}$ $\quad \dfrac{3z^5}{x^4}$

123. $\left(\dfrac{2x^2y^{-3}}{z}\right)^{-4}$ $\quad \dfrac{y^{12}z^4}{16x^8}$

124. $\left(\dfrac{b^4c^{-2}}{2d^{-3}}\right)^{-1}$ $\quad \dfrac{2c^2}{b^4d^3}$

125. $\left(\dfrac{2r^{-5}s^9}{t^{12}}\right)^{-4}$ $\quad \dfrac{r^{20}t^{48}}{16s^{36}}$

126. $\left(\dfrac{5m^{-1}n^{-3}}{p^2}\right)^{-3}$ $\quad \dfrac{m^3n^9p^6}{125}$

127. $\left(\dfrac{x^3y^{-4}z}{y^{-2}}\right)^{-6}$ $\quad \dfrac{y^{12}}{x^{18}z^6}$

128. $\left(\dfrac{3p^{-1}q^{-2}r^3}{p^2}\right)^{3}$ $\quad \dfrac{27r^9}{p^9q^6}$

129. $\left(\dfrac{a^2b^{-2}}{3a^4}\right)^{3}$ $\quad \dfrac{1}{27a^6b^6}$

130. $\left(\dfrac{x^{12}y^5}{y^{-3}z}\right)^{-4}$ $\quad \dfrac{z^4}{x^{48}y^{32}}$

Problem Solving

131. a) Does $a^{-1}b^{-1} = \dfrac{1}{ab}$? Explain your answer. yes

  b) Does $a^{-1} + b^{-1} = \dfrac{1}{a+b}$? Explain your answer. no

132. a) Does $\dfrac{x^{-1}y^2}{z} = \dfrac{y^2}{xz}$? Explain your answer. yes

  b) Does $\dfrac{x^{-1} + y^2}{z} = \dfrac{y^2}{x+z}$? Explain your answer. no

Evaluate.

133. $4^2 + 4^{-2}$ $\quad 16\tfrac{1}{16}$

134. $3^2 + 3^{-2}$ $\quad 9\tfrac{1}{9}$

135. $2^2 + 2^{-2}$ $\quad 4\tfrac{1}{4}$

136. $6^{-3} + 6^3$ $\quad 216\tfrac{1}{216}$

Evaluate.

137. $5^0 - 3^{-1}$ $\quad \tfrac{2}{3}$

138. $4^{-1} - 3^{-1}$ $\quad -\tfrac{1}{12}$

139. $2 \cdot 4^{-1} + 4 \cdot 3^{-1}$ $\quad \tfrac{11}{6}$

140. $2^{-3} - 2^3 \cdot 2^{-3}$ $\quad -\tfrac{7}{8}$

141. $2 \cdot 4^{-1} - 3^{-1}$ $\quad \tfrac{1}{6}$

142. $2 \cdot 4^{-1} - 4 \cdot 3^{-1}$ $\quad -\tfrac{5}{6}$

143. $3 \cdot 5^0 - 5 \cdot 3^{-2}$ $\quad \tfrac{22}{9}$

144. $7 \cdot 2^{-3} - 2 \cdot 4^{-1}$ $\quad \tfrac{3}{8}$

Determine the number that when placed in the shaded area makes the statement true.

145. $3^{\blacksquare} = \dfrac{1}{9}$ $\quad -2$

146. $\dfrac{1}{4^{\blacksquare}} = 16$ $\quad -2$

147. $\dfrac{1}{3^{\blacksquare}} = 9$ $\quad -2$

148. $4^{\blacksquare} = \dfrac{1}{16}$ $\quad -2$

Challenge Problems

In Exercises 149–151, determine the number (or numbers) that when placed in the shaded area (or areas) make the statement true.

149. $(x^{\blacksquare}y^3)^{-2} = \dfrac{x^4}{y^6}$ $\quad -2$

150. $(\blacksquare x^{\blacksquare}y^{-2})^3 = \dfrac{8}{x^9y^6}$ $\quad 2, -3$

151. $(x^4y^{-3})^{\blacksquare} = \dfrac{y^9}{x^{12}}$ $\quad -3$

152. For any nonzero real number a, if $a^{-1} = x$, describe the following in terms of x.

a) $-a^{-1}$ $-x$ **b)** $\dfrac{1}{a^{-1}}$ $\dfrac{1}{x}$

153. Consider $(3^{-1} + 2^{-1})^0$. We know this is equal to 1 by the zero exponent rule. As a group, determine the error in the following calculation. Explain your answer.

$$(3^{-1} + 2^{-1})^0 = (3^{-1})^0 + (2^{-1})^0$$
$$= 3^{-1(0)} + 2^{-1(0)}$$
$$= 3^0 + 2^0$$
$$= 1 + 1 = 2$$

Product rule is $(xy)^m = x^m y^m$ not $(x + y)^m$.

Group Activity

Discuss and answer Exercises 154 as a group.

154. Often problems involving exponents can be done in more than one way. Consider

$$\left(\frac{3x^2 y^3}{x}\right)^{-2}$$

a) Group member 1: Simplify this expression by first simplifying the expression within parentheses.

b) Group member 2: Simplify this expression by first using the expanded power rule.

c) Group member 3: Simplify this expression by first using the negative exponent rule.

d) Compare your answers. If you did not all get the same answers, determine why.

e) As a group, decide which method—**a)**, **b)**, or **c)**—was the easiest way to simplify this expression.

154. a)–c) $\dfrac{1}{9x^2 y^6}$

Cumulative Review Exercises

[2.6] 155. *Sailing* If a sailboat travels 3 miles in 48 minutes, how far will it sail in 80 minutes (assuming all conditions stay the same)? 5 miles

[3.1] 156. *Volume* Find the volume of a right cylinder whose radius is 5 inches and whose height is 12 inches. $\approx 942.48\ \text{in.}^3$

[3.3] 157. *Integers* The larger of two integers is 1 more than 3 times the smaller. If the sum of the two integers is 37, find the two integers. 9, 28

158. *Cost of Item* The cost of an item when increased by 20% is $150. Find the price of the item before the increase. $125

159. *Consecutive Integers* Find two consecutive integers whose sum is 75. 37, 38

4.3 SCIENTIFIC NOTATION

1 Convert numbers to and from scientific notation.

2 Recognize numbers in scientific notation with a coefficient of 1.

3 Do calculations using scientific notation.

SSM   Study Guide   CD/Video

MathPro 4/5   PH Math Tutor Center   prenhall.com/Angel

1 Convert Numbers to and from Scientific Notation

We often see, and sometimes use, very large or very small numbers. For example, in January 2001, the world population was about 6,160,000,000 people. You may have read that an influenza virus is about 0.0000001 meters in diameter. Because it is difficult to work with many zeros, we can express such numbers using

TEACHING TIP
Have students write 6000 as a
product of 6 and a number. Then
have them write it as a product
using only 6 and 10 as factors.
Have them simplify this expression using exponents to represent
repeated multiplication. Repeat
this process with 6,160,000,000.

exponents. For example, the number 6,160,000,000 could be written 6.16×10^9 and the number 0.0000001 could be written 1.0×10^{-7}.

Numbers such as 6.16×10^9 and 1.0×10^{-7} are in a form called **scientific notation**. Each number written in scientific notation is written as a number greater than or equal to 1 and less than 10 ($1 \leq a < 10$) multiplied by some power of 10. The exponent on the 10 must be an integer.

Examples of Numbers in Scientific Notation

$$1.2 \times 10^6$$
$$3.762 \times 10^3$$
$$8.07 \times 10^{-2}$$
$$1 \times 10^{-5}$$

Below we change the number 68,400 to scientific notation.

$$68,400 = 6.84 \times 10,000$$
$$= 6.84 \times 10^4 \quad \text{Note that } 10,000 = 10 \cdot 10 \cdot 10 \cdot 10 = 10^4.$$

Therefore, $68,400 = 6.84 \times 10^4$. To go from 68,400 to 6.84 the decimal point was moved four places to the left. Note that the exponent on the 10, the 4, is the same as the number of places the decimal point was moved to the left.

Following is a simplified procedure for writing a number in scientific notation.

To Write a Number in Scientific Notation

1. Move the decimal point in the original number to the right of the first nonzero digit. This will give a number greater than or equal to 1 and less than 10.

2. Count the number of places you moved the decimal point to obtain the number in step 1. If the original number was 10 or greater, the count is to be considered positive. If the original number was less than 1, the count is to be considered negative.

3. Multiply the number obtained in step 1 by 10 raised to the count (power) found in step 2.

EXAMPLE 1   Write the following numbers in scientific notation.
a) 10,700   **b)** 0.000386   **c)** 972,000   **d)** 0.0083

Solution   **a)** The original number is greater than 10; therefore, the exponent is positive. The decimal point in 10,700 belongs after the last zero.

$$10,700. = 1.07 \times 10^4$$
4 places

b) The original number is less than 1; therefore, the exponent is negative.

$$0.000386 = 3.86 \times 10^{-4}$$
4 places

c) $972,000. = 9.72 \times 10^5$ **d)** $0.0083 = 8.3 \times 10^{-3}$
5 places   *3 places*

NOW TRY EXERCISE 17

When we write a number in scientific notation, we are allowed to leave our answer with a negative exponent, as in Example 1**b)** and 1**d)**.

Now we explain how to write a number in scientific notation as a number without exponents, or in decimal form.

> ## To Convert a Number from Scientific Notation to Decimal Form
>
> 1. Observe the exponent of the power of 10.
>
> 2. a) If the exponent is positive, move the decimal point in the number (greater than or equal to 1 and less than 10) to the right the same number of places as the exponent. It may be necessary to add zeros to the number. This will result in a number greater than or equal to 10.
>
>   b) If the exponent is 0, do not move the decimal point. Drop the factor 10^0 since it equals 1. This will result in a number greater than or equal to 1 but less than 10.
>
>   c) If the exponent is negative, move the decimal point in the number to the left the same number of places as the exponent (dropping the negative sign). It may be necessary to add zeros. This will result in a number less than 1.

EXAMPLE 2 Write each number without exponents.

a) 2.9×10^4 **b)** 6.28×10^{-3} **c)** 7.95×10^8

Solution **a)** Move the decimal point four places to the right

$$2.9 \times 10^4 = 2.9 \times 10{,}000 = 29{,}000$$

TEACHING TIP
Before discussing the solution to each part in Example 2 ask, "Will the answer be greater than 1 or less than 1?"

b) Move the decimal point three places to the left.

$$6.28 \times 10^{-3} = 0.00628$$

c) Move the decimal point eight places to the right.

NOW TRY EXERCISE 39

$$7.95 \times 10^8 = 795{,}000{,}000$$

2 Recognize Numbers in Scientific Notation with a Coefficient of 1

We often hear terms like kilograms, milligrams, and gigabytes. For example, an aspirin tablet bottle may indicate that each aspirin contains 325 milligrams of aspirin. Your hard drive on your computer may hold 40 gigabytes of memory. The prefixes kilo, milli, and giga are some of the prefixes used in the *metric system*. The metric system is used in every westernized nation except the United States as the main system of measurement. The prefixes are always used with some type of base unit. The base unit may be measures like meter, m (a unit of length); gram, g (a unit of mass); liter, ℓ (a unit of volume); bits, b (a unit of computer memory); or hertz, Hz (a measure of frequency). For example, a *milli*meter is $\frac{1}{1000}$ meter. A *mega*gram is 1,000,000 grams, and so on. The following table illustrates the meaning of some prefixes.*

---

*There are other prefixes not listed. For example, centi is 10^{-2} or 0.01 time the base unit.

Prefix	Meaning	Symbol	Meaning as a Decimal Number
nano	10^{-9}	n	$\dfrac{1}{1,000,000,000}$ or 0.000000001
micro	10^{-6}	μ	$\dfrac{1}{1,000,000}$ or 0.000001
milli	10^{-3}	m	$\dfrac{1}{1000}$ or 0.001
base unit*	10^{0}		1
kilo	10^{3}	k	1000
mega	10^{6}	M	1,000,000
giga	10^{9}	G	1,000,000,000

* The base unit is not a prefix. We included this row to include 10^0 in the chart.

You will sometimes see numbers written as powers of 10, but without a numerical coefficient, as in the table above. If no numerical coefficient is indicated, the numerical coefficient is always assumed to be 1. Thus, for example, $10^{-3} = 1.0 \times 10^{-3}$ and $10^9 = 1.0 \times 10^9$. A computer hard drive that contains 40 gigabytes (40 Gb) contains about $40(1.0 \times 10^9) = 40 \times 10^9 = 40,000,000,000$ bytes. Fifty micrometers (50 μm) is $50(1.0 \times 10^{-6}) = 50 \times 10^{-6} = 0.00005$ meters. Three hundred twenty-five milligrams (325 mg) is $325(1 \times 10^{-3}) = 325 \times 10^{-3} = 0.325$ grams. Notice that in the table, each prefix represents a value that is 10^3 or 1000 times greater than the prefix above it. For example, a micrometer is 10^3 or 1000 times greater than a nanometer. A gigameter is 1000 times larger than a megameter, and so on.

In Figure 4.1 we see that the frequency of FM radio and VHF TV is about 10^8 hertz (or cycles per second). Thus, the frequency of FM radio is $10^8 = 1.0 \times 10^8 = 100,000,000$ hertz. This number, hundred million hertz, can also be expressed as 100×10^6 or 100 megahertz, 100 MHz.

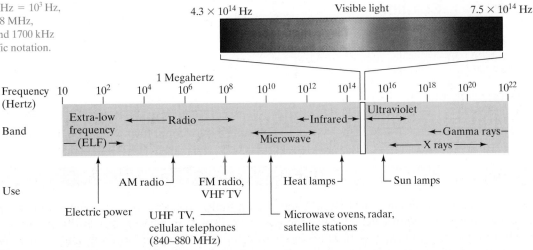

FIGURE 4.1

Now that we know how to interpret powers of ten that are given without numerical coefficients, we will work some problems using scientific notation.

EXAMPLE 3 Write each quantity without the metric prefix.

a) 52 kilograms **b)** 183 nanoseconds

Solution **a)** 52 kilograms (52 kg) = 52×10^3 grams = 52,000 grams

NOW TRY EXERCISE 45 **b)** 183 nanoseconds (183 ns) = 183×10^{-9} seconds = 0.000000183 seconds ✳

HELPFUL HINT

STUDY TIP

> Think about how often in your daily life you come across large and small quantities that may be expressed using scientific notation. This may give you more of an appreciation for scientific notation.

---

3 Do Calculations Using Scientific Notation

We can use the rules of exponents presented in Sections 4.1 and 4.2 when working with numbers written in scientific notation.

EXAMPLE 3 Multiply $(4.2 \times 10^6)(2 \times 10^{-4})$. Write the answer in decimal form.

Solution By the commutative and associative properties of multiplication we can rearrange the expression as follows.

$$(4.2 \times 10^6)(2 \times 10^{-4}) = (4.2 \times 2)(10^6 \times 10^{-4})$$
$$= 8.4 \times 10^{6+(-4)} \qquad \text{By the product rule}$$
$$= 8.4 \times 10^2 \qquad \text{Scientific notation}$$

NOW TRY EXERCISE 55
$$= 840 \qquad \text{Decimal form} \quad ✳$$

EXAMPLE 4 Divide $\dfrac{3.2 \times 10^{-6}}{5 \times 10^{-3}}$. Write the answer in scientific notation.

Solution
$$\frac{3.2 \times 10^{-6}}{5 \times 10^{-3}} = \left(\frac{3.2}{5}\right)\left(\frac{10^{-6}}{10^{-3}}\right)$$

TEACHING TIP
Point out that you can write the numerator as 32×10^{-5} to avoid dividing decimal numbers. Rework the example using 32×10^{-5} in the numerator.

$$= 0.64 \times 10^{-6-(-3)} \quad \text{By the quotient rule}$$
$$= 0.64 \times 10^{-6+3}$$
$$= 0.64 \times 10^{-3}$$
$$= 6.4 \times 10^{-4} \quad \text{Scientific notation} \quad ✳$$

NOW TRY EXERCISE 61 The answer to Example 4 in decimal form would be 0.00064.

Using Your Calculator

What will your calculator show when you multiply very large or very small numbers? The answer depends on whether your calculator has the ability to display an answer in scientific notation. On calculators without the ability to express numbers in scientific notation, you will probably get an error message because the answer will be too large or too small for the display. For example, on a calculator without scientific notation:

$$8000000 \;\boxed{\times}\; 600000 \;\boxed{=}\; \boxed{\text{Error}}$$

(continued on the next page)

On scientific calculators and graphing calculators the answer to the example on the previous page might be displayed in the following ways.

Possible displays

8000000 ☒ $\times$ ☐ 600000 ☐ $=$ ☐ | $4.8 \quad 12$ |

8000000 ☒ $\times$ ☐ 600000 ☐ $=$ ☐ | $4.8 \quad ^{12}$ |

8000000 ☒ $\times$ ☐ 600000 ☐ $=$ ☐ | $4.8E12$ |

Each answer means 4.8×10^{12}. Let's look at one more example.

Possible displays

0.0000003 ☒ $\times$ ☐ 0.004 ☐ $=$ ☐ | $1.2 \quad -9$ |

0.0000003 ☒ $\times$ ☐ 0.004 ☐ $=$ ☐ | $1.2 \quad ^{-9}$ |

0.0000003 ☒ $\times$ ☐ 0.004 ☐ $=$ ☐ | $1.2E-9$ |

Each display means 1.2×10^{-9}. On some calculators you will press the ☐ ENTER ☐ key instead of the ☐ = ☐ key. The TI-83 Plus graphing calculator displays answers using E, such as 4.8E12.

EXAMPLE 5 **Comparing Big Ships** The *Disney Magic* cruise ship gross tonnage is about 8.3×10^4 tons. The Carnival line's *Destiny* cruise ship gross tonnage is about 1.02×10^5 tons.

a) How much greater is the gross tonnage of the *Destiny* than the *Disney Magic*?
b) How many times greater is the gross tonnage of the *Destiny* than the *Disney Magic*?

Solution **a)** Understand We need to subtract 8.3×10^4 from 1.02×10^5. To add or subtract numbers in scientific notation, we generally make the exponents on the 10's the same.

Translate We can write 1.02×10^5 as 10.2×10^4. Now subtract as follows.

Carry Out
$$10.2 \times 10^4$$
$$\underline{- 8.3 \times 10^4}$$
$$1.9 \times 10^4$$

Notice that in subtraction, we did not subtract the 10^4's. This subtraction could also be done as $(10.2 \times 10^4) - (8.3 \times 10^4) = (10.2 - 8.3) \times 10^4 = 1.9 \times 10^4$

Check   We can check by writing the numbers out in decimal form.

$$102,000$$
$$\underline{-\ 85,000}$$
$$17,000 \text{ or } 1.7 \times 10^4$$

Answer   Since we obtain the same results, the difference is 1.7×10^4 (or 17,000) tons.

b) Understand   Part **b)** may seem similar to part **a)**, but it is a different question because we are asked to find the *number of times* greater rather than *how much greater*. To find the number of times greater, we perform division.

Translate   Divide the gross tonnage of the *Destiny* by the gross tonnage of the *Disney Magic*.

Carry Out
$$\frac{1.02 \times 10^5}{8.5 \times 10^4} = \frac{1.02}{8.5} \times \frac{10^5}{10^4}$$
$$= 0.12 \times 10^{5-4} \quad \text{\textit{Quotient rule of exponents}}$$
$$= 0.12 \times 10^1$$
$$= 1.2$$

Check   We can check by writing the numbers out in decimal form.

$$\frac{102,000}{85,000} = 1.2$$

Answer   Since we obtain the same result, the tonnage of the *Destiny* is 1.2 times that of the *Disney Magic*.

NOW TRY EXERCISE 75

EXAMPLE 6   **Fastest Computer**   As of January 2002, the fastest computer in the world, called *ASCI White*, located at the Lawrence Livermore National Laboratory in California, could perform a single calculation in about 0.000000000000083 second (83 quadrillionths of a second). How long would it take this computer to do 7 billion (7,000,000,000) calculations?

Solution   Understand   The computer could do 1 calculation in 1(0.000000000000083) second, 2 calculations in 2(0.000000000000083) second, 3 calculations in 3(0.000000000000083) second, and 7 billion operations in 7,000,000,000(0.000000000000083) second.

The ASCI White covers an area the size of 2 basketball courts

TEACHING TIP
Have students work in pairs to estimate how long it would take them to do the number of calculations *ASCI White* can do in 1 second. Then have them share their estimates and their process for making it.

Translate   We will multiply by converting each number to scientific notation.

$$7,000,000,000(0.000000000000083) = (7 \times 10^9)(8.3 \times 10^{-14})$$

Carry Out

$$= (7 \times 8.3)(10^9 \times 10^{-14})$$

$$= 58.1 \times 10^{-5}$$

$$= 0.000581$$

Answer   The *ASCI White* would take about 0.000581 of a second to perform 7 billion calculations.

NOW TRY EXERCISE 79

Mathematics in Action

The Air You Breathe

The Environmental Protection Agency (EPA) has identified indoor air quality as the number one environmental health concern of today. It is largely poor outdoor air quality that creates substandard indoor air quality, particularly in urban areas. Contaminants such as dust, pollen, and automobile pollutants often make their way directly into buildings. With more and more chemicals being used, people are increasingly suffering from chemical sensitivities to formaldehyde, pesticides, ozone, cleaning solvents, fiberglass, asbestos, lead, and radon. Allergies to molds are now more widespread than ever. No building or workplace is immune.

For workplaces involved with pharmaceuticals and biotechnology, the issue of contaminant-free air becomes even more crucial. One type of filter used in these setting is the HEPA (High Efficiency Particulate Air), which was originally developed to remove radioactive contaminants from the air in the development of the first atomic bomb.

The science of filtration involves the trapping of objects ranging from coal dust to viruses. The size of the particles being filtered is typically expressed in microns, where 1 micron (which is short for 1 micrometer) = 1 millionth of a meter = 10^{-6} meters = 0.000001 meters. The measurement refers to the diameter of the particle.

Here is a list of some of the types of particles for which filtration systems have to be developed; their diameter ranges is given in microns and in meters.

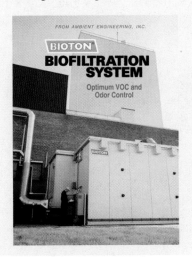

	Microns	**Meters**
Pollens	10–100	10^{-5}–10^{-4}
Tobacco smoke	0.01–1	10^{-8}–10^{-6}
Ground talc	0.4–80	4×10^{-7}–8×10^{-5}
Lung-damaging dust	0.8–6	8×10^{-7}–6×10^{-6}
Bacteria	0.5–50	5×10^{-7}–5×10^{-5}
Viruses	0.002–0.08	2×10^{-9}–8×10^{-8}

Exercise Set 4.3

Concept/Writing Exercises

1. a number greater than or equal to 1 and less than 10 multiplied by some power of 10

1. Describe the form of a number given in scientific notation.

2. a) In your own words, describe how to write a number 10 or greater in scientific notation.   Answers will vary.

b) Using the procedure described in part **a)**, write 42,100 in scientific notation. 4.21×10^4

3. a) In your own words, describe how to write a number less than 1 in scientific notation.   Answers will vary.

b) Using the procedure described in part **a)**, write 0.00568 in scientific notation. 5.68×10^{-3}

4. How many places, and in what direction, will you move the decimal point when you convert a number from scientific notation to decimal form when the exponent on the base 10 is 4? four places to the right

5. How many places, and in what direction, will you move the decimal point when you convert a number from scientific notation to decimal form when the exponent on the base 10 is −5? five places to the left

6. When changing a number to scientific notation, under what conditions will the exponent on the base 10 be positive? when the number is 10 or greater

7. When changing a number to scientific notation, under what conditions will the exponent on the base 10 be negative? when the number is less than 1

8. In writing the number 92,129 in scientific notation, will the exponent on the base 10 be positive or negative? Explain. positive

9. In writing the number 0.00734 in scientific notation, will the exponent on the base 10 be positive or negative? Explain. negative

10. Write the number 1,000,000 in scientific notation. 1×10^6

11. Write the number 0.000001 in scientific notation. 1×10^{-6}

12. a) Is 82.39×10^4 written in scientific notation? If not, how should it be written? no, 8.239×10^5

   b) Is 0.083×10^{-5} written in scientific notation? If not, how should it be written? no, 8.3×10^{-7}

Practice the Skills

Express each number in scientific notation.

13. 350,000 3.5×10^5
14. 3,610,000 3.61×10^6
15. 450 4.5×10^2
16. 0.00062 6.2×10^{-4}
17. 0.053 5.3×10^{-2}
18. 0.000726 7.26×10^{-4}
19. 19,000 1.9×10^4
20. 5,260,000,000 5.26×10^9
21. 0.00000186 1.86×10^{-6}
22. 0.0075 7.5×10^{-3}
23. 0.00000914 9.14×10^{-6}
24. 74,100 7.41×10^4
25. 220,300 2.203×10^5
26. 0.02 2.0×10^{-2}
27. 0.005104 5.104×10^{-3}
28. 416,000 4.16×10^5

Express each number in decimal form (without exponents).

29. 4.3×10^4 43,000
30. 1.63×10^{-4} 0.000163
31. 5.43×10^{-3} 0.00543
32. 6.15×10^5 615,000
33. 2.13×10^{-5} 0.0000213
34. 7.26×10^{-6} 0.00000726
35. 6.25×10^5 625,000
36. 4.6×10^1 46
37. 9×10^6 9,000,000
38. 6.475×10^1 64.75
39. 5.35×10^2 535
40. 3.14×10^{-2} 0.0314
41. 6.201×10^{-4} 0.0006201
42. 7.73×10^{-7} 0.000000773
43. 1×10^4 10,000
44. 7.13×10^{-4} 0.000713

In Exercises 45–52, write the quantity without metric prefixes. See Example 3.

45. 8 micrometers 0.000008 meter
46. 23.5 millimeters 0.0235 meter
47. 125 gigawatts 125,000,000,000 watts
48. 8.7 nanoseconds 0.0000000087 second
49. 15.3 kilometers 15,300 meters
50. 80.2 megahertz 80,200,000 hertz
51. 15 micrograms 0.000015 gram
52. 3.12 milligrams 0.00312 gram

Perform each indicated operation and express each number in decimal form (without exponents).

53. $(2 \times 10^2)(3 \times 10^5)$ 60,000,000
54. $(2 \times 10^{-3})(3 \times 10^2)$ 0.6
55. $(2.7 \times 10^{-6})(9 \times 10^4)$ 0.243
56. $(1.6 \times 10^{-2})(4 \times 10^{-3})$ 0.000064
57. $(1.3 \times 10^{-8})(1.74 \times 10^6)$ 0.02262
58. $(4 \times 10^5)(1.2 \times 10^{-4})$ 48
59. $\dfrac{8.4 \times 10^6}{2 \times 10^3}$ 4200
60. $\dfrac{6 \times 10^{-3}}{3 \times 10^1}$ 0.0002
61. $\dfrac{7.5 \times 10^6}{3 \times 10^3}$ 2500
62. $\dfrac{14 \times 10^6}{4 \times 10^8}$ 0.035
63. $\dfrac{4 \times 10^2}{8 \times 10^5}$ 0.0005
64. $\dfrac{16 \times 10^3}{8 \times 10^{-3}}$ 2,000,000

Perform each indicated operation by first converting each number to scientific notation. Write the answer in scientific notation.

65. $(700,000)(6,000,000)$ 4.2×10^{12}
66. $(0.003)(0.00015)$ 4.5×10^{-7}
67. $(0.0004)(320)$ 1.28×10^{-1}
68. $(67,000)(200,000)$ 1.34×10^{10}
69. $\dfrac{2,100,000}{7000}$ 3.0×10^2
70. $\dfrac{0.00004}{200}$ 2.0×10^{-7}
71. $\dfrac{0.00035}{0.000002}$ 1.75×10^2
72. $\dfrac{150,000}{0.0005}$ 3.0×10^8

73. List the following numbers from smallest to largest: $4.8 \times 10^5, 3.2 \times 10^{-1}, 4.6, 8.3 \times 10^{-4}$.
   $8.3 \times 10^{-4}, 3.2 \times 10^{-1}, 4.6, 4.8 \times 10^5$

74. List the following numbers from smallest to largest: $7.3 \times 10^2, 3.3 \times 10^{-4}, 1.75 \times 10^6, 5.3$.
   $3.3 \times 10^{-4}, 5.3, 7.3 \times 10^2, 1.75 \times 10^6$

10. 1×10^6 11. 1×10^{-6}

ction 4.3 • Scientific Notation • 269

Problem Solving

In Exercises 75–92, write the answer in decimal form (without exponents) unless asked to do otherwise.

75. Population In 2002, the U.S. population was about 2.81×10^8 people and the world population was about 6.20×10^9 people.

   a) How many people lived outside the United States in 2002? $\approx 5{,}919{,}000{,}000$

   b) How many times greater is the world population than the U.S. population? ≈ 22.1

76. Movies The gross ticket sales of the top five movies in the United States as of January 1, 2002, are listed below.

Movie	Year Released	Approximate U.S. Gross Ticket Sales
1. Titanic	1997	$601,000,000
2. Star Wars	1977	$461,000,000
3. Star Wars—The Phantom Menace	2001	$431,000,000
4. E.T.	1982	$400,000,000
5. Jurassic Park	1993	$357,000,000

   a) How much greater was the gross ticket sales of *Titanic* than *Jurassic Park*? Answer in scientific notation. $\$2.44 \times 10^8$

   b) How many times greater was the gross ticket sales of *Titanic* than *Jurassic Park*? ≈ 1.68

77. Niagara Falls A treaty between the United States and Canada requires that during the tourist season a minimum of 100,000 cubic feet of water per second flows over Niagara Falls (another 130,000 to 160,000 cubic feet/sec is diverted for power generation). Find the minimum volume of water that will flow over the falls in a 24-hour period during the tourist season. $8{,}640{,}000{,}000 \text{ ft}^3$

78. Big Macs On Tuesday, November 20, 2001, many newspapers had an article about Don Gorske earning a spot in *Guinness World Records*. Every day for the past 29 years Gorske has eaten one to three Big Macs. So far he has eaten 1.8×10^4 of them. The total calories consumed from all the Big Macs is about 1.06×10^7 calo-

ries. Using the information provided, estimate the number of calories in a Big Mac. ≈ 588.89

79. Computer Speed If a computer can do a calculation in 0.000002 second, how long, in seconds, would it take the computer to do 8 trillion (8,000,000,000,000) calculations? Write your answer in scientific notation.

80. Fortune Cookies Who writes the fortunes in those fortune cookies? Fortune cookies are an American invention. Steven Yang, who runs M & Y Trading Company, a San Francisco Company, prints about 90% of all the 1.02×10^9 fortunes in fortune cookies in America in any given year.

   a) How many fortunes does Yang print in a year?

   b) How many fortunes does Yang print in a day?

81. Diapers Laid end to end the 18 billion disposable diapers thrown away in the United States each year would reach the moon and back seven times (seven round trips).

   a) Write 18 billion in scientific notation. 1.8×10^{10}

   b) If the distance from Earth to the moon is 2.38×10^5 miles, what is the length of all these diapers placed end to end? Write your answer in scientific notation and as a number without exponents.

82. Jail Time The amount spent on incarceration in the United States has ballooned from about $\$6.9 \times 10^6$ in 1980 to about $\$4.5 \times 10^7$ in 2002.

   a) How much greater was the amount spent on incarceration in 2002 than in 1980? $\approx \$38{,}100{,}000$

   b) How many times greater was the amount spent on incarceration in 2002 than in 1980? (Source: *Fortune Magazine*) ≈ 6.5

83. Umpires Versus Referees The starting salary in 2001 for a major league umpire was about $\$1.05 \times 10^5$ while the starting salary for an NFL referee was $\$2.23 \times 10^4$.

   a) How much greater was the starting salary for an umpire than for a referee? $\$82{,}700$

   b) How many times greater was the starting salary for an umpire than for a referee? (Source: *Money Magazine*) ≈ 4.71

84. Astronomy The mass of Earth, Earth's moon, and the planet Jupiter are listed below.

Earth: 5,794,000,000,000,000,000,000,000 metric tons

Moon: 73,400,000,000,000,000,000 metric tons

Jupiter: 1,899,000,000,000,000,000,000,000,000 metric tons

79. $1.6 \times 10^7 \text{ sec}$ **80. a)** $\approx 918{,}000{,}000$ **b)** $\approx 2{,}515{,}068$ **81. b)** 3.332×10^6 or $3{,}332{,}000$ mi

84. a) Earth, 5.794×10^{24}; Moon, 7.34×10^{19}; Jupiter, 1.899×10^{27} **85.** 500 sec or ≈ 8.33 min **88. a)** Produced: 6.1×10^6; Imported: 9.2×10^6

a) Write the mass of Earth, the moon, and Jupiter in scientific notation.

b) How many times greater is the mass of Earth than the mass of the moon? $\approx 7.89 \times 10^4$

c) How many times greater is the mass of Jupiter than the mass of Earth? $\approx 3.28 \times 10^2$

85. *Light from the Sun* The sun is 9.3×10^7 miles from Earth. Light travels at a speed of 1.86×10^5 miles per second. How long, in both seconds and in minutes, does it take light from the sun to reach Earth?

86. *Missing Inventory* The circle graph below shows where missing inventory in 2000 went. In 2000 the total yearly loss was about $29,000,000,000 ($29 billion). How much inventory was lost to shoplifting? Write your answer in scientific notation. $\$9.483 \times 10^9$

Where Does the Missing Inventory Go?

Note: Does not add up to 100% because of rounding.

Source: 2000 National Retail Security Survey

87. *Population Growth* It took all of human history for the world's population to reach 6.16×10^9 in 2002. At current rates, the world population will double in about 54 years. Estimate the world's population in 2056. Write your answer in scientific notation. 1.232×10^{10}

88. *Sources of Oil* The following line graph shows the amount of oil produced in the United States (domestic) and the amount of oil imported into the United States.

The Flow of Oil in the U.S.

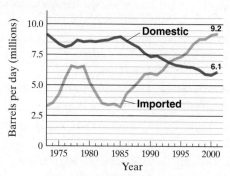

Sources: Energy Information Administration; USGS, State of Alaska

a) Write, in scientific notation, the number of barrels of oil produced in the United States in 2001 and the number of barrels of oil imported into the United States in 2001.

b) Determine, in scientific notation, the total amount of oil that is either produced or imported in the United States per day in 2001. 1.53×10^7

c) Determine, in scientific notation, the difference between the number of barrels of oil imported and the number of barrels produced in the United States per day in 2001. $\approx 3.1 \times 10^6$

89. *Digital Cameras* The following graph shows the worldwide production of digital cameras.

Digital Camera Sales Are Taking Off

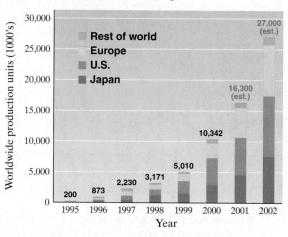

Source: Japan Camera Industrial Association + Olympus' Estimate

a) Give, in scientific notation, the worldwide production of digital cameras in 2002. 2.7×10^7

b) Olympus had the top share of digital camera sales in 2000. If Olympus produced 22% of the world's digital cameras in 2000, how many cameras did Olympus produce? Write your answer without exponents. 2,275,240

90. *Nanotechnology* The following graph shows the amount spent on nanotechnology from 1997 to 2002.

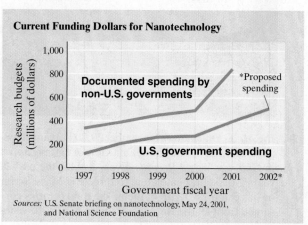

Current Funding Dollars for Nanotechnology

Sources: U.S. Senate briefing on nanotechnology, May 24, 2001, and National Science Foundation

90. a) $\approx \$1.00 \times 10^8$, $\approx \$5.00 \times 10^8$ b) $400,000,000 91. b) $\approx 5.017 \times 10^{22}$

a) Write, using scientific notation, estimates for the number of dollars spent by the U.S. government for nanotechnology in 1997 and in 2002. (For example, you may write a number like $\$1.3 \times 10^8$.)

b) Estimate the difference, in dollars, spent by the U.S. government for nanotechnology in 2002 and 1997.

c) Estimate the total amount spent, using scientific notation, for nanotechnology in 2001 including both the amount spent by the U.S. government and non-U.S. governments. $\approx \$1.25 \times 10^9$

d) Write the dollar amount determined in part c) without exponents. $1,250,000,000

91. **A Large Number** Avogadro's number, named after the nineteenth century Italian chemist Amedeo Avogadro, is roughly 6.02×10^{23}. It represents the number of atoms in 12 grams of pure carbon.

a) If this number were written out in decimal form, how many digits would it contain? 24

b) What is the number of atoms in 1 gram of pure carbon? Write your answer in scientific notation.

92. **Structure of Matter** An article in *Scientific American* states that physicists have created a *standard model* that describes the structure of matter down to 10^{-18} meters. If this number is written without exponents, how many zeroes would there be to the right of the decimal point? 17

Challenge Problems

93. **The Movie Contact** In the movie *Contact*, Jodie Foster plays an astronomer who makes the statement "There are 400 billion stars out there just in our universe alone. If only one out of a million of those had planets, and if just one out of a million of those had life, and if just one out of a million of those had intelligent life, there would be literally millions of civilizations out there." Do you believe this statement is correct? Explain your answer. Answers will vary.

94. How many times, either greater or smaller, is 10^{-12} meters than 10^{-18} meters? 1,000,000 times greater

95. How many times smaller is 1 nanosecond than one millisecond? 1,000,000

96. **Light Year** Light travels at a speed of 1.86×10^5 miles per second. A *light year* is the distance the light travels in one year. Determine the number of miles in a light year. $\approx 5.87 \times 10^{12}$ mi

Group Activity

Discuss and answer Exercise 97 as a group.

97. **A Million Versus a Billion** Do you have any idea of the difference in size between a million (1,000,000), a billion (1,000,000,000), and a trillion (1,000,000,000,000)?

a) Write a million, a billion, and a trillion in scientific notation. 1×10^6, 1×10^9, 1×10^{12}

b) Group member 1: Determine how long it would take to spend a million dollars if you spent $1000 a day. 1000 days or about 2.74 years

c) Group member 2: Repeat part b) for a billion dollars. 1,000,000 days or about 2740 years

d) Group member 3: Repeat part b) for a trillion dollars. 1,000,000,000 days or about 2,740,000 years

e) As a group, determine how many times greater a billion dollars is than a million dollars. 1000

Cumulative Review Exercises

[1.9] 98. Evaluate $4x^2 + 3x + \dfrac{x}{2}$ when $x = 0$. 0

[2.3] 99. a) If $-x = -\frac{3}{2}$, what is the value of x? $\frac{3}{2}$

b) If $5x = 0$, what is the value of x? 0

[2.5] 100. Solve the equation $2x - 3(x - 2) = x + 2$. 2

[4.1] 101. Simplify $\left(\dfrac{-2x^5 y^7}{8x^8 y^3}\right)^3$. $-\dfrac{y^{12}}{64x^9}$

4.4 ADDITION AND SUBTRACTION OF POLYNOMIALS

SSM Study Guide CD/Video

  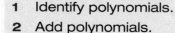
MathPro 4/5 PH Math Tutor Center prenhall.com/Angel

1 Identify polynomials.
2 Add polynomials.
3 Subtract polynomials.
4 Subtract polynomials in columns.

1 Identify Polynomials

A **polynomial in x** is an expression containing the sum of a finite number of terms of the form ax^n, for any real number a and any *whole number n*.

Examples of Polynomials	Not Polynomials	
$2x$	$4x^{1/2}$	*(Fractional exponent)*
$\frac{1}{3}x - 4$	$3x^2 + 4x^{-1} + 5$	*(Negative exponent)*
$x^2 - 2x + 1$	$4 + \frac{1}{x}$	$\left(\frac{1}{x} = x^{-1}, \text{negative exponent}\right)$

A polynomial is written in **descending order** (or **descending powers**) **of the variable** when the exponents on the variable decrease from left to right.

Example of Polynomial in Descending Order
$$2x^4 + 4x^2 - 6x + 3$$

Note in the example that the constant term 3 is last because it can be written as $3x^0$. Remember that $x^0 = 1$.

A polynomial can be in more than one variable. For example, $3xy + 2$ is a polynomial in two variables, x and y.

A polynomial with one term is called a **monomial**. A **binomial** is a two-termed polynomial. A **trinomial** is a three-termed polynomial. Polynomials containing more than three terms are not given special names. The prefix "poly" means "many." The chart that follows summarizes this information.

Type of Polynomial	Number of Terms	Examples
Monomial	One	$8, \ 4x, \ -6x^2$
Binomial	Two	$x + 5, \ x^2 - 6, \ 4y^2 - 5y$
Trinomial	Three	$x^2 - 2x + 3, \ 3z^2 - 6z + 7$

The **degree of a term** of a polynomial in one variable is the exponent on the variable in that term.

Term	Degree of Term	
$4x^2$	Second	
$2y^5$	Fifth	
$-5x$	First	*($-5x$ can be written $-5x^1$.)*
3	Zero	*(3 can be written $3x^0$.)*

TEACHING TIP
Tell students to give their answers in descending order unless otherwise instructed. Point out that this will make it easier to compare answers.

TEACHING TIP
You may want to use the definition of polynomial to show that a nonzero constant such as 5 is a polynomial.

For a polynomial in two or more variables, the degree of a term is the sum of the exponents on the variables. For example, the degree of the term $4x^2y^3$ is 5 because $2 + 3 = 5$. The degree of the term $5a^4bc^3$ is 8 because $4 + 1 + 3 = 8$. The **degree of a polynomial** is the same as that of its highest-degree term.

TEACHING TIP
Point out that polynomials are expressions not equations.

Polynomial	Degree of Polynomial	
$8x^3 + 2x^2 - 3x + 4$	Third	($8x^3$ is highest-degree term.)
$x^2 - 4$	Second	(x^2 is highest-degree term.)
$2x - 1$	First	($2x$ or $2x^1$ is highest-degree term.)
4	Zero	(4 or $4x^0$ is highest-degree term.)
$x^2y^4 + 2x + 3$	Sixth	(x^2y^4 is highest degree term.)

NOW TRY EXERCISE 53

2 Add Polynomials

In Section 2.1, we stated that like terms are terms having the same variables and the same exponents. That is, like terms may differ only in their numerical coefficients.

Examples of Like Terms

$3,$	-5
$2x$	x
$-2x^2,$	$4x^2$
$3y^2,$	$5y^2$
$3xy^2,$	$5xy^2$

To Add Polynomials

To add polynomials, combine the like terms of the polynomials.

EXAMPLE 1 Simplify $(4x^2 + 6x + 3) + (2x^2 + 5x - 1)$.

Solution Remember that $(4x^2 + 6x + 3) = 1(4x^2 + 6x + 3)$ and $(2x^2 + 5x - 1) = 1(2x^2 + 5x - 1)$. We can use the distributive property to remove the parentheses, as shown below.

TEACHING TIP
Before discussing Example 1, have students simplify $(4y + 6x + 3) + (2y + 5x - 1)$. Then have students identify the similarities between this expression and the expression in Example 1.

$$(4x^2 + 6x + 3) + (2x^2 + 5x - 1)$$
$$= 1(4x^2 + 6x + 3) + 1(2x^2 + 5x - 1)$$
$$= 4x^2 + 6x + 3 + 2x^2 + 5x - 1 \qquad \textit{Use Distributive Property.}$$
$$= \underline{4x^2 + 2x^2} + \underline{6x + 5x} + \underline{3 - 1} \qquad \textit{Rearrange terms.}$$
$$= 6x^2 + 11x + 2 \qquad \textit{Combine like terms.} \quad ✳$$

NOW TRY EXERCISE 67

In the following examples, we will not show the multiplication by 1 as was shown in Example 1.

EXAMPLE 2 Simplify $(5a^2 + 3a + b) + (a^2 - 7a + 3)$.

Solution
$$(5a^2 + 3a + b) + (a^2 - 7a + 3)$$
$$= 5a^2 + 3a + b + a^2 - 7a + 3 \qquad \textit{Remove parentheses.}$$
$$= \underline{5a^2 + a^2} + \underline{3a - 7a} + b + 3 \qquad \textit{Rearrange terms.}$$
$$= 6a^2 - 4a + b + 3 \qquad \textit{Combine like terms.} \quad ✳$$

EXAMPLE 3 Simplify $(3x^2y - 4xy + y) + (x^2y + 2xy + 3y)$.

Solution

$$(3x^2y - 4xy + y) + (x^2y + 2xy + 3y)$$
$$= 3x^2y - 4xy + y + x^2y + 2xy + 3y \qquad \textit{Remove parentheses.}$$
$$= \underbrace{3x^2y + x^2y}\ \underbrace{- 4xy + 2xy}\ \underbrace{+ y + 3y} \qquad \textit{Rearrange terms.}$$
$$= \quad 4x^2y \quad - \quad 2xy \quad + \quad 4y \qquad \textit{Combine like terms.} \ \text{✳}$$

NOW TRY EXERCISE 77

Usually, when we add polynomials, we will do so as in Examples 1 through 3. That is, we will list the polynomials horizontally. However, in Section 4.6, when we divide polynomials, there will be steps where we add polynomials in columns.

TEACHING TIP
You may wish to encourage students to translate the horizontal expression to a vertical form if that form is easier for them.

To Add Polynomials in Columns

1. Arrange polynomials in descending order, one under the other with like terms in the same columns.

2. Add the terms in each column.

EXAMPLE 4 Add $6x^2 - 2x + 2$ and $-2x^2 - x + 7$ using columns.

Solution

$$\begin{array}{r} 6x^2 - 2x + 2 \\ -2x^2 - \ x + 7 \\ \hline 4x^2 - 3x + 9 \end{array}$$

✳

EXAMPLE 5 Add $(5w^3 + 2w - 4)$ and $(2w^2 - 6w - 3)$ using columns.

Solution Since the polynomial $5w^3 + 2w - 4$ does not have a w^2 term, we will add the term $0w^2$ to the polynomial. This procedure sometimes helps in aligning like terms.

$$\begin{array}{r} 5w^3 + 0w^2 + 2w - 4 \\ 2w^2 - 6w - 3 \\ \hline 5w^3 + 2w^2 - 4w - 7 \end{array}$$

✳

3 Subtract Polynomials

Now let's subtract polynomials.

To Subtract Polynomials

1. Use the distributive property to remove parentheses. (This will have the effect of changing the sign of *every* term within the parentheses of the polynomial being subtracted.)

2. Combine like terms.

EXAMPLE 6 Simplify $(3x^2 - 2x + 5) - (x^2 - 3x + 4)$.

Solution $(3x^2 - 2x + 5)$ means $1(3x^2 - 2x + 5)$ and $(x^2 - 3x + 4)$ means $1(x^2 - 3x + 4)$. We use this information in the solution, as shown below.

$$(3x^2 - 2x + 5) - (x^2 - 3x + 4) = 1(3x^2 - 2x + 5) - 1(x^2 - 3x + 4)$$
$$= 3x^2 - 2x + 5 - x^2 + 3x - 4 \qquad \textit{Remove parentheses.}$$
$$= \underbrace{3x^2 - x^2}\ \underbrace{- 2x + 3x}\ \underbrace{+ 5 - 4} \qquad \textit{Rearrange terms.}$$

NOW TRY EXERCISE 101

$$= \quad 2x^2 \quad + \quad x \quad + \quad 1 \qquad \textit{Combine like terms.} \ \text{✳}$$

Remember from Section 2.1 that when a negative sign precedes the parentheses, the sign of every term within the parentheses is changed when the parentheses are removed. This was shown in Example 6. In Example 7, we will not show the multiplication by -1, as was done in Example 6.

EXAMPLE 7 Subtract $(-3x^2 - 5x + 3)$ from $(x^3 + 2x + 6)$.

Solution

$$(x^3 + 2x + 6) - (-3x^2 - 5x + 3)$$
$$= x^3 + 2x + 6 + 3x^2 + 5x - 3 \quad \text{Remove parentheses.}$$
$$= x^3 + 3x^2 + 2x + 5x + 6 - 3 \quad \text{Rearrange terms.}$$
$$= x^3 + 3x^2 + 7x + 3 \quad \text{Combine like terms.}$$

NOW TRY EXERCISE 107

AVOIDING COMMON ERRORS

One of the most common mistakes occurs when subtracting polynomials. When subtracting one polynomial from another, **the sign of each term in the polynomial being subtracted must be changed, not just the sign of the first term.**

CORRECT

$$6x^2 - 4x + 3 - (2x^2 - 3x + 4)$$
$$= 6x^2 - 4x + 3 - 2x^2 + 3x - 4$$
$$= 4x^2 - x - 1$$

INCORRECT

$$6x^2 - 4x + 3 - (2x^2 - 3x + 4)$$
$$= 6x^2 - 4x + 3 - 2x^2 - 3x + 4$$
$$= 4x^2 - 7x + 7$$

Do not make this mistake!

4 Subtract Polynomials in Columns

Polynomials can be subtracted as well as added using columns.

To Subtract Polynomials in Columns

1. Write *the polynomial being subtracted* below the polynomial from which it is being subtracted. List like terms in the same column.
2. *Change the sign of each term* in the polynomial being subtracted. (This step can be done mentally, if you like.)
3. Add the terms in each column.

EXAMPLE 8 Subtract $(x^2 - 4x + 6)$ from $(4x^2 + 5x + 7)$ using columns.

Solution Align like terms in columns (step 1).

$$\begin{array}{r} 4x^2 + 5x + 7 \\ -(x^2 - 4x + 6) \end{array} \quad \text{Align like terms.}$$

Change *all* signs in the second row (step 2); then add (step 3).

$$\begin{array}{r} 4x^2 + 5x + 7 \\ -x^2 + 4x - 6 \\ \hline 3x^2 + 9x + 1 \end{array} \quad \begin{array}{l} \text{Change all signs.} \\ \text{Add.} \end{array}$$

EXAMPLE 9   Subtract $(2x^2 - 6)$ from $(-3x^3 + 4x - 3)$ using columns.

Solution   To help align like terms, write each expression in descending order. If any power of x is missing, write that term with a numerical coefficient of 0.

$$-3x^3 + 4x - 3 = -3x^3 + 0x^2 + 4x - 3$$
$$2x^2 - 6 = 2x^2 + 0x - 6$$

Align like terms.

$$
\begin{array}{r}
-3x^3 + 0x^2 + 4x - 3 \\
-(2x^2 + 0x - 6) \\
\hline
\end{array}
$$

Change all signs in the second row; then add the terms in each column.

$$
\begin{array}{r}
-3x^3 + 0x^2 + 4x - 3 \\
-\ 2x^2 - 0x + 6 \\
\hline
-3x^3 - 2x^2 + 4x + 3
\end{array}
$$

NOW TRY EXERCISE 115

Note: Many of you will find that you can change the signs mentally and can therefore align and change the signs in one step.

1. an expression containing the sum of a finite number of terms of the form ax^n, for any real number a and any whole number n.
2. a) a one-termed polynomial b) a two-termed polynomial c) a three-termed polynomial 3. a) The exponent on the variable is the degree of the term. b) same as the degree of the highest-degree term in the polynomial 5. Add the exponents on the variables.
7. should be $3x + 2 - 4x + 6$ 8. Write with exponents on the variable decreasing from left to right. 9. because the exponent on the variable is 0

Exercise Set 4.4

11. b) $4x^3 + 0x^2 + 5x - 7$ 12. no, contains a negative exponent
13. no, contains a fractional exponent 14. no, $\dfrac{2}{x} = 2x^{-1}$, negative exponent

Concept/Writing Exercises

1. What is a polynomial?

2. a) What is a monomial? Make up three examples.

   b) What is a binomial? Make up three examples.

   c) What is a trinomial? Make up three examples.

3. a) Explain how to find the degree of a term in one variable.

   b) Explain how to find the degree of a polynomial in one variable.

4. Make up your own fifth-degree polynomial with three terms. Explain why it is a fifth-degree polynomial with three terms. Answers will vary.

5. Explain how to find the degree of a term in a polynomial in more than one variable.

6. Which of the following are fourth-degree terms? Explain your answer.

   a) $3xy^2$ b) $6r^2s^2$ c) $-2mn^3$ b) and c)

7. Explain why $(3x + 2) - (4x - 6) \neq 3x + 2 - 4x - 6$.

8. Explain how to write a polynomial in one variable in descending order of the variable.

9. Why is the constant term always written last when writing a polynomial in descending order?

10. Explain how to add polynomials. Combine like terms.

11. a) In your own words, describe how to add polynomials in columns. Answers will vary.

    b) How will you rewrite $4x^3 + 5x - 7$ in order to add it to $3x^3 + x^2 - 4x + 8$ using columns? Explain.

12. Is $4x^{-3} + 9$ a polynomial? Explain.

13. Is $6m^3 - 5m^{1/2}$ a polynomial? Explain.

14. Is $5x + \dfrac{2}{x}$ a polynomial? Explain.

Practice the Skills

Indicate the degree of each term.

15. x^5 fifth **16.** z^9 ninth **17.** $5a^4$ fourth **18.** $-3b^8$ eight

19. $-12n^7$ seventh **20.** $-13r^6$ sixth **21.** x^2y third **22.** a^4b^3 seventh

23. $3r^2s^8$ tenth **24.** $6m^5n^8$ thirteenth **25.** $-12p^4q^7r$ twelfth **26.** $-8x^3y^5z$ ninth

Indicate which expressions are polynomials. If the polynomial has a specific name—monomial, binomial, or trinomial—give that name.

27. $x^2 + 3$ binomial **28.** $2x^2 - 6x + 7$ trinomial **29.** 13 monomial

30. $4x^{-2}$ not polynomial **31.** $4x^3 - 8$ binomial **32.** $7x + 8$ binomial

33. $7x^3$ monomial **34.** $3x^{1/2} + 2x$ not polynomial **35.** $a^{-1} + 4$ not polynomial

36. $x^3 - 8x^2 + 8$ trinomial **37.** $6n^3 - 5n^2 + 4n - 3$ polynomial **38.** $10x^2$ monomial

39. $4 - 2b^2 - 5b$ trinomial **40.** $2x^{-2}$ not polynomial **41.** $0.6r^4 - \frac{1}{2}r^3 - 0.4r^2 - \frac{1}{3}$ polynomial

42. $\frac{2}{3}x^2 - \frac{1}{x}$ not polynomial

Express each polynomial in descending order. If the polynomial is already in descending order, so state. Give the degree of each polynomial. **55.** $-2x^4 + 3x^2 + 5x - 6$, fourth

43. $4 + 5x$ $5x + 4$, first **44.** 5 0 degree **45.** $-4 + x^2 - 2x$ $x^2 - 2x - 4$, second

46. $6x - 5$ first **47.** $x + 3x^2 - 8$ $3x^2 + x - 8$, second **48.** $4 - 3p^3$ $-3p^3 + 4$, third

49. $-x - 1$ first **50.** $2x^2 + 5x - 8$ second **51.** $2t^2 - 3t + 4$ second

52. 15 0 degree **53.** $-4 + x - 3x^2 + 4x^3$ $4x^3 - 3x^2 + x - 4$, third **54.** $1 - x^3 + 3x$ $-x^3 + 3x + 1$, third

🔒 **55.** $5x + 3x^2 - 6 - 2x^4$ **56.** $-3r - 5r^2 + 2r^4 - 6$ $2r^4 - 5r^2 - 3r - 6$, fourth

Add. **72.** $8.2n^2 - 4.8n - 0.6$ **73.** $-2x^3 - 3x^2 + 4x - 3$ **76.** $-6x^2 + xy^2$ **78.** $3x^2y + 3x - 7y + 3$

57. $(5x + 4) + (x - 5)$ $6x - 1$
58. $(5x - 6) + (2x - 3)$ $7x - 9$

59. $(-4x + 8) + (2x + 3)$ $-2x + 11$
60. $(-7x - 9) + (-2x + 9)$ $-9x$

61. $(t + 7) + (-3t - 8)$ $-2t - 1$
62. $(4x - 3) + (3x - 3)$ $7x - 6$

63. $(x^2 + 2.6x - 3) + (4x + 3.8)$ $x^2 + 6.6x + 0.8$
64. $(-4p^2 - 3p - 2) + (-p^2 - 4)$ $-5p^2 - 3p - 6$

65. $(4m - 3) + (5m^2 - 4m + 7)$ $5m^2 + 4$
66. $(-x^2 - 2x - 4) + (4x^2 + 3)$ $3x^2 - 2x - 1$

67. $(2x^2 - 3x + 5) + (-x^2 + 6x - 8)$ $x^2 + 3x - 3$
68. $(x^2 - 6x + 7) + (-x^2 + 3x + 5)$ $-3x + 12$

69. $(-x^2 - 4x + 8) + \left(5x - 2x^2 + \frac{1}{2}\right)$ $-3x^2 + x + \frac{17}{2}$
70. $(8x^2 + 3x - 5) + \left(x^2 + \frac{1}{2}x + 2\right)$ $9x^2 + \frac{7}{2}x - 3$

71. $(8x^2 + 4) + (-2.6x^2 - 5x - 2.3)$ $5.4x^2 - 5x + 1.7$
72. $(5.2n^2 - 6n + 1.7) + (3n^2 + 1.2n - 2.3)$

🔒 **73.** $(-7x^3 - 3x^2 + 4) + (4x + 5x^3 - 7)$
74. $(6x^3 - 4x^2 - 7) + (3x^2 + 3x - 3)$ $6x^3 - x^2 + 3x - 10$

75. $(8x^2 + 2xy + 4) + (-x^2 - 3xy - 8)$ $7x^2 - xy - 4$
76. $(x^2y + 6x^2 - 3xy^2) + (-x^2y - 12x^2 + 4xy^2)$

77. $(2x^2y + 2x - 3) + (3x^2y - 5x + 5)$ $5x^2y - 3x + 2$
78. $(x^2y + x - y) + (2x^2y + 2x - 6y + 3)$

Add using columns.

79. Add $3x - 6$ and $4x + 5$. $7x - 1$
80. Add $-2x + 5$ and $-3x - 5$. $-5x$

81. Add $4y^2 - 2y + 4$ and $3y^2 + 1$. $7y^2 - 2y + 5$
82. Add $9x^2 - 5x - 1$ and $-9x^2 + 6$. $-5x + 5$

🔒 **83.** Add $-x^2 - 3x + 3$ and $5x^2 + 5x - 7$. $4x^2 + 2x - 4$
84. Add $-2s^2 - s + 5$ and $3s^2 - 6s$. $s^2 - 7s + 5$

85. Add $2x^3 + 3x^2 + 6x - 9$ and $7 - 4x^2$. $2x^3 - x^2 + 6x - 2$
86. Add $-3x^3 + 3x + 9$ and $2x^2 - 4$. $-3x^3 + 2x^2 + 3x + 5$

87. Add $4n^3 - 5n^2 + n - 6$ and $-n^3 - 6n^2 - 2n + 8$.
$3n^3 - 11n^2 - n + 2$
88. Add $7x^3 + 5x - 6$ and $3x^3 - 4x^2 - x + 8$.
$10x^3 - 4x^2 + 4x + 2$

103. $2x^3 - \dfrac{23}{5}x^2 + 2x - 2$ **104.** $-x^3 - 7x^2 + \dfrac{19}{4}x - 7$

108. $-4x^2 + 8x + 13$

Subtract.

89. $(4x - 4) - (2x + 2)$ $2x - 6$

90. $(3x - 2) - (4x + 3)$ $-x - 5$

91. $(-2x - 3) - (-5x - 7)$ $3x + 4$

92. $(10x - 3) - (-2x + 7)$ $12x - 10$

93. $(-r + 5) - (2r + 5)$ $-3r$

94. $(4x + 8) - (3x + 9)$ $x - 1$

95. $(9x^2 + 7x - 5) - (3x^2 + 3.5)$ $6x^2 + 7x - 8.5$

96. $(-y^2 + 4y - 5.2) - (5y^2 + 2.1y + 7.5)$ $-6y^2 + 1.9y^2 - 12.7$

97. $(5x^2 - x - 1) - (-3x^2 - 2x - 5)$ $8x^2 + x + 4$

98. $(-a^2 + 3a + 12) - (-4a^2 - 3)$ $3a^2 + 3a + 15$

99. $(-6m^2 - 2m) - (3m^2 - 7m + 6)$ $-9m^2 + 5m - 6$

100. $(7x - 0.6) - (-2x^2 + 4x - 8)$ $2x^2 + 3x + 7.4$

101. $(8x^3 - 2x^2 - 4x + 5) - (5x^2 + 8)$ $8x^3 - 7x^2 - 4x - 3$

102. $\left(9x^3 - \dfrac{1}{5}\right) - (x^2 + 5x)$ $9x^3 - x^2 - 5x - \dfrac{1}{5}$

103. $(2x^3 - 4x^2 + 5x - 7) - \left(3x + \dfrac{3}{5}x^2 - 5\right)$

104. $(-3x^2 + 4x - 7) - \left(x^3 + 4x^2 - \dfrac{3}{4}x\right)$

105. Subtract $(7x + 4)$ from $(8x + 2)$ $x - 2$

106. Subtract $(-4x + 7)$ from $(-3x - 9)$ $x - 16$

107. Subtract $(5x - 6)$ from $(2x^2 - 4x + 8)$ $2x^2 - 9x + 14$

108. Subtract $(3x^2 - 5x - 3)$ from $(-x^2 + 3x + 10)$

109. Subtract $(4x^3 - 6x^2)$ from $(3x^3 + 5x^2 + 9x - 7)$ $-x^3 + 11x^2 + 9x - 7$

110. Subtract $(-2c^2 + 7c - 7)$ from $(-5c^3 - 6c^2 + 7)$ $-5c^3 - 4c^2 - 7c + 14$

Perform each subtraction using columns.

111. Subtract $(3x - 3)$ from $(6x + 5)$. $3x + 8$

112. Subtract $(6x + 8)$ from $(2x - 5)$. $-4x - 13$

113. Subtract $(-3d - 4)$ from $(-6d + 8)$. $-3d + 12$

114. Subtract $(-3x + 8)$ from $(6x^2 - 5x + 3)$. $6x^2 - 2x - 5$

115. Subtract $(6x^2 - 1)$ from $(7x^2 - 3x - 4)$. $x^2 - 3x - 3$

116. Subtract $(5n^3 + 7n - 9)$ from $(2n^3 - 6n + 3)$. $-3n^3 - 13n + 1$

117. Subtract $(-5m^2 + 6m)$ from $(m - 6)$. $5m^2 - 5m - 6$

118. $(5x^2 + 4)$ from $(x^2 + 4)$. $-4x^2$

119. Subtract $(x^2 + 6x - 7)$ from $(4x^3 - 6x^2 + 7x - 9)$. $4x^3 - 7x^2 + x - 2$

120. Subtract $(2x^3 + 4x^2 - 9x)$ from $(-5x^3 + 4x - 12)$. $-7x^3 - 4x^2 + 13x - 12$

Problem Solving

127. sometimes

121. Make up your own addition problem where the sum of two binomials is $-2x + 4$. **121.–124.** Answers will vary.

122. Make up your own addition problem where the sum of two trinomials is $2x^2 + 5x - 6$.

123. Make up your own subtraction problem where the difference of two trinomials is $3x + 5$

124. Make up your own subtraction problem where the difference of two trinomials is $-x^2 + 4x - 5$.

125. When two binomials are added, will the sum always, sometimes, or never be a binomial? Explain your answer and give examples to support your answer. sometimes

126. When one binomial is subtracted from another, will the difference always, sometimes, or never be a binomial? Explain your answer and give examples to support your answer. sometimes

127. When two trinomials are added, will the sum always, sometimes, or never be a trinomial? Explain your answer and give examples to support your answer.

128. When one trinomial is subtracted from another, will the difference always, sometimes, or never be a trinomial? Explain your answer and give examples to support your answer. sometimes

129. Write a fifth-degree trinomial in the variable x that has neither a third- nor second-degree term.

130. Write a sixth-degree trinomial in the variable x that has no fifth-, fourth-, or zero-degree terms.

131. Is it possible to have a fifth-degree trinomial in x that has no fourth-, third-, second-, or first-degree terms and contains no like terms? Explain. no

132. Is it possible to have a fourth-degree trinomial in x that has no third-, second-, or zero-degree terms and contains no like terms? Explain. no

129. One example is $x^5 + x^4 + x$. **130.** One example is $x^6 + x^3 + x^2$.

Write a polynomial that represents the area of each figure shown.

133. 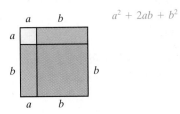 $a^2 + 2ab + b^2$

134. 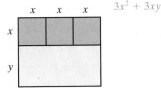 $3x^2 + 3xy$

135. 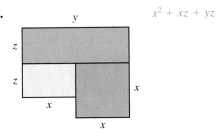 $x^2 + xz + yz$

136. 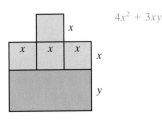 $4x^2 + 3xy$

Challenge Problems

Simplify.   **137.** $-12x + 18$ **138.** $3x^2y - 13xy + 9xy^2 + 3x$

137. $(3x^2 - 6x + 3) - (2x^2 - x - 6) - (x^2 + 7x - 9)$ **138.** $3x^2y - 6xy - 2xy + 9xy^2 - 5xy + 3x$

139. $4(x^2 + 2x - 3) - 6(2 - 4x - x^2) - 2x(x + 2)$ $8x^2 + 28x - 24$

 Group Activity

Discuss and answer Exercise 140 as a group.

140. Make up a trinomial, a binomial, and a different trinomial such that
(first trinomial) + (binomial) − (second trinomial) = 0.   Answers will vary.

Cumulative Review Exercises

[1.5] **141.** Insert either $>$, $<$, or $=$ in the shaded area to make the statement true: $|-9|$ ▓ $|-6|$. $>$

[1.6–1.8] *Indicate whether each statement is true or false.*

142. The product of two negative numbers is always a positive number.   true

143. The sum of two negative numbers is always a negative number.   true

144. The difference of two negative numbers is always a negative number.   false

145. The quotient of two negative numbers is always a negative number.   false

[4.1] **146.** Simplify $\left(\dfrac{4x^3y^5}{12x^7y^4}\right)^3$. $\dfrac{y^3}{27x^{12}}$

4.5 MULTIPLICATION OF POLYNOMIALS

 SSM
 Study Guide
 CD/Video

 MathPro 4/5
 PH Math Tutor Center
 prenhall.com/Angel

1 Multiply a monomial by a monomial.
2 Multiply a polynomial by a monomial.
3 Multiply binomials using the distributive property.
4 Multiply binomials using the FOIL method.
5 Multiply binomials using formulas for special products.
6 Multiply any two polynomials.

1 Multiply a Monomial by a Monomial

We begin our discussion of multiplication of polynomials by multiplying a monomial by a monomial. To multiply two monomials, multiply their coefficients and use the product rule of exponents to determine the exponents on the variables. Problems of this type were done in Section 4.1.

EXAMPLE 1 Multiply **a)** $(4x^2)(5x^5)$ **b)** $(-6y^4)(8y^7)$.

Solution **a)** $(4x^2)(5x^5) = 4 \cdot 5 \cdot x^2 \cdot x^5 = 20x^{2+5} = 20x^7$

b) $(-6y^4)(8y^7) = (-6)(8) \cdot y^4 \cdot y^7 = -48y^{4+7} = -48y^{11}$ ✳

EXAMPLE 2 Multiply $(6x^2y)(7x^5y^4)$.

Solution Remember that when a variable is given without an exponent we assume that the exponent on the variable is 1.

$$(6x^2y)(7x^5y^4) = 42x^{2+5}y^{1+4} = 42x^7y^5$$ ✳

EXAMPLE 3 Multiply **a)** $6xy^2z^5(-3x^4y^7z)$ **b)** $(-4x^4z^9)(-3xy^7z^3)$.

Solution **a)** $6xy^2z^5(-3x^4y^7z) = -18x^5y^9z^6$ **b)** $(-4x^4z^9)(-3xy^7z^3) = 12x^5y^7z^{12}$ ✳

NOW TRY EXERCISE 21

2 Multiply a Polynomial by a Monomial

To multiply a polynomial by a monomial, we use the distributive property presented earlier.

$$a(b + c) = ab + ac$$

The distributive property can be expanded to

$$a(b + c + d + \cdots + n) = ab + ac + ad + \cdots + an$$

EXAMPLE 4 Multiply $3x(2x^2 + 4)$.

Solution
$$3x(2x^2 + 4) = (3x)(2x^2) + (3x)(4)$$
$$= 6x^3 + 12x$$ ✳

Notice that the use of the distributive property results in monomials being multiplied by monomials. If we study Example 4, we see that the $3x$ and $2x^2$ are both monomials, as are the $3x$ and 4.

EXAMPLE 5 Multiply $-3n(4n^2 - 2n - 1)$.

Solution $$-3n(4n^2 - 2n - 1) = (-3n)(4n^2) + (-3n)(-2n) + (-3n)(-1)$$
$$= -12n^3 + 6n^2 + 3n$$

✳

EXAMPLE 6 Multiply $5x^2(4x^3 - 2x + 7)$.

Solution $$5x^2(4x^3 - 2x + 7) = (5x^2)(4x^3) + (5x^2)(-2x) + (5x^2)(7)$$

NOW TRY EXERCISE 35 $$= 20x^5 - 10x^3 + 35x^2$$

✳

EXAMPLE 7 Multiply $2x(3x^2y - 6xy + 5)$.

Solution $$2x(3x^2y - 6xy + 5) = (2x)(3x^2y) + (2x)(-6xy) + (2x)(5)$$
$$= 6x^3y - 12x^2y + 10x$$

✳

In Example 8, we perform a multiplication where the monomial is placed to the right of the polynomial. Each term of the polynomial is multiplied by the monomial, as illustrated in the example.

EXAMPLE 8 Multiply $(3x^3 - 2xy + 3)4x$.

Solution $$(3x^3 - 2xy + 3)4x = (3x^3)(4x) + (-2xy)(4x) + (3)(4x)$$

NOW TRY EXERCISE 41 $$= 12x^4 - 8x^2y + 12x$$

✳

The problem in Example 8 could be written as $4x(3x^2 - 2xy + 3)$ by the commutative property of multiplication, and then simplified as in Examples 4 through 7.

3 Multiply Binomials Using the Distributive Property

Now we will discuss multiplying a binomial by a binomial. Before we explain how to do this, consider the multiplication problem $43 \cdot 12$.

$$
\begin{array}{rcl}
& 43 & \longleftarrow \textit{Multiplicand} \\
& \underline{12} & \longleftarrow \textit{Multiplier} \\
2(4) \longrightarrow & 86 & \longleftarrow 2(3) \\
1(4) \longrightarrow & \underline{43} & \longleftarrow 1(3) \\
& 516 & \longleftarrow \textit{Product}
\end{array}
$$

Note how the 2 multiplies both the 3 and the 4, and the 1 also multiplies both the 3 and the 4. That is, every digit in the multiplier multiplies every digit in the multiplicand. We can also illustrate the multiplication process as follows.

$$
\begin{aligned}
(43)(12) &= (40 + 3)(10 + 2) \\
&= (40 + 3)(10) + (40 + 3)(2) \\
&= (40)(10) + (3)(10) + (40)(2) + (3)(2) \\
&= 400 + 30 + 80 + 6 \\
&= 516
\end{aligned}
$$

Whenever any two polynomials are multiplied, the same process must be followed. That is, **every term in one polynomial must multiply every term in the other polynomial.**

Consider multiplying $(a + b)(c + d)$. Treating $(a + b)$ as a single term and using the distributive property, we get

$$(a + b)(c + d) = (a + b)c + (a + b)d$$

Using the distributive property a second time gives

$$= ac + bc + ad + bd$$

Notice how each term of the first polynomial was multiplied by each term of the second polynomial, and all the products were added to obtain the answer.

EXAMPLE 9 Multiply $(3x + 2)(x - 5)$.

Solution
$$\begin{aligned}(3x + 2)(x - 5) &= (3x + 2)x + (3x + 2)(-5)\\ &= 3x(x) + 2(x) + 3x(-5) + 2(-5)\\ &= 3x^2 + 2x - 15x - 10\\ &= 3x^2 - 13x - 10\end{aligned}$$

Note that after performing the multiplication like terms must be combined.

EXAMPLE 10 Multiply $(x - 4)(y + 3)$.

Solution
$$\begin{aligned}(x - 4)(y + 3) &= (x - 4)y + (x - 4)3\\ &= xy - 4y + 3x - 12\end{aligned}$$

4 Multiply Binomials Using the FOIL Method

A common method used to multiply two binomials is the **FOIL method**. This procedure also results in each term of one binomial being multiplied by each term in the other binomial. Students often prefer to use this method when multiplying two binomials.

The FOIL Method

Consider $(a + b)(c + d)$.

F stands for **first**—multiply the first terms of each binomial together:

$$(a + b)(c + d) \qquad \text{product } ac$$

O stands for **outer**—multiply the two outer terms together:

$$(a + b)(c + d) \qquad \text{product } ad$$

I stands for **inner**—multiply the two inner terms together:

$$(a + b)(c + d) \qquad \text{product } bc$$

L stands for **last**—multiply the last terms together:

$$(a + b)(c + d) \qquad \text{product } bd$$

The product of the two binomials is the sum of these four products.

$$(a + b)(c + d) = ac + ad + bc + bd$$

The FOIL method is not actually a different method used to multiply binomials but rather an acronym to help students remember to correctly apply the distributive property. We could have used IFOL or any other arrangement of the four letters. However, FOIL is easier to remember than the other arrangements.

EXAMPLE 11 Using the FOIL method, multiply $(2x - 3)(x + 4)$.

Solution

$$(2x - 3)(x + 4)$$

$$
\begin{array}{cccc}
F & O & I & L \\
= (2x)(x) & + (2x)(4) & + (-3)(x) & + (-3)(4) \\
= \quad 2x^2 & + \quad 8x & - \quad 3x & - \quad 12
\end{array}
$$

$$= 2x^2 + 5x - 12$$

NOW TRY EXERCISE 45   Thus, $(2x - 3)(x + 4) = 2x^2 + 5x - 12$.

EXAMPLE 12 Multiply $(4 - 2x)(6 - 5x)$.

Solution

$$(4 - 2x)(6 - 5x)$$

$$
\begin{array}{cccc}
F & O & I & L \\
= 4(6) & + 4(-5x) & + (-2x)(6) & + (-2x)(-5x) \\
= \quad 24 & - \quad 20x & - \quad 12x & + \quad 10x^2
\end{array}
$$

$$= 10x^2 - 32x + 24$$

NOW TRY EXERCISE 63   Thus, $(4 - 2x)(6 - 5x) = 10x^2 - 32x + 24$.

EXAMPLE 13 Multiply $(2r + 3)(2r - 3)$.

Solution

$$
\begin{array}{cccc}
& F & O & I & L \\
(2r + 3)(2r - 3) = (2r)(2r) & + (2r)(-3) & + (3)(2r) & + (3)(-3) \\
= \quad 4r^2 & - \quad 6r & + \quad 6r & - \quad 9
\end{array}
$$

$$= 4r^2 - 9$$

Thus, $(2r + 3)(2r - 3) = 4r^2 - 9$.

HELPFUL HINT

STUDY TIP

Make sure you have a thorough understanding of multiplication of polynomials. In the next chapter, we will be studying factoring, which is the reverse process of multiplication of polynomials. To understand factoring, you must first understand multiplication of polynomials.

5   Multiply Binomials Using Formulas for Special Products

Example 13 illustrates a special product, the product of the sum and difference of the same two terms.

Product of the Sum and Difference of the Same Two Terms
$$(a + b)(a - b) = a^2 - b^2$$

In this special product, a represents one term and b the other term. Then $(a + b)$ is the sum of the terms and $(a - b)$ is the difference of the terms. This special product is also called the **difference of two squares formula** because the expression on the right side of the equal sign is the difference of two squares.

EXAMPLE 14   Use the rule for finding the product of the sum and difference of two quantities to multiply each expression.

a) $(x + 5)(x - 5)$ **b)** $(2x + 4)(2x - 4)$ **c)** $(3x + 2y)(3x - 2y)$

Solution   **a)** If we let $x = a$ and $5 = b$, then

$$(a + b)(a - b) = a^2 - b^2$$
$$(x + 5)(x - 5) = (x)^2 - (5)^2$$
$$= x^2 - 25$$

b)
$$(a + b)\ (a - b) = a^2 - b^2$$
$$(2x + 4)(2x - 4) = (2x)^2 - (4)^2$$
$$= 4x^2 - 16$$

c)
$$(a + b)\ (a - b) = a^2 - b^2$$
$$(3x + 2y)(3x - 2y) = (3x)^2 - (2y)^2$$
$$= 9x^2 - 4y^2$$

NOW TRY EXERCISE 77

Example 14 could also have been done using the FOIL method.

EXAMPLE 15   Using the FOIL method, find $(x + 3)^2$.

Solution $(x + 3)^2 = (x + 3)(x + 3)$

$$\begin{array}{cccc} F & O & I & L \end{array}$$
$$= x(x) + x(3) + 3(x) + (3)(3)$$
$$= x^2 + 3x + 3x + 9$$
$$= x^2 + 6x + 9$$

TEACHING TIP
Have students draw a partitioned square to represent Example 15.

Example 15 illustrates the **square of a binomial**, another special product.

Square of Binomial Formulas

$$(a + b)^2 = (a + b)(a + b) = a^2 + 2ab + b^2$$
$$(a - b)^2 = (a - b)(a - b) = a^2 - 2ab + b^2$$

To square a binomial, add the square of the first term, twice the product of the terms, and the square of the second term.

EXAMPLE 16 Use the square of a binomial formula to multiply each expression.

a) $(x + 5)^2$ **b)** $(2x - 4)^2$ **c)** $(3r + 2s)^2$ **d)** $(x - 3)(x - 3)$

Solution **a)** If we let $x = a$ and $5 = b$, then

$$(a + b)(a + b) = a^2 + 2ab + b^2$$
$$(x + 5)^2 = (x + 5)(x + 5) = (x)^2 + 2(x)(5) + (5)^2$$
$$= x^2 + 10x + 25$$

b)
$$(a - b)(a - b) = a^2 - 2ab + b^2$$
$$(2x - 4)^2 = (2x - 4)(2x - 4) = (2x)^2 - 2(2x)(4) + (4)^2$$
$$= 4x^2 - 16x + 16$$

c)
$$(a + b)(a + b) = a^2 + 2ab + b^2$$
$$(3r + 2s)^2 = (3r + 2s)(3r + 2s) = (3r)^2 + 2(3r)(2s) + (2s)^2$$
$$= 9r^2 + 12rs + 4s^2$$

d)
$$(a - b)(a - b) = a^2 - 2ab + b^2$$
$$(x - 3)(x - 3) = (x - 3)(x - 3) = (x)^2 - 2(x)(3) + (3)^2$$
$$= x^2 - 6x + 9$$

NOW TRY EXERCISE 83

Example 16 could also have been done using the FOIL method.

AVOIDING COMMON ERRORS

TEACHING TIP
Show that $(3 + 5)^2 \neq 3^2 + 5^2$.
Then show that
$(3 + 5)^2 = 3^2 + 2(3)(5) + 5^2$.

CORRECT	INCORRECT
$(a + b)^2 = a^2 + 2ab + b^2$	$(a + b)^2 = a^2 + b^2$
$(a - b)^2 = a^2 - 2ab + b^2$	$(a - b)^2 = a^2 - b^2$

Do not forget the middle term when you square a binomial.

$$(x + 2)^2 \neq x^2 + 4$$
$$(x + 2)^2 = (x + 2)(x + 2)$$
$$= x^2 + 4x + 4$$

6 Multiply Any Two Polynomials

When multiplying a binomial by a binomial, we saw that every term in the first binomial was multiplied by every term in the second binomial. When multiplying any two polynomials, each term of one polynomial must be multiplied by each term of the other polynomial. In the multiplication $(3x + 2)(4x^2 - 5x - 3)$, we use the distributive property as follows:

$$(3x + 2)(4x^2 - 5x - 3)$$
$$= 3x\,(4x^2 - 5x - 3) + 2\,(4x^2 - 5x - 3)$$
$$= 12x^3 - 15x^2 - 9x + 8x^2 - 10x - 6$$
$$= 12x^3 - 7x^2 - 19x - 6$$

Thus, $(3x + 2)(4x^2 - 5x - 3) = 12x^3 - 7x^2 - 19x - 6$.

Multiplication problems can be performed by using the distributive property, as we just illustrated. However, many students prefer to multiply a polynomial by a polynomial using a vertical procedure. On page 281, we showed that when multiplying the number 43 by the number 12, we multipy each digit in the number 43 by each digit in the number 12. Review that example now. We can follow a similar procedure when multiplying a polynomial by a polynomial, as illustrated in the following examples. We must be careful, however, to align like terms in the same columns when performing the individual multiplications.

EXAMPLE 17   Multiply $(3x + 4)(2x + 5)$.

Solution   First write the polynomials one beneath the other.

$$3x + 4$$
$$\underline{2x + 5}$$

Next, multiply each term in $(3x + 4)$ by 5.

$$3x + 4$$
$$\underline{2x + 5}$$
$$5(3x + 4) \longrightarrow 15x + 20$$

Next, multiply each term in $(3x + 4)$ by $2x$ and align like terms.

$$3x + 4$$
$$\underline{2x + 5}$$
$$15x + 20$$
$$2x(3x + 4) \longrightarrow \underline{6x^2 + 8x}$$
$$6x^2 + 23x + 20 \qquad \textit{Add like terms in columns} \quad ✳$$

The same answer for Example 17 would be obtained using the FOIL method.

EXAMPLE 18   Multiply $(3x - 2)(5x^2 + 6x - 4)$.

Solution   For convenience, we place the shorter expression on the bottom, as illustrated.

$$5x^2 + 6x - 4$$
$$\underline{3x - 2}$$
$$-10x^2 - 12x + 8 \qquad \textit{Multiply the top polynomial by} -2.$$
$$\underline{15x^3 + 18x^2 - 12x} \qquad\quad \textit{Multiply the top polynomial by } 3x; \textit{ align like terms.}$$
$$15x^3 + 8x^2 - 24x + 8 \qquad \textit{Add like terms in columns.} \qquad ✳$$

EXAMPLE 19   Multiply $x^2 - 3x + 2$ by $2x^2 - 3$.

Solution
$$x^2 - 3x + 2$$
$$\underline{2x^2 - 3}$$
$$-3x^2 + 9x - 6 \qquad \textit{Multiply the top polynomial by} -3.$$
$$\underline{2x^4 - 6x^3 + 4x^2} \qquad\qquad\quad \textit{Multiply the top polynomial by } 2x^2; \textit{ align like terms.}$$
$$2x^4 - 6x^3 + x^2 + 9x - 6 \qquad \textit{Add like terms in columns.} \qquad ✳$$

NOW TRY EXERCISE 95

EXAMPLE 20 Multiply $(3x^3 - 2x^2 + 4x + 6)(x^2 - 5x)$.

Solution

$$3x^3 - 2x^2 + 4x + 6$$
$$x^2 - 5x$$
$$\overline{-15x^4 + 10x^3 - 20x^2 - 30x} \quad \text{Multiply the top polynomial by } -5x.$$
$$3x^5 - 2x^4 + 4x^3 + 6x^2 \quad \text{Multiply the top polynomial by } x^2; \text{ align like terms.}$$
$$\overline{3x^5 - 17x^4 + 14x^3 - 14x^2 - 30x} \quad \text{Add like terms in columns.}$$

Exercise Set 4.5

Concept/Writing Exercises

2. Answers will vary. 3. first, outer, inner, last 6. because $a^2 - b^2$ is the difference of two squares 7. $(a + b)^2 = a^2 + 2ab + b^2$, $(a - b)^2 = a^2 - 2ab + b^2$
13. Answers will vary.

1. What is the name of the property used when multiplying a monomial by a polynomial? distributive property

2. Explain how to multiply a monomial by a monomial.

3. What do the letters in the acronym FOIL represent?

4. How does the FOIL method work when multiplying two binomials? every term multiplies every other term

5. When multiplying two binomials, will you get the same answer if you multiply using the order LOIF instead of FOIL? Explain your answer. yes

6. Why is the special product $(a + b)(a - b) = a^2 - b^2$ also called the difference of two squares formula?

7. Write the square of binomial formulas.

8. Use your own words to describe how to square a binomial. Answers will vary.

9. Does $(x + 5)^2 = x^2 + 5^2$? Explain. If not, what is the correct result? no, $x^2 + 10x + 25$

10. Does $(x + 3)^2 = x^2 + 3^2$? Explain. If not, what is the correct result? no, $x^2 + 6x + 9$

11. Make up a multiplication problem where a monomial in x is multiplied by a binomial in x. Determine the product. Answers will vary.

12. Make up a multiplication problem where a monomial in y is multiplied by a trinomial in y. Determine the product. Answers will vary.

13. Make up a multiplication problem where two binomials in x are multiplied. Determine the product.

14. When multiplying two polynomials, is it necessary for each term in one polynomial to multiply each term in the other polynomial? yes

Practice the Skills

Multiply.

15. $x^3 \cdot 2xy$ $2x^4y$

16. $6xy^2 \cdot 3xy^4$ $18x^2y^6$

17. $5x^3y^5(4x^2y)$ $20x^5y^6$

18. $-5x^2y^4(2x^3y^2)$ $-10x^5y^6$

19. $4x^4y^6(-7x^2y^9)$ $-28x^6y^{15}$

20. $4a^3b^7(6a^2b)$ $24a^5b^8$

21. $9xy^6 \cdot 6x^5y^8$ $54x^6y^{14}$

22. $(6m^3n^4)(3n^5)$ $18m^3n^9$

23. $(6x^2y)\left(\frac{1}{2}x^4\right)$ $3x^6y$

24. $\frac{3}{4}x(8x^2y^3)$ $6x^3y^3$

25. $(3.3x^4)(1.8x^4y^3)$ $5.94x^8y^3$

26. $(2.3x^5)(4.1x^2y^4)$ $9.43x^7y^4$

Multiply.

27. $5(x + 4)$ $5x + 20$

28. $3(x - 4)$ $3x - 12$

29. $-3x(2x - 2)$ $-6x^2 + 6x$

30. $-4p(-3p + 6)$ $12p^2 - 24p$

31. $-2(8y + 5)$ $-16y - 10$

32. $2x(x^2 + 3x - 1)$ $2x^3 + 6x^2 - 2x$

33. $-2x(x^2 - 2x + 5)$ $-2x^3 + 4x^2 - 10x$

34. $-6c(-3c^2 + 5c - 6)$ $18c^3 - 30c^2 + 36c$

35. $5x(-4x^2 + 6x - 4)$ $-20x^3 + 30x^2 - 20x$

36. $(x^2 - x + 1)x$ $x^3 - x^2 + x$

37. $0.5x^2(x^3 - 6x^2 - 1)$ $0.5x^5 - 3x^4 - 0.5x^2$

38. $2.3b^2(2b^2 - b + 3)$ $4.6b^4 - 2.3b^3 + 6.9b^2$

39. $0.3x(2xy + 5x - 6y)$ $0.6x^2y + 1.5x^2 - 1.8xy$

40. $-\frac{1}{2}x^3(2x^2 + 4x - 6y^2)$ $-x^5 - 2x^4 + 3x^3y^2$

41. $(x^2 - 4y^3 - 3)y^4$ $x^2y^4 - 4y^7 - 3y^4$

42. $\frac{1}{4}y^4(y^2 - 12y + 4x)$ $\frac{1}{4}y^6 - 3y^5 + xy^4$

Multiply.

43. $(x + 3)(x + 4)$ $x^2 + 7x + 12$

44. $(2x - 3)(x + 5)$ $2x^2 + 7x - 15$

45. $(2x + 5)(3x - 6)$ $6x^2 + 3x - 30$

46. $(4a - 1)(a + 4)$ $4a^2 + 15a - 4$

47. $(2x - 4)(2x + 4)$ $4x^2 - 16$

48. $(4 + 5w)(3 + w)$ $5w^2 + 19w + 12$

49. $(5 - 3x)(6 + 2x)$ $-6x^2 - 8x + 30$

50. $(-x + 3)(2x + 5)$ $-2x^2 + x + 15$

51. $(6x - 1)(-2x + 5)$ $-12x^2 + 32x - 5$

52. $(2n - 4)(3n - 2)$ $6n^2 - 16n + 8$

53. $(x - 2)(4x - 2)$ $4x^2 - 10x + 4$

54. $(2x + 3)(x + 5)$ $2x^2 + 13x + 15$

55. $(3k - 6)(4k - 2)$ $12k^2 - 30k + 12$

56. $(3d - 5)(4d - 1)$ $12d^2 - 23d + 5$

57. $(x - 2)(x + 2)$ $x^2 - 4$

58. $(3x - 8)(2x + 3)$ $6x^2 - 7x - 24$

59. $(2x - 3)(2x - 3)$ $4x^2 - 12x + 9$

60. $(7x + 3)(2x + 4)$ $14x^2 + 34x + 12$

61. $(6z - 4)(7 - z)$ $-6z^2 + 46z - 28$

62. $(6 - 2m)(5m - 3)$ $-10m^2 + 36m - 18$

63. $(2x + 3)(4 - 2x)$ $-4x^2 + 2x + 12$

64. $(5 - 6x)(2x - 7)$ $-12x^2 + 52x - 35$

65. $(x + y)(x - y)$ $x^2 - y^2$

66. $(z + 2y)(4z - 3)$ $4z^2 - 3z + 8yz - 6y$

67. $(2x - 3y)(3x + 2y)$ $6x^2 - 5xy - 6y^2$

68. $(2x + 3)(2y - 5)$ $4xy - 10x + 6y - 15$

69. $(3x + y)(2 + 2x)$ $6x^2 + 2xy + 6x + 2y$

70. $(2x - 0.1)(x + 2.4)$ $2x^2 + 4.7x - 0.24$

71. $(x + 0.6)(x + 0.3)$ $x^2 + 0.9x + 0.18$

72. $(3x - 6)\left(x + \dfrac{1}{3}\right)$ $3x^2 - 5x - 2$

73. $(2y - 4)\left(\dfrac{1}{2}x - 1\right)$ $xy - 2x - 2y + 4$

74. $(x + 4)\left(x - \dfrac{1}{2}\right)$ $x^2 + \dfrac{7}{2}x - 2$

Multiply using a special product formula.

75. $(x + 6)(x - 6)$ $x^2 - 36$

76. $(x + 3)^2$ $x^2 + 6x + 9$

77. $(3x - 3)(3x + 3)$ $9x^2 - 9$

78. $(r - 4)(r - 4)$ $r^2 - 8r + 16$

79. $(x + y)^2$ $x^2 + 2xy + y^2$

80. $(2x - 3)(2x + 3)$ $4x^2 - 9$

81. $(x - 0.2)^2$ $x^2 - 0.4x + 0.04$

82. $(a + 3b)(a - 3b)$ $a^2 - 9b^2$

83. $(4x + 5)(4x + 5)$ $16x^2 + 40x + 25$

84. $(5x + 4)(5x - 4)$ $25x^2 - 16$

85. $(0.4x + y)^2$ $0.16x^2 + 0.8xy + y^2$

86. $\left(x - \dfrac{1}{2}y\right)^2$ $x^2 - xy + \dfrac{1}{4}y^2$

87. $(5a - 7b)(5a + 7b)$ $25a^2 - 49b^2$

88. $(4 + 3w)(4 - 3w)$ $16 - 9w^2$

89. $(-2x + 6)(-2x - 6)$ $4x^2 - 36$

90. $(-3m + 2n)(-3m - 2n)$ $9m^2 - 4n^2$

91. $(7a + 2)^2$ $49a^2 + 28a + 4$

92. $(7s - 3t)^2$ $49s^2 - 42st + 9t^2$

Multiply.

93. $(x + 4)(3x^2 + 4x - 1)$ $3x^3 + 16x^2 + 15x - 4$

94. $(4m + 3)(4m^2 - 5m + 6)$ $16m^3 - 8m^2 + 9m + 18$

95. $(3x + 2)(4x^2 - x + 5)$ $12x^3 + 5x^2 + 13x + 10$

96. $(x - 1)(3x^2 + 3x + 2)$ $3x^3 - x - 2$

97. $(-2x^2 - 4x + 1)(7x - 3)$ $-14x^3 - 22x^2 + 19x - 3$

98. $(4x^2 + 9x - 2)(x - 2)$ $4x^3 + x^2 - 20x + 4$

99. $(-3a + 5)(2a^2 + 4a - 3)$ $-6a^3 - 2a^2 + 29a - 15$

100. $(5d^2 + 1)(3d - 2)$ $15d^3 - 10d^2 + 3d - 2$

101. $(3x^2 - 2x + 4)(2x^2 + 3x + 1)$ $6x^4 + 5x^3 + 5x^2 + 10x + 4$

102. $(x^2 - 2x + 3)(x^2 - 4)$ $x^4 - 2x^3 - x^2 + 8x - 12$

103. $(x^2 - x + 3)(x^2 - 2x)$ $x^4 - 3x^3 + 5x^2 - 6x$

104. $(6x + 4)(2x^2 + 2x - 4)$ $12x^3 + 20x^2 - 16x - 16$

105. $(a + b)(a^2 - ab + b^2)$ $a^3 + b^3$

106. $(a - b)(a^2 + ab + b^2)$ $a^3 - b^3$

Determine the cube of each expression by writing the expression as the square of an expression multiplied by another expression. For example, $(x + 3y)^3 = (x + 3y)(x + 3y)^2$.

107. $(x + 2)^3$ $x^3 + 6x^2 + 12x + 8$

108. $(b - 1)^3$ $b^3 - 3b^2 + 3b - 1$

109. $(3a - 5)^3$ $27a^3 - 135a^2 + 225a - 125$

110. $(2z + 3)^3$ $8z^3 + 36z^2 + 54z + 27$

Problem Solving   **111.** no, will always be a binomial   **113.** no

111. Will the product of a monomial and a binomial ever be a trinomial? Explain your answer.

112. Will the product of a monomial and a monomial always be a monomial? Explain your answer. yes

113. Will the product of two binomials after like terms are combined always be a trinomial? Explain your answer.

114. Will the product of any polynomial and a binomial always be a polynomial? Explain. yes

Consider the multiplications in Exercises 115 and 116. Determine the exponents to be placed in the shaded areas.

115. $3x^2(2x^{\blacksquare} - 5x^{\blacksquare} + 3x^{\blacksquare}) = 6x^8 - 15x^5 + 9x^3$.

116. $4x^3(x^{\blacksquare} + 2x^{\blacksquare} - 5x^{\blacksquare}) = 4x^7 + 8x^5 - 20x^4$. 4, 2, 1

 117. Suppose that one side of a rectangle is represented as $x + 2$ and a second side is represented as $2x + 1$.

   a) Express the area of the rectangle in terms of x.

   b) Find the area if $x = 4$ feet. 54 ft²

   c) What value of x, in feet, would result in the rectangle being a square? Explain how you determined your answer. 1 ft

118. Suppose that a rectangular solid has length $x + 5$, width $3x + 4$, and height $2x - 2$ (see the figure).

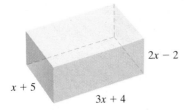

   a) Write a polynomial that represents the area of the base by multiplying the length by the width.

   b) The volume of the figure can be found by multiplying the area of the base by the height. Write a polynomial that represents the volume of the figure.

   c) Using the polynomial in part **b)**, find the volume of the figure if x is 4 feet. 864 ft³

   d) Using the binomials given for the length, width, and height, find the volume if x is 4 feet. 864 ft³

   e) Are your answers to parts **c)** and **d)** the same? If not, explain why. yes

119. Consider the figure below.

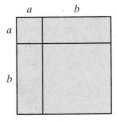

   a) Write an expression for the length of the top. $a + b$

   b) Write an expression for the length of the left side.

   c) Is this figure a square? Explain. yes

   d) Express the area of this square as the square of a binomial. $(a + b)^2$

   e) Determine the area of the square by summing the areas of the four individual pieces.

   f) Using the figure and your answer to part **e)**, complete the following.

$$(a + b)^2 = ?$$ $a^2 + 2ab + b^2$

115. 6, 3, 1 **117. a)** $(x + 2)(2x + 1)$ or $2x^2 + 5x + 2$ **118. a)** $3x^2 + 19x + 20$ **b)** $6x^3 + 32x^2 + 2x - 40$
119. **b)** $a + b$ **e)** $a^2 + 2ab + b^2$

Challenge Problems

Multiply.

120. $\left(\dfrac{1}{2}x + \dfrac{2}{3}\right)\left(\dfrac{2}{3}x - \dfrac{2}{5}\right)$ $\dfrac{1}{3}x^2 + \dfrac{11}{45}x - \dfrac{4}{15}$

121. $(2x^3 - 6x^2 + 5x - 3)(3x^3 - 6x + 4)$

  $6x^6 - 18x^5 + 3x^4 + 35x^3 - 54x^2 + 38x - 12$

Group Activity

122. Consider the trinomial $2x^2 + 7x + 3$.

   a) As a group, determine whether there is a maximum number of pairs of binomials whose product is $2x^2 + 7x + 3$. That is, how many different pairs of binomials can go in the shaded areas? yes, one

$$2x^2 + 7x + 3 = (\blacksquare)(\blacksquare)$$

   b) Individually, find a pair of binomials whose product is $2x^2 + 7x + 3$. $(x + 3)(2x + 1)$

   c) Compare your answer to part **b)** with the other members of your group. If you did not all arrive at the same answer, explain why.

Cumulative Review Exercises

[2.5] **123.** Solve the equation $4(x + 2) - 3 = 4x + 5$.
   all real numbers

[3.3] **124.** *Taxi Ride* The cost of a taxi ride is $2.00 for the first mile and $1.50 for each additional mile or part thereof. Find the maximum distance Bill Lee can ride in the taxi if he has only $20. 13 mi

[4.1] **125.** Simplify $\left(\dfrac{3xy^4}{6y^6}\right)^4$. $\dfrac{x^4}{16y^8}$

[4.1–4.2] **126.** Evaluate the following.

   a) -6^3 −216   **b)** 6^{-3} $\frac{1}{216}$

[4.4] **127.** Subtract $4x^2 - 4x - 3$ from $-x^2 - 6x + 5$.
  $-5x^2 - 2x + 8$

4.6 DIVISION OF POLYNOMIALS

SSM   Study Guide   CD/Video

MathPro 4/5   PH Math Tutor Center   prenhall.com/Angel

1 Divide a polynomial by a monomial.

2 Divide a polynomial by a binomial.

3 Check division of polynomial problems.

4 Write polynomials in descending order when dividing.

1 Divide a Polynomial by a Monomial

Now let's see how to divide polynomials. We begin by dividing a polynomial by a monomial.

> **To Divide a Polynomial by a Monomial**
>
> To divide a polynomial by a monomial, divide each term of the polynomial by the monomial.

EXAMPLE 1 Divide **a)** $\dfrac{2x + 16}{2}$ **b)** $\dfrac{10x^2 - 4x}{2x}$

Solution **a)** $\dfrac{2x + 16}{2} = \dfrac{2x}{2} + \dfrac{16}{2} = x + 8$

b) $\dfrac{10x^2 - 4x}{2x} = \dfrac{10x^2}{2x} - \dfrac{4x}{2x} = 5x - 2$

AVOIDING COMMON ERRORS

CORRECT	INCORRECT

$$\frac{x + 2}{2} = \frac{x}{2} + \frac{2}{2} = \frac{x}{2} + 1 \qquad\qquad \frac{x + \overset{1}{2}}{\underset{1}{2}} = \frac{x + 1}{1} = x + 1$$

$$\frac{x + 2}{x} = \frac{x}{x} + \frac{2}{x} = 1 + \frac{2}{x} \qquad\qquad \frac{\overset{1}{x} + 2}{\underset{1}{x}} = \frac{1 + 2}{1} = 3$$

Can you explain why the procedures on the right are not correct?

EXAMPLE 2 Divide: $\dfrac{4t^5 - 6t^4 + 8t - 3}{2t^2}$.

Solution

$$\frac{4t^5 - 6t^4 + 8t - 3}{2t^2} = \frac{4t^5}{2t^2} - \frac{6t^4}{2t^2} + \frac{8t}{2t^2} - \frac{3}{2t^2}$$

$$= 2t^3 - 3t^2 + \frac{4}{t} - \frac{3}{2t^2}$$

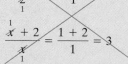

EXAMPLE 3 Divide: $\dfrac{3x^3 - 6x^2 + 4x - 1}{-3x}$.

Solution A negative sign appears in the denominator. Usually, it is easier to divide if the divisor is positive. We can multiply both numerator and denominator by -1 to get a positive denominator.

$$\frac{(-1)(3x^3 - 6x^2 + 4x - 1)}{(-1)(-3x)} = \frac{-3x^3 + 6x^2 - 4x + 1}{3x}$$

$$= \frac{-3x^3}{3x} + \frac{6x^2}{3x} - \frac{4x}{3x} + \frac{1}{3x}$$

$$= -x^2 + 2x - \frac{4}{3} + \frac{1}{3x}$$

NOW TRY EXERCISE 37

---

2 Divide a Polynomial by a Binomial

We divide a polynomial by a binomial in much the same way as we perform long division. This procedure will be explained in Example 4.

EXAMPLE 4   Divide: $\dfrac{x^2 + 6x + 8}{x + 2}$.   ← dividend   ← divisor

Solution   Rewrite the division problem in the following form:

$$x + 2 \overline{)x^2 + 6x + 8}$$

TEACHING TIP
Before explaining polynomial long division, show an example of arithmetic long division such as 4981 divided by 17.

Divide x^2 (the first term in the dividend) by x (the first term in the divisor).

$$\frac{x^2}{x} = x$$

Place the quotient, x, above the like term containing x in the dividend.

$$x + 2 \overline{)x^2 + 6x + 8} \quad \overset{x}{}$$

Next, multiply the x by $x + 2$ as you would do in long division and place the terms of the product under their like terms.

Times x

$$x + 2 \overline{)x^2 + 6x + 8}$$

Equals $x^2 + 2x$ ← $x(x + 2)$

Now subtract $x^2 + 2x$ from $x^2 + 6x$. When subtracting, remember to change the sign of the terms being subtracted and then add the like terms.

$$\begin{array}{r} x \\ x + 2 \overline{)\; x^2 + 6x + 8} \\ \underline{x^2 + 2x} \\ 4x \end{array}$$

Next, bring down the 8, the next term in the dividend.

$$\begin{array}{r} x \\ x + 2 \overline{)x^2 + 6x + 8} \\ \underline{x^2 + 2x} \\ 4x + 8 \end{array}$$

Now divide $4x$, the first term at the bottom, by x, the first term in the divisor.

$$\frac{4x}{x} = +4$$

Write the $+4$ in the quotient above the constant in the dividend.

$$
\begin{array}{r}
x + 4 \\
x + 2\overline{)\,x^2 + 6x + 8} \\
\underline{x^2 + 2x } \\
4x + 8
\end{array}
$$

Multiply the $x + 2$ by 4 and place the terms of the product under their like terms.

$$
\begin{array}{r}
\text{Times} \\
x + 4 \\
x + 2\overline{)\,x^2 + 6x + 8} \\
x^2 + 2x \\
\text{Equals} \quad 4x + 8 \\
4x + 8 \longleftarrow 4(x + 2)
\end{array}
$$

Now subtract.

$$
\begin{array}{r}
x + 4 \longleftarrow \text{Quotient} \\
x + 2\overline{)\,x^2 + 6x + 8} \\
\underline{x^2 + 2x } \\
4x + 8 \\
\underline{4x + 8 } \\
0 \longleftarrow \text{Remainder}
\end{array}
$$

Thus,

$$\frac{x^2 + 6x + 8}{x + 2} = x + 4$$

There is no remainder.

EXAMPLE 5   Divide: $\dfrac{6x^2 - 5x + 5}{2x + 3}$.

Solution

$$
\frac{6x^2}{2x} \qquad \frac{-14x}{2x}
$$

$$
\begin{array}{r}
3x - 7 \\
2x + 3\overline{)\,6x^2 - 5x + 5} \\
\underline{6x^2 + 9x} \longleftarrow 3x(2x + 3) \\
-14x + 5 \\
\underline{-14x - 21} \longleftarrow -7(2x + 3) \\
26 \longleftarrow \text{Remainder}
\end{array}
$$

When there is a remainder, as in this example, list the quotient, plus the remainder above the divisor. Thus,

NOW TRY EXERCISE 57

$$\frac{6x^2 - 5x + 5}{2x + 3} = 3x - 7 + \frac{26}{2x + 3}$$

3 Check Division of Polynomial Problems

The answer to a division problem can be checked. Consider the division problem $13 \div 5$.

$$
\begin{array}{r}
2 \\
5{\overline{\smash{\big)}\,13}} \\
\underline{10} \\
3
\end{array}
$$

Note that the divisor times the quotient, plus the remainder, equals the dividend:

$$(\text{divisor} \times \text{quotient}) + \text{remainder} = \text{dividend}$$
$$(5 \cdot 2) + 3 \stackrel{?}{=} 13$$
$$10 + 3 \stackrel{?}{=} 13$$
$$13 = 13 \qquad \textit{True}$$

This same procedure can be used to check all division problems.

To Check Division of Polynomials

$$(\text{divisor} \times \text{quotient}) + \text{remainder} = \text{dividend}$$

Let's check the answer to Example 5. The divisor is $2x + 3$, the quotient is $3x - 7$, the remainder is 26, and the dividend is $6x^2 - 5x + 5$.

Check
$$(\text{divisor} \times \text{quotient}) + \text{remainder} = \text{dividend}$$
$$(2x + 3)(3x - 7) + 26 \stackrel{?}{=} 6x^2 - 5x + 5$$
$$(6x^2 - 5x - 21) + 26 \stackrel{?}{=} 6x^2 - 5x + 5$$
$$6x^2 - 5x + 5 = 6x^2 - 5x + 5 \qquad \textit{True}$$

4 Write Polynomials in Descending Order When Dividing

When dividing a polynomial by a binomial, both the polynomial and binomial should be listed in descending order. If a given power term is missing, it is often helpful to include that term with a numerical coefficient of 0 as a placeholder. This will help in keeping like terms aligned. For example, to divide $(6x^2 + x^3 - 4)/(x - 2)$, we begin by writing $(x^3 + 6x^2 + 0x - 4)/(x - 2)$.

EXAMPLE 6 Divide $(-x + 9x^3 - 28)$ by $(3x - 4)$.

Solution First we rewrite the dividend in descending order to get $(9x^3 - x - 28) \div (3x - 4)$. Since there is no x^2 term in the dividend, we will add $0x^2$ to help align like terms.

$$\frac{9x^3}{3x} \quad \frac{12x^2}{3x} \quad \frac{15x}{3x}$$

$$\begin{array}{r} 3x^2 + 4x + 5 \\ 3x - 4\overline{)9x^3 + 0x^2 - x - 28} \\ \underline{9x^3 - 12x^2} \quad\quad\quad\quad 3x^2(3x-4)\\ 12x^2 - x \\ \underline{12x^2 - 16x} \quad\quad\quad 4x(3x-4)\\ 15x - 28 \\ \underline{15x - 20} \quad\quad 5(3x-4)\\ -8 \quad\quad Remainder \end{array}$$

Thus, $\dfrac{-x + 9x^3 - 28}{3x - 4} = 3x^2 + 4x + 5 - \dfrac{8}{3x - 4}$. Check this division yourself

NOW TRY EXERCISE 61 using the procedure just discussed.

Exercise Set 4.6

1. Divide each term in the polynomial by the monomial. **2.** (divisor)(quotient) + remainder
= polynomial **5.** in descending order **10.** $(x^2 - x + 3)(x + 1) - 6 = x^3 + 2x - 3$

Concept/Writing Exercises

1. Explain how to divide a polynomial by a monomial.

2. Explain how to check a division problem.

3. Explain why $\dfrac{y + 5}{y} \neq \dfrac{1 + 5}{1}$. Then, correctly divide the binomial by the monomial. $1 + \dfrac{5}{y}$

4. Explain why $\dfrac{2x + 8}{2} \neq \dfrac{x + 8}{1}$. Then, correctly divide the binomial by the monomial. $x + 4$

5. How should the terms of a polynomial and binomial be listed when dividing a polynomial by a binomial?

6. How would you rewrite $\dfrac{x^2 - 7}{x - 2}$ so that it is easier to complete the division? $\dfrac{x^2 + 0x - 7}{x - 2}$

7. How would you rewrite $\dfrac{x^3 - 14x + 15}{x - 3}$ so that it is easier to complete the division? $\dfrac{x^3 + 0x^2 - 14x + 15}{x - 3}$

8. Show that $\dfrac{x^2 - 3x + 7}{x + 2} = x - 5 + \dfrac{17}{x + 2}$ by checking the division. $(x - 5)(x + 2) + 17 = x^2 - 3x + 7$

9. Show that $\dfrac{x^2 + 2x - 17}{x - 3} = x + 5 - \dfrac{2}{x - 3}$ by checking the division. $(x + 5)(x - 3) - 2 = x^2 + 2x - 17$

10. Show that $\dfrac{x^3 + 2x - 3}{x + 1} = x^2 - x + 3 - \dfrac{6}{x + 1}$ by checking the division.

Rewrite each multiplication problem as a division problem. There is more than one correct answer.

11. $(x - 4)(x + 5) = x^2 + x - 20$

$\dfrac{x^2 + x - 20}{x - 4} = x + 5$ or $\dfrac{x^2 + x - 20}{x + 5} = x - 4$

12. $(x + 3)(3x - 1) = 3x^2 + 8x - 3$

$\dfrac{3x^2 + 8x - 3}{x + 3} = 3x - 1$ or $\dfrac{3x^2 + 8x - 3}{3x - 1} = x + 3$

13. $(2x + 3)(x + 1) = 2x^2 + 5x + 3$

$\dfrac{2x^2 + 5x + 3}{2x + 3} = x + 1$ or $\dfrac{2x^2 + 5x + 3}{x + 1} = 2x + 3$

14. $(2x - 3)(x + 4) = 2x^2 + 5x - 12$

$\dfrac{2x^2 + 5x - 12}{2x - 3} = x + 4$ or $\dfrac{2x^2 + 5x - 12}{x + 4} = 2x - 3$

15. $(2x + 3)(2x - 3) = 4x^2 - 9$

$\dfrac{4x^2 - 9}{2x + 3} = 2x - 3$ or $\dfrac{4x^2 - 9}{2x - 3} = 2x + 3$

16. $(3n + 4)(n - 5) = 3n^2 - 11n - 20$

$\dfrac{3n^2 - 11n - 20}{3n + 4} = n - 5$ or $\dfrac{3n^2 - 11n - 20}{n - 5} = 3n + 4$

Practice the Skills

37. $3x^2 - 2x + 6 - \dfrac{5}{2x}$ 40. $4x^3 - 2x + 5 - \dfrac{4}{3x}$ 41. $-4x^3 - x^2 + \dfrac{10}{3} + \dfrac{3}{x^2}$

Divide.

17. $\dfrac{3x + 6}{3}$ $x + 2$

18. $\dfrac{4x - 6}{2}$ $2x - 3$

19. $\dfrac{4n + 10}{2}$ $2n + 5$

20. $(-3x - 8) \div 4$ $-\dfrac{3x}{4} - 2$

21. $\dfrac{3x + 8}{2}$ $\dfrac{3}{2}x + 4$

22. $\dfrac{5x - 10}{5}$ $x - 2$

23. $\dfrac{-6x + 4}{2}$ $-3x + 2$

24. $\dfrac{-5a + 4}{-3}$ $\dfrac{5}{3}a - \dfrac{4}{3}$

25. $\dfrac{-9x - 3}{-3}$ $3x + 1$

26. $\dfrac{5x - 4}{-5}$ $-x + \dfrac{4}{5}$

27. $\dfrac{2x + 16}{4}$ $\dfrac{1}{2}x + 4$

28. $\dfrac{2p - 3}{2p}$ $1 - \dfrac{3}{2p}$

29. $\dfrac{4 - 10w}{-4}$ $-1 + \dfrac{5}{2}w$

30. $\dfrac{6 - 5x}{-3x}$ $-\dfrac{2}{x} + \dfrac{5}{3}$

31. $(3x^2 + 6x - 9) \div 3x^2$ $1 + \dfrac{2}{x} - \dfrac{3}{x^2}$

32. $\dfrac{12x^2 - 6x + 3}{3}$ $4x^2 - 2x + 1$

🔒 **33.** $\dfrac{-4x^5 + 6x + 8}{2x^2}$ $-2x^3 + \dfrac{3}{x} + \dfrac{4}{x^2}$

34. $\dfrac{6t^2 + 3t + 8}{2}$ $3t^2 + \dfrac{3}{2}t + 4$

35. $(x^5 + 3x^4 - 3) \div x^3$ $x^2 + 3x - \dfrac{3}{x^3}$

36. $(6x^2 - 7x + 9) \div 3x$ $2x - \dfrac{7}{3} + \dfrac{3}{x}$

37. $\dfrac{6x^5 - 4x^4 + 12x^3 - 5x^2}{2x^3}$

38. $\dfrac{7x^2 + 14x - 5}{-7}$ $-x^2 - 2x + \dfrac{5}{7}$

39. $\dfrac{8k^3 + 6k^2 - 8}{-4k}$ $-2k^2 - \dfrac{3}{2}k + \dfrac{2}{k}$

40. $\dfrac{-12x^4 + 6x^2 - 15x + 4}{-3x}$

41. $\dfrac{12x^5 + 3x^4 - 10x^2 - 9}{-3x^2}$

42. $\dfrac{-15m^3 - 6m^2 + 15}{-5m^3}$ $3 + \dfrac{6}{5m} - \dfrac{3}{m^3}$

Divide.

🔒 **43.** $\dfrac{x^2 + 4x + 3}{x + 1}$ $x + 3$

44. $(2x^2 + 3x - 35) \div (x + 5)$ $2x - 7$

45. $\dfrac{2x^2 - 9x - 18}{x - 6}$ $2x + 3$

46. $\dfrac{2p^2 - 7p - 15}{p - 5}$ $2p + 3$

47. $\dfrac{6x^2 + 16x + 8}{3x + 2}$ $2x + 4$

48. $\dfrac{3r^2 + 5r - 8}{r - 1}$ $3r + 8$

49. $\dfrac{x^2 - 16}{-4 + x}$ $x + 4$

50. $\dfrac{6t^2 - t - 40}{2t + 5}$ $3t - 8$

51. $(2x^2 + 7x - 18) \div (2x - 3)$

52. $\dfrac{x^2 - 36}{x - 6}$ $x + 6$

53. $(4a^2 - 25) \div (2a - 5)$ $2a + 5$

54. $\dfrac{9x^2 - 16}{3x - 4}$ $3x + 4$

55. $\dfrac{6x + 8x^2 - 25}{4x + 9}$ $2x - 3 + \dfrac{2}{4x + 9}$

56. $\dfrac{10x + 3x^2 + 6}{x + 2}$ $3x + 4 - \dfrac{2}{x + 2}$

57. $\dfrac{6x + 8x^2 - 12}{2x + 3}$ $4x - 3 - \dfrac{3}{2x + 3}$

58. $\dfrac{x^3 + 5x^2 + 2x - 8}{x + 2}$ $x^2 + 3x - 4$

59. $\dfrac{3x^3 + 18x^2 - 5x - 30}{x + 6}$ $3x^2 - 5$

60. $\dfrac{2x^3 - 3x^2 - 3x + 6}{x - 1}$

61. $\dfrac{2x^3 - 4x^2 + 12}{x - 2}$ $2x^2 + \dfrac{12}{x - 2}$

62. $\dfrac{2x^3 + 6x - 4}{x + 4}$ $2x^2 - 8x + 38 - \dfrac{156}{x + 4}$

63. $(w^3 - 8) \div (w - 3)$

64. $\dfrac{x^3 + 8}{x + 2}$ $x^2 - 2x + 4$

65. $\dfrac{x^3 - 27}{x - 3}$ $x^2 + 3x + 9$

66. $\dfrac{x^3 + 27}{x + 3}$ $x^2 - 3x + 9$

🔒 **67.** $\dfrac{4x^3 - 5x}{2x - 1}$ $2x^2 + x - 2 - \dfrac{2}{2x - 1}$

68. $\dfrac{9x^3 - x + 3}{3x - 2}$ $3x^2 + 2x + 1 + \dfrac{5}{3x - 2}$

69. $\dfrac{-m^3 - 6m^2 + 2m - 3}{m - 1}$

70. $\dfrac{-x^3 + 3x^2 + 14x + 16}{x + 3}$

71. $\dfrac{9n^3 - 6n + 4}{3n - 3}$ $3n^2 + 3n + 1 + \dfrac{7}{3n - 3}$

72. $\dfrac{4t^3 - t + 4}{t + 2}$ $4t^2 - 8t + 15 - \dfrac{26}{t + 2}$

51. $x + 5 - \dfrac{3}{2x - 3}$ **60.** $2x^2 - x - 4 + \dfrac{2}{x - 1}$ **63.** $w^2 + 3w + 9 + \dfrac{19}{w - 3}$ **69.** $-m^2 - 7m - 5 - \dfrac{8}{m - 1}$ **70.** $-x^2 + 6x - 4 + \dfrac{28}{x + 3}$

Problem Solving

73. When dividing a binomial by a monomial, must the quotient be a binomial? Explain and give an example to support your answer. no

74. When dividing a trinomial by a monomial, must the quotient be a trinomial? Explain and give an example to support your answer. no

75. If the divisor is $x + 4$, the quotient is $2x + 3$, and the remainder is 4, find the dividend (or the polynomial being divided). $2x^2 + 11x + 16$

76. If the divisor is $2x - 3$, the quotient is $3x - 1$, and the remainder is -2, find the dividend. $6x^2 - 11x + 1$

77. If a second-degree polynomial in x is divided by a first-degree polynomial in x, what will be the degree of the quotient? Explain. first degree

78. If a third-degree polynomial in x is divided by a first-degree polynomial in x, what will be the degree of the quotient? Explain. second degree

Determine the expression to be placed in the shaded area to make a true statement. Explain how you determined your answer.

79. $\dfrac{16x^4 + 20x^3 - 4x^2 + 12x}{\rule{1cm}{0.3cm}} = 4x^3 + 5x^2 - x + 3$ $4x$

80. $\dfrac{9x^5 - 6x^4 + 3x^2 + 12}{\rule{1cm}{0.3cm}} = 3x^3 - 2x^2 + 1 + \dfrac{4}{x^2}$ $3x^2$

Determine the exponents to be placed in the shaded areas to make a true statement. Explain how you determined your answer.

81. $\dfrac{8x^{\blacksquare} + 4x^{\blacksquare} - 20x^{\blacksquare} - 5x^{\blacksquare}}{2x^2} = 4x^3 + 2x - 10 - \dfrac{5}{2x}$

$5, 3, 2, 1$

82. $\dfrac{15x^{\blacksquare} + 25x^{\blacksquare} + 5x^{\blacksquare} + 10x^{\blacksquare}}{5x^2} = 3x^5 + 5x^4 + x^2 + 2$

$7, 6, 4, 2$

Challenge Problems

Divide. The quotient in Exercises 83 and 84 will contain fractions. **83.** $2x^2 - 3x + \dfrac{5}{2} - \dfrac{3}{2(2x + 3)}$

83. $\dfrac{4x^3 - 4x + 6}{2x + 3}$

84. $\dfrac{3x^3 - 5}{3x - 2}$ $x^2 + \dfrac{2}{3}x + \dfrac{4}{9} - \dfrac{37}{9(3x - 2)}$

85. $\dfrac{3x^2 + 6x - 10}{-x - 3}$ $-3x + 3 + \dfrac{1}{x + 3}$

Group Activity

Discuss and answer Exercises 86 and 87 as a group. Determine the polynomial that when substituted in the shaded area results in a true statement. Explain how you determined your answer.

86. $\dfrac{\rule{1cm}{0.3cm}}{x + 4} = x + 2 + \dfrac{2}{x + 4}$ $x^2 + 6x + 10$

87. $\dfrac{\rule{1cm}{0.3cm}}{x + 3} = x + 1 - \dfrac{1}{x + 3}$ $x^2 + 4x + 2$

Cumulative Review Exercises

[1.4] 88. Consider the set of numbers

$$\left\{2, -5, 0, \sqrt{7}, \tfrac{2}{5}, -6.3, \sqrt{3}, -\tfrac{23}{34}\right\}.$$

List those that are

a) natural numbers; 2

b) whole numbers; 2, 0

c) rational numbers; $2, -5, 0, \tfrac{2}{5}, -6.3, -\tfrac{23}{34}$

d) irrational numbers; and $\sqrt{7}, \sqrt{3}$

e) real numbers. $2, -5, 0, \sqrt{7}, \tfrac{2}{5}, -6.3, \sqrt{3}, -\tfrac{23}{34}$

[1.8] 89. a) To what is $0/1$ equal? **a)** 0 **b)** undefined

b) How do we refer to an expression like $1/0$?

[1.9] 90. Give the order of operations to be followed when evaluating a mathematical expression.

[2.5] 91. Solve the equation $2(x + 3) + 2x = x + 4$. $-\tfrac{2}{3}$

[3.1] 92. Evaluate $v = \dfrac{4}{3}\pi r^3$ when $r = 6$ ≈ 904.78

[4.2] 93. Simplify $\dfrac{x^7}{x^{-3}}$. x^{10}

90. parentheses, exponents, multiplication or division (left to right), addition or subtraction (left to right)

CHAPTER SUMMARY
Key Words and Phrases

4.1	4.3	4.4	
Base	Scientific notation	Binomial	Descending order
Exponent		Degree of a polynomial	Monomial
		Degree of a term	Polynomial
			Trinomial

IMPORTANT FACTS

Rules of Exponents

1. $x^m x^n = x^{m+n}$ — **product rule**

2. $\dfrac{x^m}{x^n} = x^{m-n}, \quad x \neq 0$ — **quotient rule**

3. $x^0 = 1, \quad x \neq 0$ — **zero exponent rule**

4. $(x^m)^n = x^{m \cdot n}$ — **power rule**

5. $\left(\dfrac{ax}{by}\right)^m = \dfrac{a^m x^m}{b^m y^m}, \quad b \neq 0, y \neq 0$ — **expanded power rule**

6. $x^{-m} = \dfrac{1}{x^m}, \quad x \neq 0$ — **negative exponent rule**

7. $\left(\dfrac{a}{b}\right)^{-m} = \left(\dfrac{b}{a}\right)^m, \quad a \neq 0, b \neq 0$ — **a fraction raised to a negative exponent rule**

Product of Sum and Difference of the Same Two Terms (also called the difference of two squares):

$$(a + b)(a - b) = a^2 - b^2$$

FOIL Method to Multiply Two Binomials (**F**irst, **O**uter, **I**nner, **L**ast)

$$(a + b)(c + d)$$

Square of a Binomial

$$(a + b)^2 = a^2 + 2ab + b^2$$
$$(a - b)^2 = a^2 - 2ab + b^2$$

Chapter Review Exercises

[4.1] Simplify.

1. $x^5 \cdot x^2$ x^7
2. $x^2 \cdot x^4$ x^6
3. $3^2 \cdot 3^3$ 243
4. $2^4 \cdot 2$ 32

5. $\dfrac{x^4}{x}$ x^3
6. $\dfrac{a^5}{a^5}$ 1
7. $\dfrac{5^5}{5^3}$ 25
8. $\dfrac{2^5}{2}$ 16

9. $\dfrac{x^6}{x^8}$ $\dfrac{1}{x^2}$
10. $\dfrac{y^4}{y}$ y^3
11. x^0 1
12. $4x^0$ 4

13. $(3x)^0$ 1
14. 6^0 1
15. $(5x)^2$ $25x^2$
16. $(3a)^3$ $27a^3$

17. $(6s)^3$ $216s^3$
18. $(-3x)^3$ $-27x^3$
19. $(2x^2)^4$ $16x^8$
20. $(-x^4)^6$ x^{24}

21. $(-m^4)^5$ $-m^{20}$
22. $\left(\dfrac{2x^3}{y}\right)^2$ $\dfrac{4x^6}{y^2}$
23. $\left(\dfrac{5y^2}{2b}\right)^2$ $\dfrac{25y^4}{4b^2}$
24. $6x^2 \cdot 4x^3$ $24x^5$

25. $\dfrac{16x^2y}{4xy^2}$ $\dfrac{4x}{y}$
26. $2x(3xy^3)^2$ $18x^3y^6$
27. $\left(\dfrac{9x^2y}{3xy}\right)^2$ $9x^2$
28. $(2x^2y)^3(3xy^4)$ $24x^7y^7$

29. $4x^2y^3(2x^3y^4)^2$ $16x^8y^{11}$
30. $3c^2(2c^4d^3)$ $6c^6d^3$
31. $\left(\dfrac{8x^4y^3}{2xy^5}\right)^2$ $\dfrac{16x^6}{y^4}$
32. $\left(\dfrac{21x^4y^3}{7y^2}\right)^3$ $27x^{12}y^3$

[4.2] Simplify.

33. x^{-4} $\dfrac{1}{x^4}$
34. 3^{-3} $\dfrac{1}{27}$
35. 5^{-2} $\dfrac{1}{25}$
36. $\dfrac{1}{z^{-2}}$ z^2

37. $\dfrac{1}{x^{-7}}$ x^7
38. $\dfrac{1}{3^{-2}}$ 9
39. $y^5 \cdot y^{-8}$ $\dfrac{1}{y^3}$
40. $x^{-2} \cdot x^{-3}$ $\dfrac{1}{x^5}$

41. $p^{-6} \cdot p^4$ $\dfrac{1}{p^2}$
42. $a^{-2} \cdot a^{-3}$ $\dfrac{1}{a^5}$
43. $\dfrac{x^3}{x^{-3}}$ x^6
44. $\dfrac{x^5}{x^{-2}}$ x^7

45. $\dfrac{x^{-3}}{x^3}$ $\dfrac{1}{x^6}$
46. $(3x^4)^{-2}$ $\dfrac{1}{9x^8}$
47. $(4x^{-3}y)^{-3}$ $\dfrac{x^9}{64y^3}$
48. $(-2m^{-3}n)^2$ $\dfrac{4n^2}{m^6}$

49. $6y^{-2} \cdot 2y^4$ $12y^2$
50. $(5y^{-3}z)^3$ $\dfrac{125z^3}{y^9}$
51. $(4x^{-2}y^3)^{-2}$ $\dfrac{x^4}{16y^6}$
52. $2x(3x^{-2})$ $\dfrac{6}{x}$

53. $(5x^{-2}y)(2x^4y)$ $10x^2y^2$
54. $4x^5(6x^{-7}y^2)$ $\dfrac{24y^2}{x^2}$
55. $4y^{-2}(3x^2y)$ $\dfrac{12x^2}{y}$
56. $\dfrac{6xy^4}{2xy^{-1}}$ $3y^5$

57. $\dfrac{12x^{-2}y^3}{3xy^2}$ $\dfrac{4y}{x^3}$
58. $\dfrac{49x^2y^{-3}}{7x^{-3}y}$ $\dfrac{7x^5}{y^4}$
59. $\dfrac{36x^4y^7}{9x^5y^{-3}}$ $\dfrac{4y^{10}}{x}$
60. $\dfrac{4x^8y^{-2}}{8x^7y^3}$ $\dfrac{x}{2y^5}$

[4.3] Express each number in scientific notation.

61. 1,720,000 1.72×10^6
62. 0.153 1.53×10^{-1}
63. 0.00763 7.63×10^{-3}
64. 47,000 4.7×10^4
65. 4820 4.82×10^3
66. 0.000314 3.14×10^{-4}

Express each number without exponents.

67. 8.4×10^{-3} 0.0084
68. 6.52×10^{-4} 0.000652
69. 9.7×10^5 970,000
70. 4.38×10^{-6} 0.00000438
71. 3.14×10^{-5} 0.0000314
72. 1.103×10^7 11,030,000

Write each of the following as a base unit without metric prefixes.

73. 6 gigameters 6,000,000,000 meters
74. 92 milliliters 0.092 liter
75. 19.2 kilograms 19,200 grams
76. 12.8 micrograms 0.0000128 gram

Perform each indicated operation and write your answer without exponents.

77. $(2.5 \times 10^2)(3.4 \times 10^{-4})$ 0.085
78. $(4.2 \times 10^{-3})(3 \times 10^5)$ 1260
79. $(3.5 \times 10^{-2})(7.0 \times 10^3)$ 245

80. $\dfrac{7.94 \times 10^6}{2 \times 10^{-2}}$ 397,000,000
81. $\dfrac{6.5 \times 10^4}{2.0 \times 10^6}$ 0.0325
82. $\dfrac{15 \times 10^{-3}}{5 \times 10^2}$ 0.00003

Convert each number to scientific notation. Then calculate. Express your answer in scientific notation.

83. $(14,000)(260,000)$ 3.64×10^9

84. $(12,500)(400,000)$ 5.0×10^9

85. $(0.00053)(40,000)$ 2.12×10^1

86. $\dfrac{250}{500,000}$ 5.0×10^{-4}

87. $\dfrac{0.000068}{0.02}$ 3.4×10^{-3}

88. $\dfrac{850,000}{0.025}$ 3.4×10^7

89. ***Milk Tank*** A milk tank holds 6.4×10^6 fluid ounces of milk. If one gallon is 1.28×10^2 fluid ounces, determine the number of gallons of milk the tank holds.
50,000 gal

90. ***Currency in Circulation*** In the United States in 2001 the total currency in \$10 bills in circulation was about \$$1.38 \times 10^{10}$ and total currency in \$5 bills in circulation was about \$$8.54 \times 10^9$.

 a) How much more was in circulation in \$10 bills than in \$5 bills? Write your answer without exponents.
 $\approx \$5,260,000,000$

 b) How many times greater is the amount in circulation in \$10 bills than in \$5 bills? ≈ 1.62

[4.4] *Indicate whether each expression is a polynomial. If the polynomial has a specific name, give that name. If the polynomial is not written in descending order, rewrite it in descending order. State the degree of each polynomial.*

91. $x^{-4} - 8$ not polynomial

92. -2 monomial, 0

93. $x^2 - 4 + 3x$ $x^2 + 3x - 4$, trinomial, second

94. $-3 - x + 4x^2$ $4x^2 - x - 3$, trinomial, second

95. $13x^3 - 4$ binomial, third

96. $4x^{1/2} - 6$ not polynomial

97. $x - 4x^2$ $-4x^2 + x$, binomial, second

98. $x^3 + x^{-2} + 3$ not polynomial

99. $2x^3 - 7 + 4x^2 - 3x$ $2x^3 + 4x^2 - 3x - 7$ polynomial, third

[4.4–4.6] *Perform each indicated operation.*

100. $(x - 5) + (2x + 4)$ $3x - 1$

101. $(2d - 3) + (5d + 7)$ $7d + 4$

102. $(-x - 10) + (-2x + 5)$ $-3x - 5$

103. $(-x^2 + 6x - 7) + (-2x^2 + 4x - 8)$ $-3x^2 + 10x - 15$

104. $(-m^2 + 5m - 8) + (6m^2 - 5m - 2)$ $5m^2 - 10$

105. $(6.2p - 4.3) + (1.9p + 7.1)$ $8.1p + 2.8$

106. $(-4x + 8) - (-2x + 6)$ $-2x + 2$

107. $(4x^2 - 9x) - (3x + 15)$ $4x^2 - 12x - 15$

108. $(5a^2 - 6a - 9) - (2a^2 - a + 12)$ $3a^2 - 5a - 21$

109. $(-2x^2 + 8x - 7) - (3x^2 + 12)$ $-5x^2 + 8x - 19$

110. $(x^2 + 7x - 3) - (x^2 + 3x - 5)$ $4x + 2$

111. $\dfrac{1}{7}x(21x + 21)$ $3x^2 + 3x$

112. $-3x(5x + 4)$ $-15x^2 - 12x$

113. $3x(2x^2 - 4x + 7)$ $6x^3 - 12x^2 + 21x$

114. $-c(2c^2 - 3c + 5)$ $-2c^3 + 3c^2 - 5c$

115. $-4z(-3z^2 - 2z - 8)$ $12z^3 + 8z^2 + 32z$

116. $(x + 4)(x + 5)$ $x^2 + 9x + 20$

117. $(3x + 6)(-4x + 1)$ $-12x^2 - 21x + 6$

118. $(-2x + 6)^2$ $4x^2 - 24x + 36$

119. $(6 - 2x)(2 + 3x)$ $-6x^2 + 14x + 12$

120. $(r + 5)(r - 5)$ $r^2 - 25$

121. $(3x + 1)(x^2 + 2x + 4)$ $3x^3 + 7x^2 + 14x + 4$

122. $(x - 1)(3x^2 + 4x - 6)$ $3x^3 + x^2 - 10x + 6$

123. $(-4x + 2)(3x^2 - x + 7)$ $-12x^3 + 10x^2 - 30x + 14$

124. $\dfrac{2x + 4}{2}$ $x + 2$

125. $\dfrac{10x + 12}{2}$ $5x + 6$

126. $\dfrac{8x^2 + 4x}{x}$ $8x + 4$

127. $\dfrac{6x^2 + 9x - 4}{3}$ $2x^2 + 3x - \dfrac{4}{3}$

128. $\dfrac{6w^2 - 5w + 3}{3w}$ $2w - \dfrac{5}{3} + \dfrac{1}{w}$

129. $\dfrac{8x^5 - 4x^4 + 3x^2 - 2}{2x}$ $4x^4 - 2x^3 + \dfrac{3}{2}x - \dfrac{1}{x}$

130. $\dfrac{8m - 4}{-2}$ $-4m + 2$

131. $\dfrac{5x^2 - 6x + 15}{3x}$ $\dfrac{5}{3}x - 2 + \dfrac{5}{x}$

132. $\dfrac{5x^3 + 10x + 2}{2x^2}$ $\dfrac{5}{2}x + \dfrac{5}{x} + \dfrac{1}{x^2}$

133. $\dfrac{x^2 + x - 12}{x - 3}$ $\;x + 4$ **134.** $\dfrac{6n^2 + 19n + 3}{6n + 1}$ $\;n + 3$ **135.** $\dfrac{5x^2 + 28x - 10}{x + 6}$ $\;5x - 2 + \dfrac{2}{x + 6}$

136. $\dfrac{4x^3 + 12x^2 + x - 12}{2x + 3}$ $\;2x^2 + 3x - 4$ **137.** $\dfrac{4x^2 - 12x + 9}{2x - 3}$ $\;2x - 3$

Chapter Practice Test

Simplify each expression.

1. $5x^4 \cdot 3x^2$ $\;15x^6$ **2.** $(3xy^2)^3$ $\;27x^3y^6$ **3.** $\dfrac{12d^5}{4d}$ $\;3d^4$ **4.** $\left(\dfrac{3x^2y}{6xy^3}\right)^3$ $\;\dfrac{x^3}{8y^6}$

5. $(2x^3y^{-2})^{-2}$ $\;\dfrac{y^4}{4x^6}$ **6.** $\dfrac{30x^6y^2}{45x^{-1}y}$ $\;\dfrac{2x^7y}{3}$ **7.** $(4x^0)(3x^2)^0$ $\;4$

Convert each number to scientific notation and then determine the answer. Express your answer in scientific notation.

8. $(175{,}000)(30{,}000)$ $\;5.25 \times 10^9$ **9.** $\dfrac{0.0008}{4000}$ $\;2.0 \times 10^{-7}$

Determine whether each expression is a polynomial. If the polynomial has a specific name, give that name.

10. $4x$ $\;$monomial   **11.** $3b + 2$ $\;$binomial   **12.** $x^{-2} + 4$ $\;$not polynomial

13. Write the polynomial $-5 + 6x^3 - 2x^2 + 5x$ in descending order, and give its degree. $\;6x^3 - 2x^2 + 5x - 5$, third degree

In Exercises 14–24, perform each indicated operation.

14. $(6x - 4) + (2x^2 - 5x - 3)$ $\;2x^2 + x - 7$ **15.** $(x^2 - 4x + 7) - (3x^2 - 8x + 7)$ $\;-2x^2 + 4x$

16. $(4x^2 - 5) - (x^2 + x - 8)$ $\;3x^2 - x + 3$ **17.** $-5d(-3d + 8)$ $\;15d^2 - 40d$

18. $(4x + 7)(2x - 3)$ $\;8x^2 + 2x - 21$ **19.** $(9 - 4c)(5 + 3c)$ $\;-12c^2 + 7c + 45$

20. $(3x - 5)(2x^2 + 4x - 5)$ $\;6x^3 + 2x^2 - 35x + 25$ **21.** $\dfrac{16x^2 + 8x - 4}{4}$ $\;4x^2 + 2x - 1$

22. $\dfrac{-12x^2 - 6x + 5}{-3x}$ $\;4x + 2 - \dfrac{5}{3x}$ **23.** $\dfrac{8x^2 - 2x - 15}{2x - 3}$ $\;4x + 5$

24. $\dfrac{12x^2 + 7x - 12}{4x + 5}$ $\;3x - 2 - \dfrac{2}{4x + 5}$

25. *Half-Life* The half-life of an element is the time it takes one half the amount of a radioactive element to decay. The half-life of carbon 14 (C^{14}) is 5730 years. The half-life of uranium 238 (U^{238}) is 4.46×10^9 years.

    a) Write the half-life of C^{14} in scientific notation.

    b) How many times longer is the half-life of U^{238} than C^{14}? $\;$**a)** 5.73×10^3 $\;$**b)** $\approx 7.78 \times 10^5$

Cumulative Review Test

Take the following test and check your answers with those that appear at the end of the test. Review any questions that you answered incorrectly. The section and objective where the material was covered is indicated after the answer.

1. Evaluate $12 + 8 \div 2^2 + 3$. 17
2. Simplify $7 - (2x - 3) + 2x - 8(1 - x)$. $8x + 2$
3. Evaluate $-4x^2 + x - 7$ when $x = -2$. -25
4. Name each indicated property.
   a) $(5 + 2) + 7 = 5 + (2 + 7)$.
   b) $7 \cdot x = x \cdot 7$. commutative property, multiplication
   c) $2(y + 9) = (y + 9)2$.
5. Solve the equation $3x + 5 = 4(x - 2)$. 13
6. Solve the equation $3(x + 2) + 3x - 5 = 4x + 1$ 0
7. Solve the inequality $3x - 11 < 5x - 2$ and graph the solution on a number line.
8. Solve the equation $3x - 2 = y - 7$ for y. $y = 3x + 5$
9. Solve $7x - 3y = 21$ for y, then find the value of y when $x = 6$. $y = (7x - 21)/3; 7$
10. Simplify $(2x^4y^3)^3(5x^2y)$. $40x^{14}y^{10}$
11. Write the polynomial $-5x + 2 - 7x^2$ in descending order and give the degree. $-7x^2 - 5x + 2$, second

Perform each indicated operation.

12. $(x^2 + 4x - 3) + (2x^2 + 5x + 1)$ $3x^2 + 9x - 2$

13. $(6a^2 + 3a + 2) - (a^2 - 3a - 3)$ $5a^2 + 6a + 5$
14. $(3y - 5)(2y + 3)$ $6y^2 - y - 15$
15. $(2x - 1)(3x^2 - 5x + 2)$ $6x^3 - 13x^2 + 9x - 2$
16. $\dfrac{10d^2 + 12d - 8}{4d}$ $\dfrac{5}{2}d + 3 - \dfrac{2}{d}$
17. $\dfrac{6x^2 + 11x - 10}{3x - 2}$ $2x + 5$
18. **Chicken Soup** At LeAnn's Grocery Store, three cans of chicken soup sell for \$1.25. Find the cost of eight cans. \$3.33
19. **Rectangle** The length of a rectangle is 2 less than 3 times the width. Find the dimensions of the rectangle if its perimeter is 28 feet. $l = 10$ ft, $w = 4$ ft
20. **Average Speed** Bob Dolan drives from Jackson, Mississippi, to Tallulah, Louisiana, a distance of 60 miles. At the same time, Nick Reide starts driving from Tallulah to Jackson along the same route. If Bob and Nick meet after 0.5 hour and Nick's average speed was 7 miles per hour greater than Bob's, find the average speed of each car. Bob, 56.5 mph; Nick, 63.5 mph

4. a) associative property, addition   c) commutative property, multiplication   7. $x > -\frac{9}{2}$,

Answers to Cumulative Review Test

1. 17; [Sec. 1.9, Obj. 5]   2. $8x + 2$; [Sec. 2.1, Obj. 6]   3. -25; [Sec. 1.9, Obj. 6]   4. a) Associative property of addition; [Sec. 1.10, Obj. 2]   b) Commutative property of multiplication; [Sec. 1.10, Obj. 1]   c) Commutative property of multiplication; [Sec. 1.10, Obj. 1]   5. 13; [Sec. 2.5, Obj. 1]   6. 0; [Sec. 2.5, Obj. 1]   7. $x > -\dfrac{9}{2}$, ;

[Sec. 2.7, Obj. 1]   8. $y = 3x + 5$; [Sec. 3.1, Obj. 3]   9. $y = \dfrac{7x - 21}{3}$, 7; [Sec. 3.1, Obj. 3]   10. $40x^{14}y^{10}$; [Sec. 4.1, Obj. 3]

11. $-7x^2 - 5x + 2$, second; [Sec. 4.4, Obj. 1]   12. $3x^2 + 9x - 2$; [Sec. 4.4, Obj. 2]   13. $5a^2 + 6a + 5$; [Sec. 4.4, Obj. 3]

14. $6y^2 - y - 15$; [Sec. 4.5, Obj. 4]   15. $6x^3 - 13x^2 + 9x - 2$; [Sec. 4.5, Obj. 6]   16. $\dfrac{5}{2}d + 3 - \dfrac{2}{d}$; [Sec. 4.6, Obj. 1]

17. $2x + 5$; [Sec. 4.6, Obj. 2]   18. \$3.33; [Sec. 2.6, Obj. 3]   19. $l = 10$ feet, $w = 4$ feet; [Sec. 3.4, Obj. 1]   20. Bob, 56.5 mph; Nick, 63.5 mph; [Sec. 3.5, Obj. 2]

Chapter 5

Factoring

Have you ever had to figure out the dimensions of a rectangular garden to be fenced in, or the size of a rectangular sign to construct, or the materials needed to put in a wood floor in a rectangular room? Often decisions are influenced by the amount of material available, or by a relationship we wish to maintain between the length and width of the sides of the rectangle, or by the desired area of the rectangle. A knowledge of quadratic equations may help you in arriving at decisions. In Example 2 on page 353, we use quadratic equations to determine the dimensions of a rectangular sign that is to have a specific area.

SSM   Study Guide   CD/Video   MathPro 4/5   PH Math Tutor Center   prenhall.com/Angel

A Look Ahead

In this chapter, we introduce *factoring polynomials*. Factoring polynomials is the reverse process of multiplying polynomials. When we factor a polynomial, we rewrite it as a product of two or more factors. In Sections 5.1 through 5.4, we discuss factoring techniques for several types of polynomials. In Section 5.5, we present some special factoring formulas, which can simplify the factoring process for some polynomials. In Section 5.6, we introduce quadratic equations and use factoring as one method for solving this type of equation. In the last section we study applications of quadratic equations.

It is essential that you have a thorough understanding of factoring, especially Sections 5.3 through 5.5, to complete Chapter 6 successfully. The factoring techniques used in this chapter are used throughout Chapter 6.

5.1 FACTORING A MONOMIAL FROM A POLYNOMIAL

SSM

Study Guide

CD/Video

MathPro 4/5

PH Math
Tutor Center

prenhall.com/Angel

1   Identify factors.

2   Determine the greatest common factor of two or more numbers.

3   Determine the greatest common factor of two or more terms.

4   Factor a monomial from a polynomial.

1 Identify Factors

In Chapter 4 you learned how to multiply polynomials. In this chapter we focus on factoring, the reverse process of multiplication. In Section 4.5 we showed that $2x(3x^2 + 4) = 6x^3 + 8x$ and determine that its factors are $2x$ and $3x^2 + 4$, and write $6x^3 + 8x = 2x(3x^2 + 4)$. To **factor an expression** means to write the expression as a product of its factors. Factoring is important because it can be used to solve equations.

TEACHING TIP
This may be a good place to reinforce the difference between a factor and a term.

If $a \cdot b = c$, then a and b are said to be *factors* of c.

$\quad 3 \cdot 5 = 15$; so 3 and 5 are factors of 15.

$\quad x^3 \cdot x^4 = x^7$; so x^3 and x^4 are factors of x^7.

$\quad x(x + 2) = x^2 + 2x$; so x and $x + 2$ are factors of $x^2 + 2x$.

$\quad (x - 1)(x + 3) = x^2 + 2x - 3$; so $x - 1$ and $x + 3$ are factors of $x^2 + 2x - 3$.

A given number or expression may have many factors. Consider the number 30.

$$1 \cdot 30 = 30, \quad 2 \cdot 15 = 30, \quad 3 \cdot 10 = 30, \quad 5 \cdot 6 = 30$$

So, the positive factors of 30 are 1, 2, 3, 5, 6, 10, 15, and 30. Factors can also be negative. Since $(-1)(-30) = 30$, -1 and -30 are also factors of 30. In fact, for each factor a of an expression, $-a$ must also be a factor. Other factors of 30 are therefore $-1, -2, -3, -5, -6, -10, -15$, and -30. When asked to list the factors of an expression that contains a positive numerical coefficient with a variable, we generally list only positive factors.

EXAMPLE 1   List the factors of $6x^3$.

Solution

$$\overbrace{\quad\quad}^{factors}$$
$$1 \cdot 6x^3 = 6x^3$$
$$2 \cdot 3x^3 = 6x^3$$
$$3 \cdot 2x^3 = 6x^3$$
$$6 \cdot x^3 = 6x^3$$

$$\overbrace{\quad\quad}^{factors}$$
$$x \cdot 6x^2 = 6x^3$$
$$2x \cdot 3x^2 = 6x^3$$
$$3x \cdot 2x^2 = 6x^3$$
$$6x \cdot x^2 = 6x^3$$

The factors of $6x^3$ are $1, 2, 3, 6, x, 2x, 3x, 6x, x^2, 2x^2, 3x^2, 6x^2, x^3, 2x^3, 3x^3$, and $6x^3$. The opposite (or negative) of each of these factors is also a factor, but these opposites are generally not listed unless specifically asked for.

TEACHING TIP
Encourage students to multiply factors to check that their product is the original expression. Then have students check if
a) $(x + 3)(x + 4)$ is the factored form of $x^2 + 7x + 12$
b) $7(x + 2)$ is the factored form of $7x + 14$
c) $2(x - 2)(3x + 1)$ is the factored form of $6x^2 - 10x - 4$

Here are examples of multiplying and factoring: Notice again that factoring is the reverse process of multiplying.

Multiplying	Factoring
$3(2x + 5) = 6x + 15$	$6x + 15 = 3(2x + 5)$
$5x(x + 4) = 5x^2 + 20x$	$5x^2 + 20x = 5x(x + 4)$
$(x + 1)(x + 3) = x^2 + 4x + 3$	$x^2 + 4x + 3 = (x + 1)(x + 3)$

2 Determine the Greatest Common Factor of Two or More Numbers

To factor a monomial from a polynomial, we make use of the *greatest common factor (GCF)*. If after studying the following material you wish to see additional material on obtaining the GCF, you may read Appendix B, where one of the topics discussed is finding the GCF.

Recall from Section 1.3 that the **greatest common factor** of two or more numbers is the greatest number that divides into all the numbers. The greatest common factor of the numbers 6 and 8 is 2. Two is the greatest number that divides into both 6 and 8. What is the GCF of 48 and 60? When the GCF of two or more numbers is not easily found, we can find it by writing each number as a product of prime numbers. A **prime number** is an integer greater than 1 that has exactly two factors, itself and one. The first 15 prime numbers are

$$2, 3, 5, 7, 11, 13, 17, 19, 23, 29, 31, 37, 41, 43, 47$$

A positive integer (other than 1) that is not prime is called **composite**. The number 1 is neither prime nor composite, it is called a **unit**. The first 15 composite numbers are

$$4, 6, 8, 9, 10, 12, 14, 15, 16, 18, 20, 21, 22, 24, 25$$

Every even number greater than 2 is a composite number since it has more than two factors, itself, 1, and 2.

To write a number as a product of prime numbers, follow the procedure illustrated in Examples 2 and 3.

EXAMPLE 2 Write 48 as a product of prime numbers.

Solution Select any two numbers whose product is 48. Two possibilities are $6 \cdot 8$ and $4 \cdot 12$, but there are other choices. Continue breaking down the factors until all the factors are prime, as illustrated in Figure 5.1.

Note that no matter how you select your initial factors,

$$48 = 2 \cdot 2 \cdot 2 \cdot 2 \cdot 3 = 2^4 \cdot 3$$

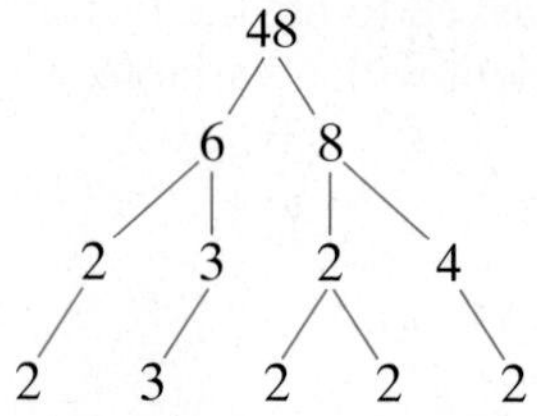

FIGURE 5.1

In Example 2 we found that $48 = 2 \cdot 2 \cdot 2 \cdot 2 \cdot 3 = 2^4 \cdot 3$. The $2 \cdot 2 \cdot 2 \cdot 2 \cdot 3$ or $2^4 \cdot 3$ may also be referred to as **prime factorizations** of 48.

EXAMPLE 3 Write 60 as a product of its prime factors.

Solution One way to find the prime factors is shown in Figure 5.2. Therefore, $60 = 2 \cdot 2 \cdot 3 \cdot 5 = 2^2 \cdot 3 \cdot 5$.

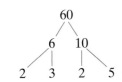

FIGURE 5.2

In the following box, we give the procedure to determine the greatest common factor of two or more numbers.

> ### To Determine the GCF of Two or More Numbers
>
> **1.** Write each number as a product of prime factors.
> **2.** Determine the prime factors common to all the numbers.
> **3.** Multiply the common factors found in step 2. The product of these factors is the GCF.

EXAMPLE 4 Determine the greatest common factor of 48 and 60.

Solution From Examples 2 and 3, we know that

Step 1
$$48 = 2 \cdot 2 \cdot 2 \cdot 2 \cdot 3 = 2^4 \cdot 3$$
$$60 = 2 \cdot 2 \cdot 3 \cdot 5 = 2^2 \cdot 3 \cdot 5$$

Step 2 Two factors of 2 and one factor of 3 are common to both numbers. The product of these factors is the GCF of 48 and 60:

Step 3
$$GCF = 2 \cdot 2 \cdot 3 = 12$$

The GCF of 48 and 60 is 12. Twelve is the greatest number that divides into both 48 and 60.

TEACHING TIP
After discussing Example 4, have students rewrite 48 and 60 as products using their GCF as one factor.

EXAMPLE 5 Determine the GCF of 18 and 24.

Solution
$$18 = 2 \cdot 3 \cdot 3 = 2 \cdot 3^2$$
$$24 = 2 \cdot 2 \cdot 2 \cdot 3 = 2^3 \cdot 3$$

One factor of 2 and one factor of 3 are common to both 18 and 24.

$$GCF = 2 \cdot 3 = 6$$

NOW TRY EXERCISE 15

3 Determine the Greatest Common Factor of Two or More Terms

The GCF of several terms containing variables is easily found. Consider the terms x^3, x^4, x^5, and x^6. The GCF of these terms is x^3, since x^3 is the largest number of x's common to all four terms. We can illustrate this by writing the terms in factored form, with x^3 as one factor.

$$x^3 = x^3 \cdot 1$$
$$x^4 = x^3 \cdot x$$
$$x^5 = x^3 \cdot x^2$$
$$x^6 = x^3 \cdot x^3$$

GCF of all four terms is x^3.

Notice that x^3 divides all four terms,

$$\frac{x^3}{x^3} = 1, \quad \text{and} \quad \frac{x^4}{x^3} = x, \quad \text{and} \quad \frac{x^5}{x^3} = x^2, \quad \text{and} \quad \frac{x^6}{x^3} = x^3.$$

EXAMPLE 6 Determine the GCF of the terms n^8, n^4, n^6, and n^7.

Solution The GCF is n^4 because n^4 is the largest number of n's common to all the terms. ✳

EXAMPLE 7 Determine the GCF of the terms x^2y^3, x^3y^2 and xy^4.

Solution The largest number of x's common to all three terms is x^1 or x. The largest number of y's common to all three terms is y^2. So the GCF of the three terms is xy^2. ✳

TEACHING TIP
After discussing Example 7, have students rewrite each expression as a product using their GCF as one factor.

To Determine the Greatest Common Factor of Two or More Terms

To determine the GCF of two or more terms, take each factor the *largest* number of times that it appears in all of the terms.

EXAMPLE 8 Determine the GCF of the terms xy, x^2y^2, and x^3.

Solution The GCF is x. The largest power of x that is common to all three terms is x^1, or x. Since the term x^3 does not contain a power of y, the GCF does not contain y. ✳

EXAMPLE 9 Determine the GCF of each group of terms.
a) $18y^2, 15y^3, 27y^5$ **b)** $-20x^2, 12x, 40x^3$ **c)** $5s^4, s^7, s^3$

Solution **a)** The GCF of 18, 15, and 27 is 3. The GCF of y^2, y^3, and y^5 is y^2. Therefore, the GCF of the three terms is $3y^2$.

b) The GCF of -20, 12, and 40 is 4. The GCF of x^2, x, and x^3 is x. Therefore, the GCF of the three terms is $4x$.

c) The GCF of 5, 1, and 1 is 1. The GCF of s^4, s^7, and s^3 is s^3. Therefore, the GCF

NOW TRY EXERCISE 33 of the three terms is $1s^3$, which we write as s^3. ✳

EXAMPLE 10 Determine the GCF of each pair of terms.
a) $x(x + 3)$ and $2(x + 3)$ **b)** $x(x - 2)$ and $x - 2$
c) $2(x + y)$ and $3x(x + y)$

Solution **a)** The GCF is $(x + 3)$.

b) $x - 2$ can be written as $1(x - 2)$. Therefore, the GCF of $x(x - 2)$ and $1(x - 2)$ is $x - 2$.

NOW TRY EXERCISE 39 **c)** The GCF is $(x + y)$. ✳

4 Factor a Monomial from a Polynomial

In section 4.5 we multiplied factors. Factoring is the reverse process of multiplying factors. As mentioned earlier, to *factor an expression* means to write the expression as a product of its factors.

To Factor a Monomial from a Polynomial
1. Determine the greatest common factor of all terms in the polynomial.
2. Write each term as the product of the GCF and its other factor.
3. Use the distributive property to factor out the GCF.

In step 3 of the process, we indicate that we use the distributive property. The distributive property is actually used in reverse. For example, if we have $4 \cdot x + 4 \cdot 2$, we use the distributive property in reverse to write $4(x + 2)$.

EXAMPLE 11 Factor $6x + 18$.

Solution The GCF is 6.

$$6x + 18 = 6 \cdot x + 6 \cdot 3 \quad \text{Write each term as a product of the GCF and some other factor.}$$
$$= 6(x + 3) \quad \text{Distributive property}$$

To check the factoring process, multiply the factors using the distributive property. If the factoring is correct, the product will be the polynomial you started with. Following is a check of the factoring in Example 11.

Check $\qquad 6(x + 3) = 6x + 18$

EXAMPLE 12 Factor $15x - 20$.

Solution The GCF is 5.

$$15x - 20 = 5 \cdot 3x - 5 \cdot 4$$
$$= 5(3x - 4)$$

Check that the factoring is correct by multiplying.

EXAMPLE 13 Factor $6y^2 + 9y^5$.

Solution The GCF is $3y^2$.

$$6y^2 + 9y^5 = 3y^2 \cdot 2 + 3y^2 \cdot 3y^3$$
$$= 3y^2(2 + 3y^3)$$

Check to see that the factoring is correct.

EXAMPLE 14 Factor $12p^3 - 24p^2 + 8p$

Solution The GCF is $4p$

$$12p^3 - 24p^2 + 8p = 4p \cdot 3p^2 - 4p \cdot 6p + 4p \cdot 2$$
$$= 4p(3p^2 - 6p + 2)$$

Check $4p(3p^2 - 6p + 2) = 12p^3 - 24p^2 + 8p$

EXAMPLE 15 Factor $35x^2 - 25x + 5$.

Solution The GCF is 5.

$$35x^2 - 25x + 5 = 5 \cdot 7x^2 - 5 \cdot 5x + 5 \cdot 1$$
$$= 5(7x^2 - 5x + 1)$$

NOW TRY EXERCISE 85

EXAMPLE 16 Factor $4x^3 + x^2 + 8x^2y$.

Solution The GCF is x^2.

$$4x^3 + x^2 + 8x^2y = x^2 \cdot 4x + x^2 \cdot 1 + x^2 \cdot 8y$$
$$= x^2(4x + 1 + 8y)$$ ✳

Notice in Examples 15 and 16 that when one of the terms is itself the GCF, we express it in factored form as the product of the term itself and 1.

EXAMPLE 17 Factor $x(5x - 2) + 7(5x - 2)$.

TEACHING TIP **Solution** The GCF of $x(5x - 2)$ and $7(5x - 2)$ is $(5x - 2)$. Factoring out the GCF gives

Before discussing Example 17,
have students factor $xy + 7y$.
Then ask, "How is the expression
in Example 17 related to this
expression?"

$$x(5x - 2) + 7(5x - 2) = (5x - 2)(x + 7)$$

Check that the factoring is correct by multiplying.   ✳

EXAMPLE 18 Factor $4x(3x - 5) - 7(3x - 5)$

Solution The GCF of $4x(3x - 5)$ and $-7(3x - 5)$ is $(3x - 5)$. Factoring out the GCF gives

$$4x(3x - 5) - 7(3x - 5) = (3x - 5)(4x - 7)$$

Recall from Section 1.10 that the commutative property of multiplication states that the order in which any two real numbers are multiplied does not matter. Therefore, $(3x - 5)(4x - 7)$ can also be written $(4x - 7)(3x - 5)$. In the book we will place

NOW TRY EXERCISE 91   the common factor on the left.   ✳

EXAMPLE 19 Factor $2x(x + 3) - 5(x + 3)$

Solution The GCF of $2x(x + 3)$ and $-5(x + 3)$ is $(x + 3)$. Factoring out the GCF gives

$$2x(x + 3) - 5(x + 3) = (x + 3)(2x - 5)$$ ✳

TEACHING TIP
After discussing Example 19, have
students verify that evaluating the
original expression for $x = 11$
gives the same result as evaluating
the factored form for $x = 11$.

Important: Whenever you are factoring a polynomial by any of the methods presented in this chapter, the first step will always be to see if there is a common factor (other than 1) to all the terms in the polynomial. If so, factor the greatest common factor from each term using the distributive property.

HELPFUL HINT

Checking a Factoring Problem
Every factoring problem may be checked by multiplying the factors. The product of the factors should be identical to the expression that was originally factored. You should check all factoring problems.

Exercise Set 5.1

Concept/Writing Exercises

1. an integer greater than 1 that has exactly two factors, itself and 1   2. a positive integer other than 1 that is not prime   3. to write an expression as the product of factors   5. the greatest number that divides into all the numbers   7. by multiplying the factors

1. What is a prime number?

2. What is a composite number?

3. What does it mean to factor an expression?

4. Is the number 1 a prime number? If not, what is it called? no; a unit

5. What is the greatest common factor of two or more numbers?

6. In your own words, explain how to factor a monomial from a polynomial.   Answers will vary.

7. How may any factoring problem be checked?

8. One student factored $5x^2 + 15x + 5$ and wrote $5(x^2 + 3x + 1)$ as the answer. A second student factored $5x^2 + 15x + 5$ and wrote $5(x^2 + 3x)$ as the answer. Which student was correct? Why was the answer given by the other student incorrect? *first student was correct*

Practice the Skills

Write each number as a product of prime numbers.

9. 56 $2^3 \cdot 7$

10. 120 $2^3 \cdot 3 \cdot 5$

11. 90 $2 \cdot 3^2 \cdot 5$

12. 540 $2^2 \cdot 3^3 \cdot 5$

13. 196 $2^2 \cdot 7^2$

14. 96 $2^5 \cdot 3$

Determine the greatest common factor for each pair of numbers.

15. 20, 24 4

16. 45, 27 9

17. 70, 98 14

18. 120, 96 24

19. 80, 126 2

20. 72, 140 4

Determine the greatest common factor for each group of terms.

21. x^5, x, x^2 x

22. y^3, y^5, y^2 y^2

23. $3x, 6x^2, 9x^3$ $3x$

24. $6p, 4p^2, 8p^3$ $2p$

25. x, y, z 1

26. a, ab, ab^2 a

27. mn, m^2n, mn^2 mn

28. $4x^2y^2, 3xy^4, 2xy^2$ xy^2

29. $x^3y^7, x^7y^{12}, x^5y^5$ x^3y^5

30. $6x, 12y, 18x^2$ 6

31. $-5, 20x, 30x^2$ 5

32. $18r^4, 6r^2s, 9s^2$ 3

33. $9x^3y^4, 8x^5y^4, 12x^4y^2$ x^2y^2

34. $16x^9y^{12}, 8x^5y^3, 20x^4y^2$ $4x^4y^2$

35. $40x^3, 27x, 30x^4y^2$ x

36. $6p^4q^3, 9p^2q^5, 9p^4q^2$ $3p^2q^2$

37. $5(x+3), 3(x+3)$ $x+3$

38. $4(x-5), 3x(x-5)$ $x-5$

39. $x^2(2x-3), 5(2x-3)$ $2x-3$

40. $x(9x-3), 9x-3$ $9x-3$

41. $3w+5, 6(3w+5)$ $3w+5$

42. $x(x+7), x+7$ $x+7$

43. $x-4, y(x-4)$ $x-4$

44. $3y(x+2), 3(x+2)$ $3(x+2)$

45. $3(x-1), 5(x-1)^2$ $x-1$

46. $5(n+2), 7(n+2)^2$ $n+2$

47. $(x+3)(x-2), (x+3)(x-4)$ $x+3$

48. $(a+4)(a-3), 5(a-3)$ $a-3$

Factor the GCF from each term in the expression.

49. $4x - 8$ $4(x-2)$

50. $4x + 2$ $2(2x+1)$

51. $15x - 5$ $5(3x-1)$

52. $12x + 15$ $3(4x+5)$

53. $6p + 12$ $6(p+2)$

54. $3t^2 - 10t$ $t(3t-10)$

55. $9x^2 - 12x$ $3x(3x-4)$

56. $24y - 6y^2$ $6y(4-y)$

57. $26p^2 - 8p$ $2p(13p-4)$

58. $8x + 16x^2$ $8x(1+2x)$

59. $3x^5 - 12x^2$ $3x^2(x^3-4)$

60. $7x^5 - 9x^4$ $x^4(7x-9)$

61. $36x^{12} + 24x^8$ $12x^8(3x^4+2)$

62. $45y^{12} + 30y^{10}$ $15y^{10}(3y^2+2)$

63. $27y^{15} - 9y^3$ $9y^3(3y^{12}-1)$

64. $30w^5 + 25w^3$ $5w^3(6w^2+5)$

65. $x + 3xy^2$ $x(1+3y^2)$

66. $4x^2y - 6x$ $2x(2xy-3)$

67. $7a^4 + 3a^2$ $a^2(7a^2+3)$

68. $3x^2y + 6x^2y^2$ $3x^2y(1+2y)$

69. $16xy^2z + 4x^3y$ $4xy(4yz+x^2)$

70. $80x^5y^3z^4 - 36x^2yz^3$ $4x^2yz^3(20x^3y^2z-9)$

71. $48m^4n^2 - 16mn^2$ $16mn^2(3m^3-1)$

72. $56xy^5z^{13} - 24y^4z^2$ $8y^4z^2(7xyz^{11}-3)$

73. $25x^2yz^3 + 25x^3yz$ $25x^2yz(z^2+x)$

74. $19x^4y^{12}z^{13} - 8x^5y^3z^9$ $x^4y^3z^9(19y^9z^4-8x)$

75. $13y^5z^3 - 11xy^2z^5$ $y^2z^3(13y^3-11xz^2)$

76. $16r^4s^5t^3 - 20r^5s^4t$ $4r^4s^4t(4st^2-5r)$

77. $8c^2 - 4c - 32$ $4(2c^2-c-8)$

78. $x^3 + 6x^2 - 4x$ $x(x^2+6x-4)$

79. $9x^2 + 18x + 3$ $3(3x^2+6x+1)$

80. $4x^2 + 8x + 24$ $4(x^2+2x+6)$

81. $4x^3 - 8x^2 + 12x$ $4x(x^2-2x+3)$

82. $12a^3 - 16a^2 - 4a$ $4a(3a^2-4a-1)$

83. $35x^2 - 15y + 10$ $5(7x^2-3y+2)$

84. $5x^3 - xy^2 + x$ $x(5x^2-y^2+1)$

85. $15p^2 - 6p + 9$ $3(5p^2-2p+3)$

86. $45y^3 - 63y^2 + 27y$ $9y(5y^2-7y+3)$

87. $9a^4 - 6a^3 + 3ab$ $3a(3a^3-2a^2+b)$

88. $40a^3b^2 + 36a^2c - 12ab^3$ $4a(10a^2b^2+9ac-3b^3)$

89. $8x^2y + 12xy^2 + 5xy$ $xy(8x+12y+5)$

90. $52x^2y^2 + 16xy^3 + 26z$ $2(26x^2y^2+8xy^3+13z)$

91. $x(x+4) + 3(x+4)$ $(x+4)(x+3)$

92. $5x(2x-5) - 8(2x-5)$ $(2x-5)(5x-8)$

93. $3b(a-2) - 4(a-2)$ $(a-2)(3b-4)$

94. $3x(7x + 1) - 2(7x + 1)$ $(7x + 1)(3x - 2)$

96. $6n(3n - 2) - 1(3n - 2)$ $(3n - 2)(6n - 1)$

98. $3x(4x - 5) + 4x - 5$ $(4x - 5)(3x + 1)$

100. $5t(t - 2) - 3(t - 2)$ $(t - 2)(5t - 3)$

95. $4x(2x + 1) + 1(2x + 1)$ $(2x + 1)(4x + 1)$

97. $5x(2x + 1) + 2x + 1$ $(2x + 1)(5x + 1)$

99. $4z(2z + 3) - 3(2z + 3)$ $(2z + 3)(4z - 3)$

Problem Solving

Factor each expression, if possible. Treat the unknown symbol as if it were a variable.

101. $3 ✱ + 6$ $3(✱ + 2)$

103. $35\Delta^3 - 7\Delta^2 + 14\Delta$ $7\Delta(5\Delta^2 - \Delta + 2)$

102. $12\nabla - 6\nabla^2$ $6\nabla(2 - \nabla)$

104. $© + 11\Delta$ Prime

Challenge Problems

105. Factor $4x^2(x - 3)^3 - 6x(x - 3)^2 + 4(x - 3)$.

106. Factor $6x^5(2x + 7) + 4x^3(2x + 7) - 2x^2(2x + 7)$.

107. Factor $x^{7/3} + 5x^{4/3} + 2x^{1/3}$ by factoring $x^{1/3}$ from all three terms. $x^{1/3}(x^2 + 5x + 2)$

108. Factor $15x^{1/2} + 5x^{-1/2}$ by factoring out $5x^{-1/2}$.

109. Factor $x^2 + 2x + 3x + 6$. (*Hint:* Factor the first two terms, then factor the last two terms, then factor the resulting two terms. We will discuss factoring problems of this type in Section 5.2.) $(x + 2)(x + 3)$

Cumulative Review Exercises

[2.1] **110.** Simplify $2x - (x - 5) + 4(3 - x)$. $-3x + 17$

[2.5] **111.** Solve the equation
$$4 + 3(x - 8) = x - 4(x + 2).$$ 2

[3.1] **112.** Solve the equation $4x - 5y = 20$ for y.

113. Find the volume of the cone shown below.

[3.3] **114.** The sum of two numbers is 41. Find the two numbers if the larger number is one less than twice the smaller number. 14, 27

[4.1] **115.** Simplify $\left(\dfrac{3x^2y^3}{2x^5y^2}\right)^2 \cdot \dfrac{9y^2}{4x^6}$

≈ 201.06 in.^{3}

105. $2(x - 3)[2x^2(x - 3)^2 - 3x(x - 3) + 2]$

106. $2x^2(2x + 7)(3x^3 + 2x - 1)$

108. $5x^{-1/2}(3x + 1)$

112. $y = \dfrac{4x - 20}{5}$ or $y = \dfrac{4}{5}x - 4$.

5.2 FACTORING BY GROUPING

1 Factor a polynomial containing four terms by grouping.

SSM   Study Guide   CD/Video

MathPro 4/5   PH Math   prenhall.com/Angel
              Tutor Center

1 Factor a Polynomial Containing Four Terms by Grouping

It may be possible to factor a polynomial containing four or more terms by removing common factors from groups of terms. This process is called **factoring by grouping**. In Sections 5.3 and 5.4 we discuss factoring trinomials. One of the methods we will use in Section 5.4 requires a knowledge of factoring by grouping. Example 1 illustrates the procedure for factoring by grouping.

EXAMPLE 1 Factor $ax + ay + bx + by$.

Solution There is no factor (other than 1) common to all four terms. However, a is common to the first two terms and b is common to the last two terms. Factor a from the first two terms and b from the last two terms.

$$a\,x + a\,y + b\,x + b\,y = a\,(x + y) + b\,(x + y)$$

This factoring gives two terms, and $(x + y)$ is common to both terms. Proceed to factor $(x + y)$ from each term, as shown below.

$$a(x + y) + b(x + y) = (x + y)(a + b)$$

Notice that when $(x + y)$ is factored out we are left with $a + b$, which becomes the other factor. Thus, $ax + ay + bx + by = (x + y)(a + b)$.　✳

TEACHING TIP
After discussing Example 1, point out that the expression could also be written as $ax + bx + ay + by$. Then have them factor this expression by grouping.

To Factor a Four-Term Polynomial Using Grouping

1. Determine whether there are any factors common to all four terms. If so, factor the greatest common factor from each of the four terms.

2. If necessary, arrange the four terms so that the first two terms have a common factor and the last two have a common factor.

3. Use the distributive property to factor each group of two terms.

4. Factor the greatest common factor from the results of step 3.

EXAMPLE 2 Factor $x^2 + 3x + 4x + 12$ by grouping.

Solution No factor is common to all four terms. However, you can factor x from the first two terms and 4 from the last two terms.

$$x^2 + 3x + 4x + 12 = x(x + 3) + 4(x + 3)$$

Notice that the expression on the right of the equal sign has two *terms* and that the *factor* $(x + 3)$ is common to both terms. Factor out the $(x + 3)$ using the distributive property.

$$x(x + 3) + 4(x + 3) = (x + 3)(x + 4)$$

Thus, $x^2 + 3x + 4x + 12 = (x + 3)(x + 4)$　✳

TEACHING TIP
After discussing Example 2, point out that a 4-term polynomial, which can be factored by grouping, can be created by multiplying 2 binomials without combining like terms. Have students create this kind of polynomial and give their polynomial to another student to factor.

NOW TRY EXERCISE 11

In Example 2, the $3x$ and $4x$ are like terms and may be combined. However, since we are explaining how to factor four terms by grouping we will not combine them. Some four-term polynomials, such as in Example 9, have no like terms that can be combined. When we discuss factoring trinomials in Section 5.4, we will sometimes start with a trinomial and rewrite it using four terms. For example, we may start with a trinomial like $2x^2 + 11x + 12$ and rewrite it as $2x^2 + 8x + 3x + 12$. We then factor the resulting four terms by grouping. This is one method that can be used to factor trinomials, as will be explained later.

EXAMPLE 3 Factor $15x^2 + 10x + 12x + 8$ by grouping.

Solution $15x^2 + 10x + 12x + 8 = 5x(3x + 2) + 4(3x + 2)$　　*Factor $5x$ from the first two terms and 4 from the last two terms.*

$$= (3x + 2)(5x + 4)$$　✳

A factoring by grouping problem can be checked by multiplying the factors using the FOIL method. If you have not made a mistake, your result will be the polynomial you began with. Here is a check of Example 3.

Check

$$
\begin{array}{cccc}
& F & O & I & L \\
(3x + 2)(5x + 4) = (3x)(5x) + (3x)(4) + (2)(5x) + (2)(4) \\
= 15x^2 + 12x + 10x + 8 \\
= 15x^2 + 10x + 12x + 8
\end{array}
$$

We are able to write $12x + 10x$ as $10x + 12x$ because of the commutative property of addition. Since this is the polynomial we started with, the factoring is correct.

HELPFUL HINT

In Example 3 when we factored $15x^2 + 10x + 12x + 8$ we obtained

$$5x(3x + 2) + 4(3x + 2)$$

When we factored out the $(3x + 2)$, to be consistent with the way we factored out common factors in Section 5.1, we placed the common factor $(3x + 2)$ on the left. That gave $(3x + 2)(5x + 4)$. We could have just as well placed the common factor on the right to obtain $(5x + 4)(3x + 2)$.

Both answers are correct since $(3x + 2)(5x + 4) = (5x + 4)(3x + 2)$ by the commutative property of multiplication.

EXAMPLE 4   Factor $15x^2 + 12x + 10x + 8$ by grouping.

Solution $15x^2 + 12x + 10x + 8 = 3x(5x + 4) + 2(5x + 4)$
$$= (5x + 4)(3x + 2)$$ ✳

Notice that Example 4 is the same as Example 3 with the two middle terms interchanged. The answers to Examples 3 and 4 are equivalent since only the order of the factors are changed. When factoring by grouping, if the two middle terms are like terms, the two like terms may be interchanged and the answer will remain the same.

EXAMPLE 5   Factor $x^2 + 4x + x + 4$ by grouping.

Solution   In the first two terms, x is the common factor. Is there a common factor in the last two terms? Yes; remember that 1 is a factor of every term. Factor 1 from the last two terms.

$$
\begin{aligned}
x^2 + 4x + x + 4 &= x^2 + 4x + 1 \cdot x + 1 \cdot 4 \\
&= x(x + 4) + 1(x + 4) \\
&= (x + 4)(x + 1)
\end{aligned}
$$

Note that $x + 4$ was expressed as $1 \cdot x + 1 \cdot 4 = 1(x + 4)$.   ✳

EXAMPLE 6   Factor $6x^2 - 3x - 2x + 1$ by grouping.

Solution   When $3x$ is factored from the first two terms, we get

$$6x^2 - 3x - 2x + 1 = 3x(2x - 1) - 2x + 1$$

What should we factor from the last two terms? We wish to factor $-2x + 1$ in such a manner that we end up with an expression that is a multiple of $(2x - 1)$.

Whenever we wish to change the sign *of each term of an expression, we can factor out a negative number from each term.* In this case, we factor out -1.

$$-2x + 1 = -1(2x - 1)$$

Now, we rewrite $-2x + 1$ as $-1(2x - 1)$.

$$3x(2x - 1) - 2x + 1 = 3x(2x - 1) - 1(2x - 1)$$

Now we factor out the common factor $(2x - 1)$.

$$3x(2x - 1) - 1(2x - 1) = (2x - 1)(3x - 1) \quad ❋$$

EXAMPLE 7 Factor $x^2 + 5x - x - 5$ by grouping.

Solution
$$
\begin{aligned}
x^2 + 5x - x - 5 &= x(x + 5) - x - 5 && \text{Factor out } x. \\
&= x(x + 5) - 1(x + 5) && \text{Factor out } -1. \\
&= (x + 5)(x - 1) && \text{Factor out } (x + 5).
\end{aligned}
$$

NOW TRY EXERCISE 19   Note that we factored -1 from $-x - 5$ to get $-1(x + 5)$. ❋

EXAMPLE 8 Factor $3x^2 - 6x - 4x + 8$ by grouping.

Solution
$$
\begin{aligned}
3x^2 - 6x - 4x + 8 &= 3x(x - 2) - 4(x - 2) \\
&= (x - 2)(3x - 4)
\end{aligned}
$$

Note: $-4x + 8 = -4(x - 2)$. ❋

HELPFUL HINT

When factoring four terms by grouping, if the coefficient of the third term is positive, as in Examples 2 through 5, you will generally factor out a positive coefficient from the last two terms. *If the coefficient of the third term is negative,* as in Examples 6 through 8, *you will generally factor out a negative coefficient from the last two terms.* The sign of the coefficient of the third term in the expression *must be included* so that the factoring results in two terms. For example,

$$2x^2 + 8x + 3x + 12 = 2x(x + 4) + 3(x + 4) = (x + 4)(2x + 3)$$
$$3x^2 - 15x - 2x + 10 = 3x(x - 5) - 2(x - 5) = (x - 5)(3x - 2)$$

In the examples illustrated so far, the two middle terms have been like terms. This need not be the case, as illustrated in Example 9.

EXAMPLE 9 Factor $xy + 3x - 2y - 6$ by grouping.

Solution This problem contains two variables, x and y. The procedure to factor here is basically the same as before. Factor x from the first two terms and -2 from the last two terms.

$$
\begin{aligned}
xy + 3x - 2y - 6 &= x(y + 3) - 2(y + 3) \\
&= (y + 3)(x - 2) && \text{Factor out } (y + 3). \quad ❋
\end{aligned}
$$

NOW TRY EXERCISE 41

EXAMPLE 10 Factor $2x^2 + 4xy + 3xy + 6y^2$.

Solution We will factor out $2x$ from the first two terms and $3y$ from the last two terms.

$$2x^2 + 4xy + 3xy + 6y^2 = 2x(x + 2y) + 3y(x + 2y)$$

Now we factor out the common factor $(x + 2y)$ from each term on the right.

$$2x(x + 2y) + 3y(x + 2y) = (x + 2y)(2x + 3y)$$

Check

$$(x + 2y)(2x + 3y) = (x)(2x) + (x)(3y) + (2y)(2x) + (2y)(3y)$$
$$= 2x^2 + 3xy + 4xy + 6y^2$$
$$= 2x^2 + 4xy + 3xy + 6y^2$$

If Example 10 were given as $2x^2 + 3xy + 4xy + 6y^2$, would the results be the same? Try it and see.

EXAMPLE 11 Factor $6r^2 - 9rs + 8rs - 12s^2$.

Solution Factor $3r$ from the first two terms and $4s$ from the last two terms.

$$6r^2 - 9rs + 8rs - 12s^2 = 3r(2r - 3s) + 4s(2r - 3s)$$
$$= (2r - 3s)(3r + 4s)$$

NOW TRY EXERCISE 31

EXAMPLE 12 Factor $3x^2 - 15x + 6x - 30$.

Solution *The first step in any factoring problem is to determine whether all the terms have a common factor. If so, we factor out that common factor.* In this polynomial, 3 is common to every term. Therefore, we begin by factoring out the 3.

$$3x^2 - 15x + 6x - 30 = 3(x^2 - 5x + 2x - 10)$$

Now we factor the expression in parentheses by grouping. We factor out x from the first two terms and 2 from the last two terms.

$$3(x^2 - 5x + 2x - 10) = 3[x(x - 5) + 2(x - 5)]$$
$$= 3[(x - 5)(x + 2)]$$
$$= 3(x - 5)(x + 2)$$

Thus, $3x^2 - 15x + 6x - 30 = 3(x - 5)(x + 2)$.

Exercise Set 5.2

Concept/Writing Exercises

1. What is the first step in any factoring by grouping problem? Factor out the GCF.

2. How can you check the solution to a factoring by grouping problem?

3. A polynomial of four terms is factored by grouping and the result is $(x - 2)(x + 4)$. Find a polynomial that was factored, and explain how you determined the answer. $x^2 + 4x - 2x - 8$

4. A polynomial of four terms is factored by grouping and the result is $(x - 2y)(x - 3)$. Find a polynomial that was factored, and explain how you determined the answer. $x^2 - 3x - 2xy + 6y$

5. In your own words, describe the steps you take to factor a polynomial of four terms by grouping.

6. What number when factored from each term in an expression changes the sign of each term in the original expression? -1

2. Multiply the factors to see if the product is the same as the original expression. 5. Answers will vary.

Practice the Skills

Factor by grouping.

7. $x^2 + 3x + 2x + 6$ $(x + 3)(x + 2)$

8. $x^2 + 7x + 3x + 21$ $(x + 7)(x + 3)$

9. $x^2 + 5x + 4x + 20$ $(x + 5)(x + 4)$

10. $x^2 - x + 3x - 3$ $(x - 1)(x + 3)$

11. $x^2 + 2x + 5x + 10$ $(x + 2)(x + 5)$

12. $x^2 - 6x + 5x - 30$ $(x - 6)(x + 5)$

13. $x^2 + 3x - 5x - 15$ $(x + 3)(x - 5)$

14. $r^2 - 4r + 6r - 24$ $(r - 4)(r + 6)$

15. $4b^2 - 10b + 10b - 25$ $(2b - 5)(2b + 5)$

16. $4x^2 - 6x + 6x - 9$ $(2x - 3)(2x + 3)$

17. $3x^2 + 9x + x + 3$ $(x + 3)(3x + 1)$

18. $a^2 + a + 3a + 3$ $(a + 1)(a + 3)$

19. $6x^2 + 3x - 2x - 1$ $(2x + 1)(3x - 1)$

20. $5x^2 + 20x - x - 4$ $(x + 4)(5x - 1)$

21. $8x^2 + 32x + x + 4$ $(x + 4)(8x + 1)$

22. $9w^2 - 6w - 6w + 4$ $(3w - 2)(3w - 2) = (3w - 2)^2$

23. $12t^2 - 8t - 3t + 2$ $(3t - 2)(4t - 1)$

24. $12x^2 + 42x - 10x - 35$ $(2x + 7)(6x - 5)$

25. $2x^2 - 4x - 3x + 6$ $(x - 2)(2x - 3)$

26. $35x^2 - 40x + 21x - 24$ $(7x - 8)(5x + 3)$

27. $6p^2 + 15p - 4p - 10$ $(2p + 5)(3p - 2)$

28. $10c^2 + 25c - 6c - 15$ $(2c + 5)(5c - 3)$

29. $x^2 + 2xy - 3xy - 6y^2$ $(x + 2y)(x - 3y)$

30. $x^2 - 3xy + 2xy - 6y^2$ $(x - 3y)(x + 2y)$

31. $3x^2 + 2xy - 9xy - 6y^2$ $(3x + 2y)(x - 3y)$

32. $3x^2 - 18xy + 4xy - 24y^2$ $(x - 6y)(3x + 4y)$

33. $10x^2 - 12xy - 25xy + 30y^2$ $(5x - 6y)(2x - 5y)$

34. $6a^2 - 3ab + 4ab - 2b^2$ $(2a - b)(3a + 2b)$

35. $x^2 + bx + ax + ab$ $(x + b)(x + a)$

36. $x^2 - bx - ax + ab$ $(x - b)(x - a)$

37. $xy + 5x - 3y - 15$ $(y + 5)(x - 3)$

38. $x^2 - 2x + ax - 2a$ $(x - 2)(x + a)$

39. $a^2 + 3a + ab + 3b$ $(a + 3)(a + b)$

40. $3x^2 - 15x - 2xy + 10y$ $(x - 5)(3x - 2y)$

41. $xy - x + 5y - 5$ $(y - 1)(x + 5)$

42. $y^2 - yb + ya - ab$ $(y - b)(y + a)$

43. $12 + 8y - 3x - 2xy$ $(3 + 2y)(4 - x)$

44. $3y - 9 - xy + 3x$ $(y - 3)(3 - x)$

45. $z^3 + 5z^2 + z + 5$ $(z + 5)(z^2 + 1)$

46. $x^3 - 3x^2 + 2x - 6$ $(x - 3)(x^2 + 2)$

47. $x^3 + 4x^2 - 3x - 12$ $(x + 4)(x^2 - 3)$

48. $y^3 - 3y + 2y^2 - 6$ $(y^2 - 3)(y + 2)$

49. $2x^2 - 12x + 8x - 48$ $2(x - 6)(x + 4)$

50. $3x^2 - 3x - 3x + 3$ $3(x - 1)(x - 1) = 3(x - 1)^2$

51. $4x^2 + 8x + 8x + 16$ $4(x + 2)(x + 2) = 4(x + 2)^2$

52. $2x^4 - 5x^3 - 6x^3 + 15x^2$ $x^2(2x - 5)(x - 3)$

53. $6x^3 + 9x^2 - 2x^2 - 3x$ $x(2x + 3)(3x - 1)$

54. $9x^3 + 6x^2 - 45x^2 - 30x$ $3x(3x + 2)(x - 5)$

55. $x^3 + 3x^2y - 2x^2y - 6xy^2$ $x(x + 3y)(x - 2y)$

56. $18x^2 + 27xy + 12xy + 18y^2$ $3(2x + 3y)(3x + 2y)$

Rearrange the terms so that the first two terms have a common factor and the last two terms have a common factor (other than 1). Then factor by grouping. There may be more than one way to arrange the factors. However, the answer should be equivalent regardless of the arrangement selected.

57. $5x + 3y + xy + 15$ $(y + 5)(x + 3)$

58. $3a + 6y + ay + 18$ $(a + 6)(y + 3)$

59. $6x + 5y + xy + 30$ $(x + 5)(y + 6)$

60. $ax - 10 - 5x + 2a$ $(a - 5)(x + 2)$

61. $ax + by + ay + bx$ $(a + b)(x + y)$

62. $ax - 21 - 3a + 7x$ $(a + 7)(x - 3)$

63. $cd - 12 - 4d + 3c$ $(d + 3)(c - 4)$

64. $ca - 2b + 2a - cb$ $(c + 2)(a - b)$

65. $ac - bd - ad + bc$ $(a + b)(c - d)$

66. $dc + 3c - ad - 3a$ $(c - a)(d + 3)$

Problem Solving

67. If you know that a polynomial with four terms is factorable by a specific arrangement of the terms, then will *any* arrangement of the terms be factorable by grouping? Explain, and support your answer with an example.
No. $xy + 2x + 5y + 10$ is factorable, $xy + 10 + 2x + 5y$ is not factorable in this arrangement.

Factor each expression, if possible. Treat the unknown symbol as if it were a variable.

68. $\heartsuit^2 + 3\heartsuit + 4\heartsuit + 12$ $(\heartsuit + 3)(\heartsuit + 4)$

69. $\odot^2 + 3\odot - 5\odot - 15$ $(\odot + 3)(\odot - 5)$

70. $\odot^2 + 3\odot + 2\odot + 7$ cannot be factored

Challenge Problems

*In Section 5.4 we will factor trinomials of the form $ax^2 + bx + c, a \neq 1$ using grouping. To do this we rewrite the middle term of the trinomial, bx, as a sum or difference of two terms. Then we factor the resulting polynomial of four terms by grouping. For Exercises 71–76, **a)** rewrite the trinomial as a polynomial of four terms by replacing the bx-term with the sum or difference given. **b)** Factor the polynomial of four terms. Note that the factors obtained are the factors of the trinomial.*

71. $3x^2 + 10x + 8, 10x = 6x + 4x$
    a) $3x^2 + 6x + 4x + 8$ **b)** $(x + 2)(3x + 4)$

72. $3x^2 + 10x + 8, 10x = 4x + 6x$
    a) $3x^2 + 4x + 6x + 8$ **b)** $(3x + 4)(x + 2)$

73. $2x^2 - 11x + 15, -11x = -6x - 5x$
    a) $2x^2 - 6x - 5x + 15$ **b)** $(x - 3)(2x - 5)$

74. $2x^2 - 11x + 15, -11x = -5x - 6x$
    a) $2x^2 - 5x - 6x + 15$ **b)** $(2x - 5)(x - 3)$

75. $4x^2 - 17x - 15, -17x = -20x + 3x$
    a) $4x^2 - 20x + 3x - 15$ **b)** $(x - 5)(4x + 3)$

76. $4x^2 - 17x - 15, -17x = 3x - 20x$
    a) $4x^2 + 3x - 20x - 15$ **b)** $(4x + 3)(x - 5)$

Factor each expression, if possible. Treat the unknown symbols as if they were variables.

77. $\star \odot + 3\star + 2\odot + 6$ $(\odot + 3)(\star + 2)$

78. $2\Delta^2 - 4\Delta\star - 8\Delta\star + 16\star^2$ $2(\Delta - 2\star)(\Delta - 4\star)$

Cumulative Review Exercises

[2.5] **79.** Solve $5 - 3(2x - 7) = 4(x + 5) - 6$. $\frac{6}{5}$

[3.5] **80.** **Special Mixture** Ed and Beatrice Petrie own a small grocery store near Leesport, Pennsylvania. The store carries a variety of bulk candy. To celebrate the fifth anniversary of the store's opening, the Petries decide to create a special candy mixture containing chocolate wafers and hard peppermint candies. The chocolate wafers sell for $6.25 per pound and the peppermint candies sell for $2.50 per pound. How many pounds of each type of candy will be needed to make a 50-pound mixture that will sell for $4.75 per pound?

[4.6] **81.** Divide $\dfrac{15x^3 - 6x^2 - 9x + 5}{3x}$.

    82. Divide $\dfrac{x^2 - 9}{x - 3}$. $x + 3$

See Exercise 80

80. 30 lb of chocolate wafers, 20 lb of peppermints

81. $5x^2 - 2x - 3 + \dfrac{5}{3x}$

5.3 FACTORING TRINOMIALS OF THE FORM $ax^2 + bx + c, a = 1$

SSM   Study Guide   CD/Video

MathPro 4/5 PH Math Tutor Center prenhall.com/Angel

1 Factor trinomials of the form $ax^2 + bx + c$, where $a = 1$.

2 Remove a common factor from a trinomial.

An Important Note Regarding Factoring Trinomials

Factoring trinomials is important in algebra, higher-level mathematics, physics, and other science courses. Because it is important, and also to be successful in Chapter 6, you should study and learn Sections 5.3 and 5.4 well.

(continued on the next page)

In this section we learn to factor trinomials of the form $ax^2 + bx + c$, where a, the numerical coefficient of the squared term, is 1. That is, we will be factoring trinomials of the form $x^2 + bx + c$. One example of this type of trinomial is $x^2 + 5x + 6$. Recall that x^2 means $1x^2$.

In Section 5.4 we will learn to factor trinomials of the form $ax^2 + bx + c$, where $a \neq 1$. One example of this type of trinomial is $2x^2 + 7x + 3$.

1 Factor Trinomials of the Form $ax^2 + bx + c$, where $a = 1$

Now we discuss how to factor trinomials of the form $ax^2 + bx + c$, where a, the numerical coefficient of the squared term, is 1. Examples of such trinomials are

$$x^2 + 7x + 12 \qquad\qquad x^2 - 2x - 24$$
$$a = 1, b = 7, c = 12 \qquad\qquad a = 1, b = -2, c = -24$$

Recall that factoring is the reverse process of multiplication. We can show with the FOIL method that

$$(x + 3)(x + 4) = x^2 + 7x + 12 \quad \text{and} \quad (x - 6)(x + 4) = x^2 - 2x - 24$$

Therefore, $x^2 + 7x + 12$ and $x^2 - 2x - 24$ factor as follows:

$$x^2 + 7x + 12 = (x + 3)(x + 4) \quad \text{and} \quad x^2 - 2x - 24 = (x - 6)(x + 4)$$

Notice that each of these trinomials when factored results in the product of two binomials in which the first term of each binomial is x and the second term is a number (including its sign). In general, when we factor a trinomial of the form $x^2 + bx + c$ we will get a pair of binomial factors as follows:

$$x^2 + bx + c = (x + \blacksquare)(x + \blacksquare)$$

Numbers go here.

If, for example, we find that the numbers that go in the shaded areas of the factors are 4 and -6, the factors are written $(x + 4)$ and $(x - 6)$. Notice that instead of listing the second factor as $(x + (-6))$, we list it as $(x - 6)$.

To determine the numbers to place in the shaded areas when factoring a trinomial of the form $x^2 + bx + c$, write down factors of the form $(x + \blacksquare)(x + \blacksquare)$ and then try different sets of factors of the constant, c, in the shaded areas of the parentheses. We multiply each pair of factors using the FOIL method, and continue until we find the pair whose sum of the products of the outer and inner terms is the same as the x-term in the trinomial. For example, to factor the trinomial $x^2 + 7x + 12$ we determine the possible factors of 12. Then we try each pair of factors until we obtain a pair whose product from the FOIL method contains $7x$, the same x-term as in the trinomial. This method for factoring is called **trial and error**. In Example 1, we factor $x^2 + 7x + 12$ by trial and error.

EXAMPLE 1 Factor $x^2 + 7x + 12$ by trial and error.

Solution Begin by listing the factors of 12 (see the left-hand column of the chart on page 318). Then list the possible factors of the trinomial, and the products of these factors. Finally, determine which, if any, of these products gives the correct middle term, $7x$.

Factors of 12	Possible Factors of Trinomial	Product of Factors
$(1)(12)$	$(x + 1)(x + 12)$	$x^2 + 13x + 12$
$(2)(6)$	$(x + 2)(x + 6)$	$x^2 + 8x + 12$
$(3)(4)$	$(x + 3)(x + 4)$	$x^2 + 7x + 12$
$(-1)(-12)$	$(x - 1)(x - 12)$	$x^2 - 13x + 12$
$(-2)(-6)$	$(x - 2)(x - 6)$	$x^2 - 8x + 12$
$(-3)(-4)$	$(x - 3)(x - 4)$	$x^2 - 7x + 12$

In the last column, we find the trinomial we are seeking in the third line. Thus,

NOW TRY EXERCISE 27

$$x^2 + 7x + 12 = (x + 3)(x + 4)$$

Now let's consider how we may more easily determine the correct factors of 12 to place in the shaded areas when factoring the trinomial in Example 1. In Section 4.5 we illustrated how the FOIL method is used to multiply two binomials. Let's multiply $(x + 3)(x + 4)$ using the FOIL method.

$$(x + 3)(x + 4) = x^2 + 4x + 3x + 12$$
$$= x^2 + 7x + 12$$

We see that $(x + 3)(x + 4) = x^2 + 7x + 12$.

Note that the *sum of the outer and inner terms is 7x and the product of the last terms is 12*. To factor $x^2 + 7x + 12$, we look for two numbers whose product is 12 and whose sum is 7. We list the factors of 12 first and then list the sum of the factors.

Factors of 12	Sum of Factors
$(1)(12) = 12$	$1 + 12 = 13$
$(2)(6) = 12$	$2 + 6 = 8$
$(3)(4) = 12$	$3 + 4 = 7$
$(-1)(-12) = 12$	$-1 + (-12) = -13$
$(-2)(-6) = 12$	$-2 + (-6) = -8$
$(-3)(-4) = 12$	$-3 + (-4) = -7$

The only factors of 12 whose sum is a positive 7 are 3 and 4. The factors of $x^2 + 7x + 12$ will therefore be $(x + 3)$ and $(x + 4)$.

$$x^2 + 7x + 12 = (x + 3)(x + 4)$$

In the previous illustration, all the possible factors of 12 were listed so that you could see them. However, when working a problem, once you find the specific factors you are seeking you need go no further.

To Factor Trinomials of the Form $ax^2 + bx + c$, where $a = 1$

1. Find two numbers whose product equals the constant, c, and whose sum equals the coefficient of the x-term, b.

2. Use the two numbers found in step 1, including their signs, to write the trinomial in factored form. The trinomial in factored form will be

$$(x + \text{one number})(x + \text{second number})$$

TEACHING TIP
Have students create trinomials from the product of 2 factors of the form $(x + c)(x + d)$ for various values of c and d. Have them identify relationships between the signs of c and d, and the coefficients of the simplified trinomial.

How do we find the two numbers mentioned in steps 1 and 2? The sign of the constant, c, is a key in finding the two numbers. *The Helpful Hint that follows is very important and useful. Study it carefully.*

HELPFUL HINT

When asked to factor a trinomial of the form $x^2 + bx + c$, first observe the sign of the constant.

a) If the constant, c, is positive, both numbers in the factors will have the same sign, either both positive or both negative. Furthermore, that common sign will be the same as the sign of the coefficient of the x-term of the trinomial being factored. That is, if b is positive, both factors will contain positive numbers, and if b is negative, both factors will contain negative numbers.

Example:

$$x^2 + 7x + 12 = (x + 3)(x + 4)$$

Both factors have positive numbers.

The coefficient, b, is positive   The constant, c, is positive   positive   positive

Example:

$$x^2 - 5x + 6 = (x - 2)(x - 3)$$

Both factors have negative numbers.

The coefficient, b, is negative   The constant, c, is positive   negative   negative

b) If the constant is negative, the two numbers in the factors will have opposite signs. That is, one number will be positive and the other number will be negative.

Example:

$$x^2 + x - 6 = (x + 3)(x - 2)$$

One factor has a positive number and the other factor has a negative number.

The coefficient, b, is positive   The constant, c, is negative   positive   negative

Example:

$$x^2 - 3x - 10 = (x + 2)(x - 5)$$

One factor has a positive number and the other factor has a negative number.

The coefficient, b, is negative   The constant, c, is negative   positive   negative

We will use this information as a starting point when factoring trinomials.

EXAMPLE 2   Consider a trinomial of the form $x^2 + bx + c$. Use the signs of b and c given below to determine the signs of the numbers in the factors.

a) b is negative and c is positive. b) b is negative and c is negative.

c) b is positive and c is negative. d) b is positive and c is positive.

Solution   In each case we look at the sign of the constant, c, first.

a) Since the constant, c, is positive, both numbers must have the same sign. Since the coefficient of the x-term, b, is negative, both factors will contain negative numbers.

b) Since the constant, c, is negative, one factor will contain a positive number and the other will contain a negative number.

c) Since the constant, c, is negative, one factor will contain a positive number and the other will contain a negative number.

d) Since the constant, c, is positive, both numbers must have the same sign. Since the coefficient of the x-term, b, is positive, both factors will contain positive numbers.

✳

EXAMPLE 3   Factor $x^2 + x - 6$.

Solution   We must find two numbers whose product is the constant, -6, and whose sum is the coefficient of the x-term, 1. Remember that x means $1x$. Since the constant is negative, one number must be positive and the other negative. Recall that the product of two numbers with unlike signs is a negative number. We now list the factors of -6 and look for the two factors whose sum is 1.

TEACHING TIP
Before working Example 3, have students (either alone or in pairs) "Name two numbers whose product is _____ and whose sum is _____."
Sample:
12, 8
24, 11
36, −13
−15, −2
−15, 2

Factors of −6	Sum of Factors
$1(-6) = -6$	$1 + (-6) = -5$
$2(-3) = -6$	$2 + (-3) = -1$
$3(-2) = -6$	$3 + (-2) = 1$
$6(-1) = -6$	$6 + (-1) = 5$

Note that the factors 1 and −6 in the top row are different from the factors −1 and 6 in the bottom row, and their sums are different.

The numbers 3 and −2 have a product of −6 and a sum of 1. Thus, the factors are $(x + 3)$ and $(x - 2)$.

$$x^2 + x - 6 = (x + 3)(x - 2)$$

The order of the factors is not crucial. Therefore, $x^2 + x - 6 = (x - 2)(x + 3)$ is also an acceptable answer.

✳

As mentioned earlier, **trinomial factoring problems can be checked by multiplying the factors using the FOIL method**. If the factoring is correct, the product obtained using the FOIL method will be identical to the original trinomial. Let's check the factors obtained in Example 3.

Check $(x + 3)(x - 2) = x^2 - 2x + 3x - 6 = x^2 + x - 6$

Since the product of the factors is identical to the original trinomial, the factoring is correct.

EXAMPLE 4 Factor $x^2 - x - 6$.

Solution The factors of -6 are illustrated in Example 3. The factors whose product is -6 and whose sum is -1 are 2 and -3.

Factors of -6	Sum of Factors
$2(-3) = -6$	$2 + (-3) = \boxed{-1}$

Therefore, $x^2 - x - 6 = (x + 2)(x - 3)$

EXAMPLE 5 Factor $x^2 - 5x + 6$.

Solution We must find two numbers whose product is 6 and whose sum is -5. Since the constant, 6, is positive, both factors must have the same sign. Since the coefficient of the x-term, -5, is negative, both numbers must be negative. Recall that the product of a negative number and a negative number is positive. We now list the negative factors of 6 and look for the pair whose sum is -5.

Factors of 6	Sum of Factors
$(-1)(-6)$	$-1 + (-6) = -7$
$(-2)(-3)$	$-2 + (-3) = \boxed{-5}$

The factors of 6 whose sum is -5 are -2 and -3.

NOW TRY EXERCISE 29

$$x^2 - 5x + 6 = (x - 2)(x - 3)$$

EXAMPLE 6 Factor $r^2 + 2r - 24$

Solution In this example the variable is r, but the factoring procedure is the same. We must find the two factors of -24 whose sum is 2. Since the constant is negative, one factor will be positive and the other factor will be negative.

Factors of -24	Sum of Factors
$(1)(-24)$	$1 + (-24) = -23$
$(2)(-12)$	$2 + (-12) = -10$
$(3)(-8)$	$3 + (-8) = -5$
$(4)(-6)$	$4 + (-6) = -2$
$(6)(-4)$	$6 + (-4) = \boxed{2}$

Since we have found the two numbers, 6 and -4, whose product is -24 and whose sum is 2, we need go no further.

$$r^2 + 2r - 24 = (r + 6)(r - 4)$$

EXAMPLE 7 Factor $x^2 - 10x + 25$

Solution We must find the factors of 25 whose sum is -10. Both factors must be negative. (Can you explain why?) The two factors whose product is 25 and whose sum is -10 are -5 and -5.

$$x^2 - 10x + 25 = (x - 5)(x - 5)$$
$$= (x - 5)^2$$

EXAMPLE 8 Factor $x^2 - 5x - 66$.

Solution We must find two numbers whose product is -66 and whose sum is -5. Since the constant is negative, one factor must be positive and the other negative. The desired factors are -11 and 6 because $(-11)(6) = -66$ and $-11 + 6 = -5$.

NOW TRY EXERCISE 51

$$x^2 - 5x - 66 = (x - 11)(x + 6)$$

EXAMPLE 9 Factor $x^2 + 7x + 18$.

Solution Let's first find the two numbers whose product is 18 and whose sum is 7. Since both the constant and the coefficient of the x-term are positive, the two numbers must also be positive.

Factors of 18	Sum of Factors
(1)(18)	$1 + 18 = 19$
(2)(9)	$2 + 9 = 11$
(3)(6)	$3 + 6 = 9$

Note that there are no two integers whose product is 18 and whose sum is 7. When two integers cannot be found to satisfy the given conditions, the trinomial cannot be factored using only integer factors. *A polynomial that cannot be factored using only integer coefficients is called a* **prime polynomial**. If you come across a polynomial that cannot be factored using only integer coefficients, as in Example 9, do not leave the answer blank. Instead, write "prime." However, before you write the answer "prime," recheck your work and make sure you have tried every possible combination.

When factoring a trinomial of the form $x^2 + bx + c$, there is at most one pair of numbers whose product is c and whose sum is b. For example, when factoring $x^2 - 12x + 32$, the two numbers whose product is 32 and whose sum is -12 are -4 and -8. No other pair of numbers will satisfy these specific conditions. Thus, the only factors of $x^2 - 12x + 32$ are $(x - 4)(x - 8)$.

A slightly different type of problem is illustrated in Example 10.

EXAMPLE 10 Factor $x^2 + 2xy + y^2$.

Solution In this problem the second term contains two variables, x and y, and the last term is not a constant. The procedure used to factor this trinomial is similar to that outlined previously. You should realize, however, that the product of the first terms of the factors we are looking for must be x^2, and the product of the last terms of the factors must be y^2.

We must find two numbers whose product is 1 (from $1y^2$) and whose sum is 2 (from $2xy$). The two numbers are 1 and 1. Thus

$$x^2 + 2xy + y^2 = (x + 1y)(x + 1y) = (x + y)(x + y) = (x + y)^2$$

EXAMPLE 11 Factor $x^2 - xy - 6y^2$.

Solution Find two numbers whose product is -6 and whose sum is -1. The numbers are -3 and 2. The last terms must be $-3y$ and $2y$ to obtain $-6y^2$.

NOW TRY EXERCISE 67

$$x^2 - xy - 6y^2 = (x - 3y)(x + 2y)$$ ✳

2 Remove a Common Factor from a Trinomial

Sometimes each term of a trinomial has a common factor. When this occurs, factor out the common factor first, as explained in Section 5.1. **The first step in any factoring problem is to factor out any factors common to all the terms in the polynomial. Whenever the numerical coefficient of the highest-degree term is not 1, you should check for a common factor.** After factoring out any common factor, you should factor the remaining trinomial further, if possible.

EXAMPLE 12 Factor $2x^2 + 2x - 12$.

Solution Since the numerical coefficient of the squared term is not 1, we check for a common factor. Because 2 is common to each term of the polynomial, we factor it out.

$$2x^2 + 2x - 12 = 2(x^2 + x - 6) \quad \textit{Factor out the common factor.}$$

TEACHING TIP
Point out that by factoring out the GCF before factoring the trinomial into binomials, the expression becomes easier to factor because there are fewer pairs of numbers whose product equals the new constant.

Now we factor the remaining trinomial $x^2 + x - 6$ into $(x + 3)(x - 2)$. Thus,

$$2x^2 + 2x - 12 = 2(x + 3)(x - 2).$$

Note that the trinomial $2x^2 + 2x - 12$ is now completely factored into *three* factors: two binomial factors, $x + 3$ and $x - 2$, and a monomial factor, 2. After 2 has been factored out, it plays no part in the factoring of the remaining trinomial. ✳

EXAMPLE 13 Factor $3n^3 + 24n^2 - 60n$.

Solution We see that $3n$ divides into each term of the polynomial and therefore is a common factor. After factoring out the $3n$, we factor the remaining trinomial.

$$3n^3 + 24n^2 - 60n = 3n(n^2 + 8n - 20) \quad \textit{Factor out the common factor.}$$
$$= 3n(n + 10)(n - 2) \quad \textit{Factor the remaining trinomial.}$$ ✳

Exercise Set 5.3

Concept/Writing Exercises

For each trinomial, determine the signs that will appear in the binomial factors. Explain how you determined your answer.

1. $x^2 + 180x + 8000$ both +
2. $x^2 - 500x + 4000$ both −
3. $x^2 + 20x - 8000$ one +, one −
4. $x^2 - 20x - 8000$ one +, one−
5. $x^2 - 240x + 8000$ both −

Write the trinomial whose factors are listed. Explain how you determined your answer.

6. $(x - 3)(x - 8)$ $x^2 - 11x + 24$
7. $(x - 2y)(x + 6y)$ $x^2 + 4xy - 12y^2$
8. $2(x - 5y)(x + y)$ $2x^2 - 8xy - 10y^2$
9. $4(a + b)(a - b)$ $4a^2 - 4b^2$
10. How can a trinomial factoring problem be checked? Multiply factors.
11. On an exam, a student factored $2x^2 - 6x + 4$ as $(2x - 4)(x - 1)$. Even though $(2x - 4)(x - 1)$ does multiply out to $2x^2 - 6x + 4$, why did his or her professor deduct points? not completely factored

13. Find the numbers whose product is c and whose sum is b. The factors are $(x + \text{first number})$ and $(x + \text{second number})$.

\ **12.** On an exam, a student factored $3x^2 + 3x - 18$ as $(3x - 6)(x + 3)$. Even though $(3x - 6)(x + 3)$ does multiply out to $3x^2 + 3x - 18$, why did his or her professor deduct points?   not completely factored

\ **13.** Explain how to determine the factors when factoring a trinomial of the form $x^2 + bx + c$.

\ **14.** In your own words, describe the trial and error method of factoring.   Answers will vary.

Practice the Skills

Factor each polynomial. If the polynomial is prime, so state.

15. $x^2 - 7x + 10$ $(x - 5)(x - 2)$
16. $x^2 + 8x + 15$ $(x + 5)(x + 3)$
17. $x^2 + 6x + 8$ $(x + 2)(x + 4)$
18. $x^2 - 3x + 2$ $(x - 2)(x - 1)$
19. $x^2 + 7x + 12$ $(x + 4)(x + 3)$
20. $x^2 - x - 12$ $(x - 4)(x + 3)$
21. $x^2 + 4x - 6$ prime
22. $y^2 - 6y + 8$ $(y - 4)(y - 2)$
23. $y^2 - 13y + 12$ $(y - 12)(y - 1)$
24. $x^2 + 3x - 28$ $(x + 7)(x - 4)$
25. $a^2 - 2a - 8$ $(a - 4)(a + 2)$
26. $p^2 + 3p - 10$ $(p + 5)(p - 2)$
27. $r^2 - 2r - 15$ $(r - 5)(r + 3)$
28. $x^2 - 6x + 8$ $(x - 4)(x - 2)$
29. $b^2 - 11b + 18$ $(b - 9)(b - 2)$
30. $x^2 + 11x - 30$ prime
31. $x^2 - 8x - 15$ prime
32. $x^2 - 10x + 9$ $(x - 9)(x - 1)$
33. $a^2 + 12a + 11$ $(a + 11)(a + 1)$
34. $x^2 + 10x + 25$ $(x + 5)^2$
35. $x^2 - 7x - 30$ $(x - 10)(x + 3)$
36. $b^2 - 9b - 36$ $(b - 12)(b + 3)$
37. $x^2 + 4x + 4$ $(x + 2)^2$
38. $x^2 - 4x + 4$ $(x - 2)^2$
39. $p^2 + 6p + 9$ $(p + 3)^2$
40. $t^2 - 6t + 9$ $(t - 3)^2$
41. $p^2 - 12p + 36$ $(p - 6)^2$
42. $x^2 - 10x - 25$ prime
43. $w^2 - 18w + 45$ $(w - 15)(w - 3)$
44. $x^2 - 11x + 10$ $(x - 10)(x - 1)$
45. $x^2 + 10x - 39$ $(x + 13)(x - 3)$
46. $x^2 - 3x + 8$ prime
47. $x^2 - x - 20$ $(x - 5)(x + 4)$
48. $t^2 - 28t - 60$ $(t - 30)(t + 2)$
49. $y^2 + 9y + 14$ $(y + 7)(y + 2)$
50. $r^2 + 14r + 48$ $(r + 6)(r + 8)$
51. $x^2 + 12x - 64$ $(x + 16)(x - 4)$
52. $x^2 - 18x + 80$ $(x - 8)(x - 10)$
53. $s^2 + 14s - 24$ prime
54. $x^2 - 13x + 36$ $(x - 4)(x - 9)$
55. $x^2 - 20x + 64$ $(x - 16)(x - 4)$
56. $x^2 + 19x + 48$ $(x + 3)(x + 16)$
57. $b^2 - 18b + 65$ $(b - 5)(b - 13)$
58. $x^2 + 5x - 24$ $(x - 3)(x + 8)$
59. $x^2 + 2 + 3x$ $(x + 2)(x + 1)$
60. $m^2 - 11 - 10m$ $(m + 1)(m - 11)$
61. $7w - 18 + w^2$ $(w + 9)(w - 2)$
62. $30 + y^2 - 13y$ $(y - 10)(y - 3)$
63. $x^2 - 8xy + 15y^2$ $(x - 3y)(x - 5y)$
64. $x^2 - 2xy + y^2$ $(x - y)^2$
65. $m^2 - 6mn + 9n^2$ $(m - 3n)^2$
66. $b^2 - 2bc - 3c^2$ $(b - 3c)(b + c)$
67. $x^2 + 8xy + 15y^2$ $(x + 5y)(x + 3y)$
68. $x^2 + 16xy - 17y^2$ $(x + 17y)(x - y)$
69. $m^2 - 5mn - 24n^2$ $(m + 3n)(m - 8n)$
70. $c^2 + 2cd - 24d^2$ $(c - 4d)(c + 6d)$

Factor completely.

71. $6x^2 - 30x + 24$ $6(x - 4)(x - 1)$
72. $2a^2 - 12a - 32$ $2(a - 8)(a + 2)$
73. $5x^2 + 20x + 15$ $5(x + 3)(x + 1)$
74. $4x^2 + 12x - 16$ $4(x + 4)(x - 1)$
75. $2x^2 - 14x + 24$ $2(x - 4)(x - 3)$
76. $3y^2 - 33y + 54$ $3(y - 2)(y - 9)$
77. $b^3 - 7b^2 + 10b$ $b(b - 5)(b - 2)$
78. $x^3 + 11x^2 - 42x$ $x(x + 14)(x - 3)$
79. $3z^3 - 21z^2 - 54z$ $3z(z - 9)(z + 2)$
80. $3x^3 - 36x^2 + 33x$ $3x(x - 11)(x - 1)$
81. $x^3 + 8x^2 + 16x$ $x(x + 4)^2$
82. $2x^3y - 12x^2y + 10xy$ $2xy(x - 5)(x - 1)$
83. $4a^2 - 24ab + 32b^2$ $4(a - 4b)(a - 2b)$
84. $3x^3 + 3x^2y - 18xy^2$ $3x(x + 3y)(x - 2y)$
85. $r^2s + 7rs^2 + 12s^3$ $s(r + 3s)(r + 4s)$
86. $3r^3 + 6r^2t - 24rt^2$ $3r(r + 4t)(r - 2t)$
87. $x^4 - 4x^3 - 21x^2$ $x^2(x - 7)(x + 3)$
88. $2z^5 + 16z^4 + 30z^3$ $2z^3(z + 3)(z + 5)$

Problem Solving

\ **89.** The first two columns in the following table describe the signs of the x-term and constant term of a trinomial of the form $x^2 + bx + c$. Determine whether the third column should contain "both positive," "both negative," or "one positive and one negative." Explain how you determined your answer.

Sign of Coefficient of x-term	Sign of Constant of Trinomial	Sign of Constant Terms in the Binomial Factors
−	+	both negative
−	−	one positive and one negative
+	−	one positive and one negative
+	+	both positive

90. a) both positive **b)** both negative **c)** and **d)** one positive and one negative

90. Assume that a trinomial of the form $x^2 + bx + c$ is factorable. Determine whether the constant terms in the factors are "both positive," "both negative," or "one positive and one negative" for the given signs of b and c. Explain your answer.

a) $b > 0, c > 0$ **b)** $b < 0, c > 0$

c) $b > 0, c < 0$ **d)** $b < 0, c < 0$

91. Write a trinomial whose binomial factors contain constant terms that sum to 5 and have a product of 4. Show the factoring of the trinomial.
$x^2 + 5x + 4 = (x + 1)(x + 4)$

92. Write a trinomial whose binomial factors contain constant terms that sum to -12 and have a product of 32 Show the factoring of the trinomial.
$x^2 - 12x + 32 = (x - 8)(x - 4)$

93. Write a trinomial whose binomial factors contain constant terms that sum to -12 and have a product of 32 Show the factoring of the trinomial.
$x^2 + 12x + 32 = (x + 8)(x + 4)$

94. Write a trinomial whose binomial factors contain constant terms that sum to 5 and have a product of -14. Show the factoring of the trinomial.
$x^2 + 5x - 14 = (x + 7)(x - 2)$

Challenge Problems

Factor.

95. $x^2 + 0.6x + 0.08$ $(x + 0.4)(x + 0.2)$

96. $x^2 - 0.5x - 0.06$ $(x - 0.6)(x + 0.1)$

97. $x^2 + \frac{2}{5}x + \frac{1}{25}$ $\left(x + \frac{1}{5}\right)\left(x + \frac{1}{5}\right)$

98. $x^2 - \frac{2}{3}x + \frac{1}{9}$ $\left(x - \frac{1}{3}\right)\left(x - \frac{1}{3}\right)$

99. $x^2 + 5x - 300$ $(x + 20)(x - 15)$

100. $x^2 - 24x - 256$ $(x + 8)(x - 32)$

Cumulative Review Exercises

[2.5] 101. Solve the equation $4(2x - 4) = 5x + 11$.   9

[3.5] 102. *Mixing Solutions* Karen Moreau, a chemist, mixes 4 liters of an 18% acid solution with 1 liter of a 26% acid solution. Find the strength of the mixture.   19.6%

[4.5] 103. Multiply $(2x^2 + 5x - 6)(x - 2)$.

[4.6] 104. Divide $3x^2 - 10x - 10$ by $x - 4$.

[5.2] 105. Factor $3x^2 + 5x - 6x - 10$ by grouping.
$(3x + 5)(x - 2)$

103. $2x^3 + x^2 - 16x + 12$ **104.** $3x + 2 - \dfrac{2}{x - 4}$

See Exercise 102.

5.4 FACTORING TRINOMIALS OF THE FORM $ax^2 + bx + c, a \neq 1$

SSM   Study Guide   CD/Video

MathPro 4/5   PH Math Tutor Center   prenhall.com/Angel

1 Factor trinomials of the form $ax^2 + bx + c, a \neq 1$, by trial and error.

2 Factor trinomials of the form $ax^2 + bx + c, a \neq 1$, by grouping.

An Important Note

In this section we discuss two methods of factoring trinomials of the form $ax^2 + bx + c$, $a \neq 1$. That is, we will be factoring trinomials whose squared term has a numerical coefficient not equal to 1, after removing any common factors. Examples of trinomials with $a \neq 1$ are

$$2x^2 + 11x + 12 \ (a = 2) \qquad 4x^2 - 3x + 1 \ (a = 4)$$

(continued on the next page)

The methods we discuss are (1) **factoring by trial and error** and (2) **factoring by grouping**. We present two different methods for factoring these trinomials because some students, and some instructors, prefer one method, while others prefer the second method. You may use either method unless your instructor asks you to use a specific method. We will use the same examples to illustrate both methods so that you can make a comparison. Each method is treated independently of the other. If your teacher asks you to use a specific method, either factoring by trial and error or factoring by grouping, you need only read the material related to that specific method. Factoring by trial and error was introduced in Section 5.3 and factoring by grouping was introduced in Section 5.2.

1 Factor Trinomials of the Form $ax^2 + bx + c$, $a \neq 1$, by Trial and Error

Let's now discuss factoring trinomials of the form $ax^2 + bx + c$, $a \neq 1$, by the trial and error method, introduced in Section 5.3. It may be helpful for you to reread that material before going any further.

Recall that factoring is the reverse of multiplying. Consider the product of the following two binomials:

$$(2x + 3)(x + 5) \overset{F\qquad O\qquad I\qquad L}{=} 2x(x) + (2x)(5) + 3(x) + 3(5)$$
$$= 2x^2 + 10x + 3x + 15$$
$$= 2x^2 + 13x + 15$$

TEACHING TIP
In polynomials with $a = 1$, the sum of the products of the outer terms and inner terms of the binomials also give the middle term; however, since the coefficients of the x's are 1, the middle term can also be found by just summing the constants in the binomials.

Notice that the product of the first terms of the binomials gives the x-squared term of the trinomial, $2x^2$. Also notice that the product of the last terms of the binomials gives the last term, or constant, of the trinomial, $+15$. Finally, notice that the sum of the products of the outer terms and inner terms of the binomials gives the middle term of the trinomial, $+13x$. When we factor a trinomial using trial and error, we make use of these important facts. Note that $2x^2 + 13x + 15$ in factored form is $(2x + 3)(x + 5)$.

$$2x^2 + 13x + 15 = (2x + 3)(x + 5)$$

When factoring a trinomial of the form $ax^2 + bx + c$ by trial and error, the product of the first terms in the binomial factors must equal the first term of the trinomial, ax^2. Also, the product of the constants in the binomial factors, including their signs, must equal the constant, c, of the trinomial.

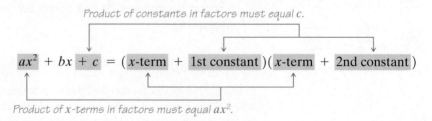

For example, when factoring the trinomial $2x^2 + 7x + 6$, each of the following pairs of factors has a product of the first terms equal to $2x^2$ and a product of the last terms equal to 6.

Trinomial	Possible Factors	Product of First Terms	Product of Last Terms
$2x^2 + 7x + 6$	$(2x + 1)(x + 6)$	$2x(x) = 2x^2$	$1(6) = 6$
	$(2x + 2)(x + 3)$	$2x(x) = 2x^2$	$2(3) = 6$
	$(2x + 3)(x + 2)$	$2x(x) = 2x^2$	$3(2) = 6$
	$(2x + 6)(x + 1)$	$2x(x) = 2x^2$	$6(1) = 6$

Each of these pairs of factors is a possible answer, but only one has the correct factors. How do we determine which is the correct factoring of the trinomial $2x^2 + 7x + 6$? The key lies in the x-term. We know that when we multiply two binomials using the FOIL method the sum of the products of the outer and inner terms gives us the x-term of the trinomial. We use this concept in reverse to determine the correct pair of factors. We need to find the pair of factors whose sum of the products of the outer and inner terms is equal to the x-term of the trinomial.

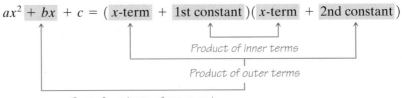

$ax^2 + bx + c = (\ x\text{-term} + 1\text{st constant})(\ x\text{-term} + 2\text{nd constant})$

Product of inner terms

Product of outer terms

Sum of products of outer and inner terms must equal bx.

Now look at the possible pairs of factors we obtained for $2x^2 + 7x + 6$ to see if any yield the correct x-term, $7x$.

Trinomial	Possible Factors	Product of the First Terms	Product of the Last Terms	Sum of the Products of Outer and Inner Terms
$2x^2 + 7x + 6$	$(2x + 1)(x + 6)$	$2x^2$	6	$2x(6) + 1(x) = 13x$
	$(2x + 2)(x + 3)$	$2x^2$	6	$2x(3) + 2(x) = 8x$
	$(2x + 3)(x + 2)$	$2x^2$	6	$2x(2) + 3(x) = 7x$
	$(2x + 6)(x + 1)$	$2x^2$	6	$2x(1) + 6(x) = 8x$

Since $(2x + 3)(x + 2)$ yields the correct x-term, $7x$, the factors of the trinomial $2x^2 + 7x + 6$ are $(2x + 3)$ and $(x + 2)$.

$$2x^2 + 7x + 6 = (2x + 3)(x + 2)$$

We can check this factoring using the FOIL method.

Check

$$\overset{F}{} \quad \overset{O}{} \quad \overset{I}{} \quad \overset{L}{}$$
$$(2x + 3)(x + 2) = 2x(x) + 2x(2) + 3(x) + 3(2)$$
$$= 2x^2 + 4x + 3x + 6$$
$$= 2x^2 + 7x + 6$$

Since we obtained the original trinomial, our factoring is correct.

Note in the preceding illustration that $(2x + 1)(x + 6)$ are different factors than $(2x + 6)(x + 1)$, because in one case 1 is paired with $2x$ and in the second case 1 is paired with x. The factors $(2x + 1)(x + 6)$ and $(x + 6)(2x + 1)$ are, however, the same set of factors with their order reversed.

HELPFUL HINT

When factoring a trinomial of the form $ax^2 + bx + c$, remember that the sign of the constant, c, and the sign of the x-term, bx, offer valuable information. When factoring a trinomial by trial and error, first check the sign of the constant. If it is positive, the signs in both factors will be the same as the sign of the x-term. If the constant is negative, one factor will contain a plus sign and the other a negative sign.

Now we outline the procedure to factor trinomials of the form $ax^2 + bx + c$, $a \neq 1$, by trial and error. Keep in mind that the more you practice, the better you will become at factoring.

To Factor Trinomials of the Form $ax^2 + bx + c$, $a \neq 1$, by Trial and Error

1. Determine whether there is any factor common to all three terms. If so, factor it out.
2. Write all pairs of factors of the coefficient of the squared term, a.
3. Write all pairs of factors of the constant term, c.
4. Try various combinations of these factors until the correct middle term, bx, is found.

TEACHING TIP
Suggest that students select a style to use when deciding the position of the factors of the coefficient of the x^2 term such as "put the larger value in the first binomial."

When factoring using this procedure, if there is more than one pair of numbers whose product is a, we generally begin with the middle-size pair. We will illustrate the procedure in Examples 1 through 8.

EXAMPLE 1   Factor $3x^2 + 20x + 12$.

Solution   We first determine that all three terms have no common factors other than 1. Since the first term is $3x^2$, one factor must contain a $3x$ and the other an x. Therefore, the factors will be of the form $(3x + \blacksquare)(x + \blacksquare)$. Now we must find the numbers to place in the shaded areas. The product of the last terms in the factors must be 12. Since the constant and the coefficient of the x-term are both positive, only the positive factors of 12 need be considered. We will list the positive factors of 12, the possible factors of the trinomial, and the sum of the products of the outer and inner terms. Once we find the factors of 12 that yield the proper sum of the products of the outer and inner terms, $20x$, we can write the answer.

Factors of 12	Possible Factors of Trinomial	Sum of the Products of the Outer and Inner Terms
1(12)	$(3x + 1)(x + 12)$	$37x$
2(6)	$(3x + 2)(x + 6)$	$20x$
3(4)	$(3x + 3)(x + 4)$	$15x$
4(3)	$(3x + 4)(x + 3)$	$13x$
6(2)	$(3x + 6)(x + 2)$	$12x$
12(1)	$(3x + 12)(x + 1)$	$15x$

Since the product of $(3x + 2)$ and $(x + 6)$ yields the correct x-term, $20x$, they are the correct factors.

$$3x^2 + 20x + 12 = (3x + 2)(x + 6)$$ ✻

In Example 1, our first factor could have been written with an x and the second with a $3x$. Had we done this, we still would have obtained the correct answer: $(x + 6)(3x + 2)$. We also could have stopped once we found the pair of factors that yielded the $20x$. Instead, we listed all the factors so that you could study them.

EXAMPLE 2 Factor $5x^2 - 7x - 6$.

Solution One factor must contain a $5x$ and the other an x. We now list the factors of -6 and look for the pair of factors that yields $-7x$.

Factors of -6	Possible Factors	Sum of the Products of the Outer and Inner Terms
$-1(6)$	$(5x - 1)(x + 6)$	$29x$
$-2(3)$	$(5x - 2)(x + 3)$	$13x$
$-3(2)$	$(5x - 3)(x + 2)$	$7x$
$-6(1)$	$(5x - 6)(x + 1)$	$-x$

Since we did not obtain the desired quantity, $-7x$, by writing the negative factor with the $5x$, we will now try listing the negative factor with the x.

Factors of -6	Possible Factors	Sum of the Products of the Outer and Inner Terms
$1(-6)$	$(5x + 1)(x - 6)$	$-29x$
$2(-3)$	$(5x + 2)(x - 3)$	$-13x$
$3(-2)$	$(5x + 3)(x - 2)$	$-7x$
$6(-1)$	$(5x + 6)(x - 1)$	x

We see that $(5x + 3)(x - 2)$ gives the $-7x$ we are looking for. Thus,

$$5x^2 - 7x - 6 = (5x + 3)(x - 2)$$

Again we listed all the possible combinations for you to study. ✻

HELPFUL HINT

In Example 2, we were asked to factor $5x^2 - 7x - 6$. When we considered the product of $-3(2)$ in the first set of possible factors, we obtained

Factors of -6	Possible Factors	Sum of the Products of the Outer and Inner Terms
$-3(2)$	$(5x - 3)(x + 2)$	$7x$

Later in the solution we tried the factors $3(-2)$ and obtained the correct answer.

$3(-2)$	$(5x + 3)(x - 2)$	$-7x$

When factoring a trinomial with a *negative constant*, if you obtain the x-term whose sign is the opposite of the one you are seeking, *reverse the signs on the constants* in the factors. This should give you the set of factors you are seeking.

TEACHING TIP
Point out that when the signs of the constants in a pair of binomial factors are both changed, the sign of the x-term in the corresponding trinomial will also change.

EXAMPLE 3   Factor $8x^2 + 33x + 4$.

Solution   There are no factors common to all three terms. Since the first term is $8x^2$, there are a number of possible combinations for the first terms in the factors. Since $8 = 8 \cdot 1$ and $8 = 4 \cdot 2$, the possible factors may be of the form $(8x \quad)(x \quad)$ or $(4x \quad)(2x \quad)$. When this situation occurs, we will generally start with the middle-size pair of factors. Thus, we begin with $(4x \quad)(2x \quad)$. If this pair does not lead to the solution, we will then try $(8x \quad)(x \quad)$. We now list the factors of the constant, 4. Since all signs are positive, we list only the positive factors of 4.

Factors of 4	Possible Factors	Sum of the Products of the Outer and Inner Terms
1(4)	$(4x + 1)(2x + 4)$	$18x$
2(2)	$(4x + 2)(2x + 2)$	$12x$
4(1)	$(4x + 4)(2x + 1)$	$12x$

Since we did not obtain the factors with $(4x \quad)(2x \quad)$, we now try $(8x \quad)(x \quad)$.

Factors of 4	Possible Factors	Sum of the Products of the Outer and Inner Terms
1(4)	$(8x + 1)(x + 4)$	$33x$
2(2)	$(8x + 2)(x + 2)$	$18x$
4(1)	$(8x + 4)(x + 1)$	$12x$

Since the product of $(8x + 1)$ and $(x + 4)$ yields the correct x-term, $33x$, they are the correct factors.

NOW TRY EXERCISE 7

$$8x^2 + 33x + 4 = (8x + 1)(x + 4)$$

EXAMPLE 4   Factor $25t^2 - 10t + 1$

Solution   The factors must be of the form $(25t \quad)(t \quad)$ or $(5t \quad)(5t \quad)$. We will start with the middle-size factors $(5t \quad)(5t \quad)$. Since the constant is positive and the coefficient of the x-term is negative, both factors must be negative.

Factors of 1	Possible Factors	Sum of the Products of the Outer and Inner Terms
$(-1)(-1)$	$(5t - 1)(5t - 1)$	$-10t$

Since we found the correct factors, we can stop.

$$25t^2 - 10t + 1 = (5t - 1)(5t - 1) = (5t - 1)^2$$

EXAMPLE 5   Factor $2x^2 + 3x + 7$.

Solution   The factors will be of the form $(2x \quad)(x \quad)$. We need only consider the positive factors of 7. Can you explain why?

Factors of 7	Possible Factors	Sum of the Products of the Outer and Inner Terms
$1(7)$	$(2x + 1)(x + 7)$	$15x$
$7(1)$	$(2x + 7)(x + 1)$	$9x$

Since we have tried all possible combinations and we have not obtained the x-term, $3x$, this trinomial *cannot be factored using only integer factors.* As explained in Section 5.3, the trinomial $2x^2 + 3x + 7$ is a *prime polynomial.* ✳

EXAMPLE 6 Factor $4x^2 + 7xy + 3y^2$.

Solution This trinomial is different from the other trinomials in that the last term is not a constant but contains y^2. Don't let this scare you. The factoring process is the same, except that the second term of both factors will contain y. We begin by considering factors of the form $(2x \quad)(2x \quad)$. If we cannot find the factors, then we try factors of the form $(4x \quad)(x \quad)$.

Factors of 3	Possible Factors	Sum of the Products of the Outer and Inner Terms
$1(3)$	$(2x + y)(2x + 3y)$	$8xy$
$3(1)$	$(2x + 3y)(2x + y)$	$8xy$
$1(3)$	$(4x + y)(x + 3y)$	$13xy$
$3(1)$	$(4x + 3y)(x + y)$	$7xy$

$$4x^2 + 7xy + 3y^2 = (4x + 3y)(x + y)$$

Check $(4x + 3y)(x + y) = 4x^2 + 4xy + 3xy + 3y^2 = 4x^2 + 7xy + 3y^2$ ✳

EXAMPLE 7 Factor $6x^2 - 13xy - 8y^2$.

Solution We begin with factors of the form $(3x \quad)(2x \quad)$. If we cannot find the solution from these, we will try $(6x \quad)(x \quad)$. Since the last term, $-8y^2$, is negative, one factor will contain a plus sign and the other will contain a minus sign.

Factors of -8	Possible Factors	Sum of the Products of the Outer and Inner Terms
$1(-8)$	$(3x + y)(2x - 8y)$	$-22xy$
$2(-4)$	$(3x + 2y)(2x - 4y)$	$-8xy$
$4(-2)$	$(3x + 4y)(2x - 2y)$	$2xy$
$8(-1)$	$(3x + 8y)(2x - y)$	$13xy \leftarrow$

We are looking for $-13xy$. When we considered $8(-1)$, we obtained $13xy$. As explained in the Helpful Hint on page 329, if we reverse the signs of the numbers in the factors, we will obtain the factors we are seeking.

$$(3x + 8y)(2x - y) \quad \text{Gives } 13xy$$
$$(3x - 8y)(2x + y) \quad \text{Gives } -13xy$$

NOW TRY EXERCISE 57   Therefore, $6x^2 - 13xy - 8y^2 = (3x - 8y)(2x + y)$. ✳

Now we will look at an example in which all the terms of the trinomial have a common factor.

EXAMPLE 8 Factor $6x^3 + 15x^2 - 36x$

Solution *The first step in any factoring problem is to determine whether all the terms contain a common factor. If so, factor out that common factor first. In this example, $3x$ is common to all three terms. We begin by factoring out the $3x$. Then we continue factoring by trial and error.*

$$6x^3 + 15x^2 - 36x = 3x(2x^2 + 5x - 12)$$
$$= 3x(2x - 3)(x + 4)$$

2 Factor Trinomials of the Form $ax^2 + bx + c$, $a \neq 1$, by Grouping

We will now discuss the use of grouping. The steps in the box that follow give the procedure for factoring trinomials by grouping.

> ### To Factor Trinomials of the Form $ax^2 + bx + c$, $a \neq 1$, by Grouping
>
> 1. Determine whether there is a factor common to all three terms. If so, factor it out.
> 2. Find two numbers whose product is equal to the product of a times c, and whose sum is equal to b.
> 3. Rewrite the middle term, bx, as the sum or difference of two terms using the numbers found in step 2.
> 4. Factor by grouping as explained in Section 5.2.

This process will be made clear in Example 9. We will rework Examples 1 through 8 here using factoring by grouping. Example 9, which follows, is the same trinomial given in Example 1. After you study this method and try some exercises, you will gain a feel for which method you prefer using.

EXAMPLE 9 Factor $3x^2 + 20x + 12$.

Solution First determine whether there is a factor common to all the terms of the polynomial. There are no common factors (other than 1) to the three terms.

$$a = 3 \quad b = 20 \quad c = 12$$

1. We must find two numbers whose product is $a \cdot c$ and whose sum is b. We must therefore find two numbers whose product equals $3 \cdot 12 = 36$ and whose sum equals 20. Only the positive factors of 36 need be considered since all signs of the trinomial are positive.

Factors of 36	Sum of Factors
(1)(36)	$1 + 36 = 37$
(2)(18)	$2 + 18 = 20$
(3)(12)	$3 + 12 = 15$
(4)(9)	$4 + 9 = 13$
(6)(6)	$6 + 6 = 12$

The desired factors are 2 and 18.

2. Rewrite $20x$ as the sum or difference of two terms using the values found in step 1. Therefore, we rewrite $20x$ as $2x + 18x$.

$$3x^2 + 20x + 12$$
$$= 3x^2 \boxed{+ 2x + 18x} + 12$$

3. Now factor by grouping. Start by factoring out a common factor from the first two terms and a common factor from the last two terms. This procedure was discussed in Section 5.2.

x is common factor   6 is common factor

$$3x^2 + 2x + 8x + 12$$
$$= x(3x + 2) + 6(3x + 2)$$
$$= (3x + 2)(x + 6)$$

※

Note that in step 2 of Example 9 we rewrote $20x$ as $2x + 18x$. Would it have made a difference if we had written $20x$ as $18x + 2x$? Let's work it out and see.

$$3x^2 + 20x + 12$$
$$= 3x^2 \boxed{+ 18x + 2x} + 12$$

$3x$ is common factor   2 is common factor

$$3x^2 + 18x + 2x + 12$$
$$= 3x(x + 6) + 2(x + 6)$$
$$= (x + 6)(3x + 2)$$

Since $(x + 6)(3x + 2) = (3x + 2)(x + 6)$, the factors are the same. We obtained the same answer by writing the $20x$ as either $2x + 18x$ or $18x + 2x$. *In general, when rewriting the middle term of the trinomial using the specific factors found, the terms may be listed in either order.* You should, however, check after you list the two terms to make sure that the sum of the terms you listed equals the middle term.

EXAMPLE 10 Factor $5x^2 - 7x - 6$.

Solution There are no common factors other than 1.

$$a = 5, \quad b = -7, \quad c = -6$$

The product of a times c is $5(-6) = -30$. We must find two numbers whose product is -30 and whose sum is -7.

Factors of -30	Sum of Factors
$(-1)(30)$	$-1 + 30 = 29$
$(-2)(15)$	$-2 + 15 = 13$
$(-3)(10)$	$-3 + 10 = 7$
$(-5)(6)$	$-5 + 6 = 1$
$(-6)(5)$	$-6 + 5 = -1$
$(-10)(3)$	$-10 + 3 = -7$
$(-15)(2)$	$-15 + 2 = -13$
$(-30)(1)$	$-30 + 1 = -29$

Rewrite the middle term of the trinomial, $-7x$, as $-10x + 3x$.

$$5x^2 - 7x - 6$$
$$= 5x^2 - 10x + 3x - 6 \quad \textit{Now factor by grouping.}$$
$$= 5x(x - 2) + 3(x - 2)$$
$$= (x - 2)(5x + 3)$$

NOW TRY EXERCISE 31

In Example 10, we could have expressed the $-7x$ as $3x - 10x$ and obtained the same answer. Try working Example 10 by rewriting $-7x$ as $3x - 10x$.

HELPFUL HINT

TEACHING TIP
Ask, "What happens to the sum of the factors when the sign of each factor is reversed?"

Notice in Example 10 that we were looking for two factors of -30 whose sum was -7. When we considered the factors -3 and 10, we obtained a sum of 7. The factors we eventually obtained that gave a sum of -7 were 3 and -10. Note that when the *constant of the trinomial is negative*, if we switch the signs of the constants in the factors, the sign of the sum of the factors changes. Thus, when trying pairs of factors to obtain the middle term, if you obtain the opposite of the coefficient you are seeking, reverse the signs in the factors. This should give you the coefficient you are seeking.

EXAMPLE 11 Factor $8x^2 + 33x + 4$.

Solution There are no common factors other than 1. We must find two numbers whose product is $8 \cdot 4$ or 32 and whose sum is 33. The numbers are 1 and 32.

Factors of 32	Sum of Factors
$(1)(32)$	$1 + 32 = 33$

Rewrite $33x$ as $32x + x$. Then factor by grouping.

$$8x^2 + 33x + 4$$
$$= 8x^2 + 32x + x + 4$$
$$= 8x(x + 4) + 1(x + 4)$$
$$= (x + 4)(8x + 1)$$

Notice in Example 11 that we rewrote $33x$ as $32x + x$ rather than $x + 32x$. We did this to reinforce factoring out 1 from the last two terms of an expression. You should obtain the same answer if you rewrite $33x$ as $x + 32x$. Try this now.

EXAMPLE 12 Factor $25t^2 - 10t + 1$

Solution There are no common factors other than 1. We must find two numbers whose product is $25 \cdot 1$ or 25 and whose sum is -10. Since the product of a times c is positive and the coefficient of the t-term is negative, both numerical factors must be negative.

Factors of 9	Sum of Factors
$(-1)(-25)$	$-1 + (-25) = -26$
$(-5)(-5)$	$-5 + (-5) = -10$

The desired factors are -5 and -5.

$$25t^2 - 10t + 1$$
$$= 25t^2 - 5t - 5t + 1 \qquad \textit{Rewrite } -10t \textit{ as } -5t - 5t$$
$$= 5t(5t - 1) - 5t + 1$$
$$= 5t(5t - 1) - 1(5t - 1) \qquad \textit{Rewrite } -5t + 1 \textit{ as } -1(5t - 1).$$
$$= (5t - 1)(5t - 1) \quad \text{or} \quad (5t - 1)^2 \qquad \text{❋}$$

HELPFUL HINT | When attempting to factor a trinomial, if there are no two integers whose product equals $a \cdot c$ and whose sum equals b, the trinomial cannot be factored.

EXAMPLE 13   Factor $2x^2 + 3x + 7$.

Solution   There are no common factors other than 1. We must find two numbers whose product is 14 and whose sum is 3. We need consider only positive factors of 14. Why?

Factors of 14	Sum of Factors
$(1)(14)$	$1 + 14 = 15$
$(2)(7)$	$2 + 7 = 9$

Since there are no factors of 14 whose sum is 3, we conclude that this trinomial cannot be factored. This is an example of a *prime polynomial*.   ❋

EXAMPLE 14   Factor $4x^2 + 7xy + 3y^2$.

Solution   There are no common factors other than 1. This trinomial contains two variables. It is factored in basically the same manner as the previous examples. Find two numbers whose product is $4 \cdot 3$ or 12 and whose sum is 7. The two numbers are 4 and 3.

$$4x^2 + 7xy + 3y^2$$
$$= 4x^2 + 4xy + 3xy + 3y^2$$
$$= 4x(x + y) + 3y(x + y)$$
$$= (x + y)(4x + 3y) \qquad \text{❋}$$

EXAMPLE 15   Factor $6x^2 - 13xy - 8y^2$.

Solution   There are no common factors other than 1. Find two numbers whose product is $6(-8)$ or -48 and whose sum is -13. Since the product is negative, one factor must be positive and the other negative. Some factors are given below.

Product of Factors	Sum of Factors
$(1)(-48)$	$1 + (-48) = -47$
$(2)(-24)$	$2 + (-24) = -22$
$(3)(-16)$	$3 + (-16) = -13$

There are many other factors, but we have found the pair we were looking for. The two numbers whose product is -48 and whose sum is -13 are 3 and -16.

$$6x^2 - 13xy - 8y^2$$
$$= 6x^2 + 3xy - 16xy - 8y^2$$
$$= 3x(2x + y) - 8y(2x + y)$$
$$= (2x + y)(3x - 8y)$$

Check $\quad (2x + y)(3x - 8y)$

$$\qquad\quad F \qquad\qquad O \qquad\quad I \qquad\quad L$$
$$= (2x)(3x) + (2x)(-8y) + (y)(3x) + (y)(-8y^2)$$
$$= 6x^2 \qquad - \quad 16xy \quad + \quad 3xy \quad - \quad 8y^2$$
$$= 6x^2 - 13xy - 8y^2$$

TEACHING TIP
After discussing Example 15, ask, "Can $6x^2 + 5xy - 18xy - 8y^2$ be factored? Explain."

If you rework Example 15 by writing $-13xy$ as $-16xy + 3xy$, what answer would you obtain? Try it now and see.

Remember that in any factoring problem our first step is to determine whether all terms in the polynomial have a common factor other than 1. If so, we use the distributive property to factor the GCF from each term. We then continue to factor the trinomial, if possible.

EXAMPLE 16 Factor $6x^3 + 15x^2 - 36x$

Solution The factor $3x$ is common to all three terms. Factor the $3x$ from each term of the polynomial.

$$6x^3 + 15x^2 - 36x = 3x(2x^2 + 5x - 12)$$

Now continue by factoring $2x^2 + 5x - 12$. The two numbers whose product is $2(-12)$ or -24 and whose sum is 5 are 8 and -3.

$$2x(2x^2 + 5x - 12)$$
$$= 2x(2x^2 + 8x - 3x - 12)$$
$$= 2x[2x(x + 4) - 3(x + 4)]$$
$$= 2x(x + 4)(2x - 3)$$

NOW TRY EXERCISE 47

HELPFUL HINT

Which Method Should You Use to Factor a Trinomial?
If your instructor asks you to use a specific method, you should use that method. If your instructor does not require a specific method, you should use the method you feel most comfortable with. You may wish to start with the trial-and-error method if there are only a few possible factors to try. If you cannot find the factors by trial and error or if there are many possible factors to consider, you may wish to use the grouping procedure. With time and practice you will learn which method you feel most comfortable with and which method gives you greater success.

TEACHING TIP
Ask, "Which method do you prefer? Why?"

Exercise Set 5.4

Concept/Writing Exercises

1. Factoring trinomials is the reverse process of multiplying binomials.

1. What is the relationship between factoring trinomials and multiplying binomials?

2. When factoring a trinomial of the form $ax^2 + bx + c$, what must the product of the first terms of the binomial factors equal? the first term of the trinomial, ax^2

3. When factoring a trinomial of the form $ax^2 + bx + c$, what must the product of the constants in the binomial factors equal? the constant, c, of the trinomial

4. Explain in your own words the procedure used to factor a trinomial of the form $ax^2 + bx + c, a \neq 1$.
Answers will vary.

Practice the Skills

Factor completely. If the polynomial is prime, so state.   **50.** $100(3x + 2)(x - 2)$ **60.** $2(3a - 4b)(2a - 3b)$

🔒 **5.** $2x^2 + 11x + 5$ $(2x + 1)(x + 5)$ **6.** $2x^2 + 9x + 4$ $(2x + 1)(x + 4)$ **7.** $3x^2 + 14x + 8$ $(3x + 2)(x + 4)$

8. $5x^2 + 13x + 6$ $(5x + 3)(x + 2)$ **9.** $5x^2 - 9x - 2$ $(5x + 1)(x - 2)$ **10.** $3y^2 + 17y + 10$ $(3y + 2)(y + 5)$

11. $3r^2 + 13r - 10$ $(3r - 2)(r + 5)$ **12.** $3x^2 - 2x - 8$ $(3x + 4)(x - 2)$ **13.** $4z^2 - 12z + 9$ $(2z - 3)^2$

14. $4n^2 - 9n + 5$ $(4n - 5)(n - 1)$ **15.** $5y^2 - y - 4$ $(5y + 4)(y - 1)$ **16.** $5m^2 - 17m + 6$ $(5m - 2)(m - 3)$

17. $5a^2 - 12a + 6$ prime   **18.** $2x^2 - x - 1$ $(2x + 1)(x - 1)$ **19.** $6z^2 + z - 12$ $(2z + 3)(3z - 4)$

20. $6y^2 - 11y + 4$ $(3y - 4)(2y - 1)$ **21.** $3x^2 + 11x + 4$ prime   **22.** $3a^2 + 7a - 20$ $(3a - 5)(a + 4)$

🔒 **23.** $5y^2 - 16y + 3$ $(5y - 1)(y - 3)$ **24.** $5x^2 + 2x + 9$ prime   **25.** $7x^2 + 43x + 6$ $(7x + 1)(x + 6)$

26. $4x^2 + 4x - 15$ $(2x + 5)(2x - 3)$ **27.** $7x^2 - 8x + 1$ $(7x - 1)(x - 1)$ **28.** $15x^2 - 19x + 6$ $(5x - 3)(3x - 2)$

29. $5b^2 - 23b + 12$ $(5b - 3)(b - 4)$ **30.** $9y^2 - 12y + 4$ $(3y - 2)^2$ **31.** $5z^2 - 6z - 8$ $(5z + 4)(z - 2)$.

32. $3z^2 - 11z - 6$ prime   🔒 **33.** $4y^2 + 5y - 6$ $(4y - 3)(y + 2)$ **34.** $4y^2 - 2y - 1$ prime

35. $10x^2 - 27x + 5$ $(5x - 1)(2x - 5)$ **36.** $6a^2 + 7a - 10$ $(6a - 5)(a + 2)$ **37.** $10d^2 - 7d - 12$ $(5d + 4)(2d - 3)$

38. $6x^2 + 13x + 3$ prime   **39.** $6x^2 - 22x - 8$ $2(3x + 1)(x - 4)$ **40.** $12x^2 - 13x - 35$ $(3x - 7)(4x + 5)$

41. $10t + 3 + 7t^2$ $(7t + 3)(t + 1)$ **42.** $n - 30 + n^2$ $(n + 6)(n - 5)$ **43.** $6x^2 + 16x + 10$ $2(3x + 5)(x + 1)$

44. $12z^2 + 32z + 20$ $4(3z + 5)(z + 1)$ 🔒 **45.** $6x^3 - 5x^2 - 4x$ $x(2x + 1)(3x - 4)$ **46.** $8x^3 + 8x^2 - 6x$ $2x(2x - 1)(2x + 3)$

47. $12x^3 + 28x^2 + 8x$ $4x(3x + 1)(x + 2)$ **48.** $18x^3 - 21x^2 - 9x$ $3x(2x - 3)(3x + 1)$ **49.** $4x^3 - 2x^2 - 12x$ $2x(2x + 3)(x - 2)$

50. $300x^2 - 400x - 400$ **51.** $36z^2 + 6z - 6$ $6(3z - 1)(2z + 1)$ **52.** $28x^2 - 28x + 7$ $7(2x - 1)^2$

53. $72 + 3r^2 - 30r$ $3(r - 4)(r - 6)$ **54.** $4p - 12 + 8p^2$ $4(2p + 3)(p - 1)$ **55.** $2x^2 + 5xy + 2y^2$ $(2x + y)(x + 2y)$

56. $8x^2 - 8xy - 6y^2$ $2(2x + y)(2x - 3y)$ **57.** $2x^2 - 7xy + 3y^2$ $(2x - y)(x - 3y)$ **58.** $15x^2 - xy - 6y^2$ $(5x + 3y)(3x - 2y)$

🔒 **59.** $12x^2 + 10xy - 8y^2$ $2(2x - y)(3x + 4y)$ **60.** $12a^2 - 34ab + 24b^2$ **61.** $6x^2 - 9xy - 27y^2$ $3(x - 3y)(2x + 3y)$

62. $24x^2 - 92x + 80$ $4(3x - 4)(2x - 5)$ **63.** $6m^2 - mn - 2n^2$ $(3m - 2n)(2m + n)$ **64.** $8m^2 + 4mn - 4n^2$ $4(2m - n)(m + n)$

65. $8x^3 + 10x^2y + 3xy^2$ $x(4x + 3y)(2x + y)$ **66.** $8a^2b + 10ab^2 + 3b^3$ $b(4a + 3b)(2a + b)$

67. $4x^4 + 8x^3y + 3x^2y^2$ $x^2(2x + y)(2x + 3y)$ **68.** $24r^2s + 30rs^2 + 9s^3$ $3s(2r + s)(4r + 3s)$

Problem Solving

Write the polynomial whose factors are listed. Explain how you determined your answer.

✎ **69.** $3x + 1, x - 7$ $3x^2 - 20x - 7$ ✎ **70.** $4x - 3, 5x - 7$ $20x^2 - 43x + 21$

✎ **71.** $5, x + 3, 2x + 1$ $10x^2 + 35x + 15$ ✎ **72.** $3, 2x + 3, x - 4$ $6x^2 - 15x - 36$

✎ **73.** $x^2, x + 1, 2x - 3$ $2x^4 - x^3 - 3x^2$ ✎ **74.** $5x^2, 3x - 7, 2x + 3$ $30x^4 - 25x^3 - 105x^2$

✎ **75. a)** If you know one binomial factor of a trinomial, explain how you can use division to find the second binomial factor of the trinomial (see Section 4.6). Trinomial divided by binomial is second factor.

b) One factor of $18x^2 + 93x + 110$ is $3x + 10$. Use division to find the second factor. $6x + 11$

✎ **76.** One factor of $30x^2 - 17x - 247$ is $6x - 19$. Find the other factor. $5x + 13$

Challenge Problems

Factor each trinomial.

✎ **77.** $18x^2 + 9x - 20$ $(6x - 5)(3x + 4)$ **78.** $8x^2 - 99x + 36$ $(8x - 3)(x - 12)$

79. $15x^2 - 124x + 160$ $(5x - 8)(3x - 20)$ **80.** $16x^2 - 62x - 45$ $(8x + 5)(2x - 9)$

81. $72x^2 - 180x - 200$ $4(6x + 5)(3x - 10)$ **82.** $72x^2 + 417x - 420$ $3(8x - 7)(3x + 20)$

83. Two factors of $6x^3 + 235x^2 + 2250x$ are x and $3x + 50$, determine the other factor. Explain how you determined your answer. $\;2x + 45$

84. Two factors of the polynomial $2x^3 + 11x^2 + 3x - 36$ are $x + 3$ and $2x - 3$. Determine the third factor. Explain how you determined your answer. $\;x + 4$

Cumulative Review Exercises

87. $12xy^2(3x^3y - 1 + 2x^4y^4)$

[1.9] **85.** Evaluate $-x^2 - 4(y + 3) + 2y^2$ when $x = -3$ and $y = -5$. $\;49$

[3.5] **86.** *Daytona 500* Ward Burton won the 2002 Daytona 500 in a time of about 3.82 hours, including cautions and pit stops. If the 500 miles were covered by circling the 2.5 mile track 200 times, find the average speed of the race. $\;\approx 130.89$ mph

[5.1] **87.** Factor $36x^4y^3 - 12xy^2 + 24x^5y^6$.

[5.3] **88.** Factor $x^2 - 15x + 54$. $\;(x - 9)(x - 6)$

See Exercise 86.

5.5 SPECIAL FACTORING FORMULAS AND A GENERAL REVIEW OF FACTORING

  SSM

 Study Guide

 CD/Video

 MathPro 4/5

 PH Math Tutor Center

prenhall.com/Angel

1 Factor the difference of two squares.

2 Factor the sum and difference of two cubes.

3 Learn the general procedure for factoring a polynomial.

TEACHING TIP
Before beginning this section, have students describe each of the following expressions in words
a) $r^2 - s^2$ b) $a^3 + b^3$ c) $x^3 - y^3$
Have them share their descriptions.

There are special formulas for certain types of factoring problems that are often used. The special formulas we focus on in this section are the *difference of two squares, the sum of two cubes, and the difference of two cubes*. There is no special formula for the sum of two squares; this is because the sum of two squares cannot be factored using the set of real numbers. *You will need to memorize the three highlighted formulas in this section* so that you can use them whenever you need them.

1 Factor the Difference of Two Squares

Let's begin with the difference of two squares. Consider the binomial $x^2 - 9$. Note that each term of the binomial can be expressed as the square of some expression.

$$x^2 - 9 = x^2 - 3^2$$

This is an example of a **difference of two squares**. To factor the difference of two squares, it is convenient to use the difference of two squares formula (which was introduced in Section 4.5).

Difference of Two Squares

$$a^2 - b^2 = (a + b)(a - b)$$

EXAMPLE 1 Factor $x^2 - 9$.

Solution If we write $x^2 - 9$ as a difference of two squares, we have $x^2 - 3^2$. Using the difference of two squares formula, where a is replaced by x and b is replaced by 3, we obtain the following:

$$a^2 - b^2 = (a + b)(a - b)$$

$$x^2 - 3^2 = (x + 3)(x - 3)$$

Thus, $x^2 - 9 = (x + 3)(x - 3)$.

EXAMPLE 2 Factor using the difference of two squares formula.

a) $x^2 - 16$ b) $25x^2 - 4$ c) $36x^2 - 49y^2$

Solution a) $x^2 - 16 = (x)^2 - (4)^2$
$= (x + 4)(x - 4)$

b) $25x^2 - 4 = (5x^2) - (2)^2$
$= (5x + 2)(5x - 2)$

c) $36x^2 - 49y^2 = (6x)^2 - (7y)^2$
$= (6x + 7y)(6x - 7y)$

EXAMPLE 3 Factor each difference of two squares.

a) $16x^4 - 9y^4$ b) $x^6 - y^4$

Solution a) Rewrite $16x^4$ as $(4x^2)^2$ and $9y^4$ as $(3y^2)^2$, then use the difference of two squares formula.

$$16x^4 - 9y^4 = (4x^2)^2 - (3y^2)^2$$
$$= (4x^2 + 3y^2)(4x^2 - 3y^2)$$

b) Rewrite x^6 as $(x^3)^2$ and y^4 as $(y^2)^2$, then use the difference of two squares formula.

$$x^6 - y^4 = (x^3)^2 - (y^2)^2$$
$$= (x^3 + y^2)(x^3 - y^2)$$

NOW TRY EXERCISE 31

EXAMPLE 4 Factor $4x^2 - 16y^2$ using the difference of two squares formula.

Solution First remove the common factor, 4.

$$4x^2 - 16y^2 = 4(x^2 - 4y^2)$$

Now use the formula for the difference of two squares.

$$4(x^2 - 4y^2) = 4[(x)^2 - (2y)^2]$$
$$= 4(x + 2y)(x - 2y)$$

NOW TRY EXERCISE 85

Notice in Example 4 that $4x^2 - 16y^2$ is the difference of two squares, $(2x)^2 - (4y)^2$. If you factor this difference of squares without first removing the common factor 4, the factoring may be more difficult. After you factor this difference of squares you will need to factor out the common factor 2 from each binomial factor, as illustrated below.

$$4x^2 - 16y^2 = (2x)^2 - (4y)^2$$
$$= (2x + 4y)(2x - 4y)$$
$$= 2(x + 2y)2(x - 2y)$$
$$= 4(x + 2y)(x - 2y)$$

We obtain the same answer as we did in Example 4. However, since we did not factor out the common factor 4 first, we had to work a little harder to obtain the answer.

EXAMPLE 5 Factor $z^4 - 16$ using the difference of two squares formula.

Solution We rewrite z^4 as $(z^2)^2$ and 16 as 4^2, then use the difference of two squares formula.

$$z^4 - 16 = (z^2)^2 - 4^2$$
$$= (z^2 + 4)(z^2 - 4)$$

Notice that the second factor, $z^2 - 4$, is also the difference of two squares. To complete the factoring, we use the difference of two squares formula again to factor $z^2 - 4$.

$$= (z^2 + 4)(z^2 - 4)$$
$$= (z^2 + 4)(z + 2)(z - 2)$$

AVOIDING COMMON ERRORS

The difference of two squares can be factored. However, a sum of two squares, where there is no common factor to the two terms, cannot be factored using real numbers.

CORRECT	INCORRECT
$a^2 - b^2 = (a + b)(a - b)$	$a^2 + b^2 = (a + b)(a + b)$

2 Factor the Sum and Difference of Two Cubes

We begin our discussion of the sum and difference of two cubes with a multiplication of polynomials problem. Consider the product of $(a + b)(a^2 - ab + b^2)$.

$$
\begin{array}{r}
a^2 - ab + b^2 \\
a + b \\
\hline
a^2b - ab^2 + b^3 \quad \leftarrow b(a^2 - ab + b^2) \\
a^3 - a^2b + ab^2 \quad\quad \leftarrow a(a^2 - ab + b^2) \\
\hline
a^3 \qquad\qquad\qquad + b^3 \quad \leftarrow \textit{Sum of terms}
\end{array}
$$

Thus, $(a + b)(a^2 - ab + b^2) = a^3 + b^3$. Since factoring is the opposite of multiplying, we may factor $a^3 + b^3$ as follows:

$$a^3 + b^3 = (a + b)(a^2 - ab + b^2)$$

We see, using the same procedure, that $a^3 - b^3 = (a - b)(a^2 + ab + b^2)$. The expression $a^3 + b^3$ is a sum of two cubes and the expression $a^3 - b^3$ is a difference of two cubes. The formulas for factoring **the sum and the difference of two cubes** follow.

Sum of Two Cubes

$$a^3 + b^3 = (a + b)(a^2 - ab + b^2)$$

TEACHING TIP
Have students work by themselves or in pairs to verify the formula for the difference of two cubes.

Difference of Two Cubes

$$a^3 - b^3 = (a - b)(a^2 + ab + b^2)$$

Note that the trinomials $a^2 - ab + b^2$ and $a^2 + ab + b^2$ cannot be factored further. Now let's solve some factoring problems using the sum and the difference of two cubes.

EXAMPLE 6 Factor $x^3 + 8$.

Solution We rewrite $x^3 + 8$ as a sum of two cubes: $x^3 + 8 = (x)^3 + (2)^3$. Using the sum of two cubes formula, if we let a correspond to x and b correspond to 2, we get

$$a^3 + b^3 = (a + b)(a^2 - a \cdot b + b^2)$$
$$x^3 + 2^3 = x^3 + 2^3 = (x + 2)[x^2 - x \cdot 2 + 2^2]$$
$$= (x + 2)(x^2 - 2x + 4)$$

You can check the factoring by multiplying $(x + 2)(x^2 - 2x + 4)$. If factored correctly, the product of the factors will equal the original expression, $x^3 + 8$. Try it and see. ✳

HELPFUL HINT

When factoring the sum or difference of two cubes remember that the sign between the terms in the *binomial factor* will be the same as the sign between the terms of the expression you are factoring. Furthermore, the sign of the ab term will be the opposite of the sign between the terms of the binomial factor. The last term in the trinomial factor will always be positive.

Consider

$$a^3 + b^3 = (a + b)(a^2 - ab + b^2)$$

same sign — opposite sign — always positive

$$a^3 - b^3 = (a - b)(a^2 + ab + b^2)$$

same sign — opposite sign — always positive

EXAMPLE 7 Factor $y^3 - 125$.

Solution We rewrite $y^3 - 125$ as a difference of two cubes: $(y)^3 - (5)^3$. Using the difference of two cubes formula, if we let a correspond to y and b correspond to 5, we get

$$a^3 - b^3 = (a - b)(a^2 + a \cdot b + b^2)$$
$$y^3 - 125 = y^3 - 5^3 = (y - 5)[y^2 + y \cdot 5 + 5^2]$$
$$= (y - 5)(y^2 + 5y + 25)$$ ✳

EXAMPLE 8 Factor $8p^3 - k^3$.

Solution We rewrite $8p^3 - k^3$ as a difference of two cubes. Since $(2p)^3 = 8p^3$, we write
$$8p^3 - k^3 = (2p)^3 - (k)^3$$
$$= (2p - k)[(2p)^2 + (2p)(k) + k^2]$$
$$= (2p - k)(4p^2 + 2pk + k^2)$$ ✳

EXAMPLE 9 Factor $8r^3 + 27s^3$.

Solution We rewrite $8r^3 + 27s^3$ as a sum of two cubes. Since $8r^3 = (2r)^3$ and $27s^3 = (3s)^3$, we write
$$8r^3 + 27s^3 = (2r)^3 + (3s)^3$$
$$= (2r + 3s)[(2r)^2 - (2r)(3s) + (3s)^2]$$
$$= (2r + 3s)(4r^2 - 6rs + 9s^2)$$ ✳

NOW TRY EXERCISE 51

AVOIDING COMMON ERRORS

Recall that $a^2 + b^2 \neq (a + b)^2$ and $a^2 - b^2 \neq (a - b)^2$. The same principle applies to the sum and difference of two cubes.

CORRECT

$$a^3 + b^3 = (a + b)(a^2 - ab + b^2)$$
$$a^3 - b^3 = (a - b)(a^2 + ab + b^2)$$

INCORRECT

$$a^3 + b^3 = (a + b)^3$$
$$a^3 - b^3 = (a - b)^3$$

Since $(a + b)^3 = (a + b)(a + b)(a + b)$, it cannot possibly equal $a^3 + b^3$. Also, since $(a - b)^3 = (a - b)(a - b)(a - b)$, it cannot possibly equal $a^3 - b^3$. At this point, we suggest you determine the products of $(a + b)(a + b)(a + b)$ and $(a - b)(a - b)(a - b)$.

It may be easier to see that, for example, $a^3 + b^3 = (a + b)(a^2 - ab + b^2)$ and not $(a + b)^3$ by substituting numbers for a and b. Suppose $a = 3$ and $b = 4$, then

$$3^3 + 4^3 = (3 + 4)[3^2 - 3(4) + 4^2]$$
$$27 + 64 = 7(13)$$
$$91 = 91$$

but $3^3 + 4^3 \neq (3 + 4)^3$

$$91 \neq 343$$

3 Learn the General Procedure for Factoring a Polynomial

In this chapter we have presented several methods of factoring. We now combine techniques from this and previous sections to give you an overview of a general factoring procedure.

Here is a general procedure for factoring any polynomial:

General Procedure for Factoring a Polynomial

1. If all the terms of the polynomial have a greatest common factor other than 1, factor it out.

2. If the polynomial has two terms (or is a binomial), determine whether it is a difference of two squares or a sum or a difference of two cubes. If so, factor using the appropriate formula.

3. If the polynomial has three terms, factor the trinomial using the methods discussed in Sections 5.3 and 5.4.

4. If the polynomial has more than three terms, try factoring by grouping.

5. As a final step, examine your factored polynomial to determine whether the terms in any factors have a common factor. If you find a common factor, factor it out at this point.

EXAMPLE 10 Factor $3x^4 - 27x^2$.

Solution First determine whether the terms have a greatest common factor other than 1. Since $3x^2$ is common to both terms, factor it out.

$$3x^4 - 27x^2 = 3x^2(x^2 - 9)$$
$$= 3x^2(x + 3)(x - 3)$$

Note that $x^2 - 9$ is a difference of two squares.

EXAMPLE 11 Factor $2m^2n^2 + 6m^2n - 36m^2$

Solution Begin by factoring the GCF, $2m^2$, from each term. Then factor the remaining trinomial.

$$2m^2n^2 + 6m^2n - 36m^2 = 2m^2(n^2 + 3n - 18)$$
$$= 2m^2(n + 6)(n - 3)$$

NOW TRY EXERCISE 81

EXAMPLE 12 Factor $10a^2b - 15ab + 20b$.

Solution $$10a^2b - 15ab + 20b = 5b(2a^2 - 3a + 4)$$

Since $2a^2 - 3a + 4$ cannot be factored, we stop here.

EXAMPLE 13 Factor $3xy + 6x + 3y + 6$

Solution Always begin by determining whether all the terms in the polynomial have a common factor. In this example, 3 is the GCF. Factor 3 from each term.

$$3xy + 6x + 3y + 6 = 3(xy + 2x + y + 2)$$

Now factor by grouping.

$$= 3[x(y + 2) + 1(y + 2)]$$
$$= 3(y + 2)(x + 1)$$

NOW TRY EXERCISE 79

In Example 13, what would happen if we forgot to factor out the common factor 3. Let's rework the problem without first factoring out the 3, and see what happens. Factor $3x$ from the first two terms, and 3 from the last two terms.

$$3xy + 6x + 3y + 6 = 3x(y + 2) + 3(y + 2)$$
$$= (y + 2)(3x + 3)$$

In step 5 of the general factoring procedure on page 342 we are reminded to examine the factored polynomial to see whether the terms in any factor have a common factor. If we study the factors, we see that the factor $3x + 3$ has a common factor of 3. If we factor out the 3 from $3x + 3$ we will obtain the same answer obtained in Example 13.

$$(y + 2)(3x + 3) = 3(y + 2)(x + 1)$$

EXAMPLE 14 Factor $12x^2 + 12x - 9$.

Solution First factor out the common factor 3. Then factor the remaining trinomial by one of the methods discussed in Section 5.4 (either by grouping or trial and error).

$$12x^2 + 12x - 9 = 3(4x^2 + 4x - 3)$$
$$= 3(2x + 3)(2x - 1)$$

EXAMPLE 15 Factor $2x^4y + 54xy$.

Solution First factor out the common factor $2xy$.

$$2x^4y + 54xy = 2xy(x^3 + 27)$$
$$= 2xy(x + 3)(x^2 - 3x + 9)$$

Note that $x^3 + 27$ is a sum of two cubes.

Exercise Set 5.5

Concept/Writing Exercises

1. a) Write the formula for factoring the difference of two squares. $a^2 - b^2 = (a + b)(a - b)$

   b) In your own words, explain how to factor the difference of two squares. Answers will vary.

2. a) Write the formula for factoring the sum of two cubes. $a^3 + b^3 = (a + b)(a^2 - ab + b^2)$

   b) In your own words, explain how to factor the sum of two cubes. Answers will vary.

3. a) Write the formula for factoring the difference of two cubes. $a^3 - b^3 = (a - b)(a^2 + ab + b^2)$

   b) In your own words, explain how to factor the difference of two cubes. Answers will vary.

4. Why is it important to memorize the special factoring formulas? to be able to use them whenever needed

5. Is there a special formula for factoring the sum of two squares? no

6. In your own words, describe the general procedure for factoring a polynomial. Answers will vary.

In Exercises 7–12, the binomial is a sum of squares. There is no formula for factoring the sum of squares. However, sometimes a common factor can be factored out from a sum of squares. Factor those polynomials that are factorable. If the polynomial is not factorable, write the word prime.

7. $x^2 + 9$ prime

8. $4y^2 + 1$ prime

9. $4a^2 + 16$ $4(a^2 + 4)$

10. $16s^2 + 64t^2$ $16(s^2 + 4t^2)$

11. $16m^2 + 36n^2$ $4(4m^2 + 9n^2)$

12. $9y^2 + 16z^2$ prime

Practice the Skills

Factor each difference of two squares.

13. $y^2 - 25$ $(y + 5)(y - 5)$

14. $x^2 - 4$ $(x + 2)(x - 2)$

15. $z^2 - 81$ $(z + 9)(z - 9)$

16. $z^2 - 64$ $(z + 8)(z - 8)$

17. $x^2 - 49$ $(x + 7)(x - 7)$

18. $x^2 - a^2$ $(x + a)(x - a)$

19. $x^2 - y^2$ $(x + y)(x - y)$

20. $16x^2 - 9$ $(4x + 3)(4x - 3)$

21. $9y^2 - 25z^2$ $(3y + 5z)(3y - 5z)$

22. $36y^2 - 25$ $(6y + 5)(6y - 5)$

23. $64a^2 - 36b^2$ $4(4a + 3b)(4a - 3b)$

24. $100x^2 - 81y^2$ $(10x + 9y)(10x - 9y)$

25. $49x^2 - 36$ $(7x + 6)(7x - 6)$

26. $y^4 - 100$ $(y^2 + 10)(y^2 - 10)$

27. $z^4 - 81x^2$ $(z^2 + 9x)(z^2 - 9x)$

28. $9x^4 - 16y^4$ $(3x^2 + 4y^2)(3x^2 - 4y^2)$

29. $9x^4 - 81y^2$ $9(x^2 + 3y)(x^2 - 3y)$

30. $4x^4 - 25y^4$ $(2x^2 + 5y^2)(2x^2 - 5y^2)$

31. $36m^4 - 49n^2$ $(6m^2 + 7n)(6m^2 - 7n)$

32. $2x^4 - 50y^2$ $2(x^2 + 5y)(x^2 - 5y)$

33. $10x^2 - 160$ $10(x + 4)(x - 4)$

34. $4x^3 - xy^2$ $x(2x + y)(2x - y)$

35. $16x^2 - 100y^4$ $4(2x + 5y^2)(2x - 5y^2)$

36. $36x^4 - 4y^2$ $4(3x^2 + y)(3x^2 - y)$

Factor each sum or difference of two cubes. 53. $(4m + 3n)(16m^2 - 12mn + 9n^2)$

37. $x^3 + y^3$ $(x + y)(x^2 - xy + y^2)$

38. $x^3 - y^3$ $(x - y)(x^2 + xy + y^2)$

39. $a^3 - b^3$ $(a - b)(a^2 + ab + b^2)$

40. $a^3 + b^3$ $(a + b)(a^2 - ab + b^2)$

41. $x^3 + 8$ $(x + 2)(x^2 - 2x + 4)$

42. $x^3 - 8$ $(x - 2)(x^2 + 2x + 4)$

43. $x^3 - 27$ $(x - 3)(x^2 + 3x + 9)$

44. $a^3 + 27$ $(a + 3)(a^2 - 3a + 9)$

45. $a^3 + 1$ $(a + 1)(a^2 - a + 1)$

46. $a^3 - 1$ $(a - 1)(a^2 + a + 1)$

47. $27x^3 - 1$ $(3x - 1)(9x^2 + 3x + 1)$

48. $27y^3 + 8$ $(3y + 2)(9y^2 - 6y + 4)$

49. $27a^3 - 125$ $(3a - 5)(9a^2 + 15a + 25)$

50. $125 + x^3$ $(5 + x)(25 - 5x + x^2)$

51. $27 - 8y^3$ $(3 - 2y)(9 + 6y + 4y^2)$

52. $8 + 27y^3$ $(2 + 3y)(4 - 6y + 9y^2)$

53. $64m^3 + 27n^3$

54. $64x^3 - 125y^3$

$(4x - 5y)(16x^2 + 20xy + 25y^2)$

55. $8a^3 - 27b^3$

$(2a - 3b)(4a^2 + 6ab + 9b^2)$

56. $27c^3 + 125d^3$

$(3c + 5d)(9c^2 - 15cd + 25d^2)$

Factor completely.

57. $2x^2 + 8x + 8$ $2(x + 2)^2$

58. $3x^2 - 9x - 12$ $3(x + 1)(x - 4)$

59. $a^2b - 25b$ $b(a + 5)(a - 5)$

60. $3x^2 - 48$ $3(x + 4)(x - 4)$

61. $3c^2 - 18c + 27$ $3(c - 3)^2$

62. $3x^2 + 9x + 12x + 36$ $3(x + 4)(x + 3)$

63. $5x^2 - 10x - 15$ $5(x - 3)(x + 1)$

64. $5x^2 - 20$ $5(x + 2)(x - 2)$

65. $3xy - 6x + 9y - 18$ $3(x + 3)(y - 2)$

66. $x^2y + 2xy - 6xy - 12y$

67. $2x^2 - 50$ $2(x + 5)(x - 5)$

68. $4a^2y - 64y^3$ $4y(a + 4y)(a - 4y)$

69. $2x^2y - 18y$ $2y(x + 3)(x - 3)$

70. $2x^3 - 72x$ $2x(x + 6)(x - 6)$

71. $3x^3y^2 + 3y^2$ $3y^2(x + 1)(x^2 - x + 1)$

66. $y(x + 2)(x - 6)$

72. $x^4 - 125x$ $x(x-5)(x^2+5x+25)$ 🔒 **73.** $2x^3 - 16$ $2(x-2)(x^2+2x+4)$ **74.** $x^3 - 27y^3$ $(x-3y)(x^2+3xy+9y^2)$

75. $18x^2 - 50$ $2(3x+5)(3x-5)$ **76.** $54a^3 - 16$ $2(3a-2)(9a^2+6a+4)$ **77.** $12x^2 + 36x + 27$ $3(2x+3)^2$

78. $12n^2 + 4n - 16$ $4(3n+4)(n-1)$ **79.** $6x^2 - 4x + 24x - 16$ $2(3x-2)(x+4)$ **80.** $2ab^2 - 4ab - 8ab + 16a$ $2a(b-2)(b-4)$

81. $2rs^2 - 10rs - 48r$ $2r(s+3)(s-8)$ **82.** $4x^4 - 26x^3 + 30x^2$ $2x^2(2x-3)(x-5)$ 🔒 **83.** $4x^2 + 5x - 6$ $(x+2)(4x-3)$

84. $12a^2 + 36a + 27$ $3(2a+3)^2$ **85.** $25b^2 - 100$ $25(b+2)(b-2)$ **86.** $3b^2 - 75c^2$ $3(b+5c)(b-5c)$

87. $a^5b^2 - 4a^3b^4$ $a^3b^2(a+2b)(a-2b)$ **88.** $12x^2 + 36x - 3x - 9$ $3(x+3)(4x-1)$ **89.** $3x^4 - 18x^3 + 27x^2$ $3x^2(x-3)^2$

90. $2a^6 + 4a^4b^2$ $2a^4(a^2+2b^2)$ **91.** $x^3 + 25x$ $x(x^2+25)$ **92.** $8y^2 - 23y - 3$ $(8y+1)(y-3)$

93. $y^4 - 16$ $(y^2+4)(y+2)(y-2)$ **94.** $16m^3 + 250$ $2(2m+5)(4m^2-10m+25)$ **95.** $36a^2 - 15ab - 6b^2$ $3(3a-2b)(4a+b)$

96. $ac + 2a + bc + 2b$ $(a+b)(c+2)$ **97.** $2ab - 3b + 4a - 6$ $(2a-3)(b+2)$ **98.** $x^3 - 100x$ $x(x+10)(x-10)$

99. $9 - 9y^4$ $9(1+y^2)(1+y)(1-y)$

Problem Solving

100. Explain why the sum of two squares, $a^2 + b^2$, cannot be factored using real numbers. Answers will vary.

101. Have you ever seen the proof that 1 is equal to 2? Here it is.

Let $a = b$, then square both sides of the equation:

$$a^2 = b^2$$
$$a^2 = b \cdot b$$
$$a^2 = ab \qquad \text{Substitute } a = b.$$
$$a^2 - b^2 = ab - b^2 \qquad \text{Subtract } b^2 \text{ from both sides of the equation.}$$
$$(a+b)(a-b) = b(a-b) \qquad \text{Factor both sides of the equation.}$$
$$\frac{(a+b)\cancel{(a-b)}}{\cancel{(a-b)}} = \frac{b\cancel{(a-b)}}{\cancel{(a-b)}} \qquad \text{Divide both sides of the equation by } (a-b) \text{ and divide out common factors.}$$
$$a + b = b$$
$$b + b = b \qquad \text{Substitute } a = b.$$
$$2b = b$$
$$\frac{\overset{1}{\cancel{2b}}}{\underset{1}{\cancel{b}}} = \frac{\overset{1}{\cancel{b}}}{\underset{1}{\cancel{b}}} \qquad \text{Divide both sides of the equation by } b.$$
$$2 = 1$$

Obviously, $2 \neq 1$. Therefore, we must have made an error somewhere. Can you find it?
cannot divide both sides of the equation by $a - b$, because it equals 0

Factor each expression. Treat the unknown symbols as if they were variables.

102. ◆✳ + 2◆ + ☺✳ + 2☺ (✳ + 2)(◆ + ☺) **103.** 2◆⁶ + 4◆⁴✳² $2◆^4(◆^2 + 2✳^2)$

104. 4◆²✳ − 6◆✳ − 20✳◆ + 30✳ 2✳(2◆ − 3)(◆ − 5)

Challenge Problems

105. Factor $x^6 + 1$. $(x^2+1)(x^4-x^2+1)$

106. Factor $x^6 - 27y^9$. $(x^2-3y^3)(x^4+3x^2y^3+9y^6)$

107. Factor $x^2 - 6x + 9 - 4y^2$. (*Hint:* Write the first three terms as the square of a binomial.) $(x-3+2y)(x-3-2y)$

108. Factor $x^2 + 10x + 25 - y^2 + 4y - 4$. (*Hint:* Group the first three terms and the last three terms.) $(x+y+3)(x-y+7)$

109. Factor $x^6 - y^6$. (*Hint:* Factor initially as the difference of two squares.) $(x-y)(x+y)(x^2+xy+y^2)(x^2-xy+y^2)$

Cumulative Review Exercises

[2.7] **110.** Solve the inequality $3x - 2(x + 4) \geq 2x - 9$ and graph the solution on a number line.

[3.1] **111.** Use the formula $A = \frac{1}{2}h(b + d)$ to find h in the following trapezoid if the area of the trapezoid is 36 square inches. 4 in.

6 in.

 h

12 in.

[3.2] **112.** Express $x + (5 - 2x) = 2$ as a statement.

[4.1] **113.** Simplify $\left(\dfrac{4x^4y}{6xy^5}\right)^3 \cdot \dfrac{8x^9}{27y^{12}}$

[4.2] **114.** Simplify $x^{-2}x^{-3}$. $\dfrac{1}{x^5}$

110. $x \leq 1$; ⟵━━⟶ 1

112. The sum of a number, and 5 decreased by twice the number is 2.

5.6 SOLVING QUADRATIC EQUATIONS USING FACTORING

  SSM
 Study Guide
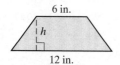 CD/Video

1 Recognize quadratic equations.
2 Solve quadratic equations using factoring.

 MathPro 4/5
 PH Math Tutor Center
 prenhall.com/Angel

1 Recognize Quadratic Equations

In this section we introduce **quadratic equations**, which are equations that contain a second-degree term and no term of a higher degree.

> ### Quadratic Equation
>
> Quadratic equations have the form
> $$ax^2 + bx + c = 0$$
> where $a, b,$ and c are real numbers, $a \neq 0$.

Examples of Quadratic Equations
$$x^2 + 4x - 12 = 0$$
$$2x^2 - 5x = 0$$
$$3x^2 - 2 = 0$$

Quadratic equations like these, in which one side of the equation is written in descending order of the variable and the other side of the equation is 0, are said to be in **standard form**.

Some quadratic equations can be solved by factoring. Two methods for solving quadratic equations that cannot be solved by factoring are given in Chapter 10. To solve a quadratic equation by factoring, we use the **zero-factor property**.

You know that if you multiply by 0, the product is 0. That is, if $a = 0$ or $b = 0$, then $ab = 0$. The reverse is also true. If a product equals 0, at least one of its factors must be 0.

> ### Zero-Factor Property
>
> If $ab = 0$, then $a = 0$ or $b = 0$.

TEACHING TIP
Before discussing the Zero-Factor Property have students investigate $ab = c$ for $c = 5, 2, 1, 0$ by asking, "What do you know about a and/or b if their product is…5? 2? 1? 0?" Then have them share their findings.

We now illustrate how the zero-factor property is used in solving equations.

EXAMPLE 1 Solve the equation $(x + 3)(x + 4) = 0$.

Solution Since the product of the factors equals 0, according to the zero-factor property, one or both factors must equal 0. Set each factor equal to 0, and solve each resulting equation.

$$x + 3 = 0 \qquad \text{or} \qquad x + 4 = 0$$
$$x + 3 - 3 = 0 - 3 \qquad x + 4 - 4 = 0 - 4$$
$$x = -3 \qquad x = -4$$

Thus, if x is either -3 or -4, the product of the factors is 0. The solutions to the equation are -3 and -4.

Check $\qquad x = -3 \qquad\qquad\qquad x = -4$

$$(x + 3)(x + 4) = 0 \qquad\qquad (x + 3)(x + 4) = 0$$
$$(-3 + 3)(-3 + 4) \overset{?}{=} 0 \qquad (-4 + 3)(-4 + 4) \overset{?}{=} 0$$
$$0(1) \overset{?}{=} 0 \qquad\qquad\qquad -1(0) \overset{?}{=} 0$$
$$0 = 0 \quad \textit{True} \qquad\qquad 0 = 0 \quad \textit{True}$$

EXAMPLE 2 Solve the equation $(5x - 3)(2x + 4) = 0$.

Solution Set each factor equal to 0 and solve for x.

$$5x - 3 = 0 \qquad \text{or} \qquad 2x + 4 = 0$$
$$5x = 3 \qquad\qquad 2x = -4$$
$$x = \frac{3}{5} \qquad\qquad x = -2$$

NOW TRY EXERCISE 11 The solutions to the equation are $\frac{3}{5}$ and -2.

2 Solve Quadratic Equations Using Factoring

Now we give a general procedure for solving quadratic equations using factoring.

To Solve a Quadratic Equation Using Factoring

1. Write the equation in standard form with the squared term having a positive coefficient. This will result in one side of the equation being 0.
2. Factor the side of the equation that is not 0.
3. Set each factor *containing a variable* equal to 0 and solve each equation.
4. Check each solution found in step 3 in the *original* equation.

EXAMPLE 3 Solve the equation $3x^2 = 12x$.

Solution To make the right side of the equation equal to 0, we subtract $12x$ from both sides of the equation. Then we factor out $3x$ from both terms. Why did we make the right side of the equation equal to 0 instead of the left side?

$$3x^2 = 12x$$
$$3x^2 - 12x = 12x - 12x$$
$$3x^2 - 12x = 0$$
$$3x(x - 4) = 0$$

Now set each factor equal to 0.

$$3x = 0 \quad \text{or} \quad x - 4 = 0$$
$$x = \frac{0}{3} \qquad\qquad x = 4$$
$$x = 0$$

The solutions to the quadratic equation are 0 and 4. Check by substituting $x = 0$, then $x = 4$ in $3x^2 = 12x$. ✳

EXAMPLE 4 Solve the equation $x^2 + 10x + 28 = 4$.

Solution To make the right side of the equation equal to 0, we subtract 4 from both sides of the equation. Then we factor and solve.

$$x^2 + 10x + 24 = 0$$
$$(x + 4)(x + 6) = 0$$

$$x + 4 = 0 \quad \text{or} \quad x + 6 = 0$$
$$x = -4 \qquad\qquad x = -6$$

The solutions are -4 and -6. We will check these values in the original equation.

Check $x = -4$ $x = -6$

$$x^2 + 10x + 28 = 4 \qquad\qquad x^2 + 10x + 28 = 4$$
$$(-4)^2 + 10(-4) + 28 \overset{?}{=} 4 \qquad (-6)^2 + 10(-6) + 28 \overset{?}{=} 4$$
$$16 - 40 + 28 \overset{?}{=} 4 \qquad\qquad 36 - 60 + 28 \overset{?}{=} 4$$
$$-24 + 28 \overset{?}{=} 4 \qquad\qquad -24 + 28 \overset{?}{=} 4$$

NOW TRY EXERCISE 23
$$4 = 4 \quad \textit{True} \qquad\qquad 4 = 4 \quad \textit{True} \quad ✳$$

EXAMPLE 5 Solve the equation $4y^2 + 5y - 20 = -11y$.

Solution Since all terms are not on the same side of the equation, add $11y$ to both sides of the equation.

$$4y^2 + 16y - 20 = 0$$

Factor out the common factor.

$$4(y^2 + 4y - 5) = 0$$

Factor the remaining trinomial.

$$4(y + 5)(y - 1) = 0$$

Now solve for y.

$$y + 5 = 0 \quad \text{or} \quad y - 1 = 0$$
$$y = -5 \qquad\qquad y = 1$$

Since 4 is a factor that does not contain a variable, we do not set it equal to 0. The solutions to the quadratic equation are -5 and 1. ✳

EXAMPLE 6 Solve the equation $-x^2 + 5x + 6 = 0$.

Solution When the squared term is negative, we generally make it positive by multiplying both sides of the equation by -1.

$$-1(-x^2 + 5x + 6) = -1 \cdot 0$$
$$x^2 - 5x - 6 = 0$$

Note that the sign of each term on the left side of the equation changed and that the right side of the equation remained 0. Why? Now proceed as before.

$$x^2 - 5x - 6 = 0$$
$$(x - 6)(x + 1) = 0$$
$$x - 6 = 0 \quad \text{or} \quad x + 1 = 0$$
$$x = 6 \qquad\qquad x = -1$$

NOW TRY EXERCISE 33 A check using the original equation will show that the solutions are 6 and −1. ✳

AVOIDING COMMON ERRORS

Be careful not to confuse factoring a polynomial with using factoring as a method to solve an equation.

CORRECT

Factor: $x^2 + 3x + 2$
$(x + 2)(x + 1)$

INCORRECT

Factor: $x^2 + 3x + 2$
$(x + 2)(x + 1)$
~~$x + 2 = 0$ or $x + 1 = 0$~~
~~$x = -2$ $x = -1$~~

Do you know what is wrong with the example on the right? It goes too far. The expression $x^2 + 3x + 2$ is a polynomial (a trinomial), not an equation. Since it is not an equation, it cannot be solved. When you are given a polynomial, you cannot just include "= 0" to change it to an equation.

Correct

Solve: $x^2 + 3x + 2 = 0$
$(x + 2)(x + 1) = 0$
$x + 2 = 0 \quad \text{or} \quad x + 1 = 0$
$x = -2 \qquad\qquad x = -1$

EXAMPLE 7 Solve the equation $x^2 = 36$

Solution Subtract 36 from both sides of the equation; then factor using the difference of two squares formula.

$$x^2 - 36 = 0$$
$$(x + 6)(x - 6) = 0$$
$$x + 6 = 0 \quad \text{or} \quad x - 6 = 0$$
$$x = -6 \qquad\qquad x = 6$$

TEACHING TIP
After the equation in Example 7 is in standard form ask, "What kind of expression is on the left side of the equation?"

The solutions are −6 and 6. ✳

EXAMPLE 8 Solve the equation $(x - 3)(x + 1) = 5$.

Solution Begin by multiplying the factors, then write the quadratic equation in standard form.

$$(x - 3)(x + 1) = 5$$
$$x^2 - 2x - 3 = 5 \quad \text{Multiply the factors}$$
$$x^2 - 2x - 8 = 0 \quad \text{Write the equation in standard form}$$
$$(x - 4)(x + 2) = 0 \quad \text{Factor}$$
$$x - 4 = 0 \quad \text{or} \quad x + 2 = 0 \quad \text{Zero-factor property}$$
$$x = 4 \qquad\qquad x = -2$$

The solutions are 4 and -2. We will check these values in the original equation.

Check $\qquad x = 4 \qquad\qquad\qquad\qquad x = -2$

$$(x - 3)(x + 1) = 5 \qquad\qquad (x - 3)(x + 1) = 5$$
$$(4 - 3)(4 + 1) \overset{?}{=} 5 \qquad\qquad (-2 - 3)(-2 + 1) \overset{?}{=} 5$$
$$1(5) \overset{?}{=} 5 \qquad\qquad (-5)(-1) \overset{?}{=} 5$$

NOW TRY EXERCISE 51

$$5 = 5 \quad \textit{True} \qquad\qquad\qquad 5 = 5 \quad \textit{True} \quad \text{✳}$$

HELPFUL HINT

In Example 8, you might have been tempted to start the problem by writing

$$x - 3 = 5 \quad \text{or} \quad x + 1 = 5.$$

This would lead to an incorrect solution. Remember, the zero-factor property only holds when one side of the equation is equal to 0. In Example 8, once we obtained $(x - 4)(x + 2) = 0$, we were able to use the zero-factor property.

Exercise Set 5.6

Concept/Writing Exercises

1. and 4. Answers will vary.   2. an equation that contains a second-degree term and no term of a higher degree   3. $ax^2 + bx + c = 0$ 5. a) The zero-factor property may only be used when one side of the equation is 0.

1. In your own words, explain the zero-factor property.
2. What is a quadratic equation?
3. What is the standard form of a quadratic equation?
4. Explain, in your own words, the procedure to use to solve a quadratic equation.
5. a) When solving the equation $(x + 1)(x - 2) = 4$, explain why we **cannot** solve the equation by first writing $x + 1 = 4$ or $x - 2 = 4$ and then solving each equation for x.

   b) Solve the equation $(x + 1)(x - 2) = 4$. $-2, 3$

6. When solving an equation such as $3(x - 4)(x + 5) = 0$, we set the factors $x - 4$ and $x + 5$ equal to 0, but we do not set the 3 equal to 0. Can you explain why?
   3 cannot equal 0

Practice the Skills

Solve.

7. $x(x + 2) = 0$ $0, -2$
8. $3x(x + 4) = 0$ $0, -4$
9. $7x(x - 8) = 0$ $0, 8$
10. $(x + 3)(x + 5) = 0$ $-3, -5$
11. $(2x + 5)(x - 3) = 0$ $-\frac{5}{2}, 3$
12. $(3x - 2)(x - 5) = 0$ $\frac{2}{3}, 5$
13. $x^2 - 9 = 0$ $3, -3$
14. $x^2 - 16 = 0$ $4, -4$
15. $x^2 - 12x = 0$ $0, 12$
16. $x^2 + 7x = 0$ $0, -7$
17. $9x^2 + 27x = 0$ $0, -3$
18. $a^2 - 4a - 12 = 0$ $-2, 6$
19. $x^2 - 8x + 16 = 0$ 4
20. $x^2 + 6x + 9 = 0$ -3
21. $x^2 + 12x = -20$ $-2, -10$
22. $3y^2 - 4 = -4y$ $\frac{2}{3}, -2$
23. $z^2 + 3z = 18$ $3, -6$
24. $3x^2 = -21x - 18$ $-1, -6$
25. $4a^2 - 4a - 48 = 0$ $4, -3$
26. $x^2 = 4x + 21$ $-3, 7$
27. $23p - 24 = -p^2$ $1, -24$
28. $3x^2 - 9x - 30 = 0$ $5, -2$
29. $33w + 90 = -3w^2$ $-5, -6$
30. $w^2 + 45 + 18w = 0$ $-3, -15$
31. $-2x - 15 = -x^2$ $5, -3$
32. $-9x + 20 = -x^2$ $4, 5$
33. $-x^2 + 29x + 30 = 0$ $30, -1$
34. $12y - 11 = y^2$ $1, 11$
35. $12 = 3n^2 + 16n$ $-6, \frac{2}{3}$
36. $z^2 + 8z = -16$ -4
37. $9p^2 = -21p - 6$ $-\frac{1}{3}, -2$
38. $2x^2 - 5 = 3x$ $-1, \frac{5}{2}$
39. $3r^2 + 13r = 10$ $\frac{2}{3}, -5$
40. $3x^2 = 7x + 20$ $-\frac{5}{3}, 4$
41. $4x^2 + 4x - 48 = 0$ $-4, 3$
42. $6x^2 + 13x + 6 = 0$ $-\frac{2}{3}, -\frac{3}{2}$
43. $8x^2 + 2x = 3$ $-\frac{3}{4}, \frac{1}{2}$
44. $2x^2 + 4x - 6 = 0$ $-3, 1$
45. $2n^2 + 36 = -18n$ $-3, -6$
46. $c^2 = 64$ $8, -8$
47. $2x^2 = 50x$ $0, 25$
48. $4x^2 - 25 = 0$ $\frac{5}{2}, -\frac{5}{2}$
49. $x^2 = 100$ $10, -10$
50. $2x^2 - 50 = 0$ $5, -5$
51. $(x - 2)(x - 1) = 12$ $-2, 5$

52. $(x + 2)(x + 5) = -2$ $-4, -3$ **53.** $(x - 1)(2x - 5) = 9$ $-\frac{1}{2}, 4$ **54.** $(3x + 2)(x + 1) = 4$ $-2, \frac{1}{3}$

55. $x(x + 5) = 6$ $-6, 1$ **56.** $2(a^2 + 9) = 15a$ $\frac{3}{2}, 6$

Problem Solving

In Exercises 57–60, create a quadratic equation with the given solutions. Explain how you determined your answers.

57. $4, -2$ $x^2 - 2x - 8 = 0$ **58.** $-3, -5$ $x^2 + 8x + 15 = 0$ **59.** $6, 0$ $x^2 - 6x = 0$ **60.** $0, -4$ $x^2 + 4x = 0$

61. The solutions to a quadratic equation are $\frac{1}{2}$ and $-\frac{1}{3}$.

    a) What factors with integer coefficients were set equal to 0 to obtain these solutions? $(2x - 1)$ and $(3x + 1)$

    b) Write a quadratic equation whose solutions are $\frac{1}{2}$ and $-\frac{1}{3}$. $6x^2 - x - 1 = 0$

62. The solutions to a quadratic equation are $\frac{2}{3}$ and $-\frac{3}{4}$.

    a) What factors with integer coefficients were set equal to 0 to obtain these solutions? $(3x - 2)$ and $(4x + 3)$

    b) Write a quadratic equation whose solutions are $\frac{2}{3}$ and $-\frac{3}{4}$. $12x^2 + x - 6 = 0$

Challenge Problems

63. Solve the equation
$(x - 3)(x - 2) = (x + 5)(2x - 3) + 21$. $0, -12$

64. Solve the equation
$(2x - 3)(x - 4) = (x - 5)(x + 3) + 7$. $4, 5$

65. Solve the equation
$x(x - 3)(x + 2) = 0$. $0, 3, -2$

66. Solve the equation
$x^3 - 10x^2 + 24x = 0$. $0, 4, 6$

Cumulative Review Exercises

[1.7] **67.** Subtract $\frac{3}{5} - \frac{4}{7}$. $\frac{1}{35}$

[2.5] **68. a)** What is the name given to an equation that has an infinite number of solutions? Identity

    b) What is the name given to an equation that has no solution? Contradiction

[2.6] **69.** ***Cyprus Gardens*** At Cyprus Gardens, there is a long line of people waiting to go through the entrance. If 160 people are admitted in 13 minutes, how many people will be admitted in 60 minutes? Assume the rate stays the same. ≈ 738

[4.1] **70.** Simplify $\left(\dfrac{2x^4y^5}{x^5y^7}\right)^3$. $\dfrac{8}{x^3y^6}$

[4.4] Identify the following as a monomial, binomial, trinomial or not polynomial. If an expression is not a polynomial, explain why.

71. $2x$ monomial   **72.** $x - 3$ binomial

73. $\dfrac{1}{x}$ not polynomial   **74.** $x^2 - 6x + 9$ trinomial

Cyprus Gardens, See Exercise 69

5.7 APPLICATIONS OF QUADRATIC EQUATIONS

SSM   Study Guide   CD/Video

MathPro 4/5   PH Math Tutor Center   prenhall.com/Angel

1 Solve applications by factoring quadratic equations.
2 Learn the Pythagorean Theorem.

1 Solve Applications by Factoring Quadratic Equations

In Section 5.6, we learned how to solve quadratic equations by factoring. In this section, we will discuss and solve application problems when it is necessary to solve quadratic equations to obtain the answer. In Example 1, we will solve a problem involving a relationship between two numbers.

EXAMPLE 1   **Number Problem**   The product of two numbers is 78. Find the two numbers if one number is 7 more than the other.

Solution   Understand and Translate   Our goal is to find the two numbers.

$$\text{Let } x = \text{smaller number}$$
$$x + 7 = \text{larger number}$$

Carry Out
$$x(x + 7) = 78$$
$$x^2 + 7x = 78$$
$$x^2 + 7x - 78 = 0$$
$$(x - 6)(x + 13) = 0$$

$$x - 6 = 0 \quad \text{or} \quad x + 13 = 0$$
$$x = 6 \qquad\qquad x = -13$$

Remember that x represents the smaller of the two numbers. This problem has two possible solutions.

	Solution 1	Solution 2
Smaller number	6	−13
Larger number	$x + 7 = 6 + 7 = 13$	$x + 7 = -13 + 7 = -6$

Thus the two possible solutions are 6 and 13, and −13 and −6.

Check	6 and 13	−13 and −6
Product of the two numbers is 78.	$6 \cdot 13 = 78$	$(-13)(-6) = 78$
One number is 7 more than the other number.	13 is 7 more than 6.	−6 is 7 more than −13.

Answer   One solution is: smaller number 6, larger number 13. A second solution is: smaller number −13, larger number −6. You must give both solutions. If the question had stated "the product of two *positive* numbers is 78," the only solution would be 6 and 13.

TEACHING TIP
After discussing Example 1, have students verify that the problem can be solved by letting
x = larger number,
$x - 7$ = smaller number.

NOW TRY EXERCISE 13

In the exercise set, we use the terms consecutive integers, consecutive even integers, and consecutive odd integers. Recall from Section 3.2 that **consecutive integers** may be represented as x and $x + 1$. **Consecutive even** or **consecutive odd integers** may be represented as x and $x + 2$. Now let us work an application problem involving geometry.

EXAMPLE 2   **Advertising**   The marketing department of a large publishing company is planning to make a large rectangular sign to advertise a new book at a convention. They want the length of the sign to be 3 feet longer than the width (Fig. 5.3). Signs at the convention may have a maximum area of 54 square feet. Find the length and width of the sign if the area is to be 54 square feet.

Solution   Understand and Translate   We need to find the length and width of the sign. We will use the formula for the area of a rectangle.

$$\text{Let } x = \text{width}$$
$$x + 3 = \text{length}$$
$$\text{area} = \text{length} \cdot \text{width}$$
$$54 = (x + 3)x$$

Carry Out
$$54 = x^2 + 3x$$
$$0 = x^2 + 3x - 54$$
$$\text{or} \quad x^2 + 3x - 54 = 0$$
$$(x - 6)(x + 9) = 0$$
$$x - 6 = 0 \quad \text{or} \quad x + 9 = 0$$
$$x = 6 \qquad\qquad x = -9$$

FIGURE 5.3

Check and Answer   Since the width of the sign cannot be a negative number, the only solution is

$$\text{width} = x = 6 \text{ feet}, \quad \text{length} = x + 3 = 6 + 3 = 9 \text{ feet}$$

The area, length $\cdot$ width, is 54 square feet, and the length is 3 feet more than the width, so the answer checks.   ✳

NOW TRY EXERCISE 21

EXAMPLE 3   **Earth's Gravitational Field**   In Earth's gravitational field, the distance, d, in feet, that an object falls t seconds after it has been released is given by the formula $d = 16t^2$. While at the top of a roller coaster, a rider's eyeglasses slide off his head and fall out of the cart. How long does it take the eyeglasses to reach the ground 64 feet below?

Solution   Understand and Translate   Substitute 64 for d in the formula and then solve for t.

$$d = 16t^2$$
$$64 = 16t^2$$

Carry Out
$$\frac{64}{16} = t^2$$
$$4 = t^2$$

Now subtract 4 from both sides of the equation and write the equation with 0 on the right side to put the quadratic equation in standard form.

$$4 - 4 = t^2 - 4$$
$$0 = t^2 - 4$$
$$\text{or} \quad t^2 - 4 = 0$$
$$(t + 2)(t - 2) = 0$$
$$t + 2 = 0 \quad \text{or} \quad t - 2 = 0$$
$$t = -2 \quad\quad\quad t = 2$$

TEACHING TIP
Have students work in pairs to write an area word problem and a gravitational word problem. Then have them give their problem to another pair to solve.

Check and Answer Since t represents the number of seconds, it must be a positive number. Thus, the only possible answer is 2 seconds. It takes 2 seconds for the eyeglasses (or any other object falling under the influence of gravity) to fall 64 feet. ✳

Mathematics in Action

Quadratic Growth

You have just been introduced to quadratic equations of the form $ax^2 + bx + c = 0$. Looks harmless enough, and the problems you learn to solve in this chapter with quadratic equations are pretty interesting.

Now, if you go to the search engine at www.google.com and search on the word "quadratic" you will get upwards of 581,000 entries. Hmm . . . why has so much been written about this simple-looking formula?

The fact is that quadratic equations can be used to describe or model patterns of growth or change in an astounding variety of settings. Here are just some of the questions that quadratic equations have been used to explore:

• What is the relationship between advancing age and cardiac disease?
• What is the rate at which a pouring of molten steel will turn solid?
• How is the growth of poultry affected by the addition of protein to their food?
• How much more complex will a computer system become if a new module is added?
• What is the pattern of world population growth?
• How fast can we expect AIDS to spread in any given country?

As your knowledge of mathematics grows, you will gain a deeper understanding of how mathematics provides models of phenomena in the real world, and how it can be used to analyze, predict, and affect those phenomena. The people engaged in these sophisticated applications of mathematics are helping to make decisions that directly affect the lives of hundreds of millions of people. And that is not a mathematical exaggeration.

2 Learn The Pythagorean Theorem

Now we will introduce the Pythagorean Theorem, which describes an important relationship between the length of the sides of a right triangle. The Pythagorean Theorem is named after Pythagoras of Samos (≈ 569 BC–475 BC) who was born in Samos, Ionia. Pythagoras is often described as the first pure

Pythagoras of Samos

mathematician. Unlike many later Greek mathematicians, relatively little is known about his life. The society he led, the Pythagorians, was half religious and half scientific. They followed a code of secrecy and did not publish any of their writings. There is fairly good agreement on the main events of Pythagoras's life, but many of the dates are disputed by scholars. Now let's discuss the Pythagorean Theorem.

A **right triangle** is a triangle that contains a right, or 90°, angle (Fig. 5.4). The two shorter sides of a right triangle are called the **legs** and the side opposite the right angle is called the **hypotenuse**. The **Pythagorean Theorem** expresses the relationship between the lengths of the legs of a right triangle and its hypotenuse.

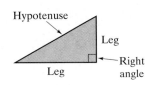

FIGURE 5.4

Pythagorean Theorem

The square of the hypotenuse of a right triangle is equal to the sum of the squares of the two legs.

$$(\text{leg})^2 + (\text{leg})^2 = (\text{hypotenuse})^2$$

If a and b represent the legs, and c represents the hypotenuse, then

$$a^2 + b^2 = c^2$$

When you use the Pythagorean Theorem, it makes no difference which leg you designate as a and which leg you designate as b, but the hypotenuse is always designated as c.

EXAMPLE 4 **Verifying Right Triangles** Determine if a right triangle can have the following sides.

a) 3 inches, 4 inches, 5 inches   **b)** 2 inches, 5 inches, 7 inches

Solution **a)** Understand To determine if a right triangle can have the sides given, we will use the Pythagorean Theorem. If the results show the Pythagorean Theorem holds true, the triangle can have the given sides. If the results are false, then the sides given cannot be those of a right triangle.

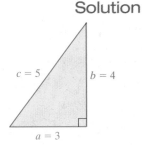

FIGURE 5.5

Translate We must always select the largest size to represent the hypotenuse, c. We will designate the length of leg a to be 3 inches and the length of leg b to be 4 inches. The length of the hypotenuse, c, will be 5 inches. See Figure 5.5.

$$a^2 + b^2 = c^2$$
$$3^2 + 4^2 \overset{?}{=} 5^2$$

Carry Out
$$9 + 16 \overset{?}{=} 25$$
$$25 = 25 \quad \textit{True}$$

Check and Answer Since using the Pythagorean Theorem results in a true statement, a right triangle can have the given sides.

b) We will work this part using the Pythagorean Theorem, as we did in part a). We will let leg a have a length of 2 inches, leg b have a length of 5 inches, and the hypotenuse, c, have a length of 7 inches.

$$a^2 + b^2 = c^2$$
$$2^2 + 5^2 \stackrel{?}{=} 7^2$$
$$4 + 25 \stackrel{?}{=} 49$$
$$29 = 49 \quad \textit{False}$$

Since 29 is not equal to 49, the Pythagorean Theorem does not hold for these lengths. Therefore, no right triangle can have sides with lengths of 2 inches, 5 inches and 7 inches.

NOW TRY EXERCISE 27

HELPFUL HINT

When drawing a right triangle, the hypotenuse, c, is always the side opposite the right angle. See Figures 5.6 a)–d).

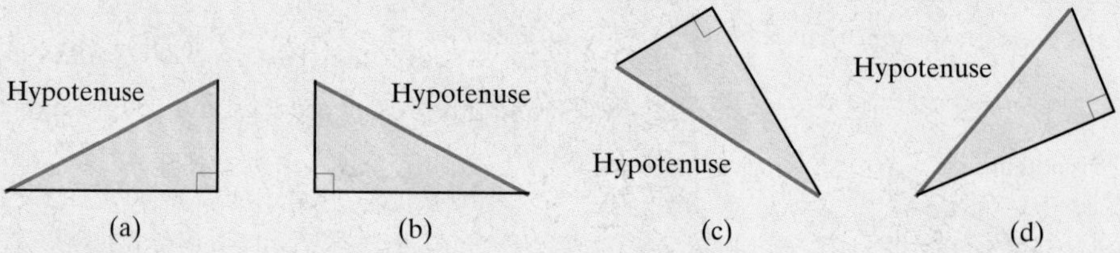

Hypotenuse (a)   Hypotenuse (b)   Hypotenuse (c)   Hypotenuse (d)

FIGURE 5.6

Notice that the hypotenuse is always the longest side of a right triangle.

EXAMPLE 5   **Using the Pythagorean Theorem**   One leg of a right triangle is 7 feet longer than the other leg. The hypotenuse is 13 feet. Find the dimensions of the right triangle.

Solution   Understand and Translate   We will first draw a diagram of the situation. See Figure 5.7.

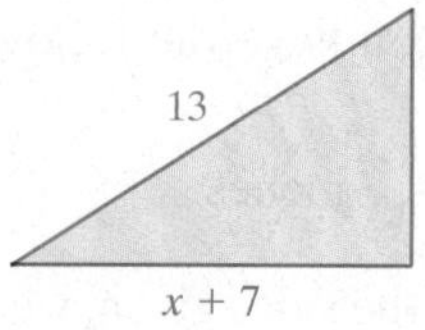

Now we will use the Pythagorean Theorem to determine the dimensions of the right triangle.

FIGURE 5.7

$$a^2 + b^2 = c^2$$
$$x^2 + (x + 7)^2 = 13^2$$

Carry Out
$$x^2 + (x^2 + 14x + 49) = 169$$
$$2x^2 + 14x - 120 = 0$$
$$2(x^2 + 7x - 60) = 0$$
$$2(x + 12)(x - 5) = 0$$

$$x + 12 = 0 \qquad \text{or} \qquad x - 5 = 0$$
$$x = -12 \qquad\qquad\qquad x = 5$$

Check and Answer   Since a length cannot be a negative number, the only answer is 5. The dimensions of the right triangle are 5 feet, $x + 7$ or 12 feet, and 13 feet. One leg is 5 feet, the other leg is 12 feet, and the hypotenuse is 13 feet.

EXAMPLE 6   **Laying Fiber Optic Cable**   Two telephone company crews are laying fiber optic cable from a specific point in the town of Stuckeyville. Crew A is laying cable going directly north, and Crew B is laying cable going directly east. See Figure 5.8a. At a certain point in time, Crew B has laid 1 mile more cable than Crew A. At that time, the distance between the two crews is two miles greater than the distance traveled by Crew A. See Figure 5.8b. Determine the distance traveled by Crew A and the distance between the crews.

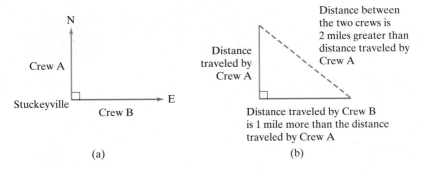

FIGURE 5.8   (a)   (b)

Solution   **Understand and Translate**   After reading the question carefully and looking at the figures, you should realize that we are working with a right triangle. Therefore we will use the Pythagorean Theorem to answer the question. To do so, we must express the distances in mathematical terms.

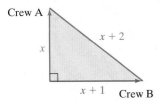

FIGURE 5.9

Let x = distance traveled by crew A,

then $x + 1$ = distance traveled by crew B,

and $x + 2$ = distance between the two crews.

Figure 5.9 illustrates this relationship. Now we use the Pythagorean Theorem. We will let x represent leg a, $x + 1$ represent leg b, and $x + 2$ represent the hypotenuse, c.

$$a^2 + b^2 = c^2$$
$$x^2 + (x + 1)^2 = (x + 2)^2$$

Carry Out   Since $(x + 1)^2 = (x + 1)(x + 1) = x^2 + 2x + 1$ and since $(x + 2)^2 = (x + 2)(x + 2) = x^2 + 4x + 4$, we write the above expression as

$$x^2 + (x^2 + 2x + 1) = x^2 + 4x + 4$$
$$2x^2 + 2x + 1 = x^2 + 4x + 4$$
$$x^2 - 2x - 3 = 0$$
$$(x - 3)(x + 1) = 0$$
$$x - 3 = 0 \quad \text{or} \quad x + 1 = 0$$
$$x = 3 \quad\quad\quad\quad x = -1$$

Check and Answer   Since a length cannot be negative, the only answer is 3. Thus the distance traveled by crew A, going north, is 3 miles. The distance between the crews is $x + 2$ or $3 + 2$ or 5 miles.

NOW TRY EXERCISE 37

Exercise Set 5.7

Concept/Writing Exercises   2. a) legs

1. What is a right triangle?   a triangle with a 90° angle

2. a) What are the smaller sides of a right triangle called?

    b) What is the longest side of a right triangle called?
      hypotenuse

3. State the Pythagorean Theorem. $a^2 + b^2 = c^2$

4. When designating the legs of a right triangle as a and b, does it make any difference which leg you call a and which leg you call b? Explain your answer.   no

Practice the Skills

In Exercises 5–8, determine the value of the question mark.

5.

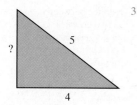

6.

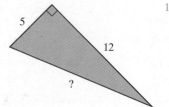

7.

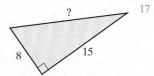

8.

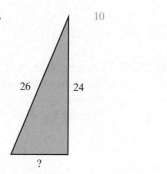

In Exercises 9–12, a *and* b *represent two legs of a right triangle and* c *represents the hypotenuse. Determine the value of the question mark (?).*

9. $a = 3, c = 5, b = ?$ 4

10. $a = 12, b = 16, c = ?$ 20

11. $a = 15, b = 36, c = ?$ 39

12. $b = 20, c = 25, a = ?$ 15

Problem Solving

Express each problem as an equation, then solve.

13. *Product of Numbers* The product of two positive integers is 117. Determine the two numbers if one is 4 more than the other.   9, 13

14. *Positive Integers* The product of two positive integers is 64. Determine the two integers if one number is 4 times the other.   4, 16

15. *Positive Integers* The product of two positive numbers is 84. Find the two numbers if one number is 2 more than twice the other.   6, 14

16. *Consecutive Integers* The product of two consecutive positive integers is 56. Find the two integers.   7, 8

17. *Consecutive Numbers* The product of two consecutive positive odd integers is 63. Determine the two integers.

18. *Consecutive Integers* The product of two consecutive positive even integers is 48. Find the two integers.   6, 8

19. *Area of Rectangle* The area of a rectangle is 36 square feet. Determine the length and width if the length is 4 times the width. $w = 3$ ft, $l = 12$ ft

17. 9, 7

20. *Area of Rectangle* The area of a rectangle is 84 square inches. Determine the length and width if the length is 2 inches less than twice the width. $w = 7$ in., $l = 12$ in.

21. *Rectangular Garden* Maureen Woolhouse has a rectangular garden whose width is 2/3 its length. If it's area is 150 square feet, determine the length and width of the garden. $w = 10$ ft, $l = 15$ ft

22. *Buying Wallpaper* Alejandro Ibanez wishes to buy a wallpaper border to go along the top of one wall in his living room. The length of the wall is 7 feet greater than its height.

  a) Find the length and height of the wall if the area of the wall is 120 square feet. $h = 8$ ft, $l = 15$ ft

  b) What is the length of the border he will need?   15 ft

  c) If the border cost $4 per linear foot, how much will the border cost? $60

23. *Square* If each side of a square is increased by 4 meters, the area becomes 49 square meters. Determine the length of a side of the original square.   3 m

24. *Sign* If the length of the sign in Example 2 is to be 2 feet longer than the width and the area is to be 35 square feet, determine the dimensions of the sign.
5 ft by 7 ft

25. *Dropped Egg* How long would it take for an egg dropped from a helicopter to fall 256 feet to the ground? See Example 3. 4 sec

26. *Falling Rock* How long would it take a rock that falls from a cliff 400 feet above the sea to hit the sea? 5 sec

In Exercises 27–30, determine if a right triangle can have the following sides where a *and* b *represent the legs and* c *represents the hypotenuse. Explain your answer.*

27. $a = 7, b = 24, c = 25$ yes

28. $a = 16, c = 20, b = 22$ no

29. $a = 9, b = 40, c = 41$ yes

30. $a = 13, b = 18, c = 28$ no

In Exercises 31–34, find the value of the question mark.

31.
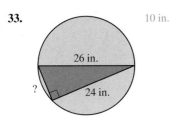
16 ft
34 ft
?
30 ft

32.
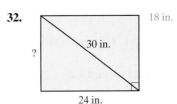
18 in.
30 in.
?
24 in.

33.
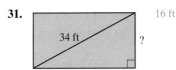
10 in.
26 in.
?
24 in.

34.
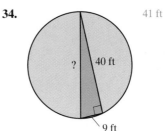
41 ft
40 ft
?
9 ft

36. 5 in., 12 in., 13 in.

35. *Triangle* One leg of a right triangle is 2 feet longer than the other leg. The hypotenuse is 10 feet. Find the lengths of the three sides of the triangle. 6 ft, 8 ft, 10 ft

36. *Triangle* One leg of a right triangle is two inches more than twice the other leg. The hypotenuse is 13 inches. Find the lengths of the three sides of the triangle.

37. *Car Ride* Alice and Bob start riding from the same point at the same time. Alice drives north and Bob drives west. Alice drives 3 miles less than 3 times the distance that Bob drives. At that instant the distance between Alice and Bob is three miles more than twice the distance Bob has driven. Find the distance between Alice and Bob. 13 miles

38. *Motorboats* Gwen and Jennifer start in motorboats at the same time. Gwen travels south while Jennifer travels west. Jennifer travels 1 mile less than twice the distance that Gwen travels. At that time the distance between Gwen and Jennifer is 1 mile more than twice the distance that Gwen traveled. Find the distance that Jennifer traveled. 15 miles

39. *Rectangular Garden* Mary Ann Tuerk has constructed a rectangular garden. The length of the garden is 3 feet more than three times its width. The diagonal of the garden is 4 feet more than three times the width. Find the length and width of the garden. $w = 7$ ft, $l = 24$ ft

40. *Height of Tree* A tree is supported by ropes. One rope goes from the top of the tree to a point on the ground. The height of the tree is 4 feet more than twice the distance between the base of the tree and the rope anchored in the ground. The length of the rope is 6 feet more than twice the distance between the base of the tree and the rope anchored in the ground. Find the height of the tree. 24 ft

41. *Video Store* A video store owner finds that her daily profit, P, is approximated by the formula $P = x^2 - 15x - 50$, where x is the number of videos she sells. How many videos must she sell in a day for her profit to be $400?

42. *Water Sprinklers* The cost, C, for manufacturing x water sprinklers is given by the formula $C = x^2 - 27x - 20$. Determine the number of water sprinklers manufactured at a cost of $70. 30

41. 30 videos

43. Sum of Numbers The sum, s, of the first n even numbers is given by the formula $s = n^2 + n$. Determine n for the given sums:

a) $s = 20$ 4   **b)** $s = 90$ 9

44. Telephone Lines For a switchboard that handles n telephone lines, the maximum number of telephone connections, C, that it can make simultaneously is given by the formula

$$C = \frac{n(n-1)}{2}.$$

a) How many telephone connections can a switchboard make simultaneously if it handles 15 lines?
105

b) How many lines does a switchboard have if it can make 55 telephone connections simultaneously?   11

Challenge Problems

45. Area Determine the area of the rectangle.

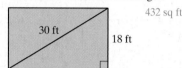

432 sq ft

30 ft

18 ft

46. Area Determine the area of the circle.

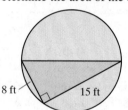

≈227 sq ft

8 ft   15 ft

47. Solve the equation $x^3 + 3x^2 - 10x = 0$. $0, 2, -5$

48. Create an equation whose solutions are 0, 3, and 5. Explain how you determined your answer.
$x^3 - 8x^2 + 15x = 0$

49. Numbers The product of two numbers is -40. Determine the numbers if their sum is 3. -5 and 8

50. Numbers The sum of two numbers is 9. The sum of the squares of the two numbers is 45. Determine the two numbers.   3 and 6

Group Activity

Discuss and solve Exercises 51 and 52 in groups.

51. Cost and Revenue The break-even point for a manufacturer occurs when its cost of production, C, is equal to its revenue, R. The cost equation for a company is $C = 2x^2 - 20x + 600$ and its revenue equation is $R = x^2 + 50x - 400$, where x is the number of units produced and sold. How many units must be produced and sold for the manufacturer to break even? There are two values.   20 and 50

52. Cannonball When a certain cannon is fired, the height, in feet, of the cannonball at time t can be found by using the formula $h = -16t^2 + 128t$.

a) Determine the height of the cannonball 3 seconds after being fired.   240 ft

b) Determine the time it takes for the cannonball to hit the ground. (*Hint:* What is the value of h at impact?)   8 sec

Cumulative Review Exercises

[3.2] **53.** Express the statement "five less than twice a number" as a mathematical expression. $2x - 5$

[4.4] **54.** Subtract $x^2 - 4x + 6$ from $3x + 2$.
$-x^2 + 7x - 4$

[4.5] **55.** Multiply $(3x^2 + 2x - 4)(2x - 1)$.
$6x^3 + x^2 - 10x + 4$

[4.6] **56.** Divide $\dfrac{6x^2 - 19x + 15}{3x - 5}$ by dividing the numerator by the denominator. $2x - 3$

[5.4] **57.** Divide $\dfrac{6x^2 - 19x + 15}{3x - 5}$ by factoring the numerator and dividing out common factors. $2x - 3$

CHAPTER SUMMARY
Key Words and Phrases

5.1
Composite number
Factor an expression
Factors
Greatest common factor
Prime factors
Prime number
Unit

5.2
Factor by grouping

5.3
Factor out the common factor
Prime polynomial
Trial-and-error method

5.4
Factor by grouping
Factor by trial and error

5.5
Difference of two squares
Difference of two cubes
Sum of two cubes

5.6
Quadratic equation
Standard form of a quadratic equation
Zero-factor property

5.7
Hypotenuse
Leg
Pythagorean Theorem
Right triangle

IMPORTANT FACTS

Difference of Two Squares
$$a^2 - b^2 = (a + b)(a - b)$$
Note: The sum of two squares, $a^2 + b^2$ cannot be factored using real numbers.

Sum of Two Cubes
$$a^3 + b^3 = (a + b)(a^2 - ab + b^2)$$

Difference of Two Cubes
$$a^3 - b^3 = (a - b)(a^2 + ab + b^2)$$

Zero-factor Property
If $a \cdot b = 0$, then $a = 0$ or $b = 0$.

General Procedure to Factor a Polynomial
1. If all the terms of the polynomial have a greatest common factor other than 1, factor it out.
2. If the polynomial has two terms, determine whether it is a difference of two squares or a sum or difference of two cubes. If so, factor using the appropriate formula.
3. If the polynomial has three terms, factor the trinomial using the methods discussed in Sections 5.3 and 5.4.
4. If the polynomial has more than three terms, try factoring by grouping.
5. As a final step, examine your factored polynomial to determine whether the terms in any factor have a common factor. If you find a common factor, factor it out at this point.

Pythagorean Theorem
If a and b represent the legs of a right triangle, and c represents the hypotenuse, then

$$a^2 + b^2 = c^2$$

Chapter Review Exercises

[5.1] *Find the greatest common factor for each set of terms.*

1. $3y^5, y^4, y^3$ y^3 **2.** $3p, 6p^2, 9p^3$ $3p$ **3.** $20a^3, 15a^2, 35a^5$ $5a^2$

4. $20x^2y^3, 25x^3y^4, 10x^5y^2z$ $5x^2y^2$ **5.** $9xyz, 12xz, 36, x^2y$ 1 **6.** $8ab, 12a^2b, 16, a^2b^2$ 1

7. $4(x-5), x-5$ $x-5$ **8.** $x(x+5), x+5$ $x+5$

15. $4x^3y^2(5+2x^6y-4x^2)$ **18.** $(5x+3)(x-2)$
19. $(x+2)(5x-2)$ **20.** $(4x-3)(2x+1)$

Factor each expression. If an expression is prime, so state.

9. $4x-12$ $4(x-3)$ **10.** $35x-5$ $5(7x-1)$ **11.** $24y^2-4y$ $4y(6y-1)$

12. $55p^3-20p^2$ $5p^2(11p-4)$ **13.** $60a^2b-36ab^2$ $12ab(5a-3b)$ **14.** $6xy-12x^2y$ $6xy(1-2x)$

15. $20x^3y^2+8x^9y^3-16x^5y^2$ **16.** $24x^2-13y^2+6xy$ prime   **17.** $14a^2b-7b-a^3$ prime

18. $x(5x+3)-2(5x+3)$ **19.** $5x(x+2)-2(x+2)$ **20.** $2x(4x-3)+4x-3$

28. $(5x-y)(x+4y)$ **29.** $(x+3y)(4x-5y)$ **30.** $(3a-5b)(2a-b)$
[5.2] *Factor by grouping.* **32.** $(x-3y)(3x+2y)$ **33.** $(a+2b)(7a-b)$

21. $x^2+6x+2x+12$ $(x+6)(x+2)$ **22.** $x^2-5x+4x-20$ $(x-5)(x+4)$ **23.** $y^2-9y-9y+81$ $(y-9)(y-9)$

24. $4a^2-4ab-a+b$ $(a-b)(4a-1)$ **25.** $3xy+3x+2y+2$ $(y+1)(3x+2)$ **26.** $x^2+3x-2xy-6y$ $(x+3)(x-2y)$

27. $2x^2+12x-x-6$ $(x+6)(2x-1)$ **28.** $5x^2-xy+20xy-4y^2$ **29.** $4x^2+12xy-5xy-15y^2$

30. $6a^2-10ab-3ab+5b^2$ **31.** $ab-a+b-1$ $(b-1)(a+1)$ **32.** $3x^2-9xy+2xy-6y^2$

33. $7a^2+14ab-ab-2b^2$ **34.** $6x^2+9x-2x-3$ $(2x+3)(3x-1)$

[5.3] *Factor completely. If an expression cannot be factored, so state.*

35. x^2-x-6 $(x-3)(x+2)$ **36.** $x^2+4x-15$ prime   **37.** $x^2-13x+42$ $(x-6)(x-7)$

38. b^2+b-20 $(b+5)(b-4)$ **39.** $n^2+3n-40$ $(n+8)(n-5)$ **40.** $x^2-15x+56$ $(x-8)(x-7)$

41. $c^2-10c-20$ prime   **42.** $x^2+11x-24$ prime   **43.** x^3-17x^2+72x $x(x-9)(x-8)$

44. x^3-3x^2-40x $x(x-8)(x+5)$ **45.** $x^2-2xy-15y^2$ $(x+3y)(x-5y)$ **46.** $4x^3+32x^2y+60xy^2$
$4x(x+5y)(x+3y)$

[5.4] *Factor completely. If an expression is prime, so state.*

47. $2x^2-x-15$ $(2x+5)(x-3)$ **48.** $3x^2-13x+4$ $(3x-1)(x-4)$ **49.** $4x^2-9x+5$ $(4x-5)(x-1)$

50. $5m^2-14m+8$ $(5m-4)(m-2)$ **51.** $9x^2+3x-2$ $(3x-1)(3x+2)$ **52.** $5x^2-32x+12$ $(5x-2)(x-6)$

53. $2t^2+14t+9$ prime   **54.** $6s^2+13s+5$ $(2s+1)(3s+5)$ **55.** $5x^2+37x-24$ $(5x-3)(x+8)$

56. $6x^2+11x-10$ $(3x-2)(2x+5)$ **57.** $12x^2+2x-4$ $2(3x+2)(2x-1)$ **58.** $9x^2-6x+1$ $(3x-1)^2$

59. $9x^3-12x^2+4x$ $x(3x-2)^2$ **60.** $18x^3+12x^2-16x$ $2x(3x+4)(3x-2)$ **61.** $16a^2-22ab-3b^2$ $(8a+b)(2a-3b)$

62. $4a^2-16ab+15b^2$ $(2a-3b)(2a-5b)$

[5.5] *Factor completely.* **70.** $(10x^2+11y^2)(10x^2-11y^2)$ **77.** $(5a+b)(25a^2-5ab+b^2)$ **80.** $3(x-4y)(x^2+4xy+16y^2)$

63. x^2-36 $(x+6)(x-6)$ **64.** x^2-100 $(x+10)(x-10)$ **65.** $4x^2-16$ $4(x+2)(x-2)$

66. $81x^2-9y^2$ $9(3x+y)(3x-y)$ **67.** $81-a^2$ $(9+a)(9-a)$ **68.** $64-x^2$ $(8+x)(8-x)$

69. $16x^4-49y^2$ $(4x^2+7y)(4x^2-7y)$ **70.** $100x^4-121y^4$ **71.** x^3-y^3 $(x-y)(x^2+xy+y^2)$

72. x^3+y^3 $(x+y)(x^2-xy+y^2)$ **73.** x^3-1 $(x-1)(x^2+x+1)$ **74.** x^3+8 $(x+2)(x^2-2x+4)$

75. a^3+27 $(a+3)(a^2-3a+9)$ **76.** b^3-64 $(b-4)(b^2+4b+16)$ **77.** $125a^3+b^3$

78. $27-8y^3$ $(3-2y)(9+6y+4y^2)$ **79.** $27x^4-75y^2$ $3(3x^2+5y)(3x^2-5y)$ **80.** $3x^3-192y^3$

[5.1–5.5] Factor completely.   **92.** $3x(2x + 3)(x + 5)$ **99.** $2x(2x + 5y)(x + 2y)$

81. $x^2 - 14x + 48$ $(x - 6)(x - 8)$

82. $3x^2 - 18x + 27$ $3(x - 3)^2$

83. $4a^2 - 64$ $4(a + 4)(a - 4)$

84. $4y^2 - 36$ $4(y + 3)(y - 3)$

85. $8x^2 + 16x - 24$ $8(x + 3)(x - 1)$

86. $x^2 - 6x - 27$ $(x - 9)(x + 3)$

87. $9x^2 - 6x + 1$ $(3x - 1)^2$

88. $4x^2 + 7x - 2$ $(4x - 1)(x + 2)$

89. $6b^3 - 6$ $6(b - 1)(b^2 + b + 1)$

90. $x^3y - 27y$ $y(x - 3)(x^2 + 3x + 9)$

91. $a^2b - 2ab - 15b$ $b(a + 3)(a - 5)$

92. $6x^3 + 30x^2 + 9x^2 + 45x$

93. $x^2 - 4xy + 3y^2$ $(x - 3y)(x - y)$

94. $3m^2 + 2mn - 8n^2$ $(3m - 4n)(m + 2n)$

95. $4x^2 - 20xy + 25y^2$ $(2x - 5y)^2$

96. $25a^2 - 49b^2$ $(5a + 7b)(5a - 7b)$

97. $xy - 7x + 2y - 14$ $(x + 2)(y - 7)$

98. $16y^5 - 25y^7$ $y^5(4 + 5y)(4 - 5y)$

99. $4x^3 + 18x^2y + 20xy^2$

100. $6x^2 + 5xy - 21y^2$ $(2x - 3y)(3x + 7y)$

101. $16x^4 - 8x^3 - 3x^2$ $x^2(4x + 1)(4x - 3)$

102. $a^4 - 1$ $(a^2 + 1)(a + 1)(a - 1)$

[5.6] Solve.

103. $x(x - 5) = 0$ $0, 5$

104. $(a - 2)(a + 6) = 0$ $2, -6$

105. $(x + 5)(4x - 3) = 0$ $-5, \frac{3}{4}$

106. $x^2 - 3x = 0$ $0, 3$

107. $5x^2 + 20x = 0$ $0, -4$

108. $6x^2 + 18x = 0$ $0, -3$

109. $r^2 + 9r + 18 = 0$ $-3, -6$

110. $x^2 - 12 = -x$ $-4, 3$

111. $x^2 - 3x = -2$ $1, 2$

112. $15x + 12 = -3x^2$ $-1, -4$

113. $x^2 - 6x + 8 = 0$ $2, 4$

114. $3p^2 + 6p = 45$ $3, -5$

115. $8x^2 - 3 = -10x$ $\frac{1}{4}, -\frac{3}{2}$

116. $2x^2 + 15x = 8$ $\frac{1}{2}, -8$

117. $4x^2 - 16 = 0$ $2, -2$

118. $49x^2 - 100 = 0$ $\frac{10}{7}, -\frac{10}{7}$

119. $8x^2 - 14x + 3 = 0$ $\frac{3}{2}, \frac{1}{4}$

120. $-48x = -12x^2 - 45$ $\frac{3}{2}, \frac{5}{2}$

[5.7] **121.** State the Pythagorean Theorem. $a^2 + b^2 = c^2$

122. What is the longest side of a right triangle called?
hypotenuse

In Exercises 123 and 124, determine the value of the question mark (?).

123.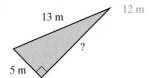
13 m   12 m   ?   5 m

124.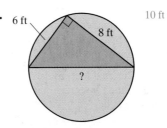
6 ft   10 ft   8 ft   ?

Express each problem as an equation, then solve.   **127.** $w = 7$ ft, $l = 9$ ft

125. *Product of Numbers* The product of two consecutive positive even integers is 48. Determine the two integers. $6, 8$

126. *Product of Numbers* The product of two positive integers is 56. Determine the integers if the larger is 6 more than twice the smaller. $4, 14$

127. *Area of Rectangle* The area of a rectangle is 63 square feet. Determine the length and width of the rectangle if the length is 2 feet greater than the width.

128. *Square* The length of each side of a square is made smaller by 4 inches. If the area of the resulting square is 25 square inches, determine the length of a side of the original square.   9 in.

129. *Right Triangle* One leg of a right triangle is 7 feet longer than the other leg. The hypotenuse is 9 feet longer than the shortest leg. Find the lengths of the three sides of the triangle.   8 ft, 15 ft, 17 ft

130. *Wading Pool* Jason has a rectangular wading pool. The length of the pool is 2 feet greater than the width of the pool. A diagonal across the pool is 4 feet greater

than the width of the pool. Find the length of the diagonal across the pool.   10 ft

131. *Falling Apple* How long would it take an apple that falls off a 16-foot tree to hit the ground?   1 sec

132. *Baking Cookies* The Pine Hills Neighborhood Association has determined that the cost, C, to make x dozen cookies can be estimated by the formula $C = x^2 - 79x + 20$. If they have $100 to be used to make the cookies, how many dozen cookies can the association make to sell at a fund-raiser?   80 dozen

Chapter Practice Test

1. Determine the greatest common factor of $9y^5$, $15y^3$, and $27y^4$. $3y^3$
2. Determine the greatest common factor of $6x^2y^3$, $9xy^2$, and $12xy^5$. $3xy^2$

Factor completely.

3. $5x^2y^3 - 15x^5y^2$ $5x^2y^2(y - 3x^3)$ 4. $8a^3b - 12a^2b^2 + 28a^2b$ $4a^2b(2a - 3b + 7)$ 5. $5x^2 - 15x + 2x - 6$ $(5x + 2)(x - 3)$
6. $a^2 - 4ab - 5ab + 20b^2$ $(a - 4b)(a - 5b)$ 7. $r^2 + 5r - 24$ $(r + 8)(r - 3)$ 8. $25a^2 - 5ab - 6b^2$ $(5a - 3b)(5a + 2b)$
9. $4x^2 - 16x - 48$ $4(x + 2)(x - 6)$ 10. $2x^3 - 3x^2 + x$ $x(2x - 1)(x - 1)$ 11. $12x^2 - xy - 6y^2$ $(3x + 2y)(4x - 3y)$
12. $x^2 - 9y^2$ $(x + 3y)(x - 3y)$ 13. $x^3 + 27$ $(x + 3)(x^2 - 3x + 9)$

Solve.

14. $(5x - 3)(x - 1) = 0$ $\frac{3}{5}, 1$ 15. $x^2 - 6x = 0$ $0, 6$ 16. $x^2 = 64$ $-8, 8$
17. $x^2 - 14x + 49 = 0$ 7 18. $x^2 + 6 = -5x$ $-2, -3$ 19. $x^2 - 7x + 12 = 0$ $3, 4$

20. **Right Triangle** Find the length of the side indicated with a question mark. 24 in.

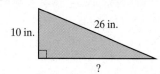

10 in.　26 in.　?

21. **Right Triangle** In a right triangle, one leg is 2 feet less than twice the length of the smaller leg. The hypotenuse is 2 feet more than twice the length of the smaller leg. Determine the hypotenuse of the triangle. 34 ft

22. **Product of Integers** The product of two positive integers is 36. Determine the two integers if the larger is 1 more than twice the smaller. 4, 9

23. **Consecutive Integers** The product of two positive consecutive odd integers is 99. Determine the integers. 9, 11

24. **Rectangle** The area of a rectangle is 24 square meters. Determine the length and width of the rectangle if its length is 2 meters greater than its width. $l = 6\,\text{m}, w = 4\,\text{m}$

25. **Fallen Object** How long would it take for an object dropped from a hot air balloon to fall 1600 feet to the ground? 10 sec

See Exercise 25

Cumulative Review Test

5. $|-4|$ 7. 19.2

Take the following test and check your answers with those that appear at the end of the test. Review any questions that you answered incorrectly. The section and objective where the material was covered is indicated after the answer.

1. Evaluate $4 - 5(2x + 4x^2 - 21)$ when $x = -3$. -41
2. Evaluate $5x^2 - 3y + 7(2 + y^2 - 4x)$ when $x = 3$ and $y = -2$. 9
3. **Hotel Room** The cost of a hotel room including a 12% state tax and 3% county tax is $103.50. Determine the cost of the room before tax. $90
4. Consider the set of numbers
$$\left\{-6, -0.2, \frac{3}{5}, \sqrt{7}, -\sqrt{2}, 7, 0, -\frac{5}{9}, 1.34\right\}.$$

List the elements that are
a) natural numbers. 7
b) rational numbers. $-6, -0.2, \frac{3}{5}, 7, 0, -\frac{5}{9}, 1.34$
c) irrational numbers. $\sqrt{7}, -\sqrt{2}$
d) real numbers. $-6, -0.2, \frac{3}{5}, \sqrt{7}, -\sqrt{2}, 7, 0, -\frac{5}{9}, 1.34$
5. Which is greater, $|-4|$ or $-|2|$? Explain your answer.
6. Solve the equation $4x - 2 = 4(x - 7) + 2x$ for x. 13
7. Solve the proportion $\frac{5}{12} = \frac{8}{x}$ for x by cross-multiplying.

8. $x \geq 5$ ⟵————⟶
                    5

8. Solve the inequality $3x - 5 \geq 10(6 - x)$, and graph the solution on a number line.

9. Solve the equation $5x - 2y = 6$ for y. $y = \frac{5}{2}x - 3$

10. *Cross-country Skiing* Two cross-country skiers follow the same trail in a local park. Brooke Stoner skis at a rate of 8 kilometers per hour and Bob Thoresen skis at a rate of 4 kilometers per hour. How long will it take Brooke to catch Bob if she leaves 15 minutes after he does? $\frac{1}{4}$ hr

11. *Acid Solution* How many liters of a 10% acid solution must be mixed with three liters of a 4% acid solution to get an 8% acid solution? $6\,\ell$

12. *Consecutive Integers* The sum of two consecutive odd integers is 96. Find the two integers. $47, 49$

13. Simplify $\left(\dfrac{3x}{5y^2}\right)^3$. $\dfrac{27x^3}{125y^6}$

14. Simplify $(2x^{-3})^{-2}(4x^{-3}y^2)^3$. $\dfrac{16y^6}{x^3}$

15. Subtract $(4x^3 - 3x^2 + 7)$ from $(x^3 - x^2 + 6x - 5)$. $-3x^3 + 2x^2 + 6x - 12$

Perform the operations indicated.

16. $(3x - 2)(x^2 + 5x - 6)$ $3x^3 + 13x^2 - 28x + 12$

17. $\dfrac{x^2 - 2x + 6}{x + 3}$ $x - 5 + \dfrac{21}{x + 3}$

18. Factor $ab + 3b - 6a - 18$ by grouping. $(a + 3)(b - 6)$

19. Factor $x^2 - 2x - 63$. $(x + 7)(x - 9)$

20. Factor $5x^3 - 125x$. $5x(x + 5)(x - 5)$

Answers to Cumulative Review Test

1. -41; [Sec. 1.9, Obj. 6] **2.** \$9; [Sec. 1.9, Obj. 6] **3.** \$90; [Sec. 3.3, Obj. 4] **4. a)** 7 **b)** $-6, -0.2, \frac{3}{5}, 7, 0, -\frac{5}{9}, 1.34$ **c)** $\sqrt{7}, -\sqrt{2}$ **d)** $-6, -0.2, \frac{3}{5}, \sqrt{7}, -\sqrt{2}, 7, 0, -\frac{5}{9}, 1.34$; [Sec. 1.4, Obj. 2] **5.** $|-4|$, since $|-4| = 4$ and $-|2| = -2$; [Sec. 1.5, Obj. 2] **6.** 13; [Sec. 2.5, Obj. 1] **7.** 19.2; [Sec. 2.6, Obj. 2] **8.** $x \geq 5$, ⟵————⟶ ; [Sec. 2.7, Obj. 1]
  5

9. $y = \frac{5}{2}x - 3$; [Sec. 3.1, Obj. 3] **10.** $\frac{1}{4}$ hour; [Sec. 3.5, Obj. 2] **11.** 6 liters; [Sec. 3.5, Obj. 4] **12.** 47, 49; [Sec. 3.3, Obj. 2]

13. $\dfrac{27x^3}{125y^6}$; [Sec. 4.1, Obj. 2] **14.** $\dfrac{16y^6}{x^3}$; [Sec. 4.2, Obj. 2] **15.** $-3x^3 + 2x^2 + 6x - 12$; [Sec. 4.4, Obj. 3]

16. $3x^3 + 13x^2 - 28x + 12$; [Sec. 4.5, Obj. 6] **17.** $x - 5 + \dfrac{21}{x + 3}$; [Sec. 4.6, Obj. 2] **18.** $(a + 3)(b - 6)$; [Sec. 5.2, Obj. 1]

19. $(x + 7)(x - 9)$; [Sec. 5.3, Obj. 1] **20.** $5x(x + 5)(x - 5)$; [Sec. 5.5, Obj. 3]

Chapter 6

Rational Expressions and Equations

Technological advances in transportation in the past 50 years have changed the way people travel and have significantly decreased traveling time. Rail travelers in Japan have benefited from the country's commitment to high-speed railroads. In Exercise 24 on page 424, we use a rational equation to determine the distance between two cities by comparing a traveler's time riding one of Japan's new bullet trains to the time the traveler would have spent making the same trip on an older train.

SSM

Study Guide

CD/Video

MathPro 4/5

PH Math Tutor Center

prenhall.com/Angel

A Look Ahead

Y ou worked with rational numbers when you worked with fractions in arithmetic. Now you will expand your knowledge to include fractions that contain variables. Fractions that contain variables are often referred to as rational expressions. The same basic procedures that you used to simplify (or reduce), add, subtract, multiply, and divide arithmetic fractions will be used with rational expressions in Sections 6.1 through 6.5. You might wish to review Section 1.3 since the material presented in this chapter builds upon the procedures presented there.

Many real-life applications involve equations that contain rational expressions. Such equations are called rational equations. We will solve rational equations in Section 6.6 and give various real-life applications of rational equations, including the use of several familiar formulas, in Sections 6.6–6.8.

To be successful in this chapter you need to have a complete understanding of factoring, which was presented in Chapter 5. Sections 5.3 and 5.4 are especially important.

6.1 SIMPLIFYING RATIONAL EXPRESSIONS

SSM · Study Guide · CD/Video

MathPro 4/5 · PH Math Tutor Center · prenhall.com/Angel

1 Determine the values for which a rational expression is defined.

2 Understand the three signs of a fraction.

3 Simplify rational expressions.

4 Factor a negative 1 from a polynomial.

1 Determine the Values for Which a Rational Expression Is Defined

We begin this chapter by defining a *rational expression*.

> A **rational expression** is an expression of the form p/q, where p and q are polynomials and $q \neq 0$.

Examples of Rational Expressions

$$\frac{4}{5}, \quad \frac{x-6}{x}, \quad \frac{x^2+2x}{x-3}, \quad \frac{a}{a^2-4}$$

The denominator of a rational expression cannot equal 0 since division by 0 is not defined. In the expression $\frac{x+3}{x}$, the value of x cannot be 0 because the denominator would then equal 0. We say that the expression $\frac{x+3}{x}$ is *defined* for all real numbers except 0. It is *undefined* when x is 0. In $\frac{x^2+4x}{x-3}$, the value of x cannot be 3 because the denominator would then be 0. What values of x cannot be used in the expression $\frac{x}{x^2-4}$? If you answered 2 and −2, you answered correctly. **Whenever we have a rational expression containing a variable in the denominator, we always assume that the value or values of the variable that make the denominator 0 are excluded.**

One method that can be used to determine the value or values of the variable that are excluded is to set the denominator equal to 0 and then solve the resulting equation for the variable.

TEACHING TIP
Have students suggest two arithmetic fractions. Use the fractions to set up an addition, subtraction, multiplication, and division problem. Create rational expressions by replacing the denominator of one fraction and the numerator of the other fraction with variable expressions. Have students work in pairs to create a list of problems that might be included in this chapter. Then make a class list. Refer to this class list throughout the chapter as appropriate.

EXAMPLE 1 Determine the value or values of the variable for which the rational expression is defined.   **a)** $\dfrac{1}{x-4}$ **b)** $\dfrac{x+1}{2x-7}$ **c)** $\dfrac{x+3}{x^2+6x-7}$

Solution **a)** We need to determine the value or values of x that make $x - 4$ equal to 0 and exclude these values. By looking at the denominator we can see that when $x = 4$, the denominator is $4 - 4$ or 0. Thus, we do not consider $x = 4$ when we consider the rational expression $\dfrac{1}{x - 4}$. This expression is defined for all real numbers except 4. We will sometimes shorten our answer and write $x \neq 4$.

b) We need to determine the value or values of x that make $2x - 7$ equal to 0 and exclude these. We can do this by setting $2x - 7$ equal to 0 and solving the equation for x.

$$2x - 7 = 0$$
$$2x = 7$$
$$x = \frac{7}{2}$$

Thus, we do not consider $x = \dfrac{7}{2}$ when we consider the rational expression $\dfrac{x + 1}{2x - 7}$. This expression is defined for all real numbers except $x = \dfrac{7}{2}$. We will sometimes shorten our answer and write $x \neq \dfrac{7}{2}$.

c) To determine the value or values that are excluded, we set the denominator equal to zero and solve the equation for the variable.

$$x^2 + 6x - 7 = 0$$
$$(x + 7)(x - 1) = 0$$
$$x + 7 = 0 \qquad \text{or} \qquad x - 1 = 0$$
$$x = -7 \qquad\qquad x = 1$$

Therefore, we do not consider the values $x = -7$ or $x = 1$ when we consider the rational expression $\dfrac{x + 3}{x^2 + 6x + 7}$. Both $x = -7$ and $x = 1$ make the denominator zero. This expression is defined for all real numbers except $x = -7$ and $x = 1$. Thus, $x \neq -7$ and $x \neq 1$.

NOW TRY EXERCISE 15

TEACHING TIP
After discussing Example 1, show students how to create a rational expression that is not defined for a certain value. Have students write a rational expression that is defined for all real numbers except

a) 5 **b)** -3 **c)** -3 and 5

2 Understand the Three Signs of a Fraction

Three **signs** are associated with any fraction: the sign of the numerator, the sign of the denominator, and the sign of the fraction.

$$\text{Sign of fraction} \longrightarrow +\frac{-a}{+b}$$

Sign of numerator

Sign of denominator

Whenever any of the three signs is omitted, we assume it to be positive. For example,

$$\frac{a}{b} \quad \text{means} \quad +\frac{+a}{+b}$$

$$\frac{-a}{b} \quad \text{means} \quad +\frac{-a}{+b}$$

$$-\frac{a}{b} \quad \text{means} \quad -\frac{+a}{+b}$$

TEACHING TIP
Point out that the value of the fraction is either positive or negative, or zero. For class consistency, suggest that students give their positive answers without any signs and their negative answer with only 1 negative sign in front of the fraction.

Changing any two of the three signs of a fraction does not change the value of a fraction. Thus,

$$\frac{-a}{b} = -\frac{a}{b} = \frac{a}{-b}$$

Generally, we do not write a fraction with a negative denominator. For example, the expression $\frac{2}{-5}$ would be written as either $\frac{-2}{5}$ or $-\frac{2}{5}$. The expression $\frac{x}{-(4-x)}$ can be written $\frac{x}{x-4}$ since $-(4-x) = -4 + x$ or $x - 4$.

3 Simplify Rational Expressions

A rational expression is **simplified** or **reduced to its lowest terms** when the numerator and denominator have no common factors other than 1. The fraction $\frac{9}{12}$ is not simplified because 9 and 12 both contain the common factor 3. When the 3 is factored out, the simplified fraction is $\frac{3}{4}$.

$$\frac{9}{12} = \frac{\overset{1}{\cancel{3}} \cdot 3}{\underset{1}{\cancel{3}} \cdot 4} = \frac{3}{4}$$

The rational expression $\frac{ab - b^2}{2b}$ is not simplified because both the numerator and denominator have a common factor, b. To simplify this expression, factor b from each term in the numerator, then divide it out.

$$\frac{ab - b^2}{2b} = \frac{\cancel{b}(a - b)}{2\cancel{b}} = \frac{a - b}{2}$$

Thus, $\frac{ab - b^2}{2b}$ becomes $\frac{a - b}{2}$ when simplified.

TEACHING TIP
Before discussing Examples 2–5 ask, "Is the numerator in factored form? Is the denominator in factored form?" Suggest that students ask themselves these questions whenever simplifying rational expressions.

To Simplify Rational Expressions

1. Factor both the numerator and denominator as completely as possible.

2. Divide out any factors common to both the numerator and denominator.

EXAMPLE 2 Simplify $\dfrac{5x^3 + 10x^2 - 25x}{10x^2}$.

Solution Factor the greatest common factor, $5x$, from each term in the numerator. Since $5x$ is a factor common to both the numerator and denominator, divide it out.

$$\frac{5x^3 + 10x^2 - 25x}{10x^2} = \frac{\cancel{5x}(x^2 + 2x - 5)}{\cancel{5x} \cdot 2x} = \frac{x^2 + 2x - 5}{2x}$$

NOW TRY EXERCISE 33

HELPFUL HINT

In Example 2 we are *simplifying* a polynomial divided by a monomial using factoring. In Section 4.6 we *divided* polynomials by monomials by writing each term in the numerator over the expression in the denominator. For example,

$$\frac{5x^3 + 10x^2 - 25x}{10x^2} = \frac{5x^3}{10x^2} + \frac{10x^2}{10x^2} - \frac{25x}{10x^2}$$

$$= \frac{x}{2} + 1 - \frac{5}{2x}$$

The answer above, $\frac{x}{2} + 1 - \frac{5}{2x}$, is equivalent to the answer obtained by factoring in Example 2, $\frac{x^2 + 2x - 5}{2x}$, as shown below.

$$\frac{x}{2} + 1 - \frac{5}{2x}$$

$$= \frac{x}{x} \cdot \frac{x}{2} + \frac{2x}{2x} - \frac{5}{2x} \qquad \textit{Write each term with LCD 2x.}$$

$$= \frac{x^2}{2x} + \frac{2x}{2x} - \frac{5}{2x}$$

$$= \frac{x^2 + 2x - 5}{2x}$$

When asked to *simplify* an expression we will factor numerators and denominators, when possible, then divide out common factors. This process was illustrated in Example 2 and will be further illustrated in Examples 3 through 5.

EXAMPLE 3 Simplify $\frac{x^2 + 2x - 3}{x + 3}$.

Solution Factor the numerator; then divide out the common factor.

$$\frac{x^2 + 2x - 3}{x + 3} = \frac{(x + 3)(x - 1)}{x + 3} = x - 1$$

EXAMPLE 4 Simplify $\frac{r^2 - 25}{r - 5}$.

Solution Factor the numerator; then divide out common factors.

$$\frac{r^2 - 25}{r - 5} = \frac{(r + 5)(r - 5)}{r - 5} = r + 5$$

EXAMPLE 5 Simplify $\frac{3x^2 - 10x - 8}{x^2 + 3x - 28}$

Solution Factor both the numerator and denominator, then divide out common factors.

$$\frac{3x^2 - 10x - 8}{x^2 + 3x - 28} = \frac{(3x + 2)(x - 4)}{(x + 7)(x - 4)} = \frac{3x + 2}{x + 7}.$$

NOW TRY EXERCISE 41

Note that $\frac{3x + 2}{x + 7}$ cannot be simplified any further.

AVOIDING COMMON ERRORS

Remember: Only common *factors* can be divided out from expressions.

CORRECT

$$\frac{\overset{5}{\cancel{20}}\,\overset{x^2}{\cancel{x^2}}}{\underset{1}{\cancel{4}}\,\underset{1}{\cancel{x}}} = 5x$$

INCORRECT

In the denominator of the example on the left, $4x$, the 4 and x are factors since they are *multiplied* together. The 4 and the x are also both factors of the numerator $20x^2$, since $20x^2$ can be written $4 \cdot x \cdot 5x$.

Some students incorrectly divide out *terms*. In the expression $\dfrac{x^2 - 20}{x - 4}$, the x and -4 are *terms* of the denominator, not factors, and therefore cannot be divided out.

4 Factor a Negative 1 from a Polynomial

Recall that when -1 is factored from a polynomial, the sign of each term in the polynomial changes.

Examples

$$-3x + 7 = -1(3x - 7) = -(3x - 7)$$
$$5 - 2x = -1(-5 + 2x) = -(2x - 5)$$
$$-2x^2 + 3x - 4 = -1(2x^2 - 3x + 4) = -(2x^2 - 3x + 4)$$

Whenever the terms in a numerator and denominator differ only in their signs (one is the opposite or additive inverse of the other), we can factor out -1 from either the numerator or denominator and then divide out the common factor. This procedure is illustrated in Example 6.

EXAMPLE 6   Simplify $\dfrac{3x - 7}{7 - 3x}$.

Solution   Since each term in the numerator differs only in sign from its like term in the denominator, we will factor -1 from each term in the denominator.

$$\frac{3x - 7}{7 - 3x} = \frac{3x - 7}{-1(-7 + 3x)}$$

$$= \frac{\cancel{3x - 7}}{-\cancel{(3x - 7)}}$$

$$= -1$$

HELPFUL HINT

In Example 6 we found that $\dfrac{3x - 7}{7 - 3x} = -1$. Note that the numerator, $3x - 7$, and the denominator, $7 - 3x$, are opposites since they differ only in sign. That is,

$$\frac{3x - 7}{7 - 3x} = \frac{3x - 7}{-(3x - 7)} = -1$$

Whenever we have the quotient of two expressions that are opposites, such as $\dfrac{a - b}{b - a}$, $a \neq b$, the quotient can be replaced by -1.

We will use the Helpful Hint given on the previous page in Example 7.

EXAMPLE 7 Simplify $\dfrac{4n^2 - 23n - 6}{6 - n}$.

Solution

$$\dfrac{4n^2 - 23n - 6}{6 - n} = \dfrac{(4n + 1)(n - 6)}{6 - n}$$ The terms in $n - 6$ differ only in sign from the terms in $6 - n$.

$$= (4n + 1)(-1)$$ Replace $\dfrac{n - 6}{6 - n}$ with -1.

$$= -(4n + 1)$$

NOW TRY EXERCISE 49   Note that $-4n - 1$ is also an acceptable answer.

1. a) and **b)** Answers will vary.   **2.** Set the denominator equal to zero and then solve the resulting equation.   **3.** The value of the variable does not make the denominator equal to 0.

Exercise Set 6.1

Concept/Writing Exercises

1. a) In your own words, define a rational expression.

  b) Give three examples of rational expressions.

2. Explain how to determine the value or values of the variable that makes a rational expression undefined.

3. In any rational expression with a variable in the denominator, what do we always assume about the variable?

4. In your own words, explain how to simplify a rational expression.   Answers will vary.

Explain why the following expressions cannot be simplified.

5. $\dfrac{2 + 3x}{4}$ **5.** and **6.** There is no factor common to both the numerator and the denominator.

6. $\dfrac{5x + 4y}{12xy}$

Explain why x *can represent any real number in the following expressions.*

7. $\dfrac{x - 6}{x^2 + 2}$ **7.** and **8.** The denominator cannot be 0.

8. $\dfrac{x + 3}{x^2 + 4}$

In Exercises 9 and 10, determine what values, if any, x *cannot represent. Explain.*

9. $\dfrac{x + 3}{x - 2}$ $x \neq 2$

10. $\dfrac{x}{(x - 4)^2}$ $x \neq 4$

11. Is $-\dfrac{x + 5}{5 - x}$ equal to -1? Explain.   no

12. Is $-\dfrac{3x + 2}{-3x - 2}$ equal to 1? Explain.   yes

Practice the Skills

Determine the value or values of the variable where each expression is defined.

13. $\dfrac{x - 3}{x}$ all real numbers except $x = 0$

14. $\dfrac{3}{r + 4}$ all real numbers except $r = -4$

15. $\dfrac{7}{4n - 12}$ all real numbers except $n = 3$

16. $\dfrac{5}{2x - 3}$ all real numbers except $x = \dfrac{3}{2}$

17. $\dfrac{x + 4}{x^2 - 4}$ all real numbers except $x = 2, x = -2$

18. $\dfrac{7}{x^2 + 4x - 5}$ all real numbers except $x = -5, x = 1$

19. $\dfrac{x - 3}{2x^2 - 9x + 9}$ all real numbers except $x = \dfrac{3}{2}, x = 3$

20. $\dfrac{x^2 + 3}{2x^2 - 13x + 15}$ all real numbers except $x = 5, x = \dfrac{3}{2}$

21. $\dfrac{x}{x^2 + 16}$ all real numbers

22. $\dfrac{6}{9 + x^2}$ all real numbers

23. $\dfrac{p + 4}{4p^2 - 25}$ all real numbers except $p = \dfrac{5}{2}, p = -\dfrac{5}{2}$

24. $\dfrac{3}{9r^2 - 16}$ all real numbers except $r = \dfrac{4}{3}, r = -\dfrac{4}{3}$

Simplify the following rational expressions which contain a monomial divided by a monomial. This material was covered in Sections 4.1 and 4.2, and it will help prepare you for the next section.

25. $\dfrac{7x^3 y}{21x^2 y^5}$ $\dfrac{x}{3y^4}$

26. $\dfrac{24x^3 y^2}{30x^4 y^5}$ $\dfrac{4}{5xy^3}$

27. $\dfrac{(2a^4 b^5)^3}{2a^{12} b^{20}}$ $\dfrac{4}{b^5}$

28. $\dfrac{(4r^2 s^3)^2}{(3r^4 s)^3}$ $\dfrac{16s^3}{27r^8}$

Simplify.

29. $\dfrac{x}{x + xy}$ $\dfrac{1}{1 + y}$

30. $\dfrac{3x}{3x + 9}$ $\dfrac{x}{x + 3}$

31. 🔒 $\dfrac{5x + 15}{x + 3}$ 5

32. $\dfrac{3x^2 + 6x}{3x^2 + 9x}$ $\dfrac{x + 2}{x + 3}$

33. $\dfrac{x^3 + 6x^2 + 3x}{2x}$ $\dfrac{x^2 + 6x + 3}{2}$

34. $\dfrac{x^2 y^2 - 2xy + 3y}{y}$ $x^2 y - 2x + 3$

35. $\dfrac{r^2 - r - 2}{r + 1}$ $r - 2$

36. $\dfrac{b - 2}{b^2 - 8b + 12}$ $\dfrac{1}{b - 6}$

37. 🔒 $\dfrac{x^2 + 2x}{x^2 + 4x + 4}$ $\dfrac{x}{x + 2}$

38. $\dfrac{x^2 + 3x - 18}{3x - 9}$ $\dfrac{x + 6}{3}$

39. $\dfrac{k^2 - 6k + 9}{k^2 - 9}$ $\dfrac{k - 3}{k + 3}$

40. $\dfrac{z^2 - 10z + 25}{z^2 - 25}$ $\dfrac{z - 5}{z + 5}$

41. $\dfrac{x^2 - 2x - 3}{x^2 - x - 6}$ $\dfrac{x + 1}{x + 2}$

42. $\dfrac{4x^2 - 12x - 40}{2x^2 - 16x + 30}$ $\dfrac{2x + 4}{x - 3}$

43. $\dfrac{2x - 3}{3 - 2x}$ -1

44. $\dfrac{8a - 6}{3 - 4a}$ -2

45. 🔒 $\dfrac{x^2 - 2x - 8}{4 - x}$ $-(x + 2)$

46. $\dfrac{7 - s}{s^2 - 12s + 35}$ $-\dfrac{1}{s - 5}$

47. $\dfrac{x^2 + 3x - 18}{-2x^2 + 6x}$ $-\dfrac{x + 6}{2x}$

48. $\dfrac{3p^2 - 13p - 10}{p - 5}$ $3p + 2$

49. $\dfrac{2x^2 + 5x - 3}{1 - 2x}$ $-(x + 3)$

50. $\dfrac{x^2 - 25}{x^2 - 3x - 10}$ $\dfrac{x + 5}{x + 2}$

51. $\dfrac{m - 2}{4m^2 - 13m + 10}$ $\dfrac{1}{4m - 5}$

52. $\dfrac{2x^2 - 11x + 15}{(x - 3)^2}$ $\dfrac{2x - 5}{x - 3}$

53. $\dfrac{x^2 - 25}{(x + 5)^2}$ $\dfrac{x - 5}{x + 5}$

54. $\dfrac{16x^2 + 24x + 9}{4x + 3}$ $4x + 3$

55. 🔒 $\dfrac{6x^2 - 13x + 6}{3x - 2}$ $2x - 3$

56. $\dfrac{6t^2 - 7t - 5}{2t + 1}$ $3t - 5$

57. $\dfrac{x^2 - 3x + 4x - 12}{x - 3}$ $x + 4$

58. $\dfrac{x^2 - 2x + 4x - 8}{2x^2 + 3x + 8x + 12}$ $\dfrac{x - 2}{2x + 3}$

59. 🔒 $\dfrac{2x^2 - 8x + 3x - 12}{2x^2 + 8x + 3x + 12}$ $\dfrac{x - 4}{x + 4}$

60. $\dfrac{x^3 + 1}{x^2 - x + 1}$ $x + 1$

61. $\dfrac{a^3 - 8}{a - 2}$ $a^2 + 2a + 4$

62. $\dfrac{x^3 - 125}{x^2 - 25}$ $\dfrac{x^2 + 5x + 25}{x + 5}$

63. $\dfrac{9s^2 - 16t^2}{3s - 4t}$ $3s + 4t$

64. $\dfrac{a + 4b}{a^2 - 16b^2}$ $\dfrac{1}{a - 4b}$

65. $\dfrac{4x + 6y}{2x^2 + xy - 3y^2}$ $\dfrac{2}{x - y}$

66. $\dfrac{3k^2 + 6kr - 9r^2}{k^2 + 5kr + 6r^2}$ $\dfrac{3(k - r)}{k + 2r}$

Problem Solving

Simplify the following expressions, if possible. Treat the unknown symbol as if it were a variable.

67. $\dfrac{3☺}{12}$ $\dfrac{☺}{4}$

68. $\dfrac{☺}{☺ + 7☺^2}$ $\dfrac{1}{1 + 7☺}$

69. $\dfrac{7\Delta}{14\Delta + 21}$ $\dfrac{\Delta}{2\Delta + 3}$

70. $\dfrac{\Delta^2 + 2\Delta}{\Delta^2 + 4\Delta + 4}$ $\dfrac{\Delta}{\Delta + 2}$

71. $\dfrac{3\Delta - 2}{2 - 3\Delta}$ -1

72. $\dfrac{(\Delta - 3)^2}{\Delta^2 - 6\Delta + 9}$ 1

Determine the denominator that will make each statement true. Explain how you obtained your answer.

73. $\dfrac{x^2 - x - 6}{\rule{1cm}{0.3cm}} = x - 3$ $x + 2$

74. $\dfrac{2x^2 + 11x + 12}{\rule{1cm}{0.3cm}} = x + 4$ $2x + 3$

Determine the numerator that will make each statement true. Explain how you obtained your answer.

75. $\dfrac{\rule{1cm}{0.3cm}}{x + 4} = x + 3$ $x^2 + 7x + 12$

76. $\dfrac{\rule{1cm}{0.3cm}}{x - 5} = 2x - 1$ $2x^2 - 11x + 5$

Challenge Problems

*In Exercises 77–79, **a)** determine the value or values that x cannot represent. **b)** Simplify the expression.*

77. $\dfrac{x + 3}{x^2 - 2x + 3x - 6}$ **a)** $x \ne -3, x \ne 2$ **b)** $\dfrac{1}{x - 2}$

78. $\dfrac{x - 4}{2x^2 - 5x - 8x + 20}$ **a)** $x \ne 4, x \ne \dfrac{5}{2}$ **b)** $\dfrac{1}{2x - 5}$

79. $\dfrac{x + 5}{2x^3 + 7x^2 - 15x}$ **a)** $x \ne 0, x \ne -5, x \ne \dfrac{3}{2}$ **b)** $\dfrac{1}{x(2x - 3)}$

Simplify. Explain how you determined your answer.

80. $\dfrac{\frac{1}{5}x^5 - \frac{2}{3}x^4}{x^4}$ $\dfrac{1}{5}x - \dfrac{2}{3}$

81. $\dfrac{\frac{1}{5}x^5 - \frac{2}{3}x^4}{\frac{1}{5}x^5 - \frac{2}{3}x^4}$ 1

82. $\dfrac{\frac{1}{5}x^5 - \frac{2}{3}x^4}{\frac{2}{3}x^4 - \frac{1}{5}x^5}$ -1

83. a) Undefined at $x = 0$, $x = 3$, and $x = -5$ **b)** $\dfrac{x - 5}{x(x - 3)}$ **g)** undefined in the original;

$-\frac{1}{4}$ in the simplified form **h)** no, only for those values for which the expression is defined

 Group Activity

Discuss and answer Exercise 83 as a group.

83. a) As a group, determine the values of the variable where the expression $\dfrac{x^2 - 25}{x^3 + 2x^2 - 15x}$ is undefined.

   b) As a group, simplify the rational expression.

   c) Group member 1: Substitute 6 in the *original expression* and evaluate. $\frac{11}{198} = \frac{1}{18}$

   d) Group member 2: Substitute 6 in the *simplified expression* from part **b)** and compare your result to that of Group member 1. $\frac{1}{18}$

   e) Group member 3: Substitute -2 in the original expression and in the simplified expression in part **b)**. Compare your answers. $-\frac{21}{30} = -\frac{7}{10}$

   f) As a group, discuss the results of your work in parts **c)**–**e)**.

   g) Now, as a group, substitute -5 in the original expression and in the simplified expression. Discuss your results.

   h) Is $\dfrac{x^2 - 25}{x^3 + 2x^2 - 15x}$ always equal to its simplified form for *any* value of x? Explain your answer.

Cumulative Review Exercises

[3.1] 84. Solve the formula $z = \dfrac{x - y}{2}$ for y. $y = x - 2z$

[3.4] 85. **Triangle** Find the measures of the three angles of a triangle if one angle is 30° greater than the smallest angle, and the third angle is 10° greater than 3 times the smallest angle. 28°, 58°, 94°

[4.1] 86. Simplify $\left(\dfrac{4x^2y^2}{9x^4y^3}\right)^2 \cdot \dfrac{16}{81x^4y^2}$

[4.4] 87. Subtract $3x^2 - 4x - 8 - (-3x^2 + 6x + 9)$.

$6x^2 - 10x - 17$

[5.3] 88. Factor $3a^2 - 30a + 72$ completely.

$3(a - 4)(a - 6)$

[5.7] 89. Find the length of the hypotenuse of the right triangle. 13 in.

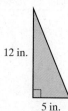

12 in.

5 in.

6.2 MULTIPLICATION AND DIVISION OF RATIONAL EXPRESSIONS

1 Multiply rational expressions.

2 Divide rational expressions.

SSM   Study Guide   CD/Video

MathPro 4/5   PH Math   prenhall.com/Angel
Tutor Center

1 Multiply Rational Expressions

In Section 1.3 we reviewed multiplication of numerical fractions. Recall that to multiply two fractions we multiply their numerators together and multiply their denominators together.

To Multiply Two Fractions

$$\frac{a}{b} \cdot \frac{c}{d} = \frac{a \cdot c}{b \cdot d}, \quad b \neq 0 \quad \text{and} \quad d \neq 0$$

EXAMPLE 1 Multiply $\left(\frac{3}{5}\right)\left(\frac{-2}{9}\right)$.

Solution First divide out common factors; then multiply.

$$\frac{\overset{1}{\cancel{3}}}{5} \cdot \frac{-2}{\underset{3}{\cancel{9}}} = \frac{1 \cdot (-2)}{5 \cdot 3} = -\frac{2}{15}$$

TEACHING TIP
Some students may prefer to write the problem as a single fraction before simplifying.

The same principles apply when multiplying rational expressions containing variables. Before multiplying, you should first divide out any factors common to both a numerator and a denominator.

To Multiply Rational Expressions

1. Factor all numerators and denominators completely.

2. Divide out common factors.

3. Multiply numerators together and multiply denominators together.

EXAMPLE 2 Multiply $\frac{3x^2}{2y} \cdot \frac{4y^3}{3x}$.

Solution This problem can be represented as

$$\frac{3xx}{2y} \cdot \frac{4yyy}{3x}$$

$$\frac{\overset{1}{\cancel{3}}\overset{1}{\cancel{x}}x}{2y} \cdot \frac{4yyy}{\underset{1}{\cancel{3}}\underset{1}{\cancel{x}}} \qquad \textit{Divide out the 3's and x's.}$$

$$\frac{\overset{1}{\cancel{3}}\overset{1}{\cancel{x}}x}{\underset{1}{\cancel{2}}\underset{1}{\cancel{y}}} \cdot \frac{\overset{2}{\cancel{4}}\overset{1}{\cancel{y}}yy}{\underset{1}{\cancel{3}}\underset{1}{\cancel{x}}} \qquad \textit{Divide both the 4 and the 2 by 2, and divide out the y's.}$$

Now we multiply the remaining numerators together and the remaining denominators together.

$$\frac{2xy^2}{1} \quad \text{or} \quad 2xy^2$$

Rather than illustrating this entire process when multiplying rational expressions, we will often proceed as follows:

$$\frac{3x^2}{2y} \cdot \frac{4y^3}{3x}$$

$$= \frac{\overset{1}{\cancel{3}}\,\overset{x}{\cancel{x^2}}}{\underset{1}{\cancel{2}}\,\underset{1}{\cancel{y}}} \cdot \frac{\overset{2}{\cancel{4}}\,\overset{y^2}{\cancel{y^3}}}{\underset{1}{\cancel{3}}\,\underset{1}{\cancel{x}}} = 2xy^2$$

EXAMPLE 3 Multiply $-\dfrac{3y^2}{2x^3} \cdot \dfrac{5x^2}{7y^2}$.

Solution

NOW TRY EXERCISE 43

$$-\frac{3\cancel{y^2}}{2\overset{}{\cancel{x^3}}} \cdot \frac{5\cancel{x^2}}{7\cancel{y^2}} = -\frac{15}{14x}$$

✳

In Example 3 when the y^2 was divided out from both the numerator and denominator we did not place a 1 above and below the y^2 factors. When a factor that appears in a numerator and denominator is factored out, we will generally not show the 1s.

EXAMPLE 4 Multiply $(x - 6) \cdot \dfrac{5}{x^3 - 6x^2}$.

Solution

$$(x - 6) \cdot \frac{5}{x^3 - 6x^2} = \frac{\cancel{x - 6}}{1} \cdot \frac{5}{x^2\cancel{(x - 6)}} = \frac{5}{x^2}$$

✳

EXAMPLE 5 Multiply $\dfrac{(x + 2)^2}{6x^2} \cdot \dfrac{3x}{x^2 - 4}$.

Solution

$$\frac{(x + 2)^2}{6x^2} \cdot \frac{3x}{x^2 - 4} = \frac{(x + 2)(x + 2)}{6x^2} \cdot \frac{3x}{(x + 2)(x - 2)}$$

NOW TRY EXERCISE 15

$$= \frac{\cancel{(x + 2)}(x + 2)}{\underset{2}{\cancel{6}}\,\underset{x}{\cancel{x^2}}} \cdot \frac{\overset{1}{\cancel{3}}\,\overset{1}{\cancel{x}}}{\cancel{(x + 2)}(x - 2)} = \frac{x + 2}{2x(x - 2)}$$

✳

In Example 5 we could have multiplied the factors in the denominator to get $\dfrac{x + 2}{2x^2 - 4x}$. This is also a correct answer. In this section we will leave rational answers with the numerator as a polynomial (in unfactored form) and the denominators in factored form, as was given in Example 5. This is consistent with how we will leave rational answers when we add and subtract rational expressions in later sections.

EXAMPLE 6 Multiply $\dfrac{a - 4}{3a} \cdot \dfrac{6a}{4 - a}$.

Solution

$$\frac{a - 4}{\underset{1}{\cancel{3a}}} \cdot \frac{\overset{2}{\cancel{6a}}}{4 - a} = \frac{2(a - 4)}{4 - a}$$

This problem is still not complete. In Section 6.1 we showed that $4 - a$ is $-1(-4 + a)$ or $-1(a - 4)$. Thus,

NOW TRY EXERCISE 13

$$\frac{2(a - 4)}{4 - a} = \frac{2\cancel{(a - 4)}}{-1\cancel{(a - 4)}} = -2$$

✳

HELPFUL HINT

When only the signs differ in a numerator and denominator in a multiplication problem, factor out -1 *from either the numerator or denominator*; then divide out the common factor.

$$\frac{a-b}{x} \cdot \frac{y}{b-a} = \frac{a-b}{x} \cdot \frac{y}{-1(a-b)} = -\frac{y}{x}$$

EXAMPLE 7 Multiply $\dfrac{3x+2}{2x-1} \cdot \dfrac{4-8x}{3x+2}$.

Solution

$$\frac{3x+2}{2x-1} \cdot \frac{4-8x}{3x+2} = \frac{3x+2}{2x-1} \cdot \frac{4(1-2x)}{3x+2} \qquad \textit{Factor.}$$

$$= \frac{3x+2}{2x-1} \cdot \frac{4(1-2x)}{3x+2}. \qquad \textit{Divide out common factors.}$$

Note that the factor $(1-2x)$ in the numerator of the second fraction differs only in sign from $2x-1$, the denominator of the first fraction. We will therefore factor -1 from each term of the $(1-2x)$ in the numerator of the second fraction.

$$= \frac{3x+2}{2x-1} \cdot \frac{4(-1)(2x-1)}{3x+2} \qquad \textit{Factor }-1\textit{ from the second numerator.}$$

$$= \frac{3x+2}{2x-1} \cdot \frac{-4(2x-1)}{3x+2} \qquad \textit{Divide out common factors.}$$

$$= \frac{-4}{1} = -4$$

EXAMPLE 8 Multiply $\dfrac{2x^2+5x-12}{6x^2-11x+3} \cdot \dfrac{3x^2+2x-1}{x^2+5x+4}$.

Solution Factor all numerators and denominators, and then divide out common factors.

$$\frac{2x^2+5x-12}{6x^2-11x+3} \cdot \frac{3x^2+2x-1}{x^2+5x+4} = \frac{(2x-3)(x+4)}{(2x-3)(3x-1)} \cdot \frac{(3x-1)(x+1)}{(x+1)(x+4)}$$

$$= \frac{(2x-3)(x+4)}{(2x-3)(3x-1)} \cdot \frac{(3x-1)(x+1)}{(x+1)(x+4)} = 1$$

EXAMPLE 9 Multiply $\dfrac{2x^3-14x^2+12x}{6y^2} \cdot \dfrac{-2y}{3x^2-3x}$.

Solution

$$\frac{2x^3-14x^2+12x}{6y^2} \cdot \frac{-2y}{3x^2-3x} = \frac{2x(x^2-7x+6)}{6y^2} \cdot \frac{-2y}{3x(x-1)}$$

$$= \frac{2x(x-6)(x-1)}{6y^2} \cdot \frac{-2y}{3x(x-1)}$$

$$= \frac{2x(x-6)(x-1)}{\underset{3}{6}y^{\underset{y}{2}}} \cdot \frac{-2y}{3x(x-1)}$$

$$= \frac{-2(x-6)}{9y} = \frac{-2x+12}{9y}$$

EXAMPLE 10   Multiply $\dfrac{x^2 - y^2}{x + y} \cdot \dfrac{x + 2y}{2x^2 - xy - y^2}$.

Solution

$$\dfrac{x^2 - y^2}{x + y} \cdot \dfrac{x + 2y}{2x^2 - xy - y^2} = \dfrac{(x + y)(x - y)}{x + y} \cdot \dfrac{x + 2y}{(2x + y)(x - y)}$$

$$= \dfrac{\cancel{(x + y)}\,\cancel{(x - y)}}{\cancel{x + y}} \cdot \dfrac{x + 2y}{(2x + y)\cancel{(x - y)}}$$

$$= \dfrac{x + 2y}{2x + y}$$

NOW TRY EXERCISE 19

2   Divide Rational Expressions

In Chapter 1 we learned that to divide one fraction by a second we invert the divisor and multiply.

To Divide Two Fractions

$$\dfrac{a}{b} \div \dfrac{c}{d} = \dfrac{a}{b} \cdot \dfrac{d}{c} = \dfrac{ad}{bc}, \quad b \neq 0, \quad d \neq 0, \quad \text{and} \quad c \neq 0$$

EXAMPLE 11   Divide. **a)** $\dfrac{2}{7} \div \dfrac{5}{7}$ **b)** $\dfrac{3}{4} \div \dfrac{5}{6}$

Solution   **a)** $\dfrac{2}{7} \div \dfrac{5}{7} = \dfrac{2}{\cancelto{1}{7}} \cdot \dfrac{\cancelto{1}{7}}{5} = \dfrac{2 \cdot 1}{1 \cdot 5} = \dfrac{2}{5}$ **b)** $\dfrac{3}{4} \div \dfrac{5}{6} = \dfrac{3}{\cancelto{2}{4}} \cdot \dfrac{\cancelto{3}{6}}{5} = \dfrac{3 \cdot 3}{2 \cdot 5} = \dfrac{9}{10}$

TEACHING TIP
After discussing division in terms of multiplication, have students rewrite Examples 1, 9, and 10 as division problems that give the same answers as the original problems.

The same principles are used to **divide rational expressions**.

To Divide Rational Expressions

Invert the divisor (the second fraction) and multiply.

EXAMPLE 12   Divide $\dfrac{8x^3}{z} \div \dfrac{5z^3}{3}$.

Solution   Invert the divisor (the second fraction), and then multiply.

NOW TRY EXERCISE 25

$$\dfrac{8x^3}{z} \div \dfrac{5z^3}{3} = \dfrac{8x^3}{z} \cdot \dfrac{3}{5z^3} = \dfrac{24x^3}{5z^4}$$

EXAMPLE 13   Divide $\dfrac{x^2 - 9}{x + 4} \div \dfrac{x - 3}{x + 4}$.

Solution $\dfrac{x^2 - 9}{x + 4} \div \dfrac{x - 3}{x + 4} = \dfrac{x^2 - 9}{x + 4} \cdot \dfrac{x + 4}{x - 3}$ *Invert the divisor and multiply.*

$$= \dfrac{(x + 3)\cancel{(x - 3)}}{\cancel{x + 4}} \cdot \dfrac{\cancel{x + 4}}{\cancel{x - 3}} = x + 3$$ *Factor, and divide out common factors.*

EXAMPLE 14 Divide $\dfrac{-1}{2x - 3} \div \dfrac{3}{3 - 2x}$.

Solution

$$\dfrac{-1}{2x - 3} \div \dfrac{3}{3 - 2x} = \dfrac{-1}{2x - 3} \cdot \dfrac{3 - 2x}{3}$$
Invert the divisor and multiply.

$$= \dfrac{-1}{2x - 3} \cdot \dfrac{-1(2x - 3)}{3}$$
Factor out −1, then divide out common factors.

$$= \dfrac{(-1)(-1)}{(1)(3)} = \dfrac{1}{3}$$

EXAMPLE 15 Divide $\dfrac{w^2 - 11w + 30}{w^2} \div (w - 5)^2$.

Solution $(w - 5)^2$ means $\dfrac{(w - 5)^2}{1}$. Factor the numerator of the first fraction, then invert the divisor and multiply.

$$\dfrac{w^2 - 11w + 30}{w^2} \div (w - 5)^2 = \dfrac{w^2 - 11w + 30}{w^2} \cdot \dfrac{1}{(w - 5)^2}$$

$$= \dfrac{(w - 6)(w - 5)}{w^2} \cdot \dfrac{1}{(w - 5)(w - 5)}$$

$$= \dfrac{w - 6}{w^2(w - 5)}$$

EXAMPLE 16 Divide $\dfrac{12x^2 - 22x + 8}{3x} \div \dfrac{3x^2 + 2x - 8}{2x^2 + 4x}$.

Solution $\dfrac{12x^2 - 22x + 8}{3x} \div \dfrac{3x^2 + 2x - 8}{2x^2 + 4x} = \dfrac{12x^2 - 22x + 8}{3x} \cdot \dfrac{2x^2 + 4x}{3x^2 + 2x - 8}$

$$= \dfrac{2(6x^2 - 11x + 4)}{3x} \cdot \dfrac{2x(x + 2)}{(3x - 4)(x + 2)}$$

$$= \dfrac{2(3x - 4)(2x - 1)}{3x} \cdot \dfrac{2x(x + 2)}{(3x - 4)(x + 2)}$$

$$= \dfrac{4(2x - 1)}{3} = \dfrac{8x - 4}{3}$$

NOW TRY EXERCISE 35

Exercise Set 6.2

Concept/Writing Exercises   **1.** and **2.** Answers will vary.

1. In your own words, explain how to multiply rational expressions.

2. In your own words, explain how to divide rational expressions.

What polynomial should be in the shaded area of the second fraction to make each statement true? Explain how you determined your answer.

✎ 3. $\dfrac{x+3}{x-4} \cdot \dfrac{}{x+3} = x+2$

$x^2 - 2x - 8$; numerator must be $(x-4)(x+2)$

✎ 4. $\dfrac{x-5}{x+2} \cdot \dfrac{}{x-5} = 2x-3$

$2x^2 + x - 6$; numerator must be $(x+2)(2x-3)$

✎ 5. $\dfrac{x-5}{x+5} \cdot \dfrac{x+5}{} = \dfrac{1}{x+3}$

$x^2 - 2x - 15$ denominator must be $(x-5)(x+3)$

✎ 6. $\dfrac{2x-1}{x-3} \cdot \dfrac{x-3}{} = \dfrac{1}{x-6}$

$2x^2 - 13x + 6$; denominator must be $(2x-1)(x-6)$

Practice the Skills

Multiply.

19. $\dfrac{x+3}{2(x+2)}$ 22. $\dfrac{t-3}{t+4}$

7. $\dfrac{5x}{4y} \cdot \dfrac{y^2}{10} \quad \dfrac{xy}{8}$

8. $\dfrac{15x^3y^2}{z} \cdot \dfrac{z}{5xy^3} \quad \dfrac{3x^2}{y}$

9. $\dfrac{16x^2}{y^4} \cdot \dfrac{5x^2}{y^2} \quad \dfrac{80x^4}{y^6}$

10. $\dfrac{7n^3}{32m} \cdot \dfrac{-4}{21m^2n^3} \quad -\dfrac{1}{24m^3}$

🔒 11. $\dfrac{6x^5y^3}{5z^3} \cdot \dfrac{6x^4}{5yz^4} \quad \dfrac{36x^9y^2}{25z^7}$

12. $\dfrac{x^2-9}{x^2-16} \cdot \dfrac{x-4}{x-3} \quad \dfrac{x+3}{x+4}$

13. $\dfrac{3x-2}{3x+2} \cdot \dfrac{4x-1}{1-4x} \quad \dfrac{-3x+2}{3x+2}$

14. $\dfrac{m-5}{2m+5} \cdot \dfrac{3m}{-m+5} \quad -\dfrac{3m}{2m+5}$

15. $\dfrac{x^2+7x+12}{x+4} \cdot \dfrac{1}{x+3} \quad 1$

16. $\dfrac{b^2+7b+12}{2b} \cdot \dfrac{b^2-4b}{b^2-b-12} \quad \dfrac{b+4}{2}$

🔒 17. $\dfrac{a}{a^2-b^2} \cdot \dfrac{a+b}{a^2+ab} \quad \dfrac{1}{a^2-b^2}$

18. $\dfrac{t^2-36}{t^2+t-30} \cdot \dfrac{t-5}{3t} \quad \dfrac{t-6}{3t}$

19. $\dfrac{6x^2-14x-12}{6x+4} \cdot \dfrac{x+3}{2x^2-2x-12}$

20. $\dfrac{2x^2-9x+9}{8x-12} \cdot \dfrac{2x}{x^2-3x} \quad \dfrac{1}{2}$

21. $\dfrac{3x^2-13x-10}{x^2-2x-15} \cdot \dfrac{x^2+x-2}{3x^2-x-2} \quad \dfrac{x+2}{x+3}$

22. $\dfrac{2t^2-t-6}{2t^2-3t-2} \cdot \dfrac{2t^2-5t-3}{2t^2+11t+12}$

23. $\dfrac{x+3}{x-3} \cdot \dfrac{x^3-27}{x^2+3x+9} \quad x+3$

24. $\dfrac{x^3+8}{x^2-x-6} \cdot \dfrac{x+3}{x^2-2x+4} \quad \dfrac{x+3}{x-3}$

Divide.

34. $\dfrac{x-15}{x+9}$ 35. $\dfrac{x-8}{x+2}$

🔒 25. $\dfrac{9x^3}{y^2} \div \dfrac{3x}{y^3} \quad 3x^2y$

26. $\dfrac{9x^3}{4} \div \dfrac{1}{16y^2} \quad 36x^3y^2$

27. $\dfrac{15xy^2}{4z} \div \dfrac{5x^2y^2}{12z^2} \quad \dfrac{9z}{x}$

28. $\dfrac{36y}{7z^2} \div \dfrac{3xy}{2z} \quad \dfrac{24}{7xz}$

29. $\dfrac{5xy}{7ab^2} \div \dfrac{6xy}{7} \quad \dfrac{5}{6ab^2}$

30. $2xz \div \dfrac{4xy}{z} \quad \dfrac{z^2}{2y}$

31. $\dfrac{10r+5}{r} \div \dfrac{2r+1}{r^2} \quad 5r$

32. $\dfrac{x-3}{4y^2} \div \dfrac{x^2-9}{2xy} \quad \dfrac{x}{2y(x+3)}$

33. $\dfrac{x^2+5x-14}{x} \div \dfrac{x-2}{x} \quad x+7$

34. $\dfrac{1}{x^2+7x-18} \div \dfrac{1}{x^2-17x+30}$

35. $\dfrac{x^2-12x+32}{x^2-6x-16} \div \dfrac{x^2-x-12}{x^2-5x-24}$

36. $\dfrac{a-b}{9a+9b} \div \dfrac{a^2-b^2}{a^2+2a+1} \quad \dfrac{a^2+2a+1}{9(a+b)^2}$

🔒 37. $\dfrac{2x^2+9x+4}{x^2+7x+12} \div \dfrac{2x^2-x-1}{(x+3)^2} \quad \dfrac{x+3}{x-1}$

38. $\dfrac{a^2-b^2}{9} \div \dfrac{3a-3b}{27x^2} \quad ax^2+bx^2$

39. $\dfrac{x^2-y^2}{x^2-2xy+y^2} \div \dfrac{x+y}{y-x} \quad -1$

40. $\dfrac{9x^2-9y^2}{6x^2y^2} \div \dfrac{3x+3y}{12x^2y^5} \quad 6xy^3-6y^4$

41. $\dfrac{5x^2-4x-1}{5x^2+6x+1} \div \dfrac{x^2-5x+4}{x^2+2x+1} \quad \dfrac{x+1}{x-4}$

42. $\dfrac{7n^2-15n+2}{n^2+n-6} \div \dfrac{n^2-3n-10}{n^2-2n-15} \quad \dfrac{7n-1}{n+2}$

Perform each indicated operation.

43. $\dfrac{9x}{6y^2} \cdot \dfrac{24x^2y^4}{9x} \quad 4x^2y^2$

44. $\dfrac{5z^3}{8} \cdot \dfrac{9x^2}{15z} \quad \dfrac{3x^2z^2}{8}$

45. $\dfrac{63a^2b^3}{16c^3} \cdot \dfrac{4c^4}{9a^3b^5} \quad \dfrac{7c}{4ab^2}$

46. $\dfrac{-2xw}{y^5} \div \dfrac{6x^2}{y^6} \quad -\dfrac{wy}{3x}$

🔒 47. $\dfrac{-xy}{a} \div \dfrac{-2ax}{6y} \quad \dfrac{3y^2}{a^2}$

48. $\dfrac{27x}{5y^2} \div 3x^2y^2 \quad \dfrac{9}{5xy^4}$

49. $\dfrac{100m^6}{21x^5y^7} \cdot \dfrac{14x^{12}y^5}{25m^5} \quad \dfrac{8mx^7}{3y^2}$

50. $\dfrac{-18x^2y}{11z^2} \cdot \dfrac{22z^3}{x^2y^5} \quad -\dfrac{36z}{y^4}$

51. $(3x+5) \cdot \dfrac{1}{6x+10} \quad \dfrac{1}{2}$

64. $\frac{q-5}{2q+3}$

52. $\frac{1}{4x-3}\cdot(20x-15)$ 5

53. $\frac{1}{4x^2y^2}\div\frac{1}{28x^3y}$ $\frac{7x}{y}$

54. $\frac{x^2y^5}{3z}\div\frac{3z}{2x}$ $\frac{2x^3y^5}{9z^2}$

55. $\frac{(4m)^2}{8n^3}\div\frac{m^6n^8}{4}$ $\frac{8}{m^4n^{11}}$

56. $\frac{3r^5s^2}{(r^2s^3)^3}\cdot\frac{6r^4}{4s}$ $\frac{9r^3}{2s^8}$

57. $\frac{r^2+5r+6}{r^2+9r+18}\cdot\frac{r^2+4r-12}{r^2-5r+6}$ $\frac{r+2}{r-3}$

58. $\frac{z^2-z-20}{z^2-3z-10}\cdot\frac{(z+2)^2}{(z+4)^2}$ $\frac{z+2}{z+4}$

59. $\frac{x^2-10x+24}{x^2-8x+12}\div\frac{x^2-7x+12}{x^2-6x+8}$ $\frac{x-4}{x-3}$

60. $\frac{p^2-5p+6}{p^2-10p+16}\div\frac{p^2+2p}{p^2-6p-16}$ $\frac{p-3}{p}$

61. $\frac{3z^2-4z-4}{z^2-4}\cdot\frac{2z^2+5z+2}{2z^2-3z-2}$ $\frac{3z+2}{z-2}$

62. $\frac{2w^2+3w-35}{w^2-7w-8}\cdot\frac{w^2-5w-24}{w^2+8w+15}$ $\frac{2w-7}{w+1}$

63. $\frac{2x^2-19x+24}{x^2-12x+32}\div\frac{2x^2+x-6}{x^2+7x+10}$ $\frac{x+5}{x-4}$

64. $\frac{q^2-11q+30}{2q^2-7q-15}\div\frac{q^2-2q-24}{q^2-q-20}$

65. $\frac{4n^2-9}{9n^2-1}\cdot\frac{3n^2-2n-1}{2n^2-5n+3}$ $\frac{2n+3}{3n-1}$

66. $\frac{2z^2+9z+9}{4z^2-9}\div\frac{(z+3)^2}{(2z-3)^2}$ $\frac{2z-3}{z+3}$

Problem Solving

Perform each indicated operation. Treat Δ and ☺ as if they were variables.

67. $\frac{6\Delta^2}{12}\cdot\frac{12}{36\Delta^5}$ $\frac{1}{6\Delta^3}$

68. $\frac{\Delta-6}{2\Delta+5}\cdot\frac{2\Delta}{-\Delta+6}$ $-\frac{2\Delta}{2\Delta+5}$

69. $\frac{\Delta-☺}{9\Delta-9☺}\div\frac{\Delta^2-☺^2}{\Delta^2+2\Delta☺+☺^2}$ $\frac{\Delta+☺}{9(\Delta-☺)}$

70. $\frac{\Delta^2-☺^2}{\Delta^2-2\Delta☺+☺^2}\div\frac{\Delta+☺}{☺-\Delta}$ -1

For each equation, fill in the shaded area with a binomial or trinomial to make the statement true. Explain how you determined your answer.

71. $\frac{\blacksquare}{x+2}=x+1$ x^2+3x+2

72. $\frac{x+3}{\blacksquare}=\frac{1}{x-3}$ x^2-9

73. $\frac{\blacksquare}{x-5}=x+2$ $x^2-3x-10$

74. $\frac{\blacksquare}{x^2-7x+10}=\frac{1}{x-5}$ $x-2$

75. $\frac{\blacksquare}{x^2-4}\cdot\frac{x+2}{x-1}=1$ x^2-3x+2

76. $\frac{x+4}{x^2+9x+20}\cdot\frac{\blacksquare}{x-2}=1$ $x^2+3x-10$

Challenge Problems

Simplify.

77. $\left(\frac{x+2}{x^2-4x-12}\cdot\frac{x^2-9x+18}{x-2}\right)\div\frac{x^2+5x+6}{x^2-4}$ $\frac{x-3}{x+3}$

78. $\left(\frac{x^2-x-6}{2x^2-9x+9}\div\frac{x^2+x-12}{x^2+3x-4}\right)\cdot\frac{2x^2-5x+3}{x^2+x-2}$ $\frac{x-1}{x-3}$

79. $\left(\frac{x^2+4x+3}{x^2-6x-16}\right)\div\left(\frac{x^2+5x+6}{x^2-9x+8}\cdot\frac{x^2-1}{x^2+4x+4}\right)$ 1

80. $\left(\frac{x^2+4x+3}{x^2-6x-16}\div\frac{x^2+5x+6}{x^2-9x+8}\right)\cdot\left(\frac{x^2-1}{x^2+4x+4}\right)$ $\frac{(x+1)^2(x-1)^2}{(x+2)^4}$

For Exercises 81 and 82, determine the polynomials that when placed in the shaded areas make the statement true. Explain how you determined your answer.

81. $\frac{\blacksquare}{\blacksquare}\cdot\frac{x^2+3x-4}{x^2-4x+3}=\frac{x-2}{x-5}$ $x^2-5x+6,\ x^2-x-20$

82. $\frac{\blacksquare}{x^2+x-2}\cdot\frac{x^2+6x+8}{\blacksquare}=\frac{x+3}{x+5}$ $x^2+2x-3,\ x^2+9x+20$

Group Activity

83. Consider the three problems that follow:

(1) $\left(\dfrac{x+2}{x-3}\right) \div \left(\dfrac{x^2-5x+6}{x-2} \cdot \dfrac{x+2}{x-3}\right)$ $\dfrac{1}{x-3}$

(2) $\left(\dfrac{x+2}{x-3} \div \dfrac{x^2-5x+6}{x-2}\right) \cdot \left(\dfrac{x+2}{x-3}\right)$ $\dfrac{(x+2)^2}{(x-3)^3}$

(3) $\left(\dfrac{x+2}{x-3}\right) \div \left(\dfrac{x^2-5x+6}{x-2}\right) \cdot \left(\dfrac{x+2}{x-3}\right)$ $\dfrac{(x+2)^2}{(x-3)^3}$

a) Without working the problem, decide as a group which of the problems will have the same answer. Explain. (2) and (3)

b) Individually, simplify each of the three problems.

c) Compare your answers to part **b)** with the other members of your group. If you did not get the same answers, determine why.

Cumulative Review Exercises

[3.6] **84.** **Tug Boat** A tug boat leaves its dock traveling at an average of 15 miles per hour towards a barge it is to pull back to the dock. On the return trip, pulling the barge, the tug averages 5 miles per hour. If the trip back to the dock took 2 hours longer than the trip out, find the time it took the tug boat to reach the barge. 1 hr

[4.5] **85.** Multiply $(4x^3y^2z^4)(5xy^3z^7)$. $20x^4y^5z^{11}$

[4.6] **86.** Divide $\dfrac{4x^3-5x}{2x-1}$. $2x^2 + x - 2 - \dfrac{2}{2x-1}$

[5.4] **87.** Factor $3x^2 - 9x - 30$. $3(x-5)(x+2)$

[5.6] **88.** Solve $3x^2 - 9x - 30 = 0$. $5, -2$

6.3 ADDITION AND SUBTRACTION OF RATIONAL EXPRESSIONS WITH A COMMON DENOMINATOR AND FINDING THE LEAST COMMON DENOMINATOR

1 Add and subtract rational expressions with a common denominator.

2 Find the least common denominator.

 SSM   Study Guide   CD/Video

 MathPro 4/5   PH Math Tutor Center   prenhall.com/Angel

1 Add and Subtract Rational Expressions with a Common Denominator

Recall that when adding (or subtracting) two arithmetic fractions with a common denominator we add (or subtract) the numerators while keeping the common denominator.

To Add or Subtract Two Fractions

$$\frac{a}{c} + \frac{b}{c} = \frac{a+b}{c}, c \neq 0 \qquad \frac{a}{c} - \frac{b}{c} = \frac{a-b}{c}, c \neq 0$$

EXAMPLE 1 **a)** Add $\dfrac{5}{16} + \dfrac{8}{16}$. **b)** Subtract $\dfrac{5}{9} - \dfrac{1}{9}$.

Solution **a)** $\dfrac{5}{16} + \dfrac{8}{16} = \dfrac{5+8}{16} = \dfrac{13}{16}$ **b)** $\dfrac{5}{9} - \dfrac{1}{9} = \dfrac{5-1}{9} = \dfrac{4}{9}$

Note in Example **1a)** that we did not simplify $\frac{8}{16}$ to $\frac{1}{2}$. The fractions are given with a common denominator, 16. If $\frac{8}{16}$ was simplified to $\frac{1}{2}$, you would lose the common denominator that is needed to add or subtract fractions.

The same principles apply when **adding or subtracting rational expressions** containing variables.

> **To Add or Subtract Rational Expressions with a Common Denominator**
>
> 1. Add or subtract the numerators.
> 2. Place the sum or difference of the numerators found in step 1 over the common denominator.
> 3. Simplify the fraction if possible.

EXAMPLE 2 Add $\dfrac{3}{x-4} + \dfrac{x+2}{x-4}$

Solution
$$\frac{3}{x-4} + \frac{x+2}{x-4} = \frac{3+(x+2)}{x-4} = \frac{x+5}{x-4}$$

EXAMPLE 3 Add $\dfrac{2x^2+5}{x+3} + \dfrac{6x-5}{x+3}$.

Solution
$$\frac{2x^2+5}{x+3} + \frac{6x-5}{x+3} = \frac{(2x^2+5)+(6x-5)}{x+3}$$
$$= \frac{2x^2+5+6x-5}{x+3}$$
$$= \frac{2x^2+6x}{x+3}.$$

Now factor $2x$ from each term in the numerator and simplify.
$$= \frac{2x\cancel{(x+3)}}{\cancel{x+3}} = 2x$$

NOW TRY EXERCISE 21

EXAMPLE 4 Add $\dfrac{x^2+3x-2}{(x+5)(x-2)} + \dfrac{4x+12}{(x+5)(x-2)}$.

Solution

$$\frac{x^2 + 3x - 2}{(x + 5)(x - 2)} + \frac{4x + 12}{(x + 5)(x - 2)} = \frac{(x^2 + 3x - 2) + (4x + 12)}{(x + 5)(x - 2)}$$ *Write as a single fraction.*

$$= \frac{x^2 + 3x - 2 + 4x + 12}{(x + 5)(x - 2)}$$ *Remove parentheses in the numerator.*

$$= \frac{x^2 + 7x + 10}{(x + 5)(x - 2)}$$ *Combine like terms.*

$$= \frac{(x + 5)(x + 2)}{(x + 5)(x - 2)}$$ *Factor, divide out common factor.*

$$= \frac{x + 2}{x - 2}$$

When subtracting rational expressions, be sure to subtract the entire numerator of the fraction being subtracted. Study the following Avoiding Common Errors box very carefully.

Consider the subtraction

$$\frac{4x}{x - 2} - \frac{2x + 1}{x - 2}$$

Many people begin problems of this type incorrectly. Here are the correct and incorrect ways of working this problem.

CORRECT

$$\frac{4x}{x - 2} - \frac{2x + 1}{x - 2} = \frac{4x - (2x + 1)}{x - 2}$$

$$= \frac{4x - 2x - 1}{x - 2}$$

$$= \frac{2x - 1}{x - 2}$$

INCORRECT

$$\frac{4x}{x - 2} - \frac{2x + 1}{x - 2} \quad \frac{4x - 2x + 1}{x - 2}$$

Note that the entire numerator of the second fraction (not just the first term) **must be subtracted**. Also note that the sign of *each* term of the numerator being subtracted will change when the parentheses are removed.

EXAMPLE 5 Subtract $\dfrac{x^2 - 2x + 3}{x^2 + 7x + 12} - \dfrac{x^2 - 4x - 5}{x^2 + 7x + 12}$.

Solution

$$\frac{x^2 - 2x + 3}{x^2 + 7x + 12} - \frac{x^2 - 4x - 5}{x^2 + 7x + 12} = \frac{(x^2 - 2x + 3) - (x^2 - 4x - 5)}{x^2 + 7x + 12}$$ *Write as a single fraction.*

$$= \frac{x^2 - 2x + 3 - x^2 + 4x + 5}{x^2 + 7x + 12}$$ *Remove parentheses.*

$$= \frac{2x + 8}{x^2 + 7x + 12}$$ *Combine like terms.*

$$= \frac{2(x + 4)}{(x + 3)(x + 4)}$$ *Factor, divide out common factor.*

$$= \frac{2}{x + 3}$$

The variable used when working with rational expressions is irrelevant. In Example 6 we work with rational expressions in variable r.

EXAMPLE 6 Subtract $\dfrac{6r}{r-5} - \dfrac{4r^2 - 17r + 15}{r-5}$

Solution
$$\dfrac{6r}{r-5} - \dfrac{4r^2 - 17r + 15}{r-5} = \dfrac{6r - (4r^2 - 17r + 15)}{r-5} \quad \text{\textit{Write as a single fraction.}}$$

$$= \dfrac{6r - 4r^2 + 17r - 15}{r-5} \quad \text{\textit{Remove parentheses.}}$$

$$= \dfrac{-4r^2 + 23r - 15}{r-5} \quad \text{\textit{Combine like terms.}}$$

$$= \dfrac{-(4r^2 - 23r + 15)}{r-5} \quad \text{\textit{Factor out} -1.}$$

$$= \dfrac{-(4r-3)\cancel{(r-5)}}{\cancel{r-5}} \quad \text{\textit{Factor, divide out common factor.}}$$

NOW TRY EXERCISE 43
$$= -(4r-3) \quad \text{or} \quad -4r + 3$$

2 Find the Least Common Denominator

To add two fractions with unlike denominators, we must first obtain a common denominator. Now we explain how to find the **least common denominator** for rational expressions. We will use this information in Section 6.4 when we add and subtract rational expressions.

EXAMPLE 7 Add $\dfrac{5}{7} + \dfrac{2}{3}$.

Solution The least common denominator (LCD) of the fractions $\frac{5}{7}$ and $\frac{2}{3}$ is 21. Twenty-one is the smallest number that is divisible by both denominators, 7 and 3. Rewrite each fraction so that its denominator is 21.

$$\dfrac{5}{7} + \dfrac{2}{3} = \boxed{\dfrac{3}{3}} \cdot \dfrac{5}{7} + \dfrac{2}{3} \cdot \boxed{\dfrac{7}{7}}$$

$$= \dfrac{15}{21} + \dfrac{14}{21} = \dfrac{29}{21} \quad \text{or} \quad 1\dfrac{8}{21}$$

To add or subtract rational expressions, we must write each expression with a common denominator.

To Find the Least Common Denominator of Rational Expressions

1. Factor each denominator completely. Any factors that occur more than once should be expressed as powers. For example, $(x-3)(x-3)$ should be expressed as $(x-3)^2$.

2. List all different factors (other than 1) that appear in any of the denominators. When the same factor appears in more than one denominator, write that factor with the highest power that appears.

3. The least common denominator is the product of all the factors listed in step 2.

EXAMPLE 8   Find the least common denominator.

$$\frac{1}{3} + \frac{1}{y}$$

Solution   The only factor (other than 1) of the first denominator is 3. The only factor (other than 1) of the second denominator is y. The LCD is therefore $3 \cdot y = 3y$.

EXAMPLE 9   Find the LCD.

$$\frac{5}{x^2} - \frac{3}{7x}$$

Solution   The factors that appear in the denominators are 7 and x. List each factor with its highest power. The LCD is the product of these factors.

Highest power of x

$$\text{LCD} = 7 \cdot x^2 = 7x^2$$

EXAMPLE 10   Find the LCD.

$$\frac{1}{18x^3y} + \frac{5}{27x^2y^3}$$

Solution   Write both 18 and 27 as products of prime factors: $18 = 2 \cdot 3^2$ and $27 = 3^3$. *If you have forgotten how to write a number as a product of prime factors, read Section 5.1 or Appendix B now.*

$$\frac{1}{18x^3y} + \frac{5}{27x^2y^3} = \frac{1}{2 \cdot 3^2 x^3 y} + \frac{5}{3^3 x^2 y^3}$$

The factors that appear are $2, 3, x,$ and y. List the highest powers of each of these factors.

NOW TRY EXERCISE 61 $$\text{LCD} = 2 \cdot 3^3 \cdot x^3 \cdot y^3 = 54x^3y^3$$

EXAMPLE 11   Find the LCD.

$$\frac{5}{x} - \frac{7y}{x + 3}$$

Solution   The factors in the denominators are x and $x + 3$. *Note that the x in the second denominator, $x + 3$, is a term, not a factor.*

$$\text{LCD} = x(x + 3)$$

EXAMPLE 12   Find the LCD.

$$\frac{7}{3x^2 - 6x} + \frac{x^2}{x^2 - 4x + 4}$$

Solution   Factor both denominators.

$$\frac{7}{3x^2 - 6x} + \frac{x^2}{x^2 - 4x + 4} = \frac{7}{3x(x - 2)} + \frac{x^2}{(x - 2)(x - 2)}$$

$$= \frac{7}{3x(x - 2)} + \frac{x^2}{(x - 2)^2}$$

The factors in the denominators are $3, x,$ and $x - 2$. List the highest power of each of these factors.

NOW TRY EXERCISE 85

$$\text{LCD} = 3 \cdot x \cdot (x - 2)^2 = 3x(x - 2)^2.$$

EXAMPLE 13 Find the LCD.

$$\frac{5x}{x^2 - x - 12} - \frac{6x^2}{x^2 - 7x + 12}$$

Solution Factor both denominators.

$$\frac{5x}{x^2 - x - 12} - \frac{6x^2}{x^2 - 7x + 12} = \frac{5x}{(x + 3)(x - 4)} - \frac{6x^2}{(x - 3)(x - 4)}$$

The factors in the denominators are $x + 3, x - 4,$ and $x - 3$.

$$\text{LCD} = (x + 3)(x - 4)(x - 3)$$

Although $x - 4$ is a common factor of each denominator, the highest power of that factor that appears in each denominator is 1.

EXAMPLE 14 Find the LCD.

$$\frac{6w}{w^2 - 14w + 45} + w + 5$$

Solution Factor the denominator of the first term.

$$\frac{6w}{w^2 - 14w + 45} + w + 5 = \frac{6w^2}{(w - 5)(w - 9)} + w + 5$$

Since the denominator of $w + 5$ is 1, the expression can be rewritten as

$$\frac{6w}{(w - 5)(w - 9)} + \frac{w + 5}{1}$$

NOW TRY EXERCISE 89 The LCD is therefore $1(w - 5)(w - 9)$ or simply $(w - 5)(w - 9)$.

Exercise Set 6.3

Concept/Writing Exercises
1. and **3.** Answers will vary. **8.** $3x(x - 3)$

1. In your own words, explain how to add or subtract rational expressions with a common denominator.

2. When subtracting rational expressions, what must happen to the sign of each term of the numerator being subtracted? signs change

3. In your own words, explain how to find the least common denominator of two rational expressions.

4. In the addition $\dfrac{1}{x} + \dfrac{1}{x + 1}$, is the least common denominator $x, x + 1,$ or $x(x + 1)$? Explain. $x(x + 1)$

Determine the LCD to be used to perform each indicated operation. Explain how you determined the LCD. Do not perform the operations.

5. $\dfrac{5}{x + 6} - \dfrac{2}{x}$ $x(x + 6)$

6. $\dfrac{2}{x - 2} + \dfrac{3}{5}$ $5(x - 2)$

7. $\dfrac{2}{x + 3} + \dfrac{1}{x} + \dfrac{1}{3}$ $3x(x + 3)$

8. $\dfrac{6}{x - 3} + \dfrac{1}{x} - \dfrac{1}{3}$

In Exercises 9–12 **a)** Explain why the expression on the left side of the equal sign is not equal to the expression on the right side of the equal sign. **b)** Show what the expression on the right side should be for it to be equal to the one on the left.

9. $\dfrac{4x - 3}{5x + 4} - \dfrac{2x - 7}{5x + 4} \neq \dfrac{4x - 3 - 2x - 7}{5x + 4}$ $\dfrac{4x - 3 - 2x + 7}{5x + 4}$

10. $\dfrac{5x}{2x - 3} - \dfrac{-3x - 7}{2x - 3} \neq \dfrac{5x + 3x - 7}{2x - 3}$ $\dfrac{5x + 3x + 7}{2x - 3}$

11. $\dfrac{6x - 2}{x^2 - 4x + 3} - \dfrac{3x^2 - 4x + 5}{x^2 - 4x + 3} \neq \dfrac{6x - 2 - 3x^2 - 4x + 5}{x^2 - 4x + 3}$ $\dfrac{6x - 2 - 3x^2 + 4x - 5}{x^2 - 4x + 3}$

12. $\dfrac{4x + 5}{x^2 - 6x} - \dfrac{-x^2 + 3x + 6}{x^2 - 6x} \neq \dfrac{4x + 5 + x^2 + 3x + 6}{x^2 - 6x}$ $\dfrac{4x + 5 + x^2 - 3x - 6}{x^2 - 6x}$

Practice the Skills

24. $\dfrac{1}{w + 1}$ **28.** $\dfrac{1}{m - 3}$ **39.** $\dfrac{x + 5}{x + 6}$ **43.** $\dfrac{x - 5}{x + 2}$ **44.** $\dfrac{5}{x - 2}$ **45.** $\dfrac{3x + 2}{x - 4}$ **48.** $\dfrac{4x + 3}{2x - 1}$

Add or subtract.

13. $\dfrac{x - 2}{7} + \dfrac{2x}{7}$ $\dfrac{3x - 2}{7}$

14. $\dfrac{2x - 7}{5} - \dfrac{6}{5}$ $\dfrac{2x - 13}{5}$

15. $\dfrac{3r + 2}{4} - \dfrac{3}{4}$ $\dfrac{3r - 1}{4}$

16. $\dfrac{3x + 6}{2} - \dfrac{x}{2}$ $x + 3$

17. $\dfrac{2}{x} + \dfrac{x + 4}{x}$ $\dfrac{x + 6}{x}$

18. $\dfrac{3x + 4}{x + 1} + \dfrac{6x + 5}{x + 1}$ 9

19. $\dfrac{n - 5}{n} - \dfrac{n + 7}{n}$ $-\dfrac{12}{n}$

20. $\dfrac{x - 6}{x} - \dfrac{x + 4}{x}$ $-\dfrac{10}{x}$

21. $\dfrac{x}{x - 1} + \dfrac{4x + 7}{x - 1}$ $\dfrac{5x + 7}{x - 1}$

22. $\dfrac{4x - 3}{x - 7} - \dfrac{2x + 8}{x - 7}$ $\dfrac{2x - 11}{x - 7}$

23. $\dfrac{4t + 7}{5t^2} - \dfrac{3t + 4}{5t^2}$ $\dfrac{t + 3}{5t^2}$

24. $\dfrac{3w + 5}{w^2 + 2w + 1} + \dfrac{-2w - 4}{w^2 + 2w + 1}$

25. $\dfrac{5x + 4}{x^2 - x - 12} + \dfrac{-4x - 1}{x^2 - x - 12}$ $\dfrac{1}{x - 4}$

26. $\dfrac{-x - 4}{x^2 - 16} + \dfrac{2(x + 4)}{x^2 - 16}$ $\dfrac{1}{x - 4}$

27. $\dfrac{x + 4}{3x + 2} - \dfrac{x + 4}{3x + 2}$ 0

28. $\dfrac{2m + 5}{(m + 4)(m - 3)} - \dfrac{m + 1}{(m + 4)(m - 3)}$

29. $\dfrac{2p - 5}{p - 5} - \dfrac{p + 5}{p - 5}$ $\dfrac{p - 10}{p - 5}$

30. $\dfrac{x^2 - 6}{3x} - \dfrac{x^2 + 4x - 5}{3x}$ $\dfrac{-4x - 1}{3x}$

31. $\dfrac{x^2 + 4x + 3}{x + 2} - \dfrac{5x + 9}{x + 2}$ $x - 3$

32. $\dfrac{-4x + 2}{3x + 6} + \dfrac{4(x - 1)}{3x + 6}$ $-\dfrac{2}{3x + 6}$

33. $\dfrac{3x + 11}{2x + 10} - \dfrac{2(x + 3)}{2x + 10}$ $\dfrac{1}{2}$

34. $\dfrac{x^2}{x + 4} - \dfrac{16}{x + 4}$ $x - 4$

35. $\dfrac{b^2 - 2b - 3}{b^2 - b - 6} + \dfrac{b - 3}{b^2 - b - 6}$ 1

36. $\dfrac{4x + 12}{3 - x} - \dfrac{3x + 15}{3 - x}$ -1

37. $\dfrac{t - 3}{t + 3} - \dfrac{-3t - 15}{t + 3}$ 4

38. $\dfrac{x^2 - 2}{x^2 + 6x - 7} - \dfrac{-4x + 19}{x^2 + 6x - 7}$ $\dfrac{x - 3}{x - 1}$

39. $\dfrac{x^2 + 2x}{(x + 6)(x - 3)} - \dfrac{15}{(x + 6)(x - 3)}$

40. $\dfrac{x^2 - 13}{x + 5} - \dfrac{12}{x + 5}$ $x - 5$

41. $\dfrac{3x^2 - 7x}{4x^2 - 8x} + \dfrac{x}{4x^2 - 8x}$ $\dfrac{3}{4}$

42. $\dfrac{x^3 - 10x^2 + 35x}{x(x - 6)} - \dfrac{x^2 + 5x}{x(x - 6)}$ $x - 5$

43. $\dfrac{3x^2 - 4x + 4}{3x^2 + 7x + 2} - \dfrac{10x + 9}{3x^2 + 7x + 2}$

44. $\dfrac{3x^2 + 15x}{x^3 + 2x^2 - 8x} + \dfrac{2x^2 + 5x}{x^3 + 2x^2 - 8x}$

45. $\dfrac{x^2 + 3x - 6}{x^2 - 5x + 4} - \dfrac{-2x^2 + 4x - 4}{x^2 - 5x + 4}$

46. $\dfrac{4x^2 + 5}{9x^2 - 64} - \dfrac{x^2 - x + 29}{9x^2 - 64}$ $\dfrac{x + 3}{3x + 8}$

47. $\dfrac{5x^2 + 40x + 8}{x^2 - 64} + \dfrac{x^2 + 9x}{x^2 - 64}$ $\dfrac{6x + 1}{x - 8}$

48. $\dfrac{20x^2 + 5x + 1}{6x^2 + x - 2} - \dfrac{8x^2 - 12x - 5}{6x^2 + x - 2}$

Find the least common denominator for each expression.

49. $\dfrac{x}{5} + \dfrac{x + 4}{5}$ 5

50. $\dfrac{2 + r}{3} - \dfrac{12}{3}$ 3

51. $\dfrac{1}{n} + \dfrac{1}{5n}$ $5n$

52. $\dfrac{1}{x + 1} - \dfrac{4}{7}$ $7(x + 1)$

53. $\dfrac{3}{5x} + \dfrac{7}{4}$ $20x$

54. $\dfrac{5}{2x} + 1$ $2x$

55. $\dfrac{6}{p} + \dfrac{3}{p^3}$ p^3

56. $\dfrac{2x}{x + 3} + \dfrac{6}{x - 4}$ $(x + 3)(x - 4)$

57. $\dfrac{m + 3}{3m - 4} + m$ $3m - 4$

58. $\dfrac{x + 4}{2x} + \dfrac{3}{7x}$ $14x$

59. $\dfrac{x}{2x + 3} + \dfrac{4}{x^2}$ $x^2(2x + 3)$

60. $\dfrac{x}{3x^2} + \dfrac{9}{7x^3}$ $21x^3$

61. $\dfrac{x + 1}{12x^2y} - \dfrac{7}{9x^3}$ $36x^3y$

62. $\dfrac{-3}{8x^2y^2} + \dfrac{6}{5x^4y^5}$ $40x^4y^5$

63. $\dfrac{4}{2r^4s^5} - \dfrac{5}{9r^3s^7}$ $18r^4s^7$

64. $\dfrac{3}{4w^5z^4} + \dfrac{2}{9wz^2}$ $\quad 36w^5z^4$

65. $\dfrac{w^2 - 7}{12w} - \dfrac{w + 3}{9(w + 5)}$ $\quad 36w(w + 5)$

66. $\dfrac{x - 3}{2x + 5} - \dfrac{6}{x - 5}$ $\quad (x - 5)(2x + 5)$

67. $\dfrac{5x - 2}{x^2 + x} - \dfrac{x^2}{x}$ $\quad x(x + 1)$

68. $\dfrac{4t}{t - 5} + \dfrac{2}{5 - t}$ $\quad t - 5 \text{ or } 5 - t$

69. $\dfrac{n}{4n - 1} + \dfrac{n - 2}{1 - 4n}$ $\quad 4n - 1 \text{ or } 1 - 4n$

70. $\dfrac{3}{-2a + 3b} - \dfrac{1}{2a - 3b}$ $\quad 2a - 3b \text{ or } -2a + 3b$

71. $\dfrac{6}{4k - 5r} - \dfrac{5}{-4k + 5r}$ $\quad 4k - 5r \text{ or } -4k + 5r$

72. $\dfrac{p}{4p^2 + 2p} - \dfrac{3}{2p + 1}$ $\quad 2p(2p + 1)$

73. $\dfrac{5}{2q^2 + 2q} - \dfrac{5}{3q}$ $\quad 6q(q + 1)$

74. $\dfrac{10}{(x + 4)(x + 2)} - \dfrac{5 - x}{x + 2}$ $\quad (x + 4)(x + 2)$

🔒 **75.** $\dfrac{21}{24x^2y} + \dfrac{x + 4}{15xy^3}$ $\quad 120x^2y^3$

76. $\dfrac{p^2 - 4}{p^2 - 25} + \dfrac{3}{p - 5}$ $\quad (p + 5)(p - 5)$

77. $\dfrac{3}{3x + 12} + \dfrac{3x + 6}{2x + 4}$ $\quad 6(x + 4)(x + 2)$

78. $6x^2 + \dfrac{9x}{x - 3}$ $\quad x - 3$

79. $\dfrac{9x + 4}{x + 6} - \dfrac{3x - 6}{x + 5}$ $\quad (x + 6)(x + 5)$

80. $\dfrac{x + 1}{x^2 + 11x + 18} - \dfrac{x^2 - 4}{x^2 - 3x - 10}$ $\quad (x + 9)(x + 2)(x - 5)$

🔒 **81.** $\dfrac{x - 2}{x^2 - 5x - 24} + \dfrac{3}{x^2 + 11x + 24}$ $\quad (x - 8)(x + 3)(x + 8)$

82. $\dfrac{4n}{n^2 - 4} - \dfrac{n - 3}{n^2 - 5n - 14}$ $\quad (n + 2)(n - 2)(n - 7)$

83. $\dfrac{7}{(a - 4)^2} - \dfrac{a + 2}{a^2 - 7a + 12}$ $\quad (a - 4)^2(a - 3)$

84. $\dfrac{3x + 5}{x^2 - 1} + \dfrac{x^2 - 8}{(x + 1)^2}$ $\quad (x + 1)^2(x - 1)$

85. $\dfrac{2x}{x^2 + 6x + 5} - \dfrac{5x^2}{x^2 + 4x + 3}$ $\quad (x + 5)(x + 1)(x + 3)$

86. $\dfrac{6x + 5}{x + 2} + \dfrac{4x}{(x + 2)^2}$ $\quad (x + 2)^2$

🔒 **87.** $\dfrac{3x - 5}{x^2 - 6x + 9} + \dfrac{3}{x - 3}$ $\quad (x - 3)^2$

88. $\dfrac{7n + 7}{(n + 5)(n + 2)} - \dfrac{3n - 5}{(n - 3)(n + 5)}$ $\quad (n + 5)(n + 2)(n - 3)$

89. $\dfrac{8x^2}{x^2 - 7x + 6} + x - 3$ $\quad (x - 6)(x - 1)$

90. $\dfrac{x - 1}{x^2 - 25} + x - 4$ $\quad (x + 5)(x - 5)$

91. $\dfrac{t - 1}{3t^2 + 10t - 8} - \dfrac{6}{3t^2 + 11t - 4}$ $\quad (3t - 2)(t + 4)(3t - 1)$

92. $\dfrac{-4x + 7}{2x^2 + 5x + 2} + \dfrac{x^2}{3x^2 + 4x - 4}$ $\quad (2x + 1)(x + 2)(3x - 2)$

93. $\dfrac{2x - 3}{4x^2 + 4x + 1} + \dfrac{x^2 - 4}{8x^2 + 10x + 3}$ $\quad (2x + 1)^2(4x + 3)$

94. $\dfrac{3x + 1}{6x^2 + 5x - 6} + \dfrac{x^2 - 5}{9x^2 - 12x + 4}$ $\quad (3x - 2)^2(2x + 3)$

Problem Solving

List the polynomial to be placed in each shaded area to make a true statement. Explain how you determined your answer.

✎ **95.** $\dfrac{x^2 - 6x + 3}{x + 3} + \dfrac{\blacksquare}{x + 3} = \dfrac{2x^2 - 5x - 6}{x + 3}$

$x^2 + x - 9$, sum of numerators must be $2x^2 - 5x - 6$.

✎ **96.** $\dfrac{4x^2 - 6x - 7}{x^2 - 4} - \dfrac{\blacksquare}{x^2 - 4} = \dfrac{2x^2 + x - 3}{x^2 - 4}$

$2x^2 - 7x - 4$, difference of numerators must be $2x^2 + x - 3$

✎ **97.** $\dfrac{-x^2 - 4x + 3}{2x + 5} + \dfrac{\blacksquare}{2x + 5} = \dfrac{5x - 7}{2x + 5}$

$x^2 + 9x - 10$, sum of numerators must be $5x - 7$

✎ **98.** $\dfrac{-3x^2 - 9}{(x + 4)(x - 2)} - \dfrac{\blacksquare}{(x + 4)(x - 2)} = \dfrac{x^2 + 3x}{(x + 4)(x - 2)}$

$-4x^2 - 3x - 9$, difference of numerators must be $x^2 + 3x$

Find the least common denominator of each expression.

99. $\dfrac{3}{☺} + \dfrac{4}{5☺}$ $\quad 5☺$

100. $\dfrac{5}{8\Delta^2☺^2} + \dfrac{6}{5\Delta^4☺^5}$ $\quad 40\Delta^4☺^5$

101. $\dfrac{8}{\Delta^2 - 9} - \dfrac{2}{\Delta + 3}$ $\quad (\Delta + 3)(\Delta - 3)$

102. $\dfrac{6}{\Delta + 3} - \dfrac{\Delta + 5}{\Delta^2 - 4\Delta + 3}$ $\quad (\Delta + 3)(\Delta - 3)(\Delta - 1)$

Challenge Problems

Perform each indicated operation.

103. $\dfrac{4x - 1}{x^2 - 25} - \dfrac{3x^2 - 8}{x^2 - 25} + \dfrac{8x - 3}{x^2 - 25}$ $\dfrac{-3x^2 + 12x + 4}{x^2 - 25}$

104. $\dfrac{x^2 - 8x + 2}{x + 7} + \dfrac{2x^2 - 5x}{x + 7} - \dfrac{3x^2 + 7x + 6}{x + 7}$ $\dfrac{-20x - 4}{x + 7}$

Find the least common denominator for each expression.

105. $\dfrac{7}{6x^5y^9} - \dfrac{9}{2x^3y} + \dfrac{4}{5x^{12}y^2}$ $30x^{12}y^9$

106. $\dfrac{12}{x - 3} - \dfrac{5}{x^2 - 9} + \dfrac{7}{x + 3}$ $(x - 3)(x + 3)$

107. $\dfrac{4}{x^2 - x - 12} + \dfrac{3}{x^2 - 6x + 8} + \dfrac{5}{x^2 + x - 6}$
$(x - 4)(x + 3)(x - 2)$

108. $\dfrac{4}{x^2 - 4} - \dfrac{11}{3x^2 + 5x - 2} + \dfrac{5}{3x^2 - 7x + 2}$
$(x - 2)(x + 2)(3x - 1)$

Cumulative Review Exercises

[1.3] **109.** Subtract $4\dfrac{3}{5} - 2\dfrac{5}{9}$. $\dfrac{92}{45}$ or $2\dfrac{2}{45}$

[2.5] **110.** Solve $6x + 4 = -(x + 2) - 3x + 4$. $-\dfrac{1}{5}$

[2.6] **111.** *Hummingbird Food* The instructions on a bottle of concentrated hummingbird food indicate that 6 ounces of the concentrate should be mixed with 1 gallon (128 ounces) of water. If you wish to mix the concentrate with only 48 ounces of water, how much concentrate should you use?   2.25 oz

[3.3] **112.** *Fitness Club* The Northern Fitness Club has two payment plans. Plan 1 is a yearly membership fee of $125 plus $2.50 per hour for use of the tennis court. Plan 2 is an annual membership fee of $300 with no charge for court time. How many hours would Malcolm Wu have to play in a year to make the cost of Plan 1 equal to the cost of Plan 2?   70 hrs

[4.3] **113.** Use scientific notation to evaluate $\dfrac{420,000,000}{0.0021}$. Leave your answer in scientific notation.
2.0×10^{11}

[5.6] **114.** Solve $2x^2 - 3 = x$. $\dfrac{3}{2}, -1$

6.4 ADDITION AND SUBTRACTION OF RATIONAL EXPRESSIONS

1   Add and subtract rational expressions.

SSM   Study Guide   CD/Video

MathPro 4/5   PH Math Tutor Center   prenhall.com/Angel

In Section 6.3 we discussed how to add and subtract rational expressions with a common denominator. Now we discuss adding and subtracting rational expressions that are not given with a common denominator.

1   Add and Subtract Rational Expressions

The method used to add and subtract rational expressions with unlike denominators is outlined in Example 1.

EXAMPLE 1 Add $\dfrac{7}{x} + \dfrac{3}{y}$.

Solution First we determine the LCD as outlined in Section 6.3.

$$\text{LCD} = xy$$

We write each fraction with the LCD. We do this by multiplying **both** the numerator and denominator of each fraction by any factors needed to obtain the LCD.

In this problem, the fraction on the left must be multiplied by y/y and the fraction on the right must be multiplied by x/x.

$$\frac{7}{x} + \frac{3}{y} = \frac{y}{y} \cdot \frac{7}{x} + \frac{3}{y} \cdot \frac{x}{x} = \frac{7y}{xy} + \frac{3x}{xy}$$

By multiplying both the numerator and denominator by the same factor, we are in effect multiplying by 1, which does not change the value of the fraction, only its appearance. Thus, the new fraction is equivalent to the original fraction.

Now we add the numerators, while leaving the LCD alone.

$$\frac{7y}{xy} + \frac{3x}{xy} = \frac{7y + 3x}{xy} \quad \text{or} \quad \frac{3x + 7y}{xy} \qquad \text{\Large ✳}$$

TEACHING TIP
Before beginning this section, have students multiply $\frac{3}{4}$ by

a) 1 **b)** $\dfrac{x}{x}$ **c)** $\dfrac{x-3}{x-3}$

Then ask, "How does the value of the fraction change with each multiplication? Explain."

TEACHING TIP
Point out that the Commutative Property of Multiplication allows us to multiply in either order. When getting the LCD, to keep the original expression intact, we place an expression to the left of the fraction on the left and an expression to the right of the fraction on the right.

To Add or Subtract Two Rational Expressions with Unlike Denominators

1. Determine the LCD.
2. Rewrite each fraction as an equivalent fraction with the LCD. This is done by multiplying both the numerator and denominator of each fraction by any factors needed to obtain the LCD.
3. Add or subtract the numerators while maintaining the LCD.
4. When possible, factor the remaining numerator and simplify the fraction.

EXAMPLE 2 Add $\dfrac{5}{4x^2y} + \dfrac{3}{14xy^3}$.

Solution The LCD is $28x^2y^3$. We must write each fraction with the denominator $28x^2y^3$. To do this, we multiply the fraction on the left by $7y^2/7y^2$ and the fraction on the right by $2x/2x$.

$$\frac{5}{4x^2y} + \frac{3}{14xy^3} = \frac{7y^2}{7y^2} \cdot \frac{5}{4x^2y} + \frac{3}{14xy^3} \cdot \frac{2x}{2x}$$

$$= \frac{35y^2}{28x^2y^3} + \frac{6x}{28x^2y^3}$$

$$= \frac{35y^2 + 6x}{28x^2y^3} \quad \text{or} \quad \frac{6x + 35y^2}{28x^2y^3} \qquad \text{\Large ✳}$$

NOW TRY EXERCISE 15

HELPFUL HINT

In Example 2 we multiplied the first fraction by $\dfrac{7y^2}{7y^2}$ and the second fraction by $\dfrac{2x}{2x}$ to get two fractions with a common denominator. How did we know what to multiply each fraction by? Many of you can determine this by observing the LCD and then determining what each denominator needs to be multiplied by to get the LCD. If this is not obvious, you can divide the LCD by the given denominator to determine what the numerator and denominator of each fraction should be multiplied by. In Example 2,

(continued on the next page)

the LCD is $28x^2y^3$. If we divide $28x^2y^3$ by each given denominator, $4x^2y$ and $14xy^3$, we can determine what the numerator and denominator of each respective fraction should be multiplied by.

$$\frac{28x^2y^3}{4x^2y} = 7y^2 \qquad \frac{28x^2y^3}{14xy^3} = 2x$$

Thus, $\dfrac{5}{4x^2y}$ should be multiplied by $\dfrac{7y^2}{7y^2}$ and $\dfrac{3}{14xy^3}$ should be multiplied by $\dfrac{2x}{2x}$ to obtain the LCD $28x^2y^3$.

EXAMPLE 3 Add $\dfrac{3}{x+2} + \dfrac{5}{x}$.

Solution We must write each fraction with the LCD, which is $x(x+2)$. To do this, we multiply the fraction on the left by x/x and the fraction on the right by $(x+2)/(x+2)$.

$$\frac{3}{x+2} + \frac{5}{x} = \frac{x}{x} \cdot \frac{3}{x+2} + \frac{5}{x} \cdot \frac{x+2}{x+2}$$

$$= \frac{3x}{x(x+2)} + \frac{5(x+2)}{x(x+2)} \qquad \text{Rewrite each fraction as an equivalent fraction with the LCD.}$$

$$= \frac{3x}{x(x+2)} + \frac{5x+10}{x(x+2)} \qquad \text{Distributive property.}$$

$$= \frac{3x + (5x+10)}{x(x+2)} \qquad \text{Write as a single fraction.}$$

$$= \frac{3x + 5x + 10}{x(x+2)} \qquad \text{Remove parentheses in the numerator.}$$

$$= \frac{8x + 10}{x(x+2)} \qquad \text{Combine like terms in the numerator.}$$

NOW TRY EXERCISE 25

HELPFUL HINT

Look at the answer to Example 3, $\dfrac{8x+10}{x(x+2)}$. Notice that the numerator could have been factored to obtain $\dfrac{2(4x+5)}{x(x+2)}$. Also notice that the denominator could have been multiplied to get $\dfrac{8x+10}{x^2+2x}$. All three of these answers are equivalent and each is correct.

In this section, when writing answers, unless there is a common factor in the numerator and denominator we will leave the numerator in unfactored form and the denominator in factored form. If both the numerator and denominator have a common factor, we will factor the numerator and simplify the fraction.

EXAMPLE 4 Subtract $\dfrac{w}{w-7} - \dfrac{3}{w-4}$.

Solution The LCD is $(w-7)(w-4)$. The fraction on the left must be multiplied by $(w-4)/(w-4)$ to obtain the LCD. The fraction on the right must be multiplied by $(w-7)/(w-7)$ to obtain the LCD.

$$\frac{w}{w-7} - \frac{3}{w-4} = \frac{w-4}{w-4} \cdot \frac{w}{w-7} - \frac{3}{w-4} \cdot \frac{w-7}{w-7}$$

$$= \frac{w(w-4)}{(w-4)(w-7)} - \frac{3(w-7)}{(w-4)(w-7)}$$
Rewrite each fraction as an equivalent fraction with the LCD.

$$= \frac{w^2 - 4w}{(w-4)(w-7)} - \frac{3w-21}{(w-4)(w-7)}$$
Distributive property.

$$= \frac{(w^2 - 4w) - (3w-21)}{(w-4)(w-7)}$$
Write as a single fraction.

$$= \frac{w^2 - 4w - 3w + 21}{(w-4)(w-7)}$$
Remove parentheses in the numerator.

$$= \frac{w^2 - 7w + 21}{(w-4)(w-7)}$$
Combine like terms in the numerator.

EXAMPLE 5 Subtract $\dfrac{x+2}{x-4} - \dfrac{x+3}{x+4}$.

Solution The LCD is $(x-4)(x+4)$.

$$\frac{x+2}{x-4} - \frac{x+3}{x+4} = \frac{x+4}{x+4} \cdot \frac{x+2}{x-4} - \frac{x+3}{x+4} \cdot \frac{x-4}{x-4}$$

$$= \frac{(x+4)(x+2)}{(x+4)(x-4)} - \frac{(x+3)(x-4)}{(x+4)(x-4)}$$
Rewrite each fraction as an equivalent fraction with the LCD.

Use the FOIL method to multiply each numerator.

$$= \frac{x^2 + 6x + 8}{(x+4)(x-4)} - \frac{x^2 - x - 12}{(x+4)(x-4)}$$

$$= \frac{(x^2 + 6x + 8) - (x^2 - x - 12)}{(x+4)(x-4)}$$
Write as a single fraction.

$$= \frac{x^2 + 6x + 8 - x^2 + x + 12}{(x+4)(x-4)}$$
Remove parentheses in the numerator.

$$= \frac{7x + 20}{(x+4)(x-4)}$$
Combine like terms in the numerator.

NOW TRY EXERCISE 37

Consider the problem

$$\frac{6}{x-2} + \frac{x+3}{2-x}$$

How do we add these rational expressions? We could write each fraction with the denominator $(x-2)(2-x)$. However, there is an easier way. Study the following Helpful Hint.

HELPFUL HINT

When adding or subtracting fractions whose denominators are opposites (and therefore differ only in signs), multiply both the numerator *and* the denominator of *either* fraction by -1. Then both fractions will have the same denominator.

$$\frac{x}{a-b} + \frac{y}{b-a} = \frac{x}{a-b} + \frac{y}{b-a} \cdot \frac{-1}{-1}$$

$$= \frac{x}{a-b} + \frac{-y}{a-b}$$

$$= \frac{x-y}{a-b}$$

EXAMPLE 6 Add $\dfrac{6}{x-2} + \dfrac{x+3}{2-x}$.

Solution Since the denominators differ only in sign, we may multiply both the numerator and the denominator of either fraction by -1. Here we will multiply the numerator and denominator of the second fraction by -1 to obtain the common denominator $x - 2$.

$$\frac{6}{x-2} + \frac{x+3}{2-x} = \frac{6}{x-2} + \frac{x+3}{2-x} \cdot \boxed{\frac{-1}{-1}}$$ *Multiply numerator and denominator by -1.*

$$= \frac{6}{x-2} + \frac{(-x-3)}{x-2}$$

$$= \frac{6 + (-x-3)}{x-2}$$ *Write as a single fraction.*

$$= \frac{6 - x - 3}{x-2}$$ *Remove parentheses in the numerator.*

$$= \frac{-x+3}{x-2}$$ *Combine like terms in the numerator.*

NOW TRY EXERCISE 31

Let's work another example where the denominators differ only in sign.

EXAMPLE 7 Subtract $\dfrac{a-5}{3a-4} - \dfrac{2a-5}{4-3a}$.

Solution The denominators of the two fractions differ only in sign. We will work this problem in a similar manner to how we worked Example 6. We will multiply both the numerator and denominator of the second fraction by -1 to obtain the common denominator $3a - 4$.

$$\frac{a-5}{3a-4} - \frac{2a-5}{4-3a} = \frac{a-5}{3a-4} - \frac{2a-5}{4-3a} \cdot \boxed{\frac{-1}{-1}}$$ *Multiply numerator and denominator by -1.*

$$= \frac{a-5}{3a-4} - \frac{(-2a+5)}{3a-4}$$

$$= \frac{(a-5) - (-2a+5)}{3a-4}$$ *Write as a single fraction.*

$$\frac{a-5+2a-5}{3a-4}$$ *Remove parentheses in the numerator.*

$$= \frac{3a-10}{3a-4}$$ *Combine like terms in the numerator.*

EXAMPLE 8 Add $\dfrac{3}{x^2+5x+6} + \dfrac{1}{3x^2+8x-3}$.

Solution $\dfrac{3}{x^2+5x+6} + \dfrac{1}{3x^2+8x-3} = \dfrac{3}{(x+2)(x+3)} + \dfrac{1}{(3x-1)(x+3)}$

The LCD is $(x+2)(x+3)(3x-1)$.

$$= \frac{3x-1}{3x-1} \cdot \frac{3}{(x+2)(x+3)} + \frac{1}{(3x-1)(x+3)} \cdot \frac{x+2}{x+2}$$

$$= \frac{9x-3}{(3x-1)(x+2)(x+3)} + \frac{x+2}{(3x-1)(x+2)(x+3)}$$

$$= \frac{(9x-3)+(x+2)}{(3x-1)(x+2)(x+3)}$$

$$= \frac{9x-3+x+2}{(3x-1)(x+2)(x+3)}$$

$$= \frac{10x-1}{(3x-1)(x+2)(x+3)}$$

NOW TRY EXERCISE 55

EXAMPLE 9   Subtract $\dfrac{5}{x^2-5x} - \dfrac{x}{5x-25}$.

Solution $\dfrac{5}{x^2-5x} - \dfrac{x}{5x-25} = \dfrac{5}{x(x-5)} - \dfrac{x}{5(x-5)}$

The LCD is $5x(x-5)$.

$$= \frac{5}{5} \cdot \frac{5}{x(x-5)} - \frac{x}{5(x-5)} \cdot \frac{x}{x}$$

$$= \frac{25}{5x(x-5)} - \frac{x^2}{5x(x-5)}$$

$$= \frac{25 - x^2}{5x(x-5)}$$

$$= \frac{(5-x)(5+x)}{5x(x-5)} \qquad \text{\textit{Factor the numerator.}}$$

$$= \frac{-1(x-5)(x+5)}{5x(x-5)} \qquad \text{\textit{5} $- x = -1(x-5)$}$$

$$= \frac{-1\cancel{(x-5)}(x+5)}{5x\cancel{(x-5)}} \qquad \text{\textit{Simplify.}}$$

$$= \frac{-1(x+5)}{5x} \quad \text{or} \quad -\frac{x+5}{5x}$$

AVOIDING COMMON ERRORS

A common error in an addition or subtraction problem is to add or subtract the numerators and the denominators. Here is one such example.

CORRECT

$$\frac{1}{x} + \frac{x}{1} = \frac{1}{x} + \frac{x}{1} \cdot \frac{x}{x}$$

$$= \frac{1}{x} + \frac{x^2}{x}$$

$$= \frac{1+x^2}{x} \text{ or } \frac{x^2+1}{x}$$

INCORRECT

$$\frac{1}{x} + \frac{x}{1} = \frac{1+x}{x+1}$$

$$\frac{1}{x} - \frac{x}{1} = \frac{1-x}{x-1}$$

(continued on the next page)

Remember that to add or subtract fractions you must first have a common denominator. Then you add or subtract the numerators while maintaining the common denominator.
   Another common mistake is to treat an addition or subtraction problem as a multiplication problem. You can divide out common factors only when *multiplying* expressions, not when adding or subtracting them.

CORRECT

$$\frac{1}{x} \cdot \frac{x}{1} = \frac{1}{\cancel{x}} \cdot \frac{\cancel{x}}{1}$$
$$= 1 \cdot 1 = 1$$

INCORRECT

$$\frac{1}{x} + \frac{x}{1} \neq \frac{1}{\cancel{x}} + \frac{\cancel{x}}{1}$$
$$= 1 + 1 = 2$$

Exercise Set 6.4

Concept/Writing Exercises

1. When adding or subtracting fractions with unlike denominators, how can you determine what each denominator should be multiplied by to get the LCD?

2. When you multiply both the numerator and denominator of a fraction by the factors needed to obtain the LCD, why are you not changing the value of the fraction? in effect, multiplying by 1

3. a) Explain in your own words a step-by-step procedure to add or subtract two rational expressions that have unlike denominators. Answers will vary.

   b) Using the procedure outlined in part **a)**, add
   $$\frac{x}{x^2 - x - 6} + \frac{3}{x^2 - 4}. \quad \frac{x^2 + x - 9}{(x+2)(x-3)(x-2)}$$

4. Explain how to add or subtract fractions whose denominators are opposites. Give an example.

1. For each fraction, divide the LCD by the denominator.
4. Multiply both the numerator and the denominator of either fraction by -1.
5. b) $\dfrac{3yz + 10}{12z^2}$

5. Consider $\dfrac{y}{4z} + \dfrac{5}{6z^2}$

   a) What is the LCD? $12z^2$

   b) Perform the indicated operation.

   c) If you mistakenly used $24z^2$ for the LCD when adding, would you eventually obtain the correct answer? Explain. yes

6. Would you use the LCD to perform the following indicated operations? Explain.

   a) $\dfrac{3}{x+3} - \dfrac{4}{x} + \dfrac{5}{3}$ yes   **b)** $\dfrac{1}{x+2} \cdot \dfrac{5}{x}$ no

   c) $x + \dfrac{2}{3}$ yes   **d)** $\dfrac{5}{x^2-9} \div \dfrac{2}{x-3}$ no

Practice the Skills

Add or subtract.

7. $\dfrac{1}{4x} + \dfrac{3}{x}$ $\dfrac{13}{4x}$

8. $\dfrac{1}{4x} + \dfrac{1}{x}$ $\dfrac{5}{4x}$

9. $\dfrac{5}{x^2} + \dfrac{3}{2x}$ $\dfrac{3x + 10}{2x^2}$

10. $2 - \dfrac{1}{x^2}$ $\dfrac{2x^2 - 1}{x^2}$

11. $3 + \dfrac{5}{x}$ $\dfrac{3x + 5}{x}$

12. $\dfrac{5}{6y} + \dfrac{3}{5y^2}$ $\dfrac{25y + 18}{30y^2}$

13. $\dfrac{2}{x^2} + \dfrac{3}{5x}$ $\dfrac{3x + 10}{5x^2}$

14. $\dfrac{3}{x} - \dfrac{5}{x^2}$ $\dfrac{3x - 5}{x^2}$

15. $\dfrac{7}{4x^2y} + \dfrac{3}{5xy^2}$ $\dfrac{35y + 12x}{20x^2y^2}$

16. $\dfrac{5}{12x^4y} - \dfrac{1}{5x^2y^3}$ $\dfrac{25y^2 - 12x^2}{60x^4y^3}$

17. $3y + \dfrac{x}{y}$ $\dfrac{3y^2 + x}{y}$

18. $x + \dfrac{x}{y}$ $\dfrac{xy + x}{y}$

19. $\dfrac{3a - 1}{2a} + \dfrac{2}{3a}$ $\dfrac{9a + 1}{6a}$

20. $\dfrac{3}{n} + 5$ $\dfrac{5n + 3}{n}$

21. $\dfrac{4x}{y} + \dfrac{2y}{xy}$ $\dfrac{4x^2 + 2y}{xy}$

22. $\dfrac{3}{5p} - \dfrac{5}{2p^2}$ $\dfrac{6p - 25}{10p^2}$

23. $\dfrac{4}{b} - \dfrac{4}{5a^2}$ $\dfrac{20a^2 - 4b}{5a^2b}$

24. $\dfrac{x - 3}{x} - \dfrac{1}{4x}$ $\dfrac{4x - 13}{4x}$

25. $\dfrac{4}{x} + \dfrac{7}{x - 3}$ $\dfrac{11x - 12}{x(x - 3)}$

26. $6 - \dfrac{3}{x - 3}$ $\dfrac{6x - 21}{x - 3}$

27. $\dfrac{9}{p + 3} + \dfrac{2}{p}$ $\dfrac{11p + 6}{p(p + 3)}$

28. $\dfrac{a}{a + b} + \dfrac{a - b}{a}$ $\dfrac{2a^2 - b^2}{a(a + b)}$

29. $\dfrac{5}{6d} - \dfrac{d}{3d + 5}$ $\dfrac{-6d^2 + 15d + 25}{6d(3d + 5)}$

30. $\dfrac{2}{x-3} - \dfrac{4}{x-1}$ $\quad \dfrac{-2x+10}{(x-3)(x-1)}$

31. $\dfrac{4}{p-3} + \dfrac{2}{3-p}$ $\quad \dfrac{2}{p-3}$

32. $\dfrac{3}{n-5} - \dfrac{1}{5-n}$ $\quad \dfrac{4}{n-5}$

🔒 **33.** $\dfrac{9}{x+7} - \dfrac{5}{-x-7}$ $\quad \dfrac{14}{x+7}$

34. $\dfrac{6}{7x-1} - \dfrac{5}{1-7x}$ $\quad \dfrac{11}{7x-1}$

35. $\dfrac{6}{a-2} + \dfrac{a}{2a-4}$ $\quad \dfrac{a+12}{2(a-2)}$

36. $\dfrac{4}{y-1} + \dfrac{3}{y+1}$ $\quad \dfrac{7y+1}{(y+1)(y-1)}$

37. $\dfrac{x+5}{x-5} - \dfrac{x-5}{x+5}$ $\quad \dfrac{20x}{(x-5)(x+5)}$

38. $\dfrac{x+7}{x+3} - \dfrac{x-3}{x+7}$ $\quad \dfrac{14x+58}{(x+3)(x+7)}$

39. $\dfrac{5}{6n+3} - \dfrac{4}{n}$ $\quad \dfrac{-19n-12}{3n(2n+1)}$

40. $\dfrac{x}{4x-4} - \dfrac{1}{3x}$ $\quad \dfrac{3x^2-4x+4}{12x(x-1)}$

41. $\dfrac{3}{2w+10} + \dfrac{5}{w+2}$ $\quad \dfrac{13w+56}{2(w+5)(w+2)}$

42. $\dfrac{5k}{4k-8} - \dfrac{k}{k+2}$ $\quad \dfrac{k^2+18k}{4(k-2)(k+2)}$

43. $\dfrac{z}{z^2-16} + \dfrac{4}{z+4}$ $\quad \dfrac{5z-16}{(z+4)(z-4)}$

44. $\dfrac{5}{(x+4)^2} + \dfrac{2}{x+4}$ $\quad \dfrac{2x+13}{(x+4)^2}$

🔒 **45.** $\dfrac{x+2}{x^2-4} - \dfrac{2}{x+2}$ $\quad \dfrac{-x+6}{(x+2)(x-2)}$

46. $\dfrac{3}{(x-2)(x+3)} + \dfrac{5}{(x+2)(x+3)}$ $\quad \dfrac{8x-4}{(x+2)(x-2)(x+3)}$

47. $\dfrac{3r+2}{r^2-10r+24} - \dfrac{2}{r-6}$ $\quad \dfrac{r+10}{(r-4)(r-6)}$

48. $\dfrac{x+3}{x^2-3x-10} - \dfrac{2}{x-5}$ $\quad \dfrac{-x-1}{(x-5)(x+2)}$

49. $\dfrac{x^2}{x^2+2x-8} - \dfrac{x-4}{x+4}$ $\quad \dfrac{6x-8}{(x+4)(x-2)}$

50. $\dfrac{x+1}{x^2-4x+4} - \dfrac{x+1}{x-2}$ $\quad \dfrac{-x^2+2x+3}{(x-2)^2}$

51. $\dfrac{x-3}{x^2+10x+25} + \dfrac{x-3}{x+5}$ $\quad \dfrac{x^2+3x-18}{(x+5)^2}$

52. $\dfrac{x}{x^2-xy} - \dfrac{y}{xy-x^2}$ $\quad \dfrac{x+y}{x(x-y)}$

53. $\dfrac{5}{a^2-9a+8} - \dfrac{3}{a^2-6a-16}$ $\quad \dfrac{2a+13}{(a-8)(a-1)(a+2)}$

54. $\dfrac{3}{a^2+2a-15} - \dfrac{1}{a^2-9}$ $\quad \dfrac{2a+4}{(a+5)(a-3)(a+3)}$

55. $\dfrac{2}{x^2+6x+9} + \dfrac{3}{x^2+x-6}$ $\quad \dfrac{5x+5}{(x+3)^2(x-2)}$

56. $\dfrac{x}{2x^2+7x-4} + \dfrac{2}{x^2-x-20}$ $\quad \dfrac{x^2-x-2}{(2x-1)(x+4)(x-5)}$

57. $\dfrac{x}{2x^2+7x+3} - \dfrac{3}{3x^2+7x-6}$ $\quad \dfrac{3x^2-8x-3}{(2x+1)(3x-2)(x+3)}$

58. $\dfrac{x}{6x^2+7x+2} + \dfrac{5}{2x^2-3x-2}$ $\quad \dfrac{x^2+13x+10}{(3x+2)(2x+1)(x-2)}$

🔒 **59.** $\dfrac{x}{4x^2+11x+6} - \dfrac{2}{8x^2+2x-3}$ $\quad \dfrac{2x^2-3x-4}{(4x+3)(x+2)(2x-1)}$

60. $\dfrac{x}{5x^2-9x-2} - \dfrac{2}{3x^2-7x+2}$ $\quad \dfrac{3x^2-11x-2}{(5x+1)(x-2)(3x-1)}$

61. $\dfrac{3w+12}{w^2+w-12} - \dfrac{2}{w-3}$ $\quad \dfrac{1}{w-3}$

62. $\dfrac{5x+10}{x^2-5x-14} - \dfrac{4}{x-7}$ $\quad \dfrac{1}{x-7}$

63. $\dfrac{3r}{2r^2-10r+12} + \dfrac{3}{r-2}$ $\quad \dfrac{9}{2(r-3)}$

64. $\dfrac{6m}{3m^2-24m+48} - \dfrac{2}{m-4}$ $\quad \dfrac{8}{(m-4)^2}$

Problem Solving

For what value(s) of x is each expression defined?

65. $\dfrac{2}{x} + 6$ $\quad$ all real numbers except $x = 0$

66. $\dfrac{2}{x-1} - \dfrac{3}{x}$ $\quad$ all real numbers except $x = 1, x = 0$

67. $\dfrac{5}{x-4} + \dfrac{7}{x+6}$ $\quad$ all real numbers except $x = 4, x = -6$

68. $\dfrac{4}{x^2-9} - \dfrac{1}{x+3}$ $\quad$ all real numbers except $x = 3, x = -3$

Add or subtract. Treat the unknown symbols as if they were variables.

69. $\dfrac{3}{\Delta-2} - \dfrac{1}{2-\Delta}$ $\quad \dfrac{4}{\Delta-2}$

70. $\dfrac{\Delta}{2\Delta^2+7\Delta-4} + \dfrac{2}{\Delta^2-\Delta-20}$ $\quad \dfrac{\Delta^2-\Delta-2}{(2\Delta-1)(\Delta+4)(\Delta-5)}$

Challenge Problems

Under what conditions is each expression defined? Explain your answers.

71. $\dfrac{5}{a+b} + \dfrac{3}{a}$ all real numbers except $a = 0$, $a = -b$

72. $\dfrac{x+2}{x+5y} - \dfrac{y-3}{2x}$ all real numbers except $x = 0$, $x = -5y$

Perform each indicated operation.

73. $\dfrac{x}{x^2 - 9} + \dfrac{3x}{x+3} + \dfrac{3x^2 - 8x}{9 - x^2}$ 0

74. $\dfrac{5x}{x^2 + x - 6} + \dfrac{x}{x+3} - \dfrac{2}{x-2}$ 1

75. $\dfrac{x+6}{4-x^2} - \dfrac{x+3}{x+2} + \dfrac{x-3}{2-x}$ $\dfrac{2x-3}{2-x}$

76. $\dfrac{3x-1}{x+2} + \dfrac{x}{x-3} - \dfrac{4}{2x+3}$ $\dfrac{8x^3 - 8x^2 - 14x + 33}{(x+2)(x-3)(2x+3)}$

77. $\dfrac{2}{x^2 - x - 6} + \dfrac{3}{x^2 - 2x - 3} + \dfrac{1}{x^2 + 3x + 2}$ $\dfrac{6x+5}{(x+2)(x-3)(x+1)}$

78. $\dfrac{3x}{x^2 - 4} + \dfrac{4}{x^3 + 8}$ $\dfrac{3x^3 - 6x^2 + 16x - 8}{(x+2)(x-2)(x^2 - 2x + 4)}$

Group Activity

79. a) $(x+y)(x+2y)(2x+y)$ **b)** $\dfrac{x^2 + 6xy + 5y^2}{(x+y)(x+2y)(2x+y)}$ **c)** $\dfrac{x+5y}{(x+2y)(2x+y)}$

Discuss and answer Exercise 79 as a group.

79. a) As a group, find the LCD of

$$\frac{x+3y}{x^2 + 3xy + 2y^2} + \frac{y-x}{2x^2 + 3xy + y^2}$$

b) As a group, perform the indicated operation, but do not simplify your answer.

c) As a group, simplify your answer.

d) Group member 1: Substitute 2 for x and 1 for y in the fraction on the left in part **a)** and evaluate. $\frac{5}{12}$

e) Group member 2: Substitute 2 for x and 1 for y in the fraction on the right in part **a)** and evaluate. $-\frac{1}{15}$

f) Group member 3: Add the numerical fractions found in parts **d)** and **e)**. $\frac{7}{20}$

g) Individually, substitute 2 for x and 1 for y in the expression obtained in part **b)** and evaluate. $\frac{21}{60} = \frac{7}{20}$

h) Individually, substitute 2 for x and 1 for y in the expression obtained in part **c)**, evaluate, and compare your answers. $\frac{7}{20}$

i) As a group, discuss what you discovered from this activity.

j) Do you think your results would have been similar for any numbers substituted for x and y (for which the denominator is not 0)? Why?   yes

Cumulative Review Exercises

[2.6] **80.** ***White Pass Railroad*** The White Pass Railroad is a narrow gauge railroad that travels slowly through the mountains of Alaska. If the train travels 22 miles in 0.8 hours, how long will it take to travel 42 miles? Assume the train travels at the same rate throughout the trip. ≈ 1.53 hr

[2.7] **81.** Solve the inequality $3(x-2) + 2 < 4(x+1)$ and graph the solution on a number line. $x > -8$,

[4.6] **82.** Divide $(8x^2 + 6x - 13) \div (2x + 3)$. $4x - 3 - \dfrac{4}{2x+3}$

[6.2] **83.** Multiply $\dfrac{x^2 + xy - 6y^2}{x^2 - xy - 2y^2} \cdot \dfrac{y^2 - x^2}{x^2 + 2xy - 3y^2}$. -1

6.5 COMPLEX FRACTIONS

SSM   Study Guide   CD/Video

MathPro 4/5   PH Math Tutor Center   prenhall.com/Angel

1   Simplify complex fractions by combining terms.
2   Simplify complex fractions using multiplication first to clear fractions.

1   Simplify Complex Fractions by Combining Terms

A **complex fraction** is one that has a fraction in its numerator or its denominator or in both its numerator and denominator.

Examples of Complex Fractions

$$\dfrac{\dfrac{3}{5}}{4} \qquad \dfrac{\dfrac{x+1}{x}}{2x} \qquad \dfrac{\dfrac{x}{y}}{x+1} \qquad \dfrac{\dfrac{a+b}{a}}{\dfrac{a-b}{b}}$$

Numerator of complex fraction $\Big\{ \dfrac{a+b}{a}$

$\longleftarrow$ Main fraction line

Denominator of complex fraction $\Big\{ \dfrac{a+b}{a}$

The expression above the main fraction line is the numerator, and the expression below the main fraction line is the denominator of the complex fraction.

There are two methods to simplify complex fractions. The first reinforces many of the concepts used in this chapter because we may need to add, subtract, multiply, and divide simpler fractions as we simplify the complex fraction. Many students prefer to use the second method because the answer may be obtained more quickly. We will give two examples using the first method and then work three examples using the second method.

Method 1—To Simplify a Complex Fraction by Combining Terms

1. Add or subtract the fractions in both the numerator and denominator of the complex fraction to obtain single fractions in both the numerator and the denominator.

2. Invert the denominator of the complex fraction and multiply the numerator by it.

3. Simplify further if possible.

EXAMPLE 1   Simplify $\dfrac{\dfrac{ab^2}{c^3}}{\dfrac{a}{bc^2}}$.

Solution   Since both numerator and denominator are already single fractions, we omit step 1 and begin with step 2. When we invert the denominator of the complex fraction and multiply it by the numerator we obtain the following.

$$\frac{\dfrac{ab^2}{c^3}}{\dfrac{a}{bc^2}} = \frac{\cancel{a}b^2}{\cancel{c}^3} \cdot \frac{b\cancel{c}^2}{\cancel{a}} = \frac{b^3}{c}$$

NOW TRY EXERCISE 11 Thus the expression simplifies to $\dfrac{b^3}{c}$. ✳

EXAMPLE 2 Simplify $\dfrac{a + \dfrac{1}{x}}{x + \dfrac{1}{a}}$.

Solution Express both the numerator and denominator of the complex fraction as single fractions. The LCD of the numerator is x and the LCD of the denominator is a.

$$\frac{a + \dfrac{1}{x}}{x + \dfrac{1}{a}} = \frac{\dfrac{x}{x} \cdot a + \dfrac{1}{x}}{\dfrac{a}{a} \cdot x + \dfrac{1}{a}} = \frac{\dfrac{ax}{x} + \dfrac{1}{x}}{\dfrac{ax}{a} + \dfrac{1}{a}} = \frac{\dfrac{ax + 1}{x}}{\dfrac{ax + 1}{a}}$$

Now invert the denominator and multiply the numerator by it.

$$= \frac{\cancel{ax + 1}}{x} \cdot \frac{a}{\cancel{ax + 1}} = \frac{a}{x}$$ ✳

In Example 4 on page 401 we will rework Example 2. However, at that time we will work the example using method 2. Most students will agree that method 2 is simpler to use for problems of this type, where the numerator or denominator consists of a sum or difference of terms. We illustrate Example 2 here to show you that method 1 works for problems of this type, and to give you more practice with method 1.

2 Simplify Complex Fractions Using Multiplication First to Clear Fractions

Here is the second method for simplifying complex fractions.

> **Method 2—To Simplify a Complex Fraction Using Multiplication First**
>
> 1. Find the least common denominator of *all* the denominators appearing in the complex fraction.
> 2. Multiply both the numerator and denominator of the complex fraction by the LCD found in step 1.
> 3. Simplify when possible.

EXAMPLE 3 Simplify $\dfrac{\dfrac{1}{3} + \dfrac{4}{5}}{\dfrac{4}{5} - \dfrac{1}{3}}$.

Solution The denominators in the complex fraction are 3 and 5. The LCD of 3 and 5 is 15. Thus 15 is the LCD of the complex fraction. Multiply both the numerator and denominator of the complex fraction by 15.

$$\frac{\dfrac{1}{3}+\dfrac{4}{5}}{\dfrac{4}{5}-\dfrac{1}{3}}=\frac{15}{15}\cdot\frac{\left(\dfrac{1}{3}+\dfrac{4}{5}\right)}{\left(\dfrac{4}{5}-\dfrac{1}{3}\right)}=\frac{15\left(\dfrac{1}{3}\right)+15\left(\dfrac{4}{5}\right)}{15\left(\dfrac{4}{5}\right)-15\left(\dfrac{1}{3}\right)}$$

Now simplify.

NOW TRY EXERCISE 9

$$=\frac{5+12}{12-5}=\frac{17}{7}$$

Now we will rework Example 2 using method 2.

EXAMPLE 4 Simplify $\dfrac{a+\dfrac{1}{x}}{x+\dfrac{1}{a}}$.

Solution The denominators in the complex fraction are x and a. Therefore, the LCD of the complex fraction is ax. Multiply both the numerator and denominator of the complex fraction by ax.

$$\frac{a+\dfrac{1}{x}}{x+\dfrac{1}{a}}=\frac{ax}{ax}\cdot\frac{\left(a+\dfrac{1}{x}\right)}{\left(x+\dfrac{1}{a}\right)}=\frac{a^2x+a}{ax^2+x}$$

$$=\frac{a\cancel{(ax+1)}}{x\cancel{(ax+1)}}=\frac{a}{x}$$

Note that the answers to Examples 2 and 4 are the same.

EXAMPLE 5 Simplify $\dfrac{x}{\dfrac{1}{x}+\dfrac{1}{y}}$.

Solution The denominators in the complex fraction are x and y. Therefore, the LCD of the complex fraction is xy. Multiply both the numerator and denominator of the complex fraction by xy.

$$\frac{x}{\dfrac{1}{x}+\dfrac{1}{y}}=\frac{xy}{xy}\cdot\frac{x}{\left(\dfrac{1}{x}+\dfrac{1}{y}\right)}$$

$$=\frac{x^2y}{xy\left(\dfrac{1}{x}\right)+xy\left(\dfrac{1}{y}\right)}$$

$$=\frac{x^2y}{y+x}$$

NOW TRY EXERCISE 17

When asked to simplify a complex fraction, you may use either method unless you are told by your instructor to use a specific method. Read the Helpful Hint that follows.

HELPFUL HINT

We have presented two methods for simplifying complex fractions. Which method should you use? Although either method can be used to simplify complex fractions, most students prefer to use method 1 when both the numerator and denominator consist of a single term, as in Example 1. When the complex fraction has a sum or difference of expressions in either the numerator or denominator, as in Examples 2, 3, 4, or 5, most students prefer to use method 2.

Exercise Set 6.5

Concept/Writing Exercises

1. What is a complex fraction?

2. What is the numerator and denominator of each complex fraction?

a) $\dfrac{5}{\dfrac{3}{x^2+5x+6}}$

b) $\dfrac{\dfrac{5}{3}}{x^2+5x+6}$

3. What is the numerator and denominator of each complex fraction?

a) $\dfrac{\dfrac{x+3}{4}}{\dfrac{7}{x^2+5x+6}}$

b) $\dfrac{\dfrac{1}{2y}+x}{\dfrac{3}{y}+x}$

4. a) Select the method you prefer to use to simplify complex fractions. Then write in your own words a step-by-step procedure for simplifying complex fractions using that method. Answers will vary.

b) Using the procedure you wrote in part **a)**, simplify the following complex fraction.

$$\dfrac{\dfrac{2}{x}-\dfrac{3}{y}}{x+\dfrac{1}{y}} \quad \dfrac{2y-3x}{x^2y+x}$$

1. A fraction that contains a fraction in its numerator or its denominator or in both.

2. a) numerator, 5; denominator, $\dfrac{3}{x^2+5x+6}$ **b)** numerator, $\dfrac{5}{3}$; denominator, x^2+5x+6

3. a) numerator, $\dfrac{x+3}{4}$; denominator, $\dfrac{7}{x^2+5x+6}$

b) numerator, $\dfrac{1}{2y}+x$; denominator, $\dfrac{3}{y}+x$

Practice the Skills

Simplify.

5. $\dfrac{4+\dfrac{2}{3}}{2+\dfrac{1}{3}}$ 2

6. $\dfrac{3+\dfrac{4}{5}}{1-\dfrac{9}{16}}$ $\dfrac{304}{35}$

7. $\dfrac{2+\dfrac{3}{8}}{1+\dfrac{1}{3}}$ $\dfrac{57}{32}$

8. $\dfrac{\dfrac{1}{4}+\dfrac{5}{6}}{\dfrac{2}{3}+\dfrac{3}{5}}$ $\dfrac{65}{76}$

9. $\dfrac{\dfrac{2}{7}-\dfrac{1}{4}}{6-\dfrac{2}{3}}$ $\dfrac{3}{448}$

10. $\dfrac{1-\dfrac{x}{y}}{x}$ $\dfrac{y-x}{xy}$

11. $\dfrac{\dfrac{xy^2}{9}}{\dfrac{3}{x^2}}$ $\dfrac{x^3y^2}{27}$

12. $\dfrac{\dfrac{12a}{b^3}}{\dfrac{b^2}{4}}$ $\dfrac{48a}{b^5}$

13. $\dfrac{\dfrac{6a^2b}{5}}{\dfrac{9ac^2}{b^2}}$ $\dfrac{2ab^3}{15c^2}$

14. $\dfrac{\dfrac{36x^4}{5y^4z^5}}{\dfrac{9xy^2}{15z^5}}$ $\dfrac{12x^3}{y^6}$

15. $\dfrac{a-\dfrac{a}{b}}{1+a}$ $\dfrac{ab-a}{1+a}$

16. $\dfrac{a+\dfrac{1}{b}}{\dfrac{a}{b}}$ $\dfrac{ab+1}{a}$

17. $\dfrac{\dfrac{9}{x} + \dfrac{3}{x^2}}{3 + \dfrac{1}{x}}$ $\dfrac{3}{x}$

18. $\dfrac{\dfrac{3}{a} + \dfrac{1}{2a}}{a + \dfrac{a}{2}}$ $\dfrac{7}{3a^2}$

19. $\dfrac{5 - \dfrac{1}{x}}{4 - \dfrac{1}{x}}$ $\dfrac{5x-1}{4x-1}$

20. $\dfrac{\dfrac{x}{x-y}}{\dfrac{x^2}{y}}$ $\dfrac{y}{x(x-y)}$

21. $\dfrac{\dfrac{m}{n} - \dfrac{n}{m}}{\dfrac{m+n}{n}}$ $\dfrac{m-n}{m}$

22. $\dfrac{1}{\dfrac{1}{x} + y}$ $\dfrac{x}{1+xy}$

23. $\dfrac{\dfrac{a^2}{b} - b}{\dfrac{b^2}{a} - a}$ $-\dfrac{a}{b}$

24. $\dfrac{\dfrac{1}{x^2} - \dfrac{3}{x}}{3 + \dfrac{1}{x^2}}$ $\dfrac{1-3x}{3x^2+1}$

25. $\dfrac{5 - \dfrac{a}{b}}{\dfrac{a}{b} - 5}$ -1

26. $\dfrac{\dfrac{x}{y} - 2}{\dfrac{-x}{y} + 2}$ -1

27. $\dfrac{\dfrac{a^2-b^2}{a}}{\dfrac{a+b}{a^3}}$ $a^2(a-b)$

28. $\dfrac{\dfrac{1}{a} + \dfrac{1}{b}}{ab}$ $\dfrac{b+a}{a^2b^2}$

29. $\dfrac{\dfrac{1}{a} - \dfrac{1}{b}}{\dfrac{1}{ab}}$ $b-a$

30. $\dfrac{\dfrac{1}{a} - \dfrac{1}{b}}{\dfrac{1}{a} + \dfrac{1}{b}}$ $\dfrac{b-a}{a+b}$

31. $\dfrac{\dfrac{a}{b} + \dfrac{1}{a}}{\dfrac{b}{a} + \dfrac{1}{a}}$ $\dfrac{a^2+b}{b(b+1)}$

32. $\dfrac{\dfrac{1}{a} + \dfrac{1}{b}}{\dfrac{1}{a}}$ $\dfrac{b+a}{b}$

33. $\dfrac{\dfrac{1}{xy}}{\dfrac{1}{x} - \dfrac{1}{y}}$ $\dfrac{1}{y-x}$

34. $\dfrac{\dfrac{4}{x^2} + \dfrac{4}{x}}{\dfrac{4}{x} + \dfrac{4}{x^2}}$ 1

35. $\dfrac{\dfrac{3}{a} + \dfrac{3}{a^2}}{\dfrac{3}{b} + \dfrac{3}{b^2}}$ $\dfrac{ab^2 + b^2}{a^2(b+1)}$

36. $\dfrac{\dfrac{x}{y} - \dfrac{1}{x}}{\dfrac{y}{x} + \dfrac{1}{y}}$ $\dfrac{x^2-y}{x+y^2}$

39. $\dfrac{x-y+3}{2x+2y-7}$

40. $\dfrac{29x+21y}{2x+8y}$

41. a) $\dfrac{\dfrac{5}{12x}}{\dfrac{8}{x^2} - \dfrac{4}{3x}}$ **b)** $\dfrac{5x}{96-16x}$

42. a) $\dfrac{\dfrac{6}{x} - \dfrac{3}{2x}}{x + \dfrac{1}{x}}$; **b)** $\dfrac{9}{2x^2+2}$

Problem Solving

For the complex fractions in Exercises 37–40,

a) *Determine which of the two methods discussed in this section you would use to simplify the fraction. Explain why.*

b) *Simplify by the method you selected in part a).*

c) *Simplify by the method you did not select in part a). If your answers to parts b) and c) are not the same, explain why.*

37. $\dfrac{5 + \dfrac{3}{5}}{\dfrac{1}{8} - 4}$ $-\dfrac{224}{155}$

38. $\dfrac{\dfrac{x+y}{x^3} - \dfrac{1}{x}}{\dfrac{x-y}{x^5} + 5}$ $\dfrac{x^3+x^2y-x^4}{x-y+5x^5}$

39. $\dfrac{\dfrac{x-y}{x+y} + \dfrac{3}{x+y}}{2 - \dfrac{7}{x+y}}$

40. $\dfrac{\dfrac{25}{x-y} + \dfrac{4}{x+y}}{\dfrac{5}{x-y} - \dfrac{3}{x+y}}$

In Exercises 41 and 42, a) write the complex fraction, and b) simplify the complex fraction.

41. The numerator of the complex fraction consists of one term: 5 is divided by $12x$. The denominator of the complex fraction consists of two terms: 4 divided by $3x$ is subtracted from 8 divided by x^2.

42. The numerator of the complex fraction consists of two terms: 3 divided by $2x$ is subtracted from 6 divided by x. The denominator of the complex fraction consists of two terms: the sum of x and the quantity 1 divided by x.

Simplify. (Hint: Refer to Section 4.2, which discusses negative exponents.)

43. $\dfrac{x^{-1} + y^{-1}}{2}$ $\dfrac{y+x}{2xy}$

44. $\dfrac{x^{-1} + y^{-1}}{x^{-1}}$ $\dfrac{y+x}{y}$

45. $\dfrac{x^{-1} + y^{-1}}{x^{-1}y^{-1}}$ $x+y$

46. $\dfrac{x^{-2} - y^{-2}}{y^{-1} - x^{-1}}$ $-\dfrac{x+y}{xy}$

Challenge Problems

47. The efficiency of a jack, E, is expressed by the formula

$E = \dfrac{\frac{1}{2}h}{h + \frac{1}{2}}$, where h is determined by the pitch of the

jack's thread. Determine the efficiency of a jack if h is

a) $\dfrac{2}{3}$ $\dfrac{2}{7}$ **b)** $\dfrac{4}{5}$ $\dfrac{4}{13}$

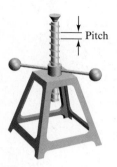 Pitch

Simplify.

48. $\dfrac{\dfrac{x}{y} + \dfrac{y}{x} + \dfrac{1}{x}}{\dfrac{x}{y} + y}$ $\dfrac{x^2 + y^2 + y}{x^2 + xy^2}$

49. $\dfrac{\dfrac{a}{b} + b - \dfrac{1}{a}}{\dfrac{a}{b^2} - \dfrac{b}{a} + \dfrac{1}{a^2}}$ $\dfrac{a^3b + a^2b^3 - ab^2}{a^3 - ab^3 + b^2}$

50. $\dfrac{x}{1 + \dfrac{x}{1 + x}}$ $\dfrac{x^2 + x}{2x + 1}$

Cumulative Review Exercises

[2.5] **51.** Solve the equation
$2x - 8(5 - x) = 9x - 3(x + 2)$. $\frac{17}{2}$

[4.4] **52.** What is a polynomial? An expression containing a finite number of terms of the form ax^n where a is a real number and n is a whole number.

[5.3] **53.** Factor $x^2 - 13x + 42$ $(x - 6)(x - 7)$

[6.4] **54.** Subtract $\dfrac{x}{3x^2 + 17x - 6} - \dfrac{2}{x^2 + 3x - 18}$.

$\dfrac{x^2 - 9x + 2}{(3x - 1)(x + 6)(x - 3)}$

6.6 SOLVING RATIONAL EQUATIONS

SSM   Study Guide   CD/Video

MathPro 4/5   PH Math Tutor Center   prenhall.com/Angel

1 Solve rational equations with integer denominators.

2 Solve rational equations where a variable appears in a denominator.

1   Solve Rational Equations with Integer Denominators

In Sections 6.1 through 6.5, we focused on how to add, subtract, multiply, and divide rational expressions. Now we are ready to solve rational equations. A **rational equation** is one that contains one or more rational (or fractional) expressions. A rational equation may be one that contains rational coefficients, such as $\dfrac{1}{2}x + \dfrac{3}{5}x = 8$ or $\dfrac{x}{2} + \dfrac{3x}{5} = 8$, or one that contains rational terms, with a variable in a denominator, such as $\dfrac{3}{x - 2} = 5$. We solved linear equations with rational coefficients in Sections 4.4 and 4.5.

The emphasis of this section will be on solving rational equations where a variable appears in a denominator. The following procedure that we will use to solve rational equations in this section is very similar to the procedure we used in Chapter 2.

To Solve Rational Equations

1. Determine the least common denominator (LCD) of all fractions in the equation.

2. Multiply **both** sides of the equation by the LCD. **This will result in every term in the equation being multiplied by the LCD**.

3. Remove any parentheses and combine like terms on each side of the equation.

4. Solve the equation using the properties discussed in earlier chapters.

5. Check your solution in the *original* equation.

The purpose of multiplying both sides of the equation by the LCD (step 2) is to eliminate all fractions from the equation. After both sides of the equation are multiplied by the LCD, the resulting equation should contain no fractions. We will omit some of the checks to save space.

Before we solve rational equations where a variable appears in a denominator, let's review how to solve equations with rational coefficients. Examples 1 and 2 illustrate the procedure to follow.

EXAMPLE 1 Solve $\dfrac{t}{4} - \dfrac{t}{5} = 1$ for t.

Solution The LCD of 4 and 5 is 20. Multiply both sides of the equation by 20.

$$\frac{t}{4} - \frac{t}{5} = 1$$

$$20\left(\frac{t}{4} - \frac{t}{5}\right) = 20 \cdot 1 \quad \textit{Multiply both sides by the LCD, 20.}$$

$$20\left(\frac{t}{4}\right) - 20\left(\frac{t}{5}\right) = 20 \quad \textit{Distributive property}$$

$$5t - 4t = 20$$

$$t = 20$$

Check

$$\frac{t}{4} - \frac{t}{5} = 1$$

$$\frac{20}{4} - \frac{20}{5} \overset{?}{=} 1$$

$$5 - 4 \overset{?}{=} 1$$

$$1 = 1 \quad \textit{True}$$

NOW TRY EXERCISE 13

EXAMPLE 2 Solve $\dfrac{x-5}{30} = \dfrac{4}{5} - \dfrac{x-1}{10}$

Solution The least common denominator is 30. Multiply both sides of the equation by 30.

$$\frac{x-5}{30} = \frac{4}{5} - \frac{x-1}{10}$$

$$30\left(\frac{x-5}{30}\right) = 30\left(\frac{4}{5} - \frac{x-1}{10}\right) \quad \textit{Multiply both sides by the LCD, 30.}$$

$$x - 5 = 30\left(\frac{4}{5}\right) - 30\left(\frac{x-1}{10}\right) \quad \textit{Distributive property}$$

$$x - 5 = 24 - 3(x - 1)$$

$$x - 5 = 24 - 3x + 3 \qquad \textit{Distributive property}$$

$$x - 5 = -3x + 27 \qquad \textit{Combine like terms}$$

$$4x - 5 = 27 \qquad \textit{3x was added to both sides}$$

$$4x = 32 \qquad \textit{5 was added to both sides}$$

$$x = 8 \qquad \textit{Both sides were divided by 4}$$

NOW TRY EXERCISE 27

A check will show that the answer is 8. We suggest you check this answer now to get practice at checking answers.   ✳

In Example 2, the equation could also have been written as

$$\frac{1}{30}(x - 5) = \frac{4}{5} - \frac{1}{10}(x - 1).$$

For additional examples of solving rational equations with integers in the denominators, review Sections 2.4 and 2.5.

2 Solve Rational Equations Where a Variable Appears in a Denominator

TEACHING TIP
Point out that when checking for extraneous solutions in rational equations, it is only necessary to check that the solution does not make any denominator in the original equation equal to zero.

Now we are ready to solve rational equations where a variable appears in a denominator. When solving a rational equation where a variable appears in any denominator, you *must* check your answer. See the following warning!

Warning **Whenever a variable appears in any denominator of a rational equation, it is necessary to check your answer in the original equation. If the answer obtained makes any denominator equal to zero, that value is not a solution to the equation. Such values are called extraneous roots or extraneous solutions.**

EXAMPLE 3 Solve the equation $3 - \dfrac{4}{x} = \dfrac{5}{2}$.

Solution Multiply both sides of the equation by the LCD, $2x$.

$$2x\left(3 - \frac{4}{x}\right) = \left(\frac{5}{2}\right) \cdot 2x \qquad \textit{Multiply both sides by the LCD, 2x}$$

$$2x(3) - 2x\left(\frac{4}{x}\right) = \left(\frac{5}{2}\right) \cdot 2x \qquad \textit{Distributive property}$$

$$6x - 8 = 5x \qquad \textit{5x was subtracted from both sides}$$

$$x - 8 = 0 \qquad \textit{8 was added to both sides}$$

$$x = 8$$

Check

$$3 - \frac{4}{x} = \frac{5}{2}$$

$$3 - \frac{4}{8} \overset{?}{=} \frac{5}{2}$$

$$3 - \frac{1}{2} \overset{?}{=} \frac{5}{2}$$

$$\frac{5}{2} = \frac{5}{2} \qquad \textit{True}$$

NOW TRY EXERCISE 51 Since 8 does check, it is the solution to the equation.   ✳

EXAMPLE 4   Solve the equation $\dfrac{p-5}{p+3} = \dfrac{1}{5}$

Solution   The LCD is $5(p+3)$. Multiply both sides of the equation by the LCD.

$$5(p+3) \cdot \dfrac{(p-5)}{p+3} = \dfrac{1}{5} \cdot 5(p+3)$$
$$5(p-5) = 1(p+3)$$
$$5p - 25 = p + 3$$
$$4p - 25 = 3$$
$$4p = 28$$
$$p = 7$$

A check will show that 7 is the solution.

TEACHING TIP
When discussing Example 4, have students compare the equation in step 2 of the solution with the original equation. Then ask, "Do you know another method for getting the equation in step 2 from the original equation?"

In Section 2.6 we illustrated that proportions of the form

$$\dfrac{a}{b} = \dfrac{c}{d}$$

can be cross-multiplied to obtain $a \cdot d = b \cdot c$. Example 4 is a proportion and can also be solved by cross-multiplying, as done in Example 5.

EXAMPLE 5   Use cross-multiplication to solve the equation $\dfrac{6}{x+3} = \dfrac{5}{x-2}$.

Solution
$$\dfrac{6}{x+3} = \dfrac{5}{x-2}$$
$$6(x-2) = 5(x+3) \quad \text{Cross multiply.}$$
$$6x - 12 = 5x + 15 \quad \text{Distributive property}$$
$$x - 12 = 15 \quad \text{5x was subtracted from both sides.}$$
$$x = 27 \quad \text{12 was added to both sides.}$$

NOW TRY EXERCISE 41   A check will show that 27 is the solution to the equation.

Now let's examine some examples that involve quadratic equations. Recall from Section 5.6 that quadratic equations have the form $ax^2 + bx + c = 0$, where $a \neq 0$.

EXAMPLE 6   Solve the equation $x + \dfrac{12}{x} = -7$.

Solution
$$x \cdot \left(x + \dfrac{12}{x}\right) = -7 \cdot x \quad \text{Multiply both sides by } x.$$
$$x(x) + x\left(\dfrac{12}{x}\right) = -7x \quad \text{Distributive property}$$
$$x^2 + 12 = -7x$$
$$x^2 + 7x + 12 = 0 \quad \text{7x was added to both sides.}$$
$$(x+3)(x+4) = 0 \quad \text{Factor.}$$
$$x + 3 = 0 \quad \text{or} \quad x + 4 = 0 \quad \text{Zero-factor property}$$
$$x = -3 \qquad\qquad x = -4$$

Check
$$x = -3$$

$$x + \frac{12}{x} = -7$$

$$-3 + \frac{12}{-3} \overset{?}{=} -7$$

$$-3 + (-4) \overset{?}{=} -7$$

$$-7 = -7 \quad \textit{True}$$

$$x = -4$$

$$x + \frac{12}{x} = -7$$

$$-4 + \frac{12}{-4} \overset{?}{=} -7$$

$$-4 + (-3) \overset{?}{=} -7$$

$$-7 = -7 \quad \textit{True}$$

NOW TRY EXERCISE 73   The solutions are -3 and -4.

EXAMPLE 7   Solve the equation $\dfrac{x^2 + x}{x - 6} = \dfrac{42}{x - 6}$.

Solution   If we try to solve this equation using cross-multiplication we will get a cubic equation, we will solve this equation by multiplying both sides of the equation by the LCM, $x - 6$.

$$\frac{x^2 + x}{x - 6} = \frac{42}{x - 6}$$

$$\cancel{x - 6} \cdot \frac{x^2 + x}{\cancel{x - 6}} = \frac{42}{\cancel{x - 6}} \cdot \cancel{x - 6} \qquad \textit{Multiply both sides by the LCM, } x - 6.$$

$$x^2 + x = 42$$

$$x^2 + x - 42 = 0 \qquad \textit{42 was subctracted from both sides.}$$

$$(x + 7)(x - 6) = 0 \qquad \textit{Factor}$$

$$x + 7 = 0 \quad \text{or} \quad x - 6 = 0 \qquad \textit{Zero-factor property}$$

$$x = -7 \qquad\qquad x = 6$$

Check

$$x = -7$$

$$\frac{x^2 + x}{x - 6} = \frac{42}{x - 6}$$

$$\frac{(-7)^2 + (-7)}{-7 - 6} \overset{?}{=} \frac{42}{-7 - 6}$$

$$\frac{49 - 7}{-13} \overset{?}{=} \frac{42}{-13}$$

$$-\frac{42}{13} = -\frac{42}{13} \quad \textit{True}$$

$$x = 6$$

$$\frac{x^2 + x}{x - 6} = \frac{42}{x - 6}$$

$$\frac{6^2 + 6}{6 - 6} \overset{?}{=} \frac{42}{6 - 6}$$

$$\frac{42}{0} = \frac{42}{0}$$

$$\uparrow \qquad \uparrow$$

Since the denominator is 0, and we cannot divide by 0, 6 is not a solution.

Since $\dfrac{42}{0}$ is not a real number, 6 is an extraneous solution. Thus, this equation has

NOW TRY EXERCISE 47   only one solution, -7.

HELPFUL HINT

Remember, when solving a rational equation in which a variable appears in a denominator, you must check *all* your answers to make sure that none is an extraneous root. Extraneous roots can usually be spotted quickly. If any of your answers make any denominator 0, that answer is an extraneous root and not a true solution.

Generally speaking, if none of your answers makes any denominator 0, and you have not made a mistake when solving the equation, the answers you obtained when solving the equation will be solutions to the equation.

EXAMPLE 8   Solve the equation $\dfrac{5w}{w^2 - 4} + \dfrac{1}{w - 2} = \dfrac{4}{w + 2}$.

Solution   First factor $w^2 - 4$.

$$\frac{5w}{(w + 2)(w - 2)} + \frac{1}{w - 2} = \frac{4}{w + 2}$$

Multiply both sides of the equation by the LCD, $(w + 2)(w - 2)$.

$$(w + 2)(w - 2)\left[\frac{5w}{(w + 2)(w - 2)} + \frac{1}{w - 2}\right] = \frac{4}{w + 2} \cdot (w + 2)(w - 2)$$

$$(w + 2)(w - 2) \cdot \frac{5w}{(w + 2)(w - 2)} + (w + 2)(w - 2) \cdot \frac{1}{w - 2} = \frac{4}{w + 2} \cdot (w + 2)(w - 2)$$

$$(w + 2)(w - 2) \cdot \frac{5w}{(w + 2)(w - 2)} + (w + 2)(w - 2) \cdot \frac{1}{w - 2} = \frac{4}{w + 2} \cdot (w + 2)(w - 2)$$

$$5w + (w + 2) = 4(w - 2)$$
$$6w + 2 = 4w - 8$$
$$2w + 2 = -8$$
$$2w = -10$$
$$w = -5$$

NOW TRY EXERCISE 65   A check will show that -5 is the solution to the equation.   ❋

HELPFUL HINT

Some students confuse adding and subtracting rational expressions with solving rational equations. When adding or subtracting rational expressions, we must rewrite each expression with a common denominator. When solving a rational equation, we multiply both sides of the equation by the LCD to eliminate fractions from the equation. Consider the following two problems. Note that the one on the right is an equation because it contains an equal sign. We will work both problems. The LCD for both problems is $x(x + 4)$.

Adding Rational Expressions	**Solving Rational Equations**
$\dfrac{x + 2}{x + 4} + \dfrac{3}{x}$	$\dfrac{x + 2}{x + 4} = \dfrac{3}{x}$

We rewrite each fraction with the LCD, $x(x + 4)$.

We eliminate fractions by multiplying both sides of the equation by the LCD, $x(x + 4)$.

$$= \frac{x}{x} \cdot \frac{x + 2}{x + 4} + \frac{3}{x} \cdot \frac{x + 4}{x + 4}$$

$$(x)(x + 4)\left(\frac{x + 2}{x + 4}\right) = \frac{3}{x}(x)(x + 4)$$

$$= \frac{x(x + 2)}{x(x + 4)} + \frac{3(x + 4)}{x(x + 4)}$$

$$x(x + 2) = 3(x + 4)$$

$$= \frac{x^2 + 2x}{x(x + 4)} + \frac{3x + 12}{x(x + 4)}$$

$$x^2 + 2x = 3x + 12$$

$$= \frac{x^2 + 2x + 3x + 12}{x(x + 4)}$$

$$x^2 - x - 12 = 0$$

$$= \frac{x^2 + 5x + 12}{x(x + 4)}$$

$$(x - 4)(x + 3) = 0$$

$$x - 4 = 0 \quad \text{or} \quad x + 3 = 0$$

$$x = 4 \qquad\qquad x = -3$$

(continued on the next page)

2. **a)** will be the same **b)** no solution **3. b) Left:** Write fractions with LCD $12(x - 1)$, combine numerators. **Right:** Multiply both sides by LCD, $12(x - 1)$, then solve.

> The numbers 4 and −3 on the right will both check and are thus solutions to the equation.
>   *Note that when adding and subtracting rational expressions we usually end up with an algebraic expression. When solving rational equations, the solution will be a numerical value or values.* The equation on the right could also be solved using cross-multiplication.

3. **c) Left:** $\dfrac{x^2 - x + 12}{12(x - 1)}$; **Right:** 4, −3   4. **b) Left:** Write fractions with LCD $6(2x + 7)$, combine numerators. **Right:** Multiply both sides by LCD $6(2x + 7)$, then solve.   **c) Left:** $\dfrac{2x^2 + 7x + 30}{6(2x + 7)}$, **Right:** $\dfrac{5}{2}$, −6   **5.** when there is a variable in the denominator

Exercise Set 6.6

Concept/Writing Exercises

1. **a)** Explain in your own words the steps to use to solve rational equations. Answers will vary.

   b) Using the procedure you wrote in part **a)**, solve the equation $\dfrac{1}{x - 1} - \dfrac{1}{x + 1} = \dfrac{3x}{x^2 - 1} \cdot \dfrac{2}{3}$.

2. **a)** Without solving the equations, determine whether the answer to the two equations that follow will be the same or different. Explain your answer.

   $$\frac{3}{x - 2} + \frac{2}{x + 2} = \frac{4x}{x^2 - 4}, \quad \frac{4x}{x^2 - 4} - \frac{2}{x + 2} = \frac{3}{x - 2}$$

   b) Determine the answer to both questions.

3. Consider the following problems.

   Simplify: Solve:

   $$\frac{x}{3} - \frac{x}{4} + \frac{1}{x - 1} \qquad \frac{x}{3} - \frac{x}{4} = \frac{1}{x - 1}$$

   a) Explain the difference between the two types of problems. Left: expression, Right: equation

   b) Explain how you would work each problem to obtain the correct answer.

   c) Find the correct answer to each problem.

4. Consider the following problems.

   Simplify: Solve:

   $$\frac{x}{2} - \frac{x}{3} + \frac{5}{2x + 7} \qquad \frac{x}{2} - \frac{x}{3} = \frac{5}{2x + 7}$$

   a) Explain the difference between the two types of problems. Left: expression, Right: equation

   b) Explain how you would work each problem to obtain the correct answer.

   c) Find the correct answer to each problem.

5. Under what conditions must you check rational equations for extraneous solutions?

6. Which of the following numbers, 3, 1, or, 0 cannot be a solution to the equation $3 - \dfrac{1}{x} = 4$? Explain.   0

7. Which of the following numbers, 0, 3, or 2 cannot be a solution to the equation $\dfrac{3}{x - 2} + 5x = 6$? Explain.   2

8. Which of the following numbers, 0, −5 or 2 cannot be a solution to the equation $4x - \dfrac{2}{x + 5} = 3$,? Explain.   −5

In Exercises 9–12, indicate whether it is necessary to check the solution obtained to see if it is extraneous. Explain your answer.

9. $\dfrac{4}{7} + \dfrac{x - 2}{5} = 5$ no

10. $\dfrac{3}{x - 2} + \dfrac{1}{x} = 6$ yes

11. $\dfrac{2}{x + 4} + 3 = 6$ yes

12. $\dfrac{z}{3} + \dfrac{1}{z} = 6$ yes

Practice the Skills

Solve each equation and check your solution. See Examples 1 and 2.

13. $\dfrac{x}{3} - \dfrac{x}{4} = 1$ 12

14. $\dfrac{t}{5} - \dfrac{t}{6} = 4$ 120

15. $\dfrac{r}{3} = \dfrac{r}{4} + \dfrac{1}{6}$ 2

16. $\dfrac{n}{5} = \dfrac{n}{6} + \dfrac{1}{3}$ 10

17. $\dfrac{z}{2} + 3 = \dfrac{z}{5}$ −10

18. $\dfrac{3w}{5} - 6 = w$ −15

19. $\dfrac{z}{6} + \dfrac{1}{3} = \dfrac{z}{5} - \dfrac{2}{3}$ 30

20. $\dfrac{m - 2}{3} = \dfrac{4}{3} + \dfrac{m}{6}$ 12

21. $d + 1 = \dfrac{3}{2}d - 1$ 4

22. $\dfrac{q}{5} + \dfrac{q}{2} = \dfrac{21}{10}$ 3

23. $k + \dfrac{1}{6} = 2k - 4$ $\dfrac{25}{6}$

24. $\dfrac{p}{4} + \dfrac{1}{2} = \dfrac{p}{3} - \dfrac{1}{4}$ 9

25. $\dfrac{n+6}{3} = \dfrac{2(n-8)}{4}$ 36

26. $\dfrac{3(x-6)}{5} = \dfrac{3(x+2)}{6}$ 46

27. $\dfrac{x-5}{15} = \dfrac{3}{5} - \dfrac{x-4}{10}$ 8

28. $\dfrac{z+4}{6} = \dfrac{3}{2} - \dfrac{2z+2}{12}$ 2

29. $\dfrac{-p+1}{4} + \dfrac{13}{20} = \dfrac{p}{5} - \dfrac{p-1}{2}$ −8

30. $\dfrac{1}{10} - \dfrac{n+1}{6} = \dfrac{1}{5} - \dfrac{n+10}{15}$ 4

31. $\dfrac{d-3}{4} + \dfrac{1}{5} = \dfrac{2d+1}{3} - \dfrac{32}{15}$ 3

32. $\dfrac{t+4}{5} = \dfrac{3}{8} + \dfrac{t+17}{40}$ 0

Solve each equation and check your solution. See Examples 3–8.

33. $2 + \dfrac{3}{x} = \dfrac{11}{4}$ 4

34. $3 - \dfrac{1}{x} = \dfrac{14}{5}$ 5

35. $6 - \dfrac{3}{x} = \dfrac{9}{2}$ 2

36. $5 + \dfrac{3}{z} = \dfrac{11}{2}$ 6

37. $\dfrac{4}{n} - \dfrac{3}{2n} = \dfrac{1}{2}$ 5

38. $\dfrac{5}{3x} + \dfrac{3}{x} = 1$ $\dfrac{14}{3}$

39. $\dfrac{x-1}{x-5} = \dfrac{4}{x-5}$ no solution

40. $\dfrac{2x+3}{x+2} = \dfrac{3}{2}$ 0

41. $\dfrac{5}{a+3} = \dfrac{4}{a+1}$ 7

42. $\dfrac{5}{x+2} = \dfrac{1}{x-4}$ $\dfrac{11}{2}$

43. $\dfrac{4}{y-3} = \dfrac{6}{y+3}$ 15

44. $\dfrac{5}{-x-6} = \dfrac{2}{x}$ $-\dfrac{12}{7}$

45. $\dfrac{2x-3}{x-4} = \dfrac{5}{x-4}$ no solution

46. $\dfrac{3}{x} + 4 = \dfrac{3}{x}$ no solution

47. $\dfrac{x^2}{x-4} = \dfrac{16}{x-4}$ −4

48. $\dfrac{x^2}{x+5} = \dfrac{25}{x+5}$ 5

49. $\dfrac{n+10}{n+2} = \dfrac{n+4}{n-3}$ 38

50. $\dfrac{x-3}{x+1} = \dfrac{x-6}{x+5}$ $\dfrac{9}{7}$

51. $\dfrac{1}{3r} = \dfrac{r}{8r+3}$ $-\dfrac{1}{3}, 3$

52. $\dfrac{1}{2r} = \dfrac{r}{r+15}$ $-\dfrac{5}{2}, 3$

53. $\dfrac{k}{k+2} = \dfrac{3}{k-2}$ 6, −1

54. $\dfrac{3a-2}{2a+2} = \dfrac{3}{a-1}$ $-\dfrac{1}{3}, 4$

55. $\dfrac{3}{r} + r = \dfrac{19}{r}$ 4, −4

56. $a + \dfrac{3}{a} = \dfrac{12}{a}$ 3, −3

57. $x + \dfrac{20}{x} = -9$ −4, −5

58. $x - \dfrac{21}{x} = 4$ 7, −3

59. $\dfrac{3y-2}{y+1} = 4 - \dfrac{y+2}{y-1}$ 4

60. $\dfrac{2b}{b+1} = 2 - \dfrac{5}{2b}$ −5

61. $\dfrac{1}{x+3} + \dfrac{1}{x-3} = \dfrac{-5}{x^2-9}$ $-\dfrac{5}{2}$

62. $\dfrac{t-1}{t-5} - \dfrac{3}{4} = \dfrac{3}{t-5}$ 1

63. $\dfrac{x}{x-3} + \dfrac{3}{2} = \dfrac{3}{x-3}$ no solution

64. $\dfrac{3x}{x^2-9} + \dfrac{1}{x-3} = \dfrac{3}{x+3}$ −12

65. $\dfrac{3}{x-5} - \dfrac{4}{x+5} = \dfrac{11}{x^2-25}$ 24

66. $\dfrac{2n^2-15}{n^2+n-6} = \dfrac{n+1}{n+3} + \dfrac{n-3}{n-2}$ 4

67. $\dfrac{y}{2y+2} + \dfrac{2y-16}{4y+4} = \dfrac{y-3}{y+1}$ no solution

68. $\dfrac{3}{x+3} + \dfrac{5}{x+4} = \dfrac{12x+19}{x^2+7x+12}$ 2

69. $\dfrac{1}{y-1} + \dfrac{1}{2} = \dfrac{2}{y^2-1}$ −3

70. $\dfrac{2y}{y+2} = \dfrac{y}{y+3} - \dfrac{3}{y^2+5y+6}$ −1

71. $\dfrac{2t}{4t+4} + \dfrac{t}{2t+2} = \dfrac{2t-3}{t+1}$ 3

72. $\dfrac{2}{x-2} - \dfrac{1}{x+1} = \dfrac{5}{x^2-x-2}$ 1

Problem Solving

In Exercises 73–78, determine the solution by observation. Explain how you determined your answer.

73. $\dfrac{3}{x-2} = \dfrac{x-2}{x-2}$ 5

74. $\dfrac{1}{2} + \dfrac{x}{2} = \dfrac{5}{2}$ 4

75. $\dfrac{x}{x-3} + \dfrac{x}{x-3} = 0$ 0

76. $\dfrac{x}{2} + \dfrac{x}{2} = x$ *x* can be any real number.

77. $\dfrac{x-2}{3} + \dfrac{x-2}{3} = \dfrac{2x-4}{3}$ *x* can be any real number.

78. $\dfrac{2}{x} - \dfrac{1}{x} = \dfrac{1}{x}$ *x* can be any real number except 0.

79. Optics A formula frequently used in optics is

$$\frac{1}{p} + \frac{1}{q} = \frac{1}{f}$$

where *p* represents the distance of the object from a mirror (or lens), *q* represents the distance of the image from the mirror (or lens), and *f* represents the focal length of the mirror (or lens). If a mirror has a focal length of 10 centimeters, how far from the mirror will the image appear when the object is 30 centimeters from the mirror? 15 cm

Challenge Problems

80. a) You would obtain a cubic equation.
82. b) 900 ohms   83. No; it is impossible for both sides of the equation to be equal.

80. a) Explain why the equation $\dfrac{x^2}{x-3} = \dfrac{9}{x-3}$ cannot be solved by cross-multiplying using the material presented in the book.

    b) Solve the equation given in part a). -3

81. Solve the equation $\dfrac{x-4}{x^2-2x} = \dfrac{-4}{x^2-4}$ -4

82. ***Electrical Resistance*** In electronics the total resistance R_T, of resistors wired in a parallel circuit is determined by the formula

    $$\frac{1}{R_T} = \frac{1}{R_1} + \frac{1}{R_2} + \frac{1}{R_3} + \cdots + \frac{1}{R_n}$$

where $R_1, R_2, R_3, \ldots, R_n$ are the resistances of the individual resistors (measured in ohms) in the circuit.

   a) Find the total resistance if two resistors, one of 200 ohms and the other of 300 ohms, are wired in a parallel circuit.   120 ohms

   b) If three identical resistors are to be wired in parallel, what should be the resistance of each resistor if the total resistance of the circuit is to be 300 ohms?

83. Can an equation of the form $\dfrac{a}{x} + 1 = \dfrac{a}{x}$ have a real number solution for any real number a? Explain your answer.

Group Activity

84. a) cross-multiply or use the LCD   c) $\dfrac{5}{x+3} = \dfrac{4}{x}$ h) The solution will be the same.

Discuss and answer Exercise 84 as a group.

84. a) As a group, discuss two different methods you can use to solve the equation $\dfrac{x+3}{5} = \dfrac{x}{4}$.

    b) Group member 1: Solve the equation by obtaining a common denominator.
    Group member 2: Solve the equation by cross-multiplying.
    Group member 3: Check the results of group member 1 and group member 2.   12

    c) Individually, create another equation by taking the reciprocal of each term in the equation in part a). Compare your results. Do you think that the reciprocal of the answer you found in part b) will be the solution to this equation? Explain.

    d) Individually, solve the equation you found in part c) and check your answer. Compare your work with the other group members. Was the conclusion you came to in part c) correct? Explain.   12

    e) As a group, solve the equation $\dfrac{1}{x} + \dfrac{1}{3} = \dfrac{2}{x}$. Check your result.   3

    f) As a group, create another equation by taking the reciprocal of each term of the equation in part e). Do you think that the reciprocal of the answer you found in part e) will be the solution to this equation? Explain. $\dfrac{x}{1} + \dfrac{3}{1} = \dfrac{x}{2}$

    g) Individually, solve the equation you found in part f) and check your answer. Compare your work with the other group members. Did your group make the correct conclusion in part f)? Explain. -6

    h) As a group, discuss the relationship between the solution to the equation $\dfrac{7}{x-9} = \dfrac{3}{x}$ and the solution to the equation $\dfrac{x-9}{7} = \dfrac{x}{3}$. Explain your answer.

Cumulative Review Exercises

[3.3] 85. ***Internet Plans*** An Internet service offers two plans for its customers. One plan includes 5 hours of use and costs $7.95 per month. Each additional minute after the 5 hours costs $0.15. The second plan costs $19.95 per month and provides unlimited Internet access. How many hours would Jake LaRue have to use the Internet monthly to make the second plan the less expensive?
more than $6\frac{1}{3}$ hours

[3.4] 86. ***Supplementary Angles*** Two angles are supplementary angles if the sum of their measures is 180°. Find the two supplementary angles if the smaller angle is 30° less than half the larger angle.
40°, 140°

87. ***Filling a Jacuzzi*** How long will it take to fill a 600-gallon Jacuzzi if water is flowing into the Jacuzzi at a rate of 8 gallons a minute?   75 min

[4.6] 88. Multiply $(3.4 \times 10^{-5})(2 \times 10^7)$. 6.8×10^2

6.7 RATIONAL EQUATIONS: APPLICATIONS AND PROBLEM SOLVING

SSM

Study Guide

CD/Video

MathPro 4/5

PH Math
Tutor Center

prenhall.com/Angel

1 Set up and solve applications containing rational expressions.

2 Set up and solve motion problems.

3 Set up and solve work problems.

1 Set Up and Solve Applications Containing Rational Expressions

Many applications of algebra involve rational equations. After we represent the application as an equation, we solve the rational equation as we did in Section 6.6. The first type of application we will consider is a **geometry problem**.

EXAMPLE 1 **A New Rug** Mary and Larry Armstrong are interested in purchasing a carpet whose area is 60 square feet. Determine the length and width if the width is 5 feet less than $\frac{3}{5}$ of the length, see Figure 6.1.

Solution Understand and Translate   Let x = length

$$\text{then } \frac{3}{5}x - 5 = \text{width}$$

$$\text{area} = \text{length} \cdot \text{width}$$

$$60 = x\left(\frac{3}{5}x - 5\right)$$

$\frac{3}{5}x - 5$

x

FIGURE 6.1

Carry Out

$$60 = \frac{3}{5}x^2 - 5x$$

$$5(60) = 5\left(\frac{3}{5}x^2 - 5x\right) \quad \textit{Multiply both sides by 5.}$$

$$300 = 3x^2 - 25x \quad \textit{Distributive property}$$

$$0 = 3x^2 - 25x - 300 \quad \textit{Subtract 300 from both sides.}$$

$$\text{or} \quad 3x^2 - 25x - 300 = 0$$

$$(3x + 20)(x - 15) = 0 \quad \textit{Factor.}$$

$$3x + 20 = 0 \quad \text{or} \quad x - 15 = 0 \quad \textit{Zero-factor property}$$

$$3x = -20 \qquad\qquad x = 15$$

$$x = -\frac{20}{3}$$

Check and Answer   Since the length of a rectangle cannot be negative, we can eliminate $-\frac{20}{3}$ as an answer to our problem.

$$\text{length} = x = 15 \text{ feet}$$

$$\text{width} = \frac{3}{5}(15) - 5 = 4 \text{ feet}$$

Check

$$a = lw$$

$$60 \stackrel{?}{=} 15(4)$$

$$60 = 60 \quad \textit{True}$$

NOW TRY EXERCISE 5   Therefore, the length is 15 feet and the width is 4 feet.

Now we will work with a problem that expresses the relationship between two numbers. Problems like this are sometimes referred to as *number problems*.

EXAMPLE 2 **Reciprocals** One number is 3 times another number. The sum of their reciprocals is $\frac{20}{9}$. Determine the numbers.

Solution Understand and Translate Let x = first number

then $3x$ = second number

The reciprocal of the first number is $\frac{1}{x}$ and the reciprocal of the second number is $\frac{1}{3x}$. The sum of their reciprocals is $\frac{20}{9}$, thus:

$$\frac{1}{x} + \frac{1}{3x} = \frac{20}{9}$$

Carry Out
$$9x\left(\frac{1}{x} + \frac{1}{3x}\right) = 9x\left(\frac{20}{9}\right)$$
Multiply both sides by the LCD, 9x.

$$9x\left(\frac{1}{x}\right) + 9x\left(\frac{1}{3x}\right) = 20x$$
Distributive property

$$9 + 3 = 20x$$

$$12 = 20x$$

$$\frac{12}{20} = x$$

$$\frac{3}{5} = x$$

Check The first number is $\frac{3}{5}$. The second number is therefore $3x = 3\left(\frac{3}{5}\right) = \frac{9}{5}$.

Let's now check if the sum of the reciprocals is $\frac{20}{9}$. The reciprocal of $\frac{3}{5}$ is $\frac{5}{3}$. The reciprocal of $\frac{9}{5}$ is $\frac{5}{9}$. The sum of the reciprocals is

$$\frac{5}{3} + \frac{5}{9} = \frac{15}{9} + \frac{5}{9} = \frac{20}{9}$$

NOW TRY EXERCISE 11 Answer Since the sum of the reciprocals is $\frac{20}{9}$, the two numbers are $\frac{3}{5}$ and $\frac{9}{5}$. ✳

2 Set Up and Solve Motion Problems

In Chapter 3 we discussed **motion problems**. Recall that

$$\text{distance} = \text{rate} \cdot \text{time}$$

If we solve this equation for time, we obtain

$$\text{time} = \frac{\text{distance}}{\text{rate}} \quad \text{or} \quad t = \frac{d}{r}$$

This equation is useful in solving motion problems when the total time of travel for two objects or the time of travel between two points is known.

EXAMPLE 3 **Canoeing** Cindy Kilborn lives near the Colorado River and loves to canoe in a section of the river where the current is not too great. One Saturday she went canoeing. She learned from a friend that the current in the river was 2 miles per hour. If it took Cindy the same amount of time to travel 10 miles downstream as 2 miles upstream, determine the speed at which Cindy's canoe would travel in still water.

Solution Understand and Translate

Let r = the canoe's speed in still water

then $r + 2$ = the canoe's speed traveling downstream (with current)

and $r - 2$ = the canoe's speed traveling upstream (against current)

Direction	Distance	Rate	Time
Downstream	10	$r + 2$	$\dfrac{10}{r + 2}$
Upstream	2	$r - 2$	$\dfrac{2}{r - 2}$

Since the time it takes to travel 10 miles downstream is the same as the time to travel 2 miles upstream, we set the times equal to each other and then solve the resulting equation.

$$\text{time downstream} = \text{time upstream}$$

$$\frac{10}{r + 2} = \frac{2}{r - 2}$$

Carry Out

$$10(r - 2) = 2(r + 2) \qquad \textit{Cross-multiply}$$
$$10r - 20 = 2r + 4$$
$$8r = 24$$
$$r = 3$$

TEACHING TIP
Before discussing the solution to Example 3, have students assume that Cindy paddles at the same rate as the current. Then ask, "How fast would she travel downstream? How fast would she travel upstream? Do you think her canoe was traveling faster or slower than the current? Explain."

Check and Answer Since this rational equation contains a variable in a denominator, the solution must be checked. A check will show that 3 satisfies the equation. Thus, the canoe would travel at 3 miles per hour in still water.

NOW TRY EXERCISE 15

EXAMPLE 4 **Patrol Route** Officer DeWolf rides his bicycle every Saturday morning in the Kahana Valley State Park in Oahu, Hawaii, as part of his patrolling duties. During the first part of the ride he is pedaling mostly uphill and his average speed is 6 miles an hour. After a certain point, he is traveling mostly downhill and averages 9 miles per hour. If the total distance he travels is 30 miles and the total time he rides is 4 hours, how long did he ride at each speed?

Solution Understand and Translate

Let d = distance traveled at 6 miles per hour

then $30 - d$ = distance traveled at 9 miles per hour

Direction	Distance	Rate	Time
Uphill	d	6	$\dfrac{d}{6}$
Downhill	$30 - d$	9	$\dfrac{30 - d}{9}$

Since the total time spent riding is 4 hours, we write

time going uphill + time going downhill = 4 hours

$$\frac{d}{6} + \frac{30 - d}{9} = 4$$

Carry Out

$$18\left(\frac{d}{6} + \frac{30 - d}{9}\right) = 18 \cdot 4 \qquad \text{Multiply both sides by the LCD, 18.}$$

$$\overset{3}{\cancel{18}}\left(\frac{d}{\cancel{6}}\right) + \overset{2}{\cancel{18}}\left(\frac{30 - d}{\cancel{9}}\right) = 72 \qquad \text{Distributive property}$$

$$3d + 2(30 - d) = 72$$

$$3d + 60 - 2d = 72$$

$$d + 60 = 72$$

$$d = 12$$

Answer The answer to the problem is not 12. Remember that the question asked us to *find the time spent* traveling at each speed. The variable d does not represent time but represents the distance traveled at 6 miles per hour. To find the time traveled and to answer the question asked, we need to evaluate $\frac{d}{6}$ and $\frac{30 - d}{9}$ for $d = 12$.

Time at 6 mph

$$\frac{d}{6} = \frac{12}{6} = 2$$

Time at 9 mph

$$\frac{30 - d}{9} = \frac{30 - 12}{9} = \frac{18}{9} = 2$$

Thus, 2 hours were spent traveling at each rate. We leave the check for you to do. ✳

EXAMPLE 5 **Distance of a Marathon** At a fund-raising marathon participants can either bike, walk, run, or rollerblade (or use any other means of nonvehicular transportation). Kimberly Clark, who rode a bike, completed the entire distance of the marathon with an average speed of 16 kilometers per hour (kph). Steve Schwartz, who jogged, completed the entire distance with an average speed of 5 kph. If Kimberly completed the race in 2.75 hours less time than Steve did, determine the distance the marathon covered.

Solution Understand and Translate Let d = the distance from the start to the finish of the marathon. Then we can construct the following table. To determine the time, we divide the distance by the rate.

Person	Distance	Rate	Time
Kimberly	d	16	$\frac{d}{16}$
Steve	d	5	$\frac{d}{5}$

We are given that Kimberly completed the marathon in 2.75 hours less time than Steve did. Therefore, to make Kimberly's and Steve's times equal, we need to subtract 2.75 hours from Steve's time (or added 2.75 hours to Kimberly's time; see the paragraph below this example). We will subtract 2.75 hours from Steve's time and use the following equation to solve the problem.

Time for Kimberly = Time for Steve −2.75 hours

$$\frac{d}{16} = \frac{d}{5} - 2.75$$

Carry Out

$$80\left(\frac{d}{16}\right) = 80\left(\frac{d}{5} - 2.75\right) \qquad \textit{Multiply both sides by the LCD, 80.}$$

$$5d = 80\left(\frac{d}{5}\right) - 80(2.75) \qquad \textit{Distributive property}$$

$$5d = 16d - 220$$

$$-11d = -220$$

$$d = 20$$

Check and Answers The distance from the starting point to the ending point of the marathon appears to be 20 kilometers. To check this answer we will determine the times it took Kimberly and Steve to complete the marathon and see if the distance between the times is 2.75 hours. To determine the times, divide the distance, 20 kilometers, by the rate.

$$\text{Kimberly's time} = \frac{20}{16} = 1.25 \text{ hours}$$

$$\text{Steve's time} = \frac{20}{5} = 4 \text{ hours}$$

NOW TRY EXERCISE 25

Since $4 - 1.25 = 2.75$ hours, the answer checks. Therefore the distance the marathon covers is 20 kilometers. ✳

In Example 5 we subtracted 2.75 hours from Steve's time to obtain an equation. We could have added 2.75 hours to Kimberly's time to obtain an equivalent equation. Rework Example 5 now by adding 2.75 hours to Kimberly's time.

3 Set Up and Solve Work Problems

Problems in which two or more machines or people work together to complete a certain task are sometimes referred to as **work problems.** Work problems often involve equations containing fractions. Generally, work problems are based on the fact that the fractional part of the work done by person 1 (or machine 1) plus the fractional part of the work done by person 2 (or machine 2) is equal to the total amount of work done by both people (or both machines). *We represent the total amount of work done by the number 1, which represents one whole job completed.*

part of task done by first person or machine	+	part of task done by second person or machine	=	1 (one whole task completed)

To determine the part of the task completed by each person or machine, we use the formula

$$\textbf{part of task completed} \; = \; \textbf{rate} \cdot \textbf{time}$$

This formula is very similar to the formula

$$\text{amount} = \text{rate} \cdot \text{time}$$

that was discussed in Section 3.5. To determine the part of the task completed, we need to determine the rate. Suppose that Paul can do a particular task in 6 hours. Then he would complete 1/6 of the task per hour. Thus, his rate is 1/6 of the task per hour. If Audrey can do a particular task in 5 minutes, her rate is 1/5 of the task per minute. In general, if a person or machine can complete a task in t units of time, the rate is $1/t$.

EXAMPLE 6 **Building a Deck** Marty Agar can build a patio deck by himself in 20 hours. His wife, Betty, by herself can build the same patio deck in 30 hours. How long will it take them to build the deck if they work together?

Solution Understand and Translate Let $t =$ the time, in hours, for Mr. and Mrs. Agar working together, to build the deck. We will construct a table to help us in finding the part of the task completed by Mr. Agar and Mrs. Agar in t hours.

Landscaper	Rate of Work (part of the task completed per hour)	Time Worked	Part of Task
Mr. Agar	$\dfrac{1}{20}$	t	$\dfrac{t}{20}$
Mrs. Agar	$\dfrac{1}{30}$	t	$\dfrac{t}{30}$

$$\left(\begin{array}{c} \text{part of the deck completed} \\ \text{by Mr Agar in } t \text{ hours} \end{array} \right) + \left(\begin{array}{c} \text{part of the deck completed} \\ \text{by Mrs Agar in } t \text{ hours} \end{array} \right) = 1 \text{ (entire deck built)}$$

$$\frac{t}{20} \qquad + \qquad \frac{t}{30} \qquad = \qquad 1$$

Carry Out   Now multiply both sides of the equation by the LCD, 60.

$$60 \left(\frac{t}{20} + \frac{t}{30} \right) = 60 \cdot 1$$

$$\overset{3}{60} \left(\frac{t}{20} \right) + \overset{2}{60} \left(\frac{t}{30} \right) = 60 \qquad \textit{Distributive property}$$

$$3t + 2t = 60$$

$$5t = 60$$

$$t = 12$$

Answer   Thus, Mr. and Mrs. Agar working together can build the deck in 12 hours. We leave the check for you.

NOW TRY EXERCISE 27

HELPFUL HINT

In Example 6, Mr. Agar could build the deck by himself in 20 hours, and Mrs. Agar could build the deck by herself in 30 hours. We determined that together they could build the deck in 12 hours. Does this answer make sense? Since you would expect the time to build the deck together to be less than the time either of them could build it alone, the answer makes sense. When solving a work problem, always examine your answer to see if it makes sense. If not, rework the problem and find your error.

EXAMPLE 7 **Storing Wine** At a winery in Napa Valley, California, one pipe can fill a tank with wine in 3 hours and another pipe can empty the tank in 5 hours. If the valves to both pipes are open, how long will it take to fill the empty tank?

Solution Understand and Translate Let t = amount of time to fill the tank with both valves open.

Pipe	Rate of Work	Time	Part of Task
Pipe filling tank	$\dfrac{1}{3}$	t	$\dfrac{t}{3}$
Pipe emptying tank	$\dfrac{1}{5}$	t	$\dfrac{t}{5}$

As one pipe is filling, the other is emptying the tank. Thus, the pipes are working against each other. Therefore, instead of adding the parts of the task, as was done in Example 6 where the people worked together, we will subtract the parts of the task.

$$\left(\begin{array}{c}\text{part of tank}\\\text{filled in } t \text{ hours}\end{array}\right) - \left(\begin{array}{c}\text{part of tank}\\\text{emptied in } t \text{ hours}\end{array}\right) = 1 \text{ (total tank filled)}$$

$$\frac{t}{3} - \frac{t}{5} = 1$$

Carry Out
$$15\left(\frac{t}{3} - \frac{t}{5}\right) = 15 \cdot 1 \qquad \begin{array}{l}\textit{Multiply both sides}\\\textit{by the LCD, 15.}\end{array}$$

$$\overset{5}{\cancel{15}}\left(\frac{t}{\cancel{3}}\right) - \overset{3}{\cancel{15}}\left(\frac{t}{\cancel{5}}\right) = 15 \qquad \textit{Distributive property}$$

$$5t - 3t = 15$$

$$2t = 15$$

$$t = 7\tfrac{1}{2}$$

TEACHING TIP
Before discussing the solution to Example 8 ask, "If Courtney worked at the same rate as Kathy, how long would it take them to clean the office building's windows together? Use this information to compare Courtney's window cleaning rate with Kathy's window cleaning rate."

Check and Answer The tank will be filled in $7\tfrac{1}{2}$ hours. This answer is reasonable because we expect it to take longer than 3 hours when the tank is being drained at the same time. ✹

EXAMPLE 8 **Window-Cleaning Service** Kathy and Courtney own a small window-cleaning service. When Kathy cleans all the windows in a specific office building by herself, it takes her 8 hours. When Kathy and Courtney work together, they can clean all the windows in the office building in 5 hours. How long does it take Courtney by herself to clean all the windows?

Solution

Let t = time for Courtney to clean the windows by herself. Then Courtney's rate is $\frac{1}{t}$. Let's make a table to help analyze the problem. Since Kathy can clean all the windows by herself in 8 hours, her rate is $\frac{1}{8}$ of the job per hour. In the table, we use the fact that together they can clean all the windows in 5 hours.

Worker	Rate of Work	Time	Part of Task
Kathy	$\frac{1}{8}$	5	$\frac{5}{8}$
Courtney	$\frac{1}{t}$	5	$\frac{5}{t}$

$$\left(\begin{array}{c}\text{part of windows}\\\text{cleaned by Kathy}\end{array}\right) + \left(\begin{array}{c}\text{part of windows}\\\text{cleaned by Courtney}\end{array}\right) = 1$$

$$\frac{5}{8} \qquad + \qquad \frac{5}{t} \qquad = 1$$

Carry Out

$$8t\left(\frac{5}{8} + \frac{5}{t}\right) = 8t \cdot 1 \quad \textit{Multiply both sides by the LCD, 8t.}$$

$$8t\left(\frac{5}{8}\right) + 8t\left(\frac{5}{t}\right) = 8t \qquad \textit{Distributive property}$$

$$5t + 40 = 8t$$

$$40 = 3t$$

$$\frac{40}{3} = t$$

$$13\frac{1}{3} = t$$

TEACHING TIP
Tell your students that in addition to checking the answer in the equation they need to make sure the answer is reasonable. If the equation was set up incorrectly, the check might work but the solution may still be incorrect.

Check and Answer   Thus, it takes Courtney $13\frac{1}{3}$ hours, or 13 hours 20 minutes, to clean all the windows by herself. This answer is reasonable because we expect it to take longer for Courtney to clean the windows by herself than it would for Kathy and Courtney working together.

NOW TRY EXERCISE 35

EXAMPLE 9   **Thank-you Notes**   Peter and Kaitlyn Kewin are handwriting thank-you notes to guests who attended their 20th wedding anniversary party. Kaitlyn by herself could write all the notes in 8 hours and Peter could write all the notes by himself in 7 hours. After Kaitlyn has been writing thank-you notes for 5 hours by herself, she must leave town on business. Peter then continues the task of writing the thank-you notes. How long will it take Peter to finish writing the remaining notes?

Solution   Understand and Translate   Let t = time it will take Peter to finish writing the notes.

Person	Rate of Work	Time	Part of Task
Kaitlyn	$\frac{1}{8}$	5	$\frac{5}{8}$
Peter	$\frac{1}{7}$	t	$\frac{t}{7}$

$$\left(\begin{array}{c}\text{part of cards written}\\\text{by Kaitlyn}\end{array}\right) + \left(\begin{array}{c}\text{part of cards written}\\\text{by Peter}\end{array}\right) = 1$$

$$\frac{5}{8} + \frac{t}{7} = 1$$

Carry Out

$$56\left(\frac{5}{8} + \frac{t}{7}\right) = 56 \cdot 1 \qquad \textit{Multiply both sides by the LCD, 56.}$$

$$56\left(\frac{5}{8}\right) + 56\left(\frac{t}{7}\right) = 56 \qquad \textit{Distributive property}$$

$$35 + 8t = 56$$

$$8t = 21$$

$$t = \frac{21}{8} \quad \text{or} \quad 2\frac{5}{8}$$

TEACHING TIP
Have students work in pairs to write two word problems that can be solved using rational equations. Then have them give their word problems to another pair to solve.

Answer Thus, it will take Peter $2\frac{5}{8}$ hours to complete the cards.

Mathematics in Action

Head Loss in Pipes

There are many real-life applications of rational equations. Some are illustrated in this chapter. One real-life application involves what is referred to as head loss. Head loss is the loss of energy of a liquid being pumped from one point in a pipe to another point. A head loss of one foot is the same loss of energy that would occur if the pipe were elevated one foot from the horizontal at the point measured. Head loss, (HL), expressed in feet, can be calculated with the rational equation:

$$HL = k \cdot \frac{L}{D} \cdot \frac{V^2}{2g}$$

where k = friction factor; L = pipe length in feet; D = pipe diameter in feet; V = flow velocity in feet per second; g = gravitational constant = 32.17 feet per second squared.

Keeping track of and controlling the pressure of water through pipes has applications wherever water has to be moved from one place to another. In nuclear power plants, huge amounts of water are pumped to cool various reactor parts. A design that did not sufficiently predict head loss could result in less-than-expected pressure for critically necessary cold water. The result could be a plant shut-down or even a life-threatening failure.

One can hardly imagine a modern home without pressurized water running to the kitchen, bath-rooms, washing machines, and outdoor hose connections. Calculations of water pressure become particularly important in the design of plumbing for septic systems. A serious health hazard could result from a flow rate that is not sufficient to move waste water to a storage or drainage location. Similarly, poorly designed swimming pool plumbing could mean that water did not circulate through filters at the proper rate, which could result in dangerously high bacteria counts.

Head loss is a significant factor at tens of thousands of agricultural and manufacturing locations where systems depend on liquids flowing through pipes in a predictable manner. Hydraulic engineers and plumbing designers use the head loss formula as an important tool to ensure adequate water flow.

Exercise Set 6.7

Concept/Writing Exercises

1. Geometric formulas, which were discussed in Section 3.1, are often rational equations. Give three formulas that are rational equations.

2. Suppose that car 1 and car 2 are traveling to the same location 60 miles away and that car 1 travels 10 miles per hour faster than car 2.

    a) Let r represent the speed of car 2. Write an expression, using time $= \dfrac{\text{distance}}{\text{rate}}$, for the time it takes car 1 to reach its destination.

    b) Now let r represent the speed of car 1. Write an expression, using time $= \dfrac{\text{distance}}{\text{rate}}$, for the time it takes car 2 to reach its destination.

3. In an equation for a work problem, one side of the equation is set equal to 1. What does the 1 represent in the problem? 1 complete task

4. Suppose Tracy Augustine can complete a particular task in 3 hours, while the same task can be completed by John Bailey in 7 hours. How would you represent the part of the task completed by Tracy in 1 hour? By John in 1 hour? Tracy, $\frac{1}{3}$; John, $\frac{1}{7}$

1. Some examples are: $A = \frac{1}{2}bh$, $A = \frac{1}{2}h(b_1 + b_2)$, $V = \frac{1}{3}\pi r^2 h$, $V = \frac{4}{3}\pi r^3$

2. a) $t = \dfrac{60}{r + 10}$ **b)** $t = \dfrac{60}{r - 10}$. **5.** $l = 10$ in., $w = 9$ in.

6. $w = 10$ in, $l = 24$ in., $h = 16$ in.

7. base $= 12$ cm, height $= 7$ cm

Practice the Skills/Problem Solving

In Exercises 5–36, solve the problem and answer the questions.

Geometry Problems, See Example 1.

5. ***Packaging Computers*** The Phillips Paper Company makes rectangular pieces of cardboard for packing computers. The sheets of cardboard are to have an area of 90 square inches, and the length of a sheet is to be 4 inches more than $\frac{2}{3}$ its width. Determine the length and width of the cardboard to be manufactured.

6. ***Carry-on Luggage*** On most airlines carry-on luggage can have a maximum width of 10 inches with a maximum volume of 3840 cubic inches, see the figure. If the height of the luggage is $\frac{2}{3}$ the length, determine the dimensions of the largest piece of carry-on luggage. Use $V = lwh$.

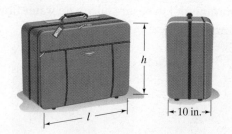

7. ***Triangles of Dough*** Pillsbury Crescent Rolls are packaged in tubes that contain perforated triangles of dough. The base of the triangular piece of dough is about 5 centimeters more than its height. Determine the base and height of a piece of dough if the area is about 42 square centimeters.

8. ***Yield Sign*** Yield right of way signs used in the United States are triangles. The area of the sign is about 558 square inches. The height of the sign is about 5 inches less than its base. Determine the length of the base of a yield right of way sign. 36 in

9. ***Triangular Garden*** A triangular area is 12 square feet. Find the base of the triangular area if the height is 1 foot less than $\frac{1}{2}$ the base. base: 8 feet

10. ***Billboard*** A billboard sign is in the shape of a trapezoid. The area of the sign is 32 square feet. If the height of the sign is $\frac{1}{4}$ the length of the sum of the 2 bases, determine the height of the sign. 4 ft

Number Problems, See Example 2.

11. *Difference of Numbers* One number is 10 times larger than another. The difference of their reciprocals is 3. Determine the two numbers. $\frac{3}{10}, 3$

12. *Sum of Numbers* One number is 3 times larger than another. The sum of their reciprocals is $\frac{4}{3}$. Determine the two numbers. 1, 3

15. 10 mph   **19.** 150 mph, 600 mph

13. *Increased Numerator* The numerator of the fraction $\frac{3}{4}$ is increased by an amount so that the value of the resulting fraction is $\frac{5}{2}$. Determine the amount by which the numerator was increased. 7

14. *Decreased Denominator* The denominator of the fraction $\frac{6}{19}$ is decreased by an amount so that the value of the resulting fraction is $\frac{3}{8}$. Determine the amount by which the denominator was decreased. 3

Motion Problems, See Example 3–5.

15. *Paddleboat Ride* In the Mississippi River near New Orleans, the Creole Queen paddleboat travels 4 miles upstream (against the current) in the same amount of time it travels 6 miles downstream (with the current). If the current of the river is 2 miles per hour, determine the speed of the Creole Queen in still water.

16. *Kayak Ride* Kathy Boothby-Sestak can paddle her kayak 6 miles per hour in still water. It takes her as long to paddle 3 miles upstream as 4 miles downstream in the Wabash River near Lafayette, Indiana. Determine the river's current. $\frac{6}{7}$ mph

17. *Trolley Ride* A trolley travels in one direction at an average of 15 miles per hour, then turns around and travels on the same track in the opposite direction at 15 miles per hour. If the total time traveling on the trolley is $2\frac{1}{2}$ hours, how far did the trolley travel in one direction? 18.75 mi

18. *Motorcycle Trip* Brandy Dawson and Jason Dodge start a motorcycle trip at the same point a little north of Fort Worth, Texas. Both are traveling to San Antonio, Texas, a distance of about 400 kilometers. Brandy rides 30 kilometers per hour faster than Jason does.

When Brandy reaches her destination, Jason has only traveled to Austin, Texas, a distance of about 250 kilometers. Determine the approximate speed of each motorcycle. 50 km/hr, 80 km/hr

19. *Jet Flight* Elenore Morales traveled 1600 miles by commercial jet from Kansas City, Missouri, to Spokane, Washington. She then traveled an additional 500 miles on a private propeller plane from Spokane to Billings, Montana. If the speed of the jet was 4 times the speed of the propeller plane and the total time in the air was 6 hours, determine the speed of each plane.

20. *Exercise Regimen* As part of his exercise regimen, Chris Barker walks a distance of 2 miles on an indoor track and then jogs at twice his walking speed for another mile. If the total time spent on the track was $\frac{3}{4}$ of an hour, determine the speeds at which he walks and jogs. walks at $3\frac{1}{3}$ mph, jogs at $6\frac{2}{3}$ mph

21. *Dover to Calais* To go from Dover, England, to Paris, France, Renee Estivez took the Hovercraft, over the English Channel, from Dover to Calais, France. Then she took the bullet train to Paris. The Hovercraft traveled at an average of 40 miles per hour, and the train traveled at an average of 120 miles per hour. If the total distance traveled was 200 miles, and the total time spent commuting (in the boat and on the train) was 2.2 hours, determine the distance she traveled on the boat and the distance she traveled on the train. boat: 32 mi, train: 168 mi

22. *Jogging and Walking* Kristen Taylor jogs a certain distance and then walks a certain distance. When she jogs she averages 5 miles per hour and when she walks she averages 2 miles per hour. If she walks and jogs a total of 3 miles in a total of 0.9 hour, how far does she jog and how far does she walk? jogs 2 mi, walks 1 mi

23. *Headwind and Tailwind* A Boeing 747 flew from San Francisco to Honolulu, a distance of 2800 miles. Flying with the wind, it averaged 600 miles per hour. When the wind changed from a tailwind to a headwind, the plane's speed dropped to 500 miles per hour. If the total

23. 3 hr at 600 mph, 2 hr at 500 mph
time of the trip was 5 hours, determine the length of time it flew at each speed.

24. *Bullet Train* Bullet trains in Japan have been known to average 240 kilometers per hour (Kph). Prior to using bullet trains, trains in Japan traveled at an average speed of 120 Kph. If a bullet train traveling from Shin-Osake to Hakata can complete its trip in 2.3 hours less time than an older train, determine the distance from Shin-Osake to Hakata. 552 km

25. *Water Skiers* At a water show a boat pulls a water skier at a speed of 30 feet per second. When it reaches the end of the lake, more skiers were added to be pulled by the boat, so the boat's speed drops to 20 feet per second. If the boat traveled the same distance with additional skiers as it did with the single skier, and the trip back with the additional skiers took 15 seconds longer than the trip with the single skier, how far, in feet, in one direction, had the boat traveled? 900 ft

26. *Cross-Country Skiing* Alana Bradley and her father Tim begin skiing the same cross-country ski trail in Elmwood Park in Sioux Falls, South Dakota, at the same time. If Alana, who averages 9 miles per hour, finishes the trail 0.25 hours sooner than her father, who averages 6 miles per hour, determine the length of the trail. 4.5 miles

Work Problems, See Examples 6–9.

27. *Wallpaper* Reynaldo and Felicia Fernandez decide to wallpaper their family room. Felicia, who has wallpapering experience, can wallpaper the room in 6 hours. Reynaldo can wallpaper the same room in 8 hours. How long will it take them to wallpaper the family room if they work together? $3\frac{3}{7}$ hr

28. *Conveyor Belt* At a salt mine, one conveyor belt requires 20 minutes to fill a large truck with ore. A second conveyor belt requires 30 minutes to fill the same truck with ore. How long would it take if both conveyor belts were working together to fill the truck with ore? 12 min

29. *Hot Tub* Pam and Loren Fornieri know that their hot tub can be filled in 2 hours and drained completely in 3 hours. Pam turned on the water and returned to the house. After 2 hours, she went out to the tub and saw, to her disappointment, that the tub was only partly filled because the drain had been left open. If the water was on and the drain was left open, how long would it take the tub to fill completely? 6 hr

See Exercise 29

30. *Filling a Tank* During a rainstorm, the rain is flowing into a large holding tank. At the rate the rain is falling, the empty tank would fill in 8 hours. At the bottom of the tank is a spigot to dispense water. Typically, it takes about 12 hours with the spigot wide open to empty the water in a full tank. If the tank is empty and the spigot has been accidentally left open, and the rain falls at the constant rate, how long would it take for the tank to fill completely? 24 hr

31. *Flowing Water* When the water is turned on and passes through a one-half inch diameter hose, a small aboveground pool can be filled in 5 hours. When the water is turned on at two spigots and passes through both the one-half inch diameter hose and a three-quarter inch diameter hose, the pool can be filled in 2 hours. How long would it take to fill the pool using only the three-quarter inch diameter hose? $3\frac{1}{3}$ hr

32. *Payroll Checks* At the NCNB Savings Bank it takes a computer 4 hours to process and print payroll checks.

When a second computer is used and the two computers work together, the checks can be processed and printed in 3 hours. How long would it take the second computer by itself to process and print the payroll checks? 12 hr

33. ***Digging a Trench*** A construction company with two backhoes has contracted to dig a long trench for drainage pipes. The larger backhoe can dig the entire trench by itself in 12 days. The smaller backhoe can dig the entire trench by itself in 15 days. The large backhoe begins working on the trench by itself, but after 5 days it is transferred to a different job and the smaller backhoe begins working on the trench. How long will it take for the smaller backhoe to complete the job?

34. ***Delivery of Condiments*** Chadwick and Melissa Wicker deliver institutional size cans of condiments to various warehouses in Pennsylvania. If Chadwick drove the entire trip, the trip would take about 10 hours. Melissa is a faster driver. If Melissa drove the entire trip, the trip would take about 8 hours. After Melissa had been driving for 4 hours, Chadwick takes over the driving. About how much longer will Chadwick drive before they reach their final destination? 5 hr

35. ***Snowstorm*** Following a severe snowstorm, Ken and Bettina Reeves must clear their driveway and sidewalk. Ken can clear the snow by himself in 4 hours, and Bettina can clear the snow by herself in 6 hours. After Bettina has been working for 3 hours, Ken is able to join her. How much longer will it take them working together to remove the rest of the snow?

36. ***Farm Cooperative*** A large farming cooperative near

Hutchinson, Kansas, owns three hay balers. The oldest baler can pick up and bale an acre of hay in 3 hours and each of the two new balers can work an acre in 2 hours.

a) How long would it take the three balers working together to pick up and bale 1 acre?

b) How long would it take for the three balers working together to pick up and bale the farm's 375 acres of hay?   **a)** $\frac{3}{4}$ hr   **b)** $281\frac{1}{4}$ hr

37. ***Skimming Oil*** A boat designed to skim oil off the surface of the water has two skimmers. One skimmer can fill the boat's holding tank in 60 hours while the second skimmer can fill the boat's holding tank in 50 hours. There is also a valve in the holding tank that is used to transfer the oil to a larger vessel. If no new oil is coming into the holding tank, a full holding tank of skimmed oil can be transferred to a larger tank in 30 hours. If both skimmers begin skimming and the valve on the holding tank is opened, how long will it take for the empty holding tank on the boat to fill?

38. ***Flower Garden*** Bob can plant a flower garden by himself in 8 hours. Mary can plant the same garden by herself in 10 hours, and Gloria can plant the same garden by herself in 12 hours. How long would it take them working together to plant the garden? ≈3.24 hr

33. $8\frac{3}{4}$ days   35. $\frac{3}{5}$ hr or 36 minutes   37. 300 hr

Challenge Problems

39. ***Reciprocal of a Number*** If 5 times a number is added to 4 times the reciprocal of the number, the answer is 12. Determine the number(s).   2 or $\frac{2}{5}$

40. ***Determine a Number*** The reciprocal of the difference of a certain number and 3 is twice the reciprocal of the difference of twice the number and 6. Determine the number(s). all real numbers except 3

41. *Picking Blueberrys* Ed and Samantha Weisman, whose parents own a fruit farm, must each pick the same number of pints of blueberries each day during the season. Ed picks an average of 8 pints per hour, while Samantha picks an average of 4 pints per hour. If Ed and Samantha begin picking blueberries at the same time, and Samantha finishes 1 hour after Ed, how many pints of blueberries must each pick? 8 pints

42. *Sorting Mail* A mail processing machine can sort a large bin of mail in 1 hour. A newer model can sort the same quantity of mail in 40 minutes. If they operate together, how long will it take them to sort the bin of mail? $\frac{2}{5}$ hr = 24 min

Cumulative Review Exercises

[2.1] **43.** Simplify $\frac{1}{2}(x + 3) - (2x + 6)$. $-\frac{3}{2}x - \frac{9}{2}$

[5.2] **44.** Factor $y^2 + 5y - y - 5$ by grouping.
  $(y - 1)(y + 5)$

[6.2] **45.** Divide $\dfrac{x^2 - 14x + 48}{x^2 - 5x - 24} \div \dfrac{2x^2 - 13x + 6}{2x^2 + 5x - 3}$. 1

[6.4] **46.** Subtract $\dfrac{x}{6x^2 - x - 15} - \dfrac{5}{9x^2 - 12x - 5}$.
  $\dfrac{3x^2 - 9x - 15}{(2x + 3)(3x - 5)(3x + 1)}$

6.8 VARIATION

SSM   Study Guide   CD/Video

MathPro 4/5   PH Math Tutor Center   prenhall.com/Angel

1   Set up and solve direct variation problems.
2   Set up and solve inverse variation problems.

Many scientific formulas are expressed as variations. Many of our daily activities also involve variation. A **variation** is an equation that relates one variable to one or more other variables using the operations of multiplication or division (or both operations). In this book we will discuss two types of variation: **direct variation** and **inverse variation.** We will first discuss direct variation.

1   Set Up and Solve Direct Variation Problems

In direct variation, the two variables both increase together or both decrease together; that is, as one variable increases, so does the other, and as one variable decreases, so does the other.

Suppose water is coming out of a garden hose for a specific amount of time and going into a bucket. The greater the flow rate of water coming out of the hose, the greater the volume of water in the bucket, and the smaller the flow rate of water coming out of the hose, the smaller the volume of water in the bucket. This is an example of direct variation. The volume of water in the bucket varies directly as the flow rate from the hose.

Consider a car traveling at 50 miles per hour. The car travels 50 miles in 1 hour, 100 miles in 2 hours, 150 miles in 3 hours. Notice that as the time increases the distance traveled increases. This is also an example of direct variation. Notice that for a constant speed, as the time increases, so does the distance traveled, and as the time decreases, so does the distance traveled.

The formula used to calculate distance is

$$\text{distance} = \text{rate} \cdot \text{time}$$

Since the rate is constant, 50 miles per hour, the formula can be written

$$d = 50t$$

We say that the distance, *d, varies directly* as the time, *t*, or that the distance is *directly proportional* to the time *t*. In the equation, the 50 is a constant referred to as a **constant of proportionality.** Now we will define direct variation.

Direct Variation

If a variable *y* varies directly as a variable *x*, then

$$y = kx$$

where *k* is the *constant of proportionality* (or the variation constant).

TEACHING TIP
You may wish to mention, that, if the direct variation is linear, *k* represents the slope of a line. You may want to come back to direct variation briefly when slope is covered.

When we are told that one quantity varies directly as another, we generally start with an equation in the form $y = kx$; however letters other than *x* and *y* may be used for the variables.

EXAMPLE 1 **Heating Up a Jacuzzi**   When the heater is turned on to warm the water in a jacuzzi, the temperature of the water, *w*, increases directly with the time, *t*, the heater is on.

a) Write the variation as an equation.

b) If the constant of proportionality, *k*, is 0.5, find the increase in temperature of the water after 30 minutes.

Solution   **a)** We are told that the water temperature varies directly with the time. Thus we set up the direct proportion, as follows.

$$w = kt$$

b) To find the increase in water temperature we will substitute 0.5 for *k*, and 30 for *t*.

$$w = kt$$
$$w = 0.5(30) = 15$$

NOW TRY EXERCISE 35   Thus, after 30 minutes the water temperature has increased by 15°.

In many variation problems you will first have to solve for the constant of proportionality before you can solve for the variable you are asked to find. To determine the constant of proportionality, substitute the values given for the variables, and solve for *k*. We illustrate this procedure in Example 2.

EXAMPLE 2 **Direct Variation Problem**   *s* varies directly as the square of *m*. If $s = 100$ when $m = 5$, find *s* when $m = 12$.

Solution   **a) Understand and Translate**   We begin by setting up the equation. Notice that we are told that *s* varies directly as the square of *m*. The square of *m* is written m^2. Therefore, the equation is $s = km^2$. Since we are not given the constant of proportionality, we find it by substituting the values we are given for the variables.

$$s = km^2$$
$$100 = k(5^2) \quad \textit{Substitute values}$$
$$100 = k(25)$$
$$100 = 25k$$
$$4 = k$$

Now that we have determined k, we can answer the question by substituting 4 for k, and 12 for m.

Carry Out
$$s = km^2$$
$$s = 4(12)^2$$
$$s = 4(144)$$
$$s = 576$$

Answer   Thus, when $m = 12$, $s = 576$.   ✳

EXAMPLE 3   **Drug Dosage**   The amount of the drug, d, given to a person is directly proportional to the person's weight, w. If a person who weighs 75 kilograms is given 150 milligrams (mg) of allopurinol, determine how many milligrams of allopurinol are given to a person who weighs 96 kg.

Solution   Understand and Translate   We are told that the amount of the drug is directly proportional to the person's weight. Thus we set up the equation
$$d = kw$$
Now we determine k by substituting the values given for d and w.

TEACHING TIP
You may wish to point out that direct variation problems like Example 3, can also be solved using proportions.

Carry Out
$$d = kw$$
$$150 = k(75)$$
$$\frac{150}{75} = k$$
$$2 = k$$

Now we proceed to find the number of milligrams of the drug to be given by substituting 2 for k and 96 for w.
$$d = kw$$
$$d = 2(96) = 192$$

Check and Answer   Since we expect the amount of the drug to be greater than 150 milligrams, our answer is reasonable. A 96 kg person should be given 192 milligrams of the drug.   ✳

NOW TRY EXERCISE 39

Example 3 could also be solved using a proportion.

2   Set Up and Solve Inverse Variation Problems

Now we will discuss inverse variation. When two quantities vary inversely, it means that as one quantity increases, the other quantity decreases, and vice versa.

Suppose you are making a casserole in the oven. Generally speaking, the higher the temperature of the oven, the shorter the period of time needed for the casserole to cook. That is, as the temperature increases, the time decreases, and vice versa. Thus, the time and temperature are inversely proportional to one another, or vary inversely with one another.

To explain inverse variation, we again use the distance formula,
$$\text{distance} = \text{rate} \cdot \text{time}.$$

If we solve the formula for time, we get time $= \dfrac{\text{distance}}{\text{rate}}$. Assume the distance is fixed at 100 miles: then
$$\text{time} = \frac{100}{\text{rate}}$$

At a rate of 100 miles per hour it would take 1 hour to cover the 100 mile distance. At 50 miles per hour it would take 2 hours to cover the distance. At 25 mph it would take 4 hours. Note that as the rate (or speed) decreases, the time increases, and vice versa. The equation above can be written

$$t = \frac{100}{r}$$

This is an example of an inverse variation in which the time and rate are inversely proportional and the constant of proportionality is 100. Now we define inverse variation.

Inverse Variation

If a variable y varies inversely as a variable x, then

$$y = \frac{k}{x} \text{ (or } xy = k)$$

where k is the constant of proportionality.

Now let's work two examples.

EXAMPLE 4   **Chartering a Sailboat**   A group of friends is going to charter a sailboat for a pleasure trip. The cost per person for chartering the boat, c, is inversely proportional to the number of people chartering the boat, n. If 8 friends decide to charter the boat, the cost per person is $40. Determine the cost per person if 14 friends decide to charter the boat.

Solution   **Understand and Translate**   We are told this an inverse variation. Therefore, we will set up an equation to represent the inverse proportion.

$$c = \frac{k}{n}$$

Since we are not given the constant of proportionality, we determine k by substituting the values given for c and n.

$$40 = \frac{k}{8}$$

$$320 = k$$

Now we can determine the answer to the question by using $k = 320$ and $n = 14$.

Carry Out
$$c = \frac{k}{n}$$

$$c = \frac{320}{14}$$

$$c = 22.86$$

Check and Answer   The cost to each person would be about $22.86 if 14 friends decided to charter the sailboat. If you multiply 22.86(14) you obtain 320.04, which is slightly more than the constant of proportionality. This is caused by the rounding off of $\frac{320}{14}$ to $22.86.

NOW TRY EXERCISE 41

EXAMPLE 5 **Speaker Loudness**   The loudness, l, of a stereo speaker, measured in decibels (dB), varies inversely as the square of the distance, d, of the listener from the speaker. Assuming that for a particular speaker the loudness is 20 dB when the listener is 6 feet from the speaker.

a) Determine an equation that expresses the relationship between the loudness and the distance.

b) Using the equation obtained in part **a)**, determine the loudness when a person is 3 feet from the speaker.

Solution   This problem is broken down into two parts. The first part asks us to find a general formula, while the second part asks us to use the formula.

a) Understand and Translate   We are told that the loudness varies inversely as the *square* of the distance. Thus we write the following equation and solve for k.

Carry Out
$$l = \frac{k}{d^2}$$

$$20 = \frac{k}{6^2}$$

$$20 = \frac{k}{36}$$

$$720 = k$$

Check and Answer   The constant of proportionality, k, is 720. Since for this speaker $k = 720$, the equation we are seeking is

$$l = \frac{720}{d^2}$$

b) Understand and Translate   In part **a)** we determined the equation used to find the loudness. We substitute 3 for d in the formula and solve for l.

$$l = \frac{720}{d^2}$$

$$l = \frac{720}{3^2}$$

Carry Out
$$l = \frac{720}{9}$$

$$l = 80$$

NOW TRY EXERCISE 49

Check and Answer   Thus at 3 feet the loudness is 80 decibels. This is reasonable because at a shorter distance (3 feet versus 6 feet) the sound will be louder. ✳

Exercise Set 6.8

Concept/Writing Exercises

1. What is the value of the constant of proportionality in the direct variation $m = 40t$?   40

2. What is the value of the constant of proportionality in the inverse variation $w = \frac{60}{r}$?   60

3. Give the general form of direct variation. $y = kx$

4. Give the general form of inverse variation. $y = \frac{k}{x}$

Practice the Skills

Determine if the following are direct or inverse variations. Explain your answer.

5. The diameter of a hose and the amount of water coming out of the hose. direct

6. The lens opening on a camera and the amount of light reaching the film. direct

7. The speed of a turtle and the length of time it takes the turtle to cross a road. inverse

8. The age of a car, up to 10 years old, and the value of a car. inverse

9. The temperature of water and the time it takes for an ice cube placed in the water to melt. inverse

10. The number of people in line at the theater and the time it takes for all people in line to purchase tickets. direct

11. The length of a roll of Scotch tape and the number of 2 inch strips that can be obtained from the roll. direct

12. A person's reading speed and the time it takes to read a novel. inverse

13. The cubic-inch displacement, in liters, and the horsepower of the engine. direct

14. The speed of a riding lawn mower and the time it takes to cut the lawn. inverse

In Exercises 15–22, find the quantity indicated.

15. x varies directly as z. Find x when $z = 11$ and $k = 3$. 33

16. x varies directly as y. Find x when $y = 12$ and $k = 6$. 72

17. x varies inversely as y. Find x when $y = 25$ and $k = 5$. $\frac{1}{5}$

18. R varies inversely as W. Find R when $W = 160$ and $k = 240$. 1.5

19. C varies directly as the square of Z. Find C when $Z = 5$ and $k = 2$. 50

20. L varies directly as the square of R. Find L when $R = 9$ and $k = 4$. 324

21. y varies inversely as the square of x. Find y when $x = 10$ and $k = 320$. 3.2

22. y varies inversely as the square of w. Find y when $w = 12$ and $k = 288$. 2

For Exercises 23–30, find the quantity indicated.

23. x varies directly as y. If $x = 9$ when $y = 18$, find x when $y = 36$. 18

24. Z varies directly as W. If $Z = 7$ when $W = 28$, find Z when $W = 140$. 35

25. C varies inversely as J. If $C = 7$ when $J = 1$ find C when $J = 2$. 3.5

26. H varies inversely as L. If $H = 15$ when $L = 50$, find H when $L = 10$. 75

27. y varies directly as the square of R. If $y = 4$ when $R = 4$, find y when $R = 8$. 16

28. A varies directly as the square of B. If $A = 245$ when $B = 7$, find A when $B = 9$. 405

29. L varies inversely as the square of P. If L is 270 when $P = 10$, find L when $P = 30$. 30

30. x varies inversely as the square of P. If $x = 10$ when $P = 6$, find x when $P = 20$. 0.9

Problem Solving

31. Assume a varies directly as b. If b is doubled, how will it affect a? Explain. doubled

32. Assume a varies directly as b^2. If b is doubled, how will it affect a? Explain. quadrupled

33. Assume y varies inversely as x. If x is doubled, how will it affect y? Explain. halved

34. Assume y varies inversely as a^2. If a is doubled, how will it affect y? Explain. quartered

In Exercises 35–54 determine the quantity you are asked to find.

35. *Distance and Speed* The distance, d, a car travels is directly proportional to the speed, s, the car is traveling. Determine the distance traveled if the constant of proportionality, k, is 2 and the speed is 40 miles per hour. 80 mi

36. *Swimming Pool* The time, t, it takes to fill an inground pool is directly proportional to the amount of water coming out of the hose, w. Determine the time it takes a hose to fill the pool if the constant of proportionality is 0.3 and the amount of water coming out of the hose is 100 gallons per hour. 30 hr

40. ≈0.73 hr   **43.** 1000 people   **44.** 16 presses

37. Reaching a Destination The time it takes to reach a certain destination, t, is inversely proportional to the speed a car is traveling, s. Determine the time it takes to reach the destination if the constant of proportionality is 100 and the speed, s, is 50 miles per hour.   2 hr

38. Light Through Water The percent of light that filters through water, l, is inversely proportional to the depth of the water, d. Determine the percent of light that filters down to a depth of 10 feet if the constant of proportionality is 300.   30

39. Speed of a Car The distance traveled in a car, d, is directly proportional to the speed of the car, s. If the distance traveled is 300 feet when the speed is 120 feet/minute, determine the distance traveled when the speed is 150 feet/minute.   375 ft

40. Cutting a Lawn The time it takes Don to cut a lawn, t, is directly proportional to the area of the lawn, A. If it takes Don 0.4 hour to cut an area of 1200 square feet, how long will it take him to cut an area of 2200 square feet?

41. Adding Fencing The time, t, it takes to fence in a large field is inversely proportional to the number of people, n, working on the fence. When 10 people are working on the fence it takes 200 hours to put up the fence. How long will it take if 15 people work on putting up the fence?   ≈133.33 hr

42. Baking a Turkey The time, t, it takes to bake a turkey is inversely proportional to the oven temperature, T. If it takes 3 hours to bake a turkey at 300°F, how long will it take to bake the turkey at 250°F?   3.6 hr

43. Basketball Game The receipts, r, at a baseball game are directly proportional to the number of people attending the game, n. If the receipts for a game are $12,000 when 800 people attend, determine how many people attend if the receipts for a game are $15,000.

44. Daily Newspaper The time, t, it takes to print a specific number of copies of a daily newspaper is inversely proportional to the number of presses it has working, n. When 6 presses are working, the newspapers are printed in 8 hours. Determine the number of presses working if the newspapers are printed in 3 hours.

45. Cost of Wedding The cost, c, of a certain wedding is directly proportional to the number of people attending, n. If the cost is $4000 when 80 people attend, how many people attend if the cost is $3000?   60 people

46. Cleaning Windows The time, t, it takes to clean all the windows in a large office building is inversely proportional to the number of teams, n, of window workers used. If 6 teams can clean all the windows in 16 days, how many teams are used if the windows are cleaned in 12 days?   8 teams

47. Area of a Circle The area of a circle, A, is directly proportional to the square of the radius of a circle, r. If the area of a circle is about 78.5 square inches when the radius is 5 inches, determine the area when the radius is 12 inches.   ≈452.16 sq in.

48. Falling Object The velocity, v, of a falling object is directly proportional to the square of the time, t, it has been in free fall. An object that has been in free fall for 2 seconds has a velocity of 64 feet per second. Determine the velocity of an object that has been falling for 8 seconds.   1024 ft/sec

49. Electrical Circuit In an electrical circuit the resistance, r, of an appliance is inversely proportional to the square of the current, c. If the resistance is 100 ohms when the current is 0.4 amps, determine the resistance if the current is 0.6 amps.   ≈44.44 ohms

50. Volume of a Cylinder For a cylinder of a specific volume, the height, h, of the cylinder is inversely proportional to the square of the radius of the cylinder, r. When the radius is 6 inches, the height is 10 inches. Determine the height when the radius is 5 inches.   14.4 in.

51. ***Finding Interest*** The amount of interest earned on an investment, I, varies directly as the interest rate, r. If the interest earned is $40 when the interest rate is 4%, find the amount of interest earned when the interest rate is 6%. $60

52. ***Brick Wall*** The time, t, required to build a brick wall varies inversely as the number of people, n, working on the wall. If it takes 8 hours for five bricklayers to build a wall, how long will it take eight bricklayers to build a wall? 5 hr

53. ***Volume of Gas*** The volume of a gas, V, varies inversely as its pressure, P. If the volume, V, is 800 cubic centimeters when the pressure is 200 millimeters (mm) of mercury, find the volume when the pressure is 25 mm of mercury. 6400 cc

54. ***Hooke's Law*** Hooke's law states that the length a spring will stretch, S, varies directly with the force (or weight), F, attached to the spring. If a spring stretches 1.4 inches when 20 pounds is attached, how far will it stretch when 10 pounds is attached? 0.7 in.

Challenge Problems

*In addition to direct variation and inverse variations, there is also joint variation and combined variation. In **joint variation** one quantity may vary directly (as the product of) two or more quantities. In **combined variation**, one quantity may vary directly with some variables and inversely with other variables. Exercise 55 is a joint variation problem, and Exercise 56 is a combined variation problem. For Exercises 55 and 56 **a)** write the variation, and **b)** find the quantity indicated.*

55. x varies jointly as y and z. If x is 72 when $y = 18$ and $z = 2$, find x when $y = 36$ and $z = 3$. **a)** $x = kyz$, **b)** 216

56. T varies directly as the square of D and inversely as F. If $T = 18$ when $D = 6$ and $F = 4$, find T when $D = 8$ and $F = 8$. **a)** $T = \dfrac{kD^2}{F}$, **b)** 16

Cumulative Review Exercises

[4.6] 57. Divide $\dfrac{8x^2 + 6x - 25}{4x + 9}$ $2x - 3 + \dfrac{2}{4x + 9}$

[5.1] 58. Factor $y(z - 2) + 3(z - 2)$ $(z - 2)(y + 3)$

[5.6] 59. Solve $3x^2 - 24 = -6x$ $-4, 2$

[6.2] 60. Multiply $\dfrac{x + 3}{x - 3} \cdot \dfrac{x^3 - 27}{x^2 + 3x + 9}$ $x + 3$

CHAPTER SUMMARY
Key Words and Phrases

6.1
Rational expression
Signs of a fraction
Simplify a rational
  expression

6.2
Divide rational
  expressions
Multiply rational
  expressions

6.3
Add rational
  expressions with a

common
  denominator
Least common
  denominator
Subtract rational
  expressions with a
  common
  denominator

6.4
Add rational
  expressions with
  unlike denominators

Subtract rational
  expressions with
  unlike denominators

6.5
Complex fraction

6.6
Extraneous root or
  extraneous solution
  to a rational equation
Rational equation

6.7
Geometry problems
Motion problems
Work problems

6.8
Constant of
  proportionality
Direct variation
Inverse variation
Variation

(continued on the next page)

IMPORTANT FACTS

For any fraction:	$-\dfrac{a}{b} = \dfrac{-a}{b} = \dfrac{a}{-b}, b \neq 0$
To add fractions:	$\dfrac{a}{c} + \dfrac{b}{c} = \dfrac{a+b}{c}, c \neq 0$
To subtract fractions:	$\dfrac{a}{c} - \dfrac{b}{c} = \dfrac{a-b}{c}, c \neq 0$
To multiply fractions:	$\dfrac{a}{b} \cdot \dfrac{c}{d} = \dfrac{ac}{bd}, b \neq 0, d \neq 0$
To divide fractions:	$\dfrac{a}{b} \div \dfrac{c}{d} = \dfrac{a}{b} \cdot \dfrac{d}{c} = \dfrac{ad}{bc}, b \neq 0, c \neq 0, d \neq 0$
Time $= \dfrac{\text{distance}}{\text{rate}}$	
Direct variation:	$y = kx$
Inverse variation:	$y = \dfrac{k}{x}$

Chapter Review Exercises

[6.1] Determine the values of the variable for which the following expressions are defined.

1. $\dfrac{5}{2x - 12}$

2. $\dfrac{2}{x^2 - 8x + 15}$

3. $\dfrac{2}{5x^2 + 4x - 1}$

1. all real numbers except $x = 6$ **2.** all real numbers except $x = 3, x = 5$ **3.** all real numbers except $x = \dfrac{1}{5}, x = -1$

Simplify.

4. $\dfrac{y}{xy - 3y}$ $\dfrac{1}{x-3}$

5. $\dfrac{x^3 + 4x^2 + 12x}{x}$ $x^2 + 4x + 12$

6. $\dfrac{9x^2 + 6xy}{3x}$ $3x + 2y$

7. $\dfrac{x^2 + 2x - 8}{x - 2}$ $x + 4$

8. $\dfrac{a^2 - 36}{a - 6}$ $a + 6$

9. $\dfrac{-2x^2 + 7x + 4}{x - 4}$ $-(2x + 1)$

10. $\dfrac{b^2 - 8b + 15}{b^2 - 3b - 10}$ $\dfrac{b-3}{b+2}$

11. $\dfrac{4x^2 - 11x - 3}{4x^2 - 7x - 2}$ $\dfrac{x-3}{x-2}$

12. $\dfrac{2x^2 - 21x + 40}{4x^2 - 4x - 15}$ $\dfrac{x-8}{2x+3}$

[6.2] Multiply.

13. $\dfrac{5a^2}{6b} \cdot \dfrac{2}{4a^2b}$ $\dfrac{5}{12b^2}$

14. $\dfrac{15x^2y^3}{3z} \cdot \dfrac{6z^3}{5xy^3}$ $6xz^2$

15. $\dfrac{40a^3b^4}{7c^3} \cdot \dfrac{14c^5}{5a^5b}$ $\dfrac{16b^3c^2}{a^2}$

16. $\dfrac{1}{x - 4} \cdot \dfrac{4 - x}{3}$ $-\dfrac{1}{3}$

17. $\dfrac{-m + 4}{15m} \cdot \dfrac{10m}{m - 4}$ $-\dfrac{2}{3}$

18. $\dfrac{a - 2}{a + 3} \cdot \dfrac{a^2 + 4a + 3}{a^2 - a - 2}$ 1

Divide.

19. $\dfrac{16x^6}{y^2} \div \dfrac{x^4}{4y}$ $\dfrac{64x^2}{y}$

20. $\dfrac{8xy^2}{z} \div \dfrac{x^4y^2}{4z^2}$ $\dfrac{32z}{x^3}$

21. $\dfrac{5a + 5b}{a^2} \div \dfrac{a^2 - b^2}{a^2}$ $\dfrac{5}{a-b}$

22. $\dfrac{1}{a^2 + 8a + 15} \div \dfrac{3}{a + 5}$ $\dfrac{1}{3(a+3)}$

23. $(t + 8) \div \dfrac{t^2 + 5t - 24}{t - 3}$ 1

24. $\dfrac{x^2 + xy - 2y^2}{4y} \div \dfrac{x + 2y}{12y^2}$ $3y(x-y)$

[6.3] Add or subtract.

25. $\dfrac{n}{n+5} - \dfrac{5}{n+5}$ $\dfrac{n-5}{n+5}$

26. $\dfrac{3x}{x+7} + \dfrac{21}{x+7}$ 3

27. $\dfrac{9x-4}{x+8} + \dfrac{76}{x+8}$ 9

28. $\dfrac{7x-3}{x^2+7x-30} - \dfrac{3x+9}{x^2+7x-30}$ $\dfrac{4}{x+10}$

29. $\dfrac{5h^2+12h-1}{h+5} - \dfrac{h^2-5h+14}{h+5}$ $4h-3$

30. $\dfrac{6x^2-4x}{2x-3} - \dfrac{-3x+12}{2x-3}$ $3x+4$

Find the least common denominator for each expression.

31. $\dfrac{a}{10} + \dfrac{4a}{3}$ 30

32. $\dfrac{8}{x+3} + \dfrac{4x}{x+3}$ $x+3$

33. $\dfrac{5}{4xy^3} - \dfrac{7}{10x^2y}$ $20x^2y^3$

34. $\dfrac{6}{x+1} - \dfrac{3x}{x}$ $x(x+1)$

35. $\dfrac{4}{n-5} + \dfrac{2n-3}{n-4}$ $(n-5)(n-4)$

36. $\dfrac{7x-12}{x^2+x} - \dfrac{4}{x+1}$ $x(x+1)$

37. $\dfrac{2r-9}{r-s} - \dfrac{6}{r^2-s^2}$ $(r+s)(r-s)$

38. $\dfrac{4x^2}{x-7} + 8x^2$ $x-7$

39. $\dfrac{19x-5}{x^2+2x-35} + \dfrac{3x-2}{x^2+9x+14}$ $(x+7)(x-5)(x+2)$

[6.4] Add or subtract.

40. $\dfrac{4}{3y^2} + \dfrac{y}{2y}$ $\dfrac{3y^2+8}{6y^2}$

41. $\dfrac{2x}{xy} + \dfrac{1}{5x}$ $\dfrac{10x+y}{5xy}$

42. $\dfrac{5x}{3xy} - \dfrac{4}{x^2}$ $\dfrac{5x^2-12y}{3x^2y}$

43. $6 - \dfrac{2}{x+2}$ $\dfrac{6x+10}{x+2}$

44. $\dfrac{x-y}{y} - \dfrac{x+y}{x}$ $\dfrac{x^2-2xy-y^2}{xy}$

45. $\dfrac{7}{x+4} + \dfrac{4}{x}$ $\dfrac{11x+16}{x(x+4)}$

46. $\dfrac{2}{3x} - \dfrac{3}{3x-6}$ $\dfrac{-x-4}{3x(x-2)}$

47. $\dfrac{3}{(z+5)} + \dfrac{7}{(z+5)^2}$ $\dfrac{3z+22}{(z+5)^2}$

48. $\dfrac{x+2}{x^2-x-6} + \dfrac{x-3}{x^2-8x+15}$ $\dfrac{2x-8}{(x-3)(x-5)}$

[6.2–6.4] Perform each indicated operation.

49. $\dfrac{x+4}{x+6} - \dfrac{x-5}{x+2}$ $\dfrac{5x+38}{(x+6)(x+2)}$

50. $3 + \dfrac{x}{x-4}$ $\dfrac{4x-12}{x-4}$

51. $\dfrac{a+2}{b} \div \dfrac{a-2}{4b^2}$ $\dfrac{4ab+8b}{a-2}$

52. $\dfrac{x+3}{x^2-9} + \dfrac{2}{x+3}$ $\dfrac{3x-3}{(x+3)(x-3)}$

53. $\dfrac{5p+10q}{p^2q} \cdot \dfrac{p^4}{p+2q}$ $\dfrac{5p^2}{q}$

54. $\dfrac{4}{(x+2)(x-3)} - \dfrac{4}{(x-2)(x+2)}$ $\dfrac{4}{(x+2)(x-3)(x-2)}$

55. $\dfrac{x+7}{x^2+9x+14} - \dfrac{x-10}{x^2-49}$ $\dfrac{8x-29}{(x+2)(x-7)(x+7)}$

56. $\dfrac{x-y}{x+y} \cdot \dfrac{xy+x^2}{x^2-y^2}$ $\dfrac{x}{x+y}$

57. $\dfrac{3x^2-27y^2}{20} \div \dfrac{(x-3y)^2}{4}$ $\dfrac{3(x+3y)}{5(x-3y)}$

58. $\dfrac{a^2-9a+20}{a-4} \cdot \dfrac{a^2-8a+15}{a^2-10a+25}$ $a-3$

59. $\dfrac{a}{a^2-1} - \dfrac{2}{3a^2-2a-5}$ $\dfrac{3a^2-7a+2}{(a+1)(a-1)(3a-5)}$

60. $\dfrac{2x^2+6x-20}{x^2-2x} \div \dfrac{x^2+7x+10}{4x^2-16}$ $\dfrac{8x-16}{x}$

[6.5] Simplify each complex fraction.

61. $\dfrac{3+\dfrac{2}{3}}{\dfrac{3}{5}}$ $\dfrac{55}{9}$

62. $\dfrac{1+\dfrac{5}{8}}{4-\dfrac{9}{16}}$ $\dfrac{26}{55}$

63. $\dfrac{\dfrac{12ab}{9c}}{\dfrac{4a}{c^2}}$ $\dfrac{bc}{3}$

64. $\dfrac{\dfrac{36x^4y^2}{9xy^5}}{4z^2}$ $\dfrac{16x^3z^2}{y^3}$

65. $\dfrac{a-\dfrac{a}{b}}{\dfrac{1+a}{b}}$ $\dfrac{ab-a}{a+1}$

66. $\dfrac{r^2+\dfrac{1}{s}}{s^2}$ $\dfrac{r^2s+1}{s^3}$

67. $\dfrac{\dfrac{4}{x}+\dfrac{2}{x^2}}{6-\dfrac{1}{x}}$ $\dfrac{4x+2}{x(6x-1)}$

68. $\dfrac{\dfrac{x}{x+y}}{\dfrac{x^2}{2x+2y}}$ $\dfrac{2}{x}$

69. $\dfrac{\dfrac{3}{x}}{\dfrac{3}{x^2}}$ x

70. $\dfrac{\dfrac{1}{a}+2}{\dfrac{1}{a}+\dfrac{1}{a}}$ $\dfrac{2a+1}{2}$

71. $\dfrac{\dfrac{1}{x^2}-\dfrac{1}{x}}{\dfrac{1}{x^2}+\dfrac{1}{x}}$ $\dfrac{-x+1}{x+1}$

72. $\dfrac{\dfrac{3x}{y}-x}{\dfrac{y}{x}-1}$ $\dfrac{3x^2-x^2y}{y(y-x)}$

[6.6] Solve.

73. $\dfrac{5}{9} = \dfrac{5}{x+3}$ 6

74. $\dfrac{x}{6} = \dfrac{x-4}{2}$ 6

75. $\dfrac{n}{5} + 12 = \dfrac{n}{2}$ 40

76. $\dfrac{3}{x} - \dfrac{1}{6} = \dfrac{1}{x}$ 12

77. $\dfrac{-4}{d} = \dfrac{3}{2} + \dfrac{4-d}{d}$ −16

78. $\dfrac{1}{x-7} + \dfrac{1}{x+7} = \dfrac{1}{x^2-49}$ $\dfrac{1}{2}$

79. $\dfrac{x-3}{x-2} + \dfrac{x+1}{x+3} = \dfrac{2x^2+x+1}{x^2+x-6}$ −6

80. $\dfrac{a}{a^2-64} + \dfrac{4}{a+8} = \dfrac{3}{a-8}$ 28

81. $\dfrac{d}{d-2} - 2 = \dfrac{2}{d-2}$ no solution

[6.7] Solve.

82. *Sandcastles* It takes John and Amy Brogan 6 hours to build a sandcastle. It takes Paul and Cindy Carter 5 hours to make the same sandcastle. How long will it take all four people together to build the sandcastle? $2\frac{8}{11}$ hr

83. *Filling a Pool* A $\frac{3}{4}$-inch-diameter hose can fill a swimming pool in 7 hours. A $\frac{5}{16}$-inch-diameter hose can siphon all the water out of a full pool in 12 hours. How long will it take to fill the pool if while one hose is filling the pool the other hose is siphoning water from the pool? $16\frac{4}{5}$ hr

84. *Sum of Numbers* One number is five times as large as another. The sum of their reciprocals is 6. Determine the numbers. $\frac{1}{5}, 1$

85. *Rollerblading and Motorcycling* Robert Johnston can travel 3 miles on his rollerblades in the same time Tran Lee can travel 8 miles on his mountain bike. If Tran's speed on his bike is 3.5 miles per hour faster than that of Robert on his rollerblades, determine Robert's and Tran's speeds. Robert, 2.1 mph; Tran, 5.6 mph

[6.8]

86. *Drug Dosage* The recommended dosage, d, of the antibiotic drug vancomycin is directly proportional to a person's weight, w. If Carmen Brown, who is 132 pounds is given 2376 milligrams, find the recommended dosage for Bill Glenn, who is 172 pounds. 3096 mg

87. *Runners' Speed* The time, t, it takes a runner to cover a specified distance is inversely proportional to the runner's speed. If Nhat Chung runs at an average of 6 miles per hour, he will finish a race in 1.4 hours. How long will it take Leif Lundgren, who runs at 5 miles per hour, to finish the same race? 1.68 hr

Chapter Practice Test

Simplify.

5. $\dfrac{x^2-6x+9}{(x+3)(x+2)}$

9. $-\dfrac{m+6}{m-5}$

14. $\dfrac{-1}{(x+4)(x-4)}$

1. $\dfrac{-6+x}{x-6}$ 1

2. $\dfrac{x^3-1}{x^2-1}$ $\dfrac{x^2+x+1}{x+1}$

Perform each indicated operation.

3. $\dfrac{15x^2y^3}{4z^2} \cdot \dfrac{8xz^3}{5xy^4}$ $\dfrac{6x^2z}{y}$

4. $\dfrac{a^2-9a+14}{a-2} \cdot \dfrac{a^2-4a-21}{(a-7)^2}$ $a+3$

5. $\dfrac{x^2-x-6}{x^2-9} \cdot \dfrac{x^2-6x+9}{x^2+4x+4}$

6. $\dfrac{x^2-1}{x+2} \cdot \dfrac{2x+4}{2-2x^2}$ −1

7. $\dfrac{x^2-4y^2}{3x+12y} \div \dfrac{x+2y}{x+4y}$ $\dfrac{x-2y}{3}$

8. $\dfrac{15}{y^2+2y-15} \div \dfrac{3}{y-3}$ $\dfrac{5}{y+5}$

9. $\dfrac{m^2+3m-18}{m-3} \div \dfrac{m^2-8m+15}{3-m}$

10. $\dfrac{4x+3}{2y} + \dfrac{2x-5}{2y}$ $\dfrac{3x-1}{y}$

11. $\dfrac{7x^2-4}{x+3} - \dfrac{6x+7}{x+3}$ $\dfrac{7x^2-6x-11}{x+3}$

12. $\dfrac{4}{xy} - \dfrac{3}{xy^3}$ $\dfrac{4y^2-3}{xy^3}$

13. $4 - \dfrac{5z}{z-5}$ $-\dfrac{z+20}{z-5}$

14. $\dfrac{x-5}{x^2-16} - \dfrac{x-2}{x^2+2x-8}$

Simplify.

15. $\dfrac{5 + \dfrac{1}{2}}{3 - \dfrac{1}{5}}$ $\dfrac{55}{28}$

16. $\dfrac{x + \dfrac{x}{y}}{\dfrac{1}{x}}$ $\dfrac{x^2 + x^2 y}{y}$

17. $\dfrac{2 + \dfrac{3}{x}}{\dfrac{2}{x} - 5}$ $\dfrac{2x + 3}{2 - 5x}$

Solve.

18. $6 + \dfrac{2}{x} = 7$ 2

19. $\dfrac{2x}{3} - \dfrac{x}{4} = x + 1$ $-\dfrac{12}{7}$

20. $\dfrac{x}{x - 8} + \dfrac{6}{x - 2} = \dfrac{x^2}{x^2 - 10x + 16}$ 12

Solve.

21. *Working Together* Mr. Jackson, on his tractor, can clear a 1-acre field in 8 hours. Mr. Hackett, on his tractor, can clear a 1-acre field in 5 hours. If they work together, how long will it take them to clear a 1-acre field? $3\frac{1}{13}$ hr

22. *Determine a Number* The sum of a positive number and its reciprocal is 2. Determine the number. 1

23. *Area of Triangle* The area of a triangle is 27 square inches. If the height is 3 inches less than 2 times the base, determine the height and base of the triangle.

24. *Exercising* LaConya Bertrell exercises for $1\frac{1}{2}$ hours each day. During the first part of her routine, she rides a bicycle and averages 10 miles per hour. For the remainder of the time, she rollerblades and averages 4 miles per hour. If the total distance she travels is 12 miles, how far did she travel on the rollerblades?

23. base, 6 in; height, 9 in   **24.** 2 mi

25. *Making Music* The wavelength of sound waves, w, is inversely proportional to the frequency, f (or pitch). If a frequency of 263 cycles per second (middle C on a piano) produces a wavelength of about 4.3 feet, determine the length of a wavelength of a frequency of 1000 cycles per second. ≈ 1.13 ft

Cumulative Review Test

1. 131   **5.** $8x^2 + 5x + 4$ **6.** $6n^3 - 23n^2 + 26n - 15$ **15.** $\dfrac{6x + 2}{(x - 5)(x + 2)(x + 3)}$

Take the following test and check your answers with those that appear at the end of the test. Review any questions that you answered incorrectly. The section and objective where the material was covered is indicated after the answer.

1. Evaluate $3x^2 - 5xy^2 + 3$ when $x = -4$ and $y = -2$.

2. Solve the equation $5z + 4 = -3(z - 7)$. $\frac{17}{8}$

3. Simplify $\left(\dfrac{8x^6 y^3}{2x^5 y^5}\right)^3$. $\dfrac{64x^3}{y^6}$

4. Solve the formula $P = 2E + 3R$ for R. $R = \dfrac{P - 2E}{3}$

5. Simplify $(6x^2 - 3x - 5) - (-2x^2 - 8x - 9)$.

6. Multiply $(3n^2 - 4n + 3)(2n - 5)$.

7. Factor $6a^2 - 6a - 5a + 5$. $(6a - 5)(a - 1)$

8. Factor $13x^2 + 26x - 39$. $13(x + 3)(x - 1)$

9. Evaluate $[7 - [3(8 \div 4)]^2 + 9 \cdot 4]^2$. 49

10. Solve $2(x + 3) \le -(x + 5) - 1$ and graph the solution on a real number line. $x \le -4$, ![number line]

11. Divide $\dfrac{4x - 34}{8}$. $\dfrac{1}{2}x - \dfrac{17}{4}$

12. Solve $2x^2 = 11x - 12$. $4, \frac{3}{2}$

13. Multiply $\dfrac{x^2 - 9}{x^2 - x - 6} \cdot \dfrac{x^2 - 2x - 8}{2x^2 - 7x - 4}$. $\dfrac{x + 3}{2x + 1}$

14. Subtract $\dfrac{r}{r + 2} - \dfrac{6}{r - 5}$. $\dfrac{r^2 - 11r - 12}{(r + 2)(r - 5)}$

15. Add $\dfrac{4}{x^2 - 3x - 10} + \dfrac{2}{x^2 + 5x + 6}$.

16. Solve the equation $\dfrac{x}{9} - \dfrac{x}{6} = \dfrac{1}{12}$. $-\dfrac{3}{2}$

17. Solve the equation $\dfrac{7}{x + 3} + \dfrac{5}{x + 2} = \dfrac{5}{x^2 + 5x + 6}$. no solution

18. **Medical Plans** A school district allows its employees to choose from two medical plans. With plan 1, the employee pays 10% of all medical bills (the school district pays the balance). With plan 2, the employee pays the school district a one-time payment of $100, then the employee pays 5% of all medical bills. What total medical bills would result in the employee paying the same amount with the two plans? $2000

19. **Bird Seed** A feed store owner wishes to make his own store-brand mixture of bird seed by mixing sunflower seed that costs $0.50 per pound with a premixed assorted seed that costs $0.15 per pound. How many pounds of each will he have to use to make a 50-pound mixture that will cost $14.50?

20 lb sunflower seed, 30 lb premixed assorted seed mix

20. **Sailing** During the first leg of a race, the sailboat *Thumper* sailed at an average speed of 6.5 miles per hour. During the second leg of the race, the winds increased and *Thumper* sailed at an average speed of 9.5 miles per hour. If the total distance sailed by *Thumper* was 12.75 miles, and the total time spent racing was 1.5 hours, determine the distance traveled by *Thumper* on each leg of the race.

1st leg: 3.25 miles; 2nd leg: 9.5 miles

Answers to Cumulative Review Test

1. 131; [Sec. 1.9, Obj. 6]   **2.** $\frac{17}{8}$; [Sec. 2.5, Obj. 1]   **3.** $\frac{64x^3}{y^6}$; [Sec. 4.1, Obj. 3]   **4.** $R = \frac{P - 2E}{3}$; [Sec. 3.1, Obj. 3]

5. $8x^2 + 5x + 4$; [Sec. 4.4, Obj. 3]   **6.** $6n^3 - 23n^2 + 26n - 15$; [Sec. 4.5, Obj. 6]   **7.** $(6a - 5)(a - 1)$; [Sec. 5.2, Obj. 1]

8. $13(x + 3)(x - 1)$; [Sec. 5.3, Obj. 2]   **9.** 49; [Sec. 1.9, Obj. 5]   **10.** $x \le -4$, ⟵————●———⟶ ; [Sec. 2.7, Obj. 1]
-4

11. $\frac{1}{2}x - \frac{17}{4}$; [Sec. 4.6, Obj. 1]   **12.** $4, \frac{3}{2}$; [Sec. 5.6, Obj. 2]   **13.** $\frac{x + 3}{2x + 1}$; [Sec. 6.2, Obj. 1]   **14.** $\frac{r^2 - 11r - 12}{(r + 2)(r - 5)}$; [Sec. 6.4, Obj. 1]

15. $\frac{6x + 2}{(x - 5)(x + 2)(x + 3)}$; [Sec. 6.4, Obj. 1]   **16.** $-\frac{3}{2}$; [Sec. 6.6, Obj. 1]   **17.** No solution; [Sec. 6.6, Obj. 2]

18. $2000; [Sec. 3.3, Obj. 4]   **19.** 20 pounds sunflower seed, 30 pounds premixed assorted seed mix; [Sec. 3.5, Obj. 4]   **20.** First leg: 3.25 miles, second leg: 9.5 miles; [Sec. 3.5, Obj. 2]

Chapter 7

Graphing Linear Equations

There are many situations in which a quantity is composed of a fixed amount and a variable amount. On pages 477 and 478, Example 5, we discuss a pottery artist who makes and sells ceramic vases at art fairs. Certain costs involved with making the vases are fixed, and others depend upon the quantity of goods made. Also, consider renting a truck when you pay a fixed amount plus a charge per mile driven. Graphs of linear equations can be used to estimate quantities such as total cost of making vases or of renting a truck. Graphs are used to display various types of information and are used a great deal in industry.

SSM   Study Guide   CD/Video   MathPro 4/5   PH Math Tutor Center   prenhall.com/Angel

A Look Ahead

In this chapter you will learn how to graph linear equations. The graphs of linear equations are straight lines. Graphing is one of the most important topics in mathematics, and each year its importance increases. Graphs are also used in many professions and industries. They are used to display information and to make projections about future trends.

In Section 7.1 we introduce the Cartesian coordinate system and explain how to plot points. In Section 7.2 we discuss two methods for graphing linear equations: by plotting points and by using the x- and y-intercepts. The slope of a line, which is a measure of its steepness, is discussed in Section 7.3. In Section 7.4 we discuss a third procedure, using slope, for graphing a linear equation.

We solved inequalities in one variable in Section 2.7. In Section 7.5 we will solve and graph linear inequalities in two variables. Graphing linear inequalities is an extension of graphing linear equations.

Functions are a unifying concept in mathematics. In Section 7.6 we give a brief and somewhat informal introduction to functions. Functions will be discussed in much more depth in later mathematics courses.

This chapter contains concepts that are central to mathematics. If you plan on taking another mathematics course, graphs and functions will probably be a significant part of that course.

TEACHING TIP
Suggest that students bring a ruler and graph paper to class. Show them a sample of the kind of graph paper they should buy.

7.1 THE CARTESIAN COORDINATE SYSTEM AND LINEAR EQUATIONS IN TWO VARIABLES

SSM   Study Guide   CD/Video

1 Plot points in the Cartesian coordinate system.
2 Determine whether an ordered pair is a solution to a linear equation.

MathPro 4/5   PH Math Tutor Center   prenhall.com/Angel

1 Plot Points in the Cartesian Coordinate System

René Descartes

Many algebraic relationships are easier to understand if we can see a picture of them. A **graph** shows the relationship between two variables in an equation. In this chapter we discuss several procedures that can be used to draw graphs using the **Cartesian (or rectangular) coordinate system**. The Cartesian coordinate system is named for its developer, the French mathematician and philosopher René Descartes (1596–1650).

The Cartesian coordinate system provides a means of locating and identifying points just as the coordinates on a map help us find cities and other locations. Consider the map of the Great Smoky Mountains (see Fig. 7.1). Can you find Cades Cove on the map? If we tell you that it is in grid A3, you can probably find it much more quickly and easily.

The Cartesian coordinate system is a grid system, like that of a map, except that it is formed by two axes (or number lines) drawn perpendicular to each other. The two intersecting axes form four **quadrants**, numbered I through IV in Fig. 7.2.

The horizontal axis is called the **x-axis**. The vertical axis is called the **y-axis**. The point of intersection of the two axes is called the **origin**. At the origin the value of x is 0 and the value of y is 0. Starting from the origin and moving to the right along the x-axis, the numbers increase (Fig. 7.3). Starting from the origin and moving to the left, the numbers decrease. Starting from the origin and moving up the y-axis, the numbers increase. Starting from the origin and moving down, the numbers decrease.

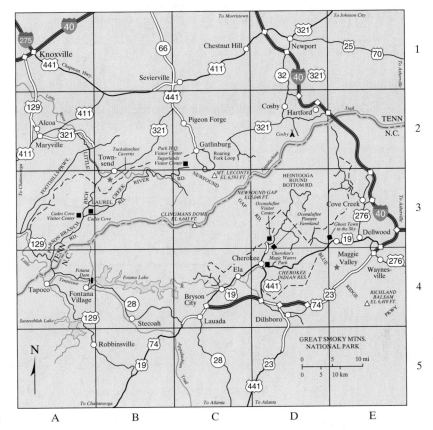

FIGURE 7.1

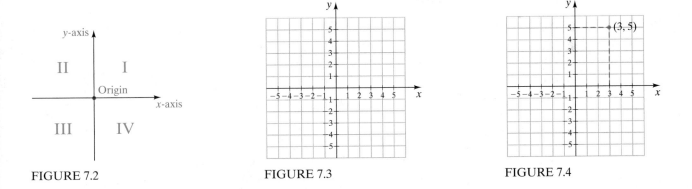

FIGURE 7.2

FIGURE 7.3

FIGURE 7.4

To locate a point, it is necessary to know both the value of x and the value of y, or the **coordinates**, of the point. When the x- and y-coordinates of a point are placed in parentheses, *with the x-coordinate listed first*, we have an **ordered pair**. In the ordered pair $(3, 5)$ the x-coordinate is 3 and the y-coordinate is 5. The point corresponding to the ordered pair $(3, 5)$ is plotted in Figure 7.4. The phrase "the point corresponding to the ordered pair $(3, 5)$" is often abbreviated "the point $(3, 5)$." For example, if we write "the point $(-1, 2)$," it means "the point corresponding to the ordered pair $(-1, 2)$."

EXAMPLE 1 Plot (or mark) each point on the same axes.

a) $A(5, 3)$ **b)** $B(2, 4)$ **c)** $C(-3, 1)$

d) $D(4, 0)$ **e)** $E(-2, -5)$ **f)** $F(0, -3)$

g) $G(0, 2)$ **h)** $H\left(6, -\frac{9}{2}\right)$ **i)** $I\left(-\frac{3}{2}, -\frac{5}{2}\right)$

Solution The first number in each ordered pair is the *x*-coordinate and the second number is the *y*-coordinate. The points are plotted in Figure 7.5.

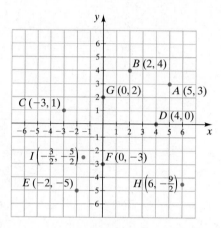

FIGURE 7.5

Note that when the *x*-coordinate is 0, as in Example 1 **f)** and 1 **g)**, the point is on the *y*-axis. When the *y*-coordinate is 0, as in Example 1 **d)**, the point is on the *x*-axis.

NOW TRY EXERCISE 29

EXAMPLE 2 List the ordered pairs for each point shown in Figure 7.6.

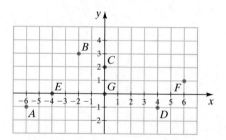

FIGURE 7.6

Solution Remember to give the *x*-value first in the ordered pair.

Point	Ordered Pair
A	$(-6, -1)$
B	$(-2, 3)$
C	$(0, 2)$
D	$(4, -1)$
E	$(-4, 0)$
F	$(6, 1)$
G	$(0, 0)$

NOW TRY EXERCISE 27

2 Determine Whether an Ordered Pair Is a Solution to a Linear Equation

In Section 7.2 we will learn to graph linear equations in two variables. Below we explain how to identify a linear equation in two variables.

> A **linear equation in two variables** is an equation that can be put in the form
>
> $$ax + by = c$$
>
> where a, b, and c are real numbers.

The graphs of equations of the form $ax + by = c$ are straight lines. For this reason such equations are called linear. Linear equations may be written in various forms, as we will show later. A linear equation in the form $ax + by = c$ is said to be in **standard form**.

Examples of Linear Equations

$$4x - 3y = 12$$
$$y = 5x + 3$$
$$x - 3y + 4 = 0$$

Note in the examples that only the equation $4x - 3y = 12$ is in standard form. However, the bottom two equations can be written in standard form, as follows:

$$y = 5x + 3 \qquad\qquad x - 3y + 4 = 0$$
$$-5x + y = 3 \qquad\qquad x - 3y = -4$$

Most of the equations we have discussed thus far have contained only one variable. Exceptions to this include formulas used in application sections. Consider the linear equation in *one* variable, $2x + 3 = 5$. What is its solution?

$$2x + 3 = 5$$
$$2x = 2$$
$$x = 1$$

This equation has only one solution, 1.

Check

$$2x + 3 = 5$$
$$2(1) + 3 \overset{?}{=} 5$$
$$5 = 5 \quad \textit{True}$$

Now consider the linear equation in *two* variables, $y = x + 1$. What is the solution? Since the equation contains two variables, its solutions must contain two numbers, one for each variable. One pair of numbers that satisfies this equation is $x = 1$ and $y = 2$. To see that this is true, we substitute both values into the equation and see that the equation checks.

Check

$$y = x + 1$$
$$2 \overset{?}{=} 1 + 1$$
$$2 = 2 \quad \textit{True}$$

We write this answer as an ordered pair by writing the x- and y-values within parentheses separated by a comma. Remember that the x-value is always listed first because the form of an ordered pair is (x, y). Therefore, one possible solution to this equation is the ordered pair $(1, 2)$. The equation $y = x + 1$ has other possible solutions. Below we show three other solutions and show their checks.

Solution	Solution	Solution
$x = 2, y = 3$	$x = -3, y = -2$	$x = -\dfrac{1}{3}, y = \dfrac{2}{3}$

Check

$$y = x + 1 \qquad\qquad y = x + 1 \qquad\qquad y = x + 1$$

$$3 \overset{?}{=} 2 + 1 \qquad\qquad -2 \overset{?}{=} -3 + 1 \qquad\qquad \frac{2}{3} \overset{?}{=} -\frac{1}{3} + 1$$

$$3 = 3 \quad \textit{True} \qquad\qquad -2 = -2 \quad \textit{True} \qquad\qquad \frac{2}{3} = \frac{2}{3} \quad \textit{True}$$

Solution Written as an Ordered Pair

$$(2, 3) \qquad\qquad (-3, -2) \qquad\qquad \left(-\frac{1}{3}, \frac{2}{3}\right)$$

TEACHING TIP
After discussing several solutions to the equation $y = x + 1$ ask, Is there a relationship between the x-and y-values? Be sure they notice that the y-value is always 1 more than the x-value. Point out that the equation concisely describes this relationship. Then have them guess another solution.

How many possible solutions does the equation $y = x + 1$ have? The equation $y = x + 1$ has an unlimited or *infinite number* of possible solutions. Since it is not possible to list all the specific solutions, the solutions are illustrated with a graph.

DEFINITION

A **graph** of an equation is an illustration of a set of points whose coordinates satisfy the equation.

Figure 7.7a shows the points $(2, 3)$, $(-3, -2)$, and $\left(-\frac{1}{3}, \frac{2}{3}\right)$ plotted in the Cartesian coordinate system. Figure 7.7b shows a straight line drawn through the three points. Arrowheads are placed at the ends of the line to show that the line continues in both directions. Every point on this line will satisfy the equation $y = x + 1$, so this graph illustrates all the solutions of $y = x + 1$. The ordered pair $(1, 2)$, which is on the line, also satisfies the equation.

In Figure 7.7b, what do you notice about the points $(2, 3)$, $(1, 2)$, $\left(-\frac{1}{3}, \frac{2}{3}\right)$, and $(-3, -2)$? You probably noticed that they are in a straight line. A set of points that are in a straight line are said to be **collinear**. *In Section 7.2 when you graph linear equations by plotting points, the points you plot should all be collinear.*

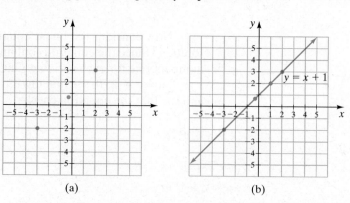

FIGURE 7.7   (a)   (b)

EXAMPLE 3 Determine whether the three points given appear to be collinear.

a) $(2, 7)$, $(0, 3)$, and $(-2, -1)$

b) $(0, 5)$, $\left(\dfrac{5}{2}, 0\right)$, and $(5, -5)$

c) $(-2, -5)$, $(0, 1)$, and $(5, 8)$

Solution   We plot the points to determine whether they appear to be collinear. The solution is shown in Figure 7.8.

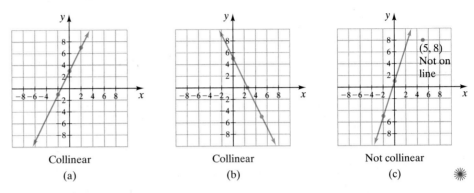

FIGURE 7.8

Collinear	Collinear	Not collinear
(a)	(b)	(c)

NOW TRY EXERCISE 33

To graph an equation, you will need to determine ordered pairs that satisfy the equation and then plot the points.

How many points do you need to graph a linear equation? As mentioned earlier, *the graph of every linear equation of the form ax + by = c will be a straight line.* Since only two points are needed to draw a straight line, only two points are needed to graph a linear equation. However, it is always a good idea to plot at least three points. See the Helpful Hint that follows.

HELPFUL HINT

Only two points are needed to graph a linear equation because the graph of every linear equation is a straight line. However, if you graph a linear equation using only two points and you have made an error in determining or plotting one of those points, your graph will be wrong and you will not know it. In Figures 7.9a and b we plot only two points to show that if only one of the two points plotted is incorrect, the graph will be wrong. In both Figures 7.9a and b we use the ordered pair $(-2, -2)$. However, in Figure 7.9a the second point is $(1, 2)$, while in Figure 7.9b the second point is $(2, 1)$. Notice how the two graphs differ.

If you use at least three points to plot your graph, as in Figure 7.7 b) on page 444, and they appear to be collinear, you probably have not made a mistake.

TEACHING TIP
Have students draw a picture to answer the following questions. "How many lines will go through any 1 point? How many lines will go through any 2 points? How many lines will go through any 3 points?"

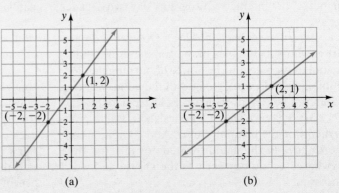

FIGURE 7.9   (a)   (b)

EXAMPLE 4

a) Determine which of the following ordered pairs satisfy the equation $2x + y = 4$.

$$(2, 0), (0, 4), (3, 3), (-1, 6)$$

b) Plot all the points that satisfy the equation on the same axes and draw a straight line through the points.

c) What does this straight line represent?

Solution   **a)** We substitute values for x and y into the equation $2x + y = 4$ and determine whether they check.

Check

$(2, 0)$
$$2x + y = 4$$
$$2(2) + 0 \overset{?}{=} 4$$
$$4 = 4 \quad \textit{True}$$

$(0, 4)$
$$2x + y = 4$$
$$2(0) + 4 \overset{?}{=} 4$$
$$4 = 4 \quad \textit{True}$$

$(3, 3)$
$$2x + y = 4$$
$$2(3) + 3 \overset{?}{=} 4$$
$$9 = 4 \quad \textit{False}$$

$(-1, 6)$
$$2x + y = 4$$
$$2(-1) + 6 \overset{?}{=} 4$$
$$4 = 4 \quad \textit{True}$$

The ordered pairs $(2, 0)$, $(0, 4)$, and $(-1, 6)$ satisfy the equation. The ordered pair $(3, 3)$ does not satisfy the equation.

b) Figure 7.10 shows the three points that satisfy the equation. A straight line drawn through the three points shows that they appear to be collinear.

c) The straight line represents all solutions of $2x + y = 4$. The coordinates of every point on this line satisfy the equation $2x + y = 4$. ✺

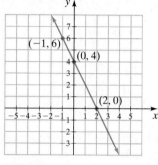

FIGURE 7.10

Mathematics in Action

René Descartes and the Computer Mouse

It is difficult these days to imagine a computer without a pointing device like a mouse or its first cousin a trackball. In the early 1950s, computers were considered marvelously rapid calculators that could be given problems, one batch at a time, and solve them in the same batched fashion. Computers were room-sized giants tended to by technicians clad in white lab coats. Implementations of concepts like "personal computer" or "interactive programs" lay two decades into the future.

Even that far back, however, Douglas C. Englebart was nursing the idea of the computer as a thinking tool to supplement the power of the human mind, a tool that would work along with the mind. Working at Stanford University with others, he created the NLS (oN Line System), which emphasized a visual environment for the computer user. At a computer conference in 1968, he demonstrated, for the first time anywhere, technology such as the mouse and word processing.

The mouse has evolved since 1968, with optical sensors replacing mechanical rollers in many models, but the underlying concept of the mouse and almost all computer pointing devices has remained the same. The position of a cursor on the screen follows the Cartesian coordinates tracked by the pointing device. No matter what direction you move the mouse, it processes the move as a combination of so many units along an x-axis and so many units along a y-axis. The circuitry of the mouse, in combination with the mouse's own software programs, "knows" where to place the cursor on the screen from instant to instant, even if you lift the mouse into the air and place it at some other spot on a mouse pad.

The story goes that René Descartes (1596–1650) thought of his coordinate system while observing a fly crawl around on the ceiling. He noticed that the path of the fly could be described by the fly's distances from each of the walls. If so, we all have that fly to thank for the way of organizing and tracking numbers that is built into the computer mouse, and for the unaccountable acts of creativity that tool has made possible.

Using Your Graphing Calculator

Some of you may have graphing calculators. In this chapter we will give a number of Using Your Graphing Calculator boxes with some information on using your calculator. Because the instructions will be general, you may need to refer to your calculator manual for more specific instructions. The keystrokes you will use will depend on the brand and model of your calculator. The keystrokes and graphing calculator screens we show in this book are from the Texas Instruments TI-83 Plus calculator.

A primary use of a graphing calculator is to graph equations. A graphing calculator *window* is the rectangular screen in which a graph is displayed. Figure 7.11 shows a calculator window with labels added. Figure 7.12 shows the meaning of the information given in Figure 7.11. These are the *standard window settings* for a graphing calculator screen.

The *x*-axis on the standard window goes from -10 (the minimum value of *x*, Xmin) to 10 (the maximum value of *x*, Xmax) with a scale of 1. Therefore, each tick mark represents 1 unit (Xscl = 1). The *y*-axis goes from -10 (the minimum value of *y*, Ymin) to 10 (the maximum value of *y*, Ymax) with a scale of 1 (Yscl = 1). The numbers below the graph in Figure 7.11 indicate, in order, the window settings: Xmin, Xmax, Xscl, Ymin, Ymax, Yscl. *When no settings are shown below a graph, always assume the standard window setting is used.* Since the window is rectangular, the distance between tick marks on the standard window is greater on the *x*-axis than on the *y*-axis.

When graphing, you will often need to change the window settings. Read your graphing calculator manual to learn how to change the window settings. On a TI-83 Plus you press the [WINDOW] key and then change the settings.

Now, turn on your calculator and press the [WINDOW] key. If necessary, adjust the window so that it looks like the window in Figure 7.12. Use the [(−)] key, if necessary, to make negative numbers. (You can also obtain the standard window settings on a TI-83 Plus by pressing the [ZOOM] key and then pressing option 6, ZStandard.) Next press the [GRAPH] key. Your screen should resemble the screen in Figure 7.11 (without the labels that were added). Now press the [WINDOW] key again. Then use the appropriate keys to change the window setting so it is the same as that shown in Figure 7.13.

Now press the [GRAPH] key again. You should get the screen shown in Figure 7.14. In Figure 7.14, the *x*-axis starts at 0 and goes to 50, and each tick mark represents 5 units (represented by the first 3 numbers under the window). The *y*-axis starts at 0 and goes to 100, and each tick mark represents 10 units (represented by the last 3 numbers under the window).

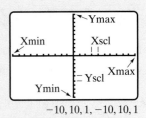

−10, 10, 1, −10, 10, 1

FIGURE 7.11

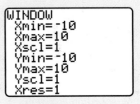

FIGURE 7.12

FIGURE 7.13

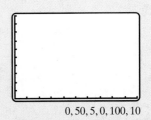

0, 50, 5, 0, 100, 10

FIGURE 7.14

Exercises

For Exercises 1 and 2, set your window to the values shown. Then use the [GRAPH] *key to show the axes formed.*

1. Xmin = −20, Xmax = 40, Xscl = 5,
   Ymin = −10, Ymax = 60, Yscl = 10   **1.** and **2.** See graphing answer section, page G1.

2. Xmin = −200, Xmax = 400, Xscl = 100,
   Ymin = −500, Ymax = 1000, Yscl = 200

(continued on the next page)

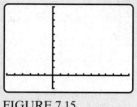

3. Consider the screen in Figure 7.15. The numbers under the window have been omitted. If Xmin = −300 and Xmax = 500, find Xscl. Explain how you determined your answer. 50

4. In Figure 7.15, if Ymin = −200 and Ymax = 1000, find Yscl. Explain how you determined your answer. 100

FIGURE 7.15

Exercise Set 7.1

Concept/Writing Exercises

4. the point of intersection of the two axes 5. Axis is singular, axes is plural. 6. Place the point directly above the −2 on the x-axis and directly to the left of 4 on the y-axis.

1. In an ordered pair, which coordinate is always listed first? the x-coordinate

2. What is another name for the Cartesian coordinate system? rectangular coordinate system

3. a) Is the *horizontal axis* the x- or y-axis in the Cartesian coordinate system? x-axis

b) Is the *vertical axis* the x- or y-axis? y-axis

4. What is the *origin* in the Cartesian coordinate system?

5. We can refer to the *x-axis* and we can refer to the *y-axis*. We can also refer to the x- and *y-axes*. Explain when we use the word *axis* and when we use the word *axes*.

6. Explain how to plot the point (−2, 4) in the Cartesian coordinate system.

7. What does the graph of a linear equation illustrate?

8. Why are arrowheads added to the ends of graphs of linear equations?

9. a) How many points are needed to graph a linear equation? two

b) Why is it always a good idea to use three or more points when graphing a linear equation? to catch errors

10. What will the graph of a linear equation look like?

11. What is the standard form of a linear equation?

12. When graphing linear equations, the points that are plotted should all be *collinear*. Explain what this means.

13. In the Cartesian coordinate system there are four quadrants. Draw the x- and y-axes and mark the four quadrants, I through IV, on your axes.

14. How many solutions does a linear equation in two variables have? an infinite number

7. an illustration of a set of points whose coordinates satisfy the equation 8. to show that the line extends in both directions
10. a straight line 11. $ax + by = c$ 12. They lie in a straight line.
13. See graphing answer section, page G1.

Practice the Skills

Indicate the quadrant in which each of the points belongs.

15. (−4, 2) II
16. (−3, 6) II
17. (5, −2) IV
18. (5, −3) IV
19. (−8, 5) II
20. (5, 30) I
21. (−16, −87) III
22. (63, −47) IV
23. (−124, −132) III
24. (75, −200) IV
25. (−8, 42) II
26. (76, −92) IV

27. List the ordered pairs corresponding to each point.

$A(3, 1)$
$B(-3, 0)$
$C(1, -3)$
$D(-2, -3)$
$E(0, 3)$
$F\left(\frac{3}{2}, -1\right)$

28. List the ordered pairs corresponding to each point.

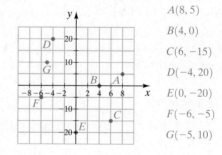

$A(8, 5)$
$B(4, 0)$
$C(6, -15)$
$D(-4, 20)$
$E(0, -20)$
$F(-6, -5)$
$G(-5, 10)$

Plot each point on the same axes.

29. $A(3, 2), B(-4, 1), C(0, -3), D(-2, 0), E(-3, -4), F\left(-4, -\frac{5}{2}\right)$ 29. and 30. See graphing answer section, page G1.

30. $A(-3, -1), B(2, 0), C(3, 2), D\left(\frac{1}{2}, -4\right), E(-4, 2), F(0, 4)$

31. $A(4, 0), B(-1, 3), C(2, 4), D(0, -2), E(-3, -3), F(2, -3)$ **31.** and **32.** See graphing answer section, page G1.

32. $A(-3, 4), B(2, 3), C(0, 3), D(-1, 0), E(-2, -2), F(2, -4)$

Plot the following points. Then determine whether they appear to be collinear.

33. $A(1, -1), B(5, 3), C(-3, -5), D(0, -2), E(2, 0)$ yes

34. $A(1, -2), B(0, -5), C(3, 1), D(-1, -8), E\left(\frac{1}{2}, -\frac{7}{2}\right)$
    (3, 1) not on line

35. $A(1, 5), B\left(-\frac{1}{2}, \frac{1}{2}\right), C(0, 2), D(-5, -3), E(-2, -4)$
    $(-5, -3)$ not on line

36. $A(1, -1), B(3, 5), C(0, -3), D(-2, -7), E(2, 1)$
    (3, 5) not on line

In Exercises 37–42, **a)** *determine which of the four ordered pairs does not satisfy the given equation.* **b)** *Plot all the points that satisfy the equation on the same axes and draw a straight line through the points.*   **37.–42. b)** See graphing answer section, page G1.

37. $y = x + 2$,   **a)** $(2, 4)$ **b)** $(-2, 0)$ **c)** $(2, 3)$ **d)** $(0, 2)$ **a)** not **c)**

38. $2x + y = -4$,   **a)** $(-2, 0)$ **b)** $(-2, 1)$ **c)** $(0, -4)$ **d)** $(-1, -2)$ **a)** not **b)**

39. $3x - 2y = 6$,   **a)** $(4, 0)$ **b)** $(2, 0)$ **c)** $\left(\frac{2}{3}, -2\right)$ **d)** $\left(\frac{4}{3}, -1\right)$ **a)** not **a)**

40. $4x - 3y = 0$,   **a)** $(3, 4)$ **b)** $(-3, -4)$ **c)** $(0, 0)$ **d)** $(3, -4)$ **a)** not **d)**

41. $\frac{1}{2}x + 4y = 4$,   **a)** $(2, -1)$ **b)** $\left(2, \frac{3}{4}\right)$ **c)** $(0, 1)$ **d)** $\left(-4, \frac{3}{2}\right)$ **a)** not **a)**

42. $y = \frac{1}{2}x + 2$,   **a)** $(0, 2)$ **b)** $(2, 0)$ **c)** $(-2, 1)$ **d)** $(4, 4)$ **a)** not **b)**

Problem Solving

Consider the linear equation $y = 3x - 4$. In Exercises 43–46, find the value of y that makes the given ordered pair a solution to the equation.

43. $(2, y)$ 2   **44.** $(-1, y)$ -7 **45.** $(0, y)$ -4 **46.** $(3, y)$ 5

Consider the linear equation $2x + 3y = 12$. In Exercises 47–50, find the value of y that makes the given ordered pair a solution to the equation.

47. $(3, y)$ 2   **48.** $(0, y)$ 4   **49.** $\left(\frac{1}{2}, y\right)$ $\frac{11}{3}$ **50.** $(-5, y)$ $\frac{22}{3}$

51. What is the value of *y* at the point where a straight line crosses the *x*-axis? Explain.   0

52. What is the value of *x* at the point where a straight line crosses the *y*-axis? Explain.   0

53. **Longitude and Latitude** Another type of coordinate system that is used to identify a location or position on Earth's surface involves *latitude* and *longitude*. On a globe, the longitudinal lines are lines that go from top to bottom; on a world map they go up and down. The latitudinal lines go around the globe, or left to right on a world map. The locations of Hurricane Georges and Tropical Storm Hermine are indicated on the map on the right.

a) Estimate the latitude and longitude of Hurricane Georges.   lat., 16°N; long., 56°W

b) Estimate the latitude and longitude of Tropical Storm Hermine.   lat., 29°N; long., 90.5°W

c) Estimate the latitude and longitude of the city of Miami.   lat., 26°N; long., 80.5°W

d) Use either a map or a globe to estimate the latitude and longitude of your college.   Answers will vary.

Source: National Weather Service

Group Activity

*In Section 7.2 we discuss how to find ordered pairs to plot when graphing linear equations. Let's see if you can draw some graphs now. Individually work parts **a)** through **c)** in Exercises 54–56.*

a) *Select any three values for x and find the corresponding values of y.*

b) *Plot the points (they should appear to be collinear).*

c) *Draw the graph.*

d) *As a group, compare your answers. You should all have the same lines.*

54. $y = x$ **55.** $y = 2x$ **56.** $y = x + 1$

54.–56. See graphing answer section, page G1.

Cumulative Review Exercises

Answer each question.   **58.** $y = \dfrac{2x - 6}{5} = \dfrac{2}{5}x - \dfrac{6}{5}$

[2.5] **57.** Solve the equation $\dfrac{1}{2}(x - 3) = \dfrac{1}{3}x + 2$.   21

[3.1] **58.** Solve the equation $2x - 5y = 6$ for y.

[4.1] **59.** Simplify $(2x^4)^3$. $8x^{12}$

[5.3] **60.** Factor $x^2 - 6x - 27$. $(x + 3)(x - 9)$

[5.6] **61.** Solve $y(y - 8) = 0$. $0, 8$

[6.4] **62.** Add $\dfrac{4}{x^2} + \dfrac{7}{3x}$. $\dfrac{7x + 12}{3x^2}$

7.2 GRAPHING LINEAR EQUATIONS

 SSM   Study Guide   CD/Video

 MathPro 4/5   PH Math Tutor Center   prenhall.com/Angel

1   Graph linear equations by plotting points.

2   Graph linear equations of the form $ax + by = 0$.

3   Graph linear equations using the x- and y-intercepts.

4   Graph horizontal and vertical lines.

5   Study applications of graphs.

In Section 7.1 we explained the Cartesian coordinate system, how to plot points, and how to recognize linear equations in two variables. Now we are ready to graph linear equations. *In this section we discuss two methods that can be used to graph linear equations: (1) graphing by plotting points and (2) graphing using the x- and y-intercepts.* In Section 7.4 we discuss graphing using the slope and the y-intercept.

1   Graph Linear Equations by Plotting Points

Graphing by plotting points is the most versatile method of graphing because we can also use it to graph second- and higher-degree equations. In Chapter 10, we will graph quadratic equations, which are second-degree equations, by plotting points.

To Graph Linear Equations by Plotting Points

1. Solve the linear equation for the variable y. That is, get the variable y by itself on the left side of the equal sign.

2. Select a value for the variable x. Substitute this value in the equation for x and find the corresponding value of y. Record the ordered pair (x, y).

3. Repeat step 2 with two different values of x. This will give you two additional ordered pairs.

(continued on the next page)

4. Plot the three ordered pairs. The three points should appear to be collinear. If they do not, recheck your work for mistakes.

5. With a straightedge, draw a straight line through the three points. Draw arrowheads on each end of the line to show that the line continues indefinitely in both directions.

In Step 1 you are asked to solve the equation for y. Although it is not necessary to do this to graph the equation, it can provide insight as to which values to select for the variable in step 2. If you have forgotten how to solve the equation for y, review Section 3.1. When selecting values in step 2, you should select integer values of x that result in integer values of y if possible. Also, you should select values of x that are small enough so that the ordered pairs obtained can be plotted on the axes. Since y is often easy to find when $x = 0$, 0 is always a good value to select for x.

EXAMPLE 1 Graph the equation $y = 3x + 6$.

Solution First we determine that this is a linear equation. Its graph must therefore be a straight line. The equation is already solved for y. We select three values for x, substitute them in the equation, and find the corresponding values for y. We will arbitrarily select the values $-2, 0,$ and 1 for x. The calculations that follow show that when $x = -2, y = 0$, when $x = 0, y = 6$, and when $x = 1, y = 9$.

x	$y = 3x + 6$	Ordered Pair
-2	$y = 3(-2) + 6 = 0$	$(-2, 0)$
0	$y = 3(0) + 6 = 6$	$(0, 6)$
1	$y = 3(1) + 6 = 9$	$(1, 9)$

x	y
-2	0
0	6
1	9

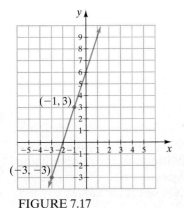

FIGURE 7.16

It is convenient to list the x- and y-values in a table. Then we plot the three ordered pairs on the same axes (Fig. 7.16).

Since the three points appear to be collinear, the graph appears correct. Connect the three points with a straight line and place arrowheads at the ends of the line to show that the line continues infinitely in both directions.

NOW TRY EXERCISE 25

To graph the equation $y = 3x + 6$, we arbitrarily used the three values $x = -2, x = 0,$ and $x = 1$. We could have selected three entirely different values and obtained exactly the same graph. When selecting values to substitute for x, use values that make the equation easy to evaluate.

The graph drawn in Example 1 represents the set of *all* ordered pairs that satisfy the equation $y = 3x + 6$. If we select any point on this line, the ordered pair represented by that point will be a solution to the equation $y = 3x + 6$. Similarly, any solution to the equation will be represented by a point on the line. Let's select some points on the line, say, $(-1, 3)$ and $(-3, -3)$, and verify that they are solutions to the equation (Fig. 7.17).

FIGURE 7.17

Check $(-1, 3)$

$$y = 3x + 6$$
$$3 \overset{?}{=} 3(-1) + 6$$
$$3 \overset{?}{=} -3 + 6$$
$$3 = 3 \quad \textit{True}$$

Check $(-3, -3)$

$$y = 3x + 6$$
$$-3 \overset{?}{=} 3(-3) + 6$$
$$-3 \overset{?}{=} -9 + 6$$
$$-3 = -3 \quad \textit{True}$$

Remember, a graph of an equation is an illustration of the set of points whose coordinates satisfy the equation.

EXAMPLE 2 Graph $3y = 5x - 6$

Solution We begin by solving the equation for y. This will help us in selecting values to use for x. To solve the equation for y, we divide both sides of the equation by 3.

$$3y = 5x - 6$$

$$y = \frac{5x - 6}{3}$$

$$y = \frac{5x}{3} - \frac{6}{3}$$

$$y = \frac{5}{3}x - 2$$

Now we can see that if we select values for x that are multiples of 3, the values we obtain for y will be integers. Let's select the values -3, 0, and 3 for x.

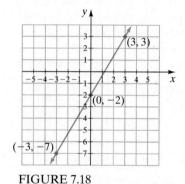

FIGURE 7.18

$$y = \frac{5}{3}x - 2$$

Let $x = -3$ $\quad y = \frac{5}{3}(-3) - 2 = -5 - 2 = -7$

Let $x = 0$ $\quad y = \frac{5}{3}(0) - 2 = -2$

Let $x = 3$ $\quad y = \frac{5}{3}(3) - 2 = 5 - 2 = 3$

x	y
-3	-7
0	-2
3	3

NOW TRY EXERCISE 31   Finally, we plot the points and draw the straight line (Figure 7.18).

2 Graph Linear Equations of the Form $ax + by = 0$

In Example 3 we graph an equation of the form $ax + by = 0$, which is a linear equation whose constant is 0.

EXAMPLE 3 Graph the equation $2x + 5y = 0$.

Solution We begin by solving the equation for y.

$$2x + 5y = 0$$

$$5y = -2x$$

$$y = -\frac{2x}{5} \quad \text{or} \quad y = -\frac{2}{5}x$$

Now we select values for x and find the corresponding values of y. Which values shall we select for x? Notice that the coefficient of the x-term is a fraction, with the denominator 5. If we select values for x that are multiples of the denominator, such as $\ldots, -15, -10, -5, 0, 5, 10, 15, \ldots$, the 5 in the denominator will divide out. This will give us integer values for y. We will arbitrarily select the values $x = -5$, $x = 0$, and $x = 5$.

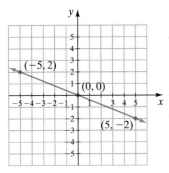

FIGURE 7.19

$$y = -\frac{2}{5}x$$

Let $x = -5$ $\qquad y = \left(-\frac{2}{5}\right)(-5) = 2$

Let $x = 0$ $\qquad y = \left(-\frac{2}{5}\right)(0) = 0$

Let $x = 5$ $\qquad y = -\frac{2}{5}(5) = -2$

x	y
-5	2
0	0
5	-2

NOW TRY EXERCISE 37 Now we plot the points and draw the graph (Fig. 7.19). ✳

The graph in Example 3 passes through the origin. The graph of every linear equation with a constant of 0 (equations of the form $ax + by = 0$) will pass through the origin.

3 Graph Linear Equations Using the *x*- and *y*-Intercepts

Now we discuss graphing linear equations using the *x*- and *y*-intercepts. The ***x*-intercept** is the point at which a graph crosses the *x*-axis and the ***y*-intercept** is the point at which the graph crosses the *y*-axis. Consider the graph in Figure 7.20 which is the graph we drew in Example 1. Note that the graph crosses the *x*-axis at -2. Therefore $(-2, 0)$ is the *x*-intercept. Since the graph crosses the *x*-axis at -2, we might say the *x*-intercept is *at* -2 (on the *x*-axis). In general, the *x*-intercept is $(x, 0)$, and the *x*-intercept is *at x* (on the *x*-axis).

Note that the graph in Figure 7.20 crosses the *y*-axis at 6. Therefore $(0, 6)$ is the *y*-intercept. Since the graph crosses the *y*-axis at 6, we might say the *y*-intercept is *at* 6 (on the *y*-axis). In general, the *y*-intercept is $(0, y)$, and the *y*-intercept is *at y* (on the *y*-axis).

Note that the graph in Figure 7.19 crosses both the *x*- and *y*-axes at the origin. Thus, both the *x*- and *y*-intercepts of this graph are $(0, 0)$.

It is often convenient to graph linear equations by finding their *x*- and *y*-intercepts. To graph an equation using the *x*- and *y*-intercepts, use the following procedure.

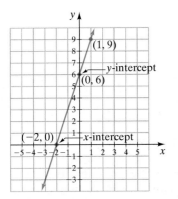

FIGURE 7.20

To Graph Linear Equations Using the *x*- and *y*-Intercepts

1. Find the *y*-intercept by setting x in the given equation equal to 0 and finding the corresponding value of y.

2. Find the *x*-intercept by setting y in the given equation equal to 0 and finding the corresponding value of x.

3. Determine a check point by selecting a nonzero value for x and finding the corresponding value of y.

4. Plot the *y*-intercept (where the graph crosses the *y*-axis), the *x*-intercept (where the graph crosses the *x*-axis), and the check point. The three points should appear to be collinear. If not, recheck your work.

5. Using a straightedge, draw a straight line through the three points. Draw an arrowhead at both ends of the line to show that the line continues indefinitely in both directions.

HELPFUL HINT

Since only two points are needed to determine a straight line, it is not absolutely necessary to determine and plot the check point in step 3. However, if you use only the x- and y-intercepts to draw your graph and one of those points is wrong, your graph will be incorrect and you will not know it. It is always a good idea to use three points when graphing a linear equation.

EXAMPLE 4   Graph the equation $3y = 6x + 12$ by plotting the x- and y-intercepts.

Solution   To find the y-intercept (where the graph crosses the y-axis), set $x = 0$ and find the corresponding value of y.

$$3y = 6x + 12$$
$$3y = 6(0) + 12$$
$$3y = 0 + 12$$
$$3y = 12$$
$$y = \frac{12}{3} = 4$$

The graph crosses the y-axis at 4. The ordered pair representing the y-intercept is $(0, 4)$. To find the x-intercept (where the graph crosses the x-axis), set $y = 0$ and find the corresponding value of x.

$$3y = 6x + 12$$
$$3(0) = 6x + 12$$
$$0 = 6x + 12$$
$$-12 = 6x$$
$$\frac{-12}{6} = x$$
$$-2 = x$$

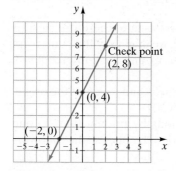

FIGURE 7.21

The graph crosses the x-axis at -2. The ordered pair representing the x-intercept is $(-2, 0)$. Now plot the intercepts (Fig. 7.21).

Before graphing the equation, select a nonzero value for x, find the corresponding value of y, and make sure that it is collinear with the x- and y-intercepts. This third point is the check point.

$$\text{Let } x = 2$$
$$3y = 6x + 12$$
$$3y = 6(2) + 12$$
$$3y = 12 + 12$$
$$3y = 24$$
$$y = \frac{24}{3} = 8$$

Plot the check point $(2, 8)$. Since the three points appear to be collinear, draw the straight line through all three points.   ✳

EXAMPLE 5 Graph the equation $2x + 5y = 12$ by finding the x- and y-intercepts.

Solution

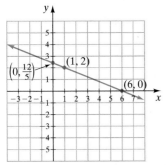

FIGURE 7.22

NOW TRY EXERCISE 45

Find y-intercept	Find x-intercept	Check Point
Let $x = 0$	Let $y = 0$	Let $x = 1$
$2x + 5y = 12$	$2x + 5y = 12$	$2x + 5y = 12$
$2(0) + 5y = 12$	$2x + 5(0) = 12$	$2(1) + 5y = 12$
$0 + 5y = 12$	$2x + 0 = 12$	$2 + 5y = 12$
$5y = 12$	$2x = 12$	$5y = 10$
$y = \dfrac{12}{5}$	$x = 6$	$y = 2$

The three ordered pairs are $\left(0, \frac{12}{5}\right)$, $(6, 0)$, and $(1, 2)$.
The three points appear to be collinear. Draw a straight line through all three points (Fig. 7.22).

EXAMPLE 6 Graph the equation $y = 20x + 60$.

Solution

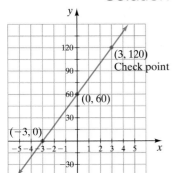

FIGURE 7.23

Find y-Intercept	Find x-Intercept	Check Point
Let $x = 0$	Let $y = 0$	Let $x = 3$
$y = 20x + 60$	$y = 20x + 60$	$y = 20x + 60$
$y = 20(0) + 60$	$0 = 20x + 60$	$y = 20(3) + 60$
$y = 60$	$-60 = 20x$	$y = 60 + 60$
	$-3 = x$	$y = 120$

The three ordered pairs are $(0, 60)$, $(-3, 0)$, and $(3, 120)$. Since the values of y are large, we let each interval on the y-axis be 15 units rather than 1 (Fig. 7.23). Sometimes you will have to use different scales on the x- and y-axes, as illustrated, to accommodate the graph. Now we plot the points and draw the graph.

When selecting the scales for your axes, you should realize that different scales will result in the same equation having a different appearance. Consider the graphs shown in Figure 7.24. Both graphs represent the same equation, $y = x$. In Figure 7.24a both the x- and y-axes have the same scale. In Figure 7.24b, the x- and y-axes do not have the same scale. Both graphs are correct in that each represents the graph of $y = x$. The difference in appearance is due to the difference in scales on the x-axis. When possible, keep the scales on the x- and y-axes the same, as in Figure 7.24a.

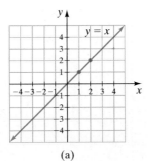

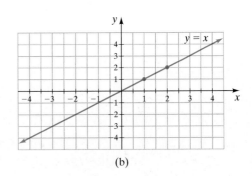

FIGURE 7.24   (a)   (b)

Using Your Graphing Calculator

To graph an equation on a graphing calculator, use the following steps.

1. Solve the equation for y, if necessary.

2. Press the $\boxed{Y=}$ key and enter the equation.

3. Press the $\boxed{GRAPH}$ key (to see the graph). You may need to adjust the window, as explained in the Using Your Graphing Calculator box on page 447.

In Example 2, when we solved the equation $2y = 4x - 12$ for y we obtained $y = 2x - 6$. If you press $\boxed{Y=}$, enter $2x - 6$ as Y_1, and then press $\boxed{GRAPH}$, you should get the graph shown in Figure 7.25.

If you do not get the graph in Figure 7.25, press $\boxed{WINDOW}$ and determine whether you have the standard window $-10, 10, 1, -10, 10, 1$. If not, change to the standard window and press the $\boxed{GRAPH}$ key again.

It is possible to graph two or more equations on your graphing calculator. If, for example, you wanted to graph both $y = 2x - 6$ and $y = -3x + 4$ on the same screen, you would begin by pressing the $\boxed{Y=}$ key. Then you would let $Y_1 = 2x - 6$ and $Y_2 = -3x + 4$. After you enter both equations and press the $\boxed{GRAPH}$ key, both equations will be graphed. Try graphing both equations on your graphing calculator now.

Exercises

Graph each equation on your graphing calculator.

1. $y = 3x - 5$
2. $y = -2x + 6$
3. $2x - 3y = 6$
4. $5x + 10y = 20$

See graphing answer section, page G2.

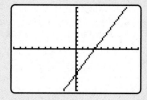

FIGURE 7.25

4 Graph Horizontal and Vertical Lines

When a linear equation contains only one variable, its graph will be either a horizontal or a vertical line, as is explained in Examples 7 and 8.

EXAMPLE 7 Graph the equation $y = 3$.

Solution This equation can be written as $y = 3 + 0x$. Thus, for any value of x selected, y will be 3. The graph of $y = 3$ is illustrated in Figure 7.26.

TEACHING TIP
Begin this objective by having students graph $x = 5$ on a number line. Be sure the number line is labeled with x. Then have students graph $x = 5$ on a 2 dimensional graph. Point out that all points with $x = 5$ should be graphed. Repeat with $y = -4$.

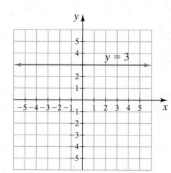

FIGURE 7.26

The graph of an equation of the form y = b is a **horizontal line** *whose y-intercept is* $(0, b)$.

EXAMPLE 8   Graph the equation $x = -2$.

Solution   This equation can be written as $x = -2 + 0y$. Thus, for any value of y selected, x will have a value of -2. The graph of $x = -2$ is illustrated in Figure 7.27.

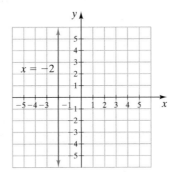

NOW TRY EXERCISE 21   FIGURE 7.27

The graph of an equation of the form x = a is a **vertical line** *whose x-intercept is* $(a, 0)$.

5 Study Applications of Graphs

Before we leave this section, let's look at an application of graphing. We will see additional applications of graphing linear equations in Sections 7.4 and 7.5.

EXAMPLE 9   **Weekly Salary**   Carol Smith recently graduated from college. She accepted a position as a sales manager trainee at a furniture store where she is paid a weekly salary plus a commission on sales. She will receive a salary of $300 per week plus a 7% commission on all sales, s.

a) Write an equation for the salary Carol will receive, R, in terms of the sales, s.

b) Graph the salary for sales of $0 up to and including $20,000.

c) From the graph, estimate Carol's salary if her weekly sales are $15,000.

d) From the graph, estimate the sales needed for Carol to earn a weekly salary of $900.

Solution   **a)** Since s is the amount of sales, a 7% commission on s dollars in sales is $0.07s$.

$$\text{salary received} = \$300 + \text{commission}$$
$$R = 300 + 0.07s$$

b) We select three values for s and find the corresponding values of R.

	$R = 300 + 0.07s$	s	R
Let $s = 0$	$R = 300 + 0.07(0) = 300$	0	300
Let $s = 10,000$	$R = 300 + 0.07(10,000) = 1000$	10,000	1000
Let $s = 20,000$	$R = 300 + 0.07(20,000) = 1700$	20,000	1700

The graph is illustrated in Figure 7.28. Notice that since we only graph the equation for values of s from 0 to 20,000, we do not place arrowheads on the ends of the graph.

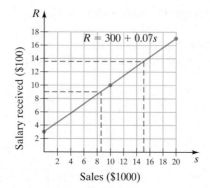

FIGURE 7.28   Sales ($1000)

c) To determine Carol's weekly salary on sales of $15,000, locate $15,000 on the sales axis. Then draw a vertical line up to where it intersects the graph, the *red* line in Figure 7.28. Now draw a horizontal line across to the salary axis. Since the horizontal line crosses the salary axis at about $1350, weekly sales of $15,000 would result in a weekly salary of about $1350. We can find the exact salary by substituting 15,000 for s in the equation $R = 300 + 0.07s$ and finding the value of R. Do this now.

d) To find the sales needed for Carol to earn a weekly salary of $900, we find $900 on the salary axis. We then draw a horizontal line from the point to the graph, as shown with the *green* line in Figure 7.28. We then draw a vertical line from the point of intersection of the graph to the sales axis. This value on the sales axis represents the sales needed for Carol to earn $900. Thus, sales of about $8600 per week would result in a salary of $900. We can find an exact answer by substituting 900 for R in the equation $R = 300 + 0.07s$ and solving the equation for s. Do this now.   ✳

NOW TRY EXERCISE 75

Using Your Graphing Calculator

Before we leave this section, let's spend a little more time discussing the graphing calculator. In the previous Using Your Graphing Calculator, we graphed $y = 2x - 6$. The graph of $y = 2x - 6$ is shown again using the standard window in Figure 7.29. After obtaining the graph, if you press the [TRACE] key, you may (depending on your calculator) obtain the graph in Figure 7.30.

Notice that the cursor, the blinking box in Figure 7.30, is on the y-intercept, that is, where $x = 0$ and $y = -6$. By pressing the left or the right arrow keys, you can move the cursor. The corresponding values of x and y change with the position of the cursor. You can magnify the part of the graph by the cursor by using the [ZOOM] key. Read your calculator manual to learn how to use the ZOOM feature and the various ZOOM options available.

Another important key on many graphing calculators is the TABLE feature. On the TI-83 Plus when you press [2nd] [GRAPH] to get the TABLE feature, you may get the display shown in Figure 7.31.

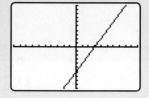

FIGURE 7.29

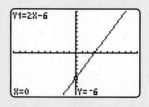

FIGURE 7.30

(continued on the next page)

If your display does not show the screen in Figure 7.31, use your up and down arrows until you get the screen shown. The table gives x-values and corresponding y-values for the graph. Notice from the table that when $x = 3$, $y = 0$ (the x-intercept) and when $x = 0$, $y = -6$ (the y-intercept).

You can change the table features by pressing $\boxed{\text{TBLSET}}$, (press $\boxed{2^{nd}}$ $\boxed{\text{WINDOW}}$ on the TI-83 Plus). For example, if you want the table to give values of x in tenths, you could change ΔTbl to 0.1 instead of the standard 1 unit.

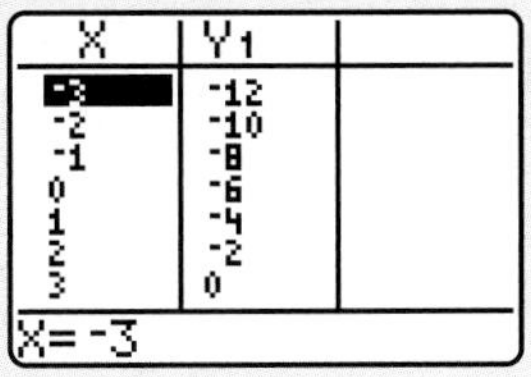

Exercises

Use the TABLE feature of your calculator to find the x- and y-intercepts of the graphs of the following equations. In Exercises 3 and 4, you will need to set ΔTbl to tenths.

1. $y = 4x - 8$ $(2, 0), (0, -8)$
2. $y = -2x - 6$ $(-3, 0), (0, -6)$
3. $4x - 5y = 10$ $(2.5, 0), (0, -2)$
4. $2y = 6x - 9$ $(1.5, 0), (0, -4.5)$

X	Y1
-3	-12
-2	-10
-1	-8
0	-6
1	-4
2	-2
3	0

X= -3

FIGURE 7.31

Exercise Set 7.2

Concept/Writing Exercises

1. x-intercept: substitute 0 for y and find corresponding value of x; y-intercept: substitute 0 for x and find corresponding value of y

1. Explain how to find the x- and y-intercepts of a line.
2. How many points are needed to graph a straight line? How many points should be used? Why? 2, 3
3. What will the graph of $y = b$ look like for any real number b? a horizontal line
4. What will the graph of $x = a$ look like for any real number a? a vertical line
5. In Example 9c and 9d, we made an estimate. Why is it sometimes not possible to obtain an exact answer from a graph?

6. In Example 9 does the salary, R, depend on the sales, s, or do the sales depend on the salary? Explain.
7. Will the equation $2x - 4y = 0$ go through the origin? Explain. yes
8. Write an equation, other than the ones given in this section, whose graph will go through the origin. Explain how you determined your answer.

5. You may not be able to read exact answers from a graph.
6. salary depends on sales   8. Answers will vary.

Practice the Skills

Find the missing coordinate in the given solutions for $3x + y = 9$.

9. $(3, ?)$ 0
10. $(-2, ?)$ 15
11. $(?, -6)$ 5
12. $(?, -3)$ 4
13. $(?, 0)$ 3
14. $\left(\frac{1}{2}, ?\right)$ $\frac{15}{2}$

Find the missing coordinate in the given solutions for $3x - 2y = 8$.

15. $(4, ?)$ 2
16. $(0, ?)$ -4
17. $(?, 0)$ $\frac{8}{3}$
18. $\left(?, -\frac{1}{2}\right)$ $\frac{7}{3}$
19. $(-4, ?)$ -10
20. $(?, -5)$ $-\frac{2}{3}$

Graph each equation. 21.–44. See graphing answer section, pages G2 and G3.

21. $x = -3$
22. $x = \frac{3}{2}$
23. $y = 4$
24. $y = -\frac{5}{3}$

Graph by plotting points. Plot at least three points for each graph.

25. $y = 3x - 1$
26. $y = -x + 3$
27. $y = 4x - 2$
28. $y = x - 4$
29. $y = -\frac{1}{2}x + 3$
30. $2y = 2x + 4$
31. $3x - 2y = 4$
32. $3x - 2y = 6$
33. $4x + 3y = -9$
34. $3x + 2y = 0$
35. $6x + 5y = 30$
36. $-2x - 3y = 6$
37. $-4x + 5y = 0$
38. $12y - 24x = 36$
39. $y = -20x + 60$
40. $2y - 50 = 100x$
41. $y = \frac{4}{3}x$
42. $y = -\frac{3}{5}x$
43. $y = \frac{1}{2}x + 4$
44. $y = -\frac{2}{5}x + 2$

Graph using the x- and y-intercepts. **45.–64.** See graphing answer section, pages G3 and G4.

45. $y = 3x + 3$

46. $y = -2x + 6$

47. $y = -4x + 2$

48. $y = -3x + 8$

49. $y = -5x + 4$

50. $y = 4x + 16$

51. $4y + 6x = 24$

52. $4x = 3y - 9$

53. $\frac{1}{2}x + 2y = 4$

54. $x + \frac{1}{2}y = 2$

55. $6x - 12y = 24$

56. $25x + 50y = 100$

57. $8y = 6x - 12$

58. $-3y - 2x = -6$

59. $-20y - 30x = 40$

60. $20x - 240 = -60y$

61. $\frac{1}{3}x + \frac{1}{4}y = 12$

62. $\frac{1}{4}x - \frac{2}{3}y = 60$

63. $\frac{1}{2}x = \frac{2}{5}y - 80$

64. $\frac{2}{3}y = \frac{5}{4}x + 120$

Write the equation represented by the given graph.

65.
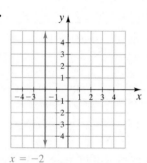
$x = -2$

66.
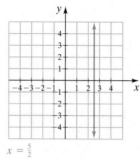
$x = \frac{5}{2}$

67.
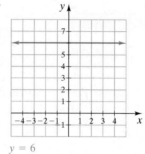
$y = 6$

68.
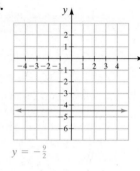
$y = -\frac{9}{2}$

Problem Solving

69. What is the value of a if the graph of $ax + 4y = 8$ is to have an x-intercept of $(2, 0)$? 4

70. What is the value of a if the graph of $ax + 8y = 12$ is to have an x-intercept of $(3, 0)$? 4

71. What is the value of b if the graph of $3x + by = 10$ is to have a y-intercept of $(0, 5)$? 2

72. What is the value of b if the graph of $4x + by = 12$ is to have a y-intercept of $(0, -3)$? −4

The bar graphs in Exercises 73 and 74 display information. State whether the graph displays a linear relationship. Explain your answer.

73. Calories Burned by Average 150–Pound Person Walking at 4.5 mph

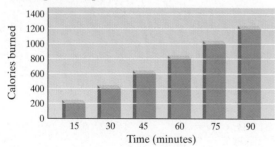

yes

74. Price of a 30-Second Super Bowl Ad

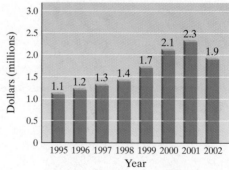

Source: NFL Research

no

Review Example 9 before working Exercises 75–80.

75. Buying Gravel Yolanda Torres needs to buy a large number of bags of gravel at Home Depot and rent a truck to transport the gravel. The truck rental is a flat $30 and each bag of gravel costs $2.

  a) Write an equation for the total cost, C, of the truck rental and for purchasing n bags of gravel.

  b) Graph the equation for up to and including 60 bags of gravel. See graphing answer section, page G4.

  c) Estimate the total cost Yolanda must pay for the truck rental and 30 bags of gravel. $90

  d) If the total cost of the truck rental and the gravel is $110, estimate how many bags of gravel Yolanda purchased. 40

76. Distance Traveled Distance traveled is calculated using the formula distance = rate · time or $d = rt$. Assume the rate of a car is a constant 60 miles per hour.

  a) Write an equation for the distance, d, in terms of time, t. $d = 60t$

  b) Graph the equation for times of 0 to 10 hours inclusive. See graphing answer section, page G4.

  c) Estimate the distance traveled in 6 hours. 360 mi

  d) If the distance traveled is 300 miles, estimate the time traveled. 5 hr

77. Truck Rental Lynn Brown needs a large truck to move some furniture. She found that the cost C, of renting a truck is $40 per day plus $1 per mile, m.

  a) Write an equation for the cost in terms of the miles driven. $C = m + 40$

  b) Graph the equation for values up to and including 100 miles.

  c) Estimate the cost of driving 50 miles in one day. $90

  d) Estimate the miles driven if the cost for one day is $60. 20 mi

78. Simple Interest Simple interest is calculated by the simple interest formula,

  interest = principal · rate · time or $I = prt$.

  Suppose the principal is $10,000 and the rate is 5%.

  a) Write an equation for simple interest in terms of time. $I = 500t$

  b) Graph the equation for times of 0 to 20 years inclusive.

  c) What is the simple interest for 10 years? $5000

  d) If the simple interest is $500, find the length of time.

79. Video Store Profit The weekly profit, P of a video rental store can be approximated by the formula $P = 1.5n - 200$, where n is the number of tapes rented weekly.

  a) Draw a graph of profit in terms of tape rentals for up to and including 1000 tapes.

  b) Estimate the weekly profit if 500 tapes are rented.

  c) Estimate the number of tapes rented if the week's profit is $1000. 800

80. Playing Tennis The cost, C, of playing tennis in the Downtown Tennis Club includes an annual $200 membership fee plus $10 per hour, h, of court time.

  a) Write an equation for the annual cost of playing tennis at the Downtown Tennis Club in terms of hours played. $C = 10h + 200$

  b) Graph the equation for up to and including 300 hours.

  c) Estimate the cost for playing 200 hours in a year.

  d) If the annual cost for playing tennis was $1200, estimate how many hours of tennis were played. 100 hr

75. **a)** $C = 2n + 30$ 77 **b)**, 78 **b)**, 79 **a)**, 80 **b)** See graphing answer section, page G5.   78. **d)** 1 yr   79. **b)** $550   80. **c)** $2200

Determine the coefficients to be placed in the shaded areas so that the graph of the equation will be a line with the x- and y-intercepts specified. Explain how you determined your answer.

81. $\blacksquare x + \blacksquare y = 20$; x-intercept at 4, y-intercept at 5
  5, 4

82. $\blacksquare x + \blacksquare y = 18$; x-intercept at −3, y-intercept at 6
  −6, 3

83. $\blacksquare x - \blacksquare y = -12$; x-intercept at −2, y-intercept at 3
  6, 4

84. $\blacksquare x - \blacksquare y = 30$; x-intercept at −5, y-intercept at −15
  −6, 2

Challenge Problems

✎ **85.** Consider the following equations: $y = 2x - 1$, $y = -x + 5$.
   a) Carefully graph both equations on the same axes.
   b) Determine the point of intersection of the two graphs.
   c) Substitute the values for x and y at the point of intersection into each of the two equations and determine whether the point of intersection satisfies each equation.
   d) Do you believe there are any other ordered pairs that satisfy both equations? Explain your answer.

(We will study equations like these, called systems of equations, in Chapter 8.) **b)** $(2, 3)$ **c)** yes **d)** no

86. In Chapter 10 we will be graphing quadratic equations. The graphs of quadratic equations are *not* straight lines. Graph the quadratic equation $y = x^2 - 4$ by selecting values for x and find the corresponding values of y, then plot the points. Make sure you plot a sufficient number of points to get an accurate graph.

85 a), 86, 87 a)–d) See graphing answer section, page G5.

Group Activity

Discuss and answer Exercise 87 as a group.   87. **e)** They appear to be parallel lines.   **f)** The constant is the *y*-intercept.

✎ **87.** Let's study the graphs of the equations $y = 2x + 4$, $y = 2x + 2$, and $y = 2x - 2$ to see how they are similar and how they differ. Each group member should start with the same axes.
   a) Group member 1: Graph $y = 2x + 4$.
   b) Group member 2: Graph $y = 2x + 2$.
   c) Group member 3: Graph $y = 2x - 2$.
   d) Now transfer all three graphs onto the same axes. (You can use one of the group members' graphs or you can construct new axes.)
   e) Explain what you notice about the three graphs.
   f) Explain what you notice about the *y*-intercepts.

Cumulative Review Exercises

[1.9] **88.** Evaluate $2[6 - (4 - 5)] \div 2 - 5^2$. -18

[2.6] **89.** *House Cleaning* According to the instructions on a bottle of concentrated household cleaner, 8 ounces of the cleaner should be mixed with 3 gallons of water. If your bucket holds only 2.5 gallons of water, how much cleaner should you use?
6.67 oz

[3.3] **90.** *Integers* The larger of two integers is 1 more than 3 times the smaller. If the sum of the two integers is 37, find the two integers. 9, 28

[6.2] **91.** Divide $\dfrac{10xy^3}{z} \div \dfrac{x^2y^2}{5z^2}$. $\dfrac{50yz}{x}$

[6.4] **92.** Add $\dfrac{3}{x - 2} + \dfrac{4}{x - 3} + 3$ $\dfrac{3x^2 - 8x + 1}{(x - 2)(x - 3)}$

[6.6] **93.** Solve $\dfrac{3}{x - 2} + \dfrac{4}{x - 3} = 3$ $5, \dfrac{7}{3}$

7.3 SLOPE OF A LINE

SSM   Study Guide   CD/Video

MathPro 4/5   PH Math Tutor Center   prenhall.com/Angel

1 Find the slope of a line.
2 Recognize positive and negative slopes.
3 Examine the slopes of horizontal and vertical lines.
4 Examine the slopes of parallel and perpendicular lines.

1 Find the Slope of a Line

In this section we discuss the *slope* of a line. In the following Helpful Hint we discuss similarities between slope as commonly used and the slope of a line.

HELPFUL HINT

Slope We often come across slopes in everyday life. A highway (or a ramp) may have grade (or slope) of 8%. A roof may have a pitch (or slope) of $\frac{6}{15}$. The slope is a measure of steepness which can be determined by dividing the vertical change, called the *rise*, by the horizontal change, called the *run*.

Suppose a road has an 8% grade. Since $8\% = \frac{8}{100}$, this means the road drops (or rises) 8 feet for each 100 feet of horizontal length. A roof pitch of $\frac{6}{15}$ means the roof drops 6 feet for each 15 feet of horizontal length.

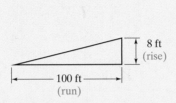

When we find the slope of a line we are also finding a ratio of the vertical change to the horizontal change. The major difference is that when we find the slope of a non-horizontal, non-vertical line, the slope can be a positive number or a negative number, as will be explained shortly.

The *slope of a line* is a measure of the *steepness* of the line. The slope of a line is an important concept in many areas of mathematics. A knowledge of slope is helpful in understanding linear equations. We now define the slope of a line.

DEFINITION

The **slope of a line** is a ratio of the vertical change to the horizontal change between any two selected points on the line.

As an example, consider the line that goes through the two points $(3, 6)$ and $(1, 2)$. (see Fig. 7.32a).

If we draw a line parallel to the *x*-axis through the point $(1, 2)$ and a line parallel to the *y*-axis through the point $(3, 6)$, the two lines intersect at $(3, 2)$, see Figure 7.32b. From the figure, we can determine the slope of the line. The vertical change (along the *y*-axis) is $6 - 2$, or 4 units. The horizontal change (along the *x*-axis) is $3 - 1$, or 2 units.

$$\text{slope} = \frac{\text{vertical change}}{\text{horizontal change}} = \frac{4}{2} = 2$$

TEACHING TIP
Draw 3 lines, one with a steep slope, one with a shallow slope, and one with a slope of zero. Have students label the lines as steep, shallow, or horizontal and explain their answer. Then point out that mathematics allows us to quantify the steepness of a line.

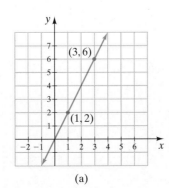

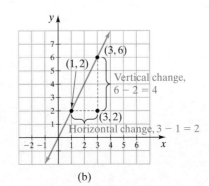

FIGURE 7.32   (a)   (b)

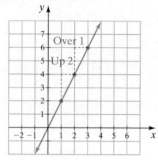

FIGURE 7.33

Thus, the slope of the line through these two points is 2. By examining the line connecting these two points, we can see that as the graph moves up 2 units on the y-axis it moves to the right 1 unit on the x-axis (Fig. 7.33).

Now we present the procedure to find the slope of a line between any two points (x_1, y_1) and (x_2, y_2). Consider Figure 7.34.

The vertical change can be found by subtracting y_1 from y_2. The horizontal change can be found by subtracting x_1 from x_2.

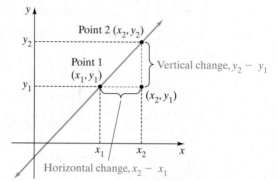

FIGURE 7.34

Slope of a Line Through the Points (x_1, y_1) and (x_2, y_2)

$$\text{slope} = \frac{\text{change in } y \text{ (vertical change)}}{\text{change in } x \text{ (horizontal change)}} = \frac{y_2 - y_1}{x_2 - x_1}$$

It makes no difference which two points are selected when finding the slope of a line. It also makes no difference which point you label (x_1, y_1) or (x_2, y_2). The Greek capital letter delta, Δ, is often used to represent the words "the change in." So, Δy is read, the change in y, and Δx is read, the change in x. Thus, the slope, which is symbolized by the letter m, is indicated as

$$m = \frac{\Delta y}{\Delta x} = \frac{y_2 - y_1}{x_2 - x_1}$$

EXAMPLE 1 Find the slope of the line through the points $(-6, -1)$ and $(3, 5)$.

Solution We will designate $(-6, -1)$ as (x_1, y_1) and $(3, 5)$ as (x_2, y_2).

$$m = \frac{y_2 - y_1}{x_2 - x_1}$$

$$= \frac{5 - (-1)}{3 - (-6)}$$

$$= \frac{5 + 1}{3 + 6} = \frac{6}{9} = \frac{2}{3}$$

Thus, the slope is $\frac{2}{3}$.

If we had designated $(3, 5)$ as (x_1, y_1) and $(-6, -1)$ as (x_2, y_2), we would have obtained the same results.

$$m = \frac{y_2 - y_1}{x_2 - x_1}$$

$$= \frac{-1 - 5}{-6 - 3} = \frac{-6}{-9} = \frac{2}{3}$$

NOW TRY EXERCISE 13

Students sometimes subtract the x's and y's in the slope formula in the wrong order. For instance, using the problem in Example 1:

$$m = \frac{\cancel{y_2} - \cancel{y_1}}{\cancel{x_1} - \cancel{x_2}} = \frac{5 - (-1)}{-6 - 3} = \frac{5 + 1}{-6 - 3} = \frac{6}{-9} = -\frac{2}{3}$$

Notice that subtracting in this incorrect order results in a negative slope, when the actual slope of the line is positive. The same sign error will occur each time subtraction is done incorrectly in this manner.

2 Recognize Positive and Negative Slopes

A straight line for which the value of y increases as x increases has a **positive slope**, see Figure 7.35a. A line with a positive slope rises as it moves from left to right. A straight line for which the value of y decreases as x increases has a **negative slope**, see Figure 7.35b. A line with a negative slope falls as it moves from left to right.

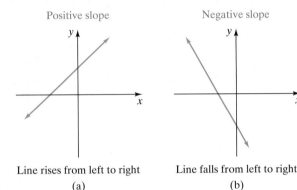

Positive slope

Negative slope

Line rises from left to right   Line falls from left to right
(a)   (b)

FIGURE 7.35

EXAMPLE 2 Consider the line in Figure 7.36.

a) Determine the slope of the line by observing the vertical change and horizontal change between the points $(1, 5)$ and $(0, 2)$.

b) Calculate the slope of the line using the two given points.

Solution **a)** The first thing you should notice is that the slope is positive since the line rises from left to right. Now determine the vertical change between the two points. The vertical change is $+3$ units. Next determine the horizontal change between the two points. The horizontal change is $+1$ unit. Since the slope is the ratio of the vertical change to the horizontal change between any two points, and since the slope is positive, the slope of the line is $\frac{3}{1}$ or 3.

b) We can use any two points on the line to determine its slope. Since we are given the ordered pairs $(1, 5)$ and $(0, 2)$, we will use them.

Let (x_2, y_2) be $(1, 5)$. Let (x_1, y_1) be $(0, 2)$.

$$m = \frac{y_2 - y_1}{x_2 - x_1} = \frac{5 - 2}{1 - 0} = \frac{3}{1} = 3$$

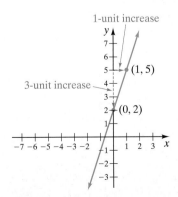

FIGURE 7.36

Note that the slope obtained in part **b)** agrees with the slope obtained in part **a)**. If we had designated $(1, 5)$ as (x_1, y_1) and $(0, 2)$ as (x_2, y_2), the slope would not have changed. Try it and see that you will still obtain a slope of 3. ✳

EXAMPLE 3   Find the slope of the line in Figure 7.37 by observing the vertical change and horizontal change between the two points shown.

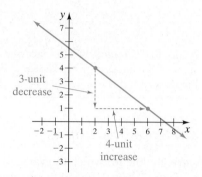

3-unit decrease

4-unit increase

FIGURE 7.37

Solution   Since the graph falls from left to right, you should realize that the line has a negative slope. The vertical change between the two given points is -3 units since it is decreasing. The horizontal change between the two given points is 4 units since it is increasing. Since the ratio of the vertical change to the horizontal change is -3 units to 4 units, the slope of this line is $\frac{-3}{4}$ or $-\frac{3}{4}$.

NOW TRY EXERCISE 29

Using the two points shown in Figure 7.37 and the definition of slope, calculate the slope of the line in Example 3. You should obtain the same answer.

EXAMPLE 4   **Golf Balls**   With technological advances in the golf equipment industry, golf balls are flying further. The graph in Figure 7.38 shows the average distance (to the nearest yard) of the top 10 longest drives in selected years over the last ten years.

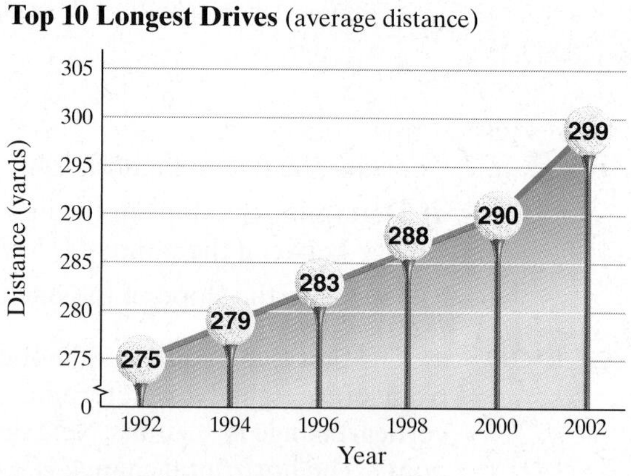

Top 10 Longest Drives (average distance)

Distance (yards)

305
300
295
290
285
280
275
0

275   279   283   288   290   299

1992   1994   1996   1998   2000   2002

Year

FIGURE 7.38   *Source:* Professional Golfers Association

The increase from 1992 to 2000 is almost linear. The red line can be used to estimate the average distance of the drives from 1992 to 2000.

a) Use the points $(1992, 275)$ and $(2000, 290)$, that are on the red line, to determine the slope of the red line.

b) Determine the slope of the green line from 2000 to 2002.

Solution   **a)** To determine the slope, divide the change in distance, on the vertical axis, by the change in years, on the horizontal axis. We will use $(2000, 290)$ as (x_2, y_2) and $(1992, 275)$ as (x_1, y_1).

$$m = \frac{y_2 - y_1}{x_2 - x_1} = \frac{\text{distance in 2000} - \text{distance in 1992}}{2000 - 1992} = \frac{290 - 275}{8} = \frac{15}{8}$$

Thus the slope of the red line is $\frac{15}{8}$.

b) Two points on the green line are $(2002, 299)$ and $(2000, 290)$.

$$m = \frac{\text{distance in 2002} - \text{distance in 2000}}{2002 - 2000} = \frac{299 - 290}{2} = \frac{9}{2}$$

Thus, the slope of the green line is $\frac{9}{2}$.

3 Examine the Slopes of Horizontal and Vertical Lines

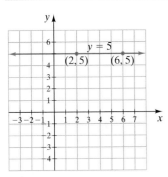

FIGURE 7.39

NOW TRY EXERCISE 37

Now we consider the slope of horizontal and vertical lines. Consider the graph of $y = 5$ (Fig. 7.39). What is its slope?

The graph is parallel to the x-axis and goes through the points $(2, 5)$ and $(6, 5)$. Arbitrarily select $(6, 5)$ as (x_2, y_2) and $(2, 5)$ as (x_1, y_1). Then the slope of the line is

$$m = \frac{y_2 - y_1}{x_2 - x_1} = \frac{5 - 5}{6 - 2} = \frac{0}{4} = 0$$

Since there is no change in y, this line has a slope of 0. Note that *any* two points on the line would yield the same slope, 0.

> Every horizontal line has a slope of 0.

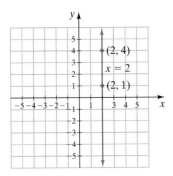

FIGURE 7.40

Now we discuss vertical lines. Consider the graph of $x = 2$ (Fig. 7.40). What is its slope?

The graph is parallel to the y-axis and goes through the points $(2, 1)$ and $(2, 4)$. Arbitrarily select $(2, 4)$ as (x_2, y_2) and $(2, 1)$ as (x_1, y_1). Then the slope of the line is

$$m = \frac{y_2 - y_1}{x_2 - x_1} = \frac{4 - 1}{2 - 2} = \frac{3}{0}$$

We learned in Section 1.8 that $\frac{3}{0}$ is undefined. Thus, we say that the slope of this line is undefined.

> The slope of any vertical line is undefined.

4 Examine the Slopes of Parallel and Perpendicular Lines

Two lines are **parallel** when they do not intersect, no matter how far they are extended. Figure 7.41 on page 468 illustrates two parallel lines.

If we compute the slope of line 1 using the given points, we obtain a slope of 3. If we compute the slope of line 2, we obtain a slope of 3. (You should compute the slopes of both lines now to verify this.) Notice both lines have the same slopes. Any two non-vertical lines that have the same slope are parallel lines.

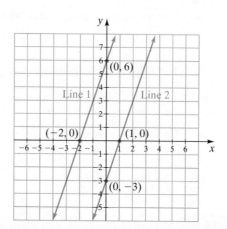

FIGURE 7.41

Parallel Lines

Two non-vertical lines with the same slope and different y-intercepts are parallel. Any two vertical lines are parallel to each other.

EXAMPLE 5 **a)** Draw a line with a slope of $\frac{1}{2}$ through the point $(2, 3)$.

b) On the same set of axes, draw a line with a slope of $\frac{1}{2}$ through the point $(-1, -3)$.

c) Are the two lines in part a) and b) parallel? Explain.

Solution **a)** Place a dot at $(2, 3)$. Because the slope is a positive $\frac{1}{2}$, from the point $(2, 3)$ move *up* 1 unit and to the *right* 2 units to get a second point. Draw a line through the two points, see the blue line in Figure 7.42.

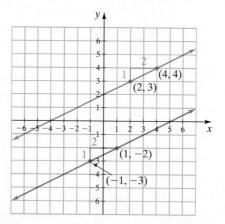

FIGURE 7.42

b) Place a dot at $(-1, -3)$. From the point $(-1, -3)$ move up 1 unit and to the right 2 units to get a second point. Draw a line through the two points; see the red line in Figure 7.42.

c) The lines appear to be parallel on the graph. Since both lines have the same slope, $\frac{1}{2}$, they are parallel lines. ✳

Now let's consider perpendicular lines. Two lines are **perpendicular** when they meet at a right ($90°$) angle. Figure 7.43 on the next page illustrates two perpendicular lines.

If we compute the slope of line 1 using the given point, we obtain a slope of $\frac{1}{2}$. If we compute the slope of line 2 using the given points, we obtain a slope of -2.

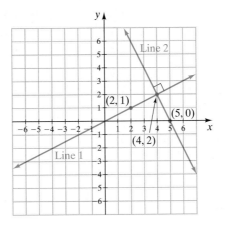

FIGURE 7.43

(You should compute the slopes of both lines now to verify this.) Notice the product of their slopes, $\frac{1}{2}(-2)$, is -1. Any two numbers whose product is -1 are said to be **negative reciprocals** of each other. In general, if m represents a number, its negative reciprocal will be $-\frac{1}{m}$ because $m\left(-\frac{1}{m}\right) = -1$. Any two lines with slopes that are negative reciprocals of each other are perpendicular lines.

Perpendicular Lines

Two lines whose slopes are negative reciprocals of each other are perpendicular lines. Any vertical line is perpendicular to any horizontal line.

EXAMPLE 6 **a)** Draw a line with a slope of -3 through the point $(2, 3)$.

b) On the same set of axes, draw a line with a slope of $\frac{1}{3}$ through the point $(-1, -3)$.

c) Are the two lines in parts **a)** and **b)** perpendicular? Explain.

Solution **a)** Place a dot at $(2, 3)$. A slope of -3 means $\frac{-3}{1}$. Because the slope is *negative*, from the point $(2, 3)$ move *down* 3 units and to the *right* 1 unit to get a second point. Draw a line through the two points; see the blue line in Figure 7.44.

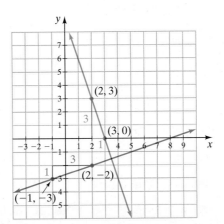

FIGURE 7.44

b) Place a dot at $(-1, -3)$. Because the slope is a *positive* $\frac{1}{3}$, from this point move *up* 1 unit and to the *right* 3 units. Draw a line through the two points; see the red line in Figure 7.44.

c) The lines appear to be perpendicular on the graph. To determine if they are perpendicular, multiply the slopes of the two lines together. If their product is -1, then the slopes are negative reciprocals, and the lines are perpendicular.

$$\text{Slope of line 1 (or } m_1) = -3, \text{ slope of line 2 (or } m_2) = \frac{1}{3}$$

$$(m_1)(m_2) = (-3)\left(\frac{1}{3}\right) = -1$$

NOW TRY EXERCISE 39   Since the slopes are negative reciprocals, the two lines are perpendicular.   ✳

EXAMPLE 7   If m_1 represents the slope of line 1 and m_2 represents the slope of line 2, determine if line 1 and line 2 are parallel, perpendicular or neither.

a) $m_1 = \frac{5}{6}, m_2 = \frac{5}{6}$ **b)** $m_1 = \frac{2}{5}, m_2 = 4$ **c)** $m_1 = \frac{3}{5}, m_2 = -\frac{5}{3}$

Solution   **a)** Since the slopes are the same, both $\frac{5}{6}$, the lines are parallel.

b) Since the slopes are not the same, the lines are not parallel. Since $m_1 \cdot m_2 = \left(\frac{2}{5}\right)(4) \neq -1$, the slopes are not negative reciprocals and the lines are not perpendicular. Thus the answer is neither.

c) Since the slopes are not the same, the lines are not parallel. Since $m_1 \cdot m_2 = \frac{3}{5}\left(-\frac{5}{3}\right) = -1$, the slopes are negative reciprocals and the lines are perpendicular.   ✳

NOW TRY EXERCISE 53

1. the ratio of the vertical change to the horizontal change between any two points on a line   **2.** Calculate the ratio of the vertical to horizontal change between any two points.   **5.** positive: lines rise from left to right; negative: lines fall from left to right   **7.** No, since we cannot divide by 0, the slope is undefined.

Exercise Set 7.3

Concept/Writing Exercises

1. Explain what is meant by the slope of a line.

2. Explain how to find the slope of a line.

3. Describe the appearance of a line that has a positive slope.   rises from left to right

4. Describe the appearance of a line that has a negative slope.   falls from left to right

5. Explain how to tell by observation whether a line has a positive slope or negative slope.

6. What is the slope of any horizontal line? Explain your answer.   0, there is no vertical change.

7. Do vertical lines have a slope? Explain.

8. What letter is used to represent the slope? m

9. If two non-vertical lines are parallel, what do we know about the slopes of the two lines?   They are the same.

10. If two non-vertical lines are perpendicular, what do we know about the slopes of the two lines?
Their slopes are negative reciprocals.

Practice the Skills

Using the slope formula, find the slope of the line through the given points.

11. $(5, 1)$ and $(7, 5)$ 2

12. $(8, -2)$ and $(6, -4)$ 1

13. $(9, 0)$ and $(5, -2)$ $\frac{1}{2}$

14. $(5, -6)$ and $(6, -5)$ 1

15. $\left(3, \frac{1}{2}\right)$ and $\left(-3, \frac{1}{2}\right)$ 0

16. $(-4, 2)$ and $(6, 5)$ $\frac{3}{10}$

17. $(-7, 6)$ and $(-2, 6)$ 0

18. $(9, 3)$ and $(5, -6)$ $\frac{9}{4}$

19. $(6, 4)$ and $(6, -2)$

20. $(-7, 5)$ and $(3, -4)$

21. $(6, 0)$ and $(-2, 3)$ $-\frac{3}{8}$

22. $(-2, 3)$ and $(-2, -1)$
undefined

23. $\left(0, \frac{3}{2}\right)$ and $\left(-\frac{3}{4}, 1\right)$ $\frac{2}{3}$

24. $(-1, 7)$ and $\left(\frac{1}{3}, -2\right)$ $-\frac{27}{4}$

19. undefined   **20.** $-\frac{9}{10}$

By observing the vertical and horizontal change of the line between the two points indicated, determine the slope of each line.

25.
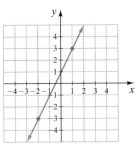
$m = 2$

26.
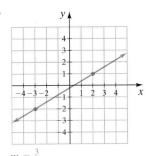
$m = \frac{3}{5}$

27.
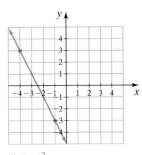
$m = -2$

28.
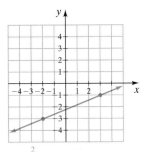
$m = \frac{2}{5}$

29.
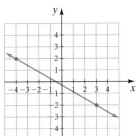
$m = -\frac{4}{7}$

30.
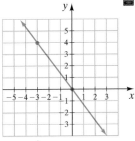
$m = -\frac{4}{3}$

🔒 **31.**
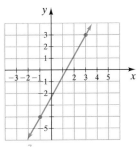
$m = \frac{7}{4}$

32.
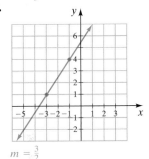
$m = \frac{3}{2}$

🔒 **33.**
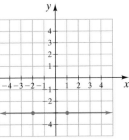
$m = 0$

34.
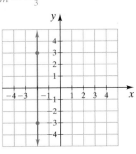
undefined

35.
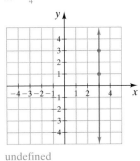
undefined

36.
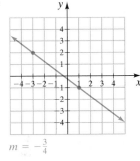
$m = -\frac{3}{4}$

37.
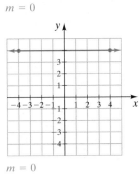
$m = 0$

38.
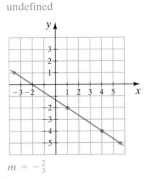
$m = -\frac{2}{3}$

39.–48. See graphing answer section, pages G5 and G6.

In Exercises 39–48, graph the line with the given slope that goes through the given point.

39. Through $(3, -1)$ with $m = 2$.

40. Through $(0, -2)$ with $m = \frac{1}{2}$

41. Through $(0, 3)$ with $m = 1$

42. Through $(-1, -2)$ with $m = -2$

43. Through $(0, 0)$ with $m = -\frac{1}{3}$

44. Through $(-3, 4)$ with $m = -\frac{2}{3}$

45. Through $(-4, 2)$ with $m = 0$

46. Through $(-2, 3)$ with $m = 0$

47. Through $(2, 3)$ with slope undefined

48. Through $(-1, -2)$ with slope undefined

49.–64. Parallel: 49, 52, 59, 61   Perpendicular: 51, 53, 58, 62, 63   Neither: 50, 54, 55, 56, 57, 60, 64

In Exercises 49–64, indicate whether distinct line 1 and line 2 are parallel, perpendicular, or neither.

49. $m_1 = 3, m_2 = 3$

50. $m_1 = \dfrac{1}{2}, m_2 = -3$

51. $m_1 = \dfrac{1}{4}, m_2 = -4$

52. $m_1 = -2, m_2 = -2$

53. $m_1 = \dfrac{2}{3}, m_2 = -\dfrac{3}{2}$

54. $m_1 = 5, m_2 = -4$

55. $m_1 = 6, m_2 = \dfrac{1}{3}$

56. $m_1 = -\dfrac{1}{4}, m_2 = -4$

57. $m_1 = \dfrac{1}{3}, m_2 = 3$

58. $m_1 = 6, m_2 = -\dfrac{1}{6}$

59. $m_1 = 0, m_2 = 0$

60. $m_1 = 0, m_2 = -1$

61. m_1 is undefined, m_2 is undefined

62. $m_1 = 0, m_2$ is undefined

63. m_1 is undefined, $m_2 = 0$

64. $m_1 = 2, m_2 = -2$

65. The slope of a given line is 2. If a line is to be drawn parallel to the given line, what will be its slope?   2

66. The slope of a given line is -4. If a line is to be drawn parallel to the given line, what will be its slope? -4

67. The slope of a given line is -4. If a line is to be drawn perpendicular to the given line, what will be its slope? $\frac{1}{4}$

68. The slope of a given line is -5. If a line is to be drawn perpendicular to the given line, what will be its slope?
$\frac{1}{5}$

Problem Solving

In Exercises 69 and 70, determine which line (the first or second) has the greater slope. Explain your answer. Notice that the scales on the x- and y-axes are different.

69.

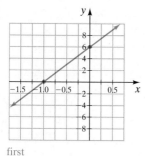

first

70.

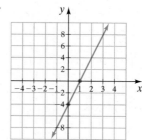

second

*In Exercises 71 and 72, find the slope of the line segments indicated in **a)** red and **b)** blue.*

71.

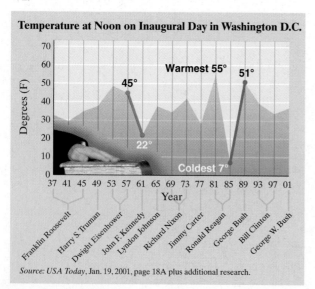

Temperature at Noon on Inaugural Day in Washington D.C.

Warmest 55°   51°   45°   22°   Coldest 7°

Degrees (F)

37 41 45 49 53 57 61 65 69 73 77 81 85 89 93 97 01
Year

Franklin Roosevelt   Harry S. Truman   Dwight Eisenhower   John F. Kennedy   Lyndon Johnson   Richard Nixon   Jimmy Carter   Ronald Reagan   George Bush   Bill Clinton   George W. Bush

Source: *USA Today*, Jan. 19, 2001, page 18A plus additional research.

72.

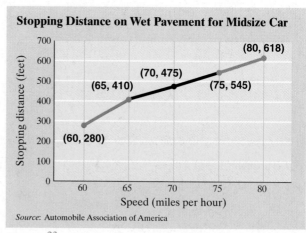

Stopping Distance on Wet Pavement for Midsize Car

(80, 618)   (70, 475)   (65, 410)   (75, 545)   (60, 280)

Stopping distance (feet)

Speed (miles per hour)

Source: Automobile Association of America

71. a) $-\dfrac{23}{4}$ **b)** 11   **72. a)** 26   **b)** 14.6

73. A given line goes through the points $(1, 6)$ and $(3, -2)$. If a line is to be drawn parallel to the given line, what will be its slope? -4

74. A given line goes through the points $(-2, 3)$ and $(4, 5)$. If a line is to be drawn parallel to the given line, what will be its slope? $\frac{1}{3}$

75. A given line goes through the points $(1, -3)$ and $(2, 5)$. If a line is to be drawn perpendicular to the given line, what will be its slope? $-\frac{1}{8}$

76. A given line goes through the points $(-3, 0)$ and $(-2, 3)$. If a line is to be drawn perpendicular to the given line, what will be its slope? $-\frac{1}{3}$

Challenge Problems

79. b) $AC, m = \frac{3}{5}$; $CB, m = -2$; $DB, m = \frac{3}{5}$; $AD, m = -2$ **c)** yes
80. a) $ab, m = 2,500,000$; $bc, m = 12,500,000$; $cd, m \approx 43,478,261$; $de, m = 100,000,000$

77. Find the slope of the line through the points $\left(\frac{1}{2}, -\frac{3}{8}\right)$ and $\left(-\frac{4}{9}, -\frac{7}{2}\right)$. $\frac{225}{68}$

78. If one point on a line is $(6, -4)$ and the slope of the line is $-\frac{5}{3}$, identify another point on the line. $(3, 1)$

79. A quadrilateral (a four-sided figure) has four vertices (the points where the sides meet). Vertex A is at $(0, 1)$, vertex B is at $(6, 2)$, vertex C is at $(5, 4)$, and vertex D is at $(1, -1)$.
  a) Graph the quadrilateral in the Cartesian coordinate system. See graphing answer section, page G6.
  b) Find the slopes of sides AC, CB, DB, and AD.
  c) Do you think this figure is a parallelogram? Explain.

80. The following graph shows the world's population estimated to the year 2016.

World Popultation Growth

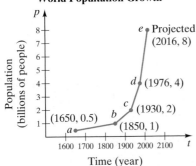

  a) Find the slope of the line segment between each pair of points, that is, ab, bc, and so on. Remember, the

second coordinate is in billions. Thus, for example, 0.5 billion is actually 500,000,000.

  b) Would you say that this graph represents a linear equation? Explain. no

81. Consider the graph below.

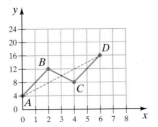

  a) Determine the slope of each of the three solid blue lines. $AB, m = 4$; $BC, m = -2$; $CD, m = 4$

  b) Determine the average of the three slopes found in part a). $[4 + (-2) + 4]/3 = 2$

  c) Determine the slope of the red dashed line from A to D. $AD, m = 2$

  d) Determine whether the slope of the red dashed line from A to D is the same as the (mean) average of the slopes of the three solid blue lines. yes

  e) Explain what this example illustrates. Answers will vary.

 **Group Activity**   **82. b)** slope of line, no units;   **c)** slope of hill, degrees

Discuss and answer Exercise 82 as a group, according to the instructions.

82. The slope of a hill and the slope of a line both measure steepness. However, there are several important differences.

  a) As a group, explain how you think the slope of a hill is determined.

  b) Is the slope of a line, graphed in the Cartesian coordinate system, measured in any specific unit?

  c) Is the slope of a hill measured in any specific unit?

Cumulative Review Exercises

[1.9] **83.** Evaluate $4x^2 + 3x + \dfrac{x}{2}$ when $x = 0$. 0

[2.3] **84. a)** If $-x = -\dfrac{3}{2}$, what is the value of x? $\dfrac{3}{2}$

     b) If $5x = 0$, what is the value of x? 0

[4.4] **85.** Subtract $(2x - 8)$ from $(4x + 7)$. $2x + 15$

[6.6] **86.** Solve the equation $\dfrac{2x}{x - 3} = 2 + \dfrac{3}{x}$ -3

[7.2] **87.** Find the x- and y-intercepts for the line whose equation is $5x - 3y = 15$.
x-intercept: $(3, 0)$; y-intercept: $(0, -5)$

7.4 SLOPE–INTERCEPT AND POINT–SLOPE FORMS OF A LINEAR EQUATION

SSM

Study Guide

CD/Video

MathPro 4/5

PH Math Tutor Center

prenhall.com/Angel

1   Write a linear equation in slope–intercept form.

2   Graph a linear equation using the slope and y-intercept.

3   Use the slope–intercept form to determine the equation of a line.

4   Use the point–slope form to determine the equation of a line.

5   Compare the three methods of graphing linear equations.

In Section 7.1 we introduced the *standard form* of a linear equation, $ax + by = c$. In this section we introduce two more forms, the slope–intercept form and the point–slope form. We begin our discussion with the slope–intercept form.

---

1   Write a Linear Equation in Slope–Intercept Form

A very important form of a linear equation is the **slope–intercept form**, $y = mx + b$. The graph of an equation of the form $y = mx + b$ will always be a straight line with a **slope of m** and a **y-intercept $(0, b)$**. For example, the graph of the equation $y = 3x - 4$ will be a straight line with a slope of 3 and a y-intercept $(0, -4)$. The graph of $y = -2x + 5$ will be a straight line with a slope of -2 and a y-intercept $(0, 5)$.

> ### Slope-Intercept Form of a Linear Equation
>
> $$y = mx + b$$
>
> where m is the slope, and $(0, b)$ is the y-intercept of the line.

$$\overset{\text{slope}}{\underset{\searrow}{}} \quad \overset{\text{y-intercept}}{\underset{\searrow}{}}$$
$$y = mx + b$$

Equations in Slope–Intercept Form	Slope	y-Intercept
$y = 4x - 5$	4	$(0, -5)$
$y = \dfrac{1}{2}x + \dfrac{3}{2}$	$\dfrac{1}{2}$	$\left(0, \dfrac{3}{2}\right)$
$y = -5x + 3$	-5	$(0, 3)$
$y = -\dfrac{2}{3}x - \dfrac{3}{5}$	$-\dfrac{2}{3}$	$\left(0, -\dfrac{3}{5}\right)$

> **To write a linear equation in slope–intercept form,** solve the equation for y.

Once the equation is solved for y, the numerical coefficient of the x-term will be the slope, and the constant term will give the y-intercept.

EXAMPLE 1   Write the equation $-3x + 4y = 8$ in slope–intercept form. State the slope and y-intercept.

Solution   To write this equation in slope–intercept form, we solve the equation for y.

$$-3x + 4y = 8$$
$$4y = 3x + 8$$
$$y = \frac{3x + 8}{4}$$
$$y = \frac{3}{4}x + \frac{8}{4}$$
$$y = \frac{3}{4}x + 2$$

NOW TRY EXERCISE 11   The slope is $\frac{3}{4}$, and the y-intercept is $(0, 2)$.

EXAMPLE 2   Determine whether the two equations represent lines that are parallel, perpendicular, or neither.

a. $2x + y = 10$
$2y = -4x + 12$

b. $3x - 2y = 4$
$6y + 4x = -6$

Solution   We learned in Section 7.3 that two lines that have the same slope are parallel lines, and two lines whose slopes are negative reciprocals are perpendicular lines. We can determine the slope of each line by solving each equation for y. The coefficient of the x term will be the slope.

a. $2x + y = 10$

$$y = \boxed{-2}x + 10$$

$2y = -4x + 12$

$$y = \frac{-4x + 12}{2}$$
$$y = \boxed{-2}x + 6$$

Since both equations have the same slope, -2, the equations represent lines that are parallel. Notice the equations represent two different lines because their y intercepts are different.

b. $3x - 2y = 4$
$$-2y = -3x + 4$$
$$y = \frac{-3x + 4}{-2}$$
$$y = \boxed{\frac{3}{2}}x - 2$$

$6y + 4x = -6$
$$6y = -4x - 6$$
$$y = \frac{-4x - 6}{6}$$
$$y = \boxed{-\frac{2}{3}}x - 1$$

The slope of one line is $\frac{3}{2}$ and the slope of the other line is $-\frac{2}{3}$. Multiplying the slopes we obtain $\left(\frac{3}{2}\right)\left(-\frac{2}{3}\right) = -1$. Since the product is -1, the slopes are negative reciprocals. Therefore the equations represent lines that are perpendicular.

NOW TRY EXERCISE 39

2   Graph a Linear Equation Using the Slope and *y*-Intercept

In Section 7.2 we discussed two methods of graphing a linear equation. They were (1) by plotting points and (2) using the *x*- and *y*-intercepts. Now we present a third method. This method makes use of the slope and the *y*-intercept. Remember that when we solve an equation for *y* we put the equation in slope–intercept form. Once it is in this form, we can determine the slope and *y*-intercept of the graph from the equation.

We graph equations using the slope and *y* intercept in a manner very similar to the way we worked Examples 5 and 6 in Section 7.3. However, when graphing using the slope–intercept form, our starting point is always the *y*-intercept. After you determine the *y*-intercept, a second point can be obtained by moving up and to the right if the slope is positive, or down and to the right if the slope is negative.

EXAMPLE 3   Write the equation $-3x + 4y = 8$ in slope–intercept form; then use the slope and *y*-intercept to graph $-3x + 4y = 8$.

Solution   In Example 1 we solved $-3x + 4y = 8$ for *y*. We found that

$$y = \frac{3}{4}x + 2$$

The slope of the line is $\frac{3}{4}$ and the *y*-intercept is $(0, 2)$. We mark the first point, the *y*-intercept at 2 on the *y*-axis (Fig. 7.45). Now we use the slope $\frac{3}{4}$, to find a second

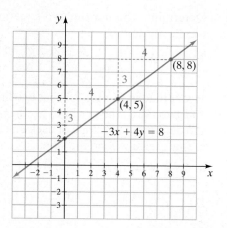

FIGURE 7.45

point. Since the slope is positive, we move 3 units up and 4 units to the right to find the second point. A second point will be at $(4, 5)$. We can continue this process to obtain a third point at $(8, 8)$. Now we draw a straight line through the three points. Notice that the line has a positive slope, which is what we expected.   ✳

EXAMPLE 4   Graph the equation $5x + 3y = 12$ by using the slope and *y*-intercept.

Solution   Solve the equation for *y*.

$$5x + 3y = 12$$
$$3y = -5x + 12$$
$$y = \frac{-5x + 12}{3}$$
$$= -\frac{5}{3}x + 4$$

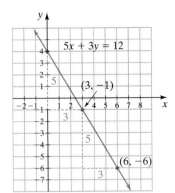

FIGURE 7.46

Thus, the slope is $-\frac{5}{3}$ and the y-intercept is $(0, 4)$. Begin by marking a point at 4 on the y-axis (Fig. 7.46). Then move 5 units down and 3 units to the right to determine the next point. Move down and to the right because the slope is negative and a line with a negative slope must fall as it goes from left to right. Finally, draw the straight line between the plotted points.

NOW TRY EXERCISE 21

3 Use the Slope–Intercept Form to Determine the Equation of a Line

Now that we know how to use the slope–intercept form of a line, we can use it to write the equation of a given line. To do so, we need to determine the slope, m, and y-intercept of the line. Once we determine these values we can write the equation in slope–intercept form, $y = mx + b$. For example, if we determine the slope of a line is -4 and the y-intercept is at 6, the equation of the line is $y = -4x + 6$.

EXAMPLE 5 Determine the equation of the line shown in Figure 7.47.

Solution The graph shows that the y-intercept is at -5. Now we need to determine the slope of the line. Since the graph falls from left to right, it has a negative slope. We can see that the vertical change is 3 units for each horizontal change of 1 unit. Thus, the slope of the line is -3. The slope can also be determined by selecting any two points on the line and calculating the slope. Let's use the point $(-2, 1)$ to represent (x_2, y_2) and the point $(0, -5)$ to represent (x_1, y_1).

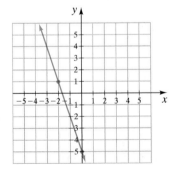

FIGURE 7.47

$$m = \frac{\Delta y}{\Delta x} = \frac{y_2 - y_1}{x_2 - x_1}$$

$$= \frac{1 - (-5)}{-2 - 0}$$

$$= \frac{1 + 5}{-2} = \frac{6}{-2} = -3$$

Again we obtain a slope of -3, Substituting -3 for m and -5 for b into the slope–intercept form of a line gives us the equation of the line in Figure 7.47, which is $y = -3x - 5$.

NOW TRY EXERCISE 29

Now let's look at an application of graphing.

EXAMPLE 6 **Artistic Vases**   Kris, a pottery artist, makes ceramic vases that he sells at art shows. His business has a fixed monthly cost (booth rental, advertising, cell phone, etc.) and a variable cost per vase made (cost of materials, cost of labor, cost for kiln use, etc.). The total monthly cost for making x vases is illustrated in the graph in Figure 7.48.

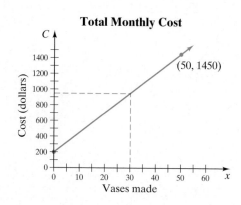

FIGURE 7.48

a) Find the equation of the total monthly cost when x vases are made.
b) Use the equation found in part **a)** to find the total monthly cost if 30 vases are made.
c) Use the graph in Figure 7.48 to see whether your answer in part **b)** appears correct.

Solution   **a)** Understand and Translate   Notice that the vertical axis is cost, C, and not y. The letters or names used on the axes do not change the way we solve the problem. We will use slope–intercept form to write the equation of the line. However, since y is replaced by C, we will use $C = mx + b$, in which b is where the graph crosses the vertical or C-axis. We first note that the graph crosses the vertical axis at 200. Thus, b is 200. Now we need to find the slope of the line. Let's use the point $(0, 200)$ as (x_1, y_1) and $(50, 1450)$ as (x_2, y_2).

Carry Out
$$m = \frac{y_2 - y_1}{x_2 - x_1}$$
$$= \frac{1450 - 200}{50 - 0} = \frac{1250}{50} = 25$$

Answer   The slope is 25. The equation in slope–intercept form is
$$C = mx + b$$
$$= 25x + 200$$

b) To find the monthly cost when 30 vases are sold, we substitute 30 for x.
$$C = 25x + 200$$
$$= 25(30) + 200$$
$$= 750 + 200 = 950$$

The monthly cost when 30 vases are made is $\$950$.

c) If we draw a vertical line up from 30 on the x-axis (the red line), we see that the corresponding cost is about $\$950$. Thus, our answer in part **b)** appears correct.   ✳

NOW TRY EXERCISE 65

4   Use the Point–Slope Form to Determine the Equation of a Line

Thus far, we have discussed the standard form of a linear equation, $ax + by = c$, and the slope–intercept form of a linear equation, $y = mx + b$. Now we will discuss another form, called the *point–slope form*.

When the slope of a line and a point on the line are known, we can use the point–slope form to determine the equation of the line. The **point–slope form** can be obtained by beginning with the slope between any selected point (x, y) and a fixed point (x_1, y_1) on a line.

$$m = \frac{y - y_1}{x - x_1} \quad \text{or} \quad \frac{m}{1} = \frac{y - y_1}{x - x_1}$$

Now cross-multiply to obtain

$$m(x - x_1) = y - y_1 \quad \text{or} \quad y - y_1 = m(x - x_1)$$

Point–Slope Form of a Linear Equation

$$y - y_1 = m(x - x_1)$$

where m is the slope of the line and (x_1, y_1) is a point on the line.

EXAMPLE 7   Write an equation, in slope–intercept form, of the line that goes through the point $(3, 2)$ and has a slope of 5.

Solution   Since we are given a point on the line and the slope of the line, we begin by writing the equation in point–slope form. The slope m is 5. The point on the line is $(3, 2)$; we will use this point for (x_1, y_1) in the formula. We substitute 5 for m, 3 for x_1, and 2 for y_1 in the point–slope form of a linear equation.

$$y - y_1 = m(x - x_1)$$
$$y - 2 = 5(x - 3) \qquad \textit{Equation in point–slope form}$$
$$y - 2 = 5x - 15 \qquad \textit{Distributive property}$$
$$y = 5x - 13 \qquad \textit{Equation in slope–intercept form}$$

NOW TRY EXERCISE 51   The graph of $y = 5x - 13$ has a slope of 5 and passes through the point $(3, 2)$. ✳

The answer to Example 7 was given in slope–intercept form. If we were asked to give the answer in standard form, two acceptable answers would be $-5x + y = -13$ and $5x - y = 13$. Your instructor may specify the form in which the equation is to be given.

In Example 8, we will work an example very similar to Example 7. However, in the solution to Example 8, the equation will contain a fraction. We solved some equations that contained fractions in Chapter 2. Recall that to simplify an equation that contains a fraction, we multiply both sides of the equation by the denominator of the fraction.

EXAMPLE 8   Write an equation, in slope–intercept form, of the line that goes through the point $(6, -2)$ and has a slope $\frac{2}{3}$.

Solution   We will begin with the point–slope form of a line, where m is $\frac{2}{3}$, 6 is x_1, and -2 is y_1.

$$y - y_1 = m(x - x_1)$$
$$y - (-2) = \frac{2}{3}(x - 6) \qquad \textit{Equation in point–slope form}$$
$$y + 2 = \frac{2}{3}(x - 6)$$
$$3(y + 2) = 3 \cdot \frac{2}{3}(x - 6) \qquad \textit{Multiply both sides by 3.}$$
$$3y + 6 = 2(x - 6) \qquad \textit{Distributive property}$$
$$3y + 6 = 2x - 12 \qquad \textit{Distributive property}$$
$$3y = 2x - 18 \qquad \textit{Subtract 6 from both sides.}$$
$$y = \frac{2x - 18}{3} \qquad \textit{Divide both sides by 3.}$$
$$y = \frac{2}{3}x - 6 \qquad \textit{Equation in slope–intercept form}$$
✳

HELPFUL HINT

We have discussed three forms of a linear equation. We summarize the three forms below. It is important that you memorize these forms.

STANDARD FORM	EXAMPLES
$ax + by = c$	$2x - 3y = 8$
	$-5x + y = -2$

SLOPE–INTERCEPT FORM	EXAMPLES
$y = mx + b$	$y = 2x - 5$
m is the slope, $(0, b)$ is the y-intercept	$y = -\dfrac{3}{2}x + 2$

POINT–SLOPE FORM	EXAMPLES
$y - y_1 = m(x - x_1)$	$y - 3 = 2(x + 4)$
m is the slope, (x_1, y_1) is a point on the line	$y + 5 = -4(x - 1)$

We now discuss how to use the point–slope form to determine the equation of a line when two points on the line are known.

EXAMPLE 9 Find an equation of the line through the points $(-1, 3)$ and $(-3, 2)$. Write the equation in slope–intercept form.

Solution To use the point–slope form, we must first find the slope of the line through the two points. To determine the slope, let's designate $(-1, 3)$ as (x_1, y_1) and $(-3, 2)$ as (x_2, y_2).

$$m = \frac{y_2 - y_1}{x_2 - x_1} = \frac{2 - 3}{-3 - (-1)} = \frac{2 - 3}{-3 + 1} = \frac{-1}{-2} = \frac{1}{2}$$

The slope is $\frac{1}{2}$. We can use either point (one at a time) in determining the equation of the line. This example will be worked out using both points to show that the solutions obtained are identical.

Using the point $(-1, 3)$ as (x_1, y_1),

$$y - y_1 = m(x - x_1)$$

$$y - 3 = \frac{1}{2}[x - (-1)]$$

$$y - 3 = \frac{1}{2}(x + 1)$$

$$2 \cdot (y - 3) = 2 \cdot \frac{1}{2}(x + 1) \qquad \textit{Multiply both sides by the LCD, 2.}$$

$$2y - 6 = x + 1$$

$$2y = x + 7$$

$$y = \frac{x + 7}{2} \quad \text{or} \quad y = \frac{1}{2}x + \frac{7}{2}$$

Using the point $(-3, 2)$ as (x_1, y_1),

$$y - y_1 = m(x - x)$$

$$y - 2 = \frac{1}{2}[x - (-3)]$$

$$y - 2 = \frac{1}{2}(x + 3)$$

$$\boxed{2} \cdot (y - 2) = \boxed{2} \cdot \frac{1}{2}(x + 3) \qquad \textit{Multiply both sides by the LCD, 2.}$$

$$2y - 4 = x + 3$$

$$2y = x + 7$$

$$y = \frac{x + 7}{2} \quad \text{or} \quad y = \frac{1}{2}x + \frac{7}{2}$$

NOW TRY EXERCISE 57 Note that the equations for the line are identical.

HELPFUL HINT

In the exercise set at the end of this section, you will be asked to write a linear equation in slope–intercept form. Even though you will eventually write the equation in slope–intercept form, you may need to start your work with the point–slope form. Below we indicate the initial form to use to solve the problem.

Begin with the **slope–intercept form** if you know:

  The slope of the line and the y-intercept

Begin with the **point–slope form** if you know:

a) The slope of the line and a point on the line, or

b) Two points on the line (first find the slope, then use the point–slope form)

---

5 Compare the Three Methods of Graphing Linear Equations

We have discussed three methods to graph a linear equation: (1) plotting points, (2) using the x- and y-intercepts, and (3) using the slope and y-intercept. In Example 10 we graph an equation using all three methods. No single method is always the easiest to use. If the equation is given in slope–intercept form, $y = mx + b$, then graphing by plotting points or by using the slope and y-intercept might be easier. If the equation is given in standard form, $ax + by = c$, then graphing using the intercepts might be easier. Unless your teacher specifies that you should graph by a specific method, you may use the method with which you feel most comfortable. Graphing by plotting points is the most versatile method since it can also be used to graph equations that are not straight lines.

EXAMPLE 10 Graph $3x - 2y = 8$ **a)** by plotting points; **b)** using the x- and y-intercepts; and **c)** using the slope and y-intercept.

Solution For parts **a)** and **c)** we will write the equation in slope–intercept form.

$$3x - 2y = 8$$

$$-2y = -3x + 8$$

$$y = \frac{-3x + 8}{-2} = \frac{3}{2}x - 4$$

a) Plotting Points   To find ordered pairs to plot, we substitute values for x and find the corresponding values of y. Three ordered pairs are indicated in the following table. Next we plot the ordered pairs and draw the graph (Fig. 7.49).

$$y = \frac{3}{2}x - 4$$

x	y
0	-4
2	-1
4	2

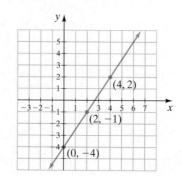

FIGURE 7.49

b) Intercepts   We find the x- and y- intercepts and a check point. Then we plot the points and draw the graph (Fig. 7.50).

$$3x - 2y = 8$$

x-Intercept	*y*-Intercept	Check Point
Let $y = 0$	Let $x = 0$	Let $x = 2$
$3x - 2y = 8$	$3x - 2y = 8$	$3x - 2y = 8$
$3x - 2(0) = 8$	$3(0) - 2y = 8$	$3(2) - 2y = 8$
$3x = 8$	$-2y = 8$	$6 - 2y = 8$
$x = \dfrac{8}{3}$	$y = -4$	$-2y = 2$
		$y = -1$

The three ordered pairs are $\left(\frac{8}{3}, 0\right)$, $(0, -4)$ and $(2, -1)$.

c) Slope and *y*-intercept   The y-intercept is $(0, -4)$, therefore we place a point at -4 on the y-axis. Since the slope is $\frac{3}{2}$, we obtain a second point by moving 3 units up and 2 units to the right. The graph is illustrated in Figure 7.51.

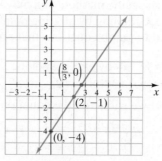

FIGURE 7.50

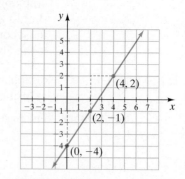

FIGURE 7.51

Notice that we get the same line by all three methods.

Using Your Graphing Calculator

In Example 10, we graphed the equation $y = \frac{3}{2}x - 4$. Let's see what the graph of $y = \frac{3}{2}x - 4$ looks like on a graphing calculator using three different window settings. Figure 7.52 shows the graph of $y = \frac{3}{2}x - 4$ using the *standard window setting*. To obtain the standard window, press the $\boxed{\text{ZOOM}}$ key and then choose option 6, ZStandard. Notice that in the standard window setting, the units on the y-axis are not as long as the units on the x-axis.

In Figure 7.53, we illustrate the graph of $y = \frac{3}{2}x - 4$ using the *square window setting*. To obtain the square window setting, press the $\boxed{\text{ZOOM}}$ key and then choose option 5, ZSquare. This setting makes the units on both axes the same size. This allows the line to be displayed at the correct orientation to the axes.

In Figure 7.54, we illustrate the graph using the *decimal window setting*. To obtain the decimal window setting, press the $\boxed{\text{ZOOM}}$ key and then choose option 4, ZDecimal. The decimal window setting also makes the units the same on both axes, but sets the increment from one pixel (dot) to the next at 0.1 on both axes.

STANDARD WINDOW SETTING
($\boxed{\text{ZOOM}}$: OPTION 6)

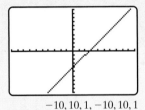

$-10, 10, 1, -10, 10, 1$

FIGURE 7.52

SQUARE WINDOW SETTING
($\boxed{\text{ZOOM}}$: OPTION 5)

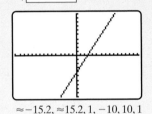

$\approx -15.2, \approx 15.2, 1, -10, 10, 1$

FIGURE 7.53

DECIMAL WINDOW SETTING
($\boxed{\text{ZOOM}}$: OPTION 4)

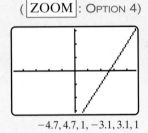

$-4.7, 4.7, 1, -3.1, 3.1, 1$

FIGURE 7.54

Exercises

*Graph each equation using the **a)** standard window, **b)** square window, and **c)** decimal window settings.*

1. $y = 4x - 6$
2. $y = -\frac{1}{5}x + 4$ See graphing answer section, page G6.

1. $y = mx + b$ 5. If the slopes are the same and the y-intercepts are different, the lines are parallel.   7. $y - y_1 = m(x - x_1)$

Exercise Set 7.4

Concept/Writing Exercises

1. Give the slope–intercept form of a linear equation.

2. When you are given an equation in a form other than slope–intercept form, how can you change it to slope–intercept form?   Solve the equation for y.

3. What is the equation of a line, in slope–intercept form, if the slope is 3 and the y-intercept is at -5? $y = 3x - 5$

4. What is the equation of a line, in slope–intercept form, if the slope is -3 and the y-intercept is at 5? $y = -3x + 5$

5. Explain how you can determine whether two equations represent parallel lines without graphing the equations.

6. Explain how you can determine whether two equations represent the same line without graphing the equations?   slopes are same and y-intercepts are same

7. Give the point–slope form of a linear equation.

8. Assume the slope of a line is 2 and the line goes through the origin. Write the equation of the line in point–slope form. $y - 0 = 2(x - 0)$

Practice the Skills

Determine the slope and y-intercept of the line represented by the given equation.

9. $y = 4x - 6$ $4, (0, -6)$
10. $y = -3x + 25$ $-3, (0, 25)$
11. $4x - 3y = 15$ $\frac{4}{3}, (0, -5)$
12. $7x = 5y + 20$ $\frac{7}{5}, (0, -4)$

Determine the slope and y-intercept of the line represented by each equation. Graph the line using the slope and y-intercept.

13. $y = x - 3$ $1, (0, -3)$

14. $y = -x + 5$ $-1, (0, 5)$ **15.** $y = 3x + 2$ $3, (0, 2)$

16. $y = 2x$ $2, (0, 0)$

17. $y = -4x$ $-4, (0, 0)$

18. $3x + y = 4$ $-3, (0, 4)$

19. $-2x + y = -3$ $2, (0, -3)$

20. $3x + 3y = 9$ $-1, (0, 3)$

21. $5x - 2y = 10$ $\frac{5}{2}, (0, -5)$

22. $-x + 2y = 8$ $\frac{1}{2}, (0, 4)$

23. $6x + 12y = 18$ $-\frac{1}{2}, \left(0, \frac{3}{2}\right)$

24. $4x = 6y + 9$ $\frac{2}{3}, \left(0, -\frac{3}{2}\right)$

25. $-6x + 2y - 8 = 0$ $3, (0, 4)$

26. $16y = 8x + 32$ $\frac{1}{2}, (0, 2)$

27. $3x = 2y - 4$ $\frac{3}{2}, (0, 2)$

28. $20x = 80y + 40$ $\frac{1}{4}, \left(0, -\frac{1}{2}\right)$

Determine the equation of each line.   **13.–28.** See graphing answer section, pages G6 and G7.

29.
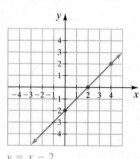
$y = x - 2$

30.
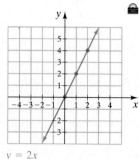
$y = 2x$

 31.
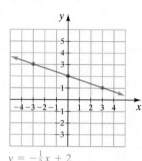
$y = -\frac{1}{3}x + 2$

32.
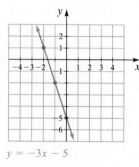
$y = -3x - 5$

33.
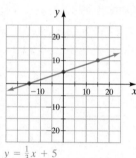
$y = \frac{1}{3}x + 5$

34.
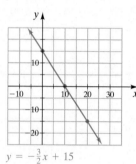
$y = -\frac{3}{2}x + 15$

35.
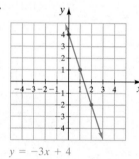
$y = -3x + 4$

36.
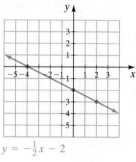
$y = -\frac{1}{2}x - 2$

Determine whether each pair of lines are parallel, perpendicular or neither.

37. $y = 4x + 2$
$y = 4x - 1$ parallel

38. $2x + 3y = 8$
$y = -\frac{2}{3}x + 5$ parallel

39. $4x + 2y = 9$
$4x = 8y + 4$ perpendicular

40. $3x - 5y = 7$
$5y + 3x = 2$ neither

 41. $3x + 5y = 9$
$6x = -10y + 9$ parallel

42. $8x + 2y = 8$
$x - 9 = 4y$ perpendicular

43. $y = \frac{1}{2}x - 4$
$2y = 6x + 9$ neither

44. $2y - 6 = -5x$
$y = -\frac{5}{2}x - 2$ parallel

45. $5y = 2x + 3$
$-10x = 4y + 8$ perpendicular

46. $3x - 9y = 12$
$-3x + 9y = 18$ parallel

47. $3x + 7y = 8$
$7x + 3y = 8$ neither

48. $5x - 6y = 10$
$-6x + 5y = 20$ neither

Problem Solving

Write the equation of each line, with the given properties, in slope–intercept form.

 49. Slope $= 3$, through $(0, 2)$ $y = 3x + 2$

50. Slope $= 4$, through $(2, 3)$ $y = 4x - 5$

51. Slope $= -3$, through $(-4, 5)$ $y = -3x - 7$

52. Slope $= -3$, through $(4, 0)$ $y = -3x + 12$

53. Slope $= \frac{1}{2}$, through $(-1, -5)$ $y = \frac{1}{2}x - \frac{9}{2}$

54. Slope $= -\frac{2}{3}$, through $(-1, -2)$ $y = -\frac{2}{3}x - \frac{8}{3}$

55. Slope $= \dfrac{2}{5}$, y-intercept is $(0, 6)$ $y = \dfrac{2}{5}x + 6$

56. Slope $= \dfrac{4}{9}$, y-intercept is $\left(0, -\dfrac{2}{3}\right)$ $y = \dfrac{4}{9}x - \dfrac{2}{3}$

57. Through $(-4, -2)$ and $(-2, 4)$ $y = 3x + 10$

58. Through $(6, 3)$ and $(5, 2)$ $y = x - 3$

59. Through $(-6, 9)$ and $(6, -9)$ $y = -\dfrac{3}{2}x$

60. Through $(3, 0)$ and $(-3, 5)$ $y = -\dfrac{5}{6}x + \dfrac{5}{2}$

61. Through $(10, 3)$ and $(0, -2)$ $y = \dfrac{1}{2}x - 2$

62. Through $(-6, -2)$ and $(5, -3)$ $y = -\dfrac{1}{11}x - \dfrac{28}{11}$

63. Slope $= 6.3$, y-intercept is $(0, -4.5)$ $y = 6.3x - 4.5$

64. Slope $= -\dfrac{5}{8}$, y-intercept is $\left(0, -\dfrac{7}{10}\right)$ $y = -\dfrac{5}{8}x - \dfrac{7}{10}$

65. ***Weight Loss Clinic*** Stacy Best owns a weight loss clinic. She charges her clients a one-time membership fee. She also charges per pound of weight lost. Therefore, the more successful she is at helping clients lose weight, the more income she will receive. The following graph shows a client's cost for losing weight.

Cost of Losing Weight

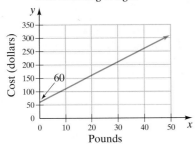

a) Find the equation that represents the cost for a client who loses x pounds. $y = 5x + 60$

b) Use the equation found in part **a)** to determine the cost for a client who loses 30 pounds. $\$210$

66. ***Submarine Submerges*** A submarine is submerged below sea level. Tom Johnson, the captain, orders the ship to dive slowly. The following graph illustrates the submarine's depth at a time t minutes after the submarine begins to dive.

Submarine's Depth

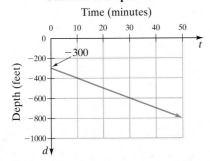

a) Find the equation that represents the depth at time t. $y = -10t - 300$

b) Use the equation found in part **a)** to find the submarine's depth after 20 minutes. -500 ft

67. Suppose that you were asked to write the equation of a line with the properties given below. Which form of a linear equation—standard form, slope–intercept form, or point–slope form—would you start with? Explain your answer.   **67. a)** slope–intercept form

a) The slope of the line and the y-intercept of the line

b) The slope and a point on the line   point–slope form

c) Two points on the line   point–slope form

68. Consider the two equations $40x - 60y = 100$ and $-40x + 60y = 80$.

a) When these equations are graphed, will the two lines have the same slope? Explain how you determined your answer.   yes

b) When these two equations are graphed, will they be parallel lines?   yes

69. Assume the slope of a line is 2 and two points on the line are $(-5, -4)$ and $(2, 10)$.

a) If you use $(-5, -4)$ as (x_1, y_1) and then $(2, 10)$ as (x_1, y_1) will the appearance of the two equations be the same in point–slope form? Explain.   no

b) Find the equation, in point–slope form, using $(-5, -4)$ as (x_1, y_1). $y + 4 = 2(x + 5)$

c) Find the equation, in point–slope form, using $(2, 10)$ as (x_1, y_1). $y - 10 = 2(x - 2)$

d) Write the equation obtained in part **b)** in slope–intercept form. $y = 2x + 6$

e) Write the equation obtained in part **c)** in slope–intercept form. $y = 2x + 6$

f) Are the equations obtained in parts **d)** and **e)** the same? If not, explain why.   yes

70. Assume the slope of a line is -3 and two points on the line are $(-1, 8)$ and $(2, -1)$.

a) If you use $(-1, 8)$ as (x_1, y_1) and then $(2, -1)$ as (x_1, y_1) will the appearance of the two equations be the same in point–slope form? Explain.   no

b) Find the equation, in point–slope form, using $(-1, 8)$ as (x_1, y_1). $y - 8 = -3(x + 1)$

c) Find the equation, in point–slope form, using $(2, -1)$ as (x_1, y_1). $y + 1 = -3(x - 2)$

d) Write the equation obtained in part **b)** in slope–intercept form. $y = -3x + 5$

e) Write the equation obtained in part **c)** in slope–intercept form. $y = -3x + 5$

f) Are the equations obtained in parts **d)** and **e)** the same? If not, explain why.   yes

Challenge Problems

71. ***Unit Conversions*** The following graph shows the approximate relationship between speed in miles per hour and feet per second.

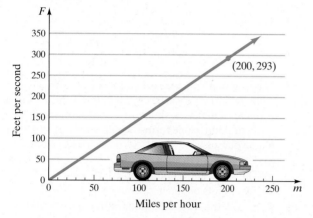

a) Determine the slope of the line. 1.465
b) Determine the equation of the line. $f = 1.465\,m$
c) At the 2002 Daytona 500, Ward Burton, the winner, had an average speed of 130.81 miles per hour. (See photo). Use the equation you obtained in part b) to determine the speed in feet per second. ≈191.64 ft/sec

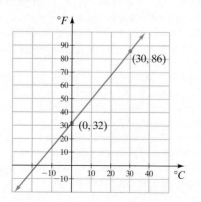

d) Use the graph to estimate a speed of 100 miles per hour in feet per second. 150 ft/sec
e) Use the graph to estimate a speed of 80 feet per second in miles per hour. 55 mph

72. ***Temperature*** The following graph shows the relationship between Fahrenheit temperature and Celsius temperature.

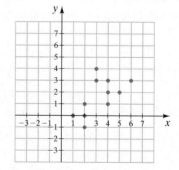

a) Determine the slope of the line. $\frac{9}{5}$
b) Determine the equation of the line in slope–intercept form. $F = \frac{9}{5}C + 32$
c) Use the equation (or formula) you obtained in part **b)** to find the Fahrenheit temperature when the Celsius temperature is 20°. 68°F
d) Use the graph to estimate the Celsius temperature when the Fahrenheit temperature is 100°. 38°C
e) Estimate the Celsius temperature that corresponds to a Fahrenheit temperature of 0°. −18°C

73. Determine the equation of the line with y-intercept at 4 that is parallel to the line whose equation is $2x + y = 6$. Explain how you determined your answer. $y = -2x + 4$

74. Will a line through the points $(60, 30)$ and $(20, 90)$ be parallel to the line with x-intercept at 2 and y-intercept at 3? Explain how you determined your answer. yes

75. Write an equation of the line parallel to the graph of $3x - 4y = 6$ that passes through the point $(-4, -1)$. $y = \frac{3}{4}x + 2$

76. Determine the equation of the straight line that intersects the greatest number of shaded points on the following graph. $y = x - 3$

Group Activity

Discuss and answer Exercise 77 as a group, according to the instructions.

77. Consider the equation $-3x + 2y = 4$.

   a) Group member 1: Explain how to graph this equation by plotting points. Then graph the equation by plotting points.

   b) Group member 2: Explain how to graph this equation using the intercepts. Then graph the equation using the intercepts.

   c) Group member 3: Explain how to graph this equation using the slope and y-intercept. Then graph the equation using the slope and y-intercept.

   d) As a group, compare your graphs. Did you all obtain the same graph? If not, determine why.
See graphing answer section, page G7.

Cumulative Review Exercises

[1.5] **78.** Insert either $>$, $<$, or $=$ in the shaded area to make the statement true: $|-4|$ ▪ $|-6|$. $<$

[2.7] **79.** Solve $2(x - 3) \geq 5x + 6$ and graph the solution on a number line. $x \leq -4$: ⟵━━━━⟶
-4

[3.1] **80.** Solve $i = prt$ for r. $r = \dfrac{i}{pt}$

[5.2] **81.** Factor $x^2 - 2xy + 3xy - 6y^2$ by grouping.
$(x - 2y)(x + 3y)$

[6.6] **82.** Solve $\dfrac{x}{3} - \dfrac{3x + 2}{6} = \dfrac{1}{2}$. -5

7.5 GRAPHING LINEAR INEQUALITIES

SSM

Study Guide

CD/Video

MathPro 4/5

PH Math Tutor Center

prenhall.com/Angel

1   Graph linear inequalities in two variables.

1   Graph Linear Inequalities in Two Variables

A **linear inequality** results when the equal sign in a linear equation is replaced with an inequality sign.

Examples of Linear Inequalities in Two Variables

$$3x + 2y > 4 \qquad -x + 3y < -2$$
$$-x + 4y \geq 3 \qquad 4x - y \leq 4$$

TEACHING TIP
Point out that the line divides the grid into 2 regions: one region above the line and one region below the line.

To Graph a Linear Inequality in Two Variables

1. Replace the inequality symbol with an equal sign.

2. Draw the graph of the equation in step 1. If the original inequality contained the symbol $\geq$ or $\leq$, draw the graph using a solid line. If the original inequality contained the symbol $>$ or $<$, draw the graph using a dashed line.

3. Select any point not on the line and determine whether this point is a solution to the original inequality. If the selected point is a solution, shade the region on the side of the line containing this point. If the selected point does not satisfy the inequality, shade the region on the side of the line not containing this point.

EXAMPLE 1 Graph the inequality $y < 2x - 4$.

Solution First we graph the equation $y = 2x - 4$ (Fig. 7.55). Since the original inequality contains an is-less-than sign, $<$, we use a dashed line when drawing the graph. The dashed line indicates that the points on this line are not solutions to the inequality $y < 2x - 4$.

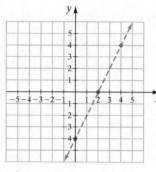

FIGURE 7.55

Next we select a point not on the line and determine whether this point satisfies the inequality. Often the easiest point to use is the origin, $(0, 0)$. In the check we will use the symbol $\overset{?}{<}$ until we determine whether the statement is true or false.

Check

$$y < 2x - 4$$
$$0 \overset{?}{<} 2(0) - 4$$
$$0 \overset{?}{<} 0 - 4$$
$$0 < -4 \qquad \textit{False}$$

Since 0 is not less than -4, the point $(0, 0)$ does not satisfy the inequality. The solution will therefore be all the points on the opposite side of the line from the point $(0, 0)$. We shade this region (Fig. 7.56).

Every point in the shaded region satisfies the given inequality. Let's check a few selected points A, B, and C.

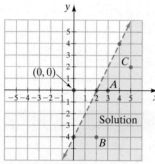

FIGURE 7.56

Point A	Point B	Point C
$(3, 0)$	$(2, -4)$	$(5, 2)$
$y < 2x - 4$	$y < 2x - 4$	$y < 2x - 4$
$0 \overset{?}{<} 2(3) - 4$	$-4 \overset{?}{<} 2(2) - 4$	$2 \overset{?}{<} 2(5) - 4$
$0 < 2$ _True_	$-4 < 0$ _True_	$2 < 6$ _True_

NOW TRY EXERCISE 11

All points in the shaded region in Figure 7.56 satisfy the inequality $y < 2x - 4$. The points in the unshaded region to the left of the dashed line would satisfy the inequality $y > 2x - 4$.

EXAMPLE 2 Graph the inequality $y \geq -\dfrac{1}{2}x$.

Solution

Graph the equation $y = -\frac{1}{2}x$. Since the inequality symbol is $\geq$, we use a solid line to indicate that the points on the line are solutions to the inequality (Fig. 7.57 on the next page). Since the point $(0, 0)$ is on the line, we cannot select it as our test point. Let's select the point $(3, 1)$.

$$y \geq -\frac{1}{2}x$$

$$1 \overset{?}{\geq} -\frac{1}{2}(3)$$

$$1 \geq -\frac{3}{2} \qquad \textit{True}$$

Since the ordered pair $(3, 1)$ satisfies the inequality, every point on the same side of the line as $(3, 1)$ will also satisfy the inequality $y \geq -\frac{1}{2}x$. We shade this region (Fig. 7.58 on the next page). Every point in the shaded region as well as every point on the line satisfies the inequality.

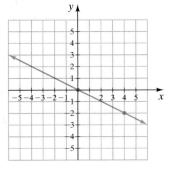

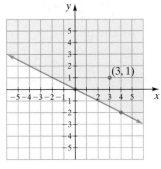

NOW TRY EXERCISE 9   FIGURE 7.57   FIGURE 7.58

TEACHING TIP
Have students graph $3x - y > 6$.
Be sure they remember to change
the inequality when multiplying or
dividing by a negative number.

In some of the exercises you may need to solve the inequality for y before graphing. For example, to graph $-2x + y < -4$ you would solve the inequality for y to obtain $y < 2x - 4$. Then you would graph the inequality $y < 2x - 4$. Note that $y < 2x - 4$ was graphed in Figure 7.56.

1. Points on the line satisfy the $=$ part of the inequality.   **2.** Points on line satisfy equation ($=$) but not any inequality that is strictly greater than or less than.

Exercise Set 7.5

Concept/Writing Exercises

1. When graphing inequalities that contain either $\leq$ or $\geq$, explain why the points on the line will be solutions to the inequality.

2. When graphing inequalities that contain either $<$ or $>$, explain why the points on the line will not be solutions to the inequality.

3. How do the graphs of $2x + 3y > 6$ and $2x + 3y < 6$ differ?   shading on opposite sides of the line

4. How do the graphs of $4x - 3y < 6$ and $4x - 3y \leq 6$ differ?   first graph, dashed line; second, solid line

Practice the Skills

Graph each inequality.   **5.–24.** See graphing answer section, pages G7 and G8.

5. $y > -3$

6. $x > 3$

7. $x \geq \dfrac{3}{2}$

8. $y < x$

9. $y \leq 3x$

10. $y > -2x$

11. $y < x - 4$

12. $y < 2x + 1$

13. $y < -3x + 4$

14. $y \geq 2x - 3$

15. $y \geq \dfrac{1}{2}x - 4$

16. $y > -\dfrac{x}{2} + 2$

17. $y > \dfrac{1}{3}x + 1$

18. $y > \dfrac{1}{2}x - 2$

19. $3x + y \leq 5$

20. $3x - 2 < y$

21. $2x + y \leq 3$

22. $3y > 2x - 3$

23. $y + 2 > -x$

24. $4x - 2y \leq 6$

Problem Solving

25. Determine whether $(4, 2)$ is a solution to each inequality.

  a) $2x + 4y < 16$ no   **b)** $2x + 4y > 16$ no

  c) $2x + 4y \geq 16$ yes   **d)** $2x + 4y \leq 16$ yes

26. Determine whether $(-3, 5)$ is a solution to each inequality.

  a) $-2x + 3y < 9$ no   **b)** $-2x + 3y > 9$ yes

  c) $-2x + 3y \geq 9$ yes   **d)** $-2x + 3y \leq 9$ no

27. If an ordered pair is not a solution to the inequality $ax + by < c$, must the ordered pair be a solution to $ax + by > c$? Explain. no

28. If an ordered pair is not a solution to the inequality $ax + by \leq c$, must the ordered pair be a solution to $ax + by > c$? Explain. yes

29. If an ordered pair is a solution to $ax + by > c$, is it possible for the ordered pair to be a solution to $ax + by \leq c$? Explain. no

30. Is it possible for an ordered pair to be a solution to both $ax + by < c$ and $ax + by > c$? Explain. no

31. Determine whether the given phrase means: less than, less than or equal to, greater than, or greater than or equal to.
    a) no more than
    b) no less than
    c) at most
    d) at least

32. Consider the two inequalities $2x + 1 > 5$ and $2x + y > 5$.
    a) How many variables does the inequality $2x + 1 > 5$ contain? 1
    b) How many variables does the inequality $2x + y > 5$ contain? 2
    c) What is the solution to $2x + 1 > 5$? Indicate the solution on a number line.
    d) Graph $2x + y > 5$.

33. Which of the following inequalities have the same graphs? Explain how you determined your answer.
    a) $2x - y > 4$
    b) $-2x + y < -4$
    c) $y < 2x - 4$
    d) $-2y + 4x < -8$

31. a) less than or equal to   b) greater than or equal to   c) less than or equal to   d) greater than or equal to   32. c) $x > 2$,

d) See graphing answers section, page G8.

33. a), b), and c)

Cumulative Review Exercises

[1.4] 34. Consider the set of numbers

$$\left\{ 2, -5, 0, \sqrt{7}, \frac{2}{5}, -6.3, \sqrt{3}, -\frac{23}{34} \right\}.$$

List those that are a) natural numbers; b) whole numbers; c) rational numbers; d) irrational numbers; e) real numbers.

34. a) 2   b) 2, 0   c) $2, -5, 0, \frac{2}{5}, -6.3, -\frac{23}{34}$ d) $\sqrt{7}, \sqrt{3}$
e) $2, -5, 0, \sqrt{7}, \frac{2}{5}, -6.3, \sqrt{3}, -\frac{23}{34}$

[2.5] 35. Solve the equation $2(x + 3) + 2x = x + 4$. $-\frac{2}{3}$

[4.6] 36. Divide $\dfrac{10x^2 - 15x + 30}{5x}$. $2x - 3 + \dfrac{6}{x}$.

[6.1] 37. Simplify $\dfrac{2x}{2x^2 + 4xy}$. $\dfrac{1}{x + 2y}$.

7.6 FUNCTIONS

SSM Study Guide CD/Video

MathPro 4/5 PH Math Tutor Center prenhall.com/Angel

1. Find the domain and range of a relation.
2. Recognize functions.
3. Evaluate functions.
4. Graph linear functions.

In this section we introduce relations and functions. As you will learn shortly, a function is a special type of relation. Functions are a common thread in mathematics courses from algebra through calculus. In this section we give an informal introduction to relations and functions.

1 Find the Domain and Range of a Relation

First we will discuss **relations**.

> **DEFINITION** A **relation** is any set of ordered pairs.

Since a relation is *any* set of points, *every graph will represent a relation.*

TEACHING TIP
You may want to describe relations in terms of nonnumeric relations such as (x, y) where x's biological daughter is y or (x, y) where x's biological mother is y.

Examples of Relations

$$\{(3, 5), (4, 6), (5, 9), (7, 12)\}$$
$$\{(1, 2), (2, 2), (3, 2), (4, 2)\}$$
$$\{(3, 2), (3, 3), (3, 4), (3, 5), (3, 6)\}$$

In the ordered pair (x, y), the x and y are called the **components of the ordered pair**. The **domain** of a relation is the set of *first components* in the set of ordered pairs. For example,

Relation	Domain
$\{(3, 5), (4, 6), (5, 9), (7, 12)\}$	$\{3, 4, 5, 7\}$
$\{(1, 2), (2, 2), (3, 2), (4, 2)\}$	$\{1, 2, 3, 4\}$
$\{(3, 2), (3, 3), (3, 4), (3, 5), (3, 6)\}$	$\{3\}$

The **range** of a relation is the set of *second components* in the set of ordered pairs. For example,

Relation	Range
$\{(3, 5), (4, 6), (5, 9), (7, 12)\}$	$\{5, 6, 9, 12\}$
$\{(1, 2), (2, 2), (3, 2), (4, 2)\}$	$\{2\}$
$\{(3, 2), (3, 3), (3, 4), (3, 5), (3, 6)\}$	$\{2, 3, 4, 5, 6\}$

In relations, the sets can contain elements other than numbers. For example,

Relation

$\{(\text{Carol, Seat 1}), (\text{Mary, Seat 2}), (\text{John, Seat 3}), (\text{Olonso, Seat 4})\}$

Domain

$\{\text{Carol, Mary, John, Olonso}\}$

Range

$\{\text{Seat 1, Seat 2, Seat 3, Seat 4}\}$

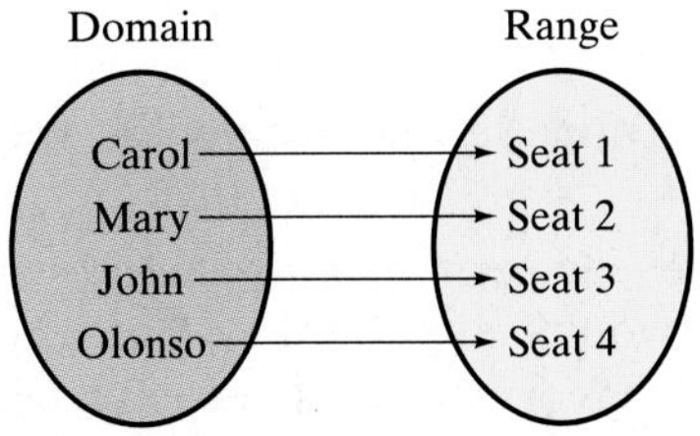

FIGURE 7.59

Figure 7.59 illustrates the relation between the person and the seat number.

2 Recognize Functions

Now we are ready to discuss functions. Consider the relation shown in Figure 7.59. Notice that each member in the domain corresponds with exactly one member of the range. That is, each person is assigned to exactly one seat. This is an example of a **function**.

DEFINITION

A **function** is a set of ordered pairs in which each first component corresponds to exactly one second component.

TEACHING TIP
Point out that the first component is often thought of as the input and the second is the output. In a function, the output depends on the input.

Since a function is a special type of relation, our discussion of domain and range applies to functions. In the definition of a function, the set of first components represents the domain of the function and the set of second components represents the range of the function.

EXAMPLE 1   Consider the relations in Figures 7.60a) through d). Which relations are functions?

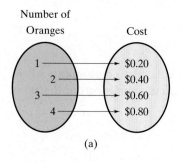

(a)

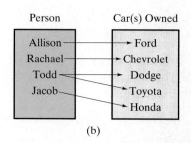

(b)

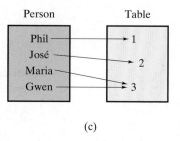

(c)

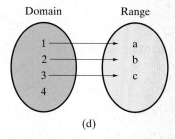

(d)

FIGURE 7.60

Solution   **a)** If we wished, we could represent the information given in the figure as the following set of ordered pairs: {(1, $0.20), (2, $0.40), (3, $0.60), (4, $0.80)}. Notice that each first component corresponds to exactly one second component. Therefore, this relation is a function.

b) If we look at Figure 7.60b, we can see that Todd does *not* correspond to exactly one car. If we were to list the set of ordered pairs to represent this relation, the set would contain the ordered pairs (Todd, Dodge) and (Todd, Toyota). Therefore, each first component does not correspond to exactly one second component and this relation is not a function.

c) Although both Maria and Gwen share a table, each person corresponds to exactly one table. If we listed the ordered pairs we would have (Phil, 1), (José, 2), (Maria, 3), (Gwen, 3). Note that each *first* component corresponds to exactly one second component. Therefore, this relation is a function.

d) Since the number 4 in the domain does not correspond to any component in the range, this relation is not a function. *Every* component in the domain must correspond to exactly one component in the range for the relation to be a function.

NOW TRY EXERCISE 19

TEACHING TIP
When discussing Example 2, point out that cost depends on the number of apples purchased. Stress that functions are useful because they allow us to predict the output when the input is known.

The functions given in Example 1a) and 1c) were determined by looking at correspondences in figures. Most functions have an infinite number of ordered pairs and are usually defined with an equation (or rule) that tells how to obtain the second component when you are given the first component. In Example 2, we determine a function from the information provided.

EXAMPLE 2   **The Cost of Apples**   Assume each apple costs $0.30. Write a function to determine the cost, c, when n apples are purchased.

Solution   When one apple is purchased, the cost is $0.30. When two apples are purchased, the cost is 2($0.30), and when n apples are purchased, the cost is $n($0.30) or $0.30n$. The function $c = 0.30n$ will give the cost, c, in dollars, when n apples are purchased. Note that for any value of n, there is exactly one value of c.

30 cents each

EXAMPLE 3   Determine whether the following sets of ordered pairs are functions.

a) $\{(4, 5), (3, 2), (-2, -3), (2, 5), (1, 6)\}$

b) $\{(4, 5), (3, 2), (-2, -3), (4, 1), (5, -2)\}$

Solution   **a)** Since each first component corresponds with exactly one second component, this set of ordered pairs is a function.

b) The ordered pairs $(4, 5)$ and $(4, 1)$ contain the same first component. Therefore, each first component does not correspond to exactly one second component, and this set of ordered pairs is not a function.

    In Figures 7.61a and 7.61b we plot the ordered pairs from Example 3**a)** and 3**b)**, respectively.

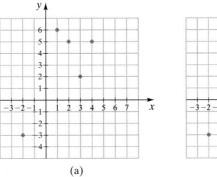

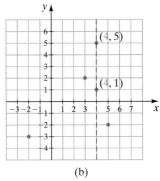

(a)

First set of ordered pairs,
Function

(b)

Second set of ordered pairs,
Not a function

FIGURE 7.61

Consider Figure 7.61a. If a vertical line is drawn through each point, no vertical line intersects more than one point. This indicates that no two ordered pairs have the same first (or x) coordinate and that each value of x in the domain corresponds to exactly one value of y in the range. Therefore, this set of points represents a function.

    Now look at Figure 7.61b. If a vertical line is drawn through each point, one vertical line passes through two points. The dashed red vertical line intersects both $(4, 5)$ and $(4, 1)$. Each element in the domain *does not* correspond to exactly one element in the range. The number 4 in the domain corresponds to two numbers, 5 and 1, in the range. Therefore, this set of ordered pairs does *not* represent a function.

    To determine whether a graph represents a function, we can use the **vertical line test** just described.

Vertical Line Test

If a vertical line can be drawn through any part of a graph and the vertical line intersects another part of the graph, then each value of x does not correspond to exactly one value of y and the graph does not represent a function.

    If a vertical line cannot be drawn to intersect the graph at more than one point, each value of x corresponds to exactly one value of y and the graph represents a function.

EXAMPLE 4   Using the vertical line test to determine which graphs represent functions.

a)   b)   c)

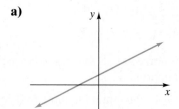 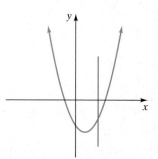 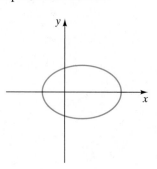

Solution

a)   b)   c)

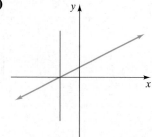 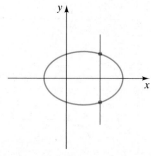

The graphs in parts **a)** and **b)** represent functions since it is not possible to draw a vertical line that intersects the graph at more than one point. The graph in part **c)** does not represent a function since a vertical line can be drawn to intersect the graph at more than one point.

NOW TRY EXERCISE 31

We see functions all around us. Consider the information provided in Table 7.1, which may apply to some salespeople. This table of values is a function because each amount of sales corresponds to exactly one income.

| TABLE 7.1 Monthly Income ||
Sales (dollars)	Income (dollars)
0	$1500
$5000	$1800
$10,000	$2100
$15,000	$2400
$20,000	$2700
$25,000	$3000

Consider the graph shown in Figure 7.62. This graph represents a function. Notice that each year corresponds with a unique percent of schools. Both graphs in Figure 7.63 represent functions. Each graph passes the vertical line test.

NOW TRY EXERCISE 53

Public Schools in the United States Connected to the Internet

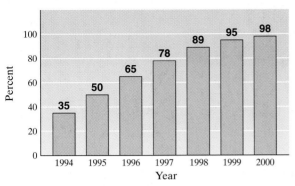

FIGURE 7.62

Life Expectancy at Birth

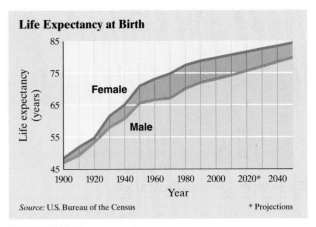

Source: U.S. Bureau of the Census   * Projections

FIGURE 7.63

3 Evaluate Functions

When a function is represented by an equation, it is often convenient to use **function notation**, $f(x)$. If we were to graph $y = x + 2$, we would see that it is a function because its graph passes the vertical line test. The value of y in the equation or function depends on the value of x. Therefore, we say that y is a function of x, and we can write $y = f(x)$. The notation $y = f(x)$ is used to show that y is a function of the variable x. If we wish, we can write

$$y = f(x) = x + 2 \quad \text{or simply} \quad f(x) = x + 2$$

The notation $f(x)$ is read "f of x" *and does not mean f times x.*

To evaluate a function for a specific value of x, we substitute that value for x everywhere the x appears in the function. For example, to evaluate the function $f(x) = x + 2$ at $x = 1$, we do the following:

$$f(x) = x + 2$$
$$f(1) = 1 + 2 = 3$$

Thus, when x is 1, $f(x)$ or y is 3.

When $x = 4$, $f(x)$ or $y = 6$, as illustrated below.

$$y = f(x) = x + 2$$
$$y = f(4) = 4 + 2 = 6$$

The notation $f(1)$ is read "f of 1" and $f(4)$ is read "f of 4."

EXAMPLE 5 For the function $f(x) = x^2 + 4x - 5$, find **a)** $f(3)$ and **b)** $f(-6)$. **c)** If $x = -1$, determine the value of y.

Solution **a)** Substitute 3 for each x in the function, and then evaluate.

TEACHING TIP
When discussing Example 5**a)**, point out that $f(3)$ is a shorthand for "Evaluate the expression that the function equals for $x = 3$."

$$f(x) = x^2 + 4x - 5$$
$$f(3) = 3^2 + 4(3) - 5$$
$$= 9 + 12 - 5 = 16$$

b)
$$f(x) = x^2 + 4x - 5$$
$$f(-6) = (-6)^2 + 4(-6) - 5$$
$$= 36 - 24 - 5 = 7$$

c) Since $y = f(x)$, we evaluate $f(x)$ at -1.

$$f(x) = x^2 + 4x - 5$$
$$f(-1) = (-1)^2 + 4(-1) - 5$$
$$= 1 - 4 - 5 = -8$$

NOW TRY EXERCISE 39   Thus, when $x = -1$, $y = -8$.   ✳

4 Graph Linear Functions

The graphs of all equations of the form $y = ax + b$ will be straight lines that are functions. Therefore, we may refer to equations of the form $y = ax + b$ as **linear functions**. Equations of the form $f(x) = ax + b$ are also linear functions since $f(x)$ is the same as y. We may graph linear functions as shown in Example 6.

EXAMPLE 6 Graph $f(x) = 2x + 4$.

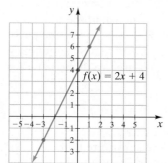

Solution Since $f(x)$ is the same as y, write $y = f(x) = 2x + 4$. Select values for x and find the corresponding values for y or $f(x)$.

$$y = f(x) = 2x + 4$$

Let $x = -3$ $y = f(-3) = 2(-3) + 4 = -2$
Let $x = 0$ $y = f(0) = 2(0) + 4 = 4$
Let $x = 1$ $y = f(1) = 2(1) + 4 = 6$

x	y
-3	-2
0	4
1	6

Now plot the points and draw the graph of the function (Fig. 7.64).   ✳

FIGURE 7.64   NOW TRY EXERCISE 45

EXAMPLE 7 **Ice Skating Rink** The weekly profit, p, of an ice skating rink is a function of the number of skaters per week, n. The function approximating the profit is $p = f(n) = 8n - 600$, where $0 \le n \le 400$.

a) Construct a graph showing the relationship between the number of skaters and the weekly profit.

b) Estimate the profit if there are 200 skaters in a given week.

Rockefeller Plaza, New York City

Solution

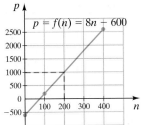

$$p = f(n) = 8n - 600$$

FIGURE 7.65

NOW TRY EXERCISE 57

a) Select values for n, and find the corresponding values for p. Then draw the graph (Fig. 7.65). Notice there are no arrowheads on the line because the function is defined only for values of n between 0 and 400 inclusive.

$$p = f(n) = 8n - 600$$

Let $n = 0$ $p = f(0) = 8(0) - 600 = -600$

Let $n = 100$ $p = f(100) = 8(100) - 600 = 200$

Let $n = 400$ $p = f(400) = 8(400) - 600 = 2600$

n	p
0	−600
100	200
400	2600

b) Using the red dashed line on the graph, we can see that if there are 200 skaters the weekly profit is $1000.

3. a set of ordered pairs in which each first component corresponds to exactly one second component   **5. a)** the set of first components in the set of ordered pairs   **b)** the set of second components in the set of ordered pairs   **6. b)** yes

Exercise Set 7.6

Concept/Writing Exercises

1. What is a relation?   any set of ordered pairs

2. Is every graph in the Cartesian coordinate system a relation? Explain.   yes

3. What is a function?

4. Is every graph in the Cartesian coordinate system a function? Explain.   no

5. a) What is the domain of a relation or function?

b) What is the range of a relation or function?

6. a) Is every relation a function?   no

b) Is every function a relation? Explain your answer.

7. If two distinct ordered pairs in a relation have the same first coordinate, can the relation be a function? Explain.   no

8. In a function is it necessary for each value of y in the range to correspond to exactly one value of x in the domain? Explain.   no

Practice the Skills

Determine which of the relations are also functions. Give the domain and range of each relation or function.

9. $\{(5, 4), (2, 2), (3, 5), (1, 3), (4, 1)\}$
function, D: $\{1, 2, 3, 4, 5\}$, R: $\{1, 2, 3, 4, 5\}$

10. $\{(2, 1), (4, 0), (3, 5), (2, 2), (5, 1)\}$
relation, D: $\{2, 3, 4, 5\}$, R: $\{0, 1, 2, 5\}$

11. $\{(5, -2), (3, 0), (3, 2), (1, 4), (2, 4), (7, 5)\}$
relation, D: $\{1, 2, 3, 5, 7\}$, R: $\{-2, 0, 2, 4, 5\}$

12. $\{(-2, 1), (1, -3), (3, 4), (4, 5), (-2, 0)\}$
relation, D: $\{-2, 1, 3, 4\}$, R: $\{-3, 0, 1, 4, 5\}$

13. $\{(5, 0), (4, -4), (0, -1), (3, 2), (1, 1)\}$
function, D: $\{0, 1, 3, 4, 5\}$, R: $\{-4, -1, 0, 1, 2\}$

14. $\{(-6, 3), (-3, 4), (0, 3), (5, 2), (3, 5), (2, 5)\}$
function, D: $\{-6, -3, 0, 2, 3, 5\}$, R: $\{2, 3, 4, 5\}$

15. $\{(3, 0), (0, -3), (1, 5), (1, 0), (1, 2)\}$
relation, D: $\{0, 1, 3\}$, R: $\{-3, 0, 2, 5\}$

16. $\{(4, -3), (3, -7), (4, -9), (3, 5)\}$
relation, D: $\{3, 4\}$, R: $\{-9, -7, -3, 5\}$

17. $\{(0, 3), (1, 3), (2, 3), (3, 3), (4, 3)\}$
function, D: $\{0, 1, 2, 3, 4\}$, R: $\{3\}$

18. $\{(3, 5), (2, 4), (1, 0), (0, 1), (-1, 5)\}$
function, D: $\{-1, 0, 1, 2, 3\}$, R: $\{0, 1, 4, 5\}$

In the figures in Exercises 19–22, the domain and range of a relation are illustrated. **a)** *Construct a set of ordered pairs that represent the relation.* **b)** *Determine whether the relation is a function. Explain your answer.*

19.

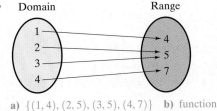

a) $\{(1, 4), (2, 5), (3, 5), (4, 7)\}$ **b)** function

20.

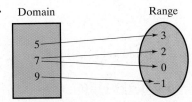

a) $\{(5, 3), (7, 2), (7, 0), (9, -1)\}$ **b)** not a function

21.

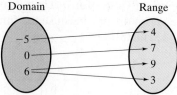

a) $\{(-5, 4), (0, 7), (6, 9), (6, 3)\}$ b) not a function

22.

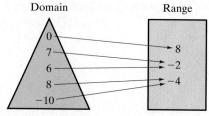

a) $\{(0, 8), (7, -2), (6, -2), (8, -4), (-10, -4)\}$ b) function

Use the vertical line test to determine whether each relation is also a function.

23.

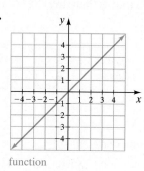

function

24.

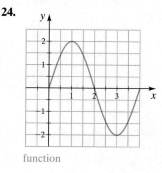

function

25.

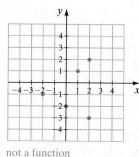

not a function

26.

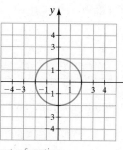

not a function

27.

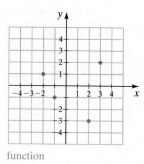

function

28.

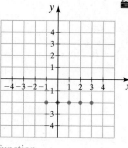

function

29.

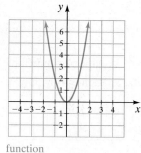

function

30.

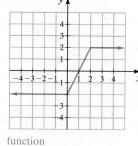

function

31.

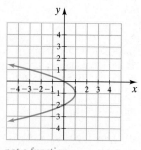

not a function

32.

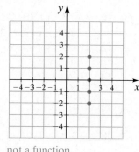

not a function

33.

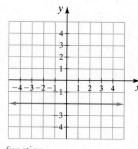

function

34.

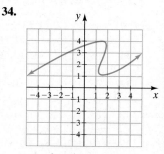

not a function

Evaluate each function at the indicated values.

35. $f(x) = 4x + 2$; find **a)** $f(3)$, **b)** $f(-1)$ a) 14 b) −2

36. $f(x) = -4x + 7$; find **a)** $f(0)$, **b)** $f(1)$ a) 7 b) 3

37. $f(x) = x^2 - 5$; find **a)** $f(6)$, **b)** $f(-2)$ a) 31 b) −1

38. $f(x) = 2x^2 + 3x - 4$; find **a)** $f(1)$, **b)** $f(-3)$ a) 1 b) 5

39. $f(x) = 3x^2 - x + 4$; find **a)** $f(0)$, **b)** $f(2)$ a) 4 b) 14

40. $f(x) = \dfrac{1}{2}x - 4$; find **a)** $f(10)$, **b)** $f(-4)$ a) 1 b) −6

41. $f(x) = \dfrac{x + 4}{2}$; find **a)** $f(2)$, **b)** $f(6)$ a) 3 b) 5

42. $f(x) = \dfrac{1}{2}x^2 + 6$; find **a)** $f(2)$, **b)** $f(-2)$ a) 8 b) 8

Graph each function. **43.–50.** See graphing answer section, pages G8 and G9.

43. $f(x) = x + 3$

44. $f(x) = -x + 4$

45. $f(x) = 2x - 1$

46. $f(x) = 4x + 2$

47. $f(x) = -2x + 4$

48. $f(x) = -x + 5$

49. $f(x) = -\frac{1}{2}x + 2$

50. $f(x) = -4x$

Problem Solving

51. If a relation consists of six ordered pairs and the domain of the relation consists of five values of x, can the relation be a function? Explain.

No, each x cannot have a unique value of y.

In exercises 53 and 54, are the graphs functions? Explain.

53. **U.S. New Car Gas Milage – National Average**

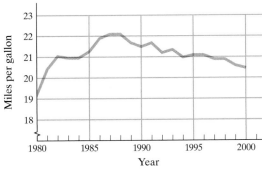

yes

52. If a relation consists of six ordered pairs and the range of the relation consists of five values of y, can the relation be a function? Explain.

Yes, each x may have a unique value of y.

55. a)–59. a) See graphing answer section, page G9.
57. b) $14,000 **58. b)** $2850

54. yes

Social Security

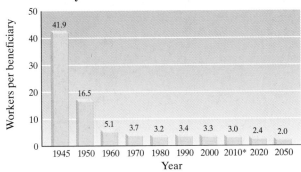

Source: Social Security Administration   * Projections for 2010 and beyond

55. *Babysitting* Jacci Cavanaugh gets paid $8 per hour for babysitting. Her weekly income, I, from babysitting can be represented using the function $I = 8h$ where h is the number of hours she babysits.
  a) Draw a graph of the function for times up to and including 60 hours.
  b) Estimate her weekly income from babysitting if she babysits for 40 hours per week. $320

56. *Bike Riding* Rachel Falk and Tony Mathias go bike riding together. If they ride at a constant rate of 8 miles per hour, the distance they travel, d, can be found by the function $d = 8t$, where t is time in hours.
  a) Draw a graph of the function for times up to and including 6 hours.
  b) Estimate the distance they will travel if they ride for 2 hours. 16 mi

57. *Highway Repair* The cost, c, in dollars, of repairing a highway can be estimated by the function $c = 2000 + 6000m$, where m is the number of miles to be repaired.
  a) Draw a graph of the function for up to and including 6 miles.
  b) Estimate the cost of repairing 2 miles of road.

58. *Vacationing* Frank Duomo plans to take a vacation in the San Diego/Los Angeles area. He determines that the cost, C, of the vacation can be estimated using the

function $C = 350n + 400$ where n is the number of days spent in the area.
  a) Draw a graph of the function for up to and including 10 days.
  b) Estimate the cost of a 7-day vacation in the area.

Hotel del Coronado, San Diego

59. *Commission* A stock broker's commission, c, on stock trades is $25 plus 2% of the sales value, s. Therefore, the broker's commission is a function of the sales, $c = 25 + 0.02s$.
  a) Draw a graph illustrating the broker's commission on sales up to and including $10,000.
  b) If the sales value of a trade is $8000, estimate the broker's commission. $185

60. **Auto Registration** A state's auto registration fee, f, is $20 plus $15 per 1000 pounds of the vehicle's gross weight. The registration fee is a function of the vehicle's weight, $f = 20 + 0.015w$, where w is the weight of the vehicle in pounds.

a) Draw a graph of the function for vehicle weights up to and including 10,000 pounds.

b) Estimate the registration fee of a vehicle whose gross weight is 4000 pounds. $80

 61. **Singing Sensation** A new singing group, Three Forks and a Spoon, sign a recording contract with the Smash Record label. Their contract provides them with a signing bonus of $10,000, plus an 8% royalty on the sales, s, of their new record, *There's Mud in*

61. b) $11,600   60. a)–62. a) See graphing answer section, page G9.

Your Eye! Their income, i, is a function of their sales, $i = 10,000 + 0.08s$.

a) Draw a graph of the function for sales of up to and including $100,000.

b) Estimate their income if their sales are $20,000.

62. **Electric Bill** A monthly electric bill, m, in dollars, consists of a $20 monthly fee plus $0.07 per kilowatt-hour, k, of electricity used. The amount of the bill is a function of the kilowatt-hours used, $m = 20 + 0.07k$.

a) Draw a graph for up to and including 3000 kilowatt-hours of electricity used in a month.

b) Estimate the bill if 1500 kilowatt-hours of electricity are used. $125

Consider the following graphs. Recall from Section 2.7 that an open circle at the end of a line segment means that the endpoint is not included in the answer. A solid circle at the end of a line segment indicates that the endpoint is included in the answer. Determine whether the following graphs are functions. Explain your answer.

63.

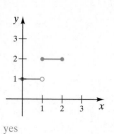

yes

64.

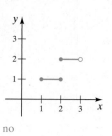

no

65.

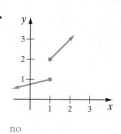

no

66.

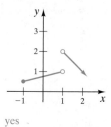

yes

Challenge Problems

67. $f(x) = \frac{1}{2}x^2 - 3x + 5$; find **a)** $f\left(\frac{1}{2}\right)$, **b)** $f\left(\frac{2}{3}\right)$, **c)** $f(0.2)$

a) $\frac{29}{8}$ b) $\frac{29}{9}$ c) 4.42

68. $f(x) = x^2 + 2x - 3$; find **a)** $f(1)$, **b)** $f(2)$, **c)** $f(a)$. Explain how you determined your answer to part **c)**.

a) 0   b) 5   c) $a^2 + 2a - 3$

Group Activity

Discuss and answer Exercises 69 and 70 as a group.

69. Submit three real-life examples (different from those already given) of a quantity that is a function of another. Write each as a function, and indicate what each variable represents.

70. In December, 2002, the cost of mailing a first class letter was 37 cents for the first ounce and 23 cents for each additional ounce. A graph showing the cost of mailing a letter first class is pictured below.

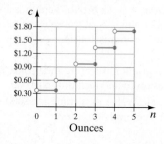

a) Does this graph represent a function? Explain your answer. Yes, each n has a unique value of c.

b) From the graph, estimate the cost of mailing a 4-ounce package first class. $1.30

c) Determine the exact cost of mailing a 4-ounce package first class. $1.34

d) From the graph, estimate the cost of mailing a 3.6-ounce package first class. $1.30

e) Determine the exact cost of mailing a 3.6-ounce package first class. $1.34

Cumulative Review Exercises

[1.3] **71.** Evaluate $\dfrac{5}{9} - \dfrac{3}{7}$. $\dfrac{8}{63}$

[2.5] **72.** Solve the equation $2x - 3(x + 2) = 8$. -14

[3.3] **73.** **Taxi Ride** The cost of a taxi ride is $2.00 for the first mile and $1.50 for each additional mile or part thereof. Find the maximum distance Andrew Collins can ride in the taxi if he has only $20. 13 mi

[5.5] **74.** Factor $25x^2 - 121y^2$. $(5x + 11y)(5x - 11y)$

[6.5] **75.** Simplify $\dfrac{\dfrac{21x}{y^2}}{\dfrac{7}{xy}} \quad \dfrac{3x^2}{y}$

[7.1] **76.** **What is a graph?** An illustration of a set of points whose coordinates satisfy an equation.

CHAPTER SUMMARY

Key Words and Phrases

7.1
Cartesian coordinate system
Collinear
Coordinates
Graph
Linear equation in two variables
Ordered pairs
Origin
Quadrants
Standard form of a linear equation

x-axis
y-axis

7.2
Horizontal line
Vertical line
x-intercept
y-intercept

7.3
Negative reciprocal
Negative slope
Parallel lines

Perpendicular lines
Positive slope
Slope of a line

7.4
Point–slope form
Slope–intercept form

7.5
Linear inequality

7.6
Components of an ordered pair

Domain
Function
Function notation
Linear function
Range
Relation
Vertical line test

IMPORTANT FACTS

To Find the x-Intercept: Set $y = 0$ and find the corresponding value of x.

To Find the y-Intercept: Set $x = 0$ and find the corresponding value of y.

Slope of Line, m, through points (x_1, y_1) and (x_2, y_2):

$$m = \frac{y_2 - y_1}{x_2 - x_1}$$

Methods of Graphing
$$y = 3x - 4$$

By Plotting Points

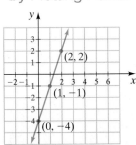

(2, 2)
(1, −1)
(0, −4)

Using Intercepts

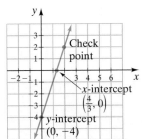

Check point
x-intercept $\left(\frac{4}{3}, 0\right)$
y-intercept $(0, -4)$

Using Slope and y-Intercept

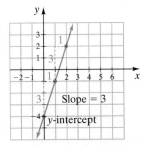

Slope = 3
y-intercept

(continued on the next page)

Standard Form of a Linear Equation: $ax + by = c.$
Slope–Intercept Form of a Linear Equation: $y = mx + b.$
Point–Slope Form of a Linear Equation: $y - y_1 = m(x - x_1).$

Review of Slope

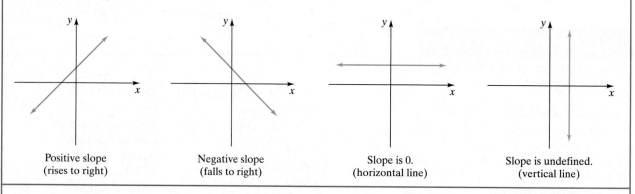

Positive slope (rises to right)   Negative slope (falls to right)   Slope is 0. (horizontal line)   Slope is undefined. (vertical line)

Vertical Line Test: If a vertical line can be drawn through any part of a graph and the vertical line intersects another part of the graph, the graph is not a function.

Chapter Review Exercises

1. See graphing answer section, page G10.

[7.1] **1.** Plot each ordered pair on the same axes.

   a) $A(5, 3)$ **b)** $B(0, 6)$ **c)** $C\left(5, \frac{1}{2}\right)$

   d) $D(-4, 3)$ **e)** $E(-6, -1)$ **f)** $F(-2, 0)$

2. Determine whether the following points are collinear.

     $(0, -4), (6, 8), (-2, 0), (4, 4)$ no

3. Which of the following ordered pairs satisfy the equation $2x + 3y = 9$?

   a) $\left(5, -\frac{1}{3}\right)$ **b)** $(0, 3)$

   c) $(-1, 4)$ **d)** $\left(2, \frac{5}{3}\right)$ a), b), d)

[7.2] **4.** Find the missing coordinate in the following solutions to $3x - 2y = 8$.

   a) $(4, ?)$ 2   **b)** $(0, ?)$ -4 **c)** $(?, 4)$ $\frac{16}{3}$ **d)** $(?, 0)$ $\frac{8}{3}$

Graph each equation using the method of your choice. 5.–14. *See graphing answer section, page G10.*

   5. $y = 4$ **6.** $x = 2$ **7.** $y = 3x$ **8.** $y = 2x - 1$

   9. $y = -2x + 5$ **10.** $y = -\frac{1}{2}x + 4$ **11.** $-2x + 3y = 6$ **12.** $-5x - 2y = 10$

   13. $25x + 50y = 100$ **14.** $\frac{2}{3}x = \frac{1}{4}y + 20$

[7.3] *Find the slope of the line through the given points.*

   15. $(3, -4)$ and $(-2, 5)$ $-\frac{9}{5}$ **16.** $(-4, -2)$ and $(8, -3)$ $-\frac{1}{12}$ **17.** $(-2, -1)$ and $(-4, 3)$ -2

18. What is the slope of a horizontal line? 0

19. What is the slope of a vertical line? undefined

20. Define the slope of a straight line. the ratio of the vertical change to the horizontal change between any two points on a line

Find the slope of each line.

21. −2

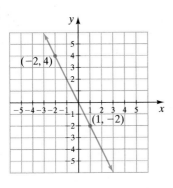

22. $\frac{1}{4}$

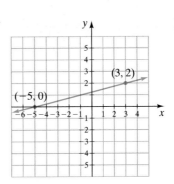

Assume that line 1 and line 2 are distinct lines. If m_1 represents the slope of line 1 and m_2 represents the slope of line 2, determine if line 1 and line 2 are parallel, perpendicular, or neither.

23. $m_1 = \frac{5}{8}, m_2 = -\frac{5}{8}$ neither

24. $m_1 = -4, m_2 = \frac{1}{4}$ perpendicular

25. The following graph shows the number of manatee deaths in Florida. Find the slope of the line segment in **a)** red and **b)** blue. **a)** 28 **b)** −49

Manatee Deaths in Florida

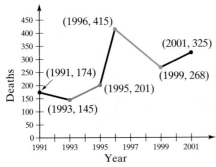

Source: *St. Petersburg Times* 1/5/02, page 46

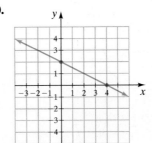

[7.4] *Determine the slope and y-intercept of the graph of each equation.*

26. $6x + 7y = 14$ $-\frac{6}{7}, (0, 2)$

27. $2x + 5 = 0$
slope is undefined, no y-intercept

28. $3y + 9 = 0$ $0, (0, -3)$

Write the equation of each line.

29. $y = 3x - 3$

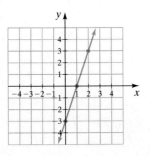

30. $y = -\frac{1}{2}x + 2$

Determine whether each pair of lines is parallel, perpendicular, or neither.

31. $y = 2x - 6$
$6y = 12x + 6$ parallel

32. $2x - 3y = 9$
$3x + 2y = 6$ perpendicular

Find the equation of each line with the given properties.

33. Slope = 3 through $(2, 4)$ $y = 3x - 2$

34. Slope = $-\dfrac{2}{3}$, through $(3, 2)$ $y = -\dfrac{2}{3}x + 4$

35. Slope = 0, through $(4, 2)$ $y = 2$

36. Slope is undefined, through $(4, 2)$ $x = 4$

37. Through $(-2, 3)$ and $(0, -4)$ $y = -\frac{7}{2}x - 4$

38. Through $(-4, -2)$ and $(-4, 3)$ $x = -4$

[7.5] *Graph each inequality.* **39.–44.** See graphing answer section, page G10.

39. $y \geq 1$

40. $x < 4$

41. $y < 3x$

42. $y > 2x + 1$

43. $-6x + y \geq 5$

44. $3y + 6 \leq x$

[7.6] *Determine which of the following relations are also functions. Give the domain and range of each.*

45. $\{(3, 2), (4, -3), (1, 5), (2, -1), (6, 4)\}$
function, D: $\{1, 2, 3, 4, 6\}$, R: $\{-3, -1, 2, 4, 5\}$

46. $\{(3, 1), (4, 2), (4, 5), (6, 1), (7, 0)\}$
not a function. D: $\{3, 4, 6, 7\}$, R: $\{0, 1, 2, 5\}$

47. $\{(3, 1), (4, 1), (5, 1), (6, 2), (3, -3)\}$
not a function, D: $\{3, 4, 5, 6\}$, R: $\{-3, 1, 2\}$

48. $\{(5, -2), (3, -2), (4, -2), (9, -2), (-2, -2)\}$
function, D: $\{-2, 3, 4, 5, 9\}$, R: $\{-2\}$

In Exercises 49 and 50, the domain and range of a relation are illustrated. ***a)*** *Construct a set of ordered pairs that represent the relation.* ***b)*** *Determine whether the relation is a function. Explain your answer.*

49.
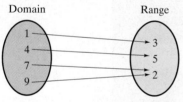
a) $\{(1, 3), (4, 5), (7, 2), (9, 2)\}$ **b)** function

50.
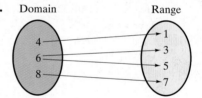
a) $\{(4, 1), (6, 3), (6, 5), (8, 7)\}$ **b)** not a function

In Exercises 51–54, ***a)*** *indicate the domain and range of the relation and* ***b)*** *indicate if the relation is a function. If it is not a function, explain why.*

51.
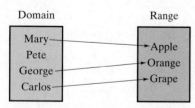
a) D: {Mary, Pete, George, Carlos}, R: {Apple, Orange, Grape}
b) not a function

52.
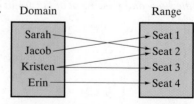
a) D: {Sarah, Jacob, Kristen, Erin}, R: {Seat 1, Seat 2, Seat 3, Seat 4}
b) not a function

53.
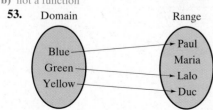
a) D: {Blue, Green, Yellow}, R: {Paul, Maria, Lalo, Duc}
b) function

54.
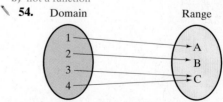
a) D: {1, 2, 3, 4}, R: {A, B, C} **b)** function

Use the vertical line test to determine whether the relation is also a function.

55.

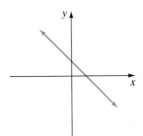

function

56.

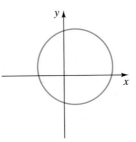

not a function

57.

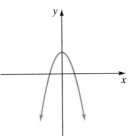

function

58.

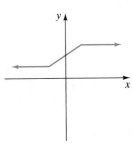

function

Evaluate each function at the indicated values.

59. $f(x) = 6x - 4$; find **a)** $f(1)$, **b)** $f(-5)$ **a)** 2   **b)** −34

60. $f(x) = -4x - 5$; find **a)** $f(-4)$, **b)** $f(8)$ **a)** 11   **b)** −37

61. $f(x) = \dfrac{1}{3}x - 5$; find **a)** $f(3)$, **b)** $f(-9)$ **a)** −4   **b)** −8

62. $f(x) = 2x^2 - 4x + 6$; find **a)** $f(3)$, **b)** $f(-5)$ **a)** 12   **b)** 76

Determine whether the following graphs are functions. Explain your answer.

63.   **Average U.S. Credit Card Debt per Household**

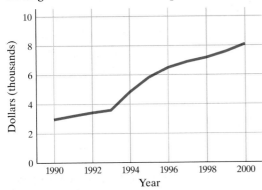

Source: Fortune Magazine, April 2, 2001

yes

64.

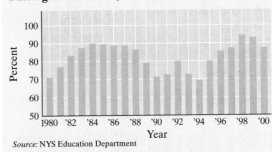

Percentage of New York State School Budgets Passing on First Vote, 1980–2000

Source: NYS Education Department

yes

Graph the following functions.   **65, 66, 67. a), 68. a)** See graphing answer section, pages G10 and G11.

65. $f(x) = 3x - 5$

66. $f(x) = -2x + 3$

67. *Stock Purchase* A discount stock broker charges $25 plus 3 cents per share of stock bought or sold. A customer's cost, c, in dollars, is a function of the number of shares, n, bought or sold, $c = 25 + 0.03n$.

   a) Draw a graph illustrating a customer's cost for up to and including 10,000 shares of stock.

   b) Estimate the cost if 1000 shares of a stock are purchased. $55

68. *Dollar Store* The monthly profit, p, of an Everything for a Dollar store can be estimated by the function $p = 4x - 1600$, where x represents the number of items sold.

   a) Draw a graph of the function for up to and including 1000 items sold.

   b) Estimate the profit if 400 items are sold. $0

Chapter Practice Test

1. an illustration of a set of points that satisfy an equation

3. a) $ax + by = c$ **c)** $y - y_1 = m(x - x_1)$

1. What is a graph?

2. In which quadrants do the following points lie?

   a) $(3, -5)$ IV

   b) $\left(-2, -\dfrac{1}{2}\right)$ III

3. a) What is the standard form of a linear equation?

   b) What is the slope–intercept form of a linear equation? $y = mx + b$

   c) What is the point–slope form of a linear equation?

4. Which of the following ordered pairs satisfy the equation $3y = 5x - 9$?

a) $(4, 2)$ **b)** $\left(\dfrac{9}{5}, 0\right)$

c) $(-2, -6)$ **d)** $(0, -3)$ b) and d)

5. Find the slope of the line through the points $(-4, 3)$ and $(2, -5)$. $-\frac{4}{3}$

6. Find the slope and y-intercept of $4x - 9y = 15$.

7. Write an equation of the graph in the accompanying figure. $y = -x - 1$

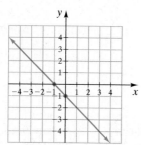

8. Graph $x = -3$

9. Graph $y = 2$.

10. Graph $y = 3x - 2$ by plotting points.

11. a) Solve the equation $2x - 4y = 8$ for y. $y = \frac{1}{2}x - 2$

b) Graph the equation by plotting points.

12. Graph $3x + 5y = 15$ using the intercepts.

13. Write, in slope–intercept form, an equation of the line with a slope of 4 passing through the point $(2, -5)$

14. Write, in slope–intercept form, an equation of the line passing through the points $(3, -1)$ and $(-4, 2)$.

15. Determine whether the following equations represent parallel lines. Explain how you determined your answer.

$$2y = 3x - 6 \quad \text{and} \quad y - \frac{3}{2}x = -5$$

16. Graph $y = 3x - 4$ using the slope and y-intercept.

17. Graph $4x - 2y = 6$ using the slope and y-intercept.

18. Define a function.

19. a) Determine whether the following relation is a function. Explain your answer. not a function

$$\{(1, 2), (3, -4), (5, 3), (1, 0), (6, 5)\}$$

b) Give the domain and range of the relation or function. D: $\{1, 3, 5, 6\}$, R: $\{-4, 0, 2, 3, 5\}$

20. Determine whether the following graphs are functions. Explain how you determined your answer.

a) **b)**

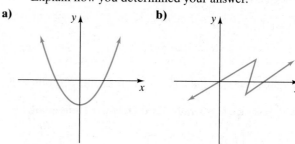

function   not a function

21. If $f(x) = 2x^2 + 3x$, find **a)** $f(2)$ and **b)** $f(-3)$

22. Graph the function $f(x) = 2x - 4$.

23. Graph $y \geq -3x + 5$.

24. Graph $y < 4x - 2$.

25. *Weekly Income* Kate Moore, a salesperson, has a weekly income, i, that can be determined by the function $i = 200 + 0.05s$, where s is her weekly sales.

a) Draw a graph of her weekly income for sales from $0 to $10,000.

b) Estimate her weekly income if her sales are $3000.

6. $\frac{4}{9}$; $\left(0, -\frac{5}{3}\right)$ 8., 9., 10., 11. b), 12., 16., 17., 22., 23., 24., 25. a) See graphing answer section, page G11.   **13.** $y = 4x - 13$
14. $y = -\frac{3}{7}x + \frac{2}{7}$ **15.** Yes, same slopes but different y-intercepts
18. A set of ordered pairs in which each first component corresponds to exactly one second component.   **21. a)** 14 **b)** 9   **25. b)** $350

Cumulative Review Test

Take the following test and check your answers with those that appear at the end of the test. Review any questions that you answered incorrectly. The section and objective where the material was covered are indicated after the answer.

1. Write the set of
a) natural numbers. $\{1, 2, 3, \ldots\}$
b) whole numbers. $\{0, 1, 2, 3, \ldots\}$

2. Name each indicated property.
a) $3(x + 2) = 3x + 3 \cdot 2$ distributive property
b) $a + b = b + a$ commutative property of addition

3. Solve the equation $2x + 5 = 3(x - 5)$. 20

4. Solve the equation $3(x - 2) - (x + 4) = 2x - 10$.
all real numbers

5. Solve the inequality $2x - 14 > 5x + 1$. Graph the solution on a number line. $x < -5$.

6. *Chicken Soup* At Tsong Hsu's Grocery Store, 3 cans of chicken soup sell for $1.50. Find the cost of 8 cans. $4.00

7. *Rectangle* The length of a rectangle is 4 more than twice the width. Find the length and width of the rectangle if its perimeter is 26 feet. w: 3 ft, l: 10 ft

8. 2 hr   **17, 18.** See graphing answer section. page G12.

8. *Running* Two runners start at the same point and run in opposite directions. One runs at 6 mph and the other runs at 8 mph. In how many hours will they be 28 miles apart?

9. Simplify $\dfrac{x^{-4}}{x^{11}} \cdot \dfrac{1}{x^{15}}$

10. Express 652.3 in scientific notation. 6.523×10^2

11. Factor $2x^2 - 12x + 10$. $2(x-5)(x-1)$

12. Factor $4a^2 + 4a - 35$ $(2a-5)(2a+7)$

13. Solve $3x^2 = 18x$. $0, 6$

14. Simplify $\dfrac{2r-7}{14-4r}$. $-\dfrac{1}{2}$

15. Multiply $\dfrac{x-2}{3x+7} \cdot \dfrac{3x}{x-2}$. $\dfrac{3x}{3x+7}$

16. Solve $\dfrac{y^2}{y-5} = \dfrac{25}{y-5}$. -5

19. $y - 2 = 3(x - 5)$

17. Graph $6x - 3y = -12$ using the intercepts.

18. Graph $y = \dfrac{2}{3}x - 3$ using the slope and y-intercept.

19. Write the equation, in point–slope form, of the line with a slope of 3 passing through the point $(5, 2)$.

20. Determine whether the following relations are functions? Explain your answer.

a)   no

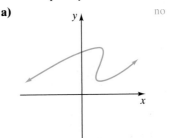

b) $\{(1, 2), (5, 3), (7, 3), (-2, 0)\}$ yes

Answers to Cumulative Review Test

1. a) $\{1, 2, 3, 4, \ldots\}$ **b)** $\{0, 1, 2, 3, \ldots\}$; [Sec. 1.4, Obj. 1] **2. a)** Distributive property; [Sec. 1.10, Obj. 3]
b) Commutative property of addition; [Sec. 1.10, Obj. 1] **3.** 20; [Sec. 2.5, Obj. 1] **4.** All real numbers; [Sec. 2.5, Obj. 3]
5. $x < -5$, ◄————⊕———► ; [Sec. 2.7, Obj. 1] **6.** \$4.00; [Sec. 2.6, Obj. 3] **7.** width: 3 feet, length: 10 feet; [Sec. 3.4, Obj. 1]
            -5

8. 2 hours; [Sec. 3.5, Obj. 2] **9.** $\dfrac{1}{x^{15}}$; [Sec. 4.2, Obj. 2] **10.** 6.523×10^2; [Sec. 4.3, Obj. 1] **11.** $2(x-5)(x-1)$;

[Sec. 5.3, Obj. 2] **12.** $(2a-5)(2a+7)$; [Sec. 5.4, Obj. 1] **13.** 0, 6; [Sec. 5.6, Obj. 2] **14.** $-\dfrac{1}{2}$; [Sec. 6.1, Obj. 4]

15. $\dfrac{3x}{3x+7}$; [Sec. 6.2, Obj. 1] **16.** -5; [Sec. 6.6, Obj. 2] **17.**

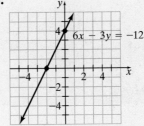

; [Sec. 7.2, Obj. 3]

18.

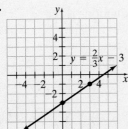

; [Sec. 7.4, Obj. 2] **19.** $y - 2 = 3(x - 5)$; [Sec. 7.4, Obj. 4] **20. a)** no **b)** yes;
[Sec. 7.6, Obj. 2]

Chapter 8

Systems of Linear Equations

8.1 Solving Systems of Equations Graphically

8.2 Solving Systems of Equations by Substitution

8.3 Solving Systems of Equations by the Addition Method

8.4 Systems of Equations: Applications and Problem Solving

8.5 Solving Systems of Linear Inequalities

Chapter Summary
Chapter Review Exercises
Chapter Practice Test
Cumulative Review Test

Almost daily we make financial decisions. Often a knowledge of systems of equations is helpful in making such decisions. For instance, in Example 6 on page 515, we explain how to use a system of equations to select a security system company, and in Exercise 35 on page 525, we use a system of equations to decide whether it is appropriate to refinance a house. As you read through this chapter, you will see many real-life applications of systems of equations.

SSM

Study Guide

CD/Video

MathPro 4/5

PH Math
Tutor Center

prenhall.com/Angel

A Look Ahead

In this chapter we learn how to express applications as systems of linear equations and how to solve systems of linear equations. The solution to a system of equations is the value or values that satisfy all equations in the system.

In this chapter we explain three procedures for solving systems of equations. In Section 8.1 we solve systems using graphs. In Section 8.2 we solve systems using substitution. In Section 8.3 we solve systems using the addition (or elimination) method. In Section 8.4 we explain how to express real-life applications as systems of linear equations. Lastly, in Section 8.5 we build on the graphical solution presented in Section 8.1, and solve systems of linear *inequalities* graphically.

To be successful with Section 8.1, solving systems of equations graphically, you need to understand the procedure for graphing straight lines presented in Sections 7.2 through 7.4. To be successful with Section 8.5, systems of inequalities, you need to understand how to graph linear inequalities, which was presented in Section 7.5.

8.1 SOLVING SYSTEMS OF EQUATIONS GRAPHICALLY

SSM   Study Guide   CD/Video

MathPro 4/5   PH Math Tutor Center   prenhall.com/Angel

1 Determine if an ordered pair is a solution to a system of equations.

2 Determine if a system of equations is consistent, inconsistent, or dependent.

3 Solve a system of equations graphically.

1 Determine If an Ordered Pair Is a Solution to a System of Equations.

When we seek a common solution to two or more linear equations, the equations are called a **system of linear equations**. An example of a system of linear equations follows:

$$\left.\begin{array}{l}(1)\ y = x + 5 \\ (2)\ y = 2x + 4\end{array}\right\} \quad \textit{System of linear equations}$$

The **solution to a system of equations** is the ordered pair or pairs that satisfy all equations in the system. The solution to the system above is $(1, 6)$.

TEACHING TIP
Point out that the solution to a system of equations is an ordered pair.

Check   In Equation (1)   In Equation (2)

$(1, 6)$ $(1, 6)$

$y = x + 5$ $y = 2x + 4$

$6 \overset{?}{=} 1 + 5$ $6 \overset{?}{=} 2(1) + 4$

$6 = 6$ *True* $6 = 6$ *True*

Because the ordered pair $(1, 6)$ satisfies *both* equations, it is a solution to the system of equations. Notice that the ordered pair $(2, 7)$ satisfies the first equation but does not satisfy the second equation.

Check   In Equation (1)   In Equation (2)

$(2, 7)$ $(2, 7)$

$y = x + 5$ $y = 2x + 4$

$7 \overset{?}{=} 2 + 5$ $7 \overset{?}{=} 2(2) + 4$

$7 = 7$ *True* $7 = 8$ *False*

Since the ordered pair $(2, 7)$ does not satisfy *both* equations, it is *not* a solution to the system of equations.

EXAMPLE 1   Determine which of the following ordered pairs satisfy the system of equations.

$$y = 2x - 8$$
$$2x + y = 4$$

a) $(1, -6)$ **b)** $(3, -2)$

Solution   **a)** Substitute 1 for x and -6 for y in each equation.

$$y = 2x - 8 \qquad\qquad 2x + y = 4$$
$$-6 \overset{?}{=} 2(1) - 8 \qquad\qquad 2(1) + (-6) \overset{?}{=} 4$$
$$-6 \overset{?}{=} 2 - 8 \qquad\qquad 2 - 6 \overset{?}{=} 4$$
$$-6 = -6 \quad \textit{True} \qquad\qquad -4 = 4 \quad \textit{False}$$

Since $(1, -6)$ does not satisfy both equations, it is not a solution to the system of equations.

b) Substitute 3 for x and -2 for y in each equation.

$$y = 2x - 8 \qquad\qquad 2x + y = 4$$
$$-2 \overset{?}{=} 2(3) - 8 \qquad\qquad 2(3) + (-2) \overset{?}{=} 4$$
$$-2 \overset{?}{=} 6 - 8 \qquad\qquad 6 - 2 \overset{?}{=} 4$$
$$-2 = -2 \quad \textit{True} \qquad\qquad 4 = 4 \quad \textit{True}$$

NOW TRY EXERCISE 13   Since $(3, -2)$ satisfies both equations, it is a solution to the system of linear equations.   ✳

In this chapter we discuss three methods for finding the solution to a system of equations: the *graphical method*, the *substitution method*, and the *addition method*. In this section we discuss the graphical method.

2   Determine If a System of Equations Is Consistent, Inconsistent, or Dependent

The **solution to a system of linear equations** is the ordered pair (or pairs) common to all lines in the system when the lines are graphed. When two lines are graphed, three situations are possible, as illustrated in Figure 8.1.

In Figure 8.1a, lines 1 and 2 are not parallel lines. They intersect at exactly one point. This system of equations has *exactly one solution*. This is an example of a **consistent system of equations**. A consistent system of equations is a system of equations that has a solution.

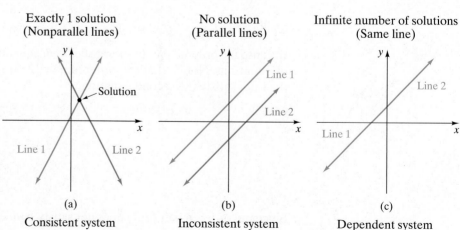

Exactly 1 solution (Nonparallel lines)	No solution (Parallel lines)	Infinite number of solutions (Same line)
(a) Consistent system	(b) Inconsistent system	(c) Dependent system

FIGURE 8.1

In Figure 8.1b, lines 1 and 2 are two different parallel lines. The lines do not intersect, and this system of equations has *no solution*. This is an example of an **inconsistent system of equations**. An inconsistent system of equations is a system of equations that has no solution.

In Figure 8.1c, lines 1 and 2 are actually the same line. In this case, every point on the line satisfies both equations and is a solution to the system of equations. This system has *an infinite number of solutions*. This is an example of a **dependent system of equations**. A dependent system of linear equations is a system of equations that has an infinite number of solutions. If a system of two linear equations is dependent, then both equations represent the same line. *Note that a dependent system is also a consistent system since it has a solution.*

TEACHING TIP
Stress this paragraph to your students.

We can determine if a system of linear equations is consistent, inconsistent, or dependent by writing each equation in slope–intercept form and comparing the slopes and y-intercepts. If the slopes of the lines are different (Fig. 8.1a), the system is consistent. If the slopes are the same but the y-intercepts are different (Fig. 8.1b), the system is inconsistent. If both the slopes and the y-intercepts are the same (Fig. 8.1c), the system is dependent.

EXAMPLE 2 Determine whether the following system has exactly one solution, no solution, or an infinite number of solutions.

$$2x + 4y = -16$$
$$-8y = 4x - 12$$

Solution Write each equation in slope–intercept form and then compare the slopes and the y-intercepts.

$$
\begin{aligned}
2x + 4y &= -16 & -8y &= 4x - 12 \\
4y &= -2x - 16 & y &= \frac{4x - 12}{-8} \\
y &= \frac{-2x - 16}{4} & y &= -\frac{4}{8}x + \frac{12}{8} \\
y &= -\frac{2}{4}x - \frac{16}{4} & y &= -\frac{1}{2}x + \frac{3}{2} \\
y &= -\frac{1}{2}x - 4
\end{aligned}
$$

Since the lines have the same slope, $-\frac{1}{2}$, and different y-intercepts, the lines are parallel. This system of equations is therefore inconsistent and has no solution. ✳

NOW TRY EXERCISE 29

3 Solve a System of Equations Graphically

Now we will see how to solve systems of equations graphically.

> **To Obtain the Solution to a System of Equations Graphically**
>
> Graph each equation and determine the point or points of intersection.

EXAMPLE 3 Solve the following system of equations graphically.

$$2x + y = 11$$
$$x + 3y = 18$$

Solution — Find the x- and y-intercepts of each graph; then draw the graphs.

$2x + y = 11$	Ordered Pair	$x + 3y = 18$	Ordered Pair
Let $x = 0$; then $y = 11$	$(0, 11)$	Let $x = 0$; then $y = 6$	$(0, 6)$
Let $y = 0$; then $x = \dfrac{11}{2}$	$\left(\dfrac{11}{2}, 0\right)$	Let $y = 0$; then $x = 18$	$(18, 0)$

The two graphs (Fig. 8.2) appear to intersect at the point $(3, 5)$. The point $(3, 5)$ may be the solution to the system of equations. To be sure, however, we must check to see that $(3, 5)$ satisfies *both* equations.

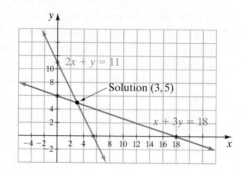

FIGURE 8.2

Check

$$2x + y = 11$$
$$2(3) + 5 \overset{?}{=} 11$$
$$11 = 11 \quad \textit{True}$$

$$x + 3y = 18$$
$$3 + 3(5) \overset{?}{=} 18$$
$$18 = 18 \quad \textit{True}$$

Since the ordered pair $(3, 5)$ checks in both equations, it is the solution to the system of equations. This system of equations is consistent. ✳

EXAMPLE 4 — Solve the following system of equations graphically.

$$2x + y = 3$$
$$4x + 2y = 12$$

Solution — Find the x- and y-intercepts of each graph; then draw the graphs.

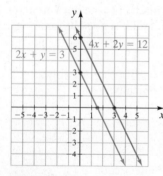

FIGURE 8.3

$2x + y = 3$	Ordered Pair	$4x + 2y = 12$	Ordered Pair
Let $x = 0$; then $y = 3$	$(0, 3)$	Let $x = 0$; then $y = 6$	$(0, 6)$
Let $y = 0$; then $x = \dfrac{3}{2}$	$\left(\dfrac{3}{2}, 0\right)$	Let $y = 0$; then $x = 3$	$(3, 0)$

The two lines (Fig. 8.3) appear to be parallel.

To show that the two lines are indeed parallel, write each equation in slope–intercept form.

$$2x + y = 3$$
$$y = -2x + 3$$

$$4x + 2y = 12$$
$$2y = -4x + 12$$
$$y = -2x + 6$$

Both equations have the same slope, -2, and different y-intercepts; thus the lines must be parallel. Since parallel lines do not intersect, this system of equations has no solution. This system of equations is inconsistent. ✳

NOW TRY EXERCISE 55

EXAMPLE 5   Solve the following system of equations graphically.

$$x - \frac{1}{2}y = 2$$
$$y = 2x - 4$$

Solution   Find the x- and y-intercepts of each graph; then draw the graphs.

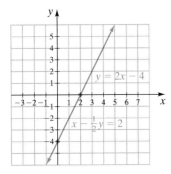

FIGURE 8.4

$x - \dfrac{1}{2}y = 2$	Ordered Pair	$y = 2x - 4$	Ordered Pair
Let $x = 0$; then $y = -4$	$(0, -4)$	Let $x = 0$; then $y = -4$	$(0, -4)$
Let $y = 0$; then $x = 2$	$(2, 0)$	Let $y = 0$; then $x = 2$	$(2, 0)$

Because the lines have the same x- and y-intercepts, both equations represent the same line (Fig. 8.4). When the equations are written in slope–intercept form, it becomes clear that the equations are identical and the system is dependent.

$$x - \frac{1}{2}y = 2 \qquad\qquad y = 2x - 4$$
$$2\left(x - \frac{1}{2}y\right) = 2(2)$$
$$2x - y = 4$$
$$-y = -2x + 4$$
$$y = 2x - 4$$

The solution to this system of equations is all the points on the line.   ✳

NOW TRY EXERCISE 53

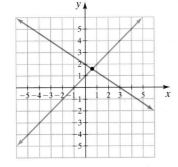

FIGURE 8.5

When graphing a system of equations, the intersection of the lines is not always easy to read on the graph. For example, can you determine the solution to the system of equations shown in Figure 8.5? You may estimate the solution to be $\left(\frac{7}{10}, \frac{3}{2}\right)$ when it may actually be $\left(\frac{4}{5}, \frac{8}{5}\right)$. The accuracy of your answer will depend on how carefully you draw the graphs and on the scale of the graph paper used. In Section 8.2 we present an algebraic method that gives exact solutions to systems of equations.

Using Your Graphing Calculator

A graphing calculator can be used to solve or check systems of equations. To solve a system of equations graphically, graph both equations as was explained in Chapter 7. The point or points of intersection of the graphs is the solution to the system of equations.

The first step in solving the system on a graphing calculator is to solve each equation for y. For example, consider the system of equations given in Example 3.

System of Equations	Equations Solved for y
$2x + y = 11$	$y = -2x + 11$
$x + 3y = 18$	$y = -\dfrac{1}{3}x + 6$

(continued on the next page)

$$\text{Let } Y_1 = -2x + 11$$
$$Y_2 = -\frac{1}{3}x + 6$$

Figure 8.6 shows the screen of a TI-83 Plus with these two equations graphed.

To find the intersection of the two graphs, you can use the $\boxed{\text{TRACE}}$ and $\boxed{\text{ZOOM}}$ keys or the TABLE feature as was explained earlier. Figure 8.7 shows a table of values for equations y_1 and y_2. Notice when $x = 3$, Y_1 and Y_2 both have the same value, 5. Therefore both graphs intersect at $(3, 5)$, and $(3, 5)$ is the solution.

Some graphing calculators have a feature that displays the intersection of graphs by pressing a sequence of keys. For example, on the TI-83 Plus if you go to CALC (which stands for calculate) by pressing $\boxed{2^{\text{nd}}}$ $\boxed{\text{TRACE}}$, you get the screen shown in Figure 8.8. Now press $\boxed{5}$, intersect. Once the *intersect* feature has been selected, the calculator will display the graph and the question

FIRST CURVE?

At this time, move the cursor along the first curve until it is close to the point of intersection. Then press $\boxed{\text{ENTER}}$. The calculator now shows

SECOND CURVE?

and has the cursor on the second curve. If the cursor is not close to the point of intersection, move it along this curve until this happens. Then press $\boxed{\text{ENTER}}$. The calculator now displays

GUESS?

Now press $\boxed{\text{ENTER}}$ and the point of intersection is displayed, see Figure 8.9. The point of intersection is $(3, 5)$.

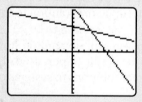

FIGURE 8.6

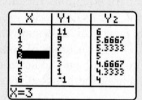

FIGURE 8.7

FIGURE 8.8

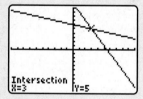

FIGURE 8.9

Exercises

Use your graphing calculator to find the solution to each system of equations. Round noninteger answers to the nearest tenth.

1. $x + 2y = -11$
   $2x - y = -2$ $(-3, -4)$

2. $x - 3y = -13$
   $-2x - 2y = 2$ $(-4, 3)$

3. $2x - y = 7.7$
   $-x - 3y = 1.4$ $(3.1, -1.5)$

4. $3x + 2y = 7.8$
   $-x + 3y = 15.0$ $(-0.6, 4.8)$

TEACHING TIP
You may want to spend a few
minutes here showing how to find
the intersection of two lines on a
graphing calculator.

In Section 3.3, Example 5, we solved a problem involving security systems using only one variable. In the example that follows, we will work that same problem using two variables and illustrate the solution in the form of a graph. Although an answer may sometimes be easier to obtain using only one variable, a graph of the situation may help you to visualize the total picture better.

EXAMPLE 6

Security Systems Meghan O'Donnell plans to install a security system in her house. She has narrowed her choices to two security dealers: Moneywell and Doile. Moneywell's system costs $3580 to install and their monitoring fee is $20 per month. Doile's equivalent system costs only $2620 to install, but their monitoring fee is $32 per month.

a) Assuming that the monthly monitoring fees do not change, in how many months would the total cost of Moneywell's and Doile's systems be the same?

b) If both dealers guarantee not to raise monthly fees for 10 years, and if Meghan plans to use the system for 10 years, which system would be the least expensive?

Solution

a) Understand and Translate   We need to determine the number of months for which both systems will have the same total cost.

$$\text{Let } n = \text{number of months}$$
$$c = \text{total cost of the security system over } n \text{ months}$$

Now we can write an equation to represent the cost of each system using the two variables c and n.

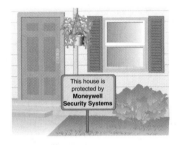

Moneywell	Doile
Total cost = $\left(\begin{array}{c}\text{initial} \\ \text{cost}\end{array}\right) + \left(\begin{array}{c}\text{fees over} \\ n \text{ months}\end{array}\right)$	Total cost = $\left(\begin{array}{c}\text{initial} \\ \text{cost}\end{array}\right) + \left(\begin{array}{c}\text{fees over} \\ n \text{ months}\end{array}\right)$
$c = 3580 + 20n$	$c = 2620 + 32n$

Thus, our system of equations is

$$c = 3580 + 20n$$
$$c = 2620 + 32n$$

Carry Out   Now let's graph each equation. Following are tables of values. Figure 8.10 shows the graphs of the equations.

$$c = 3580 + 20n$$

	$c = 3580 + 20n$
Let $n = 0$	$c = 3580 + 20(0) = 3580$
Let $n = 100$	$c = 3580 + 20(100) = 5580$
Let $n = 160$	$c = 3580 + 20(160) = 6780$

n	c
0	3580
100	5580
160	6780

$$c = 2620 + 32n$$

	$c = 2620 + 32n$
Let $n = 0$	$c = 2620 + 32(0) = 2620$
Let $n = 100$	$c = 2620 + 32(100) = 5820$
Let $n = 160$	$c = 2620 + 32(160) = 7740$

n	c
0	2620
100	5820
160	7740

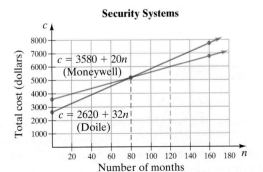

FIGURE 8.10

Check and Answer The graph (Fig. 8.10) shows that the total cost of the two security systems would be the same in 80 months. This is the same answer we obtained in Example 5 in Section 3.3.

b) Since 10 years is 120 months, we draw a dashed vertical line at $n = 120$ months and see where it intersects the two lines. Since at 120 months the Doile line is higher than the Moneywell line, the cost for the Doile system for 120 months is more than the cost of the Moneywell system. Therefore, the cost of the Moneywell system would be less expensive for 10 years.

NOW TRY EXERCISE 71

We will discuss applications of systems of equations further in Section 8.4.

HELPFUL HINT

An equation in one variable may be solved using a system of linear equations. Consider the equation $3x - 1 = x + 1$. Its solution is 1, as illustrated below.

$$3x - 1 = x + 1$$
$$2x - 1 = 1$$
$$2x = 2$$
$$x = 1$$

Let's set each side of the equation $3x - 1 = x + 1$ equal to y to obtain the following system of equations:

$$y = 3x - 1$$
$$y = x + 1$$

The graphical solution of this system of equations is illustrated in Figure 8.11.

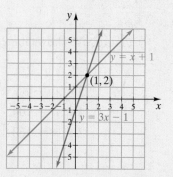

FIGURE 8.11

The solution to the system is $(1, 2)$. Notice that the x-coordinate of the solution of the system, 1, is the solution to the linear equation in one variable, $3x - 1 = x + 1$. If you have a difficult equation to solve and you have a graphing calculator, you can solve the equation on your calculator. The x-coordinate of the solution to the system will be the solution to the linear equation in one variable.

---

1. the ordered pairs that satisfy all the equations in the system

2. **a)** has a solution   **b)** has no solution   **c)** has an infinite number of solutions

3. compare their slopes and y-intercepts

Exercise Set 8.1

Concept/Writing Exercises

1. What does the solution to a system of equations represent?

2. a) What is a consistent system of equations?
  b) What is an inconsistent system of equations?

c) What is a dependent system of equations?

3. Explain how to determine without graphing if a system of linear equations has exactly one solution, no solution, or an infinite number of solutions.

4. When a dependent system of two linear equations is graphed, what will be the result? *a single line*

5. Explain why it may be difficult to obtain an exact answer to a system of equations graphically.

6. Is a dependent system of equations a consistent system or an inconsistent system? Explain. *consistent*

5. *The point of intersection can only be estimated.*

Practice the Skills

Determine which, if any, of the following ordered pairs satisfy each system of linear equations.

7. $y = 4x - 6$
$y = -2x$
  a) $(-1, -10)$ **b)** $(3, 0)$ **c)** $(1, -2)$ c)

8. $y = -4x$
$y = -2x + 8$
  a) $(0, 0)$ **b)** $(-4, 16)$ **c)** $(2, -8)$ b)

9. $y = 2x - 3$
$y = x + 5$
  a) $(8, 13)$ **b)** $(4, 5)$ **c)** $(4, 9)$ a)

10. $x + 2y = 4$
$y = 3x + 3$
  a) $(0, 2)$ **b)** $(-2, 3)$ **c)** $(4, 15)$ (none)

11. $2x + y = 9$
$5x + y = 10$
  a) $(3, 3)$ **b)** $(2, 0)$ **c)** $(4, 1)$ (none)

12. $y = 2x + 4$
$y = 2x - 1$
  a) $(0, 4)$ **b)** $(3, 10)$ **c)** $(-2, 0)$ (none)

13. $2x - 3y = 6$
$y = \dfrac{2}{3}x - 2$
  a) $(3, 0)$ **b)** $(3, -2)$ **c)** $(6, 2)$ a), c)

14. $y = -x + 5$
$2y = -2x + 10$
  a) $(6, -1)$ **b)** $(0, 5)$ **c)** $(5, -1)$ a), b)

15. $3x - 4y = 8$
$2y = \dfrac{2}{3}x - 4$
  a) $(0, -2)$ **b)** $(1, -6)$ **c)** $\left(-\dfrac{1}{3}, -\dfrac{9}{4}\right)$ a)

16. $2x + 3y = 6$
$-x + \dfrac{5}{2} = \dfrac{1}{2}y$
  a) $\left(\dfrac{1}{2}, \dfrac{5}{3}\right)$ **b)** $(2, 1)$ **c)** $\left(\dfrac{9}{4}, \dfrac{1}{2}\right)$ c)

Identify each system of linear equations (lines are labeled 1 and 2) as consistent, inconsistent, or dependent. State whether the system has exactly one solution, no solution, or an infinite number of solutions.

17.
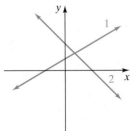
consistent—one solution

18.
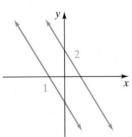
inconsistent—no solution

19.
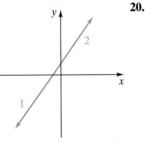
dependent—infinite number

20.
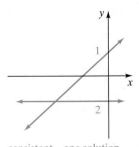
consistent—one solution

21.
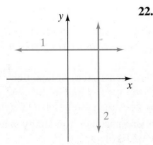
consistent—one solution

22.
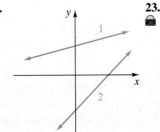
consistent—one solution

23.
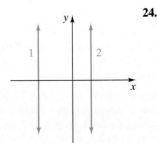
inconsistent—no solution

24.
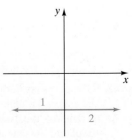
dependent—infinite number

Express each equation in slope–intercept form. Without graphing the equations, state whether the system of equations has exactly one solution, no solution, or an infinite number of solutions.

25. $y = 3x - 1$
$3y = 4x - 6$ one solution

26. $x + y = 6$
$x - y = 6$ one solution

🔒 **27.** $2y = 3x + 3$
$y = \dfrac{3}{2}x - 2$ no solution

28. $y = \dfrac{1}{2}x + 4$
$2y = x + 8$ infinite number of solutions

29. $2x = y - 6$
$4x = 4y + 5$ one solution

30. $x + 2y = 6$
$2x + y = 4$ one solution

31. $3x + 5y = -7$
$-3x - 5y = -7$ no solution

32. $x - y = 2$
$2x - 2y = -2$ no solution

33. $x = 3y + 4$
$2x - 6y = 8$ infinite number of solutions

34. $x - y = 3$
$\dfrac{1}{2}x - 2y = -6$ one solution

35. $y = \dfrac{3}{2}x + \dfrac{1}{2}$
$3x - 2y = -\dfrac{1}{2}$ no solution

36. $2y = \dfrac{7}{3}x - 9$
$4y = 6x + 9$ one solution

Determine the solution to each system of equations graphically. If the system is dependent or inconsistent, so state.

37. $y = x + 3$
$y = -x + 3$ $(0, 3)$

38. $y = 2x + 4$
$y = -3x - 6$ $(-2, 0)$

🔒 **39.** $y = 3x - 6$
$y = -x + 6$ $(3, 3)$

40. $y = 2x - 1$
$2y = 4x + 6$ inconsistent

41. $2x = 4$
$y = -3$ $(2, -3)$

42. $x + 2y = 6$
$2y = 2x - 6$ $(4, 1)$

43. $y = -x + 5$
$-x + y = 1$ $(2, 3)$

44. $2x + y = 6$
$2x - y = -2$ $(1, 4)$

45. $y = -\dfrac{1}{2}x + 4$
$x + 2y = 6$ inconsistent

46. $x + 2y = -4$
$2x - y = -3$ $(-2, -1)$

47. $x + 2y = 8$
$5x + 2y = 0$ $(-2, 5)$

48. $4x - y = 5$
$2y = 8x - 10$ dependent

49. $2x + 3y = 6$
$4x = -6y + 12$ dependent

50. $2x + 3y - 6 = 0$
$2x - 2y = 6$ $(3, 0)$

🔒 **51.** $y = 3$
$y = 2x - 3$ $(3, 3)$

52. $x = 4$
$y = 2x - 6$ $(4, 2)$

53. $x - 2y = 4$
$2x - 4y = 8$ dependent

54. $3x + y = -6$
$2x = 1 + y$ $(-1, -3)$

55. $2x + y = -2$
$6x + 3y = 6$ inconsistent

56. $y = 2x - 3$
$y = -x$ $(1, -1)$

57. $4x - 3y = 6$
$2x + 4y = 14$ $(3, 2)$

58. $2x + 6y = 6$
$y = -\dfrac{1}{3}x + 1$ dependent

59. $2x - 3y = 0$
$x + 2y = 0$ $(0, 0)$

60. $4x = 4y - 12$
$-8x + 8y = 8$ inconsistent

Problem Solving

61. and 62. lines parallel; slopes same, *y*-intercepts different   **64.** one   **65.** none

61. Given the system of equations $6x - 4y = 12$ and $12y = 18x - 24$, determine without graphing whether the graphs of the two equations will be parallel lines. Explain how you determined your answer.

62. Given the system of equations $4x - 8y = 12$ and $2x - 8 = 4y$, determine without graphing whether the graphs of the two equations will be parallel lines. Explain how you determined your answer.

63. If a system of linear equations has solutions $(4, 3)$ and $(6, 5)$, how many solutions does the system have? Explain.   an infinite number

64. If the slope of one line in a system of linear equations is 2 and the slope of the second line in the system is 3, how many solutions does the system have? Explain.

65. If two distinct lines are parallel, how many solutions does the system have? Explain.

66. If two different lines in a linear system of equations pass through the origin, must the solution to the system be $(0, 0)$? Explain. yes

67. Consider the system $x = 5$ and $y = 3$. How many solutions does the system have? What is the solution?
one, $(5, 3)$

68. A system of linear equations has $(2, 1)$ as its solution. If one line in the system is vertical and the other line is horizontal, determine the equations in the system.
$x = 2, y = 1$

In Exercises 69–72, find each solution by graphing the system of equations.

69. *Furnace Repair* Edith Hall's furnace is 10 years old and has a problem. The furnace repair man indicates that it will cost Edith $600 to repair her furnace. She can purchase a new, more efficient furnace for $1800. Her present furnace averages about $650 per year for energy cost and the new furnace would average about $450 per year.

      We can represent the total cost, c, of repair or replacement, plus energy cost over n years by the following system of equations.

    (repair) $c = 600 + 650n$
    (replacement) $c = 1800 + 450n$

Find the number of years for which the total cost of repair would equal the total cost of replacement. 6 yr

70. *Security Systems* Juan Varges is considering the two security systems discussed in Example 6. If Moneywell's system costs $4400 plus $15 per month and Doile's system costs $3400 plus $25 per month, after how many months would the total cost of the two systems be the same? 100 months or 8 yr 4 months

71. *Boat Ride* Rudy has visitors at his home and wants to take them out on a pontoon boat for a day. There are two pontoon boat rental agencies on the lake. Bob's Boat Rental charges $25 per hour for the boat rental, which includes all the gasoline used. Hopper's Rental charges $22 per hour plus a flat charge of $18 for the gasoline used. The equations that represent the total cost, c, follow. In the equations h represents the number of hours the boats are rented.

$$c = 25h$$
$$c = 22h + 18$$

Determine the number of hours the boats must be rented for the total cost to be the same. 6 hr

72. *Landscaping* The Evergreen Landscape Service charges a consultation fee of $200 plus $60 per hour for labor. The Out of Sight Landscape Service charges a consultation fee of $300 plus $40 per hour for labor. We can represent this situation with the system of equations

$$c = 200 + 60h$$
$$c = 300 + 40h$$

where c is the total cost and h is the number of hours of labor. Find the number of hours of labor for the two services to have the same total cost. 5 hr

Group Activity

Discuss and answer Exercises 73–78 as a group. Suppose that a system of three linear equations in two variables is graphed on the same axes. Find the maximum number of points where two or more of the lines can intersect if:

73. the three lines have the same slope but different y-intercepts. 0

74. the three lines have the same slope and the same y-intercept. infinite number

75. two lines have the same slope but different y-intercepts and the third line has a different slope. 2

76. the three lines have different slopes but the same y-intercept. 1

77. the three lines have different slopes but two have the same y-intercept. 3

78. the three lines have different slopes and different y-intercepts. 3

Cumulative Review Exercises

83. **a)** $(6, 0), (0, 4)$ **b)** See graphing answer section, page G12

[2.1] **79.** Simplify $3x - (x - 6) + 4(3 - x)$. $-2x + 18$

[2.5] **80.** Solve the equation $2(x + 3) - x = 5x + 2$. 1

[6.1] **81.** Simplify $\dfrac{x^2 - 9x + 14}{2 - x}$ $-(x - 7)$

[6.6] **82.** Solve the equation $\dfrac{4}{b} + 2b = \dfrac{38}{3}$. $\dfrac{1}{3}, 6$

[7.2] **83.** **a)** Find the x- and y-intercepts of $2x + 3y = 12$.
       b) Use the intercepts to graph the equation.

8.2 SOLVING SYSTEMS OF EQUATIONS BY SUBSTITUTION

SSM   Study Guide   CD/Video

MathPro 4/5   PH Math Tutor Center   prenhall.com/Angel

1   Solve systems of equations by substitution.

As we stated in Section 8.1, a graphic solution to a system of equations may be inaccurate since you must estimate the coordinates of the point of intersection. When an exact solution is necessary, the system should be solved algebraically, either by substitution or by addition of equations.

1   Solve Systems of Equations by Substitution

The procedure for solving a system of equations by **substitution** is illustrated in Example 1. The procedure for solving by addition is presented in Section 8.3. Regardless of which of the two algebraic techniques is used to solve a system of equations, our immediate goal remains the same, that is, *to obtain one equation containing only one unknown*.

EXAMPLE 1   Solve the following system of equations by substitution.

$$x + 3y = 18$$
$$2x + y = 11$$

Solution   Begin by solving for one of the variables in either of the equations. You may solve for any of the variables; however, if you solve for a variable with a numerical coefficient of 1 or -1, you may avoid working with fractions. In this system the x-term in $x + 3y = 18$ and the y-term in $2x + y = 11$; both have a numerical coefficient of 1. Let's solve for x in $x + 3y = 18$.

$$x + 3y = 18$$
$$x = -3y + 18$$

TEACHING TIP
Stress that students should solve for a variable with a coefficient of 1 or -1.

Next, substitute $-3y + 18$ for x in the *other equation*, $2x + y = 11$, and solve for the remaining variable, y.

$$2x + y = 11$$
$$2(\overbrace{-3y + 18}) + y = 11 \qquad \text{\textit{Substitution, this is now an equation in only one variable, y.}}$$
$$-6y + 36 + y = 11$$
$$-5y + 36 = 11$$
$$-5y = -25$$
$$y = 5$$

Finally, substitute $y = 5$ in the equation that is solved for x and find the value of x,

$$x = -3y + 18$$
$$x = -3(5) + 18$$
$$x = -15 + 18$$
$$x = 3$$

A check using $x = 3$ and $y = 5$ in both equations will show that the solution is the ordered pair $(3, 5)$.

The system of equations in Example 1 could also be solved by substitution by solving the equation $2x + y = 11$ for y. Do this now. You should obtain the same answer.

Note that the solution in Example 1 is identical to the graphical solution obtained in Example 3 of Section 8.1. Now we summarize the procedure for solving a system of equations by substitution.

To Solve a System of Equations by Substitution

1. Solve for a variable in either equation. (If possible, solve for a variable with a numerical coefficient of 1 or −1 to avoid working with fractions.)
2. Substitute the expression found for the variable in step 1 into the other equation.
3. Solve the equation determined in step 2 to find the value of one variable.
4. Substitute the value found in step 3 into the equation obtained in step 1 to find the value of the other variable.
5. Check by substituting both values in both original equations.

EXAMPLE 2 Solve the following system of equations by substitution.

$$2x + y = 3$$
$$4x + 2y = 12$$

Solution Solve for y in $2x + y = 3$.

$$2x + y = 3$$
$$y = -2x + 3$$

Now substitute the expression $-2x + 3$ for y in the *other equation*, $4x + 2y = 12$, and solve for x.

$$4x + 2y = 12$$
$$4x + 2(\overbrace{-2x + 3}) = 12 \quad \text{Substitution, this is now an equation in only one variable, } x.$$
$$4x - 4x + 6 = 12$$
$$6 = 12 \quad \text{False}$$

Since the statement 6 = 12 is false, the system has no solution. Therefore, the graphs of the equations will be parallel lines and the system is inconsistent because it has no solution.

TEACHING TIP
After discussing Example 2, have students solve the second equation for y and compare it to the first equation solved for y. Then ask them to identify characteristics of the original system which indicate that the graphs of the equations are parallel.

Note that the solution in Example 2 is identical to the graphical solution obtained in Example 4 of Section 8.1. Figure 8.3 on page 512 shows the parallel lines.

NOW TRY EXERCISE 21

EXAMPLE 3 Solve the following system of equations by substitution.

$$x - \frac{1}{2}y = 2$$
$$y = 2x - 4$$

Solution The equation $y = 2x - 4$ is already solved for y. Substitute $2x - 4$ for y in the other equation, $x - \frac{1}{2}y = 2$, and solve for x.

$$x - \frac{1}{2}y = 2$$
$$x - \frac{1}{2}(\overbrace{2x - 4}) = 2$$

$$x - x + 2 = 2$$
$$2 = 2 \qquad \textit{True}$$

Notice that the sum of the x terms is 0, and when simplified, x is no longer part of the equation. *Since the statement* $2 = 2$ *is true, this system has an infinite number of solutions. Therefore, the graphs of the equations represent the same line and the system is dependent.* ✳

Note that the solution in Example 3 is identical to the solution obtained graphically in Example 5 of Section 8.1. Figure 8.4 on page 513 shows that the graphs of both equations are the same line.

NOW TRY EXERCISE 13

EXAMPLE 4 Solve the following system of equations by substitution.

$$3x + 6y = 9$$
$$2x - 3y = 6$$

Solution None of the variables in either equation has a numerical coefficient of 1. However, since the numbers 3, 6, and 9 are all divisible by 3, if you solve the first equation for x, you will avoid having to work with fractions.

$$3x + 6y = 9$$
$$3x = -6y + 9$$
$$\frac{3x}{3} = \frac{-6y + 9}{3}$$
$$x = -\frac{6}{3}y + \frac{9}{3}$$
$$x = -2y + 3$$

Now substitute $-2y + 3$ for x in the other equation, $2x - 3y = 6$, and solve for the remaining variable, y.

$$2x - 3y = 6$$
$$2(\overbrace{-2y + 3}) - 3y = 6$$
$$-4y + 6 - 3y = 6$$
$$-7y + 6 = 6$$
$$-7y = 0$$
$$y = 0$$

Finally, solve for x by substituting $y = 0$ in the equation previously solved for x.

$$x = -2y + 3$$
$$x = -2(0) + 3 = 0 + 3 = 3$$

The solution is $(3, 0)$. ✳

HELPFUL HINT

Remember that a solution to a system of linear equations must contain both an x- and a y-value. Don't solve the system for one of the variables and forget to solve for the other. Write the solution as an ordered pair.

EXAMPLE 5 Solve the following system of equations by substitution.

$$4x + 4y = 3$$
$$2x = 2y + 5$$

Solution We will elect to solve for x in the second equation.

$$2x = 2y + 5$$

$$x = \frac{2y + 5}{2}$$

$$x = y + \frac{5}{2}$$

Now substitute $y + \frac{5}{2}$ for x in the other equation.

$$4x + 4y = 3$$

$$4\left(\overbrace{y + \frac{5}{2}}\right) + 4y = 3$$

$$4y + 10 + 4y = 3$$

$$8y + 10 = 3$$

$$8y = -7$$

$$y = -\frac{7}{8}$$

Finally, find the value of x.

$$x = y + \frac{5}{2}$$

$$x = -\frac{7}{8} + \frac{5}{2} = -\frac{7}{8} + \frac{20}{8} = \frac{13}{8}$$

NOW TRY EXERCISE 19   The solution is the ordered pair $\left(\frac{13}{8}, -\frac{7}{8}\right)$.   ✳

EXAMPLE 6   **Mule Trip**   In 2002 and 2003 a total of 2462 people took a mule trip to the bottom of the Grand Canyon. If the number who took the mule trip in 2003 was 372 more than the number who took the mule trip in 2002, determine the number of people who took the mule trip in 2002 and in 2003.

Solution   Understand and Translate   We are provided sufficient information to obtain two equations for our system of equations. Let x = number of people who took the mule trip in 2002 and let y = number of people who took the mule trip in 2003. Because the total for these two years was 2462, one equation is $x + y = 2462$. Since the number of people who took the trip in 2003 was 372 more than the number who took the trip in 2002, we can add 372 to the number who took the trip in 2002 to obtain the second equation, $y = x + 372$.

$$\text{system of equations} \begin{cases} x + y = 2462 \\ y = x + 372 \end{cases}$$

Carry Out   Because the second equation, $y = x + 372$, is already solved for y, we will substitute $x + 372$ for y in the first equation.

$$x + y = 2462$$

$$x + \overbrace{x + 372} = 2462$$

$$2x + 372 = 2462$$

$$2x = 2090$$

$$x = 1045$$

Check and Answer   The number of people who took the trip in 2002 was 1045. The number of people who took the trip in 2003 is therefore $x + 372 = 1045 + 372 = 1417$. Notice that $1045 + 1417 = 2462$.   ✳

1. The x in the first equation, both 6 and 12 are divisible by 3.　　**2.** The y in the first equation, both 4 and 8 are divisible by 2.

Exercise Set 8.2

Concept/Writing Exercises

1. When solving the system of equations

$$3x + 6y = 12$$
$$4x + 3y = 8$$

by substitution, which variable, in which equation, would you choose to solve for to make the solution easier? Explain your answer.

2. When solving the system of equations

$$4x + 2y = 8$$
$$3x - 9y = 8$$

by substitution, which variable, in which equation, would you choose to solve for to make the solution easier? Explain your answer.

3. When solving a system of linear equations by substitution, how will you know if the system is inconsistent?

4. When solving a system of linear equations by substitution, how will you know if the system is dependent?

3. You will obtain a false statement, such as $3 = 0$.

4. You will obtain a true statement, such as $2 = 2$.

Practice the Skills

Find the solution to each system of equations by substitution.

5. $x + 2y = 6$
　　$2x - 3y = 5$　$(4, 1)$

6. $y = x + 3$
　　$y = -x - 5$　$(-4, -1)$

7. $x + y = -2$
　　$x - y = 0$　$(-1, -1)$

8. $x + 2y = 6$　infinite number
　　$4y = 12 - 2x$　of solutions

9. $3x + y = 3$
　　$3x + y + 5 = 0$　no solution

10. $y = 2x + 4$
　　$y = -2$　$(-3, -2)$

11. $x = 3$
　　$x + y + 5 = 0$　$(3, -8)$

12. $y = 2x - 13$
　　$-4x - 7 = 9y$　$(5, -3)$

13. $x - \dfrac{1}{2}y = 6$　infinite number
　　$y = 2x - 12$　of solutions

14. $2x + 3y = 7$
　　$6x - y = 1$　$\left(\dfrac{1}{2}, 2\right)$

15. $2x + y = 11$
　　$y = 4x - 7$　$(3, 5)$

16. $y = -2x + 5$
　　$x + 3y = 0$　$(3, -1)$

17. $y = \dfrac{1}{3}x - 2$　infinite number
　　$x - 3y = 6$　of solutions

18. $x = y + 1$
　　$4x + 2y = -14$　$(-2, -3)$

19. $2x + 3y = 7$
　　$6x - 2y = 10$　$(2, 1)$

20. $3x - 3y = 4$
　　$2x + 3y = 5$　$\left(\dfrac{9}{5}, \dfrac{7}{15}\right)$

21. $3x - y = 14$
　　$6x - 2y = 10$　no solution

22. $5x - 2y = -7$
　　$5 = y - 3x$　$(-3, -4)$

23. $4x - 5y = -4$
　　$3x = 2y - 3$　$(-1, 0)$

24. $3x + 4y = 10$
　　$4x + 5y = 14$　$(6, -2)$

25. $4x + 5y = -6$
　　$x - \dfrac{5}{3}y = -2$　$\left(-\dfrac{12}{7}, \dfrac{6}{35}\right)$

26. $\dfrac{1}{2}x + y = 4$
　　$3x + \dfrac{1}{4}y = 6$　$\left(\dfrac{40}{23}, \dfrac{72}{23}\right)$

Problem Solving

27. *Positive Integers* The sum of two positive integers is 79. Find the integers if one number is 7 greater than the other. $36, 43$

28. *Positive Integers* The difference of two positive integers is 44. Find the integers if the larger number is twice the smaller number. $88, 44$

29. *Rectangle* The perimeter of a rectangle is 40 feet. Find the dimensions of the rectangle if the length is 4 feet greater than the width. $w = 8\,\text{ft}, l = 12\,\text{ft}$

30. *Rectangle* The perimeter of a rectangle is 50 feet. Find the dimensions of the rectangle if its length is four times its width. $w = 5\,\text{ft}, l = 20\,\text{ft}$

34. almonds: $20\frac{2}{3}$ lb, cashews: $41\frac{1}{3}$ lb   **35. a)** 16 months   **36. a)** 4 hr   **37. b)** 155 mile marker

31. Wooden Horse To buy a statue of a wooden horse Billy and Jean combined their money. Together they had $726. If Jean had $134 more than Billy, how much money did each have? Billy: $296, Jean: $430

32. Rodeo Attendance At a rodeo the total paid attendance was 2500. If the number of people who received a discount on the admission fee was 622 less than the number who did not, determine the number of paid attendees that paid the full fee. 1561

33. Legal Settlement After a legal settlement, the client's portion of the award was three times as much money as the attorney's portion. If the total award was $20,000, how much did the client get? $15,000

34. Mixed Nuts A grocer made a barrel of mixed nuts consisting of cashews and almonds. The barrel contained 62 pounds of nuts. There are twice as many pounds of cashews as almonds. Find the number of pounds of cashews and of almonds in the barrel.

35. Refinancing Dona Boccio is considering refinancing her house. The cost of refinancing is a one-time charge of $1280. With her reduced mortgage rate, her monthly interest and principal payments would be $794 per month. Her total cost, c, for n months could be represented by $c = 1280 + 794n$. At her current rate her mortgage payments are $874 per month and the total cost for n months can be represented by $c = 874n$.

a) Determine the number of months for which both mortgage plans would have the same total cost.

b) If Dona plans to remain in her house for 12 years, should she refinance? yes

36. Temperatures In Seattle the temperature is 82°F, but it is decreasing by 2 degrees per hour. The temperature, T, at time, t, in hours, is represented by $T = 82 - 2t$. In Spokane the temperature is 64°F, but it is increasing by 2.5 degrees per hour. The temperature, T, can be represented by $T = 64 + 2.5t$.

a) If the temperature continues decreasing and increasing at the same rate in these cities, how long will it be before both cities have the same temperature?

b) When both cities have the same temperature, what will that temperature be? 74°F

Seattle, Washington

37. Traveling by Car Jean Woody's car is at the 80 mile marker on a highway. Roberta Kieronski's car is 15 miles behind Jean's car. Jean's car is traveling at 60 miles per hour. The mile marker that Jean's car will be at in t hours can be found by the equation $m = 80 + 60t$. Roberta's car is traveling at 72 miles per hour. The mile marker that Roberta's car will be at in t hours can be found by the equation $m = 65 + 72t$.

a) Determine the time it will take for Roberta's car to catch up with Jean's car. 1.25 hr

b) At which mile marker will they be when they meet?

38. Computer Store Bill Jordan's present salary consists of a fixed weekly salary of $300 plus a $20 bonus for each computer system he sells. His weekly salary can be represented by $s = 300 + 20n$, where n is the number of computer systems he sells. He is considering another position where his weekly salary would be $400 plus a $10 bonus for each computer system he sells. The other position's weekly salary can be represented by $s = 400 + 10n$. How many computer systems would Bill need to sell in a week for his salary to be the same with both employers? 10

Challenge Problems

Answer parts **a)** *through* **d)** *on your own.*

39. Heat Transfer In a laboratory during an experiment on heat transfer, a large metal ball is heated to a temperature of 180°F. This metal ball is then placed in a gallon of oil at a temperature of 20°F. Assume that when the ball is placed in the oil it loses temperature at the rate of 10 degrees per minute while the oil's temperature rises at a rate of 6 degrees per minute.

a) Write an equation that can be used to determine the ball's temperature t minutes after being placed in the oil. $T = 180 - 10t$

b) Write an equation that can be used to determine the oil's temperature t minutes after the ball is placed in it. $T = 20 + 6t$

c) Determine how long it will take for the ball and oil to reach the same temperature. 10 min

d) When the ball and oil reach the same temperature, what will the temperature be? 80°F

Group Activity

Discuss and work Exercise 40 as a group.

40. In intermediate algebra you may solve systems containing three equations with three variables. As a group, solve the system of equations on the right. Your answer will be in the form of an **ordered triple** (x, y, z).

$$x = 4$$
$$2x - y = 6$$
$$-x + y + z = -3 \qquad (4, 2, -1)$$

43. See graphing answer section page, G12. **44.** slope: $\dfrac{3}{5}$; y-intercept is $\left(0, -\dfrac{8}{5}\right)$ **45.** $y = 2x + 4$

Cumulative Review Exercises

[3.1] **41. *Willow Tree*** The diameter of a willow tree grows about 3.5 inches per year. What is the approximate age of a willow tree whose diameter is 25 inches? 7.14 yr

[4.5] **42.** Multiply $(6x + 7)(3x - 2)$. $18x^2 + 9x - 14$

[7.2] **43.** Graph $4x - 8y = 16$ using the intercepts.

[7.4] **44.** Find the slope and y-intercept of the graph of the equation $3x - 5y = 8$.

45. Determine the equation of the following line.

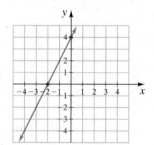

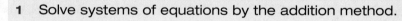

8.3 SOLVING SYSTEMS OF EQUATIONS BY THE ADDITION METHOD

1 Solve systems of equations by the addition method.

 SSM   Study Guide   CD/Video

 MathPro 4/5   PH Math Tutor Center   prenhall.com/Angel

1 Solve Systems of Equations by the Addition Method

A third, and often the easiest, method of solving a system of equations is by the **addition (or elimination) method.** *The goal of this process is to obtain two equations whose sum will be an equation containing only one variable.* Always keep in mind that our immediate goal is to obtain one equation containing only one unknown.

In the addition method, we use the fact that if $a = b$ and $c = d$, then $a + c = b + d$. Suppose we have the following two equations.

$$x - 2y = 6$$
$$3x + 2y = 2$$

If we add the left sides of the two equations we obtain $4x$ because $-2y + 2y = 0$. If we add the right sides of the equations we get $6 + 2 = 8$. Thus, the sum of the equations is $4x = 8$, which is an equation in only one variable.

$$\begin{array}{l} x - 2y = 6 \\ \underline{3x + 2y = 2} \\ 4x \qquad = 8 \end{array}$$

From the equation $4x = 8$, we determine that $x = 2$.

$$4x = 8$$
$$x = 2$$

We can now substitute 2 for x in either of the original equations to find y. We will substitute 2 for x in $x - 2y = 6$.

$$x - 2y = 6$$
$$2 - 2y = 6$$
$$-2y = 4$$
$$y = -2$$

The solution to the system is $(2, -2)$. This ordered pair will check in both equations. Check this solution now. Now let's work some examples.

EXAMPLE 1 Solve the following system of equations using the addition method.

$$x + y = 6$$
$$2x - y = 3$$

Solution Note that one equation contains $+y$ and the other contains $-y$. By adding the equations, we can eliminate the variable y and obtain one equation containing only one variable, x. When added, $+y$ and $-y$ sum to 0, and the variable y is eliminated.

TEACHING TIP
Stress that the object of this process is to obtain two equations whose sum will be an equation containing only 1 variable.

$$\begin{array}{l} x + y = 6 \\ \underline{2x - y = 3} \\ 3x \qquad = 9 \end{array}$$

Now we solve for the remaining variable, x.

$$\frac{3x}{3} = \frac{9}{3}$$
$$x = 3$$

Finally, we solve for y by substituting $x = 3$ in either of the original equations.

$$x + y = 6$$
$$3 + y = 6$$
$$y = 3$$

TEACHING TIP
After discussing Example 1, you may want to ask students what the ordered pair (3, 3) represents on the graph of the two equations.

The solution is $(3, 3)$.
We check the answer in *both* equations.

Check
$$\begin{array}{ll} x + y = 6 & 2x - y = 3 \\ 3 + 3 \overset{?}{=} 6 & 2(3) - 3 \overset{?}{=} 3 \\ 6 = 6 \quad \textit{True} & 6 - 3 \overset{?}{=} 3 \\ & 3 = 3 \quad \textit{True} \end{array}$$

Now we summarize the procedure.

> ### To Solve a System of Equations by the Addition (or Elimination) Method
>
> **1.** If necessary, rewrite each equation so that the terms containing variables appear on the left side of the equal sign and any constants appear on the right side of the equal sign.
>
> **2.** If necessary, multiply one or both equations by a constant(s) so that when the equations are added the resulting sum will contain only one variable.
>
> **3.** Add the equations. This will result in a single equation containing only one variable.
>
> **4.** Solve for the variable in the equation from step 3.
>
> **5.** a) Substitute the value found in step 4 into either of the original equations. Solve that equation to find the value of the remaining variable.
>
>   or
>
>   b) Repeat steps 2–4 to eliminate the other variable.
>
> **6.** Check the values obtained in all original equations.

In step 5 we give two methods for finding the second variable after the first variable has been found. Method a) is generally used if the value obtained for the first variable is an integer. Method b) is often preferred if the value obtained for the first variable is a fraction.

In step 2 it may be necessary to multiply one or both equations by a constant. For example, suppose we have the system of equations, labeled $(eq.\ 1)$ and $(eq.\ 2)$ as shown below.

$$3x + 2y = 6 \qquad (eq.\ 1)$$
$$x - y = 8 \qquad (eq.\ 2)$$

If we wish to multiply equation 2 by -3, we will indicate this process as follows.

$$3x + 2y = 6 \qquad\qquad\qquad\qquad 3x + 2y = 6$$
$$-3(x - y) = -3(8) \quad (eq.\ 2)\ \text{Multiplied by } -3 \quad \text{or} \quad -3x + 3y = 24$$

After performing the multiplication we will generally work with the revised system of equations, like the system given on the right.

EXAMPLE 2 Solve the following system of equations using the addition method.

$$x + 3y = 13 \qquad (eq.\ 1)$$
$$x + 4y = 18 \qquad (eq.\ 2)$$

Solution The goal of the addition process is to obtain two equations whose sum will be an equation containing only one variable. If we add these two equations, none of the variables will be eliminated. However, if we multiply either equation by -1 and then add, the terms containing x will sum to 0, and we will accomplish our goal. We will multiply $(eq.\ 1)$ by -1.

TEACHING TIP
Stress that when solving a system, *both sides* of the equation are to be multiplied by a given number.

$$-1(x + 3y) = -1 \cdot 13 \quad (eq.\ 1)\ \text{Multiplied by } -1 \quad \text{or} \quad -x - 3y = -13$$
$$x + 4y = 18 \qquad (eq.\ 2) \qquad\qquad\qquad\qquad x + 4y = 18$$

Remember that both sides of the equation must be multiplied by -1. This process changes the sign of each term in the equation being multiplied without changing the solution to the system of equations. Now add the two equations on the right.

$$
\begin{array}{rr}
-x - 3y = & -13 \\
x + 4y = & 18 \\
\hline
y = & 5
\end{array}
$$

Now we solve for x in either of the original equations.

$$
\begin{aligned}
x + 3y &= 13 \\
x + 3(5) &= 13 \\
x + 15 &= 13 \\
x &= -2
\end{aligned}
$$

NOW TRY EXERCISE 13   A check will show that the solution is $(-2, 5)$.   ✳

EXAMPLE 3   Solve the following system of equations using the addition method.

$$
\begin{aligned}
2x + y &= 11 \quad (eq.\,1) \\
x + 3y &= 18 \quad (eq.\,2)
\end{aligned}
$$

Solution   To eliminate the variable x, multiply $(eq.\,2)$ by -2 and add the two equations.

$$
\begin{array}{ll}
2x + y = 11 \quad (eq.\,1) & \qquad 2x + y = 11 \\
-2(x + 3y) = -2 \cdot 18 \quad (eq.\,2)\ \text{\scriptsize Multiplied by } -2 & \text{or} \quad -2x - 6y = -36
\end{array}
$$

Now add:

$$
\begin{array}{rr}
2x + \ y = & 11 \\
-2x - 6y = & -36 \\
\hline
-5y = & -25 \\
y = & 5
\end{array}
$$

Solve for x.

$$
\begin{aligned}
2x + y &= 11 \\
2x + 5 &= 11 \\
2x &= 6 \\
x &= 3
\end{aligned}
$$

The solution is $(3, 5)$.   ✳

TEACHING TIP
You may want to have students rework Example 3 by eliminating the variable y to show that the same answer is obtained.

       Note that the solution in Example 3 is the same as the solutions obtained graphically in Example 3 of Section 8.1 and by substitution in Example 1 of Section 8.2.

       In Example 3, we could have multiplied the $(eq.\,1)$ by -3 to eliminate the variable y. At this time, rework Example 3 by eliminating the variable y to see that you get the same answer.

EXAMPLE 4   Solve the following system of equations using the addition method.

$$
\begin{aligned}
4x &= -2y - 18 \\
-5y &= 2x + 10
\end{aligned}
$$

Solution Step 1 of the procedure indicates that we should rewrite each equation so that the terms containing variables appear on the left side of the equal sign and the constants appear on the right side of the equal sign. By adding $2y$ to both sides of the first equation and subtracting $2x$ from both sides of the second equation we obtain the following system of equations.

$$4x + 2y = -18 \quad (eq.\,1)$$
$$-2x - 5y = 10 \quad (eq.\,2)$$

We now continue to solve the system. To eliminate the variable, x, we can multiply $(eq.\,2)$ by 2 and then add.

$$4x + 2y = -18 \quad (eq.\,1) \qquad\qquad\qquad 4x + 2y = -18$$
$$\boxed{2}\,(-2x - 5y) = \boxed{2} \cdot 10 \;\; (eq.\,2) \;\text{\textit{Multiplied by 2}} \qquad \text{or} \qquad -4x - 10y = 20$$

$$
\begin{array}{rcr}
4x + 2y &=& -18 \\
-4x - 10y &=& 20 \\
\hline
-8y &=& 2 \\
y &=& -\dfrac{1}{4}
\end{array}
$$

Solve for x.

$$4x + 2y = -18$$
$$4x + 2\left(-\frac{1}{4}\right) = -18$$
$$4x - \frac{1}{2} = -18$$
$$2\left(4x - \frac{1}{2}\right) = 2(-18) \qquad \text{\textit{Multiply both sides by 2 to remove fractions.}}$$
$$8x - 1 = -36$$
$$8x = -35$$
$$x = -\frac{35}{8}$$

The solution is $\left(-\frac{35}{8}, -\frac{1}{4}\right)$.
Check the solution $\left(-\frac{35}{8}, -\frac{1}{4}\right)$ in both equations.

Check

$$4x + 2y = -18 \qquad\qquad\qquad -2x - 5y = 10$$
$$4\left(-\frac{35}{8}\right) + 2\left(-\frac{1}{4}\right) \overset{?}{=} -18 \qquad\qquad -2\left(-\frac{35}{8}\right) - 5\left(-\frac{1}{4}\right) \overset{?}{=} 10$$
$$-\frac{35}{2} - \frac{1}{2} \overset{?}{=} -18 \qquad\qquad\qquad \frac{35}{4} + \frac{5}{4} \overset{?}{=} 10$$
$$-\frac{36}{2} \overset{?}{=} -18 \qquad\qquad\qquad\qquad \frac{40}{4} \overset{?}{=} 10$$
$$-18 = -18 \quad \text{\textit{True}} \qquad\qquad\qquad 10 = 10 \quad \text{\textit{True}} \quad \text{✳}$$

TEACHING TIP
Before discussing Example 5, point out that neatness and organization will help with their success when using the addition method.

Note that the solution to Example 4 contains fractions. You should not always expect to get integers as answers.

EXAMPLE 5   Solve the following system of equations using the addition method.

$$2x + 3y = 6 \qquad (eq.\,1)$$
$$5x - 4y = -8 \qquad (eq.\,2)$$

Solution   The variable x can be eliminated by multiplying $(eq.\,1)$ by -5 and $(eq.\,2)$ by 2 and then adding the equations.

$$-5(2x + 3y) = -5 \cdot 6 \quad (eq.\,1)\text{ Multiplied by }-5 \quad \text{or} \quad -10x - 15y = -30$$
$$2(5x - 4y) = 2 \cdot (-8) \quad (eq.\,2)\text{ Multiplied by }2 \quad \text{or} \quad 10x - 8y = -16$$

$$
\begin{aligned}
-10x - 15y &= -30 \\
10x - 8y &= -16 \\
\hline
-23y &= -46 \\
y &= 2
\end{aligned}
$$

Solve for x.

$$
\begin{aligned}
2x + 3y &= 6 \\
2x + 3(2) &= 6 \\
2x + 6 &= 6 \\
2x &= 0 \\
x &= 0
\end{aligned}
$$

The solution is $(0, 2)$.

In Example 5, the same value could have been obtained for y by multiplying the $(eq.\,1)$ by 5 and $(eq.\,2)$ by -2 and then adding. Try it now and see. We could have also began by eliminating the variable y by multiplying $(eq.\,1)$ by 4 and $(eq.\,2)$ by 3.

NOW TRY EXERCISE 19

EXAMPLE 6   Solve the following system of equations using the addition method.

$$2x + y = 3 \qquad (eq.\,1)$$
$$4x + 2y = 12 \qquad (eq.\,2)$$

Solution   The variable y can be eliminated by multiplying $(eq.\,1)$ by -2 and then adding the two equations.

$$-2(2x + y) = -2 \cdot 3 \quad (eq.\,1)\text{ Multiplied by }-2 \quad \text{or} \quad -4x - 2y = -6$$
$$4x + 2y = 12 \quad (eq.\,2) \qquad\qquad\qquad\qquad 4x + 2y = 12$$

$$
\begin{aligned}
-4x - 2y &= -6 \\
4x + 2y &= 12 \\
\hline
0 &= 6 \qquad \text{False}
\end{aligned}
$$

Don't panic when both variables drop out and you see an expression like $0 = 6$. Not all systems of equations have a solution. ***Since $0 = 6$ is a false statement, this system has no solution***. The system is inconsistent. The graphs of the equations will be parallel lines.

NOW TRY EXERCISE 25

Note that the solution in Example 6 is identical to the solutions obtained by graphing in Example 4 of Section 8.1 and by substitution in Example 2 of Section 8.2.

EXAMPLE 7 Solve the following system of equations using the addition method.

$$x - \frac{1}{2}y = 2$$
$$y = 2x - 4$$

Solution First align the x- and y-terms on the left side of the equal sign by subtracting $2x$ from both sides of the second equation.

$$x - \frac{1}{2}y = 2 \qquad (eq.\,1)$$
$$-2x + y = -4 \qquad (eq.\,2)$$

Now proceed as in the previous examples. Begin by multiplying $(eq.\,1)$ by 2 to remove fractions from the equation.

$$2\left(x - \frac{1}{2}y\right) = 2 \cdot 2 \quad (eq.\,1)\text{ Multiplied by 2} \qquad \text{or} \qquad 2x - y = 4$$
$$-2x + y = -4 \quad (eq.\,2) \qquad\qquad\qquad\qquad -2x + y = -4$$

$$\begin{array}{r} 2x - y = 4 \\ -2x + y = -4 \\ \hline 0 = 0 \quad \text{True} \end{array}$$

TEACHING TIP
In Examples 6 and 7, the answer would be clear if each equation was written in slope–intercept form.

Again both variables have dropped out. Here we are left with $0 = 0$. **Since $0 = 0$ is a true statement, the system is dependent and has an infinite number of solutions.** When graphed, both equations will be the same line.

NOW TRY EXERCISE 21

The solution in Example 7 is the same as the solutions obtained by graphing in Example 5 of Section 8.1 and by substitution in Example 3 of Section 8.2.

EXAMPLE 8 Solve the following system of equations using the addition method.

$$2x + 3y = 7 \qquad (eq.\,1)$$
$$5x - 7y = -3 \qquad (eq.\,2)$$

Solution We can eliminate the variable x by multiplying $(eq.\,1)$ by -5 and $(eq.\,2)$ by 2.

$$-5(2x + 3y) = -5 \cdot 7 \quad (eq.\,1)\text{ Multiplied by }-5 \quad \text{or} \quad -10x - 15y = -35$$
$$2(5x - 7y) = 2(-3) \quad (eq.\,2)\text{ Multiplied by 2} \quad \text{or} \quad 10x - 14y = -6$$

$$\begin{array}{r} -10x - 15y = -35 \\ 10x - 14y = -6 \\ \hline -29y = -41 \\ y = \frac{41}{29} \end{array}$$

We can now find x by substituting $y = \frac{41}{29}$ into one of the original equations and solving for x. If you try this, you will see that although it can be done, the calculations are messy. An easier method of solving for x is to go back to the original equations and eliminate the variable y. We can do this by multiplying $(eq.\,1)$ by 7 and $(eq.\,2)$ by 3.

$$7\,(2x + 3y) = 7 \cdot 7 \quad (eq.\,1)\ \textit{Multiplied by 7} \qquad \text{or} \qquad 14x + 21y = 49$$
$$3\,(5x - 7y) = 3\,(-3) \quad (eq.\,2)\ \textit{Multiplied by 3} \qquad \text{or} \qquad 15x - 21y = -9$$

$$\begin{array}{r} 14x + 21y = 49 \\ 15x - 21y = -9 \\ \hline 29x \qquad\quad = 40 \\ x = \dfrac{40}{29} \end{array}$$

NOW TRY EXERCISE 35 The solution is $\left(\frac{40}{29}, \frac{41}{29}\right)$.

HELPFUL HINT

We have illustrated three methods for solving a system of linear equations: graphing, substitution, and addition. When you are given a system of equations, which method should you use to solve the system? When you need an exact solution, graphing should not be used. Of the two algebraic methods, the addition method may be easier to use if there are no numerical coefficients of 1 in the system. If one or more of the variables have a coefficient of 1, you can use either substitution or addition.

Exercise Set 8.3

Concept/Writing Exercises

1. When solving the following system of equations by the addition method, what will your first step be in solving the system? Explain your answer. Do not solve the system. multiply top equation by 2

$$-x + 3y = 4$$
$$2x + 5y = 2$$

2. When solving the following system of equations by the addition method, what will your first step be in solving the system? Explain your answer. Do not solve the system. multiply bottom equation by 2

$$2x + 4y = -8$$
$$3x - 2y = 10$$

3. When solving a system of linear equations by the addition method, how will you know if the system is inconsistent? You will obtain a false statement, such as $0 = 6$.

4. When solving a system of linear equations by the addition method, how will you know if the system is dependent? You will obtain a true statement, such as $0 = 0$.

Practice the Skills

Solve each system of equations using the addition method.

5. $\begin{aligned} x + y &= 6 \\ x - y &= 4 \end{aligned}$ $(5,1)$

6. $\begin{aligned} x - y &= 6 \\ x + y &= 8 \end{aligned}$ $(7,1)$

7. $\begin{aligned} x + y &= 5 \\ -x + y &= 1 \end{aligned}$ $(2,3)$

8. $\begin{aligned} x - y &= 2 \\ x + y &= -2 \end{aligned}$ $(0,-2)$

9. $\begin{aligned} x + 2y &= 15 \\ x - 2y &= -7 \end{aligned}$ $\left(4, \frac{11}{2}\right)$

10. $\begin{aligned} 3x + y &= 10 \\ 4x - y &= 4 \end{aligned}$ $(2,4)$

11. $\begin{aligned} 4x + y &= 6 \\ -8x - 2y &= 20 \end{aligned}$ no solution

12. $\begin{aligned} 6x + 3y &= 30 \\ 4x + 3y &= 18 \end{aligned}$ $(6,-2)$

13. $\begin{aligned} -5x + y &= 14 \\ -3x + y &= -2 \end{aligned}$ $(-8,-26)$

14. $\begin{aligned} x - y &= 2 \\ 3x - 3y &= 0 \end{aligned}$ no solution

15. $\begin{aligned} 2x + y &= -6 \\ 2x - 2y &= 3 \end{aligned}$ $\left(-\frac{3}{2}, -3\right)$

16. $\begin{aligned} -4x + 3y &= 0 \\ 5x - 6y &= 9 \end{aligned}$ $(-3,-4)$

17. $\begin{aligned} 2y &= 6x + 16 \\ y &= -3x - 4 \end{aligned}$ $(-2,2)$

18. $\begin{aligned} 2x - 3y &= 4 \\ 2x + y &= -4 \end{aligned}$ $(-1,-2)$

19. $\begin{aligned} 5x + 3y &= 12 \\ 2x - 4y &= 10 \end{aligned}$ $(3,-1)$

20. $\begin{aligned} 6x - 4y &= 9 \\ 2x - 8y &= 3 \end{aligned}$ $\left(\frac{3}{2}, 0\right)$

21. $\begin{aligned} -2y &= -4x + 8 \\ y &= 2x - 4 \end{aligned}$ infinite number of solutions

22. $\begin{aligned} 4x - 2y &= -4 \\ -3x - 4y &= -30 \end{aligned}$ $(2,6)$

23. $5x - 4y = -3$
$5y = 2x + 8$ $(1, 2)$

24. $5x + 4y = 10$
$-3x - 5y = 7$ $(6, -5)$

25. $5x - 4y = 1$
$-10x + 8y = -4$ no solution

26. $2x - 3y = 11$
$-3x = -5y - 17$ $(4, -1)$

27. $5x - 5y = 0$
$3x + 2y = 0$ $(0, 0)$

28. $4x - 2y = 8$ infinite number
$4y = 8x - 16$ of solutions.

29. $-5x + 4y = -20$
$3x - 2y = 15$ $\left(10, \dfrac{15}{2}\right)$

30. $5x = 2y - 4$
$3x - 5y = 6$ $\left(-\dfrac{32}{19}, -\dfrac{42}{19}\right)$

31. $6x = 4y + 12$
$-3y = -5x + 6$ $(-6, -12)$

32. $4x - 3y = -4$
$3x - 5y = 10$ $\left(-\dfrac{50}{11}, -\dfrac{52}{11}\right)$

33. $4x + 5y = 0$
$3x = 6y + 4$ $\left(\dfrac{20}{39}, -\dfrac{16}{39}\right)$

34. $4x - 3y = 8$
$-3x + 4y = 9$ $\left(\dfrac{59}{7}, \dfrac{60}{7}\right)$

35. $x - \dfrac{1}{2}y = 4$
$3x + y = 6$ $\left(\dfrac{14}{5}, -\dfrac{12}{5}\right)$

36. $2x - \dfrac{1}{3}y = 6$
$5x - y = 4$ $(14, 66)$

37. $3x - y = 4$
$2x - \dfrac{2}{3}y = 6$ no solution

38. $-5x + 6y = -12$
$\dfrac{5}{3}x - 4 = 2y$ infinite number of solutions

Problem Solving   **43.** $w = 3$ in., $l = 6$ in.   **45.** $w = 8$ in., $l = 10$ in.   **47.** Answers will vary.

39. *Sum of Numbers* The sum of two numbers is 16. When the second number is subtracted from the first number, the difference is 8. Find the two numbers.   12, 4

40. *Sum of Numbers* The sum of two numbers is 42. When the first number is subtracted from the second number, the difference is 6. Find the two numbers.   18, 24

41. *Sum of Numbers* The sum of a number and twice a second number is 14. When the second number is subtracted from the first number the difference is 2. Find the two numbers.   6, 4

42. *Sum of Numbers* The sum of two numbers is 9. Twice the first number subtracted from three times the second number is 7. Find the two numbers.   4, 5

43. *Rectangles* When the length of a rectangle is x inches and the width is y inches, the perimeter is 18 inches. If the length and width are both doubled, the perimeter becomes 36 inches. Find the length and width of the original rectangle.

$p = 18$ $p = 36$

44. *Perimeter of a Rectangle* When the length of a rectangle is x inches and the width is y inches, the perimeter is 28 inches. If the length is doubled and the width is tripled, the new perimeter becomes 66 inches. Find the length and width of the original rectangle. $w = 5$ in., $l = 9$ in.

45. *Photograph* A photograph has a perimeter of 36 inches. The difference between the photograph's length and width is 2 inches. Find the length and width of the photograph.

46. *Rectangular Garden* John has a large rectangular garden with a perimeter of 122 feet. The different between the garden's lenght and width is 11 feet. Determine the length and width of his garden. $w = 25$ ft, $l = 36$ ft

47. Construct a system of two equations that has no solution. Explain how you know the system has no solution.

48. Construct a system of two equations that has an infinite number of solutions. Explain how you know the system has an infinite number of solutions.   Answers will vary.

49. a) Solve the system of equations

$$4x + 2y = 1000$$
$$2x + 4y = 800 \qquad (200, 100)$$

b) If we divide all the terms in the top equation by 2 we get the following system:

$$2x + y = 500$$
$$2x + 4y = 800$$

How will the solution to this system compare to the solution in part **a)**? Explain and then check your explanation by solving this system.   same solution

50. Suppose we divided all the terms in both equations given in Exercise 49 **a)** by 2, and then solved the system. How will the solution to this system compare to the solution in part **a)**? Explain and then check your explanation by solving each system.   same solution

Challenge Problems

In Exercises 51 and 52, solve each system of equations using the addition method. (Hint: First remove all fractions by multiplying both sides of the equation by the LCD.)

51. $\dfrac{x+2}{2} - \dfrac{y+4}{3} = 4$

$\dfrac{x+y}{2} = \dfrac{1}{2} + \dfrac{x-y}{3}$ $(8, -1)$

52. $\dfrac{5}{2}x + 3y = \dfrac{9}{2} + y$

$\dfrac{1}{4}x - \dfrac{1}{2}y = 6x + 12$ $\left(-\dfrac{105}{41}, \dfrac{447}{82}\right)$

In intermediate algebra you may solve systems of three equations with three unknowns. Solve the following system.

53.
$$x + 2y - z = 2$$
$$2x - y + z = 3$$
$$2x + y + z = 7$$

Hint: Work with *one pair* of equations to get one equation in two unknowns. Then work with *a different pair*

of the original equations to get another equation in the same two unknowns. Then solve the system of two equations in two unknowns. List your answer as an *ordered triple* of the form (x, y, z). $(1, 2, 3)$

Group Activity

*Work parts **a)** and **b)** of Exercise 54 on your own. Then discuss and work parts **c)** and **d)** as a group.*

54. How difficult is it to construct a system of linear equations that has a specific solution? It is really not too difficult to do. Consider:
$$2(3) + 4(5) = 26$$
$$4(3) - 7(5) = -23$$
The system of equations
$$2x + 4y = 26$$
$$4x - 7y = -23$$
has solution $(3, 5)$.

a) Using the information provided, determine another system of equations that has $(3, 5)$ as a solution.

b) Determine a system of linear equations that has $(2, 3)$ as a solution. Answers will vary.

c) Compare your answer with the answers of the other members of your group. Answers will vary.

d) As a group, determine the number of systems of equations that have $(2, 3)$ as a solution. an infinite number

54. a) Answers will vary.

Cumulative Review Exercises

57. $2x^2y - 9xy + 4y$ **59.** $(y + c)(x - a)$

[1.9] **55.** Evaluate 5^3. 125

[2.5] **56.** Solve the equation $2(2x - 3) = 2x + 8$. 7

[4.4] **57.** Simplify
$(4x^2y - 3xy + y) - (2x^2y + 6xy - 3y)$.

[4.5] **58.** Multiply $(8a^4b^2c)(4a^2b^7c^4)$. $32a^6b^9c^5$

[5.2] **59.** Factor $xy + xc - ay - ac$ by grouping.

[7.6] **60.** If $f(x) = 2x^2 - 4$, find $f(-3)$. 14

8.4 SYSTEMS OF EQUATIONS: APPLICATIONS AND PROBLEM SOLVING

SSM   Study Guide   CD/Video

MathPro 4/5   PH Math Tutor Center   prenhall.com/Angel

1 Use systems of equations to solve application problems.

1 Use Systems of Equations to Solve Application Problems

The method you use to solve a system of equations may depend on whether you wish to see "the entire picture" or are interested in finding the exact solution. If you are interested in the trend as the variable changes, you might decide to graph

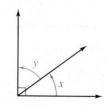

FIGURE 8.12

the equations. If you want only the solution—that is, the ordered pair common to both equations—you might use one of the two algebraic methods to find the common solution.

Many of the application problems solved in earlier chapters using only one variable can also be solved using two variables. In Example 1 we use **complementary angles**, which are two angles whose sum measures 90°. Figure 8.12 illustrates complementary angles x and y.

EXAMPLE 1 **Building a Patio** Phil Mahler is building a rectangular patio out of cement (Fig. 8.13). When he measures the two angles formed by a diagonal he finds that angle x is 16° greater than angle y. Find the two angles.

Solution **Understand and Translate** A rectangle has 4 right (or 90°) angles. Angles x and y are therefore complementary, and the sum of angles x and y is 90°. Thus one equation in the system of equations is $x + y = 90$. Since angle x is 16° greater than angle y, the second equation is $x = y + 16$.

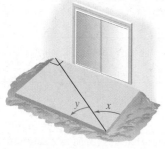

System of equations
$$\begin{cases} x + y = 90 \\ x = y + 16 \end{cases}$$

Carry Out Subtract y from each side of the second equation. Then use the addition method to solve.

$$\begin{array}{r} x + y = 90 \\ \underline{x - y = 16} \\ 2x = 106 \\ x = 53 \end{array}$$

FIGURE 8.13

Now substitute 53 for x in the first equation and solve for y.

$$x + y = 90$$
$$53 + y = 90$$
$$y = 37$$

NOW TRY EXERCISE 3

Check and Answer Angle x is 53° and angle y is 37°. Note that their sum is 90° and angle x is 16° greater than angle y. ✳

Example 1 could also have been solved using the substitution method. Try solving the system now using substitution.

EXAMPLE 2 **Fencing in an Eating Area** Marcella Laddon has opened a new restaurant that has a rectangular outdoor eating area. She wants to enclose the area with fencing but forgot its dimensions. She remembers that the perimeter of the rectangular area is 152 feet and the length is 3 times the width. Determine the dimensions of the rectangular eating area.

Solution **Understand and Translate** The formula for the perimeter of a rectangle is $P = 2l + 2w$. We will use this formula to determine one equation in the system. The other equation comes from the fact that the length is three times the width. Let w = width of the rectangular area and l = length of the rectangular area. Since the perimeter is 152 feet, one equation is $152 = 2l + 2w$. Since the length is 3 times the width, the other equation is $l = 3w$.

system of equations
$$\begin{cases} 152 = 2l + 2w \\ l = 3w \end{cases}$$

Carry Out   We will solve this system by substitution. Since $l = 3w$, substitute $3w$ for l in the equation $152 = 2l + 2w$ to obtain

$$152 = 2l + 2w$$
$$152 = 2(3w) + 2w$$
$$152 = 6w + 2w$$
$$152 = 8w$$
$$19 = w$$

Check and Answer   The width is 19 feet. Since the length is 3 times the width, the length is $3(19)$ or 57 feet. Notice that the perimeter is $2(19) + 2(57) = 152$ feet.

NOW TRY EXERCISE 7

EXAMPLE 3

Bridge Toll   Only two-axle vehicles are permitted to cross a bridge that leads to Honeymoon Island State Park. The toll for the bridge is 50 cents for motorcycles and $1.00 for cars and trucks. On Saturday, the toll booth attendant collected a total of $150, and the vehicle counter recorded 170 vehicles crossing the bridge. How many motorcycles and how many cars and trucks crossed the bridge that day?

Solution

Understand and Translate   We are provided sufficient information in the example to obtain two equations for our system of equations.

$$\text{Let } x = \text{number of motorcycles}$$
$$y = \text{number of cars and trucks}$$

Since a total of 170 vehicles crossed the bridge, one equation is $x + y = 170$. The second equation comes from the tolls collected.

tolls from motorcycles	+	tolls from cars and trucks	= 150
$0.50x$	+	$1.00y$	= 150

System of equations
$$\begin{cases} x + y = 170 \\ 0.50x + 1.00y = 150 \end{cases}$$

Carry Out   Since the first equation can be easily solved for y, we will solve this system by substitution. Solving for y in $x + y = 170$ gives $y = 170 - x$. Substitute $170 - x$ for y in the second equation and solve for x.

$$0.50x + 1.00y = 150$$
$$0.50x + 1.00(170 - x) = 150$$
$$0.50x + 170 - 1.00x = 150$$
$$170 - 0.5x = 150$$
$$-0.5x = -20$$
$$\frac{-0.5x}{-0.5} = \frac{-20}{-0.5}$$
$$x = 40$$

Answer   Forty motorcycles crossed the bridge. The total number of vehicles that crossed the bridge is 170. Therefore, $170 - 40$ or 130 cars and trucks crossed the bridge that Saturday. A check will show that the total collected for these 170 vehicles is $150.

NOW TRY EXERCISE 9

EXAMPLE 4

Horses Everywhere Many cities display statues of animals or other items for artistic, charitable, or other reasons. For example, the cities of Chicago; Beaufort, SC; Houston; Kansas City; New York; Stanford, CT; and West Orange, NJ, have displayed cows. The Tampa Bay area displayed turtles, and Boston displayed cod. Lexington, KY, and Rochester, NY, displayed horses (see photo).

Mann's Jewelers is sponsoring two horses in a city and is considering two artists to design and paint the horses. One artist, Pam, has indicated the materials needed to modify the standard horse will cost $2200, and she will charge $80 per hour to paint the horses according to specifications. The other artist, Mike, has indicated the materials he needs to modify the horse will cost $1825, and he will charge $95 per hour to paint the horses.

a) How many hours would the painting need to take for the total cost for both artists to be the same?

b) If both artists estimate the time required for painting the horse to be 20 hours, which artist would be less expensive?

Solution

a) Understand and Translate We will write equations for the total cost for both Pam and Mike.

Let n = number of hours for painting
c = total cost of horse

Pam	Mike
total cost = materials + labor	total cost = materials + labor
$c = 2200 + 80n$	$c = 1825 + 95n$

Now we have our system of equations.

System of equations
$$\begin{cases} c = 2200 + 80n \\ c = 1825 + 95n \end{cases}$$

Carry Out

$$\text{total cost for Pam} = \text{total cost for Mike}$$
$$2200 + 80n = 1825 + 95n$$
$$375 + 80n = 95n$$
$$375 = 15n$$
$$25 = n$$

Answer If 25 hours of painting is required, the cost would be equal.

b) If the painting takes 20 hours, Mike would be less expensive, as shown below.

Pam	Mike
$c = 2200 + 80n$	$c = 1825 + 95n$
$c = 2200 + 80(20)$	$c = 1825 + 95(20)$
$c = 3800$	$c = 3725$

NOW TRY EXERCISE 17

Motion Problems with Two Rates

We introduced the distance formula, distance = rate · time or $d = rt$, in Section 3.5. Now we introduce a method using two variables and a system of equations to solve motion problems that involve two rates. Often when working motion problems with two different rates it is helpful to construct a table to summarize the information given. We will do this in Example 5.

EXAMPLE 5 **Meeting for Lunch** Bob and Jim are good friends who live 420 miles apart. Sometimes they will meet for lunch at a restaurant somewhere between the towns where they live. On one occasion, they left their houses at the same time. They both arrived at the restaurant 4 hours after they began. Determine Bob's and Jim's speed if Bob drove at an average speed of 5 miles per hour faster than Jim.

Solution **a)** Understand and Translate We are asked to find Bob's and Jim's speed. We are given sufficient information in the example to obtain two equations for our system of equations.

$$\text{Let } x = \text{Bob's speed}$$
$$\text{Let } y = \text{Jim's speed}$$

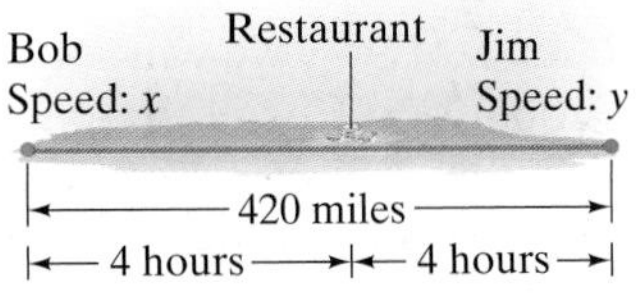

FIGURE 8.14

We will draw a sketch to help understand the problem. See Figure 8.14. We will use the formula, distance = rate · time. They both traveled for 4 hours.

Traveler	Rate	Time	Distance
Bob	x	4	$4x$
Jim	y	4	$4y$

Since the total distance covered is 420 miles, our first equation is

$$4x + 4y = 420$$

The second equation comes from the fact that Bob's speed was 5 miles per hour greater than Jim's speed. Therefore, we can add 5 miles per hour to Jim's speed to get Bob's speed.

$$x = y + 5$$

Our system of equations is

$$\begin{aligned} 4x + 4y &= 420 \\ x &= y + 5 \end{aligned}$$

Carry Out The equation $x = y + 5$ is already solved for x. Substituting $y + 5$ for x in the first equation gives

$$\begin{aligned} 4x + 4y &= 420 \\ 4(\overbrace{y + 5}) + 4y &= 420 \\ 4y + 20 + 4y &= 420 \\ 8y + 20 &= 420 \\ 8y &= 400 \\ y &= 50 \end{aligned}$$

Jim's speed is 50 miles per hour. Bob's speed is

$$\begin{aligned} x &= y + 5 \\ x &= 50 + 5 \\ x &= 55 \end{aligned}$$

Check and Answer   The answer seems reasonable. We can check to see if the equation $4x + 4y = 420$ holds true for $x = 55$ and $y = 50$.

$$4x + 4y = 420$$
$$4(55) + 4(50) \overset{?}{=} 420$$
$$220 + 200 \overset{?}{=} 420$$
$$420 = 420 \quad \textit{True}$$

NOW TRY EXERCISE 23   Thus Bob's speed is 55 miles per hour and Jim's speed is 50 miles per hour.   ✳

Mixture Problems

Mixture problems were solved with one variable in Section 3.5. Now we will solve mixture problems using two variables and systems of equations. Recall that any problem in which two or more quantities are combined to produce a different quantity, or a single quantity is separated into two or more quantities, may be considered a mixture problem. Again, we will use a table to summarize the information given.

EXAMPLE 6   **Candy Shop**   Hui Choe owns a candy and nut shop. In the shop there are both bulk and packaged candy and nuts. A customer comes into the shop and explains that he needs a large amount of a mixture of chocolate covered cherries and amaretto cordials. The customer explains that he plans to make individual packages of the mixture and give a package to each guest at a dinner party. The chocolate covered cherries sell for $7.50 per pound and the amaretto cordials sell for $6.00 per pound.

a) How many pounds of the amaretto cordials must be mixed with 12 pounds of chocolate covered cherries to obtain a mixture that sells for $6.50 per pound?

b) How many pounds of the mixture will there be?

Solution   **a)** Understand and Translate   We are asked to find the number of pounds of amaretto cordials to be mixed with 12 pounds of chocolate covered cherries. We are given sufficient information in the example to obtain two equations for our system of equations.

$$\text{Let } x = \text{number of pounds of amaretto cordials}$$
$$y = \text{number of pounds of mixture}$$

Often it is helpful to make a sketch of the situation. After we draw a sketch, we will construct a table. In our sketch we will use bins to mix the candy (Fig. 8.15).

FIGURE 8.15

	AMARETTO CORDIALS		CHOCOLATE COVERED CHERRIES		MIXTURE
Price per pound	$6.00 lb		$7.50 lb		$6.50 lb
Number of pounds	x	+	12	=	y

The value of the candy is found by multiplying the number of pounds by the price per pound.

Candy	Price	Number of Pounds	Value of Candy
Cordials	6	x	$6x$
Cherries	7.50	12	7.50(12)
Mixture	6.50	y	6.50y

Our two equations come from the following information:

$$\left(\begin{array}{c}\text{number of pounds}\\\text{of cordials}\end{array}\right) + \left(\begin{array}{c}\text{number of pounds}\\\text{of cherries}\end{array}\right) = \left(\begin{array}{c}\text{number of pounds}\\\text{of mixture}\end{array}\right)$$

$$x \qquad + \qquad 12 \qquad = \qquad y$$

$$\text{value of cordials} + \text{value of cherries} = \text{value of mixture}$$

$$6x \qquad + \qquad 7.50(12) \qquad = \qquad 6.50y$$

System of equations
$$\begin{cases} x + 12 = y \\ 6x + 7.50(12) = 6.50y \end{cases}$$

Carry Out   Since $y = x + 12$, we substitute $x + 12$ for y in the second equation and solve for x.

$$6x + 7.50(12) = 6.50y$$
$$6x + 7.50(12) = 6.50(x + 12)$$
$$6x + 90 = 6.50x + 78$$
$$90 = 0.50x + 78$$
$$12 = 0.50x$$
$$24 = x$$

Answer   Thus, 24 pounds of the amaretto cordials must be mixed with 12 pounds of the chocolate covered cherries.

NOW TRY EXERCISE 33   **b)** The total mixture will weigh 24 + 12 or 36 pounds.   ✳

Now we will work an example similar to Example 8 in Section 3.5, but this time we will use a system of equations to solve the problem.

EXAMPLE 7   **Chemistry Lab**   Eilish Main is a chemistry professor. She has a number of labs coming up shortly in which her students will need to use a 15% solution of sulfuric acid. Her assistant informs her that they are out of the 15% solution. Eilish checks the supply room and finds they have a large supply of both 5% and 30% sulfuric acid solutions. She decides to use the 5% and 30% solutions to make 60 liters of a 15% solution. How many liters of the 5% solution and the 30% solution should she mix?

Solution   Understand and Translate   Eilish will combine the 5% and 30% solutions to get 60 liters of a 15% solution of sulfuric acid. We need to determine how much of each should be mixed.

$$\text{Let } x = \text{number of liters of 5\% solution}$$
$$y = \text{number of liters of 30\% solution}$$

The problem is displayed in Figure 8.16 and the information is summarized in the following table.

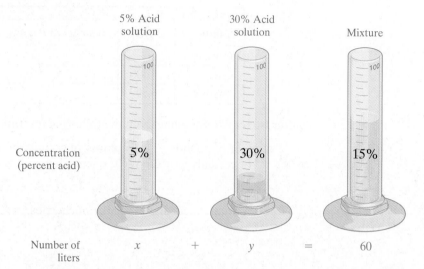

FIGURE 8.16

Solution	Number of Liters	Concentration	Acid Content
5% Solution	x	0.05	$0.05x$
30% Solution	y	0.30	$0.30y$
Mixture	60	0.15	$0.15(60)$

Because the total volume of the combination is 60 liters, we have

$$x + y = 60$$

From the table we see that

$$\left(\begin{array}{c} \text{acid content of} \\ \text{5\% solution} \end{array} \right) + \left(\begin{array}{c} \text{acid content of} \\ \text{30\% solution} \end{array} \right) = \left(\begin{array}{c} \text{acid content} \\ \text{of mixture} \end{array} \right)$$
$$0.05x \qquad\qquad + \qquad\qquad 0.30y \qquad\quad = \qquad 0.15(60)$$

System of equations
$$\begin{cases} x + y = 60 \\ 0.05x + 0.30y = 0.15(60) \end{cases}$$

Carry Out   We will solve this system by substitution. First we solve for y in the first equation.

$$x + y = 60$$
$$y = 60 - x$$

Then we substitute $60 - x$ for y in the second equation.

$$0.05x + 0.30y = 0.15(60)$$
$$0.05x + 0.30(60 - x) = 9$$
$$0.05x + 18 - 0.30x = 9$$
$$-0.25x + 18 = 9$$
$$-0.25x = -9$$
$$x = \frac{-9}{-0.25} = 36$$

Now we solve for y.

$$y = 60 - x$$
$$y = 60 - 36$$
$$y = 24$$

NOW TRY EXERCISE 31

Answer Thus, 36 liters of the 5% acid solution should be mixed with 24 liters of the 30% acid solution to obtain 60 liters of a 15% acid solution.

Mathematics in Action

Who Knew Ice Cream Was So Complicated!

Ice cream is a structurally complex mix of liquid, containing dissolved salts, sugars, and dairy proteins; tiny ice crystals; and globules of milk fat. In addition, air is whipped into the mix as it freezes and increases the volume. The amount of air is limited by a government requirement that a gallon of ice cream must weigh at least 4.5 pounds. Ice cream must also contain 10 percent milk fat. Premium ice creams go way past those minimums and can deliver close to 17 percent milk fat and weigh in at 7.5 pounds per gallon.

As producers of ice cream formulate their blends of ingredients they have to take a systems approach, balancing numerous factors as they decide how much of what to use. Will they lose the diet-conscious customer if the calories are too high? How much should calorie control depend on the amount of fat and how much on the amount of sugar? Should they use the finest natural ingredients and add more air to offset the cost?

In the end, all of these decisions are embodied in a final recipe for a product aimed at a specific market that could be non-dieting adults in households with an annual income greater than a specific amount, or lower-income families with lots of children. It can get quite complicated.

Everyone in the ice cream manufacturer's organization, from official taste testers to marketers to food engineers to cost accountants, will have an opinion. Ultimately, the decision about which ingredients to use is based on a set of interrelated equations, and on other less formal criteria. The final decision is determined by a blend of many concerns that affects the system of equations used in the decision-making process. But ultimately someone signs off on a recipe and the product heads for market.

Exercise Set 8.4

Practice the Skills and Problem Solving

In Exercises 1–38, use a system of linear equations to find the solution. Use a calculator where appropriate.

1. ***Sum of Integers*** The sum of two integers is 41. Find the numbers if one number is 7 greater than the other. 17, 24

2. ***Difference of Integers*** The difference of two integers is 25. Find the two numbers if the larger is 1 less than three times the smaller. 13, 38

3. ***Complementary Angles*** Angles A and B are complementary angles. If angle B is 18° greater than

angle A, find the measure of each angle. (See Example 1.) $A = 36°, B = 54°$

4. ***Supplementary Angles*** Two angles are **supplementary angles** when the sum of their measures is 180°. If angles A and B are supplementary angles, and angle A is four times as large as angle B, find the measure of each angle. $A = 144°, B = 36°$

5. *Supplementary Angles* If angles A and B are supplementary angles (see Exercise 4) and angle A is 24° greater than angle B, find the measure of each angle.

6. *Picture Frame* A rectangular picture frame will be made from a piece of molding 60 inches long. What dimensions will the frame have if the length is to be 6 inches greater than the width? $l = 18$ in., $w = 12$ in.

7. *American Flag* The photo shown is of a 2002 floral flag that covers 6.65 acres in Lompoc, California. The flag maintains the proper flag dimensions. The perimeter of the flag is 2260 feet. Determine the flag's length and width if the length is 350 feet greater than the width. $w = 390$ ft, $l = 740$ ft

8. *Letters* Peter Collins, an insurance agent, sent out a total of 260 envelopes by first class mail. Some weighed under 1 ounce and required a 37-cent stamp. The rest weighed 1 ounce or more but less than 2 ounces and required 60-cents postage. If his total postage cost was $133.00 determine how many envelopes were sent at each postage rate. 100 at 37¢, 160 at 60¢

9. *Farming* Celeste Nossiter plants corn and wheat on her 100-acre farm near Albuquerque, New Mexico. She estimates that her income after deducting expenses is $450 per acre of corn and $430 per acre of wheat. Find the number of acres of corn and wheat planted if her total income after expenses is $44,400.

10. *Car Wash* Countryside Hand Car Wash sells coupon books for $21 that contain 20 coupons. Each coupon allows the customer to have his or her car washed for $8.00 instead of the regular price of $11.50. After how many car washes, using a coupon with each wash, would the amount the customer saves equal the cost of the coupon book? 6

11. *Kayaking* Shane Stagg is kayaking in the St. Lawrence River. He can paddle 4.7 miles per hour with the current and 3.4 miles per hour against the current. Find the speed of the kayak in still water and the current.

12. *Airline Flights* Two commercial airplanes are flying in the same vicinity but in opposite directions. The American Airlines jet, flying with the wind, is flying at 560 miles per hour. The US Airways jet, flying against the wind, is flying at 510 miles per hour. If it were not for the wind, the two planes would be flying at identical speeds. Find the speed of the planes in still air and the speed of the wind. planes, 535 mph; wind, 25 mph

13. *Population* The population of Green Mountain is 40,000 and it is growing by 800 per year. The population of Pleasant View Valley is 66,000 and it is decreasing by 500 per year. How long will it take for both areas to have the same population? 20 yr

14. *Discount Warehouse* Sol's Club Discount Warehouse has two membership plans. Under plan A the customer pays a $50 annual membership and 85% of the manufacturer's recommended list price. Under plan B the annual membership fee is $100 and the customer pays 80% of the manufacturer's recommended list price. How much merchandise, in dollars, would one have to purchase to pay the same amount under both plans? $1000

15. *Six Flags* The Meyer family had a family reunion in Texas. They all purchased tickets to visit the Six Flags Amusement Park. The adult tickets cost $40 and the children's cost $30. If 27 tickets were purchased, and the total cost for all the adult and all the children's tickets was $930, how many adults' and how many children's tickets were purchased? 12 adults, 15 children

16. *Sales Position* Susan Summerlin, a salesperson, is considering two job offers. At the Medtec Company, Susan's salary would be $300 per week plus a 5% commission of sales. At the Genzone Company, her salary would be $200 per week plus an 8% commission of sales.

17. b) Office Copier Depot   **19.** $2500 at 8%, $5500 at 10%

a) What weekly dollar volume of sales would Susan need to make for the total income from both companies to be the same? $3333.33

b) If she expects to make sales of $4000, which company would give the greater salary?   Genzone

17. *Copier Contract* Carol Juncker just purchased a high speed copier for her office and wants a service contract on the copier. She is considering two sources for the contract. The Kate Spence Copier Sales and Service Company charges $18 a month plus 2 cents a copy. Office Copier Depot charges $25 a month but only 1.5 cents a copy.

a) Assuming the prices do not change, how many copies would Carol need to make for the monthly cost of both plans to be the same?   1400 copies

b) If Carol plans to make 2500 copies a month, which plan would be the least expensive?

18. *Financial Planner* Chris Suit is a financial planner for Walsh and Associates. His salary is a flat 40% commission of sales. As an employee he has no overhead. He is considering starting his own company. Then 100% of sales would be income to him. However, he estimates his monthly overhead for office rent, secretary, utilities, and so on, would be about $1500 per month. How much in sales would Chris's own company need to make in a month for him to make the same income he does as an employee of Walsh and Associates? $2500

19. *Savings Accounts* Mr. and Mrs. Vinny McAdams invest a total of $8000 in two savings accounts. One account yields 10% simple interest and the other 8% simple interest. Find the amount placed in each account if they receive a total of $750 in interest after 1 year. Use interest = principal · rate · time.

20. *Investments* Carol Horton invested a total of $10,000. Part of the money was placed in a savings account paying 5% simple interest. The rest was placed in a fixed annuity paying 6% simple interest. If the total interest received for the year was $540, how much had been invested in each account? $6000 in savings, $4000 in annuity

21. *Carpeting* The following bar graph shows the cost per square yard of the same carpet at two different carpet stores along with the installation cost (including matting) at each store.

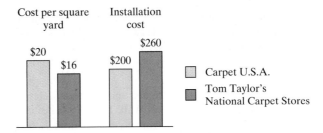

a) Find the number of square yards of carpet that Dorothy Rosene must purchase for the total cost of the carpet plus installation to be the same from both stores.   15 sq yd

b) If she needs to purchase and have installed 25 square yards of carpet, which store will be less expensive?   Tom Taylor's

22. *Self Publishing* Jorge Perez has written a book that he plans to self-publish. He has consulted two printers to determine the cost of printing his book. Each company has a one-time setup charge and a charge for each book printed. The cost information is shown on the following bar graph.

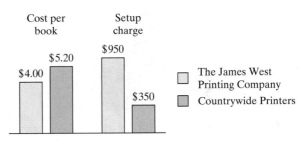

a) How many books would Jorge need to have printed for the total cost of his books to be the same from both printers?   500 copies

b) If he plans to print 1000 books, which printer would be the least expensive?   The James West Company

Review Example 5 before working Exercises 23–30.

23. *Road Trip* Dave Visser started driving from Columbus, Ohio, toward Lincoln, Nebraska—a distance of 903 miles. At the same time Alice Harra started driving to Columbus, Ohio, from Lincoln, Nebraska. If the two meet after 7 hours and Alice's speed averages 15 miles per hour greater than Dave's speed, find the speed of each car.   Dave, 57 mph; Alice, 72 mph

24. *Laying Cable* Two Lucent Technology cable crews are digging a ditch to lay fiber-optic cable. One crew, headed by John Mayleben, and a second crew, headed by Leigh Sumeral, start at opposite ends of a 30-mile stretch of land. John's crew digs at an average speed that is 0.1 miles per hour faster than Leigh's crew. If the two crews meet after 50 hours, find the average speed of each crew.   John's, 0.35 mph; Leigh's, 0.25 mph

25. **Speed Boat** During a race, Elizabeth Kell's speed boat travels 4 miles per hour faster than Melissa Suarez's boat. If Elizabeth's boat finishes the race in 3 hours and Melissa's finishes the race in 3.2 hours, find the speed of each boat. Melissa, 60 mph; Elizabeth, 64 mph

26. **Trains** Two trains are 560 miles apart on a parallel set of tracks traveling toward each other. One train is traveling 6 miles per hour faster than the other. The trains will pass each other in 4 hours. Find the speed of the trains. 67 and 73 mph

27. **Jogging** Amanda Rodriguez and Delores Melendez go jogging along the River Walk Trail in San Antonio, Texas. They start at the same point, but Amanda starts 0.3 hours before Delores does. If Amanda jogs at a rate of 5 miles per hour and Delores jogs at a rate of 8 miles per hour, how long after Delores starts will Delores catch up to Amanda? 0.5 hr

28. **Horseback Riding** Bill Leonard trots his horse Trixie east at 8 miles per hour. One half-hour later, Mary Mullaley starts at the same point and canters her horse Pegarno west at 16 miles per hour. How long after Mary starts riding will Mary and Bill be separated by 10 miles? 0.25 hr

29. **Riding on the Trail** Terri Teegarden has been rollerblading at 5 miles per hour for 0.75 hour along a trail near Pebble Beach, California. She receives a phone call on her cellular phone from her friend Randy Taylor. Randy tells her that he is where she started, at the beginning of the trail. He tells her to keep rollerblading and says that he will ride his bike to catch up with her. Randy plans to ride at 10.5 miles per hour. How long after Randy starts will they meet? ≈0.68 hr

30. **Dude Ranch** Kate and Ernie Danforth are at a dude ranch. Kate has been horseback riding for 0.5 hours on the trail at 6 miles per hour when Ernie starts riding his horse on the trail. If Ernie rides at 10 miles per hour, how long will it be before he catches up to Kate? 3/4 hr

Review Examples 6 and 7 before working Exercises 31–38.

31. **Chemist** Karl Schmid, a chemist, has a 15% hydrochloric acid solution and a 40% hydrochloric acid solution. How many liters of each should he mix to get

10 liters of a hydrochloric acid solution with a 30% acid concentration? 4 L of 15%, 6 L of 40%

32. **Pharmacist** Tamuka Williams, a pharmacist, needs 1000 milliliters of a 10% phenobarbital solution. She has only 5% and 25% phenobarbital solutions available. How many milliliters of each solution should she mix to obtain the desired solution? 750 mL of 5%, 250 mL of 25%

33. **Laying Tile** Julie Hildebrand is selecting tile for her foyer and living room. She wants to make a pattern using two different colors and types of tile. One type costs $3 per tile (per square foot) and the other type costs $5 per tile (per square foot). She needs a total of 380 tiles but does not want to spend more than $1500 on the tile. What is the maximum number of $5 tiles she can purchase? 180 tiles

34. **Melted Ice** A fruit drink is 8% juice by volume. Ice is added to the drink. After the ice melts the juice contents of the 8 ounce mixture is 6%. Find the volume of fruit drink and the volume of ice that was added. 6 oz fruit drink, 2 oz ice

35. **Dairy Farm** Wayne Froelich, a dairy farmer, has milk that is 5% butterfat and skim milk without butterfat. How much 5% milk and how much skim milk should he mix to make 100 gallons of milk that is 3.5% butterfat? 70 gal 5%, 30 gal skim

36. **Soybean Mix** Lynn Hicks wishes to mix soybean meal that is 16% protein and cornmeal that is 7% protein to get a 300-pound mixture that is 10% protein. How much of each should be used? cornmeal, 200 lb; soybean, 100 lb

37. **Juice** The All Natural Juice Company sells apple juice for 12 cents an ounce and apple drink for 6 cents an ounce. They wish to market and sell for 10 cents an ounce cans of juice drink that are part juice and part drink. How many ounces of each will be used if the juice drink is to be sold in 8-ounce cans? $2\frac{2}{3}$ oz drink, $5\frac{1}{3}$ oz juice

38. **Quiche** Pierre LaRue's recipe for quiche lorraine calls for 16 ounces (or 2 cups) of light cream, which is 20% buttermilk. It is often difficult to find light cream with 20% buttermilk at the supermarket. What is commonly found is heavy cream, which is 36% buttermilk, and half-and-half, which is 10.5% buttermilk. How many ounces of the heavy cream and how much of the half-and-half should be mixed to obtain 16 ounces of light cream that is 20% buttermilk? ≈5.96 oz heavy cream, ≈10.04 oz half-and-half

39. *Federal Taxes* As the following graph shows, more and more people are filing their federal tax returns electronically. Notice the blue bars, which represent paper returns, are decreasing at a nearly linear rate and the red bars, which represent electronic returns, are increasing at a nearly linear rate. We can represent the number of returns filed by paper, p, in millions, with the equation $p = -1.6x + 108$. We can represent the number of returns filed electronically, e, in millions, with the equation $e = 3.6x + 5$. For each equation, x represents the number of years after 1990. Thus 1991 would be 1, 1992 would be 2, and so on. Assuming this trend continues, use the given equations to estimate when the number of tax returns filed electronically will equal the number of paper tax returns filed.   19.8 years

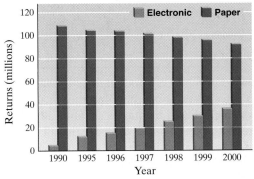

Source: Money Magazine, March 2001, page 98

40. *Cars and Trucks* The following graph shows that light trucks (pickups, SUVs and minivans) outsold cars in the United States for the first time in history in 2001. The equation representing the percent of cars, c, can be estimated with the equation $c = -1.8x + 68$. The equation representing the percent of light trucks, t, can be estimated with the equation $t = 1.6x + 34$. For both equations, x represents the number of years after 1991. Thus 1992 would be 1, 1993 would be 2, and so on.

a) Using the given equations, determine the year when the percent of sales of light trucks equals the percent of sales of cars.   2001

b) Does the graph support your answer?   yes

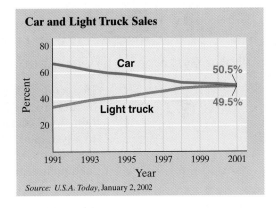

Source: U.S.A. Today, January 2, 2002

Challenge Problems

42.   200 g of first alloy, 100 g of second alloy

41. *Jogging* A brother and sister, Sean and Moira O'Donnell, jog to school daily. Sean, who is older, jogs at 9 miles per hour. Moira jogs at 5 miles per hour. When Sean reached the school, Moira is $\frac{1}{2}$ mile away. How far is the school from their house?   1.125 mi

42. *Alloys* By weight, an alloy of brass is 70% copper and 30% zinc. Another alloy of brass is 40% copper and 60% zinc. How many grams of each of these alloys must be melted and combined to obtain 300 grams of a brass alloy that is 60% copper and 40% zinc?

43. *Pressurized Tanks* Two pressurized tanks are connected by a controlled pressure valve, as shown in the figure. Initially, the internal pressure in tank 1 is 200 pounds per square inch, and the internal pressure in tank 2 is

20 pounds per square inch. The pressure valve is opened slightly to reduce the pressure in tank 1 by 2 pounds per square inch per minute. This increases the pressure in tank 2 by 2 pounds per square inch per minute. At this rate, how long will it take for the pressure to be equal in both tanks?   45 min

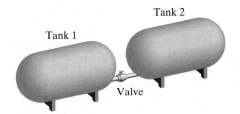

Group Activity

Discuss and work Exercise 44 as a group.

44. *Car Purchase* Debby Patterson is considering 2 cars for purchase. Car A has a list price of $16,500 and gets an average of 40 miles per gallon. Car B has a list price of $15,500 and gets an average of 20 miles per gallon. Being a conservationist, Debby wishes to purchase car A but is concerned about its greater initial cost. She plans to keep the car for many years. If she purchases car A, how many miles would she need to drive for the total cost of car A to equal the total cost of car B? Assume gasoline costs of $1.25 per gallon.   32,000 mi

45. **a)** commutative property of addition **b)** associative property of multiplication

Cumulative Review Exercises

[1.10] **45.** Name the properties illustrated.

  a) $x + 4 = 4 + x$

  b) $(3x)y = 3(xy)$

  c) $4(x + 2) = 4x + 8$ distributive property

  48. an illustration of the set of points that satisfies an equation

[3.4] **46.** *Perimeter* The perimeter of a rectangle is 22 feet. Find the length and width of the rectangle if the length is two more than twice the width.
  $w = 3\,\text{ft}, l = 8\,\text{ft}$

[6.6] **47.** Solve the equation $x + \dfrac{2}{x} = \dfrac{6}{x}$. $-2, 2$

[7.1] **48.** What is a graph?

8.5 SOLVING SYSTEMS OF LINEAR INEQUALITIES

SSM   Study Guide   CD/Video

MathPro 4/5   PH Math Tutor Center   prenhall.com/Angel

1   Solve systems of linear inequalities graphically.

1   Solve Systems of Linear Inequalities Graphically

In Section 7.5, we learned how to graph linear inequalities in two variables. In Section 8.1, we learned how to solve systems of equations graphically. In this section, we discuss how to solve systems of linear inequalities graphically. The **solution to a system of linear inequalities** is the set of points that satisfies all inequalities in the system. Although a system of linear inequalities may contain more than two inequalities, in this book, except in the Group Activity Exercises, we will consider systems with only two inequalities.

To Solve a System of Linear Inequalities Graphically

Graph each inequality on the same axes. The solution is the set of points that satisfies all the inequalities in the system.

EXAMPLE 1   Determine the solution to the following system of inequalities.

$$x + 2y \leq 6$$
$$y > 2x - 4$$

Solution   First graph the inequality $x + 2y \leq 6$ (Fig. 8.17) as explained in Section 7.5. Now, on the same axes, graph the inequality $y > 2x - 4$ (Fig. 8.18). Note that the line is dashed. Why?

TEACHING TIP
Point out that points on a dashed line are never part of the solution and points on a solid line may or may not be in the solution.

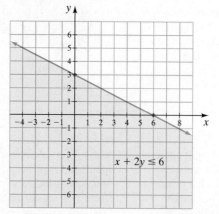

FIGURE 8.17

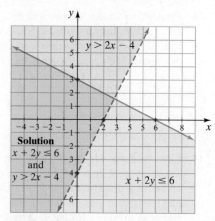

FIGURE 8.18

The solution is the set of points common to both inequalities—the part of the graph that contains both shadings (the purple color). The dashed line is not part of the solution. However, the part of the solid line that satisfies both inequalities is part of the solution.

In Example 1, the ordered pairs of each of the points in the solution (the purple colored area) satisfies both inequalities in the system. We should be able to choose any point in the shaded region, substitute its x and y values into *both* inequalities and obtain true statements. Suppose we choose the origin, which is in the shaded area. Let (x, y) be $(0, 0)$.

$$x + 2y \leq 6 \qquad\qquad y > 2x - 4$$
$$0 + 2(0) \overset{?}{\leq} 6 \qquad\qquad 0 \overset{?}{>} 2(0) - 4$$
$$0 \leq 6 \quad \textit{True} \qquad\qquad 0 > -4 \quad \textit{True}$$

As we expected, the ordered pair $(0, 0)$ checks in both inequalities.

EXAMPLE 2 Determine the solution to the system of inequalities.

$$2x + 3y > 4$$
$$2x - y \geq -6$$

Solution Graph $2x + 3y > 4$ (Fig. 8.19). Graph $2x - y \geq -6$ on the same axes (Fig. 8.20). The solution is the part of the graph with both shadings and the part of the solid line that satisfies both inequalities.

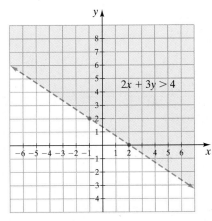

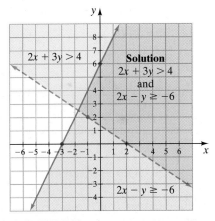

NOW TRY EXERCISE 11   FIGURE 8.19   FIGURE 8.20

EXAMPLE 3 Determine the solution to the following system of inequalities.

$$y < 2$$
$$x > -3$$

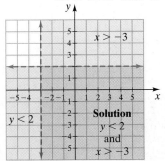

FIGURE 8.21

Solution Graph both inequalities on the same axes (Fig. 8.21). The solution is the part of the graph that contains both shadings.

NOW TRY EXERCISE 15

It is possible for a system of linear inequalities to have no solution. This can occur if the graphs of the inequalities are lines that have the same slope and are parallel to each other. For example, the system of inequalities $x + y \leq 2$ and $x + y \geq 4$ has no solution.

Exercise Set 8.5

Concept/Writing Exercises

1. If an ordered pair satisfies both inequalities in a system of linear inequalities, must that ordered pair be in the solution to the system? Explain. yes

2. If an ordered pair satisfies only one inequality in a system of linear inequalities, is it possible for that ordered pair to be in the solution to the system? Explain. no

3. Can a system of linear inequalities have no solution? Explain your answer with the use of your own example.

4. Is it possible to construct a system of two nonparallel linear inequalities that has no solution? Explain. no

   3. yes, when the lines are parallel: $x + y > 2$, $x + y < 1$

Practice the Skills

Determine the solution to each system of inequalities. 5.–24. *See graphing answer section*, pages G12 and G13.

5. $x + y > 2$
   $x - y < 2$

6. $y \leq 3x - 2$
   $y > -4x$

7. $y \leq x$
   $y < -2x + 2$

8. $2x + 3y < 6$
   $4x - 2y \geq 8$

9. $y < -2x - 3$
   $y \geq 3x + 2$

10. $x + 3y \geq 6$
    $2x - y > 4$

11. $x - 2y < 6$
    $y \leq -x + 4$

12. $y \leq 3x + 4$
    $y < 2$

13. $4x + 5y < 20$
    $x \geq -3$

14. $3x - 4y \leq 12$
    $y > -x + 4$

15. $x \leq 4$
    $y \geq -2$

16. $x \leq 0$
    $y \leq 0$

17. $x > -3$
    $y > 1$

18. $4x + 2y > 8$
    $y \leq 2$

19. $-2x + 3y \geq 6$
    $x + 4y \geq 4$

20. $2x - y \leq 3$
    $4x + 2y > 8$

21. $x - 3y > 3$
    $2x - 6y < 6$

22. $2x - y < 4$
    $4x - 2y \geq -10$

23. $2x + 4y > 6$
    $4x + 8y < 4$

24. $x - 3y \leq 6$
    $x + 3y \leq 6$

Problem Solving

25. Is it possible for a system of two linear inequalities to have only one solution? Explain. no

26. Construct a system of two linear inequalities that has no solution. Explain how you determined your answer.
    Answers will vary.

Group Activity

In more advanced mathematics courses and when working in many industries, you may need to graph more than two linear inequalities. When a system has more than two inequalities, the solution is the point or points that satisfy all inequalities in the system. As a group, determine the solutions to the systems of inequalities in Exercises 27 and 28.

27. $x + 2y \leq 6$
    $2x - y < 2$
    $y > 2$ 27., 28. See graphing answer section, page G13.

28. $x \geq 0$
    $y \geq 0$
    $y \leq 2x + 4$
    $y \leq -x + 6$

Cumulative Review Exercises

29. all real numbers, ◄———————►
    0

30. $y = \dfrac{2}{5}x - \dfrac{6}{5}$

[2.7] 29. Solve the inequality and graph the solution on a number line: $6(x - 2) < 4x - 3 + 2x$.

[3.1] 30. Solve the equation $2x - 5y = 6$ for y.

[5.6] 31. Solve the equation $4x^2 - 11x - 3 = 0$. $-\dfrac{1}{4}, 3$

[6.2] 32. Simplify $\dfrac{x^{-6}y^2}{x^3y^{-4}} \cdot \dfrac{y^6}{x^9}$.

CHAPTER SUMMARY

Key Words and Phrases

8.1
Consistent system of equations
Dependent system of equations
Graphical method to solve a system of linear equations
Inconsistent system of equations

Solution to a system of linear equations
System of linear equations

8.2
Substitution method to solve a system of linear equations

8.3
Addition (or elimination) method to solve a system of linear equations

8.4
Complementary angles
Supplementary angles

8.5
Solution to a system of linear inequalities
System of linear inequalities

IMPORTANT FACTS

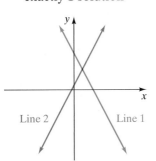

Consistent, exactly 1 solution

(a)

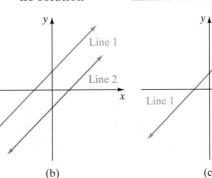

Inconsistent, no solution

(b)

Dependent, infinite number of solutions

(c)

Three methods used to solve a system of linear equations are the (1) graphical method, (2) substitution method, and (3) addition (or elimination) method. When solving a system of linear equations by either substitution or the addition method, if you get a false statement, such as $6 = 0$, the system is inconsistent and has no solution. If you get a true statement, such as $0 = 0$, the system is dependent and has an infinite number of solutions.

Chapter Review Exercises

[8.1] Determine which, if any, of the ordered pairs satisfy each system of equations.

1. $y = 4x - 2$
$2x + 3y = 8$
a) $(0, -2)$ **b)** $(-2, 4)$ **c)** $(1, 2)$ c)

2. $y = -x + 4$
$3x + 5y = 15$
a) $\left(\dfrac{5}{2}, \dfrac{3}{2}\right)$ **b)** $(0, 4)$ **c)** $\left(\dfrac{1}{2}, \dfrac{3}{5}\right)$ a)

Identify each system of linear equations as consistent, inconsistent, or dependent. State whether the system has exactly one solution, no solution, or an infinite number of solutions.

3.

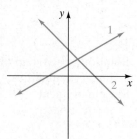

consistent, one

4.

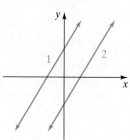

inconsistent, no solution

5.

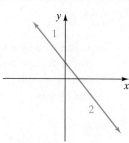

dependent, infinite number of solutions

6.

consistent, one

Write each equation in slope–intercept form. Without graphing or solving the system of equations, state whether the system of linear equations has exactly one solution, no solution, or an infinite number of solutions.

7. $x + 2y = 10$
$3x = -6y + 12$
no solution

8. $y = -3x - 6$
$2x + 5y = 8$
one solution

9. $y = \frac{1}{2}x - 4$
$x - 2y = 8$
infinite number of solutions

10. $6x = 4y - 8$
$4x = 6y + 8$
one solution

Determine the solution to each system of equations graphically.

11. $y = x - 4$
$y = 2x - 7$ $(3, -1)$

12. $x = -2$
$y = 3$ $(-2, 3)$

13. $y = 3$
$y = -2x + 5$ $(1, 3)$

14. $x + 3y = 6$
$y = 2$ $(0, 2)$

15. $x + 2y = 8$
$2x - y = -4$ $(0, 4)$

16. $y = x - 3$
$2x - 2y = 6$
infinite number of solutions

17. $3x + y = 0$
$3x - 3y = 12$ $(1, -3)$

18. $x + 2y = 4$
$\frac{1}{2}x + y = -2$
no solution

[8.2] *Find the solution to each system of equations by substitution.*

19. $y = 3x - 13$
$2x - 5y = 0$ $(5, 2)$

20. $x = 3y - 9$
$x + 2y = 1$ $(-3, 2)$

21. $2x - y = 6$
$x + 2y = 13$ $(5, 4)$

22. $x = -3y$
$x + 4y = 6$ $(-18, 6)$

23. $4x - 2y = 10$
$y = 2x + 3$ no solution

24. $2x + 4y = 8$
$4x + 8y = 16$
infinite number of solutions

25. $2x - 3y = 8$
$6x + 5y = 10$ $\left(\frac{5}{2}, -1\right)$

26. $3x - y = -5$
$x + 2y = 8$ $\left(-\frac{2}{7}, \frac{29}{7}\right)$

[8.3] *Find the solution to each system of equations using the addition method.*

27. $-x - y = 6$
$-x + y = 10$ $(-8, 2)$

28. $x + 2y = -3$
$2x - 2y = 6$ $(1, -2)$

29. $x + y = 12$
$2x + y = 5$ $(-7, 19)$

30. $4x - 3y = 8$
$2x + 5y = 8$ $\left(\frac{32}{13}, \frac{8}{13}\right)$

31. $-2x + 3y = 15$
$3x + 3y = 10$ $\left(-1, \frac{13}{3}\right)$

32. $2x + y = 9$
$-4x - 2y = 4$
no solution

33. $3x = -4y + 10$
$8y = -6x + 20$
infinite number of solutions

34. $2x - 5y = 12$
$3x - 4y = -6$
$\left(-\frac{78}{7}, -\frac{48}{7}\right)$

[8.4] *Use a system of linear equations to find the solution.*

35. *Sum of Integers* The sum of two integers is 49. Find the two numbers if the larger is 8 less than twice the smaller.

36. *Plane Flight* A plane flies 600 miles per hour with the wind and 530 miles per hour against the wind. Find the speed of the wind and the speed of the plane in still air.

37. *Truck Rental* Katz's Truck Rental charges $30 per day plus 50 cents per mile, while Willie's Truck Rental charges $40 per day plus 40 cents per mile. How far would you have to travel in one day for the total cost from both rental companies to be the same? 100 mi

35. $19, 30$ **36.** plane, 565 mph; wind, 35 mph

38. *Savings Account* Moura Hakala invested a total of $16,000. Part of the money was placed in a savings account paying 4% simple interest. The rest was placed in a savings account paying 6% simple interest. If the total interest received for the year was $760, how much had she invested in each account? $10,000 at 4%, $6000 at 6%

39. *Road Trip* Liz Wood drives from Charleston, South Carolina, to Louisville, Kentucky—a distance of 600 miles. At the same time, Mary Mayer starts driving from Louisville to Charleston along the same route. If the two meet after driving 5 hours and Mary's average speed was 6 miles per hour greater than Liz's, find the average speed of each car. Liz, 57 mph; Mary, 63 mph

40. ***Grass Seed*** Green Turf's grass seed costs 60 cents a pound and Agway's grass seed costs 45 cents a pound. How many pounds of each were used to make a 40-pound mixture that cost $20.25?

41. ***Chemist*** A chemist has a 30% acid solution and a 50% acid solution. How much of each must be mixed to get 6 liters of a 40% acid solution? 3 liters of each

40. 15 lb of Green Turf's, 25 lb of Agway's

[8.5] Determine the solution to each system of inequalities. 42.–45. see graphing answer section, page G13.

42. $2x + y > 2$
$2x - y \leq 4$

43. $2x - 3y \leq 6$
$x + 4y > 4$

44. $2x - 6y > 6$
$x > -2$

45. $x < 2$
$y \geq -3$

Chapter Practice Test

1. Determine which, if any, of the ordered pairs satisfy the system of equations.

$$x + 2y = -6$$
$$3x + 2y = -12$$

a) $(0, -6)$ **b)** $\left(-3, -\dfrac{3}{2}\right)$ **c)** $(2, -4)$ b)

Identify each system as consistent, inconsistent, or dependent. State whether the system has exactly one solution, no solution, or an infinite number of solutions.

2.

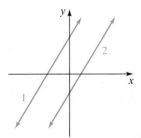

inconsistent—no solution

3.

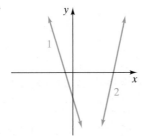

consistent—one solution

4.

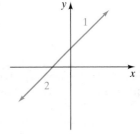

dependent—infinite number
of solutions

Write each equation in slope–intercept form. Then determine, without graphing or solving the system, whether the system of equations has exactly one solution, no solution, or an infinite number of solutions.

5. $-3y = 6x - 9$
$2x + y = 6$ no solution

6. $3x + 2y = 10$
$3x - 2y = 10$ one solution

7. $4x = 6y - 12$
$2x - 3y = -6$
infinite number of solutions

8. When solving a system of linear equations by the substitution or the addition methods, how will you know if the system is
a) inconsistent, **b)** dependent? a) You will obtain a false statement, such as 6 = 0. b) You will obtain a true statement, such as 0 = 0.

Solve each system of equations graphically.

9. $y = 2x - 4$
$y = -2x + 8$ $(3, 2)$

10. $3x - 2y = -3$
$3x + y = 6$ $(1, 3)$

11. $y = 2x + 4$
$4x - 2y = 6$ no solution

Solve each system of equations by substitution.

12. $3x + y = 8$
$x - y = 6$ $\left(\dfrac{7}{2}, -\dfrac{5}{2}\right)$

13. $3x - 4y = 8$
$4x + 2y = 18$ $(4, 1)$

14. $y = 5x - 7$
$y = 3x + 5$ $(6, 23)$

Solve each system of equations using the addition method.

15. $4x + y = -6$
$x + 3y = 4$ $(-2, 2)$

16. $3x + 2y = 12$
$-2x + 5y = 8$ $\left(\dfrac{44}{19}, \dfrac{48}{19}\right)$

17. $5x - 10y = 20$
$x = 2y + 4$
infinite number of solutions

Solve each system of equations using the method of your choice.

18. $y = 3x - 7$
$y = -2x + 8$ $(3, 2)$

19. $3x + 5y = 20$
$6x + 3y = -12$ $\left(-\dfrac{40}{7}, \dfrac{52}{7}\right)$

20. $4x - 6y = 8$
$3x + 5y = 10$ $\left(\dfrac{50}{19}, \dfrac{8}{19}\right)$

Use a system of linear equations to find the solution.

22. $13\frac{1}{3}$ lb butterscotch, $6\frac{2}{3}$ lb lemon

21. **Truck Rental** Charley's Rent a Truck Agency charges $54 per day plus 8 cents per mile to rent a certain truck. Hugh's Rent a truck charges $40 per day plus 15 cents per mile to rent the same truck. How many miles will have to be driven in one day for the cost of Charley's truck to equal the cost of Hugh's truck? 200 mi

22. **Candies** Albert's Grocery sells individually wrapped lemon candies for $6.00 a pound and individually

wrapped butterscotch candies for $4.50 a pound. How much of each must Albert mix to get 20 pounds of a mixture that he can sell for $5.00 per pound?

23. **Boat Race** During a race, Dante Hull's speed boat travels 4 miles per hour faster than Deja Rocket's speed boat. If Dante's boat finishes the race in 3 hours and Deja's finishes the race in 3.2 hours, find the speed of each boat. Deja's boat, 60 mph; Dante's boat, 64 mph

Determine the solution to each system of inequalities. **24., 25.** See graphing answer section, page G14.

24. $2x + 4y < 8$
$x - 3y \geq 6$

25. $x + 3y \geq 6$
$y < 3$

Cumulative Review Test

Take the following test and check your answers with those that appear at the end of the test. Review any questions that you answered incorrectly. The section and objective where the material was covered are indicated after the answer.

1. **Women in the Military** Women are playing an increasingly important role in the U.S. military. Women have been serving in the U.S. armed forces since 1901 when the Army Nurse Corps was established. The number of women serving in the military has grown dramatically since 1967 when the military abolished the 2% cap on women serving in the armed forces. The following graphs provide information about women and the military.

Percent of Military Branches That Are Women

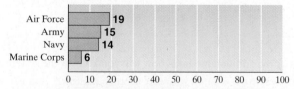

Percent of Military Positions Open to Women

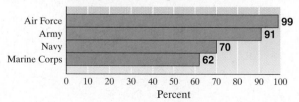

Source: Defense Department; Women's Research and Education Institute

a) Which branches of the military are composed of at least 15% women? Air Force, Army

b) Which branches of the military are composed of at least 10% women *and* have at least 70% of the military positions open to women? Air Force, Army, Navy

2. Consider the set of numbers

$$\left\{-5, -0.6, \tfrac{3}{5}, \sqrt{7}, -\sqrt{2}, 7, 0, -\tfrac{5}{9}, 1.34\right\}.$$

List the elements that are

a) natural numbers. 7

b) rational numbers. $-5, -0.6, \dfrac{3}{5}, 7, 0, -\dfrac{5}{9}, 1.34$

c) irrational numbers. $\sqrt{7}, -\sqrt{2}$

d) real numbers. $-5, -0.6, \dfrac{3}{5}, \sqrt{7}, -\sqrt{2}, 7, 0, -\dfrac{5}{9}, 1.34$

3. Simplify $-237 + 63 + (-192)$. -366

4. Simplify $8 - (3a - 2) + 4(a + 3)$. $a + 22$

5. Solve the equation $2(x - 4) + 2 = 3x - 4$. -2

6. Solve the inequality $3x - 4 \leq x + 6$. Graph the solution on a number line. $x \leq 5$;

7. Solve the formula $P = 2l + 2w$ for w. $w = \dfrac{P - 2l}{2}$

8. **Select a Salary Plan** Maria Gentile recently graduated from college and has accepted a position selling software. She is given a choice of two salary plans. Plan A is a straight 12% commission on sales. Plan B is \$350 per week plus 6% commission on sales. How much must Maria sell in a week for both plans to have the same weekly salary? $5833.33

9. *Angles of a Triangle* One angle of a triangle measures 20° greater than the smallest angle. The third angle measures 6 times the smallest angle. Find the measure of all three angles. 20°, 40°, 120°

10. Simplify $(5x^3y^2)^3(x^2y)$. $125x^{11}y^7$

11. Multiply $(x - 2y)^2$. $x^2 - 4xy + 4y^2$

12. Factor $2n^2 - 5n - 12$ $(2n + 3)(n - 4)$

13. Solve $x^2 - 3x - 40 = 0$. $-5, 8$

14. Multiply $\dfrac{x^2 - 4x - 12}{3x} \cdot \dfrac{x^2 - x}{x^2 - 7x + 6}$. $\dfrac{x + 2}{3}$

15. Solve $\dfrac{2x + 3}{3} = \dfrac{x - 2}{2}$. -12

16. Graph $2x - 4y = 8$.

16., 17. See graphing answer section, page G14.

17. Graph $\dfrac{1}{3}x + \dfrac{1}{2}y = 12$.

18. Determine if the system of equations

$$3x - y = 6$$
$$\frac{3}{2}x - 3 = \frac{1}{2}y$$

has one solution, no solution, or an infinite number of solutions. Explain how you determined your answer. infinite number of solutions

19. Solve the following system of equations graphically.

$$2x + y = 5$$
$$x - 2y = 0 \qquad (2, 1)$$

20. Solve the following system of equations using the addition method.

$$3x - 2y = 8$$
$$-6x + 3y = 2 \qquad \left(-\frac{28}{3}, -18\right)$$

Answers to Cumulative Review Test

1. a) Air Force, Army   **b)** Air Force, Army, Navy; [Sec. 1.2, Obj. 2]   **2. a)** 7   **b)** $-5, -0.6, \dfrac{3}{5}, 7, 0, -\dfrac{5}{9}, 1.34$

c) $\sqrt{7}, -\sqrt{2}$ **d)** $-5, -0.6, \dfrac{3}{5}, \sqrt{7}, -\sqrt{2}, 7, 0, -\dfrac{5}{9}, 1.34$; [Sec. 1.4, Obj. 2]   **3.** -366; [Sec. 1.7, Obj. 3]   **4.** $a + 22$;

[Sec. 2.1, Obj. 6]   **5.** -2; [Sec. 2.5, Obj. 1]   **6.** $x \le 5$; 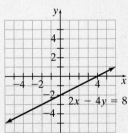;[Sec. 2.7, Obj. 1]   **7.** $w = \dfrac{P - 2l}{2}$; [Sec. 3.1, Obj. 3]

8. 5833.33 [Sec. 3.3, Obj. 4]   **9.** 20°, 40°, 120°; [Sec. 3.4, Obj. 1]   **10.** $125x^{11}y^7$; [Sec. 4.1, Obj. 3]   **11.** $x^2 - 4xy + 4y^2$;

[Sec. 4.5, Obj. 5]   **12.** $(2n + 3)(n - 4)$; [Sec. 5.4, Obj. 1]   **13.** $-5, 8$; [Sec. 5.6, Obj. 2]   **14.** $\dfrac{x + 2}{3}$; [Sec. 6.2, Obj. 1]

15. -12; [Sec. 6.6, Obj. 1]   **16.** 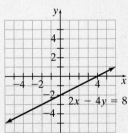; [Sec. 7.2, Obj. 3]   **17.** 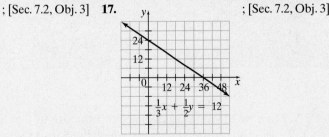; [Sec. 7.2, Obj. 3]

18. An infinite number of solutions; [Sec. 8.1, Obj. 2]   **19.** 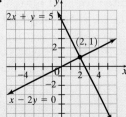; [Sec. 8.1, Obj. 3]

20. $\left(-\dfrac{28}{3}, -18\right)$; [Sec. 8.3, Obj. 1]

Chapter 9

Roots and Radicals

When we determine a distance or other items that are measured, the answer obtained is often an irrational number, which involves a radical (for example, a square root). When we obtain a radical answer, we often round it off to a decimal number. Many scientific and mathematical formulas involve square roots, including the formula for finding the velocity at which an object hits the ground when dropped from a certain height. According to legend, Galileo Galilei experimented with free falling objects using the Leaning Tower of Pisa. In Example 5 on page 596, we determine that a coconut that falls from 20 feet will hit the ground at approximately 35.78 feet per second.

SSM

Study Guide

CD/Video

MathPro 4/5

PH Math
Tutor Center

prenhall.com/Angel

A Look Ahead

In this chapter we study roots, radical expressions, and radical equations, with an emphasis on square roots. Square roots are one type of radical expression. In Sections 9.1 through 9.4, we learn how to evaluate a square root, how to simplify square root expressions, and how to add, subtract, multiply, and divide expressions that contain square roots. In Section 9.5, we discuss solving equations that contain square roots. Section 9.6, Radicals: Applications and Problem Solving, is an extension of Section 9.5 and presents some real-life uses of square roots. In Section 9.7, we discuss cube roots and higher roots. We strongly suggest that you use a scientific calculator or graphing calculator for this and the next chapter.

Radical expressions and equations play an important part in mathematics and the sciences. Many mathematical and scientific formulas involve radicals. As you will see in Chapter 10, one of the most important formulas in mathematics, the quadratic formula, contains a square root.

9.1 EVALUATING SQUARE ROOTS

SSM   Study Guide   CD/Video   MathPro 4/5   PH Math Tutor Center   prenhall.com/Angel

1   Evaluate square roots of real numbers.
2   Recognize that not all square roots represent real numbers.
3   Determine whether the square root of a real number is rational or irrational.
4   Write square roots as exponential expressions.

1 Evaluate Square Roots of Real Numbers

Earlier we discussed determining the square of a number. Recall that when we square a number we multiply the number by itself. For example,

$$\text{If } a = 5, \text{ then } a^2 = 5 \cdot 5 = 25$$
$$\text{If } a = -5, \text{ then } a^2 = (-5)(-5) = 25$$

In this chapter we consider the opposite problem. That is, if we know the square of a number, what is the number? For example,

$$\text{If } a^2 = 25, \text{ then what are the values of } a?$$

To find a in the above statement, we must find the values that when multiplied by themselves result in a product of 25. Since $5 \cdot 5 = 25$ and $(-5)(-5) = 25$, a has two values, -5 and 5.

Finding the **square root** of a given number is the reverse process of squaring a number. When we find the square root of a given number, we are determining what numbers, when multiplied by themselves, result in the given number. Every real number greater than 0 has two square roots, a positive square root and a negative square root. For example, the number 25 has two square roots, -5 and 5. We use the symbol $\sqrt{}$ to indicate the **positive** or **principal square root** of a number. For example

$$\sqrt{25} = 5 \quad \text{(The principal square root of 25 is 5.)}$$

The symbol $-\sqrt{}$ is used to denote the negative square root of a number. For example,

$$-\sqrt{25} = -5 \quad \text{(The negative square root of 25 is } -5.)$$

The negative square root is the additive inverse or opposite of the principal square root. The following boxed material summarizes this information and provides additional information about square roots.

SQUARE ROOTS

> The **positive** or **principal square root** of a positive number a is written as $\sqrt{a}$. The negative square root is written as $-\sqrt{a}$.
>
> $$\sqrt{a} = b \quad \text{if} \quad b^2 = a$$
>
> Also, the square root of 0 is 0, written $\sqrt{0} = 0$.

Note that **the principal square root of a positive number, a, is the positive number whose square equals a. Whenever we use the term square root in this book, we mean the positive or principal square root.**

Square roots are one type of radical expression that you will use in both mathematics and science.

$$\sqrt{x} \text{ is read "the square root of } x."$$

The $\sqrt{}$ is called the **radical sign**. The number or expression inside the radical sign is called the **radicand**.

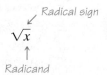

Radical sign

Radicand

The entire expression, including the radical sign and radicand, is called the **radical expression**.

Another part of a radical expression is its **index**. The index tells the "root" of the expression. Square roots have an index of 2. The index of a square root is generally not written.

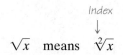

Index

$$\sqrt{x} \quad \text{means} \quad \sqrt[2]{x}$$

Other types of radical expressions have different indices. For example, $\sqrt[3]{x}$, which is read "the cube root of x," has an index of 3. Cube roots are discussed in Section 9.7.

Now let's find some square roots.

Examples

$$\sqrt{25} = 5 \qquad \text{since } 5^2 = 5 \cdot 5 = 25$$

$$\sqrt{64} = 8 \qquad \text{since } 8^2 = 8 \cdot 8 = 64$$

$$\sqrt{\frac{1}{4}} = \frac{1}{2} \qquad \text{since } \left(\frac{1}{2}\right)^2 = \left(\frac{1}{2}\right)\left(\frac{1}{2}\right) = \frac{1}{4}$$

$$\sqrt{\frac{4}{9}} = \frac{2}{3} \qquad \text{since } \left(\frac{2}{3}\right)^2 = \left(\frac{2}{3}\right)\left(\frac{2}{3}\right) = \frac{4}{9}$$

EXAMPLE 1 Evaluate. **a)** $\sqrt{81}$ **b)** $\sqrt{100}$

Solution **a)** $\sqrt{81} = 9$ since $9^2 = (9)(9) = 81$

b) $\sqrt{100} = 10$ since $(10)^2 = (10)(10) = 100$

EXAMPLE 2 Evaluate. **a)** $-\sqrt{81}$ **b)** $-\sqrt{100}$

Solution **a)** $\sqrt{81} = 9$. Now we take the opposite of both sides to get

$$-\sqrt{81} = -9$$

NOW TRY EXERCISE 21 **b)** Similarly, $-\sqrt{100} = -10$.

2 Recognize That Not All Square Roots Represent Real Numbers

You must understand that **square roots of negative numbers are not real numbers**. Consider $\sqrt{-4}$; to what is $\sqrt{-4}$ equal? To evaluate $\sqrt{-4}$, we must find some number whose square equals -4. But we know that the square of any nonzero real number must be a positive number. Therefore, no real number squared equals -4, so $\sqrt{-4}$ has no real value. Numbers like $\sqrt{-4}$, or square roots of any negative numbers, are called **imaginary numbers**. Imaginary numbers are discussed further in Section 10.5.

TEACHING TIP
Ask, "How do we know that the square of any nonzero real number must be a positive number?"

EXAMPLE 3 Indicate whether the radical expression is a real or an imaginary number.

a) $-\sqrt{16}$ **b)** $\sqrt{-16}$ **c)** $\sqrt{-37}$ **d)** $-\sqrt{37}$

Solution **a)** Real (equal to -4)   **b)** Imaginary   **c)** Imaginary   **d)** Real

HELPFUL HINT

The square root of any nonnegative number will be a real number. The square root of any negative number will be an imaginary number.

Examples of *Real* Numbers   Examples of *Imaginary* Numbers

$-\sqrt{9}, \quad -\sqrt{\frac{1}{2}}, \quad -\sqrt{6.74}, \quad -\sqrt{16}$ $\sqrt{-9}, \quad \sqrt{-\frac{1}{2}}, \quad \sqrt{-6.74}, \quad \sqrt{-16}$

Radicands are positive numbers   Radicands are negative numbers

Suppose we have an expression like $\sqrt{x}$, where x represents some number. For the radical $\sqrt{x}$ to be a real number, and not imaginary, we must assume that x is a nonnegative number.

In this chapter, unless stated otherwise, we will assume that all expressions that are radicands represent nonnegative numbers.

3 Determine Whether the Square Root of a Real Number Is Rational or Irrational

To help in our discussion of rational and irrational numbers, we will define perfect squares. The numbers $1, 4, 9, 16, 25, 36, 49, \ldots$ are called **perfect squares** because each number is *the square of a natural number*. When a perfect square is a factor of a radicand, we may refer to it as a **perfect square factor**.

TEACHING TIP
Have students graph the positive and negative square roots of perfect squares from 1–100 on a number line. Then have them estimate the location of $\sqrt{5}, -\sqrt{35}, \sqrt{50}, -\sqrt{72}, \sqrt{86}$, and $-\sqrt{95}$.

1,	2,	3,	4,	5,	6,	7,	... *Natural numbers*
1^2,	2^2,	3^2,	4^2,	5^2,	6^2,	7^2,	... *The squares of the natural numbers*
1,	4,	9,	16,	25,	36,	49,	... *Perfect squares*

What are the next two perfect squares? Note that the square root of a perfect square is an integer. That is, $\sqrt{1} = 1$, $\sqrt{4} = 2$, $\sqrt{9} = 3$, $\sqrt{16} = 4$, and so on.

Table 9.1 illustrates the first 20 perfect squares. You may wish to refer to this table when simplifying radical expressions.

	TABLE 9.1						
Perfect Square	**Square Root of Perfect Square**		**Value**	**Perfect Square**	**Square Root of Perfect Square**		**Value**
1	$\sqrt{1}$	=	1	121	$\sqrt{121}$	=	11
4	$\sqrt{4}$	=	2	144	$\sqrt{144}$	=	12
9	$\sqrt{9}$	=	3	169	$\sqrt{169}$	=	13
16	$\sqrt{16}$	=	4	196	$\sqrt{196}$	=	14
25	$\sqrt{25}$	=	5	225	$\sqrt{225}$	=	15
36	$\sqrt{36}$	=	6	256	$\sqrt{256}$	=	16
49	$\sqrt{49}$	=	7	289	$\sqrt{289}$	=	17
64	$\sqrt{64}$	=	8	324	$\sqrt{324}$	=	18
81	$\sqrt{81}$	=	9	361	$\sqrt{361}$	=	19
100	$\sqrt{100}$	=	10	400	$\sqrt{400}$	=	20

A **rational number** is one that can be written in the form $\frac{a}{b}$, where a and b are integers, and $b \neq 0$. Examples of rational numbers are $\frac{1}{2}, \frac{3}{5}, -\frac{9}{2}, 4$, and 0. All integers are rational numbers since they can be expressed with a denominator of 1. For example, $4 = \frac{4}{1}$ and $0 = \frac{0}{1}$. The square roots of perfect squares are also rational numbers since each is an integer. When a rational number is written as a decimal, it will be either a terminating or repeating decimal.

TEACHING TIP
Have students describe in their own words and give an example of
a) terminating decimal
b) repeating decimal
c) nonterminating, nonrepeating decimal

Terminating Decimals

$$\frac{1}{2} = 0.5$$

$$\frac{5}{8} = 0.625$$

$$\sqrt{4} = 2.0$$

Repeating Decimals

$$\frac{1}{3} = 0.333\ldots$$

$$\frac{4}{9} = 0.444\ldots$$

$$\frac{1}{6} = 0.1666\ldots$$

Real numbers that are not rational numbers are called **irrational numbers**. Irrational numbers when written as decimals are nonterminating, nonrepeating decimals. The square root of every positive integer that is not a perfect square is an irrational number. For example, $\sqrt{2}$ and $\sqrt{3}$ are irrational numbers. The 20 square roots listed in Table 9.1 are rational numbers. All other square roots of integers between 1 and 400 are irrational numbers. For example, since $\sqrt{50}$ is less than $\sqrt{400}$ and is not in Table 9.1, it is an irrational number. Furthermore, since $\sqrt{50}$ is between $\sqrt{49}$ and $\sqrt{64}$ in Table 9.1, the value of $\sqrt{50}$ is between 7 and 8. If you evaluate $\sqrt{50}$ on a calculator, as explained in the calculator box on the following page, you will find that $\sqrt{50} \approx 7.07$, rounded to the nearest hundredth. Notice that since $\sqrt{50}$ is just slightly larger than $\sqrt{49}$, the value of $\sqrt{50}$ is just slightly larger than 7.

Using Your Calculator

Evaluating Square Roots on a Calculator

The square root key on calculators can be used to find square roots of non-negative numbers.
   On some scientific calculators, to find the square root of 4, we press

Answer displayed

$$4 \boxed{\sqrt{x}} \; 2$$

On the TI-83 Plus graphing calculator, to find the square root of 4, we press

$$\boxed{2^{\text{nd}}} \; \boxed{x^2} \; (\; 4 \; \boxed{)} \; \boxed{\text{ENTER}} \; 2$$

To get $\sqrt{}$

Displayed by TI-83 Plus

Both calculators display the answer 2. Since 2 is an integer, $\sqrt{4}$ is a rational number.
   What would the calculator display if we evaluate $\sqrt{7}$? The display would show 2.6457513. Note that $\sqrt{7}$ is an irrational number, or a nonrepeating, nonterminating decimal. The decimal value of $\sqrt{7}$, or any other irrational number, can never be given exactly. The answers given on a calculator display are only close approximations of their value.
   Suppose we tried to evaluate $\sqrt{-4}$ on a calculator. What would the calculator give as an answer? On a scientific calculator, we press

Answer displayed

$$4 \; \boxed{^{+/-}} \; \boxed{\sqrt{x}} \; \text{Error}$$

On a TI-83 Plus graphing calculator, we press

Answer displayed

$$\boxed{2^{\text{nd}}} \; \boxed{x^2} \; (\; \boxed{(-)} \; 4 \; \boxed{)} \; \boxed{\text{ENTER}} \; \text{ERR: NONREAL ANS}$$

Both calculators would give an error message, because the square root of -4, or the square root of any other negative number, is not a real number.

Exercises

Use your calculator to evaluate each square root.

1. $\sqrt{15}$ 3.872983346   **2.** $\sqrt{151}$ 12.28820573   **3.** $\sqrt{-16}$ Error   **4.** $\sqrt{27}$ 5.196152423

Figure 9.1 shows the relationship between the different types of numbers we discussed. Notice that the rational numbers plus the irrational numbers form the real numbers, and imaginary numbers are not real numbers, and vice versa.

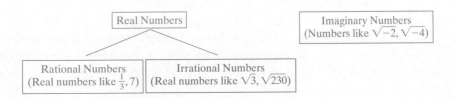

FIGURE 9.1

TEACHING TIP
Instruct students to round their
answer, rather than truncate,
when using the approximately
equal sign.

When evaluating radicals, we may use the is approximately equal to symbol, $\approx$. For example, we may write $\sqrt{2} \approx 1.414$. This is read "the square root of 2 is approximately equal to 1.414." Recall that $\sqrt{2}$ is not a perfect square, so its square root cannot be evaluated exactly.

EXAMPLE 4 Use your calculator or Table 9.1 to determine whether the following square roots are rational or irrational numbers.

    a) $\sqrt{123}$ **b)** $\sqrt{196}$ **c)** $\sqrt{326}$ **d)** $\sqrt{289}$

Solution **a)** Irrational   **b)** Rational, equal to 14

NOW TRY EXERCISE 61   **c)** Irrational   **d)** Rational, equal to 17

4 Write Square Roots as Exponential Expressions

Radical expressions can be written in exponential form. Since we are discussing square roots, we will show how to write square roots in exponential form. Writing other radicals in exponential form will be discussed in Section 9.7. We introduce this information here because your instructor may wish to use exponential form to help explain certain concepts.

Recall that the index of square roots is 2. For example,

$$\sqrt{x} \quad \text{means} \quad \sqrt[2]{x}$$

We use the index, 2, when writing square roots in exponential form. To change from an expression in square root form to an expression in exponential form, simply write the radicand of the square root to the 1/2 power, as follows:

> **Writing a Square Root in Exponential Form**
>
> $\sqrt{\blacksquare} = \blacksquare^{1/2}$ ← *Index of square root*
>
> *Radicand*

For example, $\sqrt{8}$ in exponential form is $8^{1/2}$, and $\sqrt{5ab} = (5ab)^{1/2}$. Other examples are

Square Root Form		Exponential Form
$\sqrt{25}$	$=$	$(25)^{1/2}$
$\sqrt{2x}$	$=$	$(2x)^{1/2}$
$\sqrt{15x^2y}$	$=$	$(15x^2y)^{1/2}$

EXAMPLE 5 Write each radical expression in exponential form.

    a) $\sqrt{3}$ **b)** $\sqrt{10x}$

Solution **a)** $3^{1/2}$ **b)** $(10x)^{1/2}$

NOW TRY EXERCISE 69   We can also convert an expression from exponential form to radical form. To do so, we reverse the process. For example, $(6x)^{1/2}$ can be written $\sqrt{6x}$ and $(20x^4)^{1/2}$ can be written $\sqrt{20x^4}$.

The rules of exponents presented in Sections 4.1 and 4.2 apply to rational (or fractional) exponents. For example,

$$(x^2)^{1/2} = x^{2(1/2)} = x^1 = x$$
$$(xy)^{1/2} = x^{1/2}y^{1/2}$$
and $\quad x^{1/2} \cdot x^{3/2} = x^{(1/2)+(3/2)} = x^{4/2} = x^2$

2. a) the root of the expression **b)** radicand **3.** Rational numbers can be written in the form $a/b, b \neq 0$, a and b integers. Irrational numbers are real numbers that are not rational.

4. It is nonnegative. The square root of a negative number is not a real number.

Exercise Set 9.1

Concept/Writing Exercises

5. a) rational if answer is an integer **b)** rational if number appears in Table 9.1
6. No real number when squared will be a negative number.

1. What is the principal square root of a positive real number a? positive number whose square equals a

2. a) What does the index indicate in a radical expression?

  b) What is the expression inside the radical sign called?

3. In your own words, explain the difference between a rational number and an irrational number.

4. Whenever we see an expression in a square root, what assumption do we make about the expression? Why do we make this assumption?

5. In your own words, explain how you would determine whether the square root of a positive integer less than

400 is a rational or irrational number **a)** by using a calculator, and **b)** without the use of a calculator.

6. In your own words, explain why the square root of a negative number is not a real number.

7. Is $\sqrt{25} = 5$? Explain. yes

8. Is $\sqrt{25} = -5$? Explain. no

9. Is $\sqrt{-25} = -5$? Explain. no

10. Is $-\sqrt{25} = -5$? Explain. yes

11. Is $\sqrt{\dfrac{16}{25}}$ a rational number? Explain. yes

12. Is $\sqrt{\dfrac{13}{15}}$ a rational number? Explain. no

Practice the Skills

Evaluate each square root.

13. $\sqrt{0}$ 0
14. $\sqrt{16}$ 4
15. $\sqrt{1}$ 1
16. $\sqrt{64}$ 8
17. $-\sqrt{49}$ -7
18. $\sqrt{4}$ 2
19. $\sqrt{400}$ 20
20. $\sqrt{100}$ 10
21. $-\sqrt{16}$ -4
22. $-\sqrt{36}$ -6
23. $\sqrt{144}$ 12
24. $\sqrt{49}$ 7
25. $\sqrt{169}$ 13
26. $\sqrt{225}$ 15
27. $-\sqrt{1}$ -1
28. $-\sqrt{100}$ -10
29. $\sqrt{81}$ 9
30. $-\sqrt{49}$ -7
31. $-\sqrt{121}$ -11
32. $-\sqrt{196}$ -14
33. $\sqrt{\dfrac{1}{4}}$ $\dfrac{1}{2}$
34. $\sqrt{\dfrac{25}{4}}$ $\dfrac{5}{2}$
35. $\sqrt{\dfrac{36}{49}}$ $\dfrac{6}{7}$
36. $\sqrt{\dfrac{25}{64}}$ $\dfrac{5}{8}$
37. $-\sqrt{\dfrac{25}{36}}$ $-\dfrac{5}{6}$
38. $-\sqrt{\dfrac{100}{144}}$ $-\dfrac{10}{12} = -\dfrac{5}{6}$
39. $\sqrt{\dfrac{81}{49}}$ $\dfrac{9}{7}$
40. $\sqrt{\dfrac{121}{169}}$ $\dfrac{11}{13}$

Use your calculator to evaluate each square root. Write your answer to seven decimal places.

41. $\sqrt{12}$ 3.4641016
42. $\sqrt{2}$ 1.4142136
43. $\sqrt{15}$ 3.8729833
44. $\sqrt{30}$ 5.4772256
45. $\sqrt{80}$ 8.9442719
46. $\sqrt{79}$ 8.8881944
47. $\sqrt{324}$ 18
48. $\sqrt{121}$ 11
49. $\sqrt{97}$ 9.8488578
50. $\sqrt{43}$ 6.5574385
51. $\sqrt{3}$ 1.7320508
52. $\sqrt{40}$ 6.3245553

Indicate whether each statement is true or false.

53. $\sqrt{18}$ is a rational number. false
54. $\sqrt{-25}$ is a real number. false
55. $\sqrt{25}$ is a rational number. true
56. $\sqrt{5}$ is an irrational number. true
57. $\sqrt{4}$ is an irrational number. false
58. $\sqrt{\dfrac{1}{4}}$ is a rational number. true
59. $\sqrt{\dfrac{9}{16}}$ is a rational number. true
60. $\sqrt{231}$ is a rational number. false
61. $\sqrt{125}$ is an irrational number. true
62. $\sqrt{27}$ is an irrational number. true
63. $\sqrt{(15)^2}$ is an integer. true
64. $\sqrt{(12)^2}$ is an integer. true

Write in exponential form.

65. $\sqrt{7}$ $7^{1/2}$

66. $\sqrt{31}$ $(31)^{1/2}$

67. $\sqrt{17}$ $(17)^{1/2}$

68. $\sqrt{36}$ $(36)^{1/2}$

69. $\sqrt{8x}$ $(8x)^{1/2}$

70. $\sqrt{5x}$ $(5x)^{1/2}$

🔒 **71.** $\sqrt{12x^2}$ $(12x^2)^{1/2}$

72. $\sqrt{25x^2y}$ $(25x^2y)^{1/2}$

73. $\sqrt{15ab^2}$ $(15ab^2)^{1/2}$

74. $\sqrt{34x^3y}$ $(34x^3y)^{1/2}$

75. $\sqrt{62n^3}$ $(62n^3)^{1/2}$

76. $\sqrt{36x^3y^3}$ $(36x^3y^3)^{1/2}$

Problem Solving

77. rational: $9.83, \frac{3}{5}, 0.333\ldots, 5, \sqrt{\frac{4}{49}}, \frac{3}{7}, -\sqrt{9}$; irrational: $\sqrt{\frac{5}{16}}$; imaginary: $\sqrt{-4}, -\sqrt{-16}$

🔒 **77.** Classify each number as rational, irrational, or imaginary.

$9.83, \sqrt{-4}, \frac{3}{5}, 0.333\ldots, 5, \sqrt{\frac{4}{49}}, \frac{3}{7}, \sqrt{\frac{5}{16}}, -\sqrt{9}, -\sqrt{-16}$

78. Classify each number as rational, irrational, or imaginary.

$\sqrt{5}, 8.23, \sqrt{-7}, 10, \frac{1}{3}, 0.33, \sqrt{\frac{25}{64}}, -\sqrt{90}$

79. Between what two integers is the square root of 47? Do not use your calculator or Table 9.1. Explain how you determined your answer. 6 and 7

80. Between what two integers is the square root of 88? Do not use your calculator or Table 9.1. Explain how you determined your answer. 9 and 10

81. a) Explain how you can determine without using a calculator whether 4.6 or $\sqrt{20}$ is greater.

b) Without using a calculator, determine which is greater. 4.6

82. a) Explain how you can determine without using a calculator whether 7.2 or $\sqrt{58}$ is greater.

b) Without using a calculator, determine which is greater. $\sqrt{58}$

83. Arrange the following list from smallest to largest. Do not use a calculator or Table 9.1.

$$5, -\sqrt{9}, -\sqrt{7}, 12, 2.5, -\frac{1}{2}, 4.01, \sqrt{16}$$

84. Arrange the following list from smallest to largest. Do not use a calculator or Table 9.1.

$$-\frac{1}{3}, -\sqrt{9}, 5, 0, \sqrt{9}, 8, -2, 3.25$$

78. rational: $8.23, 10, \frac{1}{3}, 0.33, \sqrt{\frac{25}{64}}$; irrational: $\sqrt{5}, -\sqrt{90}$; imaginary: $\sqrt{-7}$ **81. a)** Square 4.6 and compare to 20. **82. a)** Square 7.2 and

Challenge Problems

compare to 58. **83.** $-\sqrt{9}, -\sqrt{7}, -\frac{1}{2}, 2.5, \sqrt{16}, 4.01, 5, 12$ **84.** $-\sqrt{9}, -2, -\frac{1}{3}, 0, \sqrt{9}, 3.25, 5, 8$

We discuss the following concepts in Sections 9.2 and 9.3.

88. a) Is $\sqrt{4} \cdot \sqrt{9}$ equal to $\sqrt{4 \cdot 9}$? yes

b) Is $\sqrt{9} \cdot \sqrt{25}$ equal to $\sqrt{9 \cdot 25}$? yes

c) Using these two examples, can you guess what $\sqrt{a} \cdot \sqrt{b}$ is equal to (provided that $a \geq 0, b \geq 0$)?

d) Create your own problem like those given in parts **a)** and **b)** and see if the answer you gave in part **c)** works with your numbers. Answers will vary.

89. a) Is $\sqrt{2^2}$ equal to 2? yes

b) Is $\sqrt{5^2}$ equal to 5? yes

85. Match each number in the column on the left with the corresponding answer in the column on the right.

$\sqrt{4}$	imaginary number $\sqrt{4} = 2$,
$6^{1/2}$	2 $6^{1/2} \approx 2.45$,
$-\sqrt{9}$	≈ 2.45 $-\sqrt{9} = -3$,
$-(25)^{1/2}$	-5 $-(25)^{1/2} = -5$,
$(30)^{1/2}$	-3 $(30)^{1/2} \approx 5.48$,
$(-4)^{1/2}$	≈ 5.48 $(-4)^{1/2}$, imaginary number

86. Match each number in the column on the left with the corresponding answer in the column on the right.

$-\sqrt{36}$	10 $-\sqrt{36} = -6$,
$(40)^{1/2}$	-7 $(40)^{1/2} \approx 6.32$,
$\sqrt{100}$	imaginary number $\sqrt{100} = 10$,
$(10)^{1/2}$	≈ 6.32 $(10)^{1/2} \approx 3.16$,
$-(49^{1/2})$	≈ 3.16 $-(49^{1/2}) = -7$,
$(-16)^{1/2}$	-6 $(-16)^{1/2}$, imaginary number,

87. Is $\sqrt{0}$

a) a real number? yes

b) a positive number? no

c) a negative number? no

d) a rational number? yes

e) an irrational number? Explain your answer. no

88. c) $\sqrt{a \cdot b}$

c) Using these two examples, can you guess what $\sqrt{a^2}, a \geq 0$, is equal to? a

d) Create your own problem like those given in parts **a)** and **b)** and see if the answer you gave in part **c)** works with your numbers. Answers will vary.

90. a) Is $\dfrac{\sqrt{16}}{\sqrt{4}}$ equal to $\sqrt{\dfrac{16}{4}}$? yes

b) Is $\dfrac{\sqrt{36}}{\sqrt{9}}$ equal to $\sqrt{\dfrac{36}{9}}$? yes

c) Using these two examples, can you guess what $\dfrac{\sqrt{a}}{\sqrt{b}}$ is equal to (provided that $a \geq 0, b > 0$)? $\sqrt{\dfrac{a}{b}}$

d) Create your own problem like those given in parts **a)** and **b)** and see if the answer you gave in part **c)** works with your numbers. Answers will vary.

The rules of exponents we discussed in Chapter 4 also apply with rational exponents. Use the rules of exponents to simplify the following expressions. We will discuss problems like this in Section 9.7.

91. $\left(x^3\right)^{1/2}$ $x^{3/2}$

92. $\left(x^4\right)^{1/2}$ x^2

93. $x^{1/2} \cdot x^{5/2}$ x^3

94. $x^{3/2} \cdot x^{1/2}$ x^2

Cumulative Review Exercises

[3.3] **95.** ***Jumping on a Trampoline*** Allison jumped on the trampoline for a minute, and then her mother Elizabeth jumped on the trampoline for a minute. If Allison made twice as many jumps as her mother did, and the total number of jumps was 78, determine the number of jumps Allison made.

52 jumps

[6.6] *Solve.*

96. $\dfrac{2x}{x^2 - 4} + \dfrac{1}{x - 2} = \dfrac{2}{x + 2}$ -6

97. $\dfrac{4x}{x^2 + 6x + 9} - \dfrac{2x}{x + 3} = \dfrac{x + 1}{x + 3}$ -1

[7.3] **98.** Determine the slope of the line through the points $(-5, 3)$ and $(6, 7)$. $\dfrac{4}{11}$

[7.6] **99.** If $f(x) = x^2 - 4x - 5$, find $f(-3)$. 16

9.2 SIMPLIFYING SQUARE ROOTS

SSM

Study Guide

CD/Video

1 Use the product rule to simplify square roots containing constants.

2 Use the product rule to simplify square roots containing variables.

MathPro 4/5

PH Math Tutor Center

prenhall.com/Angel

1 Use the Product Rule to Simplify Square Roots Containing Constants

To simplify square roots in this section we will make use of the **product rule for square roots**.

TEACHING TIP
Before discussing the Product Rule for Square Roots, have students describe the rule in words.

Product Rule for Square Roots

$$\sqrt{a} \cdot \sqrt{b} = \sqrt{a \cdot b}, \quad \text{provided} \quad a \geq 0, b \geq 0 \qquad \text{Rule 1}$$

The product rule states that the product of two square roots is equal to the square root of the product. The product rule applies only when both a and b are nonnegative, since the square roots of negative numbers are not real numbers.

Examples of the Product Rule

$$\left.\begin{array}{l}\sqrt{1}\cdot\sqrt{60} = \sqrt{1\cdot 60} \\ \sqrt{2}\cdot\sqrt{30} = \sqrt{2\cdot 30} \\ \sqrt{3}\cdot\sqrt{20} = \sqrt{3\cdot 20} \\ \sqrt{4}\cdot\sqrt{15} = \sqrt{4\cdot 15} \\ \sqrt{6}\cdot\sqrt{10} = \sqrt{6\cdot 10}\end{array}\right\} = \sqrt{60}$$

Note that $\sqrt{60}$ can be factored into any of these forms.

When two square roots are placed next to one another, the square roots are to be multiplied. Thus $\sqrt{a}\,\sqrt{b}$ means $\sqrt{a}\cdot\sqrt{b}$.

To Simplify the Square Root of a Constant

1. Write the constant as a product of the largest perfect square factor and another factor.

2. Use the product rule to write the expression as a product of square roots, with each square root containing one of the factors.

3. Find the square root of the perfect square factor.

EXAMPLE 1 Simplify $\sqrt{60}$.

Solution The only perfect square factor of 60 is 4.

$$\begin{aligned}\sqrt{60} &= \sqrt{4\cdot 15} \\ &= \sqrt{4}\cdot\sqrt{15} \\ &= 2\sqrt{15}\end{aligned}$$

Since 15 is not a perfect square and has no perfect square factors, this expression cannot be simplified further. The expression $2\sqrt{15}$ is read "two times the square root of fifteen." ✳

EXAMPLE 2 Simplify $\sqrt{12}$.

Solution
$$\begin{aligned}\sqrt{12} &= \sqrt{4\cdot 3} = \sqrt{4}\cdot\sqrt{3} \\ &= 2\sqrt{3}\end{aligned}$$ ✳

EXAMPLE 3 Simplify $\sqrt{80}$.

Solution
$$\begin{aligned}\sqrt{80} &= \sqrt{16\cdot 5} = \sqrt{16}\cdot\sqrt{5} \\ &= 4\sqrt{5}\end{aligned}$$ ✳

EXAMPLE 4 Simplify $\sqrt{245}$.

Solution
$$\begin{aligned}\sqrt{245} &= \sqrt{49\cdot 5} = \sqrt{49}\cdot\sqrt{5} \\ &= 7\sqrt{5}\end{aligned}$$ ✳

HELPFUL HINT

When simplifying a square root, it is not uncommon to use a perfect square factor that is not the *largest* perfect square factor of the radicand. Let's consider Example 3 again. Four is also a perfect square factor of 80.

$$\sqrt{80} = \sqrt{4\cdot 20} = \sqrt{4}\cdot\sqrt{20} = 2\sqrt{20}$$

(continued on the next page)

Since 20 itself contains a perfect square factor of 4, the problem is not complete. Rather than starting the entire problem again, you can continue the simplification process as follows.

$$\sqrt{80} = 2\sqrt{20} = 2\sqrt{4 \cdot 5} = 2\sqrt{4} \cdot \sqrt{5} = 2 \cdot 2 \cdot \sqrt{5} = 4\sqrt{5}$$

Now the result checks with the answer in Example 3.

EXAMPLE 5 Simplify $\sqrt{140}$.

Solution
$$\sqrt{140} = \sqrt{4 \cdot 35} = \sqrt{4} \cdot \sqrt{35}$$
$$= 2\sqrt{35}$$

Although 35 can be factored into $5 \cdot 7$, neither of these factors is a perfect square. Thus, the answer cannot be simplified any further.

NOW TRY EXERCISE 21

2 Use the Product Rule to Simplify Square Roots Containing Variables

Now we will simplify square roots that contain variables in the radicand.

In Section 9.1 we noted that certain numbers were **perfect squares**. We will also refer to certain expressions that contain a variable as perfect squares. When a radical contains a variable (or number) raised to an **even exponent**, that variable (or number) and exponent together also form a perfect square. For example, in the expression $\sqrt{x^4}$, the x^4 is a perfect square since the exponent 4, is even. In the expression $\sqrt{x^5}$, the x^5 is not a perfect square since the exponent is odd. However, x^4 is a **perfect square factor** of x^5 because x^4 is a perfect square and x^4 is a factor of x^5. Note that $x^5 = x^4 \cdot x$.

To evaluate square roots when the radicand is a perfect square, we use the following rule.

$$\sqrt{a^{2 \cdot n}} = a^n, \qquad a \geq 0 \qquad \text{Rule 2}$$

This rule states that **the square root of a variable raised to an even power equals the variable raised to one-half that power**. To explain this rule, we can write the square root expression $\sqrt{a^{2n}}$ in exponential form, and then simplify as follows.

$$\sqrt{a^{2n}} = (a^{2n})^{1/2} = a^{2n(1/2)} = a^n$$

Examples of rule 2 follow. Remember that we are assuming the variables represent nonnegative real numbers.

Examples
$$\sqrt{x^2} = x$$
$$\sqrt{a^4} = a^2$$
$$\sqrt{y^{16}} = y^8$$
$$\sqrt{z^{28}} = z^{14}$$

A special case of rule 2 (when $n = 1$) is
$$\sqrt{a^2} = a, \qquad a \geq 0$$

EXAMPLE 6 Simplify. **a)** $\sqrt{x^{30}}$ **b)** $\sqrt{x^4 y^6}$ **c)** $\sqrt{a^{10} b^2}$ **d)** $\sqrt{y^{12} z^{18}}$

Solution **a)** $\sqrt{x^{30}} = x^{15}$ **b)** $\sqrt{x^4 y^6} = \sqrt{x^4}\sqrt{y^6} = x^2 y^3$

NOW TRY EXERCISE 35 **c)** $\sqrt{a^{10} b^2} = \sqrt{a^{10}}\sqrt{b^2} = a^5 b$ **d)** $\sqrt{y^{12} z^{18}} = \sqrt{y^{12}}\sqrt{z^{18}} = y^6 z^9$

To Simplify the Square Root of a Radicand Containing a Variable Raised to an Odd Power

1. Express the variable as the product of two factors, one of which has an exponent of 1 (the other will therefore be a perfect square factor).

2. Use the product rule to simplify.

The radicand of your simplified answer should not contain any perfect square factors or any variables with an exponent greater than 1.

Examples 7 and 8 illustrate the above procedure.

EXAMPLE 7 Simplify. **a)** $\sqrt{x^3}$ **b)** $\sqrt{y^7}$ **c)** $\sqrt{a^{79}}$

Solution **a)** $\sqrt{x^3} = \sqrt{x^2 \cdot x} = \sqrt{x^2} \cdot \sqrt{x}$ *(Remember that x means x^1.)*

$$= x \cdot \sqrt{x} \text{ or } x\sqrt{x}$$

b) $\sqrt{y^7} = \sqrt{y^6 \cdot y^1} = \sqrt{y^6} \cdot \sqrt{y}$

$$= y^3\sqrt{y}$$

c) $\sqrt{a^{79}} = \sqrt{a^{78} \cdot a} = \sqrt{a^{78}} \cdot \sqrt{a}$

$$= a^{39}\sqrt{a}$$

More complex radicals can be simplified using the product rule for radicals and the principles discussed in this section.

EXAMPLE 8 Simplify. **a)** $\sqrt{25x^3}$ **b)** $\sqrt{50x^2}$ **c)** $\sqrt{50x^3}$

Solution Write each expression as the product of square roots, one of which has a radicand that is a perfect square.

a) $\sqrt{25x^3} = \sqrt{25x^2} \cdot \sqrt{x} = 5x\sqrt{x}$

b) $\sqrt{50x^2} = \sqrt{25x^2} \cdot \sqrt{2} = 5x\sqrt{2}$

c) $\sqrt{50x^3} = \sqrt{25x^2} \cdot \sqrt{2x} = 5x\sqrt{2x}$

EXAMPLE 9 Simplify. **a)** $\sqrt{50x^2y}$ **b)** $\sqrt{45x^3y^4}$ **c)** $\sqrt{98a^9b^7}$

Solution **a)** $\sqrt{50x^2y} = \sqrt{25x^2} \cdot \sqrt{2y}$ *Write $\sqrt{50x^2y}$ as a product of a perfect square factor and another factor.*

$$= 5x\sqrt{2y}$$ *Simplify the perfect square factor.*

TEACHING TIP
Have students rewrite each of the following simplified expressions in an equivalent form in which the number is completely under the radical sign. $7x^2\sqrt{3x}, 9xy^4\sqrt{15xy}$

b) $\sqrt{45x^3y^4} = \sqrt{9x^2y^4} \cdot \sqrt{5x}$ *Write $\sqrt{45x^3y^4}$ as a product of a perfect square factor and another factor.*

$$= 3xy^2\sqrt{5x}$$ *Simplify the perfect square factor.*

c) $\sqrt{98a^9b^7} = \sqrt{49a^8b^6} \cdot \sqrt{2ab}$ *Write $\sqrt{98a^9b^7}$ as a product of a perfect square factor and another factor.*

NOW TRY EXERCISE 45

$$= 7a^4b^3\sqrt{2ab}$$ *Simplify the perfect square factor.*

Now let's look at an example where we use the product rule to multiply two radicals before simplifying.

EXAMPLE 10 Multiply and then simplify.

a) $\sqrt{2} \cdot \sqrt{8}$ **b)** $\sqrt{2x} \cdot \sqrt{8}$ **c)** $(\sqrt{3x})^2$

Solution **a)** $\sqrt{2} \cdot \sqrt{8} = \sqrt{2 \cdot 8} = \sqrt{16} = 4$

b) $\sqrt{2x} \cdot \sqrt{8} = \sqrt{16x} = \sqrt{16} \cdot \sqrt{x} = 4\sqrt{x}$

c) $(\sqrt{3x})^2 = \sqrt{3x} \cdot \sqrt{3x} = \sqrt{9x^2} = 3x$

EXAMPLE 11 Multiply and then simplify.

a) $\sqrt{8x^3y}\,\sqrt{4xy^5}$ **b)** $\sqrt{5ab^8}\,\sqrt{6a^5b}$

Solution **a)** $\sqrt{8x^3y}\,\sqrt{4xy^5} = \sqrt{32x^4y^6} = \sqrt{16x^4y^6} \cdot \sqrt{2}$
$$= 4x^2y^3\sqrt{2}$$

TEACHING TIP
Point out that numbers which have no perfect square factors can be written as a product of primes with no number repeated:
$30 = 2 \cdot 3 \cdot 5$.

b) $\sqrt{5ab^8}\,\sqrt{6a^5b} = \sqrt{30a^6b^9} = \sqrt{a^6b^8} \cdot \sqrt{30b}$
$$= a^3b^4\sqrt{30b}$$

NOW TRY EXERCISE 63

In part **b)**, 30 can be factored in many ways. However, none of the factors are perfect squares, so we leave the answer as given.

5. Answers will vary. 6. a), 7. a) There can be no perfect square factors or any exponents greater than 1 in the radicand.

Exercise Set 9.2

Concept/Writing Exercises

1. In your own words, state the product rule for square roots and explain what it means. *Answers will vary.*

2. a) In your own words, explain how to simplify a square root containing only a constant. *Answers will vary.*

b) Simplify $\sqrt{20}$ using the procedure you gave in part **a)**. *$2\sqrt{5}$*

3. Explain why the product rule cannot be used to simplify the problem $\sqrt{-4} \cdot \sqrt{-9}$. *Radicands cannot be negative.*

4. We learned that for $a \geq 0$, $\sqrt{a^{2 \cdot n}} = a^n$. Explain in your own words what this means. *Answers will vary.*

5. a) Explain how to simplify the square root of a radical containing a variable raised to an odd power.

b) Simplify $\sqrt{x^{13}}$ using the procedure you gave in part **a)**. *$x^6\sqrt{x}$*

6. a) Explain why $\sqrt{32x^3}$ is not a simplified expression.

b) Simplify $\sqrt{32x^3}$. *$4x\sqrt{2x}$*

7. a) Explain why $\sqrt{75x^5}$ is not a simplified expression.

b) Simplify $\sqrt{75x^5}$. *$5x^2\sqrt{3x}$*

8. Use the product rule to write $\sqrt{40}$ as four different products of factors. *$\sqrt{40} \cdot \sqrt{1}, \sqrt{20} \cdot \sqrt{2}, \sqrt{10} \cdot \sqrt{4}, \sqrt{8} \cdot \sqrt{5}$*

Determine whether the square root on the right-hand side of the equal sign is the simplified form of the square root on the left-hand side of the equal sign. If not, simplify it properly.

9. $\sqrt{75} = 5\sqrt{3}$ *yes*

10. $\sqrt{x^5} = x^2\sqrt{x}$ *yes*

11. $\sqrt{32} = 2\sqrt{8}$ *no; $4\sqrt{2}$*

12. $\sqrt{x^9} = x\sqrt{x^7}$ *no; $x^4\sqrt{x}$*

Practice the Skills

Simplify.

13. $\sqrt{12}$ *$2\sqrt{3}$*

14. $\sqrt{18}$ *$3\sqrt{2}$*

15. $\sqrt{8}$ *$2\sqrt{2}$*

16. $\sqrt{45}$ *$3\sqrt{5}$*

17. $\sqrt{96}$ *$4\sqrt{6}$*

18. $\sqrt{75}$ *$5\sqrt{3}$*

19. $\sqrt{32}$ *$4\sqrt{2}$*

20. $\sqrt{52}$ *$2\sqrt{13}$*

21. $\sqrt{160}$ *$4\sqrt{10}$*

22. $\sqrt{44}$ *$2\sqrt{11}$*

23. $\sqrt{80}$ *$4\sqrt{5}$*

24. $\sqrt{27}$ *$3\sqrt{3}$*

25. $\sqrt{72}$ *$6\sqrt{2}$*

26. $\sqrt{147}$ *$7\sqrt{3}$*

27. $\sqrt{140}$ *$2\sqrt{35}$*

28. $\sqrt{180}$ *$6\sqrt{5}$*

29. $\sqrt{243}$ *$9\sqrt{3}$*

30. $\sqrt{135}$ *$3\sqrt{15}$*

31. $\sqrt{150}$ *$5\sqrt{6}$*

32. $\sqrt{x^4}$ *x^2*

33. $\sqrt{x^6}$ *x^3*

34. $\sqrt{y^{13}}$ *$y^6\sqrt{y}$*

35. $\sqrt{x^2y^4}$ *xy^2*

36. $\sqrt{xy^2}$ *$y\sqrt{x}$*

37. $\sqrt{a^{12}b^9}$ *$a^6b^4\sqrt{b}$*

38. $\sqrt{x^4y^5z^6}$ *$x^2y^2z^3\sqrt{y}$*

39. $\sqrt{a^2b^4c}$ *$ab^2\sqrt{c}$*

40. $\sqrt{a^3b^9c^{11}}$ *$ab^4c^5\sqrt{abc}$*

41. $\sqrt{3n^3}$ *$n\sqrt{3n}$*

42. $\sqrt{12x^4y^2}$ *$2x^2y\sqrt{3}$*

43. $\sqrt{75a^3b^2}$ *$5ab\sqrt{3a}$*

44. $\sqrt{150m^4n^3}$ *$5m^2n\sqrt{6n}$*

45. $\sqrt{300a^5b^{11}}$ *$10a^2b^5\sqrt{3ab}$*

46. $\sqrt{64xyz^5}$ *$8z^2\sqrt{xyz}$*

47. $\sqrt{243x^3y^4}$ *$9xy^2\sqrt{3x}$*

48. $\sqrt{500ab^4c^3}$ *$10b^2c\sqrt{5ac}$*

49. $\sqrt{108a^2b^7c}$ *$6ab^3\sqrt{3bc}$*

50. $\sqrt{112x^6y^8}$ *$4x^3y^4\sqrt{7}$*

51. $\sqrt{180r^3s^4t^5}$ *$6rs^2t^2\sqrt{5rt}$*

52. $\sqrt{98x^4y^4z}$ *$7x^2y^2\sqrt{2z}$*

Simplify.

53. $\sqrt{5} \cdot \sqrt{5}$ 5

54. $\sqrt{8} \cdot \sqrt{8}$ 8

55. $\sqrt{24} \cdot \sqrt{5}$ $2\sqrt{30}$

56. $\sqrt{60} \cdot \sqrt{5}$ $10\sqrt{3}$

57. $\sqrt{48} \cdot \sqrt{15}$ $12\sqrt{5}$

58. $\sqrt{30} \cdot \sqrt{5}$ $5\sqrt{6}$

59. $\sqrt{3x}\,\sqrt{7x}$ $x\sqrt{21}$

60. $\sqrt{3x^3}\,\sqrt{3x}$ $3x^2$

🔒 **61.** $\sqrt{4a^2}\,\sqrt{12ab^2}$ $4ab\sqrt{3a}$

62. $\sqrt{30b^2}\,\sqrt{6b^5}$ $6b^3\sqrt{5b}$

63. $\sqrt{6xy^3}\,\sqrt{12x^2y}$ $6xy^2\sqrt{2x}$

64. $\sqrt{20xy^4}\,\sqrt{6x^5}$ $2x^3y^2\sqrt{30}$

65. $\sqrt{3r^4s^7}\,\sqrt{21r^6s^5}$ $3r^5s^6\sqrt{7}$

66. $\sqrt{20x^3y}\,\sqrt{6x^3y^5}$ $2x^3y^3\sqrt{30}$

🔒 **67.** $\sqrt{15xy^6}\,\sqrt{6xyz}$ $3xy^3\sqrt{10yz}$

68. $\sqrt{14xyz^5}\,\sqrt{3xy^2z^6}$ $xyz^5\sqrt{42yz}$

69. $\sqrt{6a^2b^4}\,\sqrt{9a^4b^6}$ $3a^3b^5\sqrt{6}$

70. $\sqrt{6a^4b^5c^6}\,\sqrt{3a^3bc^6}$ $3a^3b^3c^6\sqrt{2a}$

71. $(\sqrt{2x})^2$ $2x$

72. $(\sqrt{6x^2})^2$ $6x^2$

🔒 **73.** $\left(\sqrt{13x^4y^6}\right)^2$ $13x^4y^6$

74. $\sqrt{36x^2y^7}\,\sqrt{2x^4y}$ $6x^3y^4\sqrt{2}$

75. $(\sqrt{5a})^2(\sqrt{3a})^2$ $15a^2$

76. $(\sqrt{4ab})^2(\sqrt{3ab})^2$ $12a^2b^2$

Problem Solving

Which coefficients and exponents should be placed in the shaded areas to make a true statement? Explain how you obtained your answer.   **80.** exponent on y, 10; on x, 7   **81.** coefficient, 8; exponent on x, 12; on y, 7   **82.** coefficient, 2; exponent on x, 6; on z, 7

77. $\sqrt{25x^\blacksquare y^6} = 5x^2y^3$ 4

78. $\sqrt{\blacksquare x^4y^\blacksquare} = 4x^2y^4$ coefficient, 16; exponent on y, 8

79. $\sqrt{4x^\blacksquare y^\blacksquare} = 2x^3y^2\sqrt{y}$ exponent on x, 6; on y, 5

80. $\sqrt{3x^4y^\blacksquare} \cdot \sqrt{3x^\blacksquare y^5} = 3x^5y^7\sqrt{xy}$

81. $\sqrt{2x^\blacksquare y^5} \cdot \sqrt{\blacksquare x^3 y^\blacksquare} = 4x^7y^6\sqrt{x}$

82. $\sqrt{32x^4z^\blacksquare} \cdot \sqrt{\blacksquare x^\blacksquare z^{12}} = 8x^5z^9\sqrt{z}$

83. a) Showing all steps, simplify $\left(\sqrt{13x^3}\right)^2$. $13x^3$

    b) Showing all steps, simplify $\sqrt{(13x^3)^2}$. $13x^3$

    c) Compare your results in part **a)** and part **b)**. Are they the same?   yes

84. a) Showing all steps, simplify $\left(\sqrt{7x^4}\right)^2$. $7x^4$

    b) Showing all steps, simplify $\sqrt{(7x^4)^2}$. $7x^4$

    c) Compare your results in part **a)** and part **b)**. Are they the same?   yes

Simplify. Treat the ☺ *and* ⊗ *as if they were variables.*

85. $\sqrt{200☺^{11}}$ $10☺^5\sqrt{2☺}$

86. $\sqrt{180☺^7⊗^{16}}$ $6☺^3⊗^8\sqrt{5☺}$

87. $\sqrt{5☺^{100}} \cdot \sqrt{5☺^{36}}$ $5☺^{50}⊗^{18}$

88. $\sqrt{7☺^{10}} \cdot \sqrt{343⊗^{10}}$ $49☺^5⊗^5$

Challenge Problems

    93. rational; $\sqrt{6.25} = 2.5$ and 2.5 is a terminating decimal number

Following we illustrate two simplifications involving square roots.

$$\sqrt{x^4} = (x^4)^{1/2} = x^{4(1/2)} = x^2$$
$$\sqrt{x^{2/4}} = (x^{2/4})^{1/2} = x^{(2/4)(1/2)} = x^{1/4}$$

In Section 9.1 we indicated that the square root of an expression may be written as the expression to the $\frac{1}{2}$ power. The rules for exponents that were discussed in Section 4.1 also apply when the exponents are rational numbers. Use the two examples illustrated and the rules for exponents to simplify Exercises 89–92. We will discuss rational exponents further in Section 9.7.

89. $\sqrt{x^{2/6}}$ $x^{1/6}$

90. $\sqrt{y^{10/12}}$ $y^{5/12}$

91. $\sqrt{4x^{4/5}}$ $2x^{2/5}$

92. $\sqrt{25y^{8/3}}$ $5y^{4/3}$

93. Is $\sqrt{6.25}$ a rational or an irrational number? Explain how you determined your answer.

94. a) In Section 9.4 we will be multiplying expressions like $(\sqrt{a} + \sqrt{b})(\sqrt{a} - \sqrt{b})$ using the FOIL method. Can you find this product now? $a - b$

    b) Multiply $(\sqrt{6} + \sqrt{3})(\sqrt{6} - \sqrt{3})$.   3

95. The area of a square is found by the formula $A = s^2$. We will learn later that we can rewrite this formula as $s = \sqrt{A}$.

    a) If the area is 16 square feet, what is the length of a side?   4 ft

    b) If the area is doubled, is the length of a side doubled? Explain.   no, increased $\sqrt{2}$ or ≈ 1.414 times

    c) To double the length of a side of a square, how much must the area be increased? Explain.   4 times

96. We know that $\sqrt{a} \cdot \sqrt{b} = \sqrt{a \cdot b}$ if $a \geq 0$ and $b \geq 0$. Does $\sqrt{\dfrac{a}{b}} = \dfrac{\sqrt{a}}{\sqrt{b}}$ if $a \geq 0$ and $b > 0$? Try several pairs of values for a and b and see.   yes

97. a) Will the product of two rational numbers always be a rational number? Explain and give an example to support your answer.   yes

    b) Will the product of two irrational numbers always be an irrational number? Explain and give an example to support your answer.   no; for example, $\sqrt{2} \cdot \sqrt{2} = 2$.

 Group Activity

We learned earlier that $\sqrt{} = ^{1/2}$. For example, $\sqrt{x^6} = (x^6)^{1/2} = x^3$. We can simplify $\sqrt{x^{2n}}$ by writing the expression in exponential form, $\sqrt{x^{2n}} = (x^{2n})^{1/2} = x^{(2n)(1/2)} = x^n$. As a group, simplify the following square roots by writing the expression in exponential form. Show all the steps in the simplification process.

98. $\sqrt{x^{10a}}$ x^{5a}

99. $\sqrt{x^{8b}}$ x^{4b}

100. $\sqrt{x^{4a}y^{12b}}$ $x^{2a}y^{6b}$

101. $\sqrt{x^{8b}y^{6c}}$ $x^{4b}y^{3c}$

Cumulative Review Exercises

[3.1] **102.** *Area* Find the area of the trapezoid. Use $A = \dfrac{1}{2}h(b + d)$.

148.5 in²

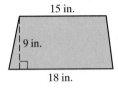

15 in.

9 in.

18 in.

103. *Volume* Find the volume of the cone. Use $V = \dfrac{1}{3}\pi r^2 h$.

≈25.13 ft³

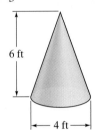

6 ft

4 ft

[6.2] **104.** Divide $\dfrac{3x^2 - 16x - 12}{3x^2 - 10x - 8} \div \dfrac{x^2 - 7x + 6}{3x^2 - 11x - 4}$.

[7.4] **105.** Write the equation $3x + 6y = 9$ in slope–intercept form and indicate the slope and the y-intercept. $y = -\frac{1}{2}x + \frac{3}{2}, m = -\frac{1}{2}, \left(0, \frac{3}{2}\right)$

[7.5] **106.** Graph $6x - 5y \geq 30$.

[8.3] **107.** Solve the following system of equations.

$$3x - 4y = 6$$
$$5x - 3y = 5 \quad \left(\frac{2}{11}, -\frac{15}{11}\right)$$

104. $\dfrac{3x + 1}{x - 1}$ **106.** See graphing answer section, page G14.

9.3 ADDING, SUBTRACTING, AND MULTIPLYING SQUARE ROOTS

SSM   Study Guide   CD/Video

MathPro 4/5   PH Math Tutor Center   prenhall.com/Angel

1 Add and subtract square roots.

2 Multiply square roots.

1 Add and Subtract Square Roots

TEACHING TIP
Use the fact that $7 = 3 + 4$ and $7x = 3x + 4x$ to lead students to $7\sqrt{3} = 3\sqrt{3} + 4\sqrt{3}$.

Like square roots are square roots having the same radicands. Like square roots are added in much the same manner that like terms are added, as illustrated below.

Examples of Adding Like Terms
$$2x + 3x = (2 + 3)x = 5x$$
$$4x + x = 4x + 1x = (4 + 1)x = 5x$$

Examples of Adding Like Square Roots
$$2\sqrt{7} + 3\sqrt{7} = (2 + 3)\sqrt{7} = 5\sqrt{7}$$
$$6\sqrt{x} + \sqrt{x} = 6\sqrt{x} + 1\sqrt{x} = (6 + 1)\sqrt{x} = 7\sqrt{x}$$

Note that adding like square roots is an application of the distributive property.

$$2\sqrt{7} + 3\sqrt{7} = (2 + 3)\sqrt{7}$$
$$= 5\sqrt{7}$$

Other Examples of Adding and Subtracting Like Square Roots

$$2\sqrt{5} - 3\sqrt{5} = (2 - 3)\sqrt{5} = -1\sqrt{5} = -\sqrt{5}$$
$$\sqrt{x} + \sqrt{x} = 1\sqrt{x} + 1\sqrt{x} = (1 + 1)\sqrt{x} = 2\sqrt{x}$$
$$6\sqrt{2} + 3\sqrt{2} - \sqrt{2} = (6 + 3 - 1)\sqrt{2} = 8\sqrt{2}$$
$$\frac{2\sqrt{3}}{5} + \frac{1\sqrt{3}}{5} = \left(\frac{2}{5} + \frac{1}{5}\right)\sqrt{3} = \frac{3}{5}\sqrt{3} \text{ or } \frac{3\sqrt{3}}{5}$$

EXAMPLE 1 Simplify if possible.

a) $4\sqrt{5} + 3\sqrt{5} - 6$ b) $\sqrt{6} - 4\sqrt{6} + 4$

c) $5\sqrt{3} + 4\sqrt{y} - 2\sqrt{3} - 6\sqrt{y}$ d) $2\sqrt{3} + 5\sqrt{2}$

Solution a) $4\sqrt{5} + 3\sqrt{5} - 6 = (4 + 3)\sqrt{5} - 6$ *Only $4\sqrt{5}$ and $3\sqrt{5}$ can be combined.*

$$= 7\sqrt{5} - 6$$ *Simplify.*

b) $\sqrt{6} - 4\sqrt{6} + 4 = (1 - 4)\sqrt{6} + 4 = -3\sqrt{6} + 4$

c) $5\sqrt{3} + 4\sqrt{y} - 2\sqrt{3} - 6\sqrt{y} = 5\sqrt{3} - 2\sqrt{3} + 4\sqrt{y} - 6\sqrt{y}$ *Place like radicals together.*

$$= 3\sqrt{3} - 2\sqrt{y}$$ *Simplify.*

d) Cannot be simplified since the radicands are different.

The answers in Example 1 could be written differently. For example, the answer to part a) could be written $-6 + 7\sqrt{5}$ because of the commutative property of addition.

EXAMPLE 2 Simplify.

a) $3\sqrt{x} - 4\sqrt{x} + 5\sqrt{x}$ b) $2\sqrt{a} + a + 4\sqrt{a}$

c) $x + \sqrt{x} + 2\sqrt{x} + 3$ d) $x\sqrt{x} + 3\sqrt{x} + x$

e) $\sqrt{xy} + 2\sqrt{xy} - \sqrt{x}$

Solution a) $3\sqrt{x} - 4\sqrt{x} + 5\sqrt{x} = (3 - 4 + 5)\sqrt{x}$ *All three terms are like radicals.*

$$= 4\sqrt{x}$$

b) $2\sqrt{a} + a + 4\sqrt{a} = a + 2\sqrt{a} + 4\sqrt{a}$ *Place like radicals together.*

$$= a + 6\sqrt{a}$$ *Simplify.*

c) $x + \sqrt{x} + 2\sqrt{x} + 3 = x + 1\sqrt{x} + 2\sqrt{x} + 3$ *$\sqrt{x}$ means $1\sqrt{x}$.*

$$= x + 3\sqrt{x} + 3$$ *Add $1\sqrt{x} + 2\sqrt{x}$ to get $3\sqrt{x}$.*

d) $x\sqrt{x} + 3\sqrt{x} + x = (x + 3)\sqrt{x} + x$ *Only $x\sqrt{x}$ and $3\sqrt{x}$ can be combined.*

e) $\sqrt{xy} + 2\sqrt{xy} - \sqrt{x} = 3\sqrt{xy} - \sqrt{x}$ *Only $\sqrt{xy}$ and $2\sqrt{xy}$ can be combined.*

NOW TRY EXERCISE 19

Unlike square roots are square roots having different radicands. It is sometimes possible to change unlike square roots into like square roots by simplifying the radicals in an expression. After simplifying, if the terms contain the same radicand they can be combined. Otherwise, they cannot. Examples 3, 4, and 5 illustrate this.

EXAMPLE 3   Simplify $\sqrt{3} + \sqrt{12}$.

Solution   Since 12 has a perfect square factor, 4, we write 12 as a product of the perfect square factor and another factor.

$$\begin{aligned}
\sqrt{3} + \sqrt{12} &= \sqrt{3} + \sqrt{4 \cdot 3} \\
&= \sqrt{3} + \sqrt{4} \cdot \sqrt{3} \quad \text{\textit{Product rule}} \\
&= \sqrt{3} + 2\sqrt{3} \quad \text{\textit{Now the terms are like terms.}} \\
&= 3\sqrt{3} \quad \text{\textit{Add like square roots.}}
\end{aligned}$$

EXAMPLE 4   Simplify $\sqrt{63} - \sqrt{28}$.

Solution   Write each radicand as a product of a perfect square factor and another factor.

$$\begin{aligned}
\sqrt{63} - \sqrt{28} &= \sqrt{9 \cdot 7} - \sqrt{4 \cdot 7} \\
&= \sqrt{9} \cdot \sqrt{7} - \sqrt{4} \cdot \sqrt{7} \quad \text{\textit{Product rule}} \\
&= 3\sqrt{7} - 2\sqrt{7} \quad \text{\textit{Now the terms are like terms.}} \\
&= \sqrt{7} \quad \text{\textit{Subtract like square roots.}}
\end{aligned}$$

EXAMPLE 5   Simplify.   **a)** $2\sqrt{8} - \sqrt{32}$ **b)** $3\sqrt{12} + 5\sqrt{27} + 2$ **c)** $\sqrt{120} - \sqrt{75}$

Solution   In each part we begin by writing each square root as a product of its largest perfect square factor and another factor. Then we simplify the perfect square factor.

a) $$\begin{aligned}
2\sqrt{8} - \sqrt{32} &= 2\sqrt{4 \cdot 2} - \sqrt{16 \cdot 2} \\
&= 2\sqrt{4}\,\sqrt{2} - \sqrt{16}\,\sqrt{2} \quad \text{\textit{Product rule}} \\
&= 2 \cdot 2\sqrt{2} - 4\sqrt{2} \quad \text{\textit{Simplify perfect square factors.}} \\
&= 4\sqrt{2} - 4\sqrt{2} \quad \text{\textit{Simplify.}} \\
&= 0
\end{aligned}$$

b) $$\begin{aligned}
3\sqrt{12} + 5\sqrt{27} + 2 &= 3\sqrt{4 \cdot 3} + 5\sqrt{9 \cdot 3} + 2 \\
&= 3\sqrt{4}\,\sqrt{3} + 5\sqrt{9}\,\sqrt{3} + 2 \quad \text{\textit{Product rule}} \\
&= 3 \cdot 2\sqrt{3} + 5 \cdot 3\sqrt{3} + 2 \quad \text{\textit{Simplify perfect square factors.}} \\
&= 6\sqrt{3} + 15\sqrt{3} + 2 \quad \text{\textit{Simplify.}} \\
&= 21\sqrt{3} + 2
\end{aligned}$$

c) $$\begin{aligned}
\sqrt{120} - \sqrt{75} &= \sqrt{4 \cdot 30} - \sqrt{25 \cdot 3} \\
&= \sqrt{4}\,\sqrt{30} - \sqrt{25}\,\sqrt{3} \quad \text{\textit{Product rule}} \\
&= 2\sqrt{30} - 5\sqrt{3} \quad \text{\textit{Simplify perfect square factors.}}
\end{aligned}$$

NOW TRY EXERCISE 27   Since 30 has no perfect square factors and since the radicands are different, the expression $2\sqrt{30} - 5\sqrt{3}$ cannot be simplified any further.

AVOIDING COMMON ERRORS   The product rule presented in Section 9.2 was $\sqrt{a} \cdot \sqrt{b} = \sqrt{a \cdot b}$. The same principle **does not apply to** addition.

INCORRECT
$$\sqrt{a} + \sqrt{b} = \sqrt{a + b}$$

For example, to evaluate $\sqrt{9} + \sqrt{16}$,

CORRECT
$$\sqrt{9} + \sqrt{16} = 3 + 4$$
$$= 7$$

INCORRECT
$$\sqrt{9} + \sqrt{16} = \sqrt{9 + 16}$$
$$= \sqrt{25}$$
$$= 5$$

2 Multiply Square Roots

Now let's discuss multiplying square roots. We introduced the product rule for radicals earlier. Now we will expand on multiplying radical expressions.

The distributive property can be used to multiply radical expressions. When the distributive property is used, each term within the parentheses is multiplied by the term preceding the parentheses. Some examples follow.

$$\sqrt{6}(\sqrt{6} + 3) = (\sqrt{6})(\sqrt{6}) + (\sqrt{6})(3) = 6 + 3\sqrt{6}$$
$$\sqrt{5}(\sqrt{x} + y) = (\sqrt{5})(\sqrt{x}) + (\sqrt{5})(y) = \sqrt{5x} + y\sqrt{5}$$

Now let's work an example.

EXAMPLE 6 Multiply **a)** $\sqrt{3}(\sqrt{6} - 2)$ **b)** $\sqrt{2x}(\sqrt{x} + \sqrt{2y})$

Solution **a)** Use the distributive property. This gives

$$\sqrt{3}(\sqrt{6} - 2) = (\sqrt{3})(\sqrt{6}) - (\sqrt{3})(2)$$
$$= \sqrt{18} - 2\sqrt{3} \qquad \textit{Product rule}$$
$$= \sqrt{9 \cdot 2} - 2\sqrt{3}$$
$$= \sqrt{9} \cdot \sqrt{2} - 2\sqrt{3} \qquad \textit{Product rule}$$
$$= 3\sqrt{2} - 2\sqrt{3}$$

b)
$$\sqrt{2x}(\sqrt{x} + \sqrt{2y}) = (\sqrt{2x})(\sqrt{x}) + (\sqrt{2x})(\sqrt{2y})$$
$$= \sqrt{2x^2} + \sqrt{4xy} \qquad \textit{Product rule}$$
$$= \sqrt{x^2} \cdot \sqrt{2} + \sqrt{4} \cdot \sqrt{xy} \qquad \textit{Product rule}$$
$$= x\sqrt{2} + 2\sqrt{xy}$$

NOW TRY EXERCISE 41

To multiply two binomials with square root terms, each term in the first binomial must be multiplied by each term in the second binomial. This can be accomplished using the FOIL method.

EXAMPLE 7 Multiply $(3 + \sqrt{5})(4 - 2\sqrt{5})$ using the FOIL method.

Solution

$$(3 + \sqrt{5})(4 - 2\sqrt{5}) = 3(4) + 3(-2\sqrt{5}) + 4\sqrt{5} + \sqrt{5}(-2\sqrt{5})$$
$$= 12 \quad - 6\sqrt{5} \quad + 4\sqrt{5} - 2\sqrt{25}$$
$$= 12 \quad\quad\quad - 2\sqrt{5} \quad\quad - 2(5)$$
$$= 12 - 2\sqrt{5} - 10$$
$$= 2 - 2\sqrt{5}$$

NOW TRY EXERCISE 55

EXAMPLE 8 Multiply using the FOIL method.
a) $(\sqrt{2} - \sqrt{7})(\sqrt{2} + \sqrt{7})$ **b)** $(x - \sqrt{y})(x + \sqrt{y})$

Solution **a)**
$$(\sqrt{2} - \sqrt{7})(\sqrt{2} + \sqrt{7}) = \sqrt{2} \cdot \sqrt{2} + \sqrt{2} \cdot \sqrt{7} + (-\sqrt{7})(\sqrt{2}) + (-\sqrt{7})(\sqrt{7})$$
$$= \sqrt{4} + \sqrt{14} - \sqrt{14} - \sqrt{49}$$
$$= \sqrt{4} - \sqrt{49}$$
$$= 2 - 7 = -5$$

b) $(x - \sqrt{y})(x + \sqrt{y}) = (x)(x) + x(\sqrt{y}) + (-\sqrt{y})(x) + (-\sqrt{y})(\sqrt{y})$

$$= x^2 + x\sqrt{y} - x\sqrt{y} - \sqrt{y^2}$$
$$= x^2 - \sqrt{y^2}$$
$$= x^2 - y$$

NOW TRY EXERCISE 73

Notice that both parts in Example 8 may be considered as the product of the sum and difference of the same two terms. In Section 4.5 we learned that $(a + b)(a - b) = a^2 - b^2$. In part **a)** if we let $a = \sqrt{2}$ and $b = \sqrt{7}$, then

$$a^2 - b^2 = (\sqrt{2})^2 - (\sqrt{7})^2 = 2 - 7 = -5.$$

In part **b)** if we let $a = x$ and $b = \sqrt{y}$, then

$$a^2 - b^2 = (x)^2 - (\sqrt{y})^2 = x^2 - y.$$

Both answers agree with the answers obtained in Example 8.

1. square roots having the same radicand; one example is $\sqrt{3}, 5\sqrt{3}$ 2. square roots having a different radicand; one example is $\sqrt{2}, \sqrt{5}$.

Exercise Set 9.3

Concept/Writing Exercises

1. What are like square roots? Give an example.

2. What are unlike square roots? Give an example.

3. Under what conditions can two square roots be added or subtracted? Only like square roots can be added or subtracted.

4. In your own words, explain how to add like square roots. Answers will vary.

Practice the Skills
8. $-2\sqrt{7} + 4$ 16. $7\sqrt{5} + 2\sqrt{x}$ 17. $\sqrt{x} + 4\sqrt{y} + x$ 19. $5m - 2\sqrt{m} + 2$ 20. $-3p - \sqrt{p}$
21. $-2\sqrt{7} - 9\sqrt{x}$ 22. $2\sqrt{5} - 2\sqrt{y}$

Simplify each expression.

5. $6\sqrt{5} - 4\sqrt{5}$ $2\sqrt{5}$

6. $8\sqrt{3} + \sqrt{3}$ $9\sqrt{3}$

7. $7\sqrt{5} - 11\sqrt{5}$ $-4\sqrt{5}$

8. $3\sqrt{7} - 4\sqrt{7} - \sqrt{7} + 4$

9. $\sqrt{7} + 4\sqrt{7} - 3\sqrt{7} + 6$ $2\sqrt{7} + 6$

10. $10\sqrt{13} - 6\sqrt{13} + 5\sqrt{13}$ $9\sqrt{13}$

11. $5\sqrt{x} + \sqrt{x}$ $6\sqrt{x}$

12. $4\sqrt{x} - 8\sqrt{x}$ $-4\sqrt{x}$

13. $-\sqrt{y} + 3\sqrt{y} - 5\sqrt{y}$ $-3\sqrt{y}$

14. $3\sqrt{y} - 6\sqrt{y} + 2$ $-3\sqrt{y} + 2$

15. $-6\sqrt{t} + 2\sqrt{t} - 6$ $-4\sqrt{t} - 6$

16. $3\sqrt{5} - \sqrt{x} + 4\sqrt{5} + 3\sqrt{x}$

17. $\sqrt{x} + \sqrt{y} + x + 3\sqrt{y}$

18. $3 + 4\sqrt{x} - 6\sqrt{x}$ $3 - 2\sqrt{x}$

19. $4 + 4\sqrt{m} - 6\sqrt{m} + 5m - 2$

20. $2\sqrt{p} - 3p - 5\sqrt{p} + 2\sqrt{p}$

21. $-3\sqrt{7} + \sqrt{7} - 2\sqrt{x} - 7\sqrt{x}$

22. $4\sqrt{5} - 2\sqrt{5} + 3\sqrt{y} - 5\sqrt{y}$

Simplify each expression.

23. $\sqrt{20} + \sqrt{18}$ $2\sqrt{5} + 3\sqrt{2}$

24. $\sqrt{8} - \sqrt{12}$ $2\sqrt{2} - 2\sqrt{3}$

25. $\sqrt{300} - \sqrt{27}$ $7\sqrt{3}$

26. $\sqrt{125} + \sqrt{20}$ $7\sqrt{5}$

27. $\sqrt{75} + \sqrt{108}$ $11\sqrt{3}$

28. $\sqrt{15} - \sqrt{135}$ $-2\sqrt{15}$

29. $4\sqrt{50} - \sqrt{72} + \sqrt{8}$ $16\sqrt{2}$

30. $-4\sqrt{99} + 3\sqrt{44} + 2\sqrt{11}$ $-4\sqrt{11}$

31. $-3\sqrt{125} + 7\sqrt{75}$ $-15\sqrt{5} + 35\sqrt{3}$

32. $5\sqrt{40} - \sqrt{50}$ $10\sqrt{10} - 5\sqrt{2}$

33. $2\sqrt{360} + 4\sqrt{160}$ $28\sqrt{10}$

34. $6\sqrt{180} - 7\sqrt{108}$ $36\sqrt{5} - 42\sqrt{3}$

35. $4\sqrt{16} - \sqrt{48}$ $16 - 4\sqrt{3}$

36. $3\sqrt{250} + 5\sqrt{160}$ $35\sqrt{10}$

39. $4\sqrt{x} - 4\sqrt{2}$ 40. $3\sqrt{5} - 3\sqrt{x}$ 43. $2\sqrt{10} - 2\sqrt{5}$ 44. $4\sqrt{6} + 4\sqrt{3}$ 45. $x + \sqrt{3x}$ 47. $6\sqrt{a} - a\sqrt{2}$ 48. $5\sqrt{d} + d\sqrt{2}$
50. $2x^2 - 3x\sqrt{y}$ 51. $12x^2 - 9x\sqrt{x}$ 52. $5y^2 + 5y\sqrt{y}$

Multiply.

37. $\sqrt{3}(4 + \sqrt{3})$ $4\sqrt{3} + 3$

38. $\sqrt{2}(6 + \sqrt{2})$ $6\sqrt{2} + 2$

39. $4(\sqrt{x} - \sqrt{2})$

40. $3(\sqrt{5} - \sqrt{x})$

41. $y(\sqrt{y} + y)$ $y\sqrt{y} + y^2$

42. $z(z - \sqrt{z})$ $z^2 - z\sqrt{z}$

43. $\sqrt{5}(\sqrt{8} - 2)$

44. $\sqrt{6}(4 + \sqrt{8})$

45. $\sqrt{x}(\sqrt{x} + \sqrt{3})$

46. $\sqrt{y}(\sqrt{y} - \sqrt{2})$ $y - \sqrt{2y}$

47. $\sqrt{a}(6 - \sqrt{2a})$

48. $\sqrt{d}(5 + \sqrt{2d})$

49. $x(x + 4\sqrt{y})$ $x^2 + 4x\sqrt{y}$

50. $x(2x - 3\sqrt{y})$

51. $3x(4x - 3\sqrt{x})$

52. $5y(y + \sqrt{y})$

Multiply.

53. $(6 + \sqrt{3})(4 - \sqrt{2})$ $24 - 6\sqrt{2} + 4\sqrt{3} - \sqrt{6}$

55. $(\sqrt{5} - 2)(\sqrt{6} + 3)$ $\sqrt{30} + 3\sqrt{5} - 2\sqrt{6} - 6$

57. $(6 - 2\sqrt{7})(8 - 2\sqrt{7})$ $76 - 28\sqrt{7}$

59. $(4 - \sqrt{x})(4 - \sqrt{x})$ $16 - 8\sqrt{x} + x$

61. $(\sqrt{3z} - 4)(\sqrt{5z} + 2)$ $z\sqrt{15} + 2\sqrt{3z} - 4\sqrt{5z} - 8$

63. $(r + 2\sqrt{s})(2r - 3\sqrt{s})$ $2r^2 + r\sqrt{s} - 6s$

65. $(x - \sqrt{2y})(2x - 2\sqrt{2y})$ $2x^2 - 4x\sqrt{2y} + 4y$

67. $(4p - 2\sqrt{3q})(p + 2\sqrt{3q})$ $4p^2 + 6p\sqrt{3q} - 12q$

64. $5\sqrt{2} - 5\sqrt{5z} - 3\sqrt{10z} + 15z$

54. $(8 + \sqrt{5})(6 - \sqrt{2})$ $48 - 8\sqrt{2} + 6\sqrt{5} - \sqrt{10}$

56. $(\sqrt{10} + 5)(\sqrt{10} - 2)$ $3\sqrt{10}$

58. $(9 + 3\sqrt{6})(9 + 3\sqrt{6})$ $135 + 54\sqrt{6}$

60. $(5 - 3\sqrt{y})(2 + \sqrt{y})$ $10 - \sqrt{y} - 3y$

62. $(x - \sqrt{z})(2x - 3\sqrt{z})$ $2x^2 - 5x\sqrt{z} + 3z$

64. $(\sqrt{5} - 3\sqrt{z})(\sqrt{10} - 5\sqrt{z})$

66. $(n - 6\sqrt{w})(2n + 5\sqrt{w})$ $2n^2 - 7n\sqrt{w} - 30w$

68. $(8n - \sqrt{6x})(2n - 5\sqrt{6x})$ $16n^2 - 42n\sqrt{6x} + 30x$

Multiply.

69. $(1 + \sqrt{3})(1 - \sqrt{3})$ -2

71. $(4 - \sqrt{2})(4 + \sqrt{2})$ 14

73. $(\sqrt{x} + 3)(\sqrt{x} - 3)$ $x - 9$

75. $(\sqrt{7} + r)(\sqrt{7} - r)$ $7 - r^2$

🔒 **77.** $(\sqrt{5x} + \sqrt{y})(\sqrt{5x} - \sqrt{y})$ $5x - y$

79. $(2\sqrt{x} + 3\sqrt{y})(2\sqrt{x} - 3\sqrt{y})$ $4x - 9y$

70. $(\sqrt{3} + 6)(\sqrt{3} - 6)$ -33

72. $(\sqrt{11} - 4)(\sqrt{11} + 4)$ -5

74. $(\sqrt{a} + 4)(\sqrt{a} - 4)$ $a - 16$

76. $(\sqrt{3} - y)(\sqrt{3} + y)$ $3 - y^2$

78. $(\sqrt{x} + \sqrt{y})(\sqrt{x} - \sqrt{y})$ $x - y$

80. $(3\sqrt{2c} + \sqrt{3d})(3\sqrt{2c} - \sqrt{3d})$ $18c - 3d$

✎ **81. a)** Is $\sqrt{10}$ twice as large as $\sqrt{5}$? no

  b) What number is twice as large as $\sqrt{5}$? Explain how you determined your answer. $2\sqrt{5}$

✎ **82. a)** Is $\sqrt{21}$ three times as large as $\sqrt{7}$? no

  b) What number is three times as large as $\sqrt{7}$? Explain how you determined your answer. $3\sqrt{7}$

Problem Solving

83. If $x > -2$, what is $\sqrt{x + 2} \cdot \sqrt{x + 2}$ equal to? $x + 2$

✎ **85.** Is the sum or difference of two rational expressions always a rational expression? Explain and give an example. yes

84. If $x > -5$, what is $\sqrt{x + 5} \cdot \sqrt{x + 5}$ equal to? $x + 5$

✎ **86.** Is the sum or difference of two irrational expressions always an irrational expression? Explain and give an example. no; for example, $\sqrt{3} - \sqrt{3} = 0$

✎ *Fill in the shaded area to make each statement true. Explain how you determined your answer.*

87. $\sqrt{\blacksquare} - \sqrt{63} = 4\sqrt{7}$ 343

88. $\sqrt{180} + \sqrt{\blacksquare} = 9\sqrt{5}$ 45

Find the perimeter and area of the following figures.

89.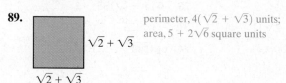
perimeter, $4(\sqrt{2} + \sqrt{3})$ units; area, $5 + 2\sqrt{6}$ square units

90.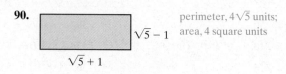
perimeter, $4\sqrt{5}$ units; area, 4 square units

91.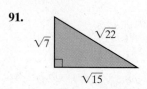
perimeter, $\sqrt{7} + \sqrt{15} + \sqrt{22}$ units; area, $\frac{1}{2}\sqrt{105}$ square units

92.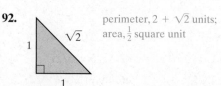
perimeter, $2 + \sqrt{2}$ units; area, $\frac{1}{2}$ square unit

In Section 5.6, we solved quadratic equations of the form $ax^2 + bx + c = 0$, by factoring. Not all quadratic equations can be solved by factoring. In Chapter 10, we will be introducing <u>another way</u> to solve quadratic equations, called the **quadratic formula**. *The quadratic formula contains the expression $\sqrt{b^2 - 4ac}$. Evaluate this expression for the following quadratic equations. Note, for example, in the quadratic equation $2x^2 - 4x - 5 = 0$ that $a = 2, b = -4,$ and $c = -5$.*

93. $x^2 + 3x + 2 = 0$ 1

94. $x^2 + 9x + 20 = 0$ 1

95. $x^2 - 14x - 5 = 0$ $\sqrt{216} = 6\sqrt{6}$

96. $x^2 + 4x - 1 = 0$ $\sqrt{20} = 2\sqrt{5}$

97. $-2x^2 + 4x + 7 = 0$ $\sqrt{72} = 6\sqrt{2}$

98. $-6x^2 - 2x + 1 = 0$ $\sqrt{28} = 2\sqrt{7}$

Challenge Problems

In Exercises 99 and 100, fill in the shaded area to make each statement true. Explain how you determined your answer.

99. $-5\sqrt{\blacksquare} + 2\sqrt{3} + 3\sqrt{27} = -9\sqrt{3}$ 48

100. $20\sqrt{2} - 4\sqrt{\blacksquare} + 4\sqrt{18} = 12\sqrt{2}$ 50

Cumulative Review Exercises

[3.5] **101.** **Sleigh Ride** Riding a sled down a hill, Jason averages 10 miles per hour. Walking up the same hill to where he started, he averages 2 miles per hour. If it took Jason 0.2 hours longer to come up the hill than to go down the hill, how long was Jason's sled ride? 0.05 hour or 3 min

[5.4] **102.** Factor $3x^2 - 12x - 96$. $3(x + 4)(x - 8)$

[6.1] **103.** Simplify $\dfrac{x - 1}{x^2 - 1}$. $\dfrac{1}{x + 1}$

[6.6] **104.** Solve the equation $x + \dfrac{24}{x} = 10$. 4, 6

[8.1] **105.** Solve the following system of equations graphically.

$$y = 2x - 2$$
$$2x + 3y = 10 \quad (2, 2)$$

9.4 DIVIDING SQUARE ROOTS

 SSM
 Study Guide
 CD/Video
 MathPro 4/5
 PH Math Tutor Center
 prenhall.com/Angel

1 Understand what it means for a square root to be simplified.

2 Use the quotient rule to simplify square roots.

3 Rationalize denominators.

4 Rationalize a denominator that contains a binomial.

1 Understand What It Means for a Square Root to be Simplified

In this section we will use a new rule, the quotient rule, to simplify square roots containing fractions. Before we do that, let's discuss what it means for a square root to be simplified.

A Square Root Is Simplified When

1. No radicand has a factor that is a perfect square.

2. No radicand contains a fraction.

3. No denominator contains a square root.

All three criteria must be met for an expression to be simplified. Let's look at some radical expressions that *are not simplified*.

TEACHING TIP

Before discussing the Quotient Rule for Square Roots, have students compare $\frac{\sqrt{64}}{\sqrt{16}}$ and $\sqrt{\frac{64}{16}}$. Then have them work in pairs to investigate the following question "Does $\frac{\sqrt{a}}{\sqrt{b}} = \sqrt{\frac{a}{b}}$? Explain."

Radical	Reason Not Simplified	
$\sqrt{8}$	Contains perfect square factor, 4. $(\sqrt{8} = \sqrt{4}\cdot\sqrt{2} = 2\sqrt{2})$	(Criteria 1)
$\sqrt{x^3}$	Contains perfect square factor, x^2. $(\sqrt{x^3} = \sqrt{x^2}\cdot\sqrt{x} = x\sqrt{x})$	
$\sqrt{\frac{1}{2}}$	Radicand contains a fraction.	(Criteria 2)
$\frac{1}{\sqrt{2}}$	Square root in the denominator.	(Criteria 3)

Before we discuss the procedure to divide and simplify square roots containing fractions, we need to understand when common factors can and cannot be divided out.

Common factors in the numerator and denominator of a fraction within a radical sign *can* be divided out. For example, the following are correct.

$$\sqrt{\frac{\overset{2}{\cancel{6}}}{\underset{1}{\cancel{3}}}} = \sqrt{\frac{2}{1}} = \sqrt{2} \quad\text{and}\quad \sqrt{\frac{\overset{x}{\cancel{x^3}}}{\underset{1}{\cancel{x^2}}}} = \sqrt{\frac{x}{1}} = \sqrt{x}$$

However, we cannot divide out common factors directly if one expression containing the common factor is within a radical sign and the other expression is not. See the following common error.

AVOIDING COMMON ERRORS

An expression under a square root *cannot* be divided by an expression not under a square root.

CORRECT	INCORRECT
$\dfrac{\sqrt{6}}{3}$ cannot be simplified any further.	$\dfrac{\sqrt{\overset{2}{\cancel{6}}}}{\underset{1}{\cancel{3}}} = \sqrt{2}$
$\dfrac{\sqrt{x^3}}{x} = \dfrac{\sqrt{x^2}\sqrt{x}}{x} = \dfrac{x\sqrt{x}}{\cancel{x}} = \sqrt{x}$	$\dfrac{\sqrt{\overset{2}{\cancel{x^3}}}}{\underset{1}{\cancel{x}}} = \sqrt{x^2} = x$

Each of the following simplifications is correct because the constant or variable divided out are not under square roots.

CORRECT	CORRECT
$\dfrac{\overset{2}{\cancel{6}}\sqrt{2}}{\underset{1}{\cancel{3}}} = 2\sqrt{2}$	$\dfrac{\cancel{x}\sqrt{2}}{\cancel{x}} = \sqrt{2}$
$\dfrac{\overset{1}{\cancel{4}}\sqrt{3}}{\underset{2}{\cancel{8}}} = \dfrac{\sqrt{3}}{2}$	$\dfrac{3\overset{x}{\cancel{x^2}}\sqrt{5}}{\cancel{x}} = 3x\sqrt{5}$

Now we will discuss the quotient rule.

2 Use the Quotient Rule to Simplify Square Roots

The quotient rule for square roots states that the quotient of two square roots is equal to the square root of the quotient.

Quotient Rule for Square Roots

$$\frac{\sqrt{a}}{\sqrt{b}} = \sqrt{\frac{a}{b}}, \quad \text{provided} \quad a \geq 0, b > 0 \qquad \text{Rule 3}$$

Examples 1 through 4 illustrate how the quotient rule is used to simplify square roots.

EXAMPLE 1 Simplify. **a)** $\sqrt{\dfrac{25}{5}}$ **b)** $\sqrt{\dfrac{64}{4}}$ **c)** $\sqrt{\dfrac{4}{9}}$

Solution When the square root contains a fraction, divide out any factor common to both the numerator and denominator. If the square root still contains a fraction, use the quotient rule to simplify.

NOW TRY EXERCISE 13

a) $\sqrt{\dfrac{25}{5}} = \sqrt{5}$ **b)** $\sqrt{\dfrac{64}{4}} = \sqrt{16} = 4$ **c)** $\sqrt{\dfrac{4}{9}} = \dfrac{\sqrt{4}}{\sqrt{9}} = \dfrac{2}{3}$

EXAMPLE 2 Simplify. **a)** $\sqrt{\dfrac{8x^2}{4}}$ **b)** $\sqrt{\dfrac{160a^2b}{4b}}$ **c)** $\sqrt{\dfrac{3x^2y^4}{27x^4}}$ **d)** $\sqrt{\dfrac{15ab^5c^2}{3a^5bc}}$

Solution First divide out any factors common to both the numerator and denominator; then simplify.

a) $\sqrt{\dfrac{8x^2}{4}} = \sqrt{2x^2}$ *Simplify the radicand.*

$\qquad\qquad = \sqrt{x^2}\,\sqrt{2}$ *Product rule*

$\qquad\qquad = x\sqrt{2}$ *Simplify.*

b) $\sqrt{\dfrac{160a^2b}{4b}} = \sqrt{40a^2}$ *Simplify the radical.*

$\qquad\qquad = \sqrt{4a^2}\,\sqrt{10}$ *Product rule*

$\qquad\qquad = 2a\sqrt{10}$ *Simplify.*

c) $\sqrt{\dfrac{3x^2y^4}{27x^4}} = \sqrt{\dfrac{y^4}{9x^2}}$ *Simplify the radicand.*

$\qquad\qquad = \dfrac{\sqrt{y^4}}{\sqrt{9x^2}}$ *Quotient rule*

$\qquad\qquad = \dfrac{y^2}{3x}$ *Simplify.*

NOW TRY EXERCISE 27

d) $\sqrt{\dfrac{15ab^5c^2}{3a^5bc}} = \sqrt{\dfrac{5b^4c}{a^4}} = \dfrac{\sqrt{5b^4c}}{\sqrt{a^4}} = \dfrac{\sqrt{b^4}\,\sqrt{5c}}{\sqrt{a^4}} = \dfrac{b^2\sqrt{5c}}{a^2}$

When you are given a fraction containing a radical expression in both the numerator and the denominator, use the quotient rule to simplify, as in Examples 3 and 4.

EXAMPLE 3   Simplify. **a)** $\dfrac{\sqrt{2}}{\sqrt{8}}$ **b)** $\dfrac{\sqrt{48}}{\sqrt{3}}$

Solution   **a)** $\dfrac{\sqrt{2}}{\sqrt{8}} = \sqrt{\dfrac{2}{8}}$ *Quotient rule*

$= \sqrt{\dfrac{1}{4}}$ *Simplify the radicand.*

$= \dfrac{\sqrt{1}}{\sqrt{4}}$ *Quotient rule*

$= \dfrac{1}{2}$ *Simplify.*

b) $\dfrac{\sqrt{48}}{\sqrt{3}} = \sqrt{\dfrac{48}{3}} = \sqrt{16} = 4$

Now we will work another example.

EXAMPLE 4   Simplify. **a)** $\dfrac{\sqrt{32x^4y^3}}{\sqrt{8xy}}$ **b)** $\dfrac{\sqrt{75x^8y^4}}{\sqrt{3x^5y^8}}$

Solution   **a)** $\dfrac{\sqrt{32x^4y^3}}{\sqrt{8xy}} = \sqrt{\dfrac{32x^4y^3}{8xy}}$ *Quotient rule*

$= \sqrt{4x^3y^2}$ *Simplify radical.*

$= \sqrt{4x^2y^2}\,\sqrt{x}$ *Product rule*

$= 2xy\sqrt{x}$ *Simplify.*

NOW TRY EXERCISE 31   **b)** $\dfrac{\sqrt{75x^8y^4}}{\sqrt{3x^5y^8}} = \sqrt{\dfrac{75x^8y^4}{3x^5y^8}} = \sqrt{\dfrac{25x^3}{y^4}} = \dfrac{\sqrt{25x^3}}{\sqrt{y^4}} = \dfrac{\sqrt{25x^2}\,\sqrt{x}}{\sqrt{y^4}} = \dfrac{5x\sqrt{x}}{y^2}$

If you observe all the answers in Examples 1 through 4 you will notice they all meet the three criteria needed for a square root to be simplified.

3   Rationalize Denominators

Suppose we are asked to simplify expressions like $\dfrac{1}{\sqrt{2}}$ or $\sqrt{\dfrac{1}{2}}$. How do we simplify them? To simplify such expressions we use a process called **rationalizing the denominator**. **To rationalize a denominator means to remove all radicals from the denominator**. We rationalize the denominator because it is easier (without a calculator) to obtain the approximate value of a number like $\dfrac{\sqrt{2}}{2}$ than a number like $\dfrac{1}{\sqrt{2}}$. Also, sometimes we may need to rationalize denominators to add or subtract radical expressions. When the denominator of a fraction contains the square root of a number that is not a perfect square, we generally simplify the expression by **rationalizing the denominator**.

> **To rationalize a denominator with a square root,** multiply *both* the numerator and the denominator of the fraction by the square root that appears in the denominator or by the square root of a number that makes the denominator a perfect square.

EXAMPLE 5 Simplify $\dfrac{1}{\sqrt{3}}$.

Solution Since $\sqrt{3} \cdot \sqrt{3} = \sqrt{9} = 3$, we multiply both the numerator and denominator by $\sqrt{3}$.

$$\frac{1}{\sqrt{3}} = \frac{1}{\sqrt{3}} \cdot \frac{\sqrt{3}}{\sqrt{3}} = \frac{\sqrt{3}}{\sqrt{9}} = \frac{\sqrt{3}}{3}$$

TEACHING TIP
Ask, "We multiply the denominator by a number that makes the denominator a perfect square, but why do we multiply the numerator by this same number?"

The answer $\dfrac{\sqrt{3}}{3}$ is simplified because it satisfies the three requirements stated earlier.

NOW TRY EXERCISE 39

In Example 5, multiplying both the numerator and denominator by $\sqrt{3}$ is equivalent to multiplying the fraction by 1, which does not change its value.

EXAMPLE 6 Simplify. **a)** $\sqrt{\dfrac{2}{3}}$ **b)** $\sqrt{\dfrac{z^2}{18}}$

Solution **a)** $\sqrt{\dfrac{2}{3}} = \dfrac{\sqrt{2}}{\sqrt{3}}$ *Quotient rule*

$\qquad\qquad = \dfrac{\sqrt{2}}{\sqrt{3}} \cdot \dfrac{\sqrt{3}}{\sqrt{3}}$ *Multiply numerator and denominator by $\sqrt{3}$.*

$\qquad\qquad = \dfrac{\sqrt{6}}{\sqrt{9}}$ *Product rule*

$\qquad\qquad = \dfrac{\sqrt{6}}{3}$ *Simplify.*

b) $\sqrt{\dfrac{z^2}{18}} = \dfrac{\sqrt{z^2}}{\sqrt{18}} = \dfrac{z}{\sqrt{9}\,\sqrt{2}} = \dfrac{z}{3\sqrt{2}}$

Now rationalize the denominator.

$$\frac{z}{3\sqrt{2}} \cdot \frac{\sqrt{2}}{\sqrt{2}} = \frac{z\sqrt{2}}{3\sqrt{4}} = \frac{z\sqrt{2}}{3 \cdot 2} = \frac{z\sqrt{2}}{6}$$

Part **b)** can also be rationalized as follows:

$$\sqrt{\frac{z^2}{18}} = \frac{\sqrt{z^2}}{\sqrt{18}} = \frac{z}{\sqrt{18}} \cdot \frac{\sqrt{2}}{\sqrt{2}} = \frac{z\sqrt{2}}{\sqrt{36}} = \frac{z\sqrt{2}}{6}$$

NOW TRY EXERCISE 63

Note that $\dfrac{z}{\sqrt{18}} \cdot \dfrac{\sqrt{18}}{\sqrt{18}}$ will also give us the same result when simplified.

Now we will discuss rationalizing a denominator where the denominator is a binomial that contains one or more radical expressions.

4 Rationalize a Denominator That Contains a Binomial

When the denominator of a rational expression is a binomial with a square root term, we again **rationalize the denominator**. We do this by multiplying both the numerator and the denominator of the fraction by the conjugate of the denominator. The **conjugate of a binomial** is a binomial having the same two terms with the sign of the second term changed.

Binomial	Its Conjugate
$5 + \sqrt{2}$	$5 - \sqrt{2}$
$\sqrt{3} - 4$	$\sqrt{3} + 4$
$2\sqrt{3} - \sqrt{5}$	$2\sqrt{3} + \sqrt{5}$
$-x + \sqrt{3}$	$-x - \sqrt{3}$

When a binomial is multiplied by its conjugate using the FOIL method, the sum of the outer and inner terms will be zero.

Now let's try some examples where we rationalize the denominator when the denominator is a binomial with one or more radical terms.

EXAMPLE 7   Simplify $\dfrac{5}{2 + \sqrt{3}}$.

Solution   To rationalize the denominator, multiply both the numerator and the denominator of the fraction by $2 - \sqrt{3}$, which is the conjugate of $2 + \sqrt{3}$.

$$\frac{5}{2 + \sqrt{3}} \cdot \frac{2 - \sqrt{3}}{2 - \sqrt{3}} = \frac{5(2 - \sqrt{3})}{(2 + \sqrt{3})(2 - \sqrt{3})} \qquad \text{\small Multiply numerator and denominator by } 2 - \sqrt{3}.$$

$$= \frac{5(2 - \sqrt{3})}{4 - 3} \qquad \text{\small Multiply factors in denominator, as discussed in Section 9.3.}$$

$$= \frac{5(2 - \sqrt{3})}{1} \qquad \text{\small Simplify.}$$

$$= 5(2 - \sqrt{3}) = 10 - 5\sqrt{3}$$

Note that $-5\sqrt{3} + 10$ is also an acceptable answer.   ✳

EXAMPLE 8   Simplify $\dfrac{6}{\sqrt{2} - \sqrt{5}}$.

Solution   Multiply both the numerator and the denominator of the fraction by $\sqrt{2} + \sqrt{5}$, the conjugate of $\sqrt{2} - \sqrt{5}$.

$$\frac{6}{\sqrt{2} - \sqrt{5}} \cdot \frac{\sqrt{2} + \sqrt{5}}{\sqrt{2} + \sqrt{5}} = \frac{6(\sqrt{2} + \sqrt{5})}{(\sqrt{2} - \sqrt{5})(\sqrt{2} + \sqrt{5})} \qquad \text{\small Multiply numerator and denominator by } \sqrt{2} + \sqrt{5}.$$

$$= \frac{6(\sqrt{2} + \sqrt{5})}{2 - 5} \qquad \text{\small Multiply factors in denominator.}$$

$$= \frac{6(\sqrt{2} + \sqrt{5})}{-3} \qquad \text{\small Simplify.}$$

$$= \frac{-\overset{2}{\cancel{6}}(\sqrt{2} + \sqrt{5})}{\cancel{3}}$$

$$= -2(\sqrt{2} + \sqrt{5}) = -2\sqrt{2} - 2\sqrt{5} \qquad ✳$$

NOW TRY EXERCISE 87

EXAMPLE 9 Simplify $\dfrac{\sqrt{3}}{2-\sqrt{6}}$.

Solution Multiply both the numerator and the denominator of the fraction by $2+\sqrt{6}$, the conjugate of $2-\sqrt{6}$.

$$\frac{\sqrt{3}}{2-\sqrt{6}}\cdot\frac{2+\sqrt{6}}{2+\sqrt{6}}=\frac{\sqrt{3}(2+\sqrt{6})}{4-6}$$

$$=\frac{2\sqrt{3}+\sqrt{3}\,\sqrt{6}}{-2}\qquad\text{Distributive property}$$

$$=\frac{2\sqrt{3}+\sqrt{18}}{-2}\qquad\text{Product rule}$$

$$=\frac{2\sqrt{3}+\sqrt{9}\,\sqrt{2}}{-2}\qquad\text{Product rule}$$

$$=\frac{2\sqrt{3}+3\sqrt{2}}{-2}\qquad\text{Simplify.}$$

$$=\frac{-2\sqrt{3}-3\sqrt{2}}{2}\qquad\text{Multiply numerator and denominator by }-1. \text{ ✳}$$

EXAMPLE 10 Simplify $\dfrac{x}{x-\sqrt{y}}$.

Solution Multiply both the numerator and the denominator of the fraction by $x+\sqrt{y}$, the conjugate of the denominator.

$$\frac{x}{x-\sqrt{y}}\cdot\frac{x+\sqrt{y}}{x+\sqrt{y}}=\frac{x(x+\sqrt{y})}{x^2-y}=\frac{x^2+x\sqrt{y}}{x^2-y}$$

NOW TRY EXERCISE 93 Remember, you cannot divide out the x^2 terms because they are not factors. ✳

7. no perfect square factors in any radicand, no radicand contains a fraction, no square roots in any denominator
9. a) $\sqrt{27}$ has a perfect square factor, 9 b) radicand contains a fraction c) square root in the denominator
10. remove all radicals from the denominator

Exercise Set 9.4

Concept/Writing Exercises

In Exercises 1–6 indicate if a common factor can be divided out from both the numerator and denominator of the fraction. If a common factor can be divided out, do so and show the simplified answer. See the Avoiding Common Errors box on page 578.

1. $\dfrac{\sqrt{5}}{5}$ cannot be simplified

2. $\dfrac{4\sqrt{3}}{2}$ yes, $2\sqrt{3}$

3. $\dfrac{x^2\sqrt{2}}{x}$ yes, $x\sqrt{2}$

4. $\dfrac{\sqrt{6}}{3}$ cannot be simplified

5. $\dfrac{\sqrt{x}}{x}$ cannot be simplified

6. $\dfrac{\sqrt{10}}{5}$ cannot be simplified

7. What are the three requirements for a square root to be considered simplified?

8. In your own words, state the quotient rule for square roots and explain what it means. Answers will vary.

9. Explain why each expression is not simplified.

a) $\sqrt{27}$ b) $\sqrt{\dfrac{1}{2}}$ c) $\dfrac{3}{\sqrt{5}}$

10. What does it mean to rationalize a denominator?

Practice the Skills

Simplify each expression.

11. $\sqrt{\dfrac{27}{3}}$ 3

12. $\sqrt{\dfrac{24}{6}}$ 2

13. $\sqrt{\dfrac{63}{7}}$ 3

14. $\sqrt{\dfrac{20}{5}}$ 2

15. $\dfrac{\sqrt{18}}{\sqrt{2}}$ 3

16. $\dfrac{\sqrt{3}}{\sqrt{27}}$ $\dfrac{1}{3}$

17. $\sqrt{\dfrac{1}{49}}$ $\dfrac{1}{7}$

18. $\sqrt{\dfrac{16}{25}}$ $\dfrac{4}{5}$

19. $\sqrt{\dfrac{81}{144}}$ $\dfrac{9}{12} = \dfrac{3}{4}$

20. $\sqrt{\dfrac{9}{121}}$ $\dfrac{3}{11}$

21. $\dfrac{\sqrt{10}}{\sqrt{1000}}$ $\dfrac{1}{10}$

22. $\sqrt{\dfrac{20}{80}}$ $\dfrac{1}{2}$

23. $\sqrt{\dfrac{48x^3}{2x}}$ $2x\sqrt{6}$

24. $\sqrt{\dfrac{9ab^4}{3b^3}}$ $\sqrt{3ab}$

25. $\sqrt{\dfrac{45x^2}{16x^2y^4}}$ $\dfrac{3\sqrt{5}}{4y^2}$

26. $\sqrt{\dfrac{50a^3b^6}{10a^3b^8}}$ $\dfrac{\sqrt{5}}{b}$

27. $\sqrt{\dfrac{16x^5y^3}{100x^7y}}$ $\dfrac{2y}{5x}$

28. $\sqrt{\dfrac{14xyz^5}{56x^3y^3z^4}}$ $\dfrac{\sqrt{z}}{2xy}$

29. $\sqrt{\dfrac{24ab}{24a^5b^3}}$ $\dfrac{1}{a^2b}$

30. $\sqrt{\dfrac{32x^3y}{32x^7y^3}}$ $\dfrac{1}{x^2y}$

31. $\dfrac{\sqrt{32n^5}}{\sqrt{8n}}$ $2n^2$

32. $\dfrac{\sqrt{24x^2y^2}}{\sqrt{6x^2y^4}}$ $\dfrac{2}{y}$

33. $\dfrac{\sqrt{81w^5z}}{\sqrt{144wz^3}}$ $\dfrac{3w^2}{4z}$

34. $\dfrac{\sqrt{108}}{\sqrt{36m^4n^6}}$ $\dfrac{\sqrt{3}}{m^2n^3}$

35. $\dfrac{\sqrt{45ab^6}}{\sqrt{9ab^4c^2}}$ $\dfrac{b\sqrt{5}}{c}$

36. $\dfrac{\sqrt{24x^2y^6}}{\sqrt{8x^4z^4}}$ $\dfrac{y^3\sqrt{3}}{xz^2}$

37. $\dfrac{\sqrt{125a^6b^8}}{\sqrt{5a^2b^2}}$ $5a^2b^3$

38. $\dfrac{\sqrt{144x^{60}y^{32}}}{\sqrt{12x^{40}y^{18}}}$ $2x^{10}y^7\sqrt{3}$

Simplify each expression.

39. $\dfrac{1}{\sqrt{5}}$ $\dfrac{\sqrt{5}}{5}$

40. $\dfrac{1}{\sqrt{7}}$ $\dfrac{\sqrt{7}}{7}$

41. $\dfrac{4}{\sqrt{8}}$ $\sqrt{2}$

42. $\dfrac{3}{\sqrt{3}}$ $\sqrt{3}$

43. $\dfrac{6}{\sqrt{12}}$ $\sqrt{3}$

44. $\dfrac{9}{\sqrt{50}}$ $\dfrac{9\sqrt{2}}{10}$

45. $\sqrt{\dfrac{1}{5}}$ $\dfrac{\sqrt{5}}{5}$

46. $\sqrt{\dfrac{2}{5}}$ $\dfrac{\sqrt{10}}{5}$

47. $\sqrt{\dfrac{2}{3}}$ $\dfrac{\sqrt{6}}{3}$

48. $\sqrt{\dfrac{7}{12}}$ $\dfrac{\sqrt{21}}{6}$

49. $\sqrt{\dfrac{5}{15}}$ $\dfrac{\sqrt{3}}{3}$

50. $\sqrt{\dfrac{5}{7}}$ $\dfrac{\sqrt{35}}{7}$

51. $\sqrt{\dfrac{3}{8}}$ $\dfrac{\sqrt{6}}{4}$

52. $\sqrt{\dfrac{5}{18}}$ $\dfrac{\sqrt{10}}{6}$

53. $\sqrt{\dfrac{3}{x}}$ $\dfrac{\sqrt{3x}}{x}$

54. $\sqrt{\dfrac{5}{a}}$ $\dfrac{\sqrt{5a}}{a}$

55. $\dfrac{\sqrt{a}}{\sqrt{b}}$ $\dfrac{\sqrt{ab}}{b}$

56. $\dfrac{\sqrt{m}}{\sqrt{n}}$ $\dfrac{\sqrt{mn}}{n}$

57. $\sqrt{\dfrac{c}{d}}$ $\dfrac{\sqrt{cd}}{d}$

58. $\sqrt{\dfrac{r}{6}}$ $\dfrac{\sqrt{6r}}{6}$

59. $\sqrt{\dfrac{3x}{5}}$ $\dfrac{\sqrt{15x}}{5}$

60. $\sqrt{\dfrac{5w}{18}}$ $\dfrac{\sqrt{10w}}{6}$

61. $\sqrt{\dfrac{x^2}{2}}$ $\dfrac{x\sqrt{2}}{2}$

62. $\sqrt{\dfrac{x^2}{3}}$ $\dfrac{x\sqrt{3}}{3}$

63. $\sqrt{\dfrac{a^2}{8}}$ $\dfrac{a\sqrt{2}}{4}$

64. $\sqrt{\dfrac{a^3}{18}}$ $\dfrac{a\sqrt{2a}}{6}$

65. $\sqrt{\dfrac{t^5}{5}}$ $\dfrac{t^2\sqrt{5t}}{5}$

66. $\sqrt{\dfrac{x^3}{11}}$ $\dfrac{x\sqrt{11x}}{11}$

67. $\sqrt{\dfrac{a^8}{14b}}$ $\dfrac{a^4\sqrt{14b}}{14b}$

68. $\sqrt{\dfrac{a^7b}{24b^2}}$ $\dfrac{a^3\sqrt{6ab}}{12b}$

69. $\sqrt{\dfrac{6c^2d^4}{30c^2d^5}}$ $\dfrac{\sqrt{5d}}{5d}$

70. $\sqrt{\dfrac{27xz^4}{6y^4}}$ $\dfrac{3z^2\sqrt{2x}}{2y^2}$

71. $\sqrt{\dfrac{50yz}{24x^4y^5z^3}}$ $\dfrac{5\sqrt{3}}{6x^2y^2z}$

72. $\dfrac{\sqrt{25x^5}}{\sqrt{100xy^5}}$ $\dfrac{x^2\sqrt{y}}{2y^3}$

73. $\dfrac{\sqrt{90x^4y}}{\sqrt{2x^5y^5}}$ $\dfrac{3\sqrt{5x}}{xy^2}$

74. $\dfrac{\sqrt{120xyz^2}}{\sqrt{9xy^2}}$ $\dfrac{2z\sqrt{30y}}{3y}$

Find the product of the given binomial and its conjugate.

75. $5 + \sqrt{3}$ 22

76. $\sqrt{6} - 2$ 2

77. $\sqrt{3} - \sqrt{6}$ -3

78. $\sqrt{7} + \sqrt{5}$ 2

79. $\sqrt{x} - y$ $x - y^2$

80. $\sqrt{x} + y$ $x - y^2$

81. $\sqrt{a} + \sqrt{b}$ $a - b$

82. $\sqrt{x} - \sqrt{y}$ $x - y$

Simplify each expression.

83. $\dfrac{2}{\sqrt{5} + 2}$ $2\sqrt{5} - 4$

84. $\dfrac{2}{4 - \sqrt{3}}$ $\dfrac{8 + 2\sqrt{3}}{13}$

85. $\dfrac{2}{\sqrt{6} - 1}$ $\dfrac{2\sqrt{6} + 2}{5}$

86. $\dfrac{4}{\sqrt{2}-7}$ $\dfrac{-4\sqrt{2}-28}{47}$

87. $\dfrac{6}{\sqrt{3}+\sqrt{5}}$ $-3\sqrt{3}+3\sqrt{5}$

88. $\dfrac{5}{\sqrt{6}+\sqrt{3}}$ $\dfrac{5\sqrt{6}-5\sqrt{3}}{3}$

89. $\dfrac{8}{\sqrt{5}-\sqrt{8}}$ $\dfrac{-8\sqrt{5}-16\sqrt{2}}{3}$

90. $\dfrac{1}{\sqrt{17}-\sqrt{8}}$ $\dfrac{\sqrt{17}+2\sqrt{2}}{9}$

91. $\dfrac{2}{\sqrt{y}+3}$ $\dfrac{2\sqrt{y}-6}{y-9}$

92. $\dfrac{5}{6-\sqrt{x}}$ $\dfrac{30+5\sqrt{x}}{36-x}$

93. $\dfrac{6}{4-\sqrt{y}}$ $\dfrac{24+6\sqrt{y}}{16-y}$

94. $\dfrac{5}{3+\sqrt{x}}$ $\dfrac{15-5\sqrt{x}}{9-x}$

95. $\dfrac{16}{\sqrt{y}+x}$ $\dfrac{16\sqrt{y}-16x}{y-x^2}$

96. $\dfrac{7}{r+\sqrt{t}}$ $\dfrac{7r-7\sqrt{t}}{r^2-t}$

97. $\dfrac{x}{\sqrt{x}+\sqrt{y}}$ $\dfrac{x\sqrt{x}-x\sqrt{y}}{x-y}$

98. $\dfrac{\sqrt{3}}{\sqrt{x}-\sqrt{3}}$ $\dfrac{\sqrt{3x}+3}{x-3}$

99. $\dfrac{\sqrt{3}}{\sqrt{3}-\sqrt{n}}$ $\dfrac{3+\sqrt{3n}}{3-n}$

100. $\dfrac{x}{\sqrt{x}-y}$ $\dfrac{x\sqrt{x}+xy}{x-y^2}$

101. $\dfrac{5\sqrt{x}}{6-\sqrt{x}}$ $\dfrac{30\sqrt{x}+5x}{36-x}$

102. $\dfrac{3\sqrt{r}}{2+\sqrt{r}}$ $\dfrac{6\sqrt{r}-3r}{4-r}$

Problem Solving

104. no; $\dfrac{\sqrt{75}}{\sqrt{3}}=\sqrt{25}=5$

103. Will the quotient of two rational numbers (denominator not equal to 0) always be a rational number? Explain and give an example to support your answer. yes

104. Will the quotient of two irrational numbers (denominator not equal to 0) always be an irrational number? Explain and give an example to support your answer.

In Exercises 105–108, if $\sqrt{5}\approx2.236$ and $\sqrt{10}\approx3.162$, find to the nearest hundredth:

105. $\sqrt{5}+\sqrt{10}$ 5.40   **106.** $\sqrt{5}\cdot\sqrt{10}$ 7.07   **107.** $\sqrt{5}/\sqrt{10}$ 0.71   **108.** $\sqrt{10}/\sqrt{5}$ 1.41

In Exercises 109–112, if $\sqrt{7}\approx2.646$ and $\sqrt{21}\approx4.583$, find to the nearest hundredth:

109. $\sqrt{7}+\sqrt{21}$ 7.23   **110.** $\sqrt{7}\cdot\sqrt{21}$ 12.13   **111.** $\sqrt{7}/\sqrt{21}$ 0.58   **112.** $\sqrt{21}/\sqrt{7}$ 1.73

In Exercises 113 and 114, find the width of each rectangle whose area and length are given. Give the answer in simplified form with integer denominator.

113. $A=24$, $l=4+\sqrt{3}$ $\dfrac{24(4-\sqrt{3})}{13}$

114. $A=\sqrt{70}$, $l=7-\sqrt{2}$ $\dfrac{7\sqrt{70}+2\sqrt{35}}{47}$

Challenge Problems

Fill in the shaded area to make the expression true. Explain how you determined your answer.

115. $\sqrt{\dfrac{\blacksquare}{4x^2}}=4x^4$ $64x^{10}$ **116.** $\dfrac{\sqrt{32x^5}}{\sqrt{\blacksquare}}=2x^2$ $8x$ **117.** $\dfrac{1}{\sqrt{\blacksquare}}=\dfrac{\sqrt{2}}{2}$ 2   **118.** $\dfrac{3x}{\sqrt{\blacksquare}}=\dfrac{3\sqrt{2x}}{2}$ $2x$

119. Simplify $\dfrac{\sqrt{x}}{1-\sqrt{3}}$ by rationalizing the denominator. $\dfrac{-\sqrt{x}-\sqrt{3x}}{2}$

120. Simplify $\dfrac{\sqrt{x}}{2+\sqrt{2}}$ by rationalizing the denominator. $\dfrac{2\sqrt{x}-\sqrt{2x}}{2}$

Group Activity

121. Consider $\dfrac{\sqrt{x}}{x + \sqrt{x}}$, where $x > 0$.

a) Individually, simplify $\dfrac{\sqrt{x}}{x + \sqrt{x}}$ by rationalizing the denominator. Compare and correct your answer if necessary.

b) Group Member 1: Substitute $x = 4$ in the original expression and in the results found in part **a)**. Evaluate each expression.   both are $\dfrac{1}{3}$

c) Group Member 2: Repeat part **b)** for $x = 6$.

d) Group Member 3: Repeat part **b)** for $x = 9$.

e) As a group, determine what relationship exists between the given expression and the rationalized expression you obtained in part **b)**.   equal

121. **a)** $\dfrac{x\sqrt{x} - x}{x^2 - x} = \dfrac{\sqrt{x} - 1}{x - 1}$ **c)** $\dfrac{\sqrt{6} - 1}{5}, \dfrac{\sqrt{6} - 1}{5}$,   **d)** $\dfrac{1}{4}$

Cumulative Review Exercises

[4.6] **122.** Divide $\dfrac{3x^2 + 4x - 25}{x + 4}$ $3x - 8 + \dfrac{7}{x + 4}$

[5.6] **123.** Solve the equation $2x^2 - x - 36 = 0$. $\dfrac{9}{2}, -4$

[6.4] **124.** Subtract $\dfrac{1}{x^2 - 4} - \dfrac{2}{x - 2}$. $\dfrac{-2x - 3}{(x + 2)(x - 2)}$

[6.7] **125.** ***Stocking Wood*** Mark DeGroat can stack a cord of wood in 20 minutes. With his wife's help, they can stack the wood in 12 minutes. How long would it take his wife, Terry, to stack the wood by herself?   30 min

See Exercise 125.

9.5 SOLVING RADICAL EQUATIONS

SSM   Study Guide   CD/Video

MathPro 4/5   PH Math Tutor Center   prenhall.com/Angel

1 Solve radical equations containing only one square root term.

2 Solve radical equations containing two square root terms.

1 Solve Radical Equations Containing Only One Square Root Term

A **radical equation** is an equation that contains a variable in a radicand. Some examples of radical equations are

$$\sqrt{x} = 3 \qquad \sqrt{x + 4} = 6 \qquad \sqrt{x - 2} = x - 6$$

In this section we will assume that all radicands represent non-negative numbers. This will allow us to write statements such as $(\sqrt{x})^2 = x$ and $(\sqrt{z - 3})^2 = z - 3$.

In the process of solving radical equations we will need to **square both sides of the equation**. An example of squaring both sides of an equation is illustrated below.

$$\sqrt{y} = 5$$
$$(\sqrt{y})^2 = 5^2 \qquad \textit{Square both sides of the equation.}$$
$$y = 25$$

Now we will give a general procedure for solving radical equations containing only one square root term.

> ## To Solve a Radical Equation Containing Only One Square Root Term
>
> 1. Use the appropriate properties to rewrite the equation with the square root term by itself on one side of the equation. We call this **isolating the square root**.
> 2. Combine like terms.
> 3. Square both sides of the equation to eliminate the square root.
> 4. Solve the equation for the variable.
> 5. Check the solution in the *original* equation for extraneous roots.

Step 3 of the procedure, squaring both sides of the equation, has the effect of removing a radical sign from the expression so that the variable can be isolated. For example, consider the following radical equation,

$$\sqrt{y-3} = 8 \qquad \textit{Given equation}$$
$$(\sqrt{y-3})^2 = 8^2 \qquad \textit{Square both sides.}$$
$$y - 3 = 64 \qquad \textit{Square root has been removed.}$$
$$y = 67 \qquad \textit{Solve for y.}$$

Now we will work some additional examples.

EXAMPLE 1 Solve the equation $\sqrt{x} = 7$.

Solution The square root containing the variable is already by itself on one side of the equation. Square both sides of the equation.

$$\sqrt{x} = 7$$
$$(\sqrt{x})^2 = 7^2 \qquad \textit{Square both sides.}$$
$$x = 49$$

Check
$$\sqrt{x} = 7$$
$$\sqrt{49} \overset{?}{=} 7$$
$$7 = 7 \qquad \textit{True}$$

EXAMPLE 2 Solve the equation $\sqrt{x-5} = 4$.

Solution The square root containing the variable is already by itself on one side of the equation. Square both sides of the equation.

$$\sqrt{x-5} = 4$$
$$(\sqrt{x-5})^2 = 4^2 \qquad \textit{Square both sides.}$$
$$x - 5 = 16 \qquad \textit{Simplify.}$$
$$x = 21$$

Check
$$\sqrt{x-5} = 4$$
$$\sqrt{21-5} \overset{?}{=} 4$$
$$\sqrt{16} \overset{?}{=} 4$$
$$4 = 4 \qquad \textit{True}$$

In the next example, the square root term containing the variable is not by itself on one side of the equation.

EXAMPLE 3 Solve the equation $\sqrt{a} + 6 = 9$.

Solution In this equation we are solving for the variable a. Since the 6 is outside the square root sign, we first subtract 6 from both sides of the equation to isolate the square root term.

$$\sqrt{a} + 6 = 9$$
$$\sqrt{a} + 6 - 6 = 9 - 6$$
$$\sqrt{a} = 3$$

Now we square both sides of the equation.

$$(\sqrt{a})^2 = 3^2$$
$$a = 9$$

NOW TRY EXERCISE 21 A check will show that 9 is the solution. ✳

HELPFUL HINT

TEACHING TIP
Before discussing Example 4, have students describe the general form of the equation. Be sure that they notice that it is a square root equal to a negative number. Then ask, "Can a square root ever equal a negative number? Now let's show there is no solution algebraically."

When you square both sides of an equation, you may introduce extraneous roots. An **extraneous root** is a number obtained when solving an equation that is not a solution to the original equation. Equations where both sides are squared in the process of finding their solutions should always be checked for extraneous roots by substituting the numbers found back in the **original** equation.

Consider the equation

$$x = 6$$

Now square both sides.

$$x^2 = 36$$

Note that the original equation $x = 6$ is true only when x is 6. However, the equation $x^2 = 36$ is true for both 6 and -6. When we squared $x = 6$, we introduced the extraneous root -6.

EXAMPLE 4 Solve the equation $\sqrt{x} = -6$.

Solution Begin by squaring both sides of the equation

$$\sqrt{x} = -6$$
$$(\sqrt{x})^2 = (-6)^2$$
$$x = 36$$

Check $\sqrt{x} = -6$
$$\sqrt{36} \overset{?}{=} -6$$
$$6 = -6 \quad \textit{False}$$

NOW TRY EXERCISE 19 Since the check results in a false statement, the number 36 is an extraneous root and is not a solution to the given equation. Thus, the equation $\sqrt{x} = -6$ has **no real solution**. ✳

In Example 4, you might have realized without working the problem that there is no solution. In the original equation, the left side is nonnegative and the right side is negative; thus they cannot possibly be equal.

EXAMPLE 5 Solve the equation $\sqrt{2x-3} = x-3$.

Solution Square both sides of the equation.

$$(\sqrt{2x-3})^2 = (x-3)^2$$
$$2x - 3 = x^2 - 6x + 9 \quad \text{\textit{Simplify on left. Write }} (x-3)^2 \text{ \textit{as} } x^2 - 6x + 9.$$

Now solve the quadratic equation as explained in Section 5.6. Make one side of the equation equal to zero by subtracting $2x$ and adding 3 to both sides of the equation. This gives the following quadratic equation.

$$0 = x^2 - 8x + 12 \quad \text{or} \quad x^2 - 8x + 12 = 0$$

Solve for x by factoring.

$$x^2 - 8x + 12 = 0$$
$$(x - 6)(x - 2) = 0 \qquad \text{\textit{Factor.}}$$
$$x - 6 = 0 \quad \text{or} \quad x - 2 = 0 \quad \text{\textit{Zero-factor property}}$$
$$x = 6 \qquad\qquad x = 2 \quad \text{\textit{Solve for x.}}$$

Check

$x = 6$	$x = 2$
$\sqrt{2x-3} = x-3$	$\sqrt{2x-3} = x-3$
$\sqrt{2(6)-3} \overset{?}{=} 6 - 3$	$\sqrt{2(2)-3} \overset{?}{=} 2 - 3$
$\sqrt{9} \overset{?}{=} 3$	$\sqrt{1} \overset{?}{=} -1$
$3 = 3$ \quad \textit{True}	$1 = -1$ \quad \textit{False}

The solution is 6. Two is not a solution to the equation.

Remember, when solving a radical equation, begin by isolating the radical, as is shown in Example 6.

EXAMPLE 6 Solve the equation $2x - 5\sqrt{x} - 3 = 0$.

Solution First rewrite the equation so that the square root containing the variable is by itself on one side of the equation.

$$2x - 5\sqrt{x} - 3 = 0$$
$$-5\sqrt{x} = -2x + 3$$
$$\text{or} \qquad 5\sqrt{x} = 2x - 3 \qquad \text{\textit{Multiplied both sides by} } -1$$

Now square both sides of the equation.

TEACHING TIP
When discussing Example 6, have students explain why the 5 gets squared when simplifying $(5\sqrt{x})^2$.

$$(5\sqrt{x})^2 = (2x-3)^2$$
$$5^2(\sqrt{x})^2 = (2x-3)^2$$
$$25x = 4x^2 - 12x + 9 \qquad \text{\textit{Simplify on left. Write} } (2x-3)^2 \text{ \textit{as} } 4x^2 - 12x + 9.$$
$$0 = 4x^2 - 37x + 9 \qquad \text{\textit{Set one side of equation equal to zero.}}$$
$$\text{or} \qquad 4x^2 - 37x + 9 = 0$$
$$(4x - 1)(x - 9) = 0 \qquad \text{\textit{Factor.}}$$
$$4x - 1 = 0 \quad \text{or} \quad x - 9 = 0 \quad \text{\textit{Zero-factor property}}$$
$$4x = 1 \qquad\qquad x = 9 \quad \text{\textit{Solve for x.}}$$
$$x = \frac{1}{4}$$

Check $x = \dfrac{1}{4}$ $x = 9$

$$2x - 5\sqrt{x} - 3 = 0 \qquad\qquad 2x - 5\sqrt{x} - 3 = 0$$

$$2\left(\dfrac{1}{4}\right) - 5\sqrt{\dfrac{1}{4}} - 3 \stackrel{?}{=} 0 \qquad\qquad 2(9) - 5\sqrt{9} - 3 \stackrel{?}{=} 0$$

$$\dfrac{1}{2} - 5\left(\dfrac{1}{2}\right) - 3 \stackrel{?}{=} 0 \qquad\qquad 18 - 5(3) - 3 \stackrel{?}{=} 0$$

$$\dfrac{1}{2} - \dfrac{5}{2} - 3 \stackrel{?}{=} 0 \qquad\qquad 18 - 15 - 3 \stackrel{?}{=} 0$$

$$-\dfrac{4}{2} - 3 \stackrel{?}{=} 0 \qquad\qquad\qquad\qquad 0 = 0 \quad \textit{True}$$

$$-2 - 3 \stackrel{?}{=} 0$$

$$-5 = 0 \quad \textit{False}$$

NOW TRY EXERCISE 39   The solution is 9; $\frac{1}{4}$ is not a solution.   ✸

2   Solve Radical Equations Containing Two Square Root Terms

Consider the radical equations

$$\sqrt{x - 2} = \sqrt{x + 7} \quad \text{and} \quad \sqrt{2x + 6} - \sqrt{3x + 5} = 0$$

These equations are different from those previously discussed because they have two square root terms containing the variable x. To solve equations of this type, rewrite the equation, when necessary, so that there is only one term containing a square root on each side of the equation. Then square both sides of the equation. Examples 7 and 8 illustrate this procedure.

EXAMPLE 7   Solve the equation $\sqrt{4x - 4} = \sqrt{3x + 6}$.

Solution   Since each side of the equation already contains one square root, it is not necessary to rewrite the equation. Square both sides of the equation, then solve for x.

$$\left(\sqrt{4x - 4}\right)^2 = \left(\sqrt{3x + 6}\right)^2$$

$$4x - 4 = 3x + 6$$

$$x - 4 = 6$$

$$x = 10$$

Check $\sqrt{4x - 4} = \sqrt{3x + 6}$

$$\sqrt{4(10) - 4} \stackrel{?}{=} \sqrt{3(10) + 6}$$

$$\sqrt{36} \stackrel{?}{=} \sqrt{36}$$

$$6 = 6 \quad \textit{True}$$

NOW TRY EXERCISE 29   The solution is 10.   ✸

EXAMPLE 8   Solve the equation $4\sqrt{r - 4} - \sqrt{10r + 14} = 0$.

Solution   Add $\sqrt{10r + 14}$ to both sides of the equation to get one square root on each side of the equation. Then square both sides of the equation.

$$4\sqrt{r - 4} - \sqrt{10r + 14} \;\boxed{+ \sqrt{10r + 14}} = 0 \;\boxed{+ \sqrt{10r + 14}}$$

$$4\sqrt{r - 4} = \sqrt{10r + 14}$$

TEACHING TIP
When discussing Example 8, be
sure students remember to square
each factor in step 3.

$$(4\sqrt{r-4})^2 = (\sqrt{10r+14})^2 \quad \text{\textit{Square both sides.}}$$
$$16(r-4) = 10r+14 \quad \text{\textit{Simplify.}}$$
$$16r-64 = 10r+14 \quad \text{\textit{Distributive property}}$$
$$6r-64 = 14 \quad \text{\textit{Subtract 10r from both sides.}}$$
$$6r = 78 \quad \text{\textit{Add 64 to both sides.}}$$
$$r = 13$$

Check $\quad 4\sqrt{r-4} - \sqrt{10r+14} = 0$

$$4\sqrt{13-4} - \sqrt{10(13)+14} \overset{?}{=} 0$$
$$4\sqrt{9} - \sqrt{130+14} \overset{?}{=} 0$$
$$4(3) - \sqrt{144} \overset{?}{=} 0$$
$$12 - 12 = 0 \quad \text{\textit{True}}$$

The solution is 13.

HELPFUL HINT

In Example 6 when we simplified $(5\sqrt{x})^2$ we obtained $25x$ and in Example 8 when we simplified $(4\sqrt{r-4})^2$ we obtained $16(r-4)$. Remember, by the power rule for exponents (discussed in Section 6.1) that when a product of factors is raised to a power, each of the factors is raised to that power. Thus, we see that

$$(5\sqrt{x})^2 = 5^2(\sqrt{x})^2 = 25x \quad \text{and} \quad (4\sqrt{r-4})^2 = 4^2(\sqrt{r-4})^2 = 16(r-4)$$

1. an equation that contains a variable in a radicand   2. a number obtained when solving an equation that is not a solution to the
original equation   4. Answers will vary.

Exercise Set 9.5

Concept/Writing Exercises

1. What is a radical equation?

2. What is an extraneous root?

3. Why is it necessary to always check solutions to radical equations?   because they may be extraneous

4. **a)** Write in your own words a step-by-step procedure for solving equations containing a single square root term.

   b) Solve the equation $\sqrt{x+1} - 1 = 1$ using the procedure you gave in part **a)**.   3

Determine whether the given value is the solution to the equation. If it is not a solution, state why.

5. $\sqrt{x} = 7, x = 49$ yes

6. $\sqrt{x} = -7, x = 49$ no

7. $-\sqrt{p} = 8, p = 64$ no

8. $-\sqrt{x} = -8, x = 64$ yes

9. $\sqrt{x+4} = 3, x = 5$ yes

10. $\sqrt{-y-5} = 3, y = 4$ no

11. $-2 = \sqrt{t-4}, t = 0$ no

12. $\sqrt{x+4} = 0, x = -4$ yes

Practice the Skills

Solve each equation. If the equation has no real solution, so state.

13. $\sqrt{x} = 4$ 16

14. $\sqrt{x} = 9$ 81

15. $\sqrt{x} = -5$ no solution

16. $\sqrt{m} = -1$ no solution

17. $\sqrt{x+5} = 3$ 4

18. $\sqrt{2x-4} = 2$ 4

19. $\sqrt{x+5} = -7$ no solution

20. $\sqrt{x-4} = 8$ 144

21. $\sqrt{z-3} = 7$ 100

22. $6 + \sqrt{w} = 9$ 9

23. $11 = 6 + \sqrt{x}$ 25

24. $5 = 7 - \sqrt{x}$ 4

25. $7 + \sqrt{n} = 3$ no solution

26. $\sqrt{3x+4} = x-2$ 7

27. $\sqrt{2x-5} = x-4$ 7

28. $\sqrt{x^2+8} = x+2$ 1

29. $\sqrt{3r-9} = \sqrt{r+3}$ 6

30. $5\sqrt{x-4} = 30$ 40

🔒 **31.** $\sqrt{4x + 4} = \sqrt{6x - 2}$ 3   **32.** $\sqrt{2x - 5} = \sqrt{x + 2}$ 7   **33.** $\sqrt{x^2 + 3} = x + 1$ 1

34. $\sqrt{4t + 8} = 2\sqrt{t}$ no solution   **35.** $\sqrt{4x - 5} = \sqrt{x + 9}$ $\frac{14}{3}$ **36.** $\sqrt{3k + 5} = 2\sqrt{k}$ 5

🔒 **37.** $3\sqrt{x} = \sqrt{x + 8}$ 1   **38.** $x - 5 = \sqrt{x^2 - 35}$ 6   **39.** $4\sqrt{x} = x + 3$ 1, 9

40. $5 + \sqrt{x - 5} = x$ 5, 6   **41.** $\sqrt{3f - 4} = 2\sqrt{3f - 2}$ no solution   **42.** $12 - 3\sqrt{2x} = 0$ 8

43. $\sqrt{x^2 - 4} = x + 2$ −2   **44.** $6\sqrt{3x - 2} = 24$ 6   🔒 **45.** $3 + \sqrt{3x - 5} = x$ 7

46. $\sqrt{4x + 5} + 5 = 2x$ 5   **47.** $\sqrt{8 - 7x} = x - 2$ no solution   **48.** $1 + \sqrt{x + 1} = x$ 3

49. $2\sqrt{3b - 5} = \sqrt{2b + 10}$ 3   **50.** $3\sqrt{s + 4} = s + 6$ 0, −3   **51.** $3\sqrt{2w + 3} = 3w$ 3

52. $4\sqrt{4c - 7} = 2c + 4$ 4, 8

Problem Solving

Exercises 53–56 show rectangles with areas. Find the value of the indicated variable.

53. $A = 24$ 7
6
$\sqrt{w + 9}$

54. 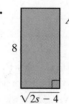 $A = 32$ 10
8
$\sqrt{2s - 4}$

55. 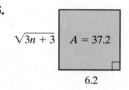 $\sqrt{3n + 3}$ $A = 37.2$ 11
6.2

56. 5
$\sqrt{3p + 1}$ $A = 29.2$
7.3

In Section 9.1, we indicated that we could write a square root expression in exponential form. For example, $\sqrt{25}$ can be written $(25)^{1/2}$. Similarly we can write certain exponential expressions as square roots. For example $(25)^{1/2}$ can be written $\sqrt{25}$, and $(x - 5)^{1/2}$ can be written $\sqrt{x - 5}$. Rewrite the following equations as equations containing square roots. Then solve each equation.

57. $(x + 3)^{1/2} = 7$
$\sqrt{x + 3} = 7; 46$

58. $(x - 6)^{1/2} = 5$
$\sqrt{x - 6} = 5; 31$

59. $(x - 2)^{1/2} = (2x - 9)^{1/2}$ **60.** $3(x - 9)^{1/2} = 21$
$\sqrt{x - 2} = \sqrt{2x - 9}; 7$ $3\sqrt{x - 9} = 21; 58$

*In Exercises 61–64, **a)** Use the FOIL method to multiply the factors on the left-hand side of the equation. **b)** Solve the equation. Remember to check your answer in the original equation.*

61. $(\sqrt{x} - 3)(\sqrt{x} + 3) = 40$ **a)** $x - 9 = 40$ **b)** 49   **62.** $(\sqrt{x} + 4)(\sqrt{x} - 4) = 20$ **a)** $x - 16 = 20$ **b)** 36

63. $(7 - \sqrt{x})(5 + \sqrt{x}) = 35$ **a)** $35 + 2\sqrt{x} - x = 35$ **b)** 0, 4   **64.** $(6 - \sqrt{x})(2 + \sqrt{x}) = 15$ **a)** $12 + 4\sqrt{x} - x = 15$ **b)** 1, 9

65. The sum of a natural number and its square root is 2. Find the number. 1

66. The product of 3 and the square root of a number is 18. Find the number. 36

Challenge Problems

Solve each equation. (Hint: You will need to square both sides of the equation, then isolate the square root containing the variable, then square both sides of the equation again.)

67. $\sqrt{x} + 2 = \sqrt{x + 16}$ 9   **68.** $\sqrt{x + 1} = 2 - \sqrt{x}$ $\frac{9}{16}$ **69.** $\sqrt{x + 7} = 5 - \sqrt{x - 8}$ 9

 ## Group Activity

Discuss and answer Exercises 70–72 as a group.

✎ **70.** Radical equations in two variables can be graphed by selecting values for x and finding the corresponding values of y, as was done in Section 7.2 when we graphed linear equations.

a) As a group, discuss what values would be appropriate to select for x if you wanted to graph $y = \sqrt{x}$. Explain your answer. Answers will vary.

71.–72. a) and b) Answers will vary.   70.–72. c) See Graphing Answer Section, page G14.   e) yes

b) Select four values for x that will result in y being a whole number. List your values for x and y in a table.

c) Individually, plot the points and sketch the graph. Compare your graphs.

d) Is the graph linear? Explain.   no

e) Is the graph a function? (See Section 7.6.) Explain.

f) Can you determine whether $y = \sqrt{x}$ has an x-intercept and a y-intercept from the graph? If so, give the intercepts.   yes, $(0,0)$

As a group, repeat parts a)–f) of exercise 70 for each the following equations.

71. $y = \sqrt{x - 2}$

71. d) no   e) yes   f) yes; x-int. $(2,0)$, no y-int.

72. $y = \sqrt{x + 4}$

72. d) no   e) yes   f) yes; x-int. $(-4,0)$, y-int. $(0,2)$

Cumulative Review Exercises

[8.1] **73.** Solve the following system of equations graphically.

$$3x - 2y = 6$$
$$y = 2x - 4 \qquad (2,0)$$

[8.2] **74.** Solve the following system of equations by substitution.

$$3x - 2y = 6$$
$$y = 2x - 4 \qquad (2,0)$$

[8.3] **75.** Solve the following system of equations using the addition method.

$$3x - 2y = 6$$
$$y = 2x - 4 \qquad (2,0)$$

[8.4] **76.** ***Boat Ride*** A ferry boat on the Hudson River can travel at a speed of 18 miles per hour with

the current and 14 miles per hour against the current. Find the speed of the ferry boat in still water and the current.   ferry: 16 mph; current, 2 mph

9.6 RADICALS: APPLICATIONS AND PROBLEM SOLVING

1 Use the Pythagorean Theorem.

2 Use the distance formula.

3 Use radicals to solve application problems.

SSM   Study Guide   CD/Video

MathPro 4/5   PH Math Tutor Center   prenhall.com/Angel

In this section we will focus on some of the many interesting applications of radicals. We will use the Pythagorean Theorem and introduce the distance formula, and then give a few additional applications of radicals.

1 Use the Pythagorean Theorem

In Section 5.7 we introduced the Pythagorean Theorem. For your convenience we state the Pythagorean Theorem again.

> ### Pythagorean Theorem
>
> The square of the hypotenuse of a right triangle is equal to the sum of the squares of the two legs.
>
> $$(\text{leg})^2 + (\text{leg})^2 = (\text{hypotenuse})^2$$

(continued on the next page)

If a and b represent the legs, and c represents the hypotenuse, then

$$a^2 + b^2 = c^2$$

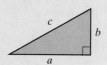

In Section 9.5, when we solved equations containing square roots, we raised both sides of the equation to the second power to eliminate the square roots. When we solve problems using the Pythagorean Theorem, we will raise both sides of the equation to the $\frac{1}{2}$ power to remove the square on one of the variables. We can do this because the rules of exponents presented in Sections 4.1 and 4.2 also apply to fractional exponents. Since lengths are positive, the values of a, b, and c in the Pythagorean Theorem must represent positive values.

EXAMPLE 1   **Right Triangle Problem**   Find the hypotenuse of a right triangle whose legs are $\sqrt{3}$ feet and 4 feet.

Solution   Draw a picture of the problem (Figure 9.2). It makes no difference which leg is called a and which leg is called b.

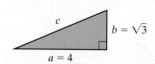

FIGURE 9.2

$$a^2 + b^2 = c^2$$
$$4^2 + (\sqrt{3})^2 = c^2 \qquad \text{\textit{Substitute values for a and b.}}$$
$$16 + 3 = c^2$$
$$19 = c^2$$
$$\sqrt{19} = \sqrt{c^2} \qquad \text{\textit{Take the square root of both sides.}}$$
$$\sqrt{19} = c \qquad \text{\textit{Since } c > 0, \sqrt{c^2} = c}$$

Thus the hypotenuse is $\sqrt{19}$ feet or about 4.36 feet.

Check $a^2 + b^2 = c^2$
$$4^2 + (\sqrt{3})^2 \stackrel{?}{=} (\sqrt{19})^2$$
$$16 + 3 \stackrel{?}{=} 19$$

NOW TRY EXERCISE 15
$$19 = 19 \quad \text{\textit{True}}$$

EXAMPLE 2   **Basketball Court**   A professional basketball court is a rectangle whose overall dimensions are 94 feet by 50 feet. Find the length of the diagonal of the court.

Solution   Understand and Translate   First, we draw the court (Fig. 9.3). We are asked to find the length of the diagonal. This length is the hypotenuse, c, of the triangle shown in Figure 9.4.

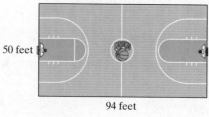

50 feet

94 feet

FIGURE 9.3

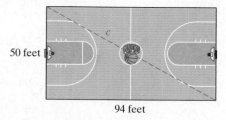

50 feet

94 feet

FIGURE 9.4

$$a^2 + b^2 = c^2$$

Carry Out $\quad (50)^2 + (94)^2 = c^2$

$$2500 + 8836 = c^2$$

$$11{,}336 = c^2$$

$$c = \sqrt{11{,}336}, \quad \text{or} \quad \text{approximately } 106.47$$

NOW TRY EXERCISE 25 $\quad$ Answer $\quad$ The length of the diagonal of the court is approximately 106.47 feet. ✳

2 Use the Distance Formula

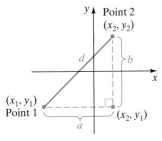

FIGURE 9.5

The **distance formula** can be used to find the distance between two points, (x_1, y_1) and (x_2, y_2), in the Cartesian coordinate system.

It can be derived using the Pythagorean Theorem. In Figure 9.5, three points are indicated. The length of side a is $x_2 - x_1$ and the length of side b is $y_2 - y_1$. The distance between points 1 and 2 can be determined using the Pythagorean Theorem. In this case we let the hypotenuse be represented by the letter d, for distance.

$$d^2 = a^2 + b^2$$

$$d^2 = (x_2 - x_1)^2 + (y_2 - y_1)^2$$

$$d = \sqrt{(x_2 - x_1)^2 + (y_2 - y_1)^2}$$

Distance Formula

$$d = \sqrt{(x_2 - x_1)^2 + (y_2 - y_1)^2}$$

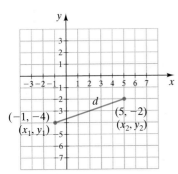

FIGURE 9.6

EXAMPLE 3 $\quad$ Find the length of the line segment between the points $(-1, -4)$ and $(5, -2)$.

Solution $\quad$ The two points are illustrated in Figure 9.6. It makes no difference which point is labeled (x_1, y_1), and which is labeled (x_2, y_2). Let $(5, -2)$ be (x_2, y_2) and $(-1, -4)$ be (x_1, y_1). Thus $x_2 = 5$, $y_2 = -2$ and $x_1 = -1$, $y_1 = -4$.

$$
\begin{aligned}
d &= \sqrt{(x_2 - x_1)^2 + (y_2 - y_1)^2} \\
&= \sqrt{[5 - (-1)]^2 + [-2 - (-4)]^2} \\
&= \sqrt{(5 + 1)^2 + (-2 + 4)^2} \\
&= \sqrt{6^2 + 2^2} \\
&= \sqrt{36 + 4} = \sqrt{40}, \quad \text{or} \quad \text{approximately } 6.32
\end{aligned}
$$

NOW TRY EXERCISE 21 $\quad$ Thus, the distance between $(-1, -4)$ and $(5, -2)$ is approximately 6.32 units. ✳

3 Use Radicals to Solve Application Problems

Radicals are often used in science and mathematics courses. Examples 4 through 6 illustrate some scientific applications that use radicals.

EXAMPLE 4 $\quad$ **Accident Investigation** $\quad$ Traffic accident investigators are trained in accident reconstruction. They use, among other things, formulas derived from the laws of physics in their work. One task they face is to try to determine the speed at which a vehicle is traveling before hitting an object. The accident investigators use the formula

$$S = 5.5\sqrt{cl}$$

to estimate the original speed of the vehicle, S, in miles per hour, where c represents the coefficient of friction between the road surface and the tire, and l represents the length of the longest skid mark measured in feet. On dry roads, the value of the coefficient of friction is between 0.69 and 0.75 for most cars. Find the speed of a car that leaves a 40-foot-long skid mark before stopping. Use $c = 0.72$.

Solution   Understand and Translate   Begin by substituting 0.72 for c and 40 for the length in the given formula.

$$S = 5.5\sqrt{cl}$$
$$= 5.5\sqrt{0.72 \times 40}$$

Carry Out $\qquad\qquad = 5.5\sqrt{28.8}$
$$\approx 29.5$$

Answer   A car leaving a 40-foot-long skid mark on a dry surface before stopping is originally traveling at approximately 29.5 miles per hour.   ✳

EXAMPLE 5

Gravity   A formula for the velocity of an object in feet per second (neglecting air resistance) after it has fallen a certain distance is

$$v = \sqrt{2gh}$$

where g is the acceleration due to gravity and h is the height the object has fallen in feet. On Earth the acceleration due to gravity, g, is approximately 32 feet per second squared. Find the velocity of a coconut after it has fallen 20 feet.

Solution   Understand and Translate   Begin by substituting 32 for g in the given equation.

$$v = \sqrt{2gh} = \sqrt{2(32)(h)} = \sqrt{64h}$$

Carry Out   At $h = 20$ feet,

$$v = \sqrt{64(20)} = \sqrt{1280} \approx 35.78$$

Answer   After a coconut has fallen 20 feet, its velocity is approximately 35.78 feet per second.   ✳

NOW TRY EXERCISE 43

EXAMPLE 6

A Pendulum   Even as far back as the 1600s, people were experimenting with ways to make clocks more accurate. A Dutch astronomer, Christiaan Huygens, is credited with being the first person to suggest using a pendulum mechanism. The *period of a pendulum* (the time required for the pendulum to make one complete swing both back and forth) depends on the length of the pendulum. Finding the length for a pendulum that would cause it to swing back and forth 60 times in one minute was a first step in creating more accurate clocks. Of course, pendulums are not used only in clocks. A detuning pendulum is found on overhead power lines to reduce the sometimes violent motion of conductors (caused by the wind). A pendulum can even be found on the playground—a simple swing, swinging freely back and forth!

The formula for the period, T (in seconds), of a pendulum is

$$T = 2\pi\sqrt{\frac{L}{32}}$$

where L is the length of the pendulum in feet (Fig. 9.7 on page 597). Find the period of the pendulum if its length is 8 feet. Use 3.14 as an approximation for π.

Solution Understand and Translate Substitute 3.14 for π and 8 for L in the formula.

$$T \approx 2(3.14)\sqrt{\frac{8}{32}}$$

Carry Out

$$\approx 6.28\sqrt{\frac{1}{4}}$$

$$\approx 6.28\left(\frac{1}{2}\right) = 3.14 \text{ seconds}$$

FIGURE 9.7

NOW TRY EXERCISE 37

Answer A pendulum 8 feet long takes about 3.14 seconds to make one complete swing.

Mathematics in Action

Body Surface Area

When we think about organs in the body, structures like the heart, lungs, and kidneys leap immediately into mind. The skin is also an organ and, in fact, it is the largest organ in your body. The measure of its extent is called the *body surface area* (BSA). One of the several alternative formulas used to calculate BSA, in square meters, is the Mosteller formula:

$$BSA = \sqrt{\frac{H \cdot W}{3600}}$$

where H is the height, in centimeters, and W is the weight, in kilograms. An alternate formula is

$$BSA = \sqrt{\frac{H \cdot W}{3131}}$$

where H is the height, in inches, and W is the weight, in pounds.

BSA represents an important measure in several areas of medicine. Doctors often use BSA rather than body weight to determine dosage for certain drugs. Also pharmacologist often use BSA rather than weight when doing drug research.

BSA plays a central role in determining burn severity. A burn is considered "critical" if it is has affected the deepest layers of the skin and covers more than 10% of the body surface area, or if it has penetrated less deeply and covers more than 30% of the body surface area.

Body surface area, but not weight, influences basal metabolic rate (BMR), the rate at which people consume energy. Two people with different shapes, and thus different BSAs, who weigh the same will have different BMRs. Even if their weight is identical, a short, heavier person will typically have a slower BMR than a tall, thin person. The tall person in this instance has a larger skin surface, loses more heat by radiation, and so must have a faster metabolism to generate the lost heat. You can find a number of BSA calculators on the Internet by entering the terms *"body surface area" calculator* (with quotations as shown) into the search engine at *www.google.com*.

Exercise Set 9.6

Concept/Writing Exercises

4. a) $d = \sqrt{(x_2 - x_1)^2 + (y_2 - y_1)^2}$ b) to find the distance between two points in the Cartesian coordinate system

1. What is a right triangle? a triangle that contains a 90° angle

2. State the Pythagorean Theorem and use your own words to describe what relationship it expresses about a right triangle. $a^2 + b^2 = c^2$

3. Can the Pythagorean Theorem be used with any triangle? Explain. no

4. a) Write the distance formula.
   b) When do we use the distance formula?

5. the two points in the coordinate plane that you are trying to find the distance between

5. What do the following ordered pairs represent in the distance formula: (x_1, y_1) and (x_2, y_2)?

6. Determine the distance between the points $(4, 0)$ and $(-3, 0)$. 7

8. the time required for the pendulum to make one complete swing back and forth.

7. Determine the distance between the points $(0, -4)$ and $(0, -10)$. 6

8. What is meant by the period of a pendulum?

11. $\sqrt{170} \approx 13.04$ **12.** $\sqrt{63} \approx 7.94$ **14.** $\sqrt{175} \approx 13.23$ **15.** $\sqrt{67} \approx 8.19$ **16.** $\sqrt{189} \approx 13.75$ **17.** $\sqrt{202} \approx 14.21$ **18.** $\sqrt{100} = 10$

Practice the Skills
19. $\sqrt{180} \approx 13.42$ **20.** $\sqrt{128} \approx 11.31$

Use the Pythagorean Theorem to find each indicated quantity. Give the exact value, then round your answer to the nearest hundredth.

9. $\sqrt{33} \approx 5.74$
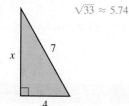

10. $\sqrt{73} \approx 8.54$
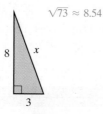

11.
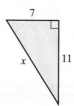

12.

13. $\sqrt{149} \approx 12.21$
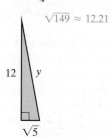

14.

15.

16.
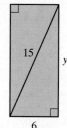

17.
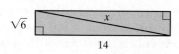

18.
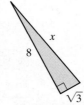

19.

20.

Use the distance formula to find the length of the line segments between each pair of points.

21. $(7, 6)$ and $(2, 9)$ $\sqrt{34} \approx 5.83$

22. $(4, -3)$ and $(-5, 6)$ $\sqrt{162} \approx 12.73$

23. $(-8, 4)$ and $(4, 11)$ $\sqrt{193} \approx 13.89$

24. $(0, 5)$ and $(-6, -4)$ $\sqrt{117} \approx 10.82$

Problem Solving
25. $\sqrt{17,240.89} \approx 131.30$ yd

25. *Football Field Dimensions* Pro Player Stadium in Miami, Florida, is home to both the Florida Marlins and the Miami Dolphins. The football field is 120 yards long from end zone to end zone. Find the length of the diagonal, to the nearest hundredth, from one end zone to the opposite corner of the other if the width of the field is 53.3 yards.

26. *Rugby Field* According to information provided by the Austin Rugby Football Club Web site, the dimensions of a rugby field are 69 meters by 100 meters. What is the length of the diagonal, to the nearest hundredth, from one corner of the field to the opposite corner of the field? $\sqrt{14,761} \approx 121.49$ yd

28. $\sqrt{232} \approx 15.23$ m

🔒 27. ***Ladder on a House*** Pedro and Vanessa are planning to paint their house. The instructions on an 8-meter extension ladder indicate that the base of the ladder should be two meters from the house when in use. If they follow the instructions, how high will the top of the ladder reach on the house? $\sqrt{60} \approx 7.75$ m

28. ***Guy Wire to a Pole*** A local cable company crew must run a guy wire to support a pole. The wire will go from the top of a 14-meter pole to a point 6 meters from the base of the pole. How long will the guy wire be?

29. ***Side of a Square*** The length of the side of a square, s, can be found by the formula $s = \sqrt{A}$, where A is the area of the square. Find the length of the side of a square boxing ring whose area is 256 square feet. 16 ft

30. ***Area of a Circle*** A formula for the area of a circle is $A = \pi r^2$, where π is approximately 3.14 and r is the radius of the circle. Find the radius of a circle whose area is 45 square inches. $\approx \sqrt{14.33} \approx 3.79$ in.

31. ***Radius of a Circle*** Find the radius of a circle whose area is 80 square feet. $\approx \sqrt{25.48} \approx 5.05$ ft

🔒 32. ***Birdbath*** A birdbath can hold water 2 inches deep, as shown in the figure. If the volume of water that can be held in the birdbath is 402 cubic inches, find the radius of the birdbath. Water volume is calculated by finding the area of the circular bottom of the birdbath and then multiplying the area by the depth of the water. Use the formula $V = \pi r^2 h$. $\approx \sqrt{64.01} \approx 8.00$ in.

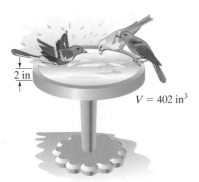

2 in

$V = 402$ in^3

33. ***Baseball*** A regulation baseball diamond is a square with 90 feet between bases. How far is second base from home plate? $\sqrt{16,200} \approx 127.28$ ft

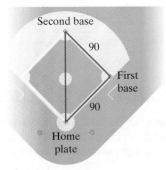

Second base

90

First base

90

Home plate

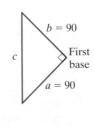

$b = 90$

c

First base

$a = 90$

34. ***Shoe Box*** The top of a shoe box is a rectangular piece of cardboard with a length of 12 inches and a width of 5 inches. Find the length of the diagonal across the top of the shoe box. $\sqrt{169} = 13$ in.

35. ***Carry-on Luggage*** The maximum dimensions of a carry-on suitcase on most airlines is length: 24 inches, height: 16 inches, and depth: 10 inches. If the suitcase is lying flat, determine the length of the diagonal across the surface of the suitcase. $\sqrt{832} \approx 28.84$ in.

36. ***Freefall*** If a pair of glasses was dropped from the Sears Tower at a height of 1431 feet (the highest occupied floor), determine the velocity of the glasses when they strike the ground. Neglect air resistance. (Use $v = \sqrt{2gh}$; refer to Example 5). $\sqrt{91,584} \approx 302.63$ ft/sec

37. ***Pendulum*** The period of a simple pendulum is determined by the length of the string, not the material composing the bob (or ball). Mikel Finn made a pendulum with a string measuring 43 inches. What is the period for this pendulum? (Use $T = 2\pi\sqrt{L/32}$; refer to Example 6.) $6.28\sqrt{0.11} \approx 2.08$ sec

38. **Foucault Pendulum** An example of a Foucault pendulum is on display in the Smithsonian Institute in Washington, D.C. It is used to demonstrate the earth's rotation. It has a cable 52 feet long and a symmetrical, hollow brass bob weighing about 240 pounds. Find the period of the pendulum. $6.28\sqrt{1.625} \approx 8.01$ sec

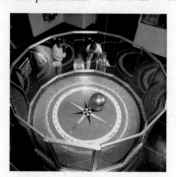

39. **Another Pendulum** The Houston Museum of Natural Science also has a Foucault pendulum on display. It has a cable 61.6 feet long and a bob weighing 180 pounds. Find the period of the pendulum. $6.28\sqrt{1.925} \approx 8.71$ sec

Earth Days *For any planet, its "year" is the time it takes for the planet to revolve once around the sun. The number of "Earth days," N, in a given planet's year, is approximated by the formula*

$$N = 0.2(\sqrt{R})^3$$

where R is the mean distance from the sun in millions of kilometers. The mean distance from the sun, in millions of kilometers, for three planets is illustrated below.

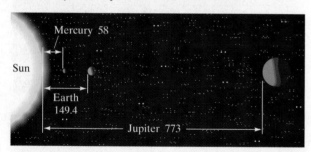

40. Find the number of Earth days in the year of Mercury.
41. Find the number of Earth days in the year of Earth.
42. Find the number of Earth days in the year of Jupiter.

Escape velocity, v_e, is the velocity needed for a spacecraft to escape a planet's gravitational field. This is found by the formula, $v_e = \sqrt{2gR}$, where g is the acceleration due to gravity of the planet and R is the radius of the planet in meters.

43. **Earth's Escape Velocity** Find the escape velocity for Earth where $g = 9.81$ meters per second squared and $R = 6{,}370{,}000$ meters.

44. **Mars's Escape Velocity** Future Mars exploration may include sending spacecraft that will land on the surface of the planet and leave after scientific work has been completed. What is the approximate velocity needed for a spacecraft to escape the gravitational field of Mars? The acceleration due to gravity on Mars is approximately 3.6835 meters per second squared. The radius of Mars is approximately 3,393,500 meters.

45. **Resultant Force** When two forces, F_1 and F_2, pull at right angles to each other as illustrated, the resultant, or the effective force, R, can be found by the formula $R = \sqrt{F_1^2 + F_2^2}$.

Two cars at a 90° angle to each other are trying to pull a third car out of the mud, as shown. If car A exerts 600 pounds of force and car B exerts 800 pounds of force, find the resulting force on the car stuck in the mud.

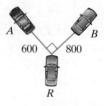

40. about 88 Earth days   41. about 365 Earth days   42. about 4298 Earth days   43. $\sqrt{19.62(6{,}370{,}000)} \approx 11{,}179.42$ m/sec
44. $\sqrt{2(3.6835)(3{,}393{,}500)} \approx 4999.99$ m/sec   45. $\sqrt{1{,}000{,}000} = 1000$ lb

Challenge Problems

46. **Rectangle** The length of a rectangle is 3 inches more than its width. If the length of the diagonal is 15 inches, find the dimensions of the rectangle. 9 in. by 12 in.

47. **Gravity on the Moon** The acceleration due to gravity on the moon is $\frac{1}{6}$ of that on Earth. If a camera falls from a rocket 100 feet above the surface of the moon, with

what velocity will it strike the moon? (Use $v = \sqrt{2gh}$; see Example 5.) ≈ 32.66 ft/sec

48. ***Length of a Pendulum*** Find the length of a pendulum if the period is 2 seconds. (Use $T = 2\pi\sqrt{L/32}$; refer to Example 6.) 3.25 ft

49. ***Diagonal of Rectangular Solid*** The length of the diagonal of a rectangular solid (see the figure on the right) is given by
$$d = \sqrt{a^2 + b^2 + c^2}$$

Find the length of the diagonal of the suitcase shown below on the right. $\sqrt{932} \approx 30.53$ in.

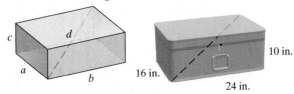

Cumulative Review Exercises

[2.7] **50.** Solve the inequality $2(x + 3) < 4x - 6$. $x > 6$

[4.2] **51.** Simplify $(4x^{-4}y^3)^{-1}$. $\dfrac{x^4}{4y^3}$

[8.3] **52.** Solve the following system of equations using the addition method.
$$3x + 4y = 12$$
$$\tfrac{1}{2}x - 2y = 8 \quad \left(7, -\tfrac{9}{4}\right)$$

[6.6] **53.** Solve the equation $5 + \dfrac{6}{x} = \dfrac{2}{3x}$. $-\dfrac{16}{15}$

9.7 HIGHER ROOTS AND RATIONAL EXPONENTS

SSM   Study Guide   CD/Video

MathPro 4/5   PH Math Tutor Center   prenhall.com/Angel

1. Evaluate cube and fourth roots.
2. Simplify cube and fourth roots.
3. Write radical expressions in exponential form.

---

1 Evaluate Cube and Fourth Roots

In this section we will use the same basic concepts used in Sections 9.1 through 9.4 to work with radicals with indexes of 3 and 4. Now we introduce **cube** and **fourth roots**.

$$\sqrt[3]{a} \text{ is read "the cube root of } a.\text{"}$$
$$\sqrt[4]{a} \text{ is read "the fourth root of } a.\text{"}$$

Note that

$$\sqrt[3]{a} = b \quad \text{if} \quad b^3 = a$$

and

$$\sqrt[4]{a} = b \quad \text{if} \quad b^4 = a, \qquad b > 0$$

Examples

$\sqrt[3]{8} = 2$	since $2^3 = 2 \cdot 2 \cdot 2 = 8$
$\sqrt[3]{-8} = -2$	since $(-2)^3 = (-2)(-2)(-2) = -8$
$\sqrt[3]{27} = 3$	since $3^3 = 3 \cdot 3 \cdot 3 = 27$
$\sqrt[4]{16} = 2$	since $2^4 = 2 \cdot 2 \cdot 2 \cdot 2 = 16$
$\sqrt[4]{81} = 3$	since $3^4 = 3 \cdot 3 \cdot 3 \cdot 3 = 81$

EXAMPLE 1 Evaluate. **a)** $\sqrt[3]{-64}$　　**b)** $\sqrt[3]{125}$

Solution **a)** To find $\sqrt[3]{-64}$, we must find the number that when cubed is -64.

$$\sqrt[3]{-64} = -4 \qquad \text{since } (-4)^3 = -64$$

b) To find $\sqrt[3]{125}$, we must find the number that when cubed is 125.

NOW TRY EXERCISE 11

$$\sqrt[3]{125} = 5 \qquad \text{since } 5^3 = 125$$

Note that the cube root of a positive number is a positive number and the cube root of a negative number is a negative number. The radicand of a fourth root (or any even root) must be a nonnegative number for the expression to be a real number. For example, $\sqrt[4]{-16}$ is not a real number because no real number raised to the fourth power can be a negative number.

It will be helpful in the explanations that follow if we define perfect cubes. A **perfect cube** is a number that is the cube of a natural number.

1,	2,	3,	4,	5,	6,	7,	8,	9,	10,	... *Natural numbers*
1^3,	2^3,	3^3,	4^3,	5^3,	6^3,	7^3,	8^3,	9^3,	10^3,	... *Cubes of natural numbers*
1,	8,	27,	64,	125,	216,	343,	512,	729,	1000,	... *Perfect cubes*

Note that $\sqrt[3]{1} = 1$, $\sqrt[3]{8} = 2$, $\sqrt[3]{27} = 3$, $\sqrt[3]{64} = 4$, and so on.

Perfect fourth powers can be expressed in a similar manner.

1,	2,	3,	4,	5,	6,	... *Natural numbers*
1^4,	2^4,	3^4,	4^4,	5^4,	6^4,	... *Fourth powers of natural numbers*
1,	16,	81,	256,	625,	1296,	... *Perfect fourth powers*

Note that $\sqrt[4]{1} = 1$, $\sqrt[4]{16} = 2$, $\sqrt[4]{81} = 3$, $\sqrt[4]{256} = 4$, and so on.

Table 9.2 illustrates some perfect cube and perfect fourth power numbers. This table may prove helpful when simplifying cube and fourth power roots.

You may wish to refer to Table 9.2 when evaluating cube and fourth roots.

TABLE 9.2

Perfect Cubes	Cube Root		Value	Perfect Fourth Powers	Fourth Root		Value
1	$\sqrt[3]{1}$	=	1	1	$\sqrt[4]{1}$	=	1
8	$\sqrt[3]{8}$	=	2	16	$\sqrt[4]{16}$	=	2
27	$\sqrt[3]{27}$	=	3	81	$\sqrt[4]{81}$	=	3
64	$\sqrt[3]{64}$	=	4	256	$\sqrt[4]{256}$	=	4
125	$\sqrt[3]{125}$	=	5	625	$\sqrt[4]{625}$	=	5
216	$\sqrt[3]{216}$	=	6	1296	$\sqrt[4]{1296}$	=	6
343	$\sqrt[3]{243}$	=	7				
512	$\sqrt[3]{512}$	=	8				

Using Your Calculator

Evaluating Cube and Higher Roots on a Scientific Calculator

If your scientific calculator contains a $\boxed{y^x}$ or $\boxed{x^y}$ key, then you can find cube and higher roots. To find cube and higher roots, you need to use both the second function key, $\boxed{2^{nd}}$, and either the $\boxed{y^x}$ or $\boxed{x^y}$ key. Some calculators use an inverse key $\boxed{INV}$, in place of the $\boxed{2^{nd}}$ key. To evaluate $\sqrt[3]{216}$ press the following keys:

$$216 \quad \boxed{2^{nd}} \quad \boxed{y^x} \quad 3 \quad \boxed{=} \quad 6$$

    ↑   ↑   ↑
   radicand   index   answer displayed

Thus $\sqrt[3]{216} = 6$. To evaluate $\sqrt[4]{158}$, press the following keys:

$$158 \quad \boxed{2^{nd}} \quad \boxed{y^x} \quad 4 \quad \boxed{=} \quad 3.54539$$

    ↑   ↑   ↑
   radicand   index   answer displayed

Thus $\sqrt[4]{158} \approx 3.54539$.

    To find the odd root of a negative number, find the odd root of that positive number and then place a negative sign before the value. For example, to find $\sqrt[3]{-64}$, find $\sqrt[3]{64}$, which is 4; then place a negative sign before the value to get -4. Thus $\sqrt[3]{-64} = -4$.

Exercises

Evaluate each of the following.

1. $\sqrt[3]{729}$ 9   2. $\sqrt[4]{16}$ 2   3. $\sqrt[4]{200}$ ≈ 3.76060

4. $\sqrt[3]{1000}$ 10   5. $\sqrt[3]{-216}$ -6 6. $\sqrt[5]{-1200}$ ≈ -4.12891

Using Your Graphing Calculator

Evaluating Cube and Higher Roots on a Graphing Calculator

    The procedure to use to find cube and higher roots depends upon your graphing calculator. Read the manual that comes with your graphing calculator for instructions.

    On a TI-83 Plus calculator you use the $\boxed{MATH}$ menu to find both cube roots and roots greater than cube roots. To find a cube root you use the $\boxed{MATH}$ menu option 4: $\sqrt[3]{}$ (.

To evaluate $\sqrt[3]{216}$ press the following keys:

$$\boxed{MATH} \quad \boxed{4} \quad 216 \quad \boxed{)} \quad \boxed{ENTER} \quad 6$$

    ↑   ↑   ↑   ↑
  MATH   TI-83 Plus   Enter   answer
  menu   then shows radicand displayed
         $\sqrt[3]{}$ (

To evaluate $\sqrt[3]{-216}$, you press the following keys:

$$\boxed{MATH} \quad \boxed{4} \quad \boxed{(-)} \quad 216 \quad \boxed{)} \quad \boxed{ENTER} \quad -6$$

(continued on the next page)

Thus $\sqrt[3]{-216} = -6$.

To evaluate roots greater than 3, you use the $\boxed{\text{MATH}}$ menu, option 5: $\sqrt[x]{}$. You enter the index before you press the $\boxed{\text{MATH}}$ key. To evaluate $\sqrt[4]{158}$ press the following keys.

$\boxed{4}$ $\boxed{\text{MATH}}$ $\boxed{5}$ 158 $\boxed{\text{ENTER}}$ 3.545392093

  ↑   ↑   ↑   ↑   ↑

Index   *MATH*   *calculator*   *Enter*   *answer*
        menu   *shows 4*$\sqrt[x]{}$ *radicand*   *displayed*

Exercises

Evaluate each of the following.

1. $\sqrt[3]{343}$ 7   **2.** $\sqrt[4]{1296}$ 6   **3.** $\sqrt[4]{76}$ ≈ 2.952591724

4. $\sqrt[3]{2744}$ 14   **5.** $\sqrt[3]{-512}$ -8 **6.** $\sqrt[5]{-1000}$ ≈ -3.981071706

2   Simplify Cube and Fourth Roots

The product rule used in simplifying square roots can be expanded to indices greater than 2.

Product Rule for Radicals

$$\sqrt[n]{a}\,\sqrt[n]{b} = \sqrt[n]{ab}, \qquad \text{for } a \geq 0, b \geq 0 \text{ and } n \text{ is even}$$

$$\sqrt[n]{a}\,\sqrt[n]{b} = \sqrt[n]{ab}, \qquad \text{when } n \text{ is odd}$$

To simplify a cube root whose radicand is a constant, write the radicand as the product of a perfect cube and another number. Then simplify, using the product rule.

EXAMPLE 2   Simplify. **a)** $\sqrt[3]{32}$ **b)** $\sqrt[3]{54}$ **c)** $\sqrt[4]{48}$

Solution   **a)** Eight is a perfect cube that is a factor of the radicand, 32. Therefore, we simplify as follows:

$$\sqrt[3]{32} = \sqrt[3]{8 \cdot 4}$$
$$= \sqrt[3]{8}\,\sqrt[3]{4} \qquad \textit{Product rule}$$
$$= 2\sqrt[3]{4} \qquad \textit{Simplify.}$$

b) Twenty-seven is a perfect cube factor of 54. Therefore we simplify as follows.

$$\sqrt[3]{54} = \sqrt[3]{27 \cdot 2} = \sqrt[3]{27}\,\sqrt[3]{2} = 3\sqrt[3]{2}$$

c) Write $\sqrt[4]{48}$ as a product of a perfect fourth power and another number, then simplify. From the listing above, we see that 16 is a perfect fourth power. Since 16 is a factor of 48, we simplify as follows:

$$\sqrt[4]{48} = \sqrt[4]{16 \cdot 3}$$
$$= \sqrt[4]{16}\,\sqrt[4]{3} \qquad \textit{Product rule}$$
$$= 2\sqrt[4]{3} \qquad \textit{Simplify.}$$ ✳

NOW TRY EXERCISE 27

3   Write Radical Expressions in Exponential Form

A radical expression can be written in **exponential form** by using the following rule.

$$\sqrt[n]{a} = a^{1/n}, \qquad a \geq 0 \text{ and } n \text{ is even} \qquad \text{Rule 4}$$

$$\sqrt[n]{a} = a^{1/n}, \qquad \text{when } n \text{ is odd}$$

Examples

$$\sqrt{8} = 8^{1/2} \qquad\qquad \sqrt{x} = x^{1/2}$$
$$\sqrt[3]{4} = 4^{1/3} \qquad\qquad \sqrt[4]{5} = 5^{1/4}$$
$$\sqrt[3]{x} = x^{1/3} \qquad\qquad \sqrt[4]{y} = y^{1/4}$$
$$\sqrt[3]{5z^2} = (5z^2)^{1/3} \qquad \sqrt[4]{3y^2} = (3y^2)^{1/4}$$

Notice $\sqrt{8} = 8^{1/2}$ and $\sqrt{x} = x^{1/2}$, which is consistent with what we learned in Section 9.1. This concept can be expanded as follows.

Power   Index

$$\sqrt[n]{a^m} = (\sqrt[n]{a})^m = a^{m/n}, \text{ for } a \geq 0 \text{ and } m \text{ and } n \text{ integers} \qquad \text{Rule 5}$$

As long as the radicand is nonnegative, we can change from one form to another.

Examples

$$\sqrt[3]{27^4} = (\sqrt[3]{27})^4 = 3^4 = 81 \qquad\qquad \sqrt[3]{x^3} = x^{3/3} = x^1 = x$$
$$8^{2/3} = (\sqrt[3]{8})^2 = 2^2 = 4 \qquad\qquad \sqrt[4]{y^8} = y^{8/4} = y^2$$

EXAMPLE 3 Write each radical in exponential form.
a) $\sqrt[3]{a^5}$ **b)** $\sqrt[4]{y^7}$ **c)** $\sqrt[4]{x^{13}}$

Solution **a)** $\sqrt[3]{a^5} = a^{5/3}$ **b)** $\sqrt[4]{y^7} = y^{7/4}$ **c)** $\sqrt[4]{x^{13}} = x^{13/4}$

EXAMPLE 4 Simplify. **a)** $\sqrt[4]{x^{12}}$ **b)** $\sqrt[3]{y^{21}}$

Solution Write each radical expression in exponential form, then simplify.

NOW TRY EXERCISE 41 **a)** $\sqrt[4]{x^{12}} = x^{12/4} = x^3$ **b)** $\sqrt[3]{y^{21}} = y^{21/3} = y^7$

The rules of exponents discussed in Sections 4.1 and 4.2 also apply when the exponents are fractions. In Example 5c) we use the negative exponent rule and in Example 8 we use the product and power rules with fractional exponents.

EXAMPLE 5 Evaluate. **a)** $8^{4/3}$ **b)** $16^{5/4}$ **c)** $8^{-2/3}$

Solution To evaluate, we write each exponential expression in radical form.
a) $8^{4/3} = (\sqrt[3]{8})^4 = 2^4 = 16$ **b)** $16^{5/4} = (\sqrt[4]{16})^5 = 2^5 = 32$

c) Recall from Section 4.2 that $x^{-m} = \dfrac{1}{x^m}$. Thus,

NOW TRY EXERCISE 67
$$8^{-2/3} = \frac{1}{8^{2/3}} = \frac{1}{(\sqrt[3]{8})^2} = \frac{1}{2^2} = \frac{1}{4}$$

AVOIDING COMMON ERRORS

Students may make mistakes simplifying expressions that contain negative exponents. Be careful when working such problems. The following is a common error.

CORRECT

$$27^{-2/3} = \frac{1}{27^{2/3}}$$

INCORRECT

$$27^{-2/3} = -27^{2/3}$$

The expression $27^{-2/3}$ simplifies to $\frac{1}{9}$. Can you show how?

EXAMPLE 6   Simplify. **a)** $\sqrt[4]{16^2}$ **b)** $\sqrt[6]{27^2}$

Solution   Write each expression in exponential form, then simplify.

a) $\sqrt[4]{16^2} = 16^{2/4} = 16^{1/2} = \sqrt{16} = 4$

NOW TRY EXERCISE 79   **b)** $\sqrt[6]{27^2} = 27^{2/6} = 27^{1/3} = \sqrt[3]{27} = 3$

EXAMPLE 7   Simplify and write the answer in radical form.

a) $\sqrt[6]{y^3}$ **b)** $\sqrt[9]{z^3}$

Solution   Write the expression in exponential form, then simplify.

a) $\sqrt[6]{y^3} = y^{3/6} = y^{1/2} = \sqrt{y}$

NOW TRY EXERCISE 51   **b)** $\sqrt[9]{z^3} = z^{3/9} = z^{1/3} = \sqrt[3]{z}$

EXAMPLE 8   Simplify. **a)** $\sqrt{x} \cdot \sqrt[4]{x}$.   **b)** $\left(\sqrt[4]{a^2}\right)^8$.

Solution   To simplify, we change each radical expression to exponential form, then apply the rules of exponents.

a) $\sqrt{x} \cdot \sqrt[4]{x} = x^{1/2} \cdot x^{1/4}$

$= x^{(1/2)+(1/4)}$

$= x^{(2/4)+(1/4)}$

NOW TRY EXERCISE 89 $= x^{3/4}$ (or $\sqrt[4]{x^3}$ in radical form)

b) $\left(\sqrt[4]{a^2}\right)^8 = \left(a^{2/4}\right)^8$

$= \left(a^{1/2}\right)^8$

$= a^4$

Using Your Calculator

In Chapter 1, page 73, we learned how to use the $\boxed{y^x}$ key or $\boxed{\wedge}$ key to raise a value to an integer exponent greater than 2. You can also use these keys if the exponent is a fraction. We demonstrate how in the following examples.

EXAMPLE 1   Evaluate $27^{2/3}$.

Solution   Scientific calculator:

$27\;\boxed{y^x}\;\boxed{(}\;2\;\boxed{\div}\;3\;\boxed{)}\;\boxed{=}\;9$ ← *Answer displayed*

Graphing calculator:

$27\;\boxed{\wedge}\;\boxed{(}\;2\;\boxed{\div}\;3\;\boxed{)}\;\boxed{\text{ENTER}}\;9$ ← *Answer displayed*

The answer is 9. Thus, $27^{2/3} = 9$.

EXAMPLE 2   Evaluate $16^{-3/4}$.

Solution   Scientific calculator:

$16\;\boxed{y^x}\;\boxed{(}\;3\;\boxed{+/-}\;\boxed{\div}\;4\;\boxed{)}\;\boxed{=}\;.125$ ← *Answer displayed*

Graphing calculator:

$16\;\boxed{\wedge}\;\boxed{(}\;\boxed{(-)}\;3\;\boxed{\div}\;4\;\boxed{)}\;\boxed{\text{ENTER}}\;.125$ ← *Answer displayed*

The answer is 0.125. Thus, $16^{-3/4} = 0.125$.

Exercises

Use your calculator to evaluate each expression.

1. $4^{3/2}$ 8   **2.** $64^{4/3}$ 256   **3.** $125^{-2/3}$ .04   **4.** $81^{-3/4}$ $\overline{.037}$

This section was meant to give you a brief introduction to roots other than square roots. If you take a course in intermediate algebra, you may study these concepts in more depth.

3. Write the radicand as the product of a perfect cube and another number.
4. Write the radicand as a product of a perfect fourth power and another number.   **5. a)** and **6. a)** Answers will vary.

Exercise Set 9.7

In this exercise set, assume all variables represent nonnegative real numbers.

Concept/Writing Exercises

1. Write how the following radicals would be read.

   a) $\sqrt{8}$ the square root of 8

   b) $\sqrt[3]{8}$ the cube root of 8

   c) $\sqrt[4]{8}$ the fourth root of 8

2. In your own words, describe a perfect cube and a perfect fourth power. Answers will vary.

3. How do you simplify a cube root whose radicand is a constant?

4. How do you simplify a fourth root whose radicand is a constant?

5. **a)** In your own words, explain how to change an expression written in radical form to exponential form.

   b) Using the procedure you wrote in part **a)**, write $\sqrt[3]{y^7}$ in exponential form. $y^{7/3}$

6. **a)** In your own words, explain how to change an expression written in exponential form to radical form.

   b) Using the procedure you wrote in part **a)**, write $x^{5/4}$ in radical form. $\sqrt[4]{x^5}$

7. Will the odd root of a positive number always be a positive number? Explain. yes

8. Can the odd root of a negative number ever be a positive number? Explain. no

Practice the Skills

Evaluate.

9. $\sqrt[3]{8}$ 2

10. $\sqrt[3]{27}$ 3

11. $\sqrt[3]{-27}$ −3

12. $\sqrt[3]{-125}$ −5

13. $\sqrt[4]{16}$ 2

14. $\sqrt[3]{216}$ 6

15. $\sqrt[4]{81}$ 3

16. $\sqrt[4]{1}$ 1

17. $\sqrt[3]{-1}$ −1

18. $\sqrt[3]{1}$ 1

19. $\sqrt[3]{-1000}$ −10

20. $\sqrt[4]{625}$ 5

Simplify.

21. $\sqrt[3]{40}$ $2\sqrt[3]{5}$

22. $\sqrt[3]{48}$ $2\sqrt[3]{6}$

23. $\sqrt[3]{16}$ $2\sqrt[3]{2}$

24. $\sqrt[3]{24}$ $2\sqrt[3]{3}$

25. $\sqrt[3]{108}$ $3\sqrt[3]{4}$

26. $\sqrt[3]{128}$ $4\sqrt[3]{2}$

27. $\sqrt[3]{32}$ $2\sqrt[3]{2}$

28. $\sqrt[3]{3000}$ $10\sqrt[3]{3}$

29. $\sqrt[4]{1250}$ $5\sqrt[4]{2}$

30. $\sqrt[3]{250}$ $5\sqrt[3]{2}$

Write each radical in exponential form.

31. $\sqrt[3]{x^7}$ $x^{7/3}$

32. $\sqrt[4]{x^3}$ $x^{3/4}$

33. $\sqrt[5]{a^2}$ $a^{2/5}$

34. $\sqrt[5]{x^{11}}$ $x^{11/5}$

35. $\sqrt[4]{y^{15}}$ $y^{15/4}$

36. $\sqrt[4]{w^{13}}$ $w^{13/4}$

37. $\sqrt[3]{y^8}$ $y^{8/3}$

38. $\sqrt[4]{x^5}$ $x^{5/4}$

Write each radical in exponential form and then simplify.

39. $\sqrt[4]{x^4}$ x

40. $\sqrt[3]{y^6}$ y^2

41. $\sqrt[3]{y^{21}}$ y^7

42. $\sqrt[4]{y^{20}}$ y^5

43. $\sqrt[3]{m^{18}}$ m^6

44. $\sqrt[3]{z^{21}}$ z^7

45. $\sqrt[4]{a^{60}}$ a^{15}

46. $\sqrt[3]{x^3}$ x

47. $\sqrt[3]{x^{15}}$ x^5

48. $\sqrt[3]{r^{30}}$ r^{10}

Write each radical in exponential form and then simplify. Write the answer in simplified radical form.

49. $\sqrt[4]{m^2}$ $\sqrt{m}$

50. $\sqrt[4]{y^2}$ $\sqrt{y}$

51. $\sqrt[6]{t^3}$ $\sqrt{t}$

52. $\sqrt[6]{p^3}$ $\sqrt{p}$

53. $\sqrt[9]{x^3}$ $\sqrt[3]{x}$

54. $\sqrt[9]{r^3}$ $\sqrt[3]{r}$

55. $\sqrt[8]{w^4}$ $\sqrt{w}$

56. $\sqrt[8]{z^4}$ $\sqrt{z}$

57. $\sqrt[6]{x^4}$ $\sqrt[3]{x^2}$

58. $\sqrt[6]{y^4}$ $\sqrt[3]{y^2}$

59. $\sqrt[8]{z^6}$ $\sqrt[4]{z^3}$

60. $\sqrt[9]{n^6}$ $\sqrt[3]{n^2}$

Write each expression in radical form and then evaluate.

61. $27^{2/3}$ 9
62. $8^{4/3}$ 16
63. $216^{2/3}$ 36
64. $64^{3/2}$ 512

65. $1^{2/3}$ 1
66. $49^{3/2}$ 343
67. $9^{3/2}$ 27
68. $64^{2/3}$ 16

69. $27^{4/3}$ 81
70. $25^{5/2}$ 3125
71. $256^{5/4}$ 1024
72. $256^{2/4}$ 16

🔒 **73.** $8^{-1/3}$ $\frac{1}{2}$
74. $16^{-3/4}$ $\frac{1}{8}$
75. $27^{-2/3}$ $\frac{1}{9}$
76. $64^{-2/3}$ $\frac{1}{16}$

Write each radical in exponential form and then simplify.

77. $\sqrt[4]{4^2}$ 2
78. $\sqrt[4]{25^2}$ 5
79. $\sqrt[6]{9^3}$ 3
80. $\sqrt[6]{16^3}$ 4

81. $\sqrt[8]{36^4}$ 6
82. $\sqrt[8]{49^4}$ 7
83. $\sqrt[4]{5^8}$ 25
84. $\sqrt[4]{3^8}$ 9

85. $\sqrt[4]{2^{12}}$ 8
86. $\sqrt[6]{3^{18}}$ 27
87. $\sqrt[5]{4^{10}}$ 16
88. $\sqrt[5]{6^{15}}$ 216

Write each radical in exponential form and then simplify. Write the answer in exponential form.

🔒 **89.** $\sqrt[4]{x}\cdot\sqrt[4]{x^3}$ x
90. $\sqrt[3]{x}\cdot\sqrt[4]{x}$ $x^{7/12}$
91. $\sqrt[4]{t^2}\cdot\sqrt[4]{t^2}$ t
92. $\sqrt[3]{x^4}\cdot\sqrt[3]{x^5}$ x^3

93. $\left(\sqrt[3]{r^4}\right)^6$ r^8
94. $\left(\sqrt[4]{x^3}\right)^4$ x^3
95. $\left(\sqrt[4]{a^2}\right)^4$ a^2
96. $\left(\sqrt[3]{x^6}\right)^2$ x^4

97. Show that $(\sqrt[3]{x})^2 = \sqrt[3]{x^2}$ for $x = 8$. both equal 4
98. Show that $(\sqrt[4]{x})^3 = \sqrt[4]{x^3}$ for $x = 16$. both equal 8

Problem Solving

The product $\sqrt[3]{2}\cdot\sqrt[3]{2}$ is not an integer, but $\sqrt[3]{2}\cdot\sqrt[3]{2^2} = \sqrt[3]{2^3} = 2^{3/3} = 2$. Use this information to determine by what radical expression you can multiply each radical so that the result is an integer.

99. $\sqrt[3]{4}$ $\sqrt[3]{4^2}$
100. $\sqrt[3]{7^2}$ $\sqrt[3]{7}$
101. $\sqrt[3]{6}$ $\sqrt[3]{6^2}$

102. $\sqrt[4]{2^3}$ $\sqrt[4]{2}$
103. $\sqrt[4]{5}$ $\sqrt[4]{5^3}$
104. $\sqrt[4]{6^2}$ $\sqrt[4]{6^2}$

Fill in the shaded area(s) in each equation to make a true statement.

105. $\sqrt[\blacksquare]{5^2}\cdot\sqrt[\blacksquare]{5} = 5$ 3
106. $\sqrt[\blacksquare]{7^2}\cdot\sqrt[\blacksquare]{7^2} = 7$ 4
107. $\sqrt[3]{6^\blacksquare}\cdot\sqrt[3]{6^2} = 6$ 1
108. $\sqrt[5]{2^3}\cdot\sqrt[5]{2^\blacksquare} = 2$ 2

Challenge Problems

Simplify.

109. $\sqrt[3]{xy}\cdot\sqrt[3]{x^2y^2}$ xy
110. $\sqrt[4]{3x^2y}\cdot\sqrt[4]{27x^6y^3}$ $3x^2y$
111. $\sqrt[4]{32} - \sqrt[4]{2}$ $\sqrt[4]{2}$
112. $\sqrt[3]{3x^3y} + \sqrt[3]{24x^3y}$ $3x\sqrt[3]{3y}$

113. a) Explain why multiplying both the numerator and denominator of $\dfrac{1}{\sqrt[3]{2}}$ by $\sqrt[3]{2^2}$ will give an integer in the denominator. $\sqrt[3]{2^3} = 2$

b) Rationalize the denominator of $\dfrac{1}{\sqrt[3]{2}}$ by multiplying both the numerator and denominator by $\sqrt[3]{2^2}$. $\dfrac{\sqrt[3]{4}}{2}$

Cumulative Review Exercises

116. $m = \dfrac{2}{3}$, y-intercept: $\left(0, -\dfrac{4}{3}\right)$

[1.9] **114.** Evaluate $-x^2 + 4xy - 6$ when $x = 2$ and $y = -4$. -42

[7.2] **116.** Determine the slope and the y-intercept of the graph of the equation $2x - 3y = 4$.

[5.4] **115.** Factor $3x^2 - 28x + 32$. $(3x - 4)(x - 8)$

[9.4] **117.** Simplify $\sqrt{\dfrac{64x^3y^7}{2x^4}}$. $\dfrac{4y^3\sqrt{2xy}}{x}$

CHAPTER SUMMARY

Key Words and Phrases

9.1
Imaginary number
Index of a radical
Irrational number
Negative square root
Perfect square
Perfect square factor
Positive or principal
  square root
Radical expression
Radical sign
Radicand
Rational number

Writing a radical
  expression in
  exponential form
Writing an expression in
  exponential form as a
  radical expression

9.2
Perfect square
Perfect square factor
Product rule for square
  roots
Simplify a square root

9.3
Like square roots
Unlike square roots

9.4
Conjugate of a binomial
Quotient rule for
  radicals
Rationalize a
  denominator
Simplify a square root

9.5
Extraneous root
Isolating the square root
Radical equation

9.6
Distance formula

9.7
Cube root
Exponential form
Fourth root
Perfect cube
Perfect fourth power

IMPORTANT FACTS

Numbers that are perfect squares: $1, 4, 9, 16, 25, 36, 49, 64, \ldots$

Numbers that are perfect cubes: $1, 8, 27, 64, 125, 216, 343, 512, \ldots$

Product rule for square roots: $\sqrt{a} \cdot \sqrt{b} = \sqrt{ab}, a \geq 0, b \geq 0$
$\sqrt{a^{2 \cdot n}} = a^n, a \geq 0$
$\sqrt{a^2} = a, a \geq 0$

Quotient rule for square roots: $\dfrac{\sqrt{a}}{\sqrt{b}} = \sqrt{\dfrac{a}{b}}, a \geq 0, b > 0$

$\sqrt[n]{a} = a^{1/n}, a \geq 0$ when n is even; $\sqrt[n]{a} = a^{1/n},$ when n is odd
$\sqrt[n]{a^m} = \left(\sqrt[n]{a}\right)^m = a^{m/n}, a \geq 0$

Pythagorean Theorem: $a^2 + b^2 = c^2$

Distance formula: $d = \sqrt{(x_2 - x_1)^2 + (y_2 - y_1)^2}$

Product rule for radicals: $\sqrt[n]{a} \cdot \sqrt[n]{b} = \sqrt[n]{ab},$ for $a \geq 0, b \geq 0,$ and n is even
$\sqrt[n]{a} \cdot \sqrt[n]{b} = \sqrt[n]{ab},$ when n is odd

Chapter Review Exercises

[9.1] Evaluate.

1. $\sqrt{81}$ 9

2. $\sqrt{49}$ 7

3. $-\sqrt{64}$ -8

Write in exponential form.

4. $\sqrt{5}$ $5^{1/2}$

5. $\sqrt{26x}$ $(26x)^{1/2}$

6. $\sqrt{13x^2y}$ $(13x^2y)^{1/2}$

[9.2] Simplify.

7. $\sqrt{32}$ $4\sqrt{2}$

8. $\sqrt{44}$ $2\sqrt{11}$

9. $\sqrt{27x^7y^4}$ $3x^3y^2\sqrt{3x}$

10. $\sqrt{125x^4y^6}$ $5x^2y^3\sqrt{5}$

11. $\sqrt{60ab^5c^4}$ $2b^2c^2\sqrt{15ab}$

12. $\sqrt{72a^2b^2c^7}$ $6abc^3\sqrt{2c}$

Simplify.

13. $\sqrt{72}\,\sqrt{20}$ $12\sqrt{10}$

14. $\sqrt{7y}\,\sqrt{7y}$ $7y$

15. $\sqrt{18x}\,\sqrt{2xy}$ $6x\sqrt{y}$

16. $\sqrt{25x^2y}\,\sqrt{3y}$ $5xy\sqrt{3}$

17. $\sqrt{12a^3b^4}\,\sqrt{3b^4}$ $6ab^4\sqrt{a}$

18. $\sqrt{5ab^3}\,\sqrt{20ab^4}$ $10ab^3\sqrt{b}$

[9.3] Simplify.

19. $5\sqrt{3} - 4\sqrt{3}$ $\sqrt{3}$

20. $4\sqrt{5} - 7\sqrt{5} - 3\sqrt{5}$ $-6\sqrt{5}$

21. $3\sqrt{x} - 5\sqrt{x}$ $-2\sqrt{x}$

22. $\sqrt{k} + 3\sqrt{k} - 4\sqrt{k}$ 0

23. $\sqrt{18} - \sqrt{27}$ $3\sqrt{2} - 3\sqrt{3}$

24. $7\sqrt{40} - 2\sqrt{10}$ $12\sqrt{10}$

25. $2\sqrt{98} - 4\sqrt{72}$ $-10\sqrt{2}$

26. $7\sqrt{50} + 2\sqrt{18} - 4\sqrt{32}$ $25\sqrt{2}$

Multiply.   **35.** $m^2 - 3m\sqrt{r} - 10r$ **36.** $3t + 5s\sqrt{t} - 2s^2$ **37.** $15m + 5\sqrt{5mn} - 2n$ **38.** $14 - 9\sqrt{7p} + 9p$

27. $\sqrt{5}(2 + \sqrt{5})$ $2\sqrt{5} + 5$

28. $\sqrt{2}(\sqrt{2} + 3)$ $2 + 3\sqrt{2}$

29. $\sqrt{y}(x - 2\sqrt{y})$ $x\sqrt{y} - 2y$

30. $5a(3a + \sqrt{2a})$ $15a^2 + 5a\sqrt{2a}$

31. $(\sqrt{3} - 2)(\sqrt{3} + 2)$ -1

32. $(9 - \sqrt{3})(9 + \sqrt{3})$ 78

33. $(x - 2\sqrt{y})(x + 2\sqrt{y})$ $x^2 - 4y$

34. $(\sqrt{c} - 2\sqrt{d})(\sqrt{c} + 2\sqrt{d})$ $c - 4d$

35. $(m + 2\sqrt{r})(m - 5\sqrt{r})$

36. $(\sqrt{t} + 2s)(3\sqrt{t} - s)$

37. $(\sqrt{5m} + 2\sqrt{n})(3\sqrt{5m} - \sqrt{n})$

38. $(\sqrt{7} - 3\sqrt{p})(2\sqrt{7} - 3\sqrt{p})$

[9.4] Simplify.

39. $\dfrac{\sqrt{32}}{\sqrt{2}}$ 4

40. $\sqrt{\dfrac{10}{490}}$ $\dfrac{1}{7}$

41. $\sqrt{\dfrac{7}{28}}$ $\dfrac{1}{2}$

42. $\dfrac{3}{\sqrt{5}}$ $\dfrac{3\sqrt{5}}{5}$

43. $\sqrt{\dfrac{n}{7}}$ $\dfrac{\sqrt{7n}}{7}$

44. $\sqrt{\dfrac{5a}{12}}$ $\dfrac{\sqrt{15a}}{6}$

45. $\sqrt{\dfrac{x^2}{3}}$ $\dfrac{x\sqrt{3}}{3}$

46. $\sqrt{\dfrac{z^4}{8}}$ $\dfrac{z^2\sqrt{2}}{4}$

47. $\sqrt{\dfrac{21x^3y^7}{3x^3y^3}}$ $y^2\sqrt{7}$

48. $\sqrt{\dfrac{30x^4y}{15x^2y^4}}$ $\dfrac{x\sqrt{2y}}{y^2}$

49. $\dfrac{\sqrt{60}}{\sqrt{27a^3b^2}}$ $\dfrac{2\sqrt{5a}}{3a^2b}$

50. $\dfrac{\sqrt{2a^4bc^4}}{\sqrt{7a^5bc^2}}$ $\dfrac{c\sqrt{14a}}{7a}$

Simplify.

51. $\dfrac{3}{1 - \sqrt{6}}$ $\dfrac{-3 - 3\sqrt{6}}{5}$

52. $\dfrac{5}{3 - \sqrt{6}}$ $\dfrac{15 + 5\sqrt{6}}{3}$

53. $\dfrac{\sqrt{2}}{2 + \sqrt{y}}$ $\dfrac{2\sqrt{2} - \sqrt{2y}}{4 - y}$

54. $\dfrac{2}{\sqrt{x} - 5}$ $\dfrac{2\sqrt{x} + 10}{x - 25}$

55. $\dfrac{\sqrt{5}}{\sqrt{x} + \sqrt{3}}$ $\dfrac{\sqrt{5x} - \sqrt{15}}{x - 3}$

56. $\dfrac{\sqrt{7}}{\sqrt{5} - x}$ $\dfrac{\sqrt{35} + x\sqrt{7}}{5 - x^2}$

[9.5] Solve.

57. $\sqrt{x} = 6$ 36

58. $\sqrt{g} = -5$ no solution

59. $\sqrt{h - 5} = 3$ 14

60. $\sqrt{3x + 1} = 5$ 8

61. $\sqrt{5x + 6} = \sqrt{4x + 8}$ 2

62. $4\sqrt{x} - x = 4$ 4

63. $\sqrt{x^2 - 3} = x - 1$ 2

64. $\sqrt{4x + 8} - \sqrt{7x - 13} = 0$ 7

65. $\sqrt{4p + 1} = 2p - 1$ 2

[9.6] Find each length indicated by x. Round answers to nearest hundredth.

66.

67.

68.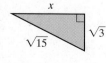

69.

67. $\sqrt{125} \approx 11.18$ **68.** $\sqrt{12} \approx 3.46$ **69.** $\sqrt{61} \approx 7.81$

71. $\sqrt{261} \approx 16.16$ in.

70. Ladder on a Cottage Adam Kurtz leans a 12-foot ladder against his grandfather's cottage. If the base of the ladder is 3 feet from the cottage how high is the ladder on the cottage? $\sqrt{135} \approx 11.62$ ft

71. Diagonal of a Rectangle Find the diagonal of a rectangle of length 15 inches and width 6 inches.

72. Find the straight-line distance between the points $(4, -3)$ and $(1, 7)$. $\sqrt{109} \approx 10.44$

73. Find the length of the line segment between the points $(6, 5)$ and $(-6, 8)$. $\sqrt{153} \approx 12.37$

74. Right-of-Way Sign An **equilateral triangle** is a triangle with three sides of the same length. A yield-right-of-way sign in the United States is an **equilateral triangle** with each side, s, measuring 36 inches. Use the formula for the area of an equilateral triangle

$$\text{Area} = \frac{s^2\sqrt{3}}{4} \quad 561.18 \text{ sq in.}$$

to find an approximation of the area of a yield sign. Round the area to the nearest hundredth.

75. Distance Seen Ken Jameston, who owns a farm outside Osceola, Iowa, enjoys climbing the ladder on the side of his silo and looking toward the horizon. After doing some research, he found that the farthest distance, in miles, he could see can be approximated by the formula $distance = \sqrt{(3/2)h}$, where h represents the height of his vantage point, measured in feet. Approximately how far can Ken see, correct to the nearest hundredth of a mile, if he climbs 40 feet? $\sqrt{60} \approx 7.75$ mi

See Exercise 75.

76. Luxor Hotel The Luxor Hotel in Las Vegas, Nevada, is the world's second largest hotel. It is a pyramid with a square base and a height of 350 feet. The length, s, of each side of the base of any pyramid with a square base can be found by the formula $s = \sqrt{(3V)/h}$, where V represents the volume, in cubic feet, of the pyramid and h represents the height, in feet. The volume of the Luxor is approximately 48,686,866.67 cubic feet. Find the length of each side of the square base of the Luxor Hotel. ≈ 646 feet

[9.7] *Evaluate.*

77. $\sqrt[3]{64}$ 4

78. $\sqrt[3]{-64}$ -4

79. $\sqrt[4]{16}$ 2

80. $\sqrt[4]{81}$ 3

Simplify.

81. $\sqrt[3]{64}$ 4

82. $\sqrt[3]{-8}$ -2

83. $\sqrt[4]{32}$ $2\sqrt[4]{2}$

84. $\sqrt[3]{48}$ $2\sqrt[3]{6}$

85. $\sqrt[3]{54}$ $3\sqrt[3]{2}$

86. $\sqrt[4]{80}$ $2\sqrt[4]{5}$

87. $\sqrt[3]{x^{21}}$ x^7

88. $\sqrt[3]{s^{30}}$ s^{10}

Evaluate.

89. $1^{2/3}$ 1

90. $25^{1/2}$ 5

91. $27^{-2/3}$ $\frac{1}{9}$

92. $64^{2/3}$ 16

93. $125^{-4/3}$ $\frac{1}{625}$

94. $49^{3/2}$ 343

Write in exponential form.

95. $\sqrt[3]{z^{11}}$ $z^{11/3}$

96. $\sqrt[3]{x^8}$ $x^{8/3}$

97. $\sqrt[4]{y^9}$ $y^{9/4}$

98. $\sqrt{x^5}$ $x^{5/2}$

99. $\sqrt{y^3}$ $y^{3/2}$

100. $\sqrt[4]{m^6}$ $m^{6/4} = m^{3/2}$

Simplify.

101. $\sqrt[3]{x} \cdot \sqrt[3]{x^2}$ x

102. $\sqrt[3]{x} \cdot \sqrt[3]{x}$ $\sqrt[3]{x^2}$

103. $\sqrt[3]{a^5} \cdot \sqrt[3]{a^7}$ a^4

104. $\sqrt[4]{x^2} \cdot \sqrt[4]{x^6}$ x^2

105. $\left(\sqrt[3]{q^3}\right)^3$ q^3

106. $\left(\sqrt[4]{d}\right)^4$ d

107. $\left(\sqrt[3]{x^8}\right)^3$ x^6

108. $\left(\sqrt[4]{x^3}\right)^8$ x^6

Chapter Practice Test

1. Write $\sqrt{3x}$ exponential form. $(3x)^{1/2}$

2. Write $x^{2/3}$ in radical form. $\sqrt[3]{x^2}$

Simplify.

3. $\sqrt{(y-4)^2}$ $y-4$

4. $\sqrt{90}$ $3\sqrt{10}$

5. $\sqrt{12x^2}$ $2x\sqrt{3}$

6. $\sqrt{50x^7y^3}$ $5x^3y\sqrt{2xy}$

7. $\sqrt{8x^2y}\,\sqrt{10xy}$ $4xy\sqrt{5x}$

8. $\sqrt{15xy^2}\,\sqrt{5x^3y^3}$ $5x^2y^2\sqrt{3y}$

9. $\sqrt{\dfrac{5}{125}}$ $\dfrac{1}{5}$

10. $\dfrac{\sqrt{7c^4d}}{\sqrt{7d^3}}$ $\dfrac{c^2}{d}$

11. $\dfrac{1}{\sqrt{6}}$ $\dfrac{\sqrt{6}}{6}$

12. $\sqrt{\dfrac{9r}{5}}$ $\dfrac{3\sqrt{5r}}{5}$

13. $\sqrt{\dfrac{40x^2y^5}{6x^3y^7}}$ $\dfrac{2\sqrt{15x}}{3xy}$

14. $\dfrac{3}{2-\sqrt{7}}$ $-2-\sqrt{7}$

15. $\dfrac{6}{\sqrt{x}-3}$ $\dfrac{6\sqrt{x}+18}{x-9}$

16. $\sqrt{48}+\sqrt{75}+2\sqrt{3}$ $11\sqrt{3}$

17. $7\sqrt{y}-3\sqrt{y}-\sqrt{y}$ $3\sqrt{y}$

Solve.

18. $\sqrt{x-8}=4$ 24

19. $2\sqrt{x-4}+4=x$ $4,8$

Solve.

20. Find the value of x in the right triangle. $\sqrt{106}\approx 10.30$

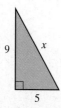

21. Find the length of the line segment between the points $(3,-2)$ and $(-4,-5)$. $\sqrt{58}\approx 7.62$

22. Evaluate $27^{-4/3}$. $1/81$

23. Simplify $\sqrt[4]{x^5}\cdot\sqrt[4]{x^7}$. x^3

24. *Side of a Square* Find the side of a square whose area is 121 square meters. 11 m

25. *Dropping Eggs* Mary Ellen Baker and her brother, Michael, were in the tree house in their backyard. They were dropping raw eggs (much to their mother's dismay)

from a ledge in the tree house 10 feet in the air. At what velocity did the eggs hit the ground? Use the formula $v=\sqrt{2gh}$ with $g=32$ feet per second squared. The velocity will be in feet per second. $\sqrt{640}\approx 25.30$ ft/sec

1. a) $-5,735,4$ **b)** $735,4$ **c)** $-5,735,0.5,4,\frac{1}{2}$ **d)** $\sqrt{12}$ **e)** all   **4.** $x>\frac{27}{11}$,

Cumulative Review Test

Take the following test and check your answers with those that appear at the end of the test. Review any questions that you answered incorrectly. The section and objective where the material was covered are indicated after the answer.

1. Consider the set of numbers

$$\left\{-5,\sqrt{12},735,0.5,4,\tfrac{1}{2}\right\}.$$

List those that are **a)** integers; **b)** whole numbers; **c)** rational numbers; **d)** irrational numbers; **e)** real numbers.

2. Evaluate $7a^2-4b^2+2ab$ when $a=-3$ and $b=2$. 35

3. Solve $-7(3-x)=4(x+2)-3x$. $\dfrac{29}{6}$

4. Solve the inequality $3(x+2)>5-4(2x-7)$ and graph the solution on a number line.

5. Factor $3x^3+x^2+6x+2$ $(3x+1)(x^2+2)$

6. Factor $2x^2-17x+21$ $(2x-3)(x-7)$

7. Use factoring to solve $r^2 - 12r = 0$ $0, 12$

8. Simplify $\dfrac{4a^3b^{-5}}{28a^8b}$. $\dfrac{1}{7a^5b^6}$

9. Graph $4x - 6y = 24$. See graphing answer section, page G15.

10. Write the equation of the graph in the accompanying figure. $y = 3x - 2$

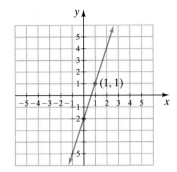

11. Find the equation of the line, in slope–intercept form, that has a slope of $\frac{2}{5}$ and goes through the point $(-5, 2)$. $y = \frac{2}{5}x + 4$

12. Solve the following system of equations.
$$-2x + 3y = 6$$
$$4x - 2y = -4 \quad (0, 2)$$

13. Simplify $\dfrac{y + 5}{8} + \dfrac{2y - 7}{8}$. $\dfrac{3y - 2}{8}$

14. Solve $\dfrac{1}{2} + \dfrac{1}{z} = 3$. $\dfrac{2}{5}$

15. Simplify $3\sqrt{11} - 4\sqrt{11}$. $-\sqrt{11}$

16. Simplify $\sqrt{\dfrac{3z}{28y}}$. $\dfrac{\sqrt{21yz}}{14y}$

17. Solve $\sqrt{x + 5} = 6$. 31

18. *Making a Fruitcake* Susan Effing is planning her first holiday meal. She has decided to make her grandmother's famous fruitcake. The recipe calls for 10 cups of flour and will make an 11-pound fruitcake. She would like to make a 3-pound fruitcake. How many cups of flour should Susan use? $2\frac{8}{11}$ cups

19. *Special Promotion* During a special promotion, the rooms in the Hyatt Hotel in Oklahoma City cost \$69. If this is a 40% discount off their regular room rate, determine the Hyatt's regular room rate. \$115

20. *Distance Between Cities* Alan Heard is in the process of getting his pilot's license. As part of his training, he flew his instructor from a small airport near Pasadena, California, to another small airport near San Diego at an average speed of 100 miles per hour. On the return trip, his average speed was 125 miles per hour. How far apart were the two airports if Alan's total flying time was 2 hours? ≈ 111.1 mi

Answers to Cumulative Review Test

1. a) $-5, 735, 4$ **b)** $735, 4$ **c)** $-5, 735, 0.5, 4, \frac{1}{2}$ **d)** $\sqrt{12}$ **e)** $-5, \sqrt{12}, 735, 0.5, 4, \frac{1}{2}$; [Sec. 1.4, Obj. 2]

2. 35; [Sec. 1.9, Obj. 6] **3.** $\frac{29}{6}$; [Sec. 2.5, Obj. 1] **4.** $x > \frac{27}{11}$, ; [Sec. 2.7, Obj. 1] **5.** $(3x + 1)(x^2 + 2)$;

[Sec. 5.2, Obj. 1] **6.** $(2x - 3)(x - 7)$; [Sec. 5.4, Obj. 1] **7.** $0, 12$; [Sec. 5.6, Obj. 2] **8.** $\dfrac{1}{7a^5b^6}$; [Sec. 4.2, Obj. 2]

9. 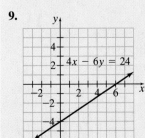 ; [Sec. 7.2, Obj. 3] **10.** $y = 3x - 2$; [Sec. 7.4, Obj. 3] **11.** $y = \dfrac{2}{5}x + 4$; [Sec. 7.4, Obj. 4]

12. $(0, 2)$; [Sec. 8.3, Obj. 1] **13.** $\dfrac{3y - 2}{8}$; [Sec. 6.3, Obj. 1] **14.** $\dfrac{2}{5}$; [Sec. 6.6, Obj. 2]

15. $-\sqrt{11}$; [Sec. 9.3, Obj. 1] **16.** $\dfrac{\sqrt{21yz}}{14y}$; [Sec. 9.4, Obj. 3] **17.** 31; [Sec. 9.5, Obj. 1]

18. $2\frac{8}{11}$ cups; [Sec. 2.6, Obj. 3] **19.** \$115; [Sec. 3.3, Obj. 4] **20.** ≈ 111.1 miles; [Sec. 6.7, Obj. 2]

Chapter 10

Quadratic Equations

We see objects projected upwards often but rarely think of the mathematics that describes the projected motion. For example, when a ball hits a bat, or a child kicks a football or a soccer ball, the ball is projected upwards. When a cannon ball is fired from a cannon, when a model rocket is launched, or when a ball is hit, an item is projected upwards. In Exercise 69 on page 637 and Exercise 63 on page 648 we describe the distance an object is above the ground at specific times after the object is projected upwards.

SSM   Study Guide   CD/Video   MathPro 4/5   PH Math Tutor Center   prenhall.com/Angel

A Look Ahead

Much of this book has dealt with solving linear equations. Quadratic equations are another important category of equations. Quadratic equations were introduced and solved by factoring in Section 5.6. You may wish to review that section now. Recall that quadratic equations are of the form $ax^2 + bx + c = 0$, where $a \neq 0$. Not every quadratic equation can be solved by factoring. In this chapter we present two additional methods to solve quadratic equations: completing the square and the quadratic formula.

The square root property is presented in Section 10.1. In Section 10.2 we solve quadratic equations by completing the square, which uses the square root property. The quadratic formula, discussed in Section 10.3, is most typically used when solving quadratic equations that cannot be factored. We graphed linear equations in Chapter 7. In Section 10.4 we graph quadratic equations. Section 10.4 is a very valuable section because it lays the groundwork for graphing cubic and higher-degree equations in other mathematics courses. In Section 10.5 we introduce the complex number system. If you take intermediate algebra or other mathematics courses you may discuss this topic in more depth.

10.1 THE SQUARE ROOT PROPERTY

1 Know that every positive real number has two square roots.

2 Solve quadratic equations using the square root property.

SSM Study Guide CD/Video

MathPro 4/5 PH Math Tutor Center prenhall.com/Angel

In Section 5.6 we solved quadratic equations by factoring. Recall that **quadratic equations** are equations of the form

$$ax^2 + bx + c = 0$$

where a, b, and c are real numbers, $a \neq 0$. A quadratic equation in this form is said to be in **standard form**. Solving quadratic equations by factoring is the preferred technique when the factors can be found quickly. To refresh your memory, below we will solve the equation $x^2 - 3x - 10 = 0$ by factoring. If you need further examples, review Section 5.6.

$$x^2 - 3x - 10 = 0$$
$$(x - 5)(x + 2) = 0$$
$$x - 5 = 0 \quad \text{or} \quad x + 2 = 0$$
$$x = 5 \qquad\qquad x = -2$$

The solutions to this equation are 5 and -2.

Not every quadratic equation can be factored easily, and many cannot be factored at all. In this chapter we give two techniques, completing the square and the quadratic formula, for solving quadratic equations that cannot be solved by factoring.

1 Know That Every Positive Real Number Has Two Square Roots

In Section 9.1 we stated that every positive number has two square roots. Thus far we have been using only the positive or principal square root. In this section we use both the positive and negative square roots of a number. For example, the positive square root of 49 is 7.

$$\sqrt{49} = 7$$

The negative square root of 49 is −7.

$$-\sqrt{49} = -7$$

Notice that $7 \cdot 7 = 49$ and $(-7)(-7) = 49$. The two square roots of 49 are $+7$ and -7. A convenient way to indicate the two square roots of a number is to use the plus or minus symbol, $\pm$. For example, the square roots of 49 can be indicated ± 7, read "plus or minus 7."

Number	Both Square Roots
64	± 8
100	± 10
3	$\pm\sqrt{3}$

The value of a number like $-\sqrt{5}$ can be found by finding the value of $\sqrt{5}$ on your calculator and then taking its opposite or negative value.

$$\sqrt{5} = 2.24 \quad \text{(rounded to the nearest hundredth)}$$
$$-\sqrt{5} = -2.24$$

Now consider the equation

$$x^2 = 49$$

We can see by substitution that this equation has two solutions, 7 and −7.

Check $x = 7$ $x = -7$

$x^2 = 49$ $x^2 = 49$

$7^2 \stackrel{?}{=} 49$ $(-7)^2 \stackrel{?}{=} 49$

$49 = 49$ *True* $49 = 49$ *True*

Therefore, the solutions to the equation $x^2 = 49$ are 7 and −7 (or ± 7).

2 Solve Quadratic Equations Using the Square Root Property

In general, for any quadratic equation of the form $x^2 = a$, we can use the **square root property** to obtain the solution.

Square Root Property

If $x^2 = a$, then $x = \sqrt{a}$ or $x = -\sqrt{a}$ (abbreviated $x = \pm\sqrt{a}$).

For example, if $x^2 = 7$, then by the square root property, $x = \sqrt{7}$ or $x = -\sqrt{7}$. We may also write $x = \pm\sqrt{7}$.

EXAMPLE 1   Solve the equation $x^2 - 25 = 0$.

Solution   Before we use the square root property we must isolate the squared variable. Add 25 to both sides of the equation to get the variable by itself on one side of the equation.

$$x^2 = 25$$

Now use the square root property to isolate the variable.

$$x = \pm\sqrt{25}$$
$$x = \pm 5$$

Check in the original equation.

Check $x = 5$ $x = -5$

$$x^2 - 25 = 0$$ $$x^2 - 25 = 0$$
$$5^2 - 25 \stackrel{?}{=} 0$$ $$(-5)^2 - 25 \stackrel{?}{=} 0$$
$$25 - 25 \stackrel{?}{=} 0$$ $$25 - 25 \stackrel{?}{=} 0$$

NOW TRY EXERCISE 7 $$0 = 0 \quad \textit{True}$$ $$0 = 0 \quad \textit{True}$$

EXAMPLE 2 Solve the equation $x^2 + 8 = 89$

Solution Begin by subtracting 8 from both sides of the equation.

$$x^2 + 8 = 89$$
$$x^2 = 81 \qquad \textit{Isolate } x^2.$$
$$x = \pm\sqrt{81} \qquad \textit{Square root property}$$
$$x = \pm 9$$

EXAMPLE 3 Solve the equation $a^2 - 7 = 0$.

Solution
$$a^2 - 7 = 0$$
$$a^2 = 7 \qquad \textit{Isolate } a^2.$$

NOW TRY EXERCISE 15
$$a = \pm\sqrt{7} \qquad \textit{Square root property}$$

EXAMPLE 4 Solve the equation $(x - 3)^2 = 4$.

Solution Begin by using the square root property.

$$(x - 3)^2 = 4$$
$$x - 3 = \pm\sqrt{4} \qquad \textit{Square root property}$$
$$x - 3 = \pm 2$$
$$x - 3 + 3 = 3 \pm 2 \qquad \textit{Add 3 to both sides.}$$
$$x = 3 \pm 2$$
$$x = 3 + 2 \quad \text{or} \quad x = 3 - 2$$
$$x = 5 \qquad\qquad x = 1$$

TEACHING TIP
Before discussing Example 4 ask,
"Is this equation more similar to
$x^2 - 3 = 4$ or $y^2 = 4$? Explain."

The solutions are 1 and 5.

EXAMPLE 5 Solve the equation $(5x + 1)^2 - 2 = 16$.

Solution We must first isolate the squared term. We do so by adding 2 to both sides of the equation.

$$(5x + 1)^2 - 2 = 16$$
$$(5x + 1)^2 = 18 \qquad \textit{Add 2 to both sides to isolate the squared term}$$
$$5x + 1 = \pm\sqrt{18} \qquad \textit{Square root property}$$
$$5x + 1 = \pm\sqrt{9}\sqrt{2} \qquad \textit{Simplify } \sqrt{18}.$$
$$5x + 1 = \pm 3\sqrt{2}$$
$$5x + 1 - 1 = -1 \pm 3\sqrt{2} \qquad \textit{Subtract 1 from both sides.}$$
$$5x = -1 \pm 3\sqrt{2}$$
$$x = \frac{-1 \pm 3\sqrt{2}}{5} \qquad \textit{Divide both sides by 5.}$$

TEACHING TIP
After discussing Example 5, have
students compare and contrast the
solutions in Examples 4 and 5.

The solutions are $\dfrac{-1 + 3\sqrt{2}}{5}$ and $\dfrac{-1 - 3\sqrt{2}}{5}$.

NOW TRY EXERCISE 43

Now let's look at one of many applications of quadratic equations.

EXAMPLE 6 **Creating Advertisements** When creating advertisements, the shape of the ad as well as its coloring and print style must be considered. When Antoinette LeMans approached an advertising firm to design a magazine ad for her company, she learned that one of the most appealing shapes is a rectangle whose length is about 1.62 times its width. Any rectangle with these proportions is called a *golden rectangle*. She decided that the ad would have the dimensions of a golden rectangle. Find the dimensions of the ad if it is to have an area of 20 square inches, see Fig. 10.1.

Solution Understand and Translate   Let x = width of rectangle, then $1.62x$ = length of rectangle

$$\text{area} = \text{length} \cdot \text{width}$$
$$20 = (1.62x)x$$

Carry Out
$$20 = 1.62x^2$$
$$\text{or} \quad 1.62x^2 = 20$$
$$x^2 = \frac{20}{1.62} \approx 12.3$$
$$x \approx \pm\sqrt{12.3} \approx \pm3.51 \text{ inches}$$

1.62x

x

FIGURE 10.1

Check and Answer   Since the width cannot be negative, the width, x, is approximately 3.51 inches. The length is about $1.62(3.51) = 5.69$ inches.

Check
$$\text{area} = \text{length} \cdot \text{width}$$
$$20 \overset{?}{=} (5.69)(3.51)$$
$$20 \approx 19.97 \qquad \text{True} \quad \textit{(There is a slight round-off error due to rounding off decimal answers.)}$$

Note that the answer is not a whole number. In many real-life situations this is the case. You should not feel uncomfortable when this occurs.

NOW TRY EXERCISE 57

1. If $x^2 = a$, then $x = \sqrt{a}$ or $x = -\sqrt{a}$.   **2.** Answers will vary.
3. The length is about 1.62 times its width.

Exercise Set 10.1

Concept/Writing Exercises

1. State the square root property.

2. In your own words, explain how to solve the quadratic equation $x^2 = 13$ using the square root property.

3. What is the relationship between the length and width of any golden rectangle?

4. If a golden rectangle has a width of 7 centimeters, what is its length?   ≈11.34 cm

5. How many different real solutions would you expect for each of the following quadratic equations? Explain.

   a) $x^2 = 9$ 2   **b)** $x^2 = 0$ 1

   c) $(x - 2)^2 = 2$ 2 **d)** $(x - 3)^2 = -2$ no real solution

6. Write the standard form of a quadratic equation.
   $ax^2 + bx + c = 0, a \neq 0$

Practice the Skills

Solve.

7. $x^2 = 64$ 8, −8

8. $x^2 = 25$ 5, −5

9. $x^2 = 81$ 9, −9

10. $x^2 = 1$ 1, −1

11. $y^2 = 169$ 13, −13

12. $z^2 = 9$ 3, −3

13. $x^2 = 100$ $10, -10$

14. $a^2 = 15$ $\sqrt{15}, -\sqrt{15}$

15. $x^2 - 20 = 0$ $2\sqrt{5}, -2\sqrt{5}$

16. $w^2 - 24 = 0$ $2\sqrt{6}, -2\sqrt{6}$

17. $3x^2 = 12$ $2, -2$

18. $7x^2 = 7$ $1, -1$

19. $2w^2 = 34$ $\sqrt{17}, -\sqrt{17}$

20. $6n^2 = 96$ $4, -4$

21. $3z^2 + 1 = 28$ $3, -3$

22. $3x^2 - 4 = 8$ $2, -2$

🔒 **23.** $9w^2 + 5 = 20$ $\sqrt{15}/3, -\sqrt{15}/3$

24. $3y^2 + 8 = 36$ $(2\sqrt{21})/3, -(2\sqrt{21})/3$

25. $16x^2 - 17 = 56$ $\sqrt{73}/4, -\sqrt{73}/4$

26. $2x^2 + 3 = 51$ $2\sqrt{6}, -2\sqrt{6}$

Solve.

27. $(x - 4)^2 = 25$ $9, -1$

28. $(y - 2)^2 = 9$ $5, -1$

29. $(a + 3)^2 = 81$ $6, -12$

30. $(x + 5)^2 = 25$ $0, -10$

31. $(x + 4)^2 = 64$ $4, -12$

32. $(x - 4)^2 = 100$ $14, -6$

33. $(r + 6)^2 = 32$ $-6 \pm 4\sqrt{2}$

34. $(x + 3)^2 = 18$ $-3 \pm 3\sqrt{2}$

35. $(d + 6)^2 = 20$ $-6 \pm 2\sqrt{5}$

36. $(k - 5)^2 = 40$ $5 \pm 2\sqrt{10}$

37. $(n - 3)^2 = 49$ $10, -4$

38. $(x - 11)^2 = 28$ $11 \pm 2\sqrt{7}$

39. $(x - 9)^2 = 100$ $19, -1$

40. $(x - 3)^2 = 15$ $3 \pm \sqrt{15}$

🔒 **41.** $(2x + 3)^2 = 18$ $(-3 \pm 3\sqrt{2})/2$

42. $(3a - 2)^2 = 30$ $(2 \pm \sqrt{30})/3$

43. $(4x + 1)^2 - 2 = 18$ $(-1 \pm 2\sqrt{5})/4$

44. $(5s - 6)^2 - 20 = 80$ $\dfrac{16}{5}, -\dfrac{4}{5}$

45. $(2p - 7)^2 + 4 = 22$ $(7 \pm 3\sqrt{2})/2$

46. $(5x + 9)^2 + 9 = 49$ $(-9 \pm 2\sqrt{10})/5$

47. $x^2 = 36$ **55.** $l \approx 1.21\sqrt{522.25}$ or ≈ 27.65 in., $w \approx \sqrt{522.25}$ or ≈ 22.85 in.

Problem Solving

56. $l \approx 0.71\sqrt{3490.14}$ or ≈ 59.08 in., $w \approx \sqrt{3490.14}$ or ≈ 41.94 in.

47. Write an equation that has the solutions 6 and -6.

48. Write an equation that has the solutions $\sqrt{7}$ and $-\sqrt{7}$. $x^2 = 7$

49. Fill in the shaded area to make a true statement. Explain how you determined your answer. The equation $x^2 - \blacksquare = 27$ has the solutions 6 and -6. 9

50. Fill in the shaded area to make a true statement. Explain how you determined your answer. The equation $x^2 + \blacksquare = 45$ has the solutions 8 and -8. -19

51. **a)** Rewrite $-3x^2 + 9x - 6 = 0$ so that the coefficient of the x^2-term is positive. $3x^2 - 9x + 6 = 0$

b) Rewrite $-3x^2 + 9x - 6 = 0$ so that the coefficient of the x^2-term is 1. $x^2 - 3x + 2 = 0$

52. **a)** Rewrite $-\dfrac{1}{2}x^2 + 3x - 4 = 0$ so that the coefficient of the x^2-term is positive. $\dfrac{1}{2}x^2 - 3x + 4 = 0$

b) Rewrite $-\dfrac{1}{2}x^2 + 3x - 4 = 0$ so that the coefficient of the x^2-term is 1. $x^2 - 6x + 8 = 0$

53. ***Product of Numbers*** The product of two positive numbers is 68. Determine the numbers if the larger number is 4.25 times the smaller number. 4 and 17

54. ***Product of Numbers*** The product of two positive numbers is 140. Determine the numbers if the larger number is 5.6 times the smaller number. 5 and 28

55. ***Newspaper Page*** The area of a newspaper page (opened up) is about 631.92 square inches. Determine the length and width of the page if its length is about 1.21 times its width.

56. ***Kitchen Table*** The area of a rectangular kitchen table is 2478 square inches. Determine the length and width of the table if its width is 0.71 times its length.

57. ***Enlarging a Garden*** Georges Marten decided to plant a rectangular garden so that it has the length-to-width ratio found in a golden rectangle. Determine the length and width of the garden if it is to have an area of 2000 square feet. width, ≈35.14 ft; length, ≈56.93 ft

58. ***Mailing Envelope*** The length of a priority mailing envelope is about 1.33 times its width. Determine the length and width of the envelope if its area is approximately 112 square inches. length, ≈12.21 in.; width, ≈9.18 in.

🔒 **59.** ***Comparing Areas*** Consider the two squares with sides x and $x + 3$ shown below.

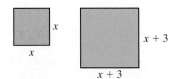

59. b) 6 in.   **c)** $\sqrt{50} \approx 7.07$ in.   **d)** 9 in.   **e)** $\sqrt{92} \approx 9.59$ in.

a) Write a quadratic expression for the area of each square.   left, x^2; right, $(x + 3)^2$

b) If the area of the square on the left is 36 square inches, what is the length of each side of the square?

c) If the area of the square on the left is 50 square inches, what is the length of each side of the square?

d) If the area of the square on the right is 81 square inches, what is the length of each side of the square?

e) If the area of the square on the right is 92 square inches, what is the length of each side of the square?

60. ***Find the Area*** Consider the figure below. If the area shaded in pink is approximately 153.94 square inches, find x (to the nearest hundredth).   4.62 in.

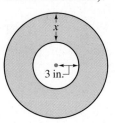

Challenge Problems

Use the square root property to solve for the indicated variable. Assume that all variables represent positive numbers. You may wish to review Section 3.1 before working these problems. List only the positive square root.

61. $A = s^2$, for s $s = \sqrt{A}$

62. $I = p^2r$, for p $p = \sqrt{\dfrac{I}{r}}$

63. $A = \pi r^2$, for r $r = \sqrt{\dfrac{A}{\pi}}$

64. $a^2 + b^2 = c^2$, for b $b = \sqrt{c^2 - a^2}$

65. $I = \dfrac{k}{d^2}$, for d $d = \sqrt{\dfrac{k}{I}}$

66. $A = p(1 + r)^2$, for r $r = \sqrt{\dfrac{A}{p}} - 1$

Cumulative Review Exercises

[5.4] **67.** Factor $4x^2 - 10x - 24$. $2(2x + 3)(x - 4)$

[6.5] **68.** Simplify $\dfrac{3 - \dfrac{1}{y}}{6 - \dfrac{1}{y}}$. $\dfrac{3y - 1}{6y - 1}$

[7.4] **69.** Determine the equation of the line illustrated on the right. $y = 4x - 1$

[9.3] **70.** Simplify $\dfrac{\sqrt{135a^4b}}{\sqrt{3a^5b^5}}$. $\dfrac{3\sqrt{5a}}{ab^2}$

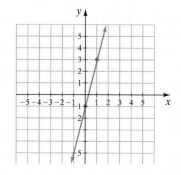

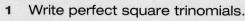

10.2 SOLVING QUADRATIC EQUATIONS BY COMPLETING THE SQUARE

1 Write perfect square trinomials.
2 Solve quadratic equations by completing the square.

SSM   Study Guide   CD/Video

MathPro 4/5   PH Math Tutor Center   prenhall.com/Angel

Quadratic equations that cannot be solved by factoring can be solved by completing the square or by the quadratic formula. In this section we focus on completing the square.

1 Write Perfect Square Trinomials

A **perfect square trinomial** is a trinomial that can be expressed as the square of a binomial. Some examples follow.

TEACHING TIP
Have students rewrite each of the perfect square trinomials with the constant term as a square. Then ask, "How is the constant term related to the coefficient of x?"

Perfect Square Trinomials		Factors		Square of a Binomial
$x^2 + 6x + 9$	$=$	$(x + 3)(x + 3)$	$=$	$(x + 3)^2$
$x^2 - 6x + 9$	$=$	$(x - 3)(x - 3)$	$=$	$(x - 3)^2$
$x^2 + 10x + 25$	$=$	$(x + 5)(x + 5)$	$=$	$(x + 5)^2$
$x^2 - 10x + 25$	$=$	$(x - 5)(x - 5)$	$=$	$(x - 5)^2$

Notice that each of the squared terms in the preceding perfect square trinomials has a numerical coefficient of 1. When the coefficient of the squared term is 1, there is an important relationship between the coefficient of the x-term and the constant. In every perfect square trinomial of this type, *the constant term is the square of one-half the coefficient of the x-term.*

Consider the perfect square trinomial $x^2 + 6x + 9$. The coefficient of the x-term is 6 and the constant is 9. Note that the constant, 9, is the square of one-half the coefficient of the x-term.

TEACHING TIP
Another method to use is to remind students that
$(x + a)^2 = x^2 + 2ax + a^2$, then
$2a = 6$
$a = 3$
$a^2 = 9$

$$x^2 + 6\,x + 9$$

$$\left[\frac{1}{2}(6)\right]^2 = 3^2 = 9$$

Consider the perfect square trinomial $x^2 - 10x + 25$. The coefficient of the x-term is -10 and the constant is 25. Note that

$$x^2 - 10\,x + 25$$

$$\left[\frac{1}{2}(-10)\right]^2 = (-5)^2 = 25$$

Consider the expression $x^2 + 8x +$ ▨. Can you determine what number must be placed in the colored box to make the trinomial a perfect square trinomial? If you answered 16, you answered correctly.

$$x^2 + 8\,x + \boxed{}$$

$$\left[\frac{1}{2}(8)\right]^2 = 4^2 = 16$$

The perfect square trinomial is $x^2 + 8x + 16$. Note that $x^2 + 8x + 16 = (x + 4)^2$.
Let's examine perfect square trinomials a little further.

Perfect Square Trinomial		Square of a Binomial
$x^2 + 6x + 9$	$=$	$(x + 3)^2$

$$\frac{1}{2}(6) = 3$$

$x^2 - 10x + 25$	$=$	$(x - 5)^2$

$$\frac{1}{2}(-10) = -5$$

Note that when a perfect square trinomial is written as the square of a binomial *the constant in the binomial is one-half the value of the coefficient of the x-term in the perfect square trinomial.*

2 Solve Quadratic Equations by Completing the Square

The procedure for solving a quadratic equation by **completing the square** is illustrated in the following example. Some of the equations that you will be solving in the examples could be easily solved by factoring, but they are being solved by completing the square as a means of illustrating the procedure on easy problems before moving on to more difficult problems.

EXAMPLE 1 Solve the equation $x^2 + 6x - 7 = 0$ by completing the square.

Solution First we make sure that the squared term has a coefficient of 1. (In Example 5 we explain what to do if the coefficient is not 1.) Next we wish to get the terms containing a variable by themselves on the left side of the equation. Therefore, we add 7 to both sides of the equation.

$$x^2 + 6x - 7 = 0$$
$$x^2 + 6x = 7$$

Now we determine one-half the numerical coefficient of the x-term. In this example, the x-term is $6x$.

$$\frac{1}{2}(6) = \boxed{3}$$

We square this number,

$$(3)^2 = (3)(3) = \boxed{9}$$

and then add this product to both sides of the equation.

$$x^2 + 6x \boxed{+ 9} = 7 \boxed{+ 9}$$

or

$$x^2 + 6x + 9 = 16$$

By following this procedure, we produce a perfect square trinomial on the left side of the equation. The expression $x^2 + 6x + 9$ is a perfect square trinomial that can be expressed as $(x + 3)^2$. Therefore,

$$x^2 + 6x + 9 = 16$$

can be written

$$(x \boxed{+ 3})^2 = 16$$

Now we use the square root property,

$$x + 3 = \pm\sqrt{16}$$
$$x + 3 = \pm 4$$

Finally, we solve for x by subtracting 3 from both sides of the equation.

$$x + 3 - 3 = -3 \pm 4$$
$$x = -3 \pm 4$$

$$x = -3 + 4 \quad \text{or} \quad x = -3 - 4$$
$$x = 1 \qquad\qquad\qquad x = -7$$

Thus, the solutions are 1 and -7. We check both solutions in the original equation.

Check $\qquad x = 1 \qquad\qquad\qquad x = -7$

$$x^2 + 6x - 7 = 0$$
$$(1)^2 + 6(1) - 7 \overset{?}{=} 0$$
$$1 + 6 - 7 \overset{?}{=} 0$$
$$0 = 0 \quad \textit{True}$$

$$x^2 + 6x - 7 = 0$$
$$(-7)^2 + 6(-7) - 7 \overset{?}{=} 0$$
$$49 - 42 - 7 \overset{?}{=} 0$$
$$0 = 0 \quad \textit{True}$$

Now let's summarize the procedure.

To Solve a Quadratic Equation by Completing the Square

1. Use the multiplication (or division) property of equality if necessary to make the numerical coefficient of the squared term equal to 1.
2. Rewrite the equation with the constant by itself on the right side of the equation.
3. Take one-half the numerical coefficient of the first-degree term, square it, and add this quantity to both sides of the equation.
4. Replace the trinomial with its equivalent squared binomial.
5. Use the square root property.
6. Solve for the variable.
7. Check your answers in the *original* equation.

EXAMPLE 2 Solve the equation $x^2 - 10x + 21 = 0$ by completing the square.

Solution
$$x^2 - 10x + 21 = 0$$
$$x^2 - 10x = -21 \quad \textit{Step 2}$$

Take half the numerical coefficient of the x-term, square it, and add this product to both sides of the equation.

$$\frac{1}{2}(-10) = -5, \quad (-5)^2 = 25$$

Now add 25 to both sides of the equation.

$$x^2 - 10x + 25 = -21 + 25 \quad \textit{Step 3}$$
$$x^2 - 10x + 25 = 4$$
$$\text{or} \qquad (x - 5)^2 = 4 \quad \textit{Step 4}$$
$$x - 5 = \pm\sqrt{4} \quad \textit{Step 5}$$
$$x - 5 = \pm 2$$
$$x = 5 \pm 2 \quad \textit{Step 6}$$
$$x = 5 + 2 \quad \text{or} \quad x = 5 - 2$$
$$x = 7 \qquad\qquad x = 3$$

NOW TRY EXERCISE 13 A check will show that the solutions are 7 and 3.

EXAMPLE 3 Solve the equation $x^2 = 3x + 18$ by completing the square.

Solution Place all terms except the constant on the left side of the equation.

$$x^2 = 3x + 18$$
$$x^2 - 3x = 18 \quad \textit{Step 2}$$

TEACHING TIP
Before discussing Example 3 ask, "How is Example 3 different from Example 2?"

Take half the numerical coefficient of the x-term, square it, and add this product to both sides of the equation.

$$\frac{1}{2}(-3) = -\frac{3}{2}, \quad \left(-\frac{3}{2}\right)^2 = \frac{9}{4}$$

$$x^2 - 3x + \frac{9}{4} = 18 + \frac{9}{4} \qquad \textit{Step 3}$$

$$\left(x - \frac{3}{2}\right)^2 = 18 + \frac{9}{4} \qquad \textit{Step 4}$$

$$\left(x - \frac{3}{2}\right)^2 = \frac{72}{4} + \frac{9}{4}$$

$$\left(x - \frac{3}{2}\right)^2 = \frac{81}{4}$$

$$x - \frac{3}{2} = \pm\sqrt{\frac{81}{4}} \qquad \textit{Step 5}$$

$$x - \frac{3}{2} = \pm\frac{9}{2}$$

$$x = \frac{3}{2} \pm \frac{9}{2} \qquad \textit{Step 6}$$

$$x = \frac{3}{2} + \frac{9}{2} \quad \text{or} \quad x = \frac{3}{2} - \frac{9}{2}$$

$$x = \frac{12}{2} = 6 \qquad\qquad x = -\frac{6}{2} = -3$$

NOW TRY EXERCISE 21 The solutions are 6 and -3. ✳

TEACHING TIP
After discussing the solution to Example 3, have students create a quadratic equation from the solutions. Then ask, "Does your quadratic equation match the original equation?"

In the following examples we will not show some of the intermediate steps.

EXAMPLE 4 Solve the equation $x^2 - 16x + 4 = 0$ by completing the square.

Solution
$$x^2 - 16x + 4 = 0$$
$$x^2 - 16x = -4 \qquad \textit{Step 2}$$
$$x^2 - 16x + 64 = -4 + 64 \qquad \textit{Step 3}$$
$$(x - 8)^2 = 60 \qquad \textit{Step 4}$$
$$x - 8 = \pm\sqrt{60} \qquad \textit{Step 5}$$
$$x - 8 = \pm\sqrt{4}\sqrt{15}$$
$$x - 8 = \pm 2\sqrt{15}$$
$$x = 8 \pm 2\sqrt{15} \qquad \textit{Step 6}$$

The solutions are $8 + 2\sqrt{15}$ and $8 - 2\sqrt{15}$. ✳

EXAMPLE 5 Solve the equation $5z^2 - 25z + 10 = 0$ by completing the square.

Solution To solve an equation by completing the square, the numerical coefficient of the squared term must be 1. Since the coefficient of the squared term is 5, we multiply both sides of the equation by $\frac{1}{5}$ (or divide every term by 5) to make the coefficient equal to 1.

$$5z^2 - 25z + 10 = 0$$
$$\frac{1}{5}(5z^2 - 25z + 10) = \frac{1}{5}(0) \qquad \textit{Step 1}$$
$$z^2 - 5z + 2 = 0$$

Now we proceed as in earlier examples.

$$z^2 - 5z = -2 \qquad \textit{Step 2}$$

$$z^2 - 5z + \frac{25}{4} = -2 + \frac{25}{4} \qquad \textit{Step 3}$$

$$\left(z - \frac{5}{2}\right)^2 = -\frac{8}{4} + \frac{25}{4} \qquad \textit{Step 4}$$

$$\left(z - \frac{5}{2}\right)^2 = \frac{17}{4}$$

$$z - \frac{5}{2} = \pm\sqrt{\frac{17}{4}} \qquad \textit{Step 5}$$

$$z - \frac{5}{2} = \pm\frac{\sqrt{17}}{2}$$

$$z = \frac{5}{2} \pm \frac{\sqrt{17}}{2} \qquad \textit{Step 6}$$

$$z = \frac{5}{2} + \frac{\sqrt{17}}{2} \quad \text{or} \quad z = \frac{5}{2} - \frac{\sqrt{17}}{2}$$

$$z = \frac{5 + \sqrt{17}}{2} \qquad\qquad z = \frac{5 - \sqrt{17}}{2}$$

TEACHING TIP
Have students explore what happens when they make a perfect square trinomial from the following equations: $x^2 - 28x + 198 = 0$, $x^2 - 2\sqrt{3}x + 3 = 0$

The solutions are $\dfrac{5 + \sqrt{17}}{2}$ and $\dfrac{5 - \sqrt{17}}{2}$.

NOW TRY EXERCISE 33

1. a) a trinomial that can be expressed as the square of a binomial **3.** constant is the square of half the coefficient **6. a)** Answers will vary.

Exercise Set 10.2

Concept/Writing Exercises

1. a) What is a perfect square trinomial?

  b) Fill in the shaded area to make a perfect square trinomial and explain how you determined your answer.

$$x^2 + 8x \quad\rule{1cm}{0.4pt}\qquad +16$$

2. Fill in the shaded area to make a perfect square trinomial and explain how you determined your answer.

$$x^2 - 10x \quad\rule{1cm}{0.4pt}\qquad +25$$

3. In a perfect square trinomial, what is the relationship between the constant and the coefficient of the x-term?

4. In a perfect square trinomial, if the coefficient of the x-term is 4, what is the constant? 4

5. In a perfect square trinomial, if the coefficient of the x-term is -12, what is the constant? 36

6. a) In your own words, explain how to solve a quadratic equation by completing the square.

  b) Compare your answer to part **a)** with the procedure given on page 623. Did you omit any steps in your explanation? Answers will vary.

Practice the Skills

Solve by completing the square.

7. $x^2 - 7x + 10 = 0$ 2, 5

8. $x^2 + 5x + 6 = 0$ $-2, -3$

9. $x^2 - 8x + 7 = 0$ 7, 1

10. $r^2 + r - 30 = 0$ 5, -6

11. $x^2 + 3x + 2 = 0$ $-1, -2$

12. $x^2 + 4x - 32 = 0$ 4, -8

13. $z^2 - 2z - 8 = 0$ 4, -2

14. $m^2 - 9m + 14 = 0$ 7, 2

15. $n^2 = -6n - 9$ -3

16. $k^2 = -9k - 18$ $-6, -3$

17. $x^2 = 2x + 15$ 5, -3

18. $x^2 = -5x - 6$ $-2, -3$

19. $x^2 + 10x + 24 = 0$ $-4, -6$

20. $-35 = n^2 - 12n$ 5, 7

21. $x^2 = 15x - 56$ 8, 7

22. $x^2 = 3x + 28$ 7, -4

23. $-32 = -p^2 + 4p$ 8, -4

24. $-x^2 - 3x + 40 = 0$ 5, -8

25. $z^2 - 4z = -2$ $2 \pm \sqrt{2}$

26. $z^2 + 2z = 6$ $-1 \pm \sqrt{7}$

27. $6w + 4 = -w^2$ $-3 \pm \sqrt{5}$

28. $g^2 - 2g = 4$ $1 \pm \sqrt{5}$

29. $m^2 + 7m + 2 = 0$ $(-7 \pm \sqrt{41})/2$

30. $x^2 + 3x - 6 = 0$ $(-3 \pm \sqrt{33})/2$

31. $2x^2 + 4x - 6 = 0$ $1, -3$

32. $2x^2 + 2x - 24 = 0$ $3, -4$

🔒 **33.** $2x^2 + 18x + 4 = 0$ $(-9 \pm \sqrt{73})/2$

34. $2x^2 = 8x + 90$ $9, -5$

35. $3h^2 - 15h = 18$ $6, -1$

36. $4x^2 = -28x + 32$ $1, -8$

🔒 **37.** $3x^2 - 11x - 4 = 0$ $4, -1/3$

38. $3x^2 - 8x + 4 = 0$ $2, 2/3$

39. $9t^2 + 6t = 6$ $(-1 \pm \sqrt{7})/3$

40. $2x^2 - x = 5$ $(1 \pm \sqrt{41})/4$

41. $x^2 - 5x = 0$ $5, 0$

42. $2x^2 - 6x = 0$ $3, 0$

43. $2x^2 = 12x$ $6, 0$

44. $3x^2 = 9x$ $3, 0$

Problem Solving

45. a) Write a perfect square trinomial that has a term of $20x$. $x^2 + 20x + 100$

b) Explain how you constructed your perfect square trinomial.   Answers will vary.

46. a) Write a perfect square trinomial that has a term of $-14x$. $x^2 - 14x + 49$

b) Explain how you constructed your perfect square trinomial.   Answers will vary.

47. *Numbers* When 3 times a number is added to the square of a number, the sum is 4. Find the number(s). $1, -4$

48. *Numbers* When 5 times a number is subtracted from 2 times the square of a number, the difference is 12. Find the number(s). $-\frac{3}{2}, 4$

49. *Numbers* If the square of 3 more than a number is 9, find the number(s). $0, -6$

50. *Numbers* If the square of 2 less than an integer is 16, find the number(s). $6, -2$

🔒 **51.** *Numbers* The product of two positive numbers is 21. Find the two numbers if the larger is 4 greater than the smaller. $3, 7$

52. *Supporting a Pole* A guy wire 20 feet long is supporting a pole as shown in the figure. Determine the height of the pole.   16 ft

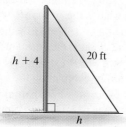

53. *Ladder Leaning on a House* A 30-foot ladder is leaning against a house, as shown. Determine the vertical distance from the ground to where the ladder rests on the house. $(9 + \sqrt{369})$ or ≈ 28.21 ft

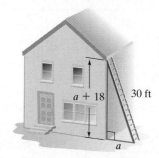

54. *Sum of Integers* The sum of the first n integers, s, can be found by the formula $s = \dfrac{n^2 + n}{2}$. Find the value of n if the sum is 28.   7

55. *Height of an Object* When an object is projected straight up from Earth with an initial velocity of 128 feet per second, its height above the ground, s, in feet, in t seconds is given by the formula $s = -16t^2 + 128t$. How long will it take the object to reach a height of 192 feet? (Therefore, $s = 192$.)   2 sec and 6 sec

56. *Height of an Object* Repeat Exercise 55 for a height of 112 feet.   1 sec and 7 sec

Challenge Problems

57. Fill in the shaded area to make a perfect square trinomial and explain how you determined your answer.

$$x^2 \quad\rule{1cm}{0.3cm}\quad + 196 \qquad +28x \text{ or } -28x$$

58. Fill in the shaded area to make a perfect square trinomial and explain how you determined your answer.

$$x^2 \quad\rule{1cm}{0.3cm}\quad + \dfrac{9}{100} \qquad +\dfrac{3}{5}x \text{ or } -\dfrac{3}{5}x$$

59. a) Solve the equation $x^2 - 14x - 1 = 0$ by completing the square. $7 \pm 5\sqrt{2}$

b) Check your solution (it will not be a rational number) by substituting the value(s) you obtained in part **a)** for each x in the equation in part **a)**.

60. a) Solve the equation $x^2 + 3x - 7 = 0$ by completing the square. $\dfrac{-3 \pm \sqrt{37}}{2}$

b) Check your solution (it will not be a rational number) by substituting the value(s) you obtained in part **a)** for each x in the equation in part **a)**.

Solve by completing the square.

61. $x^2 + \dfrac{3}{5}x - \dfrac{1}{2} = 0.$ $(-3 \pm \sqrt{59})/10$ **62.** $x^2 - \dfrac{2}{3}x - \dfrac{1}{5} = 0.$ $(5 \pm \sqrt{70})/15$ **63.** $3x^2 + \dfrac{1}{2}x = 4.$ $(-1 \pm \sqrt{193})/12$

64. $0.1x^2 + 0.2x - 0.54 = 0$ $-1 \pm \sqrt{6.4}$ **65.** $-5.26x^2 + 7.89x + 15.78 = 0$ $0.75 \pm \sqrt{3.5625}$

Cumulative Review Exercises
67. If slopes are same and *y*-intercepts are different, equations represent parallel lines.

[6.4] **66.** Simplify $\dfrac{x^2}{x^2 - x - 6} - \dfrac{x - 2}{x - 3}.$ $\dfrac{4}{(x + 2)(x - 3)}$

[7.4] **67.** Explain how you can determine whether two equations represent parallel lines without graphing the equations.

[8.2, 8.3] **68.** Solve the following system of equations.
$$3x - 4y = 6$$
$$2x + y = 8 \qquad \left(\dfrac{38}{11}, \dfrac{12}{11}\right)$$

[9.5]   **69.** Solve the equation $\sqrt{2x + 3} = 2x - 3.$ 3

10.3 SOLVING QUADRATIC EQUATIONS BY THE QUADRATIC FORMULA

SSM   Study Guide   CD/Video

MathPro 4/5   PH Math Tutor Center   prenhall.com/Angel

1 Solve quadratic equations by the quadratic formula.
2 Determine the number of solutions to a quadratic equation using the discriminant.

1   Solve Quadratic Equations by the Quadratic Formula

Another method that can be used to solve any quadratic equation is the **quadratic formula**. It is the most useful and versatile method of solving quadratic equations.

The standard form of a quadratic equation is $ax^2 + bx + c = 0$, where *a* is the coefficient of the squared term, *b* is the coefficient of the first-degree term, and *c* is the constant.

Quadratic Equation in Standard Form	Values of *a*, *b*, and *c*		
$x^2 - 5x + 6 = 0$	$a = 1,$	$b = -5,$	$c = 6$
$5x^2 + 3x = 0$	$a = 5,$	$b = 3,$	$c = 0$
$-\dfrac{1}{2}x^2 + 5 = 0$	$a = -\dfrac{1}{2},$	$b = 0$	$c = 5$

TEACHING TIP
Point out that it is necessary to label *a*, *b*, and *c* with the correct sign.

We can derive the quadratic formula by starting with a quadratic equation in standard form and completing the square, as discussed in the preceding section.

$$ax^2 + bx + c = 0$$ *Standard form of quadratic equation*

$$\dfrac{ax^2}{a} + \dfrac{b}{a}x + \dfrac{c}{a} = 0$$ *Divide both sides by a.*

$$x^2 + \dfrac{b}{a}x = -\dfrac{c}{a}$$ *Subtract c/a from both sides.*

$$x^2 + \dfrac{b}{a}x + \dfrac{b^2}{4a^2} = -\dfrac{c}{a} + \dfrac{b^2}{4a^2}$$ *Take 1/2 of b/a; that is, b/2a, and square it to get $b^2/4a^2$. Then add this expression to both sides.*

$$\left(x + \frac{b}{2a}\right)^2 = \frac{b^2}{4a^2} - \frac{c}{a}$$

Rewrite the left side of the equation as the square of a binomial.

$$\left(x + \frac{b}{2a}\right)^2 = \frac{b^2 - 4ac}{4a^2}$$

Write the right side with a common denominator.

$$x + \frac{b}{2a} = \pm\sqrt{\frac{b^2 - 4ac}{4a^2}}$$

Square root property

$$x + \frac{b}{2a} = \pm\frac{\sqrt{b^2 - 4ac}}{2a}$$

Quotient rule for radicals, $\sqrt{4a^2} = 2a$

$$x = \frac{-b}{2a} \pm \frac{\sqrt{b^2 - 4ac}}{2a}$$

Subtract $b/2a$ from both sides.

$$x = \frac{-b \pm \sqrt{b^2 - 4ac}}{2a}$$

Write with a common denominator to get the quadratic formula.

To Solve a Quadratic Equation by the Quadratic Formula

1. Write the equation in standard form, $ax^2 + bx + c = 0$, and determine the numerical values for a, b, and c.

2. Substitute the values for a, b, and c from step 1 in the quadratic formula below and then evaluate to obtain the solution.

The Quadratic Formula
$$x = \frac{-b \pm \sqrt{b^2 - 4ac}}{2a}$$

EXAMPLE 1 Use the quadratic formula to solve the equation $x^2 + 4x + 3 = 0$.

Solution In this equation $a = 1, b = 4$, and $c = 3$. Substitute these values into the quadratic formula and then evaluate.

$$x = \frac{-b \pm \sqrt{b^2 - 4ac}}{2a}$$
$$= \frac{-(4) \pm \sqrt{(4)^2 - 4(1)(3)}}{2(1)}$$
$$= \frac{-4 \pm \sqrt{16 - 12}}{2}$$
$$= \frac{-4 \pm \sqrt{4}}{2}$$
$$= \frac{-4 \pm 2}{2}$$

$$x = \frac{-4 + 2}{2} \quad \text{or} \quad x = \frac{-4 - 2}{2}$$
$$= \frac{-2}{2} = -1 \qquad\qquad = \frac{-6}{2} = -3$$

TEACHING TIP
After discussing the solution to
Example 1, have students solve it
by factoring and by completing
the square.

NOW TRY EXERCISE 29

Check

$x = -1$

$x^2 + 4x + 3 = 0$

$(-1)^2 + 4(-1) + 3 \overset{?}{=} 0$

$1 - 4 + 3 \overset{?}{=} 0$

$0 = 0$ *True*

$x = -3$

$x^2 + 4x + 3 = 0$

$(-3)^2 + 4(-3) + 3 \overset{?}{=} 0$

$9 - 12 + 3 \overset{?}{=} 0$

$0 = 0$ *True* ✳

AVOIDING COMMON ERRORS

The **entire numerator** of the quadratic formula must be divided by 2a.

CORRECT

$$x = \frac{-b \pm \sqrt{b^2 - 4ac}}{2a}$$

INCORRECT

$$x = -b \pm \frac{\sqrt{b^2 - 4ac}}{2a}$$

$$x = \frac{-b}{2a} \pm \sqrt{b^2 - 4ac}$$

EXAMPLE 2 Use the quadratic formula to solve the equation $8x^2 + 2x - 1 = 0$.

Solution

$$8x^2 + 2x - 1 = 0$$
$$a = 8, \quad b = 2, \quad c = -1$$
$$x = \frac{-b \pm \sqrt{b^2 - 4ac}}{2a}$$
$$= \frac{-(2) \pm \sqrt{(2)^2 - 4(8)(-1)}}{2(8)}$$
$$= \frac{-2 \pm \sqrt{4 + 32}}{16}$$
$$= \frac{-2 \pm \sqrt{36}}{16}$$
$$= \frac{-2 \pm 6}{16}$$

$$x = \frac{-2 + 6}{16} \quad \text{or} \quad x = \frac{-2 - 6}{16}$$
$$= \frac{4}{16} = \frac{1}{4} \qquad\qquad = \frac{-8}{16} = -\frac{1}{2}$$

Check

$x = \dfrac{1}{4}$

$8x^2 + 2x - 1 = 0$

$8\left(\dfrac{1}{4}\right)^2 + 2\left(\dfrac{1}{4}\right) - 1 \overset{?}{=} 0$

$8\left(\dfrac{1}{16}\right) + \dfrac{1}{2} - 1 \overset{?}{=} 0$

$\dfrac{1}{2} + \dfrac{1}{2} - 1 \overset{?}{=} 0$

$0 = 0$ *True*

$x = -\dfrac{1}{2}$

$8x^2 + 2x - 1 = 0$

$8\left(-\dfrac{1}{2}\right)^2 + 2\left(-\dfrac{1}{2}\right) - 1 \overset{?}{=} 0$

$8\left(\dfrac{1}{4}\right) - 1 - 1 \overset{?}{=} 0$

$2 - 1 - 1 \overset{?}{=} 0$

$0 = 0$ *True* ✳

TEACHING TIP
After discussing the solution to
Example 2, have students solve it
by factoring and by completing
the square.

EXAMPLE 3   Use the quadratic formula to solve the equation $2w^2 + 6w - 5 = 0$.

Solution   The variable in this equation is w. The procedure to solve the equation is the same.

$$a = 2, \quad b = 6, \quad c = -5$$

$$w = \frac{-b \pm \sqrt{b^2 - 4ac}}{2a}$$

$$= \frac{-(6) \pm \sqrt{(6)^2 - 4(2)(-5)}}{2(2)}$$

$$= \frac{-6 \pm \sqrt{36 + 40}}{4}$$

$$= \frac{-6 \pm \sqrt{76}}{4}$$

$$= \frac{-6 \pm \sqrt{4}\sqrt{19}}{4}$$

$$= \frac{-6 \pm 2\sqrt{19}}{4}$$

Now factor out 2 from both terms in the numerator; then divide out common sfactors as explained in Section 9.4.

$$w = \frac{\overset{1}{\cancel{2}}(-3 \pm \sqrt{19})}{\underset{2}{\cancel{4}}}$$

$$w = \frac{-3 \pm \sqrt{19}}{2}$$

Thus, the solutions are $w = \dfrac{-3 + \sqrt{19}}{2}$ and $w = \dfrac{-3 - \sqrt{19}}{2}$.   ✳

TEACHING TIP
After discussing the solution to Example 3, have students solve it by completing the square. Show that the equation cannot be solved by factoring.

Now let's try two examples where the equation is not in standard form.

EXAMPLE 4   Use the quadratic formula to solve the equation $x^2 = 6x - 4$.

Solution   First write the equation in standard form.

TEACHING TIP
Before discussing Example 4 ask, "How is the equation in Example 4 different from the equations in Examples 1–3?"

$$x^2 - 6x + 4 = 0$$ *Set one side of the equation equal to zero.*

$$a = 1, \quad b = -6, \quad c = 4$$

$$x = \frac{-b \pm \sqrt{b^2 - 4ac}}{2a}$$

$$= \frac{-(-6) \pm \sqrt{(-6)^2 - 4(1)(4)}}{2(1)}$$ *Substitute.*

$$= \frac{6 \pm \sqrt{36 - 16}}{2}$$ *Simplify.*

$$= \frac{6 \pm \sqrt{20}}{2}$$

$$= \frac{6 \pm \sqrt{4}\sqrt{5}}{2}$$ *Product rule*

$$= \frac{6 \pm 2\sqrt{5}}{2}$$

$$= \frac{\overset{1}{2}(3 \pm \sqrt{5})}{\underset{1}{2}}$$ *Factor out 2.*

$$= 3 \pm \sqrt{5}$$

NOW TRY EXERCISE 41   The solutions are $x = 3 + \sqrt{5}$ and $x = 3 - \sqrt{5}$.   ✳

AVOIDING COMMON ERRORS

Many students solve quadratic equations correctly until the last step, when they make an error. Do not make the mistake of trying to simplify an answer that cannot be simplified any further. The following are answers that cannot be simplified, along with some common errors.

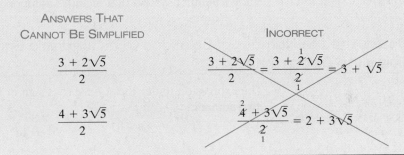

ANSWERS THAT CANNOT BE SIMPLIFIED

$$\frac{3 + 2\sqrt{5}}{2}$$

$$\frac{4 + 3\sqrt{5}}{2}$$

INCORRECT

$$\frac{3 + 2\sqrt{5}}{2} = \frac{3 + \overset{1}{2}\sqrt{5}}{\underset{1}{2}} = 3 + \sqrt{5}$$

$$\frac{\overset{2}{4} + 3\sqrt{5}}{\underset{1}{2}} = 2 + 3\sqrt{5}$$

EXAMPLE 5   Use the quadratic formula to solve the equation $t^2 = 25$.

Solution   First write the equation in standard form.

$$t^2 - 25 = 0$$ *Set one side of the equation equal to 0.*

$$a = 1, \qquad b = 0, \qquad c = -25$$

$$t = \frac{-b \pm \sqrt{b^2 - 4ac}}{2a}$$

$$= \frac{-0 \pm \sqrt{0^2 - 4(1)(-25)}}{2(1)}$$ *Substitute.*

$$= \frac{\pm\sqrt{100}}{2} = \frac{\pm 10}{2} = \pm 5$$ *Simplify and solve for t.*

Thus, the solutions are 5 and -5.   ✳

The solution to Example 5 could have been solved more quickly by using the Square Root Property discussed in Section 10.1. We worked Example 5 using the Quadratic Formula to give you more practice using the formula.

The next example illustrates a quadratic equation that has no real number solution.

EXAMPLE 6   Solve the quadratic equation $3x^2 = x - 1$.

Solution

$$3x^2 - x + 1 = 0$$

$$a = 3, \qquad b = -1, \qquad c = 1$$

$$x = \frac{-b \pm \sqrt{b^2 - 4ac}}{2a}$$

$$= \frac{-(-1) \pm \sqrt{(-1)^2 - 4(3)(1)}}{2(3)}$$

$$= \frac{1 \pm \sqrt{1 - 12}}{6}$$

$$= \frac{1 \pm \sqrt{-11}}{6}$$

Since $\sqrt{-11}$ is not a real number, we stop here. This equation has no real number solution. *When given a problem of this type, your answer should be "no real number solution." Do not leave the answer blank, and do not write 0 for the answer.* We will study numbers like $\sqrt{-11}$ and $(1 \pm \sqrt{-11})/6$ further in Section 10.5 when we study a number system that includes numbers of this type.

NOW TRY EXERCISE 47

2 Determine the Number of Solutions to a Quadratic Equation Using the Discriminant

TEACHING TIP
Before discussing the discriminant, ask students to suggest values for a, b, and c, such that

$$b^2 - 4ac > 0$$
$$b^2 - 4ac = 0$$
$$b^2 - 4ac < 0$$

Then have them write quadratic equations using those values and solve them.

The expression under the square root sign in the quadratic formula is called the **discriminant**.

$$\underbrace{b^2 - 4ac}_{Discriminant}$$

The discriminant can be used to determine the number of real solutions to a quadratic equation.

When the Discriminant Is:

1. **Greater than zero,** $b^2 - 4ac > 0$, the quadratic equation has **two distinct real number solutions.**

2. **Equal to zero,** $b^2 - 4ac = 0$, the quadratic equation has **one real number solution.**

3. **Less than zero,** $b^2 - 4ac < 0$, the quadratic equation has **no real number solution.**

We indicate this information in a shortened form in the chart below.

TEACHING TIP
Have students describe the number of solutions a quadratic equation will have if
$a = 1$ and $c < 0$
$b \neq 0$ and $c = 0$
$b = 0$, and $a, c > 0$

$b^2 - 4ac$	Number of Solutions
Positive	Two distinct real number solutions
0	One real number solution
Negative	No real number solution

EXAMPLE 7 **a)** Find the discriminant of the equation $x^2 - 12x + 36 = 0$.

b) Use the discriminant to determine the number of solutions to the equation.

c) Use the quadratic formula to find the solutions, if any exist.

Solution **a)** $a = 1,$ $b = -12,$ $c = 36$

$$b^2 - 4ac = (-12)^2 - 4(1)(36) = 144 - 144 = 0$$

b) Since the discriminant is equal to zero, there is one real number solution.

c)
$$x = \frac{-b \pm \sqrt{b^2 - 4ac}}{2a}$$

$$= \frac{-(-12) \pm \sqrt{0}}{2(1)}$$

$$= \frac{12 \pm 0}{2} = \frac{12}{2} = 6$$

NOW TRY EXERCISE 9   The only solution is 6.   ✳

EXAMPLE 8   Without actually finding the solutions, determine whether the following equations have two distinct real number solutions, one real number solution, or no real number solution.

a) $4x^2 - 4x + 1 = 0$ **b)** $2x^2 + 13x = -15$ **c)** $6p^2 = -5p - 2$

Solution   We use the discriminant of the quadratic formula to answer these equations.

a) $b^2 - 4ac = (-4)^2 - 4(4)(1) = 16 - 16 = 0$

Since the discriminant is equal to zero, this equation has one real number solution.

b) First, rewrite $2x^2 + 13x = -15$ as $2x^2 + 13x + 15 = 0.$

$$b^2 - 4ac = (13)^2 - 4(2)(15) = 169 - 120 = 49$$

Since the discriminant is positive, this equation has two distinct real number solutions.

c) First rewrite $6p^2 = -5p - 2$ as $6p^2 + 5p + 2 = 0$

$$b^2 - 4ac = (5)^2 - 4(6)(2) = 25 - 48 = -23$$

Since the discriminant is negative, this equation has no real number solution. ✳

Now let's look at one of many applications that may be solved using the quadratic formula.

EXAMPLE 9   **Building a Border**   The Johnsons have a rectangular swimming pool that measures 30 feet by 16 feet. They want to add a cement border of uniform width around all sides of the pool. How wide can they make the border if they want the area of the border to be 200 square feet?

Solution   Let's make a diagram of the pool, see Figure 10.2. Let $x =$ uniform width of the border. Then the total length of the pool and border is $2x + 30.$ The total width of the pool and border is $2x + 16.$ The area of the border can be found by subtracting the area of the pool (the smaller rectangle area) from the area of the pool and border (the larger rectangle area).

$$\text{area of pool} = l \cdot w = (30)(16) = 480$$

$$\text{area of pool and border} = l \cdot w = (2x + 30)(2x + 16)$$

$$= 4x^2 + 92x + 480$$

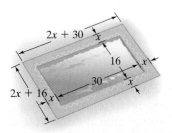

FIGURE 10.2

$$\text{area of border} = \text{area of pool and border} - \text{area of pool}$$
$$= (4x^2 + 92x + 480) - 480$$
$$= 4x^2 + 92x$$

The total area of the border is 200 square feet. Therefore,

$$\text{area of border} = 4x^2 + 92x$$
$$200 = 4x^2 + 92x$$

or $\quad 4x^2 + 92x - 200 = 0$	*Write equation in standard form.*
$4(x^2 + 23x - 50) = 0$	*Factor out 4.*
$\dfrac{1}{\cancel{4}} \cdot \cancel{4}(x^2 + 23x - 50) = \dfrac{1}{4} \cdot 0$	*Multiply both sides by $\dfrac{1}{4}$ to eliminate 4.*
$x^2 + 23x - 50 = 0$	

Now use the quadratic formula.

$$a = 1 \qquad b = 23 \qquad c = -50$$

$$x = \frac{-b \pm \sqrt{b^2 - 4ac}}{2a}$$

$$= \frac{-23 \pm \sqrt{(23)^2 - 4(1)(-50)}}{2(1)}$$

$$= \frac{-23 \pm \sqrt{529 + 200}}{2}$$

$$= \frac{-23 \pm \sqrt{729}}{2}$$

$$= \frac{-23 \pm 27}{2}$$

$$x = \frac{-23 - 27}{2} \qquad \text{or} \qquad x = \frac{-23 + 27}{2}$$

$$= \frac{-50}{2} \qquad\qquad\qquad = \frac{4}{2}$$

$$= -25 \qquad\qquad\qquad\quad = 2$$

TEACHING TIP
Have students work in pairs to write a word problem similar to Example 9. Then have them give their problem to another pair to solve.

NOW TRY EXERCISE 63

Since lengths are positive, the only possible answer is $x = 2$. Thus the uniform cement border will be 2 feet wide all around the pool. ✳

In Example 9, the answer came out to be an integer value. However, many times when working with application problems the answer comes out to be an irrational number. When this occurs in the exercise set we will round radical values to two decimal places to determine our answer.

HELPFUL HINT

Notice in Example 9 that when we had the quadratic equation $4x^2 + 92x - 200 = 0$ we factored out the common factor 4 to get

$$4x^2 + 92x - 200 = 0$$
$$4(x^2 + 23x - 50) = 0$$

We then multiplied both sides of the equation by $\frac{1}{4}$ and used the quadratic equation $x^2 + 23x - 50 = 0$, where $a = 1$, $b = 23$, and $c = -50$ in the quadratic formula. If all the terms in a quadratic equation have a common factor, it will be easier to factor it out first so that you will have smaller numbers when you use the quadratic formula. Consider the quadratic equation $4x^2 + 8x - 12 = 0$.

In this equation $a = 4$, $b = 8$, and $c = -12$. If you solve this equation with the quadratic formula, after simplification you will get the solutions -3 and 1. Try this and see. If you factor out 4 to get

$$4x^2 + 8x - 12 = 0$$
$$4(x^2 + 2x - 3) = 0$$

and then use the quadratic formula with the equation $x^2 + 2x - 3 = 0$, where $a = 1$, $b = 2$, and $c = -3$, you get the same solution. Try this and see.

Mathematics in Action

Landscape Design—Then and Now

Doing the calculations for a garden or a swimming pool border falls well within the province of a landscape designer. On the same residential property, he or she might also be planning walking paths, laying out locations for trees and shrubbery, and, if a house is not yet erected, consulting with the architects and the builder.

In days past, design work was done with paper and pen (or the closest equivalents before paper and pen existed) and with the building of physical models. In addition to extensive sketches and drawings, the landscape designer had to be prepared to man-

ually calculate all the expenses, the materials needed, and the hours of labor required to turn a vision into real flowers, grass, trees, walls, and so on. One even hears of complex mathematical formulas to determine whether a design is "balanced" or not.

Today the mathematics of landscape design has been captured in computer programs that allow the designer, professional or otherwise, to work with 2-D or 3-D visual models. Using these models, the landscaper can choose from among hundreds of plants and trees and place them onto a layout and select materials such as marble, stone, and wood to see what they look like. These programs also allow the landscaper to change the size, shape, border, and location of any element, and finally, to estimate the cost of the materials needed for a given design.

In addition to the geometric formulas needed, many additional equations are used to write the computer programs that provide the flexibility for the landscaper to move things around on the computer screen to view different designs. Although we often do not take the time to consider what makes a computer program function as it does, it is always the same thing, mathematical equations; and it takes people who understand the underlying mathematics to write those computer programs.

1. a) $b^2 - 4ac$ **b)** discriminant: greater than 0, two solutions; equal to 0, one solution; less than 0, no real solution

Exercise Set 10.3

3. $x = \dfrac{-b \pm \sqrt{b^2 - 4ac}}{2a}$ **4.** Answers will vary.   **5.** Write the equation in standard form.

6. Did not leave the -5 in the numerator   **7.** Values used for b and c are incorrect

Concept/Writing Exercises

1. a) What is the discriminant?

b) Explain how the discriminant can be used to determine the number of real number solutions a quadratic equation has.

2. How many real number solutions does a quadratic equation have if the discriminant equals

a) -4?   **b)** 0?   **c)** $\dfrac{1}{2}$?   **a)** none   **b)** one   **c)** two

3. Without looking at your notes, write down the quadratic formula. You must memorize this formula.

4. Explain in your own words why a quadratic equation will have two real number solutions when the discriminant is greater than 0, one real number solution when the discriminant is equal to 0, and no real number solution when the discriminant is less than 0. Use the quadratic formula in explaining your answer.

5. What is the first step in solving a quadratic equation using the quadratic formula?

6. A student using the quadratic formula to solve $2x^2 + 5x + 1 = 0$ wrote down the following steps. What is wrong with the student's work?

$$x = \frac{-5 \pm \sqrt{5^2 - 4(2)(1)}}{2(2)}$$
$$= -5 \pm \frac{\sqrt{25 - 8}}{4}$$
$$= -5 \pm \frac{\sqrt{17}}{4}$$

7. A student using the quadratic formula to solve $x^2 = 4x + 7$ wrote down the following steps. What is wrong with the student's work?

$$x = \frac{-4 \pm \sqrt{4^2 - 4(1)(7)}}{2(1)}$$
$$= \frac{-4 \pm \sqrt{16 - 28}}{2}$$
$$= \frac{-4 \pm \sqrt{-12}}{2}$$

8. A student using the quadratic formula to solve $x^2 + 4 = 0$ listed the following steps. What is wrong with the student's work?   Evaluated $\sqrt{-16}$ as 4

$$x = \frac{0 \pm \sqrt{0^2 - 4(1)(4)}}{2(1)}$$
$$= \frac{\pm\sqrt{-16}}{2}$$
$$= \frac{\pm 4}{2}$$
$$= \pm 2$$

9., 10., 13., 16., 18., 21., 23., 24.: two real number solutions
14., 15., 25., 26.: one real number solution
11., 12., 17., 19., 20., 22., 39.: no real number solution

Practice the Skills

Determine whether each equation has two distinct real number solutions, one real number solution, or no real number solution.

9. $x^2 + 5x - 9 = 0$ **10.** $x^2 + 2x - 5 = 0$ **11.** $2x^2 + x + 1 = 0$ **12.** $r^2 + r + 3 = 0$
13. $6n^2 + 3n - 7 = 0$ **14.** $4x^2 - 24x = -36$ **15.** $2x^2 = 16x - 32$ **16.** $5x^2 - 4x = 7$
17. $2x^2 - 7x + 8 = 0$ **18.** $z^2 = 5z + 9$ **19.** $4x = 8 + x^2$ **20.** $5x - 8 = 3x^2$
21. $x^2 + 7x - 3 = 0$ **22.** $2x^2 - 6x + 9 = 0$ **23.** $3k^2 - 9 = 0$ **24.** $6x^2 - 5x = 0$
25. $9 = -t^2 + 6t$ **26.** $-25 = w^2 + 10w$

Use the quadratic formula to solve each equation. If the equation has no real number solution, so state.

27. $x^2 - 2x - 8 = 0$ $4, -2$ **28.** $x^2 + 11x + 10 = 0$ $-1, -10$ **29.** $x^2 + 9x + 18 = 0$ $-3, -6$
30. $x^2 - 3x - 10 = 0$ $5, -2$ **31.** $x^2 - 6x = -5$ $5, 1$ **32.** $x^2 + 5x - 24 = 0$ $3, -8$
33. $30 = -z^2 - 11z$ $-5, -6$ **34.** $x^2 - 49 = 0$ $7, -7$ **35.** $m^2 = 81$ $9, -9$
36. $x^2 - 6x = 0$ $6, 0$ **37.** $t^2 - 5t = 0$ $5, 0$ **38.** $z^2 - 17z + 72 = 0$ $9, 8$
39. $2x^2 - 3x + 2 = 0$ **40.** $n^2 - 7n + 10 = 0$ $5, 2$ **41.** $2y^2 - 7y + 4 = 0$ $(7 \pm \sqrt{17})/4$

42. $15x^2 - 7x = 2$ $\frac{2}{3}, -\frac{1}{5}$

43. $6x^2 = -x + 1$ $\frac{1}{3}, -\frac{1}{2}$

44. $4r^2 + r - 3 = 0$ $\frac{3}{4}, -1$

🔒 **45.** $2x^2 = 5x + 7$ $\frac{7}{2}, -1$

46. $3w^2 - 4w + 5 = 0$

47. $2s^2 - 4s + 3 = 0$

48. $x^2 - 7x + 3 = 0$ $(7 \pm \sqrt{37})/2$

49. $4x^2 = x + 5$ $\frac{5}{4}, -1$

50. $x^2 - 3x - 1 = 0$ $(3 \pm \sqrt{13})/2$

51. $2x^2 - 7x = 9$ $\frac{9}{2}, -1$

52. $-x^2 + 2x + 15 = 0$ $-3, 5$

53. $-2x^2 + 11x - 15 = 0$ $\frac{5}{2}, 3$

54. $6y^2 + 9 = -5y$

55. $2t^2 + 4t - 30 = 0$ $3, -5$

56. $3r^2 - 27r + 54 = 0$ $3, 6$

57. $36 = -4s^2 + 40s$ $1, 9$

58. $20 = -5p^2 - 5p$

Problem Solving

46., 47., 54., 58.: no real number solution   67. 40 chairs   69. ≈1.11 sec and ≈4.52 sec

---

59. ***Product of Numbers*** The product of two consecutive positive integers is 42. Find the two consecutive integers.   6, 7

60. ***Dimensions of Rectangle*** The length of a rectangle is 3 feet longer than its width. Find the dimensions of the rectangle if its area is 28 square feet. $w = 4$ ft, $l = 7$ ft

🔒 **61.** ***Dimensions of Rectangle*** The length of a rectangle is 3 feet smaller than twice its width. Find the length and width of the rectangle if its area is 20 square feet. $w = 4$ ft, $l = 5$ ft

62. ***Rectangular Garden*** Sean McDonald's rectangular garden measures 20 feet by 30 feet. He wishes to build a uniform width brick walkway around his garden that covers an area of 336 square feet. What will be the width of the walkway?   3 ft

63. ***Swimming Pool*** Harold Goldstein and his wife Elaine recently installed a built-in rectangular swimming pool measuring 30 feet by 40 feet. They want to add a decorative tile border of uniform width around all sides of the pool. How wide can they make the tile border if they purchased enough tile to cover 296 square feet?   2 ft

64. ***Pottery Studio*** Julie Bonds is planning to plant a grass lawn of uniform width around her rectangular pottery studio, which measures 48 feet by 36 feet. How far will the lawn extend from the studio if Julie has only enough seed to plant 4000 square feet of grass?   ≈16.96 ft

65. ***Diagonals in a Polygon*** The number of diagonals, d, in a polygon with n sides is given by the formula $d = \dfrac{n^2 - 3n}{2}$. For example, a pentagon, a polygon with 5 sides, has $d = \dfrac{5^2 - 15}{2} = 5$ diagonals; see the figure.

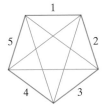

If a polygon has 27 diagonals, how many sides does it have?   9 sides

66. ***Diagonals in a Polygon*** If a polygon has 35 diagonals, how many sides does it have? See Exercise 65.   10 sides

67. ***Rocking Chairs*** The cost, c, for manufacturing x rocking chairs is given by $c = x^2 - 16x + 40$. Find the number of chairs manufactured if the cost is $1000.

68. ***Manufacturing Cost*** Repeat Exercise 67 for a cost of $2680.   60 chairs

69. ***Model Rocket*** Phil Chefetz launches a model rocket from the ground. The height, s, of the rocket above the ground at time t seconds after it is launched can be found by the formula $s = -16t^2 + 90t$. Find how long it will take for the rocket to reach a height of 80 feet.

70. ***Model Rocket*** Repeat Exercise 69 for a height of 100 feet.   ≈1.52 sec and ≈4.10 sec

71. a) $c < 9$ **b)** $c = 9$ **c)** $c > 9$ **72. a)** $c < \dfrac{9}{8}$ **b)** $c = \dfrac{9}{8}$ **c)** $c > \dfrac{9}{8}$ **73. a)** $c > -3$ **b)** $c = -3$ **c)** $c < -3$

Challenge Problems

*Find all the values of c that will result in each equation having **a)** two real number solutions, **b)** one real number solution, and **c)** no real number solution.*

71. $x^2 + 6x + c = 0$

72. $2x^2 + 3x + c = 0$

73. $-3x^2 + 6x + c = 0$

74. ***Fenced in Area*** Farmer Justina Wells wishes to form a rectangular region along a river bank by constructing fencing on three sides, as illustrated in the diagram. If she has only 400 feet of fencing and wishes to enclose an area of 15,000 square feet, find the dimensions of the rectangular region.
300 ft by 50 ft, or 100 ft by 150 ft

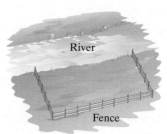
River

Fence

Group Activity

75. In Section 10.4 we will graph quadratic equations. We will learn that the graphs of quadratic equations are *parabolas*. The graph of the quadratic equation $y = x^2 - 2x - 8$ is illustrated below.

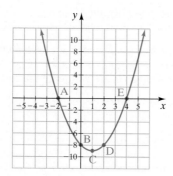

a) Each member of the group, copy the graph in your notebook.

b) Group Member 1: List the ordered pairs corresponding to points A and B. Verify that each ordered pair is a solution to the equation $y = x^2 - 2x - 8$.

c) Group Member 2: List the ordered pairs corresponding to points C and D. Verify that each ordered pair is a solution to the equation $y = x^2 - 2x - 8$.

d) Group Member 3: List the ordered pair corresponding to point E. Verify that the ordered pair is a solution to the equation $y = x^2 - 2x - 8$.

e) Individually, graph the equation $y = 2x - 3$ on the same axes that you used in part **a)**. Compare your graphs with the other members of your group.

f) The two graphs represent the system of equations

$$y = x^2 - 2x - 8$$
$$y = 2x - 3$$

As a group, estimate the points of intersection of the graphs. $(-1, -5), (5, 7)$

g) If we set the two equations equal to each other, we obtain the following quadratic equation in only the variable x.

$$x^2 - 2x - 8 = 2x - 3$$

As a group, solve this quadratic equation. Does your answer agree with the x-coordinates of the points of intersection from part **f)**? $-1, 5$; yes

h) As a group, use the values of x found in part **g)** to find the values of y in $y = x^2 - 2x - 8$ and $y = 2x - 3$. Does your answer agree with the y-coordinates of the points of intersection from part **f)**? $x = -1, y = -5; x = 5, y = 7$; yes

75. b) A, $(-2, 0)$; B, $(0, -8)$ **c)** C, $(1, -9)$; D, $(2, -8)$ **d)** E, $(4, 0)$ **e)** See graphing answer section, page G15.

Cumulative Review Exercises

[5.6, 10.2, 10.3] *Solve the following quadratic equations by **a)** factoring, **b)** completing the square, and **c)** the quadratic formula. If the equation cannot be solved by factoring, so state.*

76. $x^2 - 13x + 42 = 0$ 7, 6   **77.** $6x^2 + 11x - 35 = 0$ $\dfrac{5}{3}, -\dfrac{7}{2}$ **78.** $2x^2 + 3x - 4 = 0$ **79.** $6x^2 = 54$ 3, -3
78. cannot be solved by factoring; $(-3 \pm \sqrt{41})/4$

[6.4] 80. Subtract $\dfrac{x}{2x^2 + 7x - 4} - \dfrac{2}{x^2 - x - 20}$ $\dfrac{x^2 - 9x + 2}{(2x - 1)(x + 4)(x - 5)}$

10.4 GRAPHING QUADRATIC EQUATIONS

SSM Study Guide CD/Video

MathPro 4/5 PH Math Tutor Center prenhall.com/Angel

1. Graph quadratic equations in two variables.
2. Find the coordinates of the vertex of a parabola.
3. Use symmetry to graph quadratic equations.
4. Find the x-intercepts of the graph of a quadratic equation.

1 Graph Quadratic Equations in Two Variables

HELPFUL HINT

STUDY TIP

In this section we will graph quadratic equations. Quadratic equations can be graphed solely by plotting points, as we did in Chapter 7 when we graphed linear equations. However, there are certain things we can do to make graphing quadratic equations easier. These include finding the vertex of the graph, finding the x-intercepts of the graph, and using symmetry to draw the graph. We will discuss each of these items shortly.

TEACHING TIP
Point out that in the last section we were solving quadratic equations for the specific case when $y = 0$. Now we are graphing all the cases.

In Section 7.2 we learned how to graph linear equations. In this section we graph quadratic equations of the form

$$y = ax^2 + bx + c, \quad a \neq 0$$

The graph of every quadratic equation of this form is a **parabola**. The graph of $y = ax^2 + bx + c$ will have one of the shapes indicated in Figure 10.3. When a quadratic equation is in the form $y = ax^2 + bx + c$, the sign of a, the numerical coefficient of the squared term, will determine whether the parabola will open upward (Fig. 10.3a) or downward (Fig. 10.3b). When a is positive, the parabola will open upward, and when a is negative, the parabola will open downward.

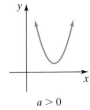
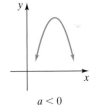

$a > 0$
Parabola opens upward
(a)

$a < 0$
Parabola opens downward
(b)

FIGURE 10.3

TEACHING TIP
Have students use a graphing calculator to investigate how the a and the c in a quadratic equation of the form $y = ax^2 + bx + c$ influence the graph of the equation.

The **vertex** is the lowest point on a parabola that opens upward or the highest point on a parabola that opens downward (Fig. 10.4). Graphs of quadratic equations of the form $y = ax^2 + bx + c$ have **symmetry** about a line through the vertex. This means that if we fold the paper along this imaginary line, called the **axis of symmetry**, the right and left sides of the graph will coincide.

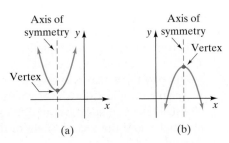

FIGURE 10.4   (a)   (b)

One method you can use to graph a quadratic equation is to plot it point by point. When determining points to plot, select values for x and determine the corresponding values for y.

EXAMPLE 1 Graph the equation $y = x^2$.

Solution Since $a = 1$, which is positive, this parabola opens upward.

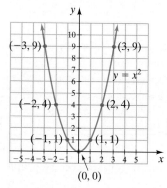

FIGURE 10.5

$$y = x^2$$

		$y = x^2$		x	y
Let $x = 3$,		$y = (3)^2 = 9$		3	9
Let $x = 2$,		$y = (2)^2 = 4$		2	4
Let $x = 1$,		$y = (1)^2 = 1$		1	1
Let $x = 0$,		$y = (0)^2 = 0$		0	0
Let $x = -1$,		$y = (-1)^2 = 1$		−1	1
Let $x = -2$,		$y = (-2)^2 = 4$		−2	4
Let $x = -3$,		$y = (-3)^2 = 9$		−3	9

Connect the points with a smooth curve (Fig. 10.5). Note how the graph is symmetric about the line $x = 0$ (or the y-axis). ✳

EXAMPLE 2 Graph the equation $y = -2x^2 + 4x + 6$.

Solution Since $a = -2$, which is negative, this parabola opens downward.

TEACHING TIP
Point out that a parabola is a smooth curve, that is, it is rounded at the vertex. It does not come to a point.

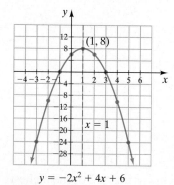

$y = -2x^2 + 4x + 6$

FIGURE 10.6

See Teaching Tip on page 641.

NOW TRY EXERCISE 23

$$y = -2x^2 + 4x + 6$$

		$y = -2x^2 + 4x + 6$		x	y
Let $x = 5$,		$y = -2(5)^2 + 4(5) + 6 = -24$		5	−24
Let $x = 4$,		$y = -2(4)^2 + 4(4) + 6 = -10$		4	−10
Let $x = 3$,		$y = -2(3)^2 + 4(3) + 6 = 0$		3	0
Let $x = 2$,		$y = -2(2)^2 + 4(2) + 6 = 6$		2	6
Let $x = 1$,		$y = -2(1)^2 + 4(1) + 6 = 8$		1	8
Let $x = 0$,		$y = -2(0)^2 + 4(0) + 6 = 6$		0	6
Let $x = -1$,		$y = -2(-1)^2 + 4(-1) + 6 = 0$		−1	0
Let $x = -2$,		$y = -2(-2)^2 + 4(-2) + 6 = -10$		−2	−10
Let $x = -3$,		$y = -2(-3)^2 + 4(-3) + 6 = -24$		−3	−24

Note how the graph (Fig. 10.6) is symmetric about the line $x = 1$, which is dashed because it is not part of the graph. The vertex of this parabola is the point $(1, 8)$. Since the y-values are large, the y-axis has been marked with 4-unit intervals to allow us to graph the points $(-3, -24)$ and $(5, -24)$. The arrows on the ends of the graph indicate that the parabola continues indefinitely. ✳

2 Find the Coordinates of the Vertex of a Parabola

When graphing quadratic equations, how do we decide what values to use for x? When the location of the vertex is unknown, this is a difficult question to answer. When the location of the vertex is known, it becomes more obvious which values to use.

In Example 2, the axis of symmetry is $x = 1$, and the x-coordinate of the vertex is also 1. For a quadratic equation in the form $y = ax^2 + bx + c$, both the axis of symmetry and the x-coordinate of the vertex can be found by using the following formula.

TEACHING TIP
Point out the duplicate y-values in the table in Example 2. Then ask, "Can you find the vertex if you know the x-values for duplicate y-values? Explain." Point out that the quadratic formula gives x-values for both occurrences of $y = 0$. Then have them average the two x-values determined by the quadratic formula to find the x-value of the vertex.

Axis of Symmetry and x-Coordinate of the Vertex

$$x = -\frac{b}{2a}$$

In the quadratic equation in Example 2, $a = -2$, $b = 4$, and $c = 6$. Substituting these values in the formula for the axis of symmetry gives

$$x = -\frac{b}{2a} = -\frac{4}{2(-2)} = -\frac{4}{-4} = 1$$

Thus, the graph is symmetric about the line $x = 1$, and the x-coordinate of the vertex is 1.

The y-coordinate of the vertex can be found by substituting the value of the x-coordinate of the vertex into the quadratic equation and solving for y.

$$y = -2x^2 + 4x + 6$$
$$= -2(1)^2 + 4(1) + 6$$
$$= -2(1) + 4 + 6$$
$$= -2 + 4 + 6$$
$$= 8$$

Thus, the vertex is at the point $(1, 8)$.

For a quadratic equation of the form $y = ax^2 + bx + c$, the y-coordinate of the vertex can also be found by the following formula.

y-Coordinate of the Vertex

$$y = \frac{4ac - b^2}{4a}$$

For Example 2,

$$y = \frac{4ac - b^2}{4a}$$
$$= \frac{4(-2)(6) - 4^2}{4(-2)}$$
$$= \frac{-48 - 16}{-8} = \frac{-64}{-8} = 8$$

You may use the method of your choice to find the y-coordinate of the vertex. Both methods result in the same value of y.

3 Use Symmetry to Graph Quadratic Equations

One method to use in selecting points to plot when graphing parabolas is to determine the axis of symmetry and the vertex of the graph. Then select nearby values of x on either side of the axis of symmetry. When graphing the equation, make use of the symmetry of the graph.

EXAMPLE 3 **a)** Find the axis of symmetry of the graph of the equation $y = x^2 + 6x + 5$.
b) Find the vertex of the graph.
c) Graph the equation.

Solution   **a)** $a = 1$, $b = 6$, $c = 5$.

$$x = -\frac{b}{2a} = -\frac{6}{2(1)} = -3$$

The parabola is symmetric about the line $x = -3$. The x-coordinate of the vertex is -3.

b) Now find the y-coordinate of the vertex. Substitute -3 for x in the quadratic equation.

$$y = x^2 + 6x + 5$$
$$y = (-3)^2 + 6(-3) + 5 = 9 - 18 + 5 = -4$$

The vertex is at the point $(-3, -4)$.

c) Since the axis of symmetry is $x = -3$, we will select values for x that are greater than or equal to -3. It is often helpful to plot each point as it is determined. If a point does not appear to lie on the parabola, check it.

$$y = x^2 + 6x + 5$$

Let $x = -2$,	$y = (-2)^2 + 6(-2) + 5 = -3$
Let $x = -1$,	$y = (-1)^2 + 6(-1) + 5 = 0$
Let $x = 0$,	$y = (0)^2 + 6(0) + 5 = 5$

x	y
-2	-3
-1	0
0	5

These points are plotted in Figure 10.7a. The entire graph of the equation is illustrated in Figure 10.7b. Note how we use symmetry to complete the graph. The points $(-2, -3)$ and $(-4, -3)$ are each 1 horizontal unit from the axis of symmetry, $x = -3$. The points $(-1, 0)$ and $(-5, 0)$ are each 2 horizontal units from the axis of symmetry, and the points $(0, 5)$ and $(-6, 5)$ are each 3 horizontal units from the axis of symmetry.

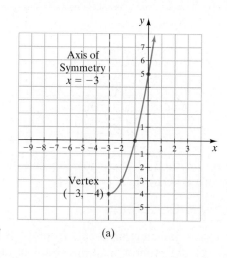

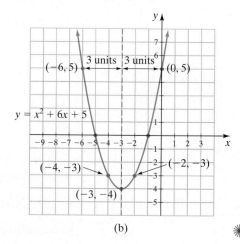

FIGURE 10.7   (a)   (b)

EXAMPLE 4   Graph the equation $y = -2x^2 + 5x - 4$.

Solution $a = -2$, $b = 5$, $c = -4$.
Since $a < 0$, this parabola will open downward.

$$\text{Axis of symmetry: } x = -\frac{b}{2a}$$

$$= -\frac{5}{2(-2)} = -\frac{5}{-4} = \frac{5}{4} \quad \left(\text{or } 1\frac{1}{4} \right)$$

Since the x-value of the vertex is a fraction, we would need to add fractions if we wished to find y by substituting $\frac{5}{4}$ for x in the given equation. Therefore, we will use the formula to find the y-coordinate of the vertex.

$$y = \frac{4ac - b^2}{4a}$$

$$= \frac{4(-2)(-4) - 5^2}{4(-2)} = \frac{32 - 25}{-8} = \frac{7}{-8} = -\frac{7}{8}$$

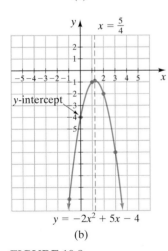

(a)

The vertex of this graph is at the point $\left(\frac{5}{4}, -\frac{7}{8} \right)$. Since the axis of symmetry is $x = \frac{5}{4}$, we will begin by selecting values of x that are greater than $\frac{5}{4}$, or $1\frac{1}{4}$.

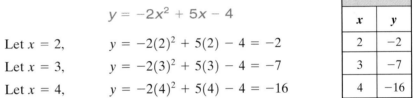

$$y = -2x^2 + 5x - 4$$

			x	y
Let $x = 2$,	$y = -2(2)^2 + 5(2) - 4 = -2$		2	-2
Let $x = 3$,	$y = -2(3)^2 + 5(3) - 4 = -7$		3	-7
Let $x = 4$,	$y = -2(4)^2 + 5(4) - 4 = -16$		4	-16

When the axis of symmetry is a fractional value, be very careful when constructing the graph. You should plot as many additional points as needed. Following we determine some values of y when x is less than $\frac{5}{4}$.

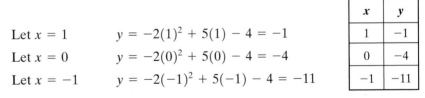

			x	y
Let $x = 1$	$y = -2(1)^2 + 5(1) - 4 = -1$		1	-1
Let $x = 0$	$y = -2(0)^2 + 5(0) - 4 = -4$		0	-4
Let $x = -1$	$y = -2(-1)^2 + 5(-1) - 4 = -11$		-1	-11

(b)

FIGURE 10.8

Figure 10.8a shows the points plotted on the right side of the axis of symmetry. Figure 10.8b shows the completed graph. The point $(4, -16)$ is not shown on the graphs. ✳

NOW TRY EXERCISE 19

When graphing, we will often have to evaluate quadratic expressions to obtain values for y. You may wish to review *Evaluating Expressions* as explained in the Using Your Calculator box in Section 1.9, page 79.

(a)

(b)

FIGURE 10.9 Not every shape that resembles a parabola is a parabola. For example, the St. Louis Arch, Figure 10.9a resembles a parabola, but it is not a parabola. However, the bridge over the Mississippi near Jefferson Barracks, Missouri, which connects Missouri and Illinois, Figure 10.9b, is a parabola.

Using Your Graphing Calculator

Let's discuss how to find the vertex of a parabola. In Example 3 we graphed $y = x^2 + 6x + 5$. This graph is shown on the window of a TI-83 Plus in Figure 10.10 using the standard window settings.

We can use the CALC (calculate) menu to find the vertex. The vertex of a parabola is also called the minimum point (if the parabola opens upward) or the maximum point (if the parabola opens downward). To obtain the CALC menu press $\boxed{2^{nd}}$ $\boxed{\text{TRACE}}$. The CALC menu is shown in Figure 10.11.

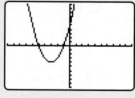

FIGURE 10.10   FIGURE 10.11

Since this graph opens upward, it has a minimum point. Scroll down the CALC menu to option 3: minimum, and press $\boxed{\text{ENTER}}$, which gives you the graph shown in Figure 10.12. Notice the cursor is at the y-intercept, $(0, 5)$. Use the left-arrow key to move the cursor to the left. Notice the values of X and Y changing as you move the cursor. The value of Y will decrease until the cursor reaches the vertex. When the cursor moves past the vertex, the value of Y will begin to increase. Move the cursor just to the left of the vertex, and press $\boxed{\text{ENTER}}$. Notice the words *Left Bound?* change to *Right Bound?*. Now move the cursor just to the right of the vertex and press $\boxed{\text{ENTER}}$. The screen now shows *Guess?*. Press $\boxed{\text{ENTER}}$ again and you get the screen shown in Figure 10.13.

Figure 10.13 shows that the minimum point of the graph occurs when $x = -3$ and $y = -4$. Thus, the vertex is at $(-3, -4)$.

If the parabola opened downward, the graph would have a maximum point. To obtain the vertex in this case we would use the CALC menu, option 4: maximum, to obtain the vertex. You then would follow the same basic procedure to find the maximum point.

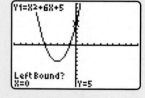

FIGURE 10.12

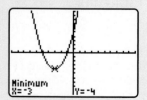

FIGURE 10.13

Exercises

Use your graphing calculator to find the vertex of the graph of each equation. You may need to adjust your window settings to show the vertex.

1. $y = x^2 - 2x - 3$
$(1, -4)$

2. $y = x^2 - 7x + 10$
$(3.5, -2.25)$

3. $y = -2x^2 - 2x + 12$
$(-0.5, 12.5)$

4. $y = 2x^2 - 7x - 15$
$(1.75, -21.125)$

4 Find the *x*-Intercepts of the Graph of a Quadratic Equation

In Example 4 the graph crossed the *y*-axis at $y = -4$. Recall from earlier sections that to find the *y*-intercept we let $x = 0$ and solve the resulting equation for *y*. The locations of the *x* and *y*-intercepts are often helpful when graphing quadratic equations.

To find the *x*-intercept when graphing straight lines in Section 7.2, we set $y = 0$ and found the corresponding value of *x*. We did this because the value of *y* where a graph crosses the *x*-axis is 0. We use the same procedure here when finding the *x*-intercepts of a quadratic equation of the form $y = ax^2 + bx + c$. To find the *x*-intercepts of the graph of $y = ax^2 + bx + c$, we set *y* equal to 0 and solve the resulting equation, $ax^2 + bx + c = 0$. The real number solutions of the equation $ax^2 + bx + c = 0$ are called the *zeros* (or *roots*) of the equation, for when they are substituted for *x* in $y = ax^2 + bx + c$, *y* has a value of zero. The zeros can be used to determine the *x*-intercepts of the graph of $y = ax^2 + bx + c$ because the zeros are the *x*-coordinates of the *x*-intercepts. To solve equations of the form $ax^2 + bx + c = 0$, we can use factoring as explained in Section 5.6; completing the square as explained in Section 10.2; or the quadratic formula as explained in Section 10.3.

In Section 10.3 we mentioned that when solving quadratic equations of the form $ax^2 + bx + c = 0$, when the discriminant, $b^2 - 4ac$, is greater than zero, there are two distinct real number solutions; when it is equal to zero, there is one real number solution; and when it is less than zero, there is no real number solution.

A quadratic equation of the form $y = ax^2 + bx + c$ will have either two distinct *x*-intercepts (Fig. 10.14a), one *x*-intercept (Fig. 10.14b), or no *x*-intercepts (Fig. 10.14c). The number of *x*-intercepts may be determined by the discriminant.

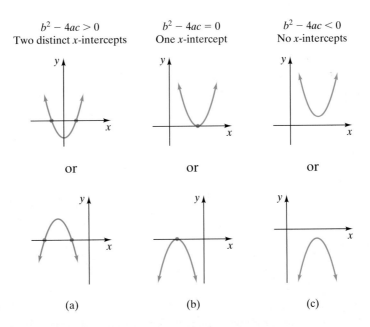

$b^2 - 4ac > 0$
Two distinct *x*-intercepts

$b^2 - 4ac = 0$
One *x*-intercept

$b^2 - 4ac < 0$
No *x*-intercepts

or

or

or

FIGURE 10.14   (a)   (b)   (c)

The *x*-intercepts can be found graphically. They may also be found algebraically by setting *y* equal to 0 and solving the resulting equation, as in Example 5.

EXAMPLE 5   **a)** Find the *x*-intercepts of the graph of the equation $y = x^2 - 6x - 7$ by factoring, by completing the square, and by the quadratic formula.

b) Graph the equation.

Solution   **a)** To find the x-intercepts algebraically we set y equal to 0 and solve the resulting equation, $x^2 - 6x - 7 = 0$. We will solve this equation by all three algebraic methods.

Method 1:   Factoring.

$$x^2 - 6x - 7 = 0$$
$$(x - 7)(x + 1) = 0$$
$$x - 7 = 0 \quad \text{or} \quad x + 1 = 0$$
$$x = 7 \qquad\qquad x = -1$$

Method 2:   Completing the square.

$$x^2 - 6x - 7 = 0$$
$$x^2 - 6x = 7$$
$$x^2 - 6x + \boxed{9} = 7 + \boxed{9}$$
$$(x - 3)^2 = 16$$
$$x - 3 = \pm 4$$
$$x = 3 \pm 4$$
$$x = 3 + 4 \quad \text{or} \quad x = 3 - 4$$
$$x = 7 \qquad\qquad x = -1$$

Method 3:   Quadratic formula.

$$x^2 - 6x - 7 = 0$$
$$a = 1, \qquad b = -6, \qquad c = -7$$
$$x = \frac{-b \pm \sqrt{b^2 - 4ac}}{2a}$$
$$= \frac{-(-6) \pm \sqrt{(-6)^2 - 4(1)(-7)}}{2(1)}$$
$$= \frac{6 \pm \sqrt{36 + 28}}{2}$$
$$= \frac{6 \pm \sqrt{64}}{2}$$
$$= \frac{6 \pm 8}{2}$$

$$x = \frac{6 + 8}{2} \quad \text{or} \quad x = \frac{6 - 8}{2}$$
$$= \frac{14}{2} = 7 \qquad\qquad = \frac{-2}{2} = -1$$

Note that the same solutions, 7 and -1, were obtained by all three methods. The graph of the equation $y = x^2 - 6x - 7$ will have two distinct x-intercepts. The graph will cross the x-axis at 7 and -1. The x-intercepts are $(7, 0)$ and $(-1, 0)$.

b) Since $a > 0$, this parabola opens upward.

$$\text{axis of symmetry:} \quad x = -\frac{b}{2a} = -\frac{-6}{2(1)} = \frac{6}{2} = 3$$

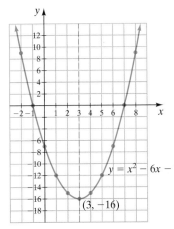

FIGURE 10.15

$$y = x^2 - 6x - 7$$

Let $x = 3$,	$y = 3^2 - 6(3) - 7 = -16$	
Let $x = 4$,	$y = 4^2 - 6(4) - 7 = -15$	
Let $x = 5$,	$y = 5^2 - 6(5) - 7 = -12$	
Let $x = 6$,	$y = 6^2 - 6(6) - 7 = -7$	
Let $x = 7$,	$y = 7^2 - 6(7) - 7 = 0$	
Let $x = 8$,	$y = 8^2 - 6(8) - 7 = 9$	

x	y
3	−16
4	−15
5	−12
6	−7
7	0
8	9

The vertex is at $(3, -16)$. Again we use symmetry to complete the graph (Fig. 10.15). The x-intercepts $(7, 0)$ and $(-1, 0)$ agree with the answer obtained in part **a)**.

NOW TRY EXERCISE 31

Using Your Graphing Calculator

The x-intercepts of a parabola can be found using a graphing calculator. Read the manual that comes with your graphing calculator to find out how to determine the x-intercepts of a parabola. On the TI-83 Plus, you can use the CALC (calculate) menu with option 2: *zero* (see Figure 10.11).

Exercise Set 10.4

2. highest point on a parabola that opens downward, lowest point on a parabola that opens upward

5. **a)** where the graph crosses the x-axis **b)** Let $y = 0$ and solve for x. **6.** If you fold the paper along the line, the right and left sides of the graph will coincide.

Concept/Writing Exercises

1. What do we call the graph of a quadratic equation of the form $y = ax^2 + bx + c, a \neq 0$? a parabola

2. What is the vertex of a parabola?

3. Use your own words to explain how to find the coordinates of the vertex of a parabola. Answers will vary.

4. What determines whether the graph of a quadratic equation of the form $y = ax^2 + bx + c, a \neq 0$, is a parabola that opens upward or downward? Explain your answer. the sign of a; $a > 0$, upward; $a < 0$, downward

5. **a)** What are the x-intercepts of a graph?

   b) How can you find the x-intercepts of a graph algebraically?

6. What does it mean when we say that graphs of quadratic equations of the form $y = ax^2 + bx + c$ have symmetry about an imaginary vertical line through the vertex?

7. **a)** When graphing a quadratic equation of the form $y = ax^2 + bx + c$, what is the equation of the vertical line about which the parabola will be symmetric? $x = -\dfrac{b}{2a}$

   b) What is this vertical line called? axis of symmetry

8. How many x-intercepts will the graph of a quadratic equation have if the discriminant has a value of

   a) 25 two **b)** −2 none **c)** 0 one

Practice the Skills

Indicate the axis of symmetry, the coordinates of the vertex, and whether the parabola opens upward or downward.

9. $y = x^2 + 4x - 3$ $x = -2, (-2, -7)$, upward

10. $y = x^2 + 2x - 8$ $x = -1, (-1, -9)$, upward

11. $y = -x^2 + 3x - 4$ $x = \frac{3}{2}, \left(\frac{3}{2}, -\frac{7}{4}\right)$, downward

12. $y = 3x^2 + 6x - 9$ $x = -1, (-1, -12)$, upward

13. $y = -3x^2 + 2x + 8$ $x = \frac{1}{3}, \left(\frac{1}{3}, \frac{25}{3}\right)$, downward

14. $y = x^2 + 3x - 6$ $x = -\frac{3}{2}, \left(-\frac{3}{2}, -\frac{33}{4}\right)$, upward

15. $y = 4x^2 + 8x + 3$ $x = -1, (-1, -1)$ upward

16. $y = 3x^2 - 2x + 2$ $x = \frac{1}{3}, \left(\frac{1}{3}, \frac{5}{3}\right)$, upward

17. $y = 2x^2 + 3x + 5$ $x = -\frac{3}{4}, \left(-\frac{3}{4}, \frac{31}{8}\right)$ upward

18. $y = -x^2 + x + 8$ $x = \frac{1}{2}, \left(\frac{1}{2}, \frac{33}{4}\right)$, downward

19. $y = -5x^2 + 6x - 1$ $x = \frac{3}{5}, \left(\frac{3}{5}, \frac{4}{5}\right)$, downward

20. $y = -2x^2 - 6x - 5$ $x = -\frac{3}{2}, \left(-\frac{3}{2}, -\frac{1}{2}\right)$, downward

Graph each quadratic equation and determine the x-intercepts, if they exist. **21.–48.** See graphing answer section, pages G15 and G16.

21. $y = x^2 - 1$ **22.** $y = x^2 + 3$ **23.** $y = -x^2 + 5$ **24.** $y = -x^2 - 1$

25. $y = x^2 + 4x + 3$ **26.** $y = 2x^2 + 2$ **27.** $y = x^2 + 4x + 4$ **28.** $y = x^2 + 2x - 15$

29. $y = -x^2 - 5x - 4$ **30.** $y = x^2 - 4x - 5$ **31.** $y = x^2 + 5x - 6$ **32.** $y = -x^2 - 5x + 6$

33. $y = x^2 + 5x - 14$ **34.** $y = 2x^2 + 4x + 2$ **35.** $y = x^2 - 6x + 9$ **36.** $y = x^2 - 4x + 4$

37. $y = x^2 - 6x$ **38.** $y = -x^2 + 5x$ **39.** $y = 4x^2 + 12x + 9$ **40.** $y = x^2 - 2x + 1$

41. $y = -x^2 + 7x - 10$ **42.** $y = x^2 + x + 1$ **43.** $y = x^2 - 2x - 16$ **44.** $y = 2x^2 + 3x - 2$

45. $y = -2x^2 + 3x - 2$ **46.** $y = -4x^2 - 6x + 4$ **47.** $y = 2x^2 - x - 15$ **48.** $y = 6x^2 + 10x - 4$

Using the discriminant, determine the number of x-intercepts the graph of each equation will have. Do not graph the equation.

49. $y = 4x^2 - 2x - 16$ two **50.** $y = -x^2 - 5$ none **51.** $y = 4x^2 - 6x - 7$ two **52.** $y = x^2 - 6x + 9$ one

53. $y = x^2 - 20x + 100$ one **54.** $y = -4.3x^2 + 5.7x$ two **55.** $y = 5.7x^2 + 2x - 3.9$ two **56.** $y = 5x^2 - 13.2x + 9.3$ none

Problem Solving

57. none; the vertex is below the x-axis, the parabola opens downward **58.** two; the vertex is below the x-axis, the parabola opens upward **59.** one; the vertex of the parabola is on the x-axis

The graph of a quadratic equation of the form $y = ax^2 + bx + c$ is a parabola. The value of a (the coefficient of the squared term in the equation) and the vertex of the parabola, are given. Determine the number of x-intercepts the parabola will have. Explain how you determined your answer. **60.** none; the vertex is below the x-axis, the parabola opens downward

57. $a = -2$, vertex at $(0, -3)$

58. $a = 5$, vertex at $(4, -3)$

59. $a = -3$, vertex at $(-4, 0)$

60. $a = -1$, vertex at $(2, -4)$

61. yes; if y is set to 0, both equations have the same solutions 5 and −3

62. a) The graphs will be symmetric about the x-axis.
b) See graphing answer section, page G16.

61. Will the equations below have the same x-intercepts when graphed? Explain how you determined your answer.

$$y = x^2 - 2x - 15 \quad \text{and} \quad y = -x^2 + 2x + 15$$

62. a) How will the graphs of the following equations compare? Explain how you determined your answer.

$$y = x^2 - 2x - 8 \quad \text{and} \quad y = -x^2 + 2x + 8$$

b) Graph $y = x^2 - 2x - 8$ and $y = -x^2 + 2x + 8$ on the same axes.

63. *Height above Ground* An object is projected upward from the ground. The height of the object above the ground, in feet, at time t, in seconds, is illustrated in the following graph.

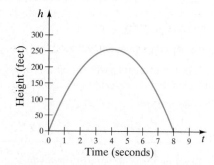

Time (seconds)

a) Estimate the maximum height the object will obtain. ≈255 ft

b) How long will it take for the object to reach its maximum height? 4 sec

c) Estimate how long will it take for the object to strike the ground. 8 sec

d) Estimate the object's height at 2 seconds and at 5 seconds. ≈190 ft, ≈240 ft

64. *Height above Ground* An object is projected straight upward from a platform that is 100 feet above the ground. The height of the object above the ground, in feet, at time t, in seconds, is illustrated in the following graph.

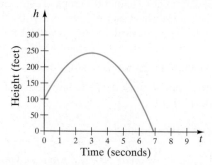

Time (seconds)

a) Estimate the maximum height the object will obtain. ≈245 ft

b) How long will it take for the object to reach its maximum height? 3 sec

c) Approximately how long will it take for the object to strike the ground? ≈7 sec

d) Estimate the object's height above the ground at 2 seconds and at 5 seconds. ≈230 ft, ≈180 ft

65. *Maximum Area* An area is to be fenced in along a river as shown in the figure. Only 180 feet of fencing is available. The fenced-in area is $A = $ length $\cdot$ width or $A = x(180 - 2x) = -2x^2 + 180x$.

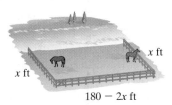

x ft

x ft

$180 - 2x$ ft

a) Graph $A = -2x^2 + 180x$.

b) Using the graph, estimate the value of x that will yield the maximum area.   45 ft

c) Estimate the maximum area.   4050 ft^2

66. *Maximum Area* Read Exercise 65. If 260 feet of fencing is available, the fenced-in area is $A = x(260 - 2x) = -2x^2 + 260x$.

a) Graph $A = -2x^2 + 260x$.

b) Using the graph, estimate the value of x that yields the maximum area.   65 ft

c) Estimate the maximum area.   8450 ft^2

Challenge Problems

67. a) Graph the quadratic equation $y = -x^2 + 6x$.

b) On the same axes, graph the quadratic equation $y = x^2 - 2x$.

c) Estimate the points of intersection of the graphs. The points represent the solution to the system of equations. $(0, 0), (4, 8)$

68. a) Graph the quadratic equation $y = x^2 + 2x - 3$.

b) On the same axes, graph the quadratic equation $y = -x^2 + 1$.

c) Estimate the points of intersection of the graphs. The points represent the solution to the system of equations. $(-2, -3), (1, 0)$

65. a), **66. a)**, **67.** and **68. a)** and **b)** See graphing answer section, page G17.

Cumulative Review Exercises

[6.4] **69.** Subtract $\dfrac{3}{x + 3} - \dfrac{x - 2}{x - 4}$. $\dfrac{-x^2 + 2x - 6}{(x + 3)(x - 4)}$

[6.6] **70.** Solve the equation $\dfrac{1}{3}(x + 6) = 3 - \dfrac{1}{4}(x - 5)$.

$\dfrac{27}{7}$

[8.3] **71.** Solve the system of equations

$$5x + 4y = 10$$
$$3x + 5y = -7 \quad (6, -5)$$

[9.5] **72.** Solve the equation $3\sqrt{2x} - 12 = 0$ 8

10.5 COMPLEX NUMBERS

SSM   Study Guide   CD/Video

MathPro 4/5   PH Math Tutor Center   prenhall.com/Angel

1 Write complex numbers using *i*.

2 Add and subtract complex numbers.

3 Solve quadratic equations with complex number solutions.

1 Write Complex Numbers Using *i*

In Section 10.3 we stated that the square roots of negative numbers, such as $\sqrt{-11}$, are not real numbers. Numbers like $\sqrt{-11}$ are called **imaginary numbers**. Such numbers are called imaginary because when they were introduced many mathematicians refused to believe that they existed. Although they do not belong to the set of real numbers, the imaginary numbers do exist and are very useful in mathematics and science.

Every imaginary number has $\sqrt{-1}$ as a factor. The $\sqrt{-1}$, called the **imaginary unit**, is often denoted by the letter *i*.

$$i = \sqrt{-1}$$

TEACHING TIP
Stress that i is a shorthand for $\sqrt{-1}$.

We can therefore write

$$\sqrt{-4} = \sqrt{4}\,\sqrt{-1} = 2\sqrt{-1} = 2i$$

$$\sqrt{-9} = \sqrt{9}\,\sqrt{-1} = 3\sqrt{-1} = 3i$$

$$\sqrt{-11} = \sqrt{11}\,\sqrt{-1} = \sqrt{11}i \quad \text{or} \quad i\sqrt{11}$$

We will generally write $i\sqrt{11}$ rather than $\sqrt{11}i$ to avoid confusion with $\sqrt{11i}$.

To help in writing square roots of negative numbers using i, we give the following rule.

> For any positive real number n,
> $$\sqrt{-n} = i\sqrt{n}$$

Examples

$$\sqrt{-4} = i\sqrt{4} = 2i \qquad \sqrt{-3} = i\sqrt{3}$$

$$\sqrt{-25} = i\sqrt{25} = 5i \qquad \sqrt{-10} = i\sqrt{10}$$

The real number system is a part of a larger number system, called the *complex number system*. Now we will discuss **complex numbers**.

DEFINITION

> Every number of the form
> $$a + bi$$
> where a and b are real numbers, is a **complex number**.

Every real number and every imaginary number are also complex numbers. A complex number has two parts: a real part, a, and an imaginary part, b.

Real part ⟶ ⟵ Imaginary part
$$a + bi$$

If $b = 0$, the complex number is a real number. If $a = 0$, the complex number is a *pure imaginary number*.

Examples of Complex Numbers

$3 + 4i$	$a = 3, \; b = 4$	
$5 - i\sqrt{3}$	$a = 5, \; b = -\sqrt{3}$	
5	$a = 5, \; b = 0$	(real number, $b = 0$)
$2i$	$a = 0, \; b = 2$	(imaginary number, $a = 0$)
$-i\sqrt{7}$	$a = 0, \; b = -\sqrt{7}$	(imaginary number, $a = 0$)

We stated that all real numbers and imaginary numbers are also complex numbers. The relationship between the various sets of numbers is illustrated in Figure 10.16. As the figure illustrates, the real numbers include both the rational numbers and the irrational numbers, and the complex numbers include both the real numbers and the non-real numbers.

Complex Numbers		
Real Numbers		Nonreal Numbers
Rational numbers $\frac{1}{2}, -\frac{3}{5}, \frac{9}{4}$ Integers $-4, -9,$ Whole numbers $0, 4, 12$	Irrational numbers $\sqrt{2}, \sqrt{3}$ $-\sqrt{7}, \pi$	$\sqrt{-4}$ $2 + 3i$ $6 - 4i$ $\sqrt{2} + i\sqrt{3}$ $i\sqrt{5}$ $6i$

FIGURE 10.16

EXAMPLE 1 Write each of the following complex numbers in the form $a + bi$.
a) $5 + \sqrt{-4}$ **b)** $3 - \sqrt{-12}$

Solution **a)** $5 + \sqrt{-4} = 5 + \sqrt{4}\,\sqrt{-1}.$
$$= 5 + 2i$$

b) $3 - \sqrt{-12} = 3 - \sqrt{12}\,\sqrt{-1}$
$$= 3 - \sqrt{4}\,\sqrt{3}\,\sqrt{-1}$$

NOW TRY EXERCISE 21
$$= 3 - 2\sqrt{3}i \quad \text{or} \quad 3 - 2i\sqrt{3}$$

In Section 10.3, Example 6, we obtained the answers $\dfrac{1 + \sqrt{-11}}{6}$ and $\dfrac{1 - \sqrt{-11}}{6}$. We indicated that since $\sqrt{-11}$ is not a real number we would stop there. Now we can rewrite these answers as $\dfrac{1 + i\sqrt{11}}{6}$ and $\dfrac{1 - i\sqrt{11}}{6}$, or $\dfrac{1 \pm i\sqrt{11}}{6}$.

2 Add and Subtract Complex Numbers

Complex numbers can be added, subtracted, multiplied, and divided. To add (or subtract) complex numbers, add (or subtract) their real parts and add (or subtract) their complex parts.

EXAMPLE 2 Add each of the following
a) $3 + 6i$ and $-2 - 3i$ **b)** $-4 - 5i$ and $-3 - 2i$

Solution **a)** $(\,3\; + 6i\,) + (-2\; - 3i\,) = (\,3 - 2\,) + (\,6i - 3i\,)$
$$= 1 + 3i$$

b) $(-4\; - 5i\,) + (-3\; - 2i\,) = (-4 - 3\,) + (-5i - 2i\,)$
$$= -7 - 7i$$

Complex numbers can also be multiplied and divided, but we will not multiply and divide complex numbers in this text.

3 Solve Quadratic Equations with Complex Number Solutions

Before we leave this section let's solve a quadratic equation that does not have real numbers as solutions. We will write the answers as complex numbers.

EXAMPLE 3 Solve the equation $2x^2 - 4x + 7 = 0$.

Solution We will use the quadratic formula.

$$a = 2, \quad b = -4, \quad c = 7$$

$$x = \frac{-b \pm \sqrt{b^2 - 4ac}}{2a} \qquad \text{Quadratic formula}$$

$$= \frac{-(-4) \pm \sqrt{(-4)^2 - 4(2)(7)}}{2(2)}$$

$$= \frac{4 \pm \sqrt{16 - 56}}{4}$$

$$= \frac{4 \pm \sqrt{-40}}{4} \qquad \text{These are complex numbers.}$$

$$= \frac{4 \pm \sqrt{4}\,\sqrt{10}\,\sqrt{-1}}{4}$$

$$= \frac{4 \pm 2i\sqrt{10}}{4} \qquad \text{Write complex numbers using } i.$$

$$= \frac{\overset{1}{2}(2 \pm i\sqrt{10})}{\underset{2}{4}} \qquad \text{Factor numerator, divide out common factors.}$$

$$= \frac{2 \pm i\sqrt{10}}{2}$$

The solutions, $\dfrac{2 + i\sqrt{10}}{2}$ and $\dfrac{2 - i\sqrt{10}}{2}$, are complex numbers.

NOW TRY EXERCISE 51

Exercise Set 10.5

Concept/Writing Exercises

1. Is every real number a complex number? Explain.   Yes

2. Are there any complex numbers that are real numbers? If so, how many complex numbers are real numbers? Explain.   yes, an infinite number of them

3. What is the value of i? $\sqrt{-1}$

4. a) Is $\sqrt{-11}$ a member of the real number system? Explain.   No
   b) Is $\sqrt{-11}$ a member of the complex number system? Explain.   Yes

5. What is the general form of a complex number? $a + bi$

6. Explain how to add or subtract complex numbers.   add real parts, add imaginary parts

Practice the Skills

Write each imaginary number in terms of i.

7. $\sqrt{-4}$ $2i$

8. $\sqrt{-9}$ $3i$

9. $\sqrt{-100}$ $10i$

10. $\sqrt{-64}$ $8i$

11. $\sqrt{-15}$ $i\sqrt{15}$

12. $\sqrt{-10}$ $i\sqrt{10}$

13. $\sqrt{-23}$ $i\sqrt{23}$

14. $\sqrt{-13}$ $i\sqrt{13}$

15. $\sqrt{-8}$ $2i\sqrt{2}$

16. $\sqrt{-18}$ $3i\sqrt{2}$

17. $\sqrt{-20}$ $2i\sqrt{5}$

18. $\sqrt{-98}$ $7i\sqrt{2}$

Write each complex number in the form of a + bi.

19. $5 + \sqrt{-4}$ $\;5 + 2i$ **20.** $7 + \sqrt{-16}$ $\;7 + 4i$ **21.** $-3 + \sqrt{-25}$ $\;-3 + 5i$ **22.** $-6 + \sqrt{-64}$ $\;-6 + 8i$

23. $-1 - \sqrt{-9}$ $\;-1 - 3i$ **24.** $-4 - \sqrt{-81}$ $\;-4 - 9i$ **25.** $-3 - \sqrt{-15}$ $\;-3 - i\sqrt{15}$ **26.** $-5 - \sqrt{-6}$ $\;-5 - i\sqrt{6}$

27. $5.2 + \sqrt{-50}$ $\;5.2 + 5i\sqrt{2}$ **28.** $6.3 - \sqrt{-32}$ $\;6.3 - 4i\sqrt{2}$ **29.** $\frac{1}{2} + \sqrt{-75}$ $\;\frac{1}{2} + 5i\sqrt{3}$ **30.** $-\frac{3}{5} + \sqrt{-80}$ $\;-\frac{3}{5} + 4i\sqrt{5}$

Add or subtract the following complex numbers.

31. $(2 + 3i) + 5$ $\;7 + 3i$ **32.** $(3 - 5i) + 6$ $\;9 - 5i$ **33.** $(4 - 3i) - 8i$ $\;4 - 11i$

34. $(12 - 5i) - 7i$ $\;12 - 12i$ **35.** $(5 + 3i) + (6 + 2i)$ $\;11 + 5i$ **36.** $(8 - 5i) + (3 - 2i)$ $\;11 - 7i$

37. $(4 - 3i) - (6 + 4i)$ $\;-2 - 7i$ **38.** $(9 - 13i) - (9 - 4i)$ $\;-9i$ **39.** $(8 - 6i) - (8 - 6i)$ $\;0$

40. $(16 - 4i) - (4 - 4i)$ $\;12$ **41.** $(45 - 3i) - (36 + i)$ $\;9 - 4i$ **42.** $(7 - 5i) - (12 + 8i)$ $\;-5 - 13i$

Solve the quadratic equation and write the solutions in terms of i.

43. $x^2 = -16$ $\;\pm 4i$ **44.** $x^2 = -25$ $\;\pm 5i$ **45.** $2x^2 = -20$ $\;\pm i\sqrt{10}$

46. $3a^2 = -36$ $\;\pm 2i\sqrt{3}$ **47.** $x^2 - 4x + 5 = 0$ $\;2 \pm i$ **48.** $x^2 + 6x + 11 = 0$ $\;-3 \pm i\sqrt{2}$

49. $2r^2 + 3r + 5 = 0$ $\;\frac{-3 \pm i\sqrt{31}}{4}$ **50.** $3w^2 - 5w + 8 = 0$ $\;\frac{5 \pm i\sqrt{71}}{6}$ **51.** $2p^2 + 4p + 9 = 0$ $\;\frac{-2 \pm i\sqrt{14}}{2}$

52. $4m^2 - 6m + 9 = 0$ $\;\frac{3 \pm 3i\sqrt{3}}{4}$ **53.** $-4w^2 + 5w - 9 = 0$ $\;\frac{5 \pm i\sqrt{119}}{8}$ **54.** $-3z^2 - 2w - 10 = 0$ $\;\frac{-1 \pm i\sqrt{29}}{3}$

Problem Solving

57. when $b^2 - 4ac < 0$

55. Under what conditions will an equation of the form $x^2 = c$ have imaginary number solutions? when $c < 0$

56. Can an equation of the form $x^2 = c$ ever have only one imaginary solution? Explain. No

57. Under what conditions will an equation of the form $ax^2 + bx + c = 0$ have non-real number solutions?

58. Is it possible for a quadratic equation to have only one non-real number solution? Explain No

Cumulative Review Exercises

[1.9] **59.** Evaluate $-\{[3(x - 4)^2 - 5] - 3x\}$ when $x = 6$ 11   [2.5] **60.** Solve the equation $\frac{1}{2}x + \frac{3}{5}x = \frac{1}{2}(x - 2)$ $-\frac{5}{3}$

[6.7] **61.** Solve the equation $\frac{w - 1}{w - 5} = \frac{3}{w - 5} + \frac{3}{4}$ 1   [9.5] **62.** Solve the equation $2\sqrt{r - 4} + 3 = 9$ 13

CHAPTER SUMMARY
Key Words and Phrases

10.1	10.2	10.4	10.5
Quadratic equation	Completing the square	Axis of symmetry	Complex number
Square root property	Perfect square trinomial	Parabola	Imaginary number
Standard form of a		Symmetry	Imaginary unit
quadratic equation	**10.3**	Vertex of a parabola	
	Discriminant		
	Quadratic formula		

(continued on the next page)

IMPORTANT FACTS

Square Root Property: If $x^2 = a$, then $x = \sqrt{a}$ or $x = -\sqrt{a}$ (or $x = \pm\sqrt{a}$).

Quadratic Formula: $x = \dfrac{-b \pm \sqrt{b^2 - 4ac}}{2a}$

Discriminant: $b^2 - 4ac$

If $b^2 - 4ac > 0$ the quadratic equation has two distinct real number solutions.

If $b^2 - 4ac = 0$ the quadratic equation has one real number solution.

If $b^2 - 4ac < 0$ the quadratic equation has no real number solution.

Coordinates of the Vertex of a Parabola:

$$\left(-\frac{b}{2a}, \frac{4ac - b^2}{4a}\right).$$

Imaginary Unit: $i = \sqrt{-1}$

Complex Numbers are numbers of the form $a + bi$ where a and b are real numbers.

Chapter Review Exercises

[10.1] Solve using the square root property.

1. $x^2 = 64$ 8, −8
2. $x^2 = 12$ $2\sqrt{3}, -2\sqrt{3}$
3. $2x^2 = 12$ $\sqrt{6}, -\sqrt{6}$
4. $x^2 + 4 = 9$ $\sqrt{5}, -\sqrt{5}$
5. $x^2 - 4 = 16$ $2\sqrt{5}, -2\sqrt{5}$
6. $2x^2 - 4 = 10$ $\sqrt{7}, -\sqrt{7}$
7. $4a^2 - 30 = 2$ $2\sqrt{2}, -2\sqrt{2}$
8. $(r - 3)^2 = 20$ $3 \pm 2\sqrt{5}$
9. $(4t - 3)^2 = 50$ $(3 \pm 5\sqrt{2})/4$
10. $(2x + 4)^2 = 30$ $(-4 \pm \sqrt{30})/2$

[10.2] Solve by completing the square.

11. $x^2 - 7x + 10 = 0$ 5, 2
12. $x^2 - 11x + 28 = 0$ 7, 4
13. $x^2 - 18x + 17 = 0$ 17, 1
14. $x^2 + x - 6 = 0$ 2, −3
15. $t^2 - 3t - 54 = 0$ 9, −6
16. $x^2 = -5x + 6$ 1, −6
17. $y^2 - 5y - 7 = 0$ $(5 \pm \sqrt{53})/2$
18. $x^2 + 2x - 5 = 0$ $-1 \pm \sqrt{6}$
19. $2x^2 - 8x = 64$ 8, −4
20. $30 = 2n^2 - 4n$ 5, −3
21. $3p^2 = -2p + 8$ $\frac{4}{3}, -2$
22. $6x^2 - 19x + 15 = 0$ $\frac{5}{3}, \frac{3}{2}$

[10.3] Determine whether each equation has two distinct real number solutions, one real number solution, or no real number solution.

23. $-4x^2 + 4x - 6 = 0$ no real solution
24. $-3x^2 + 4x = 9$ no real solution
25. $x^2 - 10x + 25 = 0$ one solution
26. $y^2 + 2y - 8 = 0$ two solutions
27. $4z^2 - 3z = 6$ two solutions
28. $3x^2 - 4x + 5 = 0$ no real solution
29. $-3x^2 - 4x + 8 = 0$ two solutions
30. $x^2 - 9x + 6 = 0$ two solutions

Use the quadratic formula to solve. If an equation has no real number solution, so state.

31. $x^2 - 10x + 16 = 0$ $2, 8$

32. $x^2 - 7x - 44 = 0$ $11, -4$

33. $x^2 = 10x - 9$ $1, 9$

34. $5x^2 - 7x = 6$ $2, -\frac{3}{5}$

35. $21 = r^2 + 4r$ $3, -7$

36. $x^2 - x + 12 = 0$ no real solution

37. $6x^2 + x - 15 = 0$ $\frac{3}{2}, -\frac{5}{3}$

38. $-2x^2 + 3x + 6 = 0$ $(3 \pm \sqrt{57})/4$

39. $2x^2 + 4x - 3 = 0$ $(-2 \pm \sqrt{10})/2$

40. $y^2 - 6y + 3 = 0$ $3 \pm \sqrt{6}$

41. $3x^2 - 4x + 6 = 0$ no real solution

42. $3x^2 - 6x - 8 = 0$ $(3 \pm \sqrt{33})/3$

43. $7x^2 - 3x = 0$ $0, \frac{3}{7}$

44. $2z^2 = 5z$ $0, \frac{5}{2}$

[10.1–10.3] Solve each quadratic equation using the method of your choice.

45. $x^2 - 10x + 24 = 0$ $4, 6$

46. $x^2 + 15x + 56 = 0$ $-7, -8$

47. $r^2 - 3r - 70 = 0$ $10, -7$

48. $x^2 + 6x = 27$ $-9, 3$

49. $x^2 - 4x - 60 = 0$ $-6, 10$

50. $x^2 - x - 42 = 0$ $7, -6$

51. $y^2 + 9y - 22 = 0$ $2, -11$

52. $t^2 = 5t$ $0, 5$

53. $x^2 = 81$ $9, -9$

54. $2x^2 + 5x = 3$ $\frac{1}{2}, -3$

55. $2x^2 = 9x - 10$ $\frac{5}{2}, 2$

56. $6x^2 + 5x = 6$ $\frac{2}{3}, -\frac{3}{2}$

57. $6n = 2n^2 - 9$ $(3 \pm 3\sqrt{3})/2$

58. $3x^2 - 11x + 10 = 0$ $2, \frac{5}{3}$

59. $-5w = 3w^2 - 8$ $1, -\frac{8}{3}$

60. $x^2 + 3x = 6$ $(-3 \pm \sqrt{33})/2$

61. $4x^2 - 9x = 0$ $0, \frac{9}{4}$

62. $3x^2 + 5x = 0$ $0, -\frac{5}{3}$

[10.4] Indicate the axis of symmetry, the coordinates of the vertex, and whether the parabola opens upward or downward.

63. $y = x^2 - 6x - 5$ $x = 3, (3, -14)$, upward

64. $y = x^2 - 12x + 6$ $x = 6, (6, -30)$, upward

65. $y = x^2 - 3x + 7$ $x = \frac{3}{2}, \left(\frac{3}{2}, \frac{19}{4}\right)$, upward

66. $y = -x^2 - 2x + 15$ $x = -1, (-1, 16)$, downward

67. $y = 3x^2 + 7x + 3$ $x = -\frac{7}{6}, \left(-\frac{7}{6}, -\frac{13}{12}\right)$, upward

68. $y = -x^2 - 5x$ $x = -\frac{5}{2}, \left(-\frac{5}{2}, \frac{25}{4}\right)$, downward

69. $y = -x^2 - 8$ $x = 0, (0, -8)$, downward

70. $y = -2x^2 - x + 20$ $x = -\frac{1}{4}\left(-\frac{1}{4}, \frac{161}{8}\right)$, downward

71. $y = -4x^2 + 8x + 5$ $x = 1, (1, 9)$, downward

72. $y = 3x^2 + 5x - 8$ $x = -\frac{5}{6}, \left(-\frac{5}{6}, -\frac{121}{12}\right)$, upward

Graph each quadratic equation and determine the x-intercepts, if they exist. If they do not exist, so state.

73.–84. See graphing answer section, pages G17 and G18.

73. $y = x^2 - 2x$

74. $y = -3x^2 + 6$

75. $y = x^2 - 2x - 15$

76. $y = -x^2 + 5x - 6$

77. $y = x^2 - x + 1$

78. $y = x^2 + 5x + 4$

79. $y = -x^2 - 6x$

80. $y = x^2 + 4x + 3$

81. $y = -x^2 + 2x - 3$

82. $y = 3x^2 - 4x - 8$

83. $y = -2x^2 + 7x - 3$

84. $y = x^2 - 5x + 4$

[10.1–10.4] Solve. **87.** width, 20 in.; length, 46 in.

85. *Product of Integers* The product of two positive consecutive even integers is 48. Find the integers. $6, 8$

86. *Product of Integers* The product of two positive integers is 88. Find the two integers if the larger one is 3 greater than the smaller. $8, 11$

87. *Rectangular Table* Samuel Jones is making a table with a rectangular top for the kitchen. He determines that the length of the table top should be 6 inches more than twice its width. What will be the dimensions of the table top if its area is to be 920 square inches?

88. *Wood Desk* Jordan and Patricia Wells recently purchased an oak desk for their son's room. To protect the rectangular writing surface of the desk, they decided to cover the top with a piece of glass. The length of the desktop is 20 inches greater than its width. Find the dimensions of the glass piece Jordan and Patricia should order if the area of the desktop is 1344 square inches. width, 28 in.; length, 48 in.

[10.5] Write each imaginary number or complex number in terms of i.

89. $\sqrt{-4}$ $2i$

90. $\sqrt{-30}$ $i\sqrt{30}$

91. $4 - \sqrt{-25}$ $4 - 5i$

92. $5 - \sqrt{-60}$ $5 - 2i\sqrt{15}$

Add or subtract the complex numbers.

93. $(4 - 6i) + (5 - 3i)$ $9 - 9i$ **94.** $(9 + 5i) - (6 - 3i)$ $3 + 8i$

Solve the quadratic equations and write the complex solutions in terms of i.

95. $3x^2 = -27$ $\pm 3i$

96. $4p^2 = -20$ $\pm i\sqrt{5}$

97. $2r^2 - 5r + 8 = 0$ $\dfrac{5 \pm i\sqrt{39}}{4}$

98. $4w^2 - 8w + 9 = 0$ $\dfrac{2 \pm i\sqrt{5}}{2}$

1. $4\sqrt{2}, -4\sqrt{2}$ **2.** $(4 \pm \sqrt{17})/2$ **6.** $(-4 \pm \sqrt{6})/2$ **8.** $x = \dfrac{-b \pm \sqrt{b^2 - 4ac}}{2a}$ **9.** Answers will vary. **14.** downward; $a < 0$

Chapter Practice Test

15. upward; $a > 0$ **16.** lowest point on a parabola that opens upward, highest point on a parabola that opens downward **17.** $(-4, 4)$ **18.** $\left(\frac{4}{3}, \frac{11}{3}\right)$

19.–21. See graphing answer section, page G18. **22.** $w = 3$ ft, $l = 10$ ft

1. Solve $x^2 - 4 = 28$ using the square root property.

2. Solve $(2p - 4)^2 = 17$ using the square root property.

3. Solve $x^2 - 6x = 40$ by completing the square. $10, -4$

4. Solve $r^2 + 7r = 44$ by completing the square. $4, -11$

5. Solve $k^2 = 13k - 42$ by the quadratic formula. $6, 7$

6. Solve $2x^2 + 5 = -8x$ by the quadratic formula.

7. Solve $16x^2 = 49$ by the method of your choice. $\frac{7}{4}, -\frac{7}{4}$

8. Write the quadratic formula.

9. Give an example of a perfect square trinomial.

10. Determine whether $-2x^2 - 4x + 2 = 0$ has two distinct real solutions, one real solution, or no real solution. Explain your answer. two real solutions

11. Determine whether $x^2 + 8x + 16 = 0$ has two distinct real solutions, one real solution, or no real solution. Explain your answer. one real solution

12. Find the equation of the axis of symmetry of the graph of $y = -x^2 - 6x + 7$. $x = -3$

13. Find the equation of the axis of symmetry of the graph of $y = 4x^2 - 16x + 9$. $x = 2$

14. Determine whether the graph of $y = -x^2 - 6x + 7$ opens upward or downward. Explain your answer.

15. Determine whether the graph of $y = 4x^2 - 8x + 9$ opens upward or downward. Explain your answer.

16. What is the vertex of the graph of a parabola?

17. Find the vertex of the graph of $y = -x^2 - 8x - 12$.

18. Find the vertex of the graph of $y = 3x^2 - 8x + 9$.

19. Graph the equation $y = x^2 + 2x - 8$ and determine the x-intercepts, if they exist.

20. Graph the equation $y = -x^2 + 6x - 9$ and determine the x-intercepts, if they exist.

21. Graph the equation $y = 2x^2 - 6x$ and determine the x-intercepts, if they exist.

22. *Wall Mural* The length of a rectangular wall mural is 1 foot greater than 3 times its width. Find the length and width of the mural if its area is 30 square feet.

23. *Consecutive Odd Integers* The product of two positive consecutive odd integers is 99. Find the larger of two integers. 11

24. *Shawn's Age* Shawn Goodwin is 4 years older than his cousin, Aaron. The product of their ages is 45. How old is Shawn? 9

25. Solve the quadratic equation $3p^2 - 2p + 6 = 0$ and write the answer in terms of i. $\dfrac{1 \pm i\sqrt{17}}{3}$

Cumulative Review Test

Take the following test and check your answers with those that appear at the end of the test. Review any questions that you answered incorrectly. The section and objective where the material was covered are indicated after the answer.

1. Evaluate $-5x^2y + 3y^2 - xy$ when $x = 4$ and $y = -3$. 279

2. Solve the equation $\dfrac{1}{2}z - \dfrac{2}{7}z = \dfrac{1}{5}(3z - 1)$. $\dfrac{14}{27}$

3. Find the length of side x. $5\frac{1}{3}$ in.

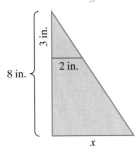

8 in.

3 in.

2 in.

x

4. Solve the inequality $2(x - 3) \le 6x - 5$ and graph the solution on a number line. $x \ge -\frac{1}{4}$,

5. Solve the formula $A = \dfrac{m + n + P}{3}$ for P. $P = 3A - m - n$

6. Simplify $(6a^4b^5)^3(3a^2b^5)^2$. $1944a^{16}b^{25}$

7. Divide $\dfrac{x^2 + 6x + 5}{x + 2}$. $x + 4 - \dfrac{3}{x + 2}$

8. Factor by grouping $2x^2 - 3xy - 4xy + 6y^2$.
$(2x - 3y)(x - 2y)$

9. Factor $6x^2 - 27x - 54$. $3(2x + 3)(x - 6)$

10. Add $\dfrac{4}{a^2 - 16} + \dfrac{2}{(a - 4)^2}$. $\dfrac{6a - 8}{(a + 4)(a - 4)^2}$

11. Solve the equation $x + \dfrac{48}{x} = 14$. 6, 8

12. Graph the equation $4x - y = 8$. See graphing answer section, page G18.

13. Solve the system of equations by the addition method.
$$5x - 3y = 12$$
$$4x - 2y = 6 \quad (-3, -9)$$

14. Simplify $\sqrt{\dfrac{2x^2y^3}{12x}} \cdot \dfrac{y\sqrt{6xy}}{6}$

15. Add $2\sqrt{28} - 3\sqrt{7} + \sqrt{63}$. $4\sqrt{7}$

16. Solve the equation $x - 2 = \sqrt{x^2 - 12}$. 4

17. Solve the equation $2x^2 + x - 8 = 0$ using the quadratic formula. $(-1 \pm \sqrt{65})/4$

18. *Fertilizer* If 4 pounds of fertilizer can fertilize 500 square feet of lawn, how many pounds of fertilizer are needed to fertilize 3200 square feet of lawn? 25.6 lb

19. *Vegetable Garden* The length of a rectangular vegetable garden is 3 feet less than three times its width. Find the width and length of the garden if its perimeter is 74 feet. $w = 10$ ft, $l = 27$ ft

20. *Jogging* Robert McCloud jogs 3 miles per hour faster than he walks. He jogs for 2 miles and then walks for 2 miles. If the total time of his outing is 1 hour, find the rate at which he walks and jogs. walks, 3 mph; jogs, 6 mph

Answers to Cumulative Review Test

1. 279; [Sec. 1.9, Obj. 6] **2.** $\dfrac{14}{27}$; [Sec. 2.5, Obj. 2] **3.** $5\frac{1}{3}$ inches; [Sec. 2.6, Obj. 5] **4.** $x \ge -\dfrac{1}{4}$;

[Sec. 2.7, Obj. 1] **5.** $P = 3A - m - n$; [Sec. 3.1, Obj. 3] **6.** $1944a^{16}b^{25}$; [Sec. 4.1, Obj. 3] **7.** $x + 4 - \dfrac{3}{x + 2}$;

[Sec. 4.6, Obj. 2] **8.** $(2x - 3y)(x - 2y)$; [Sec. 5.2, Obj. 1] **9.** $3(2x + 3)(x - 6)$; [Sec. 5.4, Obj. 1] **10.** $\dfrac{6a - 8}{(a + 4)(a - 4)^2}$;

[Sec. 6.4, Obj. 1] **11.** 6, 8; [Sec. 6.6, Obj. 2] **12.**

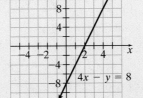

$4x - y = 8$

; [Sec. 7.2, Obj. 3] **13.** $(-3, -9)$; [Sec. 8.3, Obj. 1]

14. $\dfrac{y\sqrt{6xy}}{6}$; [Sec. 9.4, Obj. 3] **15.** $4\sqrt{7}$; [Sec. 9.3, Obj. 1] **16.** 4; [Sec. 9.5, Obj. 1] **17.** $\dfrac{-1 + \sqrt{65}}{4}, \dfrac{-1 - \sqrt{65}}{4}$;

[Sec. 10.3, Obj. 1] **18.** 25.6 pounds; [Sec. 2.6, Obj. 3] **19.** Width = 10 feet, length = 27 feet; [Sec. 3.4, Obj. 1]
20. Walks: 3 miles per hour, jogs: 6 miles per hour; [Sec. 6.7, Obj. 2]

Appendices

A Review of Decimals and Percent

B Finding the Greatest Common Factor and Least Common Denominator

C Geometry

APPENDIX A   REVIEW OF DECIMALS AND PERCENT

Decimals

To Add or Subtract Numbers Containing Decimal Points

1. Align the numbers by the decimal points.
2. Add or subtract the numbers as if they were whole numbers.
3. Place the decimal point in the sum or difference directly below the decimal points in the numbers being added or subtracted.

EXAMPLE 1 Add $4.6 + 13.813 + 9.02$.

Solution

$$\begin{array}{r} 4.600 \\ 13.813 \\ +\ 9.020 \\ \hline 27.433 \end{array}$$

EXAMPLE 2 Subtract 3.062 from 34.9.

Solution

$$\begin{array}{r} 34.900 \\ -\ 3.062 \\ \hline 31.838 \end{array}$$

To Multiply Numbers Containing Decimal Points

1. Multiply as if the factors were whole numbers.
2. Determine the total number of digits to the right of the decimal points in the factors.
3. Place the decimal point in the product so that the product contains the same number of digits to the right of the decimal as the total found in step 2. For example, if there are a total of three digits to the right of the decimal points in the factors, there must be three digits to the right of the decimal point in the product.

EXAMPLE 3 Multiply 2.34×1.9.

Solution

$$\begin{array}{r} 2.34 \quad \longleftarrow \textit{two digits to the right of the decimal point} \\ \times\ 1.9 \quad \longleftarrow \textit{one digit to the right of the decimal point} \\ \hline 2106 \\ 234 \\ \hline 4.446 \quad \longleftarrow \textit{three digits to the right of the decimal point in the product} \end{array}$$

EXAMPLE 4 Multiply 2.13 × 0.02.

Solution

$$2.13 \longleftarrow \text{\textit{two digits to the right of the decimal point}}$$
$$\underline{\times\ 0.02} \longleftarrow \text{\textit{two digits to the right of the decimal point}}$$
$$0.0426 \longleftarrow \text{\textit{four digits to the right of the decimal point in the product}}$$

Note that it was necessary to add a zero preceding the digit 4 in the answer in order to have four digits to the right of the decimal point.　☀

To Divide Numbers Containing Decimal Points

1. Multiply both the dividend and divisor by a power of 10 that will make the divisor a whole number.
2. Divide as if working with whole numbers.
3. Place the decimal point in the quotient directly above the decimal point in the dividend.

To make the divisor a whole number, multiply *both* the dividend and divisor by 10 if the divisor is given in tenths, by 100 if the divisor is given in hundredths, by 1000 if the divisor is given in thousandths, and so on. Multiplying both the numerator and denominator by the same nonzero number is the same as multiplying the fraction by 1. Therefore, the value of the fraction is unchanged.

EXAMPLE 5 Divide $\dfrac{1.956}{0.12}$.

Solution Since the divisor, 0.12, is twelve-hundredths, we multiply both the divisor and dividend by 100.

$$\frac{1.956}{0.12} \times \frac{100}{100} = \frac{195.6}{12.}$$

Now we divide.

$$
\begin{array}{r}
16.3 \\
12\overline{)195.6} \\
\underline{12} \\
75 \\
\underline{72} \\
36 \\
\underline{36} \\
0
\end{array}
$$

The decimal point in the answer is placed directly above the decimal point in the dividend. Thus, 1.956/0.12 = 16.3.　☀

EXAMPLE 6 Divide 0.26 by 10.4.

Solution First, multiply both the dividend and divisor by 10.

$$\frac{0.26}{10.4} \times \frac{10}{10} = \frac{2.6}{104.}$$

Now divide.

$$
\begin{array}{r}
0.025 \\
104\overline{)2.600} \\
\underline{2\ 08} \\
520 \\
\underline{520} \\
0
\end{array}
$$

Note that a zero had to be placed before the digit 2 in the quotient.

$$\frac{0.26}{10.4} = 0.025$$

Percent

The word *percent* means "per hundred." The symbol % means percent. One percent means "one per hundred," or

$$1\% = \frac{1}{100} \quad \text{or} \quad 1\% = 0.01$$

EXAMPLE 7 Convert 16% to a decimal.

Solution Since 1% = 0.01,

$$16\% = 16(0.01) = 0.16$$

EXAMPLE 8 Convert 4.7% to a decimal.

Solution $4.7\% = 4.7(0.01) = 0.047.$

EXAMPLE 9 Convert 1.14 to a percent.

Solution To change a decimal number to a percent, we multiply the number by 100%.

$$1.14 = 1.14 \times 100\% = 114\%$$

Often, you will need to find an amount that is a certain percent of a number. For example, when you purchase an item in a state or county that has a sales tax you must often pay a percent of the item's price as the sales tax. Examples 10 and 11 show how to find a certain percent of a number.

EXAMPLE 10 Find 32% of 300.

Solution To find a percent of a number, use multiplication. Change 32% to a decimal number, then multiply by 300.

$$(0.32)(300) = 96.$$

Thus, 32% of 300 is 96.

EXAMPLE 11 Johnson County charges an 8% sales tax.
a) Find the sales tax on a stereo system that cost $580.
b) Find the total cost of the system, including tax.

Solution **a)** The sales tax is 8% of 580.

$$(0.08)(580) = 46.40$$

The sales tax is $46.40
b) The total cost is the purchase price plus the sales tax:

$$\text{total cost} = \$580 + \$46.40 = \$626.40$$

APPENDIX B   FINDING THE GREATEST COMMON FACTOR AND LEAST COMMON DENOMINATOR

Prime Factorization

In Section 1.3 we mentioned that to simplify fractions you can divide both the numerator and denominator by the *greatest common factor* (GCF). One method to find the GCF is to use *prime factorization*. Prime factorization is the process of writing a given number as a product of prime numbers. *Prime numbers* are natural numbers, excluding 1, that can be divided by only themselves and 1. The first ten prime numbers are 2, 3, 5, 7, 11, 13, 17, 19, 23, and 29. Can you find the next prime number? If you answered 31, you answered correctly.

To write a number as a product of primes, we can use a *tree diagram*. Begin by selecting any two numbers whose product is the given number. Then continue factoring each of these numbers into prime numbers, as shown in Example 1.

EXAMPLE 1   Determine the prime factorization of the number 120.

Solution   We will use three different tree diagrams to illustrate the prime factorization of 120.

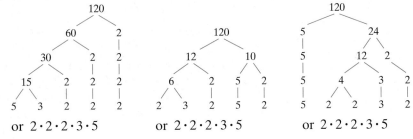

$$\text{or } 2 \cdot 2 \cdot 2 \cdot 3 \cdot 5 \qquad \text{or } 2 \cdot 2 \cdot 2 \cdot 3 \cdot 5 \qquad \text{or } 2 \cdot 2 \cdot 2 \cdot 3 \cdot 5$$

Note that no matter how you start, if you do not make a mistake, you find that the prime factorization of 120 is $2 \cdot 2 \cdot 2 \cdot 3 \cdot 5$. There are other ways 120 can be factored but all will lead to the prime factorization $2 \cdot 2 \cdot 2 \cdot 3 \cdot 5$. ✳

Greatest Common Factor

The greatest common factor (GCF) of two natural numbers is the greatest integer that is a factor of both numbers. We use the GCF when simplifying fractions.

> **To Find the Greatest Common Factor of a Given Numerator and Denominator**
>
> **1.** Write both the numerator and the denominator as a product of primes.
> **2.** Determine all the prime factors that are common to both prime factorizations.
> **3.** Multiply the prime factors found in step 2 to obtain the GCF.

EXAMPLE 2   Consider the fraction $\frac{108}{156}$.
a) Find the GCF of 108 and 156.   **b)** Simplify $\frac{108}{156}$.

Solution   **a)** First determine the prime factorizations of both 108 and 156.

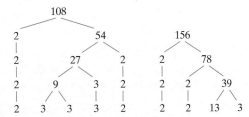

There are two 2s and one 3 common to both prime factorizations; thus

$$\text{GCF} = 2 \cdot 2 \cdot 3 = 12$$

The greatest common factor of 108 and 156 is 12. Twelve is the greatest integer that divides into both 108 and 156.

b) To simplify $\frac{108}{156}$, we divide both the numerator and denominator by the GCF, 12.

$$\frac{108 \div 12}{156 \div 12} = \frac{9}{13}$$

Thus, $\frac{108}{156}$ simplifies to $\frac{9}{13}$.

Least Common Denominator

When adding two or more fractions, you must write each fraction with a common denominator. The best denominator to use is the *least common denominator* (LCD). The LCD is the smallest number that each denominator divides into. Sometimes the least common denominator is referred to as the *least common multiple* of the denominators.

> **To Find the Least Common Denominator of Two or More Fractions**
>
> **1.** Write each denominator as a product of prime numbers.
> **2.** For each prime number, determine the maximum number of times that prime number appears in any of the prime factorizations.
> **3.** Multiply all the prime numbers, found in step 2. Include each prime number the maximum number of times it appears in any of the prime factorizations. The product of all these prime numbers will be the LCD.

Example 3 illustrates the procedure to determine the LCD.

EXAMPLE 3 Consider $\frac{7}{108} + \frac{5}{156}$.

a) Determine the least common denominator.
b) Add the fractions.

Solution **a)** We found in Example 2 that

$$108 = 2 \cdot 2 \cdot 3 \cdot 3 \cdot 3 \quad \text{and} \quad 156 = 2 \cdot 2 \cdot 3 \cdot 13$$

We can see that the maximum number of 2s that appear in either prime factorization is two (there are two 2s in both factorizations), the maximum number of 3s is three, and the maximum number of 13s is one. Multiply as follows:

$$2 \cdot 2 \cdot 3 \cdot 3 \cdot 3 \cdot 13 = 1404$$

Thus, the least common denominator is 1404. This is the smallest number that both 108 and 156 divide into.

b) To add the fractions, we need to write both fractions with a common denominator. The best common denominator to use is the LCD. Since $1404 \div 108 = 13$, we will multiply $\frac{7}{108}$ by $\frac{13}{13}$. Since $1404 \div 156 = 9$, we will multiply $\frac{5}{156}$ by $\frac{9}{9}$.

$$\frac{7}{108} \cdot \frac{13}{13} + \frac{5}{156} \cdot \frac{9}{9} = \frac{91}{1404} + \frac{45}{1404} = \frac{136}{1404} = \frac{34}{351}$$

Thus, $\frac{7}{108} + \frac{5}{156} = \frac{34}{351}$.

APPENDIX C   GEOMETRY

This appendix introduces or reviews important geometric concepts. Table C.1 gives the names and descriptions of various types of angles.

Angles

TABLE C.1	
Angle	**Sketch of Angle**
An **acute angle** is an angle whose measure is between 0° and 90°.	
A **right angle** is an angle whose measure is 90°.	
An **obtuse angle** is an angle whose measure is between 90° and 180°.	
A **straight angle** is an angle whose measure is 180°.	
Two angles are **complementary angles** when the sum of their measures is 90°. Each angle is the complement of the other. Angles A and B are complementary angles.	60° A B 30°
Two angles are **supplementary angles** when the sum of their measures is 180°. Each angle is the supplement of the other. Angles A and B are supplementary angles.	130° A B 50°

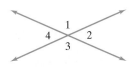

FIGURE C.1

When two lines intersect, four angles are formed as shown in Figure C.1 The pair of opposite angles formed by the intersecting lines are called **vertical angles**.

Angles 1 and 3 are vertical angles. Angles 2 and 4 are also vertical angles. *Vertical angles have equal measures.* Thus, angle 1, symbolized by $\angle 1$, is equal to angle 3, symbolized by $\angle 3$. We can write $\angle 1 = \angle 3$. Similarly, $\angle 2 = \angle 4$.

Parallel and Perpendicular Lines

Parallel lines are two lines in the same plane that do not intersect (Fig. C.2). **Perpendicular lines** are lines that intersect at right angles (Fig. C.3).

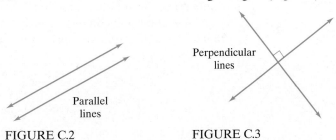

Parallel lines

Perpendicular lines

FIGURE C.2   FIGURE C.3

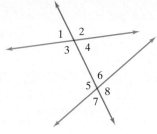

FIGURE C.4

A **transversal** is a line that intersects two or more lines at different points. When a transversal line intersects two other lines, eight angles are formed, as illustrated in Figure C.4. Some of these angles are given special names.

Interior angles: 3, 4, 5, 6

Exterior angles: 1, 2, 7, 8

Pairs of corresponding angles: 1 and 5; 2 and 6; 3 and 7; 4 and 8

Pairs of alternate interior angles: 3 and 6; 4 and 5

Pairs of alternate exterior angles: 1 and 8; 2 and 7

Parallel Lines Cut by a Transversal

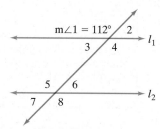

When two parallel lines are cut by a transversal:

1. Corresponding angles are equal ($\angle 1 = \angle 5$, $\angle 2 = \angle 6$, $\angle 3 = \angle 7$, $\angle 4 = \angle 8$).
2. Alternate interior angles are equal ($\angle 3 = \angle 6$, $\angle 4 = \angle 5$).
3. Alternate exterior angles are equal ($\angle 1 = \angle 8$, $\angle 2 = \angle 7$).

EXAMPLE 1 If line 1 and line 2 are parallel lines and $m\angle 1 = 112°$, find the measure of angles 2 through 8.

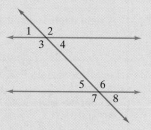

Solution Angles 1 and 2 are supplementary. So $m\angle 2$ is $180° - 112° = 68°$. The measures of angles 1 and 4 are equal since they are vertical angles. Thus, $m\angle 4 = 112°$. Angles 1 and 5 are corresponding angles. Thus, $m\angle 5 = 112°$. It is equal to its vertical angle, $\angle 8$, so $m\angle 8 = 112°$. The measures of angles 2, 3, 6, and 7 are all equal and measure 68°.   ✳

Polygons

A **polygon** is a closed figure in a plane determined by three or more line segments. Some polygons are illustrated in Figure C.5.

A **regular polygon** has sides that are all the same length, and interior angles that all have the same measure. In Figure C.5, (b) and (d) are regular polygons.

FIGURE C.5   (a)   (b)   (c)   (d)

Sum of the Interior Angles of a Polygon

The sum of the interior angles of a polygon can be found by the formula

$$\text{Sum} = (n - 2)180°$$

where n is the number of sides of the polygon.

EXAMPLE 2 Find the sum of the measures of the interior angles of **a)** a triangle; **b)** a quadrilateral (4 sides); **c)** an octagon (8 sides).

Solution **a)** Since $n = 3$, we write

$$\text{Sum} = (n - 2)180°$$
$$= (3 - 2)180° = 1(180°) = 180°$$

The sum of the measures of the interior angles in a triangle is 180°.

b) $$\text{Sum} = (n - 2)180°$$
$$= (4 - 2)180° = 2(180°) = 360°$$

The sum of the measures of the interior angles in a quadrilateral is 360°.

c) $$\text{Sum} = (n - 2)(180°) = (8 - 2)180° = 6(180°) = 1080°$$

The sum of the measures of the interior angles in an octagon is 1080°.

Now we will briefly define several types of triangles in Table C.2.

Triangles

TABLE C.2	
Triangle	**Sketch of Triangle**
An **acute triangle** is one that has three acute angles (angles of less than 90°).	
An **obtuse triangle** has one obtuse angle (an angle greater than 90°).	
A **right triangle** has one right angle (an angle equal to 90°). The longest side of a right triangle is opposite the right angle and is called the **hypotenuse**. The other two sides are called the **legs**.	
An **isosceles triangle** has two sides of equal length. The angles opposite the equal sides have the same measure.	
An **equilateral triangle** has three sides of equal length. It also has three equal angles that measure 60° each.	

When two sides of a *right triangle* are known, the third side can be found using the **Pythagorean Theorem**, $a^2 + b^2 = c^2$, where a and b are the legs and c is the hypotenuse of the triangle. (See Section 5.7 for examples.)

Congruent and Similar Figures

If two triangles are **congruent**, it means that the two triangles are identical in size and shape. Two congruent triangles could be placed one on top of the other if we were able to move and rearrange them.

Two Triangles Are Congruent If Any One of the Following Statements Is True.

1. Two angles of one triangle are equal to two corresponding angles of the other triangle, and the lengths of the sides between each pair of angles are equal. This method of showing that triangles are congruent is called the *angle, side, angle* method.

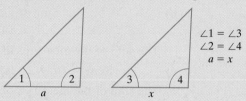

$\angle 1 = \angle 3$
$\angle 2 = \angle 4$
$a = x$

2. Corresponding sides of both triangles are equal. This is called the *side, side, side* method.

3. Two corresponding pairs of sides are equal, and the angle between them is equal. This is referred to as the *side, angle, side* method.

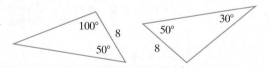

EXAMPLE 3 Determine whether the two triangles are congruent.

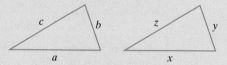

Solution The unknown angle in the figure on the right must measure 100° since the sum of the angles of a triangle is 180°. Both triangles have the same two angles (100° and 50°), with the same length side between them, 8 units. Thus, these two triangles are congruent by the angle, side, angle method.

Two triangles are **similar** if all three pairs of corresponding angles are equal and corresponding sides are in proportion. Similar figures do not have to be the same size but must have the same general shape.

Two Triangles Are Similar If Any One of the Following Statements Is True

1. Two angles of one triangle equal two angles of the other triangle.

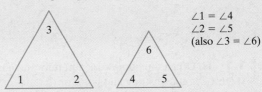

$\angle 1 = \angle 4$
$\angle 2 = \angle 5$
(also $\angle 3 = \angle 6$)

2. Corresponding sides of the two triangles are proportional.

$$\frac{a}{x} = \frac{b}{y} = \frac{c}{z}$$

3. Two pairs of corresponding sides are proportional, and the angles between them are equal.

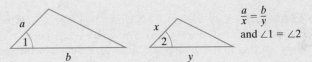

$$\frac{a}{x} = \frac{b}{y}$$
and $\angle 1 = \angle 2$

EXAMPLE 4   Are the triangles ABC and $AB'C'$ similar?

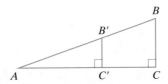

Solution   Angle A is common to both triangles. Since angle C and angle C' are equal (both 90°), then $\angle B$ and $\angle B'$ must be equal. Since the three angles of triangle ABC equal the three angles of triangle $AB'C'$, the two triangles are similar.   ✳

Answers

CHAPTER 1

Exercise Set 1.1
11. Do all the homework carefully and completely and preview the new material that is to be covered in class. **13.** At least 2 hours of study and homework time for each hour of class time is generally recommended. **15. a)** You need to do the homework in order to practice what was presented in class. **b)** When you miss class, you miss important information; therefore, it is important that you attend class regularly. **17.** Answers will vary.

Exercise Set 1.2
1. Understand, translate, calculate, check, state answer. **3.** Substitute smaller or larger numbers. **5.** Rank the data. The median is the value in the middle. **7.** Mean **9.** He actually missed a B by 10 points. **11. a)** 76.4 **b)** 74 **13. a)** $163.37 **b)** $153.85 **15. a)** 11.593 **b)** 11.68 **17.** $470 **19. a)** $1169 **b)** $17,869 **21. a)** $992 **b)** $42 **23.** $\approx$19,236 women **25.** $\approx$18.49 miles per gallon **27.** $153 **29. a)** $5191.80 **b)** $9120.25 **31. a)** 4106.25 gallons **b)** $\approx$$21.35 **33.** $6.20 **35. a)** $7420 **b)** 22 hours **37.** On the 3 on the right **39.** $183.84 **41. a)** 48 **b)** He cannot get a C. **43.** Answers will vary.

Exercise Set 1.3
1. a) Variables are letters that represent numbers. **b)** $x, y,$ and z **3.** $5(x), (5)x, (5)(x), 5x, 5 \cdot x$ **5.** Divide out the common factors. **7. a)** The smallest number that is divisible by the two denominators **b)** Answers will vary. **9.** Part b) shows simplification **11.** Part a) is incorrect. **13.** c) **15.** Divide out common factors, then multiply the numerators, and multiply the denominators **17.** Write fractions with a common denominator, then add or subtract the numerators while keeping the common denominator. **19.** Yes, the numerator and denominator have no common factors (other than 1). **21.** $\frac{1}{4}$ **23.** $\frac{2}{3}$ **25.** 1 **27.** $\frac{9}{19}$ **29.** $\frac{5}{33}$ **31.** Simplified **33.** $\frac{13}{5}$ **35.** $\frac{43}{15}$ **37.** $\frac{19}{4}$ **39.** $\frac{89}{19}$ **41.** $1\frac{3}{4}$ **43.** $3\frac{3}{4}$ **45.** $5\frac{1}{2}$ **47.** $4\frac{4}{7}$ **49.** $\frac{8}{15}$ **51.** $\frac{1}{9}$ **53.** $\frac{3}{2}$ or $1\frac{1}{2}$ **55.** $\frac{1}{2}$ **57.** $\frac{5}{16}$ **59.** 6 **61.** $\frac{77}{40}$ or $1\frac{37}{40}$ **63.** $\frac{43}{10}$ or $4\frac{3}{10}$ **65.** 1 **67.** $\frac{1}{3}$ **69.** $\frac{9}{17}$ **71.** $\frac{6}{5}$ or $1\frac{1}{5}$ **73.** $\frac{1}{9}$ **75.** $\frac{7}{24}$ **77.** $\frac{13}{36}$ **79.** $\frac{13}{45}$ **81.** $\frac{47}{15}$ or $3\frac{2}{15}$ **83.** $\frac{65}{12}$ or $5\frac{5}{12}$ **85.** $8\frac{15}{16}$ inches **87.** $\frac{11}{36}$ **89.** $\frac{11}{50}$ **91.** $7\frac{5}{12}$ tons **93.** 11 ft, $11\frac{1}{4}$ inches or $143\frac{1}{4}$ inches or $\approx$11.94 ft **95.** $1\frac{9}{16}$ inches **97.** $1\frac{3}{8}$ cups **99.** 40 times **101.** $1\frac{1}{2}$ inches **103.** 6 strips **105. a)** Yes **b)** $\frac{5}{8}$ inches **c)** $52\frac{3}{4}$ inches **107. a)** $\frac{* + ?}{a}$ **b)** $\frac{\odot - \square}{?}$ **c)** $\frac{\triangle + 4}{\square}$ **d)** $\frac{x - 2}{3}$ **e)** $\frac{8}{x}$ **109.** 270 pills **111.** Answers will vary. **112.** 16 **113.** 15 **114.** Variables are letters used to represent numbers.

Exercise Set 1.4
1. A set is a collection of elements. **3.** Answers will vary **5.** The set of natural numbers does not contain the number 0. The set of whole numbers contains the number 0. **7. a)** A number that can be written as a quotient of two integers, denominator not 0. **b)** Every integer can be written with a denominator of 1. **9. a)** Yes **b)** No **c)** No **d)** Yes **11.** $\{\ldots, -3, -2, -1, 0, 1, 2, 3, \ldots\}$ **13.** $\{0, 1, 2, 3, 4, \ldots\}$ **15.** $\{\ldots, -3, -2, -1\}$ **17.** True **19.** True **21.** False **23.** False **25.** True **27.** True **29.** False **31.** False **33.** True **35.** False **37.** True **39.** True **41.** True **43.** True **45.** False **47.** True **49. a)** 0 **b)** $0, 2\frac{1}{2}$ **c)** $0, 2\frac{1}{2}$ **51. a)** $3, 77$ **b)** $0, 3, 77$ **c)** $0, -2, 3, 77$ **d)** $-\frac{5}{7}, 0, -2, 3, 6\frac{1}{4}, 1.63, 77$ **e)** $\sqrt{7}, -\sqrt{3}$ **f)** $-\frac{5}{7}, 0, -2, 3, 6\frac{1}{4}, \sqrt{7}, -\sqrt{3}, 1.63, 77$ **53.** Answers will vary; three examples are 0, 1, and 2 **55.** Answers will vary; three examples are $\sqrt{2}, \sqrt{3}$, and $\sqrt{5}$ **57.** Answers will vary; three examples are $-\frac{2}{3}, \frac{1}{2}$, and 6.3 **59.** Answers will vary; three examples are $-13, -5$, and -1. **61.** Answers will vary; three examples are $\sqrt{2}, \sqrt{3}$, and $-\sqrt{5}$. **63.** Answers will vary; three examples are $-7, 1$, and 5. **65.** 87 **67. a)** $\{1, 3, 4, 5, 8\}$ **b)** $\{2, 5, 6, 7, 8\}$ **c)** $\{5, 8\}$ **d)** $\{1, 2, 3, 4, 5, 6, 7, 8\}$ **69. a)** Set B continues beyond 4. **b)** 4 **c)** An infinite number of elements **d)** An infinite set **71. a)** An infinite number **b)** An infinite number **73.** $\frac{27}{5}$ **74.** $5\frac{1}{3}$ **75.** $\frac{49}{40}$ or $1\frac{9}{40}$ **76.** $\frac{70}{27}$ or $2\frac{16}{27}$

Exercise Set 1.5

1. a)
-6 -5 -4 -3 -2 -1 0 1 2 3 4 5 6

b)
-6 -5 -4 -3 -2 -1 0 1 2 3 4 5 6
c) Greater than **d)** $-4 < -2$ **e)** $-2 > -4$ **3. a)** 4 is 4 units
from 0 on the number line. **b)** -4 is 4 units from 0 on the number line. **c)** 0 is 0 units from 0 on the number line.
5. Yes **7.** No, $-3 > -4$ but $|-3| < |-4|$. **9.** No, $|-4| > |-3|$ but $-4 < -3$. **11.** 7 **13.** 15 **15.** 0 **17.** -5 **19.** -21
21. > **23.** < **25.** > **27.** < **29.** > **31.** > **33.** < **35.** > **37.** < **39.** > **41.** > **43.** > **45.** > **47.** < **49.** > **51.** <
53. < **55.** > **57.** > **59.** > **61.** < **63.** < **65.** < **67.** = **69.** = **71.** < **73.** < **75.** $-|-8|, 0.38, \frac{4}{9}, \frac{3}{5}, |-6|$
77. $\frac{5}{12}, 0.6, \frac{2}{3}, \frac{19}{25}, |-2.6|$ **79.** 4, -4 **81.** Answers will vary; one example is $4\frac{1}{2}$, 5, and 5.5. **83.** Answers will vary; one example
is $-3, -4$, and -5. **85.** Answers will vary; one example is 4, 5, and 6. **87.** Answers will vary; one example is 3, 4, and 5.
89. a) Does not include the endpoints **b)** Answers will vary; one example is 4.1, 5, and $5\frac{1}{2}$. **c)** No **d)** Yes **e)** True
91. a) All the time **b)** March, 2000 **c)** January–March, 1999 **d)** March, 2000–May, 2001, except June 2000
93. Greater than **95.** Yes, 0 **98.** $\frac{31}{24}$ or $1\frac{7}{24}$ **99.** $\{0, 1, 2, 3, \ldots\}$ **100.** $\{1, 2, 3, 4, \ldots\}$ **101. a)** 5 **b)** 5, 0 **c)** 5, -2, 0
d) $5, -2, 0, \frac{1}{3}, -\frac{5}{9}, 2.3$ **e)** $\sqrt{3}$ **f)** $5, -2, 0, \frac{1}{3}, \sqrt{3}, -\frac{5}{9}, 2.3$

Exercise Set 1.6

1. Addition, subtraction, multiplication, and division **3. a)** No **b)** $\frac{2}{3}$ **5.** Either
7. Answers will vary. **9. a)** Answers will vary. **b)** -82 **c)** Answers will vary. **11.** Correct **13.** -9 **15.** 28 **17.** 0 **19.** $-\frac{5}{3}$
21. $-2\frac{3}{5}$ **23.** -3.72 **25.** 11 **27.** 1 **29.** -6 **31.** 0 **33.** 0 **35.** -10 **37.** 0 **39.** -13 **41.** 0 **43.** -6 **45.** 9 **47.** -64
49. -51 **51.** -21 **53.** 5 **55.** 2.3 **57.** -15.1 **59.** -20 **61.** -39 **63.** 91 **65.** -140 **67.** -105 **69.** -38.5 **71.** -4.6
73. $\frac{26}{35}$ **75.** $\frac{107}{84}$ **77.** $\frac{4}{55}$ **79.** $-\frac{26}{45}$ **81.** $\frac{3}{10}$ **83.** $-\frac{16}{15}$ **85.** $-\frac{13}{15}$ **87.** $-\frac{5}{72}$ **89.** $-\frac{43}{60}$ **91.** $-\frac{23}{21}$ **93. a)** Positive **b)** 390
c) Yes **95. a)** Negative **b)** -373 **c)** Yes **97. a)** Negative **b)** -452 **c)** Yes **99. a)** Negative **b)** -1300 **c)** Yes
101. a) Negative **b)** -112 **c)** Yes **103. a)** Negative **b)** -3880 **c)** Yes **105. a)** Positive **b)** 1111 **c)** Yes
107. a) Negative **b)** -2050 **c)** Yes **109.** True **111.** True **113.** False **115.** $277 **117.** 21 yards **119.** 61 feet
121. 13,796 feet **123. a)** $-$3 billion **b)** 1999: $0.4 billion; 2000: $-$0.3 billion; 2001: $-$3 billion, net loss: $-$2.9 billion
125. -22 **127.** 20 **129.** 0. **131.** $\frac{11}{30}$ **133.** 55 **135.** 1 **136.** $\frac{43}{16}$ or $2\frac{11}{16}$ **137.** $\{1, 2, 3, 4, \ldots\}$ **138.** > **139.** >

Exercise Set 1.7

1. 2 $-$ 7 **3.** □ $-$ * **5. a)** Answers will vary. **b)** $5 + (-14)$ **c)** -9 **7. a)** $a - b$ **b)** $7 - 9$
c) -2 **9. a)** $3 + 6 - 5$ **b)** 4 **11.** Correct **13.** 7 **15.** -1 **17.** -6 **19.** -1 **21.** -8 **23.** -7 **25.** 0 **27.** -4 **29.** 2 **31.** 9
33. 5 **35.** -20 **37.** 0 **39.** 4 **41.** -4 **43.** -11 **45.** -27 **47.** -150 **49.** -82 **51.** 140 **53.** -7.4 **55.** -3.93 **57.** -29
59. -16 **61.** -2 **63.** 13 **65.** -11 **67.** 6.1 **69.** $-\frac{1}{30}$ **71.** $\frac{17}{45}$ **73.** $-\frac{11}{12}$ **75.** $-\frac{5}{12}$ **77.** $\frac{1}{4}$ **79.** $-\frac{17}{24}$ **81.** $\frac{13}{16}$ **83.** $-\frac{1}{24}$
85. $-\frac{13}{63}$ **87.** $-\frac{7}{60}$ **89. a)** Positive **b)** 99 **c)** Yes **91. a)** Negative **b)** -619 **c)** Yes **93. a)** Positive **b)** 1588 **c)** Yes
95. a) Positive **b)** 196 **c)** Yes **97. a)** Negative **b)** -448 **c)** Yes **99. a)** Positive **b)** 116.1 **c)** Yes
101. a) Negative **b)** -69 **c)** Yes **103. a)** Negative **b)** -1670 **c)** Yes **105. a)** Zero **b)** 0 **c)** Yes **107.** 4
109. -6 **111.** -15 **113.** -2 **115.** 13 **117.** -5 **119.** -32 **121.** -4 **123.** 9 **125.** -12 **127.** 12 **129.** -18 **131. a)** 43
b) 143 **133.** 16 feet below sea level **135.** Dropped $100°$ F **137. a)** 276 **b)** 8 strokes less **c)** ≈ 273.33 **139.** -5
141. a) 7 units **b)** $5 - (-2)$ **143. a)** 9 feet **b)** -3 feet **144.** $\{\ldots, -3, -2, -1, 0, 1, 2, 3, \ldots\}$ **145.** The set of rational
numbers together with the set of irrational numbers form the set of real numbers. **146.** > **147.** < **148.** $\frac{17}{40}$

Exercise Set 1.8

1. Like signs: product is positive. Unlike signs: product is negative. **3.** Even number of
negatives, product is positive, odd number of negatives, product is negative **5.** $-\frac{a}{b}$ or $\frac{-a}{b}$ **7. a)** With $3 - 5$ you subtract,
but with $3(-5)$ you multiply. **b)** $-2, -15$ **9. a)** With $x - y$ you subtract, but with $x(-y)$ you multiply. **b)** 7 **c)** 10
d) -3 **11.** Negative **13.** Positive **15.** Negative **17.** 20 **19.** -18 **21.** -8 **23.** 0 **25.** 42 **27.** -56 **29.** 30 **31.** 0

A3 • Answers

33. -84 **35.** -72 **37.** 140 **39.** 0 **41.** $-\dfrac{3}{10}$ **43.** $\dfrac{7}{27}$ **45.** 4 **47.** $-\dfrac{5}{16}$ **49.** 2 **51.** 4 **53.** 4 **55.** -18 **57.** 19 **59.** -10
61. -6 **63.** -9 **65.** 8 **67.** 0 **69.** -3 **71.** -4 **73.** $-\dfrac{2}{5}$ **75.** $\dfrac{5}{36}$ **77.** 1 **79.** $-\dfrac{144}{5}$ **81.** -32 **83.** -20 **85.** -14 **87.** -9
89. -20 **91.** -1 **93.** 0 **95.** Undefined **97.** 0 **99.** Undefined **101. a)** Negative **b)** -3496 **c)** Yes **103. a)** Negative **b)** -16 **c)** Yes **105. a)** Negative **b)** -9 **c)** Yes **107. a)** Positive **b)** 6174 **c)** Yes **109. a)** Zero **b)** 0 **c)** Yes **111. a)** Undefined **b)** Undefined **c)** Yes **113. a)** Positive **b)** 3.2 **c)** Yes **115. a)** Positive **b)** 226.8 **c)** Yes **117.** False **119.** False **121.** True **123.** False **125.** False **127.** True **129.** 60 yard loss or -60 yards **131. a)** $150
b) $-$300 **133.** The total loss is $4\dfrac{1}{2}$ points. **135. a)** ≈ 2.075 **b)** ≈ 8.3 **137. a)** 102 to 128 beats per minutes
b) Answers will vary. **139.** -8 **141.** 1 **143.** Positive **146.** $\dfrac{25}{7}$ or $3\dfrac{4}{7}$ **147.** -2 **148.** -3 **149.** 3 **150.** 5

Exercise Set 1.9

1. Base; exponent **3. a)** 1 **b)** $1,3,2,1$ **5. a)** $5x$ **b)** x^5 **7.** Parentheses, exponents, multiplication or division, then addition or subtraction **9.** No **11. a)** 1 **b)** 10 **c)** b **13. a)** Answers will vary.
b) -180 **15. a)** Answers will vary. **b)** -91 **17.** 25 **19.** 1 **21.** -25 **23.** 9 **25.** -1 **27.** 64 **29.** 36 **31.** 4 **33.** 256
35. -16 **37.** $\dfrac{9}{16}$ **39.** $-\dfrac{1}{32}$ **41.** 225 **43.** 576 **45. a)** Positive **b)** 343 **c)** Yes **47. a)** Positive **b)** 1296 **c)** Yes
49. a) Negative **b)** -243 **c)** Yes **51. a)** Positive **b)** 625 **c)** Yes **53. a)** Positive **b)** 447.7456 **c)** Yes
55. a) Negative **b)** -0.765625 **c)** Yes **57.** 15 **59.** 8 **61.** 13 **63.** 0 **65.** 16 **67.** 29 **69.** -19 **71.** -77 **73.** $\dfrac{83}{100}$
75. -2 **77.** 10 **79.** 576 **81.** $\dfrac{1}{2}$ **83.** 169 **85.** 90.74 **87.** 36.75 **89.** $\dfrac{5}{8}$ **91.** $\dfrac{1}{4}$ **93.** $\dfrac{49}{30}$ **95.** $\dfrac{1}{17}$ **97.** $\dfrac{32}{53}$ **99.** 9 **101.** -4
103. a) 9 **b)** -9 **c)** 9 **105. a)** 16 **b)** -16 **c)** 16 **107. a)** 36 **b)** -36 **c)** 36 **109. a)** $\dfrac{1}{9}$ **b)** $-\dfrac{1}{9}$ **c)** $\dfrac{1}{9}$ **111.** 4
113. 18 **115.** 3 **117.** -1 **119.** -4 **121.** $\dfrac{15}{4}$ **123.** 994 **125.** -1 **127.** -5 **129.** 9 **131.** -3 **133.** $[(6 \cdot 3) - 4] - 2; 12$
135. $9\{[18 \div 3) + 9] - 8\}; 63$ **137.** $\left(\dfrac{4}{5} + \dfrac{3}{7}\right) \cdot \dfrac{2}{3}; \dfrac{86}{105}$ **139.** All real numbers **141.** $1.12 **143.** $16,050 **145. a)** .08
b) .16 **147.** 1.71 inches **149.** $12 - (4 - 6) + 10$ **155. a)** 4 **b)**

Occupants	Number of Houses
1	3
2	5
3	4
4	6
5	2

c) 59 **d)** 2.95 people per house **156.** $6.40 **157.** 144 **158.** $\dfrac{10}{3}$ or $3\dfrac{1}{3}$

Exercise Set 1.10

1. The commutative property of addition states that the sum of two numbers is the same regardless of the order in which they are added; $3 + 4 = 4 + 3$. **3.** The associative property of addition states that the sum of three numbers is the same regardless of the way the numbers are grouped; $(2 + 3) + 4 = 2 + (3 + 4)$
5. a) Answers will vary **b)** 15 **c)** 44 **7.** The associative property involves changing parentheses, and uses only one operation whereas the distributive property uses two operations, multiplication and addition. **9.** 0 **11. a)** -6 **b)** $\dfrac{1}{6}$
13. a) 3 **b)** $-\dfrac{1}{3}$ **15. a)** $-x$ **b)** $\dfrac{1}{x}$ **17. a)** -1.6 **b)** $\dfrac{1}{1.6}$ or 0.625 **19. a)** $-\dfrac{1}{5}$ **b)** 5 **21. a)** $\dfrac{3}{5}$ **b)** $-\dfrac{5}{3}$ **23.** Associative property of addition **25.** Distributive property **27.** Commutative property of multiplication **29.** Associative property of multiplication **31.** Distributive property **33.** Identity property of addition **35.** Inverse property of multiplication **37.** $6 + x$ **39.** $(-6 \cdot 4) \cdot 2$ **41.** $1 \cdot x + 1 \cdot y$ or $x + y$ **43.** $y \cdot x$ **45.** $3y + 4x$ **47.** $a + (b + 3)$ **49.** $3x + (4 + 6)$
51. $(m + n)3$ **53.** $4x + 4y + 12$ **55.** 0 **57.** $\dfrac{5}{2}n$ **59.** Yes **61.** No **63.** No **65.** Yes **67.** No **69.** No **71.** The $(3 + 4)$ is treated as one value. **73.** Commutative property of addition **75.** No; Associative property of addition **77.** $\dfrac{49}{15}$ or $3\dfrac{4}{15}$
78. $\dfrac{23}{16}$ or $1\dfrac{7}{16}$ **79.** 45 **80.** -25

Chapter Review Exercises
1. 80 **2.** $551.25 **3.** Less than **4.** $30 **5. a)** 78.4 **b)** 79 **6. a)** 80° F
b) 79°F **7. a)** $8.33 **b)** $73.20 **8. a)** 17.1 billion barrels **b)** $\approx$12.5 **c)** 1028.8 billion barrels **9.** $\frac{1}{2}$ **10.** $\frac{9}{25}$ **11.** $\frac{25}{36}$
12. $\frac{7}{6}$ or $1\frac{1}{6}$ **13.** $\frac{19}{72}$ **14.** $\frac{17}{15}$ or $1\frac{2}{15}$ **15.** $\{1, 2, 3, \ldots\}$ **16.** $\{0, 1, 2, 3, \ldots\}$ **17.** $\{\ldots, -3, -2, -1, 0, 1, 2, 3, \ldots\}$ **18.** The set of
all numbers which can be expressed as the quotient of two integers, denominator not zero **19. a)** 3, 426 **b)** 3, 0, 426
c) 3, -5, -12, 0, 426 **d)** 3, -5, -12, 0, $\frac{1}{2}$, -0.62, 426, $-3\frac{1}{4}$ **e)** $\sqrt{7}$ **f)** 3, -5, -12, 0, $\frac{1}{2}$, -0.62, $\sqrt{7}$, 426, $-3\frac{1}{4}$ **20. a)** 1
b) 1 **c)** -8, -9 **d)** -8, -9, 1 **e)** -2.3, -8, -9, $1\frac{1}{2}$, 1, $-\frac{3}{17}$ **f)** -2.3, -8, -9, $1\frac{1}{2}$, $\sqrt{2}$, $-\sqrt{2}$, 1, $-\frac{3}{17}$ **21.** $<$ **22.** $>$
23. $<$ **24.** $>$ **25.** $<$ **26.** $>$ **27.** $<$ **28.** $=$ **29.** -14 **30.** 0 **31.** -3 **32.** -6 **33.** -6 **34.** 2 **35.** 8 **36.** 0 **37.** 14
38. -5 **39.** 4 **40.** -12 **41.** $\frac{7}{12}$ **42.** $\frac{11}{10}$ **43.** $-\frac{7}{36}$ **44.** $-\frac{19}{56}$ **45.** $-\frac{5}{4}$ **46.** $-\frac{37}{84}$ **47.** $-\frac{7}{90}$ **48.** $\frac{7}{12}$ **49.** 8 **50.** -4 **51.** -12
52. -7 **53.** 6 **54.** 11 **55.** -27 **56.** 40 **57.** -120 **58.** $-\frac{6}{35}$ **59.** $-\frac{6}{11}$ **60.** $\frac{15}{56}$ **61.** 0 **62.** 144 **63.** -5 **64.** -6 **65.** -4
66. 0 **67.** -10 **68.** 9 **69.** $\frac{56}{27}$ **70.** $-\frac{35}{9}$ **71.** 0 **72.** 0 **73.** Undefined **74.** Undefined **75.** Undefined **76.** 0 **77.** 25
78. -8 **79.** 1 **80.** 3 **81.** -6 **82.** 18 **83.** 6 **84.** -4 **85.** 10 **86.** 1 **87.** 15 **88.** -4 **89.** 49 **90.** 729 **91.** 81 **92.** -27
93. -1 **94.** -32 **95.** $\frac{16}{25}$ **96.** $\frac{8}{125}$ **97.** 500 **98.** 4 **99.** 12 **100.** 512 **101.** 23 **102.** 32 **103.** 22 **104.** -17 **105.** -39
106. -4 **107.** $\frac{9}{7}$ **108.** 0 **109.** -60 **110.** 10 **111.** 20 **112.** 20 **113.** 14 **114.** 9 **115.** -4 **116.** 50 **117.** 24 **118.** 26
119. 45 **120.** 0 **121.** -3 **122.** -11 **123.** -3 **124.** 39 **125.** -215 **126.** 353.6 **127.** -2.88 **128.** 117.8 **129.** 729 **130.**
-74.088 **131.** Associative property of addition **132.** Commutative property of multiplication **133.** Distributive property
134. Commutative property of multiplication **135.** Commutative property of addition **136.** Associative property of
addition **137.** Identity property of addition **138.** Inverse property of multiplication

Chapter Practice Test
1. a) $10.65 **b)** $0.23 **c)** $10.88 **d)** $39.12 **2. a)** $\approx$ $4400 **b)** $\approx$ $1400
3. a) $\approx$108.5 million **b)** Half were above and half were below this age **4. a)** 42 **b)** 42, 0 **c)** -6, 42, 0, -7, -1
d) -6, 42, $-3\frac{1}{2}$, 0, 6.52, $\frac{5}{9}$, -7, -1 **e)** $\sqrt{5}$ **f)** -6, 42, $-3\frac{1}{2}$, 0, 6.52, $\sqrt{5}$, $\frac{5}{9}$, -7, -1 **5.** $>$ **6.** $>$ **7.** -15 **8.** -11 **9.** -14
10. 8 **11.** -24 **12.** $\frac{16}{63}$ **13.** -3 **14.** $-\frac{53}{56}$ **15.** 12 **16.** $\frac{27}{125}$ **17.** $2^2 5^2 y^2 z^3$ **18.** $2 \cdot 2 \cdot 3 \cdot 3 \cdot 3xxxxyy$ **19.** 72 **20.** 10 **21.** 11
22. 1 **23.** Commutative property of addition **24.** Distributive property **25.** Associative property of addition

CHAPTER 2

Exercise Set 2.1
1. a) Terms are the parts that are added. **b)** $3x$, $-4y$, and -5 **c)** $6xy$, $3x$, $-y$, and -9
3. a) The parts that are multiplied **b)** They are multiplied together. **c)** They are multiplied together. **5. a)** Numerical
coefficient or coefficient **b)** 4 **c)** 1 **d)** -1 **d)** $\frac{3}{5}$ **e)** $\frac{4}{7}$ **7. a)** The signs of all the terms inside the parentheses change
when the parentheses are removed. **b)** $-x + 8$ **9.** $8x$ **11.** $-x$ **13.** $5y + 3$ **15.** $3x$ **17.** $-6x + 7$ **19.** $-5w + 5$
21. $-2x$ **23.** 0 **25.** $-2x + 11$ **27.** $-2r - 8$ **29.** $-5x + y + 2$ **31.** $2x - 3$ **33.** $b + \frac{23}{5}$ **35.** $0.8n + 6.42$
37. $\frac{1}{2}a + 3b + 1$ **39.** $2x^2 + 4x + 8y^2$ **41.** $x^2 + y$ **43.** $-3x - 5y$ **45.** $-3n^2 - 2n + 13$ **47.** $21.72x - 7.11$
49. $-\frac{23}{20}x - 5$ **51.** $5w^3 + 2w^2 + w + 3$ **53.** $-7z^3 - z^2 + 2z$ **55.** $6x^2 - 6xy + 3y^2$ **57.** $4a^2 + 3ab + b^2$ **59.** $5x + 10$
61. $5x + 20$ **63.** $-2x + 8$ **65.** $-x + 2$ **67.** $x - 4$ **69.** $\frac{4}{5}s - 4$ **71.** $-0.9x - 1.5$ **73.** $r - 4$ **75.** $1.4x + 0.35$ **77.** $x - y$
79. $-2x - 4y + 8$ **81.** $3.41x - 5.72y + 3.08$ **83.** $6x - 4y + \frac{1}{2}$ **85.** $x + 3y - 9$ **87.** $3x - 6y - 12$ **89.** $2x - 15$
91. $2x + 1$ **93.** $14x + 18$ **95.** $4x - 2y + 3$ **97.** $5c$ **99.** $7x + 3$ **101.** $x - 9$ **103.** $2x - 2$ **105.** $-4s - 6$ **107.** $2x + 3$
109. 0 **111.** $y + 4$ **113.** $3x - 5$ **115.** $x + 15$ **117.** $0.2x - 4y - 2.8$ **119.** $-6x + 7y$ **121.** $\frac{3}{2}x + \frac{7}{2}$ **123.** $2\square + 3\ominus$

125. $2x + 3y + 2\triangle$ **127.** $1, 2, 3, 4, 6, 12$ **129.** $2\triangle + 2\square$ **131.** $22x^2 - 25y^2 - 4x + 3$ **133.** $6x^2 + 5y^2 + 3x + 7y$ **135.** 7
136. -16 **137.** -1 **138.** Answers will vary. **139.** -12

Exercise Set 2.2
1. An equation is a statement that shows two algebraic expressions are equal. **3.** Substitute the value in the equation and then determine if it results in a true statement. **5.** Equivalent equations are two or more equations with the same solution. **7.** Add 4 to both sides of the equation to get the variable by itself. **9.** One example is $x + 2 = 1$. **11.** Subtraction is defined in terms of addition. **13.** Yes **15.** No **17.** Yes **19.** Yes **21.** No **23.** Yes **25.** 4
27. -7 **29.** -9 **31.** 43 **33.** 15 **35.** 11 **37.** -4 **39.** -5 **41.** -13 **43.** -30 **45.** 0 **47.** -1 **49.** -4 **51.** -20 **53.** 0
55. 17 **57.** -26 **59.** 28 **61.** -46.1 **63.** 46.5 **65.** -8.23 **67.** 5.57 **69.** No, the equation is equivalent to $1 = 2$, a false statement. **71.** $x = \square + \triangle$ **73.** $\square = \odot - \triangle$ **76.** 18 **77.** -8 **78.** $2x - 13$ **79.** $7t - 19$

Exercise Set 2.3
1. Answers will vary. **3. a)** $x = -a$ **b)** $x = -5$ **c)** $x = 5$ **5.** Divide by -2 to isolate the variable. **7.** Multiply both sides by 3. **9.** 3 **11.** 8 **13.** -3 **15.** -8 **17.** 5 **19.** -3 **21.** $-\dfrac{7}{3}$ **23.** 11 **25.** -10 **27.** 39
29. $-\dfrac{1}{3}$ **31.** 6 **33.** $\dfrac{26}{43}$ **35.** 2 **37.** -1 **39.** $-\dfrac{3}{40}$ **41.** -60 **43.** 240 **45.** -35 **47.** 20 **49.** -50 **51.** 0 **53.** 0 **55.** 22.5
57. 6 **59.** -20.2 **61.** $-\dfrac{7}{48}$ **63.** 9 **65.** In $5 + x = 10$, 5 is added to the variable, whereas in $5x = 10$, 5 is multiplied by the variable. **b)** $x = 5$ **c)** $x = 2$ **67.** Multiply by $\dfrac{3}{2}$; 6 **69.** Multiply by $\dfrac{7}{3}$; $\dfrac{28}{15}$ **71. a)** $\square$ **b)** Divide both sides of the equation by $\triangle$. **c)** $\square = \dfrac{\odot}{\triangle}$ **73.** -4 **74.** 0 **75.** 6 **76.** $-11x + 38$ **77.** -57

Exercise Set 2.4
1. No, there is an x on both sides of the equation. **3.** $x = \dfrac{1}{3}$ **5.** $x = -\dfrac{1}{2}$ **7.** $x = \dfrac{3}{5}$
9. Solve **11. a)** Answers will vary. **b)** Answers will vary. **13. a)** Answers will vary. **b)** $x = -2$ **15.** 2 **17.** -4 **19.** 5
21. $\dfrac{12}{5}$ **23.** -6 **25.** 3 **27.** $\dfrac{11}{3}$ **29.** $-\dfrac{19}{16}$ **31.** -10 **33.** 2 **35.** $-\dfrac{51}{5}$ **37.** 3 **39.** 6.8 **41.** 15 **43.** 12 **45.** 58 **47.** 60
49. -13 **51.** -11 **53.** 0 **55.** $\dfrac{19}{8}$ **57.** 0 **59.** 0 **61.** -1 **63.** 6 **65.** -6 **67.** 3 **69.** -4 **71.** 4 **73.** 5 **75.** 0.8 **77.** -1
79. $\dfrac{2}{7}$ **81.** -2.6 **83.** $-\dfrac{14}{5}$ **85.** 23 **87.** $-\dfrac{1}{15}$ **89.** -2 **91.** $-\dfrac{16}{21}$ **93.** 10 **95.** $-\dfrac{1}{5}$ **97.** $-\dfrac{23}{3}$ **99.** 2 **101.** $\dfrac{25}{3}$ **103.** $\dfrac{88}{135}$
105. a) You will not have to work with fractions **b)** $x = 3$ **107.** $\dfrac{35}{6}$ **109.** -4 **113.** 64 **114.** -47 **115.** Isolate the variable on one side of the equation. **116.** Divide both sides of the equation by -4 to isolate the variable.

Exercise Set 2.5
1. Answers will vary. **3. a)** An identity is an equation that is true for infinitely many values of the variable. **b)** All real numbers **5.** Both sides of the equation are identical. **7.** You will obtain a false statement.
9. a) Answers will vary. **b)** $x = 21$ **11.** 3 **13.** 1 **15.** $\dfrac{3}{5}$ **17.** 3 **19.** 2 **21.** No Solution **23.** 0.1 **25.** 3.2 **27.** 3
29. $-\dfrac{17}{7}$ **31.** No solution **33.** $\dfrac{34}{5}$ **35.** $\dfrac{7}{9}$ **37.** 5 **39.** 16 **41.** $\dfrac{3}{4}$ **43.** $\dfrac{5}{2}$ **45.** 25 **47.** All real numbers **49.** 23 **51.** 0
53. All real numbers **55.** $\dfrac{21}{20}$ **57.** 14 **59.** $\dfrac{5}{2}$ **61.** 2 **63.** $-\dfrac{5}{3}$ **65.** 0 **67.** 16 **69.** $-\dfrac{12}{5}$ **71.** 10 **73.** $\dfrac{10}{3}$ **75.** 30 **77.** 4
79. a) One example is $x + x + 1 = x + 2$. **b)** It has a single solution. **c)** $x = 1$ **81. a)** One example is $x + x + 1 = 2x + 1$. **b)** Both sides simplify to the same expression. **c)** All real numbers **83. a)** One example is $x + x + 1 = 2x + 2$. **b)** It simplifies to a false statement. **c)** No solution **85.** $* = 6$ **87.** All real numbers
89. $x = -4$ **91. a)** 4 **b)** 7 **c)** 0 **92.** ≈ 0.131687243 **93.** Factors are expressions that are multiplied; terms are expressions that are added. **94.** $7x - 10$ **95.** $\dfrac{10}{7}$ **96.** -3

Exercise Set 2.6
1. A ratio is a quotient of two quantities. **3.** c to d, $c:d$, $\dfrac{c}{d}$ **5.** Need a given ratio and one of the two parts of a second ratio **7.** No, their corresponding angles must be equal but their corresponding sides only need to be in proportion. **9.** Yes **11.** No **13.** $2:3$ **15.** $1:2$ **17.** $8:1$ **19.** $7:4$ **21.** $1:3$ **23.** $6:1$ **25.** $13:32$ **27.** $8:1$
29. a) $199:140$ **b)** $\approx 1.42:1$ **31. a)** $1.13:0.38$ **b)** $\approx 2.97:1$ **33. a)** $434:174$ or $217:87$ **b)** $374:434$ or $187:217$

35. a) 40:32 or 5:4 **b)** 15:11 **37.** 12 **39.** 45 **41.** −100 **43.** −2 **45.** −54 **47.** 6 **49.** 32 inches **51.** 15.75 inches
53. 19.5 inches **55.** 25 loads **57.** 361.1 miles **59.** 1.5 feet **61.** 24 teaspoons **63.** ≈0.43 feet **65.** 1.25 inches
67. ≈9.49 feet **69.** 0.55 milliliter **71.** 23 minutes **73.** ≈339 children **75.** 6.5 feet **77.** 2.9 square yards **79.** 20 inches
81. ≈23 **83.** \$0.85 **85.** 4 points **87.** Yes, her ratio is 2.12:1. **89.** It must increase **91.** ≈41,667 miles **93.** 0.625 cubic
centimeters **97.** Commutative property of addition **98.** Associative property of multiplication **99.** Distributive property
100. $x = \dfrac{3}{4}$ **101.** All real numbers.

Exercise Set 2.7
1. >: is greater than; ≥: is greater than or equal to; <: is less than; ≤: is less than or equal to
3. a) False **b)** True **5.** When multiplying or dividing by a negative number **7.** All real numbers **9.** No solution

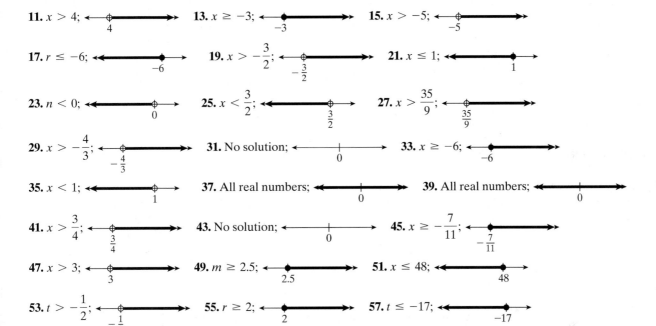

11. $x > 4$; **13.** $x \geq -3$; **15.** $x > -5$;

17. $r \leq -6$; **19.** $x > -\dfrac{3}{2}$; **21.** $x \leq 1$;

23. $n < 0$; **25.** $x < \dfrac{3}{2}$; **27.** $x > \dfrac{35}{9}$;

29. $x > -\dfrac{4}{3}$; **31.** No solution; **33.** $x \geq -6$;

35. $x < 1$; **37.** All real numbers; **39.** All real numbers;

41. $x > \dfrac{3}{4}$; **43.** No solution; **45.** $x \geq -\dfrac{7}{11}$;

47. $x > 3$; **49.** $m \geq 2.5$; **51.** $x \leq 48$;

53. $t > -\dfrac{1}{2}$; **55.** $r \geq 2$; **57.** $t \leq -17$;

59. a) May, September, June, August, and July **b)** January, February, December, March, November, and April
c) January, February, and December **d)** June, August, and July **61.** ≠ **63.** We do not know that y is positive. If y is
negative, we must reverse the sign of the inequality. **65.** $x > 4$ **66.** −9 **67.** −25 **68.** $\dfrac{14}{5}$ **69.** 500 kilowatt-hours

Chapter Review Exercises
1. $3x + 12$ **2.** $3x - 6$ **3.** $-2x - 8$ **4.** $-x - 2$ **5.** $-m - 3$ **6.** $-16 + 4x$
7. $25 - 5p$ **8.** $24x - 30$ **9.** $-25x + 25$ **10.** $-4x + 12$ **11.** $x + 2$ **12.** $-3 - 2y$ **13.** $-x - 2y + z$
14. $-6a + 15b - 21$ **15.** $4x$ **16.** $-3y + 8$ **17.** $5x + 1$ **18.** $-3x + 3y$ **19.** $8m + 8n$ **20.** $9x + 3y + 2$
21. $4x + 3y + 6$ **22.** 3 **23.** $-12x^2 + 3$ **24.** 0 **25.** $5x + 7$ **26.** $-10b + 12$ **27.** 0 **28.** $4x - 4$ **29.** $22x - 42$
30. $6x^2 - 3x + y$ **31.** $-\dfrac{7}{20}d + 7$ **32.** 3 **33.** $\dfrac{1}{6}x + 2$ **34.** $-\dfrac{7}{12}n$ **35.** 1 **36.** −13 **37.** 11 **38.** −27 **39.** 2 **40.** $\dfrac{11}{2}$
41. −6 **42.** −3 **43.** 12 **44.** 4 **45.** 2 **46.** −3 **47.** $\dfrac{3}{2}$ **48.** −3 **49.** $-\dfrac{1}{2}$ **50.** −1 **51.** 0 **52.** $\dfrac{1}{5}$ **53.** 2 **54.** −5 **55.** −35.5
56. 0.6 **57.** $-\dfrac{21}{4}$ **58.** $\dfrac{78}{7}$ **59.** $-\dfrac{3}{2}$ **60.** $-\dfrac{6}{11}$ **61.** 3 **62.** $\dfrac{10}{7}$ **63.** 0 **64.** −1 **65.** No solution **66.** 10 **67.** All real
numbers **68.** −4 **69.** No solution **70.** All real numbers **71.** $\dfrac{17}{3}$ **72.** $-\dfrac{20}{7}$ **73.** No solution **74.** 6 **75.** 52 **76.** 32
77. 7 **78.** −18 **79.** 3:5 **80.** 5:12 **81.** 1:1 **82.** 2 **83.** 20 **84.** 9 **85.** $\dfrac{135}{4}$ **86.** −10 **87.** −16 **88.** 36 **89.** 90
90. 40 inches **91.** 1 foot **92.** $x \geq 2$; **93.** $a < 1$;

94. $r \geq -2$; **95.** No solution; **96.** All real numbers;

97. $x < -3$; -3 **98.** $x \le \dfrac{9}{5}$; $\dfrac{9}{5}$ **99.** $x > \dfrac{8}{5}$; $\dfrac{8}{5}$

100. $x > -12$; -12 **101.** $t \ge -3$; -3 **102.** 6.3 hours   **103.** 240 calories   **104.** 440 pages

105. $6\dfrac{1}{3}$ inches   **106.** 15.75 feet   **107.** $\approx\$0.109$ **108.** 192 bottles

Chapter Practice Test

1. $6x - 12$ **2.** $-x - 3y + 4$ **3.** $-3x + 4$ **4.** $-x + 10$ **5.** $-5x - y - 6$

6. $7a - 8b - 3$ **7.** $2x^2 + 6x - 1$ **8.** $x = 3$ **9.** $x = 8$ **10.** $x = \dfrac{1}{2}$ **11.** $x = 0$ **12.** $x = -\dfrac{1}{7}$ **13.** No solution   **14.** All real

numbers   **15.** $x = -45$ **16.** $x = 0$ **17. a)** Conditional equation   **b)** Contradiction   **c)** Identity

18. $x > -7$; -7 **19.** $x \le 12$; 12 **20.** No solution; 0

21. All real numbers; 0 **22.** $x = \dfrac{32}{3}$ feet or $10\dfrac{2}{3}$ feet   **23.** 150 gallons   **24.** 50,000 gallons

25. 175 minutes or 2 hours 55 minutes

CHAPTER 3

Exercise Set 3.1

1. A formula is an equation used to express a relationship mathematically.   **3.** $i = prt$; i is the amount of interest earned or owed, p is the amount invested or borrowed, r is the rate in decimal form, and t is the length of time invested.   **5.** The diameter of a circle is 2 times its radius.   **7.** When you multiply a unit by the same unit, you get a square unit.   **9.** 24   **11.** 49   **13.** 26   **15.** 78.54   **17.** 2   **19.** 15   **21.** 56   **23.** 59   **25.** 6.00   **27.** 127.03   **29.** 179.2

31. 12 square inches   **33.** ≈ 452.39 cubic centimeters   **35.** 16.5 square feet   **37.** $s = \dfrac{P}{4}$ **39.** $t = \dfrac{d}{r}$ **41.** $l = \dfrac{V}{wh}$

43. $b = \dfrac{2A}{h}$ **45.** $w = \dfrac{P - 2l}{2}$ **47.** $t = \dfrac{-m + 5}{2}$ **49.** $b = y - mx$ **51.** $x = \dfrac{y - b}{m}$ **53.** $y = \dfrac{-ax + c}{b}$ **55.** $h = \dfrac{V}{\pi r^2}$

57. $m = 2A - d$ **59.** $w = \dfrac{2R - I}{3}$ **61. a)** $y = -3x + 5$ **b)** -1 **63. a)** $y = \dfrac{2x + 4}{3}$ **b)** 8   **65. a)** $y = \dfrac{3x - 12}{5}$ **b)** 0

67. a) $y = \dfrac{3x - 10}{5}$ **b)** $\dfrac{2}{5}$ **69. a)** $y = \dfrac{x - 5}{2}$ **b)** $-\dfrac{5}{2}$ **71. a)** $y = \dfrac{-x + 8}{2}$ **b)** 6   **73. a)** $y = \dfrac{-x - 5}{3}$ **b)** $-\dfrac{11}{3}$

75. a) $y = \dfrac{30x + 13}{15}$ **b)** $\dfrac{133}{15}$ **77.** 35   **79.** $C = 10°$ **81.** $F = 77°$ **83.** $P = 40$ **85.** $K = 4$ **87.** The area is 4 times as large.   **89.** 42   **91.** \$1440   **93.** \$5000   **95.** 25 feet   **97.** 558 square inches   **99.** ≈ 7.07 square feet   **101.** 3 square feet

103. ≈ 150.80 cubic feet   **105.** 14,000 square feet   **107.** ≈ 11.78 cubic inches   **109. a)** $B = \dfrac{703w}{h^2}$ **b)** ≈ 23.91

111. a) $V = 18x^3 - 3x^2$ **b)** 6027 cubic centimeters   **c)** $S = 54x^2 - 8x$ **d)** 2590 square centimeters   **113.** 0   **114.** $3:2$

115. $\dfrac{3}{25} = \dfrac{x}{13,500}$; 1620 minutes or 27 hours   **116.** $x \le -17$

Exercise Set 3.2

1. Added to, more than, increased by, and sum indicate addition.   **3.** Multiplied by, product of, twice, and three times indicate multiplication.   **5.** The cost is increased by 25% of the cost, so the expression should be

$c + 0.25c$.   **7.** $n + 7$ **9.** $4x$ **11.** $\dfrac{x}{2}$ **13.** $h + 0.8$ **15.** $p - 0.08p$ **17.** $\dfrac{1}{10}n - 5$ **19.** $\dfrac{8}{9}m + 16,000$ **21.** $45 + 0.40x$

23. $25x$ **25.** $16x + y$ **27.** $n + 0.04n$ **29.** $p - 0.02p$ **31.** $220 + 80x$ **33.** $275x + 25y$ **35.** Three less than a number **37.** One more than 4 times a number   **39.** Seven less than 6 times a number   **41.** Four times a number, decreased by 2 **43.** Three times a number, subtracted from 2   **45.** Twice the difference between a number and 1   **47.** $s + 5$ **49.** $b - 6$

51. $600 - a$ **53.** $100 - m$ **55.** $\dfrac{2}{3}m - 6$ **57.** $2r - 673$ **59.** $2p - 2.7$ **61.** $3n - 15$ **63.** $2n - 67,109$ **65.** $s + 0.20s$

67. $s + 0.15s$ **69.** $f - 0.12f$ **71.** $c + 0.07c$ **73.** $p - 0.50p$ **75.** $x + 4x = 20$ **77.** $x + (x + 1) = 41$ **79.** $2x - 8 = 12$

81. $\frac{1}{5}(x + 10) = 150$ **83.** $s + (2s - 4) = 890$ **85.** $c + 0.07c = 32,600$ **87.** $c + 0.15c = 42.50$ **89.** $1.28f - f = 16,762$
91. $s + (s + 0.55) = 2.45$ **93.** Two more than a number is 5. **95.** Three times a number, decreased by 1, is 4 more than twice the number. **97.** Four times the difference between a number and 1 is 6. **99.** Six more than 5 times a number is the difference between 6 times the number and 1. **101.** The sum of a number and the number increased by 4 is 8. **103.** The sum of twice a number and the number increased by 3 is 5. **105.** Answers will vary. **107. a)** $86,400d + 3600h + 60m + s$
b) 368,125 seconds **111.** 4 **112.** 2.52 **113.** $x > \frac{7}{2}$; **114.** 15 **115.** $y = \frac{3x - 6}{2}$ or $y = \frac{3}{2}x - 3$; 6

Exercise Set 3.3 **1.** Understand, Translate, Carry out, Check, Answer **3.** 42, 43 **5.** 51, 53 **7.** 8, 19 **9.** 25, 42
11. 73.6 years **13.** 2180 **15.** 65 **17.** 22.2 hours **19.** ≈ 62.7 weeks **21.** 11.75 years **23.** ≈ 13.3 weeks **25.** 140 miles
27. 18,100 copies **29.** 90 minutes above the 500 minutes **31.** 2 years **33.** 6 years **35.** 5000 pages **37.** ≈ 28.6 months
39. \$62,000 **41.** \$261.68 **43.** \$23,230.77 **45.** ≈ 2682.93 square feet **47.** \$25,000 **49.** \$39,387.76 **51.** \$6000 **53.** \$7500
55. \$50,000 **57.** \$300 **59.** \$77,777.78 **61. a)** $\frac{74 + 88 + 76 + x}{4} = 80$ **b)** 82 **63. a)** 0.75 year, or 9 months **b)** \$375
65. $\frac{17}{12}$ **66.** Associative property of addition **67.** Commutative property of multiplication **68.** Distributive property
69. 56 pounds **70.** $b = \frac{2A}{h}$

Exercise Set 3.4 **1.** The area remains the same. **3.** The volume is eight times as great. **5.** The area is nine times as great. **7.** An isosceles triangle is a triangle with 2 equal sides. **9.** $180°$ **11.** $46°, 46°, 88°$ **13.** 9.5 inches
15. $A = 67°, B = 23°$ **17.** $A = 47°, B = 133°$ **19.** $88°$ **21.** $50°, 60°, 70°$ **23.** Length is 16 feet, width is 8 feet. **25.** Length is 78 feet, width is 36 feet. **27.** Smaller angles are $50°$; larger angles are $130°$ **29.** $63°, 73°, 140°, 84°$ **31.** Width is 4 feet; height is 7 feet **33.** Width is 2.5 feet; height is 5 feet **35.** Width is 10 feet; length is 14 feet **37.** $ac + ad + bc + bd$
39. $<$ **40.** $>$ **41.** -8 **42.** $-2x - 4y + 6$ **43.** $y = \frac{-2x + 9}{3}$ or $y = -\frac{2}{3}x + 3$; 1

Exercise Set 3.5 **1.** 50 mph **3.** 2.4 centimeters **5.** 14 hours **7.** 250 cubic centimeters per hour **9.** 1320 seconds or 22 minutes **11.** ≈ 189.75 mph **13.** 2.4 hours **15.** 70 mph **17.** 5 minutes **19.** 35 mph, 40 mph **21. a)** 2.4 miles **b)** 112.0 miles **c)** 26.2 miles **d)** 140.6 miles **e)** 9.45 hours **23.** *Apollo:* 5 mph; *Pythagoras:* 9 mph
25. a) 1.25 hours **b)** 43.75 miles **27. a)** 2 seconds **b)** 100 feet **29.** 1.5 hours **31.** 570 mph; 600 mph **33.** Bridge: 1.25 feet per day; road: 2.45 feet per day **35.** \$2400 at 5%; \$7000 at 7% **37.** \$2400 at 6%; \$3600 at 4% **39.** \$2000 at 4%; \$8000 at 5%
41. November **43.** 8 hours at Home Depot; 10 hours at veterinary clinic **45.** 80 of the \$1550 systems **47. a)** 32 shares of GE; 160 shares of PepsiCo **b)** \$32 **49.** 2.86 pounds of Family, 7.14 pounds of Spot Filler **51.** 300 gallons regular, 200 gallons premium plus **53.** \$2.74 per pound **55.** $\approx 11.1\%$ **57.** $1\frac{2}{3}$ liters **59.** 4 ounces **61.** 3.25% **63.** 6250 gallons **65.** 0.5 gallon
67. ≈ 5.74 hours **71. a)** $\frac{22}{13}$ or $1\frac{9}{13}$ **b)** $\frac{35}{8}$ or $4\frac{3}{8}$ **72.** All real numbers **73.** $\frac{3}{4}$ or 0.75 **74.** $x \leq \frac{1}{4}$

Review Exercises **1.** 37.70 **2.** 18 **3.** 48 **4.** 25 **5.** 21 **6.** 5000 **7. a)** $y = x - 2$ **b)** 8 **8. a)** $y = -2x - 3$
b) -27 **9. a)** $y = \frac{5}{2}x - 8$ **b)** -3 **10. a)** $y = \frac{2}{3}x - 4$ **b)** -8 **11.** $w = \frac{A}{l}$ **12.** $h = \frac{2A}{b}$ **13.** $t = \frac{i}{pr}$
14. $w = \frac{P - 2l}{2}$ **15.** $h = \frac{V}{\pi r^2}$ **16.** $h = \frac{3V}{B}$ **17.** \$108 **18.** 6 inches **19.** The sum of a number and the number increased by 5 is 9. **20.** The sum of a number and twice the number decreased by 1 is 10. **21.** 33 and 41 **22.** 118 and 119 **23.** 38 and 7
24. \$21,738.32 **25.** 19 months **26.** \$2000 **27.** \$650 **28.** ≈ 4.76 months **29.** $45°, 55°, 80°$ **30.** $30°, 40°, 150°, 140°$ **31.** Width: 15.5 feet; length: 19.5 feet **32.** Width: 50 feet; length: 80 feet **33.** 30 gallons per hour **34.** 6.5 mph **35.** 2 hours **36.** 4 hours
37. ≈ 8.8 feet per second **38.** \$4000 at 8%; \$8000 at $7\frac{1}{4}\%$ **39.** \$700 at 3%; \$3300 at 3.5% **40.** ≈ 0.67 gallon **41.** 9 smaller and 21 larger chimes **42.** 1.2 liters of 10%; 0.8 liters of 5% **43.** 103 and 105 **44.** \$450 **45.** \$12,000 **46.** $42°, 50°, 88°$
47. 8 years **48.** $70°, 70°, 110°, 110°$ **49.** 500 copies **50. a)** 10 minutes **b)** 600 feet **51.** 60 pounds of \$3.50; 20 pounds of \$4.10 **52.** Older brother: 55 mph; younger brother: 60 mph **53.** 0.4 liters

Practice Test **1.** 9% **2.** 18 feet **3.** 145 **4.** 85 **5.** ≈ 7.96 **6. a)** $y = \frac{4}{3}x - 3$ **b)** 13 **7.** $R = \frac{P}{I}$ **8.** $a = 3A - b$
9. 28 square feet **10.** ≈ 3106.86 square feet **11.** $500 - n$ **12.** $2f + 6000$ **13.** The sum of a number and the number increased by 4 is 9. **14.** 56 and 102 **15.** 21 and 22 **16.** $2500 **17.** $32.79 **18.** Peter: $40,000; Julie $80,000 **19.** 6 times
20. ≈ 33.3 months **21.** 15 inches, 30 inches, and 30 inches **22.** Width is 6 feet, length is 8 feet **23.** Harlene 0.3 foot per minute; Ellis 0.5 foot per minute **24.** Jelly beans: ≈ 1.91 pounds; kits: ≈ 1.09 pounds **25.** 20 liters

CHAPTER 4

Exercise Set 4.1 **1.** In the expression c^r, c is the base, r is the exponent. **3. a)** $\frac{x^m}{x^n} = x^{m-n}$, $x \neq 0$ **b)** Answers will vary. **5. a)** $(x^m)^n = x^{m \cdot n}$ **b)** Answers will vary. **7.** $x^0 \neq 1$ when $x = 0$ **9.** x^9 **11.** z^5 **13.** $3^5 = 243$ **15.** y^5 **17.** z^8
19. y^7 **21.** 6 **23.** x^7 **25.** $3^3 = 27$ **27.** $\frac{1}{y^2}$ **29.** 1 **31.** $\frac{1}{a^4}$ **33.** 1 **35.** 3 **37.** 4 **39.** -3 **41.** $5x^3y$ **43.** $-5r$ **45.** x^8 **47.** x^{25}
49. x^3 **51.** x^{12} **53.** n^{18} **55.** $1.69x^2$ **57.** $-27x^9$ **59.** $27a^6b^{12}$ **61.** $\frac{x^2}{9}$ **63.** $\frac{y^4}{x^4}$ **65.** $\frac{216}{x^3}$ **67.** $\frac{27x^3}{y^3}$ **69.** $\frac{16p^2}{25}$ **71.** $\frac{16y^{12}}{x^4}$
73. $\frac{x^5}{y^2}$ **75.** $\frac{5x^2}{y^2}$ **77.** $\frac{1}{9a^2b^3}$ **79.** $\frac{7}{3x^5y^3}$ **81.** $-\frac{3y^2}{x^3}$ **83.** $-\frac{2}{x^3y^2z^5}$ **85.** $\frac{8}{x^6}$ **87.** $27y^9$ **89.** 1 **91.** $\frac{x^4}{y^4}$ **93.** $\frac{z^{24}}{16y^{28}}$ **95.** $64x^3y^{12}$
97. $25x^2y^8$ **99.** $3ab^4$ **101.** $-6x^2y^2$ **103.** $15x^3y^6$ **105.** $-9p^6q^3$ **107.** $25r^6s^4$ **109.** x^2 **111.** x^{12} **113.** $6.25x^6$ **115.** $\frac{x^6}{y^4}$
117. $-\frac{m^{12}}{n^9}$ **119.** $-216x^9y^6$ **121.** $-8x^{12}y^6z^3$ **123.** $729r^{12}s^{15}$ **125.** $108x^5y^7$ **127.** $53.29x^4y^8$ **129.** $x^{11}y^{13}$ **131.** x^8z^8
133. Cannot be simplified **135.** Cannot be simplified **137.** $6z^2$ **139.** Cannot be simplified **141.** 32 **143.** 1 **145.** The sign will be negative because a negative number with an odd exponent will be negative. This is because $(-1)^m = -1$ when m is odd.
147. $7x^2$ **149.** $3x^2 + 4xy$ **151.** $\frac{9x^2}{8y^3}$ **154.** 13 **155.** $x + 10$ **156.** All real numbers **157.** 4 inches, 4 inches, 9 inches, 9 inches
b) $w = \frac{P - 2l}{2}$

Exercise Set 4.2 **1.** Answers will vary. **3.** No, it is not simplified because of the negative exponent; $\frac{x^5}{y^3}$
5. The given simplification is not correct since $(y^4)^{-3} = \frac{1}{y^{12}}$. **7. a)** The numerator has one term, x^5y^2. **b)** The factors of the numerator are x^5 and y^2. **9.** The sign of the exponent changes when a factor is moved from the numerator to the denominator of a fraction. **11.** $\frac{1}{x^6}$ **13.** $\frac{1}{5}$ **15.** x^3 **17.** x **19.** 36 **21.** $\frac{1}{x^6}$ **23.** $\frac{1}{y^{20}}$ **25.** $\frac{1}{x^8}$ **27.** 64 **29.** y^2 **31.** x^2 **33.** 9
35. $\frac{1}{r}$ **37.** p^3 **39.** $\frac{1}{x^4}$ **41.** 27 **43.** $\frac{1}{27}$ **45.** z^9 **47.** p^{24} **49.** y^6 **51.** $\frac{1}{x^4}$ **53.** $\frac{1}{x^{15}}$ **55.** $-\frac{1}{16}$ **57.** $-\frac{1}{16}$ **59.** $-\frac{1}{64}$ **61.** $\frac{1}{36}$
63. $\frac{1}{x^{10}}$ **65.** n^2 **67.** 1 **69.** 1 **71.** 64 **73.** x^8 **75.** 1 **77.** $\frac{1}{4}$ **79.** $\frac{1}{36}$ **81.** x^3 **83.** $\frac{1}{16}$ **85.** 125 **87.** $\frac{1}{9}$ **89.** 1 **91.** $\frac{1}{36x^4}$
93. $\frac{3y^2}{x^2}$ **95.** 4 **97.** $\frac{64}{125}$ **99.** $\frac{y^2}{x^4}$ **101.** $-\frac{s^4}{r^{16}}$ **103.** $\frac{7}{a^3b^4}$ **105.** $\frac{y^9}{x^{15}}$ **107.** $\frac{20}{y^5}$ **109.** -8 **111.** $12x^5$ **113.** $\frac{20z^5}{y}$ **115.** $3c^5$
117. $\frac{4}{x^2}$ **119.** $\frac{x^4}{2y^5}$ **121.** $\frac{8x^6}{y^2}$ **123.** $\frac{y^{12}z^4}{16x^8}$ **125.** $\frac{r^{20}t^{48}}{16s^{36}}$ **127.** $\frac{y^{12}}{x^{18}z^6}$ **129.** $\frac{1}{27a^6b^6}$ **131. a)** Yes **b)** No **133.** $16\frac{1}{16}$ **135.** $4\frac{1}{4}$
137. $\frac{2}{3}$ **139.** $\frac{11}{6}$ **141.** $\frac{1}{6}$ **143.** $\frac{22}{9}$ **145.** -2 **147.** -2 **149.** -2 **151.** -3 **153.** The product rule is $(xy)^m = x^my^m$, not $(x + y)^m = x^m + y^m$ **155.** 5 miles **156.** ≈ 942.48 cubic inches **157.** 9, 28 **158.** $125 **159.** 37, 38

Exercise Set 4.3 **1.** A number in scientific notation is written as a number greater than or equal to 1 and less than 10 that is multiplied by some power of 10. **3. a)** Answers will vary. **b)** 5.68×10^{-3} **5.** 5 places to the left
7. The exponent is negative when the number is less than 1. **9.** Negative since $0.00734 < 1$. **11.** 1.0×10^{-6} **13.** 3.5×10^5
15. 4.5×10^2 **17.** 5.3×10^{-2} **19.** 1.9×10^4 **21.** 1.86×10^{-6} **23.** 9.14×10^{-6} **25.** 2.203×10^5 **27.** 5.104×10^{-3}

29. 43,000 **31.** 0.00543 **33.** 0.0000213 **35.** 625,000 **37.** 9,000,000 **39.** 535 **41.** 0.0006201 **43.** 10,000 **45.** 0.000008 meter **47.** 125,000,000,000 watts **49.** 15,300 meters **51.** 0.000015 gram **53.** 60,000,000 **55.** 0.243 **57.** 0.02262 **59.** 4200 **61.** 2500 **63.** 0.0005 **65.** 4.2×10^{12} **67.** 1.28×10^{-1} **69.** 3.0×10^2 **71.** 1.75×10^2 **73.** 8.3×10^{-4}, 3.2×10^{-1}, 4.6, 4.8×10^5 **75. a)** $\approx 5,919,000,000$ **b)** ≈ 22.1 **77.** 8,640,000,000 cubic feet **79.** 1.6×10^7 seconds **81. a)** 1.8×10^{10} **b)** 3.332×10^6 or 3,332,000 miles **83. a)** \$82,700 **b)** ≈ 4.71 **85.** 500 seconds or ≈ 8.33 minutes **87.** 1.232×10^{10} **89. a)** 2.7×10^7 **b)** 2,275,240 **91. a)** 24 **b)** $\approx 5.017 \times 10^{22}$ **93.** Answers will vary.

95. 1,000,000 **98.** 0 **99. a)** $\dfrac{3}{2}$ **b)** 0 **100.** 2 **101.** $-\dfrac{y^{12}}{64x^9}$

Exercise Set 4.4

1. A polynomial is an expression containing the sum of a finite number of terms of the form ax^n, where a is a real number and n is a whole number. **3. a)** The exponent on the variable is the degree of the term. **b)** The degree of the polynomial is the same as the degree of the highest degree term in the polynomial. **5.** Add the exponents on the variables **7.** $(3x + 2) - (4x - 6) = 3x + 2 - 4x + 6$ **9.** Because the exponent on the variable in a constant term is 0. **11. a)** Answers will vary. **b)** $4x^3 + 0x^2 + 5x - 7$ **13.** No, it contains a fractional exponent. **15.** fifth **17.** fourth **19.** seventh **21.** third **23.** tenth **25.** twelfth **27.** Binomial **29.** Monomial **31.** Binomial **33.** Monomial **35.** Not a polynomial **37.** Polynomial **39.** Trinomial **41.** Polynomial **43.** $5x + 4$, first **45.** $x^2 - 2x - 4$, second **47.** $3x^2 + x - 8$, second **49.** Already in descending order, first **51.** Already in descending order, second **53.** $4x^3 - 3x^2 + x - 4$, third **55.** $-2x^4 + 3x^2 + 5x - 6$, fourth **57.** $6x - 1$ **59.** $-2x + 11$ **61.** $-2t - 1$

63. $x^2 + 6.6x + 0.8$ **65.** $5m^2 + 4$ **67.** $x^2 + 3x - 3$ **69.** $-3x^2 + x + \dfrac{17}{2}$ **71.** $5.4x^2 - 5x + 1.7$

73. $-2x^3 - 3x^2 + 4x - 3$ **75.** $7x^2 - xy - 4$ **77.** $5x^2y - 3x + 2$ **79.** $7x - 1$ **81.** $7y^2 - 2y + 5$ **83.** $4x^2 + 2x - 4$ **85.** $2x^3 - x^2 + 6x - 2$ **87.** $3n^3 - 11n^2 - n + 2$ **89.** $2x - 6$ **91.** $3x + 4$ **93.** $-3r$ **95.** $6x^2 + 7x - 8.5$

97. $8x^2 + x + 4$ **99.** $-9m^2 + 5m - 6$ **101.** $8x^3 - 7x^2 - 4x - 3$ **103.** $2x^3 - \dfrac{23}{5}x^2 + 2x - 2$ **105.** $x - 2$

107. $2x^2 - 9x + 14$ **109.** $-x^3 + 11x^2 + 9x - 7$ **111.** $3x + 8$ **113.** $-3d + 12$ **115.** $x^2 - 3x - 3$ **117.** $5m^2 - 5m - 6$ **119.** $4x^3 - 7x^2 + x - 2$ **121.** Answers will vary. **123.** Answers will vary. **125.** Sometimes **127.** Sometimes **129.** Answers will vary; one example is: $x^5 + x^4 + x$ **131.** No, all three terms would have to be degree 5 or 0. Therefore at least two of the terms could be combined. **133.** $a^2 + 2ab + b^2$ **135.** $x^2 + xz + yz$ **137.** $-12x + 18$

139. $8x^2 + 28x - 24$ **141.** $>$ **142.** True **143.** True **144.** False **145.** False **146.** $\dfrac{y^3}{27x^{12}}$

Exercise Set 4.5

1. The distributive property is used when multiplying a monomial by a polynomial. **3.** First, Outer, Inner, Last **5.** Yes **7.** $(a + b)^2 = a^2 + 2ab + b^2$; $(a - b)^2 = a^2 - 2ab + b^2$ **9.** No, $(x + 5)^2 = x^2 + 10x + 25$ **11.** Answers will vary. **13.** Answers will vary. **15.** $2x^4y$ **17.** $20x^5y^6$ **19.** $-28x^6y^{15}$ **21.** $54x^6y^{14}$ **23.** $3x^6y$ **25.** $5.94x^8y^3$ **27.** $5x + 20$ **29.** $-6x^2 + 6x$ **31.** $-16y - 10$ **33.** $-2x^3 + 4x^2 - 10x$ **35.** $-20x^3 + 30x^2 - 20x$ **37.** $0.5x^5 - 3x^4 - 0.5x^2$ **39.** $0.6x^2y + 1.5x^2 - 1.8xy$ **41.** $x^2y^4 - 4y^7 - 3y^4$ **43.** $x^2 + 7x + 12$ **45.** $6x^2 + 3x - 30$ **47.** $4x^2 - 16$ **49.** $-6x^2 - 8x + 30$ **51.** $-12x^2 + 32x - 5$ **53.** $4x^2 - 10x + 4$ **55.** $12k^2 - 30k + 12$ **57.** $x^2 - 4$ **59.** $4x^2 - 12x + 9$ **61.** $-6z^2 + 46z - 28$ **63.** $-4x^2 + 2x + 12$ **65.** $x^2 - y^2$ **67.** $6x^2 - 5xy - 6y^2$ **69.** $6x^2 + 2xy + 6x + 2y$ **71.** $x^2 + 0.9x + 0.18$ **73.** $xy - 2x - 2y + 4$ **75.** $x^2 - 36$ **77.** $9x^2 - 9$ **79.** $x^2 + 2xy + y^2$ **81.** $x^2 - 0.4x + 0.04$ **83.** $16x^2 + 40x + 25$ **85.** $0.16x^2 + 0.8xy + y^2$ **87.** $25a^2 - 49b^2$ **89.** $4x^2 - 36$ **91.** $49a^2 + 28a + 4$ **93.** $3x^3 + 16x^2 + 15x - 4$ **95.** $12x^3 + 5x^2 + 13x + 10$ **97.** $-14x^3 - 22x^2 + 19x - 3$ **99.** $-6a^3 - 2a^2 + 29a - 15$ **101.** $6x^4 + 5x^3 + 5x^2 + 10x + 4$ **103.** $x^4 - 3x^3 + 5x^2 - 6x$ **105.** $a^3 + b^3$ **107.** $x^3 + 6x^2 + 12x + 8$ **109.** $27a^3 - 135a^2 + 225a - 125$ **111.** No, it will always be a binomial. **113.** No **115.** 6, 3, 1 **117. a)** $(x + 2)(2x + 1)$ or $2x^2 + 5x + 2$ **b)** 54 square feet **c)** 1 foot **119. a)** $a + b$ **b)** $a + b$ **c)** Yes **d)** $(a + b)^2$ **e)** $a^2 + ab + ab + b^2$ $= a^2 + 2ab + b^2$ **f)** $a^2 + 2ab + b^2$ **121.** $6x^6 - 18x^5 + 3x^4 + 35x^3 - 54x^2 + 38x - 12$ **123.** All real numbers

124. 13 miles **125.** $\dfrac{x^4}{16y^8}$ **126. a)** -216 **b)** $\dfrac{1}{216}$ **127.** $-5x^2 - 2x + 8$

Exercise Set 4.6

1. To divide a polynomial by a monomial, divide each term in the polynomial by the monomial. **3.** $1 + \dfrac{5}{y}$ **5.** The terms should be listed in descending order. **7.** $\dfrac{x^3 + 0x^2 - 14x + 15}{x - 3}$

9. $(x + 5)(x - 3) - 2 = x^2 + 2x - 17$ **11.** $\dfrac{x^2 + x - 20}{x - 4} = x + 5$ or $\dfrac{x^2 + x - 20}{x + 5} = x - 4$ **13.** $\dfrac{2x^2 + 5x + 3}{2x + 3} = x + 1$ or $\dfrac{2x^2 + 5x + 3}{x + 1} = 2x + 3$ **15.** $\dfrac{4x^2 - 9}{2x + 3} = 2x - 3$ or $\dfrac{4x^2 - 9}{2x - 3} = 2x + 3$ **17.** $x + 2$ **19.** $2n + 5$ **21.** $\dfrac{3}{2}x + 4$

23. $-3x + 2$ **25.** $3x + 1$ **27.** $\frac{1}{2}x + 4$ **29.** $-1 + \frac{5}{2}w$ **31.** $1 + \frac{2}{x} - \frac{3}{x^2}$ **33.** $-2x^3 + \frac{3}{x} + \frac{4}{x^2}$ **35.** $x^2 + 3x - \frac{3}{x^3}$

37. $3x^2 - 2x + 6 - \frac{5}{2x}$ **39.** $-2k^2 - \frac{3}{2}k + \frac{2}{k}$ **41.** $-4x^3 - x^2 + \frac{10}{3} + \frac{3}{x^2}$ **43.** $x + 3$ **45.** $2x + 3$ **47.** $2x + 4$ **49.** $x + 4$

51. $x + 5 - \frac{3}{2x - 3}$ **53.** $2a + 5$ **55.** $2x - 3 + \frac{2}{4x + 9}$ **57.** $4x - 3 - \frac{3}{2x + 3}$ **59.** $3x^2 - 5$ **61.** $2x^2 + \frac{12}{x - 2}$

63. $w^2 + 3w + 9 + \frac{19}{w - 3}$ **65.** $x^2 + 3x + 9$ **67.** $2x^2 + x - 2 - \frac{2}{2x - 1}$ **69.** $-m^2 - 7m - 5 - \frac{8}{m - 1}$

71. $3n^2 + 3n + 1 + \frac{7}{3n - 3}$ **73.** No; for example $\frac{x + 2}{x} = 1 + \frac{2}{x}$ which is not a binomial. **75.** $2x^2 + 11x + 16$

77. First degree **79.** $4x$ **81.** Since the shaded areas minus 2 must equal 3, 1, 0, and -1, respectively, the shaded areas are 5,

3, 2, and 1, respectively. **83.** $2x^2 - 3x + \frac{5}{2} - \frac{3}{2(2x + 3)}$ **85.** $-3x + 3 + \frac{1}{x + 3}$ **88. a)** 2 **b)** 2, 0 **c)** $2, -5, 0, \frac{2}{5}, -6.3,$

$-\frac{23}{34}$ **d)** $\sqrt{7}, \sqrt{3}$ **e)** $2, -5, 0, \sqrt{7}, \frac{2}{5}, -6.3, \sqrt{3}, -\frac{23}{34}$ **89. a)** 0 **b)** Undefined **90.** Parentheses, exponents,

multiplication or division from left to right, addition or subtraction from left to right **91.** $-\frac{2}{3}$ **92.** ≈ 904.78 **93.** x^{10}

Chapter Review Exercises **1.** x^7 **2.** x^6 **3.** 243 **4.** 32 **5.** x^3 **6.** 1 **7.** 25 **8.** 16 **9.** $\frac{1}{x^2}$ **10.** y^3 **11.** 1

12. 4 **13.** 1 **14.** 1 **15.** $25x^2$ **16.** $27a^3$ **17.** $216s^3$ **18.** $-27x^3$ **19.** $16x^8$ **20.** x^{24} **21.** $-m^{20}$ **22.** $\frac{4x^6}{y^2}$ **23.** $\frac{25y^4}{4b^2}$ **24.** $24x^5$

25. $\frac{4x}{y}$ **26.** $18x^3y^6$ **27.** $9x^2$ **28.** $24x^7y^7$ **29.** $16x^8y^{11}$ **30.** $6c^6d^3$ **31.** $\frac{16x^6}{y^4}$ **32.** $27x^{12}y^3$ **33.** $\frac{1}{x^4}$ **34.** $\frac{1}{27}$ **35.** $\frac{1}{25}$ **36.** z^2

37. x^7 **38.** 9 **39.** $\frac{1}{y^3}$ **40.** $\frac{1}{x^5}$ **41.** $\frac{1}{p^2}$ **42.** $\frac{1}{a^5}$ **43.** x^6 **44.** x^7 **45.** $\frac{1}{x^6}$ **46.** $\frac{1}{9x^8}$ **47.** $\frac{x^9}{64y^3}$ **48.** $\frac{4n^2}{m^6}$ **49.** $12y^2$ **50.** $\frac{125z^3}{y^9}$

51. $\frac{x^4}{16y^6}$ **52.** $\frac{6}{x}$ **53.** $10x^2y^2$ **54.** $\frac{24y^2}{x^2}$ **55.** $\frac{12x^2}{y}$ **56.** $3y^5$ **57.** $\frac{4y}{x^3}$ **58.** $\frac{7x^5}{y^4}$ **59.** $\frac{4y^{10}}{x}$ **60.** $\frac{x}{2y^5}$ **61.** 1.72×10^6

62. 1.53×10^{-1} **63.** 7.63×10^{-3} **64.** 4.7×10^4 **65.** 4.82×10^3 **66.** 3.14×10^{-4} **67.** 0.0084 **68.** 0.000652 **69.** 970,000
70. 0.00000438 **71.** 0.0000314 **72.** 11,030,000 **73.** 6,000,000,000 meters **74.** 0.092 liter **75.** 19,200 grams
76. 0.0000128 gram **77.** 0.085 **78.** 1260 **79.** 245 **80.** 379,000,000 **81.** 0.0325 **82.** 0.00003 **83.** 3.64×10^9
84. 5.0×10^9 **85.** 2.12×10^1 **86.** 5.0×10^{-4} **87.** 3.4×10^{-3} **88.** 3.4×10^7 **89.** 50,000 gallons **90. a)** $\approx \$5,260,000,000$
b) ≈ 1.62 **91.** Not a polynomial **92.** Monomial, zero **93.** $x^2 + 3x - 4$, trinomial, second **94.** $4x^2 - x - 3$, trinomial,
second **95.** Binomial, third **96.** Not a polynomial **97.** $-4x^2 + x$, binomial, second **98.** Not a polynomial
99. $2x^3 + 4x^2 - 3x - 7$, polynomial, third **100.** $3x - 1$ **101.** $7d + 4$ **102.** $-3x - 5$ **103.** $-3x^2 + 10x - 15$
104. $5m^2 - 10$ **105.** $8.1p + 2.8$ **106.** $-2x + 2$ **107.** $4x^2 - 12x - 15$ **108.** $3a^2 - 5a - 21$ **109.** $-5x^2 + 8x - 19$
110. $4x + 2$ **111.** $3x^2 + 3x$ **112.** $-15x^2 - 12x$ **113.** $6x^3 - 12x^2 + 21x$ **114.** $-2c^3 + 3c^2 - 5c$ **115.** $12z^3 + 8z^2 + 32z$
116. $x^2 + 9x + 20$ **117.** $-12x^2 - 21x + 6$ **118.** $4x^2 - 24x + 36$ **119.** $-6x^2 + 14x + 12$ **120.** $r^2 - 25$
121. $3x^3 + 7x^2 + 14x + 4$ **122.** $3x^3 + x^2 - 10x + 6$

123. $-12x^3 + 10x^2 - 30x + 14$ **124.** $x + 2$ **125.** $5x + 6$ **126.** $8x + 4$ **127.** $2x^2 + 3x - \frac{4}{3}$ **128.** $2w - \frac{5}{3} + \frac{1}{w}$

129. $4x^4 - 2x^3 + \frac{3}{2}x - \frac{1}{x}$ **130.** $-4m + 2$ **131.** $\frac{5}{3}x - 2 + \frac{5}{x}$ **132.** $\frac{5}{2}x + \frac{5}{x} + \frac{1}{x^2}$ **133.** $x + 4$ **134.** $n + 3$

135. $5x - 2 + \frac{2}{x + 6}$ **136.** $2x^2 + 3x - 4$ **137.** $2x - 3$

Chapter Practice Test **1.** $15x^6$ **2.** $27x^3y^6$ **3.** $3d^4$ **4.** $\frac{x^3}{8y^6}$ **5.** $\frac{y^4}{4x^6}$ **6.** $\frac{2x^7y}{3}$ **7.** 4 **8.** 5.25×10^9

9. 2.0×10^{-7} **10.** Monomial **11.** Binomial **12.** Not a polynomial **13.** $6x^3 - 2x^2 + 5x - 5$, third degree
14. $2x^2 + x - 7$ **15.** $-2x^2 + 4x$ **16.** $3x^2 - x + 3$ **17.** $15d^2 - 40d$ **18.** $8x^2 + 2x - 21$ **19.** $-12c^2 + 7c + 45$

20. $6x^3 + 2x^2 - 35x + 25$ **21.** $4x^2 + 2x - 1$ **22.** $4x + 2 - \frac{5}{3x}$ **23.** $4x + 5$ **24.** $3x - 2 - \frac{2}{4x + 5}$ **25. a)** 5.73×10^3

b) $\approx 7.78 \times 10^5$

CHAPTER 5

Exercise Set 5.1 **1.** A prime number is an integer greater than 1 that has exactly two factors, itself and 1.
3. To factor an expression means to write the expression as the product of its factors. **5.** The greatest common factor is the
greatest number that divides into all the numbers. **7.** A factoring problem may be checked by multiplying the factors.
9. $2^3 \cdot 7$ **11.** $2 \cdot 3^2 \cdot 5$ **13.** $2^2 \cdot 7^2$ **15.** 4 **17.** 14 **19.** 2 **21.** x **23.** $3x$ **25.** 1 **27.** mn **29.** $x^3 y^5$ **31.** 5 **33.** $x^2 y^2$ **35.** x
37. $x + 3$ **39.** $2x - 3$ **41.** $3w + 5$ **43.** $x - 4$ **45.** $x - 1$ **47.** $x + 3$ **49.** $4(x - 2)$ **51.** $5(3x - 1)$ **53.** $6(p + 2)$
55. $3x(3x - 4)$ **57.** $2p(13p - 4)$ **59.** $3x^2(x^3 - 4)$ **61.** $12x^8(3x^4 + 2)$ **63.** $9y^3(3y^{12} - 1)$ **65.** $x(1 + 3y^2)$
67. $a^2(7a^2 + 3)$ **69.** $4xy(4yz + x^2)$ **71.** $16mn^2(3m^3 - 1)$ **73.** $25x^2 yz(z^2 + x)$ **75.** $y^2 z^3(13y^3 - 11xz^2)$
77. $4(2c^2 - c - 8)$ **79.** $3(3x^2 + 6x + 1)$ **81.** $4x(x^2 - 2x + 3)$ **83.** $5(7x^2 - 3y + 2)$ **85.** $3(5p^2 - 2p + 3)$
87. $3a(3a^3 - 2a^2 + b)$ **89.** $xy(8x + 12y + 5)$ **91.** $(x + 4)(x + 3)$ **93.** $(a - 2)(3b - 4)$ **95.** $(2x + 1)(4x + 1)$
97. $(2x + 1)(5x + 1)$ **99.** $(2z + 3)(4z - 3)$ **101.** $3(\divideontimes + 2)$ **103.** $7\Delta(5\Delta^2 - \Delta + 2)$
105. $2(x - 3)[2x^2(x - 3)^2 - 3x(x - 3) + 2]$ **107.** $x^{1/3}(x^2 + 5x + 2)$ **109.** $(x + 2)(x + 3)$ **110.** $-3x + 17$ **111.** 2
112. $y = \dfrac{4x - 20}{5}$ or $y = \dfrac{4}{5}x - 4$ **113.** ≈ 201.06 cubic inches **114.** 14, 27 **115.** $\dfrac{9y^2}{4x^6}$

Exercise Set 5.2 **1.** The first step is to factor out a common factor, if one exists. **3.** $x^2 + 4x - 2x - 8$; found
by multiplying the factors. **5.** Answers will vary. **7.** $(x + 3)(x + 2)$ **9.** $(x + 5)(x + 4)$ **11.** $(x + 2)(x + 5)$
13. $(x + 3)(x - 5)$ **15.** $(2b - 5)(2b + 5)$ **17.** $(x + 3)(3x + 1)$ **19.** $(2x + 1)(3x - 1)$ **21.** $(x + 4)(8x + 1)$
23. $(3t - 2)(4t - 1)$ **25.** $(x - 2)(2x - 3)$ **27.** $(2p + 5)(3p - 2)$ **29.** $(x + 2y)(x - 3y)$ **31.** $(3x + 2y)(x - 3y)$
33. $(5x - 6y)(2x - 5y)$ **35.** $(x + b)(x + a)$ **37.** $(y + 5)(x - 3)$ **39.** $(a + 3)(a + b)$ **41.** $(y - 1)(x + 5)$
43. $(3 + 2y)(4 - x)$ **45.** $(z + 5)(z^2 + 1)$ **47.** $(x + 4)(x^2 - 3)$ **49.** $2(x - 6)(x + 4)$
51. $4(x + 2)(x + 2) = 4(x + 2)^2$ **53.** $x(2x + 3)(3x - 1)$ **55.** $x(x + 3y)(x - 2y)$ **57.** $(y + 5)(x + 3)$
59. $(x + 5)(y + 6)$ **61.** $(a + b)(x + y)$ **63.** $(d + 3)(c - 4)$ **65.** $(a + b)(c - d)$ **67.** No; $xy + 2x + 5y + 10$
is factorable; $xy + 10 + 2x + 5y$ is not factorable in this arrangement. **69.** $(\odot + 3)(\odot - 5)$ **71. a)** $3x^2 + 6x + 4x + 8$
b) $(x + 2)(3x + 4)$ **73. a)** $2x^2 - 6x - 5x + 15$ **b)** $(x - 3)(2x - 5)$ **75. a)** $4x^2 - 20x + 3x - 15$
b) $(x - 5)(4x + 3)$ **77.** $(\odot + 3)(\star + 2)$ **79.** $\dfrac{6}{5}$ **80.** 30 pounds of chocolate wafers, 20 pounds of peppermints.
81. $5x^2 - 2x - 3 + \dfrac{5}{3x}$ **82.** $x + 3$

Exercise Set 5.3 **1.** Since 8000 is positive, both signs will be the same. Since 180 is positive, both signs will be
positive. **3.** Since -8000 is negative, one sign will be positive, the other will be negative. **5.** Since 8000 is positive, both
signs will be the same. Since -240 is negative, both signs will be negative. **7.** The trinomial $x^2 + 4xy - 12y^2$ is obtained
by multiplying the factors using the FOIL method. **9.** The trinomial $4a^2 - 4b^2$ is obtained by multiplying all the factors
and combining like terms. **11.** It is not completely factored. It factors to $2(x - 2)(x - 1)$. **13.** Find two numbers whose
product is c, and whose sum is b. The factors are $(x + \text{first number})$ and $(x + \text{second number})$. **15.** $(x - 5)(x - 2)$
17. $(x + 2)(x + 4)$ **19.** $(x + 4)(x + 3)$ **21.** Prime **23.** $(y - 12)(y - 1)$ **25.** $(a - 4)(a + 2)$ **27.** $(r - 5)(r + 3)$
29. $(b - 9)(b - 2)$ **31.** Prime **33.** $(a + 11)(a + 1)$ **35.** $(x - 10)(x + 3)$ **37.** $(x + 2)^2$ **39.** $(p + 3)^2$ **41.** $(p - 6)^2$
43. $(w - 15)(w - 3)$ **45.** $(x + 13)(x - 3)$ **47.** $(x - 5)(x + 4)$ **49.** $(y + 7)(y + 2)$ **51.** $(x + 16)(x - 4)$ **53.** Prime
55. $(x - 16)(x - 4)$ **57.** $(b - 5)(b - 13)$ **59.** $(x + 2)(x + 1)$ **61.** $(w + 9)(w - 2)$ **63.** $(x - 3y)(x - 5y)$
65. $(m - 3n)^2$ **67.** $(x + 5y)(x + 3y)$ **69.** $(m + 3n)(m - 8n)$ **71.** $6(x - 4)(x - 1)$ **73.** $5(x + 3)(x + 1)$
75. $2(x - 4)(x - 3)$ **77.** $b(b - 5)(b - 2)$ **79.** $3z(z - 9)(z + 2)$ **81.** $x(x + 4)^2$ **83.** $4(a - 4b)(a - 2b)$
85. $s(r + 3s)(r + 4s)$ **87.** $x^2(x - 7)(x + 3)$ **89.** Both negative; one positive and one negative; one positive and
one negative; both positive **91.** $x^2 + 5x + 4 = (x + 1)(x + 4)$ **93.** $x^2 + 12x + 32 = (x + 8)(x + 4)$
95. $(x + 0.4)(x + 0.2)$ **97.** $\left(x + \dfrac{1}{5}\right)\left(x + \dfrac{1}{5}\right) = \left(x + \dfrac{1}{5}\right)^2$ **99.** $(x + 20)(x - 15)$ **101.** 9 **102.** 19.6%
103. $2x^3 + x^2 - 16x + 12$ **104.** $3x + 2 - \dfrac{2}{x - 4}$ **105.** $(3x + 5)(x - 2)$

Exercise Set 5.4 **1.** Factoring trinomials is the reverse process of multiplying binomials. **3.** The constant, c, of
the trinomial **5.** $(2x + 1)(x + 5)$ **7.** $(3x + 2)(x + 4)$ **9.** $(5x + 1)(x - 2)$ **11.** $(3r - 2)(r + 5)$ **13.** $(2z - 3)^2$
15. $(5y + 4)(y - 1)$ **17.** Prime **19.** $(2z + 3)(3z - 4)$ **21.** Prime **23.** $(5y - 1)(y - 3)$ **25.** $(7x + 1)(x + 6)$
27. $(7x - 1)(x - 1)$ **29.** $(5b - 3)(b - 4)$ **31.** $(5z + 4)(z - 2)$ **33.** $(4y - 3)(y + 2)$ **35.** $(5x - 1)(2x - 5)$
37. $(5d + 4)(2d - 3)$ **39.** $2(3x + 1)(x - 4)$ **41.** $(7t + 3)(t + 1)$ **43.** $2(3x + 5)(x + 1)$ **45.** $x(2x + 1)(3x - 4)$
47. $4x(3x + 1)(x + 2)$ **49.** $2x(2x + 3)(x - 2)$ **51.** $6(3z - 1)(2z + 1)$ **53.** $3(r - 4)(r - 6)$ **55.** $(2x + y)(x + 2y)$

57. $(2x - y)(x - 3y)$　**59.** $2(2x - y)(3x + 4y)$　**61.** $3(x - 3y)(2x + 3y)$　**63.** $(3m - 2n)(2m + n)$
65. $x(4x + 3y)(2x + y)$　**67.** $x^2(2x + y)(2x + 3y)$　**69.** $3x^2 - 20x - 7$; obtained by multiplying the factors.
71. $10x^2 + 35x + 15$; obtained by multiplying the factors.　**73.** $2x^4 - x^3 - 3x^2$; obtained by multiplying the factors.
75. a) Dividing the trinomial by the binomial gives the second factor.　**b)** $6x + 11$　**77.** $(6x - 5)(3x + 4)$
79. $(5x - 8)(3x - 20)$　**81.** $4(6x + 5)(3x - 10)$　**83.** $2x + 45$, the product of the three first terms must equal $6x^3$, and the product of the constants must equal 2250.　**85.** 49　**86.** ≈ 130.89 mph　**87.** $12xy^2(3x^3y - 1 + 2x^4y^4)$　**88.** $(x - 9)(x - 6)$

Exercise Set 5.5　**1. a)** $a^2 - b^2 = (a + b)(a - b)$　**b)** Answers will vary.
3. a) $a^3 - b^3 = (a - b)(a^2 + ab + b^2)$　**b)** Answers will vary.　**5.** No　**7.** Prime　**9.** $4(a^2 + 4)$　**11.** $4(4m^2 + 9n^2)$
13. $(y + 5)(y - 5)$　**15.** $(z + 9)(z - 9)$　**17.** $(x + 7)(x - 7)$　**19.** $(x + y)(x - y)$　**21.** $(3y + 5z)(3y - 5z)$
23. $4(4a + 3b)(4a - 3b)$　**25.** $(7x + 6)(7x - 6)$　**27.** $(z^2 + 9x)(z^2 - 9x)$　**29.** $9(x^2 + 3y)(x^2 - 3y)$
31. $(6m^2 + 7n)(6m^2 - 7n)$　**33.** $10(x + 4)(x - 4)$　**35.** $4(2x + 5y^2)(2x - 5y^2)$　**37.** $(x + y)(x^2 - xy + y^2)$
39. $(a - b)(a^2 + ab + b^2)$　**41.** $(x + 2)(x^2 - 2x + 4)$　**43.** $(x - 3)(x^2 + 3x + 9)$　**45.** $(a + 1)(a^2 - a + 1)$
47. $(3x - 1)(9x^2 + 3x + 1)$　**49.** $(3a - 5)(9a^2 + 15a + 25)$　**51.** $(3 - 2y)(9 + 6y + 4y^2)$
53. $(4m + 3n)(16m^2 - 12mn + 9n^2)$　**55.** $(2a - 3b)(4a^2 + 6ab + 9b^2)$　**57.** $2(x + 2)^2$　**59.** $b(a + 5)(a - 5)$
61. $3(c - 3)^2$　**63.** $5(x - 3)(x + 1)$　**65.** $3(x + 3)(y - 2)$　**67.** $2(x + 5)(x - 5)$　**69.** $2y(x + 3)(x - 3)$
71. $3y^2(x + 1)(x^2 - x + 1)$　**73.** $2(x - 2)(x^2 + 2x + 4)$　**75.** $2(3x + 5)(3x - 5)$　**77.** $3(2x + 3)^2$
79. $2(3x - 2)(x + 4)$　**81.** $2r(s + 3)(s - 8)$　**83.** $(x + 2)(4x - 3)$　**85.** $25(b + 2)(b - 2)$　**87.** $a^3b^2(a + 2b)(a - 2b)$
89. $3x^2(x - 3)^2$　**91.** $x(x^2 + 25)$　**93.** $(y^2 + 4)(y + 2)(y - 2)$　**95.** $3(3a - 2b)(4a + b)$　**97.** $(2a - 3)(b + 2)$
99. $9(1 + y^2)(1 + y)(1 - y)$　**101.** You cannot divide both sides of the equation by $a - b$ because it equals 0.
103. $2\blacklozenge^4(\blacklozenge^2 + 2\ast^2)$　**105.** $(x^2 + 1)(x^4 - x^2 + 1)$　**107.** $(x - 3 + 2y)(x - 3 - 2y)$
109. $(x + y)(x - y)(x^2 - xy + y^2)(x^2 + xy + y^2)$　**110.** $x \le 1$; ⟵——●——⟶ **111.** 4 inches　**112.** The sum of a

number, and 5 decreased by twice the number is 2.　**113.** $\dfrac{8x^9}{27y^{12}}$　**114.** $\dfrac{1}{x^5}$

Exercise Set 5.6　**1.** Answers will vary.　**3.** $ax^2 + bx + c = 0$　**5. a)** The zero-factor property may only be used

when one side of the equation is equal to 0　**b)** $-2, 3$　**7.** $0, -2$　**9.** $0, 8$　**11.** $-\dfrac{5}{2}, 3$　**13.** $3, -3$　**15.** $0, 12$　**17.** $0, -3$　**19.** 4

21. $-2, -10$　**23.** $3, -6$　**25.** $4, -3$　**27.** $1, -24$　**29.** $-5, -6$　**31.** $5, -3$　**33.** $30, -1$　**35.** $-6, \dfrac{2}{3}$　**37.** $-\dfrac{1}{3}, -2$　**39.** $\dfrac{2}{3}, -5$

41. $-4, 3$　**43.** $-\dfrac{3}{4}, \dfrac{1}{2}$　**45.** $-3, -6$　**47.** $0, 25$　**49.** $10, -10$　**51.** $-2, 5$　**53.** $-\dfrac{1}{2}, 4$　**55.** $-6, 1$　**57.** $x^2 - 2x - 8 = 0$ (other

answers are possible)　**59.** $x^2 - 6x = 0$ (other answers are possible)　**61. a)** $(2x - 1)$ and $(3x + 1)$

b) $6x^2 - x - 1 = 0$　**63.** $0, -12$　**65.** $0, 3, -2$　**67.** $\dfrac{1}{35}$　**68. a)** Identity　**b)** Contradiction　**69.** ≈ 738　**70.** $\dfrac{8}{x^3y^6}$

71. Monomial　**72.** Binomial　**73.** Not polynomial　**74.** Trinomial

Exercise Set 5.7　**1.** A triangle with a 90° angle　**3.** $(log)^2 + (log)^2 =$ (hypotenuse)^2 or $a^2 + b^2 = c^2$　**5.** 3
7. 17　**9.** 4　**11.** 39　**13.** 9, 13　**15.** 6, 14　**17.** 9, 7　**19.** Width: 3 feet; length: 12 feet　**21.** Width: 10 feet; length: 15 feet
23. 3 meters　**25.** 4 seconds　**27.** yes　**29.** yes　**31.** 16 feet　**33.** 10 inches　**35.** 6 feet, 8 feet, 10 feet　**37.** 13 miles
39. Width: 7 feet; length: 24 feet　**41.** 30 videos　**43. a)** 4　**b)** 9　**45.** 432 square feet　**47.** 0, 2, −5　**49.** −5 and 8
53. $2x - 5$　**54.** $-x^2 + 7x - 4$　**55.** $6x^3 + x^2 - 10x + 4$　**56.** $2x - 3$　**57.** $2x - 3$

Chapter Review Exercises　**1.** y^3　**2.** $3p$　**3.** $5a^2$　**4.** $5x^2y^2$　**5.** 1　**6.** 1　**7.** $x - 5$　**8.** $x + 5$　**9.** $4(x - 3)$
10. $5(7x - 1)$　**11.** $4y(6y - 1)$　**12.** $5p^2(11p - 4)$　**13.** $12ab(5a - 3b)$　**14.** $6xy(1 - 2x)$　**15.** $4x^3y^2(5 + 2x^6y - 4x^2)$
16. Prime　**17.** Prime　**18.** $(5x + 3)(x - 2)$　**19.** $(x + 2)(5x - 2)$　**20.** $(4x - 3)(2x + 1)$　**21.** $(x + 6)(x + 2)$
22. $(x - 5)(x + 4)$　**23.** $(y - 9)(y - 9) = (y - 9)^2$　**24.** $(a - b)(4a - 1)$　**25.** $(y + 1)(3x + 2)$　**26.** $(x + 3)(x - 2y)$
27. $(x + 6)(2x - 1)$　**28.** $(5x - y)(x + 4y)$　**29.** $(x + 3y)(4x - 5y)$　**30.** $(3a - 5b)(2a - b)$　**31.** $(b - 1)(a + 1)$
32. $(x - 3y)(3x + 2y)$　**33.** $(a + 2b)(7a - b)$　**34.** $(2x + 3)(3x - 1)$　**35.** $(x - 3)(x + 2)$　**36.** Prime
37. $(x - 6)(x - 7)$　**38.** $(b + 5)(b - 4)$　**39.** $(n + 8)(n - 5)$　**40.** $(x - 8)(x - 7)$　**41.** Prime　**42.** Prime
43. $x(x - 9)(x - 8)$　**44.** $x(x - 8)(x + 5)$　**45.** $(x + 3y)(x - 5y)$　**46.** $4x(x + 5y)(x + 3y)$　**47.** $(2x + 5)(x - 3)$
48. $(3x - 1)(x - 4)$　**49.** $(4x - 5)(x - 1)$　**50.** $(5m - 4)(m - 2)$　**51.** $(3x - 1)(3x + 2)$　**52.** $(5x - 2)(x - 6)$
53. Prime　**54.** $(2s + 1)(3s + 5)$　**55.** $(5x - 3)(x + 8)$　**56.** $(3x - 2)(2x + 5)$　**57.** $2(3x + 2)(2x - 1)$　**58.** $(3x - 1)^2$
59. $x(3x - 2)^2$　**60.** $2x(3x + 4)(3x - 2)$　**61.** $(8a + b)(2a - 3b)$　**62.** $(2a - 3b)(2a - 5b)$　**63.** $(x + 6)(x - 6)$
64. $(x + 10)(x - 10)$　**65.** $4(x + 2)(x - 2)$　**66.** $9(3x + y)(3x - y)$　**67.** $(9 + a)(9 - a)$　**68.** $(8 + x)(8 - x)$

69. $(4x^2 + 7y)(4x^2 - 7y)$ **70.** $(10x^2 + 11y^2)(10x^2 - 11y^2)$ **71.** $(x - y)(x^2 + xy + y^2)$ **72.** $(x + y)(x^2 - xy + y^2)$
73. $(x - 1)(x^2 + x + 1)$ **74.** $(x + 2)(x^2 - 2x + 4)$ **75.** $(a + 3)(a^2 - 3a + 9)$ **76.** $(b - 4)(b^2 + 4b + 16)$
77. $(5a + b)(25a^2 - 5ab + b^2)$ **78.** $(3 - 2y)(9 + 6y + 4y^2)$ **79.** $3(3x^2 + 5y)(3x^2 - 5y)$
80. $3(x - 4y)(x^2 + 4xy + 16y^2)$ **81.** $(x - 6)(x - 8)$ **82.** $3(x - 3)^2$ **83.** $4(a + 4)(a - 4)$ **84.** $4(y + 3)(y - 3)$
85. $8(x + 3)(x - 1)$ **86.** $(x - 9)(x + 3)$ **87.** $(3x - 1)^2$ **88.** $(4x - 1)(x + 2)$ **89.** $6(b - 1)(b^2 + b + 1)$
90. $y(x - 3)(x^2 + 3x + 9)$ **91.** $b(a + 3)(a - 5)$ **92.** $3x(2x + 3)(x + 5)$ **93.** $(x - 3y)(x - y)$
94. $(3m - 4n)(m + 2n)$ **95.** $(2x - 5y)^2$ **96.** $(5a + 7b)(5a - 7b)$ **97.** $(x + 2)(y - 7)$ **98.** $y^5(4 + 5y)(4 - 5y)$
99. $2x(2x + 5y)(x + 2y)$ **100.** $(2x - 3y)(3x + 7y)$ **101.** $x^2(4x + 1)(4x - 3)$ **102.** $(a^2 + 1)(a + 1)(a - 1)$ **103.** $0, 5$
104. $2, -6$ **105.** $-5, \dfrac{3}{4}$ **106.** $0, 3$ **107.** $0, -4$ **108.** $0, -3$ **109.** $-3, -6$ **110.** $-4, 3$ **111.** $1, 2$ **112.** $-1, -4$ **113.** $2, 4$
114. $3, -5$ **115.** $\dfrac{1}{4}, -\dfrac{3}{2}$ **116.** $\dfrac{1}{2}, -8$ **117.** $2, -2$ **118.** $\dfrac{10}{7}, -\dfrac{10}{7}$ **119.** $\dfrac{3}{2}, \dfrac{1}{4}$ **120.** $\dfrac{3}{2}, \dfrac{5}{2}$ **121.** $a^2 + b^2 = c^2$
122. Hypotenuse **123.** 12 meters **124.** 10 feet **125.** 6, 8 **126.** 4, 14 **127.** Width: 7 feet; length: 9 feet **128.** 9 inches
129. 8 feet, 15 feet, 17 feet **130.** 10 feet **131.** 1 second **132.** 80 dozen

Chapter Practice Test
1. $3y^3$ **2.** $3xy^2$ **3.** $5x^2y^2(y - 3x^3)$ **4.** $4a^2b(2a - 3b + 7)$ **5.** $(5x + 2)(x - 3)$
6. $(a - 4b)(a - 5b)$ **7.** $(r + 8)(r - 3)$ **8.** $(5a - 3b)(5a + 2b)$ **9.** $4(x + 2)(x - 6)$ **10.** $x(2x - 1)(x - 1)$
11. $(3x + 2y)(4x - 3y)$ **12.** $(x + 3y)(x - 3y)$ **13.** $(x + 3)(x^2 - 3x + 9)$ **14.** $\dfrac{3}{5}, 1$ **15.** $0, 6$ **16.** $-8, 8$ **17.** 7
18. $-2, -3$ **19.** $3, 4$ **20.** 24 inches **21.** 34 feet **22.** 4, 9 **23.** 9, 11 **24.** Length: 6 meters; width: 4 meters **25.** 10 seconds

CHAPTER 6

Exercise Set 6.1 **1. a–b)** Answers will vary. **3.** We assume that the value of the variable does not make the denominator equal to 0. **5.** There is no factor common to both the numerator and the denominator. **7.** The denominator cannot be 0. **9.** $x \neq 2$ **11.** No **13.** All real numbers except $x = 0$. **15.** All real numbers except $n = 3$.

17. All real numbers except $x = 2$ and $x = -2$. **19.** All real numbers except $x = \dfrac{3}{2}$ and $x = 3$. **21.** All real numbers

23. All real numbers except $p = \dfrac{5}{2}$ and $p = -\dfrac{5}{2}$ **25.** $\dfrac{x}{3y^4}$ **27.** $\dfrac{4}{b^5}$ **29.** $\dfrac{1}{1 + y}$ **31.** 5 **33.** $\dfrac{x^2 + 6x + 3}{2}$ **35.** $r - 2$

37. $\dfrac{x}{x + 2}$ **39.** $\dfrac{k - 3}{k + 3}$ **41.** $\dfrac{x + 1}{x + 2}$ **43.** -1 **45.** $-(x + 2)$ **47.** $-\dfrac{x + 6}{2x}$ **49.** $-(x + 3)$ **51.** $\dfrac{1}{4m - 5}$ **53.** $\dfrac{x - 5}{x + 5}$

55. $2x - 3$ **57.** $x + 4$ **59.** $\dfrac{x - 4}{x + 4}$ **61.** $a^2 + 2a + 4$ **63.** $3s + 4t$ **65.** $\dfrac{2}{x - y}$ **67.** $\dfrac{☺}{4}$ **69.** $\dfrac{\Delta}{2\Delta + 3}$ **71.** -1

73. $x + 2; (x + 2)(x - 3) = x^2 - x - 6$ **75.** $x^2 + 7x + 12; (x + 3)(x + 4) = x^2 + 7x + 12$

77. a) $x \neq -3, x \neq 2$ **b)** $\dfrac{1}{x - 2}$ **79. a)** $x \neq 0, x \neq -5, x \neq \dfrac{3}{2}$ **b)** $\dfrac{1}{x(2x - 3)}$

81. 1, the numerator and denominator are identical. **84.** $y = x - 2z$ **85.** $28°, 58°,$ and $94°$ **86.** $\dfrac{16}{81x^4y^2}$

87. $6x^2 - 10x - 17$ **88.** $3(a - 4)(a - 6)$ **89.** 13 inches

Exercise Set 6.2 **1.** Answers will vary. **3.** $x^2 - 2x - 8$; the numerator must be $(x - 4)(x + 2)$

5. $x^2 - 2x - 15$; the denominator must be $(x - 5)(x + 3)$ **7.** $\dfrac{xy}{8}$ **9.** $\dfrac{80x^4}{y^6}$ **11.** $\dfrac{36x^9y^2}{25z^7}$ **13.** $\dfrac{-3x + 2}{3x + 2}$ **15.** 1 **17.** $\dfrac{1}{a^2 - b^2}$

19. $\dfrac{x + 3}{2(x + 2)}$ **21.** $\dfrac{x + 2}{x + 3}$ **23.** $x + 3$ **25.** $3x^2y$ **27.** $\dfrac{9z}{x}$ **29.** $\dfrac{5}{6ab^2}$ **31.** $5r$ **33.** $x + 7$ **35.** $\dfrac{x - 8}{x + 2}$ **37.** $\dfrac{x + 3}{x - 1}$ **39.** -1

41. $\dfrac{x + 1}{x - 4}$ **43.** $4x^2y^2$ **45.** $\dfrac{7c}{4ab^2}$ **47.** $\dfrac{3y^2}{a^2}$ **49.** $\dfrac{8mx^7}{3y^2}$ **51.** $\dfrac{1}{2}$ **53.** $\dfrac{7x}{y}$ **55.** $\dfrac{8}{m^4n^{11}}$ **57.** $\dfrac{r + 2}{r - 3}$ **59.** $\dfrac{x - 4}{x - 3}$ **61.** $\dfrac{3z + 2}{z - 2}$

63. $\dfrac{x + 5}{x - 4}$ **65.** $\dfrac{2n + 3}{3n - 1}$ **67.** $\dfrac{1}{6\Delta^3}$ **69.** $\dfrac{\Delta + ☺}{9(\Delta - ☺)}$ **71.** $x^2 + 3x + 2$ **73.** $x^2 - 3x - 10$ **75.** $x^2 - 3x + 2$ **77.** $\dfrac{x - 3}{x + 3}$ **79.** 1

81. $x^2 - 5x + 6, x^2 - x - 20$ **84.** 1 hour **85.** $20x^4y^5z^{11}$ **86.** $2x^2 + x - 2 - \dfrac{2}{2x - 1}$ **87.** $3(x - 5)(x + 2)$ **88.** $5, -2$

Exercise Set 6.3

1. Answers will vary. **3.** Answers will vary. **5.** $x(x + 6)$ **7.** $3x(x + 3)$

9. a) The negative sign in $-(2x - 7)$ was not distributed. **b)** $\dfrac{4x - 3 - 2x + 7}{5x + 4}$

11. a) The negative sign in $-(3x^2 - 4x + 5)$ was not distributed. **b)** $\dfrac{6x - 2 - 3x^2 + 4x - 5}{x^2 - 4x + 3}$ **13.** $\dfrac{3x - 2}{7}$ **15.** $\dfrac{3r - 1}{4}$

17. $\dfrac{x + 6}{x}$ **19.** $-\dfrac{12}{n}$ **21.** $\dfrac{5x + 7}{x - 1}$ **23.** $\dfrac{t + 3}{5t^2}$ **25.** $\dfrac{1}{x - 4}$ **27.** 0 **29.** $\dfrac{p - 10}{p - 5}$ **31.** $x - 3$ **33.** $\dfrac{1}{2}$ **35.** 1 **37.** 4 **39.** $\dfrac{x + 5}{x + 6}$

41. $\dfrac{3}{4}$ **43.** $\dfrac{x - 5}{x + 2}$ **45.** $\dfrac{3x + 2}{x - 4}$ **47.** $\dfrac{6x + 1}{x - 8}$ **49.** 5 **51.** $5n$ **53.** $20x$ **55.** p^3 **57.** $3m - 4$ **59.** $x^2(2x + 3)$ **61.** $36x^3y$

63. $18r^4s^7$ **65.** $36w(w + 5)$ **67.** $x(x + 1)$ **69.** $4n - 1$ or $1 - 4n$ **71.** $4k - 5r$ or $-4k + 5r$ **73.** $6q(q + 1)$ **75.** $120x^2y^3$
77. $6(x + 4)(x + 2)$ **79.** $(x + 6)(x + 5)$ **81.** $(x - 8)(x + 3)(x + 8)$ **83.** $(a - 4)^2(a - 3)$ **85.** $(x + 5)(x + 1)(x + 3)$
87. $(x - 3)^2$ **89.** $(x - 6)(x - 1)$ **91.** $(3t - 2)(t + 4)(3t - 1)$ **93.** $(2x + 1)^2(4x + 3)$ **95.** $x^2 + x - 9$; the sum of the
numerators must be $2x^2 - 5x - 6$ **97.** $x^2 + 9x - 10$; the sum of the numerators must be $5x - 7$

99. 5☺ **101.** $(\Delta + 3)(\Delta - 3)$ **103.** $\dfrac{-3x^2 + 12x + 4}{x^2 - 25}$ **105.** $30x^{12}y^9$ **107.** $(x - 4)(x + 3)(x - 2)$ **109.** $\dfrac{92}{45}$ or $2\dfrac{2}{45}$

110. $-\dfrac{1}{5}$ **111.** 2.25 ounces **112.** 70 hours **113.** 2.0×10^{11} **114.** $\dfrac{3}{2}, -1$

Exercise Set 6.4

1. For each fraction, divide the LCD by the denominator. **3. a)** Answers will vary.
b) $\dfrac{x^2 + x - 9}{(x + 2)(x - 3)(x - 2)}$ **5. a)** $12z^2$ **b)** $\dfrac{3yz + 10}{12z^2}$ **c)** Yes **7.** $\dfrac{13}{4x}$ **9.** $\dfrac{3x + 10}{2x^2}$ **11.** $\dfrac{3x + 5}{x}$ **13.** $\dfrac{3x + 10}{5x^2}$

15. $\dfrac{35y + 12x}{20x^2y^2}$ **17.** $\dfrac{3y^2 + x}{y}$ **19.** $\dfrac{9a + 1}{6a}$ **21.** $\dfrac{4x^2 + 2y}{xy}$ **23.** $\dfrac{20a^2 - 4b}{5a^2b}$ **25.** $\dfrac{11x - 12}{x(x - 3)}$ **27.** $\dfrac{11p + 6}{p(p + 3)}$

29. $\dfrac{-6d^2 + 15d + 25}{6d(3d + 5)}$ **31.** $\dfrac{2}{p - 3}$ **33.** $\dfrac{14}{x + 7}$ **35.** $\dfrac{a + 12}{2(a - 2)}$ **37.** $\dfrac{20x}{(x - 5)(x + 5)}$ **39.** $\dfrac{-19n - 12}{3n(2n + 1)}$ **41.** $\dfrac{13w + 56}{2(w + 5)(w + 2)}$

43. $\dfrac{5z - 16}{(z + 4)(z - 4)}$ **45.** $\dfrac{-x + 6}{(x + 2)(x - 2)}$ **47.** $\dfrac{r + 10}{(r - 4)(r - 6)}$ **49.** $\dfrac{6x - 8}{(x + 4)(x - 2)}$ **51.** $\dfrac{x^2 + 3x - 18}{(x + 5)^2}$

53. $\dfrac{2a + 13}{(a - 8)(a - 1)(a + 2)}$ **55.** $\dfrac{5x + 5}{(x + 3)^2(x - 2)}$ **57.** $\dfrac{3x^2 - 8x - 3}{(2x + 1)(3x - 2)(x + 3)}$ **59.** $\dfrac{2x^2 - 3x - 4}{(4x + 3)(x + 2)(2x - 1)}$

61. $\dfrac{1}{w - 3}$ **63.** $\dfrac{9}{2(r - 3)}$ **65.** All real numbers except $x = 0$. **67.** All real numbers except $x = 4$ and $x = -6$.

69. $\dfrac{4}{\Delta - 2}$ **71.** All real numbers except $a = -b$ and $a = 0$. **73.** 0 **75.** $\dfrac{2x - 3}{2 - x}$ **77.** $\dfrac{6x + 5}{(x + 2)(x - 3)(x + 1)}$

80. ≈ 1.53 hours **81.** $x > -8$, **82.** $4x - 3 - \dfrac{4}{2x + 3}$ **83.** -1

Exercise Set 6.5

1. A complex fraction is a fraction whose numerator or denominator (or both) contains a fraction.
3. a) Numerator, $\dfrac{x + 3}{4}$; denominator, $\dfrac{7}{x^2 + 5x + 6}$ **b)** Numerator, $\dfrac{1}{2y} + x$; denominator, $\dfrac{3}{y} + x$ **5.** 2 **7.** $\dfrac{57}{32}$ **9.** $\dfrac{3}{448}$

11. $\dfrac{x^3y^2}{27}$ **13.** $\dfrac{2ab^3}{15c^2}$ **15.** $\dfrac{ab - a}{1 + a}$ **17.** $\dfrac{3}{x}$ **19.** $\dfrac{5x - 1}{4x - 1}$ **21.** $\dfrac{m - n}{m}$ **23.** $-\dfrac{a}{b}$ **25.** -1 **27.** $a^2(a - b)$ **29.** $b - a$

31. $\dfrac{a^2 + b}{b(b + 1)}$ **33.** $\dfrac{1}{y - x}$ **35.** $\dfrac{ab^2 + b^2}{a^2(b + 1)}$ **37. b)–c)** $-\dfrac{224}{155}$ **39. b)–c)** $\dfrac{x - y + 3}{2x + 2y - 7}$ **41. a)** $\dfrac{\dfrac{5}{12x}}{\dfrac{8}{x^2} - \dfrac{4}{3x}}$ **b)** $\dfrac{5x}{96 - 16x}$

43. $\dfrac{y + x}{2xy}$ **45.** $x + y$ **47. a)** $\dfrac{2}{7}$ **b)** $\dfrac{4}{13}$ **49.** $\dfrac{a^3b + a^2b^3 - ab^2}{a^3 - ab^3 + b^2}$ **51.** $\dfrac{17}{2}$ **52.** A polynomial is an expression containing
a finite number of terms of the form ax^n where a is a real number and n is a whole number. **53.** $(x - 6)(x - 7)$
54. $\dfrac{x^2 - 9x + 2}{(3x - 1)(x + 6)(x - 3)}$

Exercise Set 6.6   **1. a)** Answers will vary.   **b)** $\frac{2}{3}$ **3. a)** The problem on the left is an expression to be simplified while the problem on the right is an equation to be solved.   **b)** Left: Write the fractions with the LCD, $12(x-1)$, then combine numerators; Right: Multiply both sides of the equation by the LCD, $12(x-1)$, then solve.   **c)** Left: $\frac{x^2-x+12}{12(x-1)}$; Right: $4, -3$ **5.** You need to check for extraneous solutions when there is a variable in a denominator.
7. 2   **9.** No   **11.** Yes   **13.** 12   **15.** 2   **17.** -10 **19.** 30   **21.** 4   **23.** $\frac{25}{6}$ **25.** 36   **27.** 8   **29.** -8 **31.** 3   **33.** 4   **35.** 2   **37.** 5
39. No solution   **41.** 7   **43.** 15   **45.** No solution   **47.** -4 **49.** 38   **51.** $-\frac{1}{3}, 3$ **53.** 6, -1 **55.** 4, -4 **57.** $-4, -5$ **59.** 4
61. $-\frac{5}{2}$ **63.** No solution   **65.** 24   **67.** No solution   **69.** -3 **71.** 3   **73.** 5   **75.** 0   **77.** x can be any real number since the sum on the left is also $\frac{2x-4}{3}$.   **79.** 15 centimeters   **81.** -4 **83.** No, it is impossible for both sides of the equation to be equal.
85. More than $6\frac{1}{3}$ hours   **86.** $40°, 140°$ **87.** 75 minutes   **88.** 6.8×10^2

Exercise Set 6.7   **1.** Some examples are $A = \frac{1}{2}bh$, $A = \frac{1}{2}h(b_1 + b_2)$, $V = \frac{1}{3}\pi r^2 h$, and $V = \frac{4}{3}\pi r^3$
3. It represents 1 complete task.   **5.** length $= 10$ inches, width $= 9$ inches   **7.** base $= 12$ centimeters, height $= 7$ centimeters
9. Base: 8 feet   **11.** $\frac{3}{10}, 3$ **13.** 7   **15.** 10 miles per hour   **17.** 18.75 miles   **19.** 150 miles per hour, 600 miles per hour
21. boat: 32 miles, train: 168 miles   **23.** 3 hours at 600 miles per hour and 2 hours at 500 miles per hour   **25.** 900 feet
27. $3\frac{3}{7}$ hours   **29.** 6 hours   **31.** $3\frac{1}{3}$ hours   **33.** $8\frac{3}{4}$ days   **35.** $\frac{3}{5}$ hour or 36 minutes   **37.** 300 hours   **39.** 2 or $\frac{2}{5}$ **41.** 8 pints
43. $-\frac{3}{2}x - \frac{9}{2}$ **44.** $(y-1)(y+5)$ **45.** 1   **46.** $\frac{3x^2 - 9x - 15}{(2x+3)(3x-5)(3x+1)}$

Exercise Set 6.8   **1.** 40   **3.** $y = kx$ **5.** Direct   **7.** Inverse   **9.** Inverse   **11.** Direct   **13.** Direct   **15.** 33   **17.** $\frac{1}{5}$
19. 50   **21.** 3.2   **23.** 18   **25.** 3.5   **27.** 16   **29.** 30   **31.** It will be doubled   **33.** It will be halved   **35.** 80 miles   **37.** 2 hours
39. 375 feet   **41.** ≈ 133.33 hours   **43.** 1000 people   **45.** 60 people   **47.** ≈ 452.16 square inches   **49.** ≈ 44.44 ohms   **51.** \$60
53. 6400 cubic centimeters   **55. a)** $x = kyz$ **b)** 216   **57.** $2x - 3 + \frac{2}{4x+9}$ **58.** $(z-2)(y+3)$ **59.** $-4, 2$ **60.** $x + 3$

Chapter Review Exercises   **1.** All real numbers except $x = 6$ **2.** All real numbers except $x = 3$ and $x = 5$.
3. All real numbers except $x = \frac{1}{5}$ and $x = -1$.   **4.** $\frac{1}{x-3}$ **5.** $x^2 + 4x + 12$ **6.** $3x + 2y$ **7.** $x + 4$ **8.** $a + 6$
9. $-(2x+1)$ **10.** $\frac{b-3}{b+2}$ **11.** $\frac{x-3}{x-2}$ **12.** $\frac{x-8}{2x+3}$ **13.** $\frac{5}{12b^2}$ **14.** $6xz^2$ **15.** $\frac{16b^3c^2}{a^2}$ **16.** $-\frac{1}{3}$ **17.** $-\frac{2}{3}$ **18.** 1   **19.** $\frac{64x^2}{y}$
20. $\frac{32z}{x^3}$ **21.** $\frac{5}{a-b}$ **22.** $\frac{1}{3(a+3)}$ **23.** 1   **24.** $3y(x-y)$ **25.** $\frac{n-5}{n+5}$ **26.** 3   **27.** 9   **28.** $\frac{4}{x+10}$ **29.** $4h - 3$ **30.** $3x + 4$
31. 30   **32.** $x + 3$ **33.** $20x^2y^3$ **34.** $x(x+1)$ **35.** $(n-5)(n-4)$ **36.** $x(x+1)$ **37.** $(r+s)(r-s)$ **38.** $x - 7$
39. $(x+7)(x-5)(x+2)$ **40.** $\frac{3y^2+8}{6y^2}$ **41.** $\frac{10x+y}{5xy}$ **42.** $\frac{5x^2-12y}{3x^2y}$ **43.** $\frac{6x+10}{x+2}$ **44.** $\frac{x^2-2xy-y^2}{xy}$ **45.** $\frac{11x+16}{x(x+4)}$
46. $\frac{-x-4}{3x(x-2)}$ **47.** $\frac{3z+22}{(z+5)^2}$ **48.** $\frac{2x-8}{(x-3)(x-5)}$ **49.** $\frac{5x+38}{(x+6)(x+2)}$ **50.** $\frac{4x-12}{x-4}$ **51.** $\frac{4ab+8b}{a-2}$
52. $\frac{3x-3}{(x+3)(x-3)}$ **53.** $\frac{5p^2}{q}$ **54.** $\frac{4}{(x+2)(x-3)(x-2)}$ **55.** $\frac{8x-29}{(x+2)(x-7)(x+7)}$ **56.** $\frac{x}{x+y}$ **57.** $\frac{3(x+3y)}{5(x-3y)}$
58. $a - 3$ **59.** $\frac{3a^2-7a+2}{(a+1)(a-1)(3a-5)}$ **60.** $\frac{8x-16}{x}$ **61.** $\frac{55}{9}$ **62.** $\frac{26}{55}$ **63.** $\frac{bc}{3}$ **64.** $\frac{16x^3z^2}{y^3}$ **65.** $\frac{ab-a}{a+1}$ **66.** $\frac{r^2s+1}{s^3}$

67. $\dfrac{4x+2}{x(6x-1)}$ **68.** $\dfrac{2}{x}$ **69.** x **70.** $\dfrac{2a+1}{2}$ **71.** $\dfrac{-x+1}{x+1}$ **72.** $\dfrac{3x^2-x^2y}{y(y-x)}$ **73.** 6 **74.** 6 **75.** 40 **76.** 12 **77.** -16 **78.** $\dfrac{1}{2}$

79. -6 **80.** 28 **81.** No solution **82.** $2\dfrac{8}{11}$ hours **83.** $16\dfrac{4}{5}$ hours **84.** $\dfrac{1}{5}$, 1 **85.** Robert: 2.1 miles per hour; Tran: 5.6 miles

per hour **86.** 3096 milligrams **87.** 1.68 hours

Chapter Practice Test **1.** 1 **2.** $\dfrac{x^2+x+1}{x+1}$ **3.** $\dfrac{6x^2z}{y}$ **4.** $a+3$ **5.** $\dfrac{x^2-6x+9}{(x+3)(x+2)}$ **6.** -1 **7.** $\dfrac{x-2y}{3}$

8. $\dfrac{5}{y+5}$ **9.** $-\dfrac{m+6}{m-5}$ **10.** $\dfrac{3x-1}{y}$ **11.** $\dfrac{7x^2-6x-11}{x+3}$ **12.** $\dfrac{4y^2-3}{xy^3}$ **13.** $-\dfrac{z+20}{z-5}$ **14.** $\dfrac{-1}{(x+4)(x-4)}$ **15.** $\dfrac{55}{28}$

16. $\dfrac{x^2+x^2y}{y}$ **17.** $\dfrac{2x+3}{2-5x}$ **18.** 2 **19.** $-\dfrac{12}{7}$ **20.** 12 **21.** $3\dfrac{1}{13}$ hours **22.** 1 **23.** Base: 6 inches; height: 9 inches **24.** 2 miles

25. ≈ 1.13 feet

CHAPTER 7

Using Your Graphing Calculator, 7.1

1. 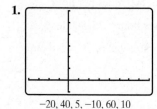   **2.** 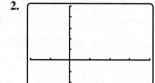   **3.** 50 **4.** 100

$-20, 40, 5, -10, 60, 10$ $-200, 400, 100, -500, 1000, 200$

Exercise Set 7.1 **1.** The x-coordinate **3. a)** x-axis **b)** y-axis **5.** Axis is singular; axes is plural.
7. An illustration of the set of points whose coordinates satisfy the equation **9. a)** Two **b)** To catch errors **11.** $ax+by=c$

13.

(Quadrants labeled II, I, III, IV)

15. II **17.** IV **19.** II
21. III **23.** III **25.** II
27. $A(3,1)$; $B(-3,0)$; $C(1,-3)$;

$D(-2,-3)$; $E(0,3)$; $F\left(\dfrac{3}{2},-1\right)$

29.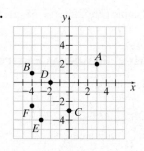

31.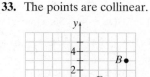

33. The points are collinear.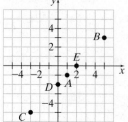

35. $(-5,-3)$ is not on the line.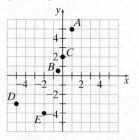

37. a) Point c) does not satisfy the equation.

b)

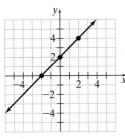

39. a) Point a) does not satisfy the equation.

b)

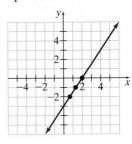

41. a) Point a) does not satisfy the equation.

b)

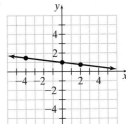

43. 2 **45.** -4 **47.** 2 **49.** $\dfrac{11}{3}$ **51.** 0 **53. a)** Latitude, $16°$ N; Longitude, $56°$ W   **b)** Latitude, $29°$ N; Longitude, $90.5°$ W

c) Latitude, $26°$ N; Longitude, $80.5°$ W   **d)** Answers will vary.   **57.** 21 **58.** $y = \dfrac{2x-6}{5} = \dfrac{2}{5}x - \dfrac{6}{5}$ **59.** $8x^{12}$

60. $(x+3)(x-9)$ **61.** $0, 8$ **62.** $\dfrac{7x+12}{3x^2}$

Using Your Graphing Calculator, 7.2

1.

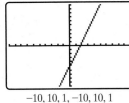

$-10, 10, 1, -10, 10, 1$

2.

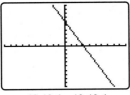

$-10, 10, 1, -10, 10, 1$

3.

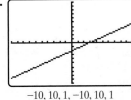

$-10, 10, 1, -10, 10, 1$

4.

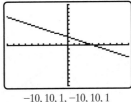

$-10, 10, 1, -10, 10, 1$

Using Your Graphing Calculator, 7.2   **1.** $(2, 0), (0, -8)$ **2.** $(-3, 0), (0, -6)$ **3.** $(2.5, 0), (0, -2)$
4. $(1.5, 0), (0, -4.5)$

Exercise Set 7.2   **1.** x-intercept: substitute 0 for y and find the corresponding value of x; y-intercept: substitute
0 for x and find the corresponding value of y **3.** A horizontal line
5. You may not be able to read exact answers from a graph.   **7.** Yes   **9.** 0 **11.** 5 **13.** 3 **15.** 2 **17.** $\dfrac{8}{3}$ **19.** -10

21.

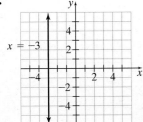

23.

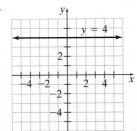

25.

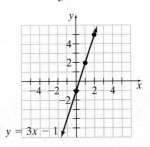

27.

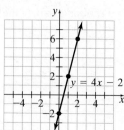

$y = 4x - 2$

29.

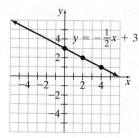

$y = -\frac{1}{2}x + 3$

31.

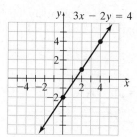

$3x - 2y = 4$

33.

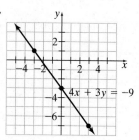

$4x + 3y = -9$

35.

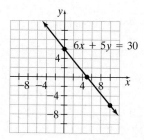

$6x + 5y = 30$

37.

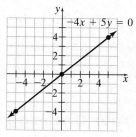

$-4x + 5y = 0$

39.

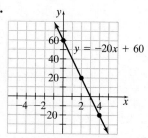

$y = -20x + 60$

41.

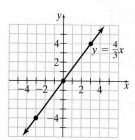

$y = \frac{4}{3}x$

43.

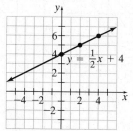

$y = \frac{1}{2}x + 4$

45.

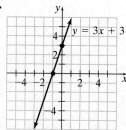

$y = 3x + 3$

47.

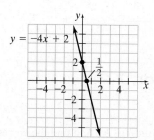

$y = -4x + 2$

49.

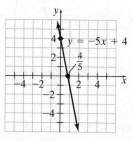

$y = -5x + 4$

51.

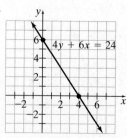

$4y + 6x = 24$

53.

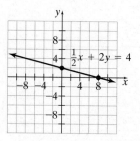

$\frac{1}{2}x + 2y = 4$

55.

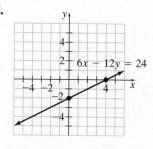

$6x - 12y = 24$

57.

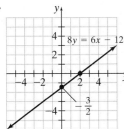

$8y = 6x - 12$

$-\dfrac{3}{2}$

59.

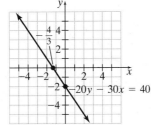

$-\dfrac{4}{3}$

$-20y - 30x = 40$

61.

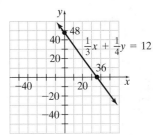

48

$\dfrac{1}{3}x + \dfrac{1}{4}y = 12$

36

63.

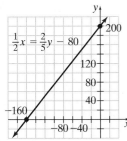

$\dfrac{1}{2}x = \dfrac{2}{5}y - 80$

200

65. $x = -2$ **67.** $y = 6$ **69.** 4 **71.** 2 **73.** Yes

75. a) $C = 2n + 30$
b)

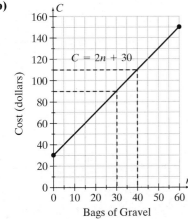

$C = 2n + 30$

Cost (dollars)

Bags of Gravel

75. c) $90 **d)** 40 bags

77. a) $C = m + 40$
b)

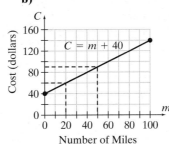

$C = m + 40$

Cost (dollars)

Number of Miles

77. c) $90 **d)** 20 miles

79. a)

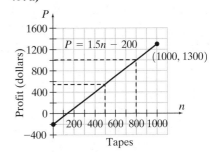

$P = 1.5n - 200$

(1000, 1300)

Profit (dollars)

Tapes

79. b) $550 **c)** 800 tapes

81. 5; 4 **83.** 6; 4 **85. a)**

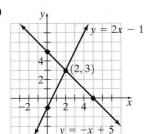

$y = 2x - 1$

$(2, 3)$

$y = -x + 5$

85. b) $(2, 3)$ **c)** Yes **d)** No **88.** -18 **89.** 6.67 ounces **90.** 9, 28

91. $\dfrac{50yz}{x}$ **92.** $\dfrac{3x^2 - 8x + 1}{(x - 2)(x - 3)}$ **93.** 5, $\dfrac{7}{3}$

Exercise Set 7.3

1. The slope of a line is the ratio of the vertical change to the horizontal change between any two points on the line. **3.** Rises from left to right **5.** Lines that rise from the left to right have a positive slope; lines that fall from left to right have a negative slope. **7.** No, since we cannot divide by 0, the slope is undefined. **9.** The slopes are the same.
11. 2 **13.** $\dfrac{1}{2}$ **15.** 0 **17.** 0 **19.** Undefined **21.** $-\dfrac{3}{8}$ **23.** $\dfrac{2}{3}$ **25.** $m = 2$ **27.** $m = -2$ **29.** $m = -\dfrac{4}{7}$ **31.** $m = \dfrac{7}{4}$ **33.** $m = 0$
35. Undefined **37.** $m = 0$

39.

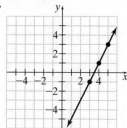

41.

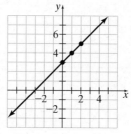

43.

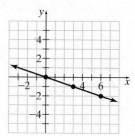

45.

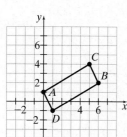

47.

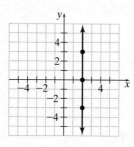

49. Parallel **51.** Perpendicular **53.** Perpendicular **55.** Neither **57.** Neither **59.** Parallel **61.** Parallel **63.** Perpendicular

65. 2 **67.** $\frac{1}{4}$ **69.** First **71. a)** $-\frac{23}{4}$ **b)** 11 **73.** -4 **75.** $-\frac{1}{8}$ **77.** $\frac{225}{68}$

79. a)

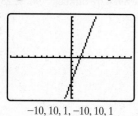

b) $AC, m = \frac{3}{5}$; $CB, m = -2$; $DB, m = \frac{3}{5}$; $AD, m = -2$ **c)** Yes, opposite sides are parallel.

81. a) $AB, m = 4$; $BC, m = -2$; $CD, m = 4$ **b)** $[4 + (-2) + 4]/3 = 2$ **c)** $AD, m = 2$ **d)** Yes

e) Answers will vary. **83.** 0 **84. a)** $\frac{3}{2}$ **b)** 0 **85.** $2x + 15$ **86.** -3 **87.** x-intercept: $(3, 0)$; y-intercept: $(0, -5)$

Using Your Graphing Calculator, 7.4

1. a)

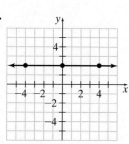

$-10, 10, 1, -10, 10, 1$

b)

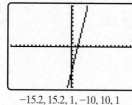

$-15.2, 15.2, 1, -10, 10, 1$

c)

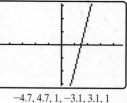

$-4.7, 4.7, 1, -3.1, 3.1, 1$

2. a)

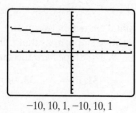

$-10, 10, 1, -10, 10, 1$

b)

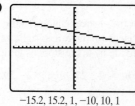

$-15.2, 15.2, 1, -10, 10, 1$

c)

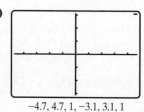

$-4.7, 4.7, 1, -3.1, 3.1, 1$

Exercise Set 7.4   **1.** $y = mx + b$ **3.** $y = 3x - 5$ **5.** Write the equations in slope-intercept form. If slopes are the same and the y-intercepts are different, the lines are parallel.   **7.** $y - y_1 = m(x - x_1)$ **9.** $4; (0, -6)$ **11.** $\frac{4}{3}; (0, -5)$

13. $1; (0, -3)$

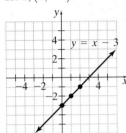

15. $3; (0, 2)$

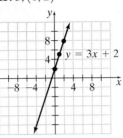

17. $-4; (0, 0)$

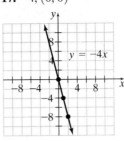

19. $2; (0, -3)$

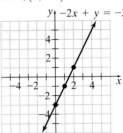

21. $\frac{5}{2}; (0, -5)$

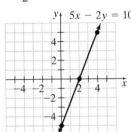

23. $-\frac{1}{2}; \left(0, \frac{3}{2}\right)$

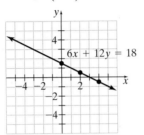

25. $3; (0, 4)$

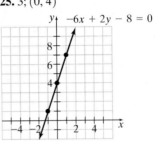

27. $\frac{3}{2}; (0, 2)$

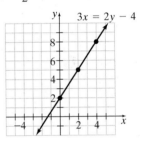

29. $y = x - 2$ **31.** $y = -\frac{1}{3}x + 2$ **33.** $y = \frac{1}{3}x + 5$ **35.** $y = -3x + 4$ **37.** Parallel   **39.** Perpendicular   **41.** Parallel

43. Neither   **45.** Perpendicular   **47.** Neither   **49.** $y = 3x + 2$ **51.** $y = -3x - 7$ **53.** $y = \frac{1}{2}x - \frac{9}{2}$ **55.** $y = \frac{2}{5}x + 6$

57. $y = 3x + 10$ **59.** $y = -\frac{3}{2}x$ **61.** $y = \frac{1}{2}x - 2$ **63.** $y = 6.3x - 4.5$ **65. a)** $y = 5x + 60$ **b)** $210

67. a) Slope-intercept form   **b)** Point-slope form   **c)** Point-slope form

69. a) No   **b)** $y + 4 = 2(x + 5)$ **c)** $y - 10 = 2(x - 2)$ **d)** $y = 2x + 6$ **e)** $y = 2x + 6$ **f)** Yes

71. a) 1.465 **b)** $f = 1.465m$ **c)** ≈ 191.64 feet per second   **d)** 150 feet per second   **e)** 55 miles per hour

73. $y = -2x + 4$ **75.** $y = \frac{3}{4}x + 2$ **78.** $<$ **79.** $x \le -4$;   **80.** $r = \dfrac{i}{pt}$ **81.** $(x - 2y)(x + 3y)$ **82.** -5

Exercise Set 7.5   **1.** Points on the line satisfy the $=$ part of the inequality   **3.** The shading is on opposite sides of the line.

5.

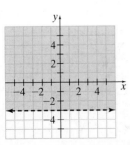

7.

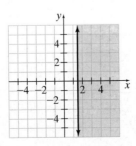

9.

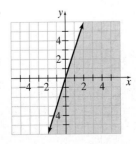

11.

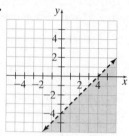

13.

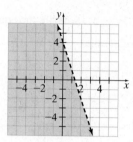

15.

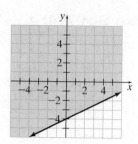

17.

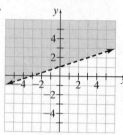

19.

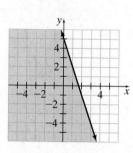

21.

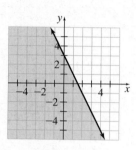

23.

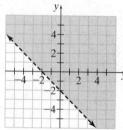

25. a) No **b)** No **c)** Yes **d)** Yes **27.** No, it could be a solution to $ax + by = c$.

29. No, the location of an ordered pair which satisfies the first inequality lies on one side of the line while an ordered pair which satisfies the other inequality lies either on the line or on the other side of the line.

31. a) Less than or equal to **b)** Greater than or equal to **c)** Less than or equal to **d)** Greater than or equal to

33. a), b), and **c)** **34. a)** 2 **b)** 2, 0 **c)** 2, $-5, 0, \frac{2}{5}, -6.3, -\frac{23}{34}$ **d)** $\sqrt{7}, \sqrt{3}$

e) 2, $-5, 0, \sqrt{7}, \frac{2}{5}, -6.3, \sqrt{3}, -\frac{23}{34}$

35. $-\frac{2}{3}$ **36.** $2x - 3 + \frac{6}{x}$ **37.** $\frac{1}{x + 2y}$

Exercise Set 7.6

1. A relation is any set of ordered pairs. **3.** A function is a set of ordered pairs in which each first component corresponds to exactly one second component **5. a)** The domain is the set of first components in the set of ordered pairs. **b)** The range is the set of second components in the set of ordered pairs. **7.** No, each x must have a unique y for it to be a function. **9.** Function; Domain: $\{1, 2, 3, 4, 5\}$; Range: $\{1, 2, 3, 4, 5\}$ **11.** Relation; Domain: $\{1, 2, 3, 5, 7\}$; Range: $\{-2, 0, 2, 4, 5\}$ **13.** Function; Domain: $\{0, 1, 3, 4, 5\}$; Range: $\{-4, -1, 0, 1, 2\}$ **15.** Relation; Domain: $\{0, 1, 3\}$; Range: $\{-3, 0, 2, 5\}$ **17.** Function; Domain: $\{0, 1, 2, 3, 4\}$; Range: $\{3\}$ **19. a)** $\{(1, 4), (2, 5), (3, 5), (4, 7)\}$ **b)** Function **21. a)** $\{(-5, 4), (0, 7), (6, 9), (6, 3)\}$ **b)** Not a function **23.** Function **25.** Not a function **27.** Function **29.** Function **31.** Not a function **33.** Function **35. a)** 14 **b)** -2 **37. a)** 31 **b)** -1 **39. a)** 4 **b)** 14 **41. a)** 3 **b)** 5

43.

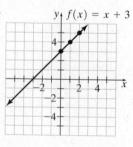

45.

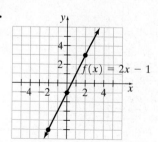

47.

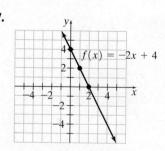

49.

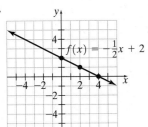

51. No, some value of x must correspond to two values of y.
53. Yes, it passes the vertical line test.

55. a)

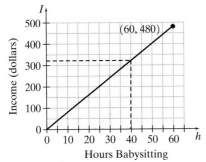

Hours Babysitting

b) $320   **57. a)**

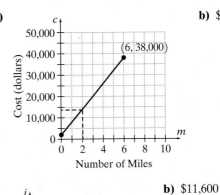

Number of Miles

b) $14,000

59. a)

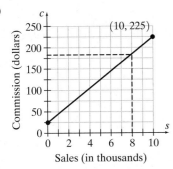

Sales (in thousands)

b) $185   **61. a)**

Sales (in thousands)

b) $11,600

63. Yes, it passes the vertical line test.   **65.** No, the vertical line $x = 1$ intersects the graph at more than one point.

67. a) $\dfrac{29}{8}$ **b)** $\dfrac{29}{9}$ **c)** 4.42   **71.** $\dfrac{8}{63}$ **72.** -14 **73.** 13 miles   **74.** $(5x + 11y)(5x - 11y)$ **75.** $\dfrac{3x^2}{y}$

76. A graph is an illustration of a set of points whose coordinates satisfy an equation.

Chapter Review Exercises   **1.**

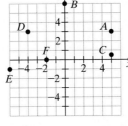

2. Not collinear   **3. a), b),** and **d)**

4. a) 2   **b)** -4 **c)** $\dfrac{16}{3}$ **d)** $\dfrac{8}{3}$

5.

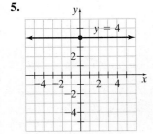

6.

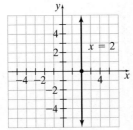

7.

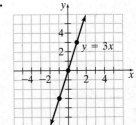

8.
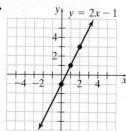

9.
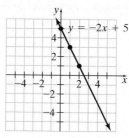

10.
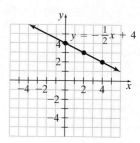

11.
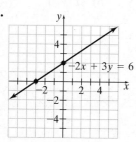

12.
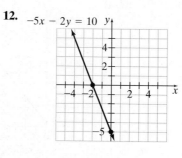

13.
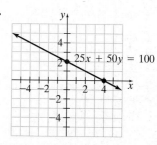

14.
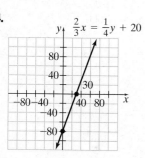

15. $-\dfrac{9}{5}$ **16.** $-\dfrac{1}{12}$ **17.** -2 **18.** 0 **19.** Undefined

20. The slope of a line is the ratio of the vertical change to the horizontal change between any two points on the line.

21. -2 **22.** $\dfrac{1}{4}$ **23.** Neither   **24.** Perpendicular

25. a) 28 **b)** -49 **26.** $m = -\dfrac{6}{7}; (0, 2)$

27. Slope is undefined; no y-intercept   **28.** $m = 0; (0, -3)$

29. $y = 3x - 3$ **30.** $y = -\dfrac{1}{2}x + 2$ **31.** Parallel

32. Perpendicular   **33.** $y = 3x - 2$ **34.** $y = -\dfrac{2}{3}x + 4$ **35.** $y = 2$ **36.** $x = 4$ **37.** $y = -\dfrac{7}{2}x - 4$ **38.** $x = -4$

39.
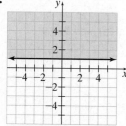

40.
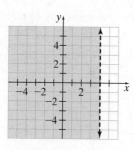

41.
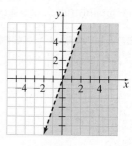

42.
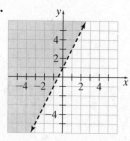

43.
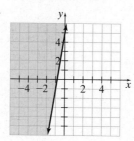

44.
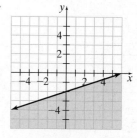

45. Function; Domain: $\{1, 2, 3, 4, 6\}$; Range: $\{-3, -1, 2, 4, 5\}$ **46.** Not a function; Domain: $\{3, 4, 6, 7\}$; Range: $\{0, 1, 2, 5\}$

47. Not a function; Domain: $\{3, 4, 5, 6\}$; Range: $\{-3, 1, 2\}$ **48.** Function; Domain: $\{-2, 3, 4, 5, 9\}$; Range: $\{-2\}$

49. a) $\{(1, 3), (4, 5), (7, 2), (9, 2)\}$ **b)** Function   **50. a)** $\{(4, 1), (6, 3), (6, 5), (8, 7)\}$ **b)** Not a function

51. a) Domain: $\{$Mary, Pete, George, Carlos$\}$; Range: $\{$Apple, Orange, Grape$\}$ **b)** Not a function

52. a) Domain: $\{$Sarah, Jacob, Kristen, Erin$\}$; Range: $\{$Seat 1, Seat 2, Seat 3, Seat 4$\}$ **b)** Not a function

53. a) Domain: $\{$Blue, Green, Yellow$\}$; Range: $\{$Paul, Maria, Lalo, Duc$\}$ **b)** Function   **54. a)** Domain: $\{1, 2, 3, 4,\}$;

Range: $\{A, B, C\}$ **b)** Function   **55.** Function   **56.** Not a function   **57.** Function   **58.** Function   **59. a)** 2   **b)** -34

60. a) 11   **b)** -37 **61. a)** -4 **b)** -8 **62. a)** 12   **b)** 76   **63.** Yes, it passes the vertical line test.

64. Yes, each year has a unique percent.   **65.**

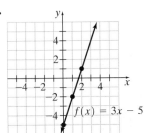

66.

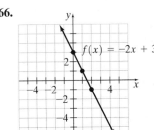

67. a)

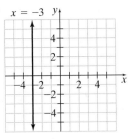

b) $55   **68. a)**

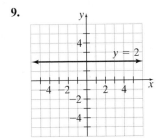

b) $0

Chapter Practice Test   **1.** A graph is an illustration of the set of points whose coordinates satisfy an equation.

2. a) IV   **b)** III   **3. a)** $ax + by = c$ **b)** $y = mx + b$ **c)** $y - y_1 = m(x - x_1)$ **4. b)** and **d)**   **5.** $-\dfrac{4}{3}$ **6.** $\dfrac{4}{9}$; $\left(0, -\dfrac{5}{3}\right)$

7. $y = -x - 1$

8.

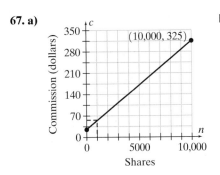

9.

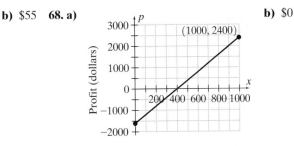

10.

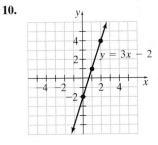

11. a) $y = \dfrac{1}{2}x - 2$

b)

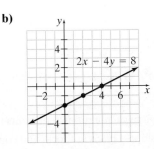

12.

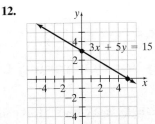

13. $y = 4x - 13$ **14.** $y = -\dfrac{3}{7}x + \dfrac{2}{7}$ **15.** The lines are parallel since they have the same slope but different y-intercepts.

16.

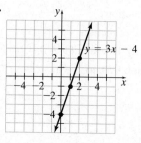

17.

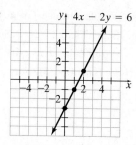

18. A set of ordered pairs in which each first component corresponds to exactly one second component.
19. a) Not a function; 1, a first component, is paired with more than one value   **b)** Domain: $\{1, 3, 5, 6\}$; Range: $\{-4, 0, 2, 3, 5\}$ **20. a)** Function; it passes the vertical line test   **b)** Not a function; a vertical line can be drawn that intersects the graph at more than one point   **21. a)** 14   **b)** 9

22.

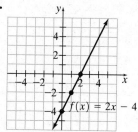

23.

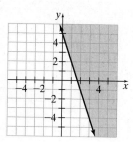

24.

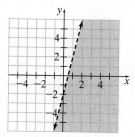

25. a)

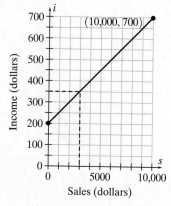

b) $350

CHAPTER 8

Using Your Graphing Calculator, 8.1   **1.** $(-3, -4)$ **2.** $(-4, 3)$ **3.** $(3.1, -1.5)$ **4.** $(-0.6, 4.8)$

Exercise Set 8.1   **1.** The solution to a system of equations represents the ordered pairs that satisfy all the equations in the system.   **3.** Write the equations in slope—intercept form and compare their slopes and y-intercepts. If the

slopes are different, the system has exactly one solution. If the slopes are the same and the *y*-intercepts are different, the system has no solution. If the slopes and *y*-intercepts are the same, the system has an infinite number of solutions.
5. The point of intersection can only be estimated. **7.** c) **9.** a) **11.** none **13.** a), c) **15.** a) **17.** Consistent; one solution
19. Dependent; infinite number of solutions **21.** Consistent; one solution **23.** Inconsistent; no solution **25.** One solution
27. No solution **29.** One solution **31.** No solution **33.** Infinite number of solutions **35.** No solution

37.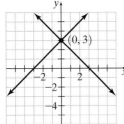

39.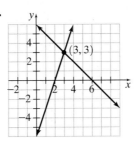

41.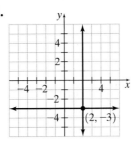

43.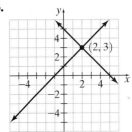

45.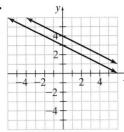
Inconsistent

47.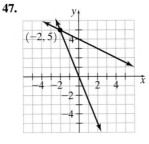

49.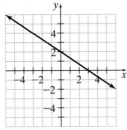
Dependent

51.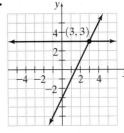

53.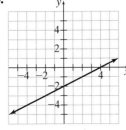
Dependent

55.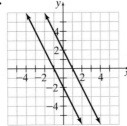
Inconsistent

57.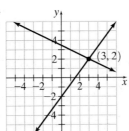

59.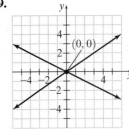

61. Lines are parallel; they have the same slope and different *y*-intercepts. **63.** The system has an infinite number of solutions. If the two lines have two points in common then they must be the same line. **65.** The system has no solution. Parallel lines do not intersect. **67.** One; (5, 3) **69.** 6 years **71.** 6 hours **79.** $-2x + 18$ **80.** 1 **81.** $-(x - 7)$

82. $\frac{1}{3}$, 6 **83. a)** $(6, 0), (0, 4)$ **b)**
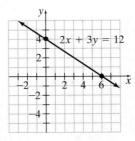

Exercise Set 8.2 **1.** The x in the first equation, since both 6 and 12 are divisible by 3. **3.** You will obtain a false statement, such as $3 = 0$. **5.** $(4, 1)$ **7.** $(-1, -1)$ **9.** No solution **11.** $(3, -8)$ **13.** Infinite number of solutions **15.** $(3, 5)$ **17.** Infinite number of solutions **19.** $(2, 1)$ **21.** No solution **23.** $(-1, 0)$ **25.** $\left(-\dfrac{12}{7}, \dfrac{6}{35}\right)$ **27.** $36, 43$ **29.** Width: 8 feet, length: 12 feet **31.** Billy: \$296, Jean: \$430 **33.** \$15,000 **35. a)** 16 months **b)** Yes **37. a)** 1.25 hours **b)** 155 mile marker **39. a)** $T = 180 - 10t$ **b)** $T = 20 + 6t$ **c)** 10 minutes **d)** $80°$ F **41.** 7.14 years **42.** $18x^2 + 9x - 14$ **43.**

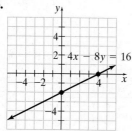

44. Slope: $\dfrac{3}{5}$; y-intercept: $\left(0, -\dfrac{8}{5}\right)$ **45.** $y = 2x + 4$

Exercise Set 8.3 **1.** Multiply the top equation by 2. **3.** You will obtain a false statement, such as $0 = 6$. **5.** $(5, 1)$ **7.** $(2, 3)$ **9.** $\left(4, \dfrac{11}{2}\right)$ **11.** No solution **13.** $(-8, -26)$ **15.** $\left(-\dfrac{3}{2}, -3\right)$ **17.** $(-2, 2)$ **19.** $(3, -1)$ **21.** Infinite number of solutions **23.** $(1, 2)$ **25.** No solution **27.** $(0, 0)$ **29.** $\left(10, \dfrac{15}{2}\right)$ **31.** $(-6, -12)$ **33.** $\left(\dfrac{20}{39}, -\dfrac{16}{39}\right)$ **35.** $\left(\dfrac{14}{5}, -\dfrac{12}{5}\right)$ **37.** No solution **39.** First number: 12, second number: 4 **41.** First number: 6, second number: 4 **43.** Width: 3 inches, length: 6 inches **45.** Width: 8 inches, length: 10 inches **47.** Answers will vary. **49. a)** $(200, 100)$ **b)** Same solution; dividing both sides of an equation by a nonzero number does not change the solution. **51.** $(8, -1)$ **53.** $(1, 2, 3)$ **55.** 125 **56.** 7 **57.** $2x^2y - 9xy + 4y$ **58.** $32a^6b^9c^5$ **59.** $(y + c)(x - a)$ **60.** 14

Exercise Set 8.4 **1.** $17, 24$ **3.** $A = 36°, B = 54°$ **5.** $A = 102°, B = 78°$ **7.** width = 390 feet, length = 740 feet **9.** 70 acres corn, 30 acres wheat **11.** kayak, 4.05 miles per hour; current, 0.65 miles per hour **13.** 20 years **15.** adults: 12, children: 15 **17. a)** 1400 copies **b)** Office Copier Depot **19.** \$2500 at 8%; \$5500 at 10% **21. a)** 15 square yards **b)** Tom Taylor's **23.** Dave: 57 miles per hour, Alice: 72 miles per hour **25.** Melissa, 60 miles per hour; Elizabeth, 64 miles per hour **27.** 0.5 hour **29.** ≈ 0.68 hour **31.** 4 liters of 15%; 6 liters of 40% **33.** 180 tiles **35.** 70 gallons 5%; 30 gallons skim **37.** $2\dfrac{2}{3}$ ounces drink, $5\dfrac{1}{3}$ ounces juice **39.** 19.8 years **41.** 1.125 miles **43.** 45 minutes **45. a)** Commutative property of addition **b)** Associative property of multiplication **c)** Distributive property **46.** width = 3 feet, length = 8 feet **47.** $-2, 2$ **48.** A graph is an illustration of the set of points that satisfies an equation.

Exercise Set 8.5 **1.** Yes, the solution to a system of linear inequalities contains all of the ordered pairs which satisfy both inequalities. **3.** Yes, when the lines are parallel. One possible system is $x + y > 2, x + y < 1$.

5.

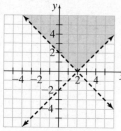

7.

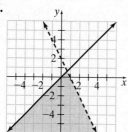

9.

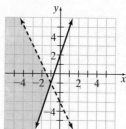

11.

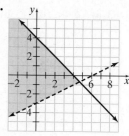

13.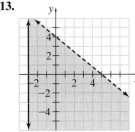

15.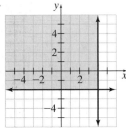

17.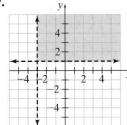

19.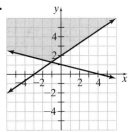

21.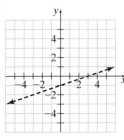

23.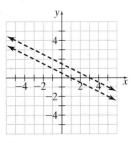

No solution   No solution

25. No, the system can have no solution or infinitely many solutions.   **29.** All real numbers; 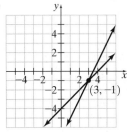 0

30. $y = \dfrac{2}{5}x - \dfrac{6}{5}$ **31.** $-\dfrac{1}{4}, 3$ **32.** $\dfrac{y^6}{x^9}$

Chapter Review Exercises   **1.** c)   **2.** a)   **3.** Consistent, one solution   **4.** Inconsistent, no solution
5. Dependent, infinite number of solutions   **6.** Consistent, one solution   **7.** No solution   **8.** One solution   **9.** Infinite number of solutions   **10.** One solution

11.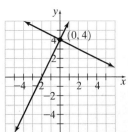

12.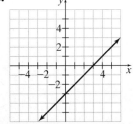

13.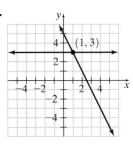

14.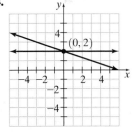

15.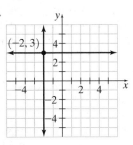

16.

17.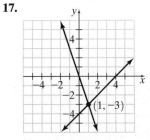

18.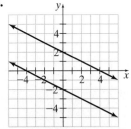

Infinite number of solutions   No solution

19. $(5, 2)$ **20.** $(-3, 2)$ **21.** $(5, 4)$ **22.** $(-18, 6)$ **23.** No solution **24.** Infinite number of solutions **25.** $\left(\frac{5}{2}, -1\right)$

26. $\left(-\frac{2}{7}, \frac{29}{7}\right)$ **27.** $(-8, 2)$ **28.** $(1, -2)$ **29.** $(-7, 19)$ **30.** $\left(\frac{32}{13}, \frac{8}{13}\right)$ **31.** $\left(-1, \frac{13}{3}\right)$ **32.** No solution **33.** Infinite

number of solutions **34.** $\left(-\frac{78}{7}, -\frac{48}{7}\right)$ **35.** 19 and 30 **36.** plane: 565 miles per hour; wind: 35 miles per hour

37. 100 miles **38.** $10,000 at 4%, $6000 at 6% **39.** Liz: 57 miles per hour; Mary: 63 miles per hour **40.** 15 pounds of Green Turf's, 25 pounds of Agway's **41.** 3 liters of each

42. 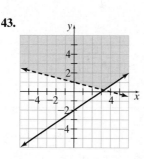 **43.** 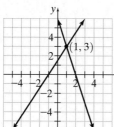 **44.** 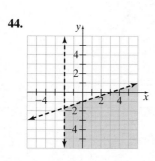 **45.**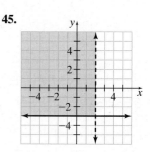

Chapter Practice Test

1. b) **2.** Inconsistent, no solution **3.** Consistent, one solution **4.** Dependent, infinite number of solutions **5.** No solution **6.** One solution **7.** Infinite number of solutions **8. a)** You will obtain a false statement, such as $6 = 0$. **b)** You will obtain a true statement, such as $0 = 0$.

9. 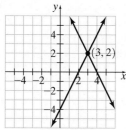 **10.** 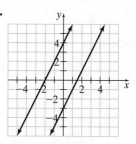 **11.**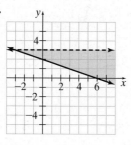

No solution

12. $\left(\frac{7}{2}, -\frac{5}{2}\right)$ **13.** $(4, 1)$ **14.** $(6, 23)$ **15.** $(-2, 2)$ **16.** $\left(\frac{44}{19}, \frac{48}{19}\right)$ **17.** Infinite number of solutions **18.** $(3, 2)$

19. $\left(-\frac{40}{7}, \frac{52}{7}\right)$ **20.** $\left(\frac{50}{19}, \frac{8}{19}\right)$ **21.** 200 miles **22.** butterscotch: $13\frac{1}{3}$ pounds; lemon: $6\frac{2}{3}$ pounds

23. Deja's boat, 60 miles per hour; Dante's boat, 64 miles per hour

24. 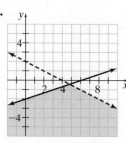 **25.**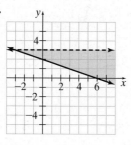

CHAPTER 9

Using Your Calculator, 9.1
1. 3.872983346 **2.** 12.28820573 **3.** Error **4.** 5.196152423

Exercise Set 9.1
1. The principal square root of a positive real number, a, is the positive number whose square equals a. **3.** Answers will vary. **5. a)** Answers will vary. **b)** Check Table 9.1 **7.** Yes, since $5^2 = 25$. **9.** No, because the square root of a negative number is not a real number. **11.** Yes, since $\sqrt{\frac{16}{25}} = \frac{4}{5}$ which is a rational number. **13.** 0 **15.** 1

17. -7 **19.** 20 **21.** -4 **23.** 12 **25.** 13 **27.** -1 **29.** 9 **31.** -11 **33.** $\frac{1}{2}$ **35.** $\frac{6}{7}$ **37.** $-\frac{5}{6}$ **39.** $\frac{9}{7}$ **41.** 3.4641016

43. 3.8729833 **45.** 8.9442719 **47.** 18 **49.** 9.8488578 **51.** 1.7320508 **53.** False **55.** True **57.** False **59.** True **61.** True

63. True **65.** $7^{1/2}$ **67.** $(17)^{1/2}$ **69.** $(8x)^{1/2}$ **71.** $(12x^2)^{1/2}$ **73.** $(15ab^2)^{1/2}$ **75.** $(62n^3)^{1/2}$ **77.** Rational: $9.83, \frac{3}{5}, 0.333\ldots, 5,$

$\sqrt{\frac{4}{49}}, \frac{3}{7}, -\sqrt{9}$; Irrational: $\sqrt{\frac{5}{16}}$; Imaginary: $\sqrt{-4}, -\sqrt{-16}$ **79.** 6 and 7 **81. a)** Square 4.6 and compare the result to 20.

b) 4.6 **83.** $-\sqrt{9}, -\sqrt{7}, -\frac{1}{2}, 2.5, \sqrt{16}, 4.01, 5, 12$ **85.** $\sqrt{4} = 2; 6^{1/2} \approx 2.45; -\sqrt{9} = -3; -(25)^{1/2} = -5; (30)^{1/2} \approx 5.48;$

$(-4)^{1/2}$, imaginary number **87. a)** Yes **b)** No **c)** No **d)** Yes **e)** No **89. a)** Yes **b)** Yes **c)** a **d)** Answers will vary. **91.** $x^{3/2}$ **93.** x^3 **95.** 52 jumps **96.** -6 **97.** -1 **98.** $\frac{4}{11}$ **99.** 16

Exercise Set 9.2
1. Answers will vary. **3.** The product rule cannot be used when radicands are negative.
5. a) Express the variable as the product of two factors, one of which has an exponent of 1. Use the product rule to simplify. **b)** $x^6\sqrt{x}$ **7. a)** There can be no perfect square factors nor any exponents greater than 1 in the radicand of a simplified square root expression. **b)** $5x^2\sqrt{3x}$ **9.** Yes **11.** No; $4\sqrt{2}$ **13.** $2\sqrt{3}$ **15.** $2\sqrt{2}$ **17.** $4\sqrt{6}$
19. $4\sqrt{2}$ **21.** $4\sqrt{10}$ **23.** $4\sqrt{5}$ **25.** $6\sqrt{2}$ **27.** $2\sqrt{35}$ **29.** $9\sqrt{3}$ **31.** $5\sqrt{6}$ **33.** x^3 **35.** xy^2 **37.** $a^6b^4\sqrt{b}$ **39.** $ab^2\sqrt{c}$
41. $n\sqrt{3n}$ **43.** $5ab\sqrt{3a}$ **45.** $10a^2b^5\sqrt{3ab}$ **47.** $9xy^2\sqrt{3x}$ **49.** $6ab^3\sqrt{3bc}$ **51.** $6rs^2t^2\sqrt{5rt}$ **53.** 5 **55.** $2\sqrt{30}$ **57.** $12\sqrt{5}$
59. $x\sqrt{21}$ **61.** $4ab\sqrt{3a}$ **63.** $6xy^2\sqrt{2x}$ **65.** $3r^5s^6\sqrt{7}$ **67.** $3xy^3\sqrt{10yz}$ **69.** $3a^3b^5\sqrt{6}$ **71.** $2x$ **73.** $13x^4y^6$ **75.** $15a^2$ **77.** 4
79. Exponent on x, 6; on y, 5 **81.** Coefficient, 8; exponent on x, 12; exponent on y, 7 **83. a)** $13x^3$ **b)** $13x^3$ **c)** Yes
85. $10\text{☺}^5\sqrt{2\text{☺}}$ **87.** $5\text{☺}^{50}\otimes^{18}$ **89.** $x^{1/6}$ **91.** $2x^{2/5}$ **93.** Rational, since $\sqrt{6.25} = 2.5$ and 2.5 is a terminating decimal
number. **95. a)** 4 feet **b)** No; the area increased $\sqrt{2}$ or ≈ 1.414 times **c)** 4 times **97. a)** Yes **b)** No, for example
$\sqrt{2} \cdot \sqrt{2} = 2$. **102.** 148.5 square inches **103.** ≈ 25.13 cubic feet **104.** $\frac{3x + 1}{x - 1}$ **105.** $y = -\frac{1}{2}x + \frac{3}{2}; m = -\frac{1}{2}; \left(0, \frac{3}{2}\right)$

106. 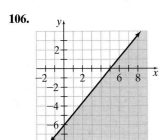 **107.** $\left(\frac{2}{11}, -\frac{15}{11}\right)$

Exercise Set 9.3
1. Like square roots are square roots having the same radicand. One example is $\sqrt{3}$ and $5\sqrt{3}$ **3.** Only like square roots can be added or subtracted. **5.** $2\sqrt{5}$ **7.** $-4\sqrt{5}$ **9.** $2\sqrt{7} + 6$ **11.** $6\sqrt{x}$ **13.** $-3\sqrt{y}$
15. $-4\sqrt{t} - 6$ **17.** $\sqrt{x} + 4\sqrt{y} + x$ **19.** $5m - 2\sqrt{m} + 2$ **21.** $-2\sqrt{7} - 9\sqrt{x}$ **23.** $2\sqrt{5} + 3\sqrt{2}$ **25.** $7\sqrt{3}$ **27.** $11\sqrt{3}$
29. $16\sqrt{2}$ **31.** $-15\sqrt{5} + 35\sqrt{3}$ **33.** $28\sqrt{10}$ **35.** $16 - 4\sqrt{3}$ **37.** $4\sqrt{3} + 3$ **39.** $4\sqrt{x} - 4\sqrt{2}$ **41.** $y\sqrt{y} + y^2$
43. $2\sqrt{10} - 2\sqrt{5}$ **45.** $x + \sqrt{3x}$ **47.** $6\sqrt{a} - a\sqrt{2}$ **49.** $x^2 + 4x\sqrt{y}$ **51.** $12x^2 - 9x\sqrt{x}$ **53.** $24 - 6\sqrt{2} + 4\sqrt{3} - \sqrt{6}$
55. $\sqrt{30} + 3\sqrt{5} - 2\sqrt{6} - 6$ **57.** $76 - 28\sqrt{7}$ **59.** $16 - 8\sqrt{x} + x$ **61.** $z\sqrt{15} + 2\sqrt{3z} - 4\sqrt{5z} - 8$
63. $2r^2 + r\sqrt{s} - 6s$ **65.** $2x^2 - 4x\sqrt{2y} + 4y$ **67.** $4p^2 + 6p\sqrt{3q} - 12q$ **69.** -2 **71.** 14 **73.** $x - 9$ **75.** $7 - r^2$
77. $5x - y$ **79.** $4x - 9y$ **81. a)** No **b)** $2\sqrt{5}$ **83.** $x + 2$ **85.** Yes **87.** 343 **89.** Perimeter: $4(\sqrt{2} + \sqrt{3})$ units;

area: $5 + 2\sqrt{6}$ square units **91.** Perimeter: $\sqrt{7} + \sqrt{15} + \sqrt{22}$ units; area: $\frac{1}{2}\sqrt{105}$ square units **93.** 1 **95.** $\sqrt{216} = 6\sqrt{6}$
97. $\sqrt{72} = 6\sqrt{2}$ **99.** 48 **101.** 0.05 hour or 3 minutes **102.** $3(x + 4)(x - 8)$ **103.** $\frac{1}{x + 1}$ **104.** 4, 6

105.

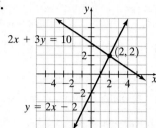

Exercise Set 9.4 **1.** Cannot be simplified **3.** Simplifies to $x\sqrt{2}$ **5.** Cannot be simplified **7.** No perfect square factors in any radicand, no radicand contains a fraction, no square roots in any denominator **9. a)** $\sqrt{27}$ has a perfect square factor, 9. **b)** The radicand contains a fraction **c)** There is a square root in the denominator. **11.** 3. **13.** 3 **15.** 3
17. $\frac{1}{7}$ **19.** $\frac{9}{12} = \frac{3}{4}$ **21.** $\frac{1}{10}$ **23.** $2x\sqrt{6}$ **25.** $\frac{3\sqrt{5}}{4y^2}$ **27.** $\frac{2y}{5x}$ **29.** $\frac{1}{a^2b}$ **31.** $2n^2$ **33.** $\frac{3w^2}{4z}$ **35.** $\frac{b\sqrt{5}}{c}$ **37.** $5a^2b^3$ **39.** $\frac{\sqrt{5}}{5}$
41. $\sqrt{2}$ **43.** $\sqrt{3}$ **45.** $\frac{\sqrt{5}}{5}$ **47.** $\frac{\sqrt{6}}{3}$ **49.** $\frac{\sqrt{3}}{3}$ **51.** $\frac{\sqrt{6}}{4}$ **53.** $\frac{\sqrt{3x}}{x}$ **55.** $\frac{\sqrt{ab}}{b}$ **57.** $\frac{\sqrt{cd}}{d}$ **59.** $\frac{\sqrt{15x}}{5}$ **61.** $\frac{x\sqrt{2}}{2}$ **63.** $\frac{a\sqrt{2}}{4}$
65. $\frac{t^2\sqrt{5t}}{5}$ **67.** $\frac{a^4\sqrt{14b}}{14b}$ **69.** $\frac{\sqrt{5d}}{5d}$ **71.** $\frac{5\sqrt{3}}{6x^2y^2z}$ **73.** $\frac{3\sqrt{5x}}{xy}$ **75.** 22 **77.** -3 **79.** $x - y^2$ **81.** $a - b$ **83.** $2\sqrt{5} - 4$
85. $\frac{2\sqrt{6} + 2}{5}$ **87.** $-3\sqrt{3} + 3\sqrt{5}$ **89.** $\frac{-8\sqrt{5} - 16\sqrt{2}}{3}$ **91.** $\frac{2\sqrt{y} - 6}{y - 9}$ **93.** $\frac{24 + 6\sqrt{y}}{16 - y}$ **95.** $\frac{16\sqrt{y} - 16x}{y - x^2}$
97. $\frac{x\sqrt{x} - x\sqrt{y}}{x - y}$ **99.** $\frac{3 + \sqrt{3n}}{3 - n}$ **101.** $\frac{30\sqrt{x} + 5x}{36 - x}$ **103.** Yes **105.** 5.40 **107.** 0.71 **109.** 7.23 **111.** 0.58
113. $\frac{24(4 - \sqrt{3})}{13}$ **115.** $64x^{10}$ **117.** 2 **119.** $\frac{-\sqrt{x} - \sqrt{3x}}{2}$ **122.** $3x - 8 + \frac{7}{x + 4}$ **123.** $\frac{9}{2}, -4$ **124.** $\frac{-2x - 3}{(x + 2)(x - 2)}$
125. 30 minutes

Exercise Set 9.5 **1.** A radical equation is an equation that contains a variable in a radicand. **3.** It is necessary to check solutions because they may be extraneous. **5.** Yes **7.** No; $-\sqrt{64} = -8$ **9.** Yes **11.** No, $\sqrt{-4} \neq -2$ **13.** 16
15. No solution **17.** 4 **19.** No solution **21.** 100 **23.** 25 **25.** No solution **27.** 7 **29.** 6 **31.** 3 **33.** 1 **35.** $\frac{14}{3}$ **37.** 1
39. 1, 9 **41.** No solution **43.** -2 **45.** 7 **47.** No solution **49.** 3 **51.** 3 **53.** 7 **55.** 11 **57.** $\sqrt{x + 3} = 7; 46$
59. $\sqrt{x - 2} = \sqrt{2x - 9}; 7$ **61. a)** $x - 9 = 40$ **b)** 49 **63. a)** $35 + 2\sqrt{x} - x = 35$ **b)** 0, 4 **65.** 1 **67.** 9 **69.** 9
73. **74.** $(2, 0)$ **75.** $(2, 0)$ **76.** Ferryboat: 16 mph; current: 2 mph

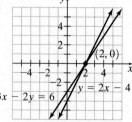

Exercise Set 9.6 **1.** A right triangle is a triangle that contains a $90°$ angle. **3.** No; only with right triangles.
5. They represent the two points in the coordinate plane that you are trying to find the distance between. **7.** 6
9. $\sqrt{33} \approx 5.74$ **11.** $\sqrt{170} \approx 13.04$ **13.** $\sqrt{149} \approx 12.21$ **15.** $\sqrt{67} \approx 8.19$ **17.** $\sqrt{202} \approx 14.21$ **19.** $\sqrt{180} \approx 13.42$
21. $\sqrt{34} \approx 5.83$ **23.** $\sqrt{193} \approx 13.89$ **25.** $\sqrt{17,240.89} \approx 131.30$ yards **27.** $\sqrt{60} \approx 7.75$ meters **29.** 16 feet

31. $\approx \sqrt{25.48} \approx 5.05$ feet **33.** $\sqrt{16{,}200} \approx 127.28$ feet **35.** $\sqrt{832} \approx 28.84$ inches **37.** $6.28\sqrt{0.11} \approx 2.08$ seconds
39. $6.28\sqrt{1.925} \approx 8.71$ seconds **41.** About 365 Earth days **43.** $\sqrt{19.62(6{,}370{,}000)} \approx 11{,}179.42$ meters per second
45. $\sqrt{1{,}000{,}000} = 1000$ pounds **47.** ≈ 32.66 feet per second **49.** $\sqrt{932} \approx 30.53$ inches **50.** $x > 6$ **51.** $\dfrac{x^4}{4y^3}$
52. $\left(7, -\dfrac{9}{4}\right)$ **53.** $-\dfrac{16}{15}$

Using Your Calculator, 9.7 (page 603) **1.** 9 **2.** 2 **3.** ≈ 3.76060 **4.** 10 **5.** -6 **6.** ≈ -4.12891

Using Your Graphing Calculator, 9.7 (page 604) **1.** 7 **2.** 6 **3.** ≈ 2.952591724 **4.** 14
5. -8 **6.** ≈ -3.981071706

Using Your Calculator, 9.7 (page 606) **1.** 8 **2.** 256 **3.** .04 **4.** $.\overline{037}$

Exercise Set 9.7 **1. a)** The square root of 8 **b)** The cube root of 8 **c)** The fourth root of 8 **3.** Write the
radicand as a product of a perfect cube and another number. **5. a)** Answers will vary. **b)** $y^{7/3}$ **7.** Yes **9.** 2 **11.** -3
13. 2 **15.** 3 **17.** -1 **19.** -10 **21.** $2\sqrt[3]{5}$ **23.** $2\sqrt[3]{2}$ **25.** $3\sqrt[3]{4}$ **27.** $2\sqrt[4]{2}$ **29.** $5\sqrt[4]{2}$ **31.** $x^{7/3}$ **33.** $a^{2/5}$ **35.** $y^{15/4}$ **37.** $y^{8/3}$
39. x **41.** y^7 **43.** m^6 **45.** a^{15} **47.** x^5 **49.** $\sqrt{m}$ **51.** $\sqrt{t}$ **53.** $\sqrt[3]{x}$ **55.** $\sqrt{w}$ **57.** $\sqrt[3]{x^2}$ **59.** $\sqrt[4]{z^3}$ **61.** 9 **63.** 36 **65.** 1
67. 27 **69.** 81 **71.** 1024 **73.** $\dfrac{1}{2}$ **75.** $\dfrac{1}{9}$ **77.** 2 **79.** 3 **81.** 6 **83.** 25 **85.** 8 **87.** 16 **89.** x **91.** t **93.** r^8 **95.** a^2

97. both equal 4 **99.** $\sqrt[3]{4^2}$ **101.** $\sqrt[3]{6^2}$ **103.** $\sqrt[4]{5^3}$ **105.** 3 **107.** 1 **109.** xy **111.** $\sqrt[4]{2}$ **113. a)** $\sqrt[3]{2^3} = 2$ **b)** $\dfrac{\sqrt[3]{4}}{2}$

114. -42 **115.** $(3x - 4)(x - 8)$ **116.** slope: $\dfrac{2}{3}$, y intercept: $\left(0, -\dfrac{4}{3}\right)$ **117.** $\dfrac{4y^3\sqrt{2xy}}{x}$

Chapter Review Exercises **1.** 9 **2.** 7 **3.** -8 **4.** $5^{1/2}$ **5.** $(26x)^{1/2}$ **6.** $(13x^2y)^{1/2}$ **7.** $4\sqrt{2}$ **8.** $2\sqrt{11}$
9. $3x^3y^2\sqrt{3x}$ **10.** $5x^2y^3\sqrt{5}$ **11.** $2b^2c^2\sqrt{15ab}$ **12.** $6abc^3\sqrt{2c}$ **13.** $12\sqrt{10}$ **14.** $7y$ **15.** $6x\sqrt{y}$ **16.** $5xy\sqrt{3}$ **17.** $6ab^4\sqrt{a}$
18. $10ab^3\sqrt{b}$ **19.** $\sqrt{3}$ **20.** $-6\sqrt{5}$ **21.** $-2\sqrt{x}$ **22.** 0 **23.** $3\sqrt{2} - 3\sqrt{3}$ **24.** $12\sqrt{10}$ **25.** $-10\sqrt{2}$ **26.** $25\sqrt{2}$
27. $2\sqrt{5} + 5$ **28.** $2 + 3\sqrt{2}$ **29.** $x\sqrt{y} - 2y$ **30.** $15a^2 + 5a\sqrt{2a}$ **31.** -1 **32.** 78 **33.** $x^2 - 4y$ **34.** $c - 4d$
35. $m^2 - 3m\sqrt{r} - 10r$ **36.** $3t + 5s\sqrt{t} - 2s^2$ **37.** $15m + 5\sqrt{5mn} - 2n$ **38.** $14 - 9\sqrt{7p} + 9p$ **39.** 4 **40.** $\dfrac{1}{7}$ **41.** $\dfrac{1}{2}$
42. $\dfrac{3\sqrt{5}}{5}$ **43.** $\dfrac{\sqrt{7n}}{7}$ **44.** $\dfrac{\sqrt{15a}}{6}$ **45.** $\dfrac{x\sqrt{3}}{3}$ **46.** $\dfrac{z^2\sqrt{2}}{4}$ **47.** $y^2\sqrt{7}$ **48.** $\dfrac{x\sqrt{2y}}{y^2}$ **49.** $\dfrac{2\sqrt{5a}}{3a^2b}$ **50.** $\dfrac{c\sqrt{14a}}{7a}$ **51.** $\dfrac{-3 - 3\sqrt{6}}{5}$
52. $\dfrac{15 + 5\sqrt{6}}{3}$ **53.** $\dfrac{2\sqrt{2} - \sqrt{2y}}{4 - y}$ **54.** $\dfrac{2\sqrt{x} + 10}{x - 25}$ **55.** $\dfrac{\sqrt{5x} - \sqrt{15}}{x - 3}$ **56.** $\dfrac{\sqrt{35} + x\sqrt{7}}{5 - x^2}$ **57.** 36 **58.** No solution **59.** 14
60. 8 **61.** 2 **62.** 4 **63.** 2 **64.** 7 **65.** 2 **66.** 26 **67.** $\sqrt{125} \approx 11.18$ **68.** $\sqrt{12} \approx 3.46$ **69.** $\sqrt{61} \approx 7.81$ **70.** $\sqrt{135} \approx 11.62$
feet **71.** $\sqrt{261} \approx 16.16$ inches **72.** $\sqrt{109} \approx 10.44$ **73.** $\sqrt{153} \approx 12.37$ **74.** 561.18 square inches **75.** $\sqrt{60} \approx 7.75$ miles
76. ≈ 646 feet **77.** 4 **78.** -4 **79.** 2 **80.** 3 **81.** 4 **82.** -2 **83.** $2\sqrt[4]{2}$ **84.** $2\sqrt[3]{6}$ **85.** $3\sqrt[3]{2}$ **86.** $2\sqrt[4]{5}$ **87.** x^7 **88.** s^{10} **89.** 1
90. 5 **91.** $\dfrac{1}{9}$ **92.** 16 **93.** $\dfrac{1}{625}$ **94.** 343 **95.** $z^{11/3}$ **96.** $x^{8/3}$ **97.** $y^{9/4}$ **98.** $x^{5/2}$ **99.** $y^{3/2}$ **100.** $m^{6/4} = m^{3/2}$ **101.** x **102.** $\sqrt[3]{x^2}$
103. a^4 **104.** x^2 **105.** q^3 **106.** d **107.** x^6 **108.** x^6

Chapter Practice Test **1.** $(3x)^{1/2}$ **2.** $\sqrt[3]{x^2}$ **3.** $y - 4$ **4.** $3\sqrt{10}$ **5.** $2x\sqrt{3}$ **6.** $5x^3y\sqrt{2xy}$ **7.** $4xy\sqrt{5x}$
8. $5x^2y^2\sqrt{3y}$ **9.** $\dfrac{1}{5}$ **10.** $\dfrac{c^2}{d}$ **11.** $\dfrac{\sqrt{6}}{6}$ **12.** $\dfrac{3\sqrt{5r}}{5}$ **13.** $\dfrac{2\sqrt{15x}}{3xy}$ **14.** $-2 - \sqrt{7}$ **15.** $\dfrac{6\sqrt{x} + 18}{x - 9}$ **16.** $11\sqrt{3}$ **17.** $3\sqrt{y}$ **18.** 24
19. 4.8 **20.** $\sqrt{106} \approx 10.30$ **21.** $\sqrt{58} \approx 7.62$ **22.** $\dfrac{1}{81}$ **23.** x^3 **24.** 11 meters **25.** $\sqrt{640} \approx 25.30$ feet per second

CHAPTER 10

Exercise Set 10.1 **1.** If $x^2 = a$, then $x = \sqrt{a}$ or $x = -\sqrt{a}$ **3.** In any golden rectangle, the length is about 1.62
times its width. **5. a)** 2 **b)** 1 **c)** 2 **d)** No real solution **7.** 8, -8 **9.** 9, -9 **11.** 13, -13 **13.** 10, -10

15. $2\sqrt{5}, -2\sqrt{5}$ **17.** $2, -2$ **19.** $\sqrt{17}, -\sqrt{17}$ **21.** $3, -3$ **23.** $\dfrac{\sqrt{15}}{3}, -\dfrac{\sqrt{15}}{3}$ **25.** $\dfrac{\sqrt{73}}{4}, -\dfrac{\sqrt{73}}{4}$ **27.** $9, -1$ **29.** $6, -12$

31. $4, -12$ **33.** $-6 + 4\sqrt{2}, -6 - 4\sqrt{2}$ **35.** $-6 + 2\sqrt{5}, -6 - 2\sqrt{5}$ **37.** $10, -4$ **39.** $19, -1$ **41.** $\dfrac{-3 + 3\sqrt{2}}{2}, \dfrac{-3 - 3\sqrt{2}}{2}$

43. $\dfrac{-1 + 2\sqrt{5}}{4}, \dfrac{-1 - 2\sqrt{5}}{4}$ **45.** $\dfrac{7 + 3\sqrt{2}}{2}, \dfrac{7 - 3\sqrt{2}}{2}$ **47.** $x^2 = 36$ **49.** 9 **51. a)** $3x^2 - 9x + 6 = 0$

b) $x^2 - 3x + 2 = 0$ **53.** $4, 17$ **55.** Width $\approx \sqrt{522.25}$ or ≈ 22.85 inches, length $\approx 1.21\sqrt{522.25}$ or ≈ 27.65 inches
57. Width ≈ 35.14 feet, length ≈ 56.93 feet   **59. a)** Left, x^2; right, $(x + 3)^2$ **b)** 6 inches   **c)** $\sqrt{50} \approx 7.07$ inches
d) 9 inches   **e)** $\sqrt{92} \approx 9.59$ inches   **61.** $s = \sqrt{A}$ **63.** $r = \sqrt{\dfrac{A}{\pi}}$ **65.** $d = \sqrt{\dfrac{k}{I}}$ **67.** $2(2x + 3)(x - 4)$ **68.** $\dfrac{3y - 1}{6y - 1}$
69. $y = 4x - 1$ **70.** $\dfrac{3\sqrt{5a}}{ab^2}$

Exercise Set 10.2   **1. a)** A perfect square trinomial is a trinomial that can be expressed as the square of a
binomial.   **b)** $+16$; The constant term is the square of half the coefficient of the x-term.   **3.** The constant is the square of
half the coefficient of the x-term.   **5.** 36 **7.** $2, 5$ **9.** $7, 1$ **11.** $-2, -1$ **13.** $4, -2$ **15.** -3 **17.** $5, -3$ **19.** $-4, -6$ **21.** $8, 7$

23. $8, -4$ **25.** $2 + \sqrt{2}, 2 - \sqrt{2}$ **27.** $-3 + \sqrt{5}, -3 - \sqrt{5}$ **29.** $\dfrac{-7 + \sqrt{41}}{2}, \dfrac{-7 - \sqrt{41}}{2}$ **31.** $1, -3$ **33.** $\dfrac{-9 + \sqrt{73}}{2},$

$\dfrac{-9 - \sqrt{73}}{2}$ **35.** $6, -1$ **37.** $4, -\dfrac{1}{3}$ **39.** $\dfrac{-1 + \sqrt{7}}{3}, \dfrac{-1 - \sqrt{7}}{3}$ **41.** $5, 0$ **43.** $6, 0$ **45. a)** $x^2 + 20x + 100$ **b)** Answers will

vary.   **47.** $1, -4$ **49.** $0, -6$ **51.** $3, 7$ **53.** $9 + \sqrt{369}$ or ≈ 28.21 feet   **55.** 2 seconds and 6 seconds   **57.** $+28x$ or $-28x$

59. $7 + 5\sqrt{2}, 7 - 5\sqrt{2}$ **61.** $\dfrac{-3 + \sqrt{59}}{10}, \dfrac{-3 - \sqrt{59}}{10}$ **63.** $\dfrac{-1 + \sqrt{193}}{12}, \dfrac{-1 - \sqrt{193}}{12}$ **65.** $0.75 + \sqrt{3.5625}, 0.75 - \sqrt{3.5625}$

66. $\dfrac{4}{(x + 2)(x - 3)}$ **67.** If the slopes are the same and the y-intercepts are different, the equations represent parallel lines.

68. $\left(\dfrac{38}{11}, \dfrac{12}{11}\right)$ **69.** 3

Exercise Set 10.3   **1. a)** $b^2 - 4ac$ **b)** Discriminant: greater than 0, two solutions; equal to 0, one solution;

less than 0, no real solution.   **3.** $x = \dfrac{-b \pm \sqrt{b^2 - 4ac}}{2a}$ **5.** The first step is to write the equation in standard form.

7. The values used for b and c are incorrect because the equation was not first put in standard form.   **9.** Two distinct real
number solutions   **11.** No real number solution   **13.** Two distinct real number solutions   **15.** One real number solution
17. No real number solution   **19.** No real number solution   **21.** Two distinct real number solutions   **23.** Two distinct real
number solutions   **25.** One real number solution   **27.** $4, -2$ **29.** $-3, -6$ **31.** $5, 1$ **33.** $-5, -6$ **35.** $9, -9$ **37.** $5, 0$

39. No real number solution   **41.** $\dfrac{7 + \sqrt{17}}{4}, \dfrac{7 - \sqrt{17}}{4}$ **43.** $\dfrac{1}{3}, -\dfrac{1}{2}$ **45.** $\dfrac{7}{2}, -1$ **47.** No real number solution   **49.** $\dfrac{5}{4}, -1$

51. $\dfrac{9}{2}, -1$ **53.** $\dfrac{5}{2}, 3$ **55.** $3, -5$ **57.** $1, 9$ **59.** $6, 7$ **61.** Width $= 4$ feet; length $= 5$ feet   **63.** 2 feet   **65.** 9 sides   **67.** 40 chairs

69. ≈ 1.11 seconds and ≈ 4.52 seconds   **71. a)** $c < 9$ **b)** $c = 9$ **c)** $c > 9$ **73. a)** $c > -3$ **b)** $c = -3$ **c)** $c < -3$

76. $7, 6$ **77.** $\dfrac{5}{3}, -\dfrac{7}{2}$ **78.** Cannot be solved by factoring; $\dfrac{-3 + \sqrt{41}}{4}, \dfrac{-3 - \sqrt{41}}{4}$ **79.** $3, -3$ **80.** $\dfrac{x^2 - 9x + 2}{(2x - 1)(x + 4)(x - 5)}$

Using Your Graphing Calculator, 10.4   **1.** $(1, -4)$ **2.** $(3.5, -2.25)$ **3.** $(-0.5, 12.5)$
4. $(1.75, -21.125)$

Exercise Set 10.4   **1.** A parabola   **3.** Answers will vary.   **5. a)** The x-intercepts are where the graph crosses

the x-axis.   **b)** The x-intercepts are found by setting $y = 0$ and solving for x.   **7. a)** $x = -\dfrac{b}{2a}$ **b)** This line is called the axis of symmetry. **9.** $x = -2$ $(-2, -7)$, upward   **11.** $x = \frac{3}{2}$, $\left(\frac{3}{2}, -\frac{7}{4}\right)$, downward   **13.** $x = \frac{1}{3}$, $\left(\frac{1}{3}, \frac{25}{3}\right)$, downward

15. $x = -1, (-1, -1)$, upward   **17.** $x = -\frac{3}{4}$, $\left(-\frac{3}{4}, \frac{31}{8}\right)$, upward   **19.** $x = \frac{3}{5}$, $\left(\frac{3}{5}, \frac{4}{5}\right)$, downward

21.

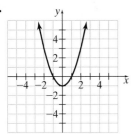

23.

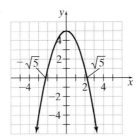

25.

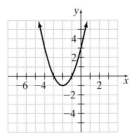

27.

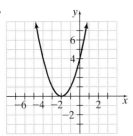

29.

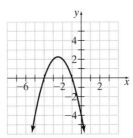

31.

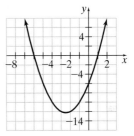

33.

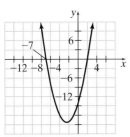

35.

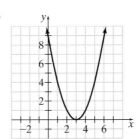

37.

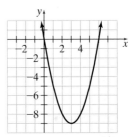

39.

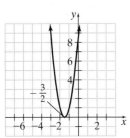

41.

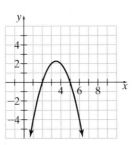

43.

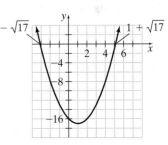

45.

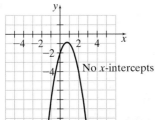

No x-intercepts

47.

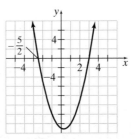

49. Two   **51.** Two   **53.** One   **55.** Two   **57.** None; the vertex is below the x-axis and the parabola opens downward   **59.** One; the vertex of the parabola is on the x-axis.   **61.** Yes; if y is set to 0, both equations have the same solutions, 5 and -3.   **63. a)** ≈ 255 feet   **b)** 4 seconds   **c)** 8 seconds   **d)** ≈ 190 feet, ≈ 240 feet

65. a)

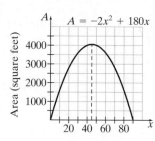

b) 45 feet   **c)** 4050 square feet

67. a)

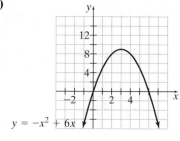

b)

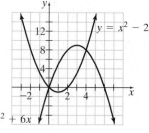

c) $(0,0), (4,8)$

69. $\dfrac{-x^2 + 2x - 6}{(x+3)(x-4)}$ **70.** $\dfrac{27}{7}$ **71.** $(6,-5)$ **73.** 8

Exercise Set 10.5

1. Yes, every real number is a complex number.   **3.** $\sqrt{-1}$ **5.** $a + bi$ **7.** $2i$ **9.** $10i$
11. $i\sqrt{15}$ **13.** $i\sqrt{23}$ **15.** $2i\sqrt{2}$ **17.** $2i\sqrt{5}$ **19.** $5 + 2i$ **21.** $-3 + 5i$ **23.** $-1 - 3i$ **25.** $-3 - i\sqrt{15}$ **27.** $5.2 + 5i\sqrt{2}$
29. $\dfrac{1}{2} + 5i\sqrt{3}$ **31.** $7 + 3i$ **33.** $4 - 11i$ **35.** $11 + 5i$ **37.** $-2 - 7i$ **39.** 0 **41.** $9 - 4i$ **43.** $4i, -4i$ **45.** $i\sqrt{10}, -i\sqrt{10}$
47. $2 + i, 2 - i$ **49.** $\dfrac{-3 + i\sqrt{31}}{4}, \dfrac{-3 - i\sqrt{31}}{4}$ **51.** $\dfrac{-2 + i\sqrt{14}}{2}, \dfrac{-2 - i\sqrt{14}}{2}$ **53.** $\dfrac{5 + i\sqrt{119}}{8}, \dfrac{5 - i\sqrt{119}}{8}$
55. When $c < 0$ **57.** When $b^2 - 4ac < 0$ **59.** 11 **60.** $-\dfrac{5}{3}$ **61.** 1 **62.** 13

Chapter Review Exercises

1. $8, -8$ **2.** $2\sqrt{3}, -2\sqrt{3}$ **3.** $\sqrt{6} - \sqrt{6}$ **4.** $\sqrt{5} - \sqrt{5}$ **5.** $2\sqrt{5}, -2\sqrt{5}$
6. $\sqrt{7}, -\sqrt{7}$ **7.** $2\sqrt{2}, -2\sqrt{2}$ **8.** $3 + 2\sqrt{5}, 3 - 2\sqrt{5}$ **9.** $\dfrac{3 + 5\sqrt{2}}{4}, \dfrac{3 - 5\sqrt{2}}{4}$ **10.** $\dfrac{-4 + \sqrt{30}}{2}, \dfrac{-4 - \sqrt{30}}{2}$ **11.** $5, 2$
12. $7, 4$ **13.** $17, 1$ **14.** $2, -3$ **15.** $9, -6$ **16.** $1, -6$ **17.** $\dfrac{5 + \sqrt{53}}{2}, \dfrac{5 - \sqrt{53}}{2}$ **18.** $-1 + \sqrt{6}, -1 - \sqrt{6}$ **19.** $8, -4$
20. $5, -3$ **21.** $\dfrac{4}{3}, -2$ **22.** $\dfrac{5}{3}, \dfrac{3}{2}$ **23.** No real solution   **24.** No real solution   **25.** One solution   **26.** Two solutions
27. Two solutions   **28.** No real solution   **29.** Two solutions   **30.** Two solutions   **31.** $8, 2$ **32.** $11, -4$ **33.** $9, 1$ **34.** $2, -\dfrac{3}{5}$
35. $3, -7$ **36.** No real solution   **37.** $\dfrac{3}{2}, -\dfrac{5}{3}$ **38.** $\dfrac{3 + \sqrt{57}}{4}, \dfrac{3 - \sqrt{57}}{4}$ **39.** $\dfrac{-2 + \sqrt{10}}{2}, \dfrac{-2 - \sqrt{10}}{2}$ **40.** $3 + \sqrt{6}, 3 - \sqrt{6}$
41. No real solution   **42.** $\dfrac{3 + \sqrt{33}}{3}, \dfrac{3 - \sqrt{33}}{3}$ **43.** $0, \dfrac{3}{7}$ **44.** $0, \dfrac{5}{2}$ **45.** $4, 6$ **46.** $-7, -8$ **47.** $10, -7$ **48.** $-9, 3$ **49.** $-6, 10$
50. $7, -6$ **51.** $2, -11$ **52.** $0, 5$ **53.** $9, -9$ **54.** $\dfrac{1}{2}, -3$ **55.** $\dfrac{5}{2}, 2$ **56.** $\dfrac{2}{3}, -\dfrac{3}{2}$ **57.** $\dfrac{3 + 3\sqrt{3}}{2}, \dfrac{3 - 3\sqrt{3}}{2}$ **58.** $2, \dfrac{5}{3}$ **59.** $1, -\dfrac{8}{3}$
60. $\dfrac{-3 + \sqrt{33}}{2}, \dfrac{-3 - \sqrt{33}}{2}$ **61.** $0, \dfrac{9}{4}$ **62.** $0, -\dfrac{5}{3}$ **63.** $x = 3, (3, -14)$, upward   **64.** $x = 6, (6, -30)$, upward   **65.** $x = \dfrac{3}{2}$,
$\left(\dfrac{3}{2}, \dfrac{19}{4}\right)$, upward   **66.** $x = -1, (-1, 16)$, downward   **67.** $x = -\dfrac{7}{6}, \left(-\dfrac{7}{6}, -\dfrac{13}{12}\right)$, upward   **68.** $x = -\dfrac{5}{2}, \left(-\dfrac{5}{2}, \dfrac{25}{4}\right)$,
downward   **69.** $x = 0, (0, -8)$, downward   **70.** $x = -\dfrac{1}{4}, \left(-\dfrac{1}{4}, \dfrac{161}{8}\right)$, downward   **71.** $x = 1, (1, 9)$, downward
72. $x = -\dfrac{5}{6}, \left(-\dfrac{5}{6}, -\dfrac{121}{12}\right)$, upward

73.

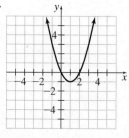

74.

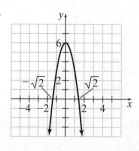

75.

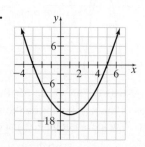

76.
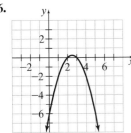

77.
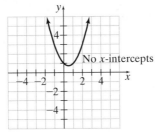
No *x*-intercepts

78.
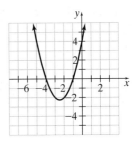

79.
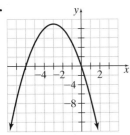

80.
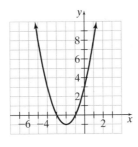

81.
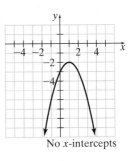
No *x*-intercepts

82.
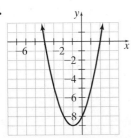

83.
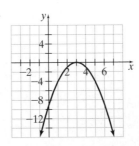

84.
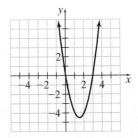

85. $6, 8$ **86.** $8, 11$ **87.** Width $= 20$ inches, length $= 46$ inches   **88.** Width $= 28$ inches, length $= 48$ inches   **89.** $2i$

90. $i\sqrt{30}$ **91.** $4 - 5i$ **92.** $5 - 2i\sqrt{15}$ **93.** $9 - 9i$ **94.** $3 + 8i$ **95.** $3i, -3i$ **96.** $i\sqrt{5}, -i\sqrt{5}$ **97.** $\dfrac{5 + i\sqrt{39}}{4}, \dfrac{5 - i\sqrt{39}}{4}$

98. $\dfrac{2 + i\sqrt{5}}{2}, \dfrac{2 - i\sqrt{5}}{2}$

Chapter Practice Test
1. $4\sqrt{2}, -4\sqrt{2}$ **2.** $\dfrac{4 + \sqrt{17}}{2}, \dfrac{4 - \sqrt{17}}{2}$ **3.** $10, -4$ **4.** $4, -11$ **5.** $6, 7$

6. $\dfrac{-4 + \sqrt{6}}{2}, \dfrac{-4 - \sqrt{6}}{2}$ **7.** $\dfrac{7}{4}, -\dfrac{7}{4}$ **8.** $x = \dfrac{-b \pm \sqrt{b^2 - 4ac}}{2a}$ **9.** Answers will vary.   **10.** Two solutions

11. One real solution   **12.** $x = -3$ **13.** $x = 2$ **14.** Downward; $a < 0$ **15.** Upward; $a > 0$ **16.** The vertex is the lowest

point on a parabola that opens upward; highest point on a parabola that opens downward.   **17.** $(-4, 4)$ **18.** $\left(\dfrac{4}{3}, \dfrac{11}{3}\right)$

19.
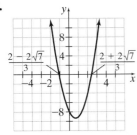

20.
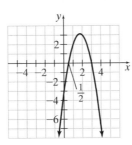

21.
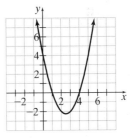

22. Width $= 3$ feet, length $= 10$ feet   **23.** 11 **24.** 9 **25.** $\dfrac{1 + i\sqrt{17}}{3}, \dfrac{1 - i\sqrt{17}}{3}$

Graphing Answer Section

Using Your Graphing Calculator, 7.1

1.
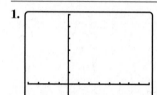
$-20, 40, 5, -10, 60, 10$

2.
$-200, 400, 100, -500, 1000, 200$

Exercise Set 7.1

13.
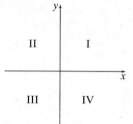

29.
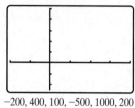

30.
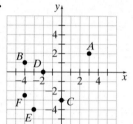

31.
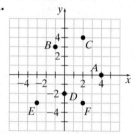

32.
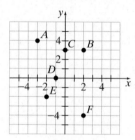

37. b)
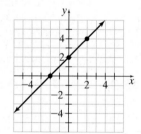

38. b)
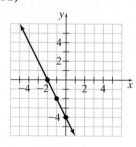

39. b)
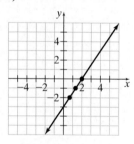

40. b)
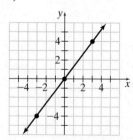

41. b)
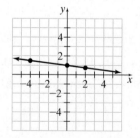

42. b)
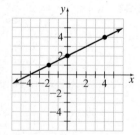

54.
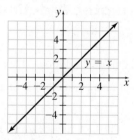

55.
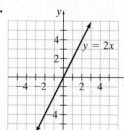

56.
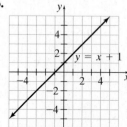

G1

Using Your Graphing Calculator, 7.2

1.
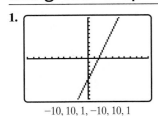
$-10, 10, 1, -10, 10, 1$

2.
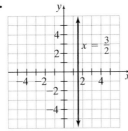
$-10, 10, 1, -10, 10, 1$

3.
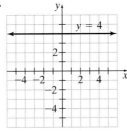
$-10, 10, 1, -10, 10, 1$

4.
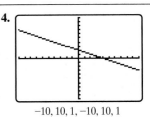
$-10, 10, 1, -10, 10, 1$

Exercise Set 7.2

21.
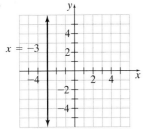

22.
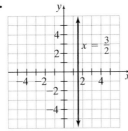

23.

24.
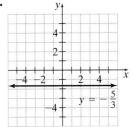

25.
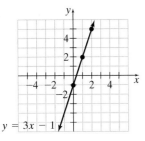

26.
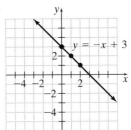

27.
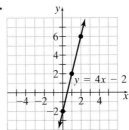

28.
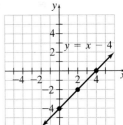

29.
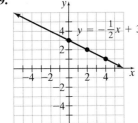

30.
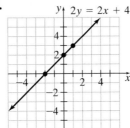

31.
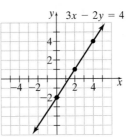

32.
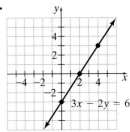

33.
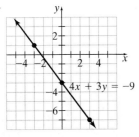

34.
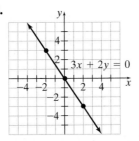

35.
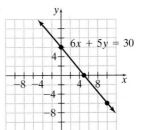

36.
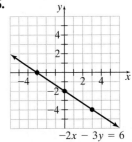

37.
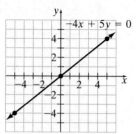
$-4x + 5y = 0$

38.
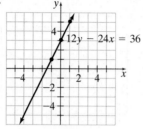
$12y - 24x = 36$

39.
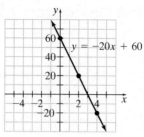
$y = -20x + 60$

40.
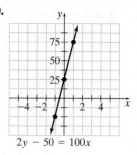
$2y - 50 = 100x$

41.
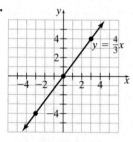
$y = \frac{4}{3}x$

42.
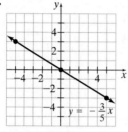
$y = -\frac{3}{5}x$

43.
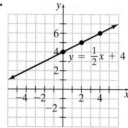
$y = \frac{1}{2}x + 4$

44.
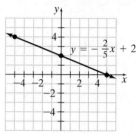
$y = -\frac{2}{5}x + 2$

45.
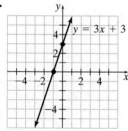
$y = 3x + 3$

46.
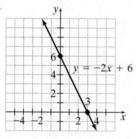
$y = -2x + 6$

47.
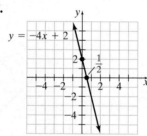
$y = -4x + 2$

48.
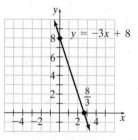
$y = -3x + 8$

49.
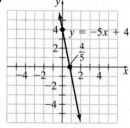
$y = -5x + 4$

50.
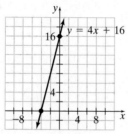
$y = 4x + 16$

51.
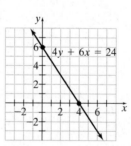
$4y + 6x = 24$

52.
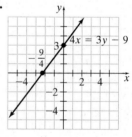
$4x = 3y - 9$

53.
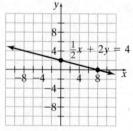
$\frac{1}{2}x + 2y = 4$

54.
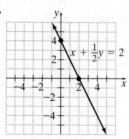
$x + \frac{1}{2}y = 2$

55.
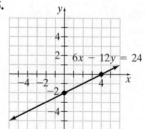
$6x - 12y = 24$

56.
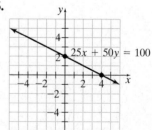
$25x + 50y = 100$

57.

$8y = 6x - 12$

$-\frac{3}{2}$

58.

$-3y - 2x = -6$

59.

$-\frac{4}{3}$

$-20y - 30x = 40$

60.

$20x - 240 = -60y$

61.

$\frac{1}{3}x + \frac{1}{4}y = 12$

48, 40, 36

62.

-90

$\frac{1}{4}x - \frac{2}{3}y = 60$

63.

$\frac{1}{2}x = \frac{2}{5}y - 80$

200, -160

64.

$\frac{2}{3}y = \frac{5}{4}x + 120$

180, -96

75. b)

$C = 2n + 30$

Cost (dollars) vs Bags of Gravel

76. b)

$d = 60t$

Distance (miles) vs Time (hours)

77. b)

$C = m + 40$

Cost (dollars) vs Number of Miles

78. b)

$I = 500t$

Interest (dollars) vs Time (years)

79. a)

$P = 1.5n - 200$

$(1000, 1300)$

Profit (dollars) vs Tapes

80. b)

$C = 10h + 200$

Cost (dollars) vs Hours

85. a)

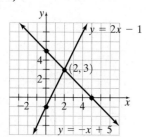

86.

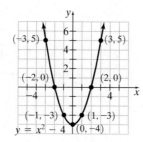

87. a–d)

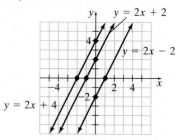

Exercise Set 7.3

39.

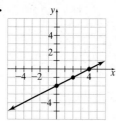

40.

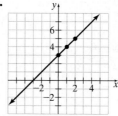

41.

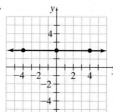

42.

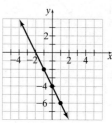

43.

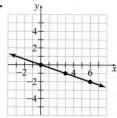

44.

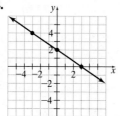

45.

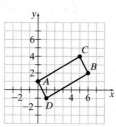

46.

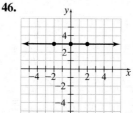

47.

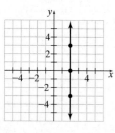

48.

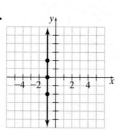

79. a)

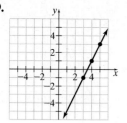

Using Your Graphing Calculator, 7.4

1. a)

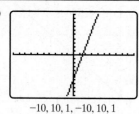

$-10, 10, 1, -10, 10, 1$

b)

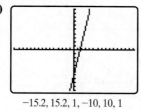

$-15.2, 15.2, 1, -10, 10, 1$

c)

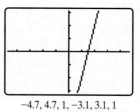

$-4.7, 4.7, 1, -3.1, 3.1, 1$

2. a)

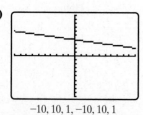

$-10, 10, 1, -10, 10, 1$

b)

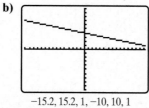

$-15.2, 15.2, 1, -10, 10, 1$

c)

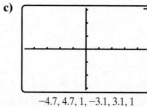

$-4.7, 4.7, 1, -3.1, 3.1, 1$

Exercise Set 7.4

13.
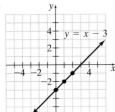
$y = x - 3$

14.
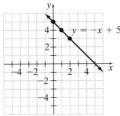
$y = -x + 5$

15.
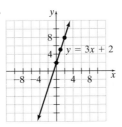
$y = 3x + 2$

16.
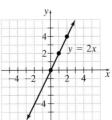
$y = 2x$

17.
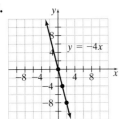
$y = -4x$

18.
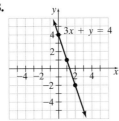
$3x + y = 4$

19.
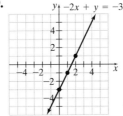
$-2x + y = -3$

20.
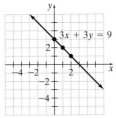
$3x + 3y = 9$

21.
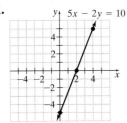
$5x - 2y = 10$

22.
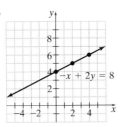
$-x + 2y = 8$

23.
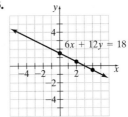
$6x + 12y = 18$

24.
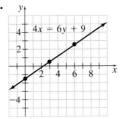
$4x = 6y + 9$

25.
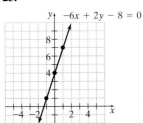
$-6x + 2y - 8 = 0$

26.
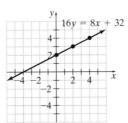
$16y = 8x + 32$

27.
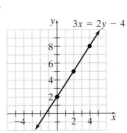
$3x = 2y - 4$

28.
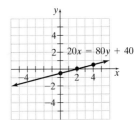
$20x = 80y + 40$

77. d)
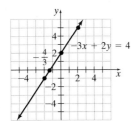
$-3x + 2y = 4$

Exercise Set 7.5

5.
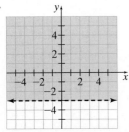

6.
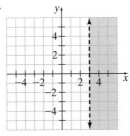

7.
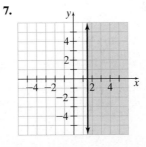

8.
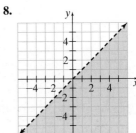

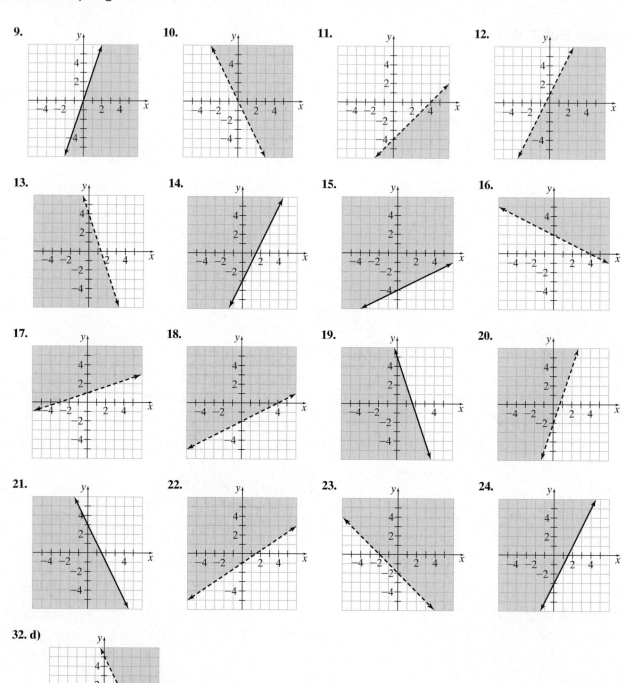

9.

10.

11.

12.

13.

14.

15.

16.

17.

18.

19.

20.

21.

22.

23.

24.

32. d)

Exercise Set 7.6

43.

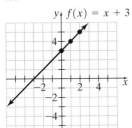

$f(x) = x + 3$

44.

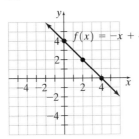

$f(x) = -x + 4$

45.

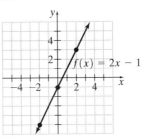

$f(x) = 2x - 1$

46.

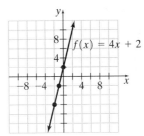

$f(x) = 4x + 2$

47.

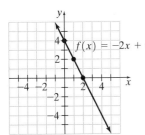

$f(x) = -2x + 4$

48.

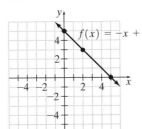

$f(x) = -x + 5$

49.

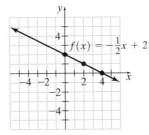

$f(x) = -\frac{1}{2}x + 2$

50.

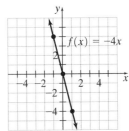

$f(x) = -4x$

55. a)

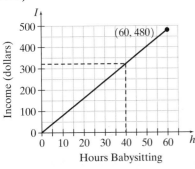

56. a)

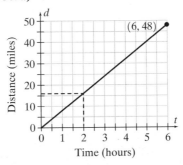

57. a)

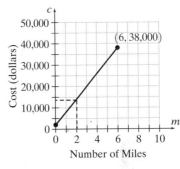

58. a)

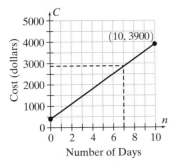

59. a)

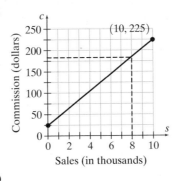

60. a)

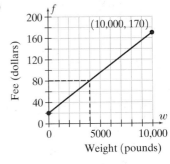

61. a)

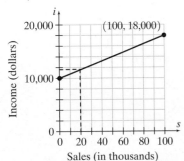

62. a)

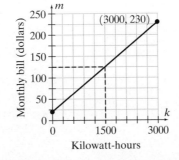

Chapter 7 Review Exercises

1.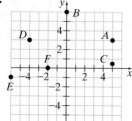

5.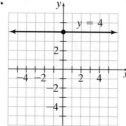

6.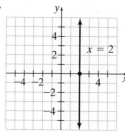

7.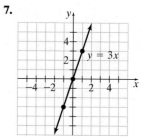

8.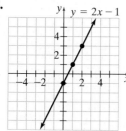

9.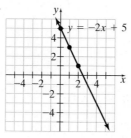

10.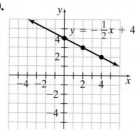

11.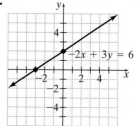

12.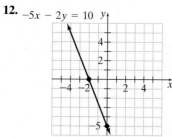

13.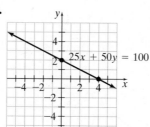

14.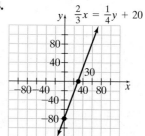

39.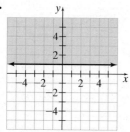

40.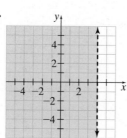

41.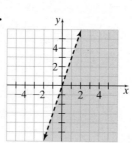

42.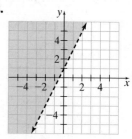

43.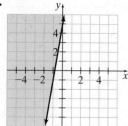

44.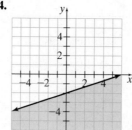

65.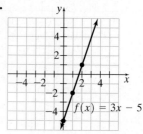

66.

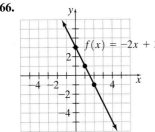

67. a)

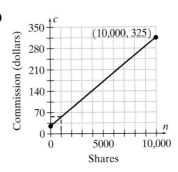

68. a)

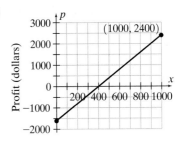

Chapter 7 Practice Test

8.

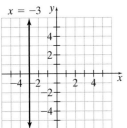

9.

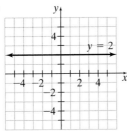

10.

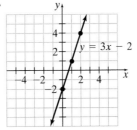

11. b)

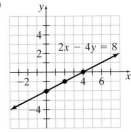

12.

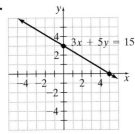

16.

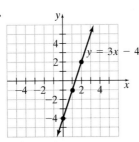

17.

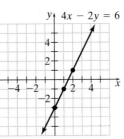

22.

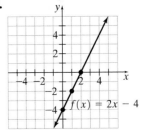

23.

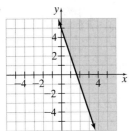

24.

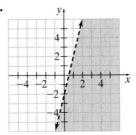

25. a)

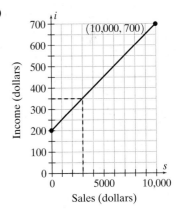

Chapter 7 Cumulative Review Test

17.

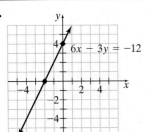

18.

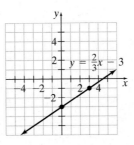

Exercise Set 8.1

83. b)

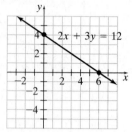

Exercise Set 8.2

43.

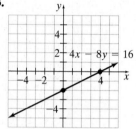

Exercise Set 8.5

5.

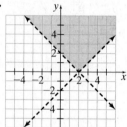

6.

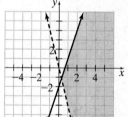

7.

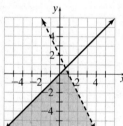

8.

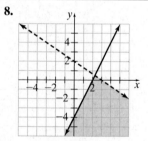

9.

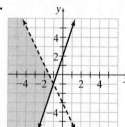

10.

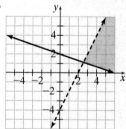

11.

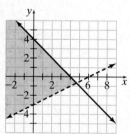

12.

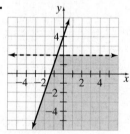

13.

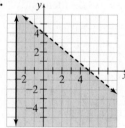

14.

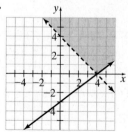

15.

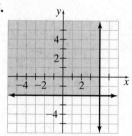

16.

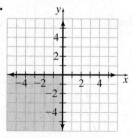

17.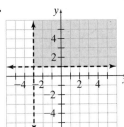

18.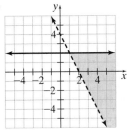

19.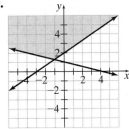

20.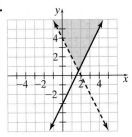

21.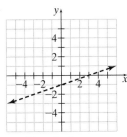

No solution

22.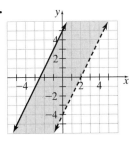

23.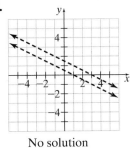

No solution

24.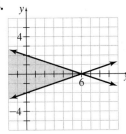

27.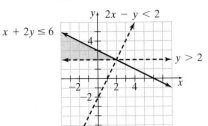

28.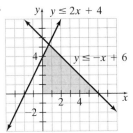

Chapter 8 Review Exercises

42.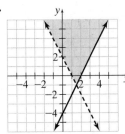

43.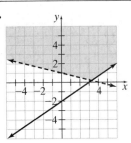

44.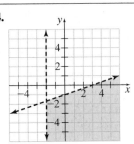

45.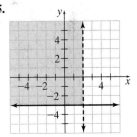

Chapter 8 Practice Test

24.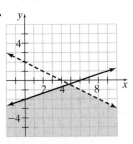

25.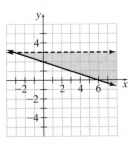

Chapter 8 Cumulative Review Test

16.

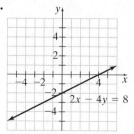

$2x - 4y = 8$

17.

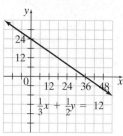

$\frac{1}{3}x + \frac{1}{2}y = 12$

Exercise Set 9.2

106.

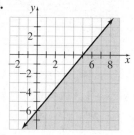

Exercise Set 9.5

70. c)

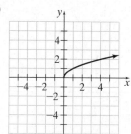

71. c)

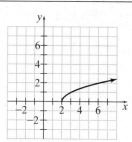

72. c)

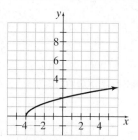

Chapter 9 Cumulative Review Test

9.

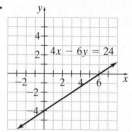

$4x - 6y = 24$

Exercise Set 10.3

75. e)

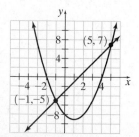

$(5, 7)$

$(-1, -5)$

Exercise Set 10.4

21.

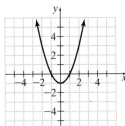

22.

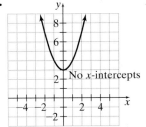
No *x*-intercepts

23.

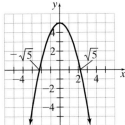
$-\sqrt{5}$ $\sqrt{5}$

24.

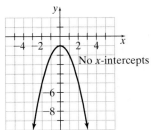
No *x*-intercepts

25.

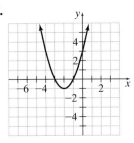

26.

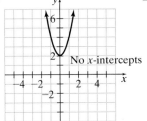
No *x*-intercepts

27.

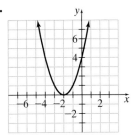

28.

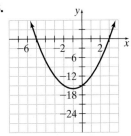

29.

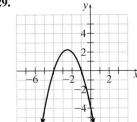

30.

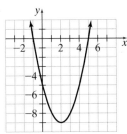

31.

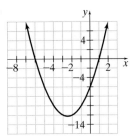

32.

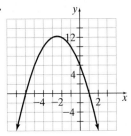

33.

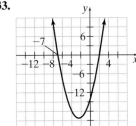

34.

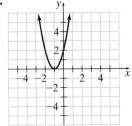

35.

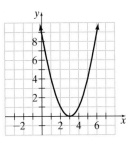

36.

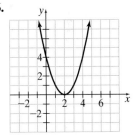

37.

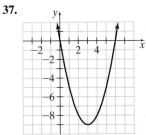

38.

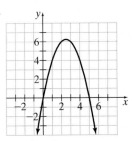

39.

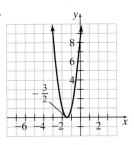
$-\dfrac{3}{2}$

40.

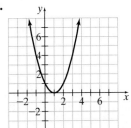

41.

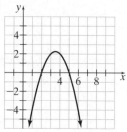

42.

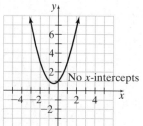

No x-intercepts

43.

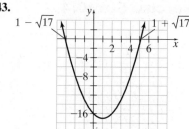

$1 - \sqrt{17}$ $1 + \sqrt{17}$

44.

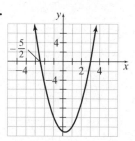

$\frac{1}{2}$

45.

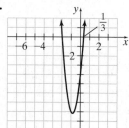

No x-intercepts

46.

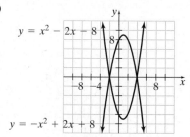

$\frac{1}{2}$

47.

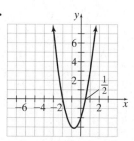

$-\frac{5}{2}$

48.

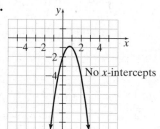

$\frac{1}{3}$

62. b)

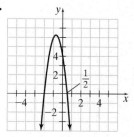

$y = x^2 - 2x - 8$

$y = -x^2 + 2x + 8$

65. a)

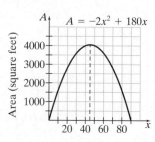

$A = -2x^2 + 180x$

66. a)

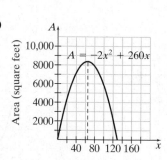

$A = -2x^2 + 260x$

67. a)

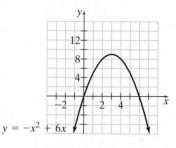

$y = -x^2 + 6x$

67. b)

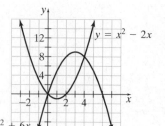

$y = x^2 - 2x$

$y = -x^2 + 6x$

68. a)

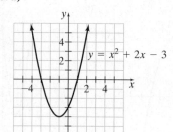

$y = x^2 + 2x - 3$

68. b)

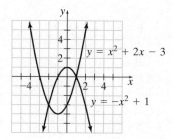

$y = x^2 + 2x - 3$

$y = -x^2 + 1$

Chapter 10 Review Exercises

73.
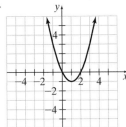

74.
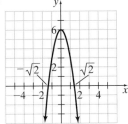

75.
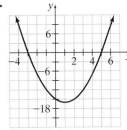

76.
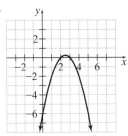

77.
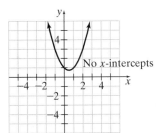
No x-intercepts

78.
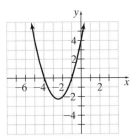

79.
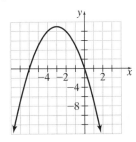

80.
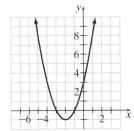

81.
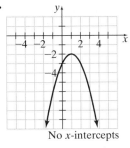
No x-intercepts

82.
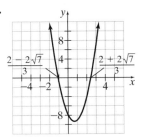

83.
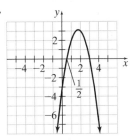

84.
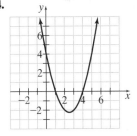
No x-intercepts

Chapter 10 Practice Test

19.
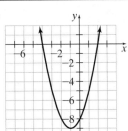

20.
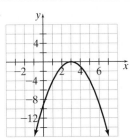

21.
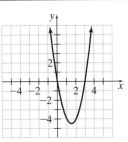

Chapter 10 Cumulative Review Test

12.
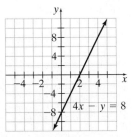

Applications Index

Index

Photo Credits

Chapter 7   Graphing Linear Equations

Linear equation in two variables: $ax + by = c$

A **graph** is an illustration of the set of points whose coordinates satisfy the equation.

Every **linear equation** of the form $ax + by = c$ will be a straight line when graphed.

To find the y-intercept (where the graph crosses the y-axis) set $x = 0$ and solve for y.

To find the x-intercept (where the graph crosses the x-axis) set $y = 0$ and solve for x.

$$\text{slope } (m) = \frac{\text{change in } y}{\text{change in } x} = \frac{y_2 - y_1}{x_2 - x_1}$$

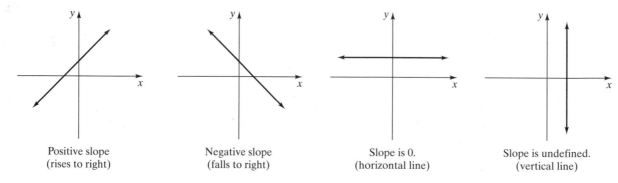

| Positive slope | Negative slope | Slope is 0. | Slope is undefined. |
| (rises to right) | (falls to right) | (horizontal line) | (vertical line) |

Linear Equations

Standard form of a linear equation: $ax + by = c$

Slope–intercept form of a linear equation: $y = mx + b$, where m is the slope and $(0, b)$ is the y-intercept.

Point–slope form of a linear equation: $y - y_1 = m(x - x_1)$, where m is the slope and (x_1, y_1) is a point on the line.

A **relation** is any set of ordered pairs.

A **function** is a set of ordered pairs in which each first component corresponds to exactly one second component.

Chapter 8   Systems of Linear Equations

The **solution** to a system of linear equations is the ordered pair or pairs that satisfy all equations in the system. A system of linear equations may have no solution, exactly one solution, or an infinite number of solutions.

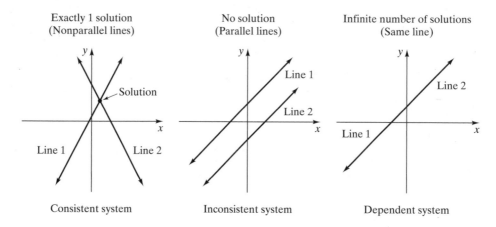

| Exactly 1 solution (Nonparallel lines) | No solution (Parallel lines) | Infinite number of solutions (Same line) |
| Consistent system | Inconsistent system | Dependent system |

A system of linear equations may be solved graphically, or algebraically by the substitution method or by the addition (or elimination) method.